Fermented Foods in Health and Disease Prevention

Fermented Foods in Health and Disease Prevention

Edited by

Juana Frias
Cristina Martinez-Villaluenga
Elena Peñas

AMSTERDAM • BOSTON • HEIDELBERG • LONDON
NEW YORK • OXFORD • PARIS • SAN DIEGO
SAN FRANCISCO • SINGAPORE • SYDNEY • TOKYO

Academic Press is an imprint of Elsevier

Academic Press is an imprint of Elsevier
125 London Wall, London EC2Y 5AS, United Kingdom
525 B Street, Suite 1800, San Diego, CA 92101-4495, United States
50 Hampshire Street, 5th Floor, Cambridge, MA 02139, United States
The Boulevard, Langford Lane, Kidlington, Oxford OX5 1GB, United Kingdom

Copyright © 2017 Elsevier Inc. All rights reserved.

No part of this publication may be reproduced or transmitted in any form or by any means, electronic or mechanical, including photocopying, recording, or any information storage and retrieval system, without permission in writing from the publisher. Details on how to seek permission, further information about the Publisher's permissions policies and our arrangements with organizations such as the Copyright Clearance Center and the Copyright Licensing Agency, can be found at our website: www.elsevier.com/permissions.

This book and the individual contributions contained in it are protected under copyright by the Publisher (other than as may be noted herein).

Notices
Knowledge and best practice in this field are constantly changing. As new research and experience broaden our understanding, changes in research methods, professional practices, or medical treatment may become necessary.

Practitioners and researchers must always rely on their own experience and knowledge in evaluating and using any information, methods, compounds, or experiments described herein. In using such information or methods they should be mindful of their own safety and the safety of others, including parties for whom they have a professional responsibility.

To the fullest extent of the law, neither the Publisher nor the authors, contributors, or editors, assume any liability for any injury and/or damage to persons or property as a matter of products liability, negligence or otherwise, or from any use or operation of any methods, products, instructions, or ideas contained in the material herein.

Library of Congress Cataloging-in-Publication Data
A catalog record for this book is available from the Library of Congress

British Library Cataloguing-in-Publication Data
A catalogue record for this book is available from the British Library

ISBN: 978-0-12-802309-9

For information on all Academic Press publications
visit our website at https://www.elsevier.com/

Publisher: Nikki Levy
Acquisition Editor: Megan Ball
Editorial Project Manager: Jaclyn A. Truesdell
Production Project Manager: Susan Li
Designer: Mark Rogers

Typeset by TNQ Books and Journals

Contents

List of Contributors xxi
Preface xxv
Acknowledgments xxvii

Section 1
Introduction

1. **Fermented Foods in Health Promotion and Disease Prevention: An Overview**
   J.R. Wilburn and E.P. Ryan

   1.1 Introduction 3
   1.2 Types of Fermented Foods and Beverages 4
   1.3 Health Benefits of Fermented Foods and Beverages 7
       1.3.1 Bioactive Compounds 7
       1.3.2 Fermented Foods Targeting Chronic Disease Control and Prevention 8
   1.4 Food Safety and Quality Control 12
   1.5 Conclusions 14
       Acknowledgments 14
       References 15

Section 2
Fermented Foods as a Source of Healthy Constituents

2. **Bioactive Peptides in Fermented Foods: Production and Evidence for Health Effects**
   C. Martinez-Villaluenga, E. Peñas and J. Frias

   2.1 Introduction 23
   2.2 Occurrence of Bioactive Peptides in Fermented Foods 30
   2.3 Production of Bioactive Peptides in Fermented Foods 32
   2.4 Strategies to Increase the Production of Bioactive Peptides in Fermented Foods 33
   2.5 Evidence for Health Effects of Bioactive Peptides Derived From Fermented Foods 35
       2.5.1 Health Benefits on the Cardiovascular System 35
       2.5.2 Health Benefits on the Nervous System 37

		2.5.3	Health Benefits on the Gastrointestinal System	38

- 2.5.3 Health Benefits on the Gastrointestinal System — 38
- 2.5.4 Health Benefits on the Immune System — 38
- 2.5.5 Health Benefits on Adipose Tissue — 39
- 2.5.6 Health Benefits on Bones — 39
- 2.6 Future Outlook — 40
- Acknowledgments — 40
- References — 41

3. Health Benefits of Exopolysaccharides in Fermented Foods

K.M. Nampoothiri, D.J. Beena, D.S. Vasanthakumari and B. Ismail

- Abbreviations — 49
- 3.1 Introduction — 49
- 3.2 Role of Exopolysaccharides in the Food Industry — 53
- 3.3 Health Benefits of Exopolysaccharides — 54
  - 3.3.1 Cholesterol Lowering Effect — 56
  - 3.3.2 Immunomodulation — 56
  - 3.3.3 Antioxidant Properties — 57
  - 3.3.4 Anticancer Properties — 58
  - 3.3.5 Interactions with Enteric Pathogens — 58
- 3.4 Conclusion — 59
- Acknowledgments — 59
- References — 59

4. Biotransformation of Phenolics by *Lactobacillus plantarum* in Fermented Foods

R. Muñoz, B. de las Rivas, F. López de Felipe, I. Reverón, L. Santamaría, M. Esteban-Torres, J.A. Curiel, H. Rodríguez and J.M. Landete

- 4.1 Phenolic Compounds in Fermented Foods — 63
- 4.2 Transformation of Phenolic Compounds by Fermentation — 65
- 4.3 *Lactobacillus plantarum* as a Model Bacteria for the Fermentation of Plant Foods — 67
- 4.4 Biotransformation of Hydroxybenzoic Acid-Derived Compounds by *Lactobacillus plantarum* — 69
- 4.5 Biotransformation of Hydroxycinnamic Acid-Derived Compounds by *Lactobacillus plantarum* — 73
- 4.6 Health Benefits Induced by the Interaction of Phenolics With *Lactobacillus plantarum* — 78
- Acknowledgments — 79
- References — 79

5. Gamma-Aminobutyric Acid-Enriched Fermented Foods

J. Quílez and M. Diana

- 5.1 Introduction — 85
- 5.2 Physiological Functions — 86
- 5.3 Mechanisms of Action — 89

5.4	Gamma-Aminobutyric Acid Production by Lactic Acid Bacteria	90
5.5	Gamma-Aminobutyric Acid-Enriched Fermented Foods	91
	5.5.1 Cereal-based Products	91
	5.5.2 Dairy Products	94
	5.5.3 Meat, Vegetables, and Legumes	95
	5.5.4 Beverages	96
5.6	Side Effects and Toxicity of Gamma-Aminobutyric Acid	96
5.7	Future Trends	97
	References	97
	Further Reading	103

6. Melatonin Synthesis in Fermented Foods
M.A. Martín-Cabrejas, Y. Aguilera, V. Benítez and R.J. Reiter

6.1	Introduction	105
6.2	Structure and Physicochemical Properties	106
6.3	Biosynthesis of Melatonin	107
6.4	Mechanisms of Action	109
6.5	Health Benefits of Melatonin	110
	6.5.1 Antioxidant Effects	110
	6.5.2 Anticancer Effects	110
	6.5.3 Antiaging Effects	111
	6.5.4 Protection Against Cardiovascular Diseases	111
	6.5.5 Antiobesity Effects	112
	6.5.6 Protection Against Alzheimer's Disease	112
	6.5.7 Effects on Bone	112
	6.5.8 Protection Against UV Radiation (UVR)	112
	6.5.9 Migraine Prevention	113
6.6	Melatonin in Plant Foods	113
6.7	Fermented Foods	115
	6.7.1 Wine	115
	6.7.2 Beer	121
	6.7.3 Orange Juice	122
	6.7.4 Other Fermented Foods	123
6.8	Conclusion	123
	References	124

7. Effect of Fermentation on Vitamin Content in Food
B. Walther and A. Schmid

7.1	Introduction	131
7.2	Folate (Vitamin B9)	132
	7.2.1 Folate Content of Fermented Food	133
	7.2.2 Folate Production by Microorganisms	136
7.3	Vitamin K	138
	7.3.1 Vitamin K Content of Fermented Food	139
	7.3.2 Enhancing Vitamin K2 Content in Fermented Food	148

7.4	Riboflavin (Vitamin B2)	148
7.5	Vitamin B12	150
7.6	Other Vitamins	152
7.7	Conclusions	152
	References	152

8. From Bacterial Genomics to Human Health
A. Benítez-Páez and Y. Sanz

8.1	Introduction	159
8.2	Fermented Milk as a Source of Beneficial Bacteria	161
8.3	How Bacterial Genomics Help in the Identification of Probiotic Health Benefits	164
8.4	Gut Ecosystem as a New Source of Beneficial Bacteria	166
	Acknowledgments	168
	References	168

Section 3
Traditional Fermented Foods

Section 3.1
Fermented Food of Animal Origin

9. Fermented Seafood Products and Health
O. Martínez-Álvarez, M.E. López-Caballero, M.C. Gómez-Guillén and P. Montero

9.1	Introduction	177
	9.1.1 Fermented Seafood Products	177
	9.1.2 Fermented Seafood Products and Health	179
9.2	Fermented Fish and Microorganisms	180
	9.2.1 Use of Microorganisms as Starters	180
	9.2.2 Antimicrobial Activity of Microorganisms From Fermented Seafood Products	183
9.3	Biological Activity in Traditional Fermented Seafood	185
	9.3.1 Antioxidant Activity	186
	9.3.2 Antihypertensive Activity	188
	9.3.3 Anticoagulant and Fibrinolytic Activity	189
	9.3.4 Other Biological Activities	189
9.4	Health Risk	190
	9.4.1 Presence of Biogenic Amines and Strategies to Mitigate Their Presence	190
	9.4.2 Strategies to Decrease Salt Content	192
	9.4.3 Strategies to Remove Heavy Metals	193
	9.4.4 Parasite Control	194
9.5	Conclusion	194
	References	195

10. Fermented Meat Sausages
R. Bou, S. Cofrades and F. Jiménez-Colmenero

10.1	Introduction	203
10.2	Composition, Nutritional Value, and Health Implications	204
	10.2.1 Proteins, Peptides, Amino Acids, and Other Nitrogen Compounds	204
	10.2.2 Fat Content, Fatty Acid Composition, and Other Lipid Compounds	207
	10.2.3 Minerals	208
	10.2.4 Vitamins and Antioxidants	209
10.3	Strategies for Optimizing the Presence of Bioactive Compounds to Improve Health and Well-Being and/or Reduce Risk of Disease	209
	10.3.1 Bioactive Peptide Formation	210
	10.3.2 Improving Fat Content	210
	10.3.3 Probiotics	211
	10.3.4 Dietary Fiber Incorporation	214
	10.3.5 Mineral Addition	216
	10.3.6 Addition of Vitamins and Antioxidants	222
	10.3.7 Reduction of Biogenic Amine Formation	222
	10.3.8 Curing Agents: Nitrite Reduction	224
10.4	Conclusion	225
	Acknowledgments	226
	References	226

Section 3.2
Dairy Fermented Foods

11. Health Effects of Cheese Components with a Focus on Bioactive Peptides
I. López-Expósito, B. Miralles, L. Amigo and B. Hernández-Ledesma

11.1	Introduction	239
11.2	Effects of Cheese Fat on Health	240
11.3	Minerals in Cheese and Their Impact on Health	244
11.4	Biological Effects of Cheese Vitamins	246
11.5	Cheese Bioactive Proteins and Peptides	246
	11.5.1 Antihypertensive Peptides	247
	11.5.2 Antioxidant Peptides	253
	11.5.3 Effects of Cheese-Derived Peptides on the Nervous System	254
	11.5.4 Antiproliferative Peptides	255
	11.5.5 Antimicrobial Peptides	256
	11.5.6 Modulatory Peptides of Mineral Absorptions: Phosphopeptides	257
11.6	Future Prospects	264
	References	265

12. Blue Cheese: Microbiota and Fungal Metabolites
J.F. Martín and M. Coton

12.1	Preparation and Maturation of Blue Cheeses	275
	12.1.1 Blue Cheese Manufacture	275
12.2	Lactic Acid Bacteria in Blue Cheeses	286
	12.2.1 Production of Undesirable Compounds	287
12.3	Filamentous Fungi and Yeasts in Cheese: Interactions Between Species	289
	12.3.1 *Penicillium roqueforti* and Related Fungi in Blue Cheese	290
	12.3.2 *Penicillium roqueforti* Strains in Local Nonindustrial Blue Cheeses	291
	12.3.3 Common Secondary Metabolites in Cheese-Associated Fungi	292
12.4	Secondary Metabolites Produced by *Penicillium roqueforti*	293
	12.4.1 Roquefortines	293
	12.4.2 PR-Toxin and Eremofortins	294
	12.4.3 Andrastins	294
	12.4.4 Mycophenolic Acid	295
	12.4.5 Agroclavine and Festuclavine	295
	12.4.6 Other *Penicillium roqueforti* Metabolites	296
12.5	Conclusions and Future Outlook	296
	References	296

13. Yogurt and Health
M.A. Fernandez, É. Picard-Deland, M. Le Barz, N. Daniel and A. Marette

13.1	Yogurt Composition	305
	13.1.1 Introduction	305
	13.1.2 Nutrient Profile	306
	13.1.3 Microorganisms	308
	13.1.4 Proteins	308
	13.1.5 Lipids	309
	13.1.6 Carbohydrates	309
	13.1.7 Vitamins and Minerals	310
13.2	Bioactive Properties of Yogurt	310
	13.2.1 Effect on Immunity	311
	13.2.2 Modulation of Gut Microbiota	312
	13.2.3 Yogurt as a Probiotic Vector	313
	13.2.4 Effect on Cholesterol Metabolism	315
	13.2.5 Effect on Lactose Intolerance	316
	13.2.6 Effect on Transit Time/Digestion	317
	13.2.7 Effect on Mineral Absorption	317
13.3	Yogurt in Disease Prevention	318
	13.3.1 Diet Quality	318
	13.3.2 Yogurt and Weight Management	319
	13.3.3 Cardiometabolic Diseases	320

		13.4	Conclusions	326

13.4 Conclusions 326
References 327

14. Kefir
H. Kesenkaş, O. Gürsoy and H. Özbaş

- 14.1 Introduction 339
- 14.2 Kefir Grains 340
- 14.3 Kefir Production 342
- 14.4 Chemical Composition of Kefir 343
- 14.5 Nutritional Characteristics of Kefir 344
- 14.6 Health Benefits of Kefir 346
  - 14.6.1 Anticarcinogenic and Antimutagenic Effects 346
  - 14.6.2 Effects on Immune System 347
  - 14.6.3 Antiinflammatory Effects 348
  - 14.6.4 Hypocholesterolemic Effects 349
  - 14.6.5 Antihypertensive Effects 350
  - 14.6.6 Antimicrobial Activity 351
  - 14.6.7 Antidiabetic Effects 351
  - 14.6.8 Kefir and Lactose Intolerance 352
  - 14.6.9 Kefir and Osteoporosis 353
- 14.7 Conclusions 353
- References 354

Section 3.3
Legume and Cereal Grains Fermented Derived Products

15. Beer and Its Role in Human Health
M.L. González-SanJosé, P.M. Rodríguez and V. Valls-Bellés

- 15.1 Introduction: Brief Notes of Brewing 365
- 15.2 Bioactive Components of Beer 365
- 15.3 Antioxidant Properties of Beer and Health Effects 371
- 15.4 Cardiovascular Diseases and Beer 373
- 15.5 Antiosteoporosis Effect of Beer 374
- 15.6 Antimutagenic and Anticarcinogenic Effects of Beer 375
- 15.7 Beer and Hydration 376
- 15.8 Effects of Beer Supplementation in Breastfeeding Mothers 377
- 15.9 Other Health Effects of Beer 377
- 15.10 Concluding Remarks 378
- References 379

16. Fermented Pulses in Nutrition and Health Promotion
J. Frias, E. Peñas and C. Martinez-Villaluenga

- 16.1 Introduction 385
- 16.2 Nutritional and Phytochemical Composition of Pulses and Their Health Benefits 390

		16.2.1	Proteins	390

16.2.1 Proteins 390
16.2.2 Carbohydrates 394
16.2.3 Lipids 395
16.2.4 Dietary Fiber 395
16.2.5 Minerals and Vitamins 395
16.2.6 Phytochemicals 396
16.3 Nutritional Changes During Fermentation of Pulse-Based Foods 397
    16.3.1 Changes in Protein and Amino Acids During Fermentation 397
    16.3.2 Changes in Starch, Carbohydrates, and Dietary Fiber During Fermentation 398
    16.3.3 Changes in Lipids During Fermentation 399
    16.3.4 Changes in Phytic Acid and Mineral Bioavailability During Fermentation 400
    16.3.5 Changes in Vitamins During Fermentation 400
    16.3.6 Changes in Phenolic Compounds and Antioxidant Activity During Fermentation 401
    16.3.7 Changes in Other Minor Bioactive Compounds During Fermentation 401
16.4 Role of Fermented Pulse Foods in Health Promotion 402
    16.4.1 Fermented Pulse Products and Weight Management 402
    16.4.2 Fermented Pulse-Products and Diabetes 403
    16.4.3 Fermented Pulse-Derived Products and Cardiovascular Diseases 403
    16.4.4 Fermented Pulse-Derived Products and Cancer 404
    16.4.5 Fermented Pulse-Derived Products in Healthy Aging and Stress 405
    16.4.6 Fermented Pulse-Derived Products as Probiotic Vehicle 405
16.5 Final Remarks 406
Acknowledgments 406
References 406

17. Nonwheat Cereal-Fermented-Derived Products

Z. Ciesarová, L. Mikušová, M. Magala, Z. Kohajdová and J. Karovičová

17.1 Introduction 417
17.2 Nutritional Aspects of Nonwheat Cereals 418
17.3 Advantages and Limitations of Fermentation Applied to Nonwheat Cereals 422
    17.3.1 Degradation of Antinutritive Compounds and Simultaneous Enhancement of Minerals Concentration 423
    17.3.2 Improvement of Nutritional Properties, Vitamins Bioavailability, Palatability, Sensorial Acceptation, and Textural Properties 423

	17.4	Fermented Nonwheat Food Products	424
		17.4.1 Fermented Nonwheat Beverages	425
		17.4.2 Fermented Nonwheat Bakery Products	426
	17.5	Health Beneficial Effects of Nonwheat Fermented Foods	427
	17.6	Conclusion	428
		Acknowledgments	428
		References	429

18. Use of Sourdough Fermentation and Nonwheat Flours for Enhancing Nutritional and Healthy Properties of Wheat-Based Foods

C.G. Rizzello, R. Coda and M. Gobbetti

	18.1	Background	433
	18.2	Use of Legumes, Minor Cereal, Pseudo-Cereal Flours, and Sourdough Fermentation for Enhancing Nutritional and Functional Properties of Wheat-Based Foods	435
		18.2.1 Legume Flours and Sourdough Fermentation	435
		18.2.2 Legume Flours in Sourdough Wheat Bread Making	439
		18.2.3 Minor Cereals and Pseudo-Cereals in Sourdough Wheat Bread Making	441
		18.2.4 Nonwheat Flours in Sourdough Gluten-Free Bread Making	442
	18.3	Use of Wheat Milling Byproducts and Sourdough Fermentation for Enhancing Nutritional and Healthy-related Properties of Wheat-Based Foods	443
	18.4	Conclusions	446
		References	446

19. *Tempeh* and Other Fermented Soybean Products Rich in Isoflavones

V. Mani and L.C. Ming

	19.1	Introduction	453
	19.2	Soybean	454
	19.3	*Tempeh*	456
		19.3.1 Nutritional Enhancements of *Tempeh*	456
		19.3.2 Antioxidant Roles of *Tempeh*	458
		19.3.3 *Tempeh* and Dementia	459
		19.3.4 *Tempeh* and Hypocholesterolemia	461
		19.3.5 *Tempeh* and Cancer Prevention	462
	19.4	Other Fermented Soybean Products With Health Benefits	464
		19.4.1 Cheonggukjang	464
		19.4.2 Doenjang	465
		19.4.3 Doubanjiang	465
		19.4.4 Douchi	465
		19.4.5 Gochujang	466

		19.4.6	Miso	466
		19.4.7	Natto	466
		19.4.8	Soy Sauce	466
		19.4.9	Tauchu (Tauco)	467
	19.5	Conclusion		467
		References		467

Section 3.4
Vegetables and Fruits Fermented Products

20. Kimchi and Its Health Benefits
K.-Y. Park, H.-Y. Kim and J.-K. Jeong

	20.1	Introduction		477
	20.2	History of Kimchi		478
	20.3	Manufacturing Kimchi		479
	20.4	Fermentation of Kimchi		480
	20.5	Health Benefits of Kimchi		482
		20.5.1	Antioxidative and Antiaging Effects	482
		20.5.2	Antimutagenic and Anticancer Effects	484
		20.5.3	Antiobesity Effects	487
		20.5.4	Other Health Benefits	489
	20.6	Safety of Kimchi		491
	20.7	Health Benefits of Kimchi LAB		492
		20.7.1	Antioxidative and Anticancer Effects	492
		20.7.2	Immune-Boosting Effects	493
		20.7.3	Antiobesity Effects	494
		20.7.4	Other Effects	494
	20.8	Conclusion		496
		References		496

21. The Naturally Fermented Sour Pickled Cucumbers
H. Zieliński, M. Surma and D. Zielińska

	21.1	Origin of Cucumber		503
	21.2	Chemical Composition and Bioactive Compounds in Cucumber		503
	21.3	Lactic Acid Fermentation of Cucumbers		504
	21.4	Factors Affecting Cucumber Fermentation		507
		21.4.1	Effects of Cucumber Size and Enzyme Activity	507
		21.4.2	Effect of Salt Concentration	507
		21.4.3	Effect of Sugar Addition	509
		21.4.4	Effect of Spice or Aromatic Herb Addition	509
		21.4.5	Effect of Chemical Preservative Addition	509
		21.4.6	Biopreservation	510
	21.5	Health Benefits of Fermented Cucumbers		510
		21.5.1	Lactic Acid Bacteria (LAB) Isolated From Fermented Cucumbers as Probiotics	510
		21.5.2	Production of Bacteriocins	511

		21.5.3	Fermented Cucumber as a Source of Starter Cultures Able to Produce Oligosaccharides	512
		21.5.4	Immune Modulation	512
		21.5.5	Macroelements From Fermented Cucumber	512
		21.5.6	Removal of Antinutrient Compounds	512
	21.6	Final Remarks		513
		References		513

22. Role of Natural Fermented Olives in Health and Disease
C.M. Peres, C. Peres and F. Xavier Malcata

22.1	General Considerations	517
22.2	Production of Traditional Fermented Olives	518
22.3	Olive Fermentation	519
22.4	Lactic Acid Bacteria of Olive Fermentation as Probiotics	520
22.5	Health Effects of Olive Fermentation Probiotics	523
22.6	Healthy Bioactive Molecules and Metabolites From Fermented Olives	527
22.7	Fermented Olives Can Modulate the Digestive Microbiota	531
22.8	Future Trends	533
	Acknowledgments	534
	References	534

23. Pulque
J.A. Gutiérrez-Uribe, L.M. Figueroa, S.T. Martín-del-Campo and A. Escalante

23.1	Introduction	543
23.2	Bioactive Constituents of Aguamiel	544
23.3	Bioactive Constituents of Pulque	546
23.4	Microorganisms in Aguamiel and Pulque	547
23.5	Conclusions	554
	References	554

24. Sauerkraut: Production, Composition, and Health Benefits
E. Peñas, C. Martinez-Villaluenga and J. Frias

	24.1	Brief History of Sauerkraut		557
	24.2	Sauerkraut Manufacture		557
	24.3	Microbial Changes During Spontaneous Sauerkraut Fermentation		559
	24.4	Inoculation of Starter Cultures During Sauerkraut Manufacture		559
	24.5	Nutritional and Phytochemical Composition of Sauerkraut		561
	24.6	Health Benefits of Sauerkraut		564
		24.6.1	Antioxidant Benefits	564
		24.6.2	Anticarcinogenic Properties	565
		24.6.3	Protection Against Oxidative DNA Damage	567

		24.6.4	Antiinflammatory Effects	568
		24.6.5	Sauerkraut as a Source of Probiotic Bacteria	568
	24.7	Concluding Remarks		569
		Acknowledgments		569
		References		569

25. Vinegars and Other Fermented Condiments

M.C. Garcia-Parrilla, M.J. Torija, A. Mas, A.B. Cerezo and A.M. Troncoso

25.1	Introduction/General Overview		577
25.2	Antimicrobial Effects		580
25.3	Bioactive Compounds and Antioxidant Activity		581
	25.3.1	Phenolic Compounds	581
	25.3.2	Phenolic Composition of Vinegars	582
	25.3.3	Antioxidant Properties of Vinegars	583
25.4	Health Effects		583
	25.4.1	Effect on Glucose Metabolism and Insulin Resistance	584
	25.4.2	Effect on Lipid Metabolism	585
	25.4.3	Other Effects and Safety Issues	586
25.5	Other Fermented Condiments		587
	References		587

26. Wine

I. Fernandes, R. Pérez-Gregorio, S. Soares, N. Mateus and V. De Freitas

26.1	Preamble		593
26.2	Disease-Protective/Preventive Effect of Wine or Its Phenolic Compounds		594
	26.2.1	Cardiovascular Diseases	597
	26.2.2	Cancer	598
	26.2.3	Metabolic Diseases	599
	26.2.4	Neurological Diseases	599
	26.2.5	Osteoporosis	601
	26.2.6	Wine and Mortality: Impact on Longevity	601
26.3	Health-Promoting Activity of Polyphenols Resulting From Their Interaction With Biological Proteins		602
	26.3.1	Phenolic Compound Journey Starts in the Oral Cavity	602
	26.3.2	Interaction With Digestive Enzymes	603
	26.3.3	Interaction With Serum Proteins	604
	26.3.4	Interaction With Platelets	605
	26.3.5	Interaction With Neurotoxic Proteins	605
	26.3.6	Interaction With Allergy Proteins	606
26.4	Bioavailability of Red Wine		606
	26.4.1	Absorption and Metabolism of Red Wine Anthocyanins and Derivatives	607
	26.4.2	Absorption of Catechins and Proanthocyanidins	609

	26.4.3	Absorption and Metabolism of Resveratrol	610
	26.4.4	Microbiota Impact on Red Wine Availability and Bioactivity	610
	References		612

Section 4
Hazardous Compounds and Their Implications in Fermented Foods

27. Biogenic Amines in Fermented Foods and Health Implications
L. Simon Sarkadi

27.1	Classification, Biosynthesis, and Metabolism of Biogenic Amines	625
27.2	Microbial Production of Biogenic Amines in Foods	627
27.3	Factors Affecting Biogenic Amine Content in Fermented Foods	628
	27.3.1 Dairy Products	628
	27.3.2 Fermented Fish Products	631
	27.3.3 Meat and Meat Products	632
	27.3.4 Alcoholic Beverages	633
	27.3.5 Fermented Plant Products	636
	27.3.6 Summary on Occurrence Data on the Major Biogenic Amines in Food and Beverages	638
27.4	Biogenic Amines and Human Health	638
27.5	Reduction of Biogenic Amines in Fermented Food	640
27.6	Conclusions	642
	References	642

28. Occurrence of Aflatoxins in Fermented Food Products
S. Shukla, D.-H. Kim, S.H. Chung and M. Kim

28.1	Introduction	653
28.2	Structures of Aflatoxins	654
28.3	Occurrence of Aflatoxins in Different Fermented Foods	656
	28.3.1 Aflatoxins in Soybean Fermented Food Products	656
	28.3.2 Aflatoxins in Dairy Fermented Food Products	657
	28.3.3 Aflatoxins in Alcoholic Beverage Products	658
28.4	Various Methods to Control Aflatoxins	659
	28.4.1 Physical Methods to Control Aflatoxins in Fermented Foods	659
	28.4.2 Biochemical Methods to Control Aflatoxins in Fermented Foods	659
	28.4.3 Microbiological Methods to Control Aflatoxins in Fermented Foods	660
28.5	Legal Validation for Aflatoxin Limits	662

28.6 General Detection Methods to Analyze Aflatoxins
in Fermented Foods 663
28.7 Rapid Detection Methods to Analyze Aflatoxins
in Fermented Foods 663
  28.7.1 Polymerase Chain Reaction (PCR)-Based Molecular
  Detection Methods 664
  28.7.2 Enzyme-Linked Immunosorbent Assay (ELISA)-Based
  Detection Methods 664
  28.7.3 Biosensor-Based Detection Methods 666
  28.7.4 Immunochromatographic Test Strip Detection Methods 666
28.8 Conclusions 667
Acknowledgments 667
References 668

29. Antibiotic Resistance Profile of Microbes From Traditional Fermented Foods
H. Abriouel, C.W. Knapp, A. Gálvez and N. Benomar

29.1 Overview of History of Antibiotic Resistance 675
29.2 Antibiotic Resistance in Traditional Fermented Foods 677
  29.2.1 Phenotypic and Genotypic Antibiotic Resistance
  Profile of Lactic Acid Bacteria 677
  29.2.2 Antibiotic Resistance Profile of Pathogens 690
29.3 New Insights Into Antibiotic Resistance 694
29.4 Conclusions 697
Acknowledgments 698
References 698

Section 5
Revalorization of Food Wastes by Fermentation into Derived Outcomes

30. Fermentation of Food Wastes for Generation of Nutraceuticals and Supplements
S. Patel and S. Shukla

30.1 Introduction 707
30.2 Functional Enhancement of Wastes by Fermentation 709
  30.2.1 Agro-Wastes 709
  30.2.2 Fruit Wastes 718
  30.2.3 Dairy Wastes 721
  30.2.4 Malting and Brewing Wastes 722
  30.2.5 Bakery Waste 722
  30.2.6 Fishery Byproducts 723
30.3 Food Wastes as Culture Medium of Functional Foods 724
30.4 Valorization of Underutilized Food Sources Through
Fermentation 724

30.5	Possible Harms and Hurdles	725
30.6	Future Directions and Barricades to Surmount	725
30.7	Conclusions	726
	References	727

Index 735

List of Contributors

H. Abriouel University of Jaén, Jaén, Spain

Y. Aguilera University Autónoma de Madrid (CIAL, CSIC-UAM, CEI UAM+CSIC), Madrid, Spain

L. Amigo Instituto de Investigación en Ciencias de la Alimentación (CSIC), Madrid, Spain

D.J. Beena National Institute for Interdisciplinary Science and Technology (NIIST), Trivandrum, Kerala, India

A. Benítez-Páez Institute of Agrochemistry and Food Technology, Spanish National Research Council (IATA-CSIC), Paterna-Valencia, Spain

V. Benítez University Autónoma de Madrid (CIAL, CSIC-UAM, CEI UAM+CSIC), Madrid, Spain

N. Benomar University of Jaén, Jaén, Spain

R. Bou Institut de Recerca i Tecnologia Agroalimentàries (IRTA), Mollens, Gerona, Spain

A.B. Cerezo University of Sevilla, Spain

S.H. Chung Korea University, Seoul, Republic of Korea

Z. Ciesarová NPPC National Agricultural and Food Centre, Food Research Institute, Bratislava, Slovak Republic

R. Coda University of Helsinki, Helsinki, Finland

S. Cofrades Institute of Food Science, Technology and Nutrition (ICTAN-CSIC), Madrid, Spain

M. Coton University of Brest, Plouzané, France

J.A. Curiel Institute of Food Science, Technology and Nutrition (ICTAN-CSIC), Madrid, Spain

N. Daniel Laval University, Québec, QC, Canada

V. De Freitas University of Porto, Porto, Portugal

B. de las Rivas Institute of Food Science, Technology and Nutrition (ICTAN-CSIC), Madrid, Spain

M. Diana Europastry S.A., Sant Cugat del Vallès, Barcelona, Spain

A. Escalante National Autonomous University of Mexico, Cuernavaca, Mexico

M. Esteban-Torres Institute of Food Science, Technology and Nutrition (ICTAN-CSIC), Madrid, Spain

I. Fernandes University of Porto, Porto, Portugal

M.A. Fernandez Laval University, Québec, QC, Canada

L.M. Figueroa Tecnologico de Monterrey, Campus Monterrey, Monterrey, Mexico

J. Frias Institute of Food Science, Technology and Nutrition (ICTAN-CSIC), Madrid, Spain

A. Gálvez University of Jaén, Jaén, Spain

M.C. Garcia-Parrilla University of Sevilla, Sevilla, Spain

M. Gobbetti University of Bari Aldo Moro, Bari, Italy

M.C. Gómez-Guillén Institute of Food Science, Technology and Nutrition (ICTAN-CSIC), Madrid, Spain

M.L. González-SanJosé University of Burgos, Burgos, Spain

O. Gürsoy Mehmet Akif Ersoy University, Burdur, Turkey

J.A. Gutiérrez-Uribe Tecnologico de Monterrey, Campus Monterrey, Monterrey, Mexico

B. Hernández-Ledesma Instituto de Investigación en Ciencias de la Alimentación (CSIC), Madrid, Spain

B. Ismail National Institute for Interdisciplinary Science and Technology (NIIST), Trivandrum, Kerala, India

J.-K. Jeong Pusan National University, Busan, Korea

F. Jiménez-Colmenero Institute of Food Science, Technology and Nutrition (ICTAN-CSIC), Madrid, Spain

J. Karovičová Slovak University of Technology, Bratislava, Slovak Republic

H. Kesenkaş Ege University, Izmir, Turkey

D.-H. Kim National Agricultural Products Quality Management Service, Gimcheon, Republic of Korea

H.-Y. Kim Pusan National University, Busan, Korea

M. Kim Yeungnam University, Gyeongsan, Republic of Korea

C.W. Knapp University of Strathclyde, Glasgow, Scotland, United Kingdom

Z. Kohajdová Slovak University of Technology, Bratislava, Slovak Republic

J.M. Landete Institute of Food Science, Technology and Nutrition (ICTAN-CSIC), Madrid, Spain

M. Le Barz Laval University, Québec, QC, Canada

F. López de Felipe Institute of Food Science, Technology and Nutrition (ICTAN-CSIC), Madrid, Spain

M.E. López-Caballero Institute of Food Science, Technology and Nutrition (ICTAN-CSIC), Madrid, Spain

I. López-Expósito Instituto de Investigación en Ciencias de la Alimentación (CSIC), Madrid, Spain

M. Magala Slovak University of Technology, Bratislava, Slovak Republic

List of Contributors **xxiii**

V. Mani Universiti Teknologi MARA (UiTM), Selangor, Malaysia; Qassim University, Buraidah, Kingdom of Saudi Arabia

A. Marette Laval University, Québec, QC, Canada

M.A. Martín-Cabrejas University Autónoma de Madrid (CIAL, CSIC-UAM, CEI UAM+CSIC), Madrid, Spain

S.T. Martín-del-Campo Tecnologico de Monterrey, Campus Querétaro, Querétaro, Mexico

O. Martínez-Álvarez Institute of Food Science, Technology and Nutrition (ICTAN-CSIC), Madrid, Spain

C. Martinez-Villaluenga Institute of Food Science, Technology and Nutrition (ICTAN-CSIC), Madrid, Spain

J.F. Martín University of León, León, Spain

A. Mas Universitat Rovira i Virgili, Tarragona, Spain

N. Mateus University of Porto, Porto, Portugal

L. Mikušová Slovak University of Technology, Bratislava, Slovak Republic

L.C. Ming University of Tasmania, Hobart, TAS, Australia

B. Miralles Instituto de Investigación en Ciencias de la Alimentación (CSIC), Madrid, Spain

P. Montero Institute of Food Science, Technology and Nutrition (ICTAN-CSIC), Madrid, Spain

R. Muñoz Institute of Food Science, Technology and Nutrition (ICTAN-CSIC), Madrid, Spain

K.M. Nampoothiri National Institute for Interdisciplinary Science and Technology, (NIIST), Trivandrum, Kerala, India

H. Özbaş Suleyman Demirel University, Isparta, Turkey

K.-Y. Park Pusan National University, Busan, Korea

S. Patel San Diego State University, San Diego, CA, United States

E. Peñas Institute of Food Science, Technology and Nutrition (ICTAN-CSIC), Madrid, Spain

C. Peres INIAV, IP, Oeiras, Portugal

C.M. Peres Instituto de Tecnologia Química e Biológica, Universidade Nova de Lisboa, Oeiras, Portugal

R. Pérez-Gregorio University of Porto, Porto, Portugal

É. Picard-Deland Laval University, Québec, QC, Canada

J. Quílez Europastry S.A., Sant Cugat del Vallès, Barcelona, Spain

R.J. Reiter UT Health Science Center, San Antonio, TX, United States

I. Reverón Institute of Food Science, Technology and Nutrition (ICTAN-CSIC), Madrid, Spain

C.G. Rizzello University of Bari Aldo Moro, Bari, Italy

H. Rodríguez Institute of Food Science, Technology and Nutrition (ICTAN-CSIC), Madrid, Spain

P.M. Rodríguez University of Burgos, Burgos, Spain

E.P. Ryan Colorado State University, Fort Collins, CO, United States

L. Santamaría Institute of Food Science, Technology and Nutrition (ICTAN-CSIC), Madrid, Spain

Y. Sanz Institute of Agrochemistry and Food Technology, Spanish National Research Council (IATA-CSIC), Paterna-Valencia, Spain

A. Schmid Agroscope, Institute for Food Sciences (IFS), Bern, Switzerland

S. Shukla Yeungnam University, Gyeongsan, Republic of Korea

L. Simon Sarkadi Szent István University, Budapest, Hungary

S. Soares University of Porto, Porto, Portugal

M. Surma University of Agriculture in Krakow, Krakow, Poland

M.J. Torija Universitat Rovira i Virgili, Tarragona, Spain

A.M. Troncoso University of Sevilla, Sevilla, Spain

V. Valls-Bellés University of Jaume I, Castellón de la Plana, Spain

D.S. Vasanthakumari National Institute for Interdisciplinary Science and Technology (NIIST), Trivandrum, Kerala, India

B. Walther Agroscope, Institute for Food Sciences (IFS), Bern, Switzerland

J.R. Wilburn Colorado State University, Fort Collins, CO, United States

F. Xavier Malcata University of Porto, Porto, Portugal; LEPABE, Laboratory for Engineering of Processes, Environment, Biotechnology and Energy, Porto, Portugal

D. Zielińska University of Warmia and Mazury in Olsztyn, Olsztyn, Poland

H. Zieliński Institute of Animal Reproduction and Food Research of the Polish Academy of Sciences, Olsztyn, Poland

Preface

Consumer preferences toward healthy food choices have driven the development of functional foods by contemporary food markets. Fermentation was historically used to preserve food, but today it is gaining tremendous attention since it provides healthy food products that go beyond basic nutrition and sensory properties. Existing scientific data show that many fermented foods have both nutritive and nonnutritive compounds that have the potential to modulate specific target functions in the body, which contribute to the health and well-being of consumers. Naturally fermented foods and beverages contain beneficial microorganisms that transform the chemical constituents of raw materials from animal/plant origin during food fermentation, thereby improving the sensory, nutritional, and health-promoting properties of the food. In particular, microorganisms in fermented foods facilitate the bioavailability of nutrients, enrich the food with bioactive compounds, confer probiotic functions, degrade toxic components and antinutritive factors, and impart biopreservative effects by producing antioxidant and antimicrobial compounds.

This book aims to provide a comprehensive overview of the nutritional, health, and safety aspects of fermented foods. These aspects include the microbial synthesis and bioavailability of nutrients and bioactive compounds, synthesis/degradation of nonnutritive and toxic compounds as well as an up-to-date compilation of studies supporting health benefits/risks associated with consumption of fermented foods. There is also a chapter that looks at the sustainable-related aspects of food waste fermentation for the production of nutraceuticals and food supplements. The core structure of the book is made up of five main sections dealing with (1) a general introduction on fermented foods and their health implications, (2) fermented foods as a source of healthy constituents, (3) traditional fermented foods from all around the world, (4) hazardous compounds and their health implications in fermented foods, and (5) revalorization of food wastes by fermentation into derived outcomes. The book covers a broad range of bioactive constituents (from peptides, exopolysaccharides, phenolic compounds, γ-aminobutyric acid, melatonin, and vitamins to probiotics) and fermented foods (from ethnic to novel products) with an emphasis on the scientific evidence of the benefits related to health promotion and disease prevention. These contributions are expected to lead to the development of new fermented foods with human health benefits and considerable market potential.

A multidisciplinary board of experts and leaders from academia, research institutions, hospitals, and food industries contributed to different areas of knowledge to this book including microbiology, molecular biology, biotechnology, food chemistry, food technology, food engineering, and nutrition to cover the current scientific evidence that supports the health effects of fermented foods. There is increased demand for natural, nutritive, and healthy food choices by consumers, and fermentation provides a wide array of bioactive constituents that contribute to several physiological functions. However, demonstration of their health benefits by human intervention trials is still very challenging despite the important progress made in the past few years.

Acknowledgments

This book would not have been possible without the admirable effort of international renowned contributors, and we would like to greatly express our gratitude for their expertise and time. We also acknowledge the Elsevier editorial team for their prompt assistance, advice, and dynamic communication.

Section 1

Introduction

Chapter 1

Fermented Foods in Health Promotion and Disease Prevention: An Overview

J.R. Wilburn, E.P. Ryan
Colorado State University, Fort Collins, CO, United States

1.1 INTRODUCTION

The word *fermentation* stems from the Latin *fermentare*, defined as "to leaven." Although the practice of food fermentation is rooted in myriad of cultural norms throughout the world, fermentation has expanded from the household to commercialized and industrial scale production systems intended for the mass marketplace. One broad definition for the practice of fermentation states that it is "the transformation of food by various bacteria, fungi, and the enzymes they produce" (Katz, 2012). It is worth noting that, in addition to the definition laid out by Katz, fermentation processes can, and do, utilize yeast for the transformation of foods and beverages (Giraffa, 2004). Fermented foods and beverages have been enzymatically altered by microorganisms in a manner that still tastes and smells desirable to humans (Steinkraus, 1997). Beyond preservation, fermentation is widely utilized to improve food palatability (Azokpota, 2015), and for purposes of producing unique and new variations of foods (Rodgers, 2008). The types of foods commonly prepared through fermentation, as well as the specific fermentation practices utilized, vary from one culture to the next, driven by availability of food sources, taste preferences, environmental conditions, raw materials, and, ultimately, new technological development (Prajapati and Nair, 2008). Fermentation can be strongly represented in the final end product, as in kimchi and kombucha, or be merely a step in the preparation, such as with chocolate and coffee. In both cases, fermentation is an important step in the development that can be responsible for creating flavor and scent patterns, as well as health properties unique to each finished product. From preservation of foods and beverages to preparation of a product for immediate consumption, fermentation is globally an important process of food and beverage production.

Fermented foods have long been thought to provide health benefits to the consumer. Some of the potential health benefits of fermented foods that have been explored in the recent past were based on an extensive body of anecdotal information, and included such benefits as: antihypertensive activity (Ferreira et al., 2007; Nakamura et al., 2013; Koyama et al., 2014; Ahren et al., 2014), blood glucose-lowering benefits (Kamiya et al., 2013; Oh et al., 2014), antidiarrheal (Kamiya et al., 2013; Parvez et al., 2006), and antithrombotic properties (Kamiya et al., 2013). The comprehensive evaluation of fermented food contents and how they may provide health benefits has led to the targeted identification of certain vitamins, minerals, amino acids, and phytochemicals (eg, phenolics, fatty acids, and saccharides) that distinguish fermented foods from their nonfermented forms (Rodgers, 2008; Rodriguez et al., 2009; Capozzi et al., 2012; Sheih et al., 2014; Xu et al., 2015). Furthermore, the evidence for bioactive components resulting from fermentation of plants and animal products is rapidly increasing with the application of new technologies, such as metabolomics (Lee et al., 2009; Yang et al., 2009; Kim et al., 2012; Liu et al., 2014). The use of high throughput, multi-omic approaches in the last decade has allowed for a more sophisticated level of microbial, gene, protein, and metabolite discovery that has potential for integration with both fermented food composition and functional assessments following ingestion. Notably, the concentrations of bioactive compounds available from fermented products may be dependent on the geographic regions from which the starting product was produced, genetic strains of bacteria utilized, the availability of specific substrates in the fermentation process, the environmental conditions, such as seasonality, and method of preparation or manufacturing process (Nikolopoulou et al., 2006; Starr et al., 2015). Addressing these variables is a major hurdle for credentialing the specific health properties associated with fermented foods. In this overview, we will distinguish between studies designed to utilize fermented food products in the form of "extracts" and with the ultimate intention of nutraceutical and/or dietary supplement utility and those attempting to ascribe health benefits to consumption of the whole fermented food or beverage. The emergence of patented, as well as advanced regulatory approaches for the commercial production of fermented foods, has been applied in an effort to achieve greater knowledge on health outcomes (reviewed in Borresen et al., 2012). It has been a challenge to evaluate traditional household use of fermented foods that date back centuries, yet it can and should be argued that this historical dietary information will be valuable for a complete understanding of the role for fermented foods and beverages in animal and human health, and even possibly for achieving food and nutritional security globally (Quave and Pieroni, 2014).

1.2 TYPES OF FERMENTED FOODS AND BEVERAGES

Fermented foods and beverages may vary by food type, timing of the fermentation process, and the intentional application of microbes utilized (Table 1.1). For example, with beer, the fermentation is intentional, and typically involves a specific yeast to create and enhance a desired flavor (Grassi et al., 2014). In the case of kimchi

TABLE 1.1 Common Fermented Foods

Fermentation in the Preparation Phase	References
Coffee	Crafack et al. (2014) and Lee et al. (2015)
Chocolate	Crafack et al. (2014)
Fermented tea leaves (Fu Zhuan and Pu-erh)	Keller et al. (2013), Lv et al. (2013), and Wang et al. (2011)
Sourdough bread	Corsetti and Luca (2007), Hansen and Schieberle (2005), and Meignen et al. (2001)
Fermentation of the Final Food Product	
Soybean products	Azokpota (2015), Bamworth and Ward (2014), and Wolff (2000)
Yogurt	Sodini et al. (2004) and Soukoulis et al. (2007)
Kimchi	Cho et al. (2006) and Mheen and Kwon (1984)
Kombucha	Chu and Chen (2006), Jayabalan et al. (2014), Sreeramulu et al. (2000), and Teoh et al. (2004)
Beer	Grassi et al. (2014)
Wine	Lee et al. (2009)

and sauerkraut, fermentation may be allowed to happen under specific environmental conditions, but without the intentional application of microbes (Cho et al., 2006). Fermentation is utilized as a starting process for products that will undergo multiple additional steps once the fermentation has been terminated, such as with coffee, chocolate, tea leaves, and sourdough breads (Corsetti and Luca, 2007; Crafack et al., 2014; Keller et al., 2013; Lee et al., 2015). Fermentation can also stand strong to signify the final product, as in the case of vinegar-tasting kombucha (Teoh et al., 2004) or pungent-smelling blue cheese (Nelson, 1970).

Coffee and chocolate represent common foods that utilize fermentation in the preparation phase, however, the final product is not typically considered a fermented beverage or food. Cocoa beans, from which chocolate is made, must first be fermented prior to processing and additions of sugar and various other flavors and ingredients that will ultimately produce the chocolate bar sold in stores (Crafack et al., 2014). Fermentation is allowed to occur naturally in humid environments, with both yeast and lactic acid bacteria (LAB) driving the conversion of the pulp surrounding the cocoa bean into ethanol and acetic acid (Crafack et al., 2014). The degree of acidification and the time for which the beans are allowed to ferment influence all the types of free amino acids, peptides, and sugars formed during the fermentation process, which are ultimately converted by Maillard reactions during the drying and roasting processes into the varying flavor and smell profiles that define the chocolate we know and eat (Crafack et al., 2014). Unlike chocolate, where the fermentation process is widely accepted as influential on the flavor and olfactory profile (Crafack et al., 2014), fermentation of raw coffee beans has traditionally been thought of as merely a method for removal of the mucilage to prepare the bean for roasting (Lee et al., 2015). Depending on whether the coffee beans are Arabica or Robusta, they are respectively fermented by wet or dry methods (Lee et al., 2015). Recent research indicates that the method of fermentation utilized is partially responsible for the difference in aroma quality of the coffee bean (Lee et al., 2015). To our knowledge, there has been limited attention to variation by fermentation during processing when investigating the health benefits of chocolate or coffee (Halliwell, 2003; George et al., 2008; Franco et al., 2013).

Fu Zhuan tea and Pu-erh tea are two examples of tea leaves that, after drying, are fermented with solid state fermentation methods to create a unique flavor profile (Wang et al., 2011; Keller et al., 2013; Lv et al., 2013). For dark teas, *Camellia sinensis* leaves are fungal fermented for several weeks from partially sun-dried leaves, producing a mellow flavor profile (Wang et al., 2011; Keller et al., 2013; Lv et al., 2013). *Aspergillus* spp. are amongst the predominant microorganisms involved in the fermentation of dark teas, though by no means the only types (Lv et al., 2013). Following fermentation, these dark teas are steeped in near-boiling water for a recommended period of several minutes prior to serving. As opposed to dark teas, where the tea leaves are fermented prior to tea brewing (Lv et al., 2013), kombucha is produced through the fermentation of already brewed tea, combined with sucrose and a gelatinous starter culture consisting of various bacteria and yeast (Sreeramulu et al., 2000; Chu and Chen, 2006; Jayabalan et al., 2014). The microbial population of the so-called "tea fungus," does not actually include any fungi, but is composed of *Acetobacter* spp. and various yeasts, such as *Brettanomyces* and *Saccharomyces* (Chu and Chen, 2006). The sugared tea is fermented in a dark and cool space until a desired sour acidic taste has been achieved, similar to that of vinegar, at which point it is bottled and may be combined with a variety of different flavors (Jayabalan et al., 2014; Sreeramulu et al., 2000).

Some food products that have not been traditionally prepared in a fermented form are finding new life through a fermented counterpart. Soybean "milk" is traditionally common in Southeast Asia and China, and has made its way into contemporary grocery stores, and utilized as an alternative to dairy milk. There has been increasing demand for and consumption of fermented soybean "milk" products, such as soybean "milk" yogurt (Bamworth and Ward, 2014). Soybeans have a rich history of preparation by fermentation in Asia (Wolff, 2000) as they are indigestible by humans in their raw form (Azokpota, 2015), and have been traditionally prepared through fermentation in order to increase their digestibility and palatability (Azokpota, 2015). These fermented food forms range from tofu to miso and many other food products in between (Wolff, 2000). The final form in which fermented soybean products are consumed and evaluated for health seems to be typically documented with clarity when compared to other fermented foods and therefore has potential to provide more mechanistic insights.

Yogurt, a popular dairy- and nondairy-derived food product, is consumed by many cultures, and results from the fermentation of the starting milk (Sodini et al., 2004; Soukoulis et al., 2007). Although different countries and cultures employ slightly different processes to ferment yogurt, the process typically involves LAB transforming milk in the presence of a small degree of heat (Sodini et al., 2004; Soukoulis et al., 2007). What differs is the application of various thickeners, the utilization of different methods of draining excess liquid, and the heat at which the yogurt is fermented (Sodini et al., 2004; Soukoulis et al., 2007).

Kimchi is a traditional spicy fermented cabbage dish in Asian culture (Cho et al., 2006). It is produced through the combination of cabbage, garlic, ginger, and hot red peppers with salt to draw out the natural fluids within the vegetables (Mheen and Kwon, 1984; Cho et al., 2006). Brine is often supplemented in the preparation, and the combination is set to ferment through the naturally occurring microbial population in a cool and anaerobic environment until appropriately fermented (Mheen and Kwon, 1984; Cho et al., 2006).

1.3 HEALTH BENEFITS OF FERMENTED FOODS AND BEVERAGES

1.3.1 Bioactive Compounds

One method by which fermented foods may exert beneficial health effects is through bioactive compounds, which are small molecules that confer a biological action, and that may result from chemical changes during the fermentation process (Martins et al., 2011). Well-known bioactive components produced and made bioavailable through fermentation include phenolic compounds, which may act as natural antioxidants and immune modulators (Dueñas et al., 2005; Martins et al., 2011). Attention to delivery of bioactives from fermented foods has also led to novel concepts for bioactive materials and packaging (Lopez-Rubio et al., 2006). Phytochemicals are commonly defined as nonnutritive substances found in food that can have disease-fighting properties. Notably though, the presence of bioactive compounds ex vivo does not necessarily speak to their activity following ingestion by an animal or human host. The acidic environment of the stomach and the many stages of digestion can lead to early conversion of certain bioactive compounds into less active, or entirely inactive forms (Stanton et al., 2005). In addition, probiotics consumed via fermented foods do not always survive the digestive systems to reach their final destination in the gut (Stanton et al., 2005). Conversely, metabolites formed through commensal gut microflora mediated fermentation can become transformed and bioactive (Stanton et al., 2005).

Fermented foods are known to produce bioactive compounds beyond those found naturally occurring in their unprocessed counterparts (Martins et al., 2011). Production of bioactive compounds in fermented foods can differentially occur depending on the specific variety or plant cultivar. In a study using different rice cultivars, rice bran fractions were fermented by *Saccharomyces boulardii*, and each of the three fermented rice brans showed significant differences in metabolite composition compared to nonfermented (Ryan et al., 2011). In this case, the fermented rice bran bioactivity for reducing the growth of human lymphomas in vitro was also dependent on the rice variety. The appearance of bioactive compounds in a fermented food that are absent in the nonfermented product has also been shown for Fu Zhuan tea. Tea leaves are fermented with the fungus *Eurotium cristatum* and showed increased amounts of dodecanamide, linoleamide, stearamide, and epicatechin gallate when compared to a nonfermented green tea (Keller et al., 2013). The relative importance for the distinct ratios and quantities of these various phytochemical components in this fermented tea are unknown. The antimicrobial activities observed by Fu Zhuan tea extracts will be essential to verify in future studies to better characterize this potential health utility that has global importance.

Red ginseng roots represent another plant containing bioactive compounds including saponins (ginsenosides) and nonsaponins, with as many as 50 ginsenosides identified to date that provide potential benefits to regulation of blood glucose and insulin levels (Oh et al., 2014). Fermentation of the ginseng roots has been demonstrated to increase the number of saponins naturally found in the plant (Oh et al., 2014). Individuals consuming 2.7 g per day of fermented red ginseng for four weeks had reduced fasting and postprandial glucose, and increased fasting and postprandial insulin levels (Oh et al., 2014).

It is important to note that the fermented form of a plant or animal food may not always be most desirable for receiving health benefits of bioactive compounds. For example, the effect of red cabbage fermentation on anthocyanin bioavailability and plasma antioxidant capacity was studied in a randomized crossover human study. A total of 18 parent anthocyanins and 12 associated metabolites were identified by HPLC-MS/MS, and revealed 10% higher bioavailability from fresh red cabbage when compared to fermented red cabbage (Wiczkowski et al., 2016). Moreover, plasma antioxidant capacity was increased following fresh cabbage consumption as compared to fermented cabbage consumption. In this case, the fermentation process reduced red cabbage anthocyanins bioavailability and human plasma antioxidant capacity.

1.3.2 Fermented Foods Targeting Chronic Disease Control and Prevention

A number of risk factors associated with the development and progression of major chronic disease conditions are inextricably linked, such as obesity, type 2 diabetes, cardiovascular disease, and cancer. These diseases share physiological aberrations and stressors, including but not limited to alterations in metabolism, oxidative stress, and inflammation (Camps and Garcia-Heredia, 2014).

Despite many dietary, nutritional, and food-based interventions that have been examined for diabetes prevention, management, and treatment, there remains an increase in the global incidence and prevalence of this disease (Fazeli Farsani et al., 2013; Goran et al., 2013; Oggioni et al., 2014; Weeratunga et al., 2014). Nearly 200 publications were identified with the search terms of 'fermented foods and diabetes' in the Entrez PubMed online database in September 2015, and while many focus on target endpoints related to glucose metabolism and insulin resistance, there are increasing numbers of studies using glycated hemoglobin (HbA1c) and other advanced glycation end-products as endpoints (Collaboration, 2015). Despite the difficulties inherent in trying to discern which of the several functional components that comprise a single fermented food or beverage may confer the potential health benefit, it is becoming well recognized that fermented products can be multifaceted in their bioactive compounds and associated mechanisms of action. Animal and human studies that assess multiple health-related endpoints at specific amounts or concentrations of fermented food intake will be useful to co-examine influences on regulation of lipid profiles, cellular energetics and metabolism, oxidative stress, as well as immune system and cognitive support. A few of the fermented foods selected for discussion herein have had information from both animal and human studies, and the clinical study designs, dosing regimens, and the inclusion of a combination of secondary endpoints are discussed. In addition, we must emphasize that there are significant challenges involved with extrapolating an effective fermented food amount and dietary intake level to humans based on available animal studies. This functional nature of fermented foods for people affected by chronic diseases may also be dependent on dietary patterns, influenced by the gut microbiome, and be related to the host nutrigenome (Daniell and Ryan, 2012).

1.3.2.1 Legumes

Fermented soybeans, commonly consumed in Korea, China, Japan, Indonesia, and Vietnam, represent an example of a major staple food that has consistently been reported for exhibiting antidiabetic effects. The fermented soybean properties are, in part, due to both quantitative and qualitative changes seen in small molecules following fermentation (reviewed in Kwon et al., 2010). Nutritional studies performed in animals and intervention studies with humans suggest that the ingestion of isoflavonoids, amino acids, and smaller bioactive peptides available in the fermented product improves glucose control and reduces insulin resistance, with strong implications for preventing or delaying the progression of type 2 diabetes. Fermentation time (short versus long term) and other host factors (animal versus human) can influence the absorption and bioavailability of the isoflavone glucosides and aglycones, yet the data supporting fermented soybeans and the beneficial effects on glucose metabolism, oxidative stress, and inflammation provided sufficient rationale to implement studies with fermented soybean extracts. Two fermented soybean compounds, isoflavones and α-galactooligosaccharides, have been shown to exert antioxidant and antiinflammatory effects, and so one study aimed to assess these effects as a dietary supplement called a fermented soy permeate (FSP). FSP was shown to reduce in vivo oxidative stress and inflammation in streptozotocin-induced type 1 diabetic rats (Malarde et al., 2015). Another three month intervention study showed that long-term ingestion of a fermented soybean-derived Touchi-extract with α-glucosidase inhibitory activity was not only safe but also effective at reducing fasting blood glucose and HbA1c in humans with borderline and mild type 2 diabetes (Fujita et al., 2001). Houji tea was used for a placebo comparison in this study as it contained steamed soybean powder that was devoid of any α-glucosidase inhibitory activity, whereby the Touchi extract was obtained by first steaming and then fermenting soybeans with koji (*Aspergillus* sp.). Finally, based on information supporting that brain insulin resistance is related to both diabetes and Alzheimer's disease, another study showed that soybeans fermented with *Bacillus lichenifomis*, when compared to a traditional method, could offer protection against cognitive dysfunction and glucose dysregulation in Alzhiemer's diseased–diabetic rats (Yang et al., 2015).

Other fermented legumes reported for health benefits include a Bambara groundnut and locust bean from Nigeria that demonstrated hypoglycemic and anticholinesterase activity in experimental diabetic rats (Ademiluyi et al., 2015). These properties could be attributed to the modulatory effect of their phytochemicals (bioactive peptides and phenolics) on α-amylase, α-glucosidase, and acetylcholinesterase activities as well as improving the in vivo antioxidant status of the diabetic rats. These findings for fermented legumes substantially advance upon the existing evidence by not just showing mitigation of hyperglycemia, but also the potential for management of diabetes-induced neurodegeneration.

Mung beans (*Vigna radiata*) also merit discussion for both fermented and nonfermented extracts (1000 mg/kg) as they were studied in normoglycemic, glucose-induced hyperglycemic, and alloxan-induced hyperglycemic mice (Yeap et al., 2012). In this case, low concentrations of the fermented mung bean extract did not show any significant difference in an antihyperglycemic effect when compared to the normal control, even though fermentation was able to improve the in vitro antioxidant and phenolic contents. The dose-dependent nature of mung bean extracts in mice is difficult to translate with regards to providing benefit in people. A human clinical trial was identified that is testing mung bean components that were included in a dietary supplement product in healthy adults, and will be evaluating changes in glucose parameters, triglycerides, and body weight as compared to a placebo (NCT02322294, https://clinicaltrials.gov/ct2/show/NCT02322294).

1.3.2.2 Specialty Foods

Blueberries fermented by tannase-active probiotic strains of the *Lactobacillus plantarum* appear to have a transient, but significant, blood pressure lowering effect (Ahren et al., 2014). Hypertension, defined by systolic blood pressure greater than 140 mmHg or diastolic blood pressure greater than 90 mmHg (Ahren et al., 2014), is a serious problem that is a major risk factor in cardiovascular disease. Research investigating a hypertension model in rats has shown that both systolic and diastolic blood pressure were significantly reduced after 2 weeks of 2 g per day of lacto-fermented blueberries, but increased back to hypertensive levels within two weeks after the intervention ended (Ahren et al., 2014). Currently, drugs that inhibit angiotensin I converting enzyme (ACE) are one of the more common antihypertensive treatments (Koyama et al., 2014). Known short-chain peptides are capable of inhibiting ACE, but they are typically degraded to inactive forms prior to reaching the target tissue (Koyama et al., 2014). Interestingly, neo-fermented buckwheat sprouts, using lactic acid fermentation methods (Nakamura et al., 2013), have been able to demonstrate blood pressure-lowering effects following a single dose (Koyama et al., 2014). Ex vivo experiments showed the mechanisms to be both an inhibition of ACE and to induce vasodilation of the aorta (Nakamura et al., 2013; Koyama et al., 2014).

A fermented papaya preparation (FPP) merits discussion for having demonstrated bioactivity in diabetic adults with defined antioxidant and immune-modulating potentials. Dickerson and colleagues reported the first experimental evidence demonstrating that FPP may improve diabetic wound outcomes by specifically influencing the response of wound-site macrophages and the subsequent angiogenic response (Dickerson et al., 2015). The notion that FPP could correct deficiencies in the respiratory burst capacity of peripheral blood mononuclear cells in type 2 diabetes mellitus patients has important implications as it was determined to be safe and without adverse effects on the hyperglycemia in patients.

Immune regulation and antiallergenic properties represents another promising area for foods and beverages fermented with specific LAB strains (Nonaka et al., 2008), yet few foods or beverages have thus far been studied in this regard. Rather, therapeutic approaches and bioengineering of LAB strains for specific mucosal adjuvant and immune regulatory properties are taking the lead (Wells and Mercenier, 2008). When LAB strains were isolated from fermented foods, researchers showed successful modulation of the immune response against the influenza virus (Kawashima et al., 2011). The relationships between these LAB strains in whole fermented foods and human disease outcomes remain to be determined.

A fungal fermentation of separated polysaccharides from rice bran is a therapeutic approach intended for activating the natural killer (NK) cells that mediate anticancer activity through a nonspecific immune response. Specifically, rice bran was digested by the carbohydrolase produced by the mycelia of *Lentinus edodes* to result in a fraction of active polysaccharides that stimulate the proliferation of the immune cell macrophages, with emphasis on the activation of NK cells and their anticancer activity (Choi et al., 2014). This double-blind, placebo controlled, parallel group study design provided 3 g/day and showed no adverse effects (study was registered with the Clinical Research Information Service

KCT0000536). Animal studies performed with this product used a dose range of 50–150 mg/kg to achieve similar effects on NK cell activity. The prospect of replacing toxic drugs with natural substances that enhance the body's natural host immune defenses against cancer has attracted considerable research, and particularly in the realm of reducing the side effects associated with chemotherapeutic agents in the prevention and treatment of cancer (Vinay et al., 2015).

While important to highlight these components of fermented foods, a majority of this research suggests that the potential for observing beneficial effects may largely depend on the individuals' characteristics, which may include, but not be limited to age, race, BMI, fat distribution, physical inactivity, gestational diabetes, and presence of cardiovascular disease risk factors. One concern regarding the interpretations of results from human studies is the dose extrapolation used following testing in animals. The percentage of total dietary intakes used in mice and rats consuming a standard diet are not directly translated to the dose in grams/day that is provided as a human supplement. Human dietary intervention studies from our group have begun to extrapolate animal food dosages to people (Borresen et al., 2014; Sheflin et al., 2015), yet these approaches have had limited applications when examining health benefits for fermented foods and beverages.

1.4 FOOD SAFETY AND QUALITY CONTROL

Although there are many potential benefits to consuming fermented foods, these benefits must be balanced against the risks of food-borne illness (Nout, 1994). Food safety and quality control are critical to the food preparation process and neither home preparation nor commercial productions of fermented foods are exempt from the potential risks posed by microbial pathogens (Nout, 1994; Bodmer et al., 1999), food spoilage (Nout, 1994), and aflatoxin production from fungi (Kinosita et al., 1968; Onilude et al., 2005; Wu et al., 2009). As new technologies begin to replace traditional tested methods of fermentation, care must be taken to ensure that foods are fermented in a safe and controlled environment to prevent potential risk of disease, illness, and death.

Contamination of fermented foods can occur during any processing stage from production of raw materials to the end-stage packaging of the product (Nout, 1994). Although highly acidic conditions and pasteurization are typically sufficient to prevent pathogenic infection of food products, high NaCl concentrations and refrigeration storage are not always sufficient at mitigating the risks of pathogen growth in food products (Nout, 1994). Some of the most common pathogenic infections seen with fermented food consumption include *Clostridium botulinum, Listeria monocytogenes,* and *Salmonella typhimurium* (Nout, 1994). In addition to pathogenic infections, some microbially derived compounds can pose health risks to consumers of fermented foods due to their potential toxicity in sufficiently high quantities, such as biogenic amines (Bodmer et al., 1999).

C. botulinum is a proteolytic microbe responsible for producing the neurotoxin that causes foodborne botulism (Peck et al., 2011). Botulism is widely considered one of the most fatal of potential foodborne illnesses, with as little as 30–100 ng of the botulinum neurotoxin being capable of conferring serious illness, and likely death, upon the consumer (Peck et al., 2011). Botulism has been implicated in foodborne deaths from fermented foods ranging from beaver tail to salmon roe and from yogurt to tofu (Peck et al., 2011). *C. botulinum* is able to form endospores that are capable of surviving harsh conditions that other microbes would be unable to withstand, such as high heat, desiccation, and UV light (Peck et al., 2011). Knowledge of ways to prevent potential contamination by *C. botulinum* is crucial to food safety with fermentation, with a standard heat cook of 3 min at 121°C as a standard in low acid foods (Stumbo et al., 1975). Botulism can be widely prevented through increased understanding of the potential hazards of improper methods of preparation and storage, when combining traditional fermented foods with modern materials.

LAB typically confer the protective effect of inhibiting foodborne pathogen contamination in the fermented foods process. The potential risk for pathogenic contamination are low for foods fermented with *L. plantarum* and other *Lactobacillus* spp. because of the combination of increased carbon dioxide enclosed medium and increased contents of organic acids, and consequential decreased pH, as well as the production of antimicrobial compounds, including nitrogen oxide and various peptides (Molin, 2008). However, the presence of LAB does not preclude pathogenic contamination of fermented foods, and both *L. monocytogenes* and *S. typhimurium* have both been observed to grow in the acidic environment of fermentation (Paramithiotis et al., 2012). *Salmonella* is especially prominent in fermented sausages, with its appearance growing in frequency with the decrease in curing agents, such as nitrates and nitrites, which have been shown to effectively inhibit *Salmonella* growth (Hospital et al., 2014).

The potential presence of biogenic amines in fermented foods is another risk associated with the production and consumption of fermented food products (Bodmer et al., 1999). Biogenic amines are nitrogenous compounds that result from decarboxylation of free amino acids often by decarboxylase-positive microorganisms. The most common biogenic amines are histamine and tyrosine (Bodmer et al., 1999). The toxic effects of biogenic amines may include such ill effects as diarrhea and vomiting (Silla Santos, 1996). Proper storage, such as refrigeration, and appropriate pH levels are important factors in the activity levels of aminodecarboxylases and the prevention of the production of biogenic amines (Silla Santos, 1996).

Aflatoxins are also carcinogenic compounds produced by fungi that pose significant risk to human health (Kinosita et al., 1968; Onilude et al., 2005). In addition to cancer risk, aflatoxins were shown to result in acute liver damage and cirrhosis (Kinosita et al., 1968; Onilude et al., 2005; Wu et al., 2009). Aflatoxins are produced primarily from species of the *Aspergillus fungi*, including *A. flavus* and *A. parasiticus* (Onilude et al., 2005). They are most common when foods are stored in facilities with humidity levels in excess of 85% humidity (Shephard, 2003). Interestingly, it appears LAB, such as *L. plantarum*, can be effective at inhibiting the formation of aflatoxin (Onilude et al., 2005). During the fermentation processes *L. plantarum* produces trace portions of phenylacids that have strong antifungal elements.

1.5 CONCLUSIONS

In addition to historical and traditional knowledge, there is a growing body of scientific literature to suggest that fermented foods exhibit promising and sustainable opportunities to target a multitude of disease conditions affecting diverse cultures and populations around the world. In addition, populations may vary, either equally or preferentially, in the process of furthering information on the health benefits associated with specific types of fermented foods and beverages consumed in the diet.

The health benefits of fermented foods may be expressed either directly through the interactions of ingested live microorganisms with the host (probiotic effect) or indirectly as the result of the ingestion of microbial metabolites synthesized during fermentation (biogenic effect). There are a number of remaining questions to be answered in future studies of fermented foods and beverages, such as determining biomarkers of fermented food health benefits, amounts of consumed fermented foods and beverages necessary to elicit a health benefit, safety of fermentation under different conditions, and the degree to which the compounds are actually accessible by the body to confer disease prevention. Metabolomics in conjunction with microbial ecology/genomics are rapidly advancing tools that have strong potential to lead in the identification of novel bioactive compounds from fermented foods. These approaches may also gain support for investigating bioavailability and disease-fighting properties of fermented foods.

Proper safety measures should be taken with regard to health promotion and disease prevention investigations. The risk of food-borne illness is not an inconsequential one, immune compromised and especially young individuals should take special considerations for the source of fermented foods. When preparing fermented foods in an at-home environment, it is critical to follow proper procedures to reduce the risk of food-borne illnesses. Fermentation can add novel and interesting flavor profiles to food that is highly regarded in the production of popular choices, such as coffee, chocolate, beer, and wine. The understanding of bioactive compounds and other interactions with fermented foods and beverages also have tremendous health potential for the consumer. Each of these topics and more will be discussed at greater length in succeeding chapters.

ACKNOWLEDGMENTS

The author thanks Erin Prapas, Nora Jean Nealon, Dustin G. Brown, and Erica C. Borresen for their assistance in conducting scientific literature searches and editorial comments.

REFERENCES

Ademiluyi, A.O., Oboh, G., Boligon, A.A., Athayde, M.L., 2015. Dietary supplementation with fermented legumes modulate hyperglycemia and acetylcholinesterase activities in Streptozotocin-induced diabetes. Pathophysiology 22, 195–201.

Ahren, I.L., Xu, J., Önning, G., Olsson, C., Ahrne, S., Molin, G., 2014. Antihypertensive activity of blueberries fermented by *Lactobacillus plantarum* DSM 15313 and effects on the gut microbiota in healthy rats. Clinical Nutrition 34, 719–726.

Azokpota, P.S., 2015. Quality aspects of alkaline-fermented foods. In: Sarkar, P.K., Nout, M.J.R. (Eds.), Handbook of Indigenous Foods Involving Alkaline Fermentation. CRC Press Taylor & Francis Group, Boca Raton, FL, pp. 315–379.

Bamworth, C.W., Ward, R.E., 2014. The Oxford Handbook of Food Fermentations. Oxford University Press, Oxford.

Bodmer, S., Imark, C., Kneubühl, M., 1999. Biogenic amines in foods: histamine and food processing. Inflammation Research 48, 296–300.

Borresen, E.C., Gundlach, K.A., Wdowik, M., Rao, S., Brown, R.J., Ryan, E.P., 2014. Feasibility of increased navy bean powder consumption for primary and secondary colorectal cancer prevention. Current Nutrition and Food Science 10, 112–119.

Borresen, E.C., Henderson, A.J., Kumar, A., Weir, T.L., Ryan, E.P., 2012. Fermented foods: patented approaches and formulations for nutritional supplementation and health promotion. Recent Patents on Food, Nutrition and Agriculture 4, 134–140.

Camps, J., Garcia-Heredia, A., 2014. Introduction: oxidation and inflammation, a molecular link between non-communicable diseases. Advances in Experimental Medicine and Biology 824, 1–4.

Capozzi, V., Russo, P., Dueñas, M.T., Lopez, P., Spano, G., 2012. Lactic acid bacteria producing B-group vitamins: a great potential for functional cereals products. Applied Microbiology and Biotechnology 96, 1383–1394.

Cho, J., Lee, D., Yang, C., Jeon, J., Kim, J., Han, H., 2006. Microbial population dynamics of kimchi, a fermented cabbage product. FEMS Microbiology Letters 257, 262–267.

Choi, J.Y., Paik, D.J., Kwon, D.Y., Park, Y., 2014. Dietary supplementation with rice bran fermented with *Lentinus edodes* increases interferon-gamma activity without causing adverse effects: a randomized, double-blind, placebo-controlled, parallel-group study. Nutrition Journal 13, 35.

Chu, S.-C., Chen, C., 2006. Effects of origins and fermentation time on the antioxidant activities of kombucha. Food Chemistry 98, 502–507.

Corsetti, A., Luca, S., 2007. Lactobacilli in sourdough fermentation. Food Research International 40, 539–558.

Crafack, M., Keul, H., Eskildsen, C.E.A., Petersen, M.A., Saerens, S., Blennow, A., Skovmand-Larsen, M., Swiegers, J.H., Petersen, G.B., Heimdal, H., Nielsen, D.S., 2014. Impact of starter cultures and fermentation techniques on the volatile aroma and sensory profile of chocolate. Food Research International 63, 306–316.

Daniell, E., Ryan, E.P., 2012. The nutrigenome and gut microbiome: chronic disease prevention with crop phytochemical diversity. In: Caliskan, M. (Ed.), The Molecular Basis of Plant Genetic Diversity. InTech, Rijeka, Croatia, pp. 357–374.

Dickerson, R., Banerjee, J., Rauckhorst, A., Pfeiffer, D.R., Gordillo, G.M., Khanna, S., Osei, K., Roy, S., 2015. Does oral supplementation of a fermented papaya preparation correct respiratory burst function of innate immune cells in type 2 diabetes mellitus patients? Antioxidants and Redox Signaling 22, 339–345.

Duenas, M., Fernandez, D., Hernandez, T., Estrella, I., Munoz, R., 2005. Bioactive phenolic compounds of cowpeas (*Vigna sinensis* L.). Modifications by fermentation with natural microflora and with *Lactobacillus plantarum* ATCC 14917. Journal of the Science of Food and Agriculture 85, 297–304.

Fazeli Farsani, S., van der Aa, M.P., van der Vorst, M.M., Knibbe, C.A., de Boer, A., 2013. Global trends in the incidence and prevalence of type 2 diabetes in children and adolescents: a systematic review and evaluation of methodological approaches. Diabetologia 56, 1471–1488.

Ferreira, I.M.P.L.V.O., Eça, R., Pinho, O., Tavares, P., Pereira, A., Roque, A.C., 2007. Development and validation of an HPLC/UV method for quantification of bioactive peptides in fermented milks. Journal of Liquid Chromatography & Related Technologies 30, 2139–2147.

Franco, R., Oñatibia-Astibia, A., Martinez-Pinilla, E., 2013. Health benefits of methylxanthines in cacao and chocolate. Nutrients 5, 4159–4173.

Fujita, H., Yamagami, T., Ohshima, K., 2001. Long-term ingestion of a fermented soybean-derived Touchi-extract with alpha-glucosidase inhibitory activity is safe and effective in humans with borderline and mild type-2 diabetes. The Journal of Nutrition 131, 2105–2108.

George, S.E., Ramalakshmi, K., Mohan Rao, L.J., 2008. A perception on health benefits of coffee. Critical Reviews in Food Science and Nutrition 48, 464–486.

Giraffa, G., 2004. Studying the dynamics of microbial populations during food fermentation. FEMS Microbiology Reviews 28, 251–260.

Goran, M.I., Ulijaszek, S.J., Ventura, E.E., 2013. High fructose corn syrup and diabetes prevalence: a global perspective. Global Public Health 8, 55–64.

Grassi, S., Amigo, J.M., Lyndgaard, C.B., Foschino, R., Casiraghi, E., 2014. Beer fermentation: monitoring of process parameters by FT-NIR and multivariate data analysis. Food Chemistry 155, 279–286.

Halliwell, B., 2003. Plasma antioxidants: health benefits of eating chocolate? Nature 426, 787 discussion 788.

Hansen, A., Schieberle, P., 2005. Generation of aroma compounds during sourdough fermentation: applied and fundamental aspects. Trends in Food Science and Technology 16, 85–94.

Hospital, X.F., Hierro, E., Fernandez, M., 2014. Effect of reducing nitrate and nitrite added to dry fermented sausages on the survival of *Salmonella Typhimurium*. Food Research International 62, 410–415.

Jayabalan, R., Malbasa, R.V., Loncar, E.S., Vitas, J.S., Sathishkumar, M., 2014. A review on kombucha tea - microbiology, composition, fermentation, beneficial effects, toxicity, and tea fungus. Comprehensive Reviews in Food Science and Food Safety 13, 538–550.

Kamiya, S., Owasawara, M., Arakawa, M., Hagimori, M., 2013. The effect of lactic acid bacteria-fermented soybean milk products on carragenan-induced tail thrombosis in rats. Bioscience of Microbiota, Food and Health 32, 101–105.

Katz, S.E., 2012. The Art of Fermentation: An In-depth Exploration of Essential Concepts and Processes from Around the World. Chelsea Green Publishing, Vermont.

Kawashima, T., Hayashi, K., Kosaka, A., Kawashima, M., Igarashi, T., Tsutsui, H., Tsuji, N.M., Nishimura, I., Hayashi, T., Obata, A., 2011. *Lactobacillus plantarum* strain YU from fermented foods activates Th1 and protective immune responses. International Immunopharmacology 11, 2017–2024.

Keller, A.C., Weir, T.L., Broeckling, C.D., Ryan, E.P., 2013. Antibacterial activity and phytochemical profile of fermented *Camellia sinensis* (fuzhuan tea). Food Research International 53, 945–949.

Kim, A.J., Choi, J.N., Kim, J., Kim, H.Y., Park, S.B., Yeo, S.H., Choi, J.H., Liu, K.H., Lee, C.H., 2012. Metabolite profiling and bioactivity of rice koji fermented by *Aspergillus* strains. Journal of Microbiology and Biotechnology 22, 100–106.

Kinosita, R., Ishiko, T., Sugiyama, S., Seto, T., Igarasi, S., Goetz, I.E., 1968. Mycotoxins in fermented food. Cancer Research 28, 2296–2311.

Koyama, M., Hattori, S., Amano, Y., Watanabe, M., Nakamura, K., 2014. Blood pressure-lowering peptides from neo-fermented buckwheat sprouts: a new approach to estimating ACE-inhibitory activity. PLoS One 9, e105802.

Kwon, D.Y., Daily 3rd, J.W., Kim, H.J., Park, S., 2010. Antidiabetic effects of fermented soybean products on type 2 diabetes. Nutrition Research 30, 1–13.

Lee, J.E., Hwang, G.S., Lee, C.H., Hong, Y.S., 2009. Metabolomics reveals alterations in both primary and secondary metabolites by wine bacteria. Journal of Agricultural and Food Chemistry 57, 10772–10783.

Lee, L.W., Cheong, M.W., Curran, P., Yu, B., Liu, S.Q., 2015. Coffee fermentation and flavor-an intricate and delicate relationship. Food Chemistry 185, 182–191.

Liu, M., Bienfait, B., Sacher, O., Gasteiger, J., Siezen, R.J., Nauta, A., Geurts, J.M., 2014. Combining chemoinformatics with bioinformatics: in silico prediction of bacterial flavor-forming pathways by a chemical systems biology approach "reverse pathway engineering". PLoS One 9, e84769.

Lopez-Rubio, A., Gavara, R., Lagaron, J.M., 2006. Bioactive packaging: turning foods into healthier foods through biomaterials. Trends in Food Science & Technology 17, 567–575.

Lv, H.P., Zhang, Y.J., Lin, Z., Liang, Y.R., 2013. Processing and chemical constituents of Pu-erh tea: a review. Food Research International 53, 608–618.

Malarde, L., Groussard, C., Lefeuvre-Orfila, L., Vincent, S., Efstathiou, T., Gratas-Delamarche, A., 2015. Fermented soy permeate reduces cytokine level and oxidative stress in streptozotocin-induced diabetic rats. Journal of Medicinal Food 18, 67–75.

Martins, S., Mussatto, S.I., Martinez-Avila, G., Montanez-Saenz, J., Aguilar, C.N., Teixeira, J.A., 2011. Bioactive phenolic compounds: production and extraction by solid-state fermentation. A review. Biotechnology Advances 29, 365–373.

Meignen, B., O, B., Gelinas, P., Infantes, M., Guilois, S., Cahagnier, B., 2001. Optimization of sourdough fermentation with *Lactobacillus brevis* and baker's yeast. Food Microbiology 18, 239–245.

Mheen, T.-I., Kwon, T.-W., 1984. Effect of temperature and salt concentration on Kimchi fermentation. Korean Journal of Food Science and Technology 16, 443–450.

Molin, G., 2008. *Lactobacillus plantarum*. The role in foods and human health. In: Farnworth, E.E. (Ed.), Handbook of Fermented Functional Foods. CRC Press Taylor & Francis Group, LLC, Boca Raton, FL, pp. 353–393.

Nakamura, K., Naramoto, K., Koyama, M., 2013. Blood pressure-lowering effect of fermented buckwheat sprouts in spontaneously hypertensive rats. Journal of Functional Foods 5, 406–415.

NCD Risk Factor Collaboration, 2015. Effects of diabetes definition on global surveillance of diabetes prevalence and diagnosis: a pooled analysis of 96 population-based studies with 331,288 participants. The Lancet Diabetes and Endocrinology 3, 624–637.

Nelson, J.H., 1970. Production of Blue cheese flavor via submerged fermentation by *Penicillium roqueforti*. Journal of Agricultural and Food Chemistry 18, 567–569.

Nikolopoulou, D., Grigorakis, K., Stasini, M., Alexis, M., Iliadis, K., 2006. Effects of cultivation area and year on proximate composition and antinutrients in three different kabuli-type chickpea (*Cicer arientinum*) varieties. European Food Research and Technology 223, 737–741.

Nonaka, Y., Izumo, T., Izumi, F., Maekawa, T., Shibata, H., Nakano, A., Kishi, A., Akatani, K., Kiso, Y., 2008. Antiallergic effects of *Lactobacillus pentosus* strain S-PT84 mediated by modulation of Th1/Th2 immunobalance and induction of IL-10 production. International Archives of Allergy and Immunology 145, 249–257.

Nout, M.J.R., 1994. Fermented foods and food safety. Food Research International 27, 291–298.

Oggioni, C., Lara, J., Wells, J.C.K., Soroka, K., Siervo, M., 2014. Shifts in population dietary patterns and physical inactivity as determinants of global trends in the prevalence of diabetes: an ecological analysis. Nutrition Metabolism and Cardiovascascular Diseases 24, 1105–1111.

Oh, M.R., Park, S.H., Kim, S.Y., Back, H.I., Kim, M.G., Jeon, J.Y., Ha, K.C., Na, W.T., Cha, Y.S., Park, B.H., Park, T.S., Chae, S.W., 2014. Postprandial glucose-lowering effects of fermented red ginseng in subjects with impaired fasting glucose or type 2 diabetes: a randomized, double-blind, placebo-controlled clinical trial. BMC Complementary and Alternative Medicine 14, 237.

Onilude, A.A., Fagade, O.E., Bello, M.M., Fadahunsi, I.F., 2005. Inhibition of aflatoxin-producing aspergilli by lactic acid bacteria isolates from indigenously fermented cereal gruels. African Journal of Biotechnology 4, 1404–1408.

Paramithiotis, S., Doulgeraki, A.I., Tsilikidis, I., Nychas, G.J.E., Drosinos, E.H., 2012. Fate of *Listeria monocytogenes* and *Salmonella Typhimurium* during spontaneous cauliflower fermentation. Food Control 27, 178–183.

Parvez, S., Malik, K.A., Kang, S.A., Kim, H.Y., 2006. Probiotics and their fermented food products are beneficial for health. Journal of Applied Microbiology 100, 1171–1185.

Peck, M.W., Stringer, S.C., Carter, A.T., 2011. *Clostridium botulinum* in the post-genomic era. Food Microbiology 28, 183–191.

Prajapati, J.B., Nair, B.M., 2008. The history of fermented foods. In: Farnworth, E.R. (Ed.), Handbook of Fermented Functional Foods. CRC Press, Taylor & Francis Group, LLC, Boca Raton, FL, pp. 1–22.

Quave, C.L., Pieroni, A., 2014. Fermented foods for food security and food sovereignty in the Balkans: a case study of the Gorani people of Northeastern Albania. Journal of Ethnobiology 34, 28–43.

Rodgers, S., 2008. Novel applications of live bacteria in food services: probiotics and protective cultures. Trends in Food Science and Technology 19, 188–197.

Rodriguez, H., Curiel, J.A., Landete, J.M., de las Rivas, B., Lopez de Felipe, F., Gomez-Cordoves, C., Mancheno, J.M., Muñoz, R., 2009. Food phenolics and lactic acid bacteria. International Journal of Food Microbiology 132, 79–90.

Ryan, E.P., Heuberger, A.L., Weir, T.L., Barnett, B., Broeckling, C.D., Prenni, J.E., 2011. Rice bran fermented with *Saccharomyces boulardii* generates novel metabolite profiles with bioactivity. Journal of Agricultural and Food Chemistry 59, 1862–1870.

Sheflin, A.M., Borresen, E.C., Wdowik, M.J., Rao, S., Brown, R.J., Heuberger, A.L., Broeckling, C.D., Weir, T.L., Ryan, E.P., 2015. Pilot dietary intervention with heat-stabilized rice bran modulates stool microbiota and metabolites in healthy adults. Nutrients 7, 1282–1300.

Sheih, I.C., Fang, T.J., Wu, T.K., Chen, R.Y., 2014. Effects of fermentation on antioxidant properties and phytochemical composition of soy germ. Journal of the Science Food and Agriculture 94, 3163–3170.

Shephard, G.S., 2003. Aflatoxin and food safety: recent African perspectives. Journal of Toxicology-Toxin Reviews 22, 267–286.

Silla Santos, M.H., 1996. Biogenic amines: their importance in foods. International Journal of Food Microbiology 29, 213–231.

Sodini, I., Remeuf, F., Haddad, S., Corrieu, G., 2004. The relative effect of milk base, starter, and process on yogurt texture: a review. Critical Reviews in Food Science and Nutrition 44, 113–137.

Soukoulis, C., Panagiotidis, P., Koureli, R., Tzia, C., 2007. Industrial yogurt manufacture: monitoring of fermentation process and improvement of final product quality. Journal of Dairy Science 90, 2641–2654.

Sreeramulu, G., Zhu, Y., Knol, W., 2000. Kombucha fermentation and its antimicrobial activity. Journal of Agricultural and Food Chemistry 48, 2589–2594.

Stanton, C., Ross, R.P., Fitzgerald, G.F., Van Sinderen, D., 2005. Fermented functional foods based on probiotics and their biogenic metabolites. Current Opinion in Biotechnology 16, 198–203.

Starr, G., Petersen, M.A., Jespersen, B.M., Hansen, A.S., 2015. Variation of volatile compounds among wheat varieties and landraces. Food Chemistry 174, 527–537.

Steinkraus, K.H., 1997. Classification of fermented foods: worldwide review of household fermentation techniques. Food Control 8, 311–317.

Stumbo, C.R., Purohit, K.S., Ramakrishnan, T.V., 1975. Thermal process lethality guide for low-acid foods in metal containers. Journal of Food Science 40, 1316–1323.

Teoh, A.L., Heard, G., Cox, J., 2004. Yeast ecology of Kombucha fermentation. International Journal of Food Microbiology 95, 119–126.

Vinay, D.S., Ryan, E.P., Pawelec, G., Talib, W.H., Stagg, J., Elkord, E., Lichtor, T., Decker, W.K., Whelan, R.L., Kumara, H.M., Signori, E., Honoki, K., Georgakilas, A.G., Amin, A., Helferich, W.G., Boosani, C.S., Guha, G., Ciriolo, M.R., Chen, S., Mohammed, S.I., Azmi, A.S., Keith, W.N., Bilsland, A., Bhakta, D., Halicka, D., Fujii, H., Aquilano, K., Ashraf, S.S., Nowsheen, S., Yang, X., Choi, B.K., Kwon, B.S., 2015. Immune evasion in cancer: mechanistic basis and therapeutic strategies. Seminars in Cancer Biology pii:S1044–579X(15)00019-X.

Wang, Q., Peng, C., Gong, J., 2011. Effects of enzymatic action on the formation of theabrownin during solid state fermentation of Pu-erh tea. Journal of the Science Food and Agriculture 91, 2412–2418.

Weeratunga, P., Jayasinghe, S., Perera, Y., Jayasena, G., Jayasinghe, S., 2014. Per capita sugar consumption and prevalence of diabetes mellitus-global and regional associations. BMC Public Health 14, 186.

Wells, J.M., Mercenier, A., 2008. Mucosal delivery of therapeutic and prophylactic molecules using lactic acid bacteria. Nature Reviews Microbiology 6, 349–362.

Wiczkowski, W., Szawara-Nowak, D., Romaszko, J., 2016. The impact of red cabbage fermentation on bioavailability of anthocyanins and antioxidant capacity of human plasma. Food Chemistry 190, 730–740.

Wolff, D., 2000. Bean there: toward a soy-based history of Northeast Asia. South Atlantic Quarterly Winter 99, 241–252.

Wu, Q., Jezkova, A., Yuan, Z., Pavlikova, L., Dohnal, V., Kuca, K., 2009. Biological degradation of aflatoxins. Drug Metabolism Reviews 41, 1–7.

Xu, L., Du, B., Xu, B., 2015. A systematic, comparative study on the beneficial health components and antioxidant activities of commercially fermented soy products marketed in China. Food Chemistry 174, 202–213.

Yang, H.J., Kwon, D.Y., Kim, H.J., Kim, M.J., Jung do, Y., Kang, H.J., Kim da, S., Kang, S., Moon, N.R., Shin, B.K., Park, S., 2015. Fermenting soybeans with *Bacillus licheniformis* potentiates their capacity to improve cognitive function and glucose homeostaisis in diabetic rats with experimental Alzheimer's type dementia. European Journal of Nutrition 54, 77–88.

Yang, S.O., Kim, M.S., Liu, K.H., Auh, J.H., Kim, Y.S., Kwon, D.Y., Choi, H.K., 2009. Classification of fermented soybean paste during fermentation by ^{1}H nuclear magnetic resonance spectroscopy and principal component analysis. Bioscience Biotechnology and Biochemistry 73, 502–507.

Yeap, S.K., Mohd Ali, N., Mohd Yusof, H., Alitheen, N.B., Beh, B.K., Ho, W.Y., Koh, S.P., Long, K., 2012. Antihyperglycemic effects of fermented and nonfermented mung bean extracts on alloxan-induced-diabetic mice. Journal of Biomedicine and Biotechnology 2012:285430.

Section 2

Fermented Foods as a Source of Healthy Constituents

Chapter 2

Bioactive Peptides in Fermented Foods: Production and Evidence for Health Effects

C. Martinez-Villaluenga, E. Peñas, J. Frias
Institute of Food Science, Technology and Nutrition (ICTAN-CSIC), Madrid, Spain

2.1 INTRODUCTION

Bioactive peptides derived from fermented foods have been the subject of intensive research due to the health benefits that they may provide. Many peptides of animal and plant origin with relevant potential physiological functions in humans have been described in fermented foods. By far the most studied are the bioactive peptides present in dairy fermented products, followed by legumes, cereals, meat, and fish-derived products (Tables 2.1 and 2.2).

Bioactive peptides usually contain between 2 and 20 amino acids residues and they are encoded in the primary structure of animal and plant proteins, requiring proteolysis for their release from the precursor protein. The main proteins in milk, meat, fish, cereal, pseudocereals, and legume grains have the capacity to liberate up to 20,000 bioactive peptides (Ricci et al., 2010; Cavazos and Gonzalez et al., 2013; Lafarga et al., 2014; López-Barrios et al., 2014; Montoya-Rodríguez et al., 2015; Mora and Hayes, 2015; Sah et al., 2015). Release of peptides during the manufacture of fermented foods may occur in two ways: (1) by the microbial proteolytic system and (2) by endogenous proteolytic enzymes, both of them taking place during the fermentation and ripening processes.

Bioactive peptides might exhibit natural resistance to gastrointestinal digestion. They can penetrate through the intestinal cells under different modes of transport, such as paracellular routes (Miguel et al., 2008), passive diffusion (Satake et al., 2002), endocytosis (Ziv and Bendayan, 2000), and via the lymphatic system (Rubas and Grass, 1991). Once liberated and absorbed, bioactive peptides may exert a physiological affect on the various systems of the body (cardiovascular, digestive, endocrine, immune, and nervous systems).

TABLE 2.1 Some Examples of Bioactive Peptides Present in Fermented Foods Derived From Animal Proteins

Fermented Food	Microorganisms	Fermentation Conditions	Identified Peptides	Bioactivities	References
Fermented milk	*Enterococcus faecalis*	30°C, 48 h, 3% v/v of inoculum	LVYPFPGPIPNSLPQNIPP, LHLPLP, LHLPLPL, VLGPVRGPFP, VRGPFPIIV	Antihypertensive	Quirós et al. (2007, 2006) and Miguel et al. (2006)
	Lactococcus lactis NRRL-B-50571	30°C, 48 h, 3% v/v of inoculum	DDQNPH, LDDDLTDDI, YPSYGL, HPHPHLSFMAIPP, YDTQAIVQ, DDDLTDDIMCV, YPSYG	ACE inhibitory	Rodriguez-Figueroa et al. (2012)
	lactis NRRL-B-50572	30°C, 48 h, 3% v/v of inoculum	DVENLHLPLPLL, YPSYGL, ENGEC	ACE inhibitory	Rodriguez-Figueroa et al. (2013)
	Bb. bifidum MF 20/5	37°C, anaerobiosis, 24 h, 3% v/v of inoculum	LVYPFP	ACE inhibitory	Gonzalez-Gonzalez et al. (2013)
	Lactococcus lactis DIBCAB2	30°C, 40 h, 7 log CFU/mL	LQSW, MFPPQSVLSLSQS, PEQSLVYP, LYQEPVLGP, KPAAVRSPAQILQWQV, IHAQQK	ACE inhibitory	Nejati et al. (2013)
	Lb. helveticus and *S. cerevisiae*	Not specified	VPP, IPP	Antihypertensive, antiinflammatory, antiadipogenic, antiatherosclerotic	Fekete et al. (2015), Eisele et al. (2012), Aihara et al. (2014, 2009), and Chakrabarti and Wu (2015)
	Lb. helveticus H9	37°C, 7.5 h, 5 × 10⁷ CFU/mL	VPP, IPP	Blood pressure lowering	Chen et al. (2011)
	Lb. casei Shirota and *St. thermophilus*	37°C, 27 h, 10⁷ CFU/mL	YQEPVLGPVRGPFPIIV	ACE inhibitory, antithrombotic	Rojas-Ronquillo et al. (2012)
	Lb. helveticus LH2	37°C, Anaerobiosis, 26 h, 1% v/v of inoculum	WMHQPHQPLPPTVMFPPQ, LYQQPVLGPVR, SCDKFLDD	Immunomodulating	Tellez et al. (2010)

Product	Culture	Conditions	Crude peptide extract	Bioactivity	Reference
	Combinations of Lb. bulgaricus, St. thermophilus, Lb. acidophilus (ATCC 4356), Lactobacillus casei (ATCC 393) and Lb. paracasei ssp. paracasei (ATCC BAA52)	37°C, 20h, 1% v/v of inoculum		Antioxidant and antimutagenic	Sah et al. (2014)
	Not specified	Not specified	NLLRF, GPVRGPFPII	Anticancer	Sah et al. (2015)
	Lb. delbrueckii ssp. bulgaricus, St. thermophilus	Not specified	VPYPQ, KAVPYPQ, KVLPVPE, GVRGPFPII, IPIQY, QQPVLGPVRGPFPIIV	Antioxidant	Saabena Farvin et al. (2010)
	Lb. delbrueckii ssp. bulgaricus, St. thermophilus	Not specified	GVSKVKEAMAPKHKEMPFPKYPVEPFTESQ	Mucin stimulating, ACE inhibitory, opioid	Plaisancié et al. (2013, 2015)
	Lb. delbrueckii ssp. bulgaricus, St. thermophilus	Not specified	QEPVL, QEPV	Immunomodulating	Jiehui et al. (2014)
Dahi	Lb. delbrueckii ssp. bulgaricus, St. salivarius ssp. thermophilus and Lc. Lactis biovar diacetylactis	37°C, 8 h, 1:0.25:0.25 inoculum ratio	SKVVP	Antihypertensive	Ashar and Chand (2004)
Greek yogurt	Lb. delbrueckii subsp. bulgaricus Y10.13, St. thermophilus Y10.7 and Lb. paracasei subsp. paracasei DC412	42°C, 4 h, 2.5% v/v of inoculum	YPVEPFTE	ACE inhibitory, opioid	Papadimitriou et al. (2007)
Koumiss (fermented mare's milk)	LAB and yeasts	28°C, 52 h, 5% v/v of inoculum	YQDPRLGPTGELDPATQPIVAVHNPVIV, PKDLREN, LLAHLL, NHRNRMMDHVH	ACE inhibitory	Chen et al. (2010)

Continued

TABLE 2.1 Some Examples of Bioactive Peptides Present in Fermented Foods Derived From Animal Proteins—cont'd

Fermented Food	Microorganisms	Fermentation Conditions	Identified Peptides	Bioactivities	References
Kefir	*Lactococcus Lactis, Leuconostoc* ssp., *St. thermophilus, Lactobacillus* ssp. and kefir yeast or kefir grain microflora	25°C until pH decreased to 4.8	KAVPYPQ, NLHLPLP, SKVLPVPQ, LNVPGEIVE, YQKFPQY, SQSKVLPVPQ, VYPFPGPIPN	ACE inhibitory	Ebner et al. (2015)
			YQEPVLGPVRGPFPIIV	Antithrombotic	
			VYPFPGPIPN; ARHPHPHLSFM, YQEPVLGPVRGPFPIIV	Antioxidant	
			VLNENLLR, YQEPVLGPVRGPFPIIV	Antimicrobial	
			YQEPVLGPVRGPFPIIV, LYQEPVLGPVRGPFPIIV	Immunomodulating	
			KIEKFQSEEQQQTS7(Phospho)	Mineral binding	
			VYPFPGPIPN	Opioid	
Fermented skipjack tuna	*Aspergillus glaucus*	Not specified	LKPNM	ACE inhibitory	Ryan et al. (2011)
Fermented shrimp paste	Natural fermentation	30°C, 6 months	SV, IF	ACE inhibitory	Kleekayai et al. (2015)
Fermented blue mussel sauce	Natural fermentation	25% NaCl, 20°C, 6 months	HFGBPFH, FGHPY	Antioxidant	Rajapakse et al. (2005) and Jung et al. (2005)
Petrovac sausage	Natural fermentation	Cool smoking for 10 days and dry ripening for 90 days	Crude protein extract	Antioxidant and ACE inhibitory	Vastag et al. (2010)
Chrorizo sausages	Natural fermentation	11°C, 78% relative humidity, 1 month and 20 days	FGG, DM, RT, KPK, carnosine, anserine	Antioxidant	Broncano et al. (2012)

TABLE 2.2 Some Examples of Bioactive Peptides Present in Fermented Foods Derived From Plant Proteins

Fermented Food	Microorganisms	Fermentation Conditions	Bioactive Peptides	Bioactivities	References
Chunghookjang (fermented soybean)	*Bacillus licheniformis* B1	40°C, 72 h, 1% v/w of inoculum	LE, EW, SP, VE, VL, VT, EF	Antidiabetic	Yang et al. (2013)
Douchi (fermented soybean)	*Aspergillus aegypticus*	First: 30°C, 90% humidity, 48h, 1% v/w of inoculum Second: 37°C, 15 days, 16% salt, 10% water, and spices	Peptide containing Phe, Ile, and Gly in the ratio 1:2:5	ACE inhibitor	Zhang et al. (2006)
Fermented soy protein	*Lb. casei* spp. *pseudoplantarum*	37°C, 36 h	LIVTQ, LIVT	ACE inhibitor	S. Vallabha and Tiku (2014)
Tofuyo fermented soybean	Not specified	Not specified	WL, IFL	ACE inhibitor	Kuba et al. (2003)
Korean soybean paste	Not specified	Not specified	HHL	ACE inhibitor	Shin et al. (2001)
Fermented soy sauce	Tanekoji, rich in conidia and *Aspergillus sojae*	45°C, 5 days, 60 rpm	AW, GW, AY, SY, GY, AF, VP, AI, VG	Antihypertensive	Nakahara et al. (2011, 2010)
Fermented soy milk	*Lb. paracasei* ssp. *paracasei* NTU 101 and *Lb. plantarum* NTU 102	37°C, 24 h	Not identified	Antiosteoporotic	Chiang et al. (2012)
Navy bean fermented milk	*Lb. bulgaricus*	37°C, 2 h	Not identified	ACE inhibitory	Rui et al. (2015)
	Lb. plantarum B1-6	31°C, 3 h	Not identified	ACE inhibitory	
	Lb. plantarum 70810	37°C, 5 h	Not identified	ACE inhibitory	

Continued

TABLE 2.2 Some Examples of Bioactive Peptides Present in Fermented Foods Derived From Plant Proteins—cont'd

Fermented Food	Microorganisms	Fermentation Conditions	Bioactive Peptides	Bioactivities	References
Sourdoughs	Pool of *Lb. alimentarius* 15M, *Lb. brevis* 14G, *Lb. sanfranciscensis* strains, *Lb. hilgardii* 51B	37°C, 24 h, 5×10^7 CFU/g of dough			Coda et al., 2012
Whole wheat			MAPAAVAAAEAGSK, DNIPIVIR	Antioxidant	
Spelt			AIAGAGVLSGYDQLQILFFGK, GNQEKVLELVQR, PAGSAAGAAP, EALEAMFL, ITFAAYRR, HPVPPKKK	Antioxidant	
Rye			RLSLPAGAPVTVAVSP, LCPVHRAADL, PAEMVAAALDR, KVALMSAGSMH, DLADIPQQQRLMAGLALVVATVIFLK	Antioxidant	
Kamut			GVSNAAVVAGGH, DAQEFKR, PPGPGPCPPPPPGAAGRGGGG, HKEMQAIFDVYIMFIN	Antioxidant	
Whole wheat	Pool of *Lb. alimentarius* 15M, *Lb. brevis* 14G, *Lb. sanfranciscensis* strains, *Lb. hilgardii* 51B	dough yield 330, 37°C, 24 h, 5×10^8 CFU/g of dough	DPVAPLQRSGPEI, PVAPQLSRGLL, ELEIVMASPP, QILLPRPGQAA	ACE inhibitory	Rizzello et al. (2008)

Substrate	Microorganism	Conditions	Peptides/Fractions	Activity	Reference
Rye malt	Lb. reuteri TMW 1.106, LTH5448, LTH5795, Lb. hammesii DSM16381, Lb. rossiae 34J, Lb. plantarum FUA3002	34°C, 24h	LQP, IPP, LLP, VPP	ACE inhibitory	Hu et al. (2011)
Whole wheat	Lb. sanfranciscensis and Candida humilis	30°C, 5 h	VPFGVG	ACE inhibitory	Nakamura et al. (2007)
Bread (rye malt and whole wheat, 1:1)	Lb. reuteri TMW 1.106, yeast	rye malt sourdoughs: 37°C, 96h; whole wheat sourdough: 37°C, 24h; bread dough: 30°C, 2 h, 85% humidity	IQP, LQP, IIP, LIP, LLP, IPP, LPP, VPP	ACE inhibitory	Zhao et al. (2013)
Boza (fermented cereal beverage)	Not specified	Not specified	Not identified	ACE inhibitory	Kancabas and Karakaya (2013)
Rice wine (Huang Jiu)	Natural fermentation	25–33°C	FP, VY, LSP, WL, FR, LVQ, YW, LHV, VYP, LTF, HLL, LVR, LQQ, LHQ, LDR, YPR, LLPH	ACE inhibitory, antioxidant, hypocholesterolemic	Han and Xu (2011)
Fermented buckwheat sprouts	Lb. plantarum KT	Room temperature, 2 weeks, 25 mL/kg of inoculum	DVWY, FDART, VVG, FQ, VAE, WTFR	Vasorelaxation and ACE inhibitory	Nakamura et al. (2013b) and Koyama et al. (2014)
Grape wine	Oenococcus oeni m1	30°C, 48h	Peptide fractions	Antioxidant and ACE inhibitory	Apud et al. (2013)
	Sacharomyces cerevisiae	30°C, 121 h	Peptide fractions	Antioxidant and ACE inhibitory	Alcaide-Hidalgo et al. (2007)

Peptide bioactivity relies on inherent amino acid composition, sequence, and size. Antihypertensive, antimicrobial, immunomodulatory, anticancer, antithrombotic, opioid, and antioxidant activities are some of the biological activities attributed to peptides of fermented foods (Hebert et al., 2010). Specific amino acid sequences exhibiting more than two biological activities are also known as multifunctional peptides.

This chapter presents an overview of the identification and occurrence of bioactive peptides in fermented foods from animal and plant origin, excluding cheese peptides that are presented in Chapter 12. Recent insights and strategies for the production of bioactive peptides are reviewed to facilitate their further industrial exploitation. The chapter also addresses the current scientific evidence including human, animal, and in vitro studies that support the health benefits of the bioactive peptides from fermented foods as well as the plausible mechanisms by which they exert their physiological affect.

2.2 OCCURRENCE OF BIOACTIVE PEPTIDES IN FERMENTED FOODS

Fermented products are a well-documented source of bioactive peptides. Examples of bioactive fragments derived from animal proteins are presented in Table 2.1. Antihypertensive peptides derived from β-casein have been found in milk fermented by *Enterococcus faecalis* (Quirós et al., 2006, 2007), *Lactococcus lactis* NRRLB-50571 and NRRLB-50572 (Rodríguez-Figueroa et al., 2012, 2013), *Bifidobacterium bifidum* MF 20/5 (Gonzalez–Gonzalez et al., 2013), and *L. lactis* DIBCAB2 (Nejati et al., 2013). The most studied antihypertensive peptides are VPP and IPP and they are the bioactive components of the commercial antihypertensive dairy products Calpis and Evolus. These peptides are released from β- and κ-casein during fermentation by proteases and peptidases of *Lactobacillus helveticus* strains (Rodríguez-Figueroa et al., 2013). Recent investigations have shown that VPP and IPP also may exhibit an array of biological functions including antiinflammatory, antiadipogenic, antiatherosclerotic, and antiosteoporotic activities (Aihara et al., 2009, 2014; Eisele et al., 2012; Aihara et al., 2014; Chakrabarti and Wu, 2015; Fekete et al., 2015). Another multifunctional peptide with angiotensin I converting enzyme (ACE) inhibitory and antithrombotic activities (YQEPVLGPVRGPFPIIV) was isolated from milk fermented by a coculture of *Lactobacillus casei* Shirota and *Streptococcus thermophilus* (Rojas-Ronquillo et al., 2012). The presence of the antihypertensive peptide SKVVP, with the ability to reduce blood pressure in hypertensive subjects has also been reported in dahi, an Indian fermented product prepared by coculturing *Lactobacillus delbrueckii* ssp. *bulgaricus*, *Streptococcus salivarius* ssp. *thermophilus*, and *L. lactis* biovar *diacetylactis* (Ashar and Chand, 2004). Greek yogurt has been reported to contain some ACE inhibitory peptides derived from β-, κ-, α_{s1}-, and α_{s2}-caseins, as well as the multifunctional peptide YPVEPFTE, which displays both ACE inhibitory and opioid activities

(Papadimitriou et al., 2007). The investigation of Chen et al. (2010) indicated that fermented mare's milk (koumiss) is also rich in ACE inhibitory peptides. Asides ACE inhibitory peptides, the presence of immunomodulating (Tellez et al., 2010; Jiehui et al., 2014), antioxidant (Sabeena Farvin et al., 2010), antimutagenic (Sah et al., 2015), mucin-stimulating, and opioid sequences (Plaisancié et al., 2013, 2015) have also been reported in fermented milk products. A 2015 study describes exhaustively, the peptide profile of kefir, in which 16 newly identified peptides contained amino acid sequences previously described to exert ACE inhibitory, antimicrobial, immunomodulating, opioid, mineral binding, antioxidant, and antithrombotic activities (Ebner et al., 2015).

Fish and other aquatic organisms are also sources of bioactive peptides. ACE inhibitory amino acid sequences were found in traditional Japanese fermented skipjack tuna (*Katsuwonus pelamis*), called katsuobushi (Ryan et al., 2011), and Thai fermented shrimp pastes (Kleekayai et al., 2015). Korean researchers identified an ACE inhibitory peptide (EVMAGNLYPG) and two antioxidant peptides (HFGBPFH and FGHPY) in the sauce of fermented blue mussel (*Mytulis edulis*) (Jung et al., 2005; Rajapakse et al., 2005).

Regarding meat fermented products, there is scarce information on the occurrence of bioactive peptides. A crude peptide fraction of Petrovac sausages has been reported to possess antioxidant and ACE inhibitory activity (Vaštag et al., 2010). Several antioxidant di- and tripeptides have been identified in chorizo sausages (Broncano et al., 2012).

Soy-fermented products have shown great potential for supplying peptides with functional significance (Table 2.2). Cheonggukjang, a soybean product fermented by *Bacillus licheniformis* possess a number of dipeptides (LE, EW, SP, VE, VL, VT, and EF) that have shown insulin-sensitizing activities in animal experiments (Yang et al., 2013). Peptides isolated from douchi, fermented soy protein (LIVTQ and LIVT), tofuyo (WL and IFL), and soybean paste (HHL) have shown ACE inhibitory activity in vitro (Shin et al., 2001; Zhang et al., 2006; S Vallabha and Tiku, 2014). Tofuyo and fermented soy sauce contain a wide number of ACE inhibitory di- and tripeptides that have shown blood pressure lowering effects in spontaneously hypertensive rats (SHR) (Kuba et al., 2003, 2004; Nakahara et al., 2010, Nakahara et al., 2011). Antiosteoporotic effects of soymilk fermented with *Lactobacillus paracasei* ssp. *paracasei* NTU and *Lactobacillus plantarum* NTU 102 in Balb/c mice have been attributed to the presence of bioactive peptides as well as other active components (Chiang et al., 2012). Other legume-derived products, such as navy bean milks fermented with different lactic acid bacteria (LAB) strains have been recently developed and proven to contain ACE inhibitory peptides (Rui et al., 2015).

Sourdough has traditionally been used as a leavening agent in bread making. Some examples of antioxidant and ACE inhibitory amino acid sequences identified in sourdoughs are presented in Table 2.2. Twenty-five antioxidant peptides consisting of 8–57 amino acid residues were identified in the active fractions of sourdoughs prepared from different cereal flours including whole wheat, spelt, rye, and kamut (Coda et al., 2012). All the purified fractions ameliorated oxidative stress on mouse fibroblast cultures. In addition, several ACE inhibitory sequences have been identified in whole wheat (DPVAPLQRSGPEI, PVAPQLSRGLL, ELEIVMASPP, QILLPRPGQAA, and VPFGVG) and rye malt (LQP, IPP, LLP, and VPP) sourdoughs fermented by different LAB strains (Nakamura et al., 2007; Rizzello et al., 2008; Hu et al., 2011). Zhao et al. (2013) reported that IPP was the predominant tripeptide at 58 and 473 µmol/kg followed by LQP, IQP, and LPP in wheat and rye malt sourdoughs, respectively. Bread making processes may lead to modifications in the peptide concentrations of sourdough as a consequence of enzymatic conversions at the dough stage, and by thermal reactions during baking. Thus, concentrations of ACE inhibitory peptides in steamed bread have been reported to exceed 60 µmol/kg.

Peptide fractions of boza, a traditional Turkish beverage made by yeasts and LAB fermentation of rice, corn, wheat, and maize semolina/flour, have shown ACE inhibitory activity, although the peptides responsible for this effect have not been identified yet (Kancabaş and Karakaya, 2013). Similarly, Chinese rice wine, named Huang Jiu, contains di- and tripeptides sequences with ACE inhibitory, antioxidant, and hypocholesterolemic activities (Han and Xu, 2011).

Recently, Japanese researchers have developed a practical antihypertensive food from buckwheat sprouts by lactic acid fermentation (Nakamura et al., 2013a). Six peptides (DVWY, FDART, VVG, FQ, VAE, and WTFR) with ACE inhibitory and vasodilatory activities were identified to be responsible for the blood pressure lowering affect of fermented buckwheat sprouts in SHR (Koyama et al., 2014).

In grape wines, peptides are the least known nitrogen compounds, in spite of the fact that they are involved in the antihypertensive activity of these fermented foods (Pozo-Bayón et al., 2005). Recently, ACE inhibitory activity was reported in peptides released from red wine by *Saccharomyces cerevisiae* (Alcaide-Hidalgo et al., 2007) and *Oenococcus oeni* m1 (Apud et al., 2013). Moreover, peptide fractions of these red wines also showed antioxidant activity, suggesting that amino acid sequences released by *S. cerevisiae* during autolysis under wine conditions and *O. oeni* during malolactic fermentation could present multifunctional activities.

2.3 PRODUCTION OF BIOACTIVE PEPTIDES IN FERMENTED FOODS

Bioactive peptides can be generated by the proteolytic systems of starter and nonstarter cultures during the manufacture of fermented foods. A number of identified bioactive peptides have shown to be released by LAB. Generally, proteolytic systems of LAB contain cell envelope-associated proteinases (CEP) responsible for the first step of protein breakdown (Hebert et al., 2008). Specialized transport systems allow the uptake of the resulting oligopeptides as well as di- and tripeptides that can be further hydrolyzed to peptides and free amino acids by intracellular endopeptidases and aminopeptidases (Savijoki et al., 2006). CEP plays a key role in casein proteolysis during fermentation, contributing to the release of health benefiting peptides during food fermentation (Saavedra et al., 2013). Biochemical pathways for bioactive peptide production have been the subject of extensive research. Microarray analysis of gene expression in *L. helveticus* have proposed the biochemical pathway for the formation of peptides VPP and IPP in fermented milk (Chen et al., 2011). The proteolytic action in *L. helveticus* consists of the hydrolysis of β-casein by CEP; then the generated peptides are transferred into the cell by transport system permease protein (OppC), where endopeptidase O2 (PepO2), and aminopeptidases N and E (PepN and PepE), can act at the C- and N-terminal sequences to produce VPP and IPP.

Some bioactive peptides present in cheese are formed by the action of endogenous enzymes, such as plasmin that hydrolyzes β- and α_{s2}-caseins (Korhonen, 2009). Endogenous cathepsin D and CEP from thermophilic starters are reported to cleave α_{s1}-caseins in Emmental cheese. Moreover peptidases released from starter and nonstarter LAB strains contribute to the formation of bioactive peptides throughout the ripening period.

Release of bioactive peptides in fermented foods by molds may also occur by the action of four families of proteases including serine proteases, cysteine proteinases, aspartic proteinases, and metalloproteinases (Beermann and Hartung, 2012). A prolyl endoproteinase activity from *Aspergillus niger* has also been reported to generate potent ACE inhibitory peptides in soy sauce due to its main cleavage preference at the C-terminal side of Pro and hydroxyl Pro residues (Norris et al., 2014).

In fermented meat products, oligopeptides result from degradation of sarcoplasmic and myofibrillar proteins by endogenous muscle enzymes including cathepsins, in particular cathepsin D, and calpains (López et al., 2015). Polypeptides produced by these enzymes are further degraded by peptidases and aminopeptidases from both muscle and microbial enzymes during the latter stages of ripening. *Staphylococcus* species are recognized to have an important role on the hydrolysis of myofibrillar proteins during the ripening stages (Casaburi et al., 2008).

2.4 STRATEGIES TO INCREASE THE PRODUCTION OF BIOACTIVE PEPTIDES IN FERMENTED FOODS

The selection of microorganisms to be used in the manufacture of fermented foods is gaining importance as a strategy to produce peptides in the final product, which allow specific health claims to be made. Donkor et al. (2005) investigated the proteolytic activity of several LAB cultures and probiotic strains as a determinant of ACE inhibitory activity of soy fermented milk. A *Lactobacillus acidophilus* strain was selected as the best culture to maximize the ACE inhibitory activity of soy fermented milk. Interesting observations have also been found when selecting *Lactobacillus rhamnosus* NCDC24, *L. casei* subsp. *casei* NCDC17, and *L. paracasei* subsp. *paracasei* NCDC63 to maximize the release of antioxidant peptides in fermented milk (Ramesh et al., 2012). Ong et al. (2007) found an increased release of ACE inhibitory peptides in Cheddar cheese made with the starter lactococci when probiotic strains *L. casei* 279 and *L. casei* LAFTI L26 were added. *L. helveticus*, *L. delbrueckii* ssp. *bulgaricus*, and *L. casei* strains were selected out of 180 LAB strains, to be used as adjunct cultures, that remarkably increased the content of ACE inhibitory and Ca-binding peptides in Bulgarian white brined cheese (Dimitrov et al., 2015).

Although bifidobacteria are not considered as proteolytic as lactobacilli, investigations have demonstrated that their slow acidification rate allows the casein to remain soluble for a greater amount of time, which may represent a technological advantage for the generation of bioactive peptides. It has been demonstrated that *Bifidobacterium longum* gives rise to fermented milks with stronger ACE inhibitory activity than *L. acidophilus* or *L. casei* (Ramchandran and Shah, 2008). Similarly, *B. bifidum* MF 20/5 released a larger amount of ACE inhibitory (LVYPFP) and antioxidant (VLPVPQK) peptides in fermented milk than *L. helveticus* DSM13137, that produced few potent ACE antihypertensive peptides (IPP and VPP) and a large amount of amino acids (Gonzalez–Gonzalez et al., 2013) These results indicated that peptidases are the predominant hydrolytic enzymes in *L. helveticus* whereas protease activity was predominant in *B. bifidum*.

Comparative genomic approaches are useful to predict the proteolytic potential of the LAB strains and select microorganisms able to release bioactive peptides. Liu et al. (2010) reported that *L. helveticus* and *L. delbrueckii* subsp. *bulgaricus* have a very broad set of genes encoding for proteolytic enzymes including proteinase (Prt), endopeptidases (PepO, PepF, and PepG), proline peptidases (PepX, PepI, PepR, PepP, and PepQ), tripeptidases (PepT), dipeptidases (PepD and PepV), and aminopeptidases (PepC, PepN, PepM, PepA, and Pcp) in comparison to LAB naturally present in plant foods, such as *L. plantarum*, *O. oeni*, and *Leuconostoc mesenteroides*.

Interactions between different microorganisms in fermented foods may exert positive or negative effects for bioactive peptide production. Therefore, evaluation of cocultures of selected microorganisms for their proteolytic activity and ability to produce fermented milk enriched in ACE inhibitory peptides has been performed. For example, Chaves-Lopez et al. (2014) concluded that the most effective strain combination to enhance the release of ACE inhibitory peptides in koumiss was *Pichia kudriavzevii* KL84A, *L. plantarum* LAT3, and *E. faecalis* KL06.

The synthesis and activity of LAB proteinases are affected by fermentation conditions, such as incubation temperature, extracellular pH, agitation, and the presence of oxygen (Agyei et al., 2013). Hence, the optimization of fermentation conditions is an important step for maximizing the production yield of bioactive peptides. For example, soy sauce seasoning enriched in ACE inhibitory dipeptides (SY and GY) was produced by a combination of approaches, such as an increased soybean to wheat ratio, reduction of fermentation time from 6–24 months to 5 days, and increasing fermentation temperature from 15–30°C to 45–55°C (Nakahara et al., 2010, 2012). These processing conditions caused the inactivation of peptidases from starter cultures (*Aspergillus oryzae* and *Aspergillus sojae*) giving rise to a lower peptide degradation during soy sauce fermentation.

Other strategies to enhance the release of bioactive peptides during fermentation are focused on the improvement of proteinase yields via optimization of media-component. Supplementation of milk with peptides prior to fermentation enhanced the proteolytic activity of *L. helveticus*, *L. acidophilus*, and *L. delbrueckii* subsp. *bulgaricus* and, consequently, the ACE inhibitory activity of fermented milk (Gandhi and Shah, 2014).

2.5 EVIDENCE FOR HEALTH EFFECTS OF BIOACTIVE PEPTIDES DERIVED FROM FERMENTED FOODS

This section summarizes current scientific evidences including human, animal, and in vitro studies looking into the biological activity and mechanisms of action of peptides identified in fermented foods. A wide range of biological effects have been described for peptides present in dairy fermented foods, particularly IPP and VPP. In contrast, less information is available regarding the bioactivities of peptides present in other fermented foods, with the majority of the biological activities being described by in vitro studies. As the biological functions of peptides present in cheeses are summarized in Chapter 12, they have not been included in this section.

2.5.1 Health Benefits on the Cardiovascular System

Raised blood pressure (hypertension) is a major cardiovascular risk factor that has been considered as a key target for controlling cardiovascular diseases-related mortality and improving global health (WHO, 2014). A number of preclinical studies have proven the efficacy of fermented foods containing peptides as antihypertensive agents. For example, orally administered peptides isolated from fermented buckwheat sprouts (Nakamura et al., 2013a; Koyama et al., 2014) and sauce of fermented blue mussels (Je et al., 2005) exhibited a blood pressure lowering effect in SHR. This antihypertensive effect was related to ACE inhibition that suppressed angiotensin II-mediated vasoconstriction. Fermented soybean seasoning containing GY and SY has demonstrated antihypertensive effects in Dahl salt-sensitive rats (Nakahara et al., 2011). The antihypertensive mechanisms consisted of ACE inhibition (particularly in the lung) and a decrease of angiotensin II and aldosterone levels in blood. Peptides LHLPLP and HLPLP released from β-casein during milk fermentation with *E. faecalis* were effective at reducing systolic and diastolic blood pressure (SBP and DBP, respectively) in SHR (Miguel et al., 2006). Similarly, administration of milk fermented with *L. lactis* NRRLB-50571 and NRRLB-50572 containing ACE inhibitory peptides significantly decrease SBP, DBP, and heart rate in SHR (Rodríguez-Figueroa et al., 2013). Single oral administration of IPP and VPP peptides (0.6 and 0.3 mg/kg body weight) significantly reduced SBP by 32 and 28 mm Hg in SHR, respectively.

Human clinical trials are necessary to evaluate the efficacy of food derived bioactive peptides. Daily supplementation of the diet with fermented milk, Dahi, containing ACE inhibitory peptide SKVYP, during an eight week period, diminished SBP and DBP in Indian hypertensive subjects (Ashar and Chand, 2004). However, literature concerning antihypertensive effects of oral administration of VPP and IPP has shown a number of inconsistencies across studies in human trials (Seppo et al., 2003; Engberink et al., 2008; Boelsma and Kloek, 2010; Hove et al., 2015). A comprehensive metaanalysis of data from all relevant human studies to date (33 randomized controlled trials) have provided a more accurate estimate of the true affect of IPP and VPP ingestion on blood pressure (Fekete et al., 2015). This study concluded that the long-term ingestion of these lactotripeptides may modestly reduce blood pressure (−2.95 and −1.15 mm Hg for SBP and DBP, respectively) in comparison to antihypertensive drugs (−14.5 and −10.7 mm Hg for SBP and DBP, respectively). Moreover, subgroup analyses revealed that the country in which studies were performed had a significant influence on treatment effect. Thus, greater blood pressure reduction was reached in Japanese studies, compared to European studies. In summary, VPP and IPP ingestion represents an important strategy for reducing the risk of hypertension through lifespan, which could reduce the need for antihypertensive medication later in life.

Based on current available scientific studies, VPP and IPP may exert their antihypertensive action through an array of mechanistic pathways. ACE inhibition has been the most studied blood pressure lowering mechanism of these tripeptides as it has been demonstrated both in vitro and in vivo (Sipola et al., 2002; Jauhiainen et al., 2012). An interesting observation from a human study was that the plasma concentration of IPP and VPP is markedly lower than their effective concentration on ACE inhibition determined in vitro (Foltz et al., 2007). Activation of angiotensin converting enzyme 2 (ACE-2) is another plausible blood pressure lowering mechanism of tripeptide IPP. An ex vivo study confirmed that administration of IPP induced rat mesenteric artery vasorelaxation through ACE-2 activation (Ehlers et al., 2011). This enzyme cleaves angiotensin II to form angiotensin 1–7 that interacts with the G-protein-coupled receptor-Mas, inducing vasorelaxation (Bader, 2013).

Vascular dysfunction is another major contributor to hypertension. It is caused primarily by an increased production of proinflammatory/vasoconstrictor molecules that produce a deficiency in vasodilators, such as nitric oxide (NO) (Schulz et al., 2011). Long-term ingestion of IPP and VPP has shown to ameliorate endothelial dysfunctions and cause vasodilatation in animal models of hypertension, and mild hypertensive subjects with endothelial dysfunction (Hirota et al., 2007; Jäkälä et al., 2009; Nonaka et al., 2014). A study in cultured endothelial cells indicates that the mechanism by which these peptides modulate endothelial function is the increased production of vasodilatory molecules, such as NO (Hirota et al., 2011). These observations were consistent with in vivo investigations showing that VPP and IPP administration upregulate gene expression of endothelial NO synthase in SHR (Yamaguchi et al., 2009).

Vascular inflammation-induced peripheral resistance also contributes to hypertension and associated vascular pathologies. IPP and VPP peptides can also reduce hypertension, improving vascular pathogenesis, and hence, modulating vascular remodeling. For example, VPP and IPP ingestion offered protection against the development of atherosclerotic plaques (intima to media thickness in the aortic arch) through attenuation of inflammatory and hypertensive pathways in male apoE−/− mice (Nakamura et al., 2013b). The proposed mechanism for this atherosclerotic effect of VPP was attenuation of monocytic TH-1 cell adhesion to inflamed endothelial cells, through inhibition of the c-Jun N-terminal kinase phosphorylation pathway and, consequently, suppression of β1 and β2 integrins activation in monocytes (Aihara et al., 2009).

Thrombosis is also implicated in the physiopathology of vascular endothelial cells that lead to sudden death and acute coronary syndromes, among other disorders (Gurm and Bhatt, 2005). Thrombosis initiates with the formation of a surface where platelets can accumulate and synthesize prostaglandin H2, which is transformed to thromboxane A2, a potent stimulator of platelet aggregation. Given the central role of thromboembolism in cardiovascular disease, inhibition of thrombin has become a key therapeutic strategy.

Thrombin plays a central role in the coagulation cascade. It acts as a catalyst for converting fibrinogen to fibrin, activates coagulation factors stabilizing the fibrin cross-links (Badimon et al., 1994), enhances leukocyte chemotaxis, and induces vasoconstriction and platelet aggregation (Bar-Shavit et al., 1983a,b).

In this context, antithrombotic peptides (YQEPVLGPVRGPFPIIV) isolated from a commercial dairy product fermented by *L. casei* Shirota and *S. thermophilus* have been demonstrated to inhibit thrombin activity in vitro, suggesting their promising potential to prevent thrombosis (Rojas-Ronquillo et al., 2012).

2.5.2 Health Benefits on the Nervous System

The key mediators of stress are the sympathetic adrenomedullary system (SAM) and the hypothalamus-pituitary adrenocortical (HPA) axis. The release of catecholamines from SAM, increases heart rate, blood pressure, and blood glucose levels, whereas the activation of the HPA axis releases glucocorticoid cortisol to intensify stress adaptation (McEwen, 2008). Excessive stress and insufficient recovery, may lead to chronic heart rate and blood pressure, which can result over time in disorders, such as stroke and heart attacks. Moreover, although there is little evidence regarding the effects of life stressors on brain structure there are some data indicating that prolonged chronic stress may produce a remodeling of neuronal circuitry that is related to certain psychiatric illnesses. A randomized, controlled, double blind intervention has demonstrated that daily ingestion of yogurt containing bioactive peptides, B vitamins, and α-lactalbumin, improves stress coping in high-trait anxiety individuals (Jaatinen et al., 2014).

2.5.3 Health Benefits on the Gastrointestinal System

Intestinal mucus plays an important protective role against colonization of pathogenic microorganisms, acidic pH, luminal proteases, mechanical damage, and potential carcinogens (Gibson and Muir, 2005). The major component of mucus is gel-forming mucin, Muc2, produced by goblet cells of the epithelium in the large and small intestine. Muc4, a transmembrane-associated mucin also protects the intestine from pathogens by providing a steric barrier. Paneth cells located in the crypts of the small intestine also play a key role in intestinal protection. These cells secrete a number of antimicrobial substances including lysozyme, phospholipase A2, α-defensins, etc. Many human studies support the hypothesis that the development of a lot of intestinal disorders is linked to alterations in the mucin synthesis, secretion, and/or degradation (Mudter, 2011). Duodenal ulcerations and ulcerative colitis are associated with a decreased number of goblet cells and mucus layer (Pugh et al., 1996; Strugala et al., 2008). Based on these investigations, the strengthening of the mucus layer through diet could be extremely beneficial (Gibson and Muir, 2005). In this context, β-casein derived peptides in milk fermented by *L. delbrueckii* ssp. *bulgaricus* and *S. salivarius* ssp. *thermophilus* induce mucin secretion in intestinal goblet cells in vitro and in ex vivo models (Plaisancié et al., 2013). β-Casein (f94-123) derived peptide was identified in the most active peptide fraction of fermented milk. Daily oral administration of this peptide (0.1–100 μM), simultaneously increased mucin production, and the number of goblet cells and paneth cells in 10–18-day postnatal rats (Plaisancié et al., 2013). These results support the promising potential of bioactive peptides in fermented milk for intestinal health promotion. The mechanisms responsible for these effects were the upregulation of Muc2 and Muc4 gene expression by goblet cells, and lysozyme and defensin 5 expression by paneth cells. Interestingly, these authors found that digestion of β-casein (f94-123) by intestinal brush border enzymes could result in fragment β-casein (f117-123) that exert opposite effects, such as inhibition of expression and production of Muc2 in intestinal cells (Plaisancié et al., 2015).

2.5.4 Health Benefits on the Immune System

Inflammation is part of the host defense mechanism of the human body to external stimuli that causes injury. Inflammation involves the production of various proinflammatory mediators, such as interferon gamma (IFN-γ), tumor necrosis factor alpha (TNF-α), interleukin 6 (IL-6), and NO. Excessive production of proinflammatory mediators by immune cells results in body tissue damage and immune dysfunction. Uncontrolled and aberrant inflammatory response has often been linked to a number of chronic diseases, such as cardiovascular diseases, metabolic syndrome, obesity, and cancer (Montecucco et al., 2013). Therefore, control of excessive inflammation through diet is important to maintain health and wellness. VPP peptide has been proven to have an antiinflammatory affect on the adipose tissue of high fat diet-fed mice, via the inhibition of proinflammatory macrophages accumulation in the stromal vascular fraction (Aihara et al., 2014). The underlying mechanism by which VPP dietary supplementation attenuates adipose tissue chronic inflammation involves the inhibition of the expression of monocyte chemoattractant protein-1 and IL-6. Another study demonstrated that peptides QEPVL and QEPV derived from fermented milk, significantly activate lymphocyte proliferation both in vitro and in vivo (Jiehui et al., 2014). A further interesting observation of these studies was that QEPVL displayed antiinflammatory effects by regulation of in vitro NO release and production of proinflammatory cytokines (IL-4, IL-10, IFN-γ, and TNF-α) in Balb/c mice.

2.5.5 Health Benefits on Adipose Tissue

Insulin resistance contributes to the pathogenesis of obesity, type-2 diabetes, and metabolic syndrome. Insulin resistance results in adipose tissue dysfunction that involves adipocyte inflammation, higher levels of circulating lipids, and suppression of preadipocyte differentiation (Stehno-Bittel, 2008). A few preliminary studies on cell cultures have supported a potential insulin-mimetic action of some bioactive peptides derived from fermented foods. Research work has found that IPP and VPP peptides exert insulin-mimetic action inducing adipogenesis in murine preadipocytes (Chakrabarti and Wu, 2015). This effect involved upregulation of transcriptional regulators, such as c-Jun and C/EBPα. Moreover, both peptides suppressed secretion of TNF-α by adipocytes preventing loss of adiponectin release and the activation of proinflammatory NFκβ pathway (Chakrabarti and Wu, 2015). Similarly, Yang et al. (2013) found that the peptide fraction derived from traditionally fermented soybeans (Cheonggukjang) exerts a potent insulin-sensitizing action in 3T3-L1 adipocytes cultures via upregulation of proliferated peroxisome activator receptor gamma.

2.5.6 Health Benefits on Bones

Osteoporosis is a major skeletal disease associated with aging that is reaching epidemic proportions (Pietschmann et al., 2009). This disease is characterized by low bone mass, microarchitectural deterioration, and increased risk of fracture. Isolated VPP peptide from milk fermented by *L. helveticus* LBK-16H, has been shown to increase bone formation in vitro (Narva et al., 2004). Consistent with these results, ovariectomized rats which were fed fermented milk containing VPP, had bone loss prevented after 12 weeks of intervention (Narva et al., 2007). However, VPP given in water showed no clear affect on bone loss suggesting a protective affect of food matrix to gastrointestinal digestion. Another investigation reported that dietary supplementation with soy skim milks fermented with *L. paracasei* ssp. *paracasei* NTU 101 or *L. plantarum* NTU 102 (containing 335 and 281 mg/g peptides, respectively) can attenuate aging-induced bone loss in Balb/c mice (Chiang et al., 2012). The mechanistic explanation to this, was an increase of trabecular bone volume, mean trabecular thickness and resting area, and a reduction of bone resorption and acid phosphatase activity.

2.6 FUTURE OUTLOOK

Microbial fermentation provides a natural technology applicable for the enrichment of fermented foods in bioactive peptides from animal or plant origin. Considering prospering functional foods businesses, the generation of bioactive peptides helps to commercially exploit fermented foods. As highlighted in this chapter, strategies to increase bioactive peptide content in fermented foods are emerging as a subject of great importance for global health and functional product development. To this end, specific production strains and optimized fermentation processes are likely to be employed industrially in the future. However, challenges, such as process scalability, peptides stability, and bitter taste require interdisciplinary additional research.

Only a few commercially fermented products containing peptides as active components have launched on the market with specific health claims. Regrettably, biological function of peptides isolated from most fermented foods developed to date are supported by preliminary studies often based on in vitro approaches or animal models of disease. These studies have provided some indications for health benefits of bioactive peptides, albeit with limited validity concerning extrapolation to humans. Therefore, for a true estimation of the functional benefits of bioactive peptides, there is a need for more scientific evidence in human interventions and clinical trials, preferably including the intended fermented product to take into account the matrix effect. Such studies should directly relate the effects to in vivo parameters and meaningful markers for health and disease.

ACKNOWLEDGMENTS

This work was supported by grants AGL2013-43247-R co-funded by the Ministry of Economy and Competitiveness (MINECO, Spain) and the European Union through the FEDER programme. E. Peñas is indebted to Spanish "Ramón y Cajal" Programme for financial support.

REFERENCES

Agyei, D., Potumarthi, R., Danquah, M.K., 2013. Production of lactobacilli proteinases for the manufacture of bioactive peptides: Part I-upstream processes. In: Kim, S.-K. (Ed.), Marine Proteins and Peptides: Biological Activities and Applications. John Wiley & Sons, Oxford, pp. 207–229.

Aihara, K., Ishii, H., Yoshida, M., 2009. Casein-derived tripeptide, Val-Pro-Pro (VPP), modulates monocyte adhesion to vascular endothelium. Journal of Atherosclerosis and Thrombosis 16, 594–603.

Aihara, K., Osaka, M., Yoshida, M., 2014. Oral administration of the milk casein-derived tripeptide Val-Pro-Pro attenuates high-fat diet-induced adipose tissue inflammation in mice. British Journal of Nutrition 112, 513–519.

Alcaide-Hidalgo, J.M., Pueyo, E., Polo, M.C., Martinez-Rodriguez, A.J., 2007. Bioactive peptides released from *Saccharomyces cerevisiae* under accelerated autolysis in a wine model system. Journal of Food Science 72, M276–M279.

Apud, G.R., Vaquero, M.J.R., Rollan, G., Stivala, M.G., Fernández, P.A., 2013. Increase in antioxidant and antihypertensive peptides from Argentinean wines by *Oenococcus oeni*. International Journal of Food Microbiology 163, 166–170.

Ashar, M.N., Chand, R., 2004. Fermented milk containing ACE-inhibitory peptides reduces blood pressure in middle aged hypertensive subjects. Milchwissenschaft 59, 363–366.

Bader, M., 2013. ACE2, angiotensin-(1-7), and Mas: the other side of the coin. Pflugers Archiv European Journal of Physiology 465 (1), 79–85.

Badimon, L., Meyer, B.J., Badimon, J.J., 1994. Thrombin in arterial thrombosis. Haemostasis 24, 69–80.

Bar-Shavit, R., Kahn, A., Fenton II, J.W., Wilner, G.D., 1983a. Chemotactic response of monocytes to thrombin. Journal of Cell Biology 96, 282–285.

Bar-Shavit, R., Kahn, A., Wilner, G.D., Fenton II, J.W., 1983b. Monocyte chemotaxis: stimulation by specific exosite region in thrombin. Science 220, 728–731.

Beermann, C., Hartung, J., 2012. Current enzymatic milk fermentation procedures. European Food Research and Technology 235, 1–12.

Boelsma, E., Kloek, J., 2010. IPP-rich milk protein hydrolysate lowers blood pressure in subjects with stage 1 hypertension, a randomized controlled trial. Nutrition Journal 9, 52.

Broncano, J.M., Otte, J., Petrón, M.J., Parra, V., Timón, M.L., 2012. Isolation and identification of low molecular weight antioxidant compounds from fermented "chorizo" sausages. Meat Science 90, 494–501.

Casaburi, A., Di Monaco, R., Cavella, S., Toldrá, F., Ercolini, D., Villani, F., 2008. Proteolytic and lipolytic starter cultures and their effect on traditional fermented sausages ripening and sensory traits. Food Microbiology 25, 335–347.

Cavazos, A., Gonzalez de Mejia, E., 2013. Identification of bioactive peptides from cereal storage proteins and their potential role in prevention of chronic diseases. Comprehensive Reviews in Food Science and Food Safety 12, 364–380.

Chakrabarti, S., Wu, J., 2015. Milk-derived tripeptides IPP (Ile-Pro-Pro) and VPP (Val-Pro- Pro) promote adipocyte differentiation and inhibit inflammation in 3T3-F442A cells. PLoS One 10, 1–15.

Chaves-López, C., Serio, A., Paparella, A., Martuscelli, M., Corsetti, A., Tofalo, R., Suzzi, G., 2014. Impact of microbial cultures on proteolysis and release of bioactive peptides in fermented milk. Food Microbiology 42, 117–121.

Chen, Y., Wang, Z., Chen, X., Liu, Y., Zhang, H., Sun, T., 2010. Identification of angiotensin I-converting enzyme inhibitory peptides from koumiss, a traditional fermented mare's milk. Journal of Dairy Science 93, 884–892.

Chen, Y., Liu, W., Xue, J., Yang, J., Chen, X., Shao, Y., Kwok, L., Bilige, M., Mang, L., Zhang, H., 2011. Angiotensin-converting enzyme inhibitory activity of *Lactobacillus helveticus* strains from traditional fermented dairy foods and antihypertensive effect of fermented milk of strain H9. Journal of Dairy Science 97, 6680–6692.

Chiang, S.S., Liao, J.W., Pan, T.M., 2012. Effect of bioactive compounds in lactobacilli-fermented soy skim milk on femoral bone microstructure of aging mice. Journal of the Science of Food and Agriculture 92, 328–335.

Coda, R., Rizzello, C.G., Pinto, D., Gobbetti, M., 2012. Selected lactic acid bacteria synthesize antioxidant peptides during sourdough fermentation of cereal flours. Applied and Environmental Microbiology 78, 1087–1096.

Dimitrov, Z., Chorbadjiyska, E., Gotova, I., Pashova, K., Ilieva, S., 2015. Selected adjunct cultures remarkably increase the content of bioactive peptides in Bulgarian white brined cheese. Biotechnology and Biotechnological Equipment 29, 78–83.

Donkor, O.N., Henriksson, A., Vasiljevic, T., Shah, N.P., 2005. Probiotic strains as starter cultures improve angiotensin-converting enzyme inhibitory activity in soy yogurt. Journal of Food Science 70, M375–M381.

Ebner, J., Aşçi Arslan, A., Fedorova, M., Hoffmann, R., Küçükçetin, A., Pischetsrieder, M., 2015. Peptide profiling of bovine kefir reveals 236 unique peptides released from caseins during its production by starter culture or kefir grains. Journal of Proteomics 117, 41–57.

Ehlers, P.I., Nurmi, L., Turpeinen, A.M., Korpela, R., Vapaatalo, H., 2011. Casein-derived tripeptide Ile-Pro-Pro improves angiotensin-(1-7)- and bradykinin-induced rat mesenteric artery relaxation. Life Sciences 88, 206–211.

Eisele, T., Stressler, T., Kranz, B., Fischer, L., 2012. Quantification of dabsylated di- and tri-peptides in fermented milk. Food Chemistry 135, 2808–2813.

Engberink, M.F., Schouten, E.G., Kok, F.J., Van Mierlo, L.A.J., Brouwer, I.A., Geleijnse, J.M., 2008. Lactotripeptides show no effect on human blood pressure: results from a double-blind randomized controlled trial. Hypertension 51, 399–405.

Foltz, M., Meynen, E.E., Bianco, V., Van Platerink, C., Koning, T.M.M.G., Kloek, J., 2007. Angiotensin converting enzyme inhibitory peptides from a lactotripeptide-enriched milk beverage are absorbed intact into the circulation. Journal of Nutrition 137, 953–958.

Gandhi, A., Shah, N.P., 2014. Cell growth and proteolytic activity of *Lactobacillus acidophilus*, *Lactobacillus helveticus*, *Lactobacillus delbrueckii* ssp. *bulgaricus*, and *Streptococcus thermophilus* in milk as affected by supplementation with peptide fractions. International Journal of Food Sciences and Nutrition 65, 937–941.

Gibson, P.R., Muir, J.G., 2005. Reinforcing the mucus: a new therapeutic approach for ulcerative colitis? Gut 54, 900–903.

Gonzalez-Gonzalez, C., Gibson, T., Jauregi, P., 2013. Novel probiotic-fermented milk with angiotensin I-converting enzyme inhibitory peptides produced by *Bifidobacterium bifidum* MF 20/5. International Journal of Food Microbiology 167, 131–137.

Gurm, H.S., Bhatt, D.L., 2005. Thrombin, an ideal target for pharmacological inhibition: a review of direct thrombin inhibitors. American Heart Journal 149 (Suppl. 1), S43–S53.

Han, F.L., Xu, Y., 2011. Identification of low molecular weight peptides in Chinese rice wine (Huang Jiu) by UPLC-ESI-MS/MS. Journal of the Institute of Brewing 117 (2), 238–250.

Hebert, E.M., Mamone, G., Picariello, G., Raya, R.R., Savoy, G., Ferranti, P., Addeo, F., 2008. Characterization of the pattern of α_{s1}- and β-casein breakdown and release of a bioactive peptide by a cell envelope proteinase from *Lactobacillus delbrueckii* subsp. *lactis* CRL 581. Applied and Environmental Microbiology 74, 3682–3689.

Hebert, E.M., Saavedra, L., Ferranti, P., 2010. Bioactive peptides derived from casein and whey proteins. In: Mozzi, F., Raya, R.R., Vignolo, G.M. (Eds.), Biotechnology of Lactic Acid Bacteria: Novel Applications. John Wiley & Sons, Oxford, pp. 233–249.

Hirota, T., Nonaka, A., Matsushita, A., Uchida, N., Ohki, K., Asakura, M., Kitakaze, M., 2011. Milk casein-derived tripeptides, VPP and IPP induced NO production in cultured endothelial cells and endothelium-dependent relaxation of isolated aortic rings. Heart and Vessels 26, 549–556.

Hirota, T., Ohki, K., Kawagishi, R., Kajimoto, Y., Mizuno, S., Nakamura, Y., Kitakaze, M., 2007. Casein hydrolysate containing the antihypertensive tripeptides Val-Pro-Pro and Ile-Pro-Pro improves vascular endothelial function independent of blood pressure-lowering effects: contribution of the inhibitory action of angiotensin-converting enzyme. Hypertension Research 30, 489–496.

Hove, K.D., Brøns, C., Færch, K., Lund, S.S., Rossing, P., Vaag, A., 2015. Effects of 12 weeks of treatment with fermented milk on blood pressure, glucose metabolism and markers of cardiovascular risk in patients with type 2 diabetes: a randomised double-blind placebo-controlled study. European Journal of Endocrinology 172, 11–20.

Hu, Y., Stromeck, A., Loponen, J., Lopes-Lutz, D., Schieber, A., Gänzle, M.G., 2011. LC-MS/MS quantification of bioactive angiotensin I-converting enzyme inhibitory peptides in rye malt sourdoughs. Journal of Agricultural and Food Chemistry 59, 11983–11989.

Fekete, Á.A., Ian Givens, D., Lovegrove, J.A., 2015. Casein-derived lactotripeptides reduce systolic and diastolic blood pressure in a meta-analysis of randomised clinical trials. Nutrients 7, 659–681.

Jaatinen, N., Korpela, R., Poussa, T., Turpeinen, A., Mustonen, S., Merilahti, J., Peuhkuri, K., 2014. Effects of daily intake of yoghurt enriched with bioactive components on chronic stress responses: a double-blinded randomized controlled trial. International Journal of Food Sciences and Nutrition 65, 507–514.

Jäkälä, P., Hakala, A., Turpeinen, A.M., Korpela, R., Vapaatalo, H., 2009. Casein-derived bioactive tripeptides Ile-Pro-Pro and Val-Pro-Pro attenuate the development of hypertension and improve endothelial function in salt-loaded Goto-Kakizaki rats. Journal of Functional Foods 1, 366–374.

Jauhiainen, T., Niittynen, L., Orei, M., Järvenpää, S., Hiltunen, T.P., Rönnback, M., Vapaatalo, H., Korpela, R., 2012. Effects of long-term intake of lactotripeptides on cardiovascular risk factors in hypertensive subjects. European Journal of Clinical Nutrition 66, 843–849.

Je, J.Y., Park, P.J., Byun, H.G., Jung, W.K., Kim, S.K., 2005. Angiotensin I converting enzyme (ACE) inhibitory peptide derived from the sauce of fermented blue mussel, *Mytilus edulis*. Bioresource Technology 96, 1624–1629.

Jiehui, Z., Liuliu, M., Haihong, X., Yang, G., Yingkai, J., Lun, Z., An Li, D.X., Dongsheng, Z., Shaohui, Z., 2014. Immunomodulating effects of casein-derived peptides QEPVL and QEPV on lymphocytes *in vitro* and *in vivo*. Food and Function 5, 2061–2069.

Jung, W.K., Rajapakse, N., Kim, S.K., 2005. Antioxidative activity of a low molecular weight peptide derived from the sauce of fermented blue mussel, *Mytilus edulis*. European Food Research and Technology 220 (5–6), 535–539.

Kancabaş, A., Karakaya, S., 2013. Angiotensin-converting enzyme (ACE)-inhibitory activity of boza, a traditional fermented beverage. Journal of the Science of Food and Agriculture 93, 641–645.

Kleekayai, T., Harnedy, P.A., O'Keeffe, M.B., Poyarkov, A.A., Cunhaneves, A., Suntornsuk, W., Fitzgerald, R.J., 2015. Extraction of antioxidant and ACE inhibitory peptides from Thai traditional fermented shrimp pastes. Food Chemistry 176, 441–447.

Korhonen, H., 2009. Milk-derived bioactive peptides: from science to applications. Journal of Functional Foods 1, 177–187.

Koyama, M., Hattori, S., Amano, Y., Watanabe, M., Nakamura, K., 2014. Blood pressure-lowering peptides from neo-fermented buckwheat sprouts: a new approach to estimating ACE-inhibitory activity. PLoS One 9.

Kuba, M., Tanaka, K., Tawata, S., Takeda, Y., Yasuda, M., 2003. Angiotensin I-converting enzyme inhibitory peptides isolated from tofuyo fermented soybean food. Bioscience, Biotechnology and Biochemistry 67, 1278–1283.

Kuba, M., Shinjo, S., Yasuda, M., 2004. Antihypertensive and hypocholesterolemic effects of tofuyo in spontaneously hypertensive rats. Journal of Health Science 50, 670–673.

Lafarga, T., O'Connor, P., Hayes, M., 2014. Identification of novel dipeptidyl peptidase-IV and angiotensin-I- converting enzyme inhibitory peptides from meat proteins using in silico analysis. Peptides 59, 53–62.

Liu, M., Bayjanov, J.R., Renckens, B., Nauta, A., Siezen, R.J., 2010. The proteolytic system of lactic acid bacteria revisited: a genomic comparison. BMC Genomics 11.

López, C.M., Bru, E., Vignolo, G.M., Fadda, S.G., 2015. Identification of small peptides arising from hydrolysis of meat proteins in dry fermented sausages. Meat Science 104, 20–29.

López-Barrios, L., Gutiérrez-Uribe, J.A., Serna-Saldívar, S.O., 2014. Bioactive peptides and hydrolysates from pulses and their potential use as functional ingredients. Journal of Food Science 79, R273–R283.

McEwen, B.S., 2008. Central effects of stress hormones in health and disease: understanding the protective and damaging effects of stress and stress mediators. European Journal of Pharmacology 583, 174–185.

Miguel, M., Recio, I., Ramos, M., Delgado, M.A., Aleixandre, M.A., 2006. Antihypertensive effect of peptides obtained from *Enterococcus faecalis*-fermented milk in rats. Journal of Dairy Science 89, 3352–3359.

Miguel, M., Dávalos, A., Manso, M.A., de la Peña, G., Lasunción, M.A., López-Fandiño, R., 2008. Transepithelial transport across Caco-2 cell monolayers of antihypertensive egg-derived peptides. PepT1-mediated flux of Tyr-Pro-Ile. Molecular Nutrition and Food Research 52, 1507–1513.

Montecucco, F., Mach, F., Pende, A., 2013. Inflammation is a key pathophysiological feature of metabolic syndrome. Mediators of Inflammation:135984.

Montoya-Rodríguez, A., Gómez-Favela, M.A., Reyes-Moreno, C., Milán-Carrillo, J., González de Mejía, E., 2015. Identification of bioactive peptide sequences from amaranth (*Amaranthus hypochondriacus*) seed proteins and their potential role in the prevention of chronic diseases. Comprehensive Reviews in Food Science and Food Safety 14, 139–158.

Mora, L., Hayes, M., 2015. Cardioprotective cryptides derived from fish and other food sources: generation, application, and future markets. Journal of Agricultural and Food Chemistry 63, 1319–1331.

Mudter, J., 2011. What's new about inflammatory bowel diseases in 2011. World Journal of Gastroenterology 17, 3177.

Nakahara, T., Sano, A., Yamaguchi, H., Sugimoto, K., Chikata, H., Kinoshita, E., Uchida, R., 2010. Antihypertensive effect of peptide-enriched soy sauce-like seasoning and identification of its angiotensin I-converting enzyme inhibitory substances. Journal of Agricultural and Food Chemistry 58, 821–827.

Nakahara, T., Sugimoto, K., Sano, A., Yamaguchi, H., Katayama, H., Uchida, R., 2011. Antihypertensive mechanism of a peptide-enriched soy sauce-like seasoning: the active constituents and its suppressive effect on renin-angiotensin-aldosterone system. Journal of Food Science 76, H201–H206.

Nakahara, T., Yamaguchi, H., Uchida, R., 2012. Effect of temperature on the stability of various peptidases during peptide-enriched soy sauce fermentation. Journal of Bioscience and Bioengineering 113, 355–359.

Nakamura, T., Yoshida, A., Komatsuzaki, N., Kawasumi, T., Shima, J., 2007. Isolation and characterization of a low molecular weight peptide contained in sourdough. Journal of Agricultural and Food Chemistry 55, 4871–4876.

Nakamura, K., Naramoto, K., Koyama, M., 2013a. Blood-pressure-lowering effect of fermented buckwheat sprouts in spontaneously hypertensive rats. Journal of Functional Foods 5, 406–415.

Nakamura, T., Hirota, T., Mizushima, K., Ohki, K., Naito, Y., Yamamoto, N., Yoshikawa, T., 2013b. Milk-derived peptides, Val-Pro-Pro and Ile-Pro-Pro, attenuate atherosclerosis development in apolipoprotein e-deficient mice: a preliminary study. Journal of Medicinal Food 16, 396–403.

Narva, M., Halleen, J., Väänänen, K., Korpela, R., 2004. Effects of *Lactobacillus helveticus* fermented milk on bone cells *in vitro*. Life Sciences 75, 1727–1734.

Narva, M., Rissanen, J., Halleen, J., Vapaatalo, H., Väänänen, K., Korpela, R., 2007. Effects of bioactive peptide, valyl-prolyl-proline (VPP), and *Lactobacillus helveticus* fermented milk containing VPP on bone loss in ovariectomized rats. Annals of Nutrition and Metabolism 51, 65–74.

Nejati, F., Rizzello, C.G., Di Cagno, R., Sheikh-Zeinoddin, M., Diviccaro, A., Minervini, F., Gobbetti, M., 2013. Manufacture of a functional fermented milk enriched of Angiotensin-I Converting Enzyme (ACE)-inhibitory peptides and γ-amino butyric acid (GABA). LWT – Food Science and Technology 51, 183–189.

Nonaka, A., Nakamura, T., Hirota, T., Matsushita, A., Asakura, M., Ohki, K., Kitakaze, M., 2014. The milk-derived peptides Val-Pro-Pro and Ile-Pro-Pro attenuate arterial dysfunction in L-NAME-treated rats. Hypertension Research 37, 703–707.

Norris, R., Poyarkov, A., O'Keeffe, M.B., Fitzgerald, R.J., 2014. Characterisation of the hydrolytic specificity of *Aspergillus niger* derived prolyl endoproteinase on bovine β-casein and determination of ACE inhibitory activity. Food Chemistry 156, 29–36.

Ong, L., Henriksson, A., Shah, N.P., 2007. Angiotensin converting enzyme-inhibitory activity in cheddar cheeses made with the addition of probiotic *Lactobacillus casei* sp. Lait 87, 149–165.

Papadimitriou, C.G., Vafopoulou-Mastrojiannaki, A., Silva, S.V., Gomes, A.M., Malcata, F.X., Alichanidis, E., 2007. Identification of peptides in traditional and probiotic sheep milk yoghurt with angiotensin I-converting enzyme (ACE)-inhibitory activity. Food Chemistry 105, 647–656.

Pietschmann, P., Rauner, M., Sipos, W., Kerschan-Schindl, K., 2009. Osteoporosis: an age-related and gender-specific disease – a mini-review. Gerontology 55, 3–12.

Plaisancié, P., Claustre, J., Estienne, M., Henry, G., Boutrou, R., Paquet, A., Léonil, J., 2013. A novel bioactive peptide from yoghurts modulates expression of the gel-forming MUC2 mucin as well as population of goblet cells and paneth cells along the small intestine. Journal of Nutritional Biochemistry 24, 213–221.

Plaisancié, P., Boutrou, R., Estienne, M., Henry, G., Jardin, J., Paquet, A., Leónil, J., 2015. β-Casein(94-123)-derived peptides differently modulate production of mucins in intestinal goblet cells. Journal of Dairy Research 82, 36–46.

Pozo-Bayón, M.A., Alegría, E.G., Polo, M.C., Tenorio, C., Martín-Álvarez, P.J., Calvo De La Banda, M.T., Ruiz-Larrea, F., Moreno-Arribas, M.V., 2005. Wine volatile and amino acid composition after malolactic fermentation: effect of *Oenococcus oeni* and *Lactobacillus plantarum* starter cultures. Journal of Agricultural and Food Chemistry 53, 8729–8735.

Pugh, S., Jayaraj, A.P., Bardhan, K.D., 1996. Duodenal mucosal histology and histochemistry in active, treated and healed duodenal ulcer: correlation with duodenal prostaglandin E_2 production. Journal of Gastroenterology and Hepatology (Australia) 11, 120–124.

Quirós, A., Ramos, M., Muguerza, B., Delgado, M.A., Martín-Alvarez, P.J., Aleixandre, A., Recio, I., 2006. Determination of the antihypertensive peptide LHLPLP in fermented milk by high-performance liquid chromatography-mass spectrometry. Journal of Dairy Science 89, 4527–4535.

Quirós, A., Ramos, M., Muguerza, B., Delgado, M.A., Miguel, M., Aleixandre, A., Recio, I., 2007. Identification of novel antihypertensive peptides in milk fermented with *Enterococcus faecalis*. International Dairy Journal 17, 33–41.

Rajapakse, N., Mendis, E., Jung, W.K., Je, J.Y., Kim, S.K., 2005. Purification of a radical scavenging peptide from fermented mussel sauce and its antioxidant properties. Food Research International 38, 175–182.

Ramchandran, L., Shah, N.P., 2008. Proteolytic profiles and angiotensin-I converting enzyme and α-glucosidase inhibitory activities of selected lactic acid bacteria. Journal of Food Science 73, M75–M81.

Ramesh, V., Kumar, R., Singh, R.R.B., Kaushik, J.K., Mann, B., 2012. Comparative evaluation of selected strains of lactobacilli for the development of antioxidant activity in milk. Dairy Science and Technology 92, 179–188.

Ricci, I., Artacho, R., Olalla, M., 2010. Milk protein peptides with angiotensin I-converting enzyme inhibitory (ACEI) activity. Critical Reviews in Food Science and Nutrition 50, 390–402.

Rizzello, C.G., Cassone, A., Di Cagno, R., Gobbetti, M., 2008. Synthesis of angiotensin I-converting enzyme (ACE)-inhibitory peptides and γ-aminobutyric acid (GABA) during sourdough fermentation by selected lactic acid bacteria. Journal of Agricultural and Food Chemistry 56, 6936–6943.

Rodríguez-Figueroa, J.C., González-Córdova, A.F., Torres-Llanez, M.J., Garcia, H.S., Vallejo-Cordoba, B., 2012. Novel angiotensin I-converting enzyme inhibitory peptides produced in fermented milk by specific wild *Lactococcus lactis* strains. Journal of Dairy Science 95, 5536–5543.

Rodríguez-Figueroa, J.C., González-Córdova, A.F., Astiazaran-García, H., Vallejo-Cordoba, B., 2013. Hypotensive and heart rate-lowering effects in rats receiving milk fermented by specific *Lactococcus lactis* strains. British Journal of Nutrition 109, 827–833.

Rojas-Ronquillo, R., Cruz-Guerrero, A., Flores-Nájera, A., Rodríguez-Serrano, G., Gómez-Ruiz, L., Reyes-Grajeda, J.P., Jiménez-Guzmán, J., García-Garibay, M., 2012. Antithrombotic and angiotensin-converting enzyme inhibitory properties of peptides released from bovine casein by *Lactobacillus casei* Shirota. International Dairy Journal 26, 147–154.

Rubas, W., Grass, G.M., 1991. Gastrointestinal lymphatic absorption of peptides and proteins. Advanced Drug Delivery Reviews 7, 15–69.

Rui, X., Wen, D., Li, W., Chen, X., Jiang, M., Dong, M., 2015. Enrichment of ACE inhibitory peptides in navy bean (*Phaseolus vulgaris*) using lactic acid bacteria. Food and Function 6, 622–629.

Ryan, J.T., Ross, R.P., Bolton, D., Fitzgerald, G.F., Stanton, C., 2011. Bioactive peptides from muscle sources: meat and fish. Nutrients 3, 765–791.

Saavedra, L., Hebert, E.M., Minahk, C., Ferranti, P., 2013. An overview of "omic" analytical methods applied in bioactive peptide studies. Food Research International 54, 925–934.

Sabeena Farvin, K.H., Baron, C.P., Nielsen, N.S., Otte, J., Jacobsen, C., 2010. Antioxidant activity of yoghurt peptides: Part 2-characterisation of peptide fractions. Food Chemistry 123, 1090–1097.

Sah, B.N.P., Vasiljevic, T., McKechnie, S., Donkor, O.N., 2014. Effecto of probiotics on antioxidant and antimutagenic activities of crude peptide extract from yogurt. Food Chemisty 156, 254–270.

Sah, B.N.P., Vasiljevic, T., McKechnie, S., Donkor, O.N., 2015. Identification of anticancer peptides from bovine milk proteins and their potential roles in management of cancer: a critical review. Comprehensive Reviews in Food Science and Food Safety 14, 123–138.

Satake, M., Enjoh, M., Nakamura, Y., Takano, T., Kawamura, Y., Arai, S., Shimizu, M., 2002. Transepithelial transport of the bioactive tripeptide, Val-Pro-Pro, in human intestinal Caco-2 cell monolayers. Bioscience, Biotechnology and Biochemistry 66, 378–384.

Savijoki, K., Ingmer, H., Varmanen, P., 2006. Proteolytic systems of lactic acid bacteria. Applied Microbiology and Biotechnology 71, 394–406.

Schulz, E., Gori, T., Münzel, T., 2011. Oxidative stress and endothelial dysfunction in hypertension. Hypertension Research 34, 665–673.

Seppo, L., Jauhiainen, T., Poussa, T., Korpela, R., 2003. A fermented milk high in bioactive peptides has a blood pressure-lowering effect in hypertensive subjects. American Journal of Clinical Nutrition 77, 326–330.

Shin, Z.I., Yu, R., Park, S.A., Chung, D.K., Ahn, C.W., Nam, H.S., Kim, K.S., Lee, H.J., 2001. His-His-Leu, an angiotensin I converting enzyme inhibitory peptide derived from Korean soybean paste, exerts antihypertensive activity *in vivo*. Journal of Agricultural and Food Chemistry 49, 3004–3009.

Sipola, M., Finckenberg, P., Korpela, R., Vapaatalo, H., Nurminen, M.L., 2002. Effect of long-term intake of milk products on blood pressure in hypertensive rats. Journal of Dairy Research 69, 103–111.

Stehno-Bittel, L., 2008. Intricacies of fat. Physical Therapy 88, 1265–1278.

Strugala, V., Dettmar, P.W., Pearson, J.P., 2008. Thickness and continuity of the adherent colonic mucus barrier in active and quiescent ulcerative colitis and Crohn's disease. International Journal of Clinical Practice 62, 762–769.

Tellez, A., Corredig, M., Brovko, L.Y., Griffiths, M.W., 2010. Characterization of immune-active peptides obtained from milk fermented by *Lactobacillus helveticus*. Journal of Dairy Research 77, 129–136.

Vallabha, S.V., Tiku, P.K., 2014. Antihypertensive peptides derived from soy protein by fermentation. International Journal of Peptide Research and Therapeutics 20, 161–168.

Vaštag, T., Popović, L., Popović, S., Petrović, L., Peričin, D., 2010. Antioxidant and angiotensin-I converting enzyme inhibitory activity in the water-soluble protein extract from Petrovac Sausage (Petrovská Kolbása). Food Control 21, 1298–1302.

World Health Organization, 2014. Global Status Report on Noncommunicable Diseases. (Geneva, Switzerland).

Yamaguchi, N., Kawaguchi, K., Yamamoto, N., 2009. Study of the mechanism of antihypertensive peptides VPP and IPP in spontaneously hypertensive rats by DNA microarray analysis. European Journal of Pharmacology 620, 71–77.

Yang, H.J., Kwon, D.Y., Moon, N.R., Kim, M.J., Kang, H.J., Jung, D.Y., Park, S., 2013. Soybean fermentation with *Bacillus licheniformis* increases insulin sensitizing and insulinotropic activity. Food and Function 4, 1675–1684.

Zhang, J.-H., Tatsumi, E., Ding, C.-H., LiL, -T., 2006. Angiotensin I-converting enzyme inhibitory peptides in douchi, a Chinese traditional fermented soybean product. Food Chemistry 98, 551–557.

Zhao, C.J., Hu, Y., Schieber, A., Gänzle, M., 2013. Fate of ACE-inhibitory peptides during the bread-making process: quantification of peptides in sourdough, bread crumb, steamed bread and soda crackers. Journal of Cereal Science 57, 514–519.

Ziv, E., Bendayan, M., 2000. Intestinal absorption of peptides through the enterocytes. Microscopy Research and Technique 49, 346–352.

Chapter 3

Health Benefits of Exopolysaccharides in Fermented Foods

K.M. Nampoothiri, D.J. Beena[a], D.S. Vasanthakumari[a], B. Ismail
National Institute for Interdisciplinary Science and Technology (NIIST), Trivandrum, Kerala, India

ABBREVIATIONS

CPS Capsular polysaccharide
CVD Cardiovascular disease
DC Dendritic cells
EPS Exopolysaccharide
ETEC Enterotoxigenic *Escherichia coli*
GRAS Generally regarded as safe
LAB Lactic acid bacteria
LPS Lipopolysaccharide
MAD Malondialdehyde
ROS Reactive oxygen species
SOD Superoxide dismutase
Treg Regulatory T cells
VEGF Vascular endothelial growth factor

3.1 INTRODUCTION

Fermented foods have traditionally been used all over the world for centuries and are one of the oldest forms of food preservation. Fermentation adds flavor and generally enhances nutritive value of foods. The process also increases the number of beneficial microorganisms which are now termed as "probiotics" that helps in maintaining a healthy gut flora. These microorganisms aid in the breakdown of complex macronutrients to simple products with desired properties and they also secrete many metabolites during the process. The products of fermentation include organic acids, ethanol, B group vitamins, microbial polysaccharides, bacteriocins, etc. (Divya et al., 2012).

---

[a]These authors contributed equally to this work.

Microbial polysaccharides are high molecular weight carbohydrate polymers which are present as lipopolysaccharides (LPSs), capsular polysaccharides (CPSs), or exopolysaccharides (EPSs). LPSs are present in the outer membrane; they elicit immune responses and maintain membrane stability. CPSs are secreted and are present as a distinct surface layer (capsule) associated with the cell surface. CPSs are mainly related to the pathogenicity of the bacteria, whereas EPSs are only loosely connected to the cell surface and have diverse functions (Öner, 2013). EPSs provide protection against adverse environmental conditions, help in adhesion, cell to cell interactions, biofilm formation, and also serve as carbon and energy reserves (Mishra and Jha, 2013; Mann and Wozniak, 2012). Based on their structure and composition, polysaccharides are classified into homopolysaccharides and heteropolysaccharides. Homopolysaccharides consist of a single type of monomer (eg, dextran and bacterial cellulose), whereas heteropolysaccharides contain two or more types of monomers, usually as multiple copies of oligosaccharides (eg, xanthan and alginate). EPSs are produced by a wide range of organisms including bacteria (xanthan and dextran), fungi (pullulan), algae (alginate and carrageenan), and plants (cellulose and pectin), and have various applications in the food and pharmaceutical industries. Table 3.1 summarizes some of the bacterial and fungal origin EPSs and their linkages. The extraction and purification steps involved in the downstream processing of microbial EPSs are summarized in Fig. 3.1.

Probiotics are used as starter cultures to develop fermented functional foods. Lactic acid bacteria (LAB) are the dominant microorganisms in most of the fermented foods. Microbial polysaccharides, due to their production from cheap and plentiful raw materials and also due to their properties, overpower polysaccharides of plants and algae (Freitas et al., 2011). Certain LAB capable of producing EPSs are used to develop fermented dairy products, such as yoghurts, cheese, etc., providing a characteristic texture and flavor to these foods. The improved organoleptic properties increase the consumer acceptance. In addition to this, EPSs are retained for a long time in the gastrointestinal tract, thereby enhancing the colonization of the probiotic bacteria (Darilmaz et al., 2011). Functional EPS produced by *Lactobacillus delbrueckii* subsp. *bulgaricus* and *Streptococcus thermophilus* have been reported (De Vuyst et al., 2003; Nishimura, 2014). Functional foods containing probiotics and prebiotics termed as symbiotics are reported to have an increased beneficial effect (de Vrese and Schrezenmeir, 2008; Grosu-Tudor et al., 2013). EPS levels in fermented food depends on the interactions of the microorganisms with the food matrix as well as the fermentation conditions.

These EPSs are beneficial if their origin is from an organism that has a generally regarded as safe (GRAS) status like LAB. LAB has a long history of safe use in various fermented foods and many of them are now identified as probiotics. Bacterial EPSs have diverse applications in food, textile, and cosmetic industries. EPSs from LAB act as natural thickening agents providing suitable viscosity to food and prevent syneresis thus improving the rheological properties

TABLE 3.1 Microbial Exopolysaccharides (EPSs) and Linkages

EPS	Monomer	Linkage	Producer Organisms	References
Dextran	Glucose	α-D-glucan linked by α-(1,6)-glycosidic bonds; 1,2-, 1,3-, or 1,4-bonds are also present in some dextrans	Weissella cibaria	Tingirikari et al. (2014)
			Leuconostoc mesenteroides	Behravan et al. (2003)
Cellulose	Glucose	β-(1,4)-D-glucan	Acetobacter xylinum	Chawla et al. (2009)
			Gluconacetobacter xylinus	
Alginate	Guluronic acid	α-(1,6) and α-(1,3) glycosidic	Azotobacter vinelandii	Sabra et al. (2001)
	Mannuronic acid	Linkages, with some α-(1,3) branching	Azotobacter chroococcum	
Gellan	Glucose, rhamnose	Partially O-acetylated polymer of D-glucose-(1,4)-β-D-glucuronic acid-(1,4)-β-D-glucose-(1,4)-β-L-rhamnose tetrasaccharide units connected by α-(1,3)-glycosidic bonds	Sphingomonas paucimobilis	Bajaj et al. (2007)
	Glucuronic acid			
Curdlan	Glucose	β-(1,3)-D-glucan	Alcaligenes faecalis	Matsushita (1990)
Levan	Fructose	β-(2, 6) glycosidic bonds β-(2,1)-linked side chains	Saccharomyces cerevisiae	Franken et al. (2013)
			Bacillus subtilis	
Xanthan	Glucose, mannose, glucuronic acid	β-(1,4)-linked glucan main chain with alternating residues substituted on the 3-position with a trisaccharide chain containing two mannose and one glucuronic acid residue	Xanthomonas campestris	Morris and Harding (2009)
Pullulan	Glucose	α-(1,6)-linked α-(1,4)-D-triglucoside maltotriose units	Aureobasidium pullulans	Singh et al. (2008)

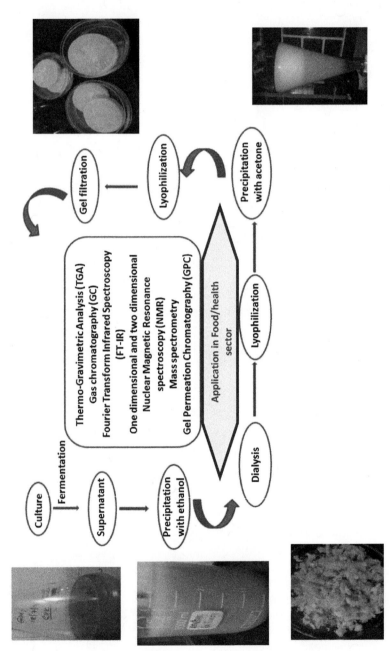

FIGURE 3.1 Steps involved in the recovery, purification, and characterization of microbial exopolysaccharides.

of food. Hence, microbial EPSs have been used extensively in the dairy and fermented food industries for improving the quality of the product. EPSs from fermented dairy foods have a potential prebiotic activity, enhancing the growth of certain microbial strains in the human gut. In addition, they have significant roles in human health (Raposo et al., 2013), including antimicrobial and immunomodulatory properties, antiviral, anticancer, cholesterol lowering, antioxidant, and antihypertensive activities.

3.2 ROLE OF EXOPOLYSACCHARIDES IN THE FOOD INDUSTRY

EPSs produced by microorganisms have unique rheological properties; this may be due to their ability to form viscous pseudoplastic fluids (Becker et al., 1998). Dextran was the first industrial polysaccharide produced by LAB; it was discovered in 1880 in sugar cane or beet root syrup. It was found that dextran is responsible for the gelation and thickening of the syrups (Crescenzi, 1995). Due to their structural differences, especially in molecular weight and glycosidic linkages, some dextrans are water soluble and others are insoluble. Dextran may be used in confectionary to improve moisture retention, viscosity, and inhibition of sugar crystallization. In ice cream, it inhibits crystal formation, and in pudding preparations it provides the desirable mouth feel (Whistler and Daniel, 1990). Xanthan, levan, alginate, emulsan, gellan, curdlan, cellulose, hyaluronic acid, succinoglycan, acetan, glucuronan, and colonic acid are some of other microbial EPSs with industrial applications. Xanthan gum produced by the plant pathogen *Xanthomonas campestris* has been described as a "benchmark" product with respect to its importance in both food and nonfood applications. Xanthan gum is included in dairy products, drinks, confectionary, food dressings, bakery products, syrups, pet foods, oil, pharmaceuticals, cosmetics, paper, paint, and textiles. A heteropolysaccharide produced by the extremely halophilic archaebacterium, *Haloferax mediterranei*, has a high viscosity at lower concentrations and exhibits pseudoplastic behavior. In addition, the polysaccharide has a wide temperature range and pH tolerance, which accounts for its wide industrial applications (Venugopal, 2011).

The rheological properties of EPSs have great importance in their functional properties and applications. It mainly depends on physicochemical characteristics of EPSs like viscosity and elasticity. Viscosity is the property of a material to resist deformation, and to recover its initial conformation after a deformation. These properties of EPS solution are attributed to its molecular weight, monosaccharide composition, chain length, and interaction with other molecules like proteins and ions. Viscosity of a solution can be divided into intrinsic and apparent viscosity. Intrinsic viscosity of EPSs increases with an increase in its molecular weight and structural rigidity. Both properties are important for the organoleptic quality, appearance, and mouth feel of a product (Patel et al., 2010).

EPSs, mainly produced by probiotic LAB, have a greater impact in improving the rheological and textural properties of fermented milk products. They help to prevent syneresis and improve the mouth feel of yogurt, curd, and cheese (Hassan et al., 2003; Ismail and Nampoothiri, 2010). This is mainly due to the pseudoplastic, non-Newtonian behavior of EPS solutions, and it means that the viscosity of a solution will decrease with an increase in shear rate. This property also accounts for its application as an emulsification and viscosity modifier in the food industry. Although EPS has no taste of its own, it increases the time the product spends in the mouth thereby enhancing the perception of taste. Yogurt manufacture remains the most important commercial application for EPSs from LAB. The microstructure of yogurt consisting of aggregates of casein changes depending on the starter cultures. The EPS producing cultures present could bind to the protein matrix and modify the structure of yoghurt (Jolly et al., 2002). The creaminess and firmness of dairy products can be improved with the incorporation of EPSs or with the incorporation of probiotic strains which could produce EPSs. Ropy cultures which can produce high molecular mass EPSs, increase the viscosity of yogurt and thus its texture and taste (Guzel-Seydim et al., 2005). EPS is also found to be effective in reducing syneresis of cooked starch. The incorporation of EPSs in starchy foods can increase the viscosity, thereby improving the texture and mouth feel of the food (Ismail and Nampoothiri, 2014). Xanthan gum is more effective in preventing syneresis of freeze–thawed sweet potato starch gel at 0.3% (w/v) concentration than guar gum, gellan gum, curdlan, locust gum, sodium alginate, and carboxymethyl cellulose. But, at 0.6% (w/v) concentration guar gum was found to be more effective than xanthan gum (Lee et al., 2002). Another major area of food processing where EPS-producing bacteria have been exploited is sourdough production. Incorporation of EPSs into the dough preparation will improve its texture by softening the gluten content of the dough and thus results in an increase in the specific volume of the product. Dextran-producing *Leuconostoc* species and *Weissella* are widely used for this purpose (Pepe et al., 2013; Wolter et al., 2014). Even though it is widely used for improving the flavor and texture of various food preparations, in some cases (eg, wine and cider fermentation) EPS production by LAB confers undesirable rheological properties to the final product (Patel et al., 2012). Table 3.2 includes some of the commercially important EPSs and their applications.

3.3 HEALTH BENEFITS OF EXOPOLYSACCHARIDES

The structural properties like molecular weight, conformation, and composition of EPSs contribute to their biological functions. Enhanced immunostimulatory activity was exhibited by high molecular weight glucans with (1,6) linkages. Structural elucidation of EPSs and the advances in immunology led to the development of polysaccharide vaccines for combating infectious diseases (Nwodo et al., 2012). Table 3.3 represents some bacterial EPSs and their potential health benefits. A few of these properties are discussed in the following sections.

TABLE 3.2 Commercially Important Exopolysaccharides (EPSs) and Their Applications

Biopolymer	Applications
Dextran	It is used in the food industry to improve moisture retention, viscosity, and inhibit sugar crystallization. In medicine, it is used as a blood plasma extender or blood flow improving agent. It is used to make resins in separation technology and also as microcarrier in tissue/cell cultures.
Xanthan gum	It is used as a viscosifier, stabilizer, emulsifier, and suspending agent in the food industry. In addition, it used in paints, pesticide and detergent formulations, pharmaceuticals, cosmetics, printing inks, and secondary and tertiary crude-oil recovery.
Levan	It is used for the production of confectionary and ice cream, as a viscosifier and stabilizer.
Curdlan	It is used as a gelling agent and immobilization matrix. Curdlan along with zidovudine (AZT), displays promising antiretroviral activity (anti AIDS-drug).
Cellulose	It is used in human medicine as a temporary artificial skin to heal burns or surgical wounds. It is also used in the food industry as a natural nondigestible fiber.
Alginate	It is used as: an immobilization matrix for viable cells and enzymes; a microencapsulation matrix for fertilizers; a hypoallergic wound-healing tissue; an antiatherosclerotic; an antiangiogenic; and as an antimetastatic agent.
Emulsan	Its main application is in crude-oil recovery. It is also used as immobilization matrix and wound healing tissue.

TABLE 3.3 Bacterial Exopolysaccharides (EPSs) and Their Health Benefits

Source of EPS	Health Benefits	References
Bacillus licheniformis	Antiviral activity and immunomodulation	Gugliandolo et al. (2014)
Leuconostoc mesenteriodes	Plasma substitute	Naessens et al. (2005)
Lactobacillus kefiranofaciens	Reduce blood pressure, cholesterol and glucose	Maeda et al. (2004a,b)
Lactobacillus brevis D7, *Lactobacillus plantarum* B2, *Lactobacillus fermentum*	Antioxidant properties	Lai et al. (2014)
Lactococcus lactis subsp. *cremoris* KVS20, *Lactobacillus delbrueckii* subsp. *bulgaricus*, *Lactobacillus plantarum* MTCC 9510	Antitumor activity	Kitazawa et al. (1993) and Ismail and Nampoothiri (2013)

3.3.1 Cholesterol Lowering Effect

Cardiovascular disease (CVD) is one of the leading causes of deaths in the world and elevated cholesterol levels are one of the major factors contributing towards CVD. Hence, it is important to lower the elevated serum cholesterol. In humans, cholesterol is synthesized de novo in the liver (700–900 mg/day) and is also obtained through diet (300–500 mg/day). Lifestyle changes, including dietary changes, have a profound effect on lowering CVDs. The role of dietary fibers in lowering serum cholesterol is well established. Some studies have also indicated the hypocholesterolemic effect of fermented milk. Nakajima et al. (1992) have demonstrated a cholesterol lowering effect in rats fed with a hypercholesterolemic diet, in combination with milk fermented with the EPS producing strain *Lactococcus lactis* subsp. *cremoris* SBT0495. HDL cholesterol and HDL:total cholesterol ratio was higher in these rats when compared to the group fed with milk fermented with a non EPS producing strain. Tok and Aslim (2010) reported that among five *L. delbrueckii* subsp. *bulgaricus* strains isolated from homemade yoghurt, those producing high amounts of EPS were able to remove more cholesterol from the medium compared to those that produce low EPS amounts. Bacterial EPSs acted in a manner similar to dietary fibres and though the exact mechanism of cholesterol lowering is not known, it was suggested that the possible mechanisms include increase excretion of bile acids and reduced absorption of cholesterol. This could result in the synthesis of new bile acids from cholesterol by the liver, thereby decreasing the level of circulating cholesterol (Akalin et al., 1997; Liu et al., 2006; Gunness and Gidley, 2010). It is also proposed that the assimilation of cholesterol by the bacterial cells in the intestine reduce the amount of cholesterol available for absorption, thus reducing the serum cholesterol level. The bacterial bile salt hydrolase activity decreases the resorption of bile acids by deconjugation and elimination of the deconjugated bile acids. The hypocholesterolemic effect was observed in humans when an oat-based product fermented with EPS-producing *Pediococcus parvulus* was administered. But, the same result could not be reproduced when individual components of the oat based product were fed to rats; it was also noted that the nonropy fermented product failed to produce a cholesterol lowering effect. Based on these observations it seemed that the hypocholesterolemic effect could be due to the synergistic effect between oats, EPS, and *P. parvulus* (Mårtensson et al., 2005; Lindström et al., 2012).

3.3.2 Immunomodulation

Several LAB strains were found to enhance host immunomodulatory activities by activating macrophages and lymphocytes. This could be a result of interactions between immune cells and bacterial cells or components like peptidoglycan, teichoic acid, and EPSs (Liu et al., 2011). EPSs from beneficial microorganisms trigger innate and adaptive immune system responses. EPSs activate dendritic cells to produce regulatory T (Treg) cells. It was hypothesized that the bacterial EPSs may be recognized by the C-type lectins on the surface of immune cells and the binding of EPSs to receptors results in changes in cell signaling molecules and transcription factors, resulting in an altered cytokine milieu in vivo, that results in immune homeostasis (Jones et al., 2014). Kefir, a drink produced by milk fermentation with a consortium of homofermentative LAB, yeasts, and acetic acid bacteria, contains an EPS known as kefiran which could elicit a gut mucosal response. Kefiran could be used as a carbon source by colon microbiota leading to the production of short chain fatty acids, such as butyric acid that may modulate the immune response. Butyrate contributes to intestinal permeability and low levels of butyrate could cause leaky gut which could alter inflammatory processes (Vinolo et al., 2011). Also the fermentation of EPSs by the microorganisms could indirectly contribute to a host immune response.

L. lactis subsp. *cremoris* KVS20, isolated from the starter culture of Scandinavian ropy sour milk (viili) produces a phosphopolysaccharide comprising of rhamnose, glucose, and galactose. This polysaccharide was found to induce IFN-1 and IL–1a production in spleen macrophages from C57BL/6 mice. The phosphate group in the polysaccharide is believed to be responsible for the immune response (Kitazawa et al., 1996). The compositional and conformational structure like the presence of phosphate groups, molecular size, structure, and overall charge of the EPS play a major role in determining its immunomodulatory effects. EPSs with a neutral charge and higher molecular mass caused a decrease in immune response while acidic and smaller EPSs caused an increased response (Welman, 2014).

3.3.3 Antioxidant Properties

Reactive oxygen species (ROS) are continuously produced in our body as a result of different metabolic activities. Accumulation of ROS in our system causes damage to DNA, proteins, lipids, and carbohydrates, resulting in disease conditions like cancer, arthritis, atherosclerosis, etc. Antioxidants play a major role in neutralizing these ROS. Recently, much attention has been paid to microbial EPSs for their potential therapeutic activities that include antioxidant properties. The antioxidant property of EPS has been determined by in vitro assays, such as 2,2-Diphenyl-1-picrylhydrazyl radical scavenging assay, reducing power assay, total antioxidant capacity determination, and inhibition of lipid peroxidation. EPSs from LAB have been reported to display antioxidant activity by both in vitro and in vivo studies. A neutral EPS from *Lactobacillus plantarum* C88 was demonstrated to have antioxidant effects that may involve scavenging of ROS, upregulation of enzymatic and nonenzymatic antioxidant activities, and reduction of lipid peroxidation (Zhang et al., 2013). Research has shown that EPSs reduced the risk of ROS accumulation by promoting the activities of superoxide dismutase (SOD) and catalase (CAT) in hepatocytes and erythrocytes of mice to some extent (Jiang et al., 2008). A novel EPS from *L. lactis* subsp. *lactis* 12 with a mean molecular weight of 6.9×10^5 Da exhibited hydroxyl and superoxide anion radical scavenging activities. It also significantly decreased the level of malondialdehyde (MAD), even while increasing SOD and CAT activities in mice in a dose dependent manner (Pan and Mei, 2010). The antitumor, antiinflammatory, and anti ulceretic properties of EPSs might be associated with its antioxidant properties.

3.3.4 Anticancer Properties

Despite advanced knowledge and developing technologies, cancer still remains as 1 of the top 10 diseases causing mortality worldwide. The role of probiotics (as well as the EPSs produced by them) in lowering the risk of cancer has been studied by various researchers. The apoptotic and antiangiogenic effects of EPSs including its effects on the c-Myc, c-Fos, and vascular endothelial growth factor (VEGF) expression was reported by Yang et al. (2005). Colon cancer is a widely occurring cancer and EPSs from probiotic strains have found to be effective against this disease. EPSs from *L. plantarum* NRRL B-4496 effectively inhibited the growth of the intestinal carcinoma cell line in vitro in a dose-dependent manner and were active in vivo against Ehrlich Ascites Carcinoma (Haroun et al., 2013). Kitazawa et al. (1993) postulated that the anticancer property of encapsulated *L. lactis* subsp. *cremoris* KVS 20 might be due to the B cell dependent mitogenic activity induced by EPSs produced by the strain. The intraperitoneal injection of *L. lactis* subsp. *cremoris* SBT 0495 resulted in the production of antibodies, which suggests that EPSs may act as an adjuvant (Madhuri and Prabhakar, 2014).

3.3.5 Interactions with Enteric Pathogens

EPSs produced by starter cultures in fermented dairy products provide protection against pathogens and toxins. EPSs can block the enteric pathogens by attaching to the gut wall surface (competitive binding) thereby preventing the adherence of pathogens or by binding to the pathogen itself (Welman, 2014). *Bifidobacterium breve* is usually found in fermented milks and cheeses. An EPS producing *B. breve* strain was found to competitively inhibit the pathogen *Citrobacter rodentium* by colonizing the gut of mice. This was not observed with a non EPS producing strain (Fanning et al., 2012). Reuteran, levan, and glucan isolated from various *Lactobacillus reuteri* strains inhibited enterotoxigenic *Escherichia coli* (ETEC) induced agglutination of porcine erythrocytes. The rapid proliferation of ETEC is by fimbria triggered attachment to specific receptors on intestinal enterocytes and subsequent secretion of toxins. EPSs decreased the adherence of ETEC strains possibly by competitive exclusion (Wang et al., 2010). EPS from *Bifidobacterium longum* BCRC14634 was shown to directly inhibit the growth of *E. coli, Pseudomonas aeruginosa, Vibrio parahaemolyticus, Staphylococcus aureus, Bacillus subtilis*, and *Bacillus cereus* in a dose dependent manner (Wu et al., 2010). It was hypothesized that EPSs produced from fermented milk products provide a barrier that retards the movement of phages (Moineau et al., 1996).

3.4 CONCLUSION

The applications of EPSs spans through various industrial sectors, such as food, textiles, cosmetics, health (medicine and pharmaceuticals), and the environment. The health benefits associated with probiotic functional foods have been partly attributed to EPSs. The recent structural–function relationship studies of EPS from LAB has paved the way for enormous research in this field. Development of EPS over secreting probiotics by metabolic engineering and strain improvement. Cost effective production of EPSs using cheap fermentation substrates, modification of fermentation conditions, metabolic engineering, exploration of new EPS producers could open up new avenue in this area.

ACKNOWLEDGMENTS

The author DJB acknowledges the Research fellowships provided by the Department of Science and Technology (DST), India, and Kerala State Council for Science, Technology and Environment (KSCSTE). The corresponding author thanks the Department of Biotechnology (DBT), New Delhi for the initial funding and the Council for Scientific and Industrial research (CSIR), New Delhi for the current grant for probiotic research under the Network Project, FUNHEALTH (CSC 0133).

REFERENCES

Akalin, A.S., Gonc, S., Duzel, S., 1997. Influence of yogurt and acidophilus yogurt on serum cholesterol levels in mice. Journal of Dairy Science 80, 2721–2725.

Bajaj, I.B., Survase, S.A., Saudagar, P.S., Singhal, R.S., 2007. Gellan gum: fermentative production, downstream processing and applications. Food Technology and Biotechnology 45, 341–354.

Becker, A., Katzen, F., Pühler, A., Ielpi, L., 1998. Xanthan gum biosynthesis and application: a biochemical/genetic perspective. Applied Microbiology and Biotechnology 50 (2), 145–152.

Behravan, J., Bazzaz, B.S., Salimi, Z., 2003. Optimization of dextran production by *Leuconostoc mesenteroides* NRRL B-512 using cheap and local sources of carbohydrate and nitrogen. Biotechnology and Applied Biochemistry 38, 267–269.

Chawla, P.R., Bajaj, I.B., Survase, A.S., Singhal, S.R., 2009. Fermentative production of microbial cellulose. Food Technology and Biotechnology 47, 107–124.

Crescenzi, V., 1995. Microbial polysaccharides of applied interest: ongoing research activities in Europe. Biotechnology Progress 11, 251–259.

Darilmaz, D.O., Aslım, B., Suludere, Z., Akca, G., 2011. Influence of gastrointestinal system conditions on adhesion of exopolysaccharide-producing *Lactobacillus delbrueckii* subsp. *bulgaricus* strains to Caco-2 cells. Brazilian Archives of Biology and Technology 54 (5), 917–926.

de Vrese, M., Schrezenmeir, J., 2008. Probiotics, prebiotics, and synbiotics. In: Stahl, U., Donalies, U.E.B., Nevoigt, E. (Eds.), Food Biotechnology. Springer Berlin Heidelberg, pp. 1–66.

De Vuyst, L., Zamfir, M., Mozzi, F., Adriany, T., Marshall, V., Degeest, B., Vaningelgem, F., 2003. Exopolysaccharide-producing *Streptococcus thermophilus* strains as functional starter cultures in the production of fermented milks. International Dairy Journal 13 (8), 707–717.

Divya, J.B., Varsha, K.K., Nampoothiri, K.M., Ismail, B., Pandey, A., 2012. Probiotic fermented foods for health benefits. Engineering in Life Sciences 12, 377–390.
Fanning, S., Hall, L.J., Cronin, M., Zomer, A., MacSharry, J., Goulding, D., van Sinderen, D., 2012. Bifidobacterial surface-exopolysaccharide facilitates commensal-host interaction through immune modulation and pathogen protection. Proceedings of the National Academy of Sciences 109 (6), 2108–2113.
Franken, J., Brandt, B.A., Tai, S.L., Bauer, F.F., 2013. Biosynthesis of levan, a bacterial extracellular polysaccharide, in the yeast *Saccharomyces cerevisiae*. PLoS One 10, e77499.
Freitas, F., Alves, V.D., Reis, M.A.M., 2011. Advances in bacterial exopolysaccharides: from production to biotechnological applications. Trends in Biotechnology 29, 388–398.
Grosu-Tudor, S.S., Zamfir, M., Van der Meulen, R., Falony, G., De Vuyst, L., 2013. Prebiotic potential of some exopolysaccharides produced by lactic acid bacteria. Romanian Biotechnological Letters 18 (5), 8666–8676.
Gugliandolo, C., Spanò, A., Lentini, V., Arena, A., Maugeri, T.L., 2014. Antiviral and immunomodulatory effects of a novel bacterial exopolysaccharide of shallow marine vent origin. Journal of Applied Microbiology 116, 1028–1034.
Gunness, P., Gidley, M.J., 2010. Mechanisms underlying the cholesterol-lowering properties of soluble dietary fibre polysaccharides. Food & Function 1 (2), 149–155.
Guzel-Seydim, Z.B., Sezgin, E., Seydim, A.C., 2005. Influences of exopolysaccharide producing cultures on the quality of plain set type yogurt. Food Control 16, 205–209.
Haroun, M.B., Refaat, M.B., El-Menoufy, A.H., Amin, A.H., El-Waseif, A.A., 2013. Structure analysis and antitumor activity of the exopolysaccharide from probiotic *Lactobacillus plantarum* NRRL B- 4496 in vitro and in vivo. Journal of Applied Sciences Research 9 (1), 425–434.
Hassan, A.N., Ipsen, I.R., Janzen, T., Qvist, K.B., 2003. Microstructure and rheology of yogurt made with cultures differing only in their ability to produce exopolysaccharides. Journal of Dairy Science 86, 1632–1638.
Ismail, B., Nampoothiri, K.M., 2013. Exposition of antitumour activity of a chemically characterized exopolysaccharide from a probiotic *Lactobacillus plantarum* MTCC 9510. Biologia : Section Cellular and Molecular Biology 68 (6), 1041–1047.
Ismail, B., Nampoothiri, K.M., 2014. Molecular characterization of an exopolysaccharide from a probiotic *Lactobacillus plantarum* MTCC 9510 and its efficacy to improve the texture of starchy food. Journal of Food Science and Technology 51 (12), 4012–4018.
Ismail, B., Nampoothiri, K.M., 2010. Exopolysaccharide production and prevention of syneresis in starch using encapsulated probiotic *Lactobacillus plantarum* MTCC 9510. Food Technology and Biotechnology 48 (4), 484–489.
Jiang, Y.H., Jiang, X.L., Wang, P., Mou, H.J., Hu, X.K., Liu, S.Q., 2008. The antitumor and antioxidative activities of polysaccharides isolated from *Isaria farinosa* B05. Microbiological Research 163 (4), 424–430.
Jolly, L., Vincent, S.J., Duboc, P., Neeser, J.R., 2002. Exploiting exopolysaccharides from lactic acid bacteria. Antonie van Leeuwenhoek 82 (1–4), 367–374.
Jones, S.E., Paynich, M.L., Knight, K.L., 2014. Exopolysaccharides: sweet success with probiotic therapeutics. Inflammation and Cell Signalling 1. http://dx.doi.org/10.14800/ics.334.
Kitazawa, H., Itoh, T., Tomioka, Y., Mizugaki, M., Yamaguchi, T., 1996. Induction of IFN-γ and IL-1α production in macrophages stimulated with phosphopolysaccharide produced by *Lactococcus lactis* ssp. *cremoris*. International Journal of Food Microbiology 31 (1), 99–106.
Kitazawa, H., Yamaguchi, T., Miura, M., Saito, T., Itoth, H., 1993. B-cell mitogen formed by slime forming encapsulated *Lactococcus lactis* subsp. *cremoris* isolated from ropy sour milk, viili. Journal of Dairy Sciences 76, 1514–1519.

Lai, Y.J., Tsai, S.H., Lee, M.Y., 2014. Isolation of exopolysaccharide producing *Lactobacillus* strains from sorghum distillery residues pickled cabbage and their antioxidant properties. Food Science and Biotechnology 23 (4), 1231–1236.

Lee, M.H., Baek, M.H., Cha, D.S., Park, H.J., Lim, S.T., 2002. Freeze–thaw stabilization of sweet potato starch gel by polysaccharide gums. Food Hydrocolloids 16 (4), 345–352.

Lindström, C., Holst, O., Nilsson, L., Öste, R., Andersson, K.E., 2012. Effects of *Pediococcus parvulus* 2.6 and its exopolysaccharide on plasma cholesterol levels and inflammatory markers in mice. AMB Express 2 (1), 66.

Liu, C.F., Tseng, K.C., Chiang, S.S., Lee, B.H., Hsu, W.H., Pan, T.M., 2011. Immunomodulatory and antioxidant potential of *Lactobacillus* exopolysaccharides. Journal of the Science of Food and Agriculture 91 (12), 2284–2291.

Liu, J.R., Wang, S.Y., Chen, M.J., Chen, H.L., Yueh, P.Y., Lin, C.W., 2006. Hypocholesterolaemic effects of milk-kefir and soyamilk-kefir in cholesterol-fed hamsters. British Journal of Nutrition 95 (05), 939–946.

Madhuri, V.K., Prabhakar, K.V., 2014. Microbial exopolysaccharides: biosynthesis and potential applications. Oriental Journal of Chemistry 30 (3), 1401–1410.

Maeda, H., Zhu, X., Omura, K., Suzuki, S., Kitamura, S., 2004a. Effects of an exopolysaccharide kefiran on lipids, blood pressure, blood glucose, and constipation. Biofactors 22, 197–200.

Maeda, H., Zhu, X., Suzuki, S., Suzuki, K., Kitamura, S., 2004b. Structural characterization and biological activities of an exopolysaccharide kefiran produced by *Lactobacillus kefiranofaciens* WT-2BT. Journal of Agricultural and Food Chemistry 52, 5533–5538.

Mann, E.E., Wozniak, D.J., 2012. Pseudomonas biofilm matrix composition and niche biology. FEMS Microbiology Reviews 36 (4), 893–916.

Mårtensson, O., Biörklund, M., Lambo, A.M., Dueñas-Chasco, M., Irastorza, A., Holst, O., Önning, G., 2005. Fermented, ropy, oat-based products reduce cholesterol levels and stimulate the bifidobacteria flora in humans. Nutrition Research 25 (5), 429–442.

Matsushita, M., 1990. Curdlan, a (1–3)-beta-D-glucan from *Alcaligenes faecalis* var. *myxogenes* IFO13140, activates the alternative complement pathway by heat treatment. Immunology Letters 26, 95–97.

Mishra, A., Jha, B., 2013. Microbial exopolysaccharides. In: Rosenberg, E., DeLong, E.F., Thompson, F., Lory, S., Stackebrandt, E. (Eds.), The Prokaryotes-Applied Bacteriology and Biotechnology. Springer, pp. 179–192.

Moineau, S., Borkaev, M., Holler, B.J., Walker, S.A., Kondo, J.K., Vedamuthu, E.R., Vandenbergh, P.A., 1996. Isolation and characterization of lactococcal bacteriophages from cultured buttermilk plants in the United States. Journal of Dairy Science 79 (12), 2104–2111.

Morris, G., Harding, S., 2009. Polysaccharides, Microbial, in Encyclopedia of Microbiology, third ed. Elsevier Inc., Philadelphia, USA, pp. 482–494.

Naessens, M., Cerdobbel, A., Soetaert, W., Vandamme, E.J., 2005. *Leuconostoc dextransucrase* and dextran: production, properties and applications. Journal of Chemical Technology and Biotechnology 80, 845–860.

Nakajima, H., Suzuki, Y., Hirota, T., 1992. Cholesterol-lowering activity of ropy fermented milk. Journal of Food Science 57, 1327–1329.

Nishimura, J., 2014. Exopolysaccharides produced from *Lactobacillus delbrueckii* subsp. *bulgaricus*. Advances in Microbiology 4 (14), 1017.

Nwodo, U.U., Green, E., Okoh, A.I., 2012. Bacterial exopolysaccharides: functionality and prospects. International Journal of Molecular Sciences 13, 14002–14015.

Öner, E.T., 2013. Microbial production of extracellular polysaccharides from biomass. In: Fang, Z. (Ed.), Pretreatment Techniques for Biofuels and Biorefineries. Springer Berlin Heidelberg, pp. 35–56.

Pan, D., Mei, X., 2010. Antioxidant activity of an exopolysaccharide purified from *Lactococcus lactis* subsp. *lactis* 12. Carbohydrate Polymers 80 (3), 908–914.

Patel, A.K., Michaud, P., Singhania, R.R., Soccol, R.C., Pandey, A., 2010. Polysaccharides from probiotics: new developments as food additives. Food Technology and Biotechnology 48 (4), 451–463.

Patel, S., Majumder, A., Goyal, A., 2012. Potentials of exopolysaccharides from lactic acid bacteria. Indian Journal of Microbiology 52 (1), 3–12.

Pepe, O., Valeria, V., Silvana, C., Massimo, F., Rachele, B., 2013. Prebiotic content of bread prepared with flour from immature wheat grain and selected dextran producing lactic acid bacteria. Applied and Environmental Microbiology 79 (12), 3779–3785.

Raposo, M.F.D.J., de Morais, R.M.S.C., Bernardo de Morais, A.M.M., 2013. Bioactivity and applications of sulphated polysaccharides from marine microalgae. Marine Drugs 11 (1), 233–252.

Sabra, W., Zeng, A.P., Deckwer, D.W., 2001. Bacterial alginate: physiology, product quality and process aspects. Applied Microbiology and Biotechnology 56, 315–325.

Singh, R.S., Saini, G.K., Kennedy, J.F., 2008. Pullulan: microbial sources, production and applications. Carbohydrate Polymers 73, 515–531.

Tingirikari, J.M., Kothari, D., Shukla, R., Goyal, A., 2014. Structural and biocompatibility properties of dextran from *Weissella cibaria* JAG8 as food additive. International Journal of Food Sciences and Nutrition 65, 686–691.

Tok, E., Aslim, B., 2010. Cholesterol removal by some lactic acid bacteria that can be used as probiotic. Microbiology and Immunology 54 (5), 257–264.

Venugopal, V., 2011. Extracellular Polysaccharides from Marine Microorganisms. Marine Polysaccharides: Food Applications. CRC Press, pp. 135–155.

Vinolo, M.A., Rodrigues, H.G., Nachbar, R.T., Curi, R., 2011. Regulation of inflammation by short chain fatty acids. Nutrients 3, 858–876.

Wang, Y., Gänzle, M.G., Schwab, C., 2010. Exopolysaccharide synthesized by *Lactobacillus reuteri* decreases the ability of enterotoxigenic *Escherichia coli* to bind to porcine erythrocytes. Applied and Environmental Microbiology 76 (14), 4863–4866.

Welman, A.D., 2014. Exopolysaccharides from fermented dairy products and health promotion. In: Holzapfel, W. (Ed.), Advances in Fermented Foods and Beverages: Improving Quality, Technologies and Health Benefits. Elsevier, pp. 23–34.

Whistler, R., Daniel, J.R., 1990. Functions of polysaccharides in foods. In: Branen, A.L., Davidson, P.M., Salminen, S. (Eds.), Food Additives. Marcel Dekker, Inc., New York, pp. 395–423.

Wolter, A., Hager, A.S., Zannini, E., Galle, S., Gänzle, M.G., Waters, D.M., Arendt, E.K., 2014. Evaluation of exopolysaccharide producing *Weissella cibaria* MG1 strain for the production of sourdough from various flours. Food Microbiology 37, 44–50.

Wu, M.H., Pan, T.M., Wu, Y.J., Chang, S.J., Chang, M.S., Hu, C.Y., 2010. Exopolysaccharide activities from probiotic bifidobacterium: immunomodulatory effects (on J774A. 1 macrophages) and antimicrobial properties. International Journal of Food Microbiology 144 (1), 104–110.

Yang, J., Zhang, W., Shi, P., Chen, J., Han, X., Wang, Y., 2005. Effects of exopolysaccharide fraction (EPSF) from a cultivated *Cordyceps sinensis* fungus on c-Myc, c-Fos, and VEGF expression in B16 melanoma-bearing mice. Pathology-Research and Practice 201 (11), 745–750.

Zhang, L., Liu, C., Li, D., Zhao, Y., Zhang, X., Zeng, X., Li, S., 2013. Antioxidant activity of an exopolysaccharide isolated from *Lactobacillus plantarum* C88. International Journal of Biological Macromolecules 54, 270–275.

Chapter 4

Biotransformation of Phenolics by *Lactobacillus plantarum* in Fermented Foods

R. Muñoz, B. de las Rivas, F. López de Felipe, I. Reverón, L. Santamaría,
M. Esteban-Torres, J.A. Curiel, H. Rodríguez, J.M. Landete
Instituto de Ciencia y Tecnología de Alimentos y Nutrición (ICTAN), CSIC, Madrid, Spain

4.1 PHENOLIC COMPOUNDS IN FERMENTED FOODS

Phenolic compounds in foods have attracted great interest since the 1990s due to growing evidence of their beneficial effect on human health. A diet rich in plant foods has been associated with a decreased risk of certain diseases, including cardiovascular disease and cancer. Phenolic compounds may play a role in the reported beneficial effects of plant foods as they have been extensively studied due to their diverse health benefits as antioxidants, and for preventing chronic inflammation, cardiovascular diseases, cancer, and diabetes (Landete, 2012).

Phenolic compounds comprise a diverse group of molecules classified as secondary metabolites in plants that have a large range of structure and functions. They can be classified into water-soluble compounds (eg, phenolic acids) and water-insoluble compounds (eg, tannins and cell wall-bound hydroxycinnamic acids). Phenolic compounds have an aromatic ring bearing one or more hydroxyl groups and their structure may vary from that of a simple phenolic molecule to that of a complex high-molecular-mass polymer (Balasundram et al., 2006). Phenolic compounds have been considered the most important, numerous, and ubiquitous group of compounds in the plant kingdom (Naczk and Shahidi, 2004).

Flavonoids account for approximately two-thirds of the dietary phenols and they are mostly present as glycosides, and partly as esters, rather than as free compounds (Haminiuk et al., 2012). In the case of the flavonoids group, when they are linked to one or more sugar molecules they are known as flavonoid glycosides, and when they are not connected to a sugar molecule they are called aglycones (Haminiuk et al., 2012). The degree of glycosylation directly affects the antioxidant capacity of flavonoids. Usually, the aglycone forms are more active than the glycoside form (Hopia and Heinonen, 1999).

The second most important group of phytochemicals comprises the phenolic acids, which account for most of the remaining third of the dietary polyphenols, and are present in fruits in a bound form (Haminiuk et al., 2012). These substances contain two distinguishing constitutive carbon frameworks: the hydroxybenzoic and hydroxycinnamic structures. Although the basic skeleton remains the same, the numbers and positions of the hydroxyl groups on the aromatic ring create the variety. In contrast to other phenolic compounds, the hydroxybenzoic and hydroxycinnamic acids present an acidic character owing to the presence of one carboxylic group in the molecule. Hydroxycinnamic acid compounds are mainly present as derivatives, having a C_6—C_3 skeleton. Ferulic acid, p-coumaric acid, and caffeic acid are some examples of this class (Fig. 4.1). Hydroxybenzoic acids (C_6—C_1) mostly occur as esters. Common phenolic acids in this last group are gallic, vanillic, and syringic acids (Fig. 4.1) (Haminiuk et al., 2012).

Tannins are the third class of polyphenols that are found among dietary phenols and are mostly present as phenolic polymers (Haminiuk et al., 2012). Tannins are astringent and bitter substances that have the ability to precipitate proteins. The two main types of tannins are condensed and hydrolyzable. Hydrolyzable tannins are complex compounds that can be degraded through pH changes, as well as by enzymatic or nonenzymatic hydrolysis into smaller fragments, mainly sugars and phenolic acids. Gallotannin or tannic acid is a type of hydrolyzable tannin found in vegetable diets which possess gallic acid as the basic unit of the polyester.

Phenolic compounds have ideal structural chemistry for free radical scavenging activities, thus exhibiting a wide range of beneficial health effects as a

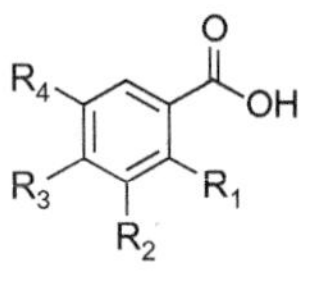

Benzoic acid	R_1	R_2	R_3	R_4
Benzoic	H	H	H	H
p-Hydroxybenzoic	H	H	OH	H
Gallic	H	OH	OH	OH
Gentisic	OH	H	H	OH
Protocatechuic	H	OH	OH	H
Salicilic	OH	H	H	H
Syringic	H	OCH_3	OH	OCH_3
Vanillic	H	OCH_3	OH	H
Veratric	H	OCH_3	OCH_3	H

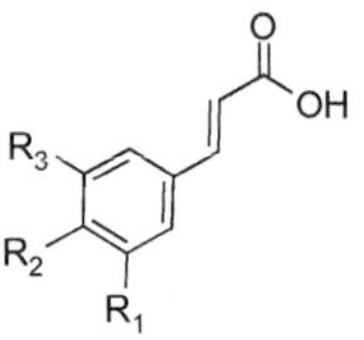

Cinnamic acid	R_1	R_2	R_3
Caffeic	OH	OH	H
Cinnamic	H	H	H
m-Coumaric	OH	H	H
p-Coumaric	H	OH	H
Ferulic	OCH_3	OH	H
Sinapic	OCH_3	OH	OCH_3

FIGURE 4.1 General structures of phenolic acids and their derivatives.

consequence of their antioxidant properties. The health effects of these compounds depend upon their bioavailability. Fruits and vegetables have most of their phytochemicals in the free or soluble conjugate forms (Acosta-Estrada et al., 2014). Bound phenolics comprise an average of 24% of the total phenolics present in these food matrices. Phenolics in the insoluble forms are covalently bound to cell wall structural components, such as cellulose, hemicellulose (eg, arabinoxylans), lignin, pectin, and structural proteins (Wong, 2006). Phenolic acids, such as hydroxycinnamic and hydroxybenzoic acids, form ether linkages with lignin through their hydroxyl groups in the aromatic ring and ester linkages with structural carbohydrates and proteins through their carboxylic group (Liu, 2007).

As natural phenolic compounds often occur as glycosides, esters, or polymers the ability to function as good antioxidants is reduced (Hur et al., 2014). For this reason, biotransformation during food processing or in the gastrointestinal tract plays an essential role in phenolic bioavailability. Bound forms of phenolic compounds can be released and absorbed in the gastrointestinal tract by large intestine microorganisms and gastrointestinal enzymes (Acosta-Estrada et al., 2014). In addition, there are several food processes that enhance the liberation of bound phenolics. These include fermentation and malting, as well as thermomechanical processes, such as extrusion, cooking, and alkaline hydrolysis (Acosta-Estrada et al., 2014). Food bioprocesses, such as fermentation or enzymatic hydrolysis of the plant sources and their products, appear to be an attractive means of enhancing functional activity of such phenolic antioxidants by increasing the concentration of free phenolics from their glycosides or other conjugates (Hur et al., 2014). Fermentation is preferred instead of the utilization of commercial enzymes to enhance the nutraceutical value of foods because it is relatively cheap (Acosta-Estrada et al., 2014).

4.2 TRANSFORMATION OF PHENOLIC COMPOUNDS BY FERMENTATION

Fermentation may be considered as a simple and valuable biotechnologically process for maintaining and/or improving the safety, nutritional, sensory, and shelf-life properties of foods. Composition of microbiota and its development are important factors influencing fermentation and final product quality (Hur et al., 2014). A greater understanding of phenolic metabolism by fermentation is necessary, because the fermentative microbiota probably plays a major role in the biological activity of many phenolic compounds. Some food phenolics are transformed during fermentation by the fermentative microbiota. This conversion is often essential for absorption and modulates the biological activity of these dietary compounds. As mentioned earlier, the bioavailability and effects of phenolic compounds greatly depend on their transformation by specific components of the fermentative microbiota via esterase, glucosidase, dehydroxylase, and decarboxylase activities (Selma et al., 2009).

TABLE 4.1 Main Lactic Acid Bacteria and Their Enzymatic Activities Useful in Food Fermentations

Species	Enzymatic Activities
Lactobacillus acidophilus	Amylase, lactate dehydrogenase, peptidase, proteinase
Lactobacillus casei	Amylase, lactate dehydrogenase, peptidase, proteinase
Lactobacillus fermentum	Amylase, lactate dehydrogenase, peptidase, proteinase
Lactobacillus lactis	Amylase, aminotransferase, decarboxylase, esterase, lactate, dehydrogenase, peptidase, proteinase
Lactobacillus plantarum	Amylase, β-glucosidase, decarboxylase, lactate, dehydrogenase, peptidase, phenolic acid decarboxylase, phenol reductase, proteinase, tannase
Lactobacillus rhamnosus	Amylase, cellulose, esterase, glucoamylase, β-glucosidase, lactate dehydrogenase, peptidase, proteinase

Great interest has emerged concerning the fermentation of cereals with lactic acid bacteria. During fermentation, the grain constituents are modified by the action of both endogenous and bacterial enzymes, including esterases, xylanases, and glycosidases, thereby affecting their structure, bioactivity, and bioavailability (Table 4.1). Cereal-based lactic acid bacteria fermentation has been shown to increase the levels of nutrients including folates, soluble dietary fiber, and total content of phenolic compounds in cereals, and to improve protein digestibility (Hole et al., 2012). Fermentation enhanced the content of bioactive compounds in kidney beans (Limón et al., 2015), lentils (Torino et al., 2013), soybean (Fernández-Orozco et al., 2007), soy milk (Martínez-Villaluenga et al., 2012a), white cabbage (Martínez-Villaluenga et al., 2009, 2012b; Peñas et al., 2010), and lupin (Fernández-Orozco et al., 2008), among other fermented foods.

Because fermentation improves antioxidative activity by increasing the release of phenolic compounds from plant-based foods, it is a useful method for increasing the supply of natural antioxidants. The fermentation-induced structural breakdown of the vegetal cell walls may also liberate and/or induce the synthesis of various bioactive compounds which could be responsible for the increase in total phenols after fermentation, and the increase in the observed antioxidant activity (Dordevic et al., 2010; Katina et al., 2007).

Microbial enzymes, such as glucosidase, amylase, cellulase, chitinase, inulinase, phytase, xylanase, tannase, esterase, invertase, or lipase produced during fermentation can hydrolyze glucosides, and break down plant cell walls or starch. These enzymes play a role in disintegrating the plant cell wall matrix and consequently facilitate flavonoid extraction. Another possible mechanism for increasing the antioxidative activity of plant-based foods by fermentation may be due to structural changes in phytochemicals. The presence of lactic acid bacteria in controlled fermentation contributes to the simple phenolic conversion and the depolymerization of high-molecular-weight phenolic compounds (Othman et al., 2009). The increase in total phenols observed after fermentation could be responsible for the increase observed in the antioxidant activity (Ng et al., 2011). As an example, the health benefits of fermented soy foods have been attributed to the antioxidant activity of specific compounds that are structurally modified or released after bacterial hydrolysis, such as the conversion of glycosylated isoflavones into their aglycones which occurs during fermentation (Hubert et al., 2008). Compared with unfermented soybean, fermented soybean foods contain more aglycones than the predominant isoflavone structures. Thus, the conversion of glucosides into their aglycone form by fermentation is a principal means of increasing antioxidative activity in plant-based foods. The isoflavones genistein and daidzein ("phytoestrogens") are much discussed because of their potential health benefits. Soybean contains high concentrations of the β-glucosides genistin and daidzin. They were reported to be hydrolyzed by β-glucosidase activities of lactic acid bacteria during soy milk fermentations. Strains of *Streptococcus thermophilus*, *L. acidophilus*, *L. delbrueckii* ssp. *bulgaricus*, *L. casei*, *Lactobacillus plantarum*, *L. fermentum*, and several *Bifidobacterium* species could increase the concentrations of genistein and daidzein in soy milk (Chien et al., 2006; Donkor and Shah, 2008; Rekha and Vijayalakshmi, 2011). Species of *Lactobacillus* and *Bifidobacterium* hydrolyzed the flavonoid β-glucosides malvidin and delphinidin, which are of interest as food additives due to their antioxidant capacities (Avila et al., 2009). All strains involved in the study could hydrolyze these glycosides. In contrast, it has been reported that the amount of flavonol glycosides, which account for up to 18% of the total phenolic compounds present in green tea, were reduced during tea fermentation and that this reduction in the flavonol glycosides might be due to oxidative degradation (Kim et al., 2011).

4.3 LACTOBACILLUS PLANTARUM AS A MODEL BACTERIA FOR THE FERMENTATION OF PLANT FOODS

The transformation observed on phenolic compounds is influenced by the microbial species present during the fermentation of plant-based foods. Fermentation has a positive influence on the total phenolic content, however, the degree of this influence depends on the microorganism. Lactic acid bacteria are widely used in food fermentation, and *L. plantarum* is the species most frequently used to ferment food products of plant origin where phenolic compounds are abundant (Rodríguez et al., 2009). Lactic acid bacteria with β-glucosidase activity (including *L. acidophilus*, *L. casei*, *L. plantarum*, *L. fermentum*, etc.) increase the isoflavone aglycone content of soymilk during fermentation (Marazza et al., 2009).

This ability may be responsible for their antioxidant activity, since the released isoflavone aglycone could act as an antioxidant.

Spontaneous and inoculated fermentation by *L. plantarum* are sufficient to improve the concentration of phenolic compounds in fermented cowpea flour; complex phenolic compounds are hydrolyzed to simpler and more biologically active compounds during fermentation (Dueñas et al., 2005). *L. plantarum* fermentation produced an increase in the bioactivity of yellow soybeans and mung beans since a conversion of glycosylated isoflavones into bioactive aglycones and an increase of the bioactive vitexin was observed in yellow and green beans, respectively (Landete et al., 2015).

L. plantarum strains degraded oleuropein, the main phenolic glucoside of olive fruit (Ciafardini et al., 1994; Marsilio et al., 1996; Zago et al., 2013). Strains of *L. plantarum* initially hydrolyze oleuropein by means of a β-glucosidase action, with formation of an aglycone, and in a second step, this derivative, by means of an esterase action, gives rise to hydroxytyrosol and elenoic acid. It has been described that *L. plantarum* strains presented hydrolytic activity against α- and β-glucopyranoside and β-galactopyranoside (Landete et al., 2014). During fermentation, the enzyme β-glucosidase catalyzes the hydrolysis of glycosidic linkages in alkyl and aryl β-glucosides, as well as glycosides containing only carbohydrate residues. This enzymatic activity might help the cleavage of intersugar linkages, releasing the corresponding glycosides that were hydrolyzed liberating the phenolic aglycon moieties (Martins et al., 2011). However, contradictory data were obtained in relation to the *L. plantarum* protein possessing this activity. A β-glucosidase, with a molecular mass of 40 kDa, was purified from *L. plantarum* extracts (Sestelo et al., 2004). However, Spano et al. (2005), by sequence similarity, described the coding region of a putative β-glucosidase of 61 kDa which was recombinantly produced (Acebrón et al., 2009). Unexpectedly, recombinant enzyme showed galactosidase activity but not glucosidase activity (Acebrón et al., 2009). Therefore, the enzyme responsible for the β-glucosidase activity observed in *L. plantarum* cells remains unknown.

The degradation of phenolic compounds present in some plant-derived foods has been studied. When the degradation of nine phenolic compounds present in olive products was tested it was found that only oleuropein and protocatechuic acid were metabolized by *L. plantarum* strains (Landete et al., 2008). Oleuropein was metabolized mainly to hydroxytyrosol, while protocatechuic acid was decarboxylated to catechol. Similarly, the *L. plantarum* capacity to degrade some phenolic compounds present in wine was studied (Landete et al., 2007). Among the 10 compounds analyzed, only some hydroxycinnamic acids, gallic acid, and methyl gallate were metabolized by the *L. plantarum* strains analyzed.

The knowledge of the metabolism of phenolic compounds in *L. plantarum* is of great interest in food science and technology, as this bacterium possesses enzymatic activities for the obtention of powerful antioxidants, such as hydroxytyrosol and pyrogallol.

4.4 BIOTRANSFORMATION OF HYDROXYBENZOIC ACID-DERIVED COMPOUNDS BY *LACTOBACILLUS PLANTARUM*

The ability of *L. plantarum* cultures to metabolize nine hydroxybenzoic acids (syringic, gallic, salicylic, benzoic, gentisic, veratric, *p*-hydroxybenzoic, protocatechuic, and vanillic acid) was studied. Among the hydroxybenzoic acids assayed, only gallic and protocatechuic acids were metabolized by both cell cultures and cell-free extracts from *L. plantarum* (Rodríguez et al., 2008a). Gallic acid was decarboxylated to pyrogallol, and protocatechuic acid was completely decarboxylated to catechol by *L. plantarum* cultures, and there was no indication of a further metabolism of pyrogallol or catechol. It is interesting to note that pyrogallol, which possesses three adjacent hydroxyl groups, is the most potent radical scavenger among simple phenols (Ordoudi and Tsimidou, 2006). Therefore, *L. plantarum* cells produce active antioxidants, such as pyrogallol, from gallic acid esters or hydrolyzable tannins.

In relation to the metabolism of complex hydroxybenzoic acid-derived compounds, such as tannins, the enzymes involved in their metabolism in *L. plantarum* have been recently described. Vegetable tannins are phenolic compounds that are present in many fermented plant foods. Tannins seem to be a double-edged sword (Chung et al., 1998). On the one hand, they are beneficial to health due to their chemopreventive activities against carcinogenesis and mutagenesis, but on the other hand, they may be involved in cancer formation, hepatoxicity, or antinutritional activity. Tannins are considered nutritionally undesirable because they precipitate proteins, inhibit digestive enzymes and affect the utilization of vitamins and minerals (Chung et al., 1998). The molar mass of tannin molecules affects the tannin characteristics directly, it has been found that the higher the molecular mass of tannin molecules, the stronger the antinutritional effects and the lower the biological activities (Chung et al., 1998). Low-molecular-weight tannins are suggested to have fewer antinutrional effects and can be more readily absorbed. Upon hydrolysis by esterase enzymes, gallotannins yield glucose and gallic acid.

L. plantarum degrades gallotannins by the action of an esterase enzyme (tannase or gallate/protocatechuate esterase) (Fig. 4.2). Tannase or tannin acyl hydrolase (EC 3.1.1.20) catalyzes the hydrolysis reaction of the ester bonds present in hydrolyzable tannins and gallic/protocatechuic acid esters. Tannase is used in the production of gallic acid for the synthesis of propylgallate, a potent antioxidant.

Osawa et al. (2000) reported for the first time tannase activity in *L. plantarum* isolates. Later, this property was confirmed in *L. plantarum* strains isolated from various food substrates (Nishitani and Osawa, 2003; Nishitani et al., 2004; Vaquero et al., 2004). It has been postulated that this enzymatic property provides an ecological advantage for this species, as it is often associated with fermentations of plant materials.

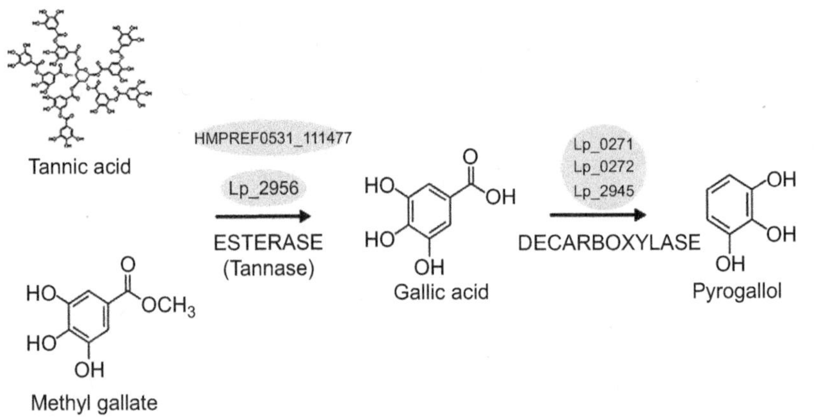

FIGURE 4.2 Biochemical pathways followed by *L. plantarum* for the degradation of hydroxybenzoic acid-derived compounds. The enzymes involved are also indicated.

Compound 1	Enzyme 1	Compound 2	Enzyme 2	Compound 3
Galloyl ester	Esterase	Gallic acid	Decarboxylase	Pyrogallol
Protocatechuoyl ester	Esterase	Protocatechuic acid	Decarboxylase	Catechol

In order to confirm tannase activity on *L. plantarum*, cell-free extracts obtained from disrupted *L. plantarum* cells were incubated in the presence of tannic acid, a complex gallotannin (Rodríguez et al., 2008b). The results obtained indicated that *L. plantarum* degrades tannic acid by depolymerization of high-molecular-weight tannins and a reduction of low-molecular-weight tannins. The biochemical pathway followed by *L. plantarum* implies that tannic acid is hydrolyzed to gallic acid and glucose, and the gallic acid formed is subsequently decarboxylated to pyrogallol (Rodríguez et al., 2008b). This metabolic transformation implies the successive action of a tannase and a gallate decarboxylase enzyme (Fig. 4.2).

Iwamoto et al. (2008) reported the identification of a gene (*tanLp1*, *tan-B_{Lp}*, or *lp_2956* in *L. plantarum* WCFS1) encoding an intracellular tannase. Controversial results were obtained about the presence of an extracellular tannase in *L. plantarum* strains since Lamia and Hamdi (2002) demonstrated that an *L. plantarum* strain produces an extracellular tannase after 24 h growth on medium containing tannic acid. As the sequences of several *L. plantarum* strains were available, a new locus (HMPREF0531_11,477) in *L. plantarum* ATCC 14917^T was annotated as "tannase" (or TanA$_{Lp}$) and encodes for an extracellular enzyme. When the presence of both tannase encoding genes was analyzed among *L. plantarum* strains, it was observed that the intracellular enzyme was present in all the analyzed strains, whereas the extracellular enzyme was only present in a few strains (Jiménez et al., 2014) (Table 4.2). In the *L. plantarum* ATCC 14917^T strain possessing both tannase genes, the expression of these genes was analyzed in the presence of a substrate (methyl gallate). The presence

TABLE 4.2 Characteristics of *L. plantarum* Hydroxybenzoyl Esterases (or Tannases)

Property	TanA$_{Lp}$	TanB$_{Lp}$
Locus	HMPREF0531_11147	lp_2956
Presence in all *L. plantarum* strains	No	Yes
Protein type	Extracellular	Intracellular
Induction by substrate (methyl gallate)	No	Yes
Molecular size (kDa)	67	51
Isoelectric point	9.94	6.17
Specific activity (U/mg)	39	408
Optimal temperature (°C)	20–30	40
Optimal pH	6	7–8
Substrates:		
Methyl gallate	Yes	Yes
Ethyl gallate	Yes	Yes
Propyl gallate	Yes	Yes
Lauryl gallate	No	Yes
Ethyl protocatechuate	Yes	Yes
Epigallocatechin gallate	Yes	Yes
Tannic acid	Yes	Yes

of the substrate did not affect the expression of *tanA$_{Lp}$*, however, it increased the expression of *tanB$_{Lp}$*. This expression behavior indicated that TanB$_{Lp}$ is a tannase inducible by tannins which is present in all *L. plantarum* strains. The obtained results indicated that the two tannases played different physiological roles in *L. plantarum* (Jiménez et al., 2014).

The *L. plantarum* genes encoding the intracellular tannase (TanB$_{Lp}$) as well as the extracellular one (TanA$_{Lp}$) were cloned and hyperexpressed in *Escherichia coli*. The recombinant tannases were further purified and biochemically characterized (Iwamoto et al., 2008; Curiel et al., 2009). The most relevant properties are summarized in Table 4.2. Both enzymes shared similar substrate spectra, hydrolysis of gallate and protocatechuic esters, except that esters with a long aliphatic alcohol were not effectively hydrolyzed by the extracellular TanA$_{Lp}$ tannase. Moreover, TanA$_{Lp}$ has a specific activity 10 times lower than the specific activity shown by $TanB_{Lp}$. In addition, the optimal temperature and pH for the intracellular enzyme (40°C and pH 7–8) differed from the values described for the extracellular $TanA_{Lp}$ tannase (20–30°C and pH 6). The biochemical properties of the extracellular $TanA_{Lp}$ enzyme are more convenient than those of $TanB_{Lp}$ for an extracellular enzyme acting on the fermentation of plant substrates. Temperatures around 20–30°C and acidic pH could be environmental conditions present on these plant fermentations.

The proposed biochemical pathway for the degradation of tannins by *L. plantarum* implies the action of a tannase and a gallate decarboxylase to decarboxylate the gallic acid formed by tannase action. The presence of gallate decarboxylase activity in *L. plantarum* has been previously reported (Osawa et al., 2000). To identify the proteins involved in the decarboxylation of gallic acid, *L. plantarum* cultures were exposed to the presence of this hydroxybenzoic acid. Cell-free extracts from these cultures showed the overproduction of a protein which was absent in the uninduced cultures (Jiménez et al., 2013). This protein was identified as Lp_2945. Homology searches among proteins in databases identified *lp_2945*-like genes located within a three-gene operon which encoded the three subunits of nonoxidative aromatic acid decarboxylases (B, C, and D subunits). *L. plantarum* is the only bacterium in which the *lpdC* (*lp_2945*) gene and the *lpdB* and *lpdD* (*lp_0271* and *lp_0272*) genes are separated in the chromosome. Experiments carried out demonstrated that LpdC is the only protein required to yield gallate decarboxylase activity, although LpdB is also essential for decarboxylase activity. Similar to tannase, which showed activity only on esters derived from gallic and protocatechuic acids, purified LpdC protein showed decarboxylase activity only against both hydroxybenzoic acids (Jiménez et al., 2013).

In contrast to tannase activity, gallate/protocatechuate decarboxylase activity is widely present among lactic acid bacteria. Apart from *Enterococcus faecalis*, among lactic acid bacteria, decarboxylation of aromatic acids has been described in *L. plantarum* and *L. brevis* (Curiel et al., 2010a). Both bacterial species decarboxylate the same hydroxybenzoic acids, gallic and protocatechuic acid, and *L. brevis* possesses gallate decarboxylase genes similar to those previously described in *L. plantarum*.

The identification of the *L. plantarum* gallate/protocatechuate decarboxylase enzyme involved in tannin degradation completes the description of the first route of degradation of a phenolic compound in lactic acid bacteria. The proposed biochemical pathway for the degradation of tannins by *L. plantarum* implies that tannins are hydrolyzed to gallic acid and glucose by an esterase (tannase) action, and the gallic acid formed is decarboxylated to pyrogallol by the action of a gallate decarboxylase enzyme (Osawa et al., 2000; Rodriguez et al., 2008a,b,c). When purified *L. plantarum* tannases were assayed against 18 phenolic acid esters, only esters derived from gallic and protocatechuic acids were hydrolyzed (Curiel et al., 2009; Jiménez et al., 2014), apparently sharing the same substrate specificity as the decarboxylase enzyme. This substrate specificity suggests a concomitant activity of tannase and gallate decarboxylase on specific phenolic substrates. This is more obvious when the chromosomal location of these genes is considered. The genes encoding inducible gallate decarboxylase (lp_2945) and tannase (lp_2956) are only 6.5 kb apart on the *L. plantarum* WCFS1 chromosome. *L. plantarum* strains do not possess appropriate mechanisms to further degrade the compounds produced by these dead-end pathways. The physiological relevance of these reactions is unknown, but in ecosystems, such as food fermentations, it could be envisioned that other organisms in a consortium will mineralize and remove these dead-end metabolites. These enzymatic activities have ecological advantages for *L. plantarum* as it is often associated with fermentation of plant materials. Therefore, *L. plantarum* plays an important role when tannins are present in food and intestine, having the capability of degrading and detoxifying harmful and antinutritional constituents into simpler and innocuous compounds.

4.5 BIOTRANSFORMATION OF HYDROXYCINNAMIC ACID-DERIVED COMPOUNDS BY *LACTOBACILLUS PLANTARUM*

The industrial use of hydroxycinnamates has attracted growing interest since they and their conjugates were shown to be bioactive molecules possessing potential antioxidant activities and health benefits. The hydroxycinnamic acids are more common than are hydroxybenzoic acids and mainly include *p*-coumaric, caffeic, ferulic, and sinapic acids (Fig. 4.1). These acids are rarely found in the free form, except in food that has undergone freezing, sterilization, or fermentation. The bound forms are glycosylated derivatives or esters. Caffeic acids, both free and esterified, are generally the most abundant phenolic acid and represent between 75% and 100% of the total hydroxycinnamic acid content of most fruits. Ferulic acid is the most abundant phenolic acid found in cereal grains.

Hydroxycinnamic acids are found covalently attached to the cell wall and the breakdown of the ester linkages between polymers is needed to release free phenolic acids. *L. plantarum* cultures were unable to hydrolyze hydroxycinnamoyl esters; however, cell-free extracts partially hydrolyzed methyl ferulate and methyl *p*-coumarate, revealing the existence of an esterase hydrolyzing these linkages (Esteban-Torres et al., 2013). In addition, the ability of *L. plantarum* cultures and cell-free extracts to metabolize seven free hydroxycinnamic acids (*p*-coumaric, *m*-coumaric, *o*-coumaric, cinnamic, caffeic, ferulic, and sinapic acid) was also studied. Among the seven hydroxycinnamic acids assayed, only *p*-coumaric and caffeic acid were metabolized by both cell cultures and cell-free extracts of *L. plantarum*. In cell extracts, the products of the decarboxylation reaction were identified (4-vinyl phenol and 4-vinyl catechol), whereas in cell cultures the products of the decarboxylation plus reduction (4-ethyl phenol and 4-ethyl catechol) of *p*-coumaric and caffeic acids were identified,

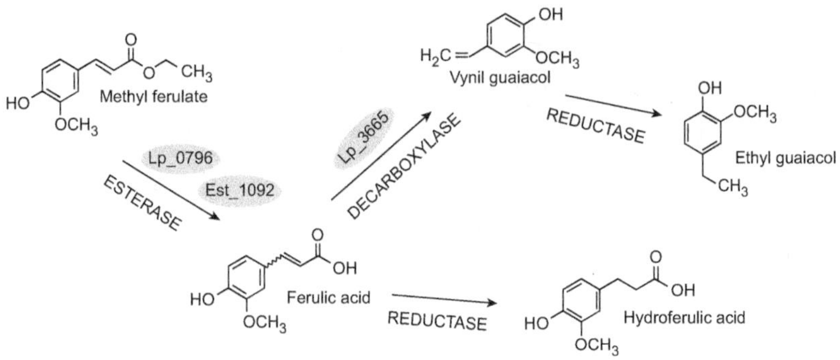

Compound 1	Enzyme 1	Compound 2	Enzyme 2	Compound 3	Enzyme 3	Compound 4
p-Coumaroyl esters	Esterase	p-Coumaric acid	Decarboxylase reductase	Vinyl phenol Phloretic acid	Reductase	Ethylphenol
Feruloyl esters	Esterase	Ferulic acid	Decarboxylase reductase	Vinyl guaiacol Hydroferulic acid	Reductase	Ethyl guaiacol
Caffeoyl esters	Esterase	Caffeic acid	Decarboxylase reductase	Vinyl catechol Hydrocaffeic acid	Reductase	Ethyl catechol
Sinapyl esters	Esterase	Sinapic acid				
m-Coumaric acid	Reductase	3-(3-hydroxyphenyl) propionic acid				

FIGURE 4.3 Biochemical pathways followed by *L. plantarum* for the degradation of hydroxycinnamic acid-derived compounds. The enzymes involved are also indicated.

respectively (Fig. 4.3). From these results it could be deduced that uninduced cell-free extracts from *L. plantarum* contained decarboxylases able to decarboxylate *p*-coumaric and caffeic acids. However, culture induction is needed to synthesize the reductase involved in the conversion of the vinyl derivatives to the corresponding ethyl derivatives. The decarboxylation of hydroxycinnamic acids originates the formation of 4-vinyl phenol and 4-vinyl guaiacol that are considered food additives and are approved as flavoring agents (JECFA, 2001). The reduction of these vinyl phenols, originates ethyl phenol and ethyl guaiacol, which are considered the most important flavor components of fermented soy sauce (Yokosuka, 1986) or, on the other hand, are considered as *off-flavor* and responsible for sensorial wine alterations (Chatonnet et al., 1992).

Unlike *p*-coumaric and caffeic acids, ferulic and *m*-coumaric acids were found to be metabolized only by *L. plantarum* cell cultures. Ferulic acid was decarboxylated to vinyl guaiacol and 3-(3-hydroxyphenyl) propionic acid was obtained from the reduction of *m*-coumaric acid (Rodríguez et al., 2008a) (Fig. 4.3).

Once the metabolism of hydroxycinnamic acid derivatives is known, there are multiple reasons for improving the understanding of the enzymes involved in this metabolism as these enzymes are involved in the formation of bioactive compounds as well as useful volatile phenol derivatives which contribute naturally to aroma in fermented foods and beverages. Similarly to hydroxybenzoic acids, the metabolism of these compounds will be explained by the degradation of complex compounds to the formation of simple phenols.

In nature, a high proportion of hydroxycinnamic acids are found esterified in plant cell walls. Feruloyl esterases, also known as ferulic acid esterases, cinnamic acid esterases, or cinnamoyl esterases, are the enzymes involved in the release of phenolic compounds, such as ferulic, *p*-coumaric, caffeic, and sinapic acid from plant cell walls. In human and rumial digestion, feuloyl esterases are important to de-esterify dietary fiber, releasing hydroxycinnamates and derivatives, which have been shown to have positive effects, such as antioxidant, antiinflammatory, and antimicrobial activities. As described earlier, *L. plantarum* WCFS1 cultures were unable to hydrolyze hydroxycinnamoyl esters; however, cell extract from this strain partially hydrolyzed some hydroxycinnamoyl esters (such as methyl ferulate and methyl *p*-coumarate). Therefore, an enzyme possessing feruloyl esterase activity should be present on *L. plantarum* WCFS1 strain. However, when cultures of additional *L. plantarum* strains were assayed for their ability to hydrolyze hydroxycinnamoyl esters, it was observed that seven out of 27 strains partially hydrolyzed methyl ferulate, methyl caffeate, and methyl *p*-coumarate. This result indicates that, contrarily to WCFS1 strain, some *L. plantarum* strains possess feruloyl esterase activity on cultures. Currently, the genome of multiple *L. plantarum* strains is available, the comparison of these genomes revealed that protein Est_1092 is present in all the strains possessing feruloyl esterase activity on cultures, such as *L. plantarum* DSM 1055 strain, and Est_1092 was absent on strains which had cultures that were unable to hydrolyze these hydroxycinnamoyl esters. Genome comparisons also allowed to identify the esterase Lp_0796 as one of the feruloyl esterases responsible for the esterase activity observed on *L. plantarum* WCFS1 cell extracts (Esteban-Torres et al., 2013).

As *L. plantarum* DSM 1055 strain possesses both feruloyl esterase enzymes, Est_1092 and Lp_0796, in order to gain insights into the specific physiological role of these esterases, the relative expression of both esterase-encoding genes under a hydroxycinnamic ester (methyl ferulate) exposure was studied. The expression levels obtained were substantially different between both esterases genes, indicating the presence of two different expression patterns. The presence of the hydroxycinnamoyl ester increased the expression of the *est_1092* gene, whereas the expression of *lp_0976* was reduced on its presence (unpublished results). Similarly to tannase enzymes, the gene expression results indicate that the two *L. plantarum* feruloyl esterase enzymes played different physiological roles in *L. plantarum* strains.

The *L. plantarum* genes encoding hydroxycinnamoyl esterases (feruloyl esterases), *est_1092* and *lp_0796*, were cloned and hyperexpressed in *E. coli*. The recombinant esterases were purified and biochemically characterized (Esteban-Torres et al., 2013, 2015). The most relevant properties of both esterases are summarized on Table 4.3. The optimal pH for activity is more acidic for Est_1092 (pH 5–6) than for Lp_0796 (pH 7–8). Both esterase enzymes exhibited high activity at low temperatures, although the optimal temperature for Est_1092

TABLE 4.3 Characteristics of *L. plantarum* Hydroxycinnamoyl Esterases (or Feruloyl Esterases)

Property	Lp_0796	Est_1092
Locus	*Lp_0796*	*JDM1_1092*
Presence in all *L. plantarum* strains	Yes	No
Protein type	Intracellular	Intracellular
Induction by substrate (methyl ferulate)	No	Yes
Molecular size (kDa)	27	33
Isoelectric point	5.60	5.42
Specific activity (U/mg)	130	30
Optimal temperature (°C)	30–37	30
Optimal pH	7–8	5–6
Substrates:		
Methyl caffeate	Yes	Yes
Methyl *p*-coumarate	Yes	Yes
Methyl ferulate	Yes	Yes
Methyl sinapinate	Yes	Yes
Methyl benzoate	No	Yes
Methyl vanillate	No	Yes
Methyl gallate	No	Yes
Ethyl gallate	No	Yes
Ethyl protocatechuate	No	Yes
Epigallocatechin gallate	No	Yes

is 30°C whereas it is 37°C for Lp_0796. Both proteins were highly inhibited by the presence of SDS on the reaction media. The most remarkable difference among both enzymes is on their substrate range. Lp_0796 efficiently hydrolyze the four model substrates for feruloyl esterases (methyl caffeate, methyl *p*-coumarate, methyl ferulate, and methyl sinapinate) although other ester substrates were less efficiently hydrolyzed (methyl vandelate or methyl benzoate). However, surprisingly, Est_1092 was not only able to hydrolyze hydroxycinnamoyl esters since all the phenolic esters assayed (including hydroxybenzoyl esters) were hydrolyzed. The hydrolytic activity on hydroxybenzoyl esters is not a common activity on feruloyl esterases, therefore Est_1092 is the first esterase described in a lactic acid bacteria which is active on both ester types and in a broad range of phenolic esters.

The biochemical pathway for the degradation of cell wall hydroxycinnamates begins with the esterase action, releasing free hydroxycinnamic acid from the naturally esterified forms. However, as mentioned earlier, some hydroxycinnamic acids could be subsequently metabolized by *L. plantarum* strains (Fig. 4.3). As early as 1997, it was described as decarboxylase activity on several hydroxycinnamic acids, eg, *p*-coumaric and caffeic. The *L. plantarum* phenolic acid decarboxylase (PAD) enzyme responsible for that decarboxylation was cloned in *E. coli* and characterized at the molecular level (Cavin et al., 1997; Gury et al., 2004). The recombinant PAD from *L. plantarum* CECT 748 is a heat-labile enzyme, showing optimal activity at 22°C. From the 19 phenolic acids assayed, PAD is able to decarboxylate exclusively *p*-coumaric, caffeic, and ferulic acids, concluding that only the acids with a *para* hydroxyl group with respect to the unsaturated side chain and with a substitution of –H, –OH, or –OCH$_3$ in position *meta* were metabolized. Kinetic parameters indicated that at high substrate concentrations, both *p*-coumaric acid and caffeic acids are much more efficiently decarboxylated than ferulic acid (Rodríguez et al., 2008c). PAD activity may confer a selective advantage upon microorganisms during growth on plants, where PAD expression could constitute a stress response induced by phenolic acid. The production of PAD enzyme rapidly degrades the phenolic acid, and thus eliminates the stress caused by it.

The decarboxylation of the hydroxycinnamic acids (*p*-coumaric, caffeic, and ferulic acids) originates the formation of 4-vinyl derivatives (4-vinyl phenol, 4-vinyl catechol, and 4-vinyl guaiacol) (Fig. 4.3) that are considered food additives and are approved as flavoring agents (JECFA, 2001). *L. plantarum* also displayed an uncharacterized vinyl reductase activity, able to reduce the vinyl derivatives into ethyl derivatives (4-ethyl phenol, 4-ethyl catechol, and 4-ethyl guayacol), which are considered the most important flavor components of fermented soy sauce (Yokosuka, 1986) or, on the other hand, are considered as *off-flavor* and responsible for sensorial wine alterations (Chatonnet et al., 1992).

In addition, *L. plantarum* strains also displayed an uncharacterized phenolic acid reductase activity, able to reduce some hydroxycinnamic acids, and to metabolize *p*-coumaric acid into phloretic acid, caffeic acid into hydrocaffeic acid, ferulic acid into hydroferulic acid, and *m*-coumaric acid into 3(3-hydroxyphenyl) propionic acid (Fig. 4.3). These reduced acids were not further degraded by *L. plantarum* strains.

Most of the metabolism of phenolic compounds in lactic acid bacteria remains unknown; however, the description of new enzymatic activities would help to uncover it. In relation to hydroxycinnamic compounds, the combined action of the esterase, decarboxylase, and reductases could allow *L. plantarum* to metabolize compounds abundant in fermented plant-derived food products, such as hydroxycinnamoyl esters abundant in plant cell walls.

4.6 HEALTH BENEFITS INDUCED BY THE INTERACTION OF PHENOLICS WITH *LACTOBACILLUS PLANTARUM*

The free radical scavenging power of diet phenolic compounds is important from a health viewpoint (Harborne and Williams, 2000), and has therefore attracted the interest of the food industry in providing the market with food products containing these antioxidant molecules. Despite the crucial role played by *L. plantarum* in vegetable fermentations, studies of the interaction between phenolic compounds and this bacterial species are scarce. Catechin is an effective scavenger of reactive oxygen species and plays a role to protect against degenerative diseases. It has been described that *L. plantarum* did not degrade catechin but this flavanol promotes quicker sugar consumption, and increased the *L. plantarum* fermentative performance (López de Felipe et al., 2010). Quercetin, which is the major contributor to the total flavonol intake in humans, also improved several key fermentation traits for the performance of *L. plantarum* in food production. These traits included accelerated fermentation of various sugars, and accelerated lactic acid production (Curiel et al., 2010b). As catequin and quercetin were not metabolized by *L. plantarum* during food fermentations, so the antioxidant properties of these flavonoids were protected against degradation, while the bacterium improved its growth performance.

Besides its successful adaptation to plant fermentations, *L. plantarum* persists and survives in the human gastrointestinal tract, where it also meets with diet phenolic compounds (Douillard and de Vos, 2014). The interaction of phenolic compounds with gut microbiota will determine to a great extent its absorption and the direct physiological effects of these metabolites. Besides direct effects on the host, other mechanisms by which phenolic compounds may benefit human health could involve effects on microbial population. In fact, some experimental data obtained in animal models show that dietary phenolics are involved in defining the distribution and abundance of microbial populations in the digestive tract. Regarding animal nutrition, tannin-rich diets are considered antinutritional as they reduce intake, and protein and carbohydrate digestibility. Some reports have evidenced that tannin-resistant microbial populations could prevent the adverse effects of diet tannins. Several studies have been undertaken to gain insight on the mechanisms used by *L. plantarum* to adapt and resist the antimicrobial action of dietary tannins. A proteomic analysis revealed that tannic acid targeted proteins involved in outstanding processes for bacterial stress resistance, and stress response at a population scale (Curiel et al., 2011). In addition, it has been described that tannins modulate selected *L. plantarum* traits linked to gastrointestinal survival (Reverón et al., 2013). To gain an insight on how the response to tannic acid influenced the molecular adaptation of *L. plantarum* to the gastrointestinal tract conditions, gene expression of selected biomarkers for gastrointestinal survival was assessed by quantitative RT-PCR. Tannin-dependent gene induction was confirmed for selected genes highly expressed in the gut or with confirmed roles in gastrointestinal survival. The obtained results indicated that this dietary constituent modulates molecular traits linked to the adaptation to intestinal environment in ways previously shown to enhance gastrointestinal survival of probiotic lactobacilli.

Similarly, it is interesting to increase the knowledge on the mechanisms involved in hydroxycinnamic acid tolerance in fermentative microorganisms. The response of *L. plantarum* to a hydroxycinnamic acid (*p*-coumaric acid) as studied by transcriptomics advances knowledge of the stress tolerance mechanisms of lactobacilli to dietary hydroxycinnamic acids and it is a step forward to decipher the role of colonic microbiota in the bioconversion of phenolic compounds (Reverón et al., 2012). Improved undestanding of the adaptative behavior of *L. plantarum* under the stress induced by *p*-coumaric acid revealed responses that would be potentially beneficial for intestinal function, such as detoxification of hydroxycinnamic acids and induction of a marked antioxidant response. As overlapping behaviors were observed with the transcriptional response of *L. plantarum* to the gut environment, contact with this phenolic acid could provide preparedness for adaptation to the gut habitat.

In summary, fermentation is an adequate technology with great potential for application on the production of active compounds from natural sources. New bioactive compounds could be found during fermentation. Production of bioactive compounds yet remains a quite unexplored potential, which could be accomplished by the elucidation of the metabolic pathways on fermentative microorganisms. *L. plantarum* is the model of lactic acid bacteria to study the metabolism of phenolic compounds. The knowledge of the metabolic pathways present in *L. plantarum* will help the elucidation of these routes on additional fermentative bacteria. Moreover, in a future when the knowledge of these transformations will be complete, fermentation with well-known microorganisms possessing specific metabolic abilities for the design of food with improved health effects can be anticipated.

ACKNOWLEDGMENTS

This work was supported by grants AGL2005-00470, AGL2008-01052, AGL2011-22745, and AGL2014-52911 from the MINECO. L. Santamaría is a recipient of an FPI predoctoral fellowship from the MINECO.

REFERENCES

Acebrón, I., Curiel, J.A., de las Rivas, B., Muñoz, R., Mancheño, J.M., 2009. Cloning, production, purification and preliminary crystallographic analysis of a glycosidase from the food lactic acid bacterium *Lactobacillus plantarum* CECT 748^T. Protein Expression and Purification 68, 177–182.

Acosta-Estrada, B.A., Gutiérrez-Uribe, J.A., Serna-Saldívar, S.O., 2014. Bound phenolics in foods, a review. Food Chemistry 152, 46–55.

Ávila, M., Hidalgo, M., Sánchez-Moreno, C., Pelaez, C., Requena, T., de Pascual-Teresa, S., 2009. Bioconversion of anthocyanin glycosides by bifidobacteria and *Lactobacillus*. Food Research International 42, 1453–1461.

Balasundram, N., Sundram, K., Samman, S., 2006. Phenolic compounds in plants and agri-industrial by-products: antioxidant activity, occurrence, and potential uses. Food Chemistry 99, 191–203.

Cavin, J.F., Barthelmebs, L., Divìes, C., 1997. Molecular characterization of an inducible *p*-coumaric acid decarboxylase from *Lactobacillus plantarum*: gene cloning, transcriptional analysis, overexpression in *Escherichia coli*, purification, and characterization. Applied and Environmental Microbiology 63, 1939–1944.

Chatonnet, P., Dubourdieu, D., Boidron, J.N., Pons, M., 1992. The origin of ethylphenols in wines. Journal of the Science of Food and Agriculture 60, 165–178.

Chien, H.L., Huang, H.Y., Chou, C.C., 2006. Transformation of isoflavone phytoestrogens during the fermentation of soymilk with lactic acid bacteria and bifidobacteria. Food Microbiology 23, 772–778.

Chung, K.-T., Wei, C.-I., Johnson, M.G., 1998. Are tannins a double-edged sword in biology and health? Trends in Food Science and Technology 9, 168–175.

Ciafardini, G., Marsilio, V., Lanza, B., Pozzi, N., 1994. Hydrolysis of oleuropein by *Lactobacillus plantarum* strain associated with olive fermentation. Applied and Environmental Microbiology 60, 4142–4147.

Curiel, J.A., Rodríguez, H., Acebrón, I., Mancheño, J.M., de las Rivas, B., Muñoz, R., 2009. Production and physiochemical properties of recombinant *Lactobacillus plantarum* tannase. Journal of Agricultural and Food Chemistry 57, 6224–6230.

Curiel, J.A., Rodríguez, H., Landete, J.M., de las Rivas, B., Muñoz, R., 2010a. Ability of *Lactobacillus brevis* strains to degrade food phenolic acids. Food Chemistry 120, 225–229.

Curiel, J.A.S., Muñoz, R., López de Felipe, F., 2010b. pH and dose-dependent effects of quercetin on the fermentation capacity of *Lactobacillus plantarum*. LWT-Food Science and Technology 43, 926–933.

Curiel, J.A., Rodríguez, H., de las Rivas, J.A., Anglade, P., Baraige, F., Zagorec, M., Champomier-Vergès, M., Muñoz, R., López de Felipe, F., 2011. Response of a *Lactobacillus plantarum* human isolate to tannic acid challenge assessed by proteomic analyses. Molecular Nutrition and Food Research 55, 1454–1465.

Donkor, O.N., Shah, N., 2008. Production of beta-glucosidase and hydrolysis of isoflavone phytoestrogens by *Lactobacillus acidophilus*, *Bifidobacterium lactis*, and *Lactobacillus casei* in soymilk. Journal of Food Science 73, M15–M20.

Dordevic, T.M., Siler-Marinkovic, S.S., Dimitrijevic-Brankovic, S.I., 2010. Effect of fermentation on antioxidant properties of some cereals and pseudo cereals. Food Chemistry 119, 957–963.

Douillard, F.P., de Vos, W.M., 2014. Functional genomics of lactic acid bacteria: from food to health. Microbial Cell Factories 13, S8.

Dueñas, M., Fernández, D., Hernández, T., Estrella, I., Muñoz, R., 2005. Bioactive phenolic compounds of cowpeas (*Vigna sinensis* L). Modificatins by fermentation with natural microflora and with *Lactobacillus plantarum*. ATCC14917. Journal of the Science of Food and Agriculture 85, 297–304.

Esteban-Torres, M., Reverón, I., Mancheño, J.M., de las Rivas, B., Muñoz, R., 2013. Characterization of a feruloyl esterase from *Lactobacillus plantarum*. Applied and Environmental Microbiology 79, 5130–5136.

Esteban-Torres, M., Landete, J.M., Reverón, I., Santamaría, L., de las Rivas, L., Muñoz, R., 2015. A *Lactobacillus plantarum* esterase active on a broad range of phenolic esters. Applied and Environmental Microbiology 81, 3235–3242.

Fernández-Orozco, R., Frias, J., Muñoz, R., Zielinski, H., Piskula, M.K., Kozlowska, H., Vidal-Valverde, C., 2007. Fermentation as a bio-process to obtain functional soybean flours. Journal of Agricultural and Food Chemistry 55, 8972–8979.

Fernández-Orozco, R., Frias, J., Muñoz, R., Zielinski, H., Piskula, M., Kozlowska, H., Vidal-Valverde, C., 2008. Effect of fermentation conditions on the antioxidant compounds and antioxidant capacity of *Lupinus angustifolius* cv. Zapaton. European Food Research and Technology 227, 979–988.

Gury, J., Barthelmebs, L., Tran, N.P., Divìes, C., Cavin, J.-F., 2004. Cloning, deletion, and characterization of PadR, the transcriptional repressor of the phenolic acid decarboxylase-encoding *padA* gene of *Lactobacillus plantarum*. Applied and Environmental Microbiology 70, 2146–2153.

Haminiuk, C.W.I., Maciel, G.M., Plata-Oviedo, M.S.V., Peralta, R.M., 2012. Phenolic compounds in fruits – an overview. International Journal of Food Science and Technology 47, 2023–2044.

Harborne, J., Williams, C., 2000. Advances in flavonoid research since 1992. Phytochemistry 55, 481–504.

Hole, A.S., Rud, I., Grimmer, S., Sigl, S., Narvhus, J., Sahlstrom, S., 2012. Improved bioavailability of dietary phenolic acids in whole grain barley and oat groats following fermentation with probiotic *Lactobacillus acidophilus*, *Lactobacillus johnsonii*, and *Lactobacillus reuteri*. Journal of Agricultural and Food Chemistry 60, 6369–6375.

Hopia, A., Heinonen, M., 1999. Antioxidant activity of flavonol aglycones and their glycosides in methyl linoleate. Journal of the American Oil Chemistry Society 76, 139–144.

Hubert, J., Berger, M., Nepveu, F., Paul, F., Daydé, J., 2008. Effects of fermentation on the phytochemical composition and antioxidant properties of soy germ. Food Chemistry 109, 709–721.

Hur, S.J., Lee, S.Y., Kim, Y.C., Choi, I., Kim, G.B., 2014. Effect of fermentation on the antioxidant activity in plant-based foods. Food Chemistry 160, 346–356.

Iwamoto, K., Tsuruta, H., Nishitani, Y., Osawa, R., 2008. Identification and cloning of a gene encoding tannase (tannin acylhydrolase) from *Lactobacillus plantarum* ATCC 14917^T. Systematic and Applied Microbiology 31, 269–277.

Jiménez, N., Curiel, J.A., Reverón, I., de las Rivas, B., Muñoz, R., 2013. Uncovering the *Lactobacillus plantarum* WCFS1 gallate decarboxylase involved in tannin degradation. Applied and Environmental Microbiology 79, 4253–4263.

Jiménez, N., Esteban-Torres, M., Mancheño, J.M., de las Rivas, B., Muñoz, R., 2014. Tannin degradation by a novel tannase enzyme present in some *Lactobacillus plantarum* strains. Applied and Environmental Microbiology 80, 2991–2997.

Joint Expert Committee on Food Additives (JECFA), 2001. Evaluation of Certain Food Additives and Contaminants. Fifty-Fifth Report of the Joint WHO/FAO Expert Committee on Food Additives WHO Technical report series no. 901. World Health Organization, Geneva.

Katina, K., Laitila, A., Juvonen, R., Liukkonen, K.H., Kariluoto, S., Piironen, V., Landberg, R., Aman, P., Poutanen, K., 2007. Bran fermentation as a means to enhance technological properties and bioactivity of rye. Food Microbiology 24, 175–186.

Kim, Y., Goodner, K.L., Park, J.D., Choi, J., Talcott, S.T., 2011. Changes in antioxidant phytochemicals and volatile composition of *Camellia sinensis* by oxidation during tea fermentation. Food Chemistry 129, 1331–1342.

Lamia, A., Hamdi, M., 2002. Culture conditions of tannase production by *Lactobacillus plantarum*. Process in Biochemistry 39, 59–65.

Landete, J.M., Rodríguez, H., de las Rivas, B., Muñoz, R., 2007. High-added value antioxidants obtained from the degradation of wine phenolics by *Lactobacillus plantarum*. Journal of Food Protection 70, 2670–2675.

Landete, J.M., Curiel, J.A., Rodríguez, H., de las Rivas, B., Muñoz, R., 2008. Study of the inhibitory activity of phenolic compounds found in olive products and their degradation by *Lactobacillus plantarum*. Food Chemistry 107, 320–326.

Landete, J.M., Curiel, J.A., Rodríguez, H., de las Rivas, B., Muñoz, R., 2014. Aryl glycosidases from *Lactobacillus plantarum* increase antioxidant activity of phenolic compounds. Journal of Functional Foods 7, 322–329.

Landete, J.M., Hernández, T., Robredo, S., Dueñas, M., de las Rivas, B., Estrella, I., Muñoz, R., 2015. Effect of soaking and fermentation on content of phenolic compounds of soybean (*Glycine max* cv. Merit) and mung beans (*Vigna radiata* [L] Wilczek). International Journal of Food Sciences and Nutrition. http://dx.doi.org/10.3109/09637486.2014.986068.

Landete, J.M., 2012. Updated knowledge about polyphenols: functions, bioavailability, metabolism, and health. Critical Reviews in Food Science and Nutrition 52, 936–948.

Limón, R.I., Peñas, E., Torino, M.I., Martínez-Villaluenga, C., Dueñas, M., Frias, J., 2015. Fermentation enhances the content of bioactive compounds in kidney bean extracts. Food Chemistry 172, 343–352.

Liu, R., 2007. Whole grain phytochemicals and health. Journal of Cereal Science 46, 207–219.

López de Felipe, F., Curiel, J.A., Muñoz, R., 2010. Improvement of the fermentation performance of *Lactobacillus plantarum* by the flavanol catechin is uncoupled from its degradation. Journal of Applied Microbiology 109, 687–697.

Marazza, J.A., Garro, M.S., Savoy de Giori, G., 2009. Aglycone production by *Lactobacillus rhamnosus* CRL981 during soymilk fermentation. Food Microbiology 26, 333–339.

Marsilio, V., Lanza, B., Pozzi, N., 1996. Progress in table olive debittering: degradation in vitro of oleuropein and its derivatives by *Lactobacillus plantarum*. Journal of AOCS 73, 593–597.

Martínez-Villaluenga, C., Peñas, E., Frias, J., Ciska, E., Honke, J., Piskula, M.K., Kozlowska, H., Vidal-Valverde, C., 2009. Influence of fermentation conditions on glucosinolates, ascorbigen, and ascorbic acid content in white cabbage (*Brassica oleracea* var. *capitala* cv. Taler) cultivated in different seasons. Journal of Food Science 74, C62–C67.

Martínez-Villaluenga, C., Torino, M.I., Martín, V., Arroyo, R., García-Mora, P., Estrella, I., Vidal-Valverde, C., Rodríguez, J.M., Frias, J., 2012a. Multifunctional properties of soy milk fermented by *Enterococcus faecium* strains isolated from raw soy milk. Journal of Agricultural and Food Chemistry 60, 10235–10244.

Martínez-Villaluenga, C., Peñas, E., Sidro, B., Ullate, M., Frias, J., Vidal-Valverde, C., 2012b. White cabbage fermentation improves ascorbigen content, antioxidant and nitric oxide production inhibitory activity in LPS-induced macrophages. LWT-Food Science and Technology 46, 77–83.

Martins, S., Mussatto, S.I., Martínez-Avila, G., Montañez-Saenz, J., Aguilar, C.N., Teixeira, J.A., 2011. Bioactive phenolic compounds: production and extraction by solid-state fermentation. A review. Biotechnology Advances 29, 365–373.

Naczk, M., Shahidi, F., 2004. Extraction and analysis of phenolics in food. Journal of Chromatography A 1054, 95–111.

Ng, C.-C., Wang, C.Y., Wang, Y.P., Tzeng, W.S., Shyu, Y.T., 2011. Lactic acid bacterial fermentation on the production of functional antioxidant herbal *Anoectochilus formosanus* Hayata. Journal of Bioscience and Bioengineeering 111, 289–293.

Nishitani, Y., Osawa, R., 2003. A novel colorimetric method to quantify tannase activity of viable bacteria. Journal of Microbiological Methods 54, 281–284.

Nishitani, Y., Sasaki, E., Fujisawa, T., Osawa, R., 2004. Genotypic analyses of lactobacilli with a range of tannase activities isolated from human feces and fermented foods. Systematic and Applied Microbiology 27, 109–117.

Ordoudi, S.A., Tsimidou, M.Z., 2006. Crocin bleaching assay (CBA) in structure-radical scavenging activity studies of selected phenolic compounds. Journal of Agricultural and Food Chemistry 54, 9347–9356.

Osawa, R., Kuroiso, K., Goto, S., Shimizu, A., 2000. Isolation of tannin-degrading lactobacilli from humans and fermented foods. Applied and Environmental Microbiology 66, 3093–3097.

Othman, N.B., Roblain, D., Chammen, N., Thonart, P., Hamdi, M., 2009. Antioxidant phenolic compounds loss during the fermentation of Chétoui olives. Food Chemistry 116, 662–669.

Peñas, E., Frias, J., Sidro, B., Vidal-Valverde, C., 2010. Chemical evaluation and sensory quality of sauerkrauts obtained by natural and induced fermentations at different NaCl levels from *Brassica oleracea* Var. *capitalia* Cv. Bronco grown in Eastern Spain: effect of storage. Journal of Agricultural and Food Chemistry 58, 3549–3557.

Rekha, C.R., Vijayalakshmi, G., 2011. Isoflavone phytoestrogens in soymilk fermented with beta-glucosidase producing probiotic lactic acid bacteria. International Journal of Food Science and Nutrition 62, 111–120.

Reverón, I., de las rivas, B., Muñoz, R., López de Felipe, F., 2012. Genome-wide transcriptomic responses of a human isolate of *Lactobacillus plantarum* exposed to *p*-coumaric acid stress. Molecular Nutrition and Food Research 56, 1848–1859.

Reverón, I., Rodríguez, H., Campos, G., Curiel, J.A., Ascaso, C., Carrascosa, A.V., Prieto, A., de las Rivas, B., Muñoz, R., López de Felipe, F., 2013. Tannic acid-dependent modulation of selected *Lactobacillus plantarum* traits linked to gastrointestinal survival. PLoS One 8, e66473.

Rodríguez, H., Landete, J.M., de las Rivas, B., Muñoz, R., 2008a. Metabolism of food phenolic acids by *Lactobacillus plantarum* CECT 748^T. Food Chemistry 107, 1393–1398.

Rodríguez, H., de las Rivas, B., Gómez-Cordovés, C., Muñoz, R., 2008b. Degradation of tannic acid by cell-free extracts of *Lactobacillus plantarum*. Food Chemistry 107, 664–670.

Rodríguez, H., Landete, J.M., Curiel, J.A., de las Rivas, B., Mancheño, J.M., Muñoz, R., 2008c. Characterization of the *p*-coumaric acid decarboxylase from *Lactobacillus plantarum* CECT 748^T. Journal of Agricultural and Food Chemistry 56, 3068–3072.

Rodríguez, H., Curiel, J.A., Landete, J.M., de las Rivas, B., López de Felipe, F., Gómez-Cordovés, C., Mancheño, J.M., Muñoz, R., 2009. Food phenolics and lactic acid bacteria. International Journal of Food Microbiology 132, 79–90.

Selma, M.V., Espin, J.C., Tomás-Barberán, F.A., 2009. Interaction between phenolics and gut microbiota role in human health. Journal of Agricultural and Food Chemistry 57, 6485–6501.

Sestelo, A.B.F., Poza, M., Villa, T.G., 2004. β-Glucosidase activity in a *Lactobacillus plantarum* wine strain. World Journal of Microbiology and Biotechnology 20, 633–637.

Spano, G., Rinaldi, A., Ugliano, M., Moio, L., Beneduce, L., Massa, S., 2005. A β-glucosidase gene isolated from wine *Lactobacillus plantarum* is regulated by abiotic stresses. Journal of Applied Microbiology 98, 855–861.

Torino, M.I., Limón, R.I., Martínez-Villaluenga, C., Mäkinen, S., Pihlanto, A., Vidal-Valverde, C., Frias, J., 2013. Antioxidant and antihypertensive properties of liquid and solid state fermented lentils. Food Chemistry 136, 1030–1037.

Vaquero, I., Marcobal, A., Muñoz, R., 2004. Tannase activity by lactic acid bacteria isolated from grape must and wine. International Journal of Food Microbiology 96, 199–204.

Wong, D.W.S., 2006. Feruloyl esterase: a key enzyme in biomass degradation. Applied Biochemistry and Biotechnology 133, 87–112.

Yokosuka, Y., 1986. Soy sauce biochemistry. Advances in Food Research 30, 196–220.

Zago, M., Lanza, B., Rossetti, L., Muzzalupo, I., Carminati, D., Giraffa, G., 2013. Selection of *Lactobacillus plantarum* strains to use as starters in fermented table olives: oleuropeinase activity and phage sensitivity. Food Microbiology 34, 81–87.

Chapter 5

Gamma-Aminobutyric Acid-Enriched Fermented Foods

J. Quílez[1], M. Diana[1]
[1]Europastry S.A., Sant Cugat del Vallès, Barcelona, Spain

5.1 INTRODUCTION

Gamma-aminobutyric acid (GABA) is a nonprotein four-carbon free amino acid. GABA is synthesized by the irreversible α-decarboxylation reaction of L-glutamic acid or its salts; the reaction is catalyzed by glutamic acid decarboxylase (GAD; EC 4.1.1.15) (Fig. 5.1) and its cofactor pyridoxal 5′ phosphate (PLP; active vitamin B6). This enzyme is found in bacteria, fungi, insects (such as cockroach, grasshopper, moth, honeybee, and fly), plants and the mammalian animal brain. However, studies have mostly focused on GABA-producing microorganisms rather than GABA in isolation. Lactic acid bacteria (LAB) and yeasts are the most important GABA producers, because they are commercially useful as starters in fermented foods.

Due to the physiological functions of GABA, development of functional foods containing high GABA concentrations has been actively pursued. On the other hand, pharmaceutical manufacturers and chemists have designed GABA-like substances (supplements or drugs) that can cross the blood–brain barrier and act as GABA in the brain. For example, Phenibut (beta-phenyl-gamma-aminobutyric acid HCl), which is a GABA analogue, can cross the blood–brain barrier because of a 6-carbon ring added to the GABA molecule; the molecule can then act as an anxiolytic, and can also enhance cognition. Picamilon, or nicotinoyl-GABA, is a dietary supplement formed by combining niacin with GABA; it is also able to cross the blood–brain barrier. Once released, GABA activates GABA receptors, potentially producing an anxiolytic effect, whilst niacin acts as a strong vasodilator.

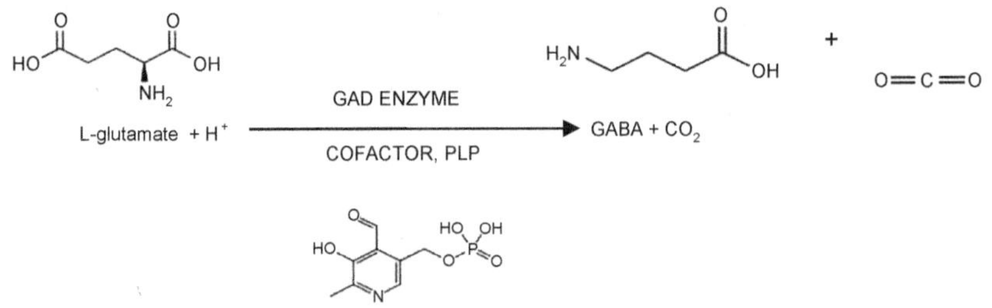

FIGURE 5.1 Decarboxylation reaction of L-glutamate to gamma-aminobutyric acid (GABA) catalyzed by glutamate decarboxylase (GAD), which is dependent on the cofactor pyridoxal-5′-phosphate (PLP) or Vitamin B_6. *Modified from Diana, M., Quílez, J., Rafecas, M., 2014d. Gamma-aminobutyric acid as a bioactive compound in food. Journal of Functional Food 10, 407–420.*

5.2 PHYSIOLOGICAL FUNCTIONS

GABA is found in almost every site in the body, giving it greater influence on a variety of physical health aspects. Although GABA's full functions and effects on the human body are not fully known, its significance in the human body is becoming increasingly indisputable.

In vertebrates, GABA is found at high levels in the brain and plays a fundamental role in inhibitory neurotransmission in several of its routes within the central nervous system, and in peripheral tissues. Alterations in GABAergic circuits are associated with Huntington's disease, Parkinson's disease, senile dementia, seizures, Alzheimer's disease, stiff person syndrome, and schizophrenia (Wong et al., 2003).

Only small amounts of GABA pass through the blood–brain barrier, however, orally ingested GABA seems to bring many physiological benefits. Blood pressure regulation is one of the most studied effects of GABA. Numerous research studies have shown that GABA can reduce elevated blood pressure in animals and humans (Hayakawa et al., 2004; Aoki et al., 2003; Wang et al., 2010; Tsai et al., 2013; Inoue et al., 2003; Pouliot-Mathieu et al., 2013; Joye et al., 2011). Other physiological functions, such as relaxation (Wong et al., 2003), sleeplessness (Wu et al., 2014), and depression (Okada et al., 2000) have been treated with GABA. Moreover, this bioactive compound could potentially protect against chronic kidney disease, ameliorate oxidative stress induced by nephrectomy (Sasaki et al., 2006), and activate liver and kidney function (Sun, 2004). Other studies have proven that GABA naturally enhanced immunity under stress conditions within 1 h of its administration in humans (Abdou et al., 2006). It may also be useful for alcohol-related disease prevention and treatment (Oh et al., 2003b). In addition, this amino acid contributes to increase the concentration of growth hormone in plasma and the rate of protein synthesis in the brain (Tujioka et al., 2009). Recent studies also indicate that it is a potent secretor of insulin, and thus could help to prevent diabetes (Adeghate and Ponery, 2002). Other authors reported that GABA tea ameliorates diabetes-induced cerebral autophagy which suggests the therapeutic potential of GABA against diabetic encephalopathy (Huang et al., 2014).

There is some evidence pointing out the chemopreventive effect of GABA against cancer. GABA could delay or inhibit the invasion and metastasis of various types of cancer cells, such as mammary gland, colon, and hepatic cancer cells (Minuk, 2000; Opolski et al., 2000). Furthermore, consumption of GABA-enhanced brown rice can inhibit leukemia cell proliferation and has a stimulatory action on cancer cell apoptosis (Oh and Oh, 2004). GABA has also been considered a potential tumor suppressor for small, airway-derived lung adenocarcinoma (Schuller et al., 2008).

Other physiological effects of GABA include antiinflammatory and fibroblast cell proliferations activities, which promote the healing of cutaneous wounds (Han et al., 2007). Besides, this amino acid is involved in maintaining cell volume homeostasis under UV radiation, in the synthesis of hyaluronic acid, and in enhancing the rate of dermal fibroblasts exposed to oxidative stress agents (Ito et al., 2007). This positive effect enables a potential novel application of GABA for dermatological purposes (Di Cagno et al., 2009). GABA can attenuate the metabolic response to ischemic incidents (Abel and McCandless, 1992). It also affects the control of asthma (Xu and Xia, 1999) and breathing (Kazemi and Hoop, 1991). GABA is found in abundance throughout the gastrointestinal (GI) tract in enteric nerves, submucosal nerve cell bodies, mucosal nerve fibers, and endocrine-like cells. Therefore, GABA largely influence GI function (Hyland and Cryan, 2010) acting as an endocrine mediator and neurotransmitter. GABA-enhanced functional foods allow for easy delivery of GABA in the GI tract (Hyland and Cryan, 2010). Thus, adding GABA into the GI tract through GABA-enriched foods may prove beneficial by taking advantage of the GI functions regulated by GABA-receptors.

There is evidence that low GABA level in plasma has an effect on the development of different mood disorders. It may even correlate with aggressiveness in patients with depression, mania, and alcoholism (Bjork et al., 2001). In contrast, normal GABA level is effective in antidepressant treatments being a target for the cure of bipolar illness (Krystal et al., 2002). There is also a relationship between GABA concentration and hormone secretion directly affecting mood behavior in women (Rapkin, 1999). Furthermore, there is clinical evidence of thyroid dysfunction (ie, hyperthyroidism or hypothyroidism) and the GABA system. In the young brain, hypothyroidism generally decreases GABA levels, whereas in the adult brain, hypothyroidism tends to increase enzyme activities and GABA levels (Wiens and Trudeau, 2006). GABA also improves oxidative stress and other thyroid functions, explaining lowered weight gain; this fact, suggests GABA as a preventer of obesity (Xie et al., 2014). Other studies have reported that GABA could improve visual function in senescent animals (Leventhal et al., 2003) and even enhance memory (Kayahara and Sugiura, 2001). Finally, GABA can control the level of quorum-sensing signal cells in eukaryotes and pathogenic bacteria (Chevrot et al., 2006).

All these physiological functions (Table 5.1) have created a strong commercial interest in developing GABA-enriched foods.

TABLE 5.1 Physiological Functions of GABA Tested on Animals and Humans

Physiological Function	Specific Function	References
Neurotransmission	Inhibitory neurotransmitter	Battaglioli et al. (2003)
Blood pressure regulator	Potent hypotensive agent	Hayakawa et al. (2004), Aoki et al. (2003), Wang et al. (2010), Tsai et al. (2013), Inoue et al. (2003), Pouliot-Mathieu et al. (2013), and Joye et al. (2011)
Brain diseases	Action on neurological disorders	Wong et al. (2003), Okada et al. (2000), and Kayahara and Sugiura (2001)
	Enhance memory	
Psychiatric diseases	Action on mood disorders	Krystal et al. (2002) and Bjork et al. (2001)
	Relaxing effect	Wong et al. (2003)
	Action on sleeplessness	Okada et al. (2000)
	Antidepressant	Okada et al. (2000) and Krystal et al. (2002)
	Prevention and treatment of alcoholism	Oh et al. (2003a,b) and Bjork et al. (2001)
Vital organs	Action on chronic kidney disease	Sasaki et al. (2006) and Sun (2004)
	Activates liver function	Sun (2004)
	Improves visual function	Leventhal et al. (2003)
	Increase rate secretion protein in brain	Tujioka et al. (2009)
Immune system	Enhance immunity	Abdou et al. (2006)
Protective against cancer	Delays and/or inhibits cancer cells proliferation	Minuk (2000) and Opolski et al. (2000)
	Stimulatory action on cancer cells apoptosis	Oh and Oh (2004)
	Potent tumor suppressor	Schuller et al. (2008)

Continued

TABLE 5.1 Physiological Functions of GABA Tested on Animals and Humans—cont'd

Physiological Function	Specific Function	References
Cell regulator	Keeps cell volume homeostasis	Warskulat et al. (2004)
	Antiinflammation and fibroblast cell proliferation	Han et al. (2007)
	Synthesis of hyaluronic acid	Ito et al. (2007)
	Enhance the rate of dermal fibroblasts	Ito et al. (2007)
	Quorum sensing signal cell-to-cell	Chevrot et al. (2006)
Protector of CVD	Attenuate the metabolic response to ischemic incidence	Abel and McCandless (1992)
Respiratory diseases	Control in asthma	Xu and Xia (1999)
	Control on breathing	Kazemi and Hoop (1991)
Hormonal regulator	Increase growth hormone	Tujioka et al. (2009)
	Regulation of progesterone	Rapkin (1999)
	Regulation of thyroid hormone	Wiens and Trudeau (2006)
	Potent secretor of insulin	Adeghate and Ponery (2002)
	Preventer of obesity	Xie et al. (2014)

Adapted from Diana, M., Quílez, J., Rafecas, M., 2014d. Gamma-aminobutyric acid as a bioactive compound in food. Journal of Functional Food 10, 407–420.

5.3 MECHANISMS OF ACTION

GABA is the most important inhibitory neurotransmitter in the brain mediating presynaptic inhibition of primary afferent fibers in the motor neuron system. It regulates brain excitability acting like a "brake" during times of runaway stress. GABA can likely benefit the body without crossing the blood–brain barrier, targeting GABA receptor sites throughout the body. GABA receptors, which recognize and bind GABA, are located in the postsynaptic membrane and throughout the human body, including the lung, liver, gastrointestinal tract, sperm, testes, mammary gland, and hepatic tumor cells (Osborn, 2011), and are categorized into three major groups: A or alpha, B or beta, and C or gamma (with subunits that determine its pharmacological activity). For example, a selected number of benzodiazepines have a tendency to bind strongly to the A1 subunit, while others bind to different A subunits. GABA-A receptors intercede in fast inhibitory synaptic transmissions. GABA regulates the seizure threshold, anxiety, panic, and response to stress through binding to A receptors (Borden et al., 1994). GABA-B receptors mediate slow inhibitory transmissions, which appear to be important in memory, mood, and pain (Meldrum and Chapman, 1999). The physiological role of GABA-C receptors have not been yet described. GABA injection in the brain does not incite any change in its concentration because the blood–brain barrier is impermeable. Thus, other effects seen after administration of GABA are due to its actions within the blood vessels or the autonomic nervous system.

In addition to neurological effects, GABA exerts effects on the endocrine system. However, the clinical significance of GABA is unclear and purely theoretical, based on clinical experience or data extrapolated from drug studies. A large-scale clinical study on a plethora of endocrine and psychoneurological conditions is needed; however, at least some benefits of GABA-enriched foods and pharmaceuticals have been established.

5.4 GAMMA-AMINOBUTYRIC ACID PRODUCTION BY LACTIC ACID BACTERIA

There are many investigations showing the GABA-producing ability of LAB isolated from fermented foods. The strains, *Lactobacillus brevis* PM17, *Lactobacillus plantarum* C48, *Lactobacillus paracasei* PF6, *Lactobacillus delbrueckii* subsp. *bulgaricus* PR1, and *Lactococcus lactis* PU1 isolated from several types of cheeses produced GABA at concentrations from 15 to 63 mg/kg in different culture mediums (Siragusa et al., 2007). Among *L. brevis* strains, *L. brevis* OPY-1 and *L. brevis* OPK-3 isolated from kimchi produced 0.825 g/L and 2.023 g/L, respectively (Park and Oh, 2005, 2007). *L. brevis* NCL912 and GABA057 isolated from Poacai produced 35.66 g GABA/L and 23.40 g GABA/L, respectively (Li et al., 2009a). In addition, *L. brevis* GABA100 produced a significant concentration of GABA in black raspberry juice (Kim et al., 2009) and *L. brevis* BJ20 synthesized 2.465 mg/L in a fermented sea tangle solution (Lee et al., 2010). *L. brevis* CGMCC1306 isolated from fresh milk produced the highest concentrations reported (76.36 g/L) in a specific medium composition (Huang et al., 2007). Other LAB strains GABA producers has been isolated from kimchi (Seok et al., 2008; Lu et al., 2008), koumiss (Sun et al., 2009), Myanmar fermented tinfoil barb (Su et al., 2011), red seaweed beverage (Ratanaburee et al., 2011), wholemeal wheat sourdough (Rizzello et al., 2008), human intestine and dental caries (Barrett et al., 2012), carrot leaves (Tamura et al., 2010) and Japanese traditional fermented fish (tuna sushi) (Komatsuzaki et al., 2005), among others.

High GABA production by LAB is related not only to their GAD activity but also to the concentration of glutamic acid in the food matrix that should be high enough. Accordingly, GABA-producing LAB can be used to develop fermented health-oriented food. There is a vast variety of GABA-enhanced food products, including cereals, sourdough and breads, cheeses, fermented sausages, teas, vegetables, legumes, dairy and soy products, alcohol beverages, and traditional Asian fermented foods.

5.5 GAMMA-AMINOBUTYRIC ACID-ENRICHED FERMENTED FOODS

GABA is naturally present in small quantities in many plant-derived foods, such as: vegetables (spinach, potatoes, cabbage, asparagus, broccoli, and tomatoes); fruits, (apples and grapes); and cereals (barley and maize) (Oh et al., 2005). In contrast, high amounts of GABA are found in fermented products.

Although the human body can produce its own supply of GABA, the production is sometimes inhibited by a lack of estrogen, zinc or vitamins, or by an excess of salicylic acid and food additives (Aoshima and Tenpaku, 1997). Indeed, GABA-enriched foods are required because the GABA content in the typical daily human diet is relatively low (Oh et al., 2005) and many of them have been tested to determine their health benefits.

Table 5.2 shows GABA-enriched food products described in the literature.

5.5.1 Cereal-based Products

Many GABA-enriched foods are cereal-derived. Numerous studies have reported the potential use of selected LAB on sourdough, resulting in GABA-enriched sourdough that could increase the GABA content in the final bread.

Rizzello et al. (2008) reported a high synthesis of GABA (258.7 mg/kg) in wholemeal wheat sourdough made using selected LAB, compared with other white wheat flours or rye flours. Another study showed 24.2 mg/100 g of GABA in bread made from a LAB fermented sourdough (98.2 mg GABA/100 mL) (Diana et al., 2014a). In this study, the maximum GABA concentration achieved in a wholemeal wheat sourdough model was 982 ppm. This sourdough was produced by *L. brevis* CECT 8183 isolated from an artisanal Spanish cheese (Diana et al., 2014b). Moreover, dough formulation based on wheat sourdough produced by *L. brevis* CECT 8183 improved peptide (<3 kDa) content, angiotensin-I converting enzyme inhibitory, and antioxidant activities of bread. Results from this study provide interesting perspectives for the development of innovative breads toward blood pressure reduction (Peñas et al., 2014). A blend of a wide variety of cereals, pseudo-cereals, and legume flours with well-characterized

TABLE 5.2 GABA-Enriched Food

Food Product	GABA Content (mg/kg)	References
Cereal-based		
Wholemeal wheat sourdough	258.7	Rizzello et al. (2008)
Whole wheat and soya sourdough	1000	Diana et al. (2014c)
Bathura sourdough bread	22.62	Coda et al. (2010)
Bread with *Yersinia* GAD supplementation	115	Lamberts et al. (2012)
Bread with *Yersinia* GAD supplementation	66	Joye et al. (2011)
Cereal bran flakes	90	Joye et al. (2011)
Cereal quinoa flakes	258	Joye et al. (2011)
Cereal malt flour flakes	435.2	Cai et al. (2012)
Fermented oats by *Aspergillus* sp.	1630	Nagaoka (2005)
Wheat germ	948	Jin et al. (2013)
Barley bran	429	Bai et al. (2008)
Foxtail millet	143	Chung et al. (2009)
Brown rice by proteolytic hydrolysis		Dai-xin et al. (2008)
Glutinous brown rice (laozao)	nr	
Diary products		
Cheese with *L. Lactis* spp. *Lactis* as starter	320	Pouliot-Mathieu et al. (2013)
Cheddar cheese with probiotic strain	6773.5	Wang et al. (2010)
Spanish artisanal cheese	980	Diana et al. (2014a)
Fermented milk from Tibet	nr	Sun et al. (2009)
Fermented milk	970	Liu et al. (2011)
Fermented milk	102	Hayakawa et al. (2004)
Fermented milk	100–120	Inoue et al. (2003)

TABLE 5.2 GABA-Enriched Food—cont'd

Food Product	GABA Content (mg/kg)	References
Fermented goat's milk	28	Minervini et al. (2009)
Fermented milk by strains isolated on old-style cheese	5000	Lacroix et al. (2013)
Skimmed milk by strains isolated from Italian cheeses	15–99.9	Siragusa et al. (2007)
Fermented skimmed milk by *L. helveticus*	113.35	Sun et al. (2009)
Fermented milk	120	Inoue et al. (2003)
Fermented milk by *L. plantarum*	77.4	Nejati et al. (2013)
Fermented milk by LAB combination	144.5	Nejati et al. (2013)
Fermented skimmed milk by *L. plantarum*	970	Liu et al. (2011)
Low fat fermented milk by LAB combination and protease	806	Hayakawa et al. (2004)
Meat, vegetables, and legumes		
Meat	100	Dai et al. (2012)
Fermented pork sausage	nr	Ratanaburee et al. (2013)
Fermented pork sausages	0.124	Li et al. (2009a)
Japanese lactic-acid fermented fish	1300	Kuda et al. (2009)
Vegetables	nr	Okita et al. (2009)
Brassica product	nr	Norimura et al. (2009)
Adzuki beans	2012	Liao et al. (2013)
Lentils	10,420	Torino et al. (2013)
Soybean nodules	0.31	Serraj et al. (1998)
Tempeh-like fermented soybean	3700	Aoki et al. (2003)
Cocoa beans	1012	Marseglia et al. (2014)
Adzuki beans	2012	Liao et al. (2013)

Continued

TABLE 5.2 GABA-Enriched Food—cont'd

Food Product	GABA Content (mg/kg)	References
Beverages		
White tea	505	Zhao et al. (2011)
Black raspberry juice	27,600	Kim et al. (2009)
Fermented grape must	9	Di Cagno et al. (2009)
Sugar cane juice-milk	3200	Hirose et al. (2008)
Fermented-pepper leaves-based	263,000	Song et al. (2014)
Fruit juice	579	Tamura et al. (2010)
Honey-based beverages	nr	Hiwatashi et al. (2010)
Rice *shochu* distillery lee	1560.5	Yokoyama et al. (2002)

GAD, glutamic acid decarboxylase; *LAB*, lactic acid bacteria; *nr*, not reported.
Adapted from Diana, M., Quílez, J., Rafecas, M., 2014d. Gamma-aminobutyric acid as a bioactive compound in food. Journal of Functional Food 10, 407–420.

GABA-producing strains have been used to make GABA-enriched sourdough and bread (504 mg/kg) (Coda et al., 2010). High GABA content was also found in Bathura sourdough bread (226.22 mg/100 g) (Bahnwar et al., 2013), followed by a GABA-enriched bread (115 ppm) made with an exogenous supplementation of recombinantly produced GAD from *Yersinia intermedia* (Lamberts et al., 2012). Another study showed the concentrations of breakfast cereal flakes by recipe and process optimization with the inclusion of bran, quinoa, or malt flour, which resulted in GABA levels of 66, 90, and 258 ppm, respectively (Joye et al., 2011). Oat fermented with a fungi strain (*Aspergillus oryzae*) produced the highest amount of GABA (435.2 μg/g) (Cai et al., 2012). GABA-enhanced fermented glutinous brown rice (*Laozao*) was developed by fermentation with a *Rhizopus* strain at 28.5°C for 48 h (Dai-xin et al., 2008).

5.5.2 Dairy Products

Cheeses and yogurt have also been studied, and many of them have shown potential in the management of hypertension. Cheese naturally enriched in GABA (16 mg of GABA/50 g of cheese) by *L. lactis* ssp. *lactis* strain, decreased blood pressure by 3.5 mmHg (Pouliot-Mathieu et al., 2013). Cheddar cheese containing a probiotic strain showed higher GABA levels than the control cheese (Wang et al., 2010). Furthermore, Diana et al. (2014c) reported high GABA concentrations in different types of artisanal Spanish cheeses, with the highest content (0.98 g/kg) being detected in two ripened cheeses made from sheep's milk. Other semiripened and fresh goat's milk cheeses showed high GABA concentrations (0.93 and 0.90 g/kg, respectively).

GABA-enriched yogurt has also been obtained by different procedures. Some commercial cheese starters were found to produce GABA, which increased during ripening in a skimmed milk culture (Nomura et al., 1998). Naturally fermented milk in Tibet had higher GABA levels than a commercial yogurt (Sun et al., 2009). Numerous studies on GABA enhancement in fermented milk have been reported, and many of such fermented products show hypotensive effects on rats (Liu et al., 2011; Hayakawa et al., 2004) and humans (Inoue et al., 2003; Nejati et al., 2013). Due to a mixed LAB starter, fermented goat's milk reached a GABA concentration of 28 mg/kg (Minervini et al., 2009) and 500 mg/100 mL was obtained in fermented milk by the action of two LAB strains isolated from an old-style cheese (Lacroix et al., 2013). A range of 15–99.9 mg/kg of GABA was obtained in a skimmed milk from several LAB-producing strains isolated from Italian cheeses (Siragusa et al., 2007). *Lactobacillus helveticus* showed 113.35 mg/L in fermented skimmed milk, and could potentially be used in the management of hypertension (Sun et al., 2009). A dose of 10–12 mg of GABA in 100 mL of fermented milk significantly decreased BP within two or four weeks in a randomized, placebo-controlled trial in mild hypertensive patients (Inoue et al., 2003). One strain of *L. plantarum* produced 77.4 mg/kg of GABA after milk fermentation and, in combination with other LAB, GABA concentration reached 144.5 mg/kg, which is a suitable dosage for a mild antihypertensive effect (Nejati et al., 2013).

The evidence of a GABA-hypotensive effect on rats has even been widely reported in milk fermented products. Systolic and diastolic BP in spontaneously hypertensive rats (SHR) significantly decreased after a diet of GABA-enriched skimmed milk (970 ppm), fermented by *L. plantarum* (Liu et al., 2011). Arterial pressure was also lower in SHRs after 8 weeks of oral administration of fresh, low-fat GABA-enriched milk (80.6 mg/100 g), fermented with five mixed LAB cultures in combination with a protease. Another nonfat fermented milk has a hypotensive effect on SHRs and in normotensive rats, at a low dose (Hayakawa et al., 2004).

5.5.3 Meat, Vegetables, and Legumes

Meat, fish, beans, and vegetables have also been used as raw materials to produce GABA-enriched fermented foods through different techniques.

Meat (Dai et al., 2012) and feedstuffs (Matsunaga et al., 2009) enriched in GABA may relieve animal stress, affecting the quality of meat. Fermented pork sausage was enriched with GABA using specific LAB starter cultures (Ratanaburee et al., 2013) and fig proteases (Li et al., 2009b) to enhance the meat value. A high content of GABA was also found in a traditional Japanese lactic-acid fermented fish (Kuda et al., 2009).

The effects of GABA-containing vegetables on the cardiac autonomic nervous system have been studied in young people (Okita et al., 2009) and the results highlighted the benefits of GABA on sympathetic nerve activity. In a fermented *Brassica* product, lactic acid fermentation produced a high GABA concentration, which suggests that this product could be used as a new functional food (Norimura et al., 2009). GABA-enhanced kimchi was found to improve long-term memory in mice (Seo et al., 2012). In this study, LAB strain *Lactobacillus sakei* B2-16 was isolated from kimchi in order to produce GABA. Seo et al. (2012) used 12% (w/v) monosodium glutamate to produce 660 mM of GABA. The culture medium included rice bran, sugar as a carbon source, and yeast extract as a nitrogen source. GABA improved the mice's memory from 132 to 48 s.

In addition, Marseglia et al. (2014) reported GABA contents ranging from 31.7 to 101.2 mg/100 g in fermented and dried cocoa beans, regarding the geographical origins.

Recently, Liao et al. (2013) proposed specific processes and fermentation conditions for GABA enrichment in *adzuki* beans, to provide a new, natural, and functional food resource. Black soybean was used to produce GABA-enhanced fermented milk as an antidepressant candidate (Ko et al., 2013). A novel tempeh-like fermented soybean with a high level of GABA was developed with a specific cultivation procedure (Aoki et al., 2003). Moreover, Torino et al. (2013) showed enhanced GABA concentration in fermented lentils.

5.5.4 Beverages

Research on different types of GABA-enriched beverages has been published. In particular, Chinese teas have been widely studied. A total of 114 samples from six types of tea were analyzed and it was concluded that white tea had higher GABA content than other types (Zhao et al., 2011). Levels of GABA can be enhanced in juices, mainly by lactic acid fermentation (Kim et al., 2009; Di Cagno et al., 2009; Song et al., 2014) or by specific fruit cultivation processes (Tamura et al., 2010). A study on the stability of GABA in fruit juices during storage at different temperatures and after radiation concluded that GABA remains stable (Shimizu and Sawai, 2008). An alcoholic Japanese beverage was found to contain a high GABA level (1.049 g GABA/L), due to the growth of *L. brevis* IFO 12005 (Yokoyama et al., 2002).

5.6 SIDE EFFECTS AND TOXICITY OF GAMMA-AMINOBUTYRIC ACID

GABA-agonist drugs can have significant side effects (from drowsiness and dizziness to addiction); however, no adverse effects have been reported for GABA dietary supplementation so far. This difference in safety profiles of synthetic and natural GABA may result from the limited capacity of the brain to retain excessive amounts of GABA. Lethal dose $(LD)_{50}$ tests conducted on natural GABA in doses of 5000 mg/kg to rats did not cause any mortality, indicating an $LD_{50} > 5000$ mg/kg in rats (Pharma Foods International Ltd., Japan).

5.7 FUTURE TRENDS

Nowadays, the pharmaceutical and food industries are promoting the use of bioactive compounds to manage diseases. However, there is a common propensity to believe that these molecules are obtained mainly synthetically. For this reason, the most important factor in the short- to medium term is to examine natural sources and investigate the effective dose of functional compounds if added both in isolation and as they occur naturally in the food matrix. The tendency of consumers to promote and maintain their health has facilitated the advancement of functional food which is growing thanks to the combination of plausible science and consumer perception.

GABA research has been intensified in recent years, establishing benefits in several physiological disorders. Nevertheless, further analysis needs to be carried on the advantages of specific GABA-enhanced functional foods in order to find the most profitable media to deliver GABA into the body.

Other questions need to be developed: why GABA has health benefits and what specific mechanisms are used to benefit the body's health at a molecular level? For example, functions of GABA-B receptors need to be discovered. In addition, little information is known about GABA-C receptors and how they influence the body. Knowing more about these receptors could bring more information about the mechanisms by which GABA-enriched foods have positive effects. The generation of GABA from glutamic acid or its salt in probiotic cells and other natural resources, such as plants or fungi is of added value to the food industry, because of the notable increased interest in natural and organic foods. High performance production, optimization through different biotechnological techniques, and the discovery of new high-GABA producing strains will remain a focus of interest in research into GABA as a health-related novel biological active compound. Testing and validation is yet another important step in increasing the prospects for clinical trials of bioactive molecules.

REFERENCES

Abdou, A.M., Higashiguchi, S., Horie, K., Kim, M., Hatta, H., Yokogoshi, H., 2006. Relaxation and immunity enhancement effects of gamma-aminobutyric acid (GABA) administration in humans. Biofactors 6 (3), 1–8.

Abel, M.S., McCandless, D.W., 1992. Elevated γ-aminobutyric acid levels attenuate the metabolic response to bilateral ischemia. Journal of Neurochemistry 58 (2), 740–744.

Adeghate, E., Ponery, A.S., 2002. GABA in the endocrine pancreas: cellular localization and function in normal and diabetic rats. Tissue Cell 34, 1–6.

Aoki, H., Uda, I., Tagami, K., Furuya, Y., Endo, Y., Fujimoto, K., 2003. The production of a new tempeh like fermented soybean containing a high level of γ- aminobutyric acid by anaerobic incubation with Rhizopus. Bioscience, Biotechnology, and Biochemistry 67 (5), 1018–1023.

Aoshima, H., Tenpaku, Y., 1997. Modulation of GABA receptors expressed in *Xenopus* oocytes by 13-L-hydroxylinoleic acid and food additives. Bioscience, Biotechnology, and Biochemistry 61 (12), 2051–2057.

Bahnwar, S., Bamnia, M., Ghosh, M., Ganguli, A., 2013. Use of *Lactococcus lactis* to enrich sourdough bread with γ-aminobutyric acid. International Journal of Food Sciences and Nutrition 64, 77–81.

Bai, Q.Y., Fan, G.J., Gu, Z.X., Cao, X.H., Gu, F.R., 2008. Effects of culture conditions on γ-aminobutyric acid accumulation during germination of foxtail millet (*Setaria italica* L.). European Food Research and Technology 228, 169–175.

Barrett, E., Ross, R.P., O'Toole, P.W., Fitzgerald, G.F., Stanton, C., 2012. γ-Aminobutyric acid production by culturable bacteria from the human intestine. Journal of Applied Microbiology 113 (2), 411–417.

Battaglioli, G., Liu, H., Martin, D.L., 2003. Kinetic differences between the isoforms of glutamate decarboxylase: implications for the regulation of GABA synthesis. Journal of Neurochemistry 86 (4), 879–887.

Bjork, J.M., Moeller, F.G., Kramer, G.L., Kram, M., Suris, A., Rush, A.J., Petty, F., 2001. Plasma GABA levels correlate with aggressiveness in relatives of patients with unipolar depressive disorder. Psychiatry Research 101 (2), 131–136.

Borden, L., Murali Dhar, T., Smith, K., 1994. Tiagabine, SK&F, 89976-A, CI-996 and NNC-711 are selective for the cloned GABA transporter GAT-1. European Journal of Pharmacology 269, 219–224.

Cai, S., Gao, F., Zhang, X., Wang, O., Wu, W., Zhu, S., Zhang, D., Zhou, F., Ji, B., 2012. Evaluation of γ- aminobutyric acid, phytate and antioxidant activity of tempeh like fermented oats (*Avena sativa* L.) prepared with different filamentous fungi. Journal of Food Science Technology 51 (10). http://dx.doi.org/10.1007/s13197-012-0748-2.

Chevrot, R., Rosen, R., Haudecoeur, E., Cirou, A., Shelp, B.J., Ron, E., Faure, D., 2006. GABA controls the level of quorum-sensing signal in *Agrobacterium tumefacien*. Proceedings of the National Academy of Sciences of the United States of America 103, 7460–7464.

Chung, H.J., Jang, S.H., Cho, H.Y., Lim, S.T., 2009. Effects of steeping and anaerobic treatment on GABA (gammaaminobutyric acid) content in germinated waxy hull-less barley. LWT – Food Science and Technology 42, 1712–1716.

Coda, R., Rizzello, C.G., Gobbetti, M., 2010. Use of sourdough fermentation and pseudo-cereals and leguminous flours for the making of a functional bread enriched of γ-aminobutyric acid (GABA). International Journal of Food Microbiology 137, 236–245.

Dai, S.F., Gao, F., Xu, X.L., Zhang, W.H., Song, S.X., Zhou, G.H., 2012. Effects of dietary glutamine and gamma-aminobutyric acid on meat colour, pH, composition, and water-holding characteristic in broilers under cyclic heat stress. British Poultry Science 53, 471–481.

Dai-xin, H., Lu, Z., Lan, J., Li-te, L., Yong-Qiang, C., 2008. Development of Laozao enriched with GABA. Food Science and Technology 1, 22–25.

Di Cagno, R., Mazzacane, F., Rizzello, C.G., Angelis, M.D.E., Giuliani, G., Meloni, M., Servi, B.D.E., Marco, G., 2009. Synthesis of γ-aminobutyric acid (GABA) by *Lactobacillus plantarum* DSM19463: functional grape must beverage and dermatological applications. Applied Microbiology and Biotechnology 86, 731–741.

Diana, M., Rafecas, M., Quílez, J., 2014a. Sourdough bread enriched with γ-aminobutyric acid (GABA). A comparison of free amino acid, biogenic amine and acrylamide content in GABA-enriched sourdough and commercial breads. Journal of Cereal Sciences 60 (3), 639–644.

Diana, M., Tres, A., Quílez, J., Llombart, M., Rafecas, M., 2014b. Spanish cheese screening and selection of lactic acid bacteria with high gamma-aminobutyric acid production. Journal of Food Science and Technology 56, 351–355.

Diana, M., Rafecas, M., Arco, C., Quílez, J., 2014c. Free amino acid profile of Spanish artisanal cheeses: importance of gamma-aminobutyric acid (GABA) and ornithine content. Food Composition and Analysis 35, 94–100.

Diana, M., Quílez, J., Rafecas, M., 2014d. Gamma-aminobutyric acid as a bioactive compound in food. Journal of Functional Food 10, 407–420.

Han, D., Kim, H.Y., Lee, H.J., Shim, I., Hahm, D.H., 2007. Wound healing activity of gamma-aminobutyric Acid (GABA) in rats. Journal of Microbiology and Biotechnology 17 (10), 1661–1669.

Hayakawa, K., Kimura, M., Kasaha, K., Matsumoto, K., Sansawa, H., Yamori, Y., 2004. Effect of a gamma-aminobutyric acid-enriched dairy product on the blood pressure of spontaneously hypertensive and normotensive Wistar-Kyoto rats. British Journal of Nutrition 92, 411–417.

Hirose, N., Ujihara, K., Teruya, R., Maeda, G., Yoshitake, H., Wada, K., Yoshimoto, M., 2008. Development of GABA-enhanced lactic acid beverage using sugar cane and its functionality. Journal of the Japanese Society for Food Science and Technology 55, 209–214.

Hiwatashi, K., Narisawa, A., Hokari, M., Toeda, K., 2010. Antihypertensive effect of honey-based beverage containing fermented rice bran in spontaneously hypertensive rats. Journal of the Japanese Society for Food Science and Technology 1, 40–43.

Huang, J., Mei, L., Sheng, Q., Yao, S., Lin, D., 2007. Purification and characterization of glutamate decarboxylase of *Lactobacillus brevis* CGMCC 1306 isolated from fresh milk. Chinese Journal of Chemical Engineering 15, 157–161.

Huang, C.-Y., Kuo, W.-W., Wang, H.-F., Lin, C.-J., Lin, Y.-M., Chen, J.-L., Kuo, C.-H., Chen, P.-K., Lin, J.-Y., 2014. GABA tea ameliorates cerebral cortex apoptosis and autophagy in streptozotocin-induced diabetic rats. Journal of Functional Foods 6, 534–544.

Hyland, N.P., Cryan, J.F., 2010. A gut feeling about GABA: focus on GABA (B) receptors. Frontiers in Pharmacology 1, 124.

Inoue, K., Shirai, T., Ochiai, H., Kasao, M., Hayakawa, K., Kimura, M., Sansawa, H., 2003. Blood-pressure-lowering effect of a novel fermented milk containing γ-aminobutyric acid (GABA) in mild hypertensives. European Journal of Clinical Nutrition 57, 490–495.

Ito, K., Tanaka, K., Nishibe, Y., Hasegawa, J., Ueno, H., 2007. GABA-synthesizing enzyme, GAD67, from dermal fibroblasts: evidence for a new skin function. Biochimica et Biophysica Acta 1770 (2), 291–296.

Jin, W., Kim, M., Kim, K., 2013. Utilization of barley or wheat bran to bioconvert glutamate to γ-aminobutyric acid (GABA). Journal of Food Science 78 (9), 1376–1382.

Joye, I.J., Lamberts, L., Brijs, K., Delcour, J.A., 2011. In situ production of γ-aminobutyric acid in breakfast cereals. Food Chemistry 129, 395–401.

Kayahara, H., Sugiura, T., 2001. Research on physiological function of GABA in recent years-improvement function of brain function and anti-hypertension. Japanese Journal of Food Development 36 (6), 4–6.

Kazemi, H., Hoop, B., 1991. Glutamic acid and gamma-aminobutyric acid neurotransmitters in central control of breathing. Journal of Applied Physiology 70 (1), 1–7.

Kim, J.Y., Lee, M.Y., Ji, G.E., Lee, Y.S., Hwang, K.T., 2009. Production of γ-aminobutyric acid in black raspberry juice during fermentation by *Lactobacillus brevis* GABA100. International Journal of Food Microbiology 130, 12–16.

Ko, C., Victor Lin, H., Tsai, G., 2013. Gamma-aminobutyric acid production in black soybean milk by *Lactobacillus brevis* FPA 3709 and the antidepressant effect of the fermented product on a forced swimming rat model. Process Biochemistry 48, 559–568.

Komatsuzaki, N., Shima, J., Kawamotoa, S., Momosed, H., Kimurab, T., 2005. Production of g-aminobutyric acid (GABA) by *Lactobacillus paracasei* isolated from traditional fermented foods. Food Microbiology 22, 497–504.

Krystal, J.H., Sanacora, G., Blumberg, H., Anand, A., Charney, D.S., Marek, G., Epperson, C.N., Goddard, A., Mason, G.F., 2002. Glutamate and GABA systems as targets for novel antidepressant and mood-stabilizing treatments. Molecular Psychiatry 7, 71–80.

Kuda, T., Tanibe, R., Mori, M., Take, H., Michihata, T., Yano, T., 2009. Microbial and chemical properties of *aji-no-susu*, a traditional fermented fish with rice product in the Noto Peninsula, Japan. Fish Science 75, 1499–1506.

Lacroix, N., St-Gelais, D., Champagne, C.P., Vuillemard, J.C., 2013. γ-aminobutyric acid-producing abilities of *lactococcal* strains isolated from old-style cheese starters. Dairy Science and Technology 9, 315–327.

Lamberts, L., Joye, I., Beliën, T., Delcour, J., 2012. Dynamics of γ-aminobutyric acid in wheat flour bread making. Food Chemistry 130, 896–901.

Lee, B.J., Kim, J.S., Kang, Y.M., Lim, J.H., Kim, Y.M., Lee, M.S., Jeong, M.H., Ahn, C.B., Je, J.Y., 2010. Antioxidant activity and γ-aminobutyric acid (GABA) content in sea tangle fermented by *Lactobacillus brevis* BJ20 isolated from traditional fermented foods. Food Chemistry 122 (1), 271–276.

Leventhal, A., Wang, Y., Pu, M., Zhou, Y., Ma, Y., 2003. GABA and its agonists improved visual cortical function in senescent monkeys. Science 300, 812–815.

Li, H., Qiu, T., Gao, D., Cao, Y., 2009a. Medium optimization for production of gamma-aminobutyric acid by *Lactobacillus brevis* NCL912. Amino Acids 38. http://dx.doi.org/10.1007/s00726-009-0355-3.

Li, J., Izumimoto, M., Yonehara, M., Hirotsu, S., Kuriki, T., Naito, I., Yamada, H., 2009b. The influence of fig proteases on the inhibition of angiotensin I-converting and GABA formation in meat. Animal Science Journal 80 (6), 691–696.

Liao, W.C., Wang, C.Y., Shyu, Y.T., Yu, R.C., Ho, K.C., 2013. Influence of preprocessing methods and fermentation of *adzuki* beans on c-aminobutyric acid (GABA) accumulation by lactic acid bacteria. Journal of Functional Foods 5 (3), 1108–1115.

Liu, C.F., Tung, Y.T., Wu, C.L., Lee, B.-H., Hsu, W.-H., Pan, T.M., 2011. Antihypertensive effects of *Lactobacillus*-fermented milk orally administered to spontaneously hypertensive rats. Journal of Agricultural and Food Chemistry 59, 4537–4543.

Lu, X., Chen, Z., Gu, Z., Han, Y., 2008. Isolation of gamma-aminobutyricacid-producing bacteria and optimization of fermentative medium. Biochemical Engineering Journal 41, 48–52.

Marseglia, A., Palla, G., Caligiani, A., 2014. Presence and variation of γ-aminobutyric acid and other free amino acids in cocoa beans from different geographical origins. Food Research International 63, 360–366.

Matsunaga, M., Saze, K., Matsunaga, T., Suzuta, Y., 2009. Feedstuff and Method for Supply of Gamma-aminobutyric Acid. US 20090181121 A1.

Meldrum, B.S., Chapman, A.G., 1999. Basic mechanisms of gabitril (tiagabine) and future potential developments. Epilepsia 40 (9), 2–6.

Minervini, F., Bilancia, M.T., Siragusa, S., Gobbetti, M., Caponio, F., 2009. Fermented goats' milk produced with selected multiple starters as a potentially functional food. Food Microbiology 26, 559–564.

Minuk, G.Y., 2000. GABA and hepatocellular carcinoma. Molecular and Cellular Biochemistry 207, 105–108.

Nejati, F., Rizzello, C., Di Cagno, R., Sheikh-Zeinoddin, M., Diviccaro, A., Minervini, F., Gobetti, M., 2013. Manufacture of a functional fermented milk enriched of angiotensin-I converting enzyme (ACE)-inhibitory peptides and gamma-amino butyric acid (gaBa). LWT Food Sciences and Technology 51 (1), 183–189.

Nomura, M., Kimoto, H., Someya, Y., Furukawa, S., 1998. Prodution of γ-aminobutyric acid by cheese starters during cheese ripening. Journal of Dairy Science 81, 1486–1491.

Norimura, N., Sonomoto, K., Maki, T., Maeda, M., Kobayashi, G., Komatsu, Y., 2009. The Enhancing Effect of Gamma-amino Butyric Acid (GABA) Concentration and Angiotensin-I Converting Enzyme (ACE) Inhibitory Capacity in Varieties of Brassica. http://agris.fao.org/aos/records/JP2009004544.

Oh, C.H., Oh, S.H., 2004. Effects of germinated brown rice extracts with enhanced levels of GABA on cancer cell proliferation and apoptosis. Journal of Medicinal Food 7 (1), 19–23.

Oh, S.H., Moon, Y.J., Oh, C.H., 2003a. γ-Aminobutyric acid (GABA) content of selected uncooked foods. Journal of Food Sciences and Nutrition 8, 75–78.

Oh, S.H., Soh, J.R., Cha, Y.S., 2003b. Germinated brown rice extract shows a nutraceutical effect in the recovery of chronic alcohol-related symptoms. Journal of Medicine Food 6, 115–121.

Oh, S.H., Choi, W.G., Lee, I.T., Oh, S., 2005. Cloning and characterization of a rice cDNA encoding glutamate decarboxylase. Journal of Biochemistry and Molecular Biology 38 (5), 595–601.

Okada, T., Sugishita, T., Murakami, T., Murai, H., Saikusa, T., Horino, T., Onoda, A., Kajmoto, O., Takahashi, R., Takahashi, T., 2000. Effect of the defatted rice germ enriched with GABA for sleeplessness depression, autonomic disorder by oral administration. Journal of the Japanese Society of Food Science and Technology 47 (8), 596–603.

Okita, Y., Nakamura, H., Kouda, K., Takahashi, I., Takaoka, T., Kimura, M., 2009. Effects of vegetable containing gamma-aminobutyric acid on the cardiac autonomic nervous system in healthy young people. Journal of Physiology and Anthropology 28, 101–107.

Opolski, A., Mazurkiewicz, M., Wietrzyk, J., Kleinrok, Z., Radzikowski, C., 2000. The role of GABA-ergic system in human mammary gland pathology and in growth of transplantable murine mammary cancer. Journal of Experimental and Clinical Cancer Research 19 (3), 383–390.

Osborn, E., 2011. GABA-enriched Functional Foods Aiding in Health and Disease Management. Lynric.org/eric.php.

Park, K.B., Oh, S.H., 2005. Production and characterization of GABArice yogurt. Food Science and Biotechnology 14, 518–522.

Park, K.B., Oh, S.H., 2007. Cloning, sequencing and expression of a novel glutamate decarboxylase gene from a newly isolated lactic acid bacterium, *Lactobacillus brevis* OPK-3. Bioresource Technology 98, 312–319.

Peñas, E., Diana, M., Frías, J., Quílez, J., Martínez-Villaluenga, C., 2014. A multistrategic approach in the development of sourdough bread targeted towards blood pressure reduction. Journal of Plant Food for Human Nutrition 70. http://dx.doi.org/10.1007/s11130-015-0469-6.

Pharma Foods International, Ltd., Kyoto, Japan (unpublished data).

Pouliot Mathieu, K., Gardner-Fortier, C., Lemieux, S., St-Gelais, D., Champagne, C.P., Vuillemard, J.-C., 2013. Effect of cheese containing gamma-aminobutyric acid-producing acid lactic bacteria on blood pressure in men. PharmaNutrition 1, 1–8.

Rapkin, A.J., 1999. Progesterone, GABA and mood disorders in women. Archives of Women and Men Health 2, 97–105.

Ratanaburee, A., Kantachote, D., Charernjiratrakul, W., Penjamras, P., Chaiyasut, C., 2011. Enhancement of γ-aminobutyric acid in a fermented red seaweed beverage by starter culture *Lactobacillus plantarum* DW12. Electronic Journal of Biotechnology 14 (3), 0717–3458.

Ratanaburee, A., Kantachote, D., Charernjiratrakul, W., Sukhoom, A., 2013. Selection of γ-aminobutyric acid-producing lactic acid bacteria and their potential as probiotics for use as starter cultures in Thai fermented sausages (Nham). International Journal of Food Science and Technology 48, 1371–1382.

Rizzello, C.G., Cassone, A., Di Cagno, R., Gobbetti, M., 2008. Synthesis of angiotensin-I converting enzyme (ACE)-inhibitory peptides and γ-aminobutyric acid (GABA). Journal of Agricultural and Food Chemistry 56 (16), 6936–6943.

Sasaki, S., Yokozawa, T., Cho, E.J., Oowada, S., Kim, M., 2006. Protective role of γ-aminobutyric acid against chronic renal failure in rats. Journal of Pharmacy and Pharmacology 58 (11), 1515–1525.

Schuller, H.M., Al-Wadei, H.A., Majidi, M., 2008. Gamma-aminobutyric acid, a potential tumor suppressor for small airway-derived lung adenocarcinoma. Carcinogenesis 29 (10), 1979–1985.

Seo, Y.C., Choi, W.Y., Kim, J., Lee, C.G., Ahn, J.H., Cho, H.Y., Lee, S.H., Cho, J.S., Joo, S.J., Lee, H.Y., 2012. Enhancement of the cognitive effects of GABA from monosodium glutamate fermentation by *Lactobacillus sakei* B2-16. Food Biotechnology 26 (1), 29–44.

Seok, J.H., Park, K.B., Kim, Y.H., Bae, M.O., Lee, M.K., Oh, S.H., 2008. Production and characterization of kimchi with enhanced levels of gamma-aminobutyric acid. Food Science and Biotechnology 17, 940–946.

Serraj, R., Shelp, B.J., Sinclair, T.R., 1998. Accumulation of g-aminobutyric acid in nodulated soybean in response to drought stress. Physiologia Plantarum 102, 79–86.

Shimizu, T., Sawai, Y., 2008. Stability of gamma-aminobutyric acid in fruit juice during storage. Food Preservation Science 34 (3), 145–149.

Siragusa, S., De Angelis, M., Di Cagno, R., Rizzello, C.G., Coda, R., Gobbetti, M., 2007. Synthesis of γ-aminobutyric acid by lactic acid bacteria isolated from a variety of Italian cheeses. Applied and Environmental Microbiology 73, 7283–7290.

Song, Y., Shin, N., Baik, S., 2014. Physicochemical and functional characteristics of a novel fermented pepper (*Capsiccum annuum* L.) leaves-based beverage using lactic acid bacteria. Food Science and Biotechnology 23 (1), 187–194.

Su, M.T., Takeshi, K., Tianyao, L., 2011. Isolation, characterization, and utilization of γ-aminobutyric acid (GABA)-producing lactic acid bacteria from Myanmar fishery products fermented with boiled rice. Fisheries Science 77 (2), 279.

Sun, T.S., Zhao, S.P., Wang, H.K., Cai, C.K., Chen, Y.F., Zhang, H.P., 2009. ACE-inhibitory activity and gamma-aminobutyric acid content of fermented skim milk by *Lactobacillus helveticus* isolated from Xinjiang koumiss in China. European Food Research and Technology 228, 607–612.

Sun, B.S., 2004. Research of Some Physiological Active Substance by Fermentation of *Monascus* Spp. (Dissertation for Master Degree). Zhejiang Industry University, China, pp. 40–55.

Tamura, T., Noda, M., Ozaki, M., Maruyama, M., Matoba, Y., Kumagai, T., Sugiyama, M., 2010. Establishment of an efficient fermentation system of gamma-aminobutyric acid by a lactic acid bacterium, *Enterococcus avium* G-15, isolated from carrot leaves. Biological and Pharmaceutical Bulletin 33 (10), 1673–2167.

Torino, M.I., Limón, R., Martínez-Villaluenga, C., Mäkinen, S., Pihlanto, A., Vidal-Valverde, C., Frías, J., 2013. Antioxidant and antihypertensive properties of liquid and solid state fermented lentils. Food Chemistry 136, 1030–1037.

Tsai, C., Chiu, T., Ho, C., Lin, P., Wu, T., 2013. Effects of anti-hypertension and intestinal microflora of spontaneously hypertensive rats fed gamma aminobutyric acid-enriched Chingshey purple sweet potato fermented milk by lactic acid bacteria. African Journal of Microbiological Research 7 (11), 932–940.

Tujioka, K., Ohsumi, M., Horie, K., Kim, M., Hayase, K., Yokogoshi, H., 2009. Dietary gamma-aminobutyric acid affects the brain protein synthesis rate in ovariectomized female rats. Journal of Nutrition and Science Vitaminology 55 (1), 75–80.

Wang, H.K., Dong, C., Chen, Y.F., Cui, L.M., Zhang, H.P., 2010. A new probiotic Cheddar cheese with high ACE-inhibitory activity and gamma-aminobutyric acid content produced with Koumiss-derived *Lactobacillus casei* Zhang. Food Technology and Biotechnology 48, 62–70.

Warskulat, U., Reinen, A., Grether-Beck, S., Krutmann, J., Häussinger, D., 2004. The osmolyte strategy of normal human keratinocytes in maintaining cell homeostasis. Journal of Investigative Dermatology 123, 516–521.

Wiens, S.C., Trudeau, V.L., 2006. Thyroid hormone and γ-aminobutyric acid (GABA) interactions in neuroendocrine systems. Comparative Biochemistry and Physiology Part A: Molecular & Integrative Physiology 144 (3), 332–344.

Wong, C.G., Bottiglieri, T., Snead, O.C., 2003. GABA, gamma-hydroxybutyric acid, and neurological disease. Annals of Neurology 54 (6), 3–12.

Wu, C., Huang, Y., Lai, X., Lai, R., Zhao, W., Zhang, M., Zhao, W., 2014. Study on quality components and sleep-promoting effect of GABA Maoyecha tea. Journal of Functional Foods 7, 180–190.

Xie, Z., Xia, S., Le, G.-W., 2014. Gamma-aminobutyric acid improves oxidative stress and function of the thyroid in high-fat diet fed mice. Journal of Functional Foods 8, 76–86.

Xu, C.W., Xia, Y.H., 1999. Clinical observations on the control acute attack of deficiency-syndrome asthma with γ-aminobutyric acid. Chinese Journal of Binzhou Medical College 22, 181.

Yokoyama, S., Hiramatsu, J., Hayakawa, K., 2002. Production of γ- aminobutyric acid from alcohol distillery lees by *Lactobacilus brevis* IFO 12005. Journal of Bioscience and Bioengineering 93 (1), 95–97.

Zhao, M., Ma, Y., Wei, Z.Z., Yuan, W.X., Li, Y.L., Zhang, C.H., Xue, X.T., Zhou, H.J., 2011. Determination and comparison of γ-aminobutyric acid (GABA) content in pu-erh and other types of Chinese tea. Journal of Agricultural and Food Chemistry 59 (8), 3641–3648.

FURTHER READING

Oh, S.H., 2003. Stimulation of gamma-aminobutyric acid synthesis activity in brown rice by a chitosan/Glugermination solution and calcium/calmodulin. Journal of Biochemistry and Molecular Biology 36, 319–325.

Chapter 6

Melatonin Synthesis in Fermented Foods

M.A. Martín-Cabrejas[1], Y. Aguilera[1], V. Benítez[1], R.J. Reiter[2]
[1]University Autónoma de Madrid, Madrid, Spain; [2]UT Health Science Center, San Antonio, TX, United States

6.1 INTRODUCTION

Melatonin (MEL) is a low-molecular weight molecule that exists in all living organisms; it exhibits many biological activities in all species from bacteria to mammals. It has also been reported in foods mainly from plant origin and proposed as a new bioactive food component, since its health benefits have been widely documented (García-Parrilla et al., 2009; Cardinali et al., 2012; Hardeland, 2013). Among other functions, MEL is a powerful antioxidant and provides relevant stimulation of the immunological system thereby enhancing resistance to infection and diseases. MEL was first identified in bovine pineal tissue in 1958 and represented exclusively as a hormone, but accumulated evidence has challenged this concept. Thus MEL is present in the earliest life forms and is found in all organisms including bacteria, algae, fungi, plants, insects, and vertebrates including humans (Handerland and Poeggeler, 2003; Posmyk and Janas, 2009). In animals, MEL is a biological modulator of several timing (circadian) processes, such as mood, sleep, sexual behavior, immunological status, etc. Since its discovery in plants in 1995 by Dubbels et al., several physiological functions have been postulated, such as a possible role in flowering, circadian rhythms and photoperiodicity, and growth regulator (Reiter et al., 2010a; Tan et al., 2011). MEL has also been detected in plant foods, including cereals, vegetables, fruits, and roots (Iriti et al., 2010). Efficient uptake of MEL from food sources should influence its very low circulating levels (~200 pg/mL at the maximum night peak and lower than 10 pg/mL during the day). Currently, the beneficial human health effects of MEL derived from the consumption of plant foods are being considered (Claustrat et al., 2005; McCune et al., 2011; Aguilera et al., 2015).

The European Food Safety Authority (EFSA) has recently accepted the link between melatonin and the alleviation of jet lag based on a list of health claims in relation to MEL (EFSA, 2010). Its recommended dose is between

0.5 and 5 mg per day. Indeed, synthetic supplements of this compound are used to treat sleeping disorders and jet lag, but, unfortunately, the EFSA did not approve melatonin for other conditions. Apart from the endogenous production of MEL, its intake increases serum melatonin levels. Reiter et al. (2005) found that rats exhibited significantly higher blood MEL levels (38.0 vs. 11.5 pg/mL) after consumption of walnuts (rich in MEL) added to rodent chow. Recently, it was observed that the consumption of fruits or fruit juices containing MEL significantly elevated serum melatonin concentration in humans (Sea-Teaw et al., 2013); the highest levels of MEL were observed at 120 min after fruit consumption (eg, pineapple 146 pg/mL, orange 151 pg/mL, and banana 140 pg/mL versus 32–48 pg/mL before juice consumption). On the other hand, endogenous MEL levels decrease with age and also in certain diseases, such as Alzheimer's and cardiovascular disease (Hardeland, 2013). These reductions have also been linked to insomnia in older patients (Leger et al., 2004) and to a higher prevalence of cancer (Bartsch and Bartsch, 2006). Therefore it is of interest to provide exogenous MEL to increase plasma levels of this important molecule in adult and aging population (Pandi-Pemural et al., 2008; Posmyk and Janas, 2009).

There is an insufficient amount of scientific data on the MEL content in foods to evaluate the adequate dietary intakes (Rodríguez-Naranjo et al., 2011a). Thus the current objective is to quantify MEL as a natural component of food intake, and for this purpose, it is necessary to identify the main sources of MEL in the diet and thereafter to document processes that would increase its level in foods.

6.2 STRUCTURE AND PHYSICOCHEMICAL PROPERTIES

Melatonin (N-acetyl-5-methoxytryptamine) is a biogenic indoleamine structurally related to other important molecules, such as tryptophan (TRP), serotonin, indole-3-acetic acid (IAA), etc. The structure of MEL has two distinguishable side chains that are added to the indole ring (Fig. 6.1); these are the methoxy group at position 5 (5-methoxy group) and the N-acetylaminoethyl group at position 3 (3-(N-acetylaminoethyl) group) (Posmyk and Janas, 2009). Theoretically, either one of these two side-chains could be potentially relocated to any one of the seven positions in the indole ring to form melatonin isomers (MIs); thus there may be 42 potential MIs. In order to distinguish them, it has been proposed that the 3-(N-acetylaminoethyl) group is designated as side chain A (A) and the methoxy group will be designated as side chain M (M). Then, an MI can be named according to the positions of A and M side chains at the seven positions on the indole ring (eg, MI A1/M6) (Tan et al., 2012).

MEL is a small molecule with low molecular weight (232.38 g/mol) and its solubility is 0.1 mg/mL in water and 8 mg/mL in ethanol. It is an amphiphilic molecule, which can freely cross-cell membranes and distribute to any aqueous or lipid compartment including the membranes, cytosol, nucleus, and mitochondria of cells (Tan et al., 2011).

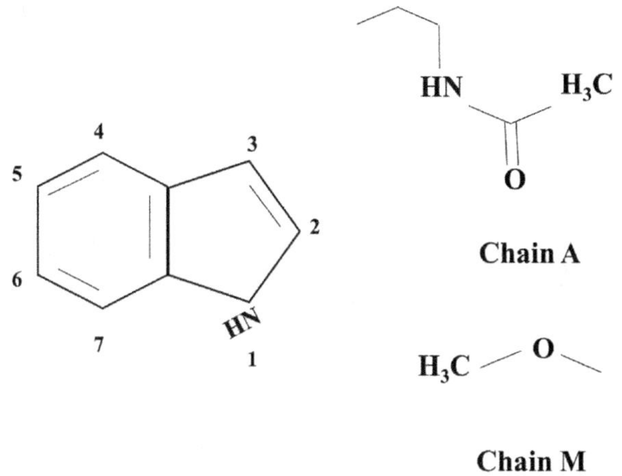

FIGURE 6.1 Chains A and M at the seven positions on the indole ring constitute the molecular structure of melatonin isomers.

6.3 BIOSYNTHESIS OF MELATONIN

MEL was originally considered as an exclusive product of the pineal gland in vertebrates, including humans. Synthesis and secretion of MEL by the pineal gland is dramatically affected by light exposure to the eyes. The main pattern observed in serum concentrations of MEL is that they are low during the daylight hours and increase to a peak during the darkness (Posmyk and Janas, 2009; Reiter et al., 2014). However, MEL production was not only restricted to this gland but also occurs in numerous other organs and tissues (retina, lacrimal glands, red blood cells, platelets, mononuclear cells, gastrointestinal tract, skin, bone narrow, etc.) (Acuña-Castroviejo et al., 2014). Most of these organs (except retina) may not synthesize melatonin in a circadian manner nor do they release melatonin into the blood in any significant amount (Reiter et al., 2014). The gastrointestinal tract contains significantly more MEL than is synthesized in the pineal gland (Claustrat et al., 2005; Tan et al., 2012).

MEL is synthesized from TRP that is taken up from the circulation and transformed to serotonin. In animals, this biosynthesis is carried out in two successive steps involving tryptophan 5-hydroxylase (TPH) and aromatic L-amino acid decarboxylase (AADC), in which TPH acts as the rate-limiting enzyme (Posmyk and Janas, 2009). Serotonin is converted to MEL by a two-step process involving the sequential activities of two enzymes, serotonin-*N*-acetyl transferase (SNAT), which is the limiting enzyme for the synthesis of MEL, and hydroxyindole-*O*-methyl transferase (HIOMT). The first step is the *N*-acetylation of serotonin to yield *N*-acetylserotonin and the second one is the transfer of a methyl group from *S*-adenosylmethionine to the 5-hydroxy position of *N*-acetylserotonin to yield MEL. In addition, MEL synthesis depends on TRP availability and other nutritional factors such folate status and vitamin B6. Once formed, pineal MEL is released into the capillaries and in higher concentrations into the cerebrospinal fluid and is then rapidly distributed throughout the body. Melatonin produced in other organs functions locally as an antioxidant, autocoid, or paracoid to regulate intracellular events (Claustrat et al., 2005; Pandi-Perumal et al., 2008; Tan et al., 2012; Reiter et al., 2014).

MEL levels in several plant tissues are as high as μg/g of tissue, values that are orders of magnitude higher than those detected in animal blood (Tan et al., 2011). The high concentrations of hemoglobin in vertebrate blood do not seem to be compatible with high MEL levels, since hemoglobin in the presence of MEL could be oxidized to methemoglobin (Tan et al., 2005). Furthermore, TRP availability is much better in organisms producing aromatic amino acids via the shikimic acid pathway than in vertebrates that are devoid of this metabolic route. These arguments are likewise valid for other organisms rich in MEL, such as dinoflagellates, euglenoids, and yeasts (Sprenger et al., 1999; Tan et al., 2011).

The biosynthetic pathway of MEL in higher plants seems to be similar to that in vertebrates (Arnao and Hernandez-Ruiz, 2006). In plants, MEL is also synthesized from TRP that is transformed into serotonin, which may be synthesized by different pathways. The two enzymes found in animals and required for TRP transformation into serotonin have been identified in plants (Fujiwara et al., 2010). However, it has been reported that the two steps of serotonin biosynthesis from TRP in a plant (rice) are reversed in plants (Park et al., 2012). The first enzymatic reaction product is not 5-hydroxytryptophan, but rather tryptamine, which is catalyzed by tryptophan decarboxylase (TDC), and consequently the resulting tryptamine is catalyzed to serotonin by tryptamine 5-hydroxylase (T5H) (Posmyk and Janas, 2009; Tan et al., 2011).

TRP in plants is also the precursor of indole-3-acetic acid (IAA) and other auxins, eg, indole-3-butyric (IBA) and *p*-hydroxyphenylacetic acids. IAA may be synthesized via several L-tryptophan dependent pathways with (a) indole-3-pyruvic acid (IPA) and indole-3-acetaldehyde (IAAld), (b) indole-3-acetamide (IAM), or (c) indole-3-acetonitrile (IAN) as intermediate products, suggesting that alternatively IAA could be also synthesized from tryptamine catalyzed by tryptamine deaminase (TDA); this intermediate is also directly connected with serotonin and melatonin biosynthesis (Fig. 6.2) (Murch and Saxena, 2002).

According to the published literature, the dominant pathway is from TRP via 5-hydroxytryptophan to serotonin and then *N*-acetylserotonin formed by SNAT and converted to MEL by HIOMT with the use of *S*-adenosyl-L-methionine (SAM) as a methyl donor (Murch and Saxena, 2002; Posmyk and Janas, 2009). In *Saccharomyces*, SNAT was characterized (Ganguly et al., 2001) and, not only MEL, but also 5-methoxytryptamine and 5-methoxytryptophol were found (Sprenger et al., 1999). Therefore the alternative pathway of serotonin via 5-methoxytryptamine to MEL is also possible (Posmyk and Janas, 2009).

The metabolism of TRP to MEL in plants is influenced by light intensity. The elevated MEL production under UV-B could be an adaptive reaction

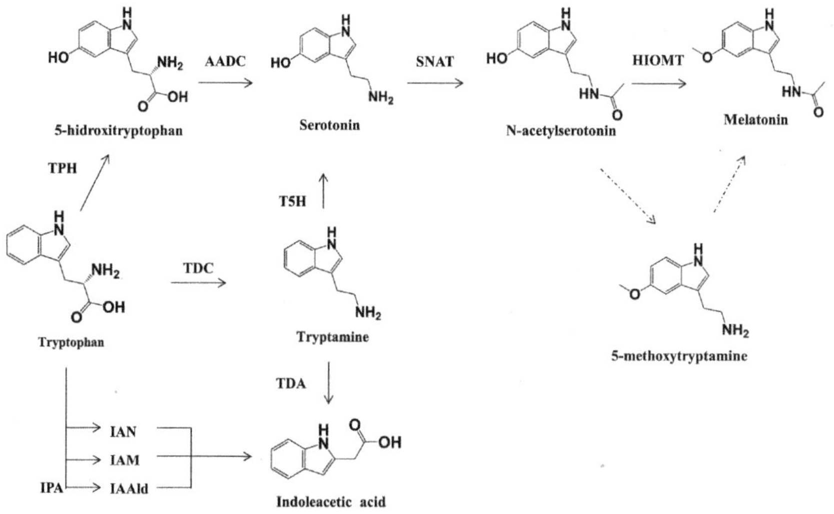

FIGURE 6.2 Biosynthetic pathway of melatonin (Enzymes involved: *TPH*, Tryptophan 5-hydroxylase; *AADC*, aromatic L-amino decarboxylase; *SNAT*, serotonin N-acetyltransferase; *HIOMT*, hydroxyindole-O-methyltransferase; *TDC*, tryptophan decarboxylase; *T5H*, tryptamine 5-hydroxylase; *TDA*, tryptaminedesaminase). With the numbers basic tryptophan dependent pathways of indoleacetic acid biosynthesis with intermediates: (1) indole-3-pyruvic acid (IPA) and indole-3-acetaldehyde(IAAld), (2) indole-3-acetamide(IAM), and (3) indole-3-acetonitrile (IAN) are marked.

of plants to tolerate this harmful radiation (Afreen et al., 2006; Paredes et al., 2009). In addition, plant roots may be able to absorb MEL (Tan et al., 2007a,b; Paredes et al., 2009). Recent studies have shown that melatonin can be taken up by leaves when MEL is sprayed onto them (Arnao and Hernández-Ruiz, 2008). Because MEL possesses both lipophilic and hydrophilic properties, it may be easy for this molecule to cross morpho- and physiological barriers with minimal difficulty, resulting in fast MEL transport into plant cells (Paredes et al., 2009). These chemical properties of melatonin are relevant in food product formulations when exogenous melatonin is added to increase its content in them.

6.4 MECHANISMS OF ACTION

MEL is involved in numerous physiological processes, some of which are receptor-mediated, whereas other functions are receptor-independent including the interactions of MEL with reactive oxygen species (ROS) and also those mediated by its bioactive metabolites (Witt-Enderby et al., 2003; Tan et al., 2011). Three MEL membrane receptors have been identified: two of them are G-protein coupled receptors (MT_1 and MT_2), while the third belongs to the family of quinone reductases. Each of the G-protein-coupled MEL receptors is classified as high affinity and can couple to multiple signal transduction cascades.

These receptors are widely distributed in the central nervous system and in the peripheral tissues, leading to widespread MEL responses in organisms. They are involved among other things in modulating circadian rhythms, constricting and dilating blood vessels, reproductive regulation, and immuno-responsiveness (Witt-Enderby et al., 2003; Reiter et al., 2014). The third binding site was originally believed to represent another membrane receptor (MT_3), but this protein turned out to be an enzyme, quinone reductase 2, and no signaling pathways are associated with this enzyme (Tan et al., 2011; Reiter et al., 2014).

MEL also acts by nonreceptor-mediated means (Reiter et al., 2014) due to its small size and highly lipophilic nature and/or its active uptake mechanism (Witt-Enderby et al., 2003). One of the nonreceptor-mediated actions of MEL is its free-radical-scavenging and antioxidant capacity (Reiter et al., 2008, 2010a; Tan et al., 2011). MEL effectively scavenges a variety of ROS and reactive nitrogen species (RNS), and protects cells, tissues, and organisms from oxidative stress (Tan et al., 2002, 2007c). This protection is due to different mechanisms. MEL is able improve the activities of several respiratory chain complexes thereby reducing electron leakage and free-radical generation. Moreover, it stimulates the activity of antioxidant enzymes, such as peroxidases (POX), glutathione reductase (GSSG-R) but also superoxide dismutase (SOD) and catalase (CAT) and inhibits prooxidative enzymes like nitric oxide synthase, myeloperoxidase, and eosinophil peroxidase (Posmyk and Janas, 2009; Reiter et al., 2014; Zhang and Zhang, 2014).

Finally, MEL can be enzymatically or nonenzymatically transformed to several biologically active metabolites including 5-methoxytryptamine (5-MT), cyclic 3-hydroxymelatonin (c3OH M), N1-acetyl-N2-formyl-5-methoxykynuramine (AFMK), and N1-acetyl-5-methoxykynuramine (AMK) (Tan et al., 2010). These metabolites also provide protection against oxidative damage by the same mechanisms describe for MEL (Zhang and Zhang, 2014).

6.5 HEALTH BENEFITS OF MELATONIN

6.5.1 Antioxidant Effects

MEL and its metabolites are potent free-radical scavengers and protect against oxidative stress by different mechanisms (Zhang and Zhang, 2014). Among other functions, MEL may provide protection against tumor growth by shielding molecules, especially DNA, from oxidative damage (Reiter, 2004; Kaur et al., 2010; Laothong et al., 2010).

6.5.2 Anticancer Effects

There is evidence that the administration of MEL reverses or inhibits tumor genesis caused by carcinogens (Brzezinski, 1997). Moreover, the findings suggest that MEL is crucial for inhibiting both cancer initiation and cancer-cell growth

(Reiter, 2004). MEL exhibits significant apoptotic, oncostatic, antiangiogenic, differentiating, and antiproliferative properties against tumors (Reiter, 2004; Di Bella et al., 2013; Kim et al., 2013; Rondanelli et al., 2013). Due to its antimitotic activity, physiologic and pharmacologic concentrations of MEL inhibit the proliferation of cultured epithelial breast cancer-cell lines (particularly MCF-7) and malignant melanoma-cell lines (M-6) in a dose-dependent manner (Brzezinski, 1997). MEL may also modulate the activity of various receptors in tumor cells (Dauchy et al., 2014). In addition, MEL enhances the immune response by increasing the production of cytokines derived from T-helper cells (interleukin-2 and interleukin-4) (Brzezinski, 1997).

MEL exerts both direct and indirect anticancer effects in synergy with other differentiating, antiproliferative, immunomodulating, and trophic molecules used on the anticancer treatment, illustrating its remarkable functional versatility (Di Bella et al., 2013; Kim et al., 2013). MEL added to conventional chemotherapy or radiotherapy may attenuate the damage to blood cells and thus make the treatment more tolerable (Bartsch and Bartsch, 2006). Nevertheless, data are still incomplete in humans; more studies must be performed in much larger groups for longer periods.

6.5.3 Antiaging Effects

MEL interacts directly or indirectly with several processes related to aging. Theories and hypotheses of aging include damage by free radicals, especially in conjunction with mitochondrial dysfunction and/or inflammatory processes ("inflamm-aging") (Boren and Gershwin, 2004; Cevenini et al., 2013); assumption of life-determining limits of energy expenditure and deterioration of the immune system (Boren and Gershwin, 2004); and telomere attrition with its consequences to reduced cell growth and progressive losses in number and proliferative potential of stem cells (Boren and Gershwin, 2004; Hardeland, 2013). Accordingly, MEL may provide protection against aging as a potent free-radical scavenger due its antioxidant properties.

6.5.4 Protection Against Cardiovascular Diseases

MEL plays an important role in the cardiovascular system (Reiter et al., 2010b). In this case, MEL has atheroprotective effects by acting on different pathogenic signaling processes. MEL reduces fatty-acid infiltration into the endothelial layer, neutralizes free radicals, reduces lipid peroxidation, modulates cholesterol clearance, and prevents electron leakage from mitochondrial respiratory chain (Favero et al., 2014). These effects result from its direct free-radical-scavenger activity, indirect antioxidant properties, and antiinflammatory actions (Reiter et al., 2010b). MEL treatment has important actions that protect against atherosclerosis-related cardiovascular disease.

6.5.5 Antiobesity Effects

MEL reduction due to aging, shiftwork, or illuminated environments during the night induces insulin resistance, glucose intolerance, sleep disturbance, and metabolic circadian disorganization, creating a state of chronodisruption and thus bringing about obesity. MEL is involved in the daily distribution of metabolic processes so that the activity/feeding phase of the day is associated with high insulin sensitivity, and the rest/fasting is synchronized to the insulin-resistant metabolic phase of the day (Cipolla-Neto et al., 2014). In addition, MEL seems to be responsible for establishing an adequate energy balance mainly by regulating energy flow to and from the stores and directly regulating the energy expenditure through the activation of brown adipose tissue and the browning process of white adipose tissue (Townsend and Tseng, 2014). The antiobesogen and the weight-reducing effects of MEL suggest that its replacement therapy might contribute to restoring a more healthy state of the organism.

6.5.6 Protection Against Alzheimer's Disease

Alzheimer's disease (AD) has multiple factors that contribute to its etiology in terms of initiation and progression. MEL production diminishes with increasing age, coincident with the onset of AD. Several hypotheses, often with overlapping features, have been formulated to explain this debilitating condition (Rosales-Corral et al., 2012). Among them are the hypotheses of the accumulation of amyloid-β, the role of neuroinflammation, insulin resistance, and the association of AD with peroxidation of brain lipids. Due to its potent antioxidant and antiinflammatory activities, MEL has many other functions that could help explain each of the hypotheses mentioned.

6.5.7 Effects on Bone

Recently, an important role for MEL in bone formation and restructuring has emerged, and studies demonstrate the multiple mechanisms for these beneficial actions (Witt-Enderby et al., 2011). It has been reported that disruption of MEL rhythms by light exposure at night, shiftwork, and disease can adversely affect bone formation (Maria and Witt-Enderby, 2014). Because of its free-radical-scavenging and antioxidant properties, MEL maintains bone health, especially, in the oral cavity. Thus this bioactive compound could be used in bone-grafting procedures, in reversing bone loss due to osteopenia and osteoporosis, and in managing periodontal disease.

6.5.8 Protection Against UV Radiation (UVR)

Many studies have documented the protective effects of MEL when applied before UVR exposure, but no effect if applied after (Dreher et al., 1999). The protection against UVR-induced skin damage was suggested to be due to MEL acting directly as an antioxidant and indirectly by regulating gene expression and inducing DNA-stabilizing effects (Scheuer et al., 2014). As these results were obtained using artificial UVR, studies using natural sunlight and evaluating possible side effects of topical MEL administration are needed.

6.5.9 Migraine Prevention

Finally, there is increasing evidence that headache disorders may be related to MEL secretion and pineal function since low MEL levels have been found in individuals with both migraine and cluster headaches. MEL may play a role in headache pathophysiology by several mechanisms (Peres, 2005). These include its antiinflammatory effect, toxic free-radical scavenging, reduction of nitric oxide synthase activity and dopamine-release inhibition, GABA and opioid analgesia potentiation, neurovascular regulation, and 5-HT modulation and its similarity in chemical structure to indomethacin (Peres, 2005). The treatment of headache disorders with MEL and other chronobiotic agents, such as MEL agonists (ramelteon and agomelatine), has yielded promising results and there is great potential for their use in headache treatment.

6.6 MELATONIN IN PLANT FOODS

MEL is detected in plant foods, including cereals, vegetables, fruits, and roots (Hattori et al., 1995; Reiter et al., 2001; Hardeland et al., 2003; Kolár and Macháčková, 2005; Stürtz et al., 2011). The content varies considerably not only between species but also among varieties of the same species and in the different organs of a given plant, usually ranging from pg/g to µg/g of tissue. The main problems concerning the reliable analysis of MEL in foods are related to sampling of different plant materials and their subsequent extraction, identification, and measurement (Cao et al., 2006; García-Parrilla et al., 2009). In general, MEL detection in foods has been hindered because of the direct adaptation of MEL methods from animals to plants. Extraction of MEL presents some difficulties in foods related to the methods from different matrices. These procedures may be a critical step in MEL measurement in foods, regardless of the subsequent quantification method used. This may include the physicochemical properties of MEL that might explain the reported variations in the MEL contents by different food extracts.

Several techniques have been used for MEL determination in plant foods. They are mainly divided into immunological assays including radioimmunoassay (RIA) and enzyme-linked immunosorbent assay (ELISA) and chromatographic techniques, such as high-performance liquid chromatography (HPLC) with electrochemical (HPLC-EC) or fluorimetric (HPLC-F) detection ($\lambda_{exc} = 280$ nm; $\lambda_{em} = 345$ nm) (Burkhardt et al., 2001; Hernández-Ruiz et al., 2005; Arnao and Hernández-Ruiz, 2008; Mercolini et al., 2008; Maldonado et al., 2009; Rodriguez-Naranjo et al., 2011a, 2012). HPLC or gas chromatography (GC) coupled to mass spectrometry (MS) identification represents a powerful tool for MEL determination in plant food samples (Kolár and Macháčková, 2005). When using plant extracts, there is often difficulty with MEL extraction and its quantification by immunodetection kits (generally developed for human samples) due to the high degree of interference. Recently, liquid chromatography–tandem mass spectrometry (LC–MS/MS) was introduced as a rapid and accurate method to identify this molecule in plant foods; this approach is very costly (Cao et al., 2006; Rodriguez-Naranjo et al., 2011b, 2012). LC–MS/MS with electrospray ionization has been identified as a quick method for the reproducible detection and quantification of MEL, especially for low levels of this compound (Cao et al., 2006; Rodriguez-Naranjo et al., 2011a; Yilmaz et al., 2014). The MEL detection limit is 5 pg/mL in plant extracts and its quantification limit is 0.02 ng/mL (Cao et al., 2006). A new method using capillary electrochromatography with immobilized carboxylic multiwalled carbon nanotubes as the stationary phase has been developed for MEL determination in complex food matrices (Stege et al., 2010). Recently, Vitalini et al. (2012) and Fernández-Pachón et al. (2014) proposed ultra-performance liquid chromatography–tandem mass spectrometry (UPLC–MS/MS) and ultra-performance liquid chromatography–high-resolution mass spectrometer (UPLC-HR-MS) as advanced techniques to confirm the identity of MEL and MIs. Nevertheless, HPLC-EC and HPLC-F appear to be efficient methodological options for MEL determination (Mercolini et al., 2008).

In general, plant tissues have higher MEL levels than those measured in animals (Reiter and Tan, 2002; Hardeland and Poeggeler, 2003). Seeds exhibit high MEL levels with considerable interspecific diversity from 189 ng/g dry weight (DW) found in *Brassica hirta* Moench (white mustard) to 2 ng/g DW detected in *Silybum marianum* (L.) Gaertn (milk thistle) (Manchester et al., 2000). Lower MEL levels have normally been reported in vegetables (0.1–0.4 ng/g DW); broccoli has the highest reported concentration, followed by red cabbage, radish, and onion (Hattori et al., 1995; Aguilera et al., 2015). Germinated seeds were also analyzed in two common legumes (*Lens culinaris* L. and *Phaseolus vulgaris* L.) and found to have moderate MEL levels (2.5 and 9.8 ng/g DW, respectively) that were higher than their respective raw seeds (Aguilera et al., 2014). Fruits exhibit lower amounts of this indoleamine than seeds (Dubbels et al., 1995; Hattori et al., 1995) with the exception of cherries (*Prunus cerasus* L.) and strawberries (*Fragaria ananassa* L.), which exhibited higher concentrations of MEL (15.2 ng/g DW) and varieties of the same species exhibit considerable differences (Burkhardt et al., 2001; Stürtz et al., 2011).

Extra-virgin olive oil is another relevant dietary source of MEL. De la Puerta et al. (2007) found significant differences when comparing refined and extra-virgin olive oils with different designations of origin (D.O.) (eg, 71 pg/mL from D.O. Bajo Aragón to 119 pg/mL D.O. Baena). Of nutritional interest, the highest MEL concentrations are found in traditional Chinese medicine herbs (12–3771 ng/g DW) commonly used to delay aging and treat diseases related to oxidative stress (Chen et al., 2003). In addition, coffee beans have also been reported to contain substantial levels of MEL (3–10 µg/g) (Ramakrishna et al., 2012), with values higher than nuts (3.5–39 ng/g) and cereals (0.4–1.8 ng/g) (Hattori et al., 1995; Manchester et al., 2000; Reiter et al., 2005).

Recent studies have also shown high concentration of MEL in beverages derived from plant products and may be considered a dietary source of this bioactive compound. They will be described in detail below (Table 6.1).

6.7 FERMENTED FOODS

While MEL levels in food vary widely, food-processing technologies may reduce the levels of this neurohormone but may also increase it such as during the fermentation process. In this regard, the fermentation is carried out by many probiotics that may be responsible at least, in part, for the MEL and MIs naturally occurring in these foods. As a result, the use of probiotics to synthesize MEL has been patented in the United States, employing the following probiotics in its industrial scale production: *Bifidobacterium* species (*breve subspec. breve, longum subspec. infantis*); *Enterococcus* species (*faecalis* TH10); *Lactobacillus* species (*brevis, acidophilus, bulgaricus, casei subspec. sakei, fermentum, helveticus subspec. jogorti, plantarum*); and Streptococcus (*thermophilus*) (Tan et al., 2012).

Enhancement of MEL and MIs levels as a result of fermentation processes has been studied (Arevalo-Villena et al., 2010; García-Moreno et al., 2013). Thus high levels of MEL and its MIs would be expected in classic Mediterranean and Oriental fermented foods with beneficial effects of the fermented products such as cheese and yogurt, kimchi, soy sauce, and homemade vinegar in which *Lactobacillus* species are used in the fermentation process. If in fact MEL levels increase during fermentation, their beneficial health effects in fermented products may be partially explained by the presence of MEL and its isomers (Tan et al., 2012). Although little information exists on the abovementioned fermented foods, it is interesting to note the research recently carried out on the enrichment of MEL content during the production of Mediterranean beverages (wine and beer). This increase was due to MEL synthesis during alcoholic fermentation by *Saccharomyces cerevisiae* (Maldonado et al., 2009; García-Moreno et al., 2013). Moreover, MEL synthesis by yeasts has also been reported during alcoholic fermentation of orange juice (Fernández-Pachón et al., 2014) and increases of MIs were detected during bread dough fermentation (Yilmaz et al., 2014).

6.7.1 Wine

Significant effort has been made to determine the bioactivity of compounds present in wine and to describe new molecules with biological activity to improve their health benefits (Bertelli, 2007). In general, studies of wine have

TABLE 6.1 Melatonin Content in Fermented Foods

Fermented Food	Melatonin (ng/mL)	Melatonin Isomers (ng/mL)	Reference
Red Wines			
Groppello	0.35–8.1	27.3	Vitalini et al. (2012) and Vitalini et al. (2011)
Melas	0.62	7.59	Vitalini et al. (2012)
Nebbiolo	0.14	14.9	Vitalini et al. (2012)
Terre di Rubinoro	0.17	26.9	Vitalini et al. (2012)
Syrah IGT	0.23	76.5	Vitalini et al. (2012)
Placido Rizzotto	0.05	57.4	Vitalini et al. (2012)
La Segreta	0.31	72.6	Vitalini et al. (2012)
Malbec	0.1–0.24	2.25	Stedge et al. (2010) and Gómez et al. (2012)
Cabernet Sauvignon	0.27–74.1	6.1	Rodriguez-Naranjo et al. (2011a, b) and Stedge et al. (2010)
Jaen Tinto	0.16	21.9	Rodriguez-Naranjo et al. (2011a)
Merlot	0.21	5.2	Rodriguez-Naranjo et al. (2011a)
Palomino Negro	0.25	16.7	Rodriguez-Naranjo et al. (2011a)
Petit Verdot	0.2–5.1	–	Rodriguez-Naranjo et al. (2011a)
Prieto Picudo	0.2–49.0	6.5	Rodriguez-Naranjo et al. (2011a)
Syrah	0.2–423.0	–	Rodriguez-Naranjo et al. (2011a, 2011b, 2012)
Tempranillo	0.14–306.9	9.3	Rodriguez-Naranjo et al. (2011a, 2011b, 2012)
Merlot	5.2–245.5	n.d.	Vitalini et al. (2011), Rodríguez-Naranjo et al. (2011b, 2012)

TABLE 6.1 Melatonin Content in Fermented Foods—cont'd

Fermented Food	Melatonin (ng/mL)	Melatonin Isomers (ng/mL)	Reference
Tintilla de Rota	18.0–322.7	n.d.	Rodríguez-Naranjo et al. (2011b, 2012)
Sangiovese	0.5	n.d.	Mercolini et al. (2008)
White Wines			
Chaudelune – Vin de glace	0.18	19.9	Vitalini et al. (2012))
Zibibbo IGT (2009) Sicilia			Vitalini et al. (2012)
Albana	0.6	n.d.	Mercolini et al. (2012)
Chardonnay	0.16	n.d.	Stedge et al. (2010)
Trebbiano	0.4	n.d.	Mercolini et al. (2008)
Palomino fino	6–390.8	n.d.	Rodriguez-Naranjo et al. (2011b, 2012)
Dessert Wines			
Recioto di Soave	0.14	7.1	Vitalini et al. (2012)
Santelmo	0.18	5.7	Vitalini et al. (2012)
Marsala	0.11	0.2	Vitalini et al. (2012)
Passito di Pantelleria	0.31	17.4	Vitalini et al. (2012)
Moscato di Pantelleria	0.29	10.1	Vitalini et al. (2012)
Pomegranate Wines			
Wonderful	5.5	n.d.	Mena et al. (2012)
Mollar de Elche	0.54	n.d.	Mena et al. (2012)
Coupage	2.91	n.d.	Mena et al. (2012)
Beers			
Commercial	0.06–0.16	n.d.	García-Moreno et al. (2013)
Volt-Damm	0.17	n.d.	Maldonado et al. (2009)

Continued

TABLE 6.1 Melatonin Content in Fermented Foods—cont'd

Fermented Food	Melatonin (ng/mL)	Melatonin Isomers (ng/mL)	Reference
Murphy's	0.14	n.d.	Maldonado et al. (2009)
Mahou Negra	0.14	n.d.	Maldonado et al. (2009)
Amstel	0.13	n.d.	Maldonado et al. (2009)
Coronita	0.13	n.d.	Maldonado et al. (2009)
Budweisser	0.12	n.d.	Maldonado et al. (2009)
Guiness	0.12	n.d.	Maldonado et al. (2009)
Cruzcampo	0.11	n.d.	Maldonado et al. (2009)
Carlsberg	0.10	n.d.	Maldonado et al. (2009)
Mahou 5 estrellas	0.10	n.d.	Maldonado et al. (2009)
Heineken	0.09	n.d.	Maldonado et al. (2009)
San Miguel	0.09	n.d.	Maldonado et al. (2009)
Mahou Clásica	0.08	n.d.	Maldonado et al. (2009)
Laiker Sin	0.07	n.d.	Maldonado et al. (2009)
San Miguel 0.0	0.06	n.d.	Maldonado et al. (2009)
Buckler Sin	0.05	n.d.	Maldonado et al. (2009)
Kaliber Sin	0.05	n.d.	Maldonado et al. (2009)
Buckler 0.0	0.05	n.d.	Maldonado et al. (2009)
Fermented Juice			
Fermented orange juice	20.0	10.5	Fernández-Pachon et al. (2014)
Bread			
Fermented dough	0.63	16.7	Yilmaz et al. (2014)

been mainly focused on polyphenols but recently MEL has been considered because of its strong antioxidant activity and its natural presence in wine (Rodriguez-Naranjo et al., 2011a). Evidence has shown that MEL in wine is not only exclusively found in grapes but is mainly generated during the process of fermentation during wine brewing (Tan et al., 2012).

Several publications have noted that microorganisms (mainly yeasts) play a role in MEL production during fermentation. Microorganisms use tryptophan and their metabolites, such as indolacetic acid, tryptophol, or biogenic amines, as a nitrogen source. The ability of yeasts to produce indoles has been confirmed and is related to the off-flavors in wine (Arevalo-Villena et al., 2010). In addition to yeasts, several lactic bacteria may also contribute to the process of malolactic fermentation of wine brewing at different stages (González-Arenzana et al., 2012), producing tryptamine from tryptophan (Arevalo-Villena et al., 2010). Thus the potential role of bacteria to enhance the MEL content in wine must be considered (Rodríguez-Naranjo et al., 2012).

Commercial yeast starters are commonly used during fermentation for obtaining standardized products; a starter for wine fermentation is *Saccharomyces cerevisiae*. Previous studies have shown that high amounts of MEL and other methoxyindoles can be synthesized by *S. cerevisiae* in a standard yeast growth and salt medium, with further production if the precursor is available (Sprenger et al., 1999).

The role of MEL in *Saccharomyces cerevisiae* has not been yet elucidated (Hardeland et al., 2003). Recently, it was reported that *S. cerevisiae* can use small aromatic molecules with an indole ring, such as tryptophol and indole-3-acetic acid as signals to modulate population growth (Dufour and Rao, 2011).

Among grapevine (*Vitis vinifera* L.) products, MEL was first detected in the berry exocarp (skin) (Iriti et al., 2006) and wine (Mercolini et al., 2008). Table 6.1 summarizes the MEL levels found in different red and white wines. Mercolini et al. (2008) detected MEL in Sangiovese red and Trebbiano white wines (0.5 and 0.4 ng/mL, respectively), followed by Stege et al. (2010) who reported its occurrence in Chardonnay, Malbec, and Cabernet Sauvignon red wines (0.16, 0.24, and 0.32 ng/mL, respectively), similar to those found by Rayne (2010). Nevertheless, Gomez et al. (2012) found MEL in grape berry exocarp, but not in experimental wine produced from these grapes.

Additional studies carried out by Rodriguez-Naranjo et al. (2011b) confirmed the absence of MEL in grapes and musts with MEL being found in five monovarietal wines (Merlot, Palomino Fino, Syrah, Tempranillo, and Tintilla de Rota), showing content ranging from 0.21 to 18 ng MEL/mL. The determinations were done during the winemaking process for the sampling steps analyzed (press for red wines and rack for white wine). In particular, MEL content was found on the fifth day and increased until the seventh day.

In general, higher MEL contents have been found in red wines than in white wines (Table 6.1), possibly for two reasons: red grapes exhibit higher MEL levels than white grapes (Iriti et al., 2010) and, additionally, in the red wine fermentation process, the grape skins are included and higher temperatures are used than with white or rose wines.

Notable differences have also been found in MEL levels for similar grape varieties. Rodriguez-Naranjo et al. (2011b, 2012) observed that the *Tempranillo* variety, grown in the same vineyard and with an identical winemaking process including the same inoculated yeasts (Actiflore F5 for red wine and Actiflore

PM for white wine), exhibited MEL levels 50-fold higher at different years (307 ng/mL in 2008, and 5.5 ng/mL in 2010). These differences might be caused by endogenous and exogenous factors between grapevines and their fermented products that change the concentration of the precursors. In this regard, genetic traits of cultivars, geographical origin, berry tissues/plant organs analyzed, differences between thin and thick-skinned grapes, phenological stages, day/night fluctuations, pathogen (mainly fungal) infections and phytosanitary treatments, agro-meteorological conditions and environmental stresses, altitude, UV radiations and high-light irradiance, and vintage and winemaking procedures may influence the final MEL content (Murch et al., 2010; Boccalandro et al., 2011; Rodríguez-Naranjo et al., 2011b; Vitallini et al., 2011).

Pomegranate wines exhibit this behavior and the evolution of MEL contents throughout the different stages of this winemaking process has been reported (Mena et al., 2012). One variety of this fruit contained some indoleamine, but its occurrence appeared primarily during the winemaking process, ranging its levels from *Wonderful* monovarietal wine (5.5 ng/mL) to *Mollar de Elche* wine (0.5 ng/mL). During the initial period of the juice fermentation stage (the first 4 days), a marked increase was noted; at this point the content of MEL increased to 7.4, 8.9, and 4.2 ng/mL for *Wonderful*, *Coupage*, and *Mollar de Elche*, respectively. However, a general reduction was found to occur at the end of the winemaking process, possibly due to the effects of yeasts and other contaminating microorganisms that might modify the metabolism of nitrogen substances and influence oxidative processes (Mena et al., 2012).

The presence of MIs in wine represents an emerging topic in the field of MEL research (Tan et al., 2012). As previously noted, two naturally occurring MIs were individually detected in red wines by different research groups (Rodríguez-Naranjo et al., 2011a; Gómez et al., 2012; Vitalini et al., 2012) and their levels in wine were approximately 20-fold higher than MEL (Gomez et al., 2012). Studies indicate that some wines only contained MEL, other wines contained a single isomer, others have both MEL and one MI (Rodríguez-Naranjo et al., 2011a), and some contain MEL and three additional isomers (Vitalini et al., 2011).

Winemaking techniques may be important for MEL extraction from grapes. The maceration time and fermentation period may possibly improve the extraction and solubility of MEL. The medium is also a determining factor in the formation of this indoleamine by *S. cerevisae*. The salt concentration of each cultivar together with other phytochemical compounds, such as polyphenols and organic acids may affect the yeasts performance with regard to MEL production (Mena et al., 2012). Two strains of *Saccharomyces cerevisiae* (ARM, QA23), ie, *S. uvarum* (S6U) and *S. cerevisiae* var. *bayanus* (Uvaferm BC), improved MEL production (Rodriguez-Naranjo et al., 2012). The presence of TRP also increases the final MEL content and accelerates its formation (Sprenger et al., 1999; Rodriguez-Naranjo et al., 2011a). The QA23 strain has shown to be the best producer of MEL when the medium contains low concentrations of reducing sugars. However, *S. uvarum* (S6U) and *S. cerevisiae* var. *bayanus* (Uvaferm BC) produced MEL in synthetic must under fermentation conditions. Thus both yeast strain and must composition seem equally important in determining the concentration of this bioactive compound in wines.

A controversy currently exists with respect to the essential role of TRP in MEL synthesis. Studies indicate the production of MEL and its MIs is independent of external TRP during the wine fermentation process because yeasts can still synthesize these indolamines without TRP in the medium (Tan et al., 2012). TRP may be synthesized intracellularly by either yeast or bacteria; both possess the shikimic acid pathway of aromatic biosynthesis (Ehammer et al., 2007). On the other hand, environmental stressors may naturally induce the production MEL and MIs (Tan et al., 2012). During the process of fermentation in winemaking, the concentration of alcohol in wine gradually increases until the yeasts disappear. To increase the tolerance for these elevated levels of alcohol in the environment, one of the biological responses of yeasts may be to upregulate the generation of MEL and, especially MIs. It seems important to evaluate the potential contribution of the *Lactobacillus* species, since it may also be essential for them to generate increased amounts of MEL and/or MIs to tolerate the gradual drop in the medium pH as a result of elevated lactic-acid production, which would act also as stressor. MEL was also reportedly produced by other species (algae, plants, and animals including humans) due to environmental stress, especially oxidative stress (Balzer et al., 1996; Antolin et al., 1997; Afreen et al., 2006; Arnao and Hernández-Ruiz, 2009; Tal et al., 2011). Thus melatonin biosynthesis in plants and plant products may be more complex than currently envisioned.

6.7.2 Beer

Beer is a beverage of low alcoholic grade produced from cereal fermentation; it is also relevant to the Mediterranean diet. Abundant evidence suggests its intake causes health benefits, such as reduction in free radicals, decreased risk factors for coronary heart disease, prevention of certain cancers, and modification of immune and inflammatory responses (Romeo et al., 2007). Beer contains approximately 400 compounds from raw or generated material during its elaboration process including ethanol, amino acids, minerals, vitamins, carbohydrates, polyphenols, aromatics compounds, and other important components that may contribute to its overall therapeutic characteristics. MEL is present in beer where it exhibits the these properties (antioxidant, oncostatic, immunoenhancing effects, etc.) (Reiter and Tan, 2003). Thus MEL may promote some of the beneficial effects of beer.

Recent studies document a direct relation between alcohol content and MEL concentration of beer (Table 6.1). The results show that all the beer analyzed did indeed contain MEL and the more melatonin present the greater its degree of alcohol (García-Moreno et al., 2013). Beer with high alcohol content (7.2%,

Volt–Damm) showed the greatest MEL concentration (169.7 pg/mL), while dealcoholized beer (0%, Buckler) contained the lowest MEL levels (0.05 pg/mL). These differences can be explained by its physicochemical characteristics. MEL is a soluble molecule in alcohol with properties of amphiphilicity but shows low solubility in water (2.4 mg/mL at 25°C). Interestingly, higher MEL concentration has been found in homemade beer (333 pg/mL MEL and 5% of alcohol) with levels threefold higher than those found in commercial brands, possibly due to thermal conditions and dilution processes being less aggressive than those used the industrial elaboration (García-Moreno et al., 2013).

To examine the MEL origin in beer, raw materials (barley, hops, and yeasts) were analyzed; high MEL levels were found in barley concentrate musts (339 pg/mL) but low MEL content was found in the analyzed hops (33 pg/mL). Of interest here is that yeast showed high MEL levels (333 pg/mL) in the second fermentation (García-Moreno et al., 2013). Autochthonous yeasts used to produce wines and beers have specific organoleptic properties, such as flavor and odor, and thus contribute positively to the final fermented products from the region and can be classified as "Denomination of Origin" (Clavijo et al., 2010). These autochthonous yeasts present distinctive features, such as different resistances to stress, different capacities to transform sugars to alcohol, and even differences in terms of greater or lesser MEL production (Santamaría et al., 2005).

Moderate consumption of beer affects the total antioxidant status (TAS) of human serum. Maldonado et al. (2009) analyzed the effects of 18 brands of beer with different percentages of alcohol content on serum MEL in healthy volunteers under basal conditions and after drinking beer. The results showed that both melatonin and TAS in human serum increased after drinking beer. Thus these promising data suggest MEL contained in beer contributes to the total antioxidative capability of human serum and moderate beer consumption can protect the organism from overall oxidative stress.

6.7.3 Orange Juice

Fresh fruits are recognized as rich sources of bioactive compounds. Among them, orange juice is known for its relevant ascorbic acid, carotenoid, and flavonoid contents. Recent studies have identified MEL in fruits including mango (699 pg/g), pineapple (302 pg/g), papaya (241 pg/g), oranges (150 pg/g), and banana (8.9 pg/g) (Sae-Teaw et al., 2013; Johns et al., 2013). Beneficial effects of MEL have been shown after orange juice consumption including increased plasma antioxidant capacity of serum (Sae-Teaw et al., 2013).

A novel commercial beverage of low alcoholic grade prepared from orange juice by controlled alcoholic fermentation was developed to determine whether MEL and IMs were synthesized (Fernández-Pachón et al., 2014). MEL content increased progressively in the soluble fraction of fermented orange juice, reaching after day 7, relevant MEL content (8.8 ng/mL). Nevertheless, the highest MEL content was observed at the end of fermentation on day 15 (22 ng/mL), a seven fold increase from day 0 (3.2 ng/mL). The major portion of MEL was identified in the supernatant, but small MEL amounts were also found in the pellet fraction. Interestingly, IMs increased gradually in the soluble fraction during orange juice fermentation from 8.3 ng/mL (day 0) to 10.5 ng/mL (day 15); however, the concentration IMs remained unchanged in the pellet fraction during alcoholic fermentation (3.3–3.6 ng/mL). The isomer levels in fermented orange juice were similar to that reported by Rodríguez-Naranjo et al. (2011a) and Gómez et al. (2012) in different types of wines after the winemaking process.

MEL levels in fermented orange juice are clearly higher than those found in other foods. The enhancement in MEL content could be due to both the occurrence of TRP or due to its de novo synthesis by yeast (Fernández-Pachón et al., 2014). This is suggested because TRP levels dropped significantly from 13.8 mg/L (day 0) up to 3.2 mg/L (day 15) during fermentation. This decrease of TRP level could be related to the simultaneous MEL synthesis during fermentation; thus MEL was inversely and significantly correlated with TRP ($r = 0.907$). In summary, the enhancement of MEL in novel fermented orange beverages could improve the health benefits of orange juice due to the antioxidant properties of MEL.

6.7.4 Other Fermented Foods

The *Saccharomyces* species are widely used in the food industry not only in wine and beer production but also in breadmaking. Because bread is widely consumed in large amounts, the formation of MEL and MIs during bread dough fermentation has been recently investigated (Yilmaz et al., 2014). The results revealed for the first time the formation of a melatonin isomer in bread dough during yeast fermentation by *Saccharomyces cerevisiae*. Formation of MEL was not significant during dough fermentation; however, MI content of nonfermented dough was 4.0 ng/g and increased to 16.7 ng/g during fermentation. Its quantities formed in dough were significantly higher than those of MEL itself. The formation rate of MI and the disappearance of TRP (58%) suggest they were also associated. This obviously indicates a potential role of TRP in the formation mechanism during yeast fermentation. MEL concentrations ranged from 0.2 to 0.6 ng/g for crumb and from 0.1 to 0.8 ng/g for crust parts. Lower amounts of MIs in crumb and crust than dough showed that the thermal process caused a remarkable degree of degradation in MIs. Compared to crumb, this effect was more pronounced in the crust part expectedly, in which higher temperatures were attained during the process (Yilmaz et al., 2014). Therefore, bread might be considered a source of MI but further investigation is needed to elucidate their biological consequences.

6.8 CONCLUSION

MEL determination in plant foods is important because in recent years attention has been paid to incorporating foods with high MEL levels as a dietary supplement. Taking into account the formation of MEL and IMs by yeasts during alcoholic fermentation, possible enhancement of these bioactive compounds in fermented foods could improve their health benefits, increasing its blood plasma levels in humans and, consequently, by enhancing its scavenging and antioxidant actions.

REFERENCES

Acuña-Castroviejo, D., Escames, G., Venegas, C., Díaz-Casado, M.E., Lima-Cabello, E., López, L.C., Rosales-Corral, S., Tan, D.X., Reiter, R.J., 2014. Extrapineal melatonin: sources, regulation, and potential functions. Cellular and Molecular Life Sciences 71, 2997–3025.

Afreen, F., Zobayed, S.M., Kozai, T., 2006. Melatonin in *Glycyrrhizauralensis*: response of plant roots to spectral quality of light and UV-B radiation. Journal of Pineal Research 41, 108–115.

Aguilera, A., Liébana, R., Herrera, T., Rebollo-Hernanz, M., Sanchez-Puelles, C., Benítez, V., Martín-Cabrejas, M.A., 2014. Effect of illumination on the content of melatonin, phenolic compounds, and antioxidant activity during germination of lentils (*Lens culinaris* L.) and kidney beans (*Phaseolus vulgaris* L.). Journal of Agricultural and Food Chemistry 62, 10736–10743.

Aguilera, A., Herrera, T., Benítez, V., Arribas, S.M., López de Pablo, A.L., Esteban, R.M., Martín-Cabrejas, M.A., 2015. Estimation of scavenging capacity of melatonin and other antioxidants: contribution and evaluation in germinated seeds. Food Chemistry 170, 203–211.

Antolin, I., Obst, B., Burkhardt, S., Hardeland, R., 1997. Antioxidative protection in a high-melatonin organism: the dinoflagellate *Gonyaulaxpolyedra* is rescued from lethal oxidative stress by strongly elevated, but physiologically possible concentrations of melatonin. Journal of Pineal Research 23, 182–190.

Arevalo-Villena, M., Bartowsky, E.J., Capone, D., Sefton, M.A., 2010. Production of indole by wine-associated microorganisms under oenological conditions. Food Microbiology 27, 685–690.

Arnao, M.B., Hernandez-Ruiz, J., 2006. The physiological function of melatonin in plants. Plant Signaling and Behavior 1, 89–95.

Arnao, M.B., Hernandez-Ruiz, J., 2008. Protective effect of melatonin against chlorophyll degradation during the senescence of barley leaves. Journal of Pineal Research 46, 58–63.

Arnao, M.B., Hernandez-Ruiz, J., 2009. Chemical stress by different agents affects the melatonin content of barley roots. Journal of Pineal Research 46, 295–299.

Balzer, I., Hardeland, R., 1996. Melatonin in algae and higher plants-possible new roles as a phytohormone and antioxidant. BotonicaActa 109, 180–183.

Bartsch, C., Bartsch, H., 2006. The antitumor activity of pineal melatonin and cancer enhancing life styles in industrialized societies. Cancer Causes Control 17, 559–571.

Bertelli, A.A.E., 2007. Wine, research and cardiovascular disease: instructions for use. Atherosclerosis 195, 242–247.

Boccalandro, H.E., Gonzales, C.V., Wunderlin, D.A., Silva, M.F., 2011. Melatonin levels, determined by LC-ESI-MS/MS, deeply fluctuate during the day in *Vitis vinifera* cv Malbec. Evidences for its antioxidant role in fruits. Journal of Pineal Research 51, 226–232.

Boren, E., Gershwin, M.E., 2004. Inflamm-aging: autoimmunity, and the immune-risk phenotype. Autoimmunity Reviews 3, 401–406.

Brzezinski, A., 1997. Melatonin in humans. The New England Journal of Medicine 336, 186–195.

Burkhardt, S., Tan, D.X., Manchester, L.C., Hardeland, R., Reiter, R.J., 2001. Detection and quantification of the antioxidant melatonin in Montmorency and Balaton tart cherries (*Prunuscerasus*). Journal of Agriculture and Food Chemistry 49, 4898–4902.

Cao, J., Murch, S.J., O'Brien, R., Saxena, P.K., 2006. Rapid method for accurate analysis of melatonin, serotonin and auxin in plant samples using liquid chromatography-tandem mass spectrometry. Journal of Chromatography A 1134, 333–337.

Cardinali, D.P., Vigo, D.E., Olivar, N., Vidal, M.F., Furio, A.M., Brusco, L.I., 2012. Therapeutic application of melatonin in mild cognitive impairment. American Journal of Neurodegenerative Disease 1, 280–291.

Cevenini, E., Monti, D., Franceschi, C., 2013. Inflamm-aging. Current Opinion in Clinical Nutrition and Metabolic Care 16, 14–20.

Cipolla-Neto, J., Amaral, F.G., Afeche, S.C., Tan, D.X., Reiter, R.J., 2014. Melatonin, energy metabolism, and obesity: a review. Journal of Pineal Research 56, 371–381.

Chen, G., Huo, Y., Tan, D.X., Liang, Z., Zhang, W., Zhang, Y., 2003. Melatonin in Chinese medicinal herbs. Life Sciences 73, 19–26.

Claustrat, B., Brun, J., Chazot, G., 2005. The basic physiology and pathophysiology of melatonin. Sleep Medicine Reviews 9, 11–24.

Clavijo, A., Caldero, N.I., Paneque, P., 2010. Diversity of *Saccharomyces* and *nonSaccharomyces* yeasts in three red grape varieties cultured in the Serranía de Ronda (Spain) vine-growing region. International Journal of Food Microbiology 143, 241–245.

Dauchy, R.T., Xiang, S., Mao, L., Brimer, S., Wren, M.A., Yuan, L., Anbalagan, M., Hauch, A., Frasch, T., Rowan, B.G., Blask, D.E., Hill, S.M., 2014. Circadian and melatonin disruption by exposure to light at night drives intrinsic resistance to tamoxifen therapy in breast cancer. Cancer Research 74, 4099–4110.

De la Puerta, C., Carrascosa-Salmoral, M.P., García-Luna, C., Lardone, P.J., Herrera, J.L., Fernández-Montesinos, R., Guerrero, J.M., Pozo, D., 2007. Melatonin is a phytochemical in olive oil. Food Chemistry 104, 609–612.

Di Bella, G., Mascia, F., Gualano, L., Di Bella, L., 2013. Melatonin anticancer effects: review. International Journal of Molecular Sciences 14, 2410–2430.

Dreher, F., Denig, N., Gabard, B., Schwindt, D.A., Maibach, H.I., 1999. Effect of topical antioxidants on UV-induced erythema formation when administered after exposure. Dermatology 198, 52–55.

Dubbels, R., Reiter, R.J., Klenke, E., Goebel, A., Schnakenberg, E., Ehlers, C., Schiwara, H.W., Schloot, W., 1995. Melatonin in edible plants identified by radioimmunoassay and by high performance liquid chromatography-mass spectrometry. Journal of Pineal Research 18, 28–31.

Dufour, N., Rao, R.P., 2011. Secondary metabolites and other small molecules as intercellular pathogenic signals. FEMS Microbiololgy Letters 314, 10–17.

European Food Safety Authority, 2010. Scientific opinion on the substantiation of health claims related to melatonin and subjective feelings of jet lag (ID1953), and reduction of sleep onset latency, and improvement of sleep quality (ID 1953) pursuant to Article 13(1) of Regulation (EC) No 1924/2006. European Food Safety Authority Journal 8, 1467.

Ehammer, H., Rauch, G., Prem, A., Kappes, B., 2007. Conservation of NADPH utilization by chorismate synthase and its implications for the evolution of the shikimate pathway. Molecular Microbiology 65, 1249–1257.

Favero, G., Rodella, L.F., Reiter, R.J., Favero, R.R., 2014. Melatonin and its atheroprotective effects: a review. Molecular and Cellular Endocrinology 382, 926–937.

Fernández-Pachón, M.S., Medina, S.G., Herrero-Martín, I., Cerrillo, G., Bernal, B., Escudero-López, F., Ferreres, F., Martín, M.C., García-Parrilla, L., Gil-Izquierdo, A., 2014. Alcoholic fermentation induces melatonin synthesis in orange juice. Journal of Pineal Research 56, 31–38.

Fujiwara, T., Maisonneuve, S., Isshiki, M., Mizutani, M., Chen, L., Wong, H.L., Kawasaki, T., Shimamoto, K., 2010. Sekiguchi lesion gene encodes a cytochrome P450 monooxygenase that catalyzes conversion of tryptamine to serotonin in rice. Journal of Biological Chemistry 285, 11308–11313.

Ganguly, S., Mummaneni, P., Steinbach, P.J., Klein, D.C., Coon, S.L., 2001. Characterization of the *Saccharomyces cerevisiae* homolog of the melatonin rhythm enzyme arylalkylamine N-acetyltransferase (EC 2.3.1.87). The Journal of Biological Chemistry 276, 47239–47247.

Garcia-Parrilla, M.C., Cantos, M., Troncoso, A.M., 2009. Analysis of melatonin in food. Journal of Food Composition and Analysis 22, 177–183.

Garcia-Moreno, H., Calvo, J.R., Maldonado, M.D., 2013. High levels of melatonin generated during the brewing process. Journal of Pineal Research 55, 26–30.

Gomez, F.J.V., Raba, J., Cerutti, S., Silva, M.F., 2012. Monitoring melatonin and its isomer in *Vitisvinifera* cv. Malbec by UHPLC-MS/MS from grape to bottle. Journal of Pineal Research 52, 349–355.

Gonzalez-Arenzana, L., Lopez, R., Santamaria, P., Tenorio, C., Lopez-Alfaro, I., 2012. Dynamics of indigenous lactic acid bacteria populations in wine fermentations from La Rioja (Spain) during three vintages. Microbiology Ecology 63, 12–19.

Hardeland, R., Poeggeler, B., 2003. Nonvertebrate melatonin. Journal of Pineal Research 34, 233–241.

Hardeland, R., 2013. Melatonin and the theories of aging: a critical appraisal of melatonin's role in antiaging mechanisms. Journal of Pineal Research 55, 325–356.

Hattori, A., Migitaka, H., Iigo, M., Itoh, M., Yamamoto, K., Ohtani, R., Hara, M., Suzuki, T., Reiter, R.J., 1995. Identification of melatonin in plants and its effects on plasma melatonin levels and binding to melatonin receptors in vertebrates. Biochemistry and Molecular Biology International 35, 27–34.

Hernández-Ruiz, J., Cano, A., Arnao, M.B., 2005. Melatonin acts as a growth-stimulating compound in some monocot species. Journal of Pineal Research 39, 137–142.

Iriti, M., Rossoni, M., Faoro, F., 2006. Melatonin content in grape: myth or panacea? Journal of Science and Food Agriculture 86, 1432–1438.

Iriti, M., Varoni, E.M., Vitalini, S., 2010. Melatonin in traditional Mediterranean diets. Journal of Pineal Research 49, 101–105.

Johns, N.P., Johns, J., Porasuphatana, S., Plaimee, P., Sae-Teaw, M., 2013. Dietary intake of melatonin from tropical fruit altered urinary excretion of 6-sulfatoxymelatonin in healthy volunteers. Journal of Agricultural and Food Chemistry 61, 913–919.

Kaur, C., Sivakumar, V., Ling, E.A., 2010. Melatonin protects periventricular white matter from damage due to hypoxia. Journal of Pineal Research 48, 185–193.

Kim, K.J., Choi, J.S., Kang, I., Kim, K.W., Jeong, C.H., Jeong, J.W., 2013. Melatonin suppresses tumor progression by reducing angiogenesis stimulated by HIF-1 in a mouse tumor model. Journal of Pineal Research 54, 264–270.

Kolár, J., Macháčková, I., 2005. Melatonin in higher plants: occurrence and possible functions. Journal of Pineal Research 39, 333–341.

Laothong, U., Pinlaor, P., Hiraku, Y., Boonsiri, P., Prakobwong, S., Khoontawad, J., Pinlaor, S., 2010. Protective effect of melatonin against *Opisthorchis viverrini*-induced oxidative and nitrosative DNA damage and liver injury in hamsters. Journal of Pineal Research 49, 271–282.

Leger, D., Laudon, M., Zisapel, N., 2004. Nocturnal 6-sulfatoxymelatonin excretion in insomnia and its relation to the response to melatonin replacement therapy. The American Journal of Medicine 116, 91–95.

Maldonado, M.D., Moreno, H., Calvo, J.R., 2009. Melatonin present in beer contributes to increase the levels of melatonin and antioxidant capacity of the human serum. Clinical Nutrition 28, 188–191.

Manchester, L.C., Tan, D.X., Reiter, R.J., Park, W., Monis, K., Qi, W., 2000. High levels of melatonin in the seeds of edible plants: possible function in germ tissue protection. Life Sciences 67, 3023–3029.

Maria, S., Witt-Enderby, P.A., 2014. Melatonin effects on bone: potential use for the prevention and treatment for osteopenia, osteoporosis, and periodontal disease and for use in bone-grafting procedures. Journal of Pineal Research 56, 115–125.

McCune, L., Kubota, C., Stendell-Hollis, N., Thomson, C., 2011. Cherries and health: a review. Critical Review of Food Science and Nutrition 51, 1–12.

Mena, P., Gil-Izquierdo, A., Moreno, D.A., Martí, N., García-Viguera, C., 2012. Assessment of the melatonin production in pomegranate wines. LWT-Food Science and Technology 47, 13–18.

Mercolini, L., Mandrioli, R., Raggi, M.A., 2012. Content of melatonin and other antioxidants in grape related foodstuffs: measurement using a MEPS-HPLC-F method. Journal of Pineal Research 55, 21–28.

Mercolini, L., Saracino, M.A., Bugamelli, F., 2008. HPLC-F analysis of melatonin and resveratrol isomers in wine using an SPE procedure. Journal of Separation Science 31, 1007–1014.

Murch, S.J., Saxena, P.K., 2002. Melatonin: a potential regulator of plant growth and development? In Vitro Cellular and Developmental Biology – Plant 38, 531–536.

Murch, S.J., Hall, B.A., Lee, C.H., Saxena, P.K., 2010. Changes in the levels of indoleamine phytochemicals during veraison and ripening of wine grapes. Journal of Pineal Research 49, 95–100.

Pandi-Pemural, S.R., Trakht, I., Srinivasan, V., Spence, D.W., Maestroni, G.J.M., Zisapel, N., Cardinali, D.P., 2008. Physiological effects of melatonin: role of melatonin receptors and signal transduction pathways. Progress in Neurobiology 85, 335–353.

Paredes, S.D., Korkmaz, A., Manchester, L.C., Tan, D.X., Reiter, R.J., 2009. Phytomelatonin: a review. Journal of Experimental Botany 60, 57–69.

Park, S., Lee, K., Kim, Y.S., Back, K., 2012. Tryptamine 5-hydroxylase-deficient Sekiguchi rice induces synthesis of 5-hydroxytryptophan and N-acetyltryptamine but decreases melatonin biosynthesis during senescence of detached rice leaves. Journal of Pineal Research 52, 211–216.

Peres, M.F.P., 2005. Melatonin, the pineal gland and their implications for headache disorders. Cephalalgia 25, 403–411.

Posmyk, M.M., Janas, K.M., 2009. Melatonin in plants. Acta Physiologia Plantarum 31, 1–11.

Ramakrishna, A., Giridhar, P., Sankar, K., Ravishankar, A., 2012. Melatonin and serotonin profiles in beans of Coffea species. Journal of Pineal Research 52, 470–476.

Rayne, S., 2010. Concentrations and Profiles of Melatonin and Serotonin in Fruits and Vegetables during Ripening: A Mini-Review. Nature Proceedings. http://dx.doi.org/10.1038/npre.2010.4722.1.

Reiter, R.J., Tan, D.X., Burkhardt, S., Manchester, L.C., 2001. Melatonin in plants. Nutrition Reviews 59, 286–290.

Reiter, R.J., Tan, D.X., 2002. Melatonin: an antioxidant in edible plants. Annals of the New York Academy of Sciences 957, 341–344.

Reiter, R.J., Tan, D.X., 2003. What constitute a physiological concentration of melatonin? Journal of Pineal Research 34, 79–80.

Reiter, R.J., 2004. Mechanisms of cancer inhibition by melatonin. Journal of Pineal Research 37, 213–214.

Reiter, R.J., Manchester, L.C., Tan, D.-X., 2005. Melatonin in walnuts: influence on levels of melatonin and total antioxidant capacity of blood. Nutrition 21, 920–924.

Reiter, R.J., Tan, D.X., Jou, M.J., Korkmaz, A., Manchester, L.C., Paredes, S.D., 2008. Biogenic amines in the reduction of oxidative stress: melatonin and its metabolites. Neuro Endocrinology Letters 29, 391–398.

Reiter, R.J., Tan, D.X., Fuentes-Broto, L., 2010a. Melatonin: a multitasking molecule. Progress in Brain Research 181, 127–151.

Reiter, R.J., Tan, D.X., Paredes, S.D., Fuentes-Broto, L., 2010b. Beneficial effects of melatonin in cardiovascular disease. Annals of Medicine 42, 276–285.

Reiter, R.J., Tan, D.X., Galano, A., 2014. Melatonin: exceeding expectations. Physiology 29, 325–333.

Rodriguez-Naranjo, M.I., Gil-Izquierdo, A., Troncoso, A.M., Cantos, E., Garcia-Parrilla, M.C., 2011a. Melatonin: a new bioactive compound present in wine. Journal of Food Composition and Analysis 24, 603–608.

Rodriguez-Naranjo, M.I., Gil-Izquierdo, A., Troncoso, A.M., Cantos, E., Garcia-Parrilla, C., 2011b. Melatonin is synthesised by yeast during alcoholic fermentation in wines. Food Chemistry 126, 1608–1613.

Rodriguez-Naranjo, M.I., Torija, M.J., Mas, A., Cantos-Villar, E., Garcia-Parrilla, M.C., 2012. Production of melatonin by *Saccharomyces* strains under growth and fermentation conditions. Journal of Pineal Research 53, 219–224.

Romeo, J., Wärnberg, J., Nova, E., Gonzalez-Gross, M., Marcos, A., 2007. Changes in the immune system after moderate beer consumption. Annals of Nutrition and Metabolism 51, 359–366.

Rondanelli, M., Faliva, M.A., Perna, S., Antoniello, N., 2013. Update on the role of melatonin in the prevention of cancer tumorigenesis and in the management of cancer correlates, such as sleep-wake and mood disturbances: review and remarks. Aging Clinical and Experimental Research 25, 499–510.

Rosales-Corral, S., Acuña-Castroviejo, D., Coto-Montes, A., Boga, J.A., Manchester, L.C., Fuentes-Broto, L., Korkmaz, A., Ma, S., Tan, D.X., Reiter, R.J., 2012. Alzheimer's disease: pathological mechanisms and the beneficial role of melatonin. Journal of Pineal Research 52, 167–202.

Sae-Teaw, M., Johns, J., Johns, N.P., Subongkot, S., 2013. Serum melatonin levels and antioxidant capacities after consumption of pineapple, orange, or banana by healthy male volunteers. Journal of Pineal Research 55, 58–64.

Santamaria, P., Garijo, P., Lopez, R., Tenorio, C., Gutierrez, A.R., 2005. Analysis of yeast population during spontaneous alcoholic fermentation: effect of the age of the winery and the practice of inoculation. International Journal of Food Microbiology 103, 49–56.

Scheuer, C., Pommergaard, H.C., Rosenberg, J., Gögenur, I., 2014. Melatonin's protective effect against UV radiation: a systematic review of clinical and experimental studies. Photodermatology, Photoimmunology & Photomedicine 30, 180–188.

Sprenger, J., Hardeland, R., Fuhrberg, B., Han, S.Z., 1999. Melatonin and other 5-methoxylated indoles in yeast: presence in high concentrations and dependence on tryptophan availability. Cytologia 64, 209–213.

Stege, P.W., Sombra, L.L., Messinga, G., Martinez, L.D., Silva, M.F., 2010. Determination of melatonin in wine and plant extracts by capillary electrochromatography with immobilized carboxylic multiwalled carbon nanotubes as stationary phase. Electrophoresis 31, 2242–2248.

Stürtz, M., Cerezo, A.B., Cantos-Villar, E., Garcia-Parrilla, M.C., 2011. Determination of the melatonin content of different varieties of tomatoes (*Lycopersiconesculentum*) and strawberries (*Fragariaananassa*). Food Chemistry 127, 1329–1334.

Tal, O., Haim, A., Harel, O., 2011. Melatonin as an antioxidant and its semi-lunar rhythm in green macroalga *Ulva* sp. Journal of Experimental Botany 62, 1903–1910.

Tan, D.X., Reiter, R.J., Manchester, L.C., Yan, M.T., El-Sawi, M., Sainz, R.M., Mayo, J.C., Kohen, R., Allegra, M., Hardeland, R., 2002. Chemical and physical properties and potential mechanisms: melatonin as a broad spectrum antioxidant and free radical scavenger. Current Topics in Medicinal Chemistry 2, 181–197.

Tan, D.X., Manchester, L.C., Sainz, R.M., Mayo, J.C., Leon, J., Hardeland, R., Poeggeler, B., Reiter, R.J., 2005. Interactions between melatonin and nicotinamide nucleotide: NADH preservation in cells and in cell-free systems by melatonin. Journal of Pineal Research 39, 185–194.

Tan, D.X., Manchester, L.C., Di Mascio, P., Martinez, G.R., Prado, F.M., Reiter, R.J., 2007a. Novel rhythms of N1-acetyl-N2-formyl-5-methoxykynuramine and its precursor melatonin in water hyacinth: importance for phytoremediation. The FASEB Journal 21, 1724–1729.

Tan, D.X., Manchester, L.C., Helton, P., Reiter, R.J., 2007b. Phytoremediative capacity of plants enriched with melatonin. Plant Signaling and Behavior 2, 514–516.

Tan, D.X., Manchester, L.C., Terron, M.P., Flores, L.J., Reiter, R.J., 2007c. One molecule, many derivatives: a never-ending interaction of melatonin with reactive oxygen and nitrogen species? Journal of Pineal Research 42, 28–42.

Tan, D.X., Hardeland, R., Manchester, L.C., Paredes, S.D., Korkmaz, A., Sainz, R.M., Mayo, J.C., Fuentes-Broto, L., Reiter, R.J., 2010. The changing biological roles of melatonin during evolution: from an antioxidant to signals of darkness, sexual selection and fitness. Biological Reviews of the Cambridge Philosophical Society 85, 607–623.

Tan, D.X., Hardeland, R., Manchester, L.C., Korkmaz, A., Rosales-Corral, S., Reiter, R.J., 2011. Functional roles of melatonin in plants, and perspectives in nutritional and agricultural science. Journal of Experimental Botany 63, 577–597.

Tan, D.X., Hardeland, R., Manchester, L.C., Rosales-Corral, S., Coto-Montes, A., Boga, J.A., Reiter, R.J., 2012. Emergence of naturally occurring melatonin isomers and their proposed nomenclature. Journal of Pineal Research 53, 113–121.

Townsend, K.L., Tseng, Y.H., 2014. Brown fat fuel utilization and thermogenesis. Trends in Endocrinology & Metabolism 25, 168–177.

Vitalini, S., Gardana, C., Zanzotto, A., 2011. From vineyard to glass: agrochemicals enhance the melatonin and total polyphenol contents and antiradical activity of red wines. Journal of Pineal Research 51, 278–285.

Vitalini, S., Gardana, C., Simonetti, P., Fico, G., Iriti, M., 2012. Melatonin, melatonin isomers and stilbenes in Italian traditional grape products and their antiradical capacity. Journal of Pineal Research 54, 322–333.

Witt-Enderby, P.A., Bennett, J., Jarzynkaa, M.J., Firestinea, S., Melan, M.A., 2003. Melatonin receptors and their regulation: biochemical and structural mechanisms. Life Sciences 72, 2183–2198.

Witt-Enderby, P., Clafshenkel, W.P., Kotlarczyk, M., et al., 2011. Melatonin in bone health. In: Watson, R.R. (Ed.), Melatonin in the Promotion of Health. CRC Press, Boca Raton, Florida, pp. 261–270.

Yılmaz, C., Kocadagl, T., Gökmen, V., 2014. Formation of melatonin and its isomer during bread dough fermentation and effect of baking. Journal of Agricultural of Food Chemistry 62, 2900–2905.

Zhang, H.M., Zhang, Y., 2014. Melatonin: a well-documented antioxidant with conditional prooxidant actions. Journal of Pineal Research 57, 131–146.

Chapter 7

Effect of Fermentation on Vitamin Content in Food

B. Walther, A. Schmid
Agroscope, Institute for Food Sciences (IFS), Bern, Switzerland

7.1 INTRODUCTION

Fermentation alters a food in several ways. Fermentation increases shelf life and microbiological safety and may also make a food more digestible and more palatable. In addition, fermentation can be used to improve the nutritional qualities of a food since microorganisms increase specific beneficial substances, such as macronutrients, micronutrients (eg, vitamins), or non-nutritive compounds.

Vitamins are organic compounds that mostly cannot be synthesized by the human organism or only in inadequate amounts and are thus essential for humans. Vitamins are not building elements of organs or tissue but are necessary for the maintenance of biochemical reactions in the cell. Most vitamins must be obtained from the diet. Although they are present in many foods, vitamin deficiencies still exist in many countries and population groups. The reason for this is not always insufficient food intake but may also be an unbalanced diet. Furthermore, food processing and cooking may destroy or remove some of the vitamins normally present in foods. Therefore, fortification of specific foods with certain vitamins and minerals has been instituted by many countries. However, not all countries adopt national fortification programs due to concerns about possible unwanted side effects. Ingestion of diverse fortified foods may result in excessive intake or an imbalance of essential nutrients with a negative impact on health. Moreover, fortification is not always well accepted by the consumer. In this context, the microbial production of vitamins provides a very attractive approach for improving the nutritional composition of fermented foods. It is a more natural and economically interesting alternative to fortification with chemically synthesized vitamins and may contribute to cover the vitamin supply in populations in which it is low (LeBlanc et al., 2011; Caplice and Fitzgerald, 1999).

This chapter reviews the vitamin content of various fermented foods and provides strategies for increasing the production of vitamins in fermented foods. The focus is folic acid and vitamin K, but attention is also given to other vitamins.

7.2 FOLATE (VITAMIN B9)

Folate is a generic term used to describe a family of compounds with the activity of folic acid, including naturally occurring food folate and synthetically produced folic acid (pteroylglutamic acid). The vitamin belongs to the B vitamin group and is heat as well as light sensitive and water-soluble. Common structures of the folate family embrace a pteridine ring system, p-aminobenzoic acid and one or more glutamic acid moieties. The number of glutamic acid moieties distinguishes pteroylmonoglutamate from pteroylpolyglutamates. Pteroylpolyglutamates are mainly found naturally in plant and animal foods (Stahl and Heseker, 2007a).

Folate acts as a coenzyme for single-carbon transfers involved in the synthesis, interconversion, and modification of nucleotides, amino acids and other key cellular components. This includes purine and pyrimidine synthesis, histidine catabolism, interconversion of serine and glycine, conversion of homocysteine to methionine, generation of formate, and methylation of transfer RNA in mitochondrial protein synthesis (Office of Dietary Supplements, 2012). In humans, folate is absorbed in the small intestine in the form of pteroylmonoglutamate and is then metabolized to the biologically active forms tetrahydrofolate and derivatives. 5-Methyltetrahydrofolate is the folate derivative normally present in the circulation and in body stores (Stahl and Heseker, 2007a; Iyer and Tomar 2009). Body stores are relatively small; thus, temporary inadequate dietary intake can lead to short-term deficiency. Folate deficiency is characterized by megaloblastic anemia; symptoms include weakness, fatigue, difficulty concentrating, irritability, headache, heart palpitations, and shortness of breath. Furthermore, folate deficiency may cause changes in the skin, hair, or fingernail pigmentation; elevated blood concentrations of homocysteine; and soreness and shallow ulcerations in the tongue and oral mucosa. Folate deficiency during pregnancy increases the risk of neural tube defects in the infant. In addition, an increased risk of preterm delivery, low infant birth weight, and fetal growth retardation are associated with inadequate folate status. Inadequate folate intake has also been associated with several health disorders, such as Alzheimer's disease, poor cognitive performance, coronary heart disease, osteoporosis, colorectal cancer, breast cancer, and hearing loss (Office of Dietary Supplements, 2012; Iyer and Tomar 2009; Bailey and Gregory, 2006).

A wide variety of foods contain folate naturally, including dark green leafy vegetables, fruits, nuts, beans, peas, dairy products, poultry and meat, eggs, seafood, and grains. Spinach, liver, yeast, asparagus, and Brussels sprouts are among the foods with the highest levels of folate. The recommended dietary allowance for healthy male and female adults is 400-μg food folate per day. Higher intakes are recommended during pregnancy (600 μg/day) and lactation (500 μg/day) (https://fnic.nal.usda.gov/sites/fnic.nal.usda.gov/files/uploads/RDA_AI_vitamins_elements.pdf).

7.2.1 Folate Content of Fermented Food

Food fermentation can result in increased folate concentrations due to folate synthesis from starter cultures. For example, milk contains only 2–5 μg/100 mL of folate and therefore is not a rich source. However, if milk is processed into yogurt, the folate content may increase to more than 20 μg/100 mL depending on the strains used for the fermentation (Wouters et al., 2002). Other foods with increased folate content due to fermentation with lactic acid bacteria (LAB) are fermented vegetables (Jägerstad et al., 2004) and corn flour (Murdock and Fields, 1984). Sauerkraut, for example, is a traditional fermented vegetable produced by spontaneous fermentation. Commercial canned products of sauerkraut contain folate in the range of 5–21 μg/100 g (Jägerstad et al., 2004). Fermenting rye and wheat dough to produce bread frequently leads to increases in folate concentration. However, in this case, folate is mainly produced by yeasts whereas LAB isolated from sourdoughs do not produce folate (Kariluoto et al., 2006; Jägerstad et al., 2005). Fungal fermentation (*Rhizopus oligosporus*) of soybeans in order to produce tempeh also increases folate concentrations (Arcot et al., 2002; Murata et al., 1970). An analysis of 120 European beers yielded folate concentrations in the range of 3–18 μg/100 mL (Jägerstad et al., 2005). Samples taken at different processing stages indicate that high values may be due to a secondary fermentation step, as is the case in Bavarian wheat beers. In general, folate and alcohol content are related, and both are probably attributed to the amount of brewing raw material used (Jägerstad et al., 2005). Generally, folate increases up to twofold in bread making and fermenting vegetables, up to sevenfold in beer brewing, and up to 20-fold in milk products (depending on the starter culture) (Jägerstad et al., 2005). Table 7.1 lists folate concentrations of fermented foods reported in the literature, and Table 7.2 indicates folate concentrations of some fermented foods taken from US, French, and Dutch food databases. Differences in folate concentrations found between food databases may be due to variations in food samples (eg, different strains used for fermentation and processing conditions) or due to different analytical methods.

Although fermentation can increase folate concentrations in food, other food processing techniques can lead to a substantial loss of folate in fermented food, mostly by leakage, oxidation, or both. This is the case especially in heat treatments, such as boiling and canning (Hawkes and Villota, 1989; Arcot et al., 2002). In fermented milk, Rao and co-workers (Rao et al., 1984) did not document an appreciable change in folate content after storage at 5°C for 5 days, and neither did Wigertz and coauthors (Wigertz et al., 1997) in several fermented milks refrigerated for 2 weeks. Storing yogurt at 4°C for 28 days also did not result in decreased folate concentrations (Laino et al., 2013). In contrast,

TABLE 7.1 Total Folate Concentrations in Common Fermented Foods

Food	Total Folate (µg/100 g or 100 mL)	Method of Analysis	References
Yogurt	13	HPLC	Müller (1993)
Plain yogurt	5	HPLC	Wigertz et al. (1997)
Plain yogurt, 1.9% fat	11.8 ± 2.8	Microb. assay	Hoppner and Lampi (1990)
Plain yogurt, 9.5% fat	4.9 ± 1.6	Microb. assay	Hoppner and Lampi (1990)
Flavored yogurt, various	3.7–13.9	Microb. assay	Hoppner and Lampi (1990)
Plain yogurt, various	3.2 ± 3.9	Microb. assay	Kneifel et al. (1991)
Plain yogurt	3.9	Microb. assay	Reddy et al. (1976)
Kefir	1.4 ± 1.3	Microb. assay	Kneifel et al. (1991)
Sour cream	7	HPLC	Müller (1993)
Hard cheese, various	12–21	HPLC	Wigertz et al. (1997)
Emmental cheese	7	HPLC	Müller (1993)
Sauerkraut, canned	5–21	HPLC	Jägerstad et al. (2004)
Beer, various	3–18	HPLC	Jägerstad et al. (2005)

HPLC, high performance liquid chromatography; *Microb. assay*, microbiological assay.

Reddy et al. (1976) found a 29% decrease in folate in cultured yogurt during storage at 5°C for 16 days. Furthermore, a significant loss of folate (14.3 versus 10.8-µg 5-methyltetrahydrofolate per 100 g) was described in hard cheese after a storage period of 24 weeks (Wigertz et al., 1997). In beer, the long-term stability of folate was good with little loss over 6 months, although an initial loss during packaging occurred (Jägerstad et al., 2005).

Fermentation conditions, such as incubation temperature, length of incubation, and medium used affect folate concentrations. The maximum folate concentration in yogurt was attained with incubation at 42°C (Reddy et al., 1976; Laino et al., 2013). Folate content increased 10-fold in the first 3 h of incubation

TABLE 7.2 Folate Concentrations (μg/100 g) in Various Fermented Foods According to US, Dutch and French Food Databases

Food	US[a]	France[b]	The Netherlands[c]
Butter, unsalted	3	2	traces
Cheese			
Brie	65	58	38
Camembert	62	53.3	83
Cheddar	26	102	38
Edam	16	16	11.7
Feta	32	23	na
Gouda	21	43	25
Gruyère	10	12	12
Mozzarella	7	19	15
Parmesan	6	12	12
Sour cream (crème fraiche)	7	23.5	7
Buttermilk	5	7.8	7.9
Yogurt, plain, whole milk	7	20	12.6
Salami	2	3.6	2.5
Tofu	15	17.3	na
Sauerkraut, canned	24	4.5	7

na, not available.
[a]http://ndb.nal.usda.gov.
[b]https://pro.anses.fr/TableCIQUAL/.
[c]http://nevo-online.rivm.nl.

followed by smaller increases at 4 and 5 h (Reddy et al., 1976). According to Laino et al.'s study, the best conditions for increasing folate concentrations during milk fermentation are 6 h of incubation at 42°C (Laino et al., 2013). Furthermore, an investigation by Sybesma et al. (2003), showed that the amount of folate produced under controlled growth conditions increased threefold when the pH was increased from 5.5 to 7.5.

Quantitative analysis of folate in food is a difficult task since there are multiple forms and generally low levels. Furthermore, the absolute content of folate in food is only conditionally meaningful because bioavailability differs greatly. On the one hand, different kinetics and bioavailability of the various folate vitamers (compounds with similar molecular structure, each showing vitamin-activity) as well as the entrapment of folates in the food matrix influence bioavailability. On the other hand, substances in certain plant foods (eg, yeast nucleic acid) reduce folate bioavailability by inhibiting the enzyme pteroylpolyglutamate hydrolase (also referred to as folate conjugase), which is responsible for the hydrolysis of folate polyglutamates. Furthermore, folate-binding proteins from milk may increase folate absorption in the small intestine. In general, the bioavailability of pteroylmonoglutamate is significantly greater than that of pteroylpolyglutamate (Stahl and Heseker, 2007a; Iyer and Tomar 2009; Mönch et al., 2015; Rosenberg and Godwin, 1971). A recent investigation found the highest bioavailability for folate in spinach, followed by wheat germ; the lowest was in Camembert cheese. The result emphasizes that folate bioavailability depends on the type of food (Mönch et al., 2015).

7.2.2 Folate Production by Microorganisms

De novo folate production is found in green plants, fungi, certain protozoa, and bacteria (Iyer and Tomar 2009). Many studies have documented that LAB synthesize folate. However, large differences have been found, not only between different species but also between different strains (Hugenholtz et al., 2002). For example, the industrial starter cultures *Lactococcus lactis*, *Streptococcus thermophilus*, and *Leuconostoc* species have the ability to synthesize folate whereas many *Lactobacilli* species consume folate. In addition to these starter bacteria, *Lactobacillus acidophilus*, *Lactobacillus reuteri*, *Bifidobacterium longum*, and some *Propionibacteria* also synthetize folate (Iyer and Tomar 2009; LeBlanc et al., 2011). Not all folate-producing bacteria are equally suitable for biofortification, however, some assemble folate intracellularly, and it is not excreted into food. This folate is less bioavailable than folate produced extracellularly. The majority of folate produced by *L. lactis*, *Leuconostoc* spp. and *Bifidobacteria* spp., but not that produced by *S. thermophilus*, is retained intracellularly. The last has a strain-specific ability for extracellular folate production (Iyer and Tomar 2009, 2011; Sybesma et al., 2003).

Two LAB species are traditionally used to produce yogurt: *S. thermophilus* and *Lactobacillus delbrueckii* subsp. *bulgaricus*. A screening of common yogurt starter bacteria and probiotic bacteria (species of the genera *Lactobacillus, Streptococcus, Bifidobacterium*, and *Enterocccus*) for their ability to produce or utilize folate proved *S. thermophilus* strains are the best folate producers, increasing the folate level in skim milk approximately fourfold (from 11.5 ng to between 40 and 50 ng/g) (Crittenden et al., 2003). Rao and co-workers (Rao et al., 1984) documented for skim milk fermented with *S. thermophilus*, maximum folate levels of 25 ng/mL compared with 7.4 ng/mL in unfermented skim milk. Other studies support the notion of *S. thermophilus* being a good folate producer (Jägerstad et al., 2005; Sybesma et al., 2003; Holasova et al., 2004; Friend et al., 1983). However, substantial differences in the amount of folate produced by individual strains were documented. Jägerstad et al. (2005)

compared three *S. thermophilus* strains and found folate levels in the product differed by a factor of 7.

The second LAB in yogurt production, *L. bulgaricus*, is predominantly known for not producing but consuming folate (LeBlanc et al., 2011). Friend et al. (1983) fermented skim milk for 24 h with three strains of *L. bulgaricus* and reported decreased folate concentrations in all three products. Of the three *L. bulgaricus* strains investigated by Crittenden et al. (2003), two did not significantly influence folate levels, and one depleted the available folate in milk. A study by Rao et al. (1984) also documented decreased levels of folate in skim milk fermented with *L. bulgaricus* compared with unfermented skim milk, and Lin and Young (2000) described folate accumulation by *L. bulgaricus* during the first 6 h of growth followed by a decrease to approximately the initial levels in the following 12 h. However, four *L. bulgaricus* strains with the ability to synthesize folate were identified in starter cultures isolated from artisanal Argentinean yogurts (Laino et al., 2012). Not only *L. bulgaricus* but also *Lactobacilli* in general are thought to deplete or to not influence folate concentrations. For instance, *L. helveticus* biotype jugurti had no affect on folate content and *L. rhamnosus*, *L. reuteri*, *L. casei*, *L. johnsonii*, and *L. acidophilus* have been proven to consume folate (Crittenden et al., 2003). Sybesma et al. (2003) also concluded, after having investigated several *Lactobacillus* strains, that most, with the exception of *Lactobacillus plantarum*, did not produce folate. Alm (1982) documented in several fermented milk products higher folate concentrations than in milk, but this was not the case in acidophilus milk. Furthermore, *L. acidophilus*, *L. brevis*, *L. plantarum*, and *L. sanfranciscensis* grown in nonsterile rye flour and water at 30°C for 19 h did not increase folate concentrations (Kariluoto et al., 2006). In contrast, Rao et al. (1984) noticed a folate increase in skim milk fermented with *L. acidophilus* (21.2 ng/mL compared to 7.4 ng/mL in unfermented skim milk after 24 h), as did Lin and Young (2000) in reconstituted milk fermented with two different *L. acidophilus* strains (folate increase of 31.5 and 43.2 ng/mL after 6 h). These different findings may reflect strain differences or the effects of different culture conditions.

Several studies document that individual strains and species of *Bifidobacterium* differ in folate production. Three *B. longum* strains grown in ultra-heat treated milk with 1.5% fat produced folate (maximum increase of 4.8-ng 5-methylthetrahydrofolate per g after 12 h fermentation), but two *B. bifidum* strains did not increase the folate concentration (Holasova et al., 2004). Congruently, Lin and Young (2000) found that two *B. longum* strains increased folate levels in reconstituted milk by 53.3 and 77.0 ng/mL within 6 h, and Crittenden et al. (2003) identified seven *Bifidobacterium* strains (*B. lactis*, *B. animalis*, *B. infantis*, and *B. breve*) as folate producers. However, among 76 wild-type *Bifidobacterium* strains, only 17 strains have the ability to produce folate (between 0.6 and 82 ng/mL in 48 h on a folate-free synthetic medium) (Pompei et al., 2007). Among the 17 strains, four *B. adolescentis* and two *B. pseudocatenulatum* strains synthesized significantly higher folate concentrations (≥41 ng/mL). Thus, folate production seems to be an uncommon characteristic of *Bifidobacterium* but can be found in single strains of these species.

Other LAB have been reported to produce folate. Strains of *L. lactis* subsp. *cremoris* and *L. lactis* subsp. *lactis* produced folate in the range of 57–291 ng/mL, and *Leuconostoc lactis* and *Leuconostoc paramesenteroides* synthesize folate, although to a lesser extent (45 and 44 ng/mL) (Sybesma et al., 2003). However, three *Propionibacterium freudenreichii* subsp. *shermanii* strains did not produce folate during milk fermentation (Holasova et al., 2004).

7.3 VITAMIN K

Vitamin K is a fat-soluble vitamin, coincidentally discovered in 1929 by the Danish nutritional biochemist Henrik Dam as part of his experiments on sterol metabolism and associated with blood coagulation. In the following decade, the two main K vitamers, phylloquinone (PK) also called K1 and the group of menaquinones (MKs) also called K2, were isolated and characterized. PK is primarily found in green leafy vegetables (eg, spinach, kale, and cabbage) and plant oils, whereas MKs are of bacterial origin and are mainly present in meat, dairy, and fermented food products (Ferland, 2012). Vitamin K is stable to air, but decomposed by sunlight and alkalis (IARC Working Group on the Evaluation of Carcinogenic Risks to Humans, 2000).

Vitamin K deficiency in adults is associated with hypoprothrombinemia as a beneficial effect of anticoagulant drug intake in patients with thromboembolic disease or at risk of it and, therefore, not perceived as a threat to health. In infants, however, spontaneous vitamin K deficiency is a well-known hazard, especially in those who are exclusively breastfed. In the 1950s, routine vitamin K prophylaxis was introduced to prevent life-threatening bleeding in newborns (Shearer, 1995; Shearer et al., 2012). In recent years, the relationship between vitamin K and chronic diseases, such as osteoporosis, cardiovascular disease, and cancer, and the possible health roles of osteocalcin and matrix Gla protein has become a topic of interest for researchers (Booth, 2009). Dietary recommendations for vitamin K intake are based on the current knowledge of PK and prevention of insufficient blood coagulation and range from 50 to 120 µg per day for 19-year old adults and older. These recommendations are generally presented as adequate intake or estimated values, and no tolerable upper intake level has been established for vitamin K (Walther et al., 2013; Ferland, 2012).

Since the main topic of this book is fermented food, bacterial MKs are the focus of this chapter. A common 2-methyl-1,4-naphtoquinone ring, also known as menadione or vitamin K3, characterizes all compounds with vitamin K activity. The structure of the different vitamers vary at the 3-position. PK holds a phytyl side chain located at the 3-position, while unsaturated isoprenyl side chains of different lengths are attached at the same locus in MKs (Ferland, 2012). The chain length generally ranges from 4 to 13 prenyl units depending on the organism by which the chain is synthesized. The MKs are classified according to the number of prenyl units. The number of units is given in a suffix (-n), for example, MK-7 for a chain length of seven prenyl units. To complicate the picture, some bacteria produce isoprenologues with one or more unsaturated prenyl units. The additional hydrogen atoms are indicated by the prefix dihydro-, tetrahydro-, and so on, abbreviated as MK-n(H2), MK-n(H4), etc. (Shearer and Newman, 2008). An exception builds MK-4 that is not synthesized by bacteria but produced in humans and animals by tissue-specific conversion of PK and/or menadione (Okano et al., 2008).

Two biochemical pathways are described for MK synthesis by bacteria (Bentley and Meganathan, 1982; Nowicka and Kruk, 2010; Hiratsuka et al., 2008). LAB, usually used in the food industry for fermentation, synthesizes vitamin K in three steps with a series of enzymes encoded by the *men* genes. First, the naphthoquinone ring is synthesized from chorismate, a product of the shikimate pathway. Then the isoprenoid side chain is synthesized over the mevalonate pathway and joined to the naphthoquinone ring to form dimethylmenaquinone (DMK). Finally, subsequent methylation of DMK completes MK biosynthesis (Bentley and Meganathan, 1982). Studies have shown that microorganisms, such as *Helicobacter pylori*, *Campylobacter jejuni*, and *Lactobacilli* do not have orthologs of the *men* genes and synthesize MKs in an alternative pathway via futalosine (Kurosu and Begari, 2010). Generally, gram-negative bacteria produce MK and DMK, whereas gram-positive bacteria primarily produce MK. Some bacteria species exclusively produce DMK, because they lack the required methylase (Collins and Jones, 1981).

MKs function in respiratory and photosynthetic electron transport chains of bacteria, may exhibit antioxidant properties (Nowicka and Kruk, 2010) and are involved in active molecule transport across the cell membrane as well as in sporulation in *Bacillus subtilis* (Farrand and Taber, 1974; Hojo et al., 2007). Most vitamin K–producing bacteria live in an anaerobic environment, such as the gut. The most important representatives among these bacteria are *Bacteroides* species that mainly produce MK-10 and MK-11, *Enterobacteria* (MK-8), *Veillonella* species (MK-7), and *Eubacterium lentum* (MK-6) (Shearer, 1995). The importance of the contribution of the gut MKs was discussed by Suttie (1995), and he concluded that diet is a much more important source of functionally available vitamin K than the gut microbiome. This conclusion has also been supported by more data showing that a short-term decrease in dietary vitamin K intake was not compensated by intestinal MKs (Booth et al., 2001; Urano et al., 2015; Iwamoto et al., 2014; Theuwissen et al., 2014).

7.3.1 Vitamin K Content of Fermented Food

MKs are primarily found in dairy products, meat, and fermented food. However, the vitamin K content of many food products is unknown. From the more than 70 national food composition databases listed by the Food and Agriculture

Organization of the United Nations (http://www.fao.org/infoods/infoods/tables-and-databases/en/), only 12 databases include the vitamin K content of individual food items. Most list only PK (France, United Kingdom, and Germany) or total vitamin K (Finland). Five countries (Sweden, Canada, Denmark, Japan, and Estonia) do not specify whether the values include only PK or MK as well. The only three databases that distinguish between PK and MK are the ones for the United States, the Netherlands and Turkey. Whereas the Netherlands and Turkey do not further specify vitamin K2 values, the United States Department of Agriculture (USDA) National Nutrient Database for Standard Reference, Release 28 (http://www.ars.usda.gov/Services/docs.htm?docid=8964) publishes data for PK, dihydrophylloquinone and MK-4 for 546 food items. Table 7.3 shows data from the US, Dutch and Turkish databases for selected fermented food items. The large variation in these values between the databases may be due to various reasons. First, the values given in the US database are those for MK-4. However, the Dutch database includes several types of MK, ranging from MK-4 to MK-10. For the data from the Turkish database, no further information is available concerning the precise definition of vitamin K2. Second, the concentration of MK-4 is higher in countries where animals are supplemented with menadione as practiced in the United States (Elder et al., 2006) and the Netherlands (EFSA FEEDAP Panel, 2014). No information on animal supplementation in Turkey is

TABLE 7.3 Comparison of Vitamin K2 Concentrations in Commonly Consumed Animal (Fermented) Products Published in Three National Food Composition Databases

	United States[a] (µg/100 g)	The Netherlands[b] (µg/100 g)	Turkey[c] (µg/100 g)
Yogurt whole	na	0.5–0.9	1.8
Butter	1.7	15	8.4
Cheese			
Parmesan	7.1	76.5	na
Cheddar	8.6	10.2	na
Cheese, semisoft	na	na	2.7
Cheese, fresh and white	5.8	25.2	6.8
Cheese mozzarella	4.1	10.8	na

na, not available.
[a] http://ndb.nal.usda.gov/.
[b] http://www.rivm.nl/.
[c] http://www.turkomp.gov.tr.

available. Third, the conditions during processing as well as the bacterial strains used in the production of fermented food determine the concentration and forms of MK in the products (Walther et al., 2013).

Recent analysis of 62 fermented dairy products resulted in MK concentrations that range widely from undetectable values for a milk fermented with thermophilic strains up to 1.1 µg/g in a French soft cheese (Manoury et al., 2013). However, compared to vitamin K content in vegetables, where more than 1.0 µg/g of PK was found, the concentration of 0.1–0.8 µg/g in cheese and most other fermented food seems moderate. Walther et al. (2013) published an overview of the distribution of various MKs and their concentrations in various fermented foods. These results were amended by recent published data on quantitative measurement of vitamin K in various fermented dairy products by Manoury et al. (2013) as well as by findings in fermented plant products. Values for MK-4 to MK-10 are available. MK-4 is found in all reported fermented products except buckwheat, hikiwari natto and black bean natto. In sour milk and buttermilk as well as in curd, hard and soft cheese, MK-8 and MK-9 mainly account for the total concentration of vitamin K followed by MK-6 and MK-7 where reported. Fermented plant products are characterized by a high concentration of MK-7 (up to 10 µg/g) (Table 7.4).

Until now, no data about the stability and changes in vitamin K concentrations during ripening of cheese and other fermented food have been available.

The investigation of the bioavailability and bioactivity of vitamin K is in its infancy, and the results are still fragmentary. Studies comparing plasma PK, MK-4, and MK-9 concentrations after consumption of equivalent doses of these forms demonstrated differences in the absorption and transport of vitamin K depending on the vitamer. PK peaked postprandially at more than twice the relative concentration of MK-4 or MK-9 in plasma, which could be explained by better absorption of PK compared to MK-4 and MK-9, faster uptake of MK into tissue or both. Schurgers and Vermeer (2000) reported that MK-9, similar to MK-7, has a much longer plasma half-life than PK or MK-4. However, in addition to absorption and kinetics, transport differs between PK and MK. Whereas PK remains concentrated in triglyceride-rich lipoproteins during the postprandial and fasting states (Shearer and Newman, 2008), MK-4 and MK-9 are redistributed to low-density lipoproteins during and after the postprandial period (Schurgers and Vermeer, 2000, 2002; Sato et al., 2012; Schurgers et al., 2007; Novotny et al., 2010). The bioavailability of other long-chained MKs as well as the absorption and transport of MK from different food sources are unknown (Walther et al., 2013). Schurgers and Vermeer (2000) concluded in their review that absorption of MK from cheese and natto is very high (almost complete) whereas only 5–15% of PK is available from vegetables. Among all the MKs, MK-7 is said to be the longest-lasting, most bioactive and most bioavailable form of vitamin K (Shearer et al., 2012).

TABLE 7.4 Concentrations of and Variations in Menaquinones With Different Chain Lengths in Fermented Foods (µg/100g; Mean ± SD or Range)

Food	Menaquinone Content (µg/100g; Mean ± SD or Range)								Source
	MK-4	MK-5	MK-6	MK-7	MK-8	MK-9	MK9 (4H)	MK-10	
Fermented Milk									
Whole milk, sour	0.6 ± 0.02	0.3 ± 0.002	0.2 ± 0.03	0.4 ± 0.04	2.0 ± 0.1	4.7 ± 0.2	nr	nd	Koivu-Tikkanen et al. (2000)
Buttermilk[a]	0.2–0.3	0.1–0.2	0–0.2	0.1–0.3	0.5–0.6	1.2–1.6	nr	nr	Schurgers and Vermeer (2000)
Mesophilic	nr	nr	4.2	5.0	25.9	100.8	nr	8.5	Manoury et al. (2013)
Yogurt									
Whole	0.4–1.0	nr	nr	nr	nr	0–2.0	nr	nr	Vermeer et al. (1998)
	0.5–0.7	0–0.2	nd	nd	nd	nd	nr	nr	Schurgers and Vermeer (2000)
	1 ± 0.1	nr	nr	0.1 ± 0.2	nr	nr	nr	nr	Kamao et al. (2007)

									References
Plain	0.4±0.03	0.1±0.006	nd	nd	nd	nd	nr	nd	Koivu-Tikkanen et al. (2000)
Skimmed	nd	nd	nd	nd	0–0.2	nd	nr	nr	Schurgers and Vermeer (2000)
Cheese									
Curd	0.3–0.6	0–0.2	0.1–0.3	0.2–0.5	4.8–5.4	18.1–19.2	nr	nr	Schurgers and Vermeer (2000)
Hard	2–10	nr	nr	nr	nr	40–70	nr	nr	Vermeer et al. (1998)
	4.2–6.6	1.3–1.7	0.6–1	1.1–1.5	14.9–18.2	45.3–54.9	nr	nr	Schurgers and Vermeer (2000)
Semi-hard	nr	nr	18.9	10.9	39.1	175.3	nr	47	Manoury et al. (2013)
Soft	3.3–3.9	0.2–0.4	0.5–0.7	0.9–1.1	10.7–12.2	35.1–42.7	nr	nr	Schurgers and Vermeer (2000)
	nr	nr	16.8	11.8	70	277.3	nr	29.2	Manoury et al. (2013)

Continued

TABLE 7.4 Concentrations of and Variations in Menaquinones With Different Chain Lengths in Fermented Foods (μg/100 g; Mean ± SD or Range)—cont'd

Food	Menaquinone Content (μg/100 g; Mean ± SD or Range)									Source
	MK-4	MK-5	MK-6	MK-7	MK-8	MK-9	MK9 (4H)	MK-10		
Processed	5 ± 2	nr	nr	0.3 ± 0.1	nr	nr	nr	nr		Kamao et al. (2007)
Blue cheese	nr	nr	48.8	124	76.8	193.3	nr	28.9		Manoury et al. (2013)
Caerphilly	nr	nr	15.8 ± 1.3	nd	78.9 ± 2.2	324.3 ± 8	nr	nd		Manoury et al. (2013)
Cheddar	10.2	nr	nr	nr	nr	nr	nr	nr		Elder et al. (2006)
Cheddar	nr	nr	22.2	20.6	31.5	129.3	nr	51.7		Manoury et al. (2013)
Cheshire	nr	nr	15.7 ± 0.7	nd	57.9 ± 1.7	241.7 ± 4	nr	nd		(Manoury et al., 2013)
Swiss	6.2–8.8	nr	nr	nr	nr	nr	nr	nr		Elder et al. 2006)
Emmental	5.1–5.2	nr	nr	nr	nr	nr	22.2–31.4	nr		Hojo et al. (2007)

Effect of Fermentation on Vitamin Content in Food Chapter | 7 145

	Manoury et al. (2013)	Koivu-Tikkanen et al. 2000	Koivu-Tikkanen et al. (2000)	Manoury et al. (2013)	Hojo et al. (2007)	Hojo et al. (2007)	Hojo et al. (2007)	Hojo et al. (2007)
Emmental	nr	nr	nd	nd	nd	nd	nr	40.1
Aged 90d	5.2±0.1	nd	trace	trace	nd	nd	nr	nd
Aged 180d	6.1±0.5	nd	trace	nd	nd	nd	nr	nd
Leicester	nr	nr	20±1.1	21.5±1.3	47.6±2.4	162.4±2.8	nr	43.8±2.2
Jarlsberg	6	nr	nr	nr	nr	nr	65.2	nr
Gruyère	3.3–5.8	nr	nr	nr	nr	nr	nd	nr
Comte	5.1–5.3	nr	nr	nr	nr	nr	5.2–6	nr
Appenzeller	4.3–5.2	nr	nr	nr	nr	nr	nd–2	nr
Raclette	2.6	nr	nr	nr	nr	nr	4.7	nr
Edam	3.3±0.2	1±0.1	0.6±0.1	1.3±0.1	10.5±0.8	30±2.6	nr	0.9±0.1
Mozzarella	3.1–4	nr	nr	nr	nr	nr	nr	nr

Continued

TABLE 7.4 Concentrations of and Variations in Menaquinones With Different Chain Lengths in Fermented Foods (µg/100g; Mean ± SD or Range)—cont'd

Food	Menaquinone Content (µg/100g; Mean ± SD or Range)									Source
	MK-4	MK-5	MK-6	MK-7	MK-8	MK-9	MK9 (4H)	MK-10		
Other Dairy										
Cream	8 ± 3	nr	nr	nd	nr	nr	nr	nr		Kamao et al. (2007)
Whipping cream	5.2–5.6	nd	nd	nd	nd	nd	nr	nr		Schurgers and Vermeer (2000)
Butter	13.5–15.9	nd	nd	nd	nd	nd	nr	nr		Schurgers and Vermeer (2000)
	21 ± 7	nr	nr	nd	nr	nr	nr	nr		Kamao et al. (2007)
Meat										
Salami	8.2–10.1	nd	nd	nd	nd	nd	nr	nr		Schurgers and Vermeer (2000)

Bread									
Buckwheat	nd	nd	nd	1–1.2	nd	nd	nr	nr	Schurgers and Vermeer (2000)
Plant Products									
Sauerkraut	0.3–0.5	0.6–1	1.4–1.6	0.1–0.3	0.6–0.9	0.9–1.3	nr	nr	Schurgers and Vermeer (2000)
Natto	nd	7.1–7.8	12.7–14.8	882–1034	78.3–89.8	nd	nr	nr	Schurgers and Vermeer (2000)
	2±3	nr	nr	939±753	nr	nr	nr	nr	Kamao et al. (2007)
Hikiwari natto (chopped natto)	nd	nr	nr	827±194	nr	nr	nr	nr	Kamao et al. (2007)
Black bean natto	nd	nr	nr	796±93	nr	nr	nr	nr	Kamao et al. (2007)

MK, Menaquinone; nd, not detectable; nr, not reported.
[a] μg/100 mL.

7.3.2 Enhancing Vitamin K2 Content in Fermented Food

Propionibacterium strains mainly produce MK-9 and tetrahydromenaquinone-9 (MK-9 (4H)) whereas *B. subtilis natto* produces large amounts of MK-7. *L. lactis* strains are known to produce MK-7 to MK-9 but no MK-9 (4H) (Hojo et al., 2007). A large study that measured vitamin K2 concentrations in 62 fermented dairy products confirmed earlier findings that thermophilic species, such as *S. thermophilus*, *L. delbrueckii*, and *Bifidobacterium* do not produce vitamin K2. In fermented dairy products that use only thermophilic starter cultures, no long-chained MKs (MK6–MK-10) have been detected. In contrast, all products that contain mesophilic LAB species and especially *Lactococcus* spp. as starter cultures contain high concentrations of vitamin K, more than 0.1 µg/g (Manoury et al., 2013).

In addition to bacterial strain selection with high production of MKs (Table 7.5), gene mutation, as well as process optimization can enhance the bacterial growth and vitamin K–producing capacity of distinct strains. A menadione-resistant mutant of a *B. subtilis* strain, isolated from the traditional Japanese food natto, produced 30% more MK than the wild strain. When this *B. subtilis* mutant is cultured on soybean extract containing, glycerol and yeast extract, the concentration reached a maximum of 35 µg/g MK after four days. These results were observed in a pH-controlled fermentation in which the decrease in pH to 5.5 after inoculation with a subsequent increase to 7.7–8.0 seems to be important for a high MK concentration (Sato et al., 2001).

7.4 RIBOFLAVIN (VITAMIN B2)

Riboflavin, also called vitamin B2, is a water-soluble vitamin. Plants and microorganisms synthesize riboflavin unlike animals and humans who have to consume this essential nutrient with their diet. The most important sources for humans are milk and dairy products, fresh meat, liver, leafy vegetables, whole grains, and egg whites. The dietary reference intake for adult men (19–50 years) is 1.3 mg of riboflavin daily and 1.1 mg for adult women. Riboflavin is stable against heat and oxygen exposure but readily destroyed by ultraviolet (UV) rays and sunlight (Northrop-Clewes and Thurnham, 2012).

Riboflavin serves as a precursor for the coenzymes flavin adenine dinucleotide (FAD) and flavin mononucleotide (FMN) and, therefore, plays a key role in energy metabolism, especially metabolism of fats, ketone bodies, carbohydrates, and proteins, as well as drug metabolism. Riboflavin is involved in supporting the immune and nervous system, forming red blood cells, producing cells, and activating folate and pyridoxine. Riboflavin also has a powerful antioxidant potential derived from its role as a precursor to FMN and FAD. Consequently, riboflavin deficiency is associated with increased lipid peroxidation. Because flavin coenzymes are involved in the metabolism of folic acid, pyridoxine, vitamin K, and niacin, a profound deficiency of riboflavin has consequences not only for enzyme systems that require flavin coenzymes but also

TABLE 7.5 Menaquinone-Producing Bacterial Species Commonly Used in Industrial Food Fermentation (Furuichi et al., 2006; Morishita et al., 1999; Walther et al., 2013)

Species/Subspecies	Food Use
Lactococcus lactis ssp. *lactis*	Cheese, buttermilk, sour cream, cottage cheese, cream cheese, kefir
Lactococcus lactis ssp. *cremoris*	Cheese, buttermilk, sour cream, cottage cheese, cream cheese, kefir, yogurt
Lactococcus plantarum	No information
Lactococcus raffinolactis	No information
Leuconostoc lactis	Cheese
Leuconostoc mesenteroides ssp. *cremoris*	No information
Leuconostoc mesenteroides ssp. *dextranicum*	No information
Brevibacterium linens	Cheese
Brochonthrix thermosphacta	Meat
Hafnia alvei	Cheese
Staphylococcus xylosus	Dairy, sausage
Staphylococcus equorum	Dairy, meat
Arthrobacter nicotianae	Cheese
Bacillus subtilis "natto"	Natto (fermented soybean)
Propionibacterium shermanii	Cheese
Propionibacterium freudenreichii	Cheese

many other enzyme systems linked with the aforesaid vitamins. Riboflavin deficiency is mostly related to other B-vitamin deficiencies and manifests in various symptoms, such as stomatitis including painful red tongue with sore throat, chapped and fissured lips (cheilosis), and inflammation of the corners of the mouth. Additional signs of deficiency are itchy, watery and red eyes, sensitivity to light and dermatitis, corneal vascularization, anemia, and brain dysfunction (Rivlin, 2006).

In 1970, van Veen and Steinkraus reported an increase of riboflavin in fermented food that was confirmed in 1984 by Murdock and Fields, who observed that the riboflavin content in fermented cornmeal was increased compared to the unfermented control. The changes in vitamin content mostly occur in the beginning of fermentation. The riboflavin concentration in cornmeal increased from 1.4 μg/g in the control to 2.9 μg/g after 1 day and to 4 μg/g after 2 days of fermentation. Similar results were produced in the fermentation of soybeans to tempeh and the fermentation of lentils, black gram, and rice by *Rhizopus* species and *Klebsiella pneumonia*, respectively (Ghosh and Chattopadhyay, 2011; van der Riet et al., 1987). In fermented dairy products, the riboflavin level can vary depending on the processing technology and the microorganisms utilized for fermentation. Whereas some yogurt starter cultures decrease riboflavin concentrations, others significantly increase the level of this vitamin compared to unfermented milk (Gulko and Kruglova, 1966; LeBlanc et al., 2011). In the past 15 years, biotechnological processes have replaced the traditional production of riboflavin with chemical processes (Stahmann et al., 2000). The screening of strains for riboflavin production resulted in a considerable number of species, such as *B. subtilis*, *Ashbya gossypii*, *Candida famata*, *Corynebacterium ammoniagenes*, and several LAB. Among them, *Lactobacilli*, *Leuconostoc*, *Lactococci*, and *Propionibacterium* were the most promising. In other studies, mutants of isolated *L. lactis*, *L. plantarum*, and *Lactobacillus fermentum*, showed overexpression of the riboflavin biosynthesis genes. The *Lactococcus* species led to an approximately two- and threefold increase in vitamin B2 content in sourdough bread and a diet supplemented with *L. fermentum* or milk fermented with *L. fermentum* eliminated most physiological manifestations in riboflavin-depleted rats (Burgess et al., 2006; LeBlanc et al., 2011; Russo et al., 2014).

7.5 VITAMIN B12

Vitamin B12 (cobalamin) is a very complex molecule that contains a corrin ring with cobalt as the central atom. Vitamin B12 is the name used for several corrinoids that have the biological activity of cyanocobalamin. The vitamin is water soluble and sensitive to light as well as to oxidizing and reducing agents. Vitamin B12 is exclusively synthesized by microorganisms and is naturally found predominantly in animal-based food (meat, milk, fish, and eggs) but not in plant food. The vitamin B12 in animal tissue comes from bacteria. The recommended daily amount of vitamin B12 for adults is 2.4 μg/day (https://fnic.nal.usda.gov/sites/fnic.nal.usda.gov/files/uploads/RDA_AI_vitamins_elements.pdf). Vitamin B12 is a cofactor for the two enzymes methionine synthase and L-methylmalonyl-CoA mutase and thus plays an essential role in amino acid and fatty acid metabolism, and in DNA and hemoglobin synthesis. The main deficiency syndrome of vitamin B12 is pernicious anemia, which is characterized by the two primary symptoms megaloblastic anemia and neuropathy (Gille and Schmid, 2014; Office of Dietary Supplements, 2011; Stahl and Heseker, 2007b).

Several studies showed that fermentation of milk leads to decreased concentrations of vitamin B12. Alm (1982) documented vitamin B12 losses in all fermented products analyzed (buttermilk, yogurt, kefir, ropy milk, acidophilus milk, bifidus milk). Vitamin B12 concentration in yogurt decreased progressively over an incubation period of 5 h (Reddy et al., 1976). The depletion continued when the fermented milks were stored at 4°C for 14 days and decreases of 40–60% were seen overall (Arkbage et al., 2003). These losses are predominantly attributable to the addition of LAB since they require vitamin B12 for growth. Various studies have documented that the traditional yogurt starter cultures, *S. thermophilus* and *L. bulgaricus*, either do not influence or decrease vitamin B12 content (Friend et al., 1983; Kneifel et al., 1989; Rao et al., 1984). In cheese, lower vitamin B12 levels compared to milk are mainly due to the removal of the whey fraction, whereas the fermentation with starter cultures does not significantly affect vitamin B12 content (Arkbage et al., 2003). However, Karlin (1969) documented increases in vitamin B12 content in Gruyère cheese, which may be due to the application of propionic bacteria. Lactic acid fermentation of cornmeal produced significantly higher vitamin B12 concentrations compared to the nonfermented control (3.51 μg/100 g versus 0.06 μg/100 g after 96 h fermentation) (Murdock and Fields, 1984).

The following genera have been reported to produce vitamin B12: *Aerobacter, Agrobacterium, Alcaligenes, Azotobacter, Bacillus, Clostridium, Corynebacterium, Flavobacterium, Micromonospora, Mycobacterium, Nocardia, Propionibacterium, Protaminobacter, Proteus, Pseudomonas, Rhizobium, Salmonella, Serratia, Streptomyces, Streptococcus*, and *Xanthomonas* (Martens et al., 2002; Perlman, 1959). The dairy *Propionibacterium* strains have long been known for their vitamin B12 production capacity. In contrast to other microorganisms used for industrial vitamin B12 production, such as *Pseudomonas denitrificans* and *Bacillus megaterium, Propionibacteria* have the advantage that they are generally recognized as safe (GRAS status) and thus can be applied to food (Hugenholtz et al., 2002; Martens et al., 2002). In Swiss-type cheeses, *Propionibacteria* are added for the characteristic flavor and eye formation. Furthermore, they have been used as adjunct cultures for increasing the vitamin B12 concentrations in several fermented foods, such as kefir (Van Wyk et al., 2011), yogurt (Amrutha, 2010), kimchi (Ro et al., 1979), and fermented vegetables (Babuchowski et al., 1999). *Propionibacterium freudenreichii* subsp. *shermanii* strains have been shown to produce vitamin B12 in concentrations ranging from 0.7 to 31.7 μg/mL (Poonam et al., 2012). Cerna and Hrabova (1977) reported that the addition of *P. freudenreichii* subsp. *shermanii* to starter cultures of various milk beverages, increased the vitamin B12 content in these products. Furthermore, several *Propionibacterium* strains were investigated for their vitamin B12 production capacity on a whey medium, and the amounts of vitamin B12 produced varied (Staniszewski and Kujawski, 2007). LAB might also be able to synthesize vitamin B12. *L. reuteri* was shown to produce a cobalamin-like compound and to revert vitamin B12 deficiency in an experimental model of female mice. The latter demonstrates the bioavailability of the vitamin produced by this strain (LeBlanc et al., 2011; Molina et al., 2009; Taranto et al., 2003).

7.6 OTHER VITAMINS

A few reports have shown that in addition to riboflavin, folate, and vitamin B12, other B group vitamins can be produced by fermentation. Increased levels of thiamine, pyridoxine, niacin, and pantothenic acid have been reported to be the result of LAB fermentation in yogurt, buttermilk, cheese, and other fermented products (Kneifel et al., 1989, 1991, 1992; Shahani and Chandan, 1979). However, other studies did not confirm these findings (Alm, 1982; Friend et al., 1983; Kneifel, 1989; Reddy et al., 1976). The differences found by the investigators may reflect variations in the metabolic activity of the fermenting microorganisms. Hou et al. (2000) showed *B. longum* had a positive influence on thiamine concentration in milk (11% increase after 48 h of fermentation). In soy milk fermented with pure and mixed cultures of *S. thermophilus*, *L. helveticus*, and *B. longum*, no changes in pyridoxine concentrations and only slight (not statistically significant) increases in thiamine were documented (Champagne et al., 2010). During lactic acid fermentation of cornmeal up to 96 h, the concentration of several B vitamins was analyzed. In all fermented samples, decreased concentrations of pyridoxine were found. At 24 h, significantly less thiamine and choline were contained in the fermented samples, but the concentrations were restored at later time points. Samples fermented for 96 h showed significantly increased concentrations of pantothenic acid (1.34 versus 0.93 mg/100 g) (Murdock and Fields, 1984).

7.7 CONCLUSIONS

It is difficult to draw a general conclusion about the vitamin content of fermented foods and the vitamin-synthesizing capacity of microorganisms because substantial variations depending on the strain used, the processing conditions applied, and the food matrix chosen have been documented. However, with a judicious selection of microbial species and cultivation conditions, concentrations of specific vitamins, such as riboflavin, vitamin B_{12}, folate, and vitamin K can significantly and naturally be increased in fermented food. Genetic modification of microorganisms promises an additional increase in their vitamin-synthesizing capacity. Using fermentation to increase vitamin concentrations in food is a natural, cost-effective, and better accepted alternative to current vitamin fortification programs. Fermented foods with increased vitamin content may assist in covering the vitamin requirements in populations with unbalanced diets and thus contribute to improving human health.

REFERENCES

Alm, L., 1982. Effect of fermentation on B-vitamin content of milk in Sweden. Journal of Dairy Science 65, 353–359.

Amrutha, N., 2010. Studies on the Vitamin B12 Enrichment of Yoghurt Using *Lactobacillus plantarum* and *Propionibacterium freudenreichii*. CFTRI, India.

Arcot, J., Wong, S., Shrestha, A.K., 2002. Comparison of folate losses in soybean during the preparation of tempeh and soymilk. Journal of the Science of Food and Agriculture 82 (12), 1365–1368.

Arkbage, K., Witthöft, C., Fondèn, R., Jägerstad, M., 2003. Retention of vitamin B12 during manufacture of six fermented dairy products using a validated radio protein-binding assay. International Dairy Journal 13 (2–3), 101–109.

Babuchowski, A., Laniewska-Moroz, L., Warminska-Radyko, I., 1999. Propionibacteria in fermented vegetables. Le Lait 79, 113–124.

Bailey, L.B., Gregory, J.F., 2006. Folate. In: Bowman, B.A., Russell, R.M. (Eds.), Present Knowledge in Nutrition, vol. 1. ninth ed. ILSI, Washington, DC, pp. 278–301.

Bentley, R., Meganathan, R., 1982. Biosynthesis of vitamin K (menaquinone) in bacteria. Microbiological Reviews 46 (3), 241–280.

Booth, S.L., Lichtenstein, A.H., O'Brien-Morse, M., McKeown, N.M., Wood, R.J., Saltzman, E., Gundberg, C.M., 2001. Effects of a hydrogenated form of vitamin K on bone formation and resorption. American Journal of Clinical Nutrition 74 (6), 783–790.

Booth, S.L., 2009. Roles for vitamin K beyond coagulation. Annual Review of Nutrition 29, 89–110.

Burgess, C.M., Smid, E.J., Rutten, G., van, S.D., 2006. A general method for selection of riboflavin-overproducing food grade micro-organisms. Microbial Cell Factories 5, 24.

Caplice, E., Fitzgerald, G.F., 1999. Food fermentations: role of microorganisms in food production and preservation. International Journal of Food Microbiology 50 (1–2), 131–149.

Cerna, J., Hrabova, H., 1977. Biologic enrichment of fermented milk beverages with vitamin B12 and folic acid. Milchwissenschaft – Milk Science International 32 (5), 274–277.

Champagne, C.P., Tompkins, T.A., Buckley, N.D., Green-Johnson, J.M., 2010. Effect of fermentation by pure and mixed cultures of *Streptococcus thermophilus* and *Lactobacillus helveticus* on isoflavone and B-vitamin content of a fermented soy beverage. Food Microbiology 27 (7), 968–972.

Collins, M.D., Jones, D., 1981. Distribution of isoprenoid quinone structural types in bacteria and their taxonomic implication. Microbiological Reviews 45 (2), 316–354.

Crittenden, R.G., Martinez, N.R., Playne, M.J., 2003. Synthesis and utilisation of folate by yoghurt starter cultures and probiotic bacteria. International Journal of Food Microbiology 80, 217–222.

Elder, S.J., Haytowitz, D.B., Howe, J.R., Peterson, J.W., Booth, S.L., 2006. Vitamin K contents of meat, dairy, and fast food in the US diet. Journal of Agricultural and Food Chemistry 54 (2), 463–467.

EFSA FEEDAP Panel (EFSA Panel on Additives and Products or Substances used in Animal Feed), 2014. Scientific opinion on the safety and efficacy of vitamin K3 (menadione sodium bisulphite and menadione nicotinamide bisulphite) as a feed additive for all animal species. EFSA Journal 12 (1), 1–29.

Farrand, S.K., Taber, H.W., 1974. Changes in menaquinone concentration during growth and early sporulation in *Bacillus subtilis*. Journal of Bacteriology 117 (1), 324–326.

Ferland, G., 2012. The discovery of vitamin K and its clinical applications. Annals of Nutrition and Metabolism 61 (3), 213–218.

Friend, B.A., Fiedler, J.M., Shahani, K.M., 1983. Influence of culture selection on the flavor, anti microbial activity, beta-galactosidase and B-vitamins of yogurt. Milchwissenschaft – Milk Science International 38 (3), 133–136.

Furuichi, K., Hojo, K., Katakura, Y., Ninomiya, K., Shioya, S., 2006. Aerobic culture of *Propionibacterium freudenreichii* ET-3 can increase production ratio of 1,4-dihydroxy-2-naphthoic acid to menaquinone. The Society for Biotechnology, Japan 101 (6), 464–470.

Ghosh, D., Chattopadhyay, P., 2011. Preparation of idli batter, its properties and nutritional improvement during fermentation. Journal of Food Science and Technology 48 (5), 610–615.

Gille, D., Schmid, A., 2014. Vitamin B12 in milch-und Fleischprodukten. Alimenta 10 (16), 32–33.

Gulko, L., Kruglova, L., 1966. Content of vitamins of the "B" complex in fermented milks. In: Proceedings Internationaler Milchwirtschaftskongress, pp. 689–693.

Hawkes, J.G., Villota, R., 1989. Folates in foods: reactivity, stability during processing, and nutritional implications. Food Science and Nutrition 28, 439–526.

Hiratsuka, T., Furihata, K., Ishikawa, J., Yamashita, H., Itoh, N., Seto, H., Dairi, T., 2008. An alternative menaquinone biosynthetic pathway operating in microorganisms. Science 321 (5896), 1670–1673.

Hojo, K., Watanabe, R., Mori, T., Taketomo, N., 2007. Quantitative measurement of tetrahydromenaquinone-9 in cheese fermented by propionibacteria. Journal of Dairy Science 90 (9), 4078–4083.

Holasova, M., Fiedlerova, V., Roubal, P., Pechacova, M., 2004. Biosynthesis of folates by lactic acid bacteria and propionibacteria in fermented milk. Czech Journal of Food Sciences 22 (5), 175–181.

Hoppner, K., Lampi, B., 1990. Total folate, pantothenic acid and biotin content of yogurt products. Canadian Institute of Food Science and Technology Journal 23 (4/5), 223–225.

Hou, J.W., Yu, R.C., Chou, C.C., 2000. Changes in some components of soymilk during fermentation with bifidobacteria. Food Research International 33 (5), 393–397.

Hugenholtz, J., Hunik, J., Santos, H., Smid, E., 2002. Nutraceutical production by propionibacteria. Le Lait 82 (1), 103–112.

IARC Working Group on the Evaluation of Carcinogenic Risks to Humans, 2000. Vitamin K substances. In: Some Antiviral and Antineoplastic Drugs, and Other Pharmaceutical Agents, IARC Monograph, vol. 76, pp. 417–486 (Lyon, France).

Iwamoto, J., Takada, T., Sato, Y., 2014. Vitamin K nutritional status and undercarboxylated osteocalcin in postmenopausal osteoporotic women treated with bisphosphonates. Asia Pacific Journal of Clinical Nutrition 23 (2), 256–262.

Iyer, R., Tomar, S., 2009. Folate: a functional food constituent. Journal of Food Science 74 (9), R114–R122.

Iyer, R., Tomar, S.K., 2011. Dietary effect of folate-rich fermented milk produced by *Streptococcus thermophilus* strains on hemoglobin level. Nutrition 27 (10), 994–997.

Jägerstad, M., Jastrebova, J., Svensson, U., 2004. Folates in fermented vegetables—a pilot study. LWT—Food Science and Technology 37 (6), 603–611.

Jägerstad, M., Piironen, V., Walker, C., Ros, G., Carnovale, E., Holasova, M., Nau, H., 2005. Increasing natural-food folates through bioprocessing and biotechnology. Trends in Food Science & Technology 16 (6–7), 298–306.

Kamao, M., Suhara, Y., Tsugawa, N., Uwano, M., Yamaguchi, N., Uenishi, K., Ishida, H., Sasaki, S., Okano, T., 2007. Vitamin K content of foods and dietary vitamin K intake in Japanese young women. Journal of Nutritional Science and Vitaminology 53 (6), 464–470.

Kariluoto, S., Aittamaa, M., Korhola, M., Salovaara, H., Vahteristo, L., Piironen, V., 2006. Effects of yeasts and bacteria on the levels of folates in rye sourdoughs 13. International Journal of Food Microbiology 106 (2), 137–143.

Karlin, R., 1969. Sur la teneur en folates des laits de grand mélange. Effets de divers traitements thermiques sur les taux de folates, B12 et B6 de ces laits. Internationale Zeitschrift für Vitaminforschung 39, 359–371.

Kneifel, W., Holub, S., Wirthmann, M., 1989. Monitoring of B-complex vitamins in yogurt during fermentation. Journal of Dairy Research 56, 651–656.

Kneifel, W., Erhard, F., Jaros, D., 1991. Production and utilization of some water-soluble vitamins by yogurt and yogurt-related starter cultures. Milchwissenschaft – Milk Science International 46 (11), 685–690.

Kneifel, W., Kaufmann, M., Fleischer, A., Ulberth, F., 1992. Screening of commercially available mesophilic dairy starter cultures: biochemical, sensory, and microbiological properties. Journal of Dairy Science 75 (11), 3158–3166.

Kneifel, W., 1989. Abbau und synthese von vitaminen während der fermentation von sauermilchprodukten. Milchwirtschaftliche Berichte Wolfpassing Rotholz 99, 110–115.

Koivu-Tikkanen, T.J., Ollilainen, V., Piironen, V.I., 2000. Determination of phylloquinone and menaquinones in animal products with fluorescence detection after postcolumn reduction with metallic zinc. Journal of Agricultural and Food Chemistry 48 (12), 6325–6331.

Kurosu, M., Begari, E., 2010. Vitamin K2 in electron transport system: are enzymes involved in vitamin K2 biosynthesis promising drug targets? Molecules 15 (3), 1531–1553.

Laino, J.E., LeBlanc, J.G., Savoy de, G.G., 2012. Production of natural folates by lactic acid bacteria starter cultures isolated from artisanal Argentinean yogurts. Canadian Journal of Microbiology 58 (5), 581–588.

Laino, J.E., Juarez del Valle, M., Savoy de Giori, G., LeBlanc, J.G.J., 2013. Development of a high folate concentration yogurt naturally bio-enriched using selected lactic acid bacteria. LWT—Food Science and Technology 54 (1), 1–5.

LeBlanc, J.G., Laino, J.E., del Valle, M.J., Vannini, V., van, S.D., Taranto, M.P., de Valdez, G.F., de Giori, G.S., Sesma, F., 2011. B-group vitamin production by lactic acid bacteria–current knowledge and potential applications. Journal of Applied Microbiology 111 (6), 1297–1309.

Lin, M.Y., Young, C.M., 2000. Folate levels in cultures of lactic acid bacteria. International Dairy Journal 10 (5–6), 409–413.

Manoury, E., Jourdon, K., Boyaval, P., Fourcassie, P., 2013. Quantitative measurement of vitamin K2 (menaquinones) in various fermented dairy products using a reliable high-performance liquid chromatography method. Journal of Dairy Science 96 (3), 1335–1346.

Martens, J.H., Barg, H., Warren, M.J., Jahn, D., 2002. Microbial production of vitamin B12. Applied Microbiology and Biotechnology 58 (3), 275–285.

Molina, V.C., Medici, M., Taranto, M.P., de Valdez, G.F., 2009. *Lactobacillus reuteri* CRL 1098 prevents side effects produced by a nutritional vitamin B deficiency. Journal of Applied Microbiology 106, 467–473.

Mönch, S., Netzel, M., Netzel, G., Ott, U., Frank, T., Rychlik, M., 2015. Folate bioavailability from foods rich in folates assessed in a short term human study using stable isotope dilution assays. Food & Function 6 (1), 241–247.

Morishita, T., Tamura, N., Makino, T., Kudo, S., 1999. Production of menaquinones by lactic acid bacteria. Journal of Dairy Science 82 (9), 1897–1903.

Müller, H., 1993. Die Bestimmung der Folsäure-Gehalte von Lebensmitteln tierischer Herkunft mit Hilfe der Hochleistungsflüssigchromatographie (HPLC). Zeitschrift für Lebensmittel-Untersuchung und-Forschung 196, 518–521.

Murata, K., Miyamoto, T., Kokufu, E., Sanke, Y., 1970. Studies on the nutritional value of tempeh. III. Changes in biotin and folic acid contents during tempeh fermentation. The Journal of Vitaminology 16, 281–284.

Murdock, F.A., Fields, M.L., 1984. B-Vitamin content of natural lactic acid fermented cornmeal. Journal of Food Science 49 (2), 373–375.

Northrop-Clewes, C.A., Thurnham, D.I., 2012. The discovery and characterization of riboflavin. Annals of Nutrition and Metabolism 61 (3), 224–230.

Novotny, J.A., Kurilich, A.C., Britz, S.J., Baer, D.J., Clevidence, B.A., 2010. Vitamin K absorption and kinetics in human subjects after consumption of 13C-labelled phylloquinone from kale. British Journal of Nutrition 104 (06), 858–862.

Nowicka, B., Kruk, J., 2010. Occurrence, biosynthesis and function of isoprenoid quinones. Biochimica et Biophysica Acta (BBA)—Bioenergetics 1797 (9), 1587–1605.
Office of Dietary Supplements, U.N.I.O.H., June 24, 2011. Vitamin B12-Dietary Supplement Fact Sheet. http://ods.od.nih.gov/factsheets/VitaminB12-HealthProfessional/.
Office of Dietary Supplements, U.N.I.O.H., December 14, 2012. Folate - Dietary Supplement Fact Sheet. http://ods.od.nih.gov/factsheets/Folate-HealthProfessional/.
Okano, T., Shimomura, Y., Yamane, M., Suhara, Y., Kamao, M., Sugiura, M., Nakagawa, K., 2008. Conversion of phylloquinone (Vitamin K1) into menaquinone-4 (Vitamin K2) in mice: two possible routes for menaquinone-4 accumulation in cerebra of mice. Journal of Biological Chemistry 283 (17), 11270–11279.
Perlman, D., 1959. Microbial synthesis of cobamides. Advances in Applied Microbiology 1, 87–122.
Pompei, A., Cordisco, L., Amaretti, A., Zanoni, S., Matteuzzi, D., Rossi, M., 2007. Folate production by bifidobacteria as a potential probiotic property. Applied and Environmental Microbiology 73 (1), 179–185.
Poonam, Pophaly, S.D., Tomar, S.K., De, S., Singh, R., 2012. Multifaceted attributes of dairy propionibacteria: a review. World Journal of Microbiology and Biotechnology 28 (11), 3081–3095.
Rao, D.R., Reddy, A.V., Pulusani, S.R., Cornwell, P.E., 1984. Biosynthesis and utilization of folic acid and vitamin B12 by lactic cultures in skim milk. Journal of Dairy Science 67, 1174–2269.
Reddy, K.P., Shahani, K.M., Kulkarni, S.M., 1976. B-complex vitamins in cultured and acidified yogurt. Journal of Dairy Science 59, 191–195.
Rivlin, R.S., 2006. Riboflavin. In: Bowman, B.A., Russell, R.M. (Eds.), Present Knowledge in Nutrition, ninth ed. ILSI Press, Washington, DC, pp. 250–259.
Ro, S.L., Burn, M., Sandine, W.E., 1979. Vitamin B12 and ascorbic acid in kimchi inoculated with *Propionibacterium freudenreichii* subsp. *shermanii*. Journal of Food Science 44 (873), 877.
Rosenberg, I.H., Godwin, H.A., 1971. Inhibition of intestinal γ–glutamyl carboxypeptidase by yeast nucleic acid: an explanation of variability in utilization of dietary polyglutamyl folate. Journal of Clinical Investigation 50, 78a.
Russo, P., Capozzi, V., Arena, M.P., Spadaccino, G., Duenas, M.T., Lopez, P., Fiocco, D., Spano, G., 2014. Riboflavin-overproducing strains of *Lactobacillus fermentum* for riboflavin-enriched bread. Applied Microbiology and Biotechnology 98 (8), 3691–3700.
van der Riet, W.B., Wight, A.W., Cilliers, J.J.L., Datel, J.M., 1987. Food chemical analysis of tempeh prepared from South African-grown soybeans. Food Chemistry 25 (3), 197–206.
Sato, T., Yamada, Y., Ohtani, Y., Mitsui, N., Murasawa, H., Araki, S., 2001. Efficient production of menaquinone (vitamin K2) by a menadione-resistant mutant of *Bacillus subtilis*. Journal of Industrial Microbiology and Biotechnology 26 (3), 115–120.
Sato, T., Schurgers, L., Uenishi, K., 2012. Comparison of menaquinone-4 and menaquinone-7 bioavailability in healthy women. Nutrition Journal 11 (1), 93–101.
Schurgers, L.J., Vermeer, C., 2000. Determination of phylloquinone and menaquinones in food–effect of food matrix on circulating vitamin K concentrations. Haemostasis 30 (6), 298–307.
Schurgers, L.J., Vermeer, C., 2002. Differential lipoprotein transport pathways of K-vitamins in healthy subjects. Biochimica et Biophysica Acta (BBA)—General Subjects 1570 (1), 27–32.
Schurgers, L.J., Teunissen, K.J., Hamulyak, K., Knapen, M.H., Vik, H., Vermeer, C., 2007. Vitamin K-containing dietary supplements: comparison of synthetic vitamin K1 and natto-derived menaquinone-7. Blood 109 (8), 3279–3283.
Shahani, K.M., Chandan, R.C., 1979. Nutritional and healthful aspects of cultured and culture containing dairy foods. Journal of Dairy Science 62, 1685–1694.

Shearer, M.J., Newman, P., 2008. Metabolism and cell biology of vitamin K. Thrombosis and Haemostasis 100 (4), 530–547.
Shearer, M.J., Fu, X., Booth, S.L., 2012. Vitamin K nutrition, metabolism, and requirements: current concepts and future research. Advances in Nutrition 3 (2), 182–195.
Shearer, M.J., 1995. Vitamin K. The Lancet 345 (8944), 229–234.
Stahl, A., Heseker, H., 2007a. Folat. Ernährungs-Umschau 54 (6), 336–343.
Stahl, A., Heseker, H., 2007b. Vitamin B12 (cobalamine). Ernährungs-Umschau 54 (10), 594–601.
Stahmann, K.-P., Revuelta, J.L., Seulberger, H., 2000. Three biotechnical processes using *Ashbya gossypii, Candida famata*, or *Bacillus subtilis* compete with chemical riboflavin production. Applied Microbiology and Biotechnology 53 (5), 509–516.
Staniszewski, M., Kujawski, M., 2007. The dynamics of vitamin B12 synthesis by some strains of propionic fermentation bacteria 16. Milchwissenschaft-Milk Science International 62 (4), 428–430.
Suttie, J.W., 1995. The importance of menaquinones in human nutrition. Annual Review of Nutrition 15, 399–417.
Sybesma, W., Starrenburg, M., Tijsseling, L., Hoefnagel, M.H., Hugenholtz, J., 2003. Effects of cultivation conditions on folate production by lactic acid bacteria. Applied and Environmental Microbiology 69 (8), 4542–4548.
Taranto, M.P., Vera, J.L., Hugenholtz, J., de Valdez, G.F., Sesma, F., 2003. *Lactobacillus reuteri* CRL1098 produces cobalamin. Journal of Bacteriology 185, 5643–5647.
Theuwissen, E., Magdeleyns, E.J., Braam, L.A.J.L., Teunissen, K.J., Knapen, M.H., Binnekamp, I.A.G., van Summeren, M.J.H., Vermeer, C., 2014. Vitamin K status in healthy volunteers. Food & Function 5, 229–234.
Urano, A., Hotta, M., Ohwada, R., Araki, M., 2015. Vitamin K deficiency evaluated by serum levels of undercarboxylated osteocalcin in patients with anorexia nervosa with bone loss. Clinical Nutrition 34 (3), 443–448.
Van Veen, A.G., Steinkraus, K.H., 1970. Nutritive value and wholesomeness of fermented foods. Journal of Agricultural and Food Chemistry 18 (4), 576–578.
Vermeer, C., Knapen, M.H., Schurgers, L.J., 1998. Vitamin K and metabolic bone disease. Journal of Clinical Pathology 51 (6), 424–426.
Van Wyk, J., Witthuhn, R.C., Britz, T.J., 2011. Optimisation of vitamin B12 and folate production by *Propionibacterium freudenreichii* strains in kefir. International Dairy Journal 21, 69–74.
Walther, B., Karl, J.P., Booth, S.L., Boyaval, P., 2013. Menaquinones, bacteria, and the food supply: the relevance of dairy and fermented food products to vitamin K requirements. Advances in Nutrition 4 (4), 463–473.
Wigertz, K., Svensson, U.K., Jagerstad, M., 1997. Folate and folate-binding protein content in dairy products. Journal of Dairy Research 64 (2), 239–252.
Wouters, J.T.M., Ayad, E.H.E., Hugenholtz, J., Smit, G., 2002. Microbes from raw milk for fermented dairy products. International Dairy Journal 12 (2–3), 91–109.

Chapter 8

From Bacterial Genomics to Human Health

A. Benítez-Páez, Y. Sanz
Institute of Agrochemistry and Food Technology, Spanish National Research Council (IATA-CSIC), Paterna-Valencia, Spain

8.1 INTRODUCTION

The first documented benefits of fermented milk dates back to the times of Elie Metchnikoff and Chalmers (1908), who established associations between the consumption of those products and longevity in rural populations of Bulgaria. These effects were further attributed to lactic-acid bacteria (LAB) acting as fermenters as well as to their metabolic activities, which may partly contribute to improve the digestibility and nutritional value of fermented milk. By that time, bifidobacteria were also isolated from healthy breast-fed infant feces by Henry Tissier, who suggested that they could prevent infections by displacing bacteria-causing colitis in breast-fed infants (reviewed in Sanz et al., 2016). These historical hints were considered the basis for the "probiotic" concept but whose definition and use was not harmonized until relatively recently. This term was first used by Lilly and Stillwell in 1965 to describe substances produced by bacteria that, unlike antibiotics, stimulate the growth of beneficial bacteria. Further, Roy Fuller defined probiotics as "live microbial feed supplements which beneficially affect the host animal by improving its intestinal microbial balance" in 1989. However, it was not until 2001 when an expert consultation group working under the umbrella of the Food and Agriculture Organization of the United Nations (FAO) and World Health Organization (WHO) agreed on a probiotic definition that was generally accepted by the scientific community. From this moment on, probiotics were defined as "live microorganisms that, when administered in adequate amounts, confer a health benefit on the host" (FAO/WHO, 2006). In 2002, a similar expert working group continued this task to generate the first guidelines for the evaluation of probiotics in food (FAO/WHO, 2006). The two documents generated by this consultation group constituted the first attempt to harmonize the definition as well as the criteria needed to for a microorganism to be considered probiotic. This definition implies that probiotics should be alive by the end of the shelf-life of the product and exert a measurable benefit on the host health at a defined dose. These consensus documents also highlight the need to identify microorganisms at the species and strain level by appropriate molecular methods and to evaluate their efficacy preferentially in randomized, double-blind, placebo-controlled trials in humans. Thirteen years after the definition of probiotics, another expert group working under the umbrella of the International Scientific Association for Probiotics and Prebiotics (ISAPP) has reviewed the probiotic concept in light of recent scientific advances (Hill et al., 2014). According to this last review, the original probiotic definition was maintained but it was also concluded that the term probiotic (generally attributed to strains of LAB and bifidobacteria) could also be applied to other commensal bacteria, which are inhabitants of our gut and may be identified as beneficial for health according to the corresponding safety and efficacy trials (Hill et al., 2014).

Undoubtedly, the probiotic definition has been useful to avoid the misuse of this term in scientific and commercial communications and to standardize the minimal requirements that this category of food or food ingredients should comply with. However, this definition is now too vague and does not enable consumers to identify different probiotic products and to inform them of their benefits, which generally vary from product to product. Furthermore, current regulations stipulate how the benefits of food and food ingredients, including probiotics, should be substantiated (Sanz et al., 2016). In fact, since the implementation of the new European Regulation of the European Parliament and the Council on Nutrition and Health Claims made on food on July 1, 2007 (EC No. 1924/2006) the situation has dramatically changed. According to this new legislative framework, the messages about the healthy properties of food or food ingredients should refer to a specific strain or strain combination (not to the general term probiotic) and to a specific (nongeneric) health benefit previously proven in humans and authorized by the European Commission and the Parliament on the basis of a scientific assessment carried out by the Panel on Dietetic Products, Nutrition, and Allergies (NDA) of the European Food Safety Authority (EFSA).

In spite of the large number of publications reporting health benefits of probiotics or fermented milk containing probiotics, only one health claim related to live yogurt cultures and its ability to improve lactose digestion has been approved under the current EU regulation of health claims made on food. Therefore further research should be reorientated, taking into account the evaluation criteria applied by authoritative bodies to improve the quality of studies that should prove the efficacy and mechanisms of action to make possible the substantiation of other potential beneficial effects. Further advances in the understanding of the complexity of the gut microbiota and the effector strains that impact human health could also help to select a next generation of probiotics with enhanced efficacy (Neef and Sanz, 2013). Here, we review the role of fermented milk as a source of live bacteria that may bring health benefits, the role of genomics in uncovering critical traits of current and future probiotic candidates, and the information provided by gut ecological studies to identify new probiotics and their functions.

8.2 FERMENTED MILK AS A SOURCE OF BENEFICIAL BACTERIA

Most of the currently commercialized probiotics for human consumption are still LAB and bifidobacteria, selected from biological samples and most often derived from fermented food due to their long history of safe use in food and their feasible production and technologic adaptation (Bernardeau et al., 2008; Gotteland et al., 2014). Fermented milk also constitutes one of the most attractive food matrices to deliver beneficial live bacteria since it generally allows their growth or at least their survival. These products are also perceived as healthy food easily accepted by consumers of different age ranges. LAB are the prototype probiotics used in yogurt and fermented milk production including mainly *Lactobacillus* spp. and *Streptococcus thermophilus*. Indeed, the definition of yogurt implies the use of *Lactobacillus delbrueckii* subsp. *bulgaricus* and *S. thermophilus* as starter cultures. *Lactobacillus* is the most numerous group of LAB, containing close to 200 different species (Bull et al., 2013). Among them, the species *Lactobacillus acidophilus* has been the most commonly used for producing fermented milk other than yogurt (Bull et al., 2013). However, the range of *Lactobacillus* species used in dairy industry has been expanded to include *Lactobacillus rhamnosus*, *Lactobacillus reuteri*, *Lactobacillus plantarum*, *Lactobacillus casei*, *Lactobacillus johnsonii*, and *Lactobacillus paracasei*, which are the most frequent species present in commercial fermented milk today. In addition, *Bifidobacterium* spp. has also been recently incorporated into fermented milk formulations. For technical reasons, fermented milk was initially produced by inoculating a sole LAB species to facilitate the fermentation process; today, several strains are mixed to initiate the fermentation process or added at later stages to ensure their viability together with other ingredients that act as prebiotic fibers to improve survival and potential health effects (Donkor and Shah, 2008).

While LAB was once used only to produce safe fermented milk products with desirable taste and flavor, today the health benefits potentially brought by this food are a priority, in part due to both marketing and consumers demands. This has led to investigations in this area and thus scientific evidence to document such benefits. Some of the nutritional or health benefits attributed to yogurt and fermented milk and the bacteria responsible for those fermentations are briefly described in the next sections and illustrated in Fig. 8.1.

The starter cultures of yogurt have been acknowledged for their ability to provide both nutritional and health benefits. For example, yogurt starter cultures are known to alleviate the symptoms of lactose maldigestion when ingested in yogurt. The yogurt starter cultures produce the enzyme (β-galactosidase), which

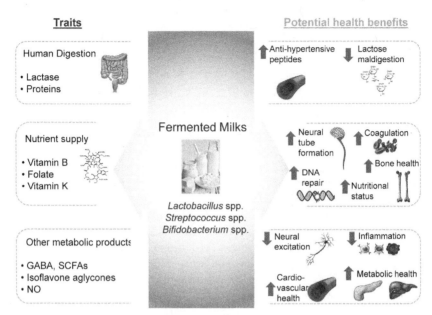

FIGURE 8.1 Schematic representation of the traits and the corresponding potential health benefits reported for LAB and bifidobacteria supplied in yogurt and fermented milks.

facilitates lactose hydrolysis and compensates for the decline in human lactase activity over time. Furthermore, the yogurt acts as a buffer, maintaining adequate pH for lactase activity and also slowing the gastrointestinal transit time, thus facilitating lactose digestion in the small intestine (Savaiano, 2014). In addition, yogurt starter cultures and other LAB and bifidobacteria involved in milk fermentation can provide micronutrients (eg, vitamins) essential to the human host (see Chapter 7 for more information). Folate is one of those micronutrients, which is indispensable for DNA synthesis and repair, and its deficiency is well-known to be associated with defects in fetal neural tube formation. For example, the addition of mixed starter cultures of *Streptococcus thermophilus* and *Bifidobacterium animalis* to fermented milk has been shown to increase up the milk folate content up to sixfold. Similar results were also obtained with the combination of *S. thermophilus* and *Lactobacillus delbrueckii* subsp. *bulgaricus*. In both cases, *S. thermophilus* appeared to be the major contributor to folate production, while the role of partner species remained unclear as to whether they were also folate producers or only inducers of folate synthesis by *S. thermophilus* (Crittenden et al., 2003; Laino et al., 2012). Bifidobacteria can also contribute to synthesizing B-group vitamins since transcriptomic analysis revealed the expression of those bifidobacterial genes to be involved in biosynthesis of several of those B-vitamins and folate (LeBlanc et al., 2013). Strains of LAB used for production of fermented milk are also known to be able to synthesize vitamin K, a cofactor involved in γ-carboxylation reactions associated with coagulation. Vitamin K comprises a group of fat-soluble and structurally related molecules, of which vitamin K_1 (or phytomenadione) is primarily found in oils from green leafy vegetables. In particular, some bacterial species are able to produce menaquinones (MKs), ie, vitamin K_2, although in lower proportions than vitamin K_1 produced by plants. *Lactococcus lactis* subsp *cremoris*, *Lactococcus lactis* subsp *lactis*, and *Leuconostoc lactis* have been showed to be quinone producers, producing up to 123 μg of MKs/L of fermented skim milk- or soy milk-based medium (Morishita et al., 1999). The supply of vitamin K via consumption of fermented milk by LAB could bring health benefits given its role as an enzyme cofactor for coagulation and its association with cardiovascular and bone health in epidemiological studies (reviewed in Walther et al., 2013).

Even though the prototype nutrients produced primarily by LAB and bifidobacteria during milk fermentation are vitamins and enzyme cofactors, the enrichment of fermented milk with other bioactive metabolites is also being investigated for the potential health benefits of these products. Production of γ-aminobutyric acid (GABA) by *L. plantarum*, *L. casei*, *L. acidophilus*, *L. bulgaricus*, *S. thermophilus*, and *Bifidobacterium longum* has been documented in previous studies (Tsai et al., 2006; Shan et al., 2015). Enrichment of dairy products with GABA could provide additional health benefits since it acts as neurotransmitter, reducing activation of the central nervous system and, thereby, stress and anxiety in adults (see Chapter 6 for more information). Moreover, recent studies have demonstrated that oral administration of GABA to mice with high-fat diet (HFD)-induced obesity protects against weight gain and improves glucose tolerance and insulin sensitivity and inhibits the obesity-related inflammation via upregulation of regulatory T cells, probably through activation of peripheral GABA receptors (Tian et al., 2011). Nevertheless, the extent to which GABA concentrations produced during milk fermentation by LAB could exert such effects in humans remains to be proven.

In experimental models, it has also been demonstrated that soy milk fermented with some LAB strains (*L. plantarum* TWK10 or *S. thermophilus* BCRC 14085) showed a greater transformation of glucoside isoflavones to the bioavailable forms (aglycones) via their glucosidase activities. These forms stimulated NO production via activation of endothelial NO synthase activity in human umbilical vein endothelial cells, which could contribute to regulating blood pressure and to cardiovascular protection (Cheng et al., 2013). In this regard, peptides generated during milk protein fermentation by LAB with angiotensin I-converting enzyme inhibitory activity has also been identified in yogurt and other fermented milk (Tsai et al., 2006; Shakerian et al., 2015), which could also exert antihypertensive effects (see Chapter 2 for more information).

In summary, fermented milk are sources of bacteria that facilitate nutrient digestion (lactose and proteins) and can theoretically provide vitamins/cofactors and other products derived from their metabolic activity (eg, neuroactive and antiinflammatory mediators, antihypertensive peptides, etc.), theoretically with the potential mental and cardiometabolic health effects.

In general, the compounds generated during milk fermentation and the bacterial strains contained in dairy products could exert synergic and complementary effects on human health, but their respective roles have been scarcely investigated. In human studies, the intake of milk beverages fermented by single or multiple LAB have been related to mild improvement of symptoms associated with chronic intestinal conditions, such as ulcerative colitis or irritable bowel syndrome (Simren et al., 2010; de Vrese et al., 2011; Ishikawa et al., 2011; Sondergaard et al., 2011). A few human intervention studies have also reported the ability of milk fermented products to restore the serum lipid profile in women (Andrade and Borges, 2009; Sadrzadeh-Yeganeh et al., 2010) and to reduce abdominal adiposity (Kadooka et al., 2010). Immunoregulatory and immune-stimulating effects have also been attributed to LAB and bifidobacteria, which when provided in the form of fermented milk could help to reduce incidence of infections as reported in a few studies in infants and the elderly (Sanz, 2016). Those beneficial properties could be attributed to both living organisms and/or to the fermented milk itself. In any case, currently there is still limited evidence to firmly support their efficacy in humans and further well-designed randomized control-trials should be conducted to make general recommendations regarding the benefits of their consumption (de Vrese et al., 2011).

8.3 HOW BACTERIAL GENOMICS HELP IN THE IDENTIFICATION OF PROBIOTIC HEALTH BENEFITS

Bacterial genomics have rapidly evolved in the last decade due to the second generation of DNA sequencing techniques commonly known as next-generation sequencing (NGS). Massive and parallel DNA sequencing has permitted the exploration of microbial diversity in-depth—to date, more than 45,000 bacterial genome sequencing projects have been publicly declared (Reddy et al., 2014). This vast genomic information available in public databases is currently poorly approached essentially due to a bottleneck in data processing and a poor rate of genome completion. Some studies have shown central features from bacterial genomes to be useful for understanding the potential beneficial functions of so-called classical or new probiotic candidates. Information retrieved from bacterial and eukaryote genome-sequencing projects has permitted deciphering a great variety of metabolic circuits where major molecular processes carried out inside cells are described in detail including enzymes reactions, whole metabolic and signaling pathways, end-products, secondary metabolites, and cofactors. All the above metabolic information rescued from genome analyses and integrated in a coherent manner is well exemplified in the KEGG database, ie, the Kyoto Encyclopedia of Genes and Genomes (Kanehisa et al., 2013). The KEGG is one of the most popular and useful databases regarding functional annotation and integration of metabolic and genomic information where dozens of pathways are hierarchically classified for comprehensiveness. Once a bacterial genome is fully or partially sequenced, functional analysis is required to determine if such a microorganism contains new functional traits of interest for their potential health effects via bioactive molecule production or their ability to eliminate toxic compounds from its ecological niche.

Classically, bacterial genomics applied to probiotic science is used to find elements potentially harmful for humans, including antibiotic resistance genes and mobile elements that could be involved in their horizontal transference. This is possible due to large compilations of genetic signatures well-known to be associated with virulence, pathogenicity, and resistance (Chen et al., 2012; Yoon et al., 2014). Currently, other important genomic features of probiotic strains are being investigated to understand their ability to produce beneficial molecules for human health, eg, vitamin production. As described previously, vitamins are key for all living cells where they act as enzyme cofactors in hundreds of pivotal metabolic reactions and are primarily obtained from exogenous sources. In recent years, components of gut microbiota have also been acknowledged for their ability to act as micromanufactories that can supply essential nutrients and other nonnutritional bioactive compounds (LeBlanc et al., 2013). Thus far, the ability of probiotic strains to produce group K and B-vitamins such as thiamine (B_1), riboflavin (B_2), niacin (B_3), pyridoxine (B_6), biotin (B_7), folate (B_9), and cobalamin (B_{12}) has been primarily studied given the well-known genetics of these biosynthetic pathways. In addition, genetic circuits encoding proteins/enzymes synthesizing or regulating the production of biomolecules, which impact immunity and gut physiology, such as short-chain fatty acids (SCFAs) and poly-unsaturated fatty acids (PUFAs), are also being investigated (Patterson et al., 2014).

Anaerobic fermentation of dietary fibers (oligo and polysaccharide) by certain gut microbiota components generates a variety of SCFAs, which depending on the types and concentrations, may exert health benefits on the host. For example, butyrate has been reported to be mainly utilized as an energy source by enterocytes, to exert an antiinflammatory role, and to contribute to the production of peptide hormones (GLP-1, GLP-2, etc.) that improve gut barrier function and glucose metabolism and induce satiety (den Besten et al., 2013). A recent intervention study in humans with inulin and propionate also showed that increased propionate intestinal levels reduces energy intake (Chambers et al., 2015). However, animal studies have associated excessive propionate production with aberrations in lipid metabolism, membrane fluidity, and neuroinflammation, which is involved in the pathogenesis of autism spectrum disorders (Thomas et al., 2012). Members of the phyla bacteroidetes and firmicutes are the main producers of SCFAs and can use different redox pathways to synthesize such products from a common substrate phosphoenolpyruvate (PEP) (den Besten et al., 2013). As the molecular machinery needed to produce SCFAs seems to be generalized among gut bacteria, genomics may contribute to understanding additional aspects of these metabolic circuits. Therefore potential probiotic strains exhibiting genomic variants at coding or regulatory regions of genes responsible for SCFAs could be further considered in order to better select those that increase the production of specific SCFAs or their relative proportions fitting more specifically the host physiological requirements. Furthermore, the investigation of gene expression patterns when bacterial strains are growing in the presence of different fermentable substrates in vitro represents an additional tool to select probiotic candidates. Overexpression of genes related to the production of SCFAs or other metabolic pathways as a result of the preferential utilization of certain carbon sources could be of interest for particular product formulations with pro- and prebiotics and for understanding the role of diet–microbe interactions on the generation of healthy metabolites (Yang et al., 2013). As a result, the genomes of LAB and bifidobacteria are recurrently investigated for sugar transporters and glycosidase gene content (Turroni et al., 2010). The number of such genes as well as their level of expression depending on the carbon source may help determine whether a probiotic candidate has good fitness based on its capability to metabolize a wide variety of complex carbon sources from human, animal, or vegetal origin. This will also help determine both its competitiveness and persistence in the host intestine.

Microbial genomics has also enabled the identification of the role of certain genetic elements of gut bacteria in regulating the host immune responses, beyond the effects likely exerted by metabolites, such as butyrate (Schell et al., 2002; Le Marechal et al., 2014). This type of analysis can also help to identify the presence of certain bacterial genes with potential adverse effects on human health. In this context, particular bacterial species or strains found in the human gut have been shown to activate or aggravate inflammatory response through different surface molecules recognized by the host immune system (Lakhdari et al., 2010; Deutsch et al., 2012). The presence of genes associated with antibiotic resistance in the genome of potential probiotic species, which can be transferred horizontally, could also constitute a serious safety issue. Those strains may become reservoirs of antibiotic resistance and facilitate their acquisition by intestinal pathogens (Ammor et al., 2008a,b; van Hoek et al., 2008). Therefore, genome sequencing and genome-wide studies on LAB used as starter cultures for manufacturing of dairy product or as probiotics is desirable to gain insight into the safety level of different microbial species and strains used for human nutrition. In addition, implementation of this type of genomic analysis as a working routine in characterization of LAB will help to obtain more information to better understand their traits and potential benefits to humans.

8.4 GUT ECOSYSTEM AS A NEW SOURCE OF BENEFICIAL BACTERIA

An increasing number of comparative studies of gut microbiota composition in health and disease population groups by high-throughput sequencing approaches is facilitating the identification of different bacterial taxa (eg, phylum, genus, or species) that are increased or reduced in a specific condition and, therefore, could be potentially associated with health. Nevertheless, one of the biggest challenges that remains in this area is how to progress from those associations established in observational studies between specific taxa and health or disease to cause–effect relationships. In other words, how to pinpoint those bacterial or bacterial combinations that alone or in conjunction are responsible for bringing benefits to the human host and their dependencies. This is indeed a complex question due to the complexity of the gut microbiota, the multiplicity and potential overlapping functions, and the biological variability across individuals. Furthermore, we are still unable to recover an important fraction of the bacteria inhabiting the human gut using classical microbiological methods, thus limiting the possibility of directly assessing their beneficial effects at species and strain level.

In this regard, most of the advances thus far have been carried by preclinical evaluation of bacterial species associated in human observational studies with a healthy phenotype. Examples of those potential next-generation probiotics are *Akkermansia muciniphila*, *Bacteroides uniformis*, and *Faecalibacterium prausnitzii*. *A. muciniphila* was recently identified as a mucus-degrading bacterium, which may represent between 3% and 5% of the total intestinal bacterial population of healthy human subjects. In a few observational studies the abundance of *A. muciniphila* was inversely correlated to body weight or to body weight gain and, therefore, its role in mouse models of obesity has recently been evaluated. In those studies, the administration of live *A. muciniphila* MucT (ATTC BAA-835) to HFD-fed mice normalized diet-induced metabolic endotoxemia, fat-mass gain, adipose tissue, and insulin resistance, while these effects were not detected with the heat-killed bacterium. From a mechanistic point of view, it has been suggested that this bacterium could exert those beneficial effects by restoring obesity-decreased mucus thickness, which could help reduce LPS (lipopolysaccharide) translocation to the bloodstream (endotoxemia). This bacterial strain also increased the intestinal endocannabinoid content (acylglycerols), which could also contribute to controlling inflammation, gut permeability, and gut peptide secretion (Everard et al., 2011).

B. uniformis CECT 7771 has also been recently evaluated in an obesity-based mouse model of a large number of observational studies reporting associations between an increased abundance of Bacteroidetes or *Bacteroides* spp. and a lean phenotype of body weight loss under dietary intervention (Sanz et al., 2013). This particular strain was also selected given the association established between its increased prevalence in breast-fed infants compared with formula-fed infants and the role attributed to breastfeeding in protecting from obesity (Sanchez et al., 2011; Gauffin Cano et al., 2012). In our preclinical trial, the administration of *B. uniformis* CECT 7771 ameliorated HFD-induced metabolic and immune dysfunction was associated with intestinal dysbiosis. In particular, *B. uniformis* CECT 7771 reduced body weight gain, liver steatosis and liver cholesterol and triglyceride concentrations, and increased small adipocyte numbers in HFD-fed mice. This strain also reduced serum cholesterol, triglyceride, glucose, insulin and leptin levels and fat absorption, and improved oral tolerance to glucose in obese mice. *B. uniformis* CECT 7771 also improved immune function of macrophages and dendritic cells, which are impaired in obesity (Gauffin Cano et al., 2012).

F. prausnitzii has also been reported to be reduced in many of the studies conducted on patients with chronic inflammatory bowel disease (including Crohn's disease and ulcerative colitis), although not in all. As a consequence, strains belonging to this species have been evaluated as potential probiotic candidates for chronic bowel inflammatory conditions in preclinical models. In this context, administration of several strains of *F. prausnitzii* has shown to protect against 2,4,6-trinitrobenzenesulfonic acid (TNBS)-induced colitis. Different mechanisms of action have been proposed to explain the antiinflammatory roles of these strains including the secretion of antiinflammatory compounds, the effects of its extracellular polymeric matrix, and the induction of Foxp3$^+$ regulatory T cells (Treg) and IL (interleukin)10 production (Qiu et al., 2013; Martin et al., 2015; Rossi et al., 2015).

Even though those preclinical trials are promising, further studies in humans are necessary to prove the efficacy and safety of the new probiotic candidates as well as production trials to evaluate the feasibility of their commercial exploitation in the form of food or supplements. An example in this field is pasteurized fermented milk by *Bacteroides xylanisolvens* DSM 23964, whose safety and tolerability have been evaluated in humans recently, presumably because of its possible commercial exploitation as a healthy product (Ulsemer et al., 2012). Furthermore, more intensive efforts are needed to capture the diversity of potential healthy synthetic microbiota that better cover the loss of functions that can be associated with different disorders in humans. The feasibility of the commercial exploitation of this next generation of probiotics as such or in the form of food/supplements (eg, fermented milk) still needs to be proven.

ACKNOWLEDGMENTS

This work was supported by the grant AGL2014-52101-P from the Spanish Ministry of Economy and Competitiveness (MINECO, Spain) and the grant 613979 (MyNewGut) from the European Union's Seventh Framework Program.

REFERENCES

Ammor, M.S., Florez, A.B., Alvarez-Martin, P., Margolles, A., Mayo, B., 2008a. Analysis of tetracycline resistance tet(W) genes and their flanking sequences in intestinal *Bifidobacterium* species. The Journal of Antimicrobial and Chemotheraphy 62, 688–693.

Ammor, M.S., Florez, A.B., van Hoek, A.H., de Los Reyes-Gavilan, C.G., Aarts, H.J., Margolles, A., Mayo, B., 2008b. Molecular characterization of intrinsic and acquired antibiotic resistance in lactic acid bacteria and bifidobacteria. Journal of Molecular Microbiology and Biotechnology 14, 6–15.

Andrade, S., Borges, N., 2009. Effect of fermented milk containing *Lactobacillus acidophilus* and *Bifidobacterium longum* on plasma lipids of women with normal or moderately elevated cholesterol. The Journal of Dairy Research 76, 469–474.

Bernardeau, M., Vernoux, J.P., Henri-Dubernet, S., Gueguen, M., 2008. Safety assessment of dairy microorganisms: the *Lactobacillus* genus. International Journal of Food Microbiology 126, 278–285.

Bull, M., Plummer, S., Marchesi, J., Mahenthiralingam, E., 2013. The life history of *Lactobacillus acidophilus* as a probiotic: a tale of revisionary taxonomy, misidentification and commercial success. FEMS Microbiology Letters 349, 77–87.

Chambers, E.S., Viardot, A., Psichas, A., Morrison, D.J., Murphy, K.G., Zac-Varghese, S.E., MacDougall, K., Preston, T., Tedford, C., Finlayson, G.S., Blundell, J.E., Bell, J.D., Thomas, E.L., Mt-Isa, S., Ashby, D., Gibson, G.R., Kolida, S., Dhillo, W.S., Bloom, S.R., Morley, W., Clegg, S., Frost, G., 2015. Effects of targeted delivery of propionate to the human colon on appetite regulation, body weight maintenance and adiposity in overweight adults. Gut 64, 1744–1754.

Chen, L., Xiong, Z., Sun, L., Yang, J., Jin, Q., 2012. VFDB 2012 update: toward the genetic diversity and molecular evolution of bacterial virulence factors. Nucleic Acids Research 40, D641–D645.

Cheng, C.P., Tsai, S.W., Chiu, C.P., Pan, T.M., Tsai, T.Y., 2013. The effect of probiotic-fermented soy milk on enhancing the NO-mediated vascular relaxation factors. Journal of the Science of Food and Agriculture 93, 1219–1225.

Crittenden, R.G., Martinez, N.R., Playne, M.J., 2003. Synthesis and utilisation of folate by yogurt starter cultures and probiotic bacteria. International Journal of Food and Microbiology 80, 217–222.

de Vrese, M., Kristen, H., Rautenberg, P., Laue, C., Schrezenmeir, J., 2011. Probiotic lactobacilli and bifidobacteria in a fermented milk product with added fruit preparation reduce antibiotic associated diarrhea and *Helicobacter pylori* activity. Journal of Dairy Research 78, 396–403.

den Besten, G., van Eunen, K., Groen, A.K., Venema, K., Reijngoud, D.J., Bakker, B.M., 2013. The role of short-chain fatty acids in the interplay between diet, gut microbiota, and host energy metabolism. Journal of Lipid Research 54, 2325–2340.

Deutsch, S.M., Parayre, S., Bouchoux, A., Guyomarc'h, F., Dewulf, J., Dols-Lafargue, M., Bagliniere, F., Cousin, F.J., Falentin, H., Jan, G., Foligne, B., 2012. Contribution of surface β-glucan polysaccharide to physicochemical and immunomodulatory properties of *Propionibacterium freudenreichii*. Applied and Environmental Microbiology 78, 1765–1775.

Donkor, O.N., Shah, N.P., 2008. Production of beta-glucosidase and hydrolysis of isoflavone phytoestrogens by *Lactobacillus acidophilus*, *Bifidobacterium lactis*, and *Lactobacillus casei* in soymilk. Journal of Food Science 73, M15–M20.

Everard, A., Lazarevic, V., Derrien, M., Girard, M., Muccioli, G.G., Neyrinck, A.M., Possemiers, S., Van Holle, A., Francois, P., de Vos, W.M., Delzenne, N.M., Schrenzel, J., Cani, P.D., 2011. Responses of gut microbiota and glucose and lipid metabolism to prebiotics in genetic obese and diet-induced leptin-resistant mice. Diabetes 60, 2775–2786.

FAO/WHO, 2006. Probiotics in food: health and nutritional properties and guidelines for evaluation. FAO Food Nutr Pap 85.

Gauffin Cano, P., Santacruz, A., Moya, A., Sanz, Y., 2012. *Bacteroides uniformis* CECT 7771 ameliorates metabolic and immunological dysfunction in mice with high-fat-diet induced obesity. PLoS One 7, e41079.

Gotteland, M., Cires, M.J., Carvallo, C., Vega, N., Ramirez, M.A., Morales, P., Rivas, P., Astudillo, F., Navarrete, P., Dubos, C., Figueroa, A., Troncoso, M., Ulloa, C., Mizgier, M.L., Carrasco-Pozo, C., Speisky, H., Brunser, O., Figueroa, G., 2014. Probiotic screening and safety evaluation of *Lactobacillus* strains from plants, artisanal goat cheese, human stools, and breast milk. Journal of Medicinal Food 17, 487–495.

Hill, C., Guarner, F., Reid, G., Gibson, G.R., Merenstein, D.J., Pot, B., Morelli, L., Canani, R.B., Flint, H.J., Salminen, S., Calder, P.C., Sanders, M.E., 2014. Expert consensus document. The International Scientific Association for Probiotics and Prebiotics consensus statement on the scope and appropriate use of the term probiotic. Nature Reviews. Gastroenterology and Hepatology 11, 506–514.

Ishikawa, H., Matsumoto, S., Ohashi, Y., Imaoka, A., Setoyama, H., Umesaki, Y., Tanaka, R., Otani, T., 2011. Beneficial effects of probiotic *Bifidobacterium* and galacto-oligosaccharide in patients with ulcerative colitis: a randomized controlled study. Digestion 84, 128–133.

Kadooka, Y., Sato, M., Imaizumi, K., Ogawa, A., Ikuyama, K., Akai, Y., Okano, M., Kagoshima, M., Tsuchida, T., 2010. Regulation of abdominal adiposity by probiotics (*Lactobacillus gasseri* SBT2055) in adults with obese tendencies in a randomized controlled trial. European Journal of Clinical Nutrition 64, 636–643.

Kanehisa, M., Goto, S., Sato, Y., Kawashima, M., Furumichi, M., Tanabe, M., 2013. Data, information, knowledge and principle: back to metabolism in KEGG. Nucleic Acids Research 42, D199–D205.

Laino, J.E., Leblanc, J.G., Savoy de Giori, G., 2012. Production of natural folates by lactic acid bacteria starter cultures isolated from artisanal Argentinean yogurts. Canadian Journal of Microbiology 58, 581–588.

Lakhdari, O., Cultrone, A., Tap, J., Gloux, K., Bernard, F., Ehrlich, S.D., Lefevre, F., Dore, J., Blottiere, H.M., 2010. Functional metagenomics: a high throughput screening method to decipher microbiota-driven NF-κB modulation in the human gut. PLoS One 5.

Le Marechal, C., Peton, V., Ple, C., Vroland, C., Jardin, J., Briard-Bion, V., Durant, G., Chuat, V., Loux, V., Foligne, B., Deutsch, S.M., Falentin, H., Jan, G., 2014. Surface proteins of *Propionibacterium freudenreichii* are involved in its antiinflammatory properties. Journal of Proteomics 113, 447–461.

LeBlanc, J.G., Milani, C., de Giori, G.S., Sesma, F., van Sinderen, D., Ventura, M., 2013. Bacteria as vitamin suppliers to their host: a gut microbiota perspective. Current Opinion in Biotechnology 24, 160–168.

Martin, R., Miquel, S., Chain, F., Natividad, J.M., Jury, J., Lu, J., Sokol, H., Theodorou, V., Bercik, P., Verdu, E.F., Langella, P., Bermudez-Humaran, L.G., 2015. *Faecalibacterium prausnitzii* prevents physiological damages in a chronic low-grade inflammation murine model. BMC Microbiology 15, 67.

Metchnikoff, É., Chalmers, M.P., 1908. The Prolongation of Life: Optimistic Studies. G. P. Putman's Sons, New York, p. 384.

Morishita, T., Tamura, N., Makino, T., Kudo, S., 1999. Production of menaquinones by lactic acid bacteria. Journal of Dairy Science 82, 1897–1903.

Neef, A., Sanz, Y., 2013. Future for probiotic science in functional food and dietary supplement development. Current Opinion in Clinical Nutrition and Metabolic Care 16, 679–687.

Patterson, E., Cryan, J.F., Fitzgerald, G.F., Ross, R.P., Dinan, T.G., Stanton, C., 2014. Gut microbiota, the pharmabiotics they produce and host health. The Proceedings of the Nutrition Society. 73, 477–489.

Qiu, X., Zhang, M., Yang, X., Hong, N., Yu, C., 2013. *Faecalibacterium prausnitzii* upregulates regulatory T cells and antiinflammatory cytokines in treating TNBS-induced colitis. Journal of Crohn's and Colitis 7, e558–e568.

Reddy, T.B., Thomas, A.D., Stamatis, D., Bertsch, J., Isbandi, M., Jansson, J., Mallajosyula, J., Pagani, I., Lobos, E.A., Kyrpides, N.C., 2014. The Genomes OnLine Database (GOLD) v.5: a metadata management system based on a four level (meta)genome project classification. Nucleic Acids Research 43, D1099–D1106.

Rossi, O., Khan, M.T., Schwarzer, M., Hudcovic, T., Srutkova, D., Duncan, S.H., Stolte, E.H., Kozakova, H., Flint, H.J., Samsom, J.N., Harmsen, H.J., Wells, J.M., 2015. *Faecalibacterium prausnitzii* strain HTF-F and its extracellular polymeric matrix attenuate clinical parameters in DSS-induced colitis. PLoS One 10, e0123013.

Sadrzadeh-Yeganeh, H., Elmadfa, I., Djazayery, A., Jalali, M., Heshmat, R., Chamary, M., 2010. The effects of probiotic and conventional yogurt on lipid profile in women. The British Journal of Nutrition 103, 1778–1783.

Sanchez, E., De Palma, G., Capilla, A., Nova, E., Pozo, T., Castillejo, G., Varea, V., Marcos, A., Garrote, J.A., Polanco, I., Lopez, A., Ribes-Koninckx, C., Garcia-Novo, M.D., Calvo, C., Ortigosa, L., Palau, F., Sanz, Y., 2011. Influence of environmental and genetic factors linked to celiac disease risk on infant gut colonization by Bacteroides species. Applied and Environmental Microbiology 77, 5316–5323.

Sanz, Y., Rastmanesh, R., Agostoni, C., 2013. Understanding the role of gut microbes and probiotics in obesity: how far are we? Pharmacological Research 69, 144–155.

Sanz, Y., Portune, K., Gomez, E.M., Benitez-Paez, A., 2016. Targeting the microbiota: considerations for developing probiotics as functional foods. In: Hyland, N., Stanton, C. (Eds.), The Gut-brain Axis: Dietary, Probiotic, and Prebiotic Interventions on the Microbiota. Academic Press, Oxford, UK, p. 640.

Sanz, Y., 2016. Bifidobacteria in foods: health effects. In: Caballero, B., Finglas, P., Toldra, F. (Eds.), The Encyclopedia of Food and Health. Academic Press, Oxford, pp. 388–394.

Savaiano, D.A., 2014. Lactose digestion from yogurt: mechanism and relevance. The American Journal of Clinical Nutrition 99, 1251S–1255S.

Schell, M.A., Karmirantzou, M., Snel, B., Vilanova, D., Berger, B., Pessi, G., Zwahlen, M.C., Desiere, F., Bork, P., Delley, M., Pridmore, R.D., Arigoni, F., 2002. The genome sequence of *Bifidobacterium longum* reflects its adaptation to the human gastrointestinal tract. The Proceedings of the National Academy of Sciences of the United States of America 99, 14422–14427.

Shakerian, M., Razavi, S.H., Ziai, S.A., Khodaiyan, F., Yarmand, M.S., Moayedi, A., 2015. Proteolytic and ACE-inhibitory activities of probiotic yogurt containing nonviable bacteria as affected by different levels of fat, inulin and starter culture. Journal of Food Science and Technology 52, 2428–2433.

Shan, Y., Man, C.X., Han, X., Li, L., Guo, Y., Deng, Y., Li, T., Zhang, L.W., Jiang, Y.J., 2015. Evaluation of improved γ-aminobutyric acid production in yogurt using *Lactobacillus plantarum* NDC75017. Journal of Dairy Science 98, 2138–2149.

Simren, M., Ohman, L., Olsson, J., Svensson, U., Ohlson, K., Posserud, I., Strid, H., 2010. Clinical trial: the effects of a fermented milk containing three probiotic bacteria in patients with irritable bowel syndrome - a randomized, double-blind, controlled study. Alimentary Pharmacology and Therapeutics 31, 218–227.

Sondergaard, B., Olsson, J., Ohlson, K., Svensson, U., Bytzer, P., Ekesbo, R., 2011. Effects of probiotic fermented milk on symptoms and intestinal flora in patients with irritable bowel syndrome: a randomized, placebo-controlled trial. Scandinavian Journal of Gastroenterology 46, 663–672.

Thomas, R.H., Meeking, M.M., Mepham, J.R., Tichenoff, L., Possmayer, F., Liu, S., MacFabe, D.F., 2012. The enteric bacterial metabolite propionic acid alters brain and plasma phospholipid molecular species: further development of a rodent model of autism spectrum disorders. Journal of Neuroinflammation 9, 153.

Tian, J., Dang, H.N., Yong, J., Chui, W.S., Dizon, M.P., Yaw, C.K., Kaufman, D.L., 2011. Oral treatment with γ-aminobutyric acid improves glucose tolerance and insulin sensitivity by inhibiting inflammation in high fat diet-fed mice. PLoS One 6, e25338.

Tsai, J.S., Lin, Y.S., Pan, B.S., Chen, T.J., 2006. Antihypertensive peptides and γ-aminobutyric acid from prozyme 6 facilitated lactic acid bacteria fermentation of soymilk. Process Biochemistry 41, 1282–1288.

Turroni, F., van Sinderen, D., Ventura, M., 2010. Genomics and ecological overview of the genus *Bifidobacterium*. International Journal of Food Microbiology 149, 37–44.

Ulsemer, P., Toutounian, K., Kressel, G., Schmidt, J., Karsten, U., Hahn, A., Goletz, S., 2012. Safety and tolerance of *Bacteroides xylanisolvens* DSM 23964 in healthy adults. Beneficial Microbes 3, 99–111.

van Hoek, A.H., Mayrhofer, S., Domig, K.J., Florez, A.B., Ammor, M.S., Mayo, B., Aarts, H.J., 2008. Mosaic tetracycline resistance genes and their flanking regions in *Bifidobacterium thermophilum* and *Lactobacillus johnsonii*. Antimicrobial Agents and Chemotheraphy 52, 248–252.

Walther, B., Karl, J.P., Booth, S.L., Boyaval, P., 2013. Menaquinones, bacteria, and the food supply: the relevance of dairy and fermented food products to vitamin K requirements. Advance in Nutrition: An International Review Journal 4, 463–473.

Yang, J., Martinez, I., Walter, J., Keshavarzian, A., Rose, D.J., 2013. In vitro characterization of the impact of selected dietary fibers on fecal microbiota composition and short chain fatty acid production. Anaerobe 23, 74–81.

Yoon, S.H., Park, Y.K., Kim, J.F., 2014. PAIDB v2.0: exploration and analysis of pathogenicity and resistance islands. Nucleic Acids Research 43, D624–D630.

Section 3

Traditional Fermented Foods

Section 3.1

Fermented Food of Animal Origin

Chapter 9

Fermented Seafood Products and Health

O. Martínez-Álvarez, M.E. López-Caballero, M.C. Gómez-Guillén, P. Montero
Institute of Food Science, Technology and Nutrition (ICTAN-CSIC), Madrid, Spain

9.1 INTRODUCTION

The origins of fermented fish in the human diet go back many thousands of years. There is evidence of the consumption of fermented fish during the Yayoi period (300 BC to AD 300) in Japan, in ancient Greece (the fish sauce *aimeteon*), and also in the Roman era (the fish sauce *garum*) (Ishige, 1993; Beddows, 1997). Today, many varieties of fermented seafood products are available and widely consumed around the world. Unlike fresh fish and seafood products, which are often too expensive in developing countries, traditional fermented seafood is comparatively cheap and well known. Fermented seafood products are used mainly used as flavor enhancers or as condiments. They can be found in almost every part of the world, although mainly in Asia, Africa, and Europe, where each nation has its own type of fermented products. Different examples are given in Table 9.1.

9.1.1 Fermented Seafood Products

Fermentation can be a natural (uncontrolled) or a controlled process in which starter cultures are added to the raw material (eg, marine fish, shellfish, and crustaceans). Fish fermentation also involves salting and ripening and occasionally smoking, marinating, and drying (Beddows, 1997). Depending on the proportion of salt added, the products can also be classified into high-salt (more than 20% of total weight), low-salt (6–8%), and no-salt products. During salting, the high concentration of salt reduces water activity and prevents microbial spoilage and the proliferation of pathogens in favor of halophilic and halotolerant microorganisms (Lopetcharat et al., 2001). During ripening, fish is subjected to endogenous proteases from the fish digestive system, enzymes from bacteria, yeasts, or molds, and occasionally exogenous proteolytic enzymes. The enzymes split proteins and

TABLE 9.1 Different Examples of Fermented Seafood Products Around the World

Name	Type	Raw Material	Preservatives	Country
Kapi	Paste	Shrimp	Salt + drying	Thailand
Kung-jom	Paste	Shrimp	Salt	Thailand
Nuoc-nam	Sauce	Anchovy	Salt	Vietnam, Cambodia
Nam-pla	Sauce	Anchovy	Salt	Thailand
Prahok	Paste	Mudfish	Salt	Cambodia
Cincaluk	Sauce	Shrimp, krill	Salt	Malaysia
Bagoong	Paste	Fish, krill,	Salt	Philippines
Feseekh	Whole	Boury fish	Salt	Egypt
Yeet	Whole	Sea snails or shellfish	Drying	Senegal
Gravlax	Whole (slices)	Salmon	Salt and sugar	Sweden
Surströmming	Whole	Lean herring	Salt	Sweden
Rakfisk	Whole (slices)	Trout	Salt	Norway

fats in the fish musculature into peptides, amino acids, ammonia, and free fatty acids that create a characteristic taste. In addition, the microorganisms produce metabolites (eg, organic acids, alcohols, esters, ketones, aldehydes, and other compounds) that contribute to the characteristic aroma and odor of fermented seafood products (Junus et al., 2010). Moreover, the texture of the salted fish becomes soft and tender.

There are many variations on the general scheme of preparation of fermented products. In the Southeast Asian subregion, fish fermentation normally lasts for several months up to one year. The final product is (1) a fish sauce (used as liquid with more or less viscosity), such as *nuoc-mam* (Vietnam) or *nam-pla* (Thailand); (2) a product consisting of small pieces of meat in a brown viscous paste, such as *ngapi* (Burma), *prahoc* (Kampuchea), or *bagoong* (Philippines), grounded (comminuted by either pounding or grinding) or ungrounded (Ishige, 1993); or (3) fish lactofermented with a source of carbohydrates (boiled/steamed rice, vegetables, and even sucrose), such as *budu* (Malaysia) and *mahyawa* (Gulf countries). The carbohydrates accelerate the fermenting process and also impart a distinctive flavor to the final product.

In Africa and Europe, different processing methods are uses in fish fermentation from region to region including fermentation with salting and drying, fermentation and drying without salting, and fermentation with salting but without drying (Anihouvi et al., 2012). In Europe, fish fermentation lasts from a few days to 1–2 years and in some cases additives, such as sugars and spices are added to the brine. The fermented products are commercialized as a whole (eg, *gravlaks* in Norway), or as a fish sauce. In Africa, the period for fish fermentation is short compared to Asia and Europe (from a few days to several weeks) and the product is not transformed into a paste or sauce.

9.1.2 Fermented Seafood Products and Health

Although fermentation has traditionally been used to preserve fresh fish, especially in tropical climates, today it is used to enhance nutritive value, improve appearance and taste, destroy undesirable factors, and also to reduce the energy needed for cooking (Paredes-López and Harry, 1988).

Fermentation can also help produce healthy compounds. However, there are no comprehensive studies on the effects on human health of fermented seafood products. Some seafood sauces contain taurine, which has been linked with several beneficial physiological actions including regulation of the oxidative stress, osmoregulation, antihypertension, neuromodulation, and hypoglycemic action (Saisithi et al., 1966; Park et al., 2005). Maillard reaction products (MRPs) exhibiting strong antioxidant activity can be formed by reaction of amino acids and peptides with reducing sugars during fermentation (Peralta et al., 2008). High coenzyme Q-content (a biomolecule that plays an important role in the electron-transport chain and as an antioxidant) has been found in *jeotgal*, a fermented product from Korea (Pyo and Oh, 2011). Moreover, the protein breakdown that occurs during fish fermentation liberates the amino acids and peptides with demonstrated bioactive properties as antioxidant (Faithong et al., 2010; Peralta et al., 2008; Binsan et al., 2008; Kleekayai et al., 2014), antihypertensive (Fujita et al., 2001; Je et al., 2005a; Itou et al., 2007), antiproliferative (Lee et al., 2004), hypoglycemic (Ichimura et al., 2003), immune-stimulating (Chowka and Fahey, 1998; Thongthai and Gildberg, 2005; Duarte et al., 2006), and anticoagulant properties (Kim et al., 2004b). Other biologically active molecules have been found in fermented seafood products, such as S-adenosyl-L-methionine, with interesting therapeutic applications in the treatment of depression, osteoarthritis, and liver disease (Lee et al., 2008) and bacteriocin-like compounds. Some lactic-acid bacteria (LAB) strains used to ferment seafood have also demonstrated healthy effects (Kim et al., 2002; Thapa et al., 2004). These LAB strains can be used for the design of probiotic formulations or to produce new fermented seafood products (Wong and Mine, 2004; Singh et al., 2014; Hwang et al., 2007).

Furthermore, different studies support the nutritional quality of fermented seafood products. They contain protein, amino acids, and polypeptides that contribute to health just like fish protein. They are also good sources of minerals (mainly calcium and iron), some B-group vitamins, peptides, and polyunsaturated fatty acids that may survive oxidation during prolonged fermentation because of the presence of natural antioxidants as free amino acids (FAAs) and some MRPs (Peralta et al., 2008).

However, the beneficial aspects of consumption of fermented seafood products can be limited due to the occurrence of several compounds associated with health risks, such as sodium chloride, histamine and others biogenic amines, and N-nitroso compounds (Ke et al., 2002; Junus et al., 2010). Although no one consumes large quantities of these products on a daily basis, improvement in the quality of these products must be achieved. To ensure safety, temperatures during fermentation should be kept below 10°C until the final pH is reached. Using good quality fish, washing the raw material before fermentation, and removing undesirable compounds are also necessary to reduce or eliminate health risks. The high amount of salt used in most fermented products should be reduced by electrodialysis (Chindapan et al., 2009) or by partial substitution of NaCl by KCl or sugar (Wongkhalaung, 2004). The use of starter cultures could also reduce the amount of salt used (Yin et al., 2002) to inhibit the growth of spoilage and pathogenic microorganisms and to degrade biogenic amines (Yin et al., 2002; Mah and Hwang, 2009). Biogenic amines could also be eliminated by using enzymes (diamine oxidase), yeasts (Wang et al., 2014), or bentonite (Shozen et al., 2012). Food-processing techniques as irradiation should be applied to degrade undesired compounds, such as histamine molecules (Kim et al., 2004b, 2005).

This chapter presents and discusses recent advances in the search for bioactive components in fermented seafood products with an interest in human health and emphasis on the technological approaches used to eliminate undesirable compounds associated with health risks.

9.2 FERMENTED FISH AND MICROORGANISMS

For fermented products to reach the industrial era they must reach a level of quality and safety without losing their character (Lan et al., 2013). In this context, innovation plays an important role since the microbiota of certain traditional products can be used to design new products with better nutritional quality. Fermented products are a source of high biological value material, sometimes even higher than unfermented source materials (Yuan et al., 2014). This section covers some aspects related to the use of microorganisms from fermented products in their role as starter cultures, antimicrobials, and/or probiotics.

9.2.1 Use of Microorganisms as Starters

Many of the microorganisms naturally present in fermented seafood products can be used as starters in traditional or commercial products. These microorganisms improve the acceptability of the product and meet some requirements, such as not favoring the production of biogenic amines. Other required properties are to be sensitive to antibiotics and/or to exert antimicrobial activity against some target pathogenic microorganisms.

The variety and diversity of traditional and commercial fermented products causes that the microorganisms related to these products are very innumerous. Many of them belong to the group of **LAB**, which includes the genus *Lactobacillus*, with wide metabolic activity and the ability to colonize several habitats (Giraffa et al., 2010). LAB including *Lactobacillus plantarum* and *Lactobacillus pentosus* were the predominant flora in *som-fak* (prepared with a mixture of fish, salt, rice, sucrose, and garlic) (Paludan-Müller et al., 2002). Zhao et al. (2014) identified *Lactococcus lactis* subsp. *lactis* as an excellent starter among the isolated microbiota from air-dried and fermented *Megalobrama amblycephala* (fermented fish from China). Out of the 61 cultures of LAB isolated from *chouguiyu* (a traditional Chinese fermented-fish product), *Lactobacillus sakei* was the predominant species during fermentation, while species of the genera *Lactococcus*, *Vagococcus*, *Enterococcus*, *Macrococcuss*, and *Staphylococcus* were also identified (Dai et al., 2013). *L. plantarum* and *Pediococcus pentosaceus* ZY40 GY23 were also used as starters in grass carp sausages (Nie et al., 2014). *Lactobacillus* and *Pediococcus* were also predominant in fermented *narezushi* (modern Japanese sushi) (Koyanagi et al., 2011). Different mixed starter cultures composed of *L. plantarum*-15, *Staphylococcus xylosus*-12, *P. pentosaceus*-ATCC33316, and *Lactobacillus casei* subsp. casei-1001 have been successfully used to improve the characteristics and functionality of minced silver carp muscle, in terms of flavor, digestibility, and nutritional value (Hu et al., 2008).

The isolation and characterization of **halophilic bacteria** or strongly halo-tolerant microorganisms could help standardize the quality of fermented seafood products, since the use of autochthonous cultures (ie, genus *Pediococcus*) to produce fish-type products could increase processing rates and product consistency, improve sensory characteristics and microbiological quality, and shorten fermentation time (Speranza et al., 2015). Zeng et al. (2015) showed that inoculation with autochthonous starter cultures (*Lactobacillus plantarum* 120, *L. plantarum* 145, and *Pediococcus pentosaceus* 220) reduced fermentation time and improved the quality of whole carp.

While the genus *Lactobacillus* dominates the market, it is not unusual to find **other species** in fermented products, as is the case for *Staphylococcus* spp., a major microorganism found in *cincaluk* (Malaysian fermented shrimp) (Hajar and Hamid, 2013). *Enterococcus faecium* and *Enterococcus faecalis*, bacteriocin-producing strains with no amino-acid decarboxylase activity, were isolated from raw seafood (shrimp and mussel) and Thai fermented seafood products including fermented shrimp (*kung-jom*), mussel (*hoi-dong*), and fish (*pla-jom*) (Nanasombat et al., 2012). In addition to *Lactobacillus*, Asiedu and Sanni (2002) reported the presence of *Pediococcus, Bacillus, Pseudomonas,*

Klebsiella, Staphylococcus, Debaromyces, and *Candida* genus in *enam nesetaakye,* a fermented fish-carbohydrate meal that also contains yam, onions, ginger, and salt. The coagulase-negative *Staphylococcus aureus* and *Micrococcus varians* are the predominant flora in *seedal* (traditional semifermented carp of Northeast India) (Muzaddadi and Basu, 2012). Yongsawatdigul et al. (2007) used several starter cultures prepared from *Virgibacillus* sp. SK33, *Virgibacillus* sp. SK37, and *Staphylococcus* sp. SK1-1-5 to produce fermented anchovy sauce, and reported that *Staphylococcus* can be used to accelerate fish-sauce fermentation without affecting the sensory characteristics of the final product.

Molds are also used as starters. For example, *Actinomucor elegans* XH-22 has been used to prepare *pehtze* surimi (Zhou et al., 2014), and koji molds (containing live *Aspergillus oryzae*) and douchi starter cultures have been used to ferment silver carp fish (Kasankala et al., 2012). Commercial molds starters (eg, SP-01, NY, M1 and kome miso) have also been used to obtain fish pastes commercially valuable from a nutritional and sensory point of view (Giri et al., 2009). The halo-tolerant starter microorganisms *Zygosaccharomyces rouxii, Candida versatilis, Tetragenococcus halophilus,* and barley koji (barley steamed and molded with *Aspergillus oryzae*) have also been used to obtain Chum salmon sauce mash (Yoshikawa et al., 2010).

In many cases, the microbiota of fermented seafood products evolve over time. Such is the case for *ngari* (fermented-fish product from India), whose flora differs greatly from the initial one and suffers a drastic bacterial community structural change during the sixth month of fermentation. Devi et al. (2015) identified *Tetragenococcus halophilus, Lactobacillus pobuzihii, Staphylococcus carnosus,* and *Bacillus indicus* are microorganisms that could effectively be used to design starter cultures. Kiyohara et al. (2012) showed that the *Lactobacillus* population drastically increased during fermentation of *narezushi* (fermentation of salted fish with rice), while the populations of other bacteria remained unchanged. Out of a total of 762 LAB isolated during the fermentation process of *plaa-som* (Thai fermented fish with rice, salt, and garlic), *Lactobacillus plantarum* dominated the final stages of fermentation, but in previous steps, *Lactococcus garvieae, Streptococcus bovis, Weissella cibaria, Pediococcus pentosaceus, L. plantarum,* and *Lactobacillus fermentum* were mainly identified (Kopermsub and Yunchalard, 2010).

In the search for the ideal starters, the susceptibility of microorganisms from fermented food products to antibiotics has been studied. Sornplang et al. (2011) isolated 10 LAB strains from *pla-chom* and three of them were sensitive to four antibiotics (penicillin, ampicillin, erythromycin, and tetracycline). These bacteria could therefore be candidates for starter cultures.

The study of microorganisms from fermented seafood has not been limited to the search for starters. Researchers are also interested in their use as **probiotics**. Kanno et al. (2012) found five isolates of LAB with the ability to ferment lactose and to scavenge DPPH$^-$ and/or O_2^- radicals in *saba-narezushi* (fermented chub mackerel with rice). Among these isolates, *Lactobacillus* *plantarum* 7FM10 and *Leuconostoc mesenteroides* 1RM3 showed a synergistic effect when grown together and could be used as probiotics starters. Kuda et al. (2014) isolated LAB with bile, salt, and acid resistance from the intestines of fish and fermented-fish samples, among which *L. plantarum, Lactococcus lactis,* and *Pediococcus pentosaceus* were identified. The activity of these strains showed their potential as starters or functional components with antiinflammatory and antioxidant properties. *Aji-narezushi* (salted and fermented horse mackerel with rice) and *kaburazushi* (salted and fermented turnip with yellow tail and malted rice) are sources of lactose, acidophilic, and antioxidative LAB (Kuda et al., 2010). Thapa et al. (2004) demonstrated wide diversity in the microbiota of *ngari, hentak,* and *tungtap* (fermented seafood products from Northeast India). They identified LAB (*L. lactis* subsp. *cremoris, L. plantarum, Enterococcus faecium, Lactobacillus fructosus, Lactobacillus amylophilus, Lactobacillus coryniformis* subsp. *Torquens,* and *L. plantarum*) that mostly showed a high degree of hydrophobicity (indicating their "probiotic" character). They also identified endospore-forming rods (*Bacillus subtilis* and *Bacillus pumilus*), aerobic coccal strains (*Micrococcus*), and yeasts (*Candida* and *Saccharomycopsis*). These strains were isolated and unable to produce biogenic amines (tyramine, cadaverine, histidine, and putrescine) in screening medium.

9.2.2 Antimicrobial Activity of Microorganisms From Fermented Seafood Products

The **antimicrobial activity** of many microorganisms used as starters in fermented seafood products is of great importance for ensuring the safety of products. This activity is mainly ascribed to the production of bacteriocins. Several strains of *L. plantarum* have been isolated from *suan-yu* (a Chinese traditional low-salt fermented whole fish) and mostly produced bacteriocins against *L. monocytogenes, Staphylococcus aureus,* and *Escherichia coli*. The isolates also mostly showed negative decarboxylase activities (Zeng et al., 2014). Abbasiliasi et al. (2010) reported the ability of a crude extract from *Lactobacillus paracasei* LA07 isolated from *budu* to inhibit the growth of indicator microorganisms (*Bacillus cereus, Lactococcus lactis, S. aureus, Salmonella enterica, L. monocytogenes, and E. coli*). The antimicrobial activity was completely inactivated when the crude extract was treated with proteinase K and α-amylase, confirming its proteinaceous nature and possibly glycosylated composition.

L. lactis subsp. *lactis* CWBI B1410 was used as a nisin-producing starter culture to improve the quality of *guedj* (a traditional Senegalese fermented-fish product). The presence of the bacteria and glucose contributed to the decrease in pH and reduction of enterobacteria in the final product (Diop et al., 2009). The same species (HKBT-9 strain) isolated from *hukuti maas* produced a bacteriocin active against foodborne pathogens and with specificity toward *Aeromonas* spp. (Kumar et al., 2014). *L. lactis* subsp. *lactis* USC-39, *Enterococcus faecium* USC-46, and *Enterococcus mundtii* USC-51 isolated from farmed turbot produced acidic and heat-resistant bacteriocins able to inhibit the growth of both *L. monocytogenes* and *Staphylococcus aureus*. These bacteriocins could be applied as biopreservatives in fermented and/or heated food products (Campos et al., 2006).

Rattanachaikunsopon and Phumkhachorn (2008) reported the ability of *Lactococcus lactis* subsp. *lactis* TFF 221 isolated from *kung jom* to produce nisin. The TFF 221 nisin showed antimicrobial activity against LAB but also against foodborne pathogens. This nisin could be used to improve the food safety of fermented products, especially in countries with problems of poor food hygiene. TFF 221 nisin was very similar to nisin A produced by *L. lactis* subsp. *lactis* NCDO 2111.

Pringsulaka et al. (2012) isolated six strains from Thai fermented-fish products showing antimicrobial activity against *Weissella confusa* N31, and only the isolate N23 (*Weissella cibaria* AB494716.1) was identified. Complete inactivation of the bacteriocin produced by *W. cibaria* N23 was observed after treatment with several enzymes (trypsin, actinase, etc.) and not with catalase. Wilaipun et al. (2008) described that *Staphylococcus hominis* KQU-131, isolated from Thai fermented marine fish, produced a heat-stable bacteriocin nukacin KQU-131, a variant of nukacin ISK-1, a type-A (II) lantibiotic. *Weissella cibaria* 110, isolated from the Thai fermented-fish product *plaa-som*, produced a bacteriocin (weissellicin 110) active against some gram-positive bacteria (Srionnual et al., 2007).

In clinical applications, bacteriocins isolated from fermented fish have been described as alternatives to the treatment of bovine mastitis and systemic bacterial infection eg, lacticin NK34 (produced by *Lactococcus lactis* isolated from *jeotgal*) (Kim et al., 2010) and pediocin SA131 (produced by *Pediococcus pentosaceus* SA131 isolated from *jeotgal*) (Lee et al., 2010).

In other cases, the inhibition of microorganisms occurs by mechanisms other than the production of bacteriocins (or at least, the different works do not mention as such). *Bacillus subtilis* JKK238 from *oh-jeot* (a fermented shrimp product from Korea) produces a wide variety of antifungal lipopeptide isomers from the iturin, fengycin, and surfactin families simultaneously (Yoon et al., 2005). Desniar et al. (2013) characterized 23 LAB isolates from *bekasam* (Indonesian fermented-fish product that tastes sour) with the potency to inhibit growth of the pathogenic bacteria *E. coli*, *Salmonella typhimurium* ATCC 14,028, *Bacillus cereus*, *Staphylococcus aureus*, and *L. monocytogenes*. Nevertheless, the neutralized supernatant of the LAB cultures did not inhibit pathogenic bacteria, which indicated that organic acids were responsible for the inhibition. Moreover, Saithong et al. (2010) isolated *Lactobacillus plantarum* IFRPD P15 and *Lactobacillus reuteri* IFRPD P17 from *plaa-som*. The former was dominant as an acidity producer, whereas the latter showed an ability to suppress and eliminate pathogenic bacteria, such as *E. coli* within 24 h.

The starter culture isolated from *ngari* consisted of three species of *Bacillus* and three species of *Micrococcus*, and no growth was detected for coliforms, *Salmonella*, and *E. coli* (Sarojnalini and Suchitra, 2009). In *ngari* (considered for its therapeutic value in stomach ulcers), *Enterococcus faecium* BDU7 has been isolated and characterized and its exopolysaccharide (EPS) had significant antioxidant properties, so this organism can be considered an efficient probiotic candidate (Abdhul et al., 2014). Hajar and Hamid (2013) reported that the major isolates from *cincaluk* showed Gram-positive cocci morphology and belonged to the group *Staphyloccoccus* spp. These isolates moderately inhibited the growth of several pathogenic strains, ie, *E. coli, Staphylococcus aureus, Salmonella typhimurium,* and *Bacillus subtilis,* which were used as indicator bacteria. These authors identified *Staphylococcus piscifermentans,* a rare strain previously reported to be specifically isolated from fish sources.

Pseudoalteromonas alienates KCCM 11,207P was isolated from traditional fermented Korean food with activity against *Vibrio harveyi,* a pathogenic microorganism associated with economic losses in aquaculture (Morya et al., 2014). Evaluation of the survival and potential growth of *Salmonella enterica* serovar Weltevreden, *S. enterica* serovar *Enteritidis, Vibrio cholera,* and *V. parahaemolyticus* in *som-fak* (Thai fermented-fish product) revealed that *S. Weltevreden* can grow independent of the inhibitory substances normally present in this type of product (sodium chloride, garlic, and lactic acid). Moreover, the use of garlic and *Lactobacillus plantarum* 509 was enough to prevent growth of the inoculated *Salmonella* in *som-fak* (Bernbom et al., 2009). LAB microbiota isolated from *budu* was identified and their activity was tested against *E. coli, Staphylococcus aureus, Salmonella thypi, Bacillus subtilis,* and *Listeria* monocytogenes (Yusra et al., 2013). In this work, *Bacillus cereus* strain HVR22 showed the highest antimicrobial activity.

Sometimes the inhibition of pathogenic microorganisms by native microbiota from fermented fish occurs as the result of several mechanisms combined. Østergaard et al. (1998) screened Thai fermented-fish products for LAB capable of inhibiting *Listeria* sp. Out of 4150 assumed LAB, 44 were further characterized and 43 strains inhibited *L. monocytogenes.* The strains inhibited other Gram-positive LAB probably due to production of bacteriocins. All 44 strains inhibited both *Vibrio cholerae* and *V. parahaemolyticus* and 37 also inhibited mesophilic fish spoilage bacterium (*Aeromonas* sp.). Inhibition of Gram-negative bacteria was attributed to production of lactic acid. Most strains were identified as *Lactobacillus* spp. Only 4 out of 44 isolates could degrade and ferment complex carbohydrates, such as rice, potatoes, and maize starch. This indicates that other types of bacteria may be responsible for the rapid spontaneous fermentation of the products or that some unknown factors ensure rapid fermentation.

9.3 BIOLOGICAL ACTIVITY IN TRADITIONAL FERMENTED SEAFOOD

Extensive protein breakdown occurring in fermented seafood products has been associated with the potential to promote consumer health since these products constitute a source of biologically active substances; however, in vivo studies and clinical trials are needed to validate the more common in vitro findings.

9.3.1 Antioxidant Activity

Uncontrolled free radicals are involved in cellular damage leading to corporal aging process and to a number of pathologies including atherosclerosis, arthritis, carcinogenesis, diabetes, neurodegenerative disorders, etc. (Halliwell, 1994). Antioxidants are molecules that can safely interact with free radicals and terminate the chain reaction before vital molecules are damaged. Noticeable free-radical-scavenging activity, as well as reduction of power and ion-chelating abilities, have been found in several traditional high-salt-fermented seafood products (Faithong et al., 2010; Peralta et al., 2008; Binsan et al., 2008; Kleekayai et al., 2014; Jung et al., 2005; Park et al., 1999). The large amount of small peptides released during fermentation is considered the main cause of their high antioxidative properties in a concentration-dependent manner (Binsan et al., 2008; Faithong et al., 2010). Compounds with masses less than 500 Da were predominantly present in water-soluble extracts of two fermented shrimp pastes (*kapi*), with ABTS radical (2,20-azinobis-(3-ethylbenzothiazoline-6-sulfonic acid)) scavenging capacity, expressed as EC_{50} values, of 60.9 and 46.4 µg protein/mL. These values were relatively close to the EC_{50} (15.3 µM) value reported for ascorbic acid, a well-recognized potent antioxidant compound (Kleekayakai et al., 2015).

Amino acids and peptides may function as primary antioxidants. However, upon prolonged fermentation they can also interact with reducing sugars present in fermented seafood to form desirable MRPs with strong antioxidant activity (Peralta et al., 2008). A close correlation between antioxidant activity and browning intensity was shown in *kapi* during the fermentation period (Faithong and Benjakul, 2014). The presumable presence of MRPs, as revealed by increased browning intensity and fluorescence, has also been noted in *mungoong*, a traditional Thai fishery product (Binsan et al., 2008). The thermally induced development of MRPs in food has been associated with decreased digestibility and toxic and mutagenic undesirable effects. However, their presence in traditionally fermented shrimp processed at ambient temperature would presumably be less toxic (Peralta et al., 2008). The accumulation of both antioxidant peptides and MRPs may retard lipid oxidation in fermented shrimp paste, as indicated by a gradual decrease in thiobarbituric reactive substances (TBARs) (Faithong and Benjakul, 2014). In this connection, important polyunsaturated fatty acids, such as eicosapentaenoic (EPA) and docosahexaenoic (DHA) in the shrimp paste, were not substantially damaged during long-term fermentation (Peralta et al., 2008). However, excessive fermentation may cause the breakdown of peptides and amino acids, as well as conversion of nitrogen compounds into ammonia, which is mainly responsible for fishy off-odor. A significant decrease in the radical-scavenging capacity and reduced browning intensity at the latest stages of fermentation in shrimp paste have been attributed to further peptide hydrolysis or polymerization, as well as MRPs interacting with peptides and other compounds (Peralta et al., 2008; Faithong and Benjakul, 2014). A similar effect was reported in blue-mussel sauce fermented for 12 months, where maximum antioxidant activity was found at month 6 (Jung et al., 2005).

A number of antioxidative peptides from traditional fermented seafood products have been isolated and characterized, following interest in recent years to obtain nonhazardous natural antioxidants. These peptides are able to exert different mechanisms of actions in free radical-mediated oxidative processes. The hepta-peptide sequence HFGNPFH (MW 962 kDa) in fermented blue-mussel sauce (Rajapakse et al., 2005) was found to be an effective radical scavenger, with IC_{50} values for DPPH, superoxide, hydroxyl, and carbon-centered radicals of 96, 21, 34, and 52 µM, respectively. The high scavenging properties were largely attributed to the presence of aromatic amino-acid and Histidine residues, as well as to the specific positioning of Phe–His in the peptide sequence. The presence of Histidine at the N-terminal also contributed greatly to the effective metal-ion chelation ability. This peptide showed lipid peroxidation inhibition higher than α-tocopherol and a ferric-ion chelating effect higher than citrate, which was also able to enhance the viability of cultured human-lung fibroblasts oxidized with t-BHP (Rajapakse et al., 2005). Another peptide sequence (FGHPY, MW 620 Da) was also identified in fermented blue-mussel sauce at the fermentation step where maximum antioxidant activity was recorded (6-month fermentation) (Jung et al., 2005). Although an IC_{50} value was not provided, the peptide showed high radical-scavenging capacity, as it scavenged 89.5% of hydroxyl radical at 64.8 µM and also proved to be more effective than α-tocopherol at inhibiting lipid peroxidation. A cyclic dipeptide, cyclo(His–Pro) (CHP), has been found in many processed foods, including dried shrimp and fish sauce, at much higher levels than previously reported in human plasma (Hilton et al., 1992). The authors suggested that the presence of this cyclic peptide in food is due to thermal manipulation of hydrolyzed proteins rather than released from endogenous proteolysis. There is evidence in the literature clearly demonstrating that dietary CHP can be absorbed from the GI tract. Originally discovered in the brain, CHP exerts multiple biological activities in the central nervous system, which have been related to antioxidant protection at the cellular level (Minelli et al., 2008). The dipeptide Trp–Pro, with EC_{50} 17.5 µM, was found to be a main contributor to the high radical-scavenging capacity of *kapi* (Kleekayakai et al., 2015). This Trp-containing dipeptide sequence may exhibit hypoglycemic activity via dipeptidyl peptidase IV (DPP-IV) inhibition (Nongonierma and FitzGerald, 2013), which could be of great therapeutic interest for patients with type 2 diabetes mellitus.

Coenzyme Q (CoQs) or ubiquinones are endogenous lipophilic enzyme cofactors, produced in all living cells in humans, that participate in the electron-transport chain for oxidative phosphorylation. The reduced form of CoQ10, ubiquinol-10, which represents more than 90% of the CoQ10 present in human serum and biological tissues, is a powerful antioxidant (Kaikkonen et al., 2002). CoQ10 deficiency has been implicated in several clinical disorders, including cardiovascular disease (Overvad et al., 1999) or neurological disorders (Shults and Haas, 2005). Noticeable high content of CoQ10 has been reported in *jeotgal*, with values ranging from 297 to 316 mg/kg in the different samples tested (Pyo and Oh, 2011). The CoQ10 content in the fermented products was considerably higher than in the unfermented ones, which was attributed to the contribution of the diverse microflora during the fermentation process.

Taurine, another biologically active compound that may regulate the oxidative stress, has been found in blue-mussel fermented sauce, accounting for around 10% of total FAA content (Park et al., 2005), as well as in salt-fermented shrimp paste (Peralta et al., 2008). Taurine has been associated with beneficial physiological actions, as it serves as a regulator of mitochondrial protein synthesis, thereby enhancing electron-transport chain activity and protecting the mitochondria against excessive superoxide generation (Jong et al., 2012). The muscle tissue of crustaceans and mollusks contains high levels of taurine (Carr et al., 1996), and it has also been found at high concentration in oyster cooker effluents (Kim et al., 2000). In the course of fermentation, taurine and other abundant FAAs tend to accumulate; however, at the very late stages a significant decrease may occur, due to extensive degradation and possible involvement in MRPs development (Peralta et al., 2008).

9.3.2 Antihypertensive Activity

Many fermented seafood products, especially long-term fermented-fish or shellfish sauces having high concentration of sodium chloride (20–25%), may not claim to be functional foods. However, they can be useful as a source of antihypertensive peptides (Ichimura et al., 2003). Noticeable ACE-inhibitory activity has been reported in a number of seafood sauces, such as salmon, sardine, and anchovy sauce, with IC_{50} values of 4.1, 2.7, and 17.5 µL/mL, respectively (Okamoto et al., 1995). The IC_{50} for fermented blue-mussel sauce was 1.01 mg/mL (Je et al., 2005a), whereas the IC_{50} of 2.45 mg/mL was found in the case of oyster sauce (Je et al., 2005b). Peptides with different molecular weights were purified from both mussel- and oyster-fermented sauces showing competitive inhibition against ACE, with IC_{50} values of 2.98 µM (in mussel sauce, MW 6.5 kDa) and 0.147 mM (in oyster sauce, MW 593 Da), both effectively reducing blood pressure in spontaneously hypertensive rats (SHRs) after oral administration (Je et al., 2005a,b).

Different types of fermented fish were also noted for ACE-inhibition capacity, eg, *rakfisk*, a Norwegian fermented trout muscle (Sørensen et al., 2004), *katsuobushi*, Japanese traditional dried bonito (Fujita et al., 2001), and *narezushi* and *heshiko*, Japanese fermented chub mackerel products made with low (5% NaCl) and high-salt content (≥20% NaCl), respectively (Itou et al., 2007). The oral administration of the *narezushi* water extract, with IC_{50} 0.06 mg/mL of ACE inhibition, effectively reduced the systolic blood pressure in SHRs (Itou et al., 2007). *Ngari* has also been proposed to be useful in alleviating hypertension at the expense of its noticeable ACE-inhibitory activity (Phadke et al., 2014). Potent ACE inhibition has been also reported for aqueous extracts of *kapi*, where compounds with molecular masses less than 500 Da were predominantly present (Kleekayai et al., 2014).

A number of peptide sequences with strong ACE-inhibitory activity have been identified in many of the products mentioned above. Several proline-containing ACE-inhibitory dipeptides were isolated in fish sauces from anchovy, sardine, or bonito (Ichimura et al., 2003). Due to the unique structure of Pro as iminoacid, peptide bonds containing Pro residues are often resistant to hydrolysis by common peptidases; this may be the reason some Proline-containing dipeptides survive even after long-term fermentation. In particular, Arg–Pro, Lys–Pro, and Ala–Pro from anchovy sauce showed the highest ACE-inhibitory activity, with IC_{50} of 22, 21, and 29 µM, respectively. The oral administration of Lys–Pro reduced the blood pressure of SHRs (Ichimura et al., 2003). Two dipeptides identified in *kapi*, Ser–Val and Ile–Phe, also exhibited ACE-inhibition capacity, with IC_{50} of 60.7 and 70.0 µM, respectively (Kleekayai et al., 2015). Naturally occurring ACE inhibitors raise the possibility that hypertension could be modulated through dietary intake. ACE-inhibitory peptides with weak in vitro activity may act as Pro drugs and produce a strong antihypertensive effect in vivo perhaps due to further endogenous enzymatic cleavage.

9.3.3 Anticoagulant and Fibrinolytic Activity

Management of heart disease and stroke has historically relied on the use of anticoagulant and antiplatelet drugs to prevent the formation of thrombus by an imbalance in hemostasis. A stable fibrin-clotting inhibitor, presumably a protein with molecular mass between 1 and 5 kDa, has been found in a salt-fermented anchovy sauce. The bioactive substance may act as an anticoagulant against fibrin clotting by protecting fibrinogen from the thrombin action (Kim et al., 2004b). Thrombolytic agents (fibrinolytic enzymes) can lyse preexisting thrombus. These enzymes can also prevent damage if the clot is removed soon after it occurs (Mine et al., 2005). Fibrinolytic enzymes can be found in a variety of fermented foods. In particular, fermented shrimp paste has shown strong fibrinolytic activity related to the presence of a novel fibrinolytic enzyme produced during the natural fermentation stage (Wong and Mine, 2004). The enzyme, with apparent molecular weight of 18 kDa and N-terminal amino-acid sequence of DPYEEPGPCENLQVA, was specific to fibrin and fibrinogen, without breaking down other blood proteins that are not involved in the blood-clotting cascade, such as bovine serum albumin, immunoglobulin G, thrombin, or hemoglobin. High fibrinolytic activity has been also reported in fermented small cyprinid fish (*Puntius sophore*), a traditional fermented product from Northeast India (Singh et al., 2014) as well as in traditional fermented shrimp, anchovy, and yellow corvine Korean *jeotgal* products (Hwang et al., 2007).

9.3.4 Other Biological Activities

Other biologically active molecules have also been found in fermented fish-derived products. Thus, *S*-adenosyl-L-methionine (SAM) can be produced by LAB (*Leuconostoc* and *Lactobacillus*) in *kimchi* containing shrimp *jeotgal* (Lee et al., 2008). SAM is a naturally occurring molecule distributed in body tissues and fluids that is involved in a number of biochemical reactions of enzymatic *trans*-methylation, contributing to the synthesis, activation, and/or metabolism of certain hormones, neurotransmitters, proteins, nucleic acids, and phospholipids. SAM could be considered an interesting bioactive material with therapeutic interest for the treatment of several disorders, such as liver disease or depression (Friedel et al., 1989).

A hydrophobic peptide fraction isolated from anchovy sauce was proposed to be highly feasible for the prevention of cancer, since it was shown to possess a strong antiproliferative effect against human lymphoma cell (U937) by inducing apoptosis ($IC_{50} = 31$ µg/mL). This peptide fraction presented a molecular weight of 440.9 Da and only comprised Ala and Phe residues at the ratio of approximately 4:1 (Lee et al., 2004).

Development of immune modulators from natural sources for diet supplementation is an active area of research, since most immune-modulatory pharmaceuticals are not suitable for chronic or preventive use. Small peptides with ACE-inhibitory activity might also exert an indirect immune-stimulating effect by enhancing the activity of bradykinin, which may mediate immune-stimulation and neurotransmission (Paegelow and Werner, 1986). Small-size peptides from commercial Thai anchovy fish sauce have shown the ability to stimulate the proliferation of human white blood cells at low concentration (5 µg/mL, Thongthai and Gildberg, 2005). The commercial dietary supplement Seacure, derived from fermented pacific whiting, served as an immune-modulating agent in a murine model and did not induce inflammatory immune response and morphological changes of the small intestine (Duarte et al., 2006). Free glutamine at high concentration and glutamine-containing di- and tripeptides, as well as several fatty-acid constituents, were the main compounds responsible for the biological activity of this commercial supplement (Fitzgerald et al., 2005).

9.4 HEALTH RISK

Fermented products from seafood can be a highly nutritious food due to its composition and its high digestibility, further accentuated during fermentation by release of healthy peptides from native proteins. However, other compounds associated with health risk, such as heavy metals, microorganisms, or other contaminating compounds may be present in the raw material used. High amounts of salt are also usually added to ferment fish and biogenic amines can be formed during the fermentation process. In recent years, different techniques or procedures have been reported to decrease or eliminate those factors that may negatively affect human health after consumption of fermented products.

9.4.1 Presence of Biogenic Amines and Strategies to Mitigate Their Presence

The biosynthesis of biogenic amines in fermented foods mostly proceeds through amino-acid decarboxylation carried out by LAB. Different articles cite the presence of biogenic amines, especially histamine, in different commercial fermented-fish products from Korea (Moon et al., 2010), Europe, Turkey (Köse et al., 2012), and China (Jiang et al., 2014). Although the recommended limits are not always exceeded, the levels of biogenic amines in fermented products are sometimes very high.

It is known that the presence of biogenic amines in diverse foods may have adverse effects in humans after consumption. Some of the hazardous effects described are rash, migraine, hypertension, and hypotension (Ten Brink et al., 1990). Moreover, they can be potential precursors of carcinogenic nitrosamines (Karovičová and Kohajdová, 2005). For this reason the presence of biogenic amines in various foodstuffs has been limited by different countries and/or organizations. For example, the Food and Drug Administration (FDA) recommends that the content of histamine, tyramine, and total biogenic amines in fish and seafood products be less than 50 mg/kg (FDA, 2002), 100 mg/kg, and 1000 mg/kg respectively, and less than 500 mg/kg in fish sauce (Brillantes and Samosorn, 2001). The European Union (EU) indicates that acceptable levels of histamine should be less than 400 mg/kg for fish sauce produced by fermentation of fishery products (Commission Regulation (EU) No 1019/2013). In Canada, Finland, Switzerland, and South Africa, the maximum amount of biogenic amines permitted in fish and seafood products is 100 mg/kg (Lange and Witmann, 2002; Auerswald et al., 2006). In Turkey and Australia, the limit for histamine in processed seafood products is 200 mg/kg (Köse et al., 2012; Auerswald et al., 2006).

The acid pH derived from LAB growth favors the activity of cathepsins present in muscle. As a result, some amino acids released from native proteins are used as substrate for microbial decarboxylase to produce amines. High temperatures also favor the bacterial activity (Gökoglu et al., 2004). The tendency to incorporate less salt content and/or carbohydrates in some fermented products as sauce or paste contributes to the increase in biogenic amines content (Köse et al., 2012). The quality of the fish used as raw material, the salting technique, and the time from capture until salt is added are key factors used to control the presence of amines in fish-sauce products (Brillantes et al., 2002).

Biogenic amines are heat stable and not destroyed even at autoclaving temperatures, which has encouraged the development of nonthermal strategies to eliminate or prevent the formation of biogenic amines in fermented fish and seafood. Thus physical methods such as the use of bentonite have been used for adsorption of histamine (Shozen et al., 2012). Irradiation has also been used for degrading histamine molecules, but it has the disadvantage that it can generate free-radical compounds (Kim et al., 2004a, 2005). Other strategies are based on the use of starter cultures with negative decarboxylase activity for controlling the growth of amine-positive bacteria. In this field, Hu et al. (2007) inoculated silver carp sausages with different mixed starter cultures of *Lactobacillus plantarum*-15, *Staphylococcus xylosus*-12, *Pediococcus pentosaceus*-ATCC33316, and *Lactobacillus casei* subsp. casei-1.001 and found inhibition of the growth of contaminant microorganisms present in the raw materials and also suppression of accumulations of histamine, cadaverine, putrescine, tryptamine, and tyramine. The use of LAB-producing bacteriocins is also of interest as an alternative to inhibit the growth of amine-producing bacteria in fermented products, as described earlier in this chapter. The use of halophilic microorganisms in fermented products to degrade biogenic amines has also gained attention in recent years. Biogenic amines can be degraded through oxidative deamination mainly catalyzed by amine oxidases (monoamine and diamine oxidases) with the production of ammonia, aldehyde, and hydrogen peroxide (Zaman et al., 2011). It has been observed that *S. xylosus* N°.0538 behaves as a protective culture as it shows a great ability to degrade histamine and tyramine. This strain can reduce levels of biogenic amines in 16% and may also produce bacteriocin-like substances with very high inhibitory activity against the amines producer *Bacillus licheniformis* (Mah and Hwang, 2009). *Bacillus amyloliquefaciens* FS-05 and *Staphylococcus carnosus* FS-19, isolated from fish sauces, were able to degrade histamine up to 59.9% and 29.1%, respectively, from its initial concentration (Zaman et al., 2010). These authors also isolated *Staphylococcus intermedius* FS-20 and *Bacillus subtilis* FS-12 from fish sauces and found that they were able to degrade putrescine and cadaverine up to 30.4% and 28.9%, respectively. Most of these isolates tolerated salt concentrations up to 15% and temperatures up to 45°C. The histamine-degrading strain *Natrinema gari*, isolated from fish sauce made in Thailand, showed high histamine-degrading activity at pH 6.5 to 8 in the presence of 3.5–5 M NaCl and 40–55°C (Tapingkae et al., 2010).

The use of new starter cultures for fermenting seafood products would thus substantially improve the safety and quality of these products, as previously reported. However, little research exists on the effect of starter cultures on the sensory attributes. Xu et al. (2010) used *Pediococcus pentosaceus* to ferment silver carp sausages and found a decrease in levels of salt- and water-soluble proteins and an increase in the amount of insoluble proteins, which could affect the sensorial properties of the product. The use of *L. plantarum* 120, *Lactobacillus plantarum* 145, to produce *suan-yu* led to low pH and rancidity (TBARS), inhibition of the growth of the main microflora, and decrease in total volatile base nitrogen (TVB-N) and FAAs. Moreover, the fermented products showed great acceptability from a sensory point of view (Zeng et al., 2015). The use of the yeast strain Omer Kodak M8 in fermented-fish sauce was effective at degrading biogenic amines (histamine and tyramine) and also improved the smell and taste of the final product (Wang et al., 2014). The taste of the fish sauce was improved due to enhancement of umami taste, cheese- and meat-like aroma, whereas fishy, ammonia, and rancid odors decreased.

9.4.2 Strategies to Decrease Salt Content

The high concentration of salt is an important health issue that limits the consumption of most traditional fermented products. Nonetheless, it should be noted that some salted and fermented products such as paste and sauces are used as condiments and thus are consumed in low quantities.

The use of lower amounts of salts in the manufacture of fermented products as well as the extraction of salt without affecting the product characteristics have been proposed to reduce the salt content in these foods. The use of lower amounts of salt must be accompanied by fast fermentation processing, either by incorporation of fish juice (Yu et al., 2014), koji (Giri et al., 2010; Xu et al., 2008), enzymes, etc. Xu et al. (2008) prepared a low-salt fermented sauce from squid byproducts (heads, viscera, skin, and fins), 8% salt, koji (20%), and Flavourzyme (0.2%) and found that the nutritional properties were improved without apparent deterioration. Moreover, the product did not show particularly strong or unpleasant aroma. Giri et al. (2010) obtained low-salt miso fish soups from the muscle of mackerel (washed or unwashed) and found that the use of washed muscle reduced the levels of some aldehydes, ketones, and nitrogen compounds, thereby providing pleasant aroma in the finished products by reduction of strong fishy odors.

The partial replacement of NaCl by KCl has been proposed to reduce the amount of NaCl in fermented products. Sanceda et al. (2003) partially substituted the initial quantity of NaCl by KCl to obtain fermented sardine sauce and found that the substitution of 25% NaCl by KCl did not significantly affect the acceptability of the final product.

Electrodialysis is an interesting strategy used to desalt fermented seafood products. Chindapan et al. (2009) desalted a Thai commercial fish sauce containing 25% salt by electrodialysis and observed that this technique reduced the final salt concentration by 7.1–13.7% with small loss of nitrogen compounds (14.5–17.1%), while also maintaining acceptable sensory properties. Studies carried out in Korean traditional sauces of fermented blue mussel have shown that the effectiveness in removing salt can be much higher and the initial 25% of NaCl can be decreased by up to 1% (Park et al., 2005). These authors also observed that the products with lower salt content contained a higher amount of certain amino acids, such as glycine, alanine, proline, aspartic acid, and glutamic acid, which influence the taste of the final product. Others methods based on particle size have been proposed to reduce the salt content in fermented products. However, they would not be suitable because compounds such as FAA could also be eliminated.

9.4.3 Strategies to Remove Heavy Metals

Fermented seafood products may contain heavy metals (eg, cadmium, arsenic, mercury, or lead) due to accumulation of these metals in internal organs, mainly in viscera. The removal of heavy metals would improve their quality, though studies in this field are limited. Several research works have described the use of bacteria, tannin, and cation-chelating resin to remove heavy metals from fermented products. Thus strains of halotolerant bacteria (genus *Staphylococcus* and *Halobacillus*) isolated from fermented shrimp and crabs have been used to reduce 30% of Cd content in fermented squid-gut sauce (Kawasaki et al., 2008). Tannin and cation-chelating resin have also been used to decrease the amount of Cd in fish sauce 13- and 16-fold times, respectively (Sasaki et al., 2013, 2015). The use of tannin or cation-chelating resin did not affect the biological activity of the final product (angiotensin I-converting enzyme inhibitory and antioxidant activities). Moreover, major nutritional components were maintained, including FAA and peptides.

9.4.4 Parasite Control

There are concerns about the presence of nematodes and other species of worms in fermented seafood products. The presence of Anisakis L3 larvae in fish and cephalopods has long been recognized as a potential human health risk (Oh et al., 2014). The FDA recommends freezing (times/temperatures) and thermal processing to kill Anisakis larvae in raw and undercooked seafood for seafood safety. However, the fermented seafood products are not subjected to thermal processing. Moreover, recommendations are general and fish and seafood are very diverse, so the specific conditions for inactivation of larvae by freezing can vary depending on the raw material used. Nonetheless, it is worth noting that the use of certain amounts of salt to prepare a fermented product should be enough to kill Anisakis larvae. Oh et al. (2014) studied the specific guidelines for manufacture of salt-fermented squid and Pollock tripe with regards to freezing and salting conditions in order to inactivate the Anisakis larvae. They found that larvae were inactivated after 48 h at -20°C, after 24 h at −40°C, and also after immersion in 15% NaCl for 7 days.

Other trematodes can also be found in fermented seafood products. Thus species in the genus *Opisthorchis* have been detected in low-salt fermented cyprinids from Thailand. Salt concentrations higher than 13.6% during manufacture would be effective for inactivation of the parasite, and irradiation could also be an effective control measure (Wilkinson, 1997).

9.5 CONCLUSION

A wide range of biological activities and beneficial LAB has recently been reported in traditional fermented seafood products, which may help improve consumers' opinion about the nutritional and healthy properties of these products. Much attention has also been paid to the health risk of consumption of fermented seafood products and different methodologies have been developed to increase their quality. Future research on fermented seafood products should be focused on the study of the healthy properties of some of the numerous products found in local markets around the world, as well as on new methods to improve quality. In conclusion, fermented seafood products can be healthy after adequate processing and can also be a valuable source of bioactive molecules and LAB and used as functional ingredients in foods.

REFERENCES

Abbasiliasi, S., Nagasundara Ramanan, R., Azmi Tengku Ibrahim, T., Shuhaimi, M., Rosfarizan, M., Ariff, A.B., 2010. Partial characterization of antimicrobial compound produced by *Lactobacillus paracasei* LA07, a strain isolated from Budu. Minerva Biotecnologica 22 (3–4), 75–82.

Abdhul, K., Ganesh, M., Shanmughapriya, S., Kanagavel, M., Anbarasu, K., Natarajaseenivasan, K., 2014. Antioxidant activity of exopolysaccharide from probiotic strain *Enterococcus faecium* (BDU7) from Ngari. International Journal of Biological Macromolecules 70, 450–454.

Anihouvi, V.B., Kindossi, J.M., Hounhouigan, J.D., 2012. Processing and quality characteristics of some major fermented fish products from Africa: a critical review. International Research Journal of Biological Science 1 (7), 72–84.

Asiedu, M., Sanni, A.I., 2002. Chemical composition and microbiological changes during spontaneous and starter culture fermentation of Enam Ne-Setaakye, a West African fermented fish-carbohydrate product. European Food Research and Technology 215 (1), 8–12.

Auerswald, L., Morren, C., Lopata, A.L., 2006. Histamine levels in seventeen species of fresh and processed South African seafood. Food Chemistry 98, 231–239.

Beddows, C.G., 1997. Fermented fish and fish products. In: Wood, B.B. (Ed.), Microbiology of Fermented Foods. Springer, US, pp. 416–440.

Bernbom, N., Ng, Y.Y., Paludan-Müller, C., Gram, L., 2009. Survival and growth of *Salmonella* and *Vibrio* in som-fak, a Thai low-salt garlic containing fermented fish product. International Journal of Food Microbiology 134 (3), 223–229.

Binsan, W., Benjakul, S., Visessanguan, W., Roytrakul, S., Tanaka, M., Kishimura, H., 2008. Antioxidative activity of Mungoong, an extract paste, from the cephalothorax of white shrimp (*Litopenaeus vannamei*). Food Chemistry 106 (1), 185–193.

Brillantes, S., Paknoi, S., Totakien, A., 2002. Histamine formation in fish sauce production. Journal of Food Science 67 (6), 2090–2094.

Brilliantes, S., Samosorn, W., 2001. Determination of histamine in fish sauce from Thailand using a solid phase extraction and high-performance liquid chromatography. Fisheries Science 67, 1163–1168.

Campos, C.A., Rodríguez, O., Calo-Mata, P., Prado, M., Barros-Velázquez, J., 2006. Preliminary characterization of bacteriocins from *Lactococcus lactis*, *Enterococcus faecium* and *Enterococcus mundtii* strains isolated from turbot (*Psetta maxima*). Food Research International 39 (3), 356–364.

Carr, W.E.S., Netherton III, J.C., Gleeson, R.A., Derby, C.D., 1996. Stimulants of feeding behavior in fish: analyses of tissues of diverse marine organisms. Biological Bulletin 190 (2), 149–160.

Chindapan, N., Devahastin, S., Chiewchan, N., 2009. Electrodialysis. Desalination of fish sauce: electrodialysis performance and product quality. Journal of Food Science 74, 363–371.

Chowka, P.B., Fahey, P., 1998. History and clinical applications of a promising new protein supplement. Townsend Letter for Doctors & Patients 6, 99–103.

Dai, Z., Li, Y., Wu, J., Zhao, Q., 2013. Diversity of lactic acid bacteria during fermentation of a traditional Chinese fish product, Chouguiyu (stinky mandarinfish). Journal of Food Science 78 (11), M1778–M1783.

Desniar, A., Rusmana, I., Suwanto, A., Mubarik, D.N.R., 2013. Characterization of lactic acid bacteria isolated from an Indonesian fermented fish (bekasam) and their antimicrobial activity against pathogenic bacteria. Emirates Journal of Food and Agriculture 25 (6), 489–494.

Devi, K.R., Manab, D., Jeyaram, K., 2015. Bacterial dynamics during yearlong spontaneous fermentation for production of ngari, a dry fermented fish product of Northeast India. International Journal of Food Microbiology 199, 62–71.

Diop, M.B., Dubois-Dauphin, R., Destain, J., Tine, E., Thonart, P., 2009. Use of a nisin-producing starter culture of *Lactococcus lactis* subsp. *lactis* to improve traditional fish fermentation in Senegal. Journal of Food Protection 72 (9), 1930–1934.

Duarte, J., Vinderola, G., Ritz, B., Perdigón, G., Matar, C., 2006. Immunomodulating capacity of commercial fish protein hydrolysate for diet supplementation. Immunobiology 211 (5), 341–350.

Commission Regulation (EU) No 1019/2013 (OJ L282, p46, 24/10/2013) of 23 October 2013 amending Annex I to Regulation (EC) No 2073/2005 as regards histamine in fishery products.

Faithong, N., Benjakul, S., 2014. Changes in antioxidant activities and physicochemical properties of Kapi, a fermented shrimp paste, during fermentation. Journal of Food Science and Technology 51 (10), 2463–2471.

Faithong, N., Benjakul, S., Phatcharat, S., Binsan, W., 2010. Chemical composition and antioxidative activity of Thai traditional fermented shrimp and krill products. Food Chemistry 119 (1), 133–140.

Fitzgerald, A.J., Rai, P.S., Marchbank, T., Taylor, G.W., Ghosh, S., Ritz, B.W., Playford, R.J., 2005. Reparative properties of a commercial fish protein hydrolysate preparation. Gut 54 (6), 775–781.

FDA, 2002. FDA's Evaluation of the Seafood HACCP Program for Fiscal Years 2000/2001. Department of Health and Human Services, Public Health Service, Food and Drug Administration, Center for Food Safety and Applied Nutrition, Office of Seafood, Washington, DC.

Friedel, H.A., Goa, K.L., Benfield, P., 1989. S-adenosyl-L-methionine. Drugs 38 (3), 389–416.

Fujita, H., Yamagami, T., Ohshima, K., 2001. Effects of an ace-inhibitory agent, katsuobushi oligopeptide, in the spontaneously hypertensive rat and in borderline and mildly hypertensive subjects. Nutrition Research 21 (8), 1149–1158.

Giraffa, G., Chanishvili, N., Widyastuti, Y., 2010. Importance of lactobacilli in food and feed biotechnology. Research in Microbiology 161 (6), 480–487.

Giri, A., Osako, K., Ohshima, T., 2009. Effect of raw materials on the extractive components and taste aspects of fermented fish paste: sakana miso. Fisheries Science 75 (3), 785–796.

Giri, A., Osako, K., Okamoto, A., Ohshima, T., 2010. Olfactometric characterization of aroma active compounds in fermented fish paste in comparison with fish sauce, fermented soy paste and sauce products. Food Research International 43, 1027–1040.

Gökoglu, N., Yerlikaya, P., Cengiz, E., 2004. Changes in biogenic amine contents and sensory quality of sardine (*Sardina pilchardus*) stored at 4°C and 20 °C. Journal of Food Quality 27, 221–231.

Hajar, S., Hamid, T.H.T.A., 2013. Isolation of lactic acid bacteria strain *Staphylococcus piscifermentans* from Malaysian traditional fermented shrimp cincaluk. International Food Research Journal 20 (1), 125–129.

Halliwell, B., 1994. Free radicals, antioxidants, and human disease: curiosity, cause, or consequence? The Lancet 344, 721–724.

Hilton, C.W., Prasad, C., Vo, P., Mouton, C., 1992. Food contains the bioactive peptide, cyclo(His-Pro). Journal of Clinical Endocrinology and Metabolism 75 (2), 375–378.

Hu, Y., Xia, W., Liu, X., 2007. Changes in biogenic amines in fermented silver carp sausages inoculated with mixed starter cultures. Food Chemistry 104 (1), 188–195.

Hu, Y., Xia, W., Ge, C., 2008. Characterization of fermented silver carp sausages inoculated with mixed starter culture. LWT—Food Science and Technology 41, 730–738.

Hwang, K.-J., Choi, K.-H., Kim, M.-J., Park, C.-S., Cha, J., 2007. Purification and characterization of a new fibrinolytic enzyme of *Bacillus licheniformis* KJ-31, isolated from Korean traditional Jeot-gal. Journal of Microbiology and Biotechnology 17 (9), 1469–1476.

Ichimura, T., Hu, J., Aita, D.Q., Maruyama, S., 2003. Angiotensin I-converting enzyme inhibitory activity and insulin secretion stimulative activity of fermented fish sauce. Journal of Bioscience and Bioengineering 96 (5), 496–499.
Ishige, N., 1993. Cultural aspects of fermented fish products in Asia. In: Lee, C.-H., Steinkraus, K.H., Reilly, P.J.A. (Eds.), Fish Fermentation Technology. United Nations University Press, Tokyo, pp. 13–32.
Itou, K., Nagahashi, R., Saitou, M., Akahane, Y., 2007. Antihypertensive effect of narezushi, a fermented mackerel product, on spontaneously hypertensive rats. Fisheries Science 73 (6), 1344–1352.
Je, J.-Y., Park, P.-J., Byun, H.-G., Jung, W.-K., Kim, S.-K., 2005a. Angiotensin I converting enzyme (ACE) inhibitory peptide derived from the sauce of fermented blue mussel, *Mytilus edulis*. Bioresource Technology 96 (14), 1624–1629.
Je, J.-Y., Park, J.-Y., Jung, W.-K., Park, P.-J., Kim, S.-K., 2005b. Isolation of angiotensin I converting enzyme (ACE) inhibitor from fermented oyster sauce, *Crassostrea gigas*. Food Chemistry 90 (4), 809–814.
Jiang, W., Xu, Y., Li, C., Dong, X., Wang, D., 2014. Biogenic amines in commercially produced Yulu, a Chinese fermented fish sauce. Food Additives & Contaminants, Part B 7 (1), 25–29.
Jong, C.J., Azuma, J., Schaffer, S., 2012. Mechanism underlying the antioxidant activity of taurine: prevention of mitochondrial oxidant production. Amino Acids 42 (6), 2223–2232.
Jung, W.-K., Rajapakse, N., Kim, S.-K., 2005. Antioxidative activity of a low molecular weight peptide derived from the sauce of fermented blue mussel, *Mytilus edulis*. European Food Research and Technology 220 (5–6), 535–539.
Junus, S., Kasipathy, K., Namrata, T., 2010. Fermented fish products. In: Tamang, J.P., Kailasapathy, K. (Eds.), Fermented Food and Beverages of the World. CRC Press, Boca Ratón, Florida, pp. 289–307.
Kaikkonen, J., Tuomainen, T.-P., Nyyssönen, K., Salonen, J.T., 2002. Coenzyme Q10: absorption, antioxidative properties, determinants, and plasma levels. Free Radical Research 36 (4), 389–397.
Kanno, T., Kuda, T., An, C., Takahashi, H., Kimura, B., 2012. Radical scavenging capacities of saba-narezushi, Japanese fermented chub mackerel, and its lactic acid bacteria. LWT—Food Science and Technology 47 (1), 25–30.
Karovičová, J., Kohajdová, Z., 2005. Biogenic amines in food. Chemical Papers 59 (1), 70–79.
Kasankala, L.M., Xiong, Y.L., Chen, J., 2012. Enzymatic activity and flavor compound production in fermented silver carp fish paste inoculated with douchi starter culture. Journal of Agricultural and Food Chemistry 60 (1), 226–233.
Kawasaki, K.I., Matsuoka, T., Satomi, M., Ando, M., Tukamasa, Y., Kawasaki, S., 2008. Reduction of cadmium in fermented squid gut sauce using cadmium-adsorbing bacteria isolated from food sources. Journal of Food Agriculture and Environment 6 (1), 45–49.
Ke, L., Yu, P., Zhang, Z.X., 2002. Novel epidemiologic evidence for the association between fermented fish sauce and esophageal cancer in South China. International Journal of Cancer 99, 424–426.
Kim, J.H., Ahn, H.J., Jo, C., Park, H.J., Chung, Y.J., Byun, M.W., 2004a. Radiolysis of biogenic amines in model system by gamma irradiation. Food Control 15, 405–408.
Kim, D.C., Chae, H.J., In, M.-J., 2004b. Existence of stable fibrin-clotting inhibitor in salt-fermented anchovy sauce. Journal of Food Composition and Analysis 17 (1), 113–118.
Kim, D.S., Baek, H.H., Ahn, C.B., Byun, D.S., Jung, K.J., Lee, H.G., Cadwallader, K.R., Kim, H.R., 2000. Development and characterization of a flavoring agent from oyster cooker effluent. Journal of Agricultural and Food Chemistry 48 (10), 4839–4843.

Kim, J.H., Kim, D.H., Ahn, H.J., Park, H.J., Byun, M.W., 2005. Reduction of the biogenic amine contents in low salt-fermented soybean paste by gamma irradiation. Food Control 16, 43–49.

Kim, K.P., Rhee, C.H., Park, H.D., 2002. Degradation of cholesterol by *Bacillus subtilis* SFF34 isolated from Korean traditional fermented flatfish. Letters in Applied Microbiology 35 (6), 468–472.

Kim, S.Y., Shin, S., Koo, H.C., Youn, J.H., Paik, H.D., Park, Y.H., 2010. In vitro antimicrobial effect and in vivo preventive and therapeutic effects of partially purified lantibiotic lacticin NK34 against infection by *Staphylococcus* species isolated from bovine mastitis. Journal of Dairy Science 93 (8), 3610–3615.

Kiyohara, M., Koyanagi, T., Matsui, H., Yamamoto, K., Take, H., Katsuyama, Y., Tsuji, A., Miyamae, H., Kondo, T., Nakamura, S., Katayama, T., Kumagai, H., 2012. Changes in microbiota population during fermentation of narezushi as revealed by pyrosequencing analysis. Bioscience, Biotechnology, and Biochemistry 76 (1), 48–52.

Kleekayai, T., Harnedy, P.A., O'Keeffe, M.B., Poyarkov, A.A., Cunhaneves, A., Suntornsuk, W., Fitzgerald, R.J., 2015. Extraction of antioxidant and ACE inhibitory peptides from Thai traditional fermented shrimp pastes. Food Chemistry 176, 441–447.

Kleekayai, T., Saetae, D., Wattanachaiyingyong, O., Tachibana, S., Yasuda, M., Suntornsuk, W., 2014. Characterization and in vitro biological activities of Thai traditional fermented shrimp pastes. Journal of Food Science and Technology 52 (3), 1839–1848.

Kopermsub, P., Yunchalard, S., 2010. Identification of lactic acid bacteria associated with the production of plaa-som, a traditional fermented fish product of Thailand. International Journal of Food Microbiology 138 (3), 200–204.

Köse, S., Koral, S., Tufan, B., Pompe, M., ScavniÇar, A., KoÇar, D., 2012. Biogenic amines contents of commercially processed traditional fish products originating from European countries and Turkey. European Food Research and Technology 235, 669–683.

Koyanagi, T., Kiyohara, M., Matsui, H., Yamamoto, K., Kondo, T., Katayama, T., Kumagai, H., 2011. Pyrosequencing survey of the microbial diversity of "narezushi", an archetype of modern Japanese sushi. Letters in Applied Microbiology 53 (6), 635–640.

Kuda, T., Kaneko, N., Yano, T., Mori, M., 2010. Induction of superoxide anion radical scavenging capacity in Japanese white radish juice and milk by *Lactobacillus plantarum* isolated from aji-narezushi and kaburazushi. Food Chemistry 120 (2), 517–522.

Kuda, T., Kawahara, M., Nemoto, M., Takahashi, H., Kimura, B., 2014. In vitro antioxidant and antiinflammation properties of lactic acid bacteria isolated from fish intestines and fermented fish from the Sanriku Satoumi region in Japan. Food Research International 64, 248–255.

Kumar, M., Kumar, J.A., Ghosh, M., Ganguli, A., 2014. Characterization and optimization of an antiaeromonas bacteriocin produced by *Lactococcus lactis* isolated from hukuti maas, an indigenous fermented fish product. Journal of Food Processing and Preservation 38, 935–947.

Lan, C.H., Son, C.K., Ha, H.P., Florence, H., Binh, L.T., Mai, L.T., Tram, N.T.H., Khanh, T.T.M., Phu, T.V., Dominique, V., Yves, W., 2013. Tropical traditional fermented food, a field full of promise. Examples from the Tropical Bioresources and Biotechnology programme and other related French-Vietnamese programmes on fermented food. International Journal of Food Science and Technology 48, 1115–1126.

Lange, J., Witmann, C., 2002. Enzyme sensor array for the determination of biogenic amines in food samples. Analytical and Bioanalytical Chemistry 372 (2), 276–283.

Lee, M.-K., Lee, J.-K., Son, J.-A., Kang, M.-H., Koo, K.-H., Suh, J.-W., 2008. S-adenosyl-L-methionine (SAM) production by lactic acid bacteria strains isolated from different fermented kimchi products. Food Science and Biotechnology 17 (4), 857–860.

Lee, N.-K., Pario, Y.-L., Park, Y.-H., Kims, J.-M., Nams, H.-M., Jungs, S.-C., Paik, H.-D., 2010. Purification and characterization of pediocin SA131 produced by *Pediococcus pentosaceus* SA131 against bovine mastitis pathogens. Milchwissenschaft 65 (1), 19–21.

Lee, Y.G., Lee, K.W., Kim, J.Y., Kim, K.H., Lee, H.J., 2004. Induction of apoptosis in a human lymphoma cell line by hydrophobic peptide fraction separated from anchovy sauce. BioFactors 21, 63–67.

Lopetcharat, K., Choi, Y.J., Park, J.W., Daeschel, M.A., 2001. Fish sauce products and manufacturing: a review. Food Reviews International 17, 65–88.

Mah, J.H., Hwang, H.J., 2009. Inhibition of biogenic amine formation in a salted and fermented anchovy by *Staphylococcus xylosus* as a protective culture. Food Control 20, 796–801.

Mine, Y., Kwan Wong, A.H., Jiang, B., 2005. Fibrinolytic enzymes in Asian traditional fermented foods. Food Research International 38 (3), 243–250.

Minelli, A., Bellezza, I., Grottelli, S., Galli, F., 2008. Focus on cyclo(His-Pro): history and perspectives as antioxidant peptide. Amino Acids 35 (2), 283–289.

Moon, J.S., Kim, Y., Jang, K.I., Han, N.S., Cho, K.J., Yang, S.J., Yoon, G.M., Kim, S.Y., 2010. Analysis of biogenic amines in fermented fish products consumed in Korea. Food Science and Biotechnology 19 (6), 1692–1698.

Morya, V.K., Choi, W., Kim, E., 2014. Isolation and characterization of *Pseudoalteromonas* sp. from fermented Korean food, as an antagonist to *Vibrio harveyi*. Applied Microbiology and Biotechnology 98, 1389–1395.

Muzaddadi, A.U., Basu, S., 2012. An accelerated process for fermented fish (seedal) production in Northeast region of India. Indian Journal of Animal Sciences 82 (1), 98–106.

Nanasombat, S., Phunpruch, S., Jaichalad, T., 2012. Screening and identification of lactic acid bacteria from raw seafoods and Thai fermented seafood products for their potential use as starter cultures. Songklanakarin Journal of Science and Technology 34 (3), 255–262.

Nie, X., Zhang, Q., Lin, S., 2014. Biogenic amine accumulation in silver carp sausage inoculated with *Lactobacillus plantarum* plus *Saccharomyces cerevisiae*. Food Chemistry 153, 432–436.

Nongonierma, A.B., FitzGerald, R.J., 2013. Dipeptidyl peptidase IV inhibitory and antioxidative properties of milk protein-derived dipeptides and hydrolysates. Peptides 39, 157–163.

Oh, S., Zhang, C.-Y., Kim, T.-I., Hong, S.-J., Ju, I.-S., Lee, S.-h., Kim, S.-H., Cho, J.-I., Ha, S.-D., 2014. Inactivation of *Anisakis larvae* in salt-fermented squid and pollock tripe by freezing, salting, and combined treatment with chlorine and ultrasound. Food Control 40, 46–49.

Okamoto, A., Hanagata, H., Matsumoto, E., Kawamura, Y., Koizumi, Y., Yanagida, F., 1995. Angiotensin I converting enzyme inhibitory activities of various fermented foods. Bioscience, Biotechnology, and Biochemistry 59 (6), 1147–1149.

Østergaard, A., Ben Embarek, P.K., Yamprayoon, J., Wedell-Neergaard, C., Huss, H.H., Gram, L., 1998. Fermentation and spoilage of som fak, a Thai low-salt fish product. Tropical Science 38 (2), 105–112.

Overvad, K., Diamant, B., Holm, L., Holmer, G., Mortensen, S.A., Stender, S., 1999. Coenzyme Q10 in health and disease. European Journal of Clinical Nutrition 53, 764–770.

Paegelow, I., Werner, H., 1986. Immunomodulation by some oligopeptides. Methods and Findings in Experimental and Clinical Pharmacology 8 (2), 91–95.

Paludan-Müller, C., Valyasevi, R., Huss, H.H., Gram, L., 2002. Genotypic and phenotypic characterization of garlic-fermenting lactic acid bacteria isolated from som-fak, a Thai low-salt fermented fish product. Journal of Applied Microbiology 92 (2), 307–314.

Paredes-López, O., Harry, G.I., 1988. Food biotechnology review: traditional solid state fermentations of plant raw materials–application, nutritional significance and future prospects. Critical Review in Food Science and Nutrition 27, 159–187.

Park, D.C., Yoon, M.S., Cho, E.J., Kim, H.S., Ryu, B.H., 1999. Antioxidant effects of fermented anchovy. Korean Journal of Food Science and Technology 31, 1378–1385.

Park, P., Je, J., Kim, S., 2005. Amino acid changes in the Korean traditional fermentation process for blue mussel (*Mytilus edulis*). Journal of Food Biochemistry 29 (1), 108–116.

Peralta, E.M., Hatate, H., Kawabe, D., Kuwahara, R., Wakamatsu, S., Yuki, T., Murata, H., 2008. Improving antioxidant activity and nutritional components of Philippine salt-fermented shrimp paste through prolonged fermentation. Food Chemistry 111 (1), 72–77.

Phadke, G., Elavarasan, K., Shamasundar, B.A., 2014. Angiotensin-I converting enzyme (ACE) inhibitory activity and antioxidant activity of fermented fish product ngari as influenced by fermentation period. International Journal of Pharma and Bio Sciences 5 (2), P134–P142.

Pringsulaka, O., Thongngam, N., Suwannasai, N., Atthakor, W., Pothivejkul, K., Rangsiruji, A., 2012. Partial characterization of bacteriocins produced by lactic acid bacteria isolated from Thai fermented meat and fish products. Food Control 23 (2), 547–551.

Pyo, Y.-H., Oh, H.-J., 2011. Ubiquinone contents in Korean fermented foods and average daily intakes. Journal of Food Composition and Analysis 24 (8), 1123–1129.

Rajapakse, N., Mendis, E., Jung, W.-K., Je, J.-Y., Kim, S.-K., 2005. Purification of a radical scavenging peptide from fermented mussel sauce and its antioxidant properties. Food Research International 38 (2), 175–182.

Rattanachaikunsopon, P., Phumkhachorn, P., 2008. Incidence of nisin Z production in *Lactococcus lactis* subsp. *lactis* TFF 221 isolated from Thai fermented foods. Journal of Food Protection 71 (10), 2024–2029.

Saisithi, P., Kasemsarn, R.O., Liston, J., Dollar, A.M., 1966. Microbiology and chemistry of fermented fish. Journal of Food Science 31, 105–110.

Saithong, P., Panthavee, W., Boonyaratanakornkit, M., Sikkhamondhol, C., 2010. Use of a starter culture of lactic acid bacteria in *plaa-som*, a Thai fermented fish. Journal of Bioscience and Bioengineering 110 (5), 553–557.

Sanceda, N., Suzuki, E., Kurata, T., 2003. Quality and sensory acceptance of fish sauce partially substituting sodium chloride or natural salt with potassium chloride during the fermentation process. International Journal of Food Science and Technology 38, 435–443.

Sarojnalini, Ch., Suchitra, T., 2009. Microbial profile of starter culture fermented fish product "Ngari" of Manipur. Indian Journal of Fisheries 56 (2), 123–127.

Sasaki, T., Araki, R., Michihata, T., Kozawa, M., Tokuda, K., Koyanagi, T., Enomoto, T., 2015. Removal of cadmium from fish sauce using chelate resin. Food Chemistry 173, 375–381.

Sasaki, T., Michihata, T., Katsuyama, Y., Take, H., Nakamura, S., Aburatani, M., Tokuda, K., Koyanagi, T., Taniguchi, H., Enomoto, T., 2013. Effective removal of cadmium from fish sauce using tannin. Journal of Agricultural and Food Chemistry 61, 1184–1188.

Shozen, K., Mori, M., Harada, Y., Yokoi, K., Satomi, M., Funatsu, Y., 2012. Histamine reduction in fish sauce with bentonite. Journal of the Japanese Society for Food Science and Technology-Nippon Shokuhin Kagaku Kogaku Kaishi 1, 17–21.

Shults, C.W., Haas, R., 2005. Clinical trials of coenzyme Q10 in neurological disorders. BioFactors 25, 117–126.

Singh, T.A., Devi, K.R., Ahmed, G., Jeyaram, K., 2014. Microbial and endogenous origin of fibrinolytic activity in traditional fermented foods of Northeast India. Food Research International 55, 356–362.

Sørensen, R., Kildal, E., Stepaniak, L., Pripp, A.H., Sørhaug, T., 2004. Screening for peptides from fish and cheese inhibitory to prolyl endopeptidase. Nahrung-Food 48 (1), 53–56.

Sornplang, P., Leelavatcharamas, V., Sukon, P., Yowarach, S., 2011. Antibiotic resistance of lactic acid bacteria isolated from a fermented fish product, *pla-chom*. Research Journal of Microbiology 6 (12), 898–903.

Speranza, B., Racioppo, A., Bevilacqua, A., Beneduce, L., Sinigaglia, M., Corbo, M.R., 2015. Selection of autochthonous strains as starter cultures for fermented fish products. Journal of Food Science 80 (1), M151–M160.

Srionnual, S., Yanagida, F., Lin, L.-H., Hsiao, K.-N., Chen, Y.-S., 2007. Weissellicin 110, a newly discovered bacteriocin from *Weissella cibaria* 110, isolated from *plaa-som*, a fermented fish product from Thailand. Applied and Environmental Microbiology 73 (7), 2247–2250.

Tapingkae, W., Tanasupawat, S., Parkin, K.L., Benjakul, S., Visessanguan, W., 2010. Degradation of histamine by extremely halophilic archaea isolated from high salt-fermented fishery products. Enzyme and Microbial Technology 46, 92–99.

Ten Brink, B., Damink, C., Joosten, H.M., Huis in't Veld, J.H., 1990. Occurrence and formation of biologically active amines in foods. International Journal Food Microbiology 11, 73–84.

Thapa, N., Pal, J., Tamang, J.P., 2004. Microbial diversity in ngari, hentak, and tungkap, fermented fish products of North-East India. World Journal of Microbiology and Biotechnology 20, 599–607.

Thongthai, C., Gildberg, A., 2005. Asian fish sauce as a source of nutrition. In: Shi, J., Sahidi, F., Ho, C.T. (Eds.), Asian Functional Foods. Marcel Dekker/CRC Press, Boca Raton, FL, pp. 215–265.

Wang, H., Fu, X., Wu, W., Wu, Y., Ren, J., Lin, Q., Li, Z., 2014. Effect of omer kodak yeast on the degrading of biogenic amine in fish sauce. Journal of Chinese Institute of Food Science and Technology 14 (8), 137–141.

Wilaipun, P., Zendo, T., Okuda, K.-I., Nakayama, J., Sonomoto, K., 2008. Identification of the nukacin KQU-131, a new type-A(II) lantibiotic produced by *Staphylococcus hominis* KQU-131 isolated from Thai fermented fish product (Pla-ra). Bioscience, Biotechnology and Biochemistry 72 (8), 2232–2235.

Wilkinson, V.M., 1997. Food Irradiation: A Reference Guide. Taylor & Francis, Great Britain.

Wong, A.H.K., Mine, Y., 2004. Novel fibrinolytic enzyme in fermented shrimp paste, a traditional Asian fermented seasoning. Journal of Agricultural and Food Chemistry 52 (4), 980–986.

Wongkhalaung, C., 2004. Industrialization of Thai fish sauce (nam pla). In: Steinkraus, K.H. (Ed.), Industrialization of Indigenous Fermented Foods, Revised and Expanded. Marcell Dekker, Inc., New York.

Xu, W., Yu, G., Xue, C., Xue, Y., Ren, Y., 2008. Biochemical changes associated with fast fermentation of squid processing by-products for low salt fish sauce. Food Chemistry 107 (4), 1597–1604.

Xu, Y., Xia, W., Yang, F., Kim, J.M., Nie, X., 2010. Effect of fermentation temperature on the microbial and physicochemical properties of silver carp sausages inoculated with *Pediococcus pentosaceus*. Food Chemistry 118, 512–518.

Yin, L.-J., Pan, C.L., Jiang, S.T., 2002. New technology for producing paste-like fish products using lactic acid bacteria fermentation. Journal of Food Science 67, 3114–3118.

Yongsawatdigul, J., Rodtong, S., Raksakulthai, N., 2007. Acceleration of Thai fish sauce fermentation using proteinases and bacterial starter cultures. Journal of Food Science 72 (9), M382–M390.

Yoon, S.-H., Kim, J.-B., Lim, Y.-H., Hong, S.-R., Song, J.-K., Kim, S.-S., Kwon, S.-W., Park, I.-C., Kim, S.-J., Yeo, Y.-S., Koo, B.-S., 2005. Isolation and characterization of three kinds of lipopeptides produced by *Bacillus subtilis* JKK238 from Jeot-Kal of Korean traditional fermented fishes. Korean Journal of Microbiology and Biotechnology 33 (4), 295–301.

Yoshikawa, S., Kurihara, H., Kawai, Y., Yamazaki, K., Tanaka, A., Nishikiori, T., Ohta, T., 2010. Effect of halotolerant starter microorganisms on chemical characteristics of fermented chum salmon (*Oncorhynchus keta*) sauce. Journal of Agricultural and Food Chemistry 58 (10), 6410–6417.

Yu, X., Mao, X., Pei Liu, S.H., Wang, Y., Xue, C., 2014. Biochemical properties of fish sauce prepared using low salt, solid state fermentation with anchovy by-products. Food Science Biotechnology 23 (5), 1497–1506.

Yuan, K., Mao, T., Cao, X., Wei, Y., 2014. Research on solid-state fermentation by combined strains with fish meal wastewater and soybean meal. Chinese Journal of Environmental Engineering 8 (7), 3041–3046.

Yusra, A., Azima, F., Novelina, N., Periadnadi, M., 2013. Antimicrobial activity of lactic acid bacteria isolated from Budu of West Sumatera to food biopreservatives. Pakistan Journal of Nutrition 12 (7), 628–635.

Zaman, M.Z., Bakar, F.A., Selamat, J., Bakar, J., 2010. Occurrence of biogenic amines and amines degrading bacteria in fish sauce. Czech Journal of Food Sciences 28, 440–449.

Zaman, M.J., Bakar, F.A., Selemat, J., Bakar, J., 2011. Novel starter cultures to inhibit biogenic amines accumulation during fish sauce fermentation. International Journal of Food Microbiology 145, 84–91.

Zeng, X., Xia, W., Jiang, Q., Guan, L., 2015. Biochemical and sensory characteristics of whole carp inoculated with autochthonous starter cultures. Journal of Aquatic Food Product Technology 24 (1), 52–67.

Zeng, X., Xia, W., Wang, J., Jiang, Q., Xu, Y., Qiu, Y., Wang, H., 2014. Technological properties of *Lacto bacillus plantarum* strains isolated from Chinese traditional low salt fermented whole fish. Food Control 40 (1), 351–358.

Zhao, J.-R., Cheng, S.-M., Hong, W.-D., Wang, R.-L., Li, C.-H., 2014. Isolation, identification, and fermentation properties of lactic acid bacteria from air dried and fermented *Megalobrama amblycephala*. Modern Food Science and Technology 30 (8), 100–105.

Zhou, X.-X., Zhao, D.-D., Liu, J.-H., Lu, F., Ding, Y.-T., 2014. Physical, chemical and microbiological characteristics of fermented surimi with *Actinomucor elegans*. LWT—Food Science and Technology 59 (1), 335–341.

Chapter 10

Fermented Meat Sausages

R. Bou[1,2], S. Cofrades[2], F. Jiménez-Colmenero[2]
[1]*Institut de Recerca i Tecnologia Agroalimentàries (IRTA). Finca Camps i Armet, Monells, Girona, Spain;* [2]*Institute of Food Science, Technology and Nutrition (ICTAN-CSIC), Madrid, Spain*

10.1 INTRODUCTION

It is believed that fermented sausages first appeared in the Mediterranean as they were known to the ancient Greeks and Romans (Toldrá, 2004). They have been consumed for centuries in many different parts of the world and are an important food source (Liu et al., 2011; Lucke, 1994). Fermented sausages can globally be defined as products manufactured with selected minced meat and fat, with or without offal, mixed with salt, seasonings, spices, and authorized additives that are then stuffed into casings, acidified through lactic-acid bacteria (fermentation), ripened, dried, and sometimes smoked. Regional customs, environmental conditions, and traditional recipes, among other factors, have given rise to a wide variety of fermented sausages and there are almost as many types of sausages as geographical regions or even manufacturers. Despite this variety, their production processes require both fermentation and ripening–drying (Toldrá, 2007). Classification of fermented sausages are based on criteria, such as acidity, degree of mincing, use and type of starter cultures, spices and seasonings, diameter and type of casing, degree of drying, smoking, etc. However, the addition of curing agents (nitrate and/or nitrite) to provide the characteristic red color and the action of microorganisms to produce lactic acid is a common trait. During fermentation and ripening various complex physical and chemical reactions determine the characteristic tangy flavor, texture, cured meat color, and increased shelf-life of this type of sausage (Campbell-Platt and Cook, 1995; Toldrá, 2007). Fermented sausages are ripened to different moisture levels depending on processing conditions (eg, air drying, smoking) and the specific variety of sausage. Fermented sausages with prolonged ripening and drying lead to low moisture content and consequently, more concentrated flavor greater concentration of nutrients, and firmer texture. These sausages are also known as dry fermented or dry-cured sausages and are considered shelf-stable even at higher temperatures. Typical examples of this type of product are the different versions of Italian Salami, Spanish Salchichón, and French Saucisson. Cervelat and summer sausages are examples of semidry fermented sausages. They have a relatively low pH and are very popular in Northern European and in North America.

10.2 COMPOSITION, NUTRITIONAL VALUE, AND HEALTH IMPLICATIONS

In 2011, the World Health Organization reported that nearly two-thirds of the deaths that occurred in the global population were due to noncommunicable diseases, mostly cardiovascular diseases (CVD), cancer, diabetes, and chronic lung disease (WHO, 2011). However, the incidence of these diseases can be reduced by changing behavioral risk factors (eg, tobacco and alcohol consumption) and by promoting a healthy diet and physical activity.

Today, dietary patterns are shifting toward increased consumption of energy-dense diets that are high in fat, particularly saturated fat, and low in unrefined carbohydrates (WHO/FAO Expert Consultation, 2003). It has been shown consumption is heavily and disproportionately concentrated in industrial countries where noncommunicable diseases such as obesity are prevalent (Bruinsma, 2003). In this context, the excessive consumption of energy-dense foods and the consumption of red and processed meats have been associated with some of the most important chronic diseases affecting the western world, such as CVD and cancer (Cross et al., 2010; Wang and Jiang, 2012; Ward et al., 2012). Thus meat and processed meat products are now viewed negatively by consumers. However, in a healthy and balanced diet, they provide different nutrients needed to meet metabolic requirements and are therefore important for proper nutrition (Olmedilla-Alonso et al., 2013). However, meat and meat products also contribute to the intake of fat, saturated fatty acids (SFA), cholesterol, salt, and other substances that can have negative health implications, depending on a variety of factors and pathophysiological circumstances. Thus, it is essential to be familiar with the composition, nutritional value, and health implications of fermented sausages and the availability of different strategies to optimize (increase or reduce) the presence of bioactive compounds to produce healthier fermented sausage.

10.2.1 Proteins, Peptides, Amino Acids, and Other Nitrogen Compounds

Fermented sausages with protein content between 15% and 26% (Table 10.1) constitute a primary source of this macronutrient necessary for body growth, maintenance, and repair as well as supply of energy. In addition, animal proteins are considered to have high biological value since a high percentage of the amino acids found in meat are essential (Toldrá, 2004).

During processing and as a consequence of proteolysis mechanisms (endogenous and microbial enzymes), an important amount of free amino acids and peptides as well as ammonia are generated contributing to their characteristic flavor (Campbell-Platt and Cook, 1995; Toldrá, 1998). The amount of hydrolyzed compounds can exceed 1% dry weight (Candogan et al., 2009; Díaz et al., 1997). The amount of free essential amino acids, such as threonine, valine,

TABLE 10.1 Proximate Composition and Energy, Fatty Acid (FA) Categories, Cholesterol, Mineral, and Vitamin Content in 100 g of Product

		Raw Meat		Fermented Sausage					
	Units	Pork, Fillet	Bacon, Fat Only	Salami	Salchichón	Pepperoni	Chorizo	Cervelat
Water	g	73.6	12.8	33.7	34.1	29.6	43.9	46.4
Protein	g	22.0	4.8	20.9	25.8	22.9	22	14.7
Fat	g	6.5	80.9	39.2	38.1	43.5	32.1	32.1
Energy	kcal	147	747	438	454	494	385	363
Saturated FA	g	2.30	31.50	14.60	12.30	14.86	12.06	(10.80)
Monounsaturated FA	g	2.30	36.10	17.70	15.93	17.16	13.92	(13.20)
Polyunsaturated FA	g	1.30	8.90	4.40	5.83	3.43	4.26	(2.40)
Trans FA	g	Tr	N	0.15	NA	1.63	NA	N
Cholesterol	mg	63.0	198.0	83.0	72	105	72	65.0
Na	mg	53	560	1800	1060	1761	1060	1100
K	mg	400	75	320	207	279	207	210
Ca	mg	5	3	11	15	22	21	35
Mg	mg	24	4	18	10	21	11	13
P	mg	230	38	170	260	177	160	150
Fe	mg	0.70	0.70	1.30	2.4	1.62	2.4	3.10
Cu	mg	0.08	0.06	0.12	NA	0.11	NA	N
Zn	mg	1.6	0.6	3.0	1.7	2.5	1.2	N
Cl	mg	54	810	3270	NA	NA	NA	N
Mn	mg	0.02	Tr	0.04	NA	0.60	NA	N

Continued

TABLE 10.1 Proximate Composition and Energy, Fatty Acid (FA) Categories, Cholesterol, Mineral, and Vitamin Content in 100 g of Product—cont'd

		Raw Meat			Fermented Sausage			
	Units	Pork, Fillet	Bacon, Fat Only	Salami	Salchichón	Pepperoni	Chorizo	Cervelat
Se	µg	14	(1)	(7)	26.1	35	21.1	N
I	µg	3	(11)	(15)	NA	NA	NA	N
Carotenoids	µg	Tr	Tr	Tr	Tr	0	Tr	Tr
Retinol eq.	µg	Tr	Tr	Tr	Tr	0	Tr	Tr
Vitamin D	µg	0.5	Tr	N	Tr	0.2	Tr	0.5
Vitamin E	mg	Tr	0.11	0.23	0.28	0	0.28	0.12
Vitamin K	mg	NA	NA	1.11	NA	5.8	NA	
Thiamine	mg	1.16	N	0.60	0.2	0.36	0.3	0.14
Vitamin B$_2$	mg	0.32	N	0.23	0.21	0.32	0.13	0.16
Niacin	mg	NA	NA	NA	10	4.6	7.1	
Vitamin B$_6$	mg	0.50	N	0.36	0.15	0.35	0.15	0.14
Vitamin B$_{12}$	µg	1.0	Tr	2.0	1	1.7	1	3.0
Folate	µg	1	Tr	3	3	5	1	4
Pantothenic	mg	1.49	N	1.66	NA	1.48	NA	0.40
Biotin	µg	1.0	Tr	7.0	NA	NA	NA	N
Vitamin C	mg	0	0	N	0	0.7	0	N

Nutrient values within brackets are estimated. "Tr" stands for trace values, "N" for nutrients present in significant quantities with no reliable information, and "NA" for not available. Data obtained from Food Standards Agency, 2010. McCance and Widdowson's the Composition of Foods Integrated Dataset (CoF IDS). Retrieved from: http://tna.europarchive.org/20110116113217/http://www.food.gov.uk/science/dietarysurveys/dietsurveys/; USDA, December 7, 2011. USDA National Nutrient Database for Standard Reference. National Agricultural Library. Retrieved from: http://ndb.nal.usda.gov/; Moreiras, O., Carbajal, A., Cabrera, L., Cuadrado, C., 2006. Tabla de Composición de Alimentos, tenth ed. Ediciones Pirámide, Madrid.

leucine, and lysine increases during ripening (Beriain et al., 2000; Campbell-Platt and Cook, 1995). The addition of certain proteases (Díaz et al., 1997) and/or starter cultures (Petron et al., 2013) may contribute to an increase in proteolysis and thus to higher amounts of free amino acids. Meat and fermented sausages are important sources of taurine (2-aminoethane sulfonic acid), which has been reported to act as an antioxidant in vivo in addition to other various biological functions (Marcinkiewicz and Kontny, 2014; Reig et al., 2013). Taurine can be synthesized from methionine and cysteine by humans but its biosynthetic capacity is low (Ito et al., 2012). Thus the presence of this amino acid and others may be of interest when diets are not well balanced or in the case of population groups with special needs or requirements (eg, children, elderly, people with diseases) (Wu, 2013).

Raw meats also contain carnosine (β-alanyl-1-histidine) and anserine (N-β-alanyl-1-methyl-1-histidine). These histidyl dipeptides are the most abundant endogenous antioxidants in meats (Arihara, 2006; Chan and Decker, 1994). The concentration of carnosine is much higher than anserine in pork meat but concentrations vary significantly depending on animal species and muscle type (Chan and Decker, 1994). Carnosine was found to increase during ripening of dry-cured Salchichón (Beriain et al., 2000). Glutathione is a tripeptide (γ-Glu-Cys-Gly) affecting animal energy metabolism and also acts as an antioxidant in vivo (Decker and Xu, 1998). Carnitine, biosynthesized from lysine and methionine, is also related with energy production and other biological functions including neuroprotective and glucose-lowering effects (Khan et al., 2011; Zammit et al., 2009). However, it has also been reported that gut microbiota can transform carnitine into trimethylamine, thus enhancing the risk of CVD (Koeth et al., 2013).

10.2.2 Fat Content, Fatty Acid Composition, and Other Lipid Compounds

Fat is the other main ingredient used in fermented sausages with important implications from a sensorial, technological, and health point of view. Depending on the type of sausage and its formulation, fat content ranges from 30% to 45% (Table 10.1). On average, the fatty-acid composition of lipids in fermented sausages includes 35–40% SFA, 45–50% monounsaturated fatty acids (MUFA), and 7–15% polyunsaturated fatty acids (PUFA) (Table 10.1). Although the association between saturated fats and the risk of CVD has recently been called into question (Siri-Tarino et al., 2010), it is still widely accepted that excessive consumption of fat, SFA, and cholesterol has negative effects on health, depending on a variety of factors and physiological circumstances (EFSA, 2010; Institute of Medicine, 2002). Low-density lipoprotein (LDL) cholesterol levels have been reported to increase with the consumption of SFA (Baum et al., 2012). Conversely, diets rich in MUFA lower total and LDL cholesterol, whereas high-density lipoprotein (HDL) cholesterol levels remain the same (Kris-Etherton,

1999). Moreover, in a randomized trial, the incidence of major CVD events was reduced among those individuals with diets rich in oleic acid (18:1 n-9) (Estruch et al., 2013). Furthermore, it is well known that n-3 fatty acids play an important role in the modulation and prevention of some diseases (including coronary disease), while also exhibiting other positive health effects (antithrombotic, antiinflammatory, etc.) (Connor, 2000). High proportions of PUFA are not necessarily healthy in the absence of a balanced n-6/n-3 ratio. Excessive amounts of n-6 PUFA and very high n-6/n-3 PUFA ratios promote pathogenesis of many sorts of diseases including CVD, cancer, and inflammatory and autoimmune diseases, whereas increased levels of n-3 PUFA (and low n-6/n-3 PUFA ratios) exhibit suppressive effects (Simopoulos, 2002).

Conjugated linoleic acid (CLA), present in meat and adipose tissues of ruminants, has attracted considerable interest over the last two decades for its array of attributed beneficial biological effects including body fat reduction, improved bone mass, prevention of CVD and cancer, and modulation of immune and inflammatory responses (Dilzer and Park, 2012). In fermented sausages, cholesterol normally ranges between 50 and 110 mg/100 g (Table 10.1). These levels are relatively high considering the recommendation of less than 300 mg/day (WHO/FAO Expert Consultation, 2003). However, with the aim of controlling hypercholesterolemia, several national and international organizations recommended a reduction in the intake of SFA (EFSA, 2010; Institute of Medicine, 2002). The presence of cholesterol oxides in products of animal origin has been reported to have a number of adverse effects in humans (Bjorkhem et al., 2002). However, the content of cholesterol oxidation products in dry-cured products and dry fermented sausages is considered to be low (Demeyer et al., 2000; Zanardi et al., 2004).

10.2.3 Minerals

Meat products are an important source of iron and zinc and also contain considerable concentrations of phosphorus and potassium and significant amounts of other elements, such as magnesium and selenium, which are important for human health (Olmedilla-Alonso et al., 2013). The iron (Fe) content in fermented sausages typically ranges between 1 and 2.5 mg/100 g (Table 10.1), which is similar to that of red meats and thus is considered an important source. This fact should be kept in mind since Fe deficiency is one of the most prevalent health issues around the world (Benoist et al., 2008). Low Fe intake and Fe levels in young children and women of childbearing age are common in developed countries (Elmadfa and Freisling, 2009). Moreover, 50% or more of Fe in meat and meat products is present in the more bioavailable heme Fe form. The increased bioavailability of Fe in meats, typically 15–25% compared with 1–7% for Fe from plant sources that do not contain heme Fe, is an important aspect to consider (Higgs, 2000). In addition, the absorption of nonheme Fe from other foods consumed in combination with meat is enhanced due to the so-called "meat factor."

Zinc (Zn) is also present in relatively high amounts (1.6–4.0 mg/100 g) in meats and its bioavailability is also high (20–40%) (Higgs, 2000). This element is needed for regulation of gene expression, protein synthesis, normal cell growth, and cell differentiation (Institute of Medicine, 2001). Several studies carried out in developed countries have shown that elderly people are at greater risk of inadequate Zn intake, which has been related to impaired immune response in this group (Girodon et al., 1999). Fe and Zn deficiency also occurs simultaneously, especially in adolescents who do not eat meat (Sandstead et al., 2000). Therefore, reduced consumption of meat and meat products can cause serious nutritional problems given that meat and meat products contribute more to total Zn intake than to total Fe intake (Baghurst, 2007).

10.2.4 Vitamins and Antioxidants

Meat and meat products are rich in certain vitamins (Table 10.1). In fact, they are considered as an excellent source of B-group vitamins (Olmedilla-Alonso et al., 2013). Among these, the most noteworthy present in pork meat and dry-cured meat products are thiamine (B_1) ranging from 0.2 to 1.2 mg/100 g, riboflavin (B_2) 0.1–0.4 mg/100 g, niacin (B_3) 3–10 mg/100 g, pyridoxine and related forms (B_6) 0.14–1.5 mg/100 g, and cobalamin (vitamin B_{12}) 0.4–3.0 µg/100 g (Table 10.1). Thus in many cases the consumption of 100 g of meat provides over one-third of the daily requirement (Institute of Medicine, 1998). Lipid-soluble vitamin content in fermented sausages is mainly determined by the amount of fatty tissue. Nevertheless, levels of vitamins, such as D and α-tocopherol (vitamin E) are generally low. However, it is possible to find moderate levels of α-tocopherol in fermented sausages because, as it cannot be synthesized by animal tissue, levels depend on the animal's diet (Magrinya et al., 2009). In fact, tocopherol-enriched fermented sausages can be achieved by strategically altering animal feed (Zanardi et al., 2000).

10.3 STRATEGIES FOR OPTIMIZING THE PRESENCE OF BIOACTIVE COMPOUNDS TO IMPROVE HEALTH AND WELL-BEING AND/OR REDUCE RISK OF DISEASE

As previously noted, fermented sausages are highly nutritious foods providing valuable amounts of protein, fatty acids, vitamins, minerals, and energy (Table 10.1), although they also contain components that can have negative health effects depending on a variety of factors. Different strategies have been employed to effectively optimize (increase or reduce) amounts of bioactive compounds in meat and meat products, although those based on reformulation and modified processing are the ones best suited to the development of healthier fermented sausage. Reformulation, ie, changes in the ingredients (raw meat and others) used to manufacture these meat products, offer a wide range of options to remove, reduce, increase, add, and/or replace different components, including those with health implications (Arihara, 2006; Jiménez-Colmenero et al., 2001).

The following is an overview of the bioactive compounds that have been considered for the development of healthier fermented sausages. The role of these substances is analyzed individually but in many cases the optimization process affects more than one of them simultaneously.

10.3.1 Bioactive Peptide Formation

Bioactive peptides are sequences that are inactive fragments within the precursor protein but once released after proteolysis (aging, fermentation, proteases, etc.) they can spark different types of bioactivity. Various meat and meat product extracts have been reported to have antihypertensive, antioxidant, prebiotic, and other biological effects (Arihara and Ohata, 2008; Toldrá and Reig, 2011). Although these compounds are vital to the development of functional fermented products, there are few studies regarding the specific effect or isolation and purification of individual bioactive peptides on fermented sausages. It would stand to reason that the most interesting bioactive peptides in fermented sausages are those exhibiting angiotensin I-converting enzyme (ACE) inhibitory activity. The presence of ACE inhibitors in fermented sausages may help to compensate for the high amounts of sodium in these products and thus diminish the risk of hypertension. Arihara and Ohata (2008) described ACE inhibitory activities in extracts of several fermented meat products. Vastag et al. (2010) reported different hydrolyzates during the ripening of petrovská kolbása (traditional dry fermented sausage) with antioxidant and ACE inhibitory properties. This may be related to the microbial growth of certain cultures, such as *Lactobacillus sakei* and *Lactobacillus curvatus*, which have been reported to produce different ACE inhibitory peptides (Castellano et al., 2013).

The production of antioxidant peptides is also interesting. Broncano et al. (2012) reported the isolation and identification of low molecular weight antioxidant compounds from fermented "chorizo" sausages. In addition, a number of microorganisms are able to produce bacteriocins during the production of fermented sausages, enabling them to be used as a bioprotective strategy (Barbosa et al., 2015; Fontana et al., 2015). In addition to bacteriocins, lactic-acid bacteria (LAB) have been reported to form different organic acids (mainly lactic and acetic acid), hydrogen peroxide, and antifungal peptides (Reis et al., 2012). All these properties are potentially important in terms of the growth of undesirable microbiota.

10.3.2 Improving Fat Content

Three main goals have been identified in terms of improving the fat content of fermented sausage: reduction of total fat (lower energy content), reduction of cholesterol, and modification of fatty-acid profiles.

Fat reduction in fermented sausages is usually based on two main criteria: the use of leaner meat as raw material together with other ingredients with little or no caloric content. Most of these ingredients are dietary fibers that have been added alone or combined with other ingredients to reduce the fat content of different products. Inulin, cereal fiber (oat and wheat) and fruit (peach, orange), ι- and κ-carrageenan, short-chain fructooligosaccharides, and soy fiber have been added to dry fermented products to reduce the caloric value and contribute to ensuring desired product characteristics.

Strategies for improving the fatty-acid profile generally entail replacement of the animal fat normally present in the fermented sausage with other fats more in line with health recommendations—ie, with lower proportions of SFA and larger proportions of MUFAs, n-3 PUFAs (especially long-chain) or CLA, enhanced n-6/n-3 PUFA and PUFA/SFA ratios, and when possible, reduced cholesterol. The type of vegetable oil used affects the fatty-acid composition of reformulated meat products. While vegetable oils are rich in MUFA and PUFA (α-linolenic acid, 18:3 n-3) and are cholesterol-free, fish and algae oils are sources of long-chain n-3 PUFAs (eicosapentaenoic-EPA, 20:5 n-3 and docosahexaenoic-DHA 22:6 n-3 acids). Many studies have reported new formulations for healthy fermented sausages using different lipid materials (mainly oils) of vegetable (olive, linseed, sunflower, soy, canola) and marine origin (fish, algae, etc.) as partial or total substitutes for animal fat (Table 10.2). Although the addition of these individual lipids of vegetal or marine origin improves the fatty-acid profile of meat products, better lipid composition from a health point of view can be achieved using healthier combinations of them as animal fat replacers. The procedures used to incorporate natural or processed plant and marine lipids into fermented sausages range from direct addition as liquid oils or as solid-like interesterified oils, to incorporation in encapsulated form as part of plant ingredients stabilized in different matrices (eg, konjac). But the most widely assayed method is pre-emulsification, mainly with soy protein isolate. The improvement in lipid profile achieved by these reformulation processes depends on the type and amount of oil that is incorporated. In general, the addition of oil raises MUFA and PUFA content as well as PUFA/SFA ratios and reduces cholesterol level and n-6/n-3 PUFA ratios (Table 10.2), bringing values closer to the recommended goals. One of the potential problems associated with these modifications is an acceleration of oxidative processes due to the increase in unsaturated fatty acids, mainly PUFA, which are more susceptible to oxidation. Lipid oxidation in healthier fermented sausages varies depending on the nature of the product, the type, amount, and method employed to incorporate nonmeat fat and the antioxidative system used to control rancidity (Jiménez-Colmenero, 2007).

10.3.3 Probiotics

Probiotics are live microorganisms present in food that, when ingested in sufficient quantities, provide health benefits for consumers (Ashwell, 2002). Although fermented dairy products are the most typical probiotic foods, meat products also serve as an excellent medium for probiotic growth. Meat products

TABLE 10.2 Examples of Fermented Meat Products With Improved Fatty-Acid Composition by Formulation With Different Nonmeat Fats

Product	Lipid Material/ Procedure of Incorporation	Main Lipid Content Changes	References
Fermented sausage	Olive oil/liquid and pre-emulsified with isolated soy protein (ISP)	Cholesterol reduction. Increased MUFA content	Bloukas et al. (1997)
Dry fermented sausages (chorizo)	Olive oil/ pre-emulsified with ISP	Fat and cholesterol reduction. Increased MUFA and PUFA contents	Muguerza et al. (2001)
Turkish dry fermented sausages	Palm and cottonseed oils/interesterified	Decreased SFA/ UFA ratio. Increased PUFA/ SFA ratio	Vural (2003)
Dry fermented sausages (salami)	Extra-virgin olive oil/ premixed with sodium caseinate (SC)	–	Severini et al. (2003)
Turkish dry fermented sausages (sucuk)	Olive oil/ pre-emulsified with ISP	Fat and cholesterol reduction. Increased oleic and linoleic acid content	Kayaardi and Gok (2004)
Dry fermented sausages (chorizo)	Soy oil/pre-emulsified with ISP	Cholesterol SFA and MUFA reduction. Increased PUFA content	Muguerza et al. (2003)
Dry fermented sausages	Fish oil extract	Increased DHA and EPA content	Muguerza et al. (2004)
Dry fermented sausages	Deodorized fish oil/ pre-emulsified with ISP	Increased DHA and EPA content	Valencia et al. (2006)
Dry fermented sausages	Pre-emulsified linseed fish oil/ISP	Increased MUFA/ PUFA content	Ansorena and Astiasaran (2004)
Dutch style fermented sausage	Flaxseed oil/ encapsulated, flaxseed oil/pre-emulsified with ISP or SC, canola oil/pre-emulsified with ISP, fish oil/ encapsulated	Increased MUFA/ PUFA content. n-6/n-3 ratio reduction	Pelser et al. (2007)

TABLE 10.2 Examples of Fermented Meat Products With Improved Fatty-Acid Composition by Formulation With Different Nonmeat Fats—cont'd

Product	Lipid Material/ Procedure of Incorporation	Main Lipid Content Changes	References
Turkish dry fermented sausages (sucuk)	Hazelnut oil/ pre-emulsified with whey protein powder (WPP)	Cholesterol reduction. Increased MUFA, PUFA, and MUFA+PUFA/SFA ratio	Yildiz-Turp and Serdaroglu (2008)
Turkish dry fermented sausages (sucuk)	Walnut paste	Fat reduction. Improved lipid profile	Ercoskun and Demirci-Ercoskun (2010)
Dry fermented sausages (chorizo)	Linseed and algae oils (3:2, w/w)/ pre-emulsified with ISP	Increased PUFA content (α-linolenic, EPA and DHA). n-6/n-3 ratio reduction	García-Iñiguez de Ciriano et al. (2013)
Dry fermented sausages (chorizo)	Linseed oil/ pre-emulsified with ISP	SFA and MUFA reduction. Increased PUFA content, n-6/n-3 ratio reduction	García-Iñiguez de Ciriano et al. (2013)
Dutch style fermented sausage	Fish oil/pure, pre-emulsified with ISP or encapsulated	SFA reduction. Increased PUFA content, n-6/n-3 ratio reduction	Josquin et al. (2012)
Dry fermented sausages (chorizo)	Olive, linseed, and fish oils combination/ stabilized in konjac matrix	Fat reduction (up to 60%). SFA reduction. Increased PUFA, n-6/n-3 ratio reduction	Jimenez-Colmenero et al. (2013)
Nonacid fermented sausages (Fuet)	Diacylglycerols (DAGs) and sunflower oil/liquid	Fat reduction (more than 70%)	Mora-Gallego et al. (2013)
Dry fermented chicken sausages	Corn oil/ pre-emulsified with ISP and with inulin	Fat reduction	Menegas et al. (2013)

Continued

TABLE 10.2 Examples of Fermented Meat Products With Improved Fatty-Acid Composition by Formulation With Different Nonmeat Fats—cont'd

Product	Lipid Material/ Procedure of Incorporation	Main Lipid Content Changes	References
Dry fermented sausages	Sunflower oil/liquid	Fat reduction (up to 50%). Increased PUFA content	Fuentes et al. (2014)
Nonacid fermented sausages (fuet)	Sunflower oil/liquid	Fat reduction (50%)	Mora-Gallego et al. (2014)
Dry-ripened venison sausages	Olive oil/pre-emulsified with SPI	SFA reduction. Increased MUFA content. SFA/UFA ratio reduction	Utrilla et al. (2014)

processed by fermentation without further heating are excellent vehicles for probiotics. Fermented sausage containing viable probiotic bacteria can have positive health effects for consumers, providing protection against gastrointestinal infections and exhibiting antimicrobial activity, improvement in lactose metabolism, lowering of blood serum cholesterol levels, immune-system stimulation, antimutagenic properties, anticarcinogenic activities, antidiarrheal properties, improvement in inflammatory bowel disease, etc. (Rivera-Espinoza and Gallardo-Navarro, 2010; Vuyst et al., 2008). Table 10.3 shows examples of fermented sausages with different probiotic strains, mainly LAB and bifidobacteria. Most of this research has focused on the survival of the species added to the meat matrix and its influence on the technological and sensory characteristics of the final product. There have been a number of reviews on probiotics in fermented meat products (Ammor and Mayo, 2007; Khan et al., 2011; Rouhi et al., 2013; Tyopponen et al., 2003). Some functional probiotic fermented meat products are available commercially (Salami in Germany and meat spread in Japan) that have been reported to contain human intestinal LAB, such as *Lactobacillus acidophilus, L. casei, Bifidobacterium* spp., and *L. rhamnosus* (Arihara, 2006).

10.3.4 Dietary Fiber Incorporation

Although dietary fiber has been defined in several ways, one widely accepted definition suggests that dietary fiber consists of the remnants of the edible part of plants and analogous carbohydrates that are resistant to digestion and absorption in the human small intestine with complete or partial fermentation in the human large intestine (Prosky, 1999). Dietary fiber has been associated

TABLE 10.3 Examples of Fermented Meat Products With Specific Probiotic Bacteria

Product	Probiotic Strain	References
Fermented meat	Lactobacillus acidophilus, Lactobacillus crispatus, Lactobacillus amylovorus, Lactobacillus gallinarum, Lactobacillus gasseri, Lactobacillus johnsonii	Arihara et al. (1998)
Fermented dry sausage	Lactobacillus casei, Bifidobacterium lactis	Andersen (1998)
Fermented sausage	Lactobacillus rhamnosus, Lactobacillus paracasei	Sameshima et al. (1998)
Fermented dry sausage	L. rhamnosus (GG, E–97800 or LC-705)	Erkkila et al. (2001)
Probiotic sausage	L. paracasei LTH 2579	Jahreis et al. (2002)
Greek dry fermented sausage	L. plantarum, Lactobacillus curvatus	Papamanoli et al. (2002)
Scandinavian-type fermented sausage	L. plantarum, (MF1291 and MF1298), Lactobacillus pentosus (MF1300)	Klingberg et al. (2005)
Greek dry fermented sausage	L. plantarum	Pennacchia et al. (2006)
Dry fermented sausage	Lactobacillus reuteri	Muthukumarasamy and Holley (2006)
Dry fermented sausage	L. casei/paracasei	Rebucci et al. (2007)
Iberian dry fermented sausage (salchichón)	Lactobacillus fermentum (HL57), Pediococcus acidilactici (SP979)	Ruiz-Moyano et al. (2011b)
Iberian dry fermented sausage (sachichón)	L. reuteri (PL519)	Ruiz-Moyano et al. (2011a)
Dry fermented sausage	L. casei (LOCK 0900)	Wojciak et al. (2012)
Italian salami sausage	L. acidophilus, Bifidobacterium lactis	Nogueira Ruiz et al. (2014)
Low-acid fermented sausages (fuets)	L. rhamnosus (CTC1679)	Rubio et al. (2014b)
Low-acid fermented sausages (fuets)	L. casei (CTC1677), L. casei (CTC1678), L. rhamnosus (CTC1679)	Rubio et al. (2014a)
Dry fermented sausages	L. casei (ATCC 393)	Sidira et al. (2014)

Adapted from Jiménez-Colmenero et al. (2012).

with various health benefits, including maintenance of gut health (by facilitating excretion), prevention of carcinogenesis, reduced risk of coronary heart disease (hypocholesterolemic effects), prevention of type 2 diabetes (ability of the fiber to reduce the glycemic response), and reduction of obesity (by providing the sensation of satiety) (Verma and Banerjee, 2010). Due to these health benefits, interest in fiber-rich products has grown in recent years. In addition, dietary fiber has technological properties that are useful for food processing, and since it is easily recoverable from industrial processing its use in value-added products provides a means of upgrading coproducts (Jiménez-Colmenero and Delgado-Pando, 2013).

Dietary fiber and fiber-rich ingredients from cereals (oats, rice, wheat, etc.), fruits (apple, peach, lemon, orange, etc.), legumes (soy, peas, etc.), roots (carrot, sugarbeet, konjac, etc.), tubers (potato), and seaweed (red and brown algae) are used as additives/ingredients in the manufacture of fermented sausages (Table 10.4). The use of purified carbohydrate-based compounds, such as inulin, short-chain fructooligasaccharides, and carrageenans (ι, κ) has been shown to enhance nutritional quality while maintaining sensorial properties of the product. Thus the use of dietary fiber for its technological properties and health benefits opens up interesting possibilities in functional meat product development (Jiménez-Colmenero, 2007; Verma and Banerjee, 2010). Note that this dietary fiber has been added as a fat replacer and/or as a functional ingredient. As previously reported, fat reduction strategies are of particular interest since the fat content of fermented sausages is usually high.

10.3.5 Mineral Addition

Depending on the mineral in question and its potential health implications, two different types of interventions have been assayed; one by reducing the presence of those with negative health effects such as sodium and the other by raising the levels of those with beneficial effects, such as calcium, magnesium, selenium, and iodine (Table 10.5).

10.3.5.1 Strategies for Sodium Reduction

One important drawback of fermented sausages is their sodium content, which is notoriously high as a result of the addition of salt (NaCl) in sausage formulations (Table 10.1). A recent study showed that 87% of the Spanish population exceeded the daily recommended allowance of sodium (AESAN, 2009). Sodium reduction is especially challenging for the meat industry as meat products (not only fermented sausages) comprise one of the major sources of dietary sodium intake, approximately 20–30% according to the WHO (WHO, 2010) and over 26% in Spain (AESAN, 2009).

There is strong evidence showing that salt (sodium) consumption is now the number one factor underlying high blood pressure and hence CVD. Reduction in salt intake worldwide would bring about a major improvement in public

TABLE 10.4 Examples of Fermented Sausages Formulated With Different Dietary Fibers and Fiber-Rich Materials Incorporated

Product	Ingredients	Fiber Content (g/100g Product)	References
Dry fermented sausages	Inulin	9.3–16.9	Mendoza et al. (2001)
Dry fermented sausages	Wheat, oat, peach fiber	2.0–4.0	Garcia et al. (2002)
Dry fermented sausages	Apple, orange fiber	2.0	Garcia et al. (2002)
Dry fermented sausage (salchichón)	Orange fiber	1.0–2.0[b]	Fernandez-Lopez et al. (2008)
Fermented sausages	I-, K-carrageenan	1.0–3.0[b]	Koutsopoulos et al. (2008)
Dry fermented sausage (sobrassada)	Carrot fiber	3.0–12.0[b]	Eim et al. 2008, 2013
Dry fermented sausages (salchichón)	sc-FOS[a]	2.9–8.7	Salazar et al. (2009)
Dry fermented sausages (chorizo)	Inulin	3–10[b]	Beriain et al. (2011)
Turkish dry fermented sausage (sucuk)	Orange fiber	2.0–4.0[b]	Yalinkilic et al. (2012)
Dry fermented sausage	Konjac flour	0.7[b]	Ruiz-Capillas et al. (2012)
Fermented sausages	Soy fiber	1.0[b]	Bastianello Campagnol et al. (2013)
Dry fermented chicken sausage	Inulin	6.8[b]	Menegas et al. (2013)
Dry fermented sausage (longaniza de Pascua)	Tiger Nut Fiber	1–2[b]	Sanchez-Zapata et al. (2013)

[a]Short-chain fructooligosaccharides.
[b]% of added ingredient.
Adapted from Jiménez-Colmenero, F., Delgado-Pando, G., 2013. Fiber-enriched meat products. In: Delcour, J.A., Poutanen, K. (Eds.), Fiber-Rich and Wholegrain Foods: Improving Quality, vol. 237. Woodhead Publishing Series in Food Science, Technology and Nutrition. pp. 329–347.

TABLE 10.5 Examples of Fermented Sausages With Improved Mineral Composition

Product	Mineral Source	Improvement	References
Reduced sodium:		% NaCl reduction:	
Dry fermented sausages	KCl	50	Ibañez et al. (1995)
Dry fermented sausages (chorizo)	Mixtures: KCl and $CaCl_2$	60	Gimeno et al. (2001)
Fermented sausages	KCl, K-lactate or glycine	60–100	Gou et al. (1996)
Fermented sausages	Mixtures: KCl, K-lactate and glycine	40	Gelabert et al. (2003)
Small caliber fermented sausages	Mixtures: KCl and K-lactate	50	Guardia et al. (2008)
Italian dry fermented sausage (salami)	Mixture: KCl, $MgCl_2$, and $CaCl_2$	50	Zanardi et al. (2010)
Fermented sausages	KCl or mixture: KCl + yeast extract	25–50	Bastianello Campagnol et al. (2011)
Dry fermented sausages (chorizo)	Calcium ascorbate	38–50	García-Iñiguez de Ciriano et al. (2013)
Fermented sausages	KCl, disodium guanylate (IMP)/ disodium inosinate (DMP) and mixture: lysine + IMP/DMP	50	Bastianello Campagnol et al. (2012)
Slow fermented sausages	KCl	16	Corral et al. (2013)
Dry fermented sausages	KCl, *Debaryomyces hansenii* yeast	17–20	Corral et al. (2014)
Nonacid fermented sausages (fuet)	Mixture: KCl and K-lactate	30–35	Mora-Gallego et al. (2014)
Dry fermented sausages	KCl, $CaCl_2$, and mixture: KCl and $CaCl_2$	50	Dos Santos et al. (2015)

TABLE 10.5 Examples of Fermented Sausages With Improved Mineral Composition—cont'd

Product	Mineral Source	Improvement	References
Calcium-enriched:		Ca content in mg/100 g:	
Dry fermented sausages	Calcium chloride	182	Gimeno et al. (1998)
Dry fermented sausage (chorizo)	Calcium chloride	–	Gimeno et al. (2001)
Fermented pork sausage (nhams)	Commercial or hen egg shell calcium lactate	150–450	Daengprok et al. (2002)
Dry fermented sausages	Calcium chloride	–	Flores et al. (2005)
Dry fermented sausages	Calcium lactate, gluconate and citrate	292–304	Selgas et al. (2009)
Italian dry fermented sausage (salami)	Calcium chloride	–	Zanardi et al. (2010)
Dry fermented sausage (chorizo)	Calcium chloride and calcium alginate	–	Beriain et al. (2011)
Dry fermented sausages (chorizo)	Calcium ascorbate	400–526	García-Iñiguez de Ciriano et al. (2013)
Dry fermented sausage	Calcium citrate-malate and calcium lactate	301–311	Soto et al. (2014)
Dry fermented sausages	Calcium chloride	–	Dos Santos et al. (2015)
Magnesium-enriched:		Mg content in mg/100 g	
Dry fermented sausages	Magnesium chloride	182	Gimeno et al. (1998)
Italian dry fermented sausage (salami)	Magnesium chloride	–	Zanardi et al. (2010)

Continued

TABLE 10.5 Examples of Fermented Sausages With Improved Mineral Composition—cont'd

Product	Mineral Source	Improvement	References
Selenium-enriched:		Se content in µg/100 g	
Dry fermented sausages (chorizo)	Selenium yeast	364	García-Iñiguez de Ciriano et al. (2010)
Iodine-enriched:		I content in µg/100 g	
Dry fermented sausages (chorizo)	Iodized salt	208	García-Iñiguez de Ciriano et al. (2010)

health and, therefore, a gradual and sustained reduction in the amount of salt added to food by the food industry is needed (He and MacGregor, 2009). However, sodium chloride (NaCl) is a multifunctional ingredient that plays an essential role in inhibiting microbial growth and ensuring pleasant texture and good taste in processed meat. This multipurpose use represents an important challenge for the meat industry. The strategies used to reduce it are generally based on using alternative ingredients, such as salt substitutes, flavor enhancers, masking agents, and optimization of the physical form of salt (Desmond, 2006) as well as modification of processing techniques (Jiménez-Colmenero et al., 2012). All are designed to maintain the characteristic properties of the product and reduce the adverse effects that salt reduction has on shelf-life, food safety, product texture, yield, and taste. A large number of salt reduction (and even salt-free) studies have been carried out with regard to meat products in recent years. Some of these have targeted fermented sausages since these foods are consumed worldwide and are among the meat products with the highest NaCl content, reaching as high as 6% (Bastianello Campagnol et al., 2011).

Because of the technological contribution of NaCl, the feasibility of direct salt reduction in meat products is limited. The most typical and versatile strategy is the replacement of NaCl by potassium chloride (KCl) alone or in combination with other compounds (Table 10.5). KCl has been used in combination with $CaCl_2$ or $MgCl_2$, although depending on the replacement ratio of each salt, some undesirable sensory effects like bitterness and metallic aftertaste may emerge. This means that careful studies need to be conducted to identify the correct proportions for each salt (Toldrá and Reig, 2011). KCl has also been combined with potassium lactate, glycine, yeast extract, lysine, etc., as a salt-reduction strategy. The partial substitution of NaCl by calcium ascorbate also allows reduction in sodium content (Table 10.5).

10.3.5.2 Calcium, Magnesium, Selenium, and Iodine Enrichment

Calcium (Ca) is one of the most important nutrients in the human diet. Inadequate Ca intake has been associated with increased incidence of osteoporosis, hypertension, and colon cancer (Weaver, 2014). Since meat is a rather poor source of Ca, fortification of meat products with Ca may be an excellent opportunity to enhance Ca intake. Various technological strategies based on the addition of different Ca salts have been used to achieve Ca-enriched fermented sausage (Table 10.5). In fact, Ca salts have been used in some cases as part of the technological strategy to reduce NaCl (Beriain et al., 2011; Flores et al., 2005), whereas in other cases they were specifically designed to increase nutritional value (Daengprok et al., 2002; Gimeno et al., 1998; Selgas et al., 2009), reaching levels that ranged between 150 and 526 mg/100 g (Table 10.5). A priority for functional food design should be not only to know the final characteristics of the enriched food but also to evaluate its Ca bioavailability. With this aim, the Ca bioavailability of dry-fermented calcium-enriched sausage using citrate-malate and Ca lactate has been studied by means of in vitro gastrointestinal digestion followed by an assay with a Caco2 cell culture (Soto et al., 2014). The percentage of Ca available and Ca transported through the cell membrane in dry fermented sausages was less than 10% of the Ca added to the product, much lower than that of fresh meat products, bologna sausages, and milk.

In addition to Ca, there are other elements of interest, such as magnesium (Mg), selenium (Se), and iodine (I) that have been used in the development of healthier fermented sausages. Fermented sausages contain Mg levels in the vicinity of 2 mg/100 g. Mg is responsible for the activation of more than 300 enzymes, helping to maintain muscle and nerve function in the human body (Fayardi, 2012). Magnesium deficiency is linked to various pathological conditions. Partial NaCl replacement strategies in dry fermented sausage based on the use of different salts, including $MgCl_2$, would significantly boost Mg levels to the point that these products could be considered as a viable source of this mineral (Gimeno et al., 1998). Pork meat contains approximately 10–14 μg of Se per 100 g (Higgs, 2000), while fermented sausages contain up to 35 μg/100 g (Table 10.1). As an essential component of glutathione peroxidase and other selenoproteins, the dietary intake of Se may favor cognitive and immune-system functions. It also has antiviral properties and reduces the risk of coronary disease, cancer, etc. (Rayman, 2008; The British Nutrition Foundation, 2001). Nutritional and technological strategies have been used to produce Se-enriched meat and meat products (Jiménez-Colmenero et al., 2012). Apart from taking organic Se as a dietary supplement, this element can also be added directly to fermented sausages (García-Iñiguez de Ciriano et al., 2010). Iodine deficiency is a major health problem in Europe and other areas of the world (WHO, 2007), goiter being its main clinical manifestation. Other consequences include brain damage and irreversible mental retardation (García-Iñiguez de Ciriano et al., 2010). Iodine fortification of meat products can be achieved by reformulation processes. With this aim, modified dry fermented sausages were made with iodized salt, with the new formulation supplying approximately 208 µg I/100 g of product (Table 10.5).

10.3.6 Addition of Vitamins and Antioxidants

Different vitamins and antioxidants have been incorporated into fermented sausages in order to make them healthier or to reduce the formation of unhealthy components such as those resulting from lipid oxidation. Tocopherol analogs are commonly used antioxidants in a wide range of foods pursuing a dual objective. First, they contribute to product stability by limiting the onset of oxidation and rancidity. Different undesirable compounds with negative health implications (cytotoxic and mutagenic) may be formed during these oxidation processes (Dobarganes and Marquez-Ruiz, 2003). Secondly, the dietary intake of antioxidants plays a crucial role in preventing oxidative damage and reducing the risk of various diseases. In addition to tocols (tocopherols and tocotrienols), other antioxidant compounds, such as vitamin C, carotenoids, flavonoids, and phenolic compounds may be used. Some of these have been incorporated into fermented sausages, either as specific preparations or as complex nonmeat ingredients (extract, powder, etc.) (Table 10.6). In addition, the antioxidant nutrient profile of meat products has also been improved by adding vitamin E or vitamin C (Table 10.6). Carotenoids have been associated with antioxidant activity and reduction of the risk of CVD, cancer, age-related macular degeneration, and cataracts (Krinsky and Johnson, 2005). Lycopene has been used in fermented-sausage reformulation (Table 10.6). Some plant derivatives rich in natural antioxidants (eg, flavonoids and phenolic compounds) have also been used to enhance the oxidative stability of different types of fermented sausages (Table 10.6). Some of these plant-derived compounds also have antiinflammatory and anticancer activities and exhibit antimicrobial activity (Zhang et al., 2010).

Different vitamins have been incorporated into fermented sausages to make them healthier. Folic acid is associated with the prevention of several diseases, such as neural tube defects in birth and some cancers and CVDs. Deficiencies of this vitamin have been reported in some societies or specific consumer groups. For example, in European countries over 90% of women of childbearing age fail to consume optimal levels of dietary folate (Biesalski, 2005). Some frequently consumed products have been fortified with this vitamin over the last several years to provide required levels. This is the case for dry fermented sausages manufactured with concentrations of folic acid (0.88–3.37 mg/100 g), assuring its daily recommended allowance (Galán et al., 2011).

10.3.7 Reduction of Biogenic Amine Formation

Biogenic amines are low molecular weight nitrogenous compounds found in a wide range of foods including meat and meat products. They are mainly

TABLE 10.6 Examples of Fermented Sausages With Vitamin and/or Antioxidant Improvement

Product	Vitamin and Antioxidant Type or Source	References
Dry fermented sausage (chorizo)	Paprika and garlic	Aguirrezabal et al. (2000)
Fermented sausage	Freeze-dried kimchi-powder (paprika, garlic, and ginger)	Lee and Kunz (2005)
Dry fermented sausage (lukanka)	Mixture: ascorbyl palmitate, tocopherols, esters of citric acid	Balev et al. (2005)
	Rosemary, routine, sodium erythorbate, and L-ascorbic acid	
Dry fermented sausage (sucuk)	Green tea extract and *Thymbraspicata* oil	Bozkurt (2006)
Dry fermented sausage (sucuk)	Sesame and *Thymbraspicata* oil	Bozkurt (2007)
Dutch style fermented sausages	*Geranium macrorrhizum*, *Potentilla fruticosa* and *Rosmarinus officinalis* extracts	Miliauskas et al. (2007)
Dry fermented sausage (salchichón)	Lycopene (dry tomato peel)	Calvo et al. (2008)
Dry fermented sausage (sucuk)	Extract of *Urticadioica*, *Hibiscus sabdariffa*	Karabacak and Bozkurt (2008)
Fermented dry-cured sausages	Tocopherol extract	Magrinya et al. (2009)
Dry fermented sausages (chorizo)	*Borago officinalis* extract	García-Iñiguez de Ciriano et al. (2009)
Dry fermented sausages (chorizo)	*Melissa officinalis* L. extract	García-Iñiguez de Ciriano et al. (2010)
Dry fermented sausages (salchichón)	Folic acid	Galán et al. (2011)
Fermented Italian-type sausages	Dried extract of mate leaves (*Ilex paraguariensis* St. Hil)	Beal et al. (2011)
Dry-fermented sausage (sucuk)	Rosemary powder, rosemary extract and α-tocopherol, and their combinations	Gok et al. (2011)

Continued

TABLE 10.6 Examples of Fermented Sausages With Vitamin and/or Antioxidant Improvement—cont'd

Product	Vitamin and Antioxidant Type or Source	References
Dry fermented sausage (chorizo)	Grape seed and chestnut extract	Lorenzo et al. (2013)
Fermented paprika sausages	Lycopene and rosemary extract	Rohlik et al. (2013)
Fermented dry sausages	Extract of *Kitaibelia vitifolia*	Kurcubic et al. (2014)

formed by the decarboxylation of free amino acids due to the action of the amino-acid decarboxylase, an enzyme of microbial origin. Consumption of foods with high concentrations of biogenic amines (tyramine and histamine being the most biologically active) is a potential public health concern due to their physiological and toxicological effects. Fermented sausages may contain high concentrations of biogenic amines since their accumulation is mainly related to the action of decarboxylase-positive bacteria and meat enzymes during fermentation and ripening (Ruiz-Capillas and Jiménez-Colmenero, 2004). It is thus important to take the necessary measures to reduce biogenic amine formation in fermented sausages. Most of such measures aim to inhibit the growth of biogenic amine-producing microorganisms and/or to select a suitable starter with low or no amino-acid decarboxylase activity, which also helps to develop the required sensorial attributes of each product (Ruiz-Capillas et al., 2011). The ability of microorganisms to produce biogenic amines must be considered as a criterion when selecting probiotic strains since they have potential for different biogenic amine formation (Priyadarshani and Rakshit, 2011). Natural antioxidants such as green tea extract and *Thymbraspicata* oil reduce the formation of biogenic amines (eg, putrescine, histamine, and tyramine) in Turkish dry-fermented sausage (Bozkurt, 2006). Spermine and spermidine belong to a different group of amines that are endogenous in meats and meat products. These polyamines were reported to have a number of positive health attributes (Kalac, 2014; Larque et al., 2007) and antioxidant activity (Decker and Xu, 1998).

10.3.8 Curing Agents: Nitrite Reduction

Nitrate and nitrite are added to meat products because of the role they play in enhancing color and flavor, microbial growth control, and antioxidant activity. However, nitrite is recognized as a potentially toxic compound in cured meat, mainly associated with the formation of carcinogens by reaction with some biogenic amines and formation of *N*-nitrosamines in food or after ingestion. To limit these harmful effects, nitrite reduction (or even omission) in meat formulation and/or the use of strategies to reduce damage arising from nitrosylation reactions have been proposed (Demeyer et al., 2008; Pegg and Shahidi, 2000). The addition of other sources of nitrite (eg, nitrate-rich vegetable powders) is an alternative approach to replacing pure nitrite or nitrate (Sebranek and Bacus, 2007), but those strategies do not necessarily contribute to a reduction in residual nitrate and/or nitrite levels (Magrinya et al., 2009). Nitrite reduction by glucono-δ-lactone and ascorbic acid in Turkish-type fermented sausage (Sucuk) was reported by Yilmaz and Zorba (2010). The ethanol extract from the aerial parts (stems, leaves, and flowers) of *Kitaibelia vitifolia* were also used as a nitrite substitute to formulate nonnitrite-added dry-fermented sausages because of their antioxidant and antimicrobial properties (Kurcubic et al., 2014).

10.4 CONCLUSION

Fermented sausages are especially popular due to their sensorial characteristics. They form part of a food group characterized by high-density nutrients, making them particularly attractive for certain situations and groups of people such as growing children. However, worldwide health problems like obesity, hypertension, and CVD are associated with excessive caloric intake, high saturated fat content, cholesterol, and sodium. Therefore due to their characteristics, fermented products must be consumed moderately as part of a healthy diet.

While the consumption of meat (particularly red meat) and meat products is related to the incidence of diseases such as colon cancer, there is no conclusive evidence that fermented sausages are particularly dangerous. Thus it is important to note that biogenic amine content can be reduced by good manufacturing practices. The same is true of some cancer-causing compounds that can appear as a consequence of thermal and/or smoking processes. In addition, nitrosamine levels are very low, especially in products eaten raw, and levels in their precursors (nitrates and nitrites) are lower than in many vegetables. The addition of nitrite is necessary to ensure sensorial characteristics and safety. It also controls oxidation in the product that could otherwise reach unhealthy levels. Despite these negative aspects, fermented sausages provide an array of essential nutrients, such as vitamins and minerals (sometimes with better absorption rates). They also contain trace compounds, which, due to their importance in the human body, may also be of great interest. Peptides with their inhibitory effect on ACE can help to offset the relatively high Na content of these products.

Thus considering all factors and the most common health risks, the elimination of these products from one's diet for food safety reasons could be more detrimental than the benefit obtained from their regular consumption. Nevertheless, different strategies are being employed to develop healthier sausages. These strategies are designed to improve certain aspects of these products and limit others. The most efficient ones are based on the reformulation of these products with the aim of producing sausages that are more appropriate for general consumption and/or target a sector of the population with specific needs.

ACKNOWLEDGMENTS

The authors acknowledge the following projects enabling this research: AGL 2011-29644-C02-01 (MINECO) and Intramural CSIC: 201470E056. Ricard Bou thanks the Spanish Ministry of Economy and Competitiveness for the Ramón y Cajal program contract.

REFERENCES

AESAN, 2009. Plan de reducción del consumo de sal La Granja de San Ildefonso, Spain. http://www.naos.aesan.msps.es/naos/ficheros/estrategia/Memoria_Plan_de_reduccion_del_consumo_de_sal_-_Jornadas_de_debate.pdf.

Aguirrezabal, M.M., Mateo, J., Dominguez, M.C., Zumalacarregui, J.M., 2000. The effect of paprika, garlic and salt on rancidity in dry sausages. Meat Science 54, 77–81.

Ammor, M.S., Mayo, B., 2007. Selection criteria for lactic acid bacteria to be used as functional starter cultures in dry sausage production: an update. Meat Science 76, 138–146.

Andersen, L., 1998. Fermented dry sausages produced with the admixture of probiotic cultures. In: Paper Presented at the 44th International Congress of Meat Science and Technology.

Ansorena, D., Astiasaran, I., 2004. The use of linseed oil improves nutritional quality of the lipid fraction of dry-fermented sausages. Food Chemistry 87, 69–74.

Arihara, K., 2006. Strategies for designing novel functional meat products. Meat Science 74, 219–229.

Arihara, K., Ohata, M., 2008. Bioactive compounds in meat. In: Toldrá, F. (Ed.), Meat Biotechnology. Springer Science+Business Media LLC, New York, pp. 231–249.

Arihara, K., Ota, H., Itoh, M., Kondo, Y., Sameshima, T., Yamanaka, H., et al., 1998. *Lactobacillus acidophilus* group lactic acid bacteria applied to meat fermentation. Journal of Food Science 63, 544–547.

Ashwell, M., 2002. Concepts of Functional Foods. International Life Science Institute, Brussels, Belgium.

Baghurst, K., 2007. Red meat and food guides. Nutrition & Dietetics 64, S140–S142.

Balev, D., Vulkova, T., Dragoev, S., Zlatanov, M., Bahtchevanska, S., 2005. A comparative study on the effect of some antioxidants on the lipid and pigment oxidation in dry-fermented sausages. International Journal of Food Science and Technology 40, 977–983.

Barbosa, M.D., Todorov, S.D., Ivanova, I., Chobert, J.M., Haertle, T., Franco, B., 2015. Improving safety of salami by application of bacteriocins produced by an autochthonous *Lactobacillus curvatus* isolate. Food Microbiology 46, 254–262.

Bastianello Campagnol, P.C., dos Santos, B.A., Terra, N.N., Rodrigues Pollonio, M.A., 2012. Lysine, disodium guanylate and disodium inosinate as flavor enhancers in low-sodium fermented sausages. Meat Science 91, 334–338.

Bastianello Campagnol, P.C., dos Santos, B.A., Wagner, R., Terra, N.N., Rodrigues Pollonio, M.A., 2011. The effect of yeast extract addition on quality of fermented sausages at low NaCl content. Meat Science 87, 290–298.

Bastianello Campagnol, P.C., Dos Santos, B.A., Wagner, R., Terra, N.N., Rodrigues Pollonio, M.A., 2013. The effect of soy fiber addition on the quality of fermented sausages at low-fat content. Journal of Food Quality 36, 41–50.

Baum, S.J., Kris-Etherton, P.M., Willett, W.C., Lichtenstein, A.H., Rudel, L.L., Maki, K.C., et al., 2012. Fatty acids in cardiovascular health and disease: a comprehensive update. Journal of Clinical Lipidology 6, 216–234.

Beal, P., Faion, A.M., Cichoski, A.J., Cansian, R.L., Valduga, A.T., de Oliveira, D., et al., 2011. Oxidative stability of fermented Italian-type sausages using mate leaves (*Ilex paraguariensis* St. Hil) extract as natural antioxidant. International Journal of Food Sciences and Nutrition 62, 703–710.

Benoist, B., McLean, E., Cogswell, M., Egli, I., Wojdyla, D., 2008. Worldwide Prevalence of Anaemia 1993–2005. WHO Global Database on Anaemia. (Geneva, Switzerland) http://whqlibdoc.who.int/publications/2008/9789241596657_eng.pdf.

Beriain, M.J., Gomez, I., Petri, E., Insausti, K., Sarries, M.V., 2011. The effects of olive oil emulsified alginate on the physicochemical, sensory, microbial, and fatty acid profiles of low-salt, inulin-enriched sausages. Meat Science 88, 189–197.

Beriain, M.J., Lizaso, G., Chasco, J., 2000. Free amino acids and proteolysis involved in 'salchichon' processing. Food Control 11, 41–47.

Biesalski, H.K., 2005. Meat as a component of a healthy diet—are there any risks or benefits if meat is avoided in the diet? Meat Science 70, 509–524.

Bjorkhem, I., Meaney, S., Diczfalusy, U., 2002. Oxysterols in human circulation: which role do they have? Current Opinion in Lipidology 13, 247–253.

Bloukas, J.G., Paneras, E.D., Fournitzis, G.C., 1997. Effect of replacing pork backfat with olive oil on processing and quality characteristics of fermented sausages. Meat Science 45, 133–144.

Bozkurt, H., 2006. Utilization of natural antioxidants: green tea extract and *Thymbra spicata* oil in Turkish dry-fermented sausage. Meat Science 73, 442–450.

Bozkurt, H., 2007. Comparison of the effects of sesame and *Thymbra spicata* oil during the manufacturing of Turkish dry-fermented sausage. Food Control 18, 149–156.

Broncano, J.M., Otte, J., Petron, M.J., Parra, V., Timon, M.L., 2012. Isolation and identification of low molecular weight antioxidant compounds from fermented "chorizo" sausages. Meat Science 90, 494–501.

Bruinsma, J., 2003. World Agriculture: Towards 2015/2030. http://www.fao.org/fileadmin/user_upload/esag/docs/y4252e.pdf.

Calvo, M.M., Garcia, M.L., Selgas, M.D., 2008. Dry fermented sausages enriched with lycopene from tomato peel. Meat Science 80, 167–172.

Campbell-Platt, G., Cook, P.E., 1995. Fermented Meats, first ed. Blackie Academic & Professional, London, New York.

Candogan, K., Wardlaw, F.B., Acton, J.C., 2009. Effect of starter culture on proteolytic changes during processing of fermented beef sausages. Food Chemistry 116, 731–737.

Castellano, P., Aristoy, M.C., Sentandreu, M.A., Vignolo, G., Toldrá, F., 2013. Peptides with angiotensin I converting enzyme (ACE) inhibitory activity generated from porcine skeletal muscle proteins by the action of meat-borne *Lactobacillus*. Journal of Proteomics 89, 183–190.

Connor, W.E., 2000. Importance of n-3 fatty acids in health and disease. American Journal of Clinical Nutrition 71, 171S–175S.

Corral, S., Salvador, A., Belloch, C., Flores, M., 2014. Effect of fat and salt reduction on the sensory quality of slow fermented sausages inoculated with *Debaryomyces hansenii* yeast. Food Control 45, 1–7.

Corral, S., Salvador, A., Flores, M., 2013. Salt reduction in slow fermented sausages affects the generation of aroma active compounds. Meat Science 93, 776–785.

Cross, A.J., Ferrucci, L.M., Risch, A., Graubard, B.I., Ward, M.H., Park, Y., et al., 2010. A large prospective study of meat consumption and colorectal cancer risk: an investigation of potential mechanisms underlying this association. Cancer Research 70, 2406–2414.

Chan, K.M., Decker, E.A., 1994. Endogenous skeletal-muscle antioxidants. Critical Reviews in Food Science and Nutrition 34, 403–426.

Daengprok, W., Garnjanagoonchorn, W., Mine, Y., 2002. Fermented pork sausage fortified with commercial or hen eggshell calcium lactate. Meat Science 62, 199–204.

Decker, E.A., Xu, Z.M., 1998. Minimizing rancidity in muscle foods. Food Technology 52, 54–59.

Demeyer, D., Honikel, K., De Smet, S., 2008. The world cancer research fund report 2007: a challenge for the meat processing industry. Meat Science 80, 953–959.

Demeyer, D., Raemaekers, M., Rizzo, A., Holck, A., De Smedt, A., ten Brink, B., et al., 2000. Control of bioflavour and safety in fermented sausages: first results of a European project. Food Research International 33, 171–180.

Desmond, E., 2006. Reducing salt: a challenge for the meat industry. Meat Science 74, 188–196.

Díaz, O., Fernandez, M., García de Fernando, G.D., de la Hoz, L., Ordoñez, J.A., 1997. Proteolysis in dry fermented sausages: the effect of selected exogenous proteases. Meat Science 46, 115–128.

Dilzer, A., Park, Y., 2012. Implication of conjugated linoleic acid (CLA) in human health. Critical Reviews in Food Science and Nutrition 52, 488–513.

Dobarganes, C., Marquez-Ruiz, G., 2003. Oxidized fats in foods. Current Opinion in Clinical Nutrition and Metabolic Care 6, 157–163.

Dos Santos, B.A., Campagnol, P.C.B., Cavalcanti, R.N., Pacheco, M.T.B., Netto, F.M., Motta, E.M.P., et al., 2015. Impact of sodium chloride replacement by salt substitutes on the proteolysis and rheological properties of dry fermented sausages. Journal of Food Engineering 151, 16–24.

EFSA, 2010. Scientific Opinion on Dietary Reference Values for Fats, Including Saturated Fatty Acids, Polyunsaturated Fatty Acids, Monounsaturated Fatty Acids, Trans Fatty Acids, and Cholesterol Parma, Italy. http://www.efsa.europa.eu/en/efsajournal/pub/1461.htm.

Eim, V.S., Simal, S., Rossello, C., Femenia, A., 2008. Effects of addition of carrot dietary fibre on the ripening process of a dry fermented sausage (sobrassada). Meat Science 80, 173–182.

Eim, V.S., Simal, S., Rossello, C., Femenia, A., Bon, J., 2013. Optimisation of the addition of carrot dietary fibre to a dry fermented sausage (sobrassada) using artificial neural networks. Meat Science 94, 341–348.

Elmadfa, I., Freisling, H., 2009. Nutritional status in Europe: methods and results. Nutrition Reviews 67, S130–S134.

Ercoskun, H., Demirci-Ercoskun, T., 2010. Walnut as fat replacer and functional component in sucuk. Journal of Food Quality 33, 646–659.

Erkkila, S., Suihko, M.L., Eerola, S., Petaja, E., Mattila-Sandholm, T., 2001. Dry sausage fermented by *Lactobacillus rhamnosus* strains. International Journal of Food Microbiology 64, 205–210.

Estruch, R., Ros, E., Salas-Salvado, J., Covas, M.-I., Corella, D., Aros, F., et al., 2013. Primary prevention of cardiovascular disease with a mediterranean diet. New England Journal of Medicine 368, 1279–1290.

Fayardi, Q., 2012. The magnificent effect of magnesium to human health: a critical review. International Journal of Applied Science and Technology 2, 118–126.

Fernandez-Lopez, J., Sendra, E., Sayas-Barbera, E., Navarro, C., Perez-Alvarez, J.A., 2008. Physicochemical and microbiological profiles of "salchichon" (Spanish dry-fermented sausage) enriched with orange fiber. Meat Science 80, 410–417.

Flores, M., Nieto, P., Ferrer, J.M., Flores, J., 2005. Effect of calcium chloride on the volatile pattern and sensory acceptance of dry-fermented sausages. European Food Research and Technology 221, 624–630.

Fontana, C., Cocconcelli, P.S., Vignolo, G., Saavedra, L., 2015. Occurrence of antilisterial structural bacteriocins genes in meat borne lactic acid bacteria. Food Control 47, 53–59.

Food Standards Agency, 2010. McCance and Widdowson's the Composition of Foods Integrated Dataset (CoF IDS). Retrieved: http://tna.europarchive.org/20110116113217/http:/www.food.gov.uk/science/dietarysurveys/dietsurveys/.

Fuentes, V., Estevez, M., Ventanas, J., Ventanas, S., 2014. Impact of lipid content and composition on lipid oxidation and protein carbonylation in experimental fermented sausages. Food Chemistry 147, 70–77.

Galán, I., García, M.L., Selgas, M.D., 2011. Effects of ionising irradiation on quality and sensory attributes of ready-to-eat dry fermented sausages enriched with folic acid. International Journal of Food Science and Technology 46, 469–477.

García-Iñiguez de Ciriano, M., Berasategi, I., Navarro-Blasco, I., Astiasaran, I., Ansorena, D., 2013. Reduction of sodium and increment of calcium and omega-3 polyunsaturated fatty acids in dry fermented sausages: effects on the mineral content, lipid profile and sensory quality. Journal of the Science of Food and Agriculture 93, 876–881.

García-Iñiguez de Ciriano, M., García-Herreros, C., Larequi, E., Valencia, I., Ansorena, D., Astiasarán, I., 2009. Use of natural antioxidants from lyophilized water extracts of *Borago officinalis* in dry fermented sausages enriched in omega-3 PUFA. Meat Science 83, 271–277.

García-Iñiguez de Ciriano, M., Larequi, E., Rehecho, S., Calvo, M.I., Cavero, R.Y., Navarro-Blasco, I., et al., 2010. Selenium, iodine, omega-3 PUFA and natural antioxidant from *Melissa officinalis* L.: a combination of components from healthier dry fermented sausages formulation. Meat Science 85, 274–279.

Garcia, M.L., Dominguez, R., Galvez, M.D., Casas, C., Selgas, M.D., 2002. Utilization of cereal and fruit fibres in low fat dry fermented sausages. Meat Science 60, 227–236.

Gelabert, J., Gou, P., Guerrero, L., Arnau, J., 2003. Effect of sodium chloride replacement on some characteristics of fermented sausages. Meat Science 65, 833–839.

Gimeno, O., Astiasaran, I., Bello, J., 1998. A mixture of potassium, magnesium, and calcium chlorides as a partial replacement of sodium chloride in dry fermented sausages. Journal of Agricultural and Food Chemistry 46, 4372–4375.

Gimeno, O., Astiasaran, I., Bello, J., 2001. Calcium ascorbate as a potential partial substitute for NaCl in dry fermented sausages: effect on colour, texture and hygienic quality at different concentrations. Meat Science 57, 23–29.

Girodon, F., Galan, P., Monget, A.L., Boutron-Ruault, M.C., Brunet-Lecomte, P., Preziosi, P., et al., 1999. Impact of trace elements and vitamin supplementation on immunity and infections in institutionalized elderly patients—a randomized controlled trial. Archives of Internal Medicine 159, 748–754.

Gok, V., Obuz, E., Sahim, M.E., Serteser, A., 2011. The effects of some natural antioxidants on the color, chemical and microbiological properties of sucuk (Turkish dry-fermented sausage) during ripening and storage periods. Journal of Food Processing and Preservation 35, 677–690.

Gou, P., Guerrero, L., Gelabert, J., Arnau, J., 1996. Potassium chloride, potassium lactate and glycine as sodium chloride substitutes in fermented sausages and in dry-cured pork loin. Meat Science 42, 37–48.

Guardia, M.D., Guerrero, L., Gelabert, J., Gou, P., Arnau, J., 2008. Sensory characterisation and consumer acceptability of small calibre fermented sausages with 50% substitution of NaCl by mixtures of KCl and potassium lactate. Meat Science 80, 1225–1230.

He, F.J., MacGregor, G.A., 2009. A comprehensive review on salt and health and current experience of worldwide salt reduction programmes. Journal of Human Hypertension 23, 363–384.

Higgs, J.D., 2000. The changing nature of red meat: 20 years of improving nutritional quality. Trends in Food Science & Technology 11, 85–95.

Ibañez, C., Quintanilla, L., Irigoyen, A., Garciajalon, I., Cid, C., Astiasaran, I., et al., 1995. Partial replacemente of sodium chloride with potassium chloride in dry fermented sausages: influence on carbohydrate fermetnation and the nitrosation process. Meat Science 40, 45–53.

Institute of Medicine, 2001. Dietary Reference Intakes for Vitamin A, Vitamin K, Arsenic, Boron, Chromium, Copper, Iodine, Iron, Manganese, Molybdenum, Nickel, Silicon, Vanadium, and Zinc. National Academy Press, Washington, DC.

Institute of Medicine, 2002. Dietary Reference Intakes for Energy, Carbohydrates, Fiber, Protein and Amino Acids (Macronutrients). National Academy Press, Washington, DC.

Institute of Medicine, 1998. Dietary Reference Intakes for Thiamin, Riboflavin, Niacin, Vitamin B6, Folate, Vitamin B12, Pantothenic Acid, Biotin, and Choline. National Academy Press, Washington, DC.

Ito, T., Schaffer, S.W., Azuma, J., 2012. The potential usefulness of taurine on diabetes mellitus and its complications. Amino Acids 42, 1529–1539.

Jahreis, G., Vogelsang, H., Kiessling, G., Schubert, R., Bunte, C., Hammes, W.P., 2002. Influence of probiotic sausage (*Lactobacillus paracasei*) on blood lipids and immunological parameters of healthy volunteers. Food Research International 35, 133–138.

Jiménez-Colmenero, F., 2007. Healthier lipid formulation approaches in meat-based functional foods. Technological options for replacement of meat fats by non-meat fats. Trends in Food Science & Technology 18, 567–578.

Jiménez-Colmenero, F., Carballo, J., Cofrades, S., 2001. Healthier meat and meat products: their role as functional foods. Meat Science 59, 5–13.

Jiménez-Colmenero, F., Delgado-Pando, G., 2013. Fibre-enriched meat products. In: Delcour, J.A., Poutanen, K. (Eds.), Fibre-Rich and Wholegrain Foods: Improving Quality, vol. 237. Woodhead Publishing Series in Food Science, Technology, and Nutrition, pp. 329–347.

Jiménez-Colmenero, F., Herrero, A., Cofrades, S., Ruiz-Capillas, C., 2012. Meat and functional foods. In: Hui, Y.H. (Ed.), Handbook of Meat and Meat Processing. CRC Press, pp. 225–248.

Jimenez-Colmenero, F., Triki, M., Herrero, A.M., Rodriguez-Salas, L., Ruiz-Capillas, C., 2013. Healthy oil combination stabilized in a konjac matrix as pork fat replacement in low-fat, PUFA-enriched, dry fermented sausages. LWT—Food Science and Technology 51, 158–163.

Josquin, N.M., Linssen, J.P.H., Houben, J.H., 2012. Quality characteristics of Dutch-style fermented sausages manufactured with partial replacement of pork back-fat with pure, preemulsified or encapsulated fish oil. Meat Science 90, 81–86.

Kalac, P., 2014. Health effects and occurrence of dietary polyamines: a review for the period 2005–mid 2013. Food Chemistry 161, 27–39.

Karabacak, S., Bozkurt, H., 2008. Effects of *Urtica dioica* and *Hibiscus sabdariffa* on the quality and safety of sucuk (Turkish dry-fermented sausage). Meat Science 78, 288–296.

Kayaardi, S., Gok, V., 2004. Effect of replacing beef fat with olive oil on quality characteristics of Turkish soudjouk (sucuk). Meat Science 66, 249–257.

Khan, M.I., Arshad, M.S., Anjum, F.M., Sameen, A., Aneeq ur, R., Gill, W.T., 2011. Meat as a functional food with special reference to probiotic sausages. Food Research International 44, 3125–3133.

Klingberg, T.D., Axelsson, L., Naterstad, K., Elsser, D., Budde, B.B., 2005. Identification of potential probiotic starter cultures for Scandinavian-type fermented sausages. International Journal of Food Microbiology 105, 419–431.

Koeth, R.A., Wang, Z.E., Levison, B.S., Buffa, J.A., Org, E., Sheehy, B.T., et al., 2013. Intestinal microbiota metabolism of L-carnitine, a nutrient in red meat, promotes atherosclerosis. Nature Medicine 19, 576–585.

Koutsopoulos, D.A., Koutsimanis, G.E., Bloukas, J.G., 2008. Effect of carrageenan level and packaging during ripening on processing and quality characteristics of low-fat fermented sausages produced with olive oil. Meat Science 79, 188–197.

Krinsky, N.I., Johnson, E.J., 2005. Carotenoid actions and their relation to health and disease. Molecular Aspects of Medicine 26, 459–516.

Kris-Etherton, P.M., 1999. AHA Science Advisory: monounsaturated fatty acids and risk of cardiovascular disease. Journal of Nutrition 129, 2280–2284.

Kurcubic, V.S., Maskovic, P.Z., Vujic, J.M., Vranic, D.V., Veskovic-Moracanin, S.M., Okanovic, D.G., et al., 2014. Antioxidant and antimicrobial activity of *Kitaibelia vitifolia* extract as alternative to the added nitrite in fermented dry sausage. Meat Science 97, 459–467.

Larque, E., Sabater-Molina, M., Zamora, S., 2007. Biological significance of dietary polyamines. Nutrition 23, 87–95.

Lee, J.Y., Kunz, B., 2005. The antioxidant properties of baechu-kimchi and freeze-dried kimchi-powder in fermented sausages. Meat Science 69, 741–747.

Liu, S.N., Han, Y., Zhou, Z.J., 2011. Lactic acid bacteria in traditional fermented Chinese foods. Food Research International 44, 643–651.

Lorenzo, J.M., Gonzalez-Rodriguez, R.M., Sanchez, M., Amado, I.R., Franco, D., 2013. Effects of natural (grape seed and chestnut extract) and synthetic antioxidants (buthylatedhydroxytoluene, BHT) on the physical, chemical, microbiological and sensory characteristics of dry cured sausage "chorizo". Food Research International 54, 611–620.

Lucke, F.K., 1994. Fermented meat-products. Food Research International 27, 299–307.

Magrinya, N., Bou, R., Tres, A., Rius, N., Codony, R., Guardiola, F., 2009. Effect of tocopherol extract, *Staphylococcus carnosus* culture, and celery concentrate addition on quality parameters of organic and conventional dry-cured sausages. Journal of Agricultural and Food Chemistry 57, 8963–8972.

Marcinkiewicz, J., Kontny, E., 2014. Taurine and inflammatory diseases. Amino Acids 46, 7–20.

Mendoza, E., Garcia, M.L., Casas, C., Selgas, M.D., 2001. Inulin as fat substitute in low fat, dry fermented sausages. Meat Science 57, 387–393.

Menegas, L.Z., Pimentel, T.C., Garcia, S., Prudencio, S.H., 2013. Dry-fermented chicken sausage produced with inulin and corn oil: physicochemical, microbiological, and textural characteristics and acceptability during storage. Meat Science 93, 501–506.

Miliauskas, G., Mulder, E., Linssen, J.P.H., Houben, J.H., van Beek, T.A., Venskutonis, P.R., 2007. Evaluation of antioxidative properties of *Geranium macrorrhizum* and *Potentilla fruticosa* extracts in Dutch style fermented sausages. Meat Science 77, 703–708.

Mora-Gallego, H., Serra, X., Dolors Guardia, M., Arnau, J., 2014. Effect of reducing and replacing pork fat on the physicochemical, instrumental and sensory characteristics throughout storage time of small caliber nonacid fermented sausages with reduced sodium content. Meat Science 97, 62–68.

Mora-Gallego, H., Serra, X., Dolors Guardia, M., Miklos, R., Lametsch, R., Arnau, J., 2013. Effect of the type of fat on the physicochemical, instrumental and sensory characteristics of reduced fat nonacid fermented sausages. Meat Science 93, 668–674.

Moreiras, O., Carbajal, A., Cabrera, L., Cuadrado, C., 2006. Tabla de Composición de Alimentos, tenth ed. Ediciones Pirámide, Madrid.

Muguerza, E., Ansorena, D., Astiasaran, I., 2003. Improvement of nutritional properties of Chorizo de Pamplona by replacement of pork backfat with soy oil. Meat Science 65, 1361–1367.

Muguerza, E., Ansorena, D., Astiasaran, I., 2004. Functional dry fermented sausages manufactured with high levels of n-3 fatty acids: nutritional benefits and evaluation of oxidation. Journal of the Science of Food and Agriculture 84, 1061–1068.

Muguerza, E., Gimeno, O., Ansorena, D., Bloukas, J.G., Astiasaran, I., 2001. Effect of replacing pork backfat with preemulsified olive oil on lipid fraction and sensory quality of Chorizo de Pamplona—a traditional Spanish fermented sausage. Meat Science 59, 251–258.

Muthukumarasamy, P., Holley, R.A., 2006. Microbiological and sensory quality of dry fermented sausages containing alginate-microencapsulated *Lactobacillus reuteri*. International Journal of Food Microbiology 111, 164–169.

Nogueira Ruiz, J., Villanueva, N.D.M., Favaro-Trindade, C.S., Contreras-Castillo, C.J., 2014. Physicochemical, microbiological and sensory assessments of Italian salami sausages with probiotic potential. Scientia Agricola 71, 204–211.

Olmedilla-Alonso, B., Jiménez-Colmenero, F., Sánchez-Muniz, F.J., 2013. Development and assessment of healthy properties of meat and meat products designed as functional foods. Meat Science 95, 919–930.

Papamanoli, E., Kotzekidou, P., Tzanetakis, N., Litopoulou-Tzanetaki, E., 2002. Characterization of Micrococcaceae isolated from dry fermented sausage. Food Microbiology 19, 441–449.

Pegg, R.B., Shahidi, F., 2000. Nitrite Curing of Meat: The *N*-Nitrosamine Problem and Nitrite Alternatives. Food & Nutrition Press, Inc., Trumbull, CT.

Pelser, W.M., Linssen, J.P.H., Legger, A., Houben, J.H., 2007. Lipid oxidation in n-3 fatty acid enriched Dutch style fermented sausages. Meat Science 75, 1–11.

Pennacchia, C., Vaughan, E.E., Villani, F., 2006. Potential probiotic *Lactobacillus* strains from fermented sausages: further investigations on their probiotic properties. Meat Science 73, 90–101.

Petron, M.J., Broncano, J.M., Otte, J., Martin, L., Timon, M.L., 2013. Effect of commercial proteases on shelf-life extension of Iberian dry-cured sausage. LWT—Food Science and Technology 53, 191–197.

Priyadarshani, W.M.D., Rakshit, S.K., 2011. Screening selected strains of probiotic lactic acid bacteria for their ability to produce biogenic amines (histamine and tyramine). International Journal of Food Science and Technology 46, 2062–2069.

Prosky, L., 1999. What is fibre? Current controversies. Trends in Food Science and Technology 10, 271–275.

Rayman, M.P., 2008. Food-chain selenium and human health: emphasis on intake. British Journal of Nutrition 100, 254–268.

Rebucci, R., Sangalli, L., Fava, M., Bersani, C., Cantoni, C., Baldi, A., 2007. Evaluation of functional aspects in *Lactobacillus* strains isolated from dry fermented sausages. Journal of Food Quality 30, 187–201.

Reig, M., Aristoy, M.C., Toldrá, F., 2013. Variability in the contents of pork meat nutrients and how it may affect food composition databases. Food Chemistry 140, 478–482.

Reis, J.A., Paula, A.T., Casarotti, S.N., Penna, A.L.B., 2012. Lactic acid bacteria antimicrobial compounds: characteristics and applications. Food Engineering Reviews 4, 124–140.

Rivera-Espinoza, Y., Gallardo-Navarro, Y., 2010. Nondairy probiotic products. Food Microbiology 27, 1–11.

Rohlik, B.A., Pipek, P., Panek, J., 2013. The effect of natural antioxidants on the colour and lipid stability of paprika salami. Czech Journal of Food Sciences 31, 307–312.

Rouhi, M., Sohrabvandi, S., Mortazavian, A.M., 2013. Probiotic fermented sausage: viability of probiotic microorganisms and sensory characteristics. Critical Reviews in Food Science and Nutrition 53, 331–348.

Rubio, R., Jofre, A., Aymerich, T., Guardia, M.D., Garriga, M., 2014a. Nutritionally enhanced fermented sausages as a vehicle for potential probiotic lactobacilli delivery. Meat Science 96, 937–942.

Rubio, R., Martin, B., Aymerich, T., Garriga, M., 2014b. The potential probiotic *Lactobacillus rhamnosus* CTC1679 survives the passage through the gastrointestinal tract and its use as starter culture results in safe nutritionally enhanced fermented sausages. International Journal of Food Microbiology 186, 55–60.

Ruiz-Capillas, C., Herrero, A.M., Jiménez-Colmenero, F., 2011. Reduction of biogenic amines in meat and meat products. In: Rai, M., Chikindas, M.L. (Eds.), Natural Antimicrobials in Food Safety and Quality. CAB International, Oxforshire, UK, pp. 154–166.

Ruiz-Capillas, C., Jiménez-Colmenero, F., 2004. Biogenic amines in meat and meat products. Critical Reviews in Food Science and Nutrition 44, 489–499.

Ruiz-Capillas, C., Triki, M., Herrero, A.M., Rodriguez-Salas, L., Jimenez-Colmenero, F., 2012. Konjac gel as pork backfat replacer in dry fermented sausages: processing and quality characteristics. Meat Science 92, 144–150.

Ruiz-Moyano, S., Martin, A., Benito, M.J., Aranda, E., Casquete, R., de Guia Cordoba, M., 2011a. Implantation ability of the potential probiotic strain, *Lactobacillus reuteri* PL519, in "salchichon," a traditional Iberian dry fermented sausage. Journal of Food Science 76, M268–M275.

Ruiz-Moyano, S., Martin, A., Benito, M.J., Hernandez, A., Casquete, R., de Guia Cordoba, M., 2011b. Application of *Lactobacillus fermentum* HL57 and *Pediococcus acidilactici* SP979 as potential probiotics in the manufacture of traditional Iberian dry-fermented sausages. Food Microbiology 28, 839–847.

Salazar, P., Garcia, M.L., Selgas, M.D., 2009. Short-chain fructooligosaccharides as potential functional ingredient in dry fermented sausages with different fat levels. International Journal of Food Science and Technology 44, 1100–1107.

Sameshima, T., Magome, C., Takeshita, K., Arihara, K., Itoh, M., Kondo, Y., 1998. Effect of intestinal *Lactobacillus* starter cultures on the behaviour of *Staphylococcus aureus* in fermented sausage. International Journal of Food Microbiology 41, 1–7.

Sanchez-Zapata, E., Zunino, V., Perez-Alvarez, J.A., Fernandez-Lopez, J., 2013. Effect of tiger nut fibre addition on the quality and safety of a dry-cured pork sausage "Chorizo" during the dry-curing process. Meat Science 95, 562–568.

Sandstead, H.H., Frederickson, C.J., Penland, J.G., 2000. History of zinc as related to brain function. Journal of Nutrition 130, 496S–502S.

Sebranek, J.G., Bacus, J.N., 2007. Cured meat products without direct addition of nitrate or nitrite: what are the issues? Meat Science 77, 136–147.

Selgas, M.D., Salazar, P., Garcia, M.L., 2009. Usefulness of calcium lactate, citrate and gluconate for calcium enrichment of dry fermented sausages. Meat Science 82, 478–480.

Severini, C., De Pilli, T., Baiano, A., 2003. Partial substitution of pork backfat with extra-virgin olive oil in 'salami' products: effects on chemical, physical and sensorial quality. Meat Science 64, 323–331.

Sidira, M., Galanis, A., Nikolaou, A., Kanellaki, M., Kourkoutas, Y., 2014. Evaluation of *Lactobacillus casei* ATCC 393 protective effect against spoilage of probiotic dry-fermented sausages. Food Control 42, 315–320.

Simopoulos, A.P., 2002. The importance of the ratio of omega-6/omega-3 essential fatty acids. Biomedicine & Pharmacotherapy 56, 365–379.

Siri-Tarino, P.W., Sun, Q., Hu, F.B., Krauss, R.M., 2010. Meta-analysis of prospective cohort studies evaluating the association of saturated fat with cardiovascular disease. American Journal of Clinical Nutrition 91, 535–546.

Soto, A.M., Morales, P., Haza, A.I., Garcia, M.L., Selgas, M.D., 2014. Bioavailability of calcium from enriched meat products using Caco-2 cells. Food Research International 55, 263–270.

The British Nutrition Foundation, 2001. Selenium and Health. London, UK. http://www.nutrition.org.uk/publications/briefingpapers/selenium-and-health.html.

Toldrá, F., 1998. Proteolysis and lipolysis in flavour development of dry-cured meat products. Meat Science 49, S101–S110.

Toldrá, F., 2007. Handbook of Fermented Meat and Poultry. Blackwell Publishing, Oxford, UK.

Toldrá, F., 2004. Dry-Cured Meat Products. Food and Nutrition Press, Inc., Trumbull, CT.

Toldrá, F., Reig, M., 2011. Innovations for healthier processed meats. Trends in Food Science & Technology 22, 517–522.

Tyopponen, S., Petaja, E., Mattila-Sandholm, T., 2003. Bioprotectives and probiotics for dry sausages. International Journal of Food Microbiology 83, 233–244.

USDA, December 7, 2011. USDA National Nutrient Database for Standard Reference. National Agricultural Library. Retrieved from: http://ndb.nal.usda.gov/.

Utrilla, M.C., Garcia Ruiz, A., Soriano, A., 2014. Effect of partial replacement of pork meat with an olive oil organogel on the physicochemical and sensory quality of dry-ripened venison sausages. Meat Science 97, 575–582.

Valencia, I., Ansorena, D., Astiasaran, I., 2006. Nutritional and sensory properties of dry fermented sausages enriched with n-3 PUFAs. Meat Science 72, 727–733.

Vastag, Z., Popovic, L., Popovic, S., Petrovic, L., Pericin, D., 2010. Antioxidant and angiotensin-I converting enzyme inhibitory activity in the water-soluble protein extract from petrovac sausage (petrovska kolbasa). Food Control 21, 1298–1302.

Verma, A.K., Banerjee, R., 2010. Dietary fibre as functional ingredient in meat products: a novel approach for healthy living—a review. Journal of Food Science and Technology-Mysore 47, 247–257.

Vural, H., 2003. Effect of replacing beef fat and tail fat with interesterified plant oil on quality characteristics of Turkish semi-dry fermented sausages. European Food Research and Technology 217, 100–103.

Vuyst, L.D., Falony, G., Leroy, F., 2008. Probiotics in fermented sausages. Meat Science 80, 75–78.

Wang, C., Jiang, H., 2012. Meat intake and risk of bladder cancer: a meta-analysis. Medical Oncology 29, 848–855.

Ward, M.H., Cross, A.J., Abnet, C.C., Sinha, R., Markin, R.S., Weisenburger, D.D., 2012. Heme iron from meat and risk of adenocarcinoma of the esophagus and stomach. European Journal of Cancer Prevention 21, 134–138.

Weaver, C.M., 2014. Calcium supplementation: is protecting against osteoporosis counter to protecting against cardiovascular disease? Current Osteoporosis Reports 12, 211–218.

WHO, 2007. Iodine Deficiency in Europe: A Continuing Public Health Problem. http://www.who.int/nutrition/publications/VMNIS_Iodine_deficiency_in_Europe.pdf.

WHO, 2010. Strategies to Monitor and Evaluate Population Sodium Consumption and Sources of Sodium in the Diet. http://whqlibdoc.who.int/publications/2011/9789241501699_eng.pdf.

WHO, 2011. Global Status Report on Noncommunicable Diseases 2010. WHO Library Cataloguing-in-Publication Data, Geneva, Switzerland.

WHO/FAO Expert Consultation, 2003. Diet, Nutrition and the Prevention of Chronic Diseases. http://www.who.int/dietphysicalactivity/publications/trs916/en/.

Wojciak, K.M., Dolatowski, Z.J., Kolozyn-Krajewska, D., Trzaskowska, M., 2012. The effect of the *Lactobacillus casei* lock 0900 probiotic strain on the quality of dry-fermented sausage during chilling storage. Journal of Food Quality 35, 353–365.

Wu, G., 2013. Functional amino acids in nutrition and health. Amino Acids 45, 407–411.

Yalinkilic, B., Kaban, G., Kaya, M., 2012. The effects of different levels of orange fiber and fat on microbiological, physical, chemical and sensorial properties of sucuk. Food Microbiology 29, 255–259.

Yildiz-Turp, G., Serdaroglu, M., 2008. Effect of replacing beef fat with hazelnut oil on quality characteristics of sucuk—a Turkish fermented sausage. Meat Science 78, 447–454.

Yilmaz, M.T., Zorba, O., 2010. Response surface methodology study on the possibility of nitrite reduction by glucono-delta-lactone and ascorbic acid in Turkish-type fermented sausage (sucuk). Journal of Muscle Foods 21, 15–30.

Zammit, V.A., Ramsay, R.R., Bonomini, M., Arduini, A., 2009. Carnitine, mitochondrial function and therapy. Advanced Drug Delivery Reviews 61, 1353–1362.

Zanardi, E., Ghidini, S., Battaglia, A., Chizzolini, R., 2004. Lipolysis and lipid oxidation in fermented sausages depending on different processing conditions and different antioxidants. Meat Science 66, 415–423.

Zanardi, E., Ghidini, S., Conter, M., Ianieri, A., 2010. Mineral composition of Italian salami and effect of NaCl partial replacement on compositional, physicochemical and sensory parameters. Meat Science 86, 742–747.

Zanardi, E., Novelli, E., Chizzolini, R., Ghiretti, G.P., 2000. Oxidative stability of lipids and cholesterol in salame Milano, coppa and Parma ham: dietary supplementation with vitamin E and oleic acid. Meat Science 55, 169–175.

Zhang, W., Xiao, S., Samaraweera, H., Lee, E.J., Ahn, D.U., 2010. Improving functional value of meat products. Meat Science 86, 15–31.

Section 3.2

Dairy Fermented Foods

Chapter 11

Health Effects of Cheese Components with a Focus on Bioactive Peptides

I. López-Expósito, B. Miralles, L. Amigo, B. Hernández-Ledesma
Instituto de Investigación en Ciencias de la Alimentación (CSIC), Madrid, Spain

11.1 INTRODUCTION

Cheese is a nutritious and versatile dairy food with a long history in the human diet. The high fat and protein content in cheese makes it an energy-rich and nutritious food for all ages, and today a wide variety of cheese types are available. Europe is the biggest producer of cheese, with an average production of 8998000 American Ton during the period 2010–2012. Average per capita consumption of cheese in the EU was reported to be 17.2 kg during the period 2010–2012. France, Iceland, Luxembourg, Germany, and Finland were the biggest consumers, with more than 23.5 kg/capita during 2012 (Bulletin of the International Dairy Federation, 2013).

Cheese is a rich source of essential nutrients, in particular, proteins, fat, vitamins, and minerals. In addition to their nutritional role, cheese components have also shown to exert important health benefits (Walther et al., 2008). Calcium, present in large amounts in cheese, has been shown to have a positive effect on various disorders such as osteoporosis (Heaney, 2000) or dental caries (Kato et al., 2002). In addition to calcium, some fatty acids present in cheese such as conjugated linoleic acid (CLA) have shown to display anticarcinogenic and antiatherogenic properties (Lee et al., 2005; Battacharaya et al., 2006). Bioactive peptides are also present in cheese. They are produced during cheese ripening, where casein is broken down by proteases and peptidases from milk, rennet, starter culture, and secondary microbial flora. Some of the generated peptides can survive gastrointestinal digestion or serve as precursors from the final peptide form, which is responsible for a wide array of biological activities (López-Expósito et al., 2012).

This chapter presents an update on the health effects of cheese components focusing on the most commonly studied bioactive peptides activities. The degree of evidence reached and the bioavailability studies performed in each case are discussed.

11.2 EFFECTS OF CHEESE FAT ON HEALTH

The physical, organoleptic, and nutritional properties that milk lipids impart to dairy products make them one of the most important components of milk. However, their high saturated fatty-acid and cholesterol content has been associated with an increased incidence of obesity, metabolic disorders, and cardiovascular diseases (CVD), as noted by many nutritionists. Recent studies indicate that there are no demonstrated evidences to recommend the consumption of low-fat dairy foods in healthy people because of the biological activities demonstrated for several compounds present in milk fat (Fontecha et al., 2011).

CLA refers to a group of polyunsaturated fatty acids that exist as positional and stereo isomers of conjugated dienoic octadecadienoate (C18:2). The fatty-acid profile and CLA content of cheese (Table 11.1) are influenced by environmental factors, farming practices, and genetic and physiological aspects related to the animals (Collomb et al., 2006). Moreover, factors involved in the cheesemaking process, such as heat treatment of milk and/or curd, addition of starter cultures, and ripening have an important effect on CLA levels in cheese. It has been shown that certain lactic-acid bacteria used in milk fermentation have the ability to generate CLA (Prandini et al., 2007). Therefore recent studies have focused on increasing the CLA content of cheese by supplementing the animal's diet with vegetable oils (rich in polyunsaturated fatty acids, Nudda et al., 2014) and/or by adding CLA dairy starters producers during cheese manufacture (de Lima Alves et al., 2011). Rumenic acid, which corresponds to the CLA isomer C18:2 *cis* 9, *trans* 11, is the most abundant isomer, accounting for around 80–90% of the total CLA (Parodi, 1999). A high number of *in vitro* studies and animal models have shown that it is responsible for the anticarcinogenic and antiatherogenic properties attributed to CLA (Lee et al., 2005; Battacharaya et al., 2006). In addition to rumenic acid, other CLA isomers have a positive impact on human health. For example, the C18:2 *trans* 10, *cis* 12, has been reported to promote weight loss, decreasing glucose levels and insulin resistance in obese men with metabolic syndrome (Riserus et al., 2002), while isomer C18:2 *trans* 9, *trans* 11 has been shown to inhibit the growth of human colon cancer cells (Coakley et al., 2006).

Although a notable number of clinical studies have evaluated the biological properties of purified rumenic acid or CLA isomer mixtures on CVD risk factors (Wang et al., 2012), relatively little information from dietary intervention studies with CLA-enriched cheeses is available. Sofi et al. (2010) observed beneficial effects of consumption by healthy subjects of a sheep cheese naturally rich in rumenic acid on several atherosclerotic biomarkers, in comparison with a commercially available cheese. Pintus et al. (2013) reported improvement of the lipid profile and reduction of endocannabinoid synthesis in the plasma

TABLE 11.1 Fatty-Acid Composition (g/100 g) and cis-9, trans-11 CLA Content (mg/g fat) of Different Cheeses From Cow, Sheep, and Goat Milk

Type of Milk	Cheese Type	SFA[a] Content (%)	MUFA[b] Content (%)	PUFA[c] Content (%)	cis-9, trans-11 CLA[d] (mg/g Fat)	References
Cow	Alpine	65.23	31.19	3.57	4.79 ± 1.07	Prandini et al. (2007)
	Asiago	n.d.	n.d.	n.d.	5.91 ± 1.95	Cicognini et al. (2014)
	Belpaese	n.d.	n.d.	n.d.	5.22 ± 1.97	Cicognini et al. (2014)
	Caciotta	n.d.	n.d.	n.d.	5.98 ± 1.18	Cicognini et al. (2014)
	Caciocavallao	n.d.	n.d.	n.d.	5.35 ± 2.77	Cicognini et al. (2014)
	Camembert	58.41	25.70	3.20	9.50 ± 0.20	Domagała et al. (2010)
	Caprino fresh	66.87	24.45	3.44	6.10 ± 2.65	Prandini et al. (2011)
	Cheddar	n.d.	n.d.	n.d.	5.86 ± 0.89	Cicognini et al. (2014)
	Crescenza	n.d.	n.d.	n.d.	4.76 ± 0.79	Cicognini et al. (2014)
	Dziurawiec	58.27	21.72	2.55	9.00 ± 0.90	Domagała et al. (2010)
	Emmentaler	57.88	25.36	3.06	7.40 ± 0.10	Domagała et al. (2010)
	Swiss Emmental	66.67	29.85	3.48	7.66 ± 1.42	Prandini et al. (2007)
	Fontina Valdostana	67.26	28.57	4.17	8.11 ± 1.53	Prandini et al. (2007)
	Gorgonzola	64.10	25.79	3.92	5.16 ± 0.29	Prandini et al. (2011)
	Gouda	56.38	24.89	3.45	12.00 ± 0.70	Domagała et al. (2010)
	Gouda	n.d.	n.d.	n.d.	4.32 ± 1.43	Cicognini et al. (2014)

Continued

TABLE 11.1 Fatty-Acid Composition (g/100g) and cis-9, trans-11 CLA Content (mg/g fat) of Different Cheeses From Cow, Sheep, and Goat Milk—cont'd

Type of Milk	Cheese Type	SFA[a] Content (%)	MUFA[b] Content (%)	PUFA[c] Content (%)	cis-9, trans-11 CLA[d] (mg/g Fat)	References
	Grana/Parmigiano	68.52	27.90	3.58	3.85 ± 0.51	Prandini et al. (2007)
	Gruyere	n.d.	n.d.	n.d.	10.21 ± 2.47	Cicognini et al. (2014)
	Mazdamer	56.05	26.51	3.20	9.80 ± 0.30	Domagała et al. (2010)
	Montasio	n.d.	n.d.	n.d.	4.78 ± 1.23	Cicognini et al. (2014)
	Mozzarella	67.04	29.20	3.75	0.24 ± 0.01[e]	Caroprese et al. (2013)
	Parmigiano	59.77	22.77	3.60	4.40 ± 0.20	Domagała et al. (2010)
	Philadelphia	n.d.	n.d.	n.d.	5.41 ± 1.15	Cicognini et al. (2014)
	Prato	n.d.	n.d.	n.d.	3.79 ± 0.50	Cortes Nunes and Guedes Torres (2010)
	Provola	n.d.	n.d.	n.d.	6.96 ± 5.12	Cicognini et al. (2014)
	Provolone	n.d.	n.d.	n.d.	4.92 ± 0.23	Cicognini et al. (2014)
	Radamer	57.74	25.86	2.94	7.10 ± 0.50	Domagała et al. (2010)
	Ricotta	n.d.	n.d.	n.d.	5.53 ± 1.08	Cicognini et al. (2014)
	Robiola	n.d.	n.d.	n.d.	4.77 ± 1.26	Cicognini et al. (2014)
	Salami	56.92	25.89	3.33	11.80 ± 0.70	Domagała et al. (2010)
	Scamorza	n.d.	n.d.	n.d.	4.93 ± 0.31	Cicognini et al. (2014)
	Stracchino	n.d.	n.d.	n.d.	5.01 ± 1.26	Cicognini et al. (2014)

Sheep	Pecorino	67.69	26.83	5.48	7.77 ± 2.25	Prandini et al. (2007)
	Pecorino hard	65.94	23.80	4.93	10.90 ± 0.03	Prandini et al. (2011)
	Roquefort	70.10	21.11	3.92	8.81 ± 3.84	Prandini et al. (2011)
	Feta	70.20	21.00	4.70	9.24 ± 4.80	Zlatanos et al. (2002)
	Feta	n.d.	n.d.	n.d.	8.50 ± 1.95	Cicognini et al. (2014)
	Rokpol	55.47	27.55	3.71	13.10 ± 0.10	Domagała et al. (2010)
Goat	Caprino fresh	66.92	23.89	4.08	6.80 ± 1.26	Prandini et al. (2011)
	Caprino hard	67.06	23.71	4.23	7.06 ± 3.42	Prandini et al. (2011)

^aSaturated fatty acids.
^bMonounsaturated fatty acids.
^cPolyunsaturated fatty acids.
^dConjugated linoleic acid.
^eExpressed as g of isomer/100 g fatty acids.
n.d., not detected.

of hypercholesterolemic subjects after consumption of sheep cheese naturally enriched in α-linolenic, CLA, and vaccenic acid. If these beneficial effects are confirmed by further studies, this will likely have important repercussions for both human nutrition and the food industry.

Short-chain fatty acids such as butyric acid (C4) and caproic (C6) and medium-chain fatty acids such as caprilic (C8) and capric acids (C10) represent 10–12% of total saturated fatty acids. Apart from its antimicrobial properties (Huang et al., 2011), the butyric acid present in milk fat in concentrations ranging from to 2–5% has been described as an anticarcinogenic agent inhibiting growth of a wide range of human cancer-cell lines (Blank-Porat et al., 2007). Furthermore, synergisms between butyric acid and other dietary compounds and common drugs in reducing cancer cell growth have also been shown (Parodi, 2006). However, no human studies have been conducted to confirm these effects after consumption of cheese.

11.3 MINERALS IN CHEESE AND THEIR IMPACT ON HEALTH

The specific amount of minerals in cheese differs according to the manufacturing procedure. For example, the method of coagulation used and the acidity of the curd will have an effect on the salt concentration. Ripened cheeses constitute an abundant source of mineral compounds. Calcium, phosphorus, sodium, and chloride are found in high amounts, while elements such as zinc and potassium are found at much lower concentration (Table 11.2). Note the significantly higher daily value of calcium and phosphorus with regard to the rest of the minerals shown. Research has shown that an insufficient intake of calcium, in particular, raises the risk of obesity (Parikh and Yanovski, 2003), hyperlipidemia, and insulin-resistance syndrome (Teegarden, 2003; Zemel, 2004). In contrast, a diet rich in dairy calcium intake has shown on enhance weight reduction in type 2 diabetic patients (Shahar et al., 2007).

A considerable body of research has been published on the cariostatic effect of cheese (Kashket and DePaola, 2002). Assays on human subjects have suggested a caries-protective effect in childhood populations (Ohlund et al., 2007). Similarly, Gedalia et al. (1991) showed that the consumption of hard cheese led to significant rehardening of softened enamel surfaces, while a cheese with *Lactobacillus rhamnosus* as probiotic diminished caries-associated salivary microbial counts in young adults (Ahola et al., 2002). Although more research is needed to elucidate the precise mechanisms for the cariostatic effects of cheese, there is enough evidence to support the consumption of cheese as an anticaries measure (European Food Safety Agency, 2008). The most plausible mechanism is related to the mineralization potential of the casein-calcium phosphate of cheese, to the stimulation of saliva flow induced by its texture, the buffering effects of cheese proteins on acid formation in dental plaque, and the inhibition of cariogenic bacteria (O'Brien and O'Connor, 2004).

TABLE 11.2 Levels of Minerals of Different International Cheeses, mg/Serving Portion of 30g

	Calcium	Magnesium	Phosphorus	Sodium	Potassium	Iron	References
Camembert	116	6.0	104	253	56	0.10	Gaucheron (2011)
% DV[a]	11	2	10	10	1	1	
Cheddar	216	8.4	154	186	29	0.20	Gaucheron (2011)
% DV[a]	20	2	15	7	1	1	
Feta	147	5.7	101	335	19	0.19	Gaucheron (2011)
% DV[a]	14	2	10	12	1	1	
Gouda	210	8.7	164	246	36	0.07	Gaucheron (2011)
% DV[a]	20	2	16	9	1	1	
Gruyère	303	10.8	182	101	24	0.05	Gaucheron (2011)
% DV[a]	29	3	19	4	1	1	
Idiazabal	295	14.3	188	166	33	0.19	Cámara-Martos et al. (2013)
% DV[a]	29	4	19	7	1	1	
Manchego	278	13.2	177	197	31	0.13	Cámara-Martos et al. (2013)
% DV[a]	27	4	18	8	1	1	
Parmigiano	413	15.3	242	558	32	0.29	Gaucheron (2011)
% DV[a]	39	4	23	22	1	1	
Roquefort	199	9.0	118	543	27	0.17	Gaucheron (2011)
% DV[a]	20	2	14	22	1	1	
Tetilla	243	9.5	154	162	31	0.08	Cámara-Martos et al. (2013)
% DV[a]	24	2	15	6	1	1	
Torta del Casar	158	9.9	117	161	34	0.16	Cámara-Martos et al. (2013)
% DV[a]	15	3	12	6	1	1	

[a]*Percentage of daily value.*
% DV, based on energy and nutrient recommendations for a general 2000-calorie diet.

One particular characteristic of cheese's calcium is its high bioavailability. The suggested factors responsible for it are the prevention of the formation of insoluble calcium salts in the intestine by casein phosphopeptides (CPPs) (Kitts et al., 1992) and the optimal calcium:phosphorus ratio (Cámara-Martos et al., 2013). In ripened cheeses, calcium availability is determined by fat content, which is lower in full-fat cheeses (13.85%) than in semifat cheeses (24.13%) (Ibeagha-Awemu et al., 2009). On the other hand, recent studies suggest that the use of probiotic cultures in the production of cheese increases the availability of calcium, magnesium, and phosphorus. The stimulating effect of probiotic cultures on the availability of mineral compounds can be attributed to intensified enzymatic conversion, mainly proteolysis and lipolysis (Aljewicz et al., 2014).

11.4 BIOLOGICAL EFFECTS OF CHEESE VITAMINS

As with other nutrients, the vitamin content is highly variable between types of cheese, depending on fat content. Most of the fat-soluble vitamins in milk are retained in cheese fat. Absorption of vitamin A and other fat-soluble vitamins such as vitamins D, E, and K are increased in the body by the presence of fats in the diet, making cheese a good vehicle to deliver them. It has been reported that cheese can provide one-third of the recommended intake of vitamin A, a vitamin with an important role in the normal functioning of the immune system, in gene-expression regulation, and in sustaining low-light vision (Russell, 2000).

With regard to the water-soluble vitamins, such as riboflavin, vitamin B12 (niacin), and folate, their content in cheese is lower than in milk. However, the loss of some B vitamins is compensated by the microbial synthesis during ripening. Semihard cheese can provide the average male and female adult with 80% of the recommended intake of vitamin B12. This vitamin along with folic acid help to reduce high levels of homocysteine in blood, an amino acid that has been linked to CVD (Quinlivan et al., 2002). Note that ovine milk is richer than cow's milk in most vitamins, and therefore 90 g of sheep milk cheese could completely cover the daily requirements for riboflavin (Recio et al., 2009).

11.5 CHEESE BIOACTIVE PROTEINS AND PEPTIDES

The content of proteins present in cheese depends on the variety, ranging from 4% (cream cheese) to 40% (Parmesan cheese). Proteins are one of the major nutrients responsible for the nutritional, physicochemical, functional, and sensory aspects of cheese. In addition, cheese protein is almost 100% digestible, as the ripening phase of cheese manufacture involves a progressive breakdown of casein by enzymes present in milk and those derived from rennet or microorganisms, releasing both amino acids and peptides. Interestingly, lysine is one of the major amino acids found in cheese with high bioavailability because of the absence of Maillard reactions (de la Fuente and Juárez, 2001). The type and quantity of milk protein-derived peptides is affected by factors influencing proteolysis, such as pH, type of enzymes, salt-to-moisture ratio, humidity, and storage time and temperature (Park and Jin, 1998). In this sense, different varieties of cheeses are likely to exhibit peptides with variable bioactivities, such as antihypertensive, antioxidant, opioid, antimicrobial, antiproliferative, and mineral absorption modulatory effects, among others (López-Expósito et al., 2012).

11.5.1 Antihypertensive Peptides

Hypertension is a chronic condition affecting up to 30% of the adult population in developed countries and can lead to more serious diseases such as stroke and coronary heart disease (Chen et al., 2009). One of the many physiological pathways that control blood pressure in the human body is the renin-angiotensin-aldosterone system. A key component of this system is the angiotensin I-converting enzyme (ACE), which catalyzes the conversion of angiotensin I into the potent vasoconstrictor angiotensin II and, simultaneously, the degradation of the vasodilators bradykinin and kallidin. These two reactions cause a contraction of blood vessels and the release of aldosterone, increasing sodium concentration and blood pressure. Thus, inhibiting this enzyme can lead to an antihypertensive effect. Most hypertensive patients are treated worldwide with ACE-inhibitory drugs such as Captopril or Enalapril (Nussberger, 2007). However, it has been shown that small reductions in average blood pressure may be obtained by nutritional measures such as reducing sodium intake and increasing daily consumption of food-derived ACE-inhibitory peptides (Sieber et al., 2010). Because of its content of sequences released during milk fermentation and ripening, cheese has been studied extensively as a potential source of ACE-inhibitory peptides. As shown in Table 11.3, the water- and/or ethanol-soluble extracts (WSE, ESE) of many cheese varieties have been obtained and evaluated for their ACE-inhibitory properties. The influence of the addition of certain probiotic strains (Ong and Shah, 2008; Wang et al., 2011) or simulated gastrointestinal digestion (Parrot et al., 2003) on the ACE-inhibitory activity of different kind of cheeses has also been studied. In some studies, different fractionation steps have been performed to purify novel bioactive fragments in which their ACE-inhibitory activities have been evaluated. As an example, this strategy has been applied to isolated peptides from Manchego cheese (Gómez-Ruiz et al., 2004a,b) and other Spanish cheeses such as Cabrales, Mahon, Idiazabal, and Roncal (Gómez-Ruiz et al., 2006). These authors applied tandem mass spectrometry (MS/MS) to identify and quantify sequences contained in those cheese varieties. Some of them, derived from α_{s1}- and β-casein, showed moderate ACE-inhibitory activity, indicating their potential contribution to the activity reported for the cheese WSE (Gómez-Ruiz et al., 2006). The MS/MS technique has also been found useful to confirm the presence of peptides previously identified in other milk products obtained by enzymatic hydrolysis or fermentation. Antihypertensive β-casein peptides VPP and IPP have been found in different cheese

TABLE 11.3 Angiotensin 1-Converting Enzyme Inhibitory Activity of Different Cheese Varieties and Peptides Identified as Potentially Responsible for the Observed Effects

Type of Cheese	Analyzed Sample	ACE Inhibitory Activity (%)	IC_{50} (mg/mL)	Identified Peptides	References
Appenzeller:					Bütikofer et al. (2007)
Classic, 3 months old	WSE[a]		17.6	VPP; IPP	
Surchoix, 4 months old	WSE[a]		8.0	VPP; IPP	
Extra, 6–7 months old	WSE[a]		16.8	VPP; IPP	
1/4 fat, 3 months old	WSE[a]		7.1	VPP; IPP	
1/4 fat, 7–10 months old	WSE[a]		4.2	VPP; IPP	
Brie	WSE[a]		25.9	VPP; IPP	Bütikofer et al. (2007)
	WSE[a]			VPP; IPP; RYLG; RYLGY; HLPLP; LHLPLP	Stuknytè et al. (2015)
Cabrales	WSE[a]	74.7		PP; MPI; MPL; EVVR; REL; DKIHPF; FPGPIH; PGPIH; GPIH; PVEP; PVEPF	Gómez-Ruiz et al. (2006)
	Fraction <1 kDa	76.1			
Caprino	WSE[a]			VPP; IPP; RYLG; RYLGY; AYFYPEL; LHLPLP	Stuknytè et al. (2015)
Cheddar	WSE[a]			VPP; IPP; RYLG; RYLGY; HLPLP; AYFYPEL; LHLPLP	Stuknytè et al. (2015)
+ *Lb. casei* LAFTI® L26 (36 weeks)	WSE[a]		0.25	RPKHPIKHQ; RPKHPIK; RPKHPI; DKIHPF; FVAPPEVF; KKYKVPQLE; YQEPVLGPVRGPFPIIV	Ong et al. (2007)
+ *Lb. acidophilus* LAFTI® L10 (24 weeks)	WSE[a]		0.13	ARHPHPH; RPKHPIKHQ; RPKHPIK; RPKHPI; FVAPPEVF; YQEPVLGPVRGPFPIIV	Ong and Shah (2008)
Crescenza	FPLC fraction from WSE[a]	37		LVYPFPGPIHNSLPQ	Smacchi and Gobbetti (1998)

Edam	WSE[a]		13.3	VPP; IPP	Bütikofer et al. (2007)
Emmental	WSE[a]			VPP; IPP	Stuknytė et al. (2015)
Classic, 4 months old	WSE[a]		10.5	VPP; IPP	Bütikofer et al. (2007)
Organic, 4 months old	WSE[a]		7.5	VPP; IPP	
Reserved, 8 months old	WSE[a]		9.9	VPP; IPP	
Cave-aged, 12 months old	WSE[a]		7.1	VPP; IPP	
Feta	WSE[a]		14.3	VPP; IPP	Bütikofer et al. (2007)
Fontina	WSE[a]			VPP; IPP; HLPLP	Stuknytė et al. (2015)
Gamalost	WSE[a]	~77	~2.5	SLVYPFPGPIPNSLPQNIPPLTQTPVVVPPFLQPEVM; VYPFPGPIPNSLPQNIPPLTQTPVVVPPFLQPE; HKEMPFPKYPVEPFT; SLTLTDVENLHLPLL; NLHLPLPLL; HLPLLLQ; WMHQPHQPLPPTVMFPPQSVL; FLLYQEPVLGPVRGPFPIIV	Qureshi et al. (2013)
	Fraction <1 kDa	~70	~0.5		
Goat cheese	WSE[a]	72.8		QP; PP; RPK; EEL; EEI; DKIHP; PLTQTP; VVVPP; GVPK; TDVEK; VRGP	Gómez-Ruiz et al. (2006)
	Fraction <1 kDa	59.4			
Gouda	WSE[a]			VPP; IPP	Stuknytė et al. (2015)
8-months old	WSE[a]	75.7[b]		RPKHPIKHQ; RPKHPIKHQGLPQ; YPFPGPIPN; MPFPKYPVQPF	Saito et al. (2000)
Old	WSE[a]		2.0	VPP; IPP	Bütikofer et al. (2007)
Gorgonzola	WSE[a]		21.8	VPP; IPP	Bütikofer et al. (2007)
	WSE[a]			VPP; IPP; AYFYPE; HLPLP; AYFYPEL; LHLPLP	Stuknytė et al. (2015)

Continued

TABLE 11.3 Angiotensin I-Converting Enzyme Inhibitory Activity of Different Cheese Varieties and Peptides Identified as Potentially Responsible for the Observed Effects—cont'd

Type of Cheese	Analyzed Sample	ACE Inhibitory Activity		Identified Peptides	References
		(%)	IC$_{50}$ (mg/mL)		
Grana Padano	WSE[a]			VPP; IPP; RYLG; HLPLP; LHLPLP	Stuknytė et al. (2015)
Graubünden, full fat	WSE[a]		9.2	VPP; IPP	Bütikofer et al. (2007)
Graubünden, ½ fat	WSE[a]		8.9	VPP; IPP	
Graubünden, ¼ fat	WSE[a]		4.8	VPP; IPP	
Gruyere, 5 months old	WSE[a]		14.9	VPP; IPP	Bütikofer et al. (2007)
Gruyere, 8 months old	WSE[a]		18.7	VPP; IPP	
Gruyere, 10 months old	WSE[a]		14.2	VPP; IPP	
Gruyere, 12 months old	WSE[a]		29.4	VPP; IPP	
Hobelkäse	WSE[a]		2.6	VPP; IPP	Bütikofer et al. (2007)
Idiazabal	WSE[a]	87.5		QP; FP; NINE; PSE; ERYL; PQL; EIVPK; AWY; DKIHP; DKIHPF	Gómez-Ruiz et al. (2006)
	Fraction <1 kDa WSE[a] and ESE[c]	68.5		QNALIVRYTR; KALPMHIRLAF; EGPKLVAS; PVGLVQPASATLYDY; KVGTKCCAKP; REKVLASS; PKIDAMREKVLA; PRKEKLCTTS; YQEPVLGPVRG; NAGPFTPTVNREQLSTS	Sagardia et al. (2013)
Küsnachter	WSE[a]		19.6	VPP; IPP	Bütikofer et al. (2007)
Limburger	WSE[a]		6.4	VPP; IPP	Bütikofer et al. (2007)
Maasdam	WSE[a]			VPP; IPP; HLPLP; AYFYPEL; LHLPLP	Stuknytė et al. (2015)

Mahon	WSE[a]	76.8		PK; PQ; RI; RL; PQEVL; EVLN; NENLL; ENLL; NLLRF	Gómez-Ruiz et al. (2006)
	Fraction <1 kDa	56.6			
Manchego	WSE[a]	70.6		QP; PP; FP; PFP; HPIK; HQGL; DKIHP; DKIHPF; TGPIPN; VRYL; VPSERYL; VPSERY; DVPSERYLG; KKYNVPQL; KKYNVPQ; IPY; TQOKTNAIPY; VRGPFP; PVRGPF; LEIVPK; RPKHPI; VPKVKE; REQEEL; DVPSERY; VRGPFP; LPQNILP	Gómez-Ruiz et al. (2004a,b, 2006)
	Fraction <1 kDa	65.3			
	WSE[b]		29.5	VPP; IPP	Bütikofer et al. (2007)
Mexican + *Lb. casei*	WSE[a]	95.5	5.3	YQEPVLGPVRGPFPI; YQEPVLGPVRGPFPIIV; FVAPFPEVFGK; EVLNENLLRF	Torres-Llanez et al. (2011)
Mexican + *En. faecium*	WSE[a]	96.3	10.4	YQEPVLGPVRGPFPIIV; RPKHPIKHQGLPQEV; RPKHPIKHQGLPQEVLNENLLR; FVAPFPEVFGK; EVLNENLLRF	
Münster	WSE[a]		14.1	VPP; IPP	Bütikofer et al. (2007)
Parmino	WSE[a]		28.6	VPP; IPP	Bütikofer et al. (2007)
Provolone	WSE[a]			VPP; IPP; RYLG; LHLPLP	Stuknytė et al. (2015)
Provolone picante	WSE[a]			VPP; IPP; RYLG; LHLPLP	Stuknytė et al. (2015)
Raclette, raw milk	WSE[a]		8.8	VPP; IPP	Bütikofer et al. (2007)
Raclette, pasteurized milk	WSE[a]		9.0	VPP; IPP	Bütikofer et al. (2007)
Reblonchon	WSE[a]		17.0	VPP; IPP	Bütikofer et al. (2007)
Roncal	WSE[a]	85.8		QP; PP; RPKHP; PKHP; HPIK; HQGL; DKIHP; DKIHPF; GPVR	Gómez-Ruiz et al. (2006)
	Fraction <1 kDa	70.4			
Roquefort	WSE[a]		10.2	VPP; IPP	Bütikofer et al. (2007)

Continued

TABLE 11.3 Angiotensin I-Converting Enzyme Inhibitory Activity of Different Cheese Varieties and Peptides Identified as Potentially Responsible for the Observed Effects—cont'd

Type of Cheese	Analyzed Sample	ACE Inhibitory Activity		Identified Peptides	References
		(%)	IC$_{50}$ (mg/mL)		
Roquefort-type cheese	WSE[a]		0.34	KEMPFPKYPVE; WMHQPPQPLPPTVMFPPQSVL; MHQPPQPLPPTVMFPPQSVL; HQPPQPLPPTVMFPPQSVL; YQEPVLGPVRGPFPI; QEPVLGPVRGPFPILV; QEPVLGPVRGPFPI; PVLGPVRGPFPI; LGPVRGPFPI; TDAPSFSDIPNPIGSENSGK; DIPNPIGSENSGKTTMPLW; IPNPIGSENSGKIT; NAGPFTPTVNR; YQGPIVLNPWDQVKR; YQGPIVLNPWDQVK; GPIVLNPWDQVKR; VLNPWDQVKR	Meister Meira et al. (2012)
St. Paulin	WSE[a]		23.5	VPP; IPP	Bütikofer et al. (2007)
Sbrinz	WSE[a]		16.4	VPP; IPP	Bütikofer et al. (2007)
Taleggio	WSE[a]			VPP; IPP; HLPLP	Stuknytė et al. (2015)
Tete de Moine	WSE[a]		7.1	VPP; IPP	Bütikofer et al. (2007)
Tilsit raw milk	WSE[a]		4.4	VPP; IPP	Bütikofer et al. (2007)
Tilsit pasteurized milk	WSE[a]		18.0	VPP; IPP	
Vacherin fribourgeois	WSE[a]		5.0	VPP; IPP	Bütikofer et al. (2007)
Vacherin Mont d'Or	WSE[a]		21.4	VPP; IPP	
Wangener Geissmutschli	WSE[a]		7.9	VPP; IPP	Bütikofer et al. (2007)
Winzerkäse	WSE[a]		9.9	VPP; IPP	Bütikofer et al. (2007)

[a]Water-soluble extract.
[b]Antihypertensive effect in spontaneously hypertensive rats (Dose of 6.1–7.5 mg/kg produces a decrease in Systolic blood pressure (SBP) of –24.7 mmHg.
[c]Ethanol-soluble extract.

varieties, such as Appenzeller, Brie, Caprine, Cheddar, Edam, Fontina, Gouda, Gorgonzola, and Gruyere, among others (Bütikofer et al., 2007; Stuknytè et al., 2015). Interestingly, some of these varieties contained similar concentrate ions of VPP and IPP than fermented milk products with blood-pressure-lowering capacity. Similarly, antihypertensive α_{s1}-casein-derived peptides AYFYPEL and RYLGY, identified and characterized in casein hydrolyzate by Contreras et al. (2009, 2013), were found in several cheese varieties, such as Brie, Caprino, Cheddar, Gorgonzola, and Maasdam (Stuknytè et al., 2015). Although several cheeses extracts have been characterized by their potent ACE-inhibitory activity, and the presence of antihypertensive peptides has been detected in some of them, confirmation of the antihypertensive effect of cheese in animal models or humans is still needed.

11.5.2 Antioxidant Peptides

Oxidative stress, caused by increased production of reactive-oxygen species (ROS) associated with decreased antioxidant capability of the cell, is thought to contribute to the pathogenesis of a number of human diseases including CVD, metabolic, inflammatory, neurodegenerative diseases, and cancer, in addition to accelerating the aging process (Stadtman, 2004). Natural antioxidants provide additional benefits to the endogenous defense strategies in the battle against oxidative stress (Erdmann et al., 2008), with food-derived peptides being potent antioxidants without important side effects (Sarmadi and Ismail, 2010). Among them, milk proteins have been one of the most frequently studied sources.

First studies reporting antioxidant activity of cheese-derived peptides were carried out with raw and sterilized ovine and caprine cheese-like systems coagulated with enzymes from the plant *Cynara cardunculus* (Silva et al., 2006), although sequences of those peptides were not identified. Gupta et al. (2009) assessed the antioxidant activity of peptides in WSE obtained from Cheddar cheese prepared with *Lactobacillus casei* ssp. *casei* 300, *Lactobacillus paracasei* ssp. *paracasei* 22, and without adjunct cultures by three different methods based on radical scavenging activity. In all cheese samples, changes in the antioxidant activity correlated to the rate of formation of soluble peptides (proteolysis). Later, Pritchard et al. (2010), Silva et al. (2012), and Bottesini et al. (2013) evaluated the radical scavenging activity of WSE obtained from Australian Cheddar, Brazilian artisanal "Coalho," and Parmigiano-Reggiano cheeses, respectively. Gupta et al. (2010) identified two peptides corresponding to bovine β-casein (VKEAMAPK) and α_{s1}-casein (HIQKEDVPSER) whose potent radical-scavenging activity indicated their key contribution to the antioxidant activity of Cheddar cheeses. Meister Meira et al. (2012) also identified peptides released during the manufacture of a Roquefort-type cheese with sequences sharing great homology with peptides characterized previously by their antioxidant activity. Recently, three peptides derived from α_{s1}- and β-casein were identified in a Burgos type cheese (Timón et al., 2014). One of them, corresponding to α_{s1}-casein (SDIPNPIGSENSEKTTPLLW), is a new peptide with potential antioxidant activity. These preliminary results suggest the potential of cheese as source of antioxidant peptides, although animal models and human trials confirming these effects are needed.

11.5.3 Effects of Cheese-Derived Peptides on the Nervous System

Opioid peptides are defined as opioid μ-receptor ligands with agonistic and antagonistic activities. Opioid receptors are located in the nervous, endocrine, and immune systems as well as in the gastrointestinal tract of mammals (Teschemacher, 2003). The most studied milk-derived opioid receptor ligands are those called β-casomorphins (BCMs), derived from β-casein. They are characterized by the presence of a tyrosine residue at the N-terminal and another aromatic amino acid at third or fourth position, which is an important structural motif that fits into the binding site of opioid receptors (Nagpal et al., 2011). In addition to their effects on the central nervous system, BCMs have been shown to induce intestinal mucus secretion (Plaisancié et al., 2013), increase plasma insulin level, decrease glucagon level, and to elevate the activity of superoxide dismutase and catalase (Yin et al., 2010). Furthermore, they have also been observed to prolong gastrointestinal transit time and to exert antidiarrheal action in both animals and humans (Daniel et al., 1990). Other beneficial activities attributed to BCMs can be found in Ul Haq et al. (2014).

Only a few studies deal with the presence and levels of BCMs in cheeses. BCM7 (β-casein (60-66)) and BCM5 (β-casein (60-64)) have been identified in extracts from Brie and Gouda cheeses, respectively (Jarmolowska et al., 1999; Kostyra et al., 2004). The content of agonistic BCM7 and BCM5, antagonistic casoxin-6 and casoxin-C (derived from bovine κ-casein), and lactoferroxin A (derived from lactoferrin) opioid peptides in three semihard, Edamski, Gouda, and Kasztelan, and in two ripening mold cheeses, Brie and Rokpol, was also evaluated by Sienkiewicz-Szlapka et al. (2009). It was found that the content of BCMs was higher in mold cheeses than in semihard cheeses, which contained fairly high amounts of casoxin-6. All the extracts, except that from Gouda, were found to affect the intestinal mobility in isolated rabbit ileum similar to morphine, with this effect being reversed by naloxone. Similarly, Gorgonzola, Caprino, Brie, Taleggio, Gouda, Fontina, Cheddar, and Grana Padano cheeses as well as their digests were analyzed by De Noni and Cataneo (2010) for the presence of BCM7 and BCM5. BCM7 was detected in all the cheeses evaluated with the exception of Taleggio, Caprino, and Grana Padano samples. However, BCM5 was not found in any of them.

Peptides incorporating the BCM7 sequence, which could act as precursors during further digestion processes, have also been found in cheeses such as Parmigiano Reggiano (Addeo et al., 1992), Cheddar and Jarlsberg (Stepaniak et al., 1995), and Crescenza (Smacchi and Gobetti, 1998). BCM9 (β-casein (60-68)) and a number of peptides including the BCM7 sequence were identified in Cheddar cheese by Singh et al. (1995, 1997). BCM9 was also found in Gouda cheese by Saito et al. (2000). This finding was later confirmed by Toelstede et al. (2008) who described the presence of BCM10 (β-casein (60-69)) in the WSE of a matured Gouda cheese. The caprine analog of BCM9 was also detected in the WSE of Caprino de Piemonte, an Italian goat cheese (Rizzello et al., 2005).

Relatively little is known about the actual bioavailability of this kind of peptides. BCMs have been detected in the plasma of newborn calves after milk intake (Umbach et al., 1985), and in plasma in 2- and 4-week old pups after ingesting a casein-based formula (Singh et al., 1999). As for humans, immunoreactive material to human and bovine BCM7 and BCM5 has been detected in the plasma of infants fed with human and cow milk (Kost et al., 2009; Wasilewska et al., 2011), although the method of detection employed by the authors raises many doubts regarding the cross-reactivity of the antibody with some other epitopes of antigen present in plasma. In any case, the authors speculated that the intestinal mucosa of the newborn is more permeable to the relatively large peptides due to their immature tight junction through which peptides cross, thereby escaping hydrolysis by brush-border peptidases (Sienkiewicz-Szlapka et al., 2009). It is necessary to continue exploring the role of BCMs in human health, applying *in vivo* experiments with improved diagnostic techniques to verify their presence in the blood. Furthermore, more research needs to be done to evaluate the absorption of these peptides in a complex food matrix such as cheese.

11.5.4 Antiproliferative Peptides

There are several studies demonstrating the ability of bovine milk proteins such as lactoferrin and lactoferricin to suppress the growth of cancer cells both *in vitro* and *in vivo* by serving as apoptotic inducers in tumor development (Roy et al., 2002; de Moreno de LeBlanc et al., 2005; Mader et al., 2005). However, no peptides with antiproliferative properties have been characterized. To date, the antiproliferative effects of cheese extracts are being evaluated and all the described results are based on either cheese extracts or cheese whey. The potential role of 12 commercial cheese products on the cell growth and induction of DNA fragmentation in HL-60 human promyelocytic leukemia cells was investigated by Yasuda et al. (2010). Among them, extracts from Montagnard, Pont-l'Eveque, Brie, Camembert, Danablue, and Blue cheeses revealed the highest activity. In addition, a positive correlation between the ripeness of various cheeses and their antiproliferative activity was found. It remains to be elucidated whether these active molecules from cheeses may be present at enough concentration to exert their activity as well as their bioavailability and specificity to cancer cells.

Mozzarella whey samples produced a 43% reduction in cell proliferation in Caco 2 cells (De Simone et al., 2009). The absence of this effect in the peptide extract from the original milk suggested that the release of specific bioactive compounds occurs specifically during the cheese's production process. Later, a partially RP-HPLC purified peptide subfraction was identified as responsible for the activity observed (De Simone et al., 2011), and a number of peptides, mainly derived from κ-casein glyco-caseinmacropeptide and β-casein, were identified by MALDI-TOF/MS.

11.5.5 Antimicrobial Peptides

The majority of antimicrobial peptides derived from food sources have been identified in milk, milk hydrolysates, and fermented milks (López-Expósito and Recio, 2006). However, there is almost no information on the potential of cheese as a source of antimicrobial peptides. To date, all the publications dealing with the antimicrobial properties of cheese are based on WSE, and very few of them identify the potential peptides responsible (Rizello et al., 2005; Losito et al., 2006; Lignitto et al., 2012). Because cheese's WSE also contain organic acids and salts that could contribute to the inhibition of microbial growth, it is important to perform a purification step to remove them, prior to attributing to peptides the potential antimicrobial activity observed. Recently, Théolier et al. (2014) evaluated the antibacterial activity of WSE from five commercial cheeses namely Mozzarella, Gouda, Swiss, and old and medium Cheddar. Before purification through a Sep-Pak cleanup column, WSE from Mozzarella and Gouda cheeses inhibited the growth of *Listeria ivanovii* HPB28, *L. monocytogenes* Scott A3, *Escherichia coli* MC4100, and *E. coli* O157:H7 up to 4.66 log cycles. However, after desalting, the activity against *L. monocytogenes* Scott A3 and *E. coli* O157:H7 was almost negligible, highlighting the importance of carrying out a previous purification step. The same purified WSE were shown to induce a delay in spore germination of several foodborne molds, reaching the lowest minimal inhibitory concentration values against *Penicillium camemberti* and *P. commune* strains, for Swiss, Gouda, and Mozzarella extracts. Purified WSE of Asiago d'Allevo cheese showed moderate antibacterial properties against *L. innocua* LRGIA01 growth (Lignitto et al., 2012; Thi et al., 2014). The bacterial growth inhibitory properties were affected by the ripening time with the highest activities observed after 6 months of maturation. The authors identified a large number of casein-derived peptides present in the WSE, among which they found other potent antimicrobial fragments previously described such as isracidin (α_{s1}-casein (1-23)) (Hill et al., 1974) and α_{s2}-casein (183-207) (López-Expósito et al., 2006). Similarly, Losito et al. (2006) identified more than 30 peptides with a high level of homology with N-terminal, C-terminal, or whole fragments of previously reported milk-derived antimicrobial sequences from the WSE of Pecorino Romano, Canestrato Pugliese, Crescenza, Caprino del Piemonte, and Caciocavallo cheese varieties (Rizzello et al., 2005). Recently, the antibacterial activity against *L. monocytogenes* strain 162 of Emmental de Savoie purified WSE was also reported (Thi et al., 2014), although there was no identification of the responsible peptides.

To date, there are no *in vivo* published data supporting the antimicrobial effects of cheese consumption. However, some data regarding the *in vivo* antimicrobial activity of one of the peptides identified in cheese extracts, isracidin, have been published (Lahov and Regelson, 1996). In mice, it exerts a protective effect against *L. monocytogenes*, *Streptococcus pyogenes*, and *Staphylococcus aureus*. In cows with mastitis, isracidin also obtained a success rate of over 80% in the treatment of chronic streptococcal infection.

11.5.6 Modulatory Peptides of Mineral Absorptions: Phosphopeptides

CPPs are bioactive peptides with various degrees of phosphorylation that are released *in vitro* and *in vivo* by enzymatic hydrolysis of different casein fractions (Bouhallab and Bouglé, 2004; Phelan et al., 2009). As these peptides have a high net negative charge, they efficiently bind divalent cations with the formation of soluble complexes. This property makes these peptides perfect candidates as anticariogenic compounds (Cochrane and Reynolds, 2009).

Apart from the metal-binding properties, CPPs have other bioactive properties such as antioxidant (Kitts, 2005), immunostimulatory (Kitts and Nakamura, 2006), gastric secretion regulatory (Guilloteau et al., 2009), antimicrobial (Arunachalam and Raja, 2010), and stimulatory activity of growth and differentiation of osteoblastic cells (Tulipano et al., 2010).

In recent years, several studies have isolated, characterized, and identified CPPs from different types of cheese (Table 11.4). The presence of CPPs in these cheese types has been investigated by analyzing the total WSE or using an enrichment step by precipitation. Although a number of enzymes are common to many cheese varieties, the peptide composition is unique to each cheese type and reflects a characteristic ripening process. In cheeses some of the peptides appeared to be partly dephosphorylated. For instance, Addeo et al. (1992) found the α_{s1}-casein (41-75) partly dephosphorylated, containing only three of the seven native phosphate groups in Parmigiano-Reggiano cheese. Singh et al. (1997) observed complete dephosphorylation of the peptide β-casein (8-23), which in intact β-casein contains four phosphate groups. These results suggest that phosphatase is active in these varieties of cheese, hydrolyzing the native CPPs to phosphate and partly dephosphorylated peptides.

Moreover, the presence of aminopeptidase activity has been deduced from the presence of a number of CPPs of different lengths with the loss of one or more residues from the N-terminus (Ferranti et al., 1997). In Comté cheese, the original cleavage of β- and α_{s2}-casein is due to plasmin, an alkaline milk protease that has a trypsin-like specificity cleaving K_{28}-K_{29} of β-casein and K_{21}-Q_{22}

TABLE 11.4 Phosphopeptides Derived From α_{s1}-, α_{ss}-, and β-Casein Identified in Different Cheese Types

Type of Cheese	Analyzed Sample	α_{s1}-Casein	α_{s2}-Casein	β-Casein	References
Beaufort, 6 months old	Enrichment by IMAC-Fe (III)	67SpSpEEIVPN74 75SpVEQKHIQK83 103KYKVPQLEIVPNSpAE117 104YKVPQLEIVPNSpAEERLH121 106VPQLEIVPNSpAEERLH121 106VPQLEIVPNSpAEERLHSMK124 109LEIVPNSpAEERLH121 111IVPNSpAEERLH121 111IVPNSpAEERLHSMK124 112VPNSpAEERLH121 114NSpAEERL120	137KTVDMESpTEVFTK149 137KTVDMESpTEVFTKK150 137KTVDMESpTEVFTKKT151 138KTVDMESpTEVFTKK150 141MESpTEVFTKK150 143SpTEVFTKK150	12IVESpL16 14ESpLSpSpSpEESITRIN27 15SLSpSpEESITR25 15SLSpSpEESITRIN27 16LSpSpEESITR25 16LSSpSpEESITRIN27 16LSSpEESITRIN27 17SSpEESITRIN27 29KIEKFQSpE37 29KIEKFQSpEEQQ39 29KIEKFQSpEEQQQTE42 29KIEKFQSpEEQQQTED43 29KIEKFQSpEEQQQTEDELQ46 29KIEKFQSpEEQQQTEDELQDK48 29KIEKFQSpEEQQQTEDELQDKIHPF52 29KIEKFQSpEEQQQTEDELQDKIHPFAQT55 30IEKFQSpEEQQQTEDELQDK48 30IEKFQSpEEQQQTEDELQDKIHPF52 30KFQSpEEQQQTEDELQDKIHPFAQTQ56 32KFQSpEEQQQTED43 32KFQSpEEQQQTEDELQD47 32KFQSpEEQQQTEDELQDKIHPF52 33FQSpEEQQQTEDELQDKIHPF52 33FQSpEEQQQTEDELQDKIHPFAQTQ56	Dupas et al. (2009)

Cheese					Reference
Cheddar (mature)	WSE	¹¹⁵**Sp**AEERLH¹²¹ ¹¹⁵**Sp**AEERLHSMK¹²⁴	⁶¹**Sp**AEVATEEVK⁷⁰	⁷NVPGEIVESL**SpSp**¹⁸ ⁷NVPGEIVESL**SpSpSp**¹⁹ ¹⁰GEIVE**Sp**¹⁵ ²⁹KIEKFQ**Sp**EEQQ³⁹ ²⁹KIEKFQ**Sp**EEQQ....QTQ⁵⁶ ³⁰IEKFQ**Sp**E³⁶	Singh et al. (1997)
Emmental, 2 months old	WSE		¹KNTMEHV**SpSpSp**EESII**Sp**QETYK²¹ ¹KNTMEHV**SpSpSp**EESII**Sp**QETYKQEK²⁴ ⁷V**SpSp**EESII**Sp**QETYK²¹ ¹⁶**Sp**QET¹⁹ ³⁵LCSTFCKEVVRNA....**SpSpSp**E**Sp**AEVAT⁶⁶	⁷NVPGEIVESL**SpSpSp**EESITRINK²⁸ ⁸VPGEIVESL**SpSpSp**EESITRINK²⁸ ¹¹EIVE**SpL SpSpSp**EESITRINK²⁸ ¹¹EIVE**SpL SpSpSp**EESITRINKK²⁹ ¹¹EIVE**SpL SpSpSp**EESITRINKKIEK³² ¹²IVE**SpL SpSp Sp**EESITRINK²⁸ ¹²IVESL**SpSp Sp**EESITRINK²⁸ ¹³VESL**SpSp Sp**EESITRINK²⁸ ¹⁵**SpL Sp Sp Sp**EESITR²⁵ ¹⁵**SpL Sp Sp Sp**EESITRINK²⁸ ¹⁶L**Sp Sp Sp**EESITR²⁵ ¹⁶L**Sp Sp Sp**EESITRINK²⁸ ¹⁸**Sp Sp**EESITR²⁵	Gagnaire et al. (2001)

Continued

TABLE 11.4 Phosphopeptides Derived From α_{s1}-, α_{s2}-, and β-Casein Identified in Different Cheese Types—cont'd

Type of Cheese	Analyzed Sample	α_{s1}-Casein	α_{s2}-Casein	β-Casein	References
Grana Padano, 14-months old	Enrichment by precipitation with barium nitrate	61EAEpSpSpSpEEIVPN74 61EAEpSpSpSpEEIVPNSpVEQK79 62AEpSpSpSpEEIVPN74 62AEpSpSpSpEEIVPNSpVEQK79 63EpSpSpSpEEIVPN74 63EpSpSpSpEEIVPNSpVEQK79 64pSpSpSpEEIVPN74 64SIpSpSpEEIVPN74 64pSIpSpSpEEIVPNSpVEEK79 65IpSpSpEEIVPN74 65ISpSpEEIVPN74 65IpSpSpEEIVPNSpVEEK79 67pSpEEIVPN74 67SSpEEIVPN74	7VpSpSpEESIIpQE18 7VSSpSpEESIIpQE18 7VpSpSpEESIIpQET19 7VpSpSpEESIIpQETYK21 8pSpSpEESIIpQE18	7NVPGEIVESpLSpSpSpEESITRINK28 121VESpLSpSpEESITR25 121VESpLSpSpEESITRINK28 13VESpLSpSpEESIT24 13VESpLSpSpEESITR25 13VESpLSpSpEESITRIN27 13VESpLSpSpEESITRINK28 14ESpLSpSpEESIT24 14ESpLSpSpEESITR25 14ESpLSpSpEESITRIN27 14ESpLSpSpEESITRINK28 15SLSpSpSpEES22 15SpLSpSpEESIT24 15SLSpSpSpEESIT24 15SpLSpSpEESITR25 15SpLSpSpEESITRIN27 15SLSpSpSpEESITRIN27 15SpLSpSpEESITRINK28 15SLSpSpSpEESITRINK28 16SpSpSpEES22 16LSpSpEESIT24 16LSpSpEESITR25 16LSpSpEESITRIN27 16LSpSpEESITRINK28	Ferranti et al. (1997)
Grana Padano, 2–23 months old	WSE[a]			14ESLSpSpEESITRINK28 16LSpSpEES22 16LSpSpEESITR25 16LSpSpEESITRINK28	Sforza et al. (2003)

Gruyère de Comté, 12 months old	WSE[a]		5EHVSpSpSpEESIIS?QETYK[21] 6HVSpSpSpEESIIS?QETYK[21] 7VSpSpSpEESIIS?QETYK[21]	13VESpLSpSpSpEESITRINK[28] 13VESLSpSpSpEESITRINK[28] 14ESpLSpSpSpEESITRIN[27] 14ESpLSpSpSpEESITRINK[28] 14ESLSpSpSpEESITRINK[28] 15pLSpSpSpEESITRIN[27] 15pLSpSpSpEESITRIN[28] 15SLSpSpSpEESITRINK[28] 16LSpSpSpEESITRIN[27]	Roudot-Algaron et al. (1994)
Herrgård, 10 months old	Enrichment by IMAC-Fe (III)	40LSKDIGSpEpTEDQAME[55] 42KDIGSpEpT[50] 42KDIGSpEpTED[51] 42KDIGSpEpTEDQAM[54] 43DIGSpEpTE[50] 43DIGSpEpTED[51] 44IGSpE[47] 44IGSpEpTE[50] 44IGSpEpTED[51] 44IGSpEpTEDQ[52] 44IGSpEpTEDQAM[54] 61EAEspISpSpEEIVPNSpVQEK[79] 62AESpISpSpEEINVPNSpVQEK[79] 64SpISpSpSpEEIVPN[74] 75SpVEQK[79] 75SpVEQKH[80] 75SpVEQKHIQ[82] 75SpVEQKHIQKE[84] 111IVPNSpAEER[119] 112VPNSpAE[117] 112VPNSpAEER[119] 115SpAEER[119] 115SpAEERL[120] 115SpAEERLH[121] 115SpAEERLHS[122] 115SpAEERLHSMK[124]	5EHVSpSpSpEESIIspQE[18] 5EHVSpSpSpEESIIspQETY[20] 5EHVSpSpSpEESIIspQETYK[21] 6HVSpSpSpEESIIspQE[18] 6HVSpSpSpEESIIspQETYK[21] 14IISpQE[18] 14IISpQET[19] 14IISpQETYK[21] 15ISpQET[19] 15ISpQETYK[21] 16pQETYK[21] 128LSpTSpEE[133] 129pTSEENSpKKTVD[140] 141MESpTEVF[147] 141MESpTEVFTK[149]	8VPGEIVESpLSpSpSpEESITRINK[28] 10GEIVESpLSpSpSpEESITRINK[28] 11EIVESpLSpSpSpE[21] 11EIVESpLSpSpSpEESITRINK[28] 12IVESpLSpSpSpE[21] 12IVESpLSpSpSpEE[22] 12IVESpLSpSpSpEESITR[25] 12IVESpLSpSpSpEESITRINK[28] 13VESpLSpSpSpE[21] 13VESpLSpSpSpEESITR[25] 15pLSpSpSpEESITR[25] 29KIEKFQSpEEQQTED[43] 29KIEKFQSpEEQQQTEDELQ[46] 30EKFQSpE[36] 30EKFQSpE[37] 30EKFQSpEEQQ[39] 30EKFQSpEEQQTED[43] 30EKFQSpEEQQQTEDELQ[46] 33KFQSpE[37] 32KFQSpEEQ[38] 32KFQSpEEQQQTED[43] 33FQSpE[36] 33FQSpE[37] 33FQSpEEQQQ[40]	Lund and Ardö (2004)

Continued

TABLE 11.4 Phosphopeptides Derived From α_{s1}-, α_{ss}-, and β-Casein Identified in Different Cheese Types—cont'd

Type of Cheese	Analyzed Sample	α_{s1}-Casein	α_{s2}-Casein	β-Casein	References
Parmigiano-Reggiano, 2 months old	WSE[a]		3TMEHVSpSpSpEESIISQE...PSKE33	8VPGEIVESpL16 9PGEIVESpL16 17SpSpSpES22 22SITRYNKKIEKFQSpEEQQQ....KIHPFAQ54	Addeo et al. (1992)
Parmigiano-Reggiano, 6 months old	WSE[a]	41SKNIG....SISpSpSpEEIVPNS75	4MEHVSpSpSpEESIISQE....PSKENLC36		Addeo et al. (1992)
Parmigiano-Reggiano, 15 months old	WSE[a]		4MEHVSpSpSpEESIISQE...PSKENLC36		Addeo et al. (1992)
Parmigiano-Reggiano, (Evolution 2, 3, 4, 6, 8, 10, 12, 16, 20, and 24 months old	Enrichment by acid precipitation			11EIVESpLSpSpSpEESITRINK28 12IVESpLSpSpSpEESITRINK28 13VESpLSpSpSpEESITRINK28 14ESpLSpSpSpEESITRINK28 15SpLSpSpSpEESITRINK28 15SLSpSpSpEESITRINK28 16LSpSpSpEESITR25 16LSpSpSpEESITRINK28 17SpSpSpEESITR25	Sforza et al. (2012)
Parmigiano-Reggiano, 24 months old	Enrichment by IMAC-Fe (III)	64SISpSpSpEEIVPN74 65ISpSpSpEE70 65ISpSpSpEEIVPN74 67SpSpEEIVPN74 115SpAEERLH121 115SpAEERLHSMK124	7VSpSpSpEES13	16LSpSpSpEES22	Lund and Ardö (2004)
Ragusano	WSE[a]	75SpVEQKHIQKE84 103KYKVPQLEIVPNSpAE117 104YKVPQLEIVPNSpAE117 115SpAEERLHSM123	1KNTMEHVSpSpSpEESIISpQE18 1KNTMEHVSpSpSpEESIISpQETYK21 141MESpTEVFTKK150 143SpTEVFTKK150	7NVPGEIVESpLSpSpSpEESITRINK28 15ISLSpSpSpEESITRINK28 17SpSpSpEESITRINK28 28KKIEKFQSpEEQQQTEDELQ46	Gagnaire et al. (2011)

Ragusano, 4 months old	WSF[a]	41SKDIGSp**E**Sp**T**EDQ52 41SKDIGSp**E**Sp**T**EDQAM54 41SKDIGS**pE**Sp**T**EDQAMED56 75**Sp**VEQKHIQ82 75**Sp**VEQKHIQKE84		28KKIEKFQ**Sp**EEQQQTEDELQDKIHPF52 29KIEKFQ**Sp**EEQQ39 29KIEKFQ**Sp**EEQQQ40 29KIEKFQ**Sp**EEQQQTED43 29KIEKFQ**Sp**EEQQQTEDEL45 29KIEKFQ**Sp**EEQQQTEDELQ46 29KIEKFQ**Sp**EEQQQTEDELQD47 29KIEKFQ**Sp**EEQQQTEDELQDKIHP51 29KIEKFQ**Sp**EEQQQTEDELQDKIHPF52 29KIEKFQ**Sp**EEQQQTEDELQDKIHPFAQTQ56 30IEKFQ**Sp**EEQQQTEDELQDKIHPF52 32KFQ**Sp**EEQQ39 32KFQ**Sp**EEQQQ40 32KFQ**Sp**EEQQQTE42 32KFQ**Sp**EEQQQTED43 32KFQ**Sp**EEQQQTEDELQ46 32KFQ**Sp**EEQQQTEDELQD47 32KFQ**Sp**EEQQQTEDELQDKIHPF52 32KFQ**Sp**EEQQQTEDELQDKIHPFAQTQ56 33FQ**Sp**EEQQQTED43 1REL....IVE**Sp**L**Sp**Sp**E**...**SpE**EQQQTEDELQ46 11EIVE**Sp**L**Sp**Sp**E**E**Sp**EESITRINK28 12IVE**Sp**L**Sp**Sp**E**ESITRINK28 15I**Sp**L**Sp**Sp**E**ESITRINK28 17**Sp**Sp**E**ESITRINK28 29KIEKFQ**Sp**EEQQQTEDELQDKIHPF52 29KIEKFQ**Sp**EEQQQTEDELQDKIHPFAQTQ56 32KFQ**Sp**EEQQQ40 32KFQ**Sp**EEQQQTED43 32KFQ**Sp**EEQQQTEDELQDKIHPF52	Gagnaire et al. (2011)
Ragusano, 7 months old	WSF[a]	41SKDIGSp**E**Sp**T**EDQAM54 41SKDIGSp**E**Sp**T**EDQAMED56 42KDIGS**pE**Sp**T**EDQAMED56 75**Sp**VEQKHIQ82	126EQL**Sp**TSEEN**Sp**pKKT138	11EIVE**Sp**L**Sp**Sp**E**ESITRINK28 12IVE**Sp**L**Sp**Sp**E**ESITRINK28 17**Sp**Sp**E**ESITRINK28 29KIEKFQ**Sp**EEQQQTEDELQDKIHPF52	Gagnaire et al. (2011)

[a]Water-soluble extract.
Sp, Ser phosphorylated.

of α_{s2}-casein. These two peptides are further hydrolyzed by the action of another endopeptidase and of an aminopeptidase, as well as of a lysyl carboxypeptidase (Roudot-Algaron et al., 1994).

Therefore CPPs could be regarded as transient intermediate components in cheese; they either accumulate or are degraded by cheese enzymes to shorter peptides and free amino acids, including phosphorylated serine. Most CPPs come from the region (14-28) of β-casein (Sforza et al., 2003). The accumulation of a specific peptide always depends on a delicate balance between producing and degrading enzymes and the availability of the parent substrate (Sforza et al., 2012). In the case of Ragusano cheese the number of identified soluble peptides dramatically decreased after 4 and 7 months of ripening, from 123 to 47 and 25, respectively (Gagnaire et al., 2011).

It has been reported that CPPs are released in the gut and accumulate in the lower part of the small intestine (ileum) where mineral absorption takes place (FitzGerald, 1998; Zidane et al., 2012). Therefore several *in vivo* studies have investigated if CPPs could increase mineral absorption, especially Ca^{2+} absorption. Although these studies provide considerable evidence for a potential effect of CPPs to improve mineral absorption, results are still controversial (Scholtz-Ahrens and Schrezenmeir, 2000; Meisel and FitzGerald, 2003; Mills et al., 2011). This variability may be due to the diversity of the experimental approaches used to assess mineral bioavailability (Korhonen and Pihlanto, 2006). Moreover, it seems that results may depend on the food matrix (Hansen et al., 1997). On the other hand, the effect of CPPs may also be influenced by the physical status of the population (Teucher et al., 2006), the levels of CPPs (Bennett et al., 2000), and the ratio calcium:phosphorus present (Erba et al., 2001, 2002).

11.6 FUTURE PROSPECTS

Cheese is an adequate source of micronutrients and bioactives, which have generated extensive investigation during the last six decades. While knowledge of minerals is relatively complete and precise, the data for vitamins are lacking or there is discrepancy between them. However, there is no doubt about the nutritional role and health benefits of vitamins. As for lipid and protein components, beyond their nutritional role, increased scientific knowledge of biological activities has demonstrated that cheese may provide specific health benefits. Questions to be answered include the changes that may occur during processing, the role of the different components in the absorption, and the possibility of enriching cheese with micronutrients or bioactives. For example, there are some reports on the presence of bioactive peptides in duodenal effluents from human on milk protein ingestion (Boutrou et al., 2013), and even the presence in plasma of the milk bioactive peptides IPP and VPP (Foltz et al., 2007) or HLPLP (Sánchez-Rivera et al., 2014) has been demonstrated *in vivo*. However, more research is needed to evaluate the absorption of these peptides in a complex food matrix such as cheese. Further intervention studies confirming the beneficial effects of cheeses containing bioactive peptides should also be conducted.

REFERENCES

Addeo, F., Chianese, L., Salzano, A., Sacchi, R., Capuccio, U., Ferranti, P., Malorni, A., 1992. Characterization of the 12% trichloroacetic acid-insoluble oligopeptides of Parmigiano Reggiano. Journal of Dairy Research 59, 401–411.

Ahola, A.J., Yli-Knuuttila, H., Suomalainen, T., Poussa, T., Ahlström, A., Meurman, J.H., Korpela, R., 2002. Short-term consumption of probiotic-containing cheese and its effect on dental caries risk factors. Archives of Oral Biology 47, 799–804.

Aljewicz, M., Siemianowska, E., Cichosz, G., Tonska, E., 2014. The effect of probiotics (*Lactobacillus rhamnosus* HN001, *Lactobacillus paracasei* LPC-37, and *Lactobacillus acidophilus* NCFM) on the availability of minerals from Dutch-type cheese. Journal of Dairy Science 97, 4824–4831.

Arunachalam, K., Raja, R.B., 2010. Isolation and characterization of CPP (casein phosphopeptides) from fermented milk. African Journal of Food Science 4, 167–175.

Battacharaya, A., Banu, J., Rahman, M., Causey, J., Fernandes, G., 2006. Biological effects of conjugated linoleic acids in health and disease. Journal of Nutritional Biochemistry 17, 789–810.

Bennett, T., Desmond, A., Harrington, M., McDonagh, D., FitzGerald, R., Flynn, A., Cashman, K.D., 2000. The effect of high intakes of casein and casein phosphopeptides on calcium absorption in the rat. British Journal of Nutrition 83, 673–680.

Blank-Porat, D., Gruss-Fischer, T., Tarasenko, N., Malik, Z., Nudelman, A., Rephaeli, A., 2007. The anticancer prodrugs of butyric acid AN-7 and AN-9, possess antiangiogenic properties. Cancer Letters 256, 39–48.

Bottesini, C., Paolella, S., Lambertini, F., Galaverna, G., Tedeschi, T., Dossena, A., Marchelli, R., Sforza, S., 2013. Antioxidant capacity of water soluble extracts from Parmigiano-Reggiano cheese. International Journal of Food Science and Nutrition 64, 953–958.

Bouhallab, S., Bouglé, D., 2004. Biopeptides of milk: caseinophosphopeptides and mineral bioavailability. Reproduction Nutrition Development 44, 493–498.

Boutrou, R., Gaudichon, C., Dupont, D., Jardin, J., Airinei, G., Marsset-Baglieri, A., Benamouzig, R., Tomé, D., Leonil, J., 2013. Sequential release of milk protein derived bioactive peptides in the jejunum in healthy humans. The American Journal of Clinical Nutrition 97, 1314–1323.

Bulletin of the International Dairy Federation, 2013. The World Dairy Situation. No. 470/2013 470.

Bütikofer, U., Meyer, J., Sieber, R., Wechsler, D., 2007. Quantification of the angiotensin-converting enzyme inhibiting tripeptides Val-Pro-Pro and Ile-Pro-Pro in hard, semihard and soft cheeses. International Dairy Journal 17, 968–975.

Cámara-Martos, F., Moreno-Rojas, R., Pérez-Rodríguez, F., 2013. Cheese as a source of nutrients and contaminants: dietary and toxicological aspects. In: Castelli, H., Vale, L.D. (Eds.), Handbook on Cheese: Production, Chemistry and Sensory Properties. Nova Science Publishers, USA, pp. 341–370.

Caroprese, M., Sevi, A., Marino, R., Santillo, A., Tateo, A., Albenzio, M., 2013. Composition and textural properties of Mozzarella cheese naturally-enriched in polyunsaturated fatty acids. Journal of Dairy Research 80, 276–282.

Chen, Z.Y., Peng, C., Jiao, R., Wong, Y.M., Yang, N., Huang, Y., 2009. Antihypertensive nutraceuticals and functional foods. Journal of Agricultural and Food Chemistry 57, 4485–4499.

Cicognini, F.M., Rossi, F., Sigolo, S., Gallo, A., Prandini, A., 2014. Conjugated linoleic acid isomer (cis9,trans11 and trans10,cis12) content in cheeses from Italian large-scale retail trade. International Dairy Journal 34, 180–183.
Coakley, M., Johnson, M.C., McGrath, E., Rahman, S., Ross, R.P., Fitzgerald, G.F., Devery, R., Stanton, C., 2006. Intestinal bifidobacteria that produce trans-9, trans-11 conjugated linoleic acid: a fatty acid with antiproliferative activity against human colon SW480 and HT-29 cancer cells. Nutrition and Cancer 56, 95–102.
Cochrane, N.J., Reynolds, E.C., 2009. In: Wilson, M. (Ed.), Food Constituents and Oral Health. Current Status and Future Prospects. CRC Press, Boca Raton, pp. 1–543.
Collomb, M., Schmid, A., Sieber, R., Wechsler, D., Ryhänen, E.-L., 2006. Conjugated linoleic acids in milk fat: variation and physiological effects. International Dairy Journal 16, 1347–1361.
Contreras, M.M., Carrón, R., Montero, M.J., Ramos, M., Recio, I., 2009. Novel casein-derived peptides with antihypertensive activity. International Dairy Journal 19, 566–573.
Contreras, M.M., Sanchez, D., Sevilla, M.A., Recio, I., 2013. Resistance of casein derived peptides to simulated gastrointestinal digestion. International Dairy Journal 32, 71–78.
Cortes Nunes, J., Guedes Torres, A., 2010. Fatty acid and CLA composition of Brazilian dairy products, and contribution to daily intake of CLA. Journal of Food Composition and Analysis 23, 782–789.
Daniel, H., Wessendorf, A., Vohwinkel, M., Brantl, V., 1990. Effect of D-Ala2, 4, Tyr^5-β-casomorphin-5-amide on gastro-intestinal functions. In: Nyberg, F., Brantl, V. (Eds.), β-Casomorphins and Related Peptides. Fyris-Tryck AB, Uppsala, pp. 95–104.
De la Fuente, M.A., Juárez, M., 2001. Los quesos: Una fuente de nutrientes. Alimentación, Nutrición y Salud 8, 75–83.
de Lima Alves, L., Pereira dos Santos Richards, N.S., Barros Mariutti, L.R., Nogueira, G.C., Bragagnolo, N., 2011. Inulin and probiotic concentration effects on fatty and linoleic conjugated acids in cream cheeses. European Food Research Technology 233, 667–675.
De Moreno de LeBlanc, A., Matar, C., LeBlanc, N., Perdigon, G., 2005. Effects of milk fermented by *Lacto bacillus* helveticus R389 on a murine breast cancer model. Breast Cancer Research 7, R477–R486.
De Noni, I., Cattaneo, S., 2010. Occurrence of beta-casomorphins 5 and 7 in commercial dairy products and in their digests following in-vitro simulated gastro-intestinal digestion. Food Chemistry 119, 560–566.
De Simone, C., Picariello, G., Mamone, G., Stiuso, P., Dicitore, A., Vanacore, D., Chianese, L., Addeo, F., Ferranti, P., 2009. Characterization and cytomodulatory properties of peptides from Mozzarella di Bufala Campana cheese whey. Journal of Peptide Science 15, 251–258.
De Simone, C., Ferranti, P., Picariello, G., Scognamiglio, I., Dicitore, A., Addeo, F., Chianese, L., Stiuso, P., 2011. Peptides from water buffalo cheese whey induced senescence cell death via ceramide secretion in human colon adenocarcinoma cell line. Molecular Nutrition and Food Research 55, 229–238.
Domagała, J., Sady, M., Grega, T., Pustkowiak, R., Florkiewicz, A., 2010. The influence of cheese type and fat extraction method on the content of conjugated linoleic acid. Journal of Food Composition and Analysis 23, 238–243.
Dupas, C., Adt, I., Cottaz, A., Boutrou, R., Molle, D., Jardin, J., Jouvet, T., Degraeve, P., 2009. A chromatographic procedure for semiquantitative evaluation of caseinphosphopeptides in cheese. Dairy Science Technology 89, 519–529.
Erba, D., Ciappellano, S., Testolin, G., 2001. Effect of casein phosphopeptides on inhibition of calcium intestinal absorption due to phosphate. Nutrition Research 21, 649–656.

Erba, D., Ciappellano, S., Testolin, G., 2002. Effect of the ratio of casein phosphopeptides to calcium (w/w) on passive calcium transport in the distal small intestine of rats. Journal of Nutrition 18, 743–746.

Erdmann, K., Cheung, B.W.Y., Schröder, H., 2008. The possible roles of food-derived bioactive peptides in reducing the risk of cardiovascular disease. Journal of Nutritional Biochemistry 19, 643–654.

European Food Safety Agency, 2008. Scientific substantiation of a health claim pursuant related to dairy products (milk and cheese) and dental health to Article 14 of Regulation (EC) No 1924/20061. EFSA Journal 787, 1–9.

Ferranti, P., Barone, F., Chianese, L., Addeo, F., Scaloni, A., Pellegrino, L., Resmini, P., 1997. Phosphopeptides from Grana Padano cheese: nature, origin and changes during ripening. Journal of Dairy Research 64, 601–615.

FitzGerald, R.J., 1998. Potential uses of caseinophosphopeptides. International Dairy Journal 8, 451–457.

Foltz, M., Meynen, E.E., Bianco, V., van Platerink, C., Koning, T.M.M.G., Kloek, J., 2007. Angiotensin converting enzyme inhibitory peptides from a lactotripeptide-enriched milk beverage are absorbed intact into the circulation. The Journal of Nutrition 137, 953–958.

Fontecha, J., Rodriguez-Alcalá, L.M., Calvo, M.V., Juárez, M., 2011. Bioactive milk lipids. Current Nutrition & Food Science 7, 1–5.

Gagnaire, V., Molle, D., Herrouin, M., Leonil, J., 2001. Peptides identified during Emmental cheese ripening: origin and proteolytic systems involved. Journal of Agricultural and Food Chemistry 49, 4402–4413.

Gagnaire, V., Carpino, S., Pediliggieri, C., Jardin, J., Lortal, S., Licitra, G., 2011. Uncommonly through hydrolysis of peptides during ripening of Ragusano cheese revealed by tandem mass spectrometry. Journal of Agricultural and Food Chemistry 59, 12443–12452.

Gaucheron, F., 2011. Milk and dairy products: a unique micronutrient combination. Journal of the American College of Nutrition 30, 400–409.

Gedalia, I., Ionat-Bendat, D., Ben-Mosheh, S., Shapira, L., 1991. Tooth enamel softening with a cola type drink and rehardening with hard cheese or stimulated saliva. Journal of Oral Rehabilitation 18, 501–506.

Gómez-Ruiz, J.A., Ramos, M., Recio, I., 2004a. Angiotensin-converting enzyme-inhibitory activity of peptides isolated from Manchego cheese. Stability under simulated gastrointestinal digestion. International Dairy Journal 14, 1075–1080.

Gómez-Ruiz, J.A., Ramos, M., Recio, I., 2004b. Identification and formation of angiotensin-converting enzyme-inhibitory peptides in Manchego cheese by high-performance liquid chromatography–tandem mass spectrometry. Journal of Chromatography A 1054, 269–277.

Gómez-Ruiz, J.A., Taborda, G., Amigo, L., Recio, I., Ramos, M., 2006. Identification of ACE-inhibitory peptides in different Spanish cheeses by tandem mass spectrometry. European Food Research Technology 223, 595–601.

Guilloteau, P., Romé, V., Delaby, L., Mendy, F., Roger, L., Chayvialle, J.A., 2009. New role of phosphopeptides as bioactive peptides released during milk casein digestion in the young mammal: regulation of gastric secretion. Peptides 30, 2221–2227.

Gupta, A., Mann, B., Kumar, R., Sangwan, R.B., 2009. Antioxidant capacity of Cheddar cheeses at different stages of ripening. International Journal of Dairy Technology 62, 339–347.

Gupta, A., Mann, B., Kumar, R., Sangwan, R., 2010. Identification of antioxidant peptides in Cheddar cheese made with adjunct culture *Lactobacillus casei* ssp. *casei* 300. Milchwissenschaft 65, 396–399.

Hansen, M., Sandstrom, B., Jensen, M., Sorensen, S.S., 1997. Casein phosphopeptides improve zinc and calcium absorption from rice-based but not from whole-grain infant cereal. Journal of Pediatric Gastroenterology and Nutrition 24, 56–62.

Heaney, R.P., 2000. Calcium, dairy products and osteoporosis. Journal of American College of Nutrition 19, 83S–99S.

Hill, R.D., Lahov, E., Givol, D., 1974. A rennin-sensitive bond in alpha and beta casein. Journal of Dairy Research 41, 147–153.

Huang, C.B., Alimova, Y., Myers, T.M., Ebersole, J.L., 2011. Short- and medium-chain fatty acids exhibit antimicrobial activity for oral microorganisms. Archives of Oral Biology 56, 650–654.

Ibeagha-Awemu, E.M., Kgwatalala, P.M., Zhao, X., 2009. Potential for improving health: calcium bioavailability in milk and dairy products. In: Park, Y.W. (Ed.), Bioactive Components in Milk and Dairy Products. Wiley-Blackwell, Ames, pp. 363–377.

Jarmolowska, B., Kostyra, E., Krawczuck, S., Kostyra, H., 1999. β-casomorphin-7 isolated from Brie cheese. Journal of the Science of Food and Agriculture 79, 1788–1792.

Kashket, S., DePaola, D.P., 2002. Cheese consumption and the development and progression of dental caries. Nutrition Reviews 60, 97–103.

Kato, K., Takada, Y., Matsuyama, H., Kawasaki, Y., Aoe, S., Yano, H., Toba, Y., 2002. Milk calcium taken with cheese increases bone mineral strength in growing rats, Bioscience. Biotechnology and Biochemistry 66, 2342–2346.

Kitts, D.D., Nakamura, S., 2006. Calcium-enriched casein phosphopeptide stimulates release of IL-6 cytokine in human epithelial intestinal cell line. Journal of Dairy Research 73, 44–48.

Kitts, D.D., Yuan, Y.V., Nagasawa, T., Moriyama, Y., 1992. Effect of casein, casein phosphopeptides and calcium intake on ileal Ca disappearance and temporal systolic blood pressure in spontaneously hypertensive rats. British Journal of Nutrition 68, 765–781.

Kitts, D.D., 2005. Antioxidant properties of casein phosphopeptides. Trends in Food Science Technology 16, 549–554.

Korhonen, H., Pihlanto, A., 2006. Bioactive peptides: Production and functionality. International Dairy Journal 16, 945–960.

Kost, N.V., Sokolov, O.Y., Kurasova, O.B., Dmitriev, A.D., Tarakanova, J.N., Gabaeva, M., 2009. Beta-casomorphins-7 in infants on different type of feeding and different levels of psychomotor development. Peptides 30, 1854–1860.

Kostyra, E., Sienkiewicz-Sztapka, E., Jarmolowska, B., Krawczuck, S., Kostyra, H., 2004. Opioid peptides derived from milk proteins. Polish Journal of Food Nutrition Science 13, 25–35.

Lahov, E., Regelson, W., 1996. Antibacterial and immunostimulating casein-derived substances from milk: casecidin, isracidin peptides. Federal Chemistry Toxicology 34, 131–145.

Lee, K.W., Lee, H.J., Cho, H.Y., Kim, Y.J., 2005. Role of the conjugated linoleic acid in the prevention of cancer. Critical Reviews in Food Science and Nutrition 45, 135–144.

Lignitto, L., Segato, S., Balzan, S., Cavatorta, V., Oulahal, N., Sforza, S., Degraeve, P., Galaverna, G., Novelli, E., 2012. Preliminary investigation on the presence of peptides inhibiting the growth of *Listeria innocua* and *Listeria monocytogenes* in Asiago d´Allevo cheese. Dairy Science and Technology 92, 297–308.

López-Expósito, I., Recio, I., 2006. Antibacterial activity of peptides and folding variants from milk proteins. International Dairy Journal 16, 1294–1305.

López-Expósito, I., Gómez-Ruiz, J.A., Amigo, L., Recio, I., 2006. Identification of antibacterial peptides from ovine α_{s2}-casein. International Dairy Journal 16, 1072–1080.

López-Expósito, I., Amigo, L., Recio, I., 2012. A mini-review on health and nutritional aspects of cheese with a focus on bioactive peptides. Dairy Science & Technology 92, 419–438.

Losito, I., Carbonara, T., De Bari, M.D., Gobetti, M., Palmiseno, F., Rizzello, C.G., Zambonin, P.G., 2006. Identification of peptides in antimicrobial fractions of cheese extracts by electrospray ionization ion trap mass spectrometry coupled to a two-dimensional liquid chromatographic separation. Rapid Communications in Mass Spectrometry 20, 447–455.

Lund, M., Ardö, Y., 2004. Purification and identification of water soluble phosphopeptides from cheese using Fe(III) affinity chromatography and mass spectrometry. Journal of Agricultural and Food Chemistry 52, 6616–6622.

Mader, J.S., Salsman, J.S., Conrad, D.M., Hoskin, D.W., 2005. Bovine lactoferricin selectively induces apoptosis in human leukemia and carcinoma cell lines. Molecular Cancer Therapy 4, 612–624.

Meisel, H., FitzGerald, R.J., 2003. Biofunctional peptides from milk proteins: mineral binding and cytomodulatory effects. Current Pharmaceutical Design 9, 1289–1295.

Meister Meira, S.M., Daroit, D.J., Etges Helfer, V., Folmer Corrêa, A.P., Segalin, J., Carro, S., Brandelli, A., 2012. Bioactive peptides in water-soluble extracts of ovine cheeses from Southern Brazil and Uruguay. Food Research International 48, 322–329.

Mills, S., Ross, R.P., Hill, C., FitzGerald, G.F., Stanton, C., 2011. Milk intelligence: mining milk for bioactive substances associated with human health. International Dairy Journal 21, 377–401.

Nagpal, R., Behare, P., Rana, R., Kumar, A., Kumar, M., Arora, S., Morotta, F., Jain, S., Yadav, H., 2011. Bioactive peptides derived from milk proteins and their health beneficial potentials: an update. Food and Function 2, 18–27.

Nudda, N., Battacone, G., Boaventura Neto, O., Cannas, A., Francesconi, A.H.D., Atzori, A.S., Pulina, G., 2014. Feeding strategies to design the fatty acid profile of sheep milk and cheese. Revista Brasileira de Zootecnia 43, 445–456.

Nussberger, J., 2007. Blutdrucksenkende tripeptide aus der milch. Therapeutische Umschau 64, 177–179.

O'Brien, N.M., O'Connor, T.P., 2004. Nutritional aspects of cheese. In: Cheese: Chemistry, Physics and Microbiology-Volume 1: General Aspects, third ed. Elsevier Academic Press, London, pp. 573–581.

Ohlund, I., Holgerson, P.L., Backman, B., Lind, T., Hernell, O., Johansson, I., 2007. Diet intake and caries prevalence in four-year-old children living in a low prevalence country. Caries Research 41, 26–33.

Ong, L., Shah, N.P., 2008. Release and identification of angiotensin-converting enzyme-inhibitory peptides as influenced by ripening temperatures and probiotic adjuncts in Cheddar cheeses. LWT – Food Science and Technology 41, 1555–1566.

Ong, L., Henriksson, A., Shah, N.P., 2007. Angiotensin converting enzyme-inhibitory activity in Cheddar cheeses made with the addition of probiotic *Lactobacillus casei* sp. Lait 87, 149–165.

Parikh, S.J., Yanovski, J.A., 2003. Calcium intake and adiposity. American Journal of Clinical Nutrition 77, 281–287.

Park, Y.W., Jin, Y.K., 1998. Proteolytic patterns of Caciotta and Monterey Jack hard goat milk cheeses as evaluated by SDS-PAGE and densitometric analyses. Small Ruminant Research 28, 263–272.

Parodi, P.W., 1999. Conjugated linoleic acid and other anticarcinogenic agents of bovine milk fat. Journal of Dairy Science 6, 1339–1349.

Parodi, P.W., 2006. Nutritional significance of milk lipids. In: Fox, P.F., McSweeney, P.L.H. (Eds.), Advanced Dairy Chemistry, Volume 2: Lipids, third ed. Springer, EEUU. pp. 601–639.

Parrot, S., Degraeve, P., Curia, C., Martial-Gros, A., 2003. In vitro study on digestion of peptides in Emmental cheese: analytical evaluation and influence on angiotensin I converting enzyme inhibitory peptides. Nahrung 47, 87–94.

Phelan, M., Aherne, A., Fitzgerald, R., O'Brien, N., 2009. Casein-derived bioactive peptides: biological effects, industrial uses, safety aspects and regulatory status. International Dairy Journal 19, 643–654.

Pintus, S., Murru, E., Carta, G., Cordeddu, L., Batetta, B., Accossu, S., Pistis, D., Uda, S., Ghiani, M.E., Mele, M., Secchiari, P., Almerighi, G., Pintus, P., Banni, S., 2013. Sheep cheese naturally enriched in α-linolenic, conjugated linoleic and vaccenic acids improves the lipid profile and reduces anandamide in the plasma of hypercholesterolaemic subjects. British Journal of Nutrition 109, 1453–1462.

Plaisancié, P., Claustre, J., Estienne, M., Henry, G., Boutrou, R., Paquet, A., Léonil, J., 2013. A novel bioactive peptide from yoghurts modulates expression of the gel-forming MUC2 mucin as well as population of goblet cells and Paneth cells along the small intestine. Journal of Nutritional Biochemistry 24, 213–221.

Prandini, A., Sigolo, S., Tansini, G., Brogna, N., Piva, G., 2007. Different level of conjugated linoleic acid (CLA) in dairy products from Italy. Journal of Food Composition and Analysis 20, 472–479.

Prandini, A., Sigolo, S., Piva, G., 2011. A comparative study of fatty acid composition and CLA concentration in commercial cheeses. Journal of Food Composition and Analysis 24, 55–61.

Pritchard, S.R., Phillips, M., Kailasapathy, K., 2010. Identification of bioactive peptides in commercial Cheddar cheese. Food Research International 43, 1545–1548.

Quinlivan, E.P., Mc Partlin, J., McNulty, H., Ward, M., Strain, J.J., Weir, D.G., Scott, J.M., 2002. Importance of both folic acid and vitamin B12 in reduction of risk of vascular disease. Lancet 359, 227–228.

Qureshi, T.M., Vegarudm, G.E., Abrahamsen, R.K., Skeie, S., 2013. Angiotensin I-converting enzyme-inhibitory activity of the Norwegian autochthonous cheeses Gamalost and Norvegia after in vitro human gastrointestinal digestion. Journal of Dairy Science 96, 838–853.

Recio, I., de la Fuente, M.A., Juárez, M., Ramos, M., 2009. Bioactive components in sheep milk. In: Park, Y.W. (Ed.), Bioactive Components in Milk and Dairy Products. Wiley-Blackwell, USA, pp. 83–104.

Riserus, U., Brismar, K., Arner, P., Vessby, B., 2002. Treatment with dietary trans-10 cis-12 conjugated linoleic acid causes isomer-specific insulin resistance in obese men with the metabolic syndrome. Diabetes Care 25, 1516–1521.

Rizzello, C.G., Losito, I., Gobetti, M., Carbonara, T., De Bari, M.D., Zambonin, P.G., 2005. Antibacterial activity of peptides from the water-soluble extracts of Italian cheese varieties. Journal of Dairy Science 88, 2348–2360.

Roudot-Algaron, F., Le Bars, D., Kerhoas, L., Einhorn, J., Gripon, J.C., 1994. Phosphopeptides from Comté cheese: nature and origin. Journal of Food Science 59, 544–560.

Roy, M.K., Kuwabara, Y., Hara, Y., Watanabe, Y., Tamai, Y., 2002. Peptides from the N-terminal end of bovine lactoferrin induce apoptosis in human leukemic (HL-60) cells. Journal of Dairy Science 85, 2065–2074.

Russell, R.M., 2000. The vitamin A spectrum: from deficiency to toxicity. American Journal of Clinical Nutrition 71, 878–884.

Sagardia, I., Iloro, I., Elortza, F., Bald, C., 2013. Quantitative structure-activity relationship based screening of bioactive peptides identified in ripened cheese. International Dairy Journal 33, 184–190.

Saito, T., Nakamura, T., Kitazawa, H., Kawai, Y., Itoh, T., 2000. Isolation and structural analysis of antihypertensive peptides that exist naturally in Gouda cheese. Journal of Dairy Science 83, 1434–1440.

Sánchez-Rivera, L., Ares, I., Miralles, B., Gómez-Ruiz, J.Á., Recio, I., Martínez-Larrañaga, M.R., Anadón, A., Martínez, M.A., 2014. Bioavailability and kinetics of the antihypertensive casein-derived peptide HLPLP in rats. Journal of Agricultural and Food Chemistry 62, 11869–11875.

Sarmadi, B.H., Ismail, A., 2010. Antioxidative peptides from food proteins: a review. Peptides 31, 1949–1956.

Scholtz-Ahrens, K.E., Schrezenmeir, J., 2000. Effects of bioactive substances in milk on mineral and trace element metabolism with special reference to casein phosphopeptides. British Journal of Nutrition 84 (Suppl.19), S147–S153.

Sforza, S., Ferroni, L., Galaverna, G., Dossena, A., Marchelli, R., 2003. Extraction, semiquantification, and fast on-line identification of oligopeptides in Grana Padano cheese by HPLC-MS. Journal of Agricultural and Food Chemistry 51, 2130–2135.

Sforza, S., Cavatorta, V., Lambertini, F., Galaverna, G., Dossena, A., Marchelli, R., 2012. Cheese peptidomics: a detailed study on the evolution of the oligopeptide fraction in Parmigiano-Reggiano cheese from curd to 24 months of aging. Journal of Dairy Science 95, 3514–3526.

Shahar, D.R., Vardi, H., Abel, R., Fraser, D., Elhayany, A., 2007. Does dairy calcium intake enhance weight loss among overweight diabetic patients. Diabetic Care 30, 485–489.

Sieber, R., Bütikofer, U., Egger, C., Portmann, R., Walthier, B., Wechsler, D., 2010. ACE-inhibitory activity and ACE-inhibiting peptides in different cheese varieties. Dairy Science and Technology 90, 47–73.

Sienkiewicz-Szlapka, E., Jarmolowska, B., Krawczuk, S., Kostyra, E., Iwan, M., 2009. Contents of agonistic and antagonistic peptides in different cheese varieties. International Dairy Journal 19, 258–263.

Silva, S.V., Pihlanto, A., Malcata, F.X., 2006. Bioactive peptides in ovine and caprine cheese like systems prepared with proteases from *Cynara cardunculus*. Journal of Dairy Science 89, 3336–3344.

Silva, R.A., Lima, M.S.F., Viana, J.B.M., Bezerra, V.S., Pimentel, M.C.B., Porto, A.L.F., Cavalcanti, M.T.H., Lima Filho, J.L., 2012. Can artisanal "Coalho" cheese from Northeastern Brazil be used as a functional food? Food Chemistry 135, 1533–1538.

Singh, T.K., Fox, P.F., Healy, A., 1995. Water-soluble peptides in Cheddar cheese: isolation and identification of peptides in the diafiltration retentate of the water-soluble fraction. Journal of Dairy Research 62, 629–640.

Singh, T.K., Fox, P.F., Healy, A., 1997. Isolation and identification of further peptides in the diafiltration retentate of the water-soluble fraction of Cheddar cheese. Journal of Dairy Research 64, 433–443.

Singh, M., Rosen, C.L., Chang, K., Haddad, G.G., 1999. Plasma β-casomorphin-7 immunoreactive peptide increases after milk ingestion in newborn but not in adult dogs. Pediatric Research 26, 34–38.

Smacchi, E., Gobbetti, M., 1998. Peptides from several Italian cheeses inhibitory to proteolytic enzymes of lactic acid bacteria, *Pseudomonas fluorescens* ATCC 948 and to the angiotensin-I-converting enzyme. Enzyme and Microbial Technology 22, 687–694.

Sofi, F., Buccioni, A., Cesari, F., Gori, A.M., Minieri, S., Mannini, L., Casini, A., Gensini, G.F., Abbate, R., Antongiovanni, M., 2010. Effects of a dairy product (pecorino cheese) naturally rich in cis-9, trans-11 conjugated linoleic acid on lipid, inflammatory and haemorheological variables: a dietary intervention study. Nutrition, Metabolism & Cardiovascular Diseases 20, 117–124.

Stadtman, E.R., 2004. Role of oxidant species in aging. Current Medicinal Chemistry 11, 1105–1112.

Stepaniak, L., Fox, P.F., Sorhaug, T., Grabska, J., 1995. Effect of peptides from the sequence 58-72 of beta-casein on the activity of endopeptidase, aminopeptidase, and X-prolyl-dipeptidyl aminopeptidase from Lactococcus lactis spp lactis MG1363. Journal of Agricultural and Food Chemistry 43, 849–853.

Stuknytè, M., Cattaneo, S., Masotti, F., De Noni, I., 2015. Occurrence and fate of ACE-inhibitor peptides in cheeses and in their digestates following in vitro static gastrointestinal digestion. Food Chemistry 168, 27–33.

Teegarden, K., 2003. Calcium intake and reduction in weight or fat mass. Journal of Nutrition 133, 249S–251S.

Teschemacher, H., 2003. Opioid receptor ligands derived from food proteins. Current Pharmaceutical Design 9, 1331–1344.

Teucher, B., Majsak-Newman, G., Dainty, J.R., McDonagh, D., FitzGerald, R.J., Fairweather-Tait, S., 2006. Calcium absorption is not increased by caseinophosphopeptides. American Journal of Clinical Nutrition 84, 162–166.

Théolier, J., Hammami, R., Fliss, I., Jean, J., 2014. Antibacterial and antifungal activity of water soluble extracts from Mozzarella, Gouda, Swiss, and Cheddar commercial cheeses produced in Canada. Dairy Science and Technology 94, 427–438.

Thi, P.N., Dupas, C., Adt, I., Degraeve, P., Ragon, M., Missaoui, M.F., Novelli, E., Segato, S., The, D.P., Oulahal, N., 2014. Partial characterization of peptides inhibiting *Listeria* growth in two Alpine cheeses. Dairy Science and Technology 94, 61–72.

Timón, M.L., Parra, V., Otte, J., Broncano, J.M., Petrón, M.J., 2014. Identification of radical scavenging peptides (<3 kDa) from Burgos-type cheese. LWT – Food Science and Technology 57, 359–365.

Toelstede, S., Hofmann, T., 2008. Sensomics mapping and identification of the key bitter metabolites in Gouda cheese. Journal of Agricultural and Food Chemistry 56, 2795–2804.

Torres-Llanez, M.J., González-Córdova, A.F., Hernández-Mendoza, A., Garcia, H.S., Vallejo-Cordoba, B., 2011. Angiotensin-converting enzyme inhibitory activity in Mexican Fresco cheese. Journal of Dairy Science 94, 3794–3800.

Tulipano, G., Bulgari, O., Chessa, S., Nardone, A., Cocchi, D., Caroli, A., 2010. Direct effects of casein phosphopeptides on growth and differentiation of in vitro cultured osteoblastic cells (MC3T3-E1). Regulatory Peptides 160, 168–174.

Ul Haq, M.R., Kapila, R., Shandilya, U.K., Kapila, S., 2014. Impact of milk derived β-casomorphins on physiological functions and trends in research: a review. International Journal of Food Properties 17, 1726–1741.

Umbach, M., Teschemacher, H., Praetorius, K., Hirschhauser, R., Bostedt, H., 1985. Demonstration of a beta-casomorphin immunoreactive material in the plasma of newborn calves after milk intake. Regulatory Peptides 12, 223–230.

Walther, B., Schmid, A., Sieber, R., Wehrmüller, K., 2008. Cheese in nutrition and health. Dairy Science and Technology 88, 389–405.

Wang, H., Cui, L., Chen, W., Zhang, H., 2011. An application in Gouda cheese manufacture for a strain of *Lactobacillus helveticus* ND01. International Journal of Dairy Technology 64, 386–393.

Wang, Y., Jacome-Sosa, M.M., Proctor, S.D., 2012. The role of ruminant trans fat as a potential nutraceutical in the prevention of cardiovascular disease. Food Research International 46, 460–468.

Wasilewska, J., Kaczmarski, M., Kostyra, E., Iwan, M.J., 2011. Cow's-milk-induced infant apnea with increased serum content of beta-casomorphin-5. Pediatric Gastroenterology and Clinical Nutrition 52, 772–775.

Yasuda, S., Ohkura, N., Suzuki, K., Yamasaki, M., Nishiyama, K., Kobayashi, H., Hoshi, Y., Kadooka, Y., Igoshi, K., 2010. Effects of highly ripened cheeses on HL-60 human leukemia cells: antiproliferative activity and induction of apoptotic DNA damage. Journal of Dairy Science 93, 1393–1400.

Yin, H., Miao, J., Zhang, Y., 2010. Protective effect of casomorphin-7 on type 1 diabetes rats induced with streptozotocin. Peptides 31, 1725–1729.

Zemel, M.B., 2004. Role of calcium and dairy products in energy partitioning and weight management. American Journal of Clinical Nutrition 79, 907S–912S.

Zidane, F., Aurélie, M., Cakir-Kiefer, C., Miclo, L., Rahuel-Clermont, S., Girardot, J., Corbier, C., 2012. Binding of divalent metal ions to 1-25 β-caseinophosphopeptide: an isothermal titration calorimetry study. Food Chemistry 132, 391–398.

Zlatanos, S., Laskaridis, K., Feist, C., Sagredosa, A., 2002. CLA content and fatty acid composition of Greek Feta and hard cheeses. Food Chemistry 78, 471–477.

Chapter 12

Blue Cheese: Microbiota and Fungal Metabolites

J.F. Martín[1], M. Coton[2]
[1]Universidad de León, León, Spain; [2]Université de Brest, Plouzané, France

12.1 PREPARATION AND MATURATION OF BLUE CHEESES

Among the more than 1000 cheese varieties produced worldwide, blue-veined cheeses or white-blue mold rind cheeses, such as the emblematic French Roquefort, Danish Danablu, Italian Gorgonzola, English Stilton, or Spanish Cabrales cheeses, are largely manufactured. They have a long history of use (Albillos et al., 2011) and are consumed in numerous countries on all continents. Blue cheeses are made according to specific manufacturing practices and from different milk types (ewe, cow, goat, and buffalo) that can be used separately or mixed according to cheese type. These distinctive manufacturing processes provide characteristic organoleptic properties to the final product, in particular, in regards to volatile composition (distinct aromatic properties), flavor (strong character), texture, and color (Gallois and Langlois, 1990; Karlshoj and Larsen, 2005).

Specific cheese-making practices as well as defined geographical origins have led to specific denominations: Protected Designation of Origin (PDO), Protected Geographical Indication (PGI), or Traditional Specialties Guaranteed (TSG) for cheeses in general, including some blue cheeses (Table 12.1).

French Roquefort cheese is actually the oldest cheese type with a PDO denomination; it was obtained in 1925. Today, PDO is quite common among cheeses to ensure product authenticity and quality for worldwide consumers. In the case of Roquefort cheese, this status can only be obtained if certain criteria are fulfilled such as the ripening period in the famous Roquefort cheese cellars in the French village of Roquefort-sur-Soulzon. In France over 18,000 tons of Roquefort cheese were produced in 2013, which corresponds to approximately one-third of total French blue cheese production (CNIEL, 2014).

12.1.1 Blue Cheese Manufacture

During cheese manufacture and ripening, complex microbial communities are encountered (Beresford et al., 2001; McSweeny, 2004; Montel et al., 2014;

TABLE 12.1 General Overview of the Main Traditional Blue Cheese Varieties Produced in Europe

Cheese Type	Country	Designation	Milk Type	Composition					Ripening Time (month)	Estimated Production (t/year)
				Moisture Content (%)	Fat Content (%)	Protein Content (%)	NaCl Content (%)			
Cabrales	Spain	PDO	Raw cow	35.4–41.6	33.8–38.2	20.4–23.6	1.8–3.4		2–5	600
Danablu	Denmark	PGI	Thermized cow	42.7–47.3	29–31	18.5–23.9	3–3.9		3–4	nd
Gorgonzola	Italy	PDO	Pasteurized cow	42.2–49.6	29.6–31	19–22.9	1.6–2.9		>3	48,000
Roquefort	France	PDO	Raw ewe	42–44	29	20	4.1		3–9	18,000
Stilton	United Kingdom	PDO	Pasteurized cow	37–41.6	32–35.2	21–28.7	2.2–2.7		1.5–<4	8000
Valdéon	Spain	PGI	Pasteurized cow, goat	nd	34	20	3		2	nd

nd, not determined; PDO, protected designation of origin; PGI, protected geographical indication.

Irlinger and Mounier, 2009; Rattray and Eppert, 2011) and play major roles in the final product quality. A succession of various microorganisms is typically observed during blue cheese manufacture including diverse bacteria (dominant during early stages), yeasts, and molds (especially dominant during ripening). Both primary (ie, lactic acid bacteria, LAB) and secondary (ie, *Penicillium roqueforti*) starter cultures are generally used for production, although some blue cheeses are still naturally ripened and rely on indigenous microorganisms from milk or the manufacturing environment. For blue cheese manufacture, although various fabrication methods exist, all are made with the well-known mold *P. roqueforti* that, in particular, gives these cheeses their distinctive blue veins (*P. roqueforti* colonies can be seen ranging from whitish to pale green to dark blue-green, Fig. 12.1) and characteristic flavor. It should be noted that this species has a very long history of apparent safe use with technological improvements for blue cheese production. Other yeast or mold species may also be added during the starter inoculation step such as *Debaryomyces hansenii* or *Penicillium camemberti* (for white-blue mold rind cheeses).

12.1.1.1 Milk Preparation

As noted, blue cheese can be produced from different milk types, namely ewe, cow, goat, and buffalo milk. Milk can either be used individually or as mixtures according to the blue cheese variety. Raw, thermally treated, or microfiltrated milk (Table 12.1) can be used as raw materials, which will directly impact both microbial counts and diversity during cheese manufacture. This is particularly true for indigenous nonstarter lactic acid bacteria (NSLAB) that can positively influence cheese ripening during raw milk blue cheese production (Fox and McSweeney, 2004), but NSLAB can also be biogenic amine producers.

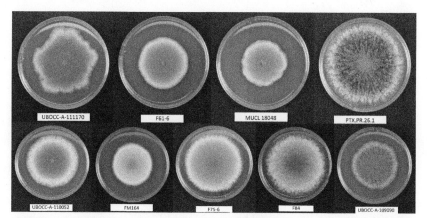

FIGURE 12.1 Different morphological aspects and pigmentations of several *Penicillium roqueforti* strains growing on potato dextrose agar medium after 7 days at 25°C. The strain codes are indicated below each Petri dish. *The photographs are courtesy of G. Gillot, LUBEM laboratory, Université de Bretagne Occidentale (UBO).*

Today industrial blue cheeses are more often produced with thermally treated milks for easier microbial and manufacturing control and some blue cheeses can also be produced from ultrafiltration retentates. Biochemical milk composition can vary and depends on milk type, in particular animal species and animal feeding habits prior to milking. Such variables can influence fat, protein, and aromatic contents (ie, taste and odor) in the final product. According to blue cheese type and denomination rules, milk standardization, which is classically performed to optimize the protein to fat ratio, may not be authorized for blue cheese manufacture.

12.1.1.2 Acidification, Coagulation, and Curd Formation

After preparation treatments, milk is then supplemented with rennet either before or after primary and secondary starter culture addition to coagulate milk. This step usually takes place at temperatures ranging between 28 and 34°C (temperatures remain below 40°C) in tanks for approximately 2.5–4 h. This ensures optimal coagulation and curd formation (Cantor et al., 2004) as well as optimal physical and biochemical characteristics for downstream operations like cutting and molding. Rennet is involved in breaking down milk caseins (Hewedi and Fox, 1984) and thereby also contributes to the overall proteolytic activity. Moreover, casein micelle breakdown leads to micelle aggregates and promotes gelation structure formation. Mesophilic and/or thermophilic primary starter cultures are often added to milk and contribute to coagulation as well as the secondary starter culture *P. roqueforti* in the form of conidia. For some blue cheese productions, producers may not supplement milk with microbial starter cultures and simply rely on indigenous species to perform spontaneous milk sugar (essentially lactose) fermentation. In either case, both indigenous or starter species present during manufacture will contribute to overall product quality.

Common lactic acid bacteria (LAB) starter cultures used for blue cheese manufacture include strains belonging to the mesophilic *Leuconostoc mesenteroides* (heterofermentative) and *Lactococcus lactis* subsp. *lactis* or *cremoris* (homofermentative) species (Cantor et al., 2004) as well as to the thermophilic bacteria *Streptococcus thermophilus* or *Lactobacillus delbrueckii* subsp. *bulgaricus* (Gobbetti et al., 1997; Cantor et al., 2004), although some species may not be systematically used. Primary starter cultures have at least two important roles during this step: the major role is acidification via lactose metabolism while the minor role (specific to some LAB species) is gas formation leading to changes in coagulated milk or curd structure. More precisely, homofermentative LAB and thermophilic strains are used to acidify milk by fermenting milk sugars (ie, lactose) to lactic acid while heterofermentative species produce not only lactic acid but also ethanol or acetate and CO_2. CO_2 production is needed to create openings in the coagulated milk or curd to facilitate air penetration and *P. roqueforti* development within the solidified or gelified structure. In general, during this phase, complete lactose fermentation is also a key step to prevent undesirable secondary microbiota development. LAB, both starter cultures or indigenous NSLAB, contribute to overall taste and aroma in the final product. This is reinforced by citrate-metabolizing primary starter cultures such as *L. lactis* subsp. *lactis* or *Leuconostoc* species. Milk contains citrate (Fox et al., 1993) and this molecule is considered to be an important precursor for flavor compounds such as diacetyl (buttery flavor) in cheese. Citrate metabolizing LAB will also produce acetate, acetoin, and CO_2 as end products that contribute to flavor and structure. Noteworthy, during this step, both acidification (pH decrease due to LAB metabolic activities) and rennet addition contribute to milk coagulation simultaneously. Within the first week of manufacture and according to the desired cheese type, pH can range from close to 4.4 to 5.1 (Cantor et al., 2004; Alonso et al., 1987; Gobbetti et al., 1997; Prieto et al., 2000). pH is considered to be an important parameter as it directly affects curd texture by influencing casein solubility (McSweeney, 2004) and will therefore have an influence on final cheese texture (ie, creating more or less soft cheeses), which is important in regards to product quality and typicity.

Concerning *P. roqueforti*, this species participates as a key secondary culture during manufacture. During this phase, conidia can be either directly added to milk, sprayed on the curd, or naturally colonize (indigenous environmental strains) cheese, which is a critical step when produced from pasteurized milk. In traditional blue cheese manufacture, *P. roqueforti* conidia can still be obtained after cultivating this fungal species on rye and wheat bread as originally performed. The action of some LAB species to create fissures, openings, and channels in the cheese curd is crucial for *P. roqueforti* growth and penetration. *Penicillium roqueforti* is involved in assimilating numerous substrates including milk sugars (lactose, glucose, galactose), utilizing lactate and citrate, and is well adapted to the pH, moderate a_w, low temperature, and low oxygen levels encountered during cheese manufacture in general but, in particular, during ripening when conditions become more hostile (Cerning et al., 1987; Vivier et al., 1992; Cantor et al., 2004). Indeed, *P. roqueforti* growth can be observed at O_2 levels as low as 0.3% (Cantor et al., 2004) and at near optimal levels between 4% and 21% O_2 (Thom and Currie, 1913; van den Tempel and Nielsen, 2000; Taniwaki et al., 2001); however, O_2–CO_2 level interactions have been shown to affect growth (Cantor et al., 2004). For a_w, *P. roqueforti* growth occurs in salt-supplemented laboratory media ranging from a_w 0.99 to 0.92 (equivalent to 0–13% w/w NaCl) (Cantor et al., 2004); *Penicillium roqueforti* growth can clearly occur at the NaCl levels encountered during cheese manufacture. Moreover, the minimal and optimal a_w values for *P. roqueforti* growth were recently determined in potato dextrose agar medium adjusted with glycerol and the minimal a_w was determined to be 0.796, while the optimal a_w was 0.98 (Rigalma, personal communication). Minimum blue cheese pH ranges from 4.6 to 4.7 for Danablu or Stilton style cheeses (Hansen, 2001; Madkor et al., 1987), while for Gorgonzola and Cabrales pH values may reach 5.15–5.30 (Gobbetti et al., 1997; Alonso et al., 1987); these values are close to the *P. roqueforti* optimal growth pH (considered to be close to pH 4 or 5). However, growth can occur in a wide pH range from 3 to 8 (Rigalma, personal communication). Finally, *P. roqueforti* addition to milk during this stage is important to obtain the blue-veined cheese aspect of the final product.

12.1.1.3 Cutting, Molding, and Salting

After milk has coagulated, mechanical stirring and cutting operations are performed to obtain the final curd and mold the cheese. Cutting operations are used to create 1.5–3 cm curds (size can be variable according to cheese type) before stirring and draining residual whey. Curds are placed in perforated molds and whey is drained. Pressure may be applied or not during this step and molds are frequently turned to facilitate drainage. *Penicillium roqueforti* spore suspensions may be injected in the molded curds during this step.

Salting is performed by immersing cheeses in brine (saturated salt solution) or by applying dry salt to the cheese surface (Guinee, 2004; Cantor et al., 2004). For example, during Roquefort type cheese production, cheeses are removed from molds and then the entire surface is salted with coarse salt (duration up to 5 days), while for Danablu type cheeses, brine salting for 2 days is carried out. This is a crucial step during blue cheese manufacture to create the characteristic NaCl gradient from the cheese surface to the core, but also to select halotolerant microorganisms for ripening and to avoid spoilage or pathogenic microbial development. This gradient will continue to settle during the ripening stages. Final salt concentrations typically range from 2% to 5% (Madkor et al., 1987; Zarmpoutis et al., 1996, 1997; Gobbetti et al., 1997; Prieto et al., 2000; Hansen and Jakobsen, 2001) and will vary according to salting times, moisture content, and cheese structure and porosity (ie, salt easily enters veins).

Before ripening, an optional step named piercing may also be carried out. This will free the CO_2 produced by some LAB species such as heterofermentative *Leuconostoc* spp. within the cheese and also favor the entry of air into the cheese core to create more favorable conditions for *P. roqueforti* growth and sporulation.

12.1.1.4 Blue Cheese Ripening

The final stage during blue cheese manufacture is ripening. This step can vary according to blue cheese type and different ripening times (from weeks to months) and storage conditions may be used (Table 12.1). Indeed, some blue cheeses are ripened in classical ripening rooms, while others are stored in traditional caves such as French Roquefort cheese. Ripening is an important and complex step involving multiple changes in biotic and abiotic factors of the microenvironment, microbial growth and communities as well as biochemical composition. In particular, pronounced gradients of NaCl, pH, a_w and gas composition (O_2 and CO_2 levels) can be observed as well as diverse microbial biochemical changes resulting from lipolysis, proteolysis, and aroma formation. Certain parameters strongly influence the biochemical activities of the various microorganisms present during this stage thereby impacting final product quality. During this period, blue cheese develops the distinctive greenish blue colored veins characteristic of *P. roqueforti* conidia and mold color can vary among shades of white, green, blue, or brown (Fig. 12.1), depending on strain and its age (Cantor et al., 2004) in the veins as well as the distinct organoleptic properties related to the final taste, texture, and aroma. In particular, *P. roqueforti* germination, growth, and sporulation occur during this phase and visible growth appears in the cheese core after approximately 2–3 weeks ripening (Fig. 12.2) as this species is well adapted to the conditions encountered.

Moreover, the characteristic blue-green veins appear after *P. roqueforti* sporulation during ripening. Ripening occurs at controlled low temperatures ranging from 8 to 12°C and high relative humidity levels ranging between 85% and 95% according to blue cheese type.

Short ripening times can be observed for some cow's milk based cheeses such as Fourme d'Ambert or Bleu d'Auvergne (from 3 to 8 weeks), while Gorgonzola cheese is usually ripened for approximately 3 months. Stilton and Cabrales are ripened for rather long periods that can reach up to 4–5 months.

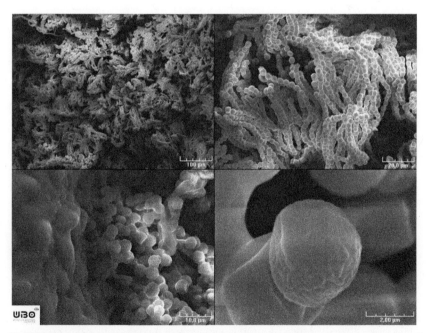

FIGURE 12.2 Electron microscopy images of *Penicillium roqueforti* growing in the veins of blue-veined cheese. The scales indicate 100 μm (upper left), 20 μm (upper right), 10 μm (lower left), and 2 μm (lower right). *The photographs are courtesy of Prof. Jérôme Mounier, Microscopy Platform from the Université de Bretagne Occidentale (UBO).*

Longer periods are also classically used for the ewe milk-based Roquefort cheese (up to 9 months). In all cases, cheese ripening may follow strict regulations according to their denomination such as observed for French PDO Roquefort cheese. Ripening is then carried out in a two-step process involving both aerobic and anaerobic conditions. In the case of French Roquefort cheese, *P. roqueforti* growth first occurs within the cheese core and on the cheese surface under optimal aerobic conditions for 15 days in ventilated Roquefort cellars. Ventilation is controlled to ensure optimal growth conditions. After this time, the second step, called "plombage" in French, involves covering the cheeses with a foil sheet to seal the cheese surface and create limiting oxygen conditions (ie, partial anaerobic conditions). In particular, this step is of interest to prevent spoilage microorganisms from developing on the cheese surface and also slows *P. roqueforti* growth and metabolic activities during the ripening period. Roquefort cheeses are then stored in cold temperature conditions as low as −2 to −4°C for at least 3 months according to PDO guidelines. Rapidly decreasing levels of O_2 creating partial anaerobic conditions within the cheese are also observed for other blue cheese types during ripening. For example, during Danablu ripening, a 50% decrease in O_2 content was observed at 4 mm below the cheese rind after 1 week, while after 13 weeks ripening O_2 was absent at levels below 0.25 mm from the rind (van den Tempel et al., 2002). In general, this anaerobic environment will be formed within a couple to a few weeks, except in the major openings or channels in direct contact with O_2.

Numerous changes will occur during ripening and are largely due to the microbial species present and their respective metabolic activities. These species are either inoculated into the milk or cheese curd during the previous steps as stated above or may spontaneously occur during manufacture. In particular, LAB (including starter cultures and NSLAB, mainly facultative heterofermentative species), yeast (predominantly *D. hansenii* and to a lesser extent *Kluyveromyces marxianus* or *Yarrowia lipolytica*, although this list is not exhaustive), and *P. roqueforti* are present but at different levels on the surface or in the core. However, *P. roqueforti* clearly plays a very important role during this step due to its dominant enzymatic activities. Important microbial activities observed during ripening are mainly involved in lipid and protein degradation as well as aroma formation (see below). In this context, the openings or veins that were formed within the cheese core and to a lesser extent the cheese surface correspond to unique zones for these diverse metabolic activities to occur.

12.1.1.5 Proteolysis

Penicillium roqueforti growth is accompanied by the production of intracellular and extracellular enzymes for protein metabolism including an aspartic protease that hydrolyzes β-caseins and $α_{S1}$-caseins, a metalloprotease with broad substrate specificity, an extracellular serine carboxypeptidase that releases acidic, basic and hydrophobic amino acids, and an extracellular metalloaminopeptidase that releases apolar amino acids not situated next to glycine residues (Grippon, 1993; Ardö, 2001; Cantor et al., 2004). Protease enzymes, in general, are known to extensively degrade milk caseins thereby participating in deacidification, textural cheese modifications, and aroma production as some precursor compounds are liberated (peptides and amino acids) (Sousa et al., 2001; Ardö, 2011). In blue cheese, proteases can also originate from other sources such as milk (eg, plasmin and protease peptones) and rennet (breaking down α_{S1}-casein to produce peptides) or other microorganisms (LAB, NSLAB, yeasts, and other mold species). Some LAB species such as *Lactococcus* sp. or *Lactobacillus* sp. starter cultures are involved in hydrolyzing the peptides produced by rennet and plasmin with limited amino acid release during the first weeks of ripening (Cantor et al., 2004). Rennet is actually considered to be the main enzyme mixture involved in casein breakdown before *P. roqueforti* growth and during outgrowth (Hewedi and Fox, 1984). However, the highest proteolytic activity in blue cheese is described to be due to *P. roqueforti* enzyme activities and is usually correlated to mold outgrowth when the mold actually becomes visible in the cheese. At this time, both extracellular and intracellular proteolytic enzymes are known to be highly active (Gripon et al., 1977; Le Bars and Grippon, 1981) and both peptides and amino acids are actively released. In some cases, these released molecules can positively contribute to flavor but also negatively via off-flavor production, such as bitterness, if not properly monitored and controlled. Indeed, *P. roqueforti* secretes aspartyl protease and metalloproteases that completely hydrolyze caseins thereby releasing different peptides including these bitter peptides (Ardö, 2011). Interestingly, *P. roqueforti* also possesses several exopeptidases and an extracellular acid carboxypeptidase that may be involved in bitter peptide breakdown as well as in degrading other peptides (Cantor et al., 2004). Finally, proteolytic activities and the levels of enzymes produced among *P. roqueforti* strains appear to be highly variable, which could be an important trait to consider for starter or adjunct culture strain selection in the future. In comparison to other cheeses, blue cheeses undergo extensive proteolysis contributing to distinct and characteristic blue cheese texture and flavor. However, amino acid catabolism probably impacts to a lesser extent final aroma content than lipid metabolism leading to methyl ketone formation, the major volatile compounds detected in this cheese type.

12.1.1.6 Lipolysis

Concerning lipolysis, this is also a major biochemical event that is very intense during blue cheese ripening and mainly due to the active growth and metabolism of *P. roqueforti* in the veins within the cheese core. Lipid metabolism is clearly a major pathway for flavor development in blue cheeses and the enzymatic activities lead to free fatty acid (FFA) and release of methyl ketone precursors involved in aroma compound formation. In blue cheeses, a 3-fold to nearly 100-fold higher FFA levels (Table 12.2) can be reached as well as higher volatile compositions in comparison to other cheese types.

TABLE 12.2 Total Free Fatty Acid (FFA) Content on Different Mold, Bacteria, and Surface-Ripened Cheese Types

Cheese Class	Cheese Name	FFA Content (mg/kg)	References
Mold-ripened	Roquefort	32,453	Woo et al. (1984)
Mold-ripened	Blue	32,320	Woo et al. (1984)
Mold-ripened	Gamonedo blue	75,685	González de Llano et al. (1992)
Mold-ripened	Camembert	5066	de la Fuente et al. (1993)
Mold-ripened	Brie	2678	Woo et al. (1984)
Bacteria-ripened	Parmesan	13,697	de la Fuente et al. (1993)
Bacteria-ripened	Cheddar	997	Woo et al. (1984)
Bacteria-ripened	Swiss	4277	Woo et al. (1984)
Bacteria-ripened	Provolone	2671	Woo et al. (1984)
Surface-ripened	Port Salut	700	Woo et al. (1984)
Surface-ripened	Munster	6260	de León-González et al. (2000)

Indeed, FFA contents can reach levels ranging from more than 30,000 mg/kg for French Roquefort (Woo et al., 1984) and even exceed 75,000 mg/kg for Spanish Gamonedo (Gonzalez de Llano et al., 1992). Interestingly, in comparison to other mold-ripened cheeses such as Camembert or Brie (involving *P. camemberti*), blue cheeses contain about 10-fold more FFA. Such high FFA levels in other cheese types would lead to rancidity (via FFA oxidation); however, in the case of blue cheese manufacture, an increase in pH as well as low oxygen levels during ripening actually prevents such problems by neutralizing FFA (Cantor et al., 2004). Moreover, high FFA concentrations and their associated flavor products (methyl ketones such as 2-heptanone or 2-nonanone) in blue cheese are associated with the distinct and piquant flavor in the final product. An FFA gradient can also be detected in blue cheeses and much lower concentrations are observed close to the cheese rind in comparison to the core. The rind corresponds to a zone with higher NaCl content and where *P. roqueforti* is less abundant thereby leading to lower lipase production and lipid metabolism. Different lipolytic agents intervene in blue cheese lipolysis including a milk lipoprotein lipase (sensitive to pasteurization therefore more active in raw milk cheeses) and yeast or, to a lesser extent, bacterial lipases; however, *P. roqueforti* lipolytic activities largely dominate. Different *P. roqueforti* lipases are known to be involved in lipid metabolism including two extracellular lipases, an acidic lipase and an alkaline lipase, and one intracellular lipase (Lamberet and Menassa, 1983; Mase et al., 1995; Menassa and Lamberet, 1982; Niki et al., 1966; Stepaniak et al., 1980). Their relative importance during ripening is still not fully clear. For *P. roqueforti*, higher amounts of long-chain fatty acids ($C_{12:0}$–$C_{18:3}$) are produced than short-chain fatty acids ($C_{4:0}$–$C_{10:0}$), and lipolytic activities tend to vary greatly among strains (Larsen and Jensen, 1999); variable FFA concentrations are also encountered in different blue cheeses (Table 12.2). During ripening, FFA are more abundant once fungal sporulation occurs, then tend to decrease toward the end of ripening following the conversion of FFA to the main aroma compounds, methyl ketones (Madkor et al., 1987; Contarini and Toppino, 1995; Prieto et al., 2000).

12.1.1.7 Aroma Compound Formation

During ripening, numerous volatile and nonvolatile compounds are produced from the various precursors previously formed or liberated in blue cheese. These compounds contribute to the unique blue cheese flavor and texture in the final product. *Penicillium roqueforti* growth throughout the cheese interior and its enzymatic activities involved in both protein and lipid metabolisms clearly influence aroma formation. The major volatile and nonvolatile compounds identified belong to the following chemical classes (in order of importance): methyl ketones, alcohols, esters, lactones, and aldehydes. Their relative proportions will determine the specific flavors and flavor intensities encountered in blue cheeses. Variations may also be strain-dependent and ripening time-dependent.

The composition of the nutritive medium used for microbial aroma production can be considered as a main factor for obtaining suitable flavors in the final product and in this context milk protein and fat concentrations are central elements. Lipolysis is considered to be the main biochemical pathway leading to blue cheese taste and aroma, which comes from FFA and the compounds produced from them. In particular, methyl ketones are by far the most abundant aroma compounds in blue cheeses and can represent up to 70% of total volatile flavor compounds (Hansen et al., 2001), followed by alcohols (15–30% of total volatiles) and to a lesser extent esters, lactones, and aldehydes (Gallois and Langlois, 1990; Cantor et al., 2004). Their concentration in cheese is directly correlated to the typical "blue cheese" note (Rothe et al., 1982, 1994) with the most abundant methyl ketones being 2-heptanone, 2-nonanone followed by 2-pentanone and 2-decanone (Madkor et al., 1987; Gallois and Langlois, 1990; de Frutos et al., 1991; Contarini and Toppino, 1995). These compounds can be related to fruity, floral, and musty odors while 2-heptanone gives spicy and characteristic "blue cheese" notes (Molimard and Spinnler, 1996; Sablé and Cottenceau, 1999). Methyl ketones are thought to be produced by *P. roqueforti* conidia and mycelium (Fan et al., 1976) via part of the β-oxidation pathway from the corresponding fatty acids (Cantor et al., 2004). Although some enzyme activities have been studied and biochemical pathways proposed, the actual biosynthetic pathways and associated genes involved in aroma production in *P. roqueforti* have not yet been described. The second most abundant aroma compounds are alcohols. Secondary alcohols such as 2-heptanol, 2-nonanol, and 2-pentanol are produced under anaerobic conditions following methyl ketone reduction (Gallois and Langlois, 1990; González de Llano et al., 1990) and are more abundant than primary alcohols. These molecules can contribute to the musty and moldy odors encountered in blue cheeses according to their relative concentrations.

Other secondary microorganisms may also participate, either directly or indirectly, to aroma formation in blue cheeses during ripening. This has been particularly studied for some yeast species found during ripening such as *Y. lipolytica*, *Kluyveromyces lactis*, or *Saccharomyces cerevisiae* in model milk or cheese systems (Hansen et al., 2001; Cantor et al., 2004; Price et al., 2014). In all cases, each yeast species could positively influence *P. roqueforti* strain aroma profiles at specific inoculum concentrations in model conditions. It should be noted that inoculum levels, strain, and chosen yeast species can all influence *P. roqueforti* aroma profiles.

12.2 LACTIC ACID BACTERIA IN BLUE CHEESES

LAB are commonly present in cheeses including blue cheese. They may be used as primary starter cultures during manufacture for both milk acidification and ripening (ie, mesophilic or thermophilic species) or be part of the secondary microbiota (starter, adjunct, or indigenous LAB species) involved in cheese ripening such as NSLAB. Many LAB species are specifically used to ensure well-defined functions within complex microbial cheese communities and also to ensure efficient rates for specific steps such as acid production, gas production, or flavor development. In this context, strain selection is essential to improve blue cheese safety (ie, strains lacking risk factors such as their potential to produce biogenic amines or undesirable metabolites) and quality (ie, sensorial properties).

In blue cheeses, both mesophilic and thermophilic LAB species are usually inoculated into milk as primary starter cultures and assure proper acidification essential for downstream steps (renneting, syneresis, microbial community selection, etc.). Although inoculated at lower counts, total LAB (lactococci and lactobacilli) can reach levels up to 10^9 colony-forming units (cfu) per gram in the core before salting, then decrease to 10^6–10^8 cfu per gram after cheese ripening (Mayo et al., 2013; Cantor et al., 2004). Variations in LAB counts can be observed according to blue cheese type, although similar patterns are typically observed with a succession of LAB species (ie, lactococci and streptococci are dominant early on while lactobacilli dominate during ripening). In the case of Spanish Valdéon cheese, lactobacilli counts are quite low early on, then levels increase to >5 log units during the early stages of ripening and remain stable, while lactococci and leuconostoc, originally dominant, decrease during ripening from nearly 8 log to 6 or 7 log units, respectively (López-Díaz et al., 2000; Diezhandino et al., 2015). In Danablu cheese made with a mesophilic starter, lactococci tend to dominate in the cheese core throughout ripening while on the cheese surface and after brining, lactococci counts greatly decrease and *Lactobacillus* species become dominant (Cantor et al., 2004). LAB counts are generally higher on the cheese surface (10^8–10^{10} cfu per gram) after brining and remain stable during ripening (Devoyod et al., 1968; Nuñez, 1978; Ordoñez et al., 1980; Gonzalez de Llano et al., 1992; Gobbetti et al., 1997; Hansen et al., 2001).

Other nonlactic acid bacteria that have been previously identified during blue cheese manufacture include *Staphylococcus* spp., *Micrococcus* spp., *Bacillus* spp., and diverse Gram negative bacteria including *Hafnia alvei*, commonly encountered in cheeses in general (Addis et al., 2001; Florez and Mayo, 2006).

12.2.1 Production of Undesirable Compounds

Among undesirable changes that may be encountered in cheese (ie, off-flavors, gas defects, discoloration, or mineral deposition) (O'Sullivan et al., 2013), biogenic amines (BA), largely associated with fermented foods in general, can pose a risk for the consumer if accumulated in high concentrations over time. BA in cheeses do not typically affect sensory qualities except in rare cases when very high levels of putrescine or cadaverine are found (ie, putrid odor). Recently, the European Food Safety Authority (EFSA) studied histamine, tyramine, cadaverine, and putrescine (decarboxylation products of histidine, tyrosine, lysine, and ornithine, respectively) levels in "at risk" food products. Data showed that cheeses were included in the main food categories containing BA along with fish sauce, fermented vegetables, and fermented sausages, while fermented beverages contained on average much lower BA quantities (EFSA, 2011). Although BA are a food safety concern, no legislation exists for BA content in cheese to date.

Cheese can contain one or more BA and their content may vary according to cheese type, ripening time, and microbial communities. The main BA detected are histamine, tyramine, cadaverine, and putrescine; however, β-phenylethylamine, tryptamine, and spermidine can also be found (Table 12.3) (Linares et al., 2011; Loizzo et al., 2013).

For BA production and accumulation, cheeses must contain free amino acids (proteolysis and ripening times can impact free amino acid levels during manufacture), BA-producing bacteria having either decarboxylase or deiminase activities, and an amino acid-BA exchange transport system as well as favorable environmental conditions including both intrinsic factors (pH, water activity, salt content) and extrinsic factors (temperature, ripening, and storage periods). In blue cheeses, total BA levels can be high reaching >1050 mg/kg, while lower levels are often observed for soft cheeses (<550 mg/kg) or semihard cheeses

TABLE 12.3 Major Cheese-Related Biogenic Amines (BA) and BA-Producing Species in Cheese

BA Category	BA	Amino Acid Precursor	Enzyme Involved in BA Biosynthesis	Examples of BA-Producing Species Found in Cheeses
Monoamines	Tyramine	Tyrosine	Tyrosine decarboxylase	*Lactobacillus* spp. (as *Lactobacillus brevis*, *Lactobacillus reuteri*), *Enterococcus* spp.
	Histamine	Histidine	Histidine decarboxylase	*Lactobacillus* spp. (as *Lactobacillus buchneri*, *L. reuteri*), *Streptococcus thermophilus*, Gram-negative bacteria
	Phenylethylamine	Phenylalanine	Phenylalanine decarboxylase	Gram-negative bacteria
	Tryptamine	Tryptophan	Tryptophan decarboxylase	Gram-negative bacteria
Diamines	Putrescine	Ornithine/Arginine via agmatine	Ornithine decarboxylase Agmatine deiminase	*Lactobacillus* spp., *Enterococcus* spp., *Lactococcus lactis*, Gram-negative bacteria
	Cadaverine	Lysine	Lysine decarboxylase	Gram-negative bacteria

(<950 mg/kg) and higher levels for hard cheeses (>1900 mg/kg) (Novella-Rodriguez et al., 2003; Fernández et al., 2007; Mayer et al., 2010).

Both Gram-positive and Gram-negative cheese-related bacteria have been shown to produce BA in cheese or using in vitro conditions and many biosynthetic pathways involved in BA production have been well studied in cheese-related species (Table 12.3) (Connil et al., 2002; Lucas et al., 2003; Coton et al., 2004; Fernández et al., 2004; Martín et al., 2005; Wolken et al., 2006; Calles-Enriquez et al., 2010; Trip et al., 2011; Linares et al., 2011; Romano et al., 2012). Gram-positive cheese-related bacteria, especially belonging to *Enterococcus*, *Lactobacillus*, *Leuconostoc*, *Streptococcus*, and *Lactococcus* are well-known BA producers. They can be encountered during cheese production as indigenous microbiota present in raw materials, in the cheese environment, or as starter cultures. In cheese, relatively high levels of histamine and tyramine and, often to a lesser extent, putrescine and cadaverine can be accumulated according to the bacterial species involved.

12.3 FILAMENTOUS FUNGI AND YEASTS IN CHEESE: INTERACTIONS BETWEEN SPECIES

Filamentous fungi and yeast populations in different cheese types play an important role in cheese quality and organoleptic characteristics due to the effect of fungal enzymes and metabolic products (eg, organic acids, methylketones, secondary alcohols, aldehydes) on cheese flavor. In addition to *P. roqueforti*, several other fungi, particularly *P. camemberti* and *Penicillium nalgiovense*, occur on the surface of Camembert and Brie cheeses (Marth and Yousef, 1991; Rousseau, 1984). *Penicillium nalgiovense*, isolated initially from a cheese surface (Samson et al., 1995), is routinely used as a surface culture in salami and other types of dry sausages (Laich et al., 1999). Several other *Penicillium* species (eg, *Penicillium brevicompactum*, *Penicillium commune*, *Penicillium verrucosum*) and some *Aspergillus* species (eg, *Aspergillus versicolor*) are sometimes found on the surface of mold-ripened cheeses, depending on the fungal microbiota in the cheese-ripening chamber environment. Several other filamentous fungi and yeasts (except technological species) have been found in some blue cheeses but are considered to be contaminants. The fungal population on the cheese is also affected by milk origin and particularly by humidity and temperature of ripening cellars. Some yeasts, particularly *Geotrichum candidum*, also play an important role in cheese ripening, particularly in Camembert cheese (Marth and Yousef, 1991; Rousseau, 1984), and others like *K. lactis* and *Yarrowia lypolytica* may also contribute to ripening of some cheeses.

In blue-veined cheeses, such as Roquefort and Danablu cheeses, *P. roqueforti* is inoculated inside the cheese (see previous sections on cheese manufacture). Growth of this fungus produces the characteristic blue/green color and texture of blue-veined cheeses and also several secondary metabolites, some of them mycotoxins, which remain inside the cheese (Fernández-Bodega et al.,

2009; Fontaine et al., 2015). Because of the interest and possible toxic effects of some of these metabolites on human health, they will be studied in detail below.

12.3.1 *Penicillium roqueforti* and Related Fungi in Blue Cheese

The blue cheese-associated fungus *P. roqueforti* Thom (Thom, 1906; Raper and Thom, 1949) was initially characterized as the prototype of filamentous fungi involved in cheese maturation (Ramirez, 1982; Engel and Teuber, 1989; Mioso et al., 2014). However, in blue cheese samples (and also in grain and grass silages) there are several closely related species that are difficult to distinguish based on morphological traits. According to metabolite patterns separated on TLC plates, early studies by Engel and Teuber (1983) divided *P. roqueforti* strains into 15 distinct groups. The production of black-green pigment in the reverse of colonies in several common culture media is characteristic of *P. roqueforti* colonies. Frisvad and Filtenborg (1989) observed that a group of strains showed pale pigmentation and produced patulin instead of the classical PR-toxin. These authors designated this class of colonies as *P. roqueforti* subspecies *carneum* (later described as *Penicillium carneum*) because they were associated with fermented meat products but were infrequent in blue cheese isolated strains. However, biosynthesis and secretion of secondary metabolites in fungi is rather variable (Martín et al., 2014a); thus this is not a reliable taxonomical criterium.

Later, using ribosomal RNA sequencing and RAPD (random amplified polymorphic DNA), Boysen et al. (1996) divided *P. roqueforti* strains into three distinct species; namely, *P. roqueforti* ("sensu stricto"), *P. carneum*, and *Penicillium paneum*. *Penicillium roqueforti* is associated with grass silage and spoiled crop grains and also with a variety of cheeses and molded bread. *Penicillium roqueforti* is resistant to short chain fatty acids, including acetic acid, and to the presence of ethanol, which is one of the reasons why it is commonly found in fermented silages where lactic acid and ethanol are formed by heterofermentative LAB. On the other hand, *P. carneum* is associated with dry sausages and related fermented food products and also, according to Nielsen et al. (2006), with some types of cheeses and bread. Finally, *P. paneum* is associated with moldy bread and silages, and is taxonomically close to *Penicillium crustosum*. However, until further comparative genome analyses are done, it is difficult to conclude whether differences in secondary metabolite spectra are merely due to different expression levels of silent or near silent clusters (Martín and Liras, 2015) or whether they respond to major genome differences.

From a genetic point of view, the differences in the 18S rRNA sequences between the three species proposed by Boysen et al. (1996) are very small. *P. carneum* differed from *P. roqueforti* in only two nucleotides in the ITS1 (internal transcribed spacer) region, whereas *P. paneum* differed from *P. roqueforti* in 12 nucleotides in the ITS1. A related fungus is *Penicillium psychrosexualis*, recently described in the Roquefortorum series by Houbraken et al. (2010). Differences between the proposed species were also detected by the Genealogical

Concordance – Phylogenetic Species Recognition (GC-PSR) method targeting multiple genetic markers (Gillot et al., 2015). Within the *P. roqueforti* species, morphological differences among isolates are often observed raising taxonomical questions. In this context, the study by Gillot et al. (2015), based on 154 isolates of worldwide origins, showed that only one species exists and that it is separated into three highly structured populations showing high genetic diversity using microsatellite markers. This population structure was even shown to reach six distinct clusters using additional targets (Ropars et al., 2014). The genome of *P. roqueforti* FM164 has been recently sequenced (Cheeseman et al., 2014) and opens new perspectives to better understand this species diversity from both genetic and metabolic points of view.

12.3.2 *Penicillium roqueforti* Strains in Local Nonindustrial Blue Cheeses

As indicated above, *P. roqueforti* predominates in the fungal biota of traditional blue-veined cheeses. There are many wild type fungal strains isolated from local blue cheese varieties in different countries of Europe, North America, Middle East, and Central Asia. An important question is whether these strains are true *P. roqueforti* or if they have some special characteristics. Since large-scale blue-veined cheese production uses selected *P. roqueforti* strains (Fernández-Bodega et al., 2009) as secondary starters, it is not surprising that other related species are not found in large amounts in these industrially produced cheeses. Homemade artisanal blue cheeses are typically produced in isolated rural regions using goat, ewe, cow, or even buffalo milk. This cheese type was actually first made in prehistorical times and was well known by Greek and Roman civilizations (Albillos et al., 2011). These "artisanal" cheeses are matured in unsterilized cellars or caves and therefore might contain other fungal species. Various morphologically different *P. roqueforti* isolates can be simultaneously observed in many blue cheeses produced worldwide (Gillot et al., 2015).

When some *P. roqueforti* strains isolated from such local varieties were compared genetically by sequencing the internal transcribed spacer sequences, ITS1 and ITS2, they all showed identical nucleotide sequences to the ATCC10110 type strain (CECT2905) (Fernández-Bodega et al., 2009). Furthermore, all studied strains showed 100% identity for the 28S rRNA D1 conserved domain and D2 (so-called partially variable) domain. Moreover, one of the main mycotoxin biosynthesis pathways, the roquefortine gene cluster, was also studied in three different *P. roqueforti* strains and found to be identical (Kosalkova et al., 2015). These results suggest that, phylogenetically, all cheese-associated *P. roqueforti* strains derive from a common lineage. Local strains could be differentiated by restriction fragments obtained by genome cleavage with the BsuRI endonuclease; these fragments presumably belong to the mitochondrial DNA (Fernández-Bodega et al., 2009). More recently, microsatellite markers have been successfully used on a very large *P. roqueforti* collection created from over 120 different blue-veined cheeses worldwide as well as other substrates and showed highly differentiated populations (Ropars et al., 2014; Gillot et al., 2015).

Some of the local cheeses matured in cellars in the north mountain region of Spain contained on their surface, in addition to *P. roqueforti*, white fungal colonies that were identified by rDNA sequencing as *P. camemberti*.

12.3.3 Common Secondary Metabolites in Cheese-Associated Fungi

Filamentous fungi produce thousands of secondary metabolites with interesting biological and pharmacological activities (Martín et al., 2014a). *Penicillium roqueforti* family strains produce about 12 different types of secondary metabolites (Frisvad et al., 2004; Hymery et al., 2014) and probably contain some additional silent gene clusters, but not all strains produce the same set of secondary metabolites, most likely due to differences in expression of some of the biosynthetic gene clusters.

Several secondary metabolites are produced by the three proposed species *P. roqueforti*, *P. carneum*, and *P. paneum* (Table 12.4).

TABLE 12.4 Secondary Metabolites Produced by *Penicillium roqueforti* and the Closely Related *Penicillium carneum* and *Penicillium paneum* Species

Metabolites Found in *P. roqueforti*	Metabolites Exclusive of *P. carneum*[a]	Metabolites Exclusive of *P. paneum*[a]
Roquefortines C, D, L	Patulin	Patulin
Hydroxyroquefortine	Penitrem A	Penitrem A
Andrastins A, B		Marcfortines A, B, C
PR-toxin	b	b
Eremofortins A, B	c	c
Mycophenolic acid		
Festuclavine		d
Agroclavine		d
Citreoisocoumarin		
Orsenillic acid		

[a]*P. carneum and P. paneum produce variable amounts of all other* P. roqueforti *secondary metabolites with the exception of eremofortins and PR toxin (see text). An early description of penicillin acid production in a* P. roqueforti *strain is now explained due to a misclassification of the producer strain.*
[b]*Do not produce PR-toxin.*
[c]*Do not produce eremofortins.*
[d]*Do not produce festuclavine and agroclavine.*

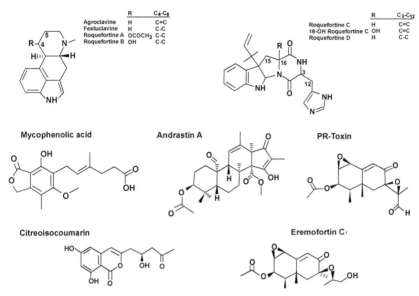

FIGURE 12.3 Chemical structures of the main *Penicillium roqueforti* secondary metabolites. Metabolites exclusively produced by the closely related species, *Penicillium carneum* and *Penicillium paneum*, are not included.

In addition, the three species produce different levels of citreoisocoumarin and orsenillic acid. The chemical structures of the metabolites produced by *P. roqueforti* and related fungi are shown in Fig. 12.3.

P. carneum strains produce penitrems and patulin (not shown in Fig. 12.3), which are not formed by *P. roqueforti*. *P. paneum* produces marcfortines A, B, and C. This chapter is focused on the secondary metabolites produced by *P. roqueforti sensus stricto*, in relation with blue-veined cheeses, but not on metabolites produced exclusively by *P. paneum* and *P. carneum*, or by contaminant fungi.

12.4 SECONDARY METABOLITES PRODUCED BY PENICILLIUM ROQUEFORTI

During the last few decades, increasing information has been reported on the ability of *P. roqueforti* to produce secondary metabolites in different culture media and inside the cheese matrix itself.

12.4.1 Roquefortines

Roquefortines are among the best-known *P. roqueforti* secondary metabolites (Ohmomo et al., 1975; Scott et al., 1976). Roquefortine C and the related roquefortines D (3, 12-dihydroroquefortine C), 16-hydroxyroquefortine C, and roquefortine L are members of the prenylated indole alkaloid family (reviewed by Martín et al., 2014b).

Roquefortines have antibacterial activity, particularly against Gram-positive bacteria (Kopp-Holtwiesche and Rehm, 1990), and may help to control the population of bacteria involved in cheese ripening. However, the biological role of roquefortines is probably more complex; some mycotoxins are known to produce distress or toxic effects on animals and may also serve as animal feed deterrents preventing soil-dwelling small animals to feed on *P. roqueforti*. Although roquefortine C is produced by *P. roqueforti* growing in a variety of solid substrates, the formation in blue cheeses does not occur in significant amounts, and there is a consensus that roquefortines in cheese do not pose a toxicity problem for humans (Nielsen et al., 2006; Fernández-Bodega et al., 2009; Hymery et al., 2014; Fontaine et al., 2015).

12.4.2 PR-Toxin and Eremofortins

Probably the most active mycotoxin produced by *P. roqueforti* is PR-toxin. This isoprenoid mycotoxin is clearly toxic for mice, rats, hamsters, and some domestic animals. It has also been shown to be cytotoxic on THP-1 human cell lines (N. Hymery, personal communication). Furthermore, PR-toxin exerts mutagenic action in vitro as shown in the Ames test. Actually, PR-toxin (perhaps in association with other secondary metabolites) is considered to be the causative agent of cow toxicosis produced by poorly conserved moldy silages (Nielsen et al., 2006). Fortunately, PR-toxin does not seem to be accumulated in large amounts in blue cheese, in contrast to what occurs with andrastins (see below) (Nielsen et al., 2005; Fernández-Bodega et al., 2009).

Recently, we studied the biosynthesis of PR-toxin and its intermediates, the eremofortins (Hidalgo et al., 2014). PR-toxin derives from the 15-carbon atoms sesquiterpene aristolochene, which is formed by aristolochene synthase (encoded by the gene *ari*1).

PR-toxin is converted in vitro and also probably in vivo to PR-amide and PR-imine by reaction of the PR-toxin carboxylic group with ammonium ions or primary amines in the culture medium or in the cells (Moreau et al., 1976; Chang et al., 1993). These derivatives appear to be less toxic than PR-toxin itself. In blue cheese, PR-toxin is probably converted to the less toxic PR-amide or PR-imine.

12.4.3 Andrastins

Andrastins are inhibitors of the *ras* oncogene protein farnesyl transferase (Overy et al., 2005; Omura et al., 1996). Since farnesylation of the human *ras* protein is essential for its biological activity causing tumor formation, inhibitors of the prenyltransferase activity are interesting for use as potential antitumor agents (Vilella et al., 2000). Indeed, andrastins were first discovered to be produced by some *Penicillium* species in a screening of antitumor agents.

The andrastins belong to the meroterpenoid class of secondary metabolites that include compounds with interesting pharmacological activities (Matsuda and Abe, 2014). Andrastins A to D were initially discovered by the 2015 Nobel Prize S. Omura et al. (1996), at the Kitasato Institute in Japan. These compounds were identified in the culture broth of *Penicillium* sp. FO4259 (Omura et al., 1996; Uchida et al., 1996a, b) and can also be produced by several other *Penicillium* species.

Interestingly Nielsen et al. (2005) and Fernández-Bodega et al. (2009) found that *P. roqueforti* produces andrastins and that andrastin A (the final product of the biosynthetic pathway) is accumulated inside blue cheeses when inoculated with *P. roqueforti* as a secondary starter. Andrastin concentrations in different blue cheeses such as Cabrales, Bejes-Tresviso, and Valdeón vary depending on the particular *P. roqueforti* strain used as starter and ripening conditions (Fernández-Bodega et al., 2009).

As *ras* prenyltransferase inhibitors, andrastins are considered to be beneficial for human health, but there are no studies that support its lack of toxicity. Blue cheese consumption (in which andrastins are present) appears to have no risk for human health, but this needs to be investigated in detail.

12.4.4 Mycophenolic Acid

An important secondary metabolite of *P. roqueforti* is mycophenolic acid (MPA). It was already described as an antibiotic active against *Bacillus anthracis* produced by a *Penicillium* sp. strain at the beginning of the 20th century, many years before the discovery of penicillin. MPA was later found to have other important biological activities. Particularly relevant is its activity as an immunosuppressant successfully used to prevent organ rejection in transplants (Meier-Kriesche et al., 2006). In addition, MPA has antitumor, antiviral, and antifungal activities and is used for psoriasis treatment (Tressler et al., 1994; Kavanagh, 1947; Borroto-Esoda et al., 2004; Nicoletti et al., 2004; Epinette et al., 1987).

While this article was in press Del Cid et al. (2016) reported the mycophenolic acid gene cluster of *P. roqueforti*, which is almost identical to that known for *P. brevicompactum*. In this context, Fontaine et al. (2015) evaluated the mycotoxin contents in 89 blue-veined cheeses and showed that 37.2% contained quantifiable amounts of MPA.

12.4.5 Agroclavine and Festuclavine

Several *P. roqueforti* strains of different origins produce the clavine type alkaloids agroclavine and festuclavine (Fig. 12.3). Clavines are also produced by *P. carneum* but not by *P. paneum* (Nielsen et al., 2006). Clavines contain a tricyclic or tetracyclic structure with small structural differences between them. Clavin alkaloids have potent effects on animal physiology (Starec et al., 2001) but clavine concentrations in blue cheese do not appear to be significant for human health. The lack of sufficient oxygen inside the cheese matrix required for biosynthetic oxygenases may limit the formation of many of these metabolites.

12.4.6 Other *Penicillium roqueforti* Metabolites

Several strains of *P. roqueforti* are also known to produce citreoisocoumarin and small amounts of orsellinic acid (Table 12.4). Very little is known about the biosynthesis of these compounds or about their formation in cheese. Although their concentration in cheese is probably very small, detailed toxicity studies are required.

12.5 CONCLUSIONS AND FUTURE OUTLOOK

Several filamentous fungi, particularly *P. roqueforti*, have a very important role in blue cheese ripening. In addition to the enzymes involved in cheese maturation, several primary and secondary metabolites are produced by *P. roqueforti* inside blue cheese (microaerobic conditions) or on the surface (aerobic conditions) and contribute to cheese flavor. Only a few of these metabolites are considered to be toxic for humans. The highest toxicity has been described for PR-toxin, but the amount of this toxin produced in cheeses does not appear to reach high levels. Other metabolites, such as andrastin A, are RAS-prenyltransferase inhibitors that were initially isolated in antitumor agent screening programs. Andrastin A is accumulated in significant amounts inside the solid cheese matrix, but there are no studies available on its toxicity or usefulness to prevent tumor formation.

There is still limited information on the biosynthesis of secondary metabolites in *P. roqueforti*. Only the biosynthesis of eremofortins, PR-toxin, and roquefortine C have been studied in some detail. The biosynthesis of some *P. roqueforti* metabolites has been studied in other fungi but not in *P. roqueforti*, and differences in the intermediates or the final products may occur, eg, *P. roqueforti* synthesizes festuclavine and agroclavine, whereas the best-known clavine producer *Aspergillus fumigatus* synthesizes festuclavine (a product of *P. roqueforti*), but this intermediate is later converted to fumigaclavine. The latter is known to contribute to the pathogenicity of *A. fumigatus* in humans but the relevance, if any, of consumption of festuclavine or agroclavine in blue cheese has not been studied in detail.

Finally, gene expression control of genes encoding putative toxic secondary metabolites is of paramount importance. For instance, the effect of oxygen, humidity, salts (eg, NaCl), amino acids, or ammonium ions may significantly affect the *P. roqueforti* metabolite levels produced in blue cheese.

REFERENCES

Addis, E., Fleet, G.H., Cox, J.M., Kolak, D., Leung, T., 2001. The growth, properties and interactions of yeasts and bacteria associated with the maturation of Camembert and blue-veined cheeses. International Journal of Food Microbiology 69, 25–36.

Albillos, S.M., García-Estrada, C., Martín, J.F., 2011. Spanish blue cheeses: functional metabolites. In: Foster, R.D. (Ed.), Cheese: Types, Nutrition and Consumption. Nova Science Publishers Inc., pp. 89–105.

Alonso, L., Juarez, J., Ramos, M., Martín-Álvarez, P.J., 1987. Overall composition, nitrogen fractions and fat characteristics of Cabrales cheese during ripening. Zeitschrift für Lebensmittel-Untersuchung und -Forschung 185, 481–486.

Ardö, Y., 2001. Cheese Ripening: General Mechanisms and Specific Cheese Varieties. Bulletin 369. International Dairy Federation, Brussels, pp. 7–12.

Ardö, Y., 2011. Blue mold cheese. In: Fuquay, J., Fow, P.F., McSweeney, P. (Eds.), Encyclopedia of Dairy Science, second ed. Elsevier Academic Press, London, pp. 767–772.

Beresford, T.P., Fitzsimons, N.A., Brennan, N.L., Cogan, T.M., 2001. Recent advances in cheese microbiology. International Dairy Journal 11, 259–274.

Borroto-Esoda, K., Myrick, F., Feng, J., Jeffrey, J., Furman, P., 2004. In vitro combination of amdoxovir and the inosine monophosphate dehydrogenase inhibitors mycophenolic acid and ribavirin demonstrates potent activity against wild-type and drug-resistant variants of human immunodeficiency virus type 1. Antimicrobial Agents Chemotherapy 48, 4387–4394.

Boysen, M., Skouboe, P., Frisvad, J., Rossen, L., 1996. Reclassification of the *Penicillium roqueforti* group into three species on the basis of molecular genetic and biochemical profiles. Microbiology 142, 541–549.

Calles-Enríquez, M., Eriksen, B.H., Andersen, P.S., Rattray, F.P., Johansen, A.H., Fernández, M., Ladero, V., Álvarez, M.A., 2010. Sequencing and transcriptional analysis of the *Streptococcus thermophilus* histamine biosynthesis gene cluster: factors that affect differential *hdcA* expression. Applied Environmental Microbiology 76, 6231–6238.

Cantor, M.D., van den Tempel, T., Hansen, T.K., Ardö, Y., 2004. Blue cheese. In: Fox, P.F., Cogan, T.M., Guinee, T.P. (Eds.), Cheese: Chemistry, Physics and Microbiology, Major Cheese Groups. Academic Press, pp. 175–198.

Cerning, J., Gripon, J.C., Lamberet, G., Lenoir, J., 1987. Les activités biochimiques des *Penicillium* utilisés en fromagerie. Lait 67, 3–39.

Chang, S.C., Lu, K.L., Yeh, S.F., 1993. Secondary metabolites resulting from degradation of PR toxin by *Penicillium roqueforti*. Applied Environmental Microbiology 59, 981–986.

Cheeseman, K., Ropars, J., Renault, P., Dupont, J., Gouzy, J., Branca, A., Abraham, A.L., Ceppi, M., Conseiller, E., Debuchy, R., Malagnac, F., Goarin, A., Silar, P., Lacoste, S., Sallet, E., Bensimon, A., Giraud, T., Brygoo, Y., 2014. Multiple recent horizontal transfers of a large genomic region in cheese making fungi. Nature Communications 5, 2876.

Contarini, G., Toppino, P.M., 1995. Lipolysis in Gorgonzola cheese during ripening. International Dairy Journal 5, 141–155.

Connil, N., Le Breton, Y., Dousset, X., Auffray, Y., Rincé, A., Prévost, H., 2002. Identification of the *Enterococcus faecalis* tyrosine decarboxylase operon involved in tyramine production. Applied Environmental Microbiology 68, 3537–3544.

Coton, M., Coton, E., Lucas, P., Lonvaud, A., 2004. Identification of the gene encoding a putative tyrosine decarboxylase of *Carnobacterium divergens* 508. Development of molecular tools for the detection of tyramine-producing bacteria. Food Microbiology 21, 125–130.

Coton, M., Delbès-Paus, C., Irlinger, F., Desmasures, N., Le Fleche, A., Stahl, V., Montel, M.-C, Coton, E., 2012. Biodiversity and assessment of potential risk factors of Gram-negative isolates associated with French cheeses. Food Microbiology 29, 88–98.

CNIEL, 2014. L'économie laitière en chiffres, Edition 2014, CNIEL.

de Frutos, M., Sanz, J., Martinez-Castro, I., 1991. Characterization of artisanal cheeses by GC and GC/MS analysis of their medium volatility (SDE) fraction. Journal of Agricultural Food Chemistry 39, 524–530.

Del-Cid, A., Gil-Durán, C., Vaca, I., Rojas-Aedo, J.F., García-Rico, R.O., Levicán, G., Chávez, R., 2016. Identification and Functional Analysis of the Mycophenolic Acid Gene Cluster of *Penicillium roqueforti*. PLoS One. 1(1), e0147047.

De Leon-Gonzalez, L.P., Wendorff, W.L., Ingham, B.H., Jaeggi, J.J., Houck, K.B., 2000. Influence of salting procedure on the composition of Muenster-type cheese. Journal of Dairy Science 83, 1396–1401.

de la Fuente, M.A., Fontecha, J., Juarez, M., 1993. Fatty acid composition of the triglyceride and free fatty acid fractions in different cows-, ewes- and goats-milk cheeses. Zeitschrift für Lebensmittel-Untersuchung und -Forschung 196, 155–158.

Devoyod, J.J., Bret, G., Auclair, J.E., 1968. La flore microbienne du fromage de Roquefort. I. Son évolution au cours de la fabrication et de l'affinage du fromage. Lait 48, 479–480.

Diezhandino, I., Fernández, D., González, L., McSweeney, P.L., Fresno, J.M., 2015. Microbiological, physico-chemical and proteolytic changes in a Spanish blue cheese during ripening (Valdeón cheese). Food Chemistry 168, 134–141.

EFSA Journal 9 (10), 2011, 2393 pp 93.

Engel, G., Teuber, M., 1983. Differentiation of *Penicillium roqueforti* strains by thin-layer chromatography of metabolites. Milchwissenschaft 38, 513–516.

Engel, G., Teuber, M., 1989. Toxic metabolites from fungal cheese starter cultures (*Penicillium camemberti* and *Penicillium roqueforti*). In: van Egmond, H.P. (Ed.), Mycotoxins in Dairy Products. Elsevier Applied Science, London, pp. 163–192.

Epinette, W.W., Parker, C.M., Jones, E.L., Greist, M.C., 1987. Mycophenolic acid for psoriasis. A review of pharmacology, long-term efficacy, and safety. Journal American Academy Dermatology 17, 962–971.

Fan, T.Y., Hwang, D.H., Kinsella, J.E., 1976. Methyl ketone formation during germination of *Penicillium roqueforti*. Journal of Agricultural Food Chemistry 24, 443–448.

Fernández-Bodega, M.A., Mauriz, E., Gómez, A., Martín, J.F., 2009. Proteolytic activity, mycotoxins and andrastin A in *Penicillium roqueforti* strains isolated from Cabrales, Valdeón and Bejes-Tresviso local varieties of blue-veined cheeses. Journal of Food Microbiology 136, 18–25.

Fernández, M., Linares, D.M., Álvarez, M.A., 2004. Sequencing of the tyrosine decarboxylase cluster of *Lactococcus lactis* IPLA 655 and the development of a PCR method for detecting tyrosine decarboxylating lactic acid bacteria. Journal of Food Protection 67, 2521–2529.

Fernández, M., Linares, D.M., Rodríguez, A., Álvarez, M.A., 2007. Factors affecting tyramine production in *Enterococcus durans* IPLA655. Applied Microbiology and Biotechnology 73, 1400–1406.

Florez, A.B., Mayo, B., 2006. Microbial diversity and succession during the manufacture and ripening of traditional, Spanish, blue-veined Cabrales cheese, as determined by PCR-DGGE. International Journal of Food Microbiology 110, 165–171.

Fontaine, K., Passeró, E., Vallones, L., Hymery, N., Coton, M., Jany, J.L., Mounier, J., Coton, E., 2015. Occurrence of roquefortine C, mycophenolic acid and aflatoxin M1 mycotoxins in blue-veined cheeses. Food Control 47, 634–640.

Fox, P.F., Law, J., McSweeney, P.L.H., Wallace, J., 1993. Biochemistry of cheese ripening. In: Fox, P.F. (Ed.), Cheese: Chemistry, Physics and Microbiology, General Aspects, vol 1. second ed. Chapman & Hall, London, pp. 389–438.

Fox, P.F., McSweeney, P.L.H., 2004. Cheese: an overview. In: Fox, P.F., Cogan, T.M., Guinee, T.P. (Eds.), Chemistry, Physics and Microbiology. Academic Press, pp. 1–18.

Frisvad, F., Smedsgaard, J., Larsen, T., Samson, R., 2004. Mycotoxins, drugs and other extrolites produced by species in *Penicillium* subgenus *Penicillium*. Studies in Mycology 49, 201–242.

Frisvad, J.C., Filtenborg, O., 1989. Terverticillate penicillia: chemotaxonomy and mycotoxin production. Mycologia 81, 837–861.

Gallois, A., Langlois, D., 1990. New results in the volatile odorous compounds of French cheeses. Lait 50, 89–106.

Gillot, G., Jany, J.-L., Coton, M., Le Floch, G., Debaets, S., Ropars, J., López-Villavicencio, M., Dupont, J., Branca, A., Giraud, T., Coton, E., 2015. Insights into *Penicillium roqueforti* morphological and genetic diversity. PLoS One 10 (6), e0129849.

Gobbetti, M., Burzigotti, R., Smacchi, E., Corsetti, A., De Angelis, M., 1997. Microbiology and biochemistry of gorgonzola cheese during ripening. International Dairy Journal 7, 519–529.

González de Llano, D., Ramos, M., Rodríguez, A., Montilla, A., Juárez, M., 1992. Microbiological and physicochemical characteristics of Gamonedo blue cheese during ripening. International Dairy Journal 2, 121–135.

Grippon, J.C., Desmazeaud, M.J., le Bars, D., Bergere, J.L., 1977. Role of proteolytic enzymes of *Streptococcus lactis, Penicillium roqueforti* and *Penicillium caseicolum* during cheese ripening. Journal of Dairy Science 60, 1532–1538.

Grippon, J.C., 1993. Mould-ripened cheeses. In: Fox, P.F. (Ed.), Cheese: Chemistry, Physics and Microbiology, Major Cheese Groups, vol. 2. second ed. Chapman & Hall, London, pp. 111–136.

Guinee, T.P., Caric, M., Kalab, M., 2004. Pasteurized processed cheese and substitute/imitation cheese products. In: Fox, P.F., McSweeney, P.L.H., Cogan, T.M., Guinee, T.P. (Eds.), Cheese: Chemistry, Physics and Microbiology, vol. 2. third ed. Elsevier, London, pp. 349–394.

Hansen, T.K., 2001. Microbial Interactions in Blue Veined Cheeses (Ph.D. thesis). The Royal Veterinary and Agricultural University, Frederiksberg, Denmark.

Hansen, T.K., Jakobsen, M., 2001. Taxonomical and technological characteristics of *Saccharomyces* spp. associated with blue veined cheese. International Journal of Food Microbiology 69, 59–68.

Hewedi, M.M., Fox, P.F., 1984. Ripening of Blue cheese: characterization of proteolysis. Milchwissenschaft 39, 198–201.

Hidalgo, P.I., Ullán, R.V., Albillos, S.M., Montero, O., Fernández-Bodega, M.Á., García-Estrada, C., Fernández-Aguado, M., Martín, J.F., 2014. Molecular characterization of the PR-toxin gene cluster in *Penicillium roqueforti* and *Penicillium chrysogenum*: cross talk of secondary metabolite pathways. Fungal Genetics Biology 62, 11–24.

Houbraken, J., Frisvad, J.C., Samson, R.A., 2010. Sex in *Penicillium* series *Roqueforti*. IMA Fungus : The Global Mycological Journal 12, 171–180.

Hymery, N., Vasseur, V., Coton, M., Mounier, J., Jany, J.-L., Barbier, G., Coton, E., 2014. Filamentous fungi and mycotoxins in cheese: a review. Comprehensive Review in Food Science and Food Safety 13, 437–456.

Irlinger, F., Mounier, J., 2009. Microbial interactions in cheese: implications for cheese quality and safety. Current Opinion in Biotechnology 20, 142–148.

Karlshøj, K., Larsen, T.O., 2005. Differentiation of species from the *Penicillium roqueforti* group by volatile metabolite profiling. Journal of Agricultural Food Chemistry 53, 708–715.

Kavanagh, F., 1947. Activities of 22 antibacterial substances against nine species of bacteria. Journal Bacteriology 54, 761–766.

Kopp-Holtwiesche, B., Rehm, H.J., 1990. Antimicrobial action of roquefortine. Journal of Environmental Pathology, Toxicology and Oncology 10, 41–44.

Kosalková, K., Domínguez-Santos, R., Coton, M., Coton, E., García-Estrada, C., Liras, P., Martín, J.F., 2015. A natural short pathway synthesizes roquefortine C but not meleagrin in three different *Penicillium roqueforti* strains. Applied Microbiology and Biotechnology 99, 7601–7612.

Laich, F., Fierro, F., Cardoza, R.E., Martín, J.F., 1999. Organization of the gene cluster for biosynthesis of penicillin in *Penicillium nalgiovense* and antibiotic production in cured dry sausages. Applied Environmental Microbiology 65, 1236–1240.

Lamberet, G., Menassa, A., 1983. Purification and properties of an acid lipase from *Penicillium roqueforti*. Journal of Dairy Research 50, 459–468.

Larsen, M.D., Jensen, K., 1999. The effects of environmental conditions on the lipolytic activity of strains of *Penicillium roqueforti*. International Journal of Food Microbiology 46, 159–166.

Le Bars, D., Grippon, J.C., 1981. Role of *Penicillium roqueforti* proteinases during blue cheese ripening. Journal of Dairy Research 48, 479–487.

Linares, D.M., Martín, M.C., Ladero, V., Álvarez, M.A., Fernández, M., 2011. Biogenic amines in dairy products. Critical Reviews in Food Science and Nutrition 51, 691–703.

Loizzo, M.R., Menichini, F., Picci, N., Puoci, F., Spizzirri, G., Restuccia, D., 2013. Technological aspects and analytical determination of biogenic amines in cheese. A review. Trends in Food Science and Technology 30, 38–55.

López-Díaz, T.M., Alonso, C., Román, C., García-López, M.L., Moreno, B., 2000. Lactic acid bacteria isolated from a hand-made blue cheese. Food Microbiology 17, 23–32.

Lucas, P., Landete, J., Coton, M., Coton, E., Lonvaud-Funel, A., 2003. The tyrosine decarboxylase operon of *Lactobacillus brevis* IOEB 9809: characterization and conservation in tyramine producing bacteria. FEMS Microbiology Letters 229, 65–71.

Madkor, S., Fox, P.F., Shalabi, S.I., Metwalli, N.H., 1987. Studies on the ripening of Stilton cheese: lipolysis. Food Chemistry 25, 13–29.

Martín, M.C., Fernández, M., Linares, D.M., Álvarez, M.A., 2005. Sequencing, characterization and transcriptional analysis of the histidine decarboxylase operon of *Lactobacillus buchneri*. Microbiology 151, 1219–1228.

Marth, E.H., Yousef, A.E., 1991. Fungi and dairy products. In: Arora, D.K., Mukerij, K.G., Marth, E.H. (Eds.), Handbook of Applied Mycology. Foods and Feeds, vol. 3. Marcel Dekker, New York, pp. 375–414.

Mase, T., Matsumiya, Y., Matsuura, A., 1995. Purification and characterization of *Penicillium roqueforti* IAM 7268 lipase. Bioscience, Biotechnology and Biochemistry 59, 329–330.

Martín, J.F., Garcia-Estrada, C., Zeilinger, S. (Eds.), 2014a. Biosynthesis and Molecular Genetics of Fungal Secondary Metabolites. Springer Verlag, New York.

Martín, J.F., Liras, P., García-Estrada, C., 2014b. Roquefortine and prenylated indole alkaloids. In: Martín, J.F., García-Estrada, C., Zeilinger, S. (Eds.), Biosynthesis and Molecular Genetics of Fungal Secondary Metabolites. Springer Verlag, New York, pp. 111–128.

Martín, J.F., Liras, P., 2015. Novel antimicrobial and other bioactive metabolites obtained from silent gene clusters. In: Demain, A.L., Sánchez, S. (Eds.), Antibiotics: Current Innovations and Future Trends. Horizon Scientific Press and Caister Academic Press, Norfolk, UK, pp. 275–292.

Matsuda, Y., Abe, I., 2014. Meroterpenoids. In: Martín, J.F., García-Estrada, C., Zeilinger, S. (Eds.), Biosynthesis and Molecular Genetics of Fungal Secondary Metabolites. Springer Verlag, New York, pp. 289–301.

Mayer, H.K., Fiechter, G., Fischer, E., 2010. A new ultra-pressure liquid chromatography method for the determination of biogenic amines in cheese. Journal of Chromatography A 1217, 3251–3257.

Mayo, B., Alonso, L., Alegría, A., 2013. Blue cheese. In: Preedy, V.R., Watson, R.R., Patel, V.B. (Eds.), Handbook of Cheese in Health: Production, Nutrition and Medical Sciences. Wageningen Academic Sciences, pp. 277–288.

McSweeney, P.L., 2004. Biochemistry of cheese ripening. International Journal of Dairy Technology 57, 127–144.

Meier-Kriesche, H.U., Li, S., Gruessner, R.W., Fung, J.J., Bustami, R.T., Barr, M.L., Leichtman, A.B., 2006. Immunosuppresion: evolution in practice and trends, 1994–2004. American Journal Transplant 6, 1111–1131.

Menassa, A., Lamberet, G., 1982. Contribution à l'étude du système lipolytique de *Penicillium roqueforti*. Caractères comparés de deux activités exocellulaires. Lait 62, 32–43.

Mioso, R., Toledo-Marante, F.J., Herrera Bravo, I., 2014. *Penicillium roqueforti*: a multifunctional cell factory of high value-added molecules. Journal of Applied Microbiology 118, 781–791.

Molimard, P., Spinnler, E., 1996. Review: compounds involved in the flavor of surface mold-ripened cheeses: origins and properties. Journal of Dairy Science 79, 169–184.

Montel, M.-C., Buchin, S., Mallet, A., Delbes-Paus, C., Vuitton, D.A., Desmasures, N., Berthier, F., 2014. Traditional cheeses: rich and diverse microbiota with associated benefits. International Journal of Food Microbiology 177, 136–154.

Moreau, S., Gaudemer, A., Lablache-Combier, A., Biguet, J., 1976. Metabolites de *Penicillium roqueforti*: PR toxine et metabolites associes. Tetrahedron Letters 11, 833–834.

Nicoletti, R., De Stefano, M., De Stefano, S., Trincone, A., Marziano, F., 2004. Identification of fungitoxic metabolites produced by some *Penicillium* isolates antagonistic to *Rhizoctonia solani*. Mycopathologia 158, 465–474.

Nielsen, K.F., Dalsgaard, P.W., Smedsgaard, J., Larsen, T.O., 2005. Andrastins A-D, *Penicillium roqueforti* metabolites consistently produced in blue-mold ripened cheese. Journal Agricultural Food Chemistry 53, 2908–2913.

Nielsen, K.F., Sumarah, M.W., Frisvad, J.C., Miller, J.D., 2006. Production of metabolites from *Penicillium roqueforti* complex. Journal Agricultural Food Chemistry 54, 3756–3763.

Niki, T., Yoshioka, Y., Ahiko, K., 1966. Proteolytic and lipolytic activities of *Penicillium roqueforti* isolated from blue cheese. In: Proceedings from the XVII International Dairy Congress, Munich, vol. D: 2, pp. 531–537.

Novella-Rodríguez, S., Veciana-Nogués, M.T., Izquerdo-Pulido, M., Vidal-Carou, M.C., 2003. Distribution of biogenic amines and polyamines in cheese. Journal of Food Science 68, 750–755.

Nuñez, M., 1978. Microflora of Cabrales cheese: changes during maturation. Journal of Dairy Research 45, 501–508.

Ohmomo, S., Sato, T., Utagawa, T., Abe, M., 1975. Production of alkaloids and related substances by fungi. Isolation of festuclavine and three new indole alkaloids, roquefortine A, B, and C from cultures of *Penicillium roqueforti*. Nippon Nogei Kagaku Kaishi 49, 615–623.

Omura, S., Inokoshi, J., Uchida, R., Shiomi, K., Masuma, R., Kawakubo, T., Tanaka, H., Iwai, Y., Kosemura, S., Yamamura, S., 1996. Andrastins A-C, new protein farnesyltransferase inhibitors produced by *Penicillium* sp. FO-3929. I. Producing strain, fermentation, isolation, and biological activities. Journal of Antibiotics (Tokyo) 49, 414–417.

Ordoñez, J.A., Masso, J.A., Marmol, M.P., Ramos, M., 1980. Contribution à l'étude du fromage "Roncal". Lait 60, 283–294.

O'Sullivan, D.J., Giblin, L., McSweeney, P.L.H., Sheehan, J.J., Cotter, P.D., 2013. Nucleic acid-based approaches to investigate microbial-related cheese quality defects. Frontiers in Microbiology 4, 1.

Overy, D.P., Nielsen, K.F., Smedsgaard, J., 2005. Roquefortine/oxaline biosynthesis pathways metabolites in *Penicillium* ser. *Corymbifera*: in planta production and implications for competitive fitness. Journal Chemistry Ecology 31, 2373–2390.

Price, E.J., Linforth, R.S., Dodd, C.E., Philips, C.A., Hewson, L., Hort, J., Gkatzionis, K., 2014. Study of the influence of yeast inoculum concentration (*Yarrowia lipolytica* and *Kluyveromyces lactis*) on blue cheese aroma development using microbiological models. Food Chemistry 145, 464–472.

Prieto, B., Franco, I., Fresno, J.M., Bernardo, A., Carballo, J., 2000. Picon Bejes-Tresviso blue cheese: an overall biochemical survey throughout the ripening process. International Dairy Journal 10, 159–167.

Ramírez, C., 1982. Manual and Atlas of the Penicillia. Elsevier, Amsterdam.

Raper, K.B., Thom, C., 1949. Manual of the Penicillia. Williams and Wilkins, Baltimore.

Rattray, F., Eppert, I., 2011. Secondary cultures. In: Fuquay, J., Fow, P.F., McSweeney, P. (Eds.), Encyclopedia or Dairy Science, second ed. Academic Press, San Diego, pp. 567–573.

Romano, A., Tri, P.H., Lonvaud-Funel, A., Lolkema, J.S., Lucas, P.M., 2012. Evidence of two functionally distinct ornithine decarboxylation systems in lactic acid bacteria. Applied Environmental Microbiology 78, 1953–1961.

Ropars, J., Lopez-Villavicencio, M., Dupont, J., Snirc, A., Gillot, G., Coton, M., Jany, J.-L., Coton, E., Giraud, T., 2014. Induction of sexual reproduction and genetic diversity in the cheese fungus *Penicillium roqueforti*. Evolutionary Applications 7 (4), 433–441.

Rothe, M., Engst, W., Erhardt, V., 1982. Studies on characterization of Blue cheese flavor. Die Nahrung 26, 591–602.

Rothe, M., Kornelson, C., Schröder, R., 1994. Key components of food flavor: a sensory study on Blue cheese flavor. In: Maarse, H., van der Heij, D.G. (Eds.), Trends in Flavour Research. Elsevier Applied Science, London, pp. 221–232.

Rousseau, M., 1984. Study of the surface flora of traditional Camembert cheese by scanning electron microscopy. Milchwissenschaft 39, 129–134.

Sablé, S., Cottenceau, G., 1999. Current knowledge of soft cheeses flavor and related compounds. Journal of Agricultural Food Chemistry 47, 4825–4836.

Samson, R.A., Hoekstra, E.S., Frisvad, J.C., Filtenbborg, O., 1995. Identification of the food-borne fungi. In: Samson, R.A., Hoekstra, E.S., Frisvad, J.C., Filtenbborg, O. (Eds.), Introduction to Food-borne Fungi, fourth ed. Centraalbureau voor Schimmelcultures, Baarn, The Netherlands, p. 158.

Scott, P.M., Merrien, M.A., Polonsky, J., 1976. Roquefortine and isofumigaclavine A, metabolites from *Penicillium roqueforti*. Experientia 32, 140–142.

Sousa, M.J., Ardö, Y., McSweeney, P.L.H., 2001. Advances in the study of proteolysis in cheese during ripening. International Dairy Journal 11, 327–345.

Starec, M., Fiserová, A., Rosina, J., Malek, J., Krsiak, M., 2001. Effect of agroclavine on NK activity in vivo under normal and stress conditions in rats. Physiological Research 50, 513–519.

Stepaniak, L., Kornacki, K., Grabska, J., Rymaszewski, J., Cichosz, G., 1980. Lipolytic and proteolytic activity of *Penicillium roqueforti*, *Penicillium candidum* and *Penicillium camemberti* strains. ACTA Scientiarum Polonorum Technologia Alimentaria 6, 155–164.

Taniwaki, M.H., Hocking, A.D., Pitt, J.I., Fleet, G.H., 2001. Growth of fungi and mycotoxin production on cheese under modified atmospheres. International Journal of Food Microbiology 68, 125–133.

Thom, C., 1906. Fungi in cheese ripening: Camembert and Roquefort. U.S.D.A. Bureau of Animal Industry Bulletin 82, 1–39.

Thom, C., Currie, J., 1913. The dominance of Roquefort mold in cheese. Journal of Biological Chemistry 15, 249–258.

Tressler, R.J., Garvin, L.J., Slate, D.L., 1994. Anti-tumor activity of mycophenolate mofetil against human and mouse tumors in vivo. International Journal of Cancer 57, 568/573.

Trip, H., Mulder, N.L., Rattray, F.P., Lolkema, J.S., 2011. HdcB, a novel enzyme catalyzing maturation of pyruvoyl-dependent histidine decarboxylase. Molecular Microbiology 79, 861–871.

Uchida, R., Shiomi, K., Inokoshi, J., Sunazuka, T., Tanaka, H., Iwai, Y., Takayanagi, H., Omura, S., 1996a. Andrastins A-C, new protein farnesyltransferasre inhibitords produced by *Penicillium* sp. FO-3929. II. Structure elucidation and biosynthesis. Journal of Antibiotics 49, 418–424.

Uchida, R., Shiomi, K., Inokoshi, J., Tanaka, H., Iwai, Y., Omura, S., 1996b. Andrastin D, novel protein farnesyltransferase inhibitor produced by Penicillium sp. FO-3929. Journal Antibiotics (Tokyo) 49, 1278–1280.

Van den Tempel, T., Nielsen, M.S., 2000. Effects of atmospheric conditions, NaCl and pH on growth and interactions between molds and yeasts related to blue cheese production. International Journal of Food Microbiology 57, 193–199.

Van den Tempel, T., Gundersen, J.K., Nielsen, M.S., 2002. The microdistribution of oxygen in Danablu cheese measured by a microsensor during ripening. International Journal of Food Microbiology 75, 157–161.

Vilella, D., Sánchez, M., Platas, G., Salazar, O., Genilloud, O., Royo, I., Cascales, C., Martín, I., Díez, T., Silverman, K.C., Lingham, R.B., Singh, S.B., Jayasuriya, H., Peláez, F., 2000. Inhibitors of farnesylation of Ras from a microbial natural products screening program. Journal Industrial Microbiology Biotechnology 25, 315–327.

Vivier, D., Rivemale, M., Reverbel, J.P., Ratomahenina, R., Galzy, P., 1992. Some observations on the physiology of *Penicillium roqueforti* Thom and *Penicillium cyclopium* Westling. Lait 72, 277–283.

Wolken, W.A., Lucas, P.M., Lonvaud-Funel, A., Lolkema, J.S., 2006. The mechanism of the tyrosine transporter TyrP supports a proton motive tyrosine decarboxylation pathway in *Lactobacillus brevis*. Journal of Bacteriology 188, 2198–2206.

Woo, A.H., Kollodge, S., Lindsay, R.C., 1984. Quantification of major free fatty acids in several cheese varieties. Journal of Dairy Science 67, 874–878.

Zarmpoutis, I.V., McSweeney, P.L.H., Beechinor, J., Fox, P.F., 1996. Proteolysis in the Irish farmhouse blue cheese. Irish Journal of Agricultural Food Research 35, 25–36.

Zarmpoutis, I.V., McSweeney, P.L.H., Fox, P.F., 1997. Proteolysis in blue-veined cheeses: an intervarietal study. Irish Journal of Agricultural Food Research 36, 219–299.

Chapter 13

Yogurt and Health

M.A. Fernandez, É. Picard-Deland, M. Le Barz, N. Daniel, A. Marette
Laval University, Québec, QC, Canada

13.1 YOGURT COMPOSITION

13.1.1 Introduction

Yogurt, as a milk-based product, offers excellent nutrient density to consumers. Its unique composition, which includes active bacterial cultures and byproducts of fermentation, offers additional health benefits beyond that of its raw material. The health benefits of yogurt include reduced risk of type 2 diabetes (Chen et al., 2014; O'Connor et al., 2014; Diaz-Lopez et al., 2015), reduced weight gain (Mozaffarian et al., 2011), prevention of and cardiovascular disease (CVD) (Astrup, 2014), and it is associated with a prudent dietary pattern (Cormier et al., 2015). The bioactive properties of yogurt and mechanisms explaining their potential effects on health will be addressed in this chapter.

Yogurt has been produced for over 10,000 years, initially to extend the shelf-life of milk (Tamime and Robinson, 2007). Historically, yogurt is a traditional fermented milk product that has been a staple food in the diets of numerous populations in Scandinavia, the Middle East, and South-East Asia (Chandan, 2006). Anecdotally, curative and medicinal properties, particularly for digestive ailments, have been attributed to yogurt. It has been produced commercially for over a century, first as a pharmaceutical product (Hartley and Denariaz, 1993). In essence, yogurt is the product of fermentation, which is milk and lactic acid bacteria combined in a closed incubated environment (Tamime and Robinson, 2007). Commercial yogurt has changed over the last 100 years, creating a widely segmented market offering a variety of consumer products including sweetened, flavored, low-fat, drinkable, probiotic, and other yogurts with functional health properties. Yogurt consumption varies worldwide, with some of the highest intake in the Netherlands and the lowest intake in China (Chandan, 2006).

Ambiguity regarding the differentiation of yogurt from other fermented milk and the labeling of yogurt as a product exists because of differences in labeling laws from country to country. For example, France and Portugal have strict regulations on products that can be labeled as yogurt, mandating the presence of *Lactobacillus delbruecki* subsp. *bulgaricus*. However, the United Kingdom permits a variety of fermented milks to be labeled as yogurts (Tamime and Robinson, 2007). According to the Codex Alimentarius internationally recognized standards, yogurt is defined as a fermented milk product characterized by the combination of two particular starter cultures during the fermentation process (ie, *L. delbruecki* subsp. *bulgaricus* and *S. thermophiles*), but alternate yogurt cultures may contain *S. thermophilus* and any *Lactobacillus* species. In general, international standards for traditional yogurts, alternate culture yogurts, and acidophilus milks stipulate that they should contain at least 2.7% milk protein, a maximum of 15% milk fat, a minimum of 0.6% acidity (lactic acid), a minimum of 10^7 colony forming units (CFU) of starter culture microorganisms per gram, and a minimum of 10^6 CFU of labeled microorganisms per gram (WHO/FAO, 2010). However, as noted previously, the definition of yogurt may vary slightly. Nevertheless, yogurt can be easily categorized into four groups according to its physical characteristics: (1) liquid/viscous yogurt; (2) semisolid, concentrated/strained; (3) solid, frozen; and (4) powder, dried (Tamime and Robinson, 2007). The present chapter will focus on traditional yogurt, ie, liquid/viscous and concentrated/strained.

13.1.2 Nutrient Profile

Yogurt is exceptionally nutrient dense and is an excellent source or various macro- and micronutrients: high-quality proteins, digestible carbohydrates, calcium, magnesium, phosphorus, vitamin B12, etc. The typical serving size for yogurt is equivalent to 187.5 mL (3/4 cup), 170–175 g, or 6 oz., although serving size may vary from country to country. An average 170 g portion of popular low-fat fruit-flavored yogurt provides 6% of potassium and magnesium, 20–24% of calcium, 26% of phosphorus, 21–25% of riboflavin, and 30% of vitamin B12 recommended daily allowances for adults (Agriculture Research Service, 2014; Institute of Medicine, 2015). In addition, plain yogurt is relatively high in protein, low in fat, and has a low glycemic index, making it an excellent snack or meal accompaniment. Table 13.1 lists the nutritional composition of various types of popular yogurts. The nutritional value of yogurt is recognized by many national health agencies that promote yogurt as a viable option for one of the recommended daily servings of dairy products. Health Canada includes yogurt in its dairy and alternatives food group of the Canadian Food Guide (Health Canada, 2011). The USDA recommends the consumption of yogurt as a low-fat dairy alternative to help Americans achieve the recommended intakes of calcium, vitamin D, and phosphorus (US Department of Agriculture and US Department of Health and Human Services, 2010). Similarly, the British Nutrition Foundation advises the consumption of low-fat fruit yogurt instead of sweets and desserts (British Nutrition Foundation, 2015).

TABLE 13.1 Nutrient Profile of Popular Yogurts

	Plain Yogurt			Flavored Yogurt	
	Nonfat (170 g)	Whole Milk (170 g)	Nonfat Artificially Sweetened (170 g)	Low-fat (170 g)	Greek Style (170 g)
Energy (kcal)	95	104	73	168	180
Protein (g)	9.74	5.9	6.56	6.77	14.0
Fat (g)	0.31	5.52	0.31	1.95	4.96
Carbohydrates (g)	13.06	7.92	12.75	31.69	19.84
Calcium (mg)	199	206	243	235	146
Phosphorus (mg)	267	162	185	185	187
Sodium (mg)	131	78	100	90	58
Potassium (mg)	434	264	301	301	223

Adapted from the USDA National Nutrient Database for Standard.

Yogurt micro- and macronutrients contribute to the growth and maintenance of muscle mass. Yogurt proteins and peptides are derived from milk, are highly digestible, and have excellent nutritional quality (Bos et al., 2000). Moreover, yogurt and dairy products are a richer source of calcium and phosphorus per kilocalorie than other foods in an adult diet (Heaney et al., 2000). These nutrients are essential for the structural integrity and development of bones, making yogurt an excellent dietary source of nutrients that help maintain bone health. There is a growing body of evidence that shows that yogurt consumption exerts beneficial effects beyond its impact on growth and development. Indeed, the bacteria present in yogurt and the bioactive compounds formed during fermentation have potential beneficial effects on health (Marsh et al., 2014).

13.1.3 Microorganisms

L. delbrueckii subsp. *bulgaricus* and *S. thermophilus* were identified as yogurt starter organisms in 1931 and defined as thermophilic lactic acid bacteria (LAB) capable of growing at 40–45°C (Orla–Jensen, 1931). Moreover, a symbiotic relationship exists between these two homofermentative LAB, enhancing their growth and promoting beneficial effects. However, industrial fermentation processes need to be well controlled, because many factors can slow the starter microorganisms' growth (incubation temperature, antibiotic or detergent residues, environment pollution, bacteriophages, etc.) (Tamime and Robinson, 2007).

Yogurt and LAB species have shown promising health effects on the immune and digestive systems (eg, constipation, diarrheal diseases, inflammatory bowel disease, *Helicobacter pylori* infection, and colon cancer) (Adolfsson et al., 2004; Parvez et al., 2006). In particular, it was shown that yogurt bacteria are involved in the improvement of lactose intolerance. A claim published by the European Food Safety Authority (EFSA) in 2011 formally approved yogurt's beneficial effects on lactose malabsorption, indicating that the dose of live starter microorganisms is at least 10^8 CFU. However, there is ongoing debate regarding the probiotic properties and viability of these yogurt starters (Morelli, 2014). There are few studies on this subject, because current research interests are focused on novel probiotic strains rather than traditional yogurt starter microorganisms. Furthermore, yogurt is often used as a vehicle for probiotic delivery. Due to the beneficial effects attributed to yogurt bacteria, more and more strains are studied as probiotics. Microorganisms are considered probiotic if they are live and confer health benefits to the host when administered in adequate amounts.

13.1.4 Proteins

Yogurt is composed of high-quality proteins called caseins (α-s1, α-s2, β-casein, κ-casein), representing around 80%, and whey proteins (β-lactoglobulin, α-lactalbumin, lactoferrin, immunoglobulins, glycomacropeptide, enzymes, and growth factors), constituting 20% of dairy proteins (McGregor and Poppitt, 2013). The LAB fermentation process leads to the release of bioactive peptides (Farnworth, 2003) that are believed to exert beneficial effects on the immune, digestive, cardiovascular, and nervous systems (Beermann and Hartung, 2013). LAB proteolytic activities capable of releasing specific peptides have now been identified and are involved in immunomodulatory, antimicrobial, antioxidative, antifungal, and anticarcinogenic activities (Beermann and Hartung, 2013). The release of branched-chain amino acids (BCAAs) during fermentation by *S. thermophilus* (Tamime and Robinson, 2007) has been specifically linked to positive effects with regards to maintenance of lean body mass, protein synthesis, and muscle function (McGregor and Poppitt, 2013). However, controversy remains over the contribution of BCAAs to health given that elevated BCAAs have been implicated in the development of insulin resistance and type 2 diabetes (Newgard et al., 2009; Wang et al., 2011). There is some evidence supporting the antihypertensive role of casokinins and lactokinins, and particularly tripeptides such as isoleucine-proline-proline and valine-proline-proline through inhibition of angiotensin I-converting enzyme (ACE) (Huth et al., 2006; Astrup, 2014).

13.1.5 Lipids

In addition to being a valuable source of energy, milk fats provide a pleasant textural component and impart richness to fat-containing yogurts (Chandan, 2006). Depending on the milk origin and the manufacturing process, the lipid content in yogurt could vary in terms of quantity, but the quality does not change significantly compared to the original milk (Boylston and Beitz, 2002; Ayerbe and Soustre, 2010). Consequently, over 95% of yogurt lipids are triglycerides (Legrand, 2008). Despite the high content of saturated fat (72%), health benefits seem to be attributed to yogurt lipids, which also contain 25% monounsaturated and 3% polyunsaturated fatty acids. These unsaturated lipids are also the vectors of fat-soluble vitamins A, D, E, and K (Tunick and Van Hekken, 2014). Moreover, dairy products, including yogurt, are a rich source of conjugated linoleic acid (CLA) (Legrand, 2008) and provide more than 80% of the dietary requirements (Institute of Medicine, 2015). The main properties of this long-chain biohydrogenated derivative of linoleic acid, particularly its anticarcinogen effect, have been well demonstrated in both human and animal models (Rodríguez-Alcalá and Fontecha, 2007). According to the type of yogurt and strains, LAB fermentation also plays a role in the improvement of free, esterified, and volatile fatty acid content and the reduction of cholesterol in milk (Tamime and Robinson, 2007).

13.1.6 Carbohydrates

The carbohydrate content of yogurt varies between 8 and 32 g per 170 g portion of yogurt (Table 13.1). Lactose is the main available carbohydrate in yogurt. Depending on the type of product and industrial additives, this disaccharide can reach up to 98% of total carbohydrates in plain yogurt. Its hydrolysis into glucose and galactose occurs mainly in the digestive tract by the intestinal brush border β-galactosidase. Yogurt is known to be better tolerated than milk by lactose-intolerant or lactose maldigestors (Webb et al., 2014). Stirred fruit yogurts contain additional unavailable carbohydrates represented by stabilizers, guar gum, locust bean gum, carrageenan, and cellulose derivatives. These longchain polysaccharides have some health effects such as slowing the orocaecal transit time, but their use is now limited due to cost. Depending on the strains, some exopolysaccharides (homo- and heteropolysaccharides) are also produced by LAB and contribute to the beneficial health effects of yogurt (Tamime and Robinson, 2007).

13.1.7 Vitamins and Minerals

Yogurt is rich in many highly bioavailable minerals and vitamins. This nutrient-dense food is an excellent source of vitamins A, B, and D and calcium, potassium, phosphorus, and zinc, which are nutrients whose intakes fall below American estimated average requirements (EAR) (Webb et al., 2014). For example, vitamin D, whose intake is below the EAR in 70–94% of the population, and calcium, whose intake is insufficient in 38–50% of the population, may be improved by adding yogurt to the diet (Hess and Slavin, 2014; Webb et al., 2014). Moreover, specific populations (ie, women, children, and seniors) are vulnerable to deficiencies of other vitamins and minerals such as calcium and vitamin D that are involved in the prevention of certain diseases like osteoporosis, cholesterol metabolism, colon cancer, and to a lesser extent metabolic syndrome (Tunick and Van Hekken, 2014). If American recommendations for dairy product intake were achieved, yogurt could provide up to 70% of calcium, 30–40% of phosphorus and vitamins A and B, and 20–30% of potassium, zinc, and choline of the recommended daily nutritional requirements (Huth et al., 2006; Webb et al., 2014). Interestingly, many of these nutrients are more concentrated and bioavailable in yogurt than milk given its acidity and fermentation process (Jacques and Wang, 2014), which mainly affects vitamin content. For example, LAB requires vitamin B (Tamime and Robinson, 2007) but some strains are also capable of synthesizing it (Buttriss, 1997).

13.2 BIOACTIVE PROPERTIES OF YOGURT

As described in the previous section, yogurt is a complex substance with multifaceted physiological functionality in the gastrointestinal tract (GIT), where it contributes to the intestinal microbiota and the development of gut-associated lymphoid tissue (Ebringer et al., 2008). In fact, it is a specific mixture of bioactive proteins, lipids, and saccharides, which represent important signals regulating the development and the maintenance of the integrity of the GIT (Clare and Swaisgood, 2000; Donovan, 2006). Given that the immune system is an important element in cancer, gastrointestinal disorders, and immunoglobulin E-mediated hypersensitivity, an immune-stimulatory mechanism induced by yogurt has been proposed and investigated by using mainly animal models and, occasionally, human subjects (Meydani and Ha, 2000; Morelli, 2014). A brief review of studies analyzing the beneficial effects of yogurt consumption on inflammation, gut microbiota modulation, and also on the use of yogurt as a probiotic vector as well as mechanisms related to host metabolism, nutrient catabolism, and improvement of lactose intolerance will be included in this section. Moreover, the matrix effect will be discussed particularly through the study of yogurt digestion and its transit time.

13.2.1 Effect on Immunity

To maintain its interaction with commensals and sustain its function as a digestive organ, the GIT environment requires the constant induction and maintenance of various classes of regulatory responses (Molloy et al., 2012) to protect the organism against pathogens. In humans, cross-sectional studies support the consumption of dairy, particularly low-fat dairy products such as yogurt, as part of a healthy diet associated with less systemic inflammation (Ebringer et al., 2008). However, conflicting results have emerged regarding the impact of dairy products on inflammation (Labonte et al., 2014). A *meta*-analysis has investigated their impact on biomarkers of inflammation by using data collected in nutritional intervention studies conducted in overweight and obese adults (Labonte et al., 2013). In one study, fat-free yogurt consumption (3 daily 6-ounce servings) was shown to reduce C-reactive protein (CRP) concentrations and increased adiponectin concentrations compared with the low-dairy control diet (Zemel et al., 2005). Improvement of key inflammatory biomarkers including CRP, interleukin-6, or tumor necrosis factor (TNF)-α after dairy product consumption (including yogurt) was shown in three out of seven studies, although the four remaining studies showed no effect (Astrup, 2014). However, Labonté et al. (2014) concluded that the experimental design does not allow for identification of yogurt intake per se as the real cause of the favorable changes observed in inflammatory biomarker concentrations after consuming the yogurt-enriched diet.

Ingestion of LAB, which are found in high quantities in yogurt (Patrick et al., 2014), can counter the effects of the proliferation of pathogenic bacterial strains by different mechanisms: production of substances (H_2O_2, lactic, and acetic acids) directly inhibiting pathogens; lowering the pH by acids produced; increasing peristalsis (Kailasapathy and Chin, 2000); detoxification by enterotoxin degradation; prevention of the synthesis of toxic amines; and a barrier effect via metabolic competition (Savadogo and Traore, 2011). Moreover, it has been demonstrated in vitro that both heat-killed bacteria and highly purified lipoteichoic acid, a protoplast component in *Lactobacillus,* mediated proinflammatory responses in macrophages (Matsuguchi et al., 2003). Another series of studies published by Japanese researchers (Makino et al., 2010) demonstrated both in vitro and in vivo that the immune modulation exerted by *L. bulgaricus* OLL1073R-1 is the result of the action of a bacterial polysaccharide, which has a marked effect on the immune system in mice (Nagai et al., 2011).

Components of nonbacterial origin, such as whey proteins, short peptides, and CLA are also believed to contribute to yogurt's beneficial properties. Antimicrobial, anticarcinogenic, immunostimulatory, and other health-promoting activities of whey proteins have been reviewed by Madureira et al. (2007). Cell-culture studies and in vivo experiments (Gomez et al., 2002) have also demonstrated that whey proteins may enhance nonspecific and specific immune responses. The long-term consumption of yogurt naturally rich in *cis*-9 or *trans*-11 CLA caused a reduction in proinflammatory parameters in healthy subjects (Sofi et al., 2010). A randomized trial demonstrated that dairy supplementation, with 1% milk or yogurt, also resulted in suppression of oxidative stress markers or lower inflammatory markers in overweight and obese subjects (Jones et al., 2013). However, some clinical studies have shown no changes in inflammatory and oxidative stress markers or cytokine gene expression after consumption of dairy products (Thompson et al., 2005; Wennersberg et al., 2009; Van Loan et al., 2011). In summary, further human as well as mechanistic studies are needed to ascertain the effects of dairy product consumption, and particularly classical yogurt, to assess inflammation-related outcomes (Da Silva and Rudkowska, 2014).

In the context of a reasonable consumption, dairy products seem to present neutral or protective effects in relation to metabolic and postprandial inflammation (Vors et al., 2015). In fact, as an example, in spite of the high content of SFA, Nestel et al. (2012) did not find significant differences in postprandial inflammatory response in a human study comparing dairy products consumption. However, a subsequent human study carried out by the same research group compared supplementation of nonfermented dairy products (butter and cream) and fermented dairy products (yogurt and cheese) and demonstrated significantly lower proinflammatory cytokine release during the postprandial step after fermented dairy product consumption (Nestel et al., 2013). In summary, there is evidence showing that yogurt intake may modulate inflammation, but this topic remains disputed with inconsistent findings.

13.2.2 Modulation of Gut Microbiota

Yogurt has been of great interest with regards to the feasibility of modulating intestinal microbiota to improve health (Alvaro et al., 2007). The effects of LAB and yogurt constituents on the metabolism and composition of the gut microbiota is an excellent example of yogurt bioactivity. In a human study it was shown that *L. bulgaricus* was present in 73% of feces samples from yogurt consumers and in only 28% of feces samples from nonyogurt consumers (Alvaro et al., 2007). However, these results do not show that the bacterial starters

(*L. bulcaricus* of *S. thermophilus*) were viable. In a subsequent study, the effects of fresh or pasteurized (after fermentation) yogurts on microbiological and immunological parameters were compared to determine whether the beneficial effects of yogurt are dependent on the viability of LAB, and no significant differences between these conditions were found (Ballesta et al., 2008). According to Garcia-Albiach et al. (2008), the main change in human microbiota observed after yogurt consumption was an increase in the concentration of LAB and *Clostridium perfringens*to the detriment of *Bacteroides*. Analysis of the predominant bacterial groups in human feces showed that *Enterobacteriaceae* were significantly lower in yogurt consumers with no significant alterations of other bacterial groups (Alvaro et al., 2007). In the yogurt-supplemented group, *Bifidobacterium* populations were positively correlated with fermented milk consumption and bacterial changes were not different after the consumption of fresh and heat-treated yogurt, suggesting that changes in microbiota do not require starter bacterial viability. Furthermore, results found by Alvaro et al. (2007) showed increased β-galactosidase activity in feces in yogurt consumers. LAB in yogurt enhances resistance against intestinal pathogens mainly via antimicrobial mechanisms (Ibeagha-Awemu et al., 2009). These mechanisms include competitive colonization, thus inhibiting the adhesion of pathogens and production of organic acids (lactic and acetic acids) bacteriocins, and other primary metabolites (Kailasapathy and Chin, 2000; Lemberg et al., 2007). Production of lactic and acetic acids lowers intestinal pH, thereby inhibiting the growth of pathogens. These organic acids also increase peristalsis, thereby indirectly removing pathogens by accelerating their rate of transit through the intestine (Kailasapathy and Chin, 2000).

Other milk components such as oligosaccharides could alter the microbiota and change host-gut microbe signals that impact host metabolism (Zivkovic et al., 2011). There are also some reports indicating that consumption of cell-free whey from milk fermented with bifidobacteria was capable of modifying the human intestinal ecosystem (Ibeagha-Awemu et al., 2009). In fact, after consumption of cell-free fermented whey for 7 days, fecal excretions of *Bacteroides fragilis*, *Clostridium perfringens*, and clostridial spores decreased, while counts of bifidobacteria increased (Romond et al., 1998), which represents one of the potential beneficial effects of yogurt consumption. More studies are needed to better understand the effect of yogurt on inflammation and gut microbiota modulation because there are few studies concerning traditional yogurt strains, ie, *S. thermophilus* and *L. bulgaricus*.

13.2.3 Yogurt as a Probiotic Vector

Although the standard yogurt bacteria, *S. thermophilus* and *L. bulgaricus*, provide some benefits to the host, their effects could be minor compared to other probiotics, because they cannot resist passage through the gastrointestinal system (Lin et al., 1991). In fact, Del Campo et al. (2005) failed to detect the presence of *S. thermophilus* or *L. delbrueckii* in the feces of 96 volunteers following a 2-week period of yogurt intake. These results have been validated in other studies (Ballesta et al., 2008). These symbiotic bacteria do not adequately survive gastric passage or colonize the gut, thus necessitating the addition of other LAB species in yogurt preparations: *L. acidophilus, Lactobacillus casei, Bifidobacterium bifidum, Bifidobacterium longum, Bifidobacterium breve, Bifidobacterium infantis*, and *Bifidobacterium lactis*, among others (Guarner et al., 2005; Mater et al., 2005; Cogan et al., 2007; Guyonnet et al., 2007). These microorganisms, combined with yogurt starters to produce fermented milks, are capable of partially resisting gastric and bile secretions in vitro and in vivo and can deliver enzymes and other substances into the intestine (Alvaro et al., 2007). LAB are able to produce bacteriocins in response to environmental stimuli, which will modulate gut microbiota populations. *L. acidophilus* has been the subject of industrial advances due to its ability to inhibit the proliferation of pathogens such as *Staphylococcus aureus, Salmonella typhimurium*, enteropathogenic *Escherichia coli,* or *Clostridium perfringens*. Several mechanisms of pathogen inhibition have been identified, and some strains of *L. acidophilus* can produce hydrogen peroxide and can secrete peptide bacteriocins (lactacin B, lactocidin, or lactacin F), exerting an antibiotic effect (Savadogo and Traore, 2011). Millette et al. (2007) also reported that a probiotic milk culture of *L. acidophilus* and *L. casei* was capable of delaying the growth of *S. aureus, Enterococcus faecium, Enterococcus faecalis,* and *Listeria innocua*. Moreover, *B. bifidum* strains have the capacity to inhibit pathogenic flora such as *Clostridium difficile* (Savadogo and Traore, 2011).

Research has also demonstrated that both the cell-wall and cytoplasmic fractions of LAB were able to stimulate macrophages to produce significant amounts of TNFα, interleukin-6, and nitric oxide (Tejada-Simon et al., 1999). Furthermore, there is evidence based on the human flora-associated mouse model (germfree mice associated with human flora) that shows that *L. casei* initiates new protein synthesis while transiting through the digestive tract. These results may explain the health benefits associated with *L. casei*-fermented milk (Oozeer et al., 2002) and that some protective effects of probiotics, such as immunostimulation, which are mediated by their own DNA rather than their viability in the GIT (Rachmilewitz et al., 2004). Sah et al. (2014) revealed that probiotics from yogurt had a statistically significant effect on the proteolytic activity and enhanced the generation of peptides with potential antioxidant and antimutagenic actions. Yogurt containing three probiotic strains (*L. acidophilus* ATCC 4356, *L. casei* ATCC 393, and *Lactobacillus paracasei* ATCC BAA52) had a high degree of hydrolysis and strong antioxidant and antimutagenic activities. In addition, probiotic *Lactobacillus* strains have been shown to increase the secretion of IgA and certain antiinflammatory cytokines and to promote the gut immunological barrier function in animal models, thus enhancing the host immune response. Sarvari et al. (2014) demonstrated that the survival of probiotic and yogurt bacteria is dependent on the strain and storage time at 4°C with an important decrease for *L. bulgaricus* viability. Korbekandi et al. (2015) incorporated *L. paracasei* ssp. *tolerans*, a probiotic species that showed beneficial effects against pathogens and inflammatory response (Maragkoudakis et al., 2006), into yogurt to verify the compatibility of this bacteria with the yogurt starters and the organoleptic acceptability of the produced probiotic yogurt. It was found that the introduction of *L. paracasei* ssp. *tolerans* into yogurt had no effect on viability of the starter cultures during cold storage or on the organoleptic properties of yogurt.

13.2.4 Effect on Cholesterol Metabolism

Dairy products containing milk fat are a dietary source of saturated fat. A high intake of saturated fat has been linked to an increased risk of CVD, which is thought to be mediated predominantly by increased blood levels of low-density lipoprotein (LDL) cholesterol and total cholesterol (TC) (Da Silva and Rudkowska, 2014). Although not without exception, existing evidence from animal and human studies suggest a moderate cholesterol-lowering action of fermented dairy products (St-Onge et al., 2000). In fact, in a study conducted by Alvaro et al. (2007), it was reported that sterol metabolism may be weakly affected by yogurt consumption. In healthy subjects, fermented milk consumption causes a large and rapid increase in postprandial triglycerides (TG) followed by rapid clearance compared to whole milk (Sanggaard et al., 2004). Several researchers have compared the effects of pasteurized milk, yogurt, and other fermented milks on serum cholesterol concentrations in animal models. One of these studies, conducted in rats fed under a cholesterol-enriched diet, compared the effects of a standard yogurt with those of a bifidus-containing yogurt on serum cholesterol concentrations (Beena and Prasad, 1997). All yogurts significantly decreased TG but the lactose-hydrolyzed and condensed whey-fortified yogurts were the most effective. All groups fed bifidus yogurt had LDL cholesterol concentrations 21–27% lower than those consuming whole milk (St-Onge et al., 2000). In another study conducted in mice by Akalin et al. (1997), values for TG and LDL cholesterol after 56 days in both the *traditional yogurt* group and the probiotic *L. acidophilus* yogurt group were reduced by 31% and 26%, respectively, when compared with the control group. High-density lipoprotein (HDL) cholesterol and TG concentrations were not affected by either treatment, but LDL cholesterol was 33% lower on day 56 than on day 28 in the *acidophilus* group and 11% lower in the yogurt group. Moreover, fecal samples showed an increase in the number of lactobacilli from 8 to 9.5 log CFU/g in the *acidophilus* group and from 8 to 8.5 log CFU/g in the yogurt group throughout the dietary intervention (Akalin et al., 1997). These findings suggest that *L. acidophilus* colonized the mouse intestinal tract more readily than *L. bulgaricus* and thus *acidophilus* yogurt was more effective in reducing serum cholesterol concentrations than yogurt (St-Onge et al., 2000). It was also concluded that specific strains of *L. acidophilus* have the ability to modify cholesterol in the gut, making it unavailable for absorption in the blood, therefore causing a lowering cholesterol effect (Gilliand et al., 1985). It is also proposed that metabolites produced during fermentation of milk might be responsible for the hypocholesterolemic action of *S. thermophilus*-fermented milk (St-Onge et al., 2000).

Whey proteins have been described to lower plasma and liver cholesterol as well as plasma TG levels in model animals fed with cholesterol-containing diets (Zhang and Beynen, 1993; Beena and Prasad, 1997; Madureira et al., 2007). A study led by Bendsen et al. (2008) showed that saturated fatty acids (SFAs), monounsaturated fatty acids (MUFAs), and polyunsaturated fatty acids (PUFAs) were excreted in larger amounts with a high-calcium diet. In fact, calcium may modify the effects on LDL cholesterol and TG (Astrup, 2014). Dairy products do not seem to exert negative effects on blood lipids as predicted by its saturated fat content. It appears that there may be other yogurt components potentially influencing saturated fats to neutralize its effect on the lipid profile. Fermentation on food-derived indigestible carbohydrates causes increased production of short-chain fatty acids, which decreases circulatory cholesterol concentrations either by inhibiting hepatic synthesis or by redistributing cholesterol from plasma to the liver. Furthermore, increased bacterial activity in the large intestine results in enhanced deconjugated bile acid synthesis, which is not well absorbed by the gut mucosa and thereby excreted. Consequently, cholesterol is utilized to a greater extent for de novo bile acid synthesis (St-Onge et al., 2000). Moreover, a study realized by Abdullah et al. (2014) suggested the existence of a gene-diet interaction modulating the impact of dairy intake on circulating cholesterol levels.

13.2.5 Effect on Lactose Intolerance

Lactose intolerance is characterized by the appearance of various symptoms such as diarrhea, nausea, bloating, borborygmi, and abdominal pain caused by the ingestion of lactose, which is malabsorbed by the gut due to lack of lactase, or β-galactosidase (Misselwitz et al., 2013). Usually, lactase hydrolyses lactose into glucose and galactose in the small intestine, but its reduced or impaired expression allows lactose to reach the large intestine, where it is cleaved into short-chain fatty acids and gas (hydrogen, carbon dioxide, and methane), leading to clinical symptoms. A β-galactosidase deficiency could be attributed to congenital lactase deficiency or intestinal disease, but it mainly depends on genetic and ethnic factors (Mattar et al., 2012). Among the many benefits of yogurt, improvement of lactose malabsorption has been well-established, supported by 14 human studies, providing the basis for the following EFSA claim: "the consumption of live cultures in yoghurt improved digestion of lactose in individuals with lactose maldigestion" (EFSA Panel on Dietetic Products, Nutrition and Allergies, 2010). Due to the presence of LAB, the fermentation process results in the partial predigestion of lactose, being 20–30% hydrolyzed to β-galactose and D-glucose, and subsequently converted to lactic acid (The Dairy

Council). Moreover, yogurt internal lactase activity seems to be protected by its buffering capacity (Savaiano, 2014), which allows it to get through the stomach and compensate for the β-galactosidase deficiency. Compared to milk, yogurt is therefore a better way for lactose-intolerant people to consume dairy products.

13.2.6 Effect on Transit Time/Digestion

It has long been established that yogurt is more digestible than milk. Its high viscosity, great osmolality, energy density (Heaney, 1998), and the complexity of its matrix (Lecerf and Legrand, 2014) is heavily involved in the intestinal transit process. Moreover, low pH induced by fermentation also allows an increase in transit time (de Vrese et al., 2001; Savaiano, 2014), whereas heat treatment during the industrial process could have adverse effects on the orocecal transit time (OCTT) (Labayen et al., 2001), which is critical in the regulation of many mechanisms. Indeed, long transit time promotes better nutrient absorption in the intestine, which is beneficial for some gastrointestinal symptoms and perturbations like lactose intolerance (Vonk et al., 2003), diarrhea, and potentially irritable bowel syndrome (IBS). In the case of lactose intolerance, the decrease of OCTT allows optimizing β-galactosidase activity in the small bowel and decreasing the osmotic load of lactose (de Vrese et al., 2001; Labayen et al., 2001). Bacterial strains found in yogurt play a role in gut microbiota and are implicated in decreasing transit time, improving the interaction between yogurt and microbiota and allowing for the elimination of pathogenic enteric bacteria as well as the reinforcement of the intestinal barrier (Adolfsson et al., 2004; Rohde et al., 2009). In order to further improve digestion, some studies have explored the introduction of newer strains in yogurt, but it seems that the beneficial effect is more related to the matrix of yogurt per se than its probiotic content (Tulk et al., 2013; Merenstein et al., 2014).

13.2.7 Effect on Mineral Absorption

Mineral absorption is closely linked to the concept of bioavailability, defined by the efficiency by which a dietary component is used systemically through normal metabolic pathways (Aggett, 2010), and it is influenced by several parameters. The fermentation process for manufacturing yogurt involves decreasing pH by LAB, which leads to the solubilization of micellar calcium phosphate (Mekmene et al., 2010) and the ionization of calcium and magnesium (Adolfsson et al., 2004). Some studies have demonstrated interactions between proteins and minerals, facilitating their absorption. It seems that bovine folate-binding protein, found in milk, may increase folate absorption (Nygren-Babol and Jagerstad, 2012) and that lactoferrin as well as some phosphopeptides improve iron absorption. Whey proteins, vitamin B12-binding protein, β-lactoglobulin, and α-lactalbumin have also been identified as having good mineral bioavailability in yogurt (Bos et al., 2000; Vegarud et al., 2000). Additionally, the aggregation of casein micelles through milk coagulation and acidification have been shown to notably contribute to the high bioavailability of calcium (Kaushik et al., 2014), because it acts as a natural delivery system whose properties permit good release to the intestine (Elzoghby et al., 2011). On the other hand, heat treatment could denature some of these proteins, which facilitate the calcium, iron, and zinc absorption in the intestine (Ebringer et al., 2008). Mineral bioavailability also depends on individual mineral reserves and hormones (CCK, GLP), which can slow the absorption of nutrients (Gueguen and Pointillart, 2000; Zemel et al., 2004). In addition, yogurt viscosity and matrix play an important role in increasing transit time, allowing better intestinal absorption (Parra et al., 2007; Lecerf and Legrand, 2014). Finally, there are some synergic effects between nutrients like vitamin D and calcium (Bronner and Pansu, 1999). It also seems that the mineral content is better absorbed and enhanced through the matrix effect (Lecerf and Legrand, 2014) since calcium from dairy sources has been reported to be more effective on weight and fat loss after a caloric restriction, as compared to calcium given as a supplement (Zemel et al., 2005).

13.3 YOGURT IN DISEASE PREVENTION

13.3.1 Diet Quality

Epidemiological studies examining dietary patterns have shown that yogurt consumption is associated with a healthy diet (Wang et al., 2012; Hess and Slavin, 2014; Wood et al., 2014; Cormier et al., 2015). This association remains positive despite cultural differences in dietary patterns and is reflected in dietary guidelines internationally. Many national health agencies recommend two to three servings of dairy products daily; yogurt is often identified as a recommended dairy product to help individuals reach daily nutrient requirements and is considered part of a healthy diet (FAO, 2013). The consumption of one serving of yogurt a day is thought to have the potential to improve diet quality of Americans by helping them meet the dietary guidelines for Americans (DGA) and in turn daily requirement for nutrients of concern (Webb et al., 2014). The Canadian Food Guide suggests 3/4 cup of low-fat yogurt (Health Canada, 2011). Even in cultures where dairy products are not traditionally consumed in great amounts, dietary guidelines recommend consuming yogurt (eg, in India (Indian Council of Medical Research, 2011)). Regardless of the type of yogurt (nonfat, low-fat, or whole fat), its regular consumption can be safely considered as a healthy component of a balanced diet, as evidenced by international dietary guideline support (FAO, 2013).

Concerns regarding high-fat and sugar content of some types of yogurt may result in classifying this dairy product as "unhealthy." For example, a study evaluating snacks of Hispanic American preschool children, considered yogurt as an "energy dense" snack and therefore unhealthy (Longley et al., 2014). This dichotomous classification of healthy/unhealthy does not take into account the nutrient density of yogurt and its contribution to dietary intake of key nutrients such as protein, calcium, and vitamin D. Nevertheless, food companies in the United States are making attempts to reduce the sugar content of yogurt (Hess and Slavin, 2014). Low-fat or nonfat yogurt is often recommended for a healthy diet (FAO, 2013); however, there is no specific evidence supporting deleterious health effects of consuming full-fat yogurt. In fact, Cormier et al. (2015) found that among yogurt consumers, normal weight subjects had significantly more servings of high-fat yogurt and significantly fewer servings of non-fat yogurt than overweight and obese subjects, even after controlling for age, sex, physical activity, and dietary patterns. This study also noted a very strong positive association between yogurt consumption and the healthy prudent dietary pattern (Cormier et al., 2015). Furthermore, there is evidence that yogurt consumers have higher potassium intakes and are less likely to have inadequate intake of vitamins B2 and B12, calcium, magnesium, and zinc (Wang et al., 2012).

Distinction between high-fat and low-fat yogurt as well as frequency of consumption has been studied against multiple health parameters and disease risk factors in clinical and epidemiological studies. These results will be discussed in the following sections.

13.3.2 Yogurt and Weight Management

In addition to better diet quality, yogurt consumption has been negatively associated with long-term weight gain and various anthropometric indices. A large body of studies have examined the role of dairy products in the maintenance of healthy weight, weight loss, and obesity-related diseases. However, few epidemiological and randomized controlled trials (RCT) have examined the specific role of yogurt among dairy products on weight loss and weight maintenance. Some prospective and cross-sectional studies have shown a positive effect of yogurt on weight change and weight circumference, while others did not find any association (Jacques and Wang, 2014). Even within the same population results are inconsistent. The relationship between yogurt consumption and reduced weight, waist-to-hip ratio, and waist circumference was found in an observational study of French Canadian adults (Cormier et al., 2015). However, in another prospective study of French Canadian adults and their offspring there was an increase in waist circumference associated with increased intake of one serving of yogurt/day from baseline to follow-up (Drapeau et al., 2004). Interestingly, a clinical trial has shown that fat-free yogurt consumption in conjunction with a reduced-calorie diet and can help obese subjects to lose weight (Zemel et al., 2005).

A prospective study with over 120,000 healthy nonobese subjects followed at 4-year intervals showed weight change that was inversely related to yogurt intake and no other dairy products, although potentially confounding lifestyle factors were not measured (Mozaffarian et al., 2011). Nevertheless, these results are in line with a study that investigated the longitudinal association between dairy consumption and changes in weight and in waist circumference; participants who consumed at least one serving per week of yogurt had a significantly lower body mass index (BMI) and waist circumference than nonconsumers (0 servings per week) (Wang et al., 2013). In a Spanish study, there was a significantly lower BMI in a group of frequent yogurt consumers (≥ 7/week) compared to the group of low yogurt consumption (0–2/week). Furthermore, central adiposity was inversely and significantly associated with total and whole-fat yogurt consumption (Sanyon-Orea et al., 2015).

Nutrients like calcium and proteins (including bioactive peptides), yogurt microorganisms influencing the gut microbiota, and the satiating properties of yogurt could have an impact on body weight maintenance (Jacques and Wang, 2014). The substitution of nutrient-poor, energy-dense items with nutrient-rich foods like yogurt can also potentially have beneficial impact on obesity (Hess and Slavin, 2014). Further longitudinal and interventional studies are needed to confirm the beneficial role of increasing yogurt intake in the prevention of obesity.

13.3.3 Cardiometabolic Diseases

13.3.3.1 Type 2 Diabetes

There is a general consensus that dairy product consumption, especially yogurt, has a moderate protective effect on type 2 diabetes risk (Tong et al., 2011; Aune et al., 2013; Gao et al., 2013; Ley et al., 2014). However, it is debatable how much more effective high-dairy diets are than other healthy diets in preventing or treating type 2 diabetes. In fact, a study by Turner et al. (2015) found that high-dairy intake led to reduced insulin sensitivity compared to a diet rich in lean red meat. Yogurt, on the other hand, has offered very strong convincing epidemiological evidence supporting its inverse relationship to type 2 diabetes. Much of this evidence comes from large American cohort studies (Chen et al., 2014). Nevertheless, similar results have been observed in various countries including the UK (O'Connor et al., 2014).

In data from the National Health and Nutrition Examination Survey (NHANES), frequent yogurt consumption was found to contribute to a healthier insulin profile in children (Zhu et al., 2015). Similarly, in a cross-sectional study of two large cohorts of American adults, yogurt consumers demonstrated lower levels of fasting glucose and insulin resistance (Wang et al., 2012). In the UK, low-fat fermented dairy products, mainly yogurt, were associated with a 24% reduced risk of type 2 diabetes, even after controlling for age, sex, BMI, sociodemographic, lifestyle, and dietary factors (O'Connor et al., 2014). Dairy consumption and its relationship to type 2 diabetes risk was further investigated in a very large prospective study with three cohorts in the United States: 41,436 men from the Health Professionals Follow-Up Study, 67,138 women from the Nurses' Health Study, and 85,884 women from the Nurses Health Study II.

This study showed that yogurt intake was significantly associated with reduced risk of type 2 diabetes across cohorts in a multivariate model with a pooled hazard ratio of 0.83 (0.75, 0.92), while other dairy foods and total dairy consumption were not (Chen et al., 2014).

A systematic review and dose–response *meta*-analysis of cohort studies evaluating the effect of different types of dairy products on type 2 diabetes risk showed a significant inverse association between intakes of total dairy, low-fat dairy, and cheese. In this *meta*-analysis, seven cohort studies were reviewed to reveal a nonlinear association between yogurt consumption and type 2 diabetes risk. While an inverse relationship was generally observed in most of the studies carried out, the summary relative risk (RR) was not significant (Aune et al., 2013). A subsequent updated *meta*-analysis of 11 dairy studies and six yogurt studies showed no appreciable effects of total dairy consumption on type 2 diabetes, but an 18% reduced risk of type 2 diabetes was observed for each serving of yogurt consumed daily (Chen et al., 2014).

The specific mechanisms by which yogurt nutrients exert their effect are not well known. Probiotic bacteria have been shown to improve lipid profiles in type 2 diabetic patients (Mohamadshahi et al., 2014). Insulinotropic effects of yogurt peptides and vitamins and minerals such as vitamin D, calcium, and magnesium may act positively to reduce type 2 diabetes risk (Chen et al., 2014). Moreover, the low glycemic load of yogurt, its protein and lipid content, texture, and acidity could also impact satiety and obesity-related mechanisms, lowering type 2 diabetes incidence. Gender differences may also play a role with differing effects of dairy product intake on type 2 diabetes risk in some cohort studies (Kirii et al., 2009; Grantham et al., 2013). Furthermore, although prospective and cross-sectional studies adjusted for established and potential type 2 diabetes risk factors, residual confounding is still possible and cannot be ignored (eg, yogurt consumption is often associated with a healthy diet and lifestyle). Experimental mechanistic and clinical studies are needed in specific population subgroups to better understand the precise effects of yogurt as an individual food or the effects of its constituents on type 2 diabetes risk. Moreover, the effects should be evaluated using standardized biomarkers (eg, HbA1c) and tests such as oral glucose tolerance tests (OGTT).

13.3.3.2 Cardiovascular Disease

As noted in Section 15.2.4, yogurt consumption, particularly probiotic yogurts, may confer beneficial effects on cholesterol metabolism, thereby showing potential protective effects on the cardiovascular system. It has also been proposed that an inverse association with CVD might be possible via a reduction in blood pressure from sustained dairy product intake (Soedamah-Muthu et al., 2012). There is evidence that yogurt and fermented milk can reduce CVD risk and CVD biomarkers (Sonestedt et al., 2011) compared to other dairy foods, but this relationship has not been consistently confirmed (Abreu et al., 2014).

With few studies on CVD that have isolated yogurt from other dairy products, the relationship is difficult to accurately assess. Nevertheless, the relationship was evaluated in adolescents from the healthy lifestyle in Europe by nutrition in adolescence (HELENA) study in which milk, yogurt, and milk/yogurt-based beverage intake was found to be inversely associated with CVD risk in girls. In both genders, waist circumference and cardiorespiratory fitness were also inversely associated with consumption of milk, yogurt, and milk/yogurt-based beverages (Bel-Serrat et al., 2014). A *meta*-analysis of 15 prospective cohort studies on stroke found a protective effect of total dairy, low-fat dairy, cheese, and fermented milk on stroke (Hu et al., 2014). Only two studies on yogurt were identified in the *meta*-analysis and no evidence of a protective relationship was noted. A subsequent *meta*-analysis of prospective cohort studies identified 22 studies (two with yogurt) examining dairy products and CVD, coronary heart disease (CHD), and stroke concluding that dairy products have a beneficial effect on CVD, and low-fat dairy and cheese, in particular, may have a protective effect on CHD (Qin et al., 2015). Similar to the previous *meta*-analysis, no beneficial or protective effects of yogurt consumption on CVD, CHD, or stroke were noted. CVD involves many metabolic pathways that could be activated by diet. More studies are needed to clarify the effect of yogurt, specific yogurt nutrients, and dairy fats on different risk factors of CVD such as blood cholesterol, blood pressure, type 2 diabetes, adiposity, and inflammation.

13.3.3.3 Hypertension

Hypertension can lead to heart failure, stroke, or renal failure and is a well-established independent risk factor for CVD. Prevention can be achieved via physical activity, weight management, moderation in alcohol consumption, increased potassium intake, and reduced sodium intake (Mancia et al., 2013; Appel et al., 2006; Mancia et al., 2009). Two to three servings of low-fat dairy are recommended in the Dietary Approaches to Stop Hypertension (DASH) diet (Appel et al., 1997). The inverse relationship between milk and milk product intake and blood pressure was recognized and backed by moderate evidence by the Dietary Guidelines Advisory Committee in 2010 (McGrane et al., 2011).

The Coronary Artery Risk Development in Young Adults (CARDIA) study tracked the dietary intake of young adults against incidence of hypertension; milk and dairy dessert consumption was inversely associated to hypertension, but not yogurt (Steffen et al., 2005). An inverse association was observed between yogurt, skim milk, and hypertension in a prospective cohort study of American women (≥ 45 years); however, the relationship was nearly attenuated in yogurt in adjusted models (Wang et al., 2008). Masala et al. (2008) noted an inverse relationship between systolic blood pressure and intake of yogurt, vegetables, and eggs. A crossover clinical trial in a small group of normotensive young adults supplemented with either low-fat or high-fat dairy products (including yogurt) failed to find any effects on blood pressure (Alonso et al., 2009). Two *meta*-analyses conducted in 2012 each with five cohorts found conflicting results. In one of the studies, no association between fermented dairy or low-fat dairy consumption and systolic blood pressure was found after adjusting for confounding factors (Heraclides et al., 2012). Conversely, Ralston et al. (2012) reported a reduction in blood pressure associated with low-fat dairy and fluid dairy products. A large *meta*-analysis performed in prospective cohorts analyzed multiple types of dairy products and found an inverse association between total dairy, low-fat dairy, and milk with hypertension (Soedamah-Muthu et al., 2012). When the five studies examining yogurt were reviewed, no association was noted. Contrary to the inverse association expected between hypertension and yogurt, a strong association between the highest group of yogurt consumers and hypertension was noted in a large cross-sectional Iranian cohort. The relationship remained consistent even after controlling for age, sex, BMI, waist circumference, total energy intake, education, smoking status, sodium, potassium, vegetables, and fiber intake (Mirmiran et al., 2015).

Research examining the relationship between yogurt and hypertension is mounting and generally supports an inverse relationship, but these findings are limited to a small number of studies. Some mechanisms of action explaining this potential relationship include the presence of bioactive peptides exerting inhibitory effects on a key blood pressure-regulating proteins such as angiotensin I-converting enzyme (ACE) (Foltz et al., 2007; Boelsma and Kloek, 2009); blood pressure regulation via nutrients in yogurt (ie, calcium, magnesium, and potassium) (Azadbakht et al., 2005), and the low saturated fat content of yogurt, all part of a dietary pattern that is protective against atherosclerotic plaque development (McGrane et al., 2011). More evidence, epidemiological and clinical, needs to be presented to draw definite conclusions about whether yogurt exerts a protective, neutral, or negative effect on hypertension.

13.3.3.4 Osteoporosis

As a major cause of fractures in older people, osteoporosis is a significant public health concern worldwide (WHO, 2004). Osteoporosis is characterized by a loss of bone mass leading to low bone-mass density and high risk of fragility fractures (Sternberg et al., 2014), which can be accelerated with age and poor diet; excess alcohol, sodium, carbonated beverage, red meat, and low calcium intake (Simmons, 2011). The International Osteoporosis Foundation recommends adequate intake of calcium and vitamin D throughout childhood and adulthood to prevent osteoporosis (IOF, 2012). Yogurt is a dietary source of key nutrients essential for bone health such as calcium, vitamin D, protein, phosphorus, and potassium (Agriculture Research Service, 2014) and therefore its intake is recommended for osteoporosis prevention (Simmons, 2011).

Increased dairy product intake is expected to alleviate the burden of hip fractures associated with poor calcium intake (Lotters et al., 2013); however, there is current debate over the contribution of high calcium intake (from diet and supplements) to kidney stone risk, myocardial infarction, and even fractures. Michaëlsson et al. (2014) found reduced fracture and mortality risk associated with intake of fermented milk products (ie, yogurt, soured milk, and cheese). However, this study has been heavily criticized for its methodological flaws (Hettinga, 2014; Schooling, 2014). A pediatric study in China investigated the effects of supplementing the diet of preschool children with 125 g/day of yogurt. The authors found a significant difference in bone-mass density in the yogurt group compared to the control group (He et al., 2005). Similarly, yogurt intake was positively associated with bone strength in a large cohort of Japanese adolescents (Uenishi, 2006). Intake of three servings of yogurt a day for 7–10 days by postmenopausal women with low calcium intake led to a decrease in N-telepeptide, a bone resorption marker (Heaney et al., 2002). A case–control study of hip-fracture patients in India identified yogurt as a protective dietary factor along with milk, paneer (fresh cheese), fish, calcium, and supplement intake (Jha et al., 2010). In the Framingham Offspring Study, subjects who consumed yogurt more than four times a week had higher trochanter bone-mass density than patients who did not consume yogurt (Sahni et al., 2013). The small number of hip fractures ($n=43$) limited the power of the results, and only a weak inverse trend was seen between yogurt consumption and hip fractures (Sahni et al., 2013).

The relationship between yogurt and bone-mass density and osteoporosis is likely complex, involving a myriad of dietary, genetic, hormonal, and lifestyle factors. Given the state of evidence available, there is no basis for recommending yogurt intake to prevent osteoporosis. Nevertheless, yogurt consumption in the context of meeting daily dairy product recommendations contributes to daily intake of nutrients, which promotes good bone health and reduces risk for fractures in older age (Rizzoli, 2014).

13.3.3.5 Cancers

A significant amount of research over the last several decades has been devoted to cancer and potential risk factors. The relationship between dietary factors, particularly dairy products, has been especially controversial. This section will review the literature investigating the relationship between yogurt and three types of cancer: colorectal, breast, and prostate.

Colorectal cancer is common in developed nations and the majority of cases can be prevented through diet and lifestyle factors (Pala et al., 2011). Dairy products have been identified as having a potential protective effect against colorectal cancer (Huncharek et al., 2009). A *meta*-analysis by Aune et al. (2012) reaffirmed the protective effect of milk and total dairy products on colorectal cancer. The same effects were not noted with cheese or other dairy products; however, yogurt is not always analyzed separately from total dairy.

In the past, study limitations such as short follow-up period or study design (Kampman et al., 1994b), quantification/reported method of fermented dairy product intake (Kampman et al., 1994a), and limited yogurt consumption have made it difficult to assess the true relationship between yogurt and colorectal cancer (Cho et al., 2004; Larsson et al., 2006; Sun et al., 2011). Pala et al. (2011) performed the first prospective study with 45,241 men and women followed for 12 years and found clear evidence between yogurt consumption and reduced colorectal cancer risk. The European Prospective Investigation into Cancer and Nutrition (EPIC) investigated 477,122 men and women with an 11-year follow-up and found reduced risk of colorectal cancer associated with milk and calcium intake; however, the inverse association with yogurt was inconsistent between categorical and linear models (Murphy et al., 2013). A systematic review *meta*-analysis investigating both fermented milk (eg, yogurt) and nonfermented milks and colorectal cancer reported a strong basis for the protective effect of nonfermented milk intake on colon cancer in men (Ralston et al., 2014). Despite the lack of strong evidence for the protective effect of yogurt on colorectal cancer, no negative effects of yogurt consumption on colorectal cancer was reported.

Breast cancer is the most common female cancer in developed nations and has been given a great deal of research attention. There is little evidence linking dietary factors such as dairy intake to breast cancer (Moorman and Terry, 2004). Nevertheless, it has been proposed that components in dairy products (milk product contaminants, fat content, and insulin-like growth factor I (IGF-I)) can promote the growth of breast cancer cells (Moorman and Terry, 2004). However, in yogurt mechanisms are thought to act in the opposite direction. In fermented milk products, milk fat has been shown to be inversely related to breast cancer (Wirfalt et al., 2005) and the IGF-I content is significantly reduced through lactic acid fermentation (ANSES, 2012). A summary of research investigating breast cancer and yogurt intake is given below.

An inverse association between yogurt intake and breast cancer and a positive association between milk fat and cheese and breast cancer was found in a French case–control study (Le et al., 1986). Results from a small Danish case–control study supported the protective effect of fermented milk products on breast cancer in combination with intake of high-fiber or low-fat diets (Van 't Veer et al., 1991). The combination of low-fat and fermented dairy products exerting a protective effect was also noted in a case–control study in Uruguay (Ronco et al., 2002). When data from eight prospective cohort studies was combined from Western Europe and North America, there was no significant effect of yogurt intake on breast cancer (Missmer et al., 2002). An inverse, but insignificant, effect of yogurt intake and breast cancer development was noted in a group of hospital-based case–control studies in Italy (Gallus et al., 2006). Similarly, the Black Women's Health Study, with 52,062 women followed for 12 years, found an insignificant inverse relationship between yogurt and breast cancer (Genkinger et al., 2013). Two case–control studies in Iran obtained differing results; Mobarakeh et al. (2014) did not observe a significant effect of high-fat yogurt on breast cancer risk, whereas Mirmiran et al. (2015) found an inverse association between fermented dairy and breast cancer.

Evidence supporting the link between calcium intake and prostate cancer has been consistent enough for the World Cancer Research Fund and the American Institute for Cancer Research to conclude a probable association between high dietary calcium intake and increased risk of prostate cancer (Lampe, 2011). Nevertheless, conflicting findings remain between specific dairy products and prostate cancer risk. Huncharek et al. (2008) performed a *meta*-analysis of 45 studies that did not support any association between dairy product intake and risk for prostate cancer. In contrast, the Aune et al. (2015) systematic review and *meta*-analysis of 32 studies confidently asserted that total risk of prostate cancer was increased with high intakes of dairy products such as milk, low-fat-milk, and cheese and dietary calcium intakes. However, the authors did not find evidence of an association between yogurt intake and prostate cancer in the six studies examined. Studies performed in Canada and the United States found no association between high yogurt intake or yogurt consumption, respectively, and increased risk for prostate cancer (Jain et al., 1999; Tseng et al., 2005). A positive relationship between yogurt consumption and prostate cancer was found in three other studies (Gallus et al., 2006; Kesse et al., 2006; Kurahashi et al., 2008). More yogurt-specific research needs to be conducted to validate the results of Aune et al., 2015.

13.4 CONCLUSIONS

The popularity, acceptability, and generally perceived healthy image of yogurt all make it an ideal snack or meal accompaniment in many cultures around in the world. Its nutrient density and relatively low calorie content further make it an excellent substitution for nutrient-poor energy-dense snacks, giving yogurt added value as a healthy snack in the diet, particularly for obesity-related conditions. In addition, yogurt contains important nutrients such as calcium, vitamin D, and potassium that can help its consumers meet the daily recommended values of key dietary nutrients. Beyond the obvious nutritional value of yogurt, further GIT, metabolic, and cardiovascular health benefits may be conferred via specific peptides/amino acids that can act on type 2 diabetes and cardiometabolic diseases. A unique yogurt matrix with potentially greater bioavailability and metabolic properties combined with live microorganisms and bioactive fermentation products truly make yogurt an exceptional food item. While the functional health benefits of yogurt are plentiful, more randomized control trials and hypothesis-driven mechanistic-based experimental studies are needed to validate the results obtained by epidemiological studies before yogurt consumption can be advocated for specific disease prevention guidelines. Nevertheless, yogurt's contribution to a healthy balanced diet and lifestyle is fully justified.

REFERENCES

Abdullah, M., Cyr, A., M.-È, L., Lépine, M.-C., Couture, P., Eck, P., Lamarche, B., Jones, P., 2014. The impact of dairy consumption on circulating cholesterol levels is modulated by common single nucleotide polymorphisms in cholesterol synthesis- and transport-related genes (1038.4). The FASEB Journal 28 (1 Suppl.).

Abreu, S., Moreira, P., Moreira, C., Mota, J., Moreira-Silva, I., Santos, P.-C., Santos, R., 2014. Intake of milk, but not total dairy, yogurt, or cheese, is negatively associated with the clustering of cardiometabolic risk factors in adolescents. Nutrition Research 34 (1), 48–57.

Adolfsson, O., Meydani, S.N., Russell, R.M., 2004. Yogurt and gut function. American Journal of Clinical Nutrition 80 (2), 245–256.

Aggett, P., 2010. Population reference intakes and micronutrient bioavailability: a European perspective. American Journal of Clinical Nutrition 91 (5), 1433S–1437S.

Agriculture Research Service, USDA, 2014. USDA National Nutrient Database for Standard Reference, Agriculture Research Service, U.S. Department of Agriculture.

Akalin, A.S., Gonc, S., Duzel, S., 1997. Influence of yogurt and acidophilus yogurt on serum cholesterol levels in mice. Journal of Dairy Science 80 (11), 2721–2725.

Alonso, A., Zozaya, C., Vazquez, Z., Alfredo Martinez, J., Martinez-Gonzalez, M.A., 2009. The effect of low-fat versus whole-fat dairy product intake on blood pressure and weight in young normotensive adults. Journal of Human Nutrition and Dietetics 22 (4), 336–342.

Alvaro, E., Andrieux, C., Rochet, V., Rigottier-Gois, L., Lepercq, P., Sutren, M., Galan, P., Duval, Y., Juste, C., Dor, J.l, 2007. Composition and metabolism of the intestinal microbiota in consumers and nonconsumers of yogurt. British Journal of Nutrition 97 (1), 126–133.

ANSES, 2012. AVIS de l'Agence nationale de sécurité sanitaire de l'alimentation,de l'environnement et du travail relatif à l'évaluation des risques de cancers liés aux facteurs de croissance du lait et des produits laitiers.

Appel, L.J., Brands, M.W., Daniels, S.R., Karanja, N., Elmer, P.J., Sacks, F.M., 2006. Dietary approaches to prevent and treat hypertension: a scientific statement from the American Heart Association. Hypertension 47 (2), 296–308.

Appel, L.J., Moore, T.J., Obarzanek, E., Vollmer, W.M., Svetkey, L.P., Sacks, F.M., Bray, G.A., Vogt, T.M., Cutler, J.A., Windhauser, M.M., Lin, P.H., Karanja, N., 1997. A clinical trial of the effects of dietary patterns on blood pressure. DASH Collaborative Research Group. New England Journal of Medicine 336 (16), 1117–1124.

Astrup, A., 2014. Yogurt and dairy product consumption to prevent cardiometabolic diseases: epidemiologic and experimental studies. The American Journal of Clinical Nutrition 99 (5 Suppl.), 1235S.

Aune, D., Lau, R., Chan, D.S., Vieira, R., Greenwood, D.C., Kampman, E., Norat, T., 2012. Dairy products and colorectal cancer risk: a systematic review and meta-analysis of cohort studies. Annals of Oncology 23 (1), 37–45.

Aune, D., Navarro Rosenblatt, D.A., Chan, D.S., Vieira, A.R., Vieira, R., Greenwood, D.C., Vatten, L.J., Norat, T., 2015. Dairy products, calcium, and prostate cancer risk: a systematic review and meta-analysis of cohort studies. The American Journal of Clinical Nutrition 101 (1), 87–117.

Aune, D., Norat, T., Romundstad, P., Vatten, L.J., 2013. Dairy products and the risk of type 2 diabetes: a systematic review and dose-response meta-analysis of cohort studies. American Journal of Clinical Nutrition 98 (4), 1066–1083.

Ayerbe, A., Soustre, Y., 2010. La matière grasse laitière. Impact des technologies sur les caractéristiques nutritionnelles. Sciences des Aliments 29 (1–2), 5–8.

Azadbakht, L., Mirmiran, P., Esmaillzadeh, A., Azizi, F., 2005. Dairy consumption is inversely associated with the prevalence of the metabolic syndrome in Tehranian adults. American Journal of Clinical Nutrition 82 (3), 523–530.

Ballesta, S., Velasco, C., Borobio, M.V., Arguelles, F., Perea, E.J., 2008. Fresh versus pasteurized yogurt: comparative study of the effects on microbiological and immunological parameters, and gastrointestinal comfort. Enfermedades Infecciosas y Microbiologia Clinica 26 (9), 552–557.

Beena, A., Prasad, V., 1997. Effect of yogurt and bifidus yogurt fortified with skim milk powder, condensed whey and lactose-hydrolysed condensed whey on serum cholesterol and triacylglycerol levels in rats. The Journal of Dairy Research 64 (3), 453–457.

Beermann, C., Hartung, J., 2013. Physiological properties of milk ingredients released by fermentation. Food & Function 4 (2), 185–199.

Bel-Serrat, S., Mouratidou, T., Jimenez-Pavon, D., Huybrechts, I., Cuenca-Garcia, M., Mistura, L., Gottrand, F., Gonzalez-Gross, M., Dallongeville, J., Kafatos, A., Manios, Y., Stehle, P., Kersting, M., De Henauw, S., Castillo, M., Hallstrom, L., Molnar, D., Widhalm, K., Marcos, A., Moreno, L., 2014. Is dairy consumption associated with low cardiovascular disease risk in European adolescents? Results from the HELENA study. Pediatric Obesity 9 (5), 401–410.

Bendsen, N.T., Hother, A.L., Jensen, S., Lorenzen, J., Astrup, A., 2008. Effect of dairy calcium on fecal fat excretion: a randomized crossover trial. International Journal of Obesity 32 (12), 1816–1824.

Boelsma, E., Kloek, J., 2009. Lactotripeptides and antihypertensive effects: a critical review. British Journal of Nutrition 101 (6), 776–786.

Bos, C., Gaudichon, C., Tome, D., 2000. Nutritional and physiological criteria in the assessment of milk protein quality for humans. Journal of the American College of Nutrition 19 (2 Suppl.), 191S–205S.

Boylston, T.D., Beitz, D.C., 2002. Conjugated linoleic acid and fatty acid composition of yogurt produced from milk of cows fed soy oil and conjugated linoleic acid. Journal of Food Science 67 (5), 1973–1978.

British Nutrition Foundation, 2015. Healthy Eating. Retrieved March 20, 2015, from:http://www.nutrition.org.uk/healthyliving/healthyeating.html.

Bronner, F., Pansu, D., 1999. Nutritional aspects of calcium absorption. Journal of Nutrition 129 (1), 9–12.

Buttriss, J., 1997. Nutritional properties of fermented milk products. International Journal of Dairy 50 (1), 17–21.

Chandan, R.C., 2006. Manufacturing Yogurt and Fermented Milks. Blackwell Pub., Ames, Iowa.

Chen, M., Sun, Q., Giovannucci, E., Mozaffarian, D., Manson, J.E., Willett, W.C., Hu, F.B., 2014. Dairy consumption and risk of type 2 diabetes: 3 cohorts of US adults and an updated meta-analysis. BMC Medicine 12, 215.

Cho, E., Smith-Warner, S.A., Spiegelman, D., Beeson, W.L., van den Brandt, P.A., Colditz, G.A., Folsom, A.R., Fraser, G.E., Freudenheim, J.L., Giovannucci, E., Goldbohm, R.A., Graham, S., Miller, A.B., Pietinen, P., Potter, J.D., Rohan, T.E., Terry, P., Toniolo, P., Virtanen, M.J., Willett, W.C., Wolk, A., Wu, K., Yaun, S.S., Zeleniuch-Jacquotte, A., Hunter, D.J., 2004. Dairy foods, calcium, and colorectal cancer: a pooled analysis of 10 cohort studies. Journal of the National Cancer Institute 96 (13), 1015–1022.

Clare, D.A., Swaisgood, H.E., 2000. Bioactive milk peptides: a prospectus. Journal of Dairy Science 83 (6), 1187.

Cogan, T.M., Beresford, T.P., Steele, J., Broadbent, J., Shah, N.P., Ustunol, Z., 2007. Invited review: advances in starter cultures and cultured Foods1. Journal of Dairy Science 90 (9), 4005–4021.

Cormier, H., Thifault, É., Garneau, V., Tremblay, A., Drapeau, V., Pérusse, L., Vohl, M.-C., 2015. Association between yogurt consumption, dietary patterns, and cardio-metabolic risk factors. European Journal of Nutrition 1–11.

Del Campo, R., Bravo, D., Canton, R., Ruiz-Garbajosa, P., Garcia-Albiach, R., Montesi-Libois, A., Yuste, F., Abraira, V., Baquerol, F., 2005. Scarce evidence of yogurt lactic acid bacteria in human feces after daily yogurt consumption by healthy volunteers. Applied and Environmental Microbiology 71 (1), 547–549.

Diaz-Lopez, A., Bullo, M., Martinez-Gonzalez, M.A., Corella, D., Estruch, R., Fito, M., Gomez-Gracia, E., Fiol, M., Garcia de la Corte, F.J., Ros, E., Babio, N., Serra-Majem, L., Pinto, X., Munoz, M.A., Frances, F., Buil-Cosiales, P., Salas-Salvado, J., 2015. Dairy product consumption and risk of type 2 diabetes in an elderly Spanish Mediterranean population at high cardiovascular risk. European Journal of Nutrition 55 (1), 349–360.

Donovan, S.M., 2006. Role of human milk components in gastrointestinal development: current knowledge and future needs. Journal of Pediatrics 149 (5), S49–S61.

Drapeau, V., Despres, J.P., Bouchard, C., Allard, L., Fournier, G., Leblanc, C., Tremblay, A., 2004. Modifications in food-group consumption are related to long-term body-weight changes. American Journal of Clinical Nutrition 80 (1), 29–37.

Ebringer, L., Ferencik, M., Krajcovic, J., 2008. Beneficial health effects of milk and fermented dairy products–review. Folia Microbiologica (Praha) 53 (5), 378–394.

EFSA Panel on Dietetic Products, Nutrition and Allergies, 2010. Scientific Opinion on the substantiation of health claims related to live yogurt cultures and improved lactose digestion. EFSA Journal 8 (10), 1763.

Elzoghby, A.O., El-Fotoh, W.S., Elgindy, N.A., 2011. Casein-based formulations as promising controlled release drug delivery systems. Journal of Controlled Release 153 (3), 206–216.

FAO, 2013. Milk-and-Dairy-Products-in-Human-Nutrition.

Farnworth, E.R., 2003. Handbook of Fermented Functional Foods. CRC Press, Boca Raton, FL.

Foltz, M., Meynen, E.E., Bianco, V., van Platerink, C., Koning, T.M., Kloek, J., 2007. Angiotensin converting enzyme inhibitory peptides from a lactotripeptide-enriched milk beverage are absorbed intact into the circulation. Journal of Nutrition 137 (4), 953–958.

Gallus, S., Bravi, F., Talamini, R., Negri, E., Montella, M., Ramazzotti, V., Franceschi, S., Giacosa, A., La Vecchia, C., 2006. Milk, dairy products and cancer risk (Italy). Cancer Causes Control 17 (4), 429–437.

Gao, D., Ning, N., Wang, C., Wang, Y., Li, Q., Meng, Z., Liu, Y., Li, Q., 2013. Dairy products consumption and risk of type 2 diabetes: systematic review and dose-response meta-analysis. PLoS One 8 (9).

Garcia-Albiach, R., Pozuelo de Felipe, M.J., Angulo, S., Morosini, M.I., Bravo, D., Baquero, F., del Campo, R., 2008. Molecular analysis of yogurt containing *Lactobacillus delbrueckii* subsp. bulgaricus and *Streptococcus thermophilus* in human intestinal microbiota. American Journal of Clinical Nutrition 87 (1), 91–96.

Genkinger, J.M., Makambi, K.H., Palmer, J.R., Rosenberg, L., Adams-Campbell, L.L., 2013. Consumption of dairy and meat in relation to breast cancer risk in the Black Women's Health Study. Cancer Causes Control 24 (4), 675–684.

Gilliland, S.E., Nelson, C.R., Maxwell, C., 1985. Assimilation of cholesterol by *Lactobacillus acidophilus*. Applied Environmental Microbiology 49 (2), 377–381.

Gomez, H., Ochoa, T., Herrera-Insua, I., Carlin, L., Cleary, T., 2002. Lactoferrin protects rabbits from Shigella flexneri-induced inflammatory enteritis. Infection and Immunity 70 (12), 7050–7053.

Grantham, N.M., Magliano, D.J., Hodge, A., Jowett, J., Meikle, P., Shaw, J.E., 2013. The association between dairy food intake and the incidence of diabetes in Australia: the Australian Diabetes Obesity and Lifestyle Study (AusDiab). Public Health Nutrition 16 (2), 339–345.

Guarner, F., Perdigon, G., Corthier, G.R., Salminen, S., Koletzko, B., Morelli, L., 2005. Should yoghurt cultures be considered probiotic? British Journal of Nutrition 93 (6), 783–786.

Gueguen, L., Pointillart, A., 2000. The bioavailability of dietary calcium. Journal of the American College of Nutrition 19 (2 Suppl.), 119S–136S.

Guyonnet, D., Chassany, O., Ducrotte, P., Picard, C., Mouret, M., Mercier, C.H., Matuchansky, C., 2007. Effect of a fermented milk containing Bifidobacterium animalis DN-173 010 on the health-related quality of life and symptoms in irritable bowel syndrome in adults in primary care: a multicentre, randomized, double-blind, controlled trial. Alimentary Pharmacology & Therapeutics 26 (3), 475–486.

Hartley, D.L., Denariaz, G., 1993. The role of lactic acid bacteria in yogurt fermentation. International Journal of Immunotherapy 9 (1), 3–17.

He, M., Yang, Y.X., Han, H., Men, J.H., Bian, L.H., Wang, G.D., 2005. Effects of yogurt supplementation on the growth of preschool children in Beijing suburbs. Biomedical and Environmental Sciences 18 (3), 192–197.

Health Canada, 2011. Eating Well with Canada's Food Guide. Retrieved March 10, 2015, from:http://www.hc-sc.gc.ca/fn-an/food-guide-aliment/index-eng.php.

Heaney, R.P., 1998. Excess dietary protein may not adversely affect bone. Journal of Nutrition 128 (6), 1054–1057.

Heaney, R.P., Abrams, S., Dawson-Hughes, B., Looker, A., Marcus, R., Matkovic, V., Weaver, C., 2000. Peak bone mass. Osteoporos International 11 (12), 985–1009.

Heaney, R.P., Rafferty, K., Dowell, M.S., 2002. Effect of yogurt on a urinary marker of bone resorption in postmenopausal women. Journal of the American Dietetics Association 102 (11), 1672–1674.

Heraclides, A., Mishra, G.D., Hardy, R.J., Geleijnse, J.M., Black, S., Prynne, C.J., Kuh, D., Soedamah-Muthu, S.S., 2012. Dairy intake, blood pressure and incident hypertension in a general British population: the 1946 birth cohort. European Journal of Nutrition 51 (5), 583–591.

Hess, J., Slavin, J., 2014. Snacking for a cause: nutritional insufficiencies and excesses of U.S. children, a critical review of food consumption patterns and macronutrient and micronutrient intake of U.S. Children. Nutrients 6 (11), 4750–4759.

Hettinga, K., 2014. Study used wrong assumption about galactose content of fermented dairy products. BMJ (Clinical Research Ed.) 349, g7000.

Hu, D., Huang, J., Wang, Y., Zhang, D., Qu, Y., 2014. Dairy foods and risk of stroke: a meta-analysis of prospective cohort studies. Nutrition, Metabolism and Cardiovascular Diseases 24 (5), 460–469.

Huncharek, M., Muscat, J., Kupelnick, B., 2008. Dairy products, dietary calcium and vitamin D intake as risk factors for prostate cancer: a meta-analysis of 26,769 cases from 45 observational studies. Nutrition and Cancer 60 (4), 421–441.

Huncharek, M., Muscat, J., Kupelnick, B., 2009. Colorectal cancer risk and dietary intake of calcium, vitamin D, and dairy products: a meta-analysis of 26,335 cases from 60 observational studies. Nutrition and Cancer 61 (1), 47–69.

Huth, P., Dirienzo, D., Miller, G., 2006. Major scientific advances with dairy foods in nutrition and health. Journal of Dairy Science 89 (4), 1207–1221.

Ibeagha-Awemu, E.M., Liu, J.R., Zhao, X., 2009. Bioactive Components in Yogurt Products. Bioactive Components in Milk and Dairy Products. Wiley-Blackwell, pp. 235–250.

Indian Council of Medical Research, N. I. o. N, 2011. Dietary Guidelines for Indians – A Manual. National Institute of Nutrition, Hyderabad.

Institute of Medicine, 2015. Dietary Reference Intakes: Applications in Dietary Planning. The National Academies Press, Washington (DC).

IOF, 2012. Osteoporosis & Musculoskeletal Disorders – Osteoporosis – Prevention. From: http://www.iofbonehealth.org/preventing-osteoporosis.

Jacques, P.F., Wang, H., 2014. Yogurt and weight management. The American Journal of Clinical Nutrition 99 (5), 1229S–1234S.

Jain, M.G., Hislop, G.T., Howe, G.R., Ghadirian, P., 1999. Plant foods, antioxidants, and prostate cancer risk: findings from case-control studies in Canada. Nutrition and Cancer 34 (2), 173–184.

Jha, R.M., Mithal, A., Malhotra, N., Brown, E.M., 2010. Pilot case-control investigation of risk factors for hip fractures in the urban Indian population. BMC Musculoskeletal Disorders 11, 49.

Jones, K., Eller, L., Parnell, J., Doyle-Baker, P.K., Edwards, A.L., Reimer, R., 2013. Effect of a dairy- and calcium-rich diet on weight loss and appetite during energy restriction in overweight and obese adults: a randomized trial. European Journal of Clinical Nutrition 67 (4), 371–376.

Kailasapathy, K., Chin, J., 2000. Survival and therapeutic potential of probiotic organisms with reference to *Lactobacillus acidophilus* and Bifidobacterium spp. Immunology and Cell Biology 78 (1), 80.

Kampman, E., Giovannucci, E., van 't Veer, P., Rimm, E., Stampfer, M.J., Colditz, G.A., Kok, F.J., W, W.C., 1994a. Calcium, vitamin D, dairy foods, and the occurrence of colorectal adenomas among men and women in two prospective studies. American Journal of Epidemiology 139 (1), 16–29.

Kampman, E., van 't Veer, P., Hiddink, G.J., van Aken-Schneijder, P., Kok, F.J., Hermus, R.J., 1994b. Fermented dairy products, dietary calcium and colon cancer: a case-control study in the Netherlands. International Journal of Cancer 59 (2), 170–176.

Kaushik, R., Sachdeva, B., Arora, S., Kapila, S., Wadhwa, B.K., 2014. Bioavailability of vitamin D2 and calcium from fortified milk. Food Chemistry 147, 307–311.

Kesse, E., Bertrais, S., Astorg, P., Jaouen, A., Arnault, N., Galan, P., Hercberg, S., 2006. Dairy products, calcium and phosphorus intake, and the risk of prostate cancer: results of the French prospective SU.VI.MAX (Supplementation en Vitamines et Minéraux Antioxydants) study. British Journal of Nutrition 95 (3), 539–545.

Kirii, K., Mizoue, T., Iso, H., Takahashi, Y., Kato, M., Inoue, M., Noda, M., Tsugane, S., 2009. Calcium, vitamin D and dairy intake in relation to type 2 diabetes risk in a Japanese cohort. Diabetologia 52 (12), 2542–2550.

Korbekandi, T., Abedi, D., Maracy, M., Jalali, M., Azarman, A., Iravani, S., 2015. Evaluation of probiotic yoghurt produced by *Lactobacillus paracasei* spp. Journal of Food Biosciences and Technology 5 (1), 7.

Kurahashi, N., Inoue, M., Iwasaki, M., Sasazuki, S., Tsugane, A.S., 2008. Dairy product, saturated fatty acid, and calcium intake and prostate cancer in a prospective cohort of Japanese men. Cancer Epidemiol Biomarkers Prev 17 (4), 930–937.

Labayen, I., Forga, L., Gonzalez, A., Lenoir-Wijnkoop, I., Nutr, R., Martinez, J.A., 2001. Relationship between lactose digestion, gastrointestinal transit time and symptoms in lactose malabsorbers after dairy consumption. Alimentary Pharmacology & Therapeutics 15 (4), 543–549.

Labonte, M.E., Couture, P., Richard, C., Desroches, S., Lamarche, B., 2013. Impact of dairy products on biomarkers of inflammation: a systematic review of randomized controlled nutritional intervention studies in overweight and obese adults. American Journal of Clinical Nutrition 97 (4), 706–717.

Labonte, M.E., Cyr, A., Abdullah, M.M., Lepine, M.C., Vohl, M.C., Jones, P., Couture, P., Lamarche, B., 2014. Dairy product consumption has No impact on biomarkers of inflammation among men and women with low-grade systemic inflammation. Journal of Nutrition 144 (11), 1760–1767.

Lampe, J.W., 2011. Dairy products and cancer. Journal of the American College of Nutrition 30 (5 Suppl. 1), 464S–470S.

Larsson, S.C., Bergkvist, L., Rutegard, J., Giovannucci, E., Wolk, A., 2006. Calcium and dairy food intakes are inversely associated with colorectal cancer risk in the Cohort of Swedish Men. American Journal of Clinical Nutrition 83 (3), 667–673, quiz 728–669.

Le, M.G., Moulton, L.H., Hill, C., Kramar, A., 1986. Consumption of dairy produce and alcohol in a case-control study of breast cancer. Journal of the National Cancer Institute 77 (3), 633–636.

Lecerf, J.-M., Legrand, P., 2014. Les effets des nutriments dépendent-ils des aliments qui les portent ? L'effet matrice. Cahiers de Nutrition et de Dietetique 65 (8), 1013–1018.

Legrand, P., 2008. Intérêt nutritionnel des principaux acides gras des lipides du lait. Cerin 105.

Lemberg, D.A., Ooi, C.Y., Day, A.S., 2007. Probiotics in paediatric gastrointestinal diseases. Journal of Paediatrics and Child Health 43 (5), 331–336.

Ley, S.H., Hamdy, O., Mohan, V., Hu, F.B., 2014. Prevention and management of type 2 diabetes: dietary components and nutritional strategies. The Lancet 383 (9933), 1999–2007.

Lin, M.Y., Savaiano, D., Harlander, S., 1991. Influence of nonfermented dairy products containing bacterial starter cultures on lactose maldigestion in humans. Journal of Dairy Science 74 (1), 87–95.

Longley, C.E., McArthur, L.H., Holbert, D., 2014. Rural latino parents offer preschool children few nutrient-dense snacks: a community-based study in Western Illinois. Hispanic Health Care International 12 (4), 189.

Lotters, F.J., Lenoir-Wijnkoop, I., Fardellone, P., Rizzoli, R., Rocher, E., Poley, M.J., 2013. Dairy foods and osteoporosis: an example of assessing the health-economic impact of food products. Osteoporosis International 24 (1), 139–150.

VanLoan, M.D., Keim, N.L., Adams, S.H., Souza, E., Woodhouse, L.R., Thomas, A., Witbracht, M., Gertz, E.R., Piccolo, B., Bremer, A.A., Spurlock, M., 2011. Dairy foods in a moderate energy restricted diet do not enhance central fat, weight, and intra-abdominal adipose tissue losses nor reduce adipocyte size or inflammatory markers in overweight and obese adults: a controlled feeding study. Journal of Obesity 2011.

Madureira, A.R., Pereira, C.I., Gomes, A.M.P., Pintado, M.E., Xavier Malcata, F., 2007. Bovine whey proteins – overview on their main biological properties. Food Research International 40 (10), 1197–1211.

Makino, S., Ikegami, S., Kume, A., Horiuchi, H., Sasaki, H., Orii, N., 2010. Reducing the risk of infection in the elderly by dietary intake of yoghurt fermented with *Lactobacillus delbrueckii* ssp. bulgaricus OLL1073R-1. British Journal Of Nutrition 104 (7), 998–1006.

Mancia, G., Fagard, R., Narkiewicz, K., Redon, J., Zanchetti, A., Böhm, M., Christiaens, T., Cifkova, R., De Backer, G., Dominiczak, A., Galderisi, M., Grobbee, D.E., Jaarsma, T., Kirchhof, P., Kjeldsen, S.E., Laurent, S., Manolis, A.J., Nilsson, P.M., Ruilope, L.M., Schmieder, R.E., Sirnes, P.A., Sleight, P., Viigimaa, M., Waeber, B., Zannad, F., Redon, J., Dominiczak, A., Narkiewicz, K., Nilsson, P.M., Burnier, M., Viigimaa, M., Ambrosioni, E., Caufield, M., Coca, A., Olsen, M.H., Schmieder, R.E., Tsioufis, C., van de Borne, P., Zamorano, J.L., Achenbach, S., Baumgartner, H., Bax, J.J., Bueno, H., Dean, V., Deaton, C., Erol, C., Fagard, R., Ferrari, R., Hasdai, D., Hoes, A.W., Kirchhof, P., Knuuti, J., Kolh, P., Lancellotti, P., Linhart, A., Nihoyannopoulos, P., Piepoli, M.F., Ponikowski, P., Sirnes, P.A., Tamargo, J.L., Tendera, M., Torbicki, A., Wijns, W., Windecker, S., Clement, D.L., Coca, A., Gillebert, T.C., Tendera, M., Rosei, E.A., Ambrosioni, E., Anker, S.D., Bauersachs, J., Hitij, J.B., Caufield, M., De Buyzere, M., De Geest, S., Derumeaux, G.A., Erdine, S., Farsang, C., Funck-Brentano, C., Gerc, V., Germano, G., Gielen, S., Haller, H., Hoes, A.W., Jordan, J., Kahan, T., Komajda, M., Lovic, D., Mahrholdt, H., Olsen, M.H., Ostergren, J., Parati, G., Perk, J., Polonia, J., Popescu, B.A., Reiner, Ž., Rydén, L., Sirenko, Y., Stanton, A., Struijker-Boudier, H., Tsioufis, C., van de Borne, P., Vlachopoulos, C., Volpe, M., Wood, D.A., 2013. 2013 ESH/ESC Guidelines for the management of arterial hypertension. European Heart Journal 34 (28), 2159–2219.

Mancia, G., Laurent, S., Agabiti-Rosei, E., Ambrosioni, E., Burnier, M., Caufield, M.J., Cifkova, R., Clement, D., Coca, A., Dominiczak, A., Erdine, S., Fagard, R., Farsang, C., Grassi, G., Haller, H., Heagerty, A., Kjeldsen, S.E., Kiowski, W., Mallion, J.M., Manolis, A., Narkiewicz,

K., Nilsson, P., Olsen, M.H., Rahn, K.H., Redon, J., Rodicio, J., Ruilope, L., Schmieder, R.E., Struijker-Boudier, H.A., van Zwieten, P.A., Viigimaa, M., Zanchetti, A., 2009. Reappraisal of European guidelines on hypertension management: a European Society of Hypertension Task Force document. Journal of Hypertension 27 (11), 2121–2158.

Maragkoudakis, P.A., Miaris, C., Rojez, P., Manalis, N., Magkanari, F., Kalantzopoulos, G., Tsakalidou, E., 2006. Production of traditional Greek yoghurt using Lactobacillus strains with probiotic potential as starter adjuncts. International Dairy Journal 16 (1), 52–60.

Marsh, A.J., Hill, C., Ross, R.P., Cotter, P.D., 2014. Fermented beverages with health-promoting potential: past and future perspectives. Trends in Food Science & Technology 38 (2), 113–124.

Masala, G., Bendinelli, B., Versari, D., Saieva, C., Ceroti, M., Santagiuliana, F., Caini, S., Salvini, S., Sera, F., Taddei, S., Ghiadoni, L., Palli, D., 2008. Anthropometric and dietary determinants of blood pressure in over 7000 Mediterranean women: the European Prospective Investigation into Cancer and Nutrition-Florence cohort. Journal of Hypertension 26 (11), 2112–2120.

Mater, D.D.G., Bretigny, L., Firmesse, O., Flores, M.-J., Mogenet, A., Bresson, J.-L., Corthier, G., 2005. *Streptococcus thermophilus* and *Lactobacillus delbrueckii* subsp. bulgaricus survive gastrointestinal transit of healthy volunteers consuming yogurt. FEMS Microbiology Letters 250 (2), 185–187.

Matsuguchi, T., Takagi, A., Matsuzaki, T., Nagaoka, M., Ishikawa, K., Yokokura, T., Yoshikai, Y., 2003. Lipoteichoic acids from Lactobacillus strains elicit strong tumor necrosis factor alpha-inducing activities in macrophages through toll-like receptor 2. Clinical and Diagnostic Laboratory Immunology 10 (2), 259–266.

Mattar, R., de Campos Mazo, D.F., Carrilho, F.J., 2012. Lactose intolerance: diagnosis, genetic, and clinical factors. Clinical and Experimental Gastroenterology 5, 113–121.

McGrane, M.M., Essery, E., Obbagy, J., Lyon, J., Macneil, P., Spahn, J., Van Horn, L., 2011. Dairy consumption, blood pressure, and risk of hypertension: an evidence-based review of recent literature. Current Cardiovascular Risk Reports 5 (4), 287–298.

McGregor, R., Poppitt, S., 2013. Milk protein for improved metabolic health: a review of the evidence. Nutrition & Metabolism 10 (1), 46.

Mekmene, O., Le Graet, Y., Gaucheron, F., 2010. Theoretical model for calculating ionic equilibria in milk as a function of pH: comparison to experiment. Journal of Agriculture and Food Chemistry 58 (7), 4440–4447.

Merenstein, D.J., D'Amico, F., Palese, C., Hahn, A., Sparenborg, J., Tan, T., Scott, H., Polzin, K., Kolberg, L., Roberts, R., 2014. Short-term, daily intake of yogurt containing Bifidobacterium animalis ssp. lactis Bf-6 (LMG 24384) does not affect colonic transit time in women. British Journal of Nutrition 111 (2), 279–286.

Meydani, S.N., Ha, W.K., 2000. Immunologic effects of yogurt. American Journal of Clinical Nutrition 71 (4), 861–872.

Michaëlsson, K., Wolk, A., Langenskiöld, S., Basu, S., Warensjö Lemming, E., Melhus, H., Byberg, L., 2014. Milk intake and risk of mortality and fractures in women and men: cohort studies. British Medical Journal 349, g6015.

Millette, M., Dupont, C., Archambault, D., Lacroix, M., 2007. Partial characterization of bacteriocins produced by human Lactococcus lactis and Pediococcus acidilactici isolates. Journal of Applied Microbiology 102 (1), 274–282.

Mirmiran, P., Golzarand, M., Bahadoran, Z., Mirzaei, S., Azizi, F., 2015. High-fat dairy is inversely associated with the risk of hypertension in adults: Tehran lipid and glucose study. International Dairy Journal 43, 22–26.

Misselwitz, B., Pohl, D., Frühauf, H., Fried, M., Vavricka, S.R., Fox, M., 2013. Lactose malabsorption and intolerance: pathogenesis, diagnosis and treatment. United European Gastroenterology Journal 1 (3), 151.

Missmer, S.A., Smith-Warner, S.A., Spiegelman, D., Yaun, S.S., Adami, H.O., Beeson, W.L., van den Brandt, P.A., Fraser, G.E., Freudenheim, J.L., Goldbohm, R.A., Graham, S., Kushi, L.H., Miller, A.B., Potter, J.D., Rohan, T.E., Speizer, F.E., Toniolo, P., Willett, W.C., Wolk, A., Zeleniuch-Jacquotte, A., Hunter, D.J., 2002. Meat and dairy food consumption and breast cancer: a pooled analysis of cohort studies. International Journal of Epidemiology 31 (1), 78–85.

Mobarakeh, Z.S., Mirzaei, K., Hatmi, N., Ebrahimi, M., Dabiran, S., Sotoudeh, G., 2014. Dietary habits contributing to breast cancer risk among Iranian women. Asian Pacific Journal of Cancer Prevention 15 (21), 9543.

Mohamadshahi, M., Veissi, M., Haidari, F., Javid, A.Z., Mohammadi, F., Shirbeigi, E., 2014. Effects of probiotic yogurt consumption on lipid profile in type 2 diabetic patients: a randomized controlled clinical trial. Journal of Research in Medical 19 (6), 531.

Molloy, A.P., Hutchinson, B., Toole, G.C.O., 2012. Extra-abdominal desmoid tumours: a review of the literature. Sarcoma 2012.

Moorman, P.G., Terry, P.D., 2004. Consumption of dairy products and the risk of breast cancer: a review of the literature. American Journal of Clinical Nutrition 80 (1), 5–14.

Morelli, L., 2014. Yogurt, living cultures, and gut health. American Journal of Clinical Nutrition 99 (5), 1248S–1250S.

Mozaffarian, D., Hao, T., Rimm, E.B., Willett, W.C., Hu, F.B., 2011. Changes in diet and lifestyle and long-term weight gain in women and men. New England Journal of Medicine 364 (25), 2392–2404.

Murphy, N., Norat, T., Ferrari, P., Jenab, M., Bueno-de-Mesquita, B., Skeie, G., Olsen, A., Tjonneland, A., Dahm, C.C., Overvad, K., Boutron-Ruault, M.C., Clavel-Chapelon, F., Nailler, L., Kaaks, R., Teucher, B., Boeing, H., Bergmann, M.M., Trichopoulou, A., Lagiou, P., Trichopoulos, D., Palli, D., Pala, V., Tumino, R., Vineis, P., Panico, S., Peeters, P.H., Dik, V.K., Weiderpass, E., Lund, E., Garcia, J.R., Zamora-Ros, R., Perez, M.J., Dorronsoro, M., Navarro, C., Ardanaz, E., Manjer, J., Almquist, M., Johansson, I., Palmqvist, R., Khaw, K.T., Wareham, N., Key, T.J., Crowe, F.L., Fedirko, V., Gunter, M.J., Riboli, E., 2013. Consumption of dairy products and colorectal cancer in the European prospective investigation into cancer and nutrition (EPIC). PLoS One 8 (9), e72715.

Nagai, T., Makino, S., Ikegami, S., Itoh, H., Yamada, H., 2011. Effects of oral administration of yogurt fermented with Lactobacillus delbrueckii ssp bulgaricus OLL1073R-1 and its exopolysaccharides against influenza virus infection in mice. International Immunopharmacology 11 (12), 2246–2250.

Nestel, P.J., Mellett, N., Pally, S., Wong, G., Barlow, C.K., Croft, K., Mori, T.A., Meikle, P.J., 2013. Effects of low-fat or full-fat fermented and nonfermented dairy foods on selected cardiovascular biomarkers in overweight adults. The British Journal of Nutrition 110 (12), 2242.

Nestel, P.J., Pally, S., MacIntosh, G.L., Greeve, M.A., Middleton, S., Jowett, J., Meikle, P.J., 2012. Circulating inflammatory and atherogenic biomarkers are not increased following single meals of dairy foods. European Journal of Clinical Nutrition 66 (1), 25–31.

Newgard, C., An, J., Bain, Jr, Muehlbauer, M., Stevens, R.D., Lien, L., Haqq, A., Shah, S., Arlotto, M., Slentz, C., Rochon, J., Gallup, D., Ilkayeva, O., Wenner, B., Yancy, W., Eisenson, H., Musante, G., Surwit, R., Millington, D., Butler, M.D., Svetkey, L., 2009. A branched-chain amino acid-related metabolic signature that differentiates obese and lean humans and contributes to insulin resistance. Cell Metabolism 9 (4), 311–326.

Nygren-Babol, L., Jagerstad, M., 2012. Folate-binding protein in milk: a review of biochemistry, physiology, and analytical methods. Critical Reviews in Food Science & Nutrition 52 (5), 410–425.

O'Connor, L., Lentjes, M., Luben, R., Khaw, K.-T., Wareham, N., Forouhi, N., 2014. Dietary dairy product intake and incident type 2 diabetes: a prospective study using dietary data from a 7-day food diary. Diabetologia 57 (5), 909–917.

Oozeer, R., Goupil-Feuillerat, N., Alpert, C., van de Guchte, M., Anba, J., Mengaud, J., Corthier, G., 2002. *Lactobacillus casei* is able to survive and initiate protein synthesis during its transit in the digestive tract of human flora-associated mice. Applied and Environmental Microbiology 68 (7), 3570–3574.

Orla-Jensen, S., 1931. Die Abhaingigkeit der Milchsauregarung von der Art und Weise im die Sterilisierung der Nahrboden ausgefuhrt Copenhagen Laiterie, Compt.

Pala, V., Sieri, S., Berrino, F., Vineis, P., Sacerdote, C., Palli, D., Masala, G., Panico, S., Mattiello, A., Tumino, R., Giurdanella, M.C., Agnoli, C., Grioni, S., Krogh, V., 2011. Yogurt consumption and risk of colorectal cancer in the Italian European prospective investigation into cancer and nutrition cohort. International Journal of Cancer 129 (11), 2712–2719.

Parra, M.D., Martinez de Morentin, B.E., Cobo, J.M., Lenoir-Wijnkoop, I., Martinez, J.A., 2007. Acute calcium assimilation from fresh or pasteurized yoghurt depending on the lactose digestibility status. Journal of the American College of Nutrition 26 (3), 288–294.

Parvez, S., Malik, K.A., Ah Kang, S., Kim, H.Y., 2006. Probiotics and their fermented food products are beneficial for health. Journal of Applied Microbiology 100 (6), 1171–1185.

Patrick, V., Nicolas, P., Anurag, A., Raish, O., Denis, G., Rémi, B., Jean-Michel, F., Johan, E.T.V.H.V., Lesley, A.H., Peter, J.W., Ehrlich, S.D., Sean, P.K., 2014. Changes of the human gut microbiome induced by a fermented milk product. Scientific Reports 4.

Qin, L.Q., Xu, J.Y., Han, S.F., Zhang, Z.L., Zhao, Y.Y., Szeto, I.M., 2015. Dairy consumption and risk of cardiovascular disease: an updated meta-analysis of prospective cohort studies. Asia Pacific Journal of Clinical Nutrition 24 (1), 90–100.

Rachmilewitz, D., Katakura, K., Karmeli, F., Hayashi, T., Reinus, C., Rudensky, B., Akira, S., Takeda, K., Lee, J., Takabayashi, K., Raz, E., 2004. Toll-like receptor 9 signaling mediates the antiinflammatory effects of probiotics in murine experimental colitis. Gastroenterology 126 (2), 520–528.

Ralston, R., Truby, H., Palermo, C., Walker, K., 2014. Colorectal cancer and nonfermented milk, solid cheese, and fermented milk consumption: a systematic review and meta-analysis of prospective studies. Critical Reviews in Food Science and Nutrition 54 (9), 1167–1179.

Ralston, R.A., Lee, J.H., Truby, H., Palermo, C.E., Walker, K.Z., 2012. A systematic review and meta-analysis of elevated blood pressure and consumption of dairy foods. Journal of Human Hypertension 26 (1), 3–13.

Rizzoli, R., 2014. Dairy products, yogurts, and bone health. The American Journal of Clinical Nutrition 99 (5), 1256S–1262S.

Rodríguez-Alcalá, L.M., Fontecha, J., 2007. Hot topic: fatty acid and conjugated linoleic acid (CLA) isomer composition of commercial CLA-fortified dairy products: evaluation after processing and storage. Journal of Dairy Science 90 (5), 2083–2090.

Rohde, C.L., Bartolini, V., Jones, N., 2009. The use of probiotics in the prevention and treatment of antibiotic-associated diarrhea with special interest in Clostridium difficile-associated diarrhea. Nutrition and Clinical Practica 24 (1), 33–40.

Romond, M.-B., Ais, A., Guillemot, F., Bounouader, R., Cortot, A., Romond, C., 1998. Cell-free whey from milk fermented with Bifidobacterium breve C50 used to modify the colonic microflora of healthy subjects. Journal of Dairy Science 81 (5), 1229–1235.

Ronco, A.L., De Stefani, E., Dattoli, R., 2002. Dairy foods and risk of breast cancer: a case-control study in Montevideo, Uruguay. European Journal of Cancer Prevention 11 (5), 457–463.

Silva, D., S, M., Rudkowska, I., 2014. Dairy products on metabolic health: current research and clinical implications. Maturitas 77 (3), 221–228.

Sah, B.N.P., Vasiljevic, T., McKechnie, S., Donkor, O.N., 2014. Effect of probiotics on antioxidant and antimutagenic activities of crude peptide extract from yogurt. Food Chemistry 156, 264–270.

Sahni, S., Tucker, K.L., Kiel, D.P., Quach, L., Casey, V.A., Hannan, M.T., 2013. Milk and yogurt consumption are linked with higher bone mineral density but not with hip fracture: the Framingham Offspring Study. Archives of Osteoporosis 8 (1–2), 119.

Sanggaard, K.M., Holst, J.J., Rehfeld, J.F., Sandstr, B., Raben, A., Tholstrup, T., 2004. Different effects of whole milk and a fermented milk with the same fat and lactose content on gastric emptying and postprandial lipaemia, but not on glycaemic response and appetite. British Journal of Nutrition 92 (3), 447–459.

Sayon-Orea, C., Bes-Rastrollo, M., Marti, A., Pimenta, A.M., Martin-Calvo, N., Martinez-Gonzalez, M.A., 2015. Association between yogurt consumption and the risk of metabolic syndrome over 6 years in the SUN study. BMC Public Health 15, 170.

Sarvari, F., Mortazavian, A.M., Fazeli, M.R., 2014. Biochemical characteristics and viability of probiotic and yogurt bacteria in yogurt during the fermentation and refrigerated storage. Applied Food Biotechnology 1 (1), 6.

Savadogo, A., Traore, A.S., 2011. La flore microbienne et les propriétés fonctionnelles des yaourts et laits fermentés. International Journal of Biological and Chemical Sciences 5 (5), 8.

Savaiano, D.A., 2014. Lactose digestion from yogurt: mechanism and relevance. The American Journal of Clinical Nutrition 99 (5 Suppl.), 1251S.

Schooling, C.M., 2014. Milk and mortality. BMJ 349, g6205.

Simmons, S., 2011. Osteoporosis. Nursing 41 (1), 35.

Soedamah-Muthu, S.S., Verberne, L.D., Ding, E.L., Engberink, M.F., Geleijnse, J.M., 2012. Dairy consumption and incidence of hypertension: a dose-response meta-analysis of prospective cohort studies. Hypertension 60 (5), 1131–1137.

Sofi, F., Buccioni, A., Cesari, F., Gori, A., Minieri, S., Mannini, L., Casini, A., Gensini, G., Abbate, R., Antongiovanni, M., 2010. Effects of a dairy product (pecorino cheese) naturally rich in cis-9, trans-11 conjugated linoleic acid on lipid, inflammatory and haemorheological variables: a dietary intervention study. Nutrition Metabolism and Cardiovascular Diseases 20 (2), 117–124.

Sonestedt, E., Wirfalt, E., Wallstrom, P., Gullberg, B., Orho-Melander, M., Hedblad, B., 2011. Dairy products and its association with incidence of cardiovascular disease: the Malmo diet and cancer cohort. European Journal of Epidemiology 26 (8), 609–618.

St-Onge, M.P., Farnworth, E.R., Jones, P.J., 2000. Consumption of fermented and nonfermented dairy products: effects on cholesterol concentrations and metabolism. American Journal of Clinical Nutrition 71 (3), 674–681.

Steffen, L.M., Kroenke, C.H., Yu, X., Pereira, M.A., Slattery, M.L., Van Horn, L., Gross, M.D., Jacobs Jr., D.R., 2005. Associations of plant food, dairy product, and meat intakes with 15-y incidence of elevated blood pressure in young black and white adults: the coronary artery risk development in young adults (CARDIA) Study. American Journal of Clinical Nutrition 82 (6), 1169–1177 quiz 1363–1164.

Sternberg, S.A., Levin, R., Dkaidek, S., Edelman, S., Resnick, T., Menczel, J., 2014. Frailty and osteoporosis in older women—a prospective study. Osteoporosis International 25 (2), 763–768.

Sun, Z., Wang, P.P., Roebothan, B., Cotterchio, M., Green, R., Buehler, S., Zhao, J., Squires, J., Zhu, Y., Dicks, E., Campbell, P.T., McLaughlin, J.R., Parfrey, P.S., 2011. Calcium and vitamin D and risk of colorectal cancer: results from a large population-based case-control study in Newfoundland and Labrador and Ontario. Canadian Journal of Public Health 102 (5), 382–389.

Tamime, A.Y., Robinson, R., 2007. Yoghurt - Science and Technology. Woodhead Publishing.

Tejada-Simon, M.V., Lee, J.H., Ustunol, Z., Pestka, J.J., 1999. Ingestion of yogurt containing *Lactobacillus acidophilus* and Bifidobacterium to potentiate immunoglobulin aresponses to cholera toxin in mice. Journal of Dairy Science 82 (4), 649–660.

The Dairy Council, Macronutrients in yogurt. Retrieved February 17, 2015, from: http://www.milk.co.uk/page.aspx?intPageID=82.

Thompson, W., Holdman, N., Janzow, D., Slezak, J., Morris, K., Zemel, M.B., 2005. Effect of energy-reduced diets high in dairy products and fiber on weight loss in obese adults. Obesity Research 13 (8), 1344–1353.

Tong, X., Dong, J.Y., Wu, Z.W., Li, W., Qin, L.Q., 2011. Dairy consumption and risk of type 2 diabetes mellitus: a meta-analysis of cohort studies. European Journal of Clinical Nutrition 65 (9), 1027–1031.

Tseng, M., Breslow, R.A., Graubard, B.I., Ziegler, R.G., 2005. Dairy, calcium, and vitamin D intakes and prostate cancer risk in the national health and nutrition examination epidemiologic follow-up study cohort. American Journal of Clinical Nutrition 81 (5), 1147–1154.

Tulk, H., Blonski, D.C., Murch, L., Duncan, A., Wright, A., 2013. Daily consumption of a synbiotic yogurt decreases energy intake but does not improve gastrointestinal transit time: a double-blind, randomized, crossover study in healthy adults. Nutrition Journal 12, 87.

Tunick, M.H., Van Hekken, D.L., 2014. Dairy products and health: recent insights. Journal of Agricultural and Food Chemistry 63 (43), 9381–9388 141119123139003.

Turner, K.M., Keogh, J.B., Clifton, P.M., 2015. Red meat, dairy, and insulin sensitivity: a randomized crossover intervention study. The American Journal of Clinical Nutrition 101 (6), 1173–1179.

U.S. Department of Agriculture and U.S. Department od Health and Human Services, 2010. Dietary Guidelines for Americans 2010. U.S. Government Printing Office, Washington (DC).

Uenishi, K., 2006. Prevention of osteoporosis by foods and dietary supplements. Prevention of osteoporosis by milk and dairy products. Clinical Calcium 16 (10), 1606–1614.

de Vrese, M., Stegelmann, A., Richter, B., Fenselau, S., Laue, C., Schrezenmeir, J., 2001. Probiotics–compensation for lactase insufficiency. American Journal of Clinical Nutrition 73 (2 Suppl), 421S–429S.

Van 't Veer, P., van Leer, E.M., Rietdijk, A., Kok, F.J., Schouten, E.G., Hermus, R.J., Sturmans, F., 1991. Combination of dietary factors in relation to breast-cancer occurrence. International Journal of Cancer 47 (5), 649–653.

Vegarud, G.E., Langsrud, T., Svenning, C., 2000. Mineral-binding milk proteins and peptides; occurrence, biochemical and technological characteristics. British Journal of Nutrition 84 (S1), 91–98.

Vonk, R.J., Priebe, M.G., Koetse, H.A., 2003. Lactose intolerance: analysis of underlying factors. European Journal of Clinical Investigation 33, 70–75.

Vors, C., Gayet-Boyer, C., Michalski, M.-C., 2015. Produits laitiers et inflammation métabolique: quels liens en phase postprandiale et à long terme ? Cahiers de Nutrition et de Dietetique 50 (1), 25–38.

Wang, H., Livingston, K.A., Foxb, C.S., Meig, J.B., Jacques, P.F., 2012. Yogurt consumption is associated with better diet quality and metabolic profile in American men and women. Nutrition Research 33 (1), 8.

Wang, H., T, L., Rogers, G.T., Fox, C.S., McKeown, N.M., Meigs, J.B., Jacques, P.F., 2013. Longitudinal association between dairy consumption and changes of body weight and waist circumference: the Framingham Heart Study. International Journal of Obesity (Lond) 38 (2), 299–305.

Wang, L., Manson, J.E., Buring, J.E., Lee, I.M., Sesso, H.D., 2008. Dietary intake of dairy products, calcium, and vitamin D and the risk of hypertension in middle-aged and older women. Hypertension 51 (4), 1073–1079.

Wang, T., Larson, M.G., Vasan, R., Cheng, S., Rhee, E., McCabe, E., Lewis, G., Fox, C.S., Jacques, P., Fernandez, C., O'Donnell, C.J., Carr, S.A., Mootha, V., Florez, J., Souza, A., Melander, O., Clish, C., Gerszten, R., 2011. Metabolite profiles and the risk of developing diabetes. Nature Medicine 17 (4), 448–U483.

Webb, D., Donovan, S.M., Meydani, S.N., 2014. The role of Yogurt in improving the quality of the American diet and meeting dietary guidelines. Nutrition Reviews 72 (3), 180–189.

Wennersberg, M., Smedman, A., Turpeinen, A., Retterstol, K., Tengblad, S., Lipre, E., Aro, A., Mutanen, P., Seljeflot, I., Basu, S., Pedersen, J., Mutanen, M., Vessby, B., 2009. Dairy products and metabolic effects in overweight men and women: results from a 6-mo intervention study. American Journal of Clinical Nutrition 90 (4), 960–968.

WHO, 2004. Scientific Group on the Assessment of Osteoporosis at Primary Health Care Lever. Meeting Report. Brussels, Belgium.

WHO/FAO, 2010. Codex Standard for Fermented Milks. FAO/WHO. CODEX STAN, Rome, pp. 243–2003.

Wirfalt, E., Mattisson, I., Gullberg, B., Olsson, H., Berglund, G., 2005. Fat from different foods show diverging relations with breast cancer risk in postmenopausal women. Nutrition and Cancer 53 (2), 135–143.

Wood, A.D., Strachan, A.A., Thies, F., Aucott, L.S., Reid, D.M., Hardcastle, A.C., Mavroeidi, A., Simpson, W.G., Duthie, G.G., Macdonald, H.M., 2014. Patterns of dietary intake and serum carotenoid and tocopherol status are associated with biomarkers of chronic low-grade systemic inflammation and cardiovascular risk. British Journal of Nutrition 112 (8), 1341–1352.

Zemel, M.B., Thompson, W., Milstead, A., Morris, K., Campbell, P., 2004. Calcium and dairy acceleration of weight and fat loss during energy restriction in obese adults. Obesity Research 12, 8.

Zemel, M.B., Richards, J., Mathis, S., Milstead, A., Gebhardt, L., Silva, E., 2005. Dairy augmentation of total and central fat loss in obese subjects. International Journal of Obesity (Lond) 29 (4), 391–397.

Zhang, X., Beynen, A.C., 1993. Lowering effect of dietary milk-whey protein v. casein on plasma and liver cholesterol concentrations in rats. British Journal of Nutrition 70 (01), 139–146.

Zhu, Y., Wang, H., Hollis, J., Jacques, P., 2015. The associations between yogurt consumption, diet quality, and metabolic profiles in children in the USA. European Journal of Nutrition 54 (4), 543–550.

Zivkovic, A.M., German, J.B., Lebrilla, C.B., Mills, D.A., Klaenhammer, T.R., 2011. Human milk glycobiome and its impact on the infant gastrointestinal microbiota. Proceedings of the National Academy of Sciences of the United States of America 108, 4653–4658.

Chapter 14

Kefir

H. Kesenkaş[1], O. Gürsoy[2], H. Özbaş[3]
[1]Ege University, Izmir, Turkey; [2]Mehmet Akif Ersoy University, Burdur, Turkey; [3]Suleyman Demirel University, Isparta, Turkey

14.1 INTRODUCTION

Since ancient times humans have used lactic acid bacteria (LAB), which can be found widely in nature. These bacteria have traditionally been used to produce fermented milk products such as yogurt, dahi, kishk, kefir, and koumiss. Kefir is a traditional fermented dairy beverage produced by using unique natural microbiota from kefir grains. Although there is no known record in literature about the origin of the first kefir grains or first kefir production (Guzel-Seydim et al., 2010), it has been produced for hundreds of years in the Northern Caucasus Mountains (Lopitz-Otsoa et al., 2006). According to literature, the word kefir, which is derived from the Turkish word "keyif," means "good feeling" (Leite et al., 2013a,b). It is also known in European countries by a variety of names including képhir, kiaphur, kefyr, képher, knapon, kefer, kepi, kippi, and kippe (Stepaniak and Fetlinsk, 2003; Farnworth and Mainville, 2008).

Kefir gained popularity in the second half of the 19th century in the Eastern and Central European countries, and at the end of the same century in the former Soviet Union it started to be produced industrially for the first time (Kesenkaş et al., 2013). Today, commercial production of kefir occurs in many countries like Russia, Poland, Norway, Romania, Germany, Czech Republic, Hungary, and Turkey, and its importance is increasing gradually.

Numerous reports and studies on the putative health benefits of fermented milk products including kefir have shown that beyond basic nutrition, their health benefits may include blood cholesterol-lowering effect, improved lactose utilization, stimulation of the immune system and antioxidant, antimicrobial, antimutagenic, and antitumor activities (Chen, 2005). Thus there is a growing evidence that kefir may contain bioactive ingredients, indicating that this unique fermented milk product is an important probiotic and/or functional food (Farnworth, 2005).

This chapter will review the microbiological and technological properties of kefir produced from grains, as well as the nutritional aspects of kefir, and, finally, it will summarize the studies related to the beneficial effects of kefir on human health and its role in disease prevention.

14.2 KEFIR GRAINS

The main difference between kefir and other fermented milk products is the use of either a starter culture called "kefir grains" or a percolate of the grains to ferment milk (Farnworth and Mainville, 2008). It was reported that traditional kefir grains are obtained from periodic coagulation of cow milk with calf or sheep abomasum (forth stomach) in goatskin bags. After a few weeks a water-insoluble, gel-like and spongy layer is formed on the inner surface of the bags where coagulation occurred. Finally, this layer is divided into small pieces and dried to obtain kefir grains (Varnacı, 1980; Koçak ve Gürsel, 1981).

Kefir grains range in size from 2 to 20 mm or more and resemble small cauliflower florets in shape and color (Fig. 14.1) (Stepaniak and Fetlinsk, 2003; Wszolek et al., 2006). This elastic or gelatinous biological mass is comprised of proteins, lipids, and a water-soluble capsular polysaccharide called kefiran. Several homofermentative *Lactobacillus* species, mainly *Lactobacillus kefiranofaciens,* produce a kefiran complex that surrounds yeasts and bacteria in the kefir grains (Ötleş and Çağındı, 2003; Lopitz-Ostsoa et al., 2006). Kefiran contains D-glucose and D-galactose in a 1:1 ratio and constitutes approximately 25% of dry grain weight (Pogacic et al., 2013). The grains contain 85–90% of water and the dry mass of the fresh grain consists of approximately 57% carbohydrates, 33% protein, 4% fat, and 6% ash (Stepaniak and Fetlinsk, 2003).

FIGURE 14.1 Physical appearance of fresh kefir grains.

The chemical composition of kefir grains originating from Iran contain 81.5% water, 8.6% polysaccharide, and 7.2% protein (Ghasemlou et al., 2012), while kefir grains obtained from Argentina contain 83% water, 9–10% polysaccharides, and 4–5% protein (Abraham and De Antoni, 1999; Garrote et al., 2001). In another study, Liutkevicius and Sarkinas (2004) reported that kefir grains contain 86.3% moisture, 4.5% protein, 1.2% ash, and 0.03% fat.

Kefir grains have a complex microbiological composition and consist of a mixture of lactococci, *Leunconostoc* spp., thermophilic and mesophilic lactobacilli, yeasts (lactose$^+$ and lactose$^-$), and acetic-acid bacteria (Kesenkaş and Kinik, 2010). The endogenous microbiota of kefir grains and the ratio between above-mentioned microbial groups strongly depends on the origin of the grain, the local culture, and the storage and manipulation processes (Witthuhn et al., 2004). According to recent scientific sources, this physical association arose from more than 50 various microorganism species (Pogacic et al., 2013). The most common lactobacilli isolated from kefir grains are *Lactobacillus kefiri*, *Lactobacillus kefiranofaciens*, *Lactobacillus kefirgranum*, *Lactobacillus parakefiri*, *Lactobacillus delbrueckii*, *Lactobacillus acidophilus*, *Lactobacillus brevis*, *Lactobacillus helveticus*, *Lactobacillus casei*, *Lactobacillus paracasei*, *Lactobacillus fermentum*, *Lactobacillus plantarum*, and *Lactobacillus gasseri*, whereas yeasts are represented by the species *Kluyveromyces marxianus*, *Kluyveromyces lactis*, *Saccharomyces cerevisiae*, *Torulaspora delbrueckii*, *Candida kefir*, *Pichia fermentans*, *Kazachstania unispora*, and *Kazachstania exigua* (Vardjan et al., 2013). Leite et al. (2012) reported that *Lactobacillus kefiranofaciens* and *Lactobacillus kefiri* were the major bacterial populations in Brazilian kefir grains, and the yeast community was dominated by *S. cerevisiae*. Similarly, Nalbantoglu et al. (2014) reported that the most abundant genus *Lactobacillus* was mainly represented by three species: *Lactobacillus kefiranofaciens*, *Lactobacillus buchneri*, and *Lactobacillus helveticus* in Turkish kefir grains.

The distribution of microorganisms within the matrix of a kefir grain is not constant. The rod-shaped LAB-like *Lactobacillus kefir* dominates at outer layer of kefir grain, whereas the population of *Lactobacillus kefiranofaciens* is distributed throughout the grain but mostly in the center. Although the yeasts are usually located at the core, the lactose-fermenting ones are located mainly in the peripheral layers of the grains (Stepaniak and Fetlinsk, 2003; Sarkar, 2008). Moreover, the viable counts of yeasts and bacteria are relatively equal at the intermediate zone of the grain and a progressive change according to the distance from the core has been reported (Özer, 2015).

There are several metabolites formed during kefir fermentations by the above-mentioned microorganisms, which give its unique physical, chemical, and sensory characteristics. Homofermentative LAB primarily produce lactic acid while heterofermentative ones produce lactic acid along with CO_2 from the fermentation of lactose. The citrate-positive strains of *Lactococcus lactis* produce diacetyl, acetaldehyde, ethanol, and acetate while *Leuconostoc mesenteroides* ssp. *mesenteroides* and *Leuconostoc mesenteroides* ssp. *cremoris* produce diacetyl, ethanol, and acetate by the utilization of citrate (Rattray and O'Connell, 2011). The yeasts in kefir grains are primarily responsible for the production of ethanol and CO_2 by the fermentation of lactose (Konar and Sahan, 1991). Although it comprises about 1% of the total viable counts in the grain, *Acetobacter* may play an important role in improving the taste and consistency of the product (Stepaniak and Fetlinski, 2003).

14.3 KEFIR PRODUCTION

There are several ways of making kefir: (i) kefir produced by fermented milk with grains (traditional production), (ii) the commercial process by the Russian or European method, and (iii) kefir produced by using commercial starter cultures directly inoculated into the milk (industrial production) (Fig. 14.2) (Ötles

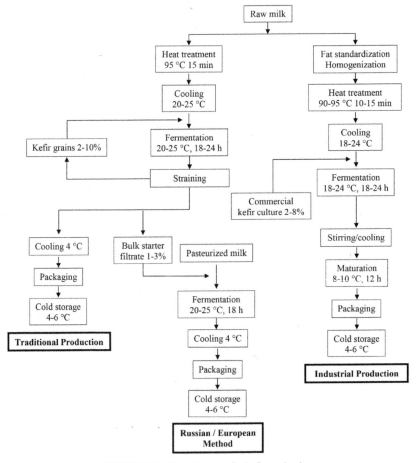

FIGURE 14.2 Flow diagram for kefir production.

and Çağındı, 2003; Farnworth, 2005; Sarkar, 2008; Rattray and O'Connel, 2011; Leite et al., 2013a,b).

The quality of raw milk is crucial for kefir production, like in other fermented milk products. The raw milk for kefir production must fulfill the following requirements: rich composition, low bacterial and somatic cell counts, and absence of pathogens or inhibitory substances like antibiotic and disinfectant residues. Although kefir can be produced with milk from different breeds (cow, goat, and sheep), full-fat, low-fat, or skimmed cow milk is preferred for industrial production (Kesenkaş and Kinik, 2010). Moreover, production of kefir from soy and oat milk has also been investigated (Liu and Lin, 2000; Kahraman, 2011; Kesenkaş et al., 2011a). During the heat treatment of milk used for kefir production, adequate temperature and time (eg, 95°C, 10–15 min) are used to denature whey proteins completely. These denatured whey proteins, which have high water-holding capacity, prevent the separation of water in the final product and increase the consistency (Kesenkaş et al., 2013). Additionally, heat treatment renders milk a better growth medium for grain microbiota due to the formation of some amino acids and other growth factors, reduces redox potential, eliminates inhibitory substances, and prevents hydrolytic rancidity through inactivation of lipase enzyme (Sarkar, 2008).

In traditional production, kefir grains (2–10%) are directly added to the heat-treated milk, but kefir production with 5% kefir grains has proven to be optimum for ethanol and volatile acid production (Sarkar, 2008). In kefir production the milk ratio is critical to obtain a product with consistent quality because it affects the pH, viscosity, final lactose concentration, CO_2 content, and the microbiological profile of the final product (Leite et al., 2013a,b). Garrote et al. (1998) recommended a ratio of 1% if a viscous and not very acidic product is desired and reported that a ratio of 10% creates an acid beverage with low viscosity and a more effervescent taste.

Following removal of kefir grains, a cooling step is carried out in traditional production but the pH is adjusted to 4.5–4.6 before cooling in industrial production. Unlike in the traditional method, the maturation step can be performed in industrial production by holding kefir at 8–10°C for up to 12 h. The pH value of kefir decreases to 4.3–4.4 at the end of maturation. The purpose of the maturation step is to allow the correct growth of microorganisms, primarily yeasts, within kefir, which contribute to the specific flavor of the product (Rattray and O'Connel, 2011; Leite et al., 2013a,b). Whey proteins also absorb higher amounts of water during the maturation period creating a more consistent and viscous product. During packaging, appropriate mechanical systems must be used to prevent the quality of the kefir.

14.4 CHEMICAL COMPOSITION OF KEFIR

The composition of kefir differs significantly due to the type and fat content of milk, the composition of grains or cultures, and the production methods. The typical composition of kefir consists of 89–90% moisture, 0.2% lipid, 3.0% protein, 6.0% sugar, 0.7% ash, and 1.0% lactic acid and alcohol (Sarkar, 2007). Kefir has been reported to contain 1.98 g/L of CO_2, 0.48% alcohol, and the amount of CO_2 increased (201.7–277.0 mL/L) with raised inoculation ratio of grains (1–10%) (Arslan, 2014). Kesenkaş et al. (2011b) reported that total solids, fat, protein, and lactose content of cow milk kefirs produced from grains or kefir culture were 11.70–11.62%, 3.05–3.10%, 4.10–4.16%, and 3.89–3.92%, respectively. The chemical composition of kefir according to literature is summarized in Table 14.1.

In addition to the major chemical components, kefir contains various flavor compounds such as acetaldehyde, acetoin, diacetyl, and fermentation metabolites like acetic acid, pyruvic acid, hippuric acid, propionic acid, and butyric acid (Guzel-Seydim et al., 2000; Beshkova, 2003). Kefir also contains several vitamins, macroelements including K, Ca, Mg, P and microelements such as Cu, Zn, Fe, Mn, Co, and Mo (Sarkar, 2007; Özer, 2015). However, continuous metabolic activities of kefir microbiota during storage and storage conditions can affect the chemical composition of kefir.

14.5 NUTRITIONAL CHARACTERISTICS OF KEFIR

The basic nutrients of kefir are similar to that of the milk from which it is produced and thus it has a high nutritional value as a source of protein and calcium (Farnworth and Mainville, 2008). Moreover, the fermentation process during kefir production can improve the nutritional value of kefir by increasing the quantity, availability, digestibility, and assimilability of nutrients.

During fermentation and storage the amino acid profile of kefir changes and the number of free amino acids like lysine, proline, cysteine, isoleucine, phenylalanine, and arginine increases (Farnworth and Mainville, 2008). Guzel-Seydim et al. (2003) reported higher amounts of threonine, serine, alanine, lysine, and ammonia in kefir samples than in milk. According to Irigoyen et al. (2012), the essential amino acid phenylalanine is the only amino acid whose content was higher in kefir than in yogurt and in a commercial fermented milk. Gul et al. (2015) showed that glutamic acid was the major amino acid in kefir samples made from cow and buffalo milk fermented by grains or starter cultures. Similar results were obtained by Grønnevik et al. (2011) in five commercial Norwegian kefir samples. The authors suggested that glutamic acid can be formed from proteolytic activity, from deamination of glutamine, or from transamination of other amino acids as the preferred amino group acceptor α-ketoglutarate is converted to glutamic acid. The degree of protein hydrolysis and, consequently, the amount of free amino acids in kefir depend on the grain-to-milk ratio and the incubation temperature (Ozer, 2015). The activities of proteolytic enzymes, especially those from yeasts, are also maintained throughout the shelf-life of kefir so the concentration of free amino acids increases during the cold storage.

Vitamins and minerals play significant roles in many metabolic functions, including growth, repair, and development of tissues, fluid and electrolyte balance, maintenance of bone health, adequate immune function, and even

TABLE 14.1 Chemical Composition of Kefir

		References
Protein		
Culture-fermented grains (cow milk)	3.54%	Şatır (2011)
Grain-fermented, 24 h	3.47%	Kök-Taş et al. (2013)
Fat		
Culture-fermented grains (cow milk)	3.46%	Şatır (2011)
Lactose		
Culture-fermented, 24 h	4.02%	Fontan et al. (2006)
Brazilian kefir, 24 h	32.4 mg/mL	Leite et al. (2013a,b)
Ethanol		
Norwegian kefir, 8 weeks	890 mg/kg	Grønnevik et al. (2011)
Grain-fermented, 24 h	0.34%	Yılmaz et al. (2006)
Grain-fermented, 24 h	111.3 mg/L	Kök-Taş et al. (2013)
Culture-fermented, 24 h	0.005%	Fontan et al. (2006)
Brazilian kefir, 28 d	1.36 mg/mL	Leite et al. (2013a,b)
Lactic Acid		
Norwegian kefir, 8 weeks	~0.8%	Grønnevik et al. (2011)
Grain-fermented, 24 h	0.89%	Kök-Taş et al. (2013)
Brazilian kefir, 24 h	7.38 mg/mL	Leite et al. (2013a,b)
Carbon dioxide		
Grain-fermented, 24 h	1.32 g/L	Yılmaz et al. (2006)
Grain-free, 24 h	0.65 g/L	Clementi et al. (1989)

genetic regulation (Chryssanthopoulos and Maridaki, 2010). In addition to the vitamins present in milk, it has also been reported that microorganisms can produce vitamins, such as cobalamin menaquinone (vitamin K_2), riboflavin (vitamin B_2), and thiamine (vitamin B_1) in fermented dairy products (O'Connor et al., 2005). Furthermore, fermentation has been reported to increase the folic-acid content in a variety of dairy products including kefir

(Kim and Oh, 2013). Kefir can contain group B vitamins such as B_1, B_2, and B_5 and vitamins A, E, C, and D in amounts of <1, <0.5, 0.3, 60, 0.11, 1, and 0.8 mg/100 g, respectively (Liutkevicius and Sarkinas, 2004; Chryssanthopoulos and Maridaki, 2010). Kneifel and Mayer (1991) reported that the vitamin content of kefir is affected by the type of milk used. Kefir enriched by >20% in pyridoxine was obtained when ewe, goat, and mare milk were used for its preparation, while folic-acid content increased by when using ewe, cow, and goat milk. Bonczar et al. (2004) determined that vitamin C content of kefir samples produced with ewe milk ranged from 1.70 to 2.27 mg/100 g during 21 days of storage. They reported that vitamin C content in kefir samples was lower than in milk after 14 days of storage. However, Magalhaes et al. (2011) did not detect vitamin C in Brazilian kefir samples.

The type of lactic acid formed in fermented dairy products has some physiological importance. Two stereoisomers of lactic acid can be found in kefir grains by the action of homofermentative and heterofermentative microorganisms. Both L(+) and D(−) isomers are absorbed in the gastrointestinal tract but differ in the proportions converted to glucose or glycogen in the body. The L(+) isomer is rapidly and completely converted to glycogen, whereas the D(−) isomer is metabolized at a slower rate and a significant quantity is excreted in urine (Uniacke-Lowe, 2011). Kefir contains almost exclusively L(+) lactic acid, whereas yogurt usually contains 45–60% L(+) lactic acid and 40–55% D(−) lactic acid (Farnworth and Mainville, 2008). Fontan et al. (2006) reported that at the end of 24 h of fermentation, the concentration of D(−) and L(+) lactic acid in kefir was 0.11% and 0.76%, respectively.

14.6 HEALTH BENEFITS OF KEFIR

14.6.1 Anticarcinogenic and Antimutagenic Effects

The anticarcinogenic role of kefir and its cell-free extracts can be related to the prevention of cancer and retardation of tumor growth by (i) induction of apoptotic mechanisms in cancer cells (Rizk et al., 2013; Ghoneum and Gimzewski, 2014), (ii) antiproliferative effect on cancer cells (Maalouf et al., 2011), (iii) activation of immune system (Liu et al., 2002), (iv) antioxidant effect (Cenesiz et al., 2008; Grishina et al., 2011), (v) inhibiting enzyme activities that convert procarcinogenic compounds to carcinogens (Sarkar, 2007), and (vi) binding of mutagens by kefir microbiota (Ahmed et al., 2013). The anticarcinogenic effect of kefir and its cell-free extracts was studied for different cancer types such as hematological cancers (leukemias and lymphoma), breast cancer, gastrointestinal system cancers (gastric and colorectal), and sarcoma (connective tissue tumor). Adult lymphoblastic leukemia is a malignancy that occurs in white blood cells and the overall cure rate in adults is only 40% (Maalouf et al., 2011). There are some promising findings concerning the curative role of cell-free extracts of kefir in adults with lymphoblastic leukemia. In an in vitro study, downregulation of tumor necrosis factor-α (TGF-α) by cell-free fraction of kefir was detected in HTLV-1 virus-infected human leukemia cell line (Rizk et al., 2009). Similar results were obtained by Maalouf et al. (2011), who found that cell-free fraction of kefir exhibited a significant antiproliferative effect on HTLV-1 negative malignant T-lymphocytes by downregulation of TGF-α and upregulation TGF-β1 mRNA expression. Ghoneum and Gimzewski (2014) examined the apoptotic effect of a kefir grain product, which was a natural mixture composed primarily of *Lactobacillus kefiri* P-IF, on human multidrug-resistant myeloid leukemia (HL60/AR) cells in vitro. Kefir-based product-induced apoptosis in HL60/AR cells and the maximum apoptosis rate was found to be 67.5% by treatment with 5 mg/mL kefir grain products for 3 days. The prospective mechanism for the apoptotic effect of kefir was reported to be the formation of piercing holes in the cellular membrane of cancer cells. De Moreno de LeBlanch et al. (2007) found that 2 days of administration of kefir and kefir cell-free fraction delayed breast tumor growth and increased the number of IgA(+) cells in the mammary glands of mice. Two days of administration of cell-free fraction of kefir also showed significant increases in apoptotic cell number and decreases in Bcl-2(+) cells, which is a type of cancer cells, in the mammary gland. Gao et al. (2013) investigated the antiproliferative effect of cell-free fraction of Tibetan kefir on human gastric cancer line SGC7901 in vitro. They found a significant antiproliferative effect when cancer cells were treated with 10 mg/mL of cell-free fraction of kefir for 24h. Cell-cycle regulation and apoptosis were also detected in gastric cancer lines treated with the cell-free kefir fraction. Khoury et al. (2014) found that kefir was able to inhibit the proliferation and induce apoptosis in HT-29 and Caco-2 colorectal cancer cells. Liu et al. (2002) reported that oral administration of kefir and soymilk kefir to mice with sarcoma, which is one of the connective tissue tumors, resulted in 64.8% and 70.9% inhibition of tumor growth respectively, compared with control groups.

UV radiation generates reactive oxygen species like singlet oxygen ($^1O_2$), superoxide (O_2^-), and hydroxyl radicals (•OH) (Nagira et al., 2002), causing oxidative DNA damage, which is involved in the cytotoxic, mutagenic, and carcinogenic changes in cells (Keibassa et al., 1997; Sinhaa and Häder, 2002). It has been suggested that kefir supernatants can significantly suppress DNA damage in different cell lines (Nagira et al., 2002; Grishina et al., 2011). Kefir extract also has the potential for stimulation of unscheduled DNA synthesis and suppressing UVC-induced apoptosis of HMV-1 cells (Nagira et al., 2002).

14.6.2 Effects on Immune System

There are numerous in vitro (Iraporda et al., 2014) and in vivo (Vinderola et al., 2006a; Hong et al., 2009; Adiloğlu et al., 2013) studies on the immunological effect of kefir and kefir fractions. Immunomodulatory effects caused by kefir and kefir products may be attributed to several components such as LAB, yeasts, exopolysaccharides, organic acids, and bioactive peptides. Kefir contains some bioactive peptides that have immunomodulatory effects (Ebner et al., 2015). Kefir supernatant may be able to promote cell-mediated immune responses against tumors and also against intracellular pathogenic infections (Hong et al., 2009). It has been demonstrated that the number of IgA+ cells on histological slices of small intestine of mice fed with kefir or pasteurized kefir significantly increased (Vinderola et al., 2005). Similar findings were obtained by Vinderola et al. (2006b), who observed the increase of IgA+ in small and large intestine lamina propria by administration of exopolysaccharide produced by *Lactobacillus kefiranofaciens*. De Moreno de LeBlanch et al. (2006) reported that kefir and kefir cell-free fraction increased interleukin 10 (IL-10) and decreased interleukin 6 (IL-6) (IL-6 is proangiogenic factor in breast tumor tissue) in mice with breast cancer. Vinderola et al. (2006a) demonstrated that supernatant of kefir induced the production of higher levels of TNFα, IL-6, and, especially, IL-10 on cells isolated from Peyer's patches. Due to the major role of IL-10 in the regulatory network of cytokines controlling mucosal tolerance, these results suggest that kefir could be a potential therapeutic fermented food product for inflammation in chronic irritable bowel disease. In an in vivo study, Adiloğlu et al. (2013) studied the serum cytokine profiles of healthy volunteers after consumption of 200 mL/day of kefir for 6 weeks and determined that oral kefir intake increased polarization of the immune response toward TH1 type and decreased TH2 type response and, accordingly, the allergic response. According to these researchers, the decrease in IL-8 level due to kefir consumption might control the inflammatory response by suppressing neutrophil chemotaxis and activation. Iraporda et al. (2014) reported that nonbacterial fraction of kefir was responsible for the modulation of certain innate immune epithelial response. Lactate from the kefir cell-free fraction inhibited the interleukin-1β, TNFα, or flagellin-induced inflammation in intestinal epithelial cells. The authors reported that lactate was a major component with immunomodulatory properties in intestinal epithelial cells. Yeasts (21 strains from *Saccharomyces, Kluyveromyces,* and *Issatchenkia*) isolated from kefir showed a high capacity to inhibit intestinal epithelial innate response triggered by different proinflammatory pathways (Romanin et al., 2010).

14.6.3 Antiinflammatory Effects

Several studies have demonstrated that kefir and its polysaccharide extract has antiinflammatory effects. Rodrigues et al. (2005a,b) studied the antiinflammatory effects of kefiran and kefir samples thawed and cultured during 15 days both in a molasses solution (50 g/L) and cow milk via cotton-induced granuloma and paw edema assays in Wistar rats. Kefir suspensions (both in molasses and milk) and kefiran in aqueous solution reduced the inflammatory process of granuloma formation in rats after 6 days of treatment. The inhibition of inflammation was more effective for the suspensions (~41% for molasses and ~44% for milk fermentation) than kefiran (~34%) in the cotton-pellet test. The authors also found that orally consumed kefir suspensions were significantly effective at decreasing the inflammation induced by carrageenan, dextran, and histamine in Wistar rats (Rodrigues et al., 2005a,b). Kwon et al. (2008) showed that intragastric administration of kefiran in an ovalbumin-induced asthma mouse model, with airway inflammation and airway hyperresponsiveness, effectively reduced ovalbumin-induced cytokine production, pulmonary eosinophilia, mucus hypersecretion, and airway hyperresponsiveness. They concluded that kefiran may have the potential to block the airway inflammation in allergic asthma patients. Similar results were obtained by Lee et al. (2007), who found that kefir (50 mg/kg administered by intragastric method) effectively suppressed allergic inflammation and airway hyperresponsiveness in a mouse model of asthma. They observed that these effects of kefir were related to decreased activity of the bronchoalveolar lavage fluid IgE and cytokines. Wang et al. (2012) recently demonstrated that kefir decreased the proinflammatory signaling proteins (TLR4 and p-NF$_\kappa$B) in ovalbumin-induced allergy BALB/c mice. The antiinflammatory activity of the isolated fraction of carbohydrates from sugary kefir was tested in vivo by Moreira et al. (2008). Researchers found that carbohydrate fraction of sugary kefir showed significant inhibitory activity by 30% and 54% for carrageenan- and dextran-induced inflammation, respectively. Yasar et al. (2013) showed an antiinflammatory effect of kefir especially in the early phase of caustic injury in an experimental corrosive esophagitis model of rats.

14.6.4 Hypocholesterolemic Effects

High serum cholesterol levels (>240 mg/dL) increase the risk of cardiovascular diseases (Gürsoy et al., 2011). Studies concerning the effects of probiotic LAB on serum cholesterol levels increased after the first observation of Mann and Spoerry (1974) about the serum cholesterol-lowering effects of fermented milk. Although some contradictory results have been observed in the past (De Roos et al., 1998; St-Onge et al., 2002), most of the results both from in vitro and in vivo studies indicated that fermented dairy products including kefir have a hypocholesterolemic effect, which can be partially attributed to LAB present in these products. The possible mechanisms for hypocholesterolemic activity of LAB have been proposed to involve (1) assimilation of cholesterol by bacterial cells (Gilliland et al., 1985), (2) inhibition of exogenous cholesterol absorption from small intestine by binding of cholesterol with bacterial cells (Noh et al., 1997), (3) suppression of bile acid reabsorption by deconjugation of bile salt hydrolase activity (De Smet et al., 1998), (4) conversion of cholesterol to coprostanol (Lye et al., 2010), and (5) production of short-chain fatty acids (Pereira et al., 2003). Vujicic et al. (1992) reported that kefir grains assimilated cholesterol in milk. The initial cholesterol concentration (12.8 mg/dL) of milk was reduced to 10.8 and 5.3 mg/dL after 24 and 48 h fermentation of milk by kefir grains, respectively. Maeda et al. (2004a) reported that kefiran from *Lactobacillus kefiranofaciens* significantly reduced the serum cholesterol levels in rats that consumed excessive dietary cholesterol. Another study with rabbits fed 30 mL/kg/body weight/day of kefir resulted in reduced levels of serum cholesterol levels (Güven and Güven, 2005). Mice fed a cholesterol-enriched diet supplemented with *Lactobacillus plantarum* MA2 isolated from Tibet kefir grains showed lower serum total cholesterol, low-density lipoprotein (LDL) cholesterol, and triglycerides levels, and no change in high-density lipoprotein (HDL) cholesterol (Wang et al., 2009). Similar results were obtained by Huang et al. (2013a) who found a significant reduction in serum total cholesterol, triglyceride, and LDL cholesterol levels as well as cholesterol and triglyceride levels in liver of rats fed a high-cholesterol diet supplemented with *Lactobacillus plantarum* Lp09 and Lp45 with bile-salt hydrolase activity isolated from kefir grains. Other *Lactobacillus* stains with cholesterol-reducing effects such as *L. plantarum* Lp27 (Huang et al., 2013b), *L. plantarum* B23, *Lactobacillus acidophilus* LA15, and *Lactobacillus kefiri* D17 (Zheng et al., 2013) were also isolated from Tibetan kefir grains. Liu et al. (2012) also demonstrated that the bile-salt hydrolase activity of *Kluyveromyces marxianus* strains K1 and M3 isolated from Tibetan kefir was the main responsible mechanism for their cholesterol-reducing activity.

14.6.5 Antihypertensive Effects

Milk proteins are a source of various biologically active peptides, which are formed enzymatically either by digestive enzymes or during milk fermentation (Kinik and Gursoy, 2002). These peptides have been shown to have different bioactivities including inhibition of angiotensin I-converting enzyme (ACE) (FitzGerald et al., 2004). The major milk-derived peptides are ACE inhibitors (Hayes et al., 2007). ACE (peptidyldipeptide hydrolase, EC 3.4.15.1) is a key enzyme involved in regulation of blood pressure (FitzGerald and Meisel, 2000). It converts angiotensin I to potent vasoconstrictor angiotensin II. Angiotensin II also takes part of the kinin–kalicrein system as it hydrolyses bradykinin, which has a vasodilator action (Quirós et al., 2005; Hernández-Ledesma et al., 2011). Quirós et al. (2005) identified two potential ACE inhibitory peptides from commercial caprine kefir. In a recent study performed by Ebner et al. (2015), 236 peptides uniquely formed during kefir production were identified. The authors found 12 peptides with ACE inhibitory activity from β-kazein and αs2-kazein. Research on the effects of kefiran in animals demonstrated that it significantly reduces blood pressure (Maeda et al., 2004a). In another study, the role of kefiran (100 and 300 mg of kefiran/kg of rat) produced by *Lactobacillus kefiranofaciens* WT-2BT on the blood pressure of rats was investigated. Researchers showed that ACE activities in the serum and thoracic aorta of kefiran-fed rats were significantly lower than those in the control group (Maeda et al., 2004b). Grønnevik et al. (2011) indicated that γ-aminobutyric acid (GABA), the decarboxylation product of glutamic acid, which has a blood pressure-lowering effect, continuously increased in Norwegian kefir for 8 weeks of storage. It appears that low molecular weight peptides, kefiran, and GABA play significant roles in the antihypertensive effect of kefir.

14.6.6 Antimicrobial Activity

Kefir has been reported to have antimicrobial activity on a wide variety of Gram-positive and Gram-negative bacteria (Powell, 2006). Pathogens inhibited by kefir, kefiran, kefir suspensions, or bacterial isolates from kefir include *Salmonella typhi* (Suriasih, 2011; Ismaiel et al., 2011), *Salmonella typhimurium* (Rodrigues et al., 2005), *Salmonella enteriditis* (Ulusoy et al., 2007; Carasi et al., 2014), *Escherichia coli* (Ulusoy et al., 2007; Suriasih, 2011), *Yersinia enterocolitica* (Gulmez and Guven, 2003), *Shigella sonnei* (Silva et al., 2009), *Bacillus cereus* (Kakisu et al., 2007; Carasi et al., 2014), *Klebsiella pneumoniae* (Ismaiel et al., 2011), *Staphylococcus aureus* (Chifiriuc et al., 2011; Carasi et al., 2014), *Staphylococcus epidermidis* (Topuz et al., 2008), *Pseudomonas aeruginosa* (Rahimzadeh et al., 2011; Carasi et al., 2014), and *Listeria monocytogenes* (Ulusoy et al., 2007; Rodrigues et al., 2005a,b). Studies have also demonstrated that kefir, kefiran, kefir grains, and water kefir possess antifungal activity against *Candida albicans* (Rodrigues et al., 2005a,b; Silva et al., 2009), *Fusarium graminearum* (Ismaiel et al., 2011), *Torulopsis glabrata*, *Microsporum nanum*, *Trichopyton mentagrophytes*, *Trichopyton rubrum* (Cevikbas et al., 1994), and *Aspergillus ochraceus* (Vélez and Peláez, 2014). The main responsible metabolites for the antimicrobial activity of kefir are undissociated organic acids (lactic and acetic acid), carbon dioxide, hydrogen peroxide, ethanol, and diacetyl (Guzel-Seydim et al., 2010). LAB isolated from kefir also produce some bacteriocins (Ryan et al., 1996; Powell et al., 2007; Miao et al., 2014), which could be partly responsible of the antimicrobial activity in a large spectrum of bacterial strains (Chifiriuc et al., 2011). These bioactive compounds may act independently or synergically to produce the antimicrobial activity. Several investigations have indicated that the antimicrobial activity of kefir and the cell-free extracts from kefir were very efficient in acidic environments (Takahashi and Kawakami, 1999; Raja et al., 2009) and ineffective at neutral (Takahashi and Kawakami, 1999) and alkaline pH (Raja et al., 2009). Past and recent studies have also shown that microorganisms and bioactive compounds of kefir have important potential beneficial effects for modulation of gastrointestinal microbiota (Marquina et al., 2002), treatment and prevention of gastrointestinal infections (Bolla et al., 2013; Chen et al., 2013; Kakisu et al., 2013), and oral health (Cogulu et al., 2010).

14.6.7 Antidiabetic Effects

Hyperglycemia, a typical symptom in type-II insulin-independent diabetes mellitus, is a condition characterized by an abnormal postprandial increase of blood glucose level (Kim et al., 2011). Digestion of dietary carbohydrates is facilitated by enteric enzymes, including α-glucosidases (Ortiz-Andradea et al., 2007). Inhibition of α-glucosidases can significantly decrease the postprandial increase of blood glucose level after food intake and can therefore be a useful strategy in the management of type-II diabetes (Kwon et al., 2006; Ortiz-Andradea et al., 2007; Kim et al., 2011). On the other hand, increased oxidative stress plays an important role in the occurrence and development of diabetes mellitus (Teruya et al., 2002). Kwon et al. (2006) showed that α-glucosidase inhibitory activity increased moderately in soy milk (supplemented with *Rhodiola crenulata* phenolic extracts) fermented by kefir culture. The authors also reported that beneficial phenolics from *Rhodiola crenulata* could be effectively mobilized through fermentation of soy milk by kefir cultures. This approach has been used by different investigations (Kwon et al., 2006; Kim et al., 2011) and could be used for complementary therapies for postprandial hyperglycemia linked to management of type-II diabetes. Teruya et al. (2002) observed that Kefram-kefir, which is a fermented milk originating from the Caucasus Mountains, increased glucose uptake in L6 myotubes of skeletal muscle cells. Results of this study showed that kefir has important potential for decreasing insulin resistance in type-II diabetic individuals. In a recent study by Punaro et al. (2014), kefir treatment (1.8 mL/day) during 8 weeks significantly reduced the progression of streptozotocin-induced hyperglycemia and oxidative stress in rats. Kefir has a low-to-moderate glycemic index and its insulinemic index is high (Kong and Hendrich, 2012). Therefore kefir consumption should be recommended for individuals with diabetes.

14.6.8 Kefir and Lactose Intolerance

Lactose is hydrolyzed by the enzyme β-galactosidase, bound to the mucosal membrane of the small intestine, producing monosaccharides that are absorbed in the intestine (Gursoy et al., 2014). Insufficient intestinal β-galactosidase activity, called lactose intolerance, does not allow hydrolysis of lactose into galactose and glucose (Alm, 2002). Lactobacilli (eg, *Lactobacillus brevis*, *Lactobacillus kefyr*, and *Lactobacillus buchneri*) in kefir grains are the main responsible microorganisms for β-galactosidase activity of kefir. Lactose-fermenting yeasts (eg, *Candida kefyr*) show lower β-galactosidase activity (De Vrese et al., 1992). To the best of our knowledge, there are only two studies that directly evaluated the effect of kefir on lactose digestion. In the first study, the effect of microbial β-galactosidase activity on intestinal lactose digestion was evaluated directly by following postprandial venous plasma galactose concentrations (De Vrese et al., 1992). Researchers observed that kefir with β-galactosidase activity resulted in 30% enhancement of the mean postprandial plasma galactose peak concentration. The second study showed that kefir reduced the perceived severity of flatulence, which is the most common reported symptom in people with lactose intolerances, by 71% in adults relative to milk (Hertzler and Clancy, 2003). On the basis of the above findings it can be concluded that kefir significantly improves lactose digestion and decrease lactose-intolerance symptoms.

14.6.9 Kefir and Osteoporosis

Osteoporosis is the most common metabolic bone disease in the world. It is characterized by low bone mass and increased fragility of bones (Wood, 2003; Huth et al., 2006). It is estimated that an osteoporotic fracture or a vertebral fracture occurs every 3 and 22 s around the world, respectively (Johnell and Kanis, 2006). Low bone mineral density is a significant risk factor for both osteoporosis and fractures (Wood, 2003; Gursoy et al., 2013). Adequate calcium intake is essential for achieving peak bone mass and for its subsequent sustenance. It is also crucial to the prevention and treatment of osteoporosis (Gursoy et al., 2013). Chen et al. (2015) studied the effects of a 12-week treatment with kefir on bone turnover markers (including bone volume, trabecular number, thickness, bone mass, bone strength, and biochemical properties) of distal femurs in ovariectomized rats. Daily oral administration of kefir in adult female ovariectomized rats prevented estrogen deficiency-induced bone loss. Kefir consumption significantly increased bone mass density, average cortical elastic moduli, harness, bone volume, thickness, and trabecular number. They also found that in an in vitro assay that kefir increased intracellular calcium uptake in Caco-2 cells through TRPV6 calcium channels. According to these results, the authors concluded that kefir has important potential as an alternative treatment for postmenopausal osteoporosis.

14.7 CONCLUSIONS

Kefir is a complex probiotic fermented milk beverage and has traditionally been regarded as a nutritious and healthy dairy food. Traditionally, kefir is produced by kefir grains, which have a complex microbiological composition in a matrix of polysaccharides and proteins. Kefir grains consist of a mixture of lactic acid bacteria, yeasts, and acetic-acid bacteria. The nutritional benefits of kefir are related to nutrients such as carbohydrates, proteins, minerals, and vitamins. The health effects caused by kefir and kefir products may be attributed to several components such as lactic acid bacteria, yeasts, exopolysaccharides, organic acids, antioxidants, and bioactive peptides. The reported health benefits of kefir include anticarcinogenic, antimutagenic, antiinflammatory, antimicrobial, antihypertensive, and antidiabetic activity, immune modulation effects, enhanced lactose utilization, and hypocholesterolemic effects. Kefir also has important potential as an alternative treatment for osteoporosis. Although there are some evidence and promising results concerning the health effects of kefir and kefir products, further clinical studies are needed to clarify those therapeutic and nutraceutical activities in human.

REFERENCES

Abraham, A.G., De Antoni, G.L., 1999. Characterization of milky kefir grains grown in milk and in soy milk. Journal of Dairy Research 66, 327–333.

Adiloğlu, A.K., Gönülateş, N., İşler, M., Şenol, A., 2013. The effect of kefir consumption on human immune system: a cytokine study. Mikrobiyoloji Bulteni 47, 273–281.

Ahmed, Z., Wang, Y., Ahmad, A., Khan, S.T., Nisa, M., Ahmad, H., Afreen, A., 2013. Kefir and health: a contemporary perspective. Critical Reviews in Food Science and Nutrition 53, 422–434.

Alm, L., 2002. In: Roginsky, H., Fuquay, J.W., Fox, P.F. (Eds.), Encyclopedia of Dairy Sciences. Academic Press, London, pp. 1533–1537.

Arslan, S., 2014. A review: chemical, microbiological and nutritional characteristics of kefir. CyTA – Journal of Food. http://dx.doi.org/10.1080/19476337.2014.981588.

Beshkovaa, D.M., Simovaa, E.D., Frengovaa, G.I., Simovb, Z.I., Dimitrovc, Zh.P., 2003. Production of volatile aroma compounds by kefir starter cultures. International Dairy Journal 13, 529–535.

Bolla, P.A., Carasi, P., de los Angeles Bolla, M., De Antoni, G.L., de los Angeles Serradell, M., 2013. Protective effect of a mixture of kefir-isolated lactic acid bacteria and yeasts in a hamster model of *Clostridium difficile* infection. Anaerobe 21, 28–33.

Bonczar, G., Regula, A., Grega, T., 2004. The vitamin C content in fermented milk beverages obtained from ewe's milk. Electronic Journal of Polish Agricultural Universities 7, #06.

Carasi, P., Diaz, M., Racedo, S.M., De Antoni, G.L., Urdaci, M.C., de los Angeles Serradell, M., 2014. Safety characterization and antimicrobial properties of kefir-isolated *Lactobacillus kefiri*. BioMed Research International. http://dx.doi.org/10.1155/2014/208974.

Cenesiz, S., Devrim, A.K., Kamber, U., Sozmen, M., 2008. The effect of kefir on glutathione (GSH), malondialdehyde (MDA) and nitric oxide (NO) levels in mice with colonic abnormal crypt formation (ACF) induced by azoxymethane (AOM). Dtsch Tierarztl Wochenschr 115, 15–19.

Cevikbas, A., Yemni, E., Ezzedenn, F.W., Yardimici, T., Cevikbas, U., Stohs, S.J., 1994. Antitumoural antibacterial and antifungal activities of kefir and kefir grain. Phytotheraphy Research 8, 78–82.

Chen, C., 2005. Antitumor Properties of Kefir: Possible Bioactive Component(s) and Mechanism(s) (Ph.D. thesis). School of Dietetics and Human Nutrition Centre for Indigenous Peoples Nutrition and Environment, McGill University, Montreal. 222 pp.

Chen, H.L., Tung, Y.T., Chuang, C.H., Tu, M.Y., Tsai, T.C., Chang, S.Y., Chen, C.M., 2015. Kefir improves bone mass and microarchitecture in an ovariectomized rat model of postmenopausal osteoporosis. Osteoporosis International 26, 589–599.

Chen, Y.P., Lee, T.Y., Hong, W.S., Hsieh, H.H., Chen, M.J., 2013. Effects of *Lactobacillus kefiranofaciens* M1 isolated from kefir grains on enterohemorrhagic *Escherichia coli* infection using mouse and intestinal cell models. Journal of Dairy Science 96, 7467–7477.

Chifiriuc, M.C., Cioaca, A.B., Lazar, V., 2011. In vitro assay of the antimicrobial activity of kephir against bacterial and fungal strains. Anaerobe 17, 433–435.

Chryssanthopoulos, C., Maridaki, M., 2010. Nutritional aspects of yogurt and functional dairy products. In: Yıldız, F. (Ed.), Development and Manufacture of Yogurt and Other Functional Dairy Products. CRC Press, Boca Raton, pp. 267–305.

Clementi, F., Gobbetti, M., Rossi, J., 1989. Carbon dioxide synthesis by immobilized yeast cells in kefir production. Milchwissenschaft 44, 70–74.

Cogulu, D., Topaloglu Ak, A., Caglar, E., Sandalli, N., Karagozlu, C., Ersin, N., Yerlikaya, O., 2010. Potential effects of a multistrain probiotic-kefir on salivary *Streptococcus mutans* and *Lactobacillus* spp. Journal of Dental Science 5, 144–149.

De Moreno de LeBlanch, A., Matar, C., Farnworth, E., Perdigon, G., 2007. Study of immune cells involved in the antitumor effect of kefir in a murine breast cancer model. Journal of Dairy Science 90, 1920–1928.

De Moreno de LeBlanch, A., Matar, C., Farnworth, E., Perdigon, G., 2006. Study of cytokines involved in the prevention of a murine experimental breast cancer by kefir. Cytokine 34, 1–8.

De Roos, N.M., Schouten, G., Katan, M.B., 1998. Yogurt enriched with *Lactobacillus acidophilus* does not lower blood lipids in healthy men and women with normal to borderline high serum cholesterol levels. European Journal of Clinical Nutrition 53, 277–280.

De Smet, I., De Boever, P., Versteaete, W., 1998. Cholesterol lowering in pigs through enhanced bacterial bile salt hydrolase activity. British Journal of Nutrition 79, 185–194.

De Vrese, M., Keller, B., Barth, C.A., 1992. Enhancement of intestinal hydrolysis of lactose by microbial β-galactosidase (EC 3.2.1.23) of kefir. British Journal of Nutrition 61, 61–75.

Ebner, J., Aşçı Arslan, A., Fedorovac, M., Hoffmannc, R., Küçükçetin, A., Pischetsriedera, M., 2015. Peptide profiling of bovine kefir reveals 236 unique peptides released from caseins during its production by starter culture or kefir grains. Journal of Proteomics 117, 41–57.

Farnworth, E.R., 2005. Kefir – a complex probiotic. Food Science and Technology Bulletin Functional Foods 2, 1–17.

Farnworth, E.R., Mainville, I., 2008. Kefir – a fermented milk product. In: Farnworth, E.R. (Ed.), Handbook of Fermented Functional Foods. CRC Press Taylor & Francis Group, Boca Raton, London, New York, pp. 89–127.

FitzGerald, R.J., Meisel, H., 2000. Milk protein-derived peptide inhibitors of angiotensin-I-converting enzyme. British Journal of Nutrition 84 (Suppl. 1), S33–S37.

FitzGerald, R.J., Murray, B.A., Walsh, D.J., 2004. Hypotensive peptides from milk proteins. Journal of Nutrition 134, 980–988.

Fontan, M.C.G., Martinez, S., Franco, I., Carballo, J., 2006. Microbiological and chemical changes during the manufacture of Kefir made from cows' milk, using a commercial starter culture. International Dairy Journal 16, 762–767.

Gao, J., Gu, F., Ruan, H., Chen, Q., He, J., He, G., 2013. Induction of apoptosis of gastric cancer cells SGC7901 in vitro by a cell-free fraction of Tibetan kefir. International Dairy Journal 30, 14–18.

Garrote, G.L., Abraham, A.G., De Antoni, G.L., 1998. Characteristics of kefir prepared with different grain: milk ratios. Journal of Dairy Research 65, 149–154.

Garrote, G.L., Abraham, A.G., De Antoni, G.L., 2001. Chemical and microbiological characterisation of kefir grains. Journal of Dairy Research 68, 639–652.

Ghasemlou, M., Khodaiyan, F., Gharibzahedi, S.M.T., 2012. Enhanced production of Iranian kefir grain biomass by optimization and empirical modeling of fermentation conditions using response surface methodology. Food Bioprocess Technology 5, 3230–3235.

Ghoneum, M., Gimzewski, J., 2014. Apoptotic effect of a novel kefir product, PFT, on multidrug-resistant myeloid leukemia cells via a hole-piercing mechanism. International Journal of Oncology 44, 830–837.

Gilliland, S.E., Nelson, C.R., Maxwell, C., 1985. Assimilation of cholesterol by *Lactobacillus acidophilus*. Applied Environmental Microbiology 49, 377–381.

Grishina, A., Kulikova, I., Alieva, L., Dodson, A., Rowland, I., Jin, J., 2011. Antigenotoxic effect of kefir and ayran supernatants on fecal water-induced DNA damage in human colon cells. Nutrition and Cancer 63, 73–79.

Grønnevik, H., Falstad, M., Narvhus, J.A., 2011. Microbiological and chemical properties of Norwegian kefir during storage. International Dairy Journal 21, 601–606.

Gul, O., Mortas, M., Atalar, I., Dervisoglu, M., Kahyaoglu, T., 2015. Manufacture and characterization of kefir made from cow and buffalo milk, using kefir grain and starter culture. Journal of Dairy Science 98, 1517–1525.

Gulmez, M., Guven, A., 2003. Behaviour of *Escherichia coli* O157: H7, *Listeria monocytogenes* 4b and *Yersinia enterocolitica* O_3 in pasteurised and nonpasteurised kefir fermented for one or two days. Food Science and Technology International 9, 365–369.

Gursoy, O., Çelik, G., Şen Gürsoy, S., 2014. Electrochemical biosensor based on surfactant doped polypyrrole (PPy) matrix for lactose determination. Journal of Applied Polymer Science 131 (9), 40200. http://dx.doi.org/10.1002/app.40200.

Gursoy, O., Ozbas, H., Yilmaz, Y., 2013. Health benefits of cheese consumption in osteoporosis. In: Preedy, V.R., Watson, R.R., Patel, V.B. (Eds.), Hand Book of Cheese in Health: Production, Nutrition and Medical Sciences. Human Health Handbooks No: 6. Wageningen Academic Publishers, the Netherlands, pp. 719–731.

Gursoy, O., Ozel, S., Ozbaş, H., Çon, A.H., 2011. Kolesterol seviyesinin *in vitro* ve *in vivo* koşullarda düşürülmesinde probiyotik mikroorganizmaların etkisi. Akademik Gıda 9, 37–45.

Guven, A., Guven, A., 2005. Hiperkolesterolemi oluşturulmuş tavşanlarda kefirin total kolesterol, trigliserit, HDL-kolesterol, LDL kolesterol ve lipit peroksidasyonu üzerine etkisi. Kafkas Üniversitesi Veteriner Fakültesi Dergisi 11, 127–131.

Guzel-Seydim, Z.B., Kok-Tas, T., Greene, A.K., 2010. Kefir and koumiss: microbiology and technology. In: Yildiz, F. (Ed.), Development and Manufacture of Yogurt and Other Functional Dairy Products. CRC Press, Boca Raton, pp. 143–163.

Guzel-Seydim, Z.B., Seydim, A.C., Greene, A.K., 2000. Organic acids and volatile flavor components evolved during refrigerated storage of kefir. Journal of Dairy Science 83, 275–277.

Guzel-Seydim, Z.B., Seydim, A.C., Greene, A.K., 2003. Comparison of amino acid profiles of milk, yogurt and Turkish kefir. Milchwissenschaft 58, 158–160.

Hayes, M., Stanton, C., Fitzgerald, G.F., Ross, R.P., 2007. Putting microbes to work: dairy fermentation, cell factories and bioactive peptides. Part II: bioactive peptide functions. Biotechnology Journal 2, 435–449.

Hernández-Ledesma, B., del Mar Contreras, M., Recio, I., 2011. Antihypertensive peptides: production, bioavailability and incorporation into foods. Advanced Colloid Interface Science 165, 23–35.

Hertzler, S.R., Clancy, S.M., 2003. Kefir improves lactose digestion and tolerance in adults with lactose maldigestion. Journal of American Dietary Association 103, 582–587.

Hong, W.-S., Chen, H.-C., Chen, Y.-P., Chen, M.-J., 2009. Effects of kefir supernatant and lactic acid bacteria isolated from kefir grain on cytokine production by macrophage. International Dairy Journal 19, 244–251.

Huang, Y., Wang, X., Wang, J., Wu, F., Sui, Y., Yang, L., Wang, Z., 2013a. *Lactobacillus plantarum* strains as potential probiotic cultures with cholesterol-lowering activity. Journal of Dairy Science 96, 2746–2753.

Huang, Y., Wu, F., Wang, X., Sui, Y., Yang, L., Wang, J., 2013b. Characterization of *Lactobacillus plantarum* Lp27 isolated from Tibetan kefir grains: a potential probiotic bacterium with cholesterol-lowering effects. Journal of Dairy Science 96, 2816–2825.

Huth, P.J., DiRienzo, D.B., Miller, G.D., 2006. Major scientific advances with dairy foods in nutrition and health. Journal of Dairy Science 89, 1207–1221.

Iraporda, C., Romanin, D.E., Rumbo, M., Garrote, G.L., Abraham, A.G., 2014. The role of lactate on the immunomodulatory properties of the nonbacterial fraction of kefir. Food Research International 62, 247–253.

Irigoyen, A., Ortigosa, M., Garcia, S., Ibanez Torre, P., 2012. Comparison of free amino acids and volatile components in three fermented milks. International Journal of Dairy Technology 65, 578–584.

Ismaiel, A.A., Ghaly, M.F., El-Naggar, A.K., 2011. Milk kefir: ultrastructure, antimicrobial activity, and efficacy on aflatoxin B1 production by *Aspergillus flavus*. Current Microbiology 62, 1602–1609.
Johnell, O., Kanis, J.A., 2006. An estimate of the worldwide prevalence and disability associated with osteoporotic fractures. Osteoporosis International 17, 1726–1733.
Kahraman, C., 2011. Production of Kefir from Bovine and Oat Milk Mixture (MSc. Thesis). School of Engineering and Sciences of İzmir Institute of Technology, Izmir. 119 pp.
Kakisu, E., Bolla, P., Abraham, A.G., de Urraza, P., de Antoni, G.L., 2013. *Lactobacillus plantarum* isolated from kefir: protection of cultured Hep-2 cells against Shigella invasion. International Dairy Journal 33, 22–26.
Kakisu, E.J., Abraham, A.G., Pérez, P.F., De Antoni, G.L., 2007. Inhibition of *Bacillus cereus* in milk fermented with kefir grains. Journal of Food Proteins 70, 2613–2616.
Kesenkaş, H., Kınık, Ö., 2010. Süt ve Süt İçecekleri. In: Ötleş, S., Akçiçek, E. (Eds.), İçecekler Beslenme Ve Sağlık. Palme Yayıncılık, Ankara, pp. 133–165.
Kesenkaş, H., Dinkçi, N., Seçkin, K., Kınık, Ö., Gönç, S., 2011a. Antioxidant properties of kefir produced from different cow and soy milk mixtures. Journal of Agricultural Science 17, 253–259.
Kesenkaş, H., Dinkçi, N., Seçkin, K., Kınık, O., Gönç, S., Ergönül, P.G., Kavas, G., 2011b. Physicochemical, microbiological and sensory characteristics of soymilk kefir. African Journal of Microbiology Research 5, 3737–3746.
Kesenkaş, H., Yerlikaya, O., Özer, E., 2013. A functional milk beverage: Kefir. Agro Food Industry Hi-Tech 24, 53–55.
Khoury, N., El-Hayek, S., Tarras, O., El-Sabban, M., El-Sibai, M., Rizk, S., 2014. Kefir exhibits antiproliferative and proapoptotic effects on colon adenocarcinoma cells with no significant effects on cell migration and invasion. International Journal of Oncology 45, 2117–2127.
Kielbassa, C., Roza, L., Epe, B., 1997. Wavelength dependence of oxidative DNA damage induced by UV and visible light. Carcinogenesis 18, 811–816.
Kim, S.H., Jo, S.H., Kwon, Y.I., Hwang, J.K., 2011. Effects of onion (*Allium cepa* L.) extract administration on intestinal α-glucosidases activities and spikes in postprandial blood glucose levels in SD rats model. International Journal of Molecular Science 12, 3757–3769.
Kim, S.H., Oh, S., 2013. Fermented milk and yoghurt. In: Park, Y.W., Haenlein, G.F.W. (Eds.), Milk and Dairy Products in Human Nutrition: Production, Composition and Health. John Wiley & Sons, Oxford, pp. 338–356.
Kinik, O., Gursoy, O., 2002. Milk proteins-derived bioactive peptides. Pamukkale University Journal of Engineering Sciences 8, 195–203.
Kneifel, W., Mayer, H.K., 1991. Vitamin profiles of kefirs made from milks of different species. International Journal of Food Science Technology 26, 423–428.
Koçak, C., Gürsel, A., 1981. Kefir. Gıda 6, 11–14.
Kök-Taş, T., Seydim, A.C., Özer, B., Guzel-Seydim, Z.B., 2013. Effects of different fermentation parameters on quality characteristics of kefir. Journal of Dairy Science 96, 780–789.
Konar, A., Şahan, N., 1991. Kefir a fermented milk. Journal of Agriculture Faculty of Çukurova University 6, 143–154.
Kong, K.L., Hendrich, S., 2012. Glycemic index, insulinemic index, and satiety index of kefir. Journal of the American College Nutrition 31, 280–287.
Kwon, O.-K., Ahn, K.-S., Lee, M.-Y., Kim, S.-Y., Park, B.-Y., Kim, M.-K., Lee, I.-Y., Oh, S.-R., Lee, H.-K., 2008. Inhibitory effect of kefiran on ovalbumin-induced lung inflammation in a murine model of asthma. Archives of Pharmacal Research 31, 1590–1596.
Kwon, Y.-I., Apostolidis, E., Shetty, K., 2006. Antidiabetes functionality of kefir culture-mediated fermented soymilk supplemented with Rhodiola extracts. Food Biotechnology 20, 13–29.

Lee, M.Y., Ahn, K.S., Kwon, O.K., Kim, M.J., Kim, M.K., Lee, I.Y., Oh, S.R., Lee, H.K., 2007. Antiinflammatory and antiallergic effects of kefir in Mouse asthma model. Immunobiology 212, 647–654.

Leite, A.M., Leite, D.C., Del Aguila, E.M., Alvares, T.S., Peixoto, R.S., Miguel, M.A., Silva, J.T., Paschoalin, V.M., 2013a. Microbiological and chemical characteristics of Brazilian kefir during fermentation and storage processes. Journal of Dairy Science 96, 4149–4159.

Leite, A.M., Mayo, B., Rachid, C.T., Peixoto, R.S., Silva, J.T., Paschoalin, V.M., Delgado, S., 2012. Assessment of the microbial diversity of Brazilian kefir grains by PCR-DGGE and pyrosequencing analysis. Food Microbiology 31, 215–221.

Leite, A.M.O., Miguel, M.A., Peixoto, R.S., Rosado, A.S., Silva, J.T., Paschoalin, V.M., 2013b. Microbiological, technological and therapeutic properties of kefir: a natural probiotic beverage. Brazilian Journal of Microbiology 44, 341–349.

Liu, H., Xie, Y., Xiong, L., Dong, R., Pan, C., Teng, G., Zhang, H., 2012. Effect and mechanism of cholesterol-lowering by Kluyveromyces from Tibetan kefir. Advances in Material Research 343–344, 1290–1298.

Liu, J.R., Lin, C.W., 2000. Production of kefir from soymilk with or without added glucose, lactose, or sucrose. Journal of Food Science 65, 716–719.

Liu, J.R., Wang, S.-Y., Lin, Y.Y., Lin, C.W., 2002. Antitumor activity of milk and soy milk kefir in tumor-bearing mice. Nutrition and Cancer 44, 182–187.

Liutkevicius, A., Sarkinas, A., 2004. Studies on the growth conditions and composition of kefir grains – as a food and forage biomass. Veterinarija ir Zootechnika 25, 64–70.

Lopitz-Otsoa, F., Rementeria, A., Elguezabal, N., Garaizar, J., 2006. Kefir: a symbiotic yeast-bacteria community with alleged healthy capabilities. Revista Iberoamericana de Micologia 23, 67–74.

Lye, H.-S., Rusul, G., Liong, M.-T., 2010. Removal of cholesterol by lactobacilli via incorporation and conversion to coprostanol. Journal of Dairy Science 93, 1383–1392.

Maalouf, K., Baydoun, E., Rizk, S., 2011. Kefir induces cell-cycle arrest and apoptosis in HTLV-I-negative malignant T-lymphocytes. Cancer Management Research 3, 39–47.

Maeda, H., Zhu, X., Omura, K., Suzuki, S., Kitamura, S., 2004a. Effects of an exopolysaccharide (kefiran) on lipids, blood pressure, blood glucose, and constipation. BioFactors 22, 197–200.

Maeda, H., Zhu, X., Suzuki, S., Suzuki, K., Kitamura, S., 2004b. Structural characterization and biological activities of an exopolysaccharide kefiran produced by *Lactobacillus kefiranofaciens* WT-2BT. Journal of Agricultural Food Chemistry 52, 5533–5538.

Magalhaes, K.T., de Melo Pereira, G.V., Campos, C.R., Dragone, G., Schwan, R.F., 2011. Brazilian kefir: structure, microbial communities and chemical composition. Brazilian Journal of Microbiology 42, 693–702.

Mann, G.V., Spoerry, A., 1974. Studies of a surfactant and cholesterolemia in the Maasai. American Journal of Clinical Nutrition 27, 464–469.

Marquina, D., Santos, A., Corpas, I., Munoz, J., Zazo, J., Peinado, J.M., 2002. Dietary influence of kefir on microbial activities in the mouse bowel. Letters on Applied Microbiology 35, 136–140.

Miao, J., Guo, H., Ou, Y., Liu, G., Fang, X., Liao, Z., Ke, C., Chen, Y., Zhao, L., Cao, Y., 2014. Kefir exhibits antiproliferative and proapoptotic effects on colon adenocarcinoma cells with no significant effects on cell migration and invasion. Food Control 42, 48–53.

Moreira, M.E.C., Dos Santos, M.H., Zolini, G.P.P., Wouters, A.T.B., Carvalho, J.C.T., Schneedorf, J.M., 2008. Antiinflammatory and cicatrizing activities of a carbohydrate fraction isolated from sugary kefir. Journal of Medicinal Food 11, 356–361.

Nagira, T., Narisawa, J., Teruya, K., Katakura, Y., Shim, S.Y., Kusumoto, K., Tokumaru, S., Tokumaru, K., Barnes, D.W., Shirahata, S., 2002. Suppression of UVC-induced cell damage and enhancement of DNA repair by the fermented milk, Kefir. Cytotechnology 40, 125–137.

Nalbantoglu, U., Cakar, A., Dogan, H., Abacı, N., Ustek, D., Sayood, K., Can, H., 2014. Metagenomic analysis of the microbial community in kefir grains. Food Microbiology 41, 42–51.
Noh, D.O., Kim, S.H., Gilliland, S.E., 1997. Incorporation of cholesterol into cellular membrane of *Lactobacillus acidophilus* ATCC 43121. Journal of Dairy Science 80, 3107–3113.
O'Connor, E.B., Barrett, E., Fitzgerald, G., Hill, C., Stanton, C., Ross, R.P., 2005. Production of vitamins, exopolysaccharides and bacteriocins by probiotic bacteria. In: Tamime, A. (Ed.), Probiotic Dairy Products. Blackwell Publishing, Oxford, pp. 167–194.
Ortiz-Andrade, R.R., García-Jiménez, S., Castillo-España, P., Ramírez-Ávila, G., Villalobos-Molinad, R., Estrada-Sotoa, S., 2007. α-Glucosidase inhibitory activity of the methanolic extract from *Tournefortia hartwegiana*: an antihyperglycemic agent. Journal of Ethnopharmacology 109, 48–53.
Otleş, S., ve Çağındı, Ö., 2003. Kefir: a probiotic dairy-composition, nutritional and therapeutic aspects. Pakistan Journal of Nutrition 2, 54–59.
Ozer, B., 2015. Microbiology and biochemistry of yogurt and other fermented milk products. In: Özer, B., Akdemir-Evrendilek, G. (Eds.), Dairy Microbiology and Biochemistry Recent Developments. CRC Press, Boca Raton, pp. 167–213.
Pereira, D.I.A., McCartney, A.L., Gibson, G.R., 2003. An *in vitro* study of the probiotic potential of a bile-salt-hydrolyzing *Lactobacillus fermentum* strain, and determination of its cholesterol-lowering properties. Applied Environmental Microbiology 69, 4743–4752.
Pogacic, T., Šinko, S., Zamberlin, S., Samaržija, D., 2013. Microbiota of kefir grains. Mljekarstvo 63, 3–14.
Powell, J.E., 2006. Bacteriocins and Bacteriocin Producers Present in Kefir and Kefir Grains (M.Sc. thesis). Department of Food Science, Faculty of AgriSciences, Stellenbosch University, South Africa. 115 pp.
Powell, J.E., Witthuhn, R.J., Todorov, S.D., Dicks, L.M.T., 2007. Characterization of bacteriocin ST8KF produced by a kefir isolate *Lactobacillus plantarum* ST8KF. International Dairy Journal 17, 190–198.
Punaro, G.R., Maciel, F.R., Rodrigues, A.M., Rogero, M.M., Bogsan, C.S.B., Oliveira, M.N., Ihara, S.S.M., Araujo, S.R.R., Sanches, T.R.C., Andrade, L.C., Higa, E.M.S., 2014. Kefir administration reduced progression of renal injury in STZ-diabetic rats by lowering oxidative stress. Nitric Oxide 37, 53–60.
Quirós, A., Hernández-Ledesma, B., Ramos, M., Amigo, L., Recio, I., 2005. Angiotensin-converting enzyme inhibitory activity of peptides derived from caprine kefir. Journal of Dairy Science 88, 3480–3487.
Rahimzadeh, G., Bahar, M.A., Mozafari, N.A., Salehi, M., 2011. Antimicrobial activity kefir on *Pseudomonas aeruginosa*. Razi Journal of Medical Sciences 18, 16.
Raja, A., Gajalakshmi, P., Raja, M.M.M., Imran, M.M., 2009. Effect of *Lactobacillus lactis cremoris* isolated from kefir against food spoilage bacteria. American Journal of Food Technology 4, 201–209.
Rattray, F.P., O'Connell, M.J., 2011. Kefir. In: Fuquay, J.W., Fox, P.F., McSweeney, P.L.H. (Eds.), Encyclopedia of Dairy Sciences, 2nd ed. Academic Press, San Diego, pp. 518–524.
Rizk, S., Maalouf, K., Baydoun, E., 2009. The antiproliferative effect of kefir cell-free fraction on HuT-102 malignant T lymphocytes. Clinical Lymphoma Myeloma 9, 198–203.
Rizk, S., Maalouf, K., Nasser, H., El-Hayek, S., 2013. The proapoptotic effect of kefir in malignant T-lymphocytes involves a p53 dependent. Clinical Lymphoma Myeloma 13 (Supplement 2), 367.
Rodrigues, K.L., Caputo, L.R.G., Carvalho, J.C.T., Evangelista, J., Schneedorf, J.M., 2005a. Antimicrobial and healing activity of kefir and kefiran extract. International Journal of Antimicrobial Agents 25, 404–408.

Rodrigues, K.L., Carvalho, J.C.T., Schneedorf, J.M., 2005b. Antiinflammatory properties of kefir and its polysaccharide extract. Inflammopharmacology 13, 485–492.

Romanin, D., Serradell, M., Maciel, D.G., Lausada, N., Garrote, G.L., Rumbo, M., 2010. Downregulation of intestinal epithelial innate response by probiotic yeasts isolated from kefir. International Journal of Food Microbiology 140, 102–108.

Ryan, M.P., Rea, M.C., Hill, C., Ross, R.P., 1996. An application in Cheddar cheese manufacture for a strain of *Lactococcus lactis* producing a novel broad-spectrum bacteriocin, lacticin 3147. Applied Environmental Microbiology 62, 612–619.

Sarkar, S., 2007. Potential of kefir as a dietetic beverage–A review. British Journal of Nutrition 109, 280–290.

Sarkar, S., 2008. Biotechnological innovations in kefir production: review. British Food Journal 110, 283–295.

Şatır, G., 2011. Effect of Kefir Fermentation on Some Functional Properties of Goat Milk (Ph.D. thesis). Süleyman Demirel University Graduate School of Applied and Natural Sciences, Department of Food Engineering, Isparta. 182 pp.

Silva, K.R., Rodrigues, S.A., Filho, L.X., Lima, A.S., 2009. Antimicrobial activity of broth fermented with kefir grains. Applied Biochemistry and Biotechnology 152, 316–325.

Sinhaa, R.P., Häder, D.-P., 2002. UV-induced DNA damage and repair: a review. Photochemical and Photobiological Sciences 1, 225–236.

Stepaniak, L., Fetliński, A., 2003. Kefir. In: Roginski, H., Fuquay, J.W., Fox, P.F. (Eds.), Encyclopedia of Dairy Science. Academic Press, London, pp. 1049–1054.

St-Onge, M.-P., Farnworth, E.R., Savard, T., Chabot, D., Mafu, A., Jones, P.J.H., 2002. Kefir consumption does not alter plasma lipid levels or cholesterol fractional synthesis rates relative to milk in hyperlipidemic men: a randomized controlled trial. BMC Complementary and Alternative Medicines 2. http://dx.doi.org/10.1186/1472-6882-2-1.

Suriasih, K., 2011. Antimicrobial activity of kefir and lactic acid bacteria isolated from kefir. Journal of ISSAAS 17, 227–273.

Takahashi, F., Kawakami, H., 1999. Antibacterial action of kefir against *Escherichia coli* O157:H7. Japanese Journal of Food Microbiology 16, 245–247.

Teruya, K., Yamashita, M., Tominaga, R., Nagira, T., Shim, S.-Y., Katakura, Y., Tokumaru, S., Tokumaru, T., Barnes, D., Shirahata, S., 2002. Fermented milk, Kefram-Kefir enhances glucose uptake into insulin-responsive muscle cells. Cytotechnology 40, 107–116.

Topuz, E., Derin, D., Can, G., Kürklü, E., Çınar, S., Akyan, F., Çevikbaş, A., Dişçi, R., Durna, Z., Şakar, B., Saglam, S., Tanyeri, H., Deniz, G., Gürer, Ü., Taş, F., Guney, N., Aydıner, A., 2008. Effect of oral administration of kefir on serum proinflammatory cytokines on 5-FU induced oral mucositis in patients with colorectal cancer. Investigational New Drugs 26, 567–572.

Ulusoy, B.H., Çolak, H., Hampikyan, H., Erkan, M.E., 2007. An in vitro study on the antibacterial effect of kefir against some food-borne pathogens. Turk Mikrobiyoloji Cemiyeti Dergisi 37, 103–107.

Uniacke-Lowe, T., 2011. Koumiss. In: Fuquay, J.W., Fox, P.F., McSweeney, P.L.H. (Eds.), Encyclopedia of Dairy Sciences, 2th ed. Academic Press, San Diego, pp. 512–517.

Vardjan, T., Lorbeg, P.M., Rogelj, I., Majhenic, A.C., 2013. Characterization and stability of lactobacilli and yeast microbiota in kefir grains. Journal of Dairy Sciences 96, 2729–2736.

Varnacı, Z., 1980. Kaybolmuş bir içecek kefir. Gıda 6, 61.

Vélez, C.A.C., Peláez, A.M.L., 2014. Inhibición del crecimiento de *Aspergillus ohraceus* mediante "Panela" Fermentada Con gránulos de kefir de agua. Revista De La Facultad De Química Farmacéutica 21, 191–200.

Vinderola, C.G., Duarte, J., Thangavel, D., Perdigon, G., Farnworth, E., Matar, C., 2005. Immunomodulating capacity of kefir. Journal of Dairy Research 72, 195–202.

Vinderola, G., Perdigon, G., Duartea, J., Farnworth, E., Matar, C., 2006b. Effects of the oral administration of the exopolysaccharide produced by *Lactobacillus kefiranofaciens* on the gut mucosal immunity. Cytokine 36, 254–260.

Vinderola, G., Perdigon, G., Duartea, J., Thangavela, D., Farnworth, E., Matar, C., 2006a. Effects of kefir fractions on innate immunity. Immunobiology 211, 149–156.

Vujicic, I.F., Vulic, M., Könyves, T., 1992. Assimilation of cholesterol in milk by kefir cultures. Biotechnology Letters 14, 847–850.

Wang, H.-F., Tseng, C.-Y., Chang, M.-H., Lin, J.A., Tsai, F.-J., Tsai, C.-H., Lu, Y.-C., Lai, C.-H., Huang, C.-Y., Tsai, C.-C., 2012. Antiinflammatory effects of probiotic *Lactobacillus paracasei* on ventricles of BALB/C mice treated with ovalbumin. Chinese Journal of Physiology 55, 37–46.

Wang, Y., Xu, N., Xi, A., Ahmed, Z., Zhang, B., Bai, X., 2009. Effects of *Lactobacillus plantarum* MA2 isolated from Tibet kefir on lipid metabolism and intestinal microflora of rats fed on high-cholesterol diet. Applied Microbiology and Biotechnology 84, 341–347.

Witthuhn, R.C., Schoeman, T., Britz, T.J., 2004. Isolation and characterization of the microbial population of different South African kefir grains. International Journal of Dairy Technology 57, 33–37.

Wood, R., 2003. Osteoporosis (Chapter 4). In: Mattila-Sandholm, T., Saarela, M. (Eds.), Functional Dairy Products. CRC Press, USA, pp. 94–107.

Wszolek, M., Kupiec-Teahan, B., Guldager, H.S., Tamime, A.Y., 2006. Production of kefir, koumiss and other related products. In: Tamime, A.Y. (Ed.), Fermented Milks. Blackwell Publishing, Oxford, pp. 174–216.

Yasar, M., Taskin, A.K., Kaya, B., Aydin, M., Ozaydin, I., Iskender, A., Erdem, H., Ankatali, H., Kandis, H., 2013. The early antiinflammatory effect of kefir in experimental corrosive esophagitis. Annali Italiani di Chirurgia 84, 681–685.

Yılmaz, L., Özcan, T., Akpinar Bayizit, A., 2006. The sensory characteristics of berry-flavoured kefir. Czech Journal of Food Science 24, 26–32.

Zheng, Y., Lu, Y., Wang, J., Yang, L., Pan, C., Huang, Y., 2013. Probiotic properties of *Lactobacillus* strains isolated from Tibetan kefir grains. PLoS One 8, e69868. http://dx.doi.org/10.1371/journal.pone.0069868.

Section 3.3

Legume and Cereal Grains Fermented Derived Products

Chapter 15

Beer and Its Role in Human Health

M.L. González-SanJosé[1], P.M. Rodríguez[1], V. Valls-Bellés[2]
[1]University of Burgos, Burgos, Spain; [2]University of Jaume I, Castellón de la Plana, Spain

15.1 INTRODUCTION: BRIEF NOTES OF BREWING

Beer is a drink with low alcohol content and is generally produced by fermenting wort aromatized with hops using selected yeasts. Wort is traditionally obtained by mashing/extraction of barley and barley malt with hot water. However, other cereals can also be used as a source of starch and enzymes other than those found in the malt can be added. The mashing/extraction process results in sweet wort that, after several more steps, including boiling in the presence of hops, is transformed into aromatized wort, or simply wort, which is itself transformed into beer upon fermentation (Ortega-Heras and González-SanJosé, 2003). Other aromatizing products in addition to hops, such as cloves and other spices, or even fruit, can be used to produce certain special beers, although these are much less widely brewed.

The composition of beer is determined by the components present in the raw materials used to make it (water, barley (malted or not), and other cereals, hops, and, in special cases, spices and fruit) together with those formed during the brewing process (malting, mashing, boiling, fermentation, and aging). Given its composition, beer is a drink with positive bioactive properties and thus has potential beneficial effects in humans, although these effects dose-dependent, especially in regards to alcohol content, as will be discussed below.

This chapter presents information concerning the major and minor components of beer, along with a brief discussion of its origins and its bioactivity and possible beneficial health effects, which are mainly related to the presence of antioxidants (Pulido et al., 2003) and the supply of minerals, vitamins, and fiber (Sendra and Carbonell, 1999).

15.2 BIOACTIVE COMPONENTS OF BEER

Although beer cannot be rated as an important source of nutrients, it contains a series of nutrient and nonnutrient components with positive physiological effects in the human body.

As noted above, beer can be considered to be a broth of cereals fermented and aromatized with hops or other spices. Numerous factors can affect the composition of beer, including those that affect the raw materials used and those that affect the brewing process. Consequently, it is difficult to provide specific "composition" information for beer as the data published in this regard must be considered to be indicative rather than absolute.

Beer is mostly water (between 90% and 95%, v/v) accompanied by other compounds such as ethanol (normally 4–6%, v/v), organic acids, nitrogenated compounds, carbohydrates, mineral salts, vitamins, phenolic compounds, melanoidins, humulones and lupulones, etc. The high water content and the presence of salts and carbohydrates contribute to the high hydrating capacity of beer, and water and carbon dioxide (approximately 0.5 g CO_2 per 100 g beer) are responsible for its high refreshing power (Piendl, 1990). The salt and carbohydrate composition of beer have led it to be proposed as an appropriate drink for partial rehydration with replacement of hydroelectrolytic losses and fast and complete replenishment of the energy deposits depleted during physical/sporting exercise (Jiménez-Pavón et al., 2009).

The quality of the water used during brewing is of prime importance and relatively soft waters are used, thus the supply of minerals, especially divalent ones, is not particularly high. Moreover, brewers tend to standardize the mineral composition of the water, markedly reducing the salt and metal content to avoid negative effects on the characteristics of the wort and resulting beer. As such, the raw materials used during brewing are the main source of the mineral salts found in beer rather than the water added to the brew.

The second most abundant components of beer are either carbohydrates or alcohol. The carbohydrates in beer come directly from the starch sources used in brewing, which are extracted and transformed by the action of malt enzymes during the mashing/extraction phases. The proportion of carbohydrates varies between 3% and 5% (w/v), although stronger beers may contain as much as 10% (w/v). Dextrins, which are formed by the enzymatic degradation of starch by the enzymes found in malt, are the most abundant carbohydrates (75–80% of the total) and contribute significantly to the dietary fiber content of beer. Although this fiber varies markedly from one type of beer to another, beer consumption may complement the fiber obtained from other foods with beneficial properties in terms of reducing cholesterol levels and the incidence of cancer. It has been reported that the dietary fiber content of Spanish beers ranges from 1.6 to 2.1 g/L, with no marked differences between normal and alcohol-free beers (Goñi et al., 2009). These values exceed those of many other drinks. Dextrins form part of the soluble fiber that are metabolized by the microbiota in the colon. These compounds are metabolized slowly, releasing glucose moieties that pass progressively into the blood, thus leading to relatively small peaks in glucose concentration. Similarly, the β-glucan and arabinoxylan content may make a positive contribution to the supply of these dietary components and thus to the positive effects associated with them, such as cholesterol control and glycemic index. The fiber in beer is characterized by high colonic fermentability (98%) with production of short-chain fatty acids such as propionic and butyric acid, which help to regulate carbohydrate and lipid metabolism and also act as an energy source and in the prevention of colorectal cancer, respectively (Saura-Calixto et al., 2002).

Other minor carbohydrates present in beer are monosaccharides such as ribose, arabinose, xylose, glucose, fructose, and galactose; disaccharides such as maltose and isomaltose; and trisaccharides such as panose, isopanose, and maltotriose. Polyalcohols such as glycerine and *myo*-inositol, whose presence is mainly associated with fermentation processes, have also been found. In addition, complex carbohydrates such as β-glucans that exert a foam-stabilizing effect may also remain.

After fermentation, the fermentable sugars formed during the mashing/extraction phases are converted into alcohol. Similarly, they are also substrates for the majority of organic acids present in beer, as many of these acids are formed during alcoholic fermentation.

The alcohol content of most common beers varies in the range 4–6% (v/v). Alcohol is a powerful bioactive agent that is highly toxic to living tissues in the pure state. Despite this fact, numerous scientific and epidemiological studies have pointed to some positive and beneficial health effects of consuming moderate amounts of alcohol (Denke, 2000; De Bree et al., 2001; Diaz et al., 2002; Di Castelnuovo et al., 2006; Romeo et al., 2007).

In addition to producing alcohol, the action of yeasts during alcoholic fermentation results in the formation of many other components that are essential for the smell and taste of beer, including other alcohols, organic acids, esters, ethers, etc. (Sendra et al., 1994). Acetic acid is the main organic acid detected in beer, ranging from 40% to 80% of total organic acids. Its concentration varies considerably depending on the type of beer and has a strong influence on the final pH of this beverage. Small quantities of lactic, formic, and succinic acids and other medium-chain fatty acids such as octadecanoic and decanoic acids are also present, along with derivatives thereof, in addition to significant quantities of carbonic acid.

Once water, carbohydrates, alcohol, and organic acids have been taken into account, the remaining components of beer can be considered to be minor as they generally account for less than 1% of the overall content.

The concentration of nitrogenated substances in beer varies between 0.2% and 0.7% (w/v), although it may exceed 1% in some special beers (Hough et al., 1982). This fraction comprises amino acids, peptides, and proteins, which are mainly derived from the cereals, although yeasts may also contribute, especially in the case of special beers. The proteins present in cereals are slowly degraded during the brewing process, especially during malting (germination) and mashing/extraction, thus resulting in amino acids and soluble peptides. The protein content of normal beer is very low as proteins extracted but not hydrolyzed during the mashing phase are removed as the beer is aged and finished. Proteins are responsible for cloudiness at low temperatures and are removed to avoid negative appearance. If this step was not performed, beer would not be clear and bright. In contrast, beer is relatively rich in amino acids, with practically all essential amino acids being present. The amino acids present in the must are used as nutrients by the yeast during fermentation and are precursors for many of the substances that affect the flavor of beer.

Beer contains small quantities of lipids from the malt and hops, with smaller amounts coming from the yeast. These lipids mainly comprise fatty acids (<1 mg/L), mono-, di-, and triglycerides (which together do not exceed 0.5 mg/L), traces of sterols, and phospholipids. As the lipid content of the raw material is somewhat higher than that of the resulting beer, it would appear that these lipids are either not extracted or are transformed or eliminated during the brewing process, thus only small quantities reach the final product.

Beer contains more than 30 minerals counting the trace elements. Minerals with physiological effects normally reach concentrations of between 0.5 and 2.0 g/L in beer, which tends to be relatively rich in magnesium and provides important quantities of phosphorus and potassium, whereas the sodium content is usually low. It has been reported that consumption of 100 mL of beer may provide 10%, 4.5%, and 4% of the daily nutritional requirements for potassium, magnesium, and phosphorus, respectively. In addition, due to its low sodium content, beer can also be considered suitable for low-sodium diets, and due to its relatively high potassium/sodium ratio of around 16, beer is a diuretic beverage (Gallen et al., 1998). Other minerals found in beer include calcium, iron, lead, copper, zinc, tin, and selenium. The recently reported ability of brewer's yeast to transform inorganic selenium into bioactive organoselenium forms (Sánchez-Martínez et al., 2012) means the selenium present in beer can be assimilated and, therefore, beer may represent a source of dietary selenium.

The minerals, and salts in general, present in beer mainly come from the cereals used in its preparation and are mainly extracted from the outer layers, or bran. Hops may also provide minerals and could be the source of metals such as copper, due to residues from pesticide treatment. Copper may also come from the typical copper vats used in traditional brewing techniques (Viñas et al., 2002).

The main salts present in beer are chlorides, sulfates, carbonates, and silicates, and the high bioavailability of silicon in beer has been noted. The bioavailable silicon in beer comes from the orthosilicic acid extracted from malt during the mashing process (Pennington, 1991; Tucker et al., 2001; Sripanyakorn et al., 2004). Indeed, it has been reported that approximately 50% of the silicon in beer is readily absorbed in humans (Sripanyakorn et al., 2004). This contrasts with the limited or complete lack of bioavailability of most of the silicon found in foodstuffs, due to its presence in the form of aluminosilicates and silica. In light of the above, beer has been proposed to be one of the most important sources of silicon in the Western diet (Jugdaohsingh et al., 2002; Pennington, 1991).

Beer contains B-group vitamins, with relatively important quantities of niacin (B3), folic acid (B9), and folates, in addition to pyridoxine (B6), riboflavin (B2), thiamine (B1), pantothenic acid (B5), biotin (B7), and cobalamine (B12). Other vitamins found in beer, with average values of around 210 mg/L, include vitamins A, D, and E. These vitamins come from the malt, and it has been reported that their content increases during germination of the barley prior to transformation into malt. Moreover, the vitamins formed are not destroyed during final roasting of the malt (Piendl, 1990). Mayer et al. (2001) reported that beer is an important source of dietary folic acid and that the bioactivity of this compound is transmitted to beer, thus moderate intake reduces blood homocysteine levels. The presence of folic acid in beer has also been associated with possible DNA repair (Bramforth, 2004). In addition, the significant vitamin content contributes to the absorption and fixation of metals from both the beer itself and other foodstuffs consumed with it.

Beer contains various different polyphenols, most of which come from malt and, to a much lesser extent, from hops. Polyphenols are of particular interest to brewers due to their influence on the colloidal stability of beer, where they are chiefly responsible for the cloudiness caused by the interaction with proteins and polysaccharides in the beer. Moreover, these compounds play a key role in the sensory characteristics (color, smell, taste) and functional properties of beer (Sendra and Carbonell, 1999). The phenols found in beer belong to various groups or types. Thus simple phenols, mainly phenolic acids such as ferulic, gallic, or syringic acid, and more complex structures such as flavonoids of various types, including both monomeric (catechins) and oligomeric flavonols (proanthocyanidins), other flavonols, such as quercetin and kaempferol, isoflavones, such as daidzein and genistein, and chalcones, such as xanthohumol, as well as lignans and ellagitannins, have been described (Gerhäuser and Becker, 2009). In addition, anthocyanins may be present in special beers prepared using red fruits. The final phenolic compound content of beer does not depend exclusively on the raw materials used in brewing as numerous other factors, the majority of which are related to the brewing technique used, may also affect the final content. Thus, for example, the phenolic content of barley is altered during the malting process, resulting in losses that vary depending on the malting conditions used. Extraction of phenols during mashing also depends on the malt/water ratio and mashing time, with simpler compounds tending to be lost with longer times. Similarly, the proanthocyanidin depolymerization processes that occur during boiling result in the release of catechins. In addition, phenolic compounds tend to be lost during the fermentation and mashing phases, in general, due either to the action of yeasts, the formation of polymers that precipitate, retention by clarifying agents, etc. (McMurrough et al., 1985; Moll et al., 1984). The above explains the marked variability in phenolic compound contents reported in the literature vary widely between beers (Gorinstein et al., 2000; Rivero et al., 2005), with values in the range of 50–1000 mg/L. Phenolic compounds make a notable contribution to the antioxidant properties of beer (Rivero et al., 2005) and are responsible for many of the beneficial health effects described for this drink, by contributing to the internal oxidant/antioxidant balance in the human body and helping to prevent various diseases.

Other components of interest in beer obtained from hops include humulones, lupulones, and essential oils. The final content of these compounds in beer varies depending on the variety and amount of hops used and the conditions under which they are applied, the time at which they are added, the use of flowers or extracts, etc. (Verzele, 1986; Stavri et al., 2004).

Beer contains specific components that, even if present in small quantities, are characteristic of this beverage, such as α–acids or humulones that make up what is known as soft hop resin. When boiled, these compounds isomerize to form isohumulones or iso-α-acids and their presence in beer explains its characteristic bitter taste. In addition to humulones, hops also contain lupulones or β-acids, which are also known as hard resins. These compounds are of less interest from a brewing point of view due to their lower flavoring power and higher instability, which may result in significant taste changes. As a result, brewers tend to avoid their presence in beer by selecting hops with low content of these resins or by using hop extracts from which they have been removed. The presence of essential hop oils is key to the flavor of beer. These oils comprise a very wide range of highly volatile substances with strong flavoring capacity. Up to 200 of the volatile compounds present in beer may be related to essential hop oils, although only around a dozen of these have a marked impact. The compounds found in beer that are of interest include humulene and humulenol mono- and diepoxides, caryophyllene and its oxidation products, and other terpenes such as linalool, geraniol derivatives, and α-terpineol (Murakami et al., 1989).

Other beer components of interest are the so-called melanoidins. These substances affect the flavor of beer but, above all, make an important contribution to its color. They are generally found in small quantities, although their concentration increases as the degree of roasting increases, reaching maximum values in dark beers/stouts. They are light- or dark-brown polymers formed in the final phases of polymerization of the initial and intermediate products of the Maillard reaction. This reaction occurs upon applying heat to media in which amino groups from amino acids, peptides, and proteins and aldehyde groups from sugars coexist. In the case of beer, they mainly come from the malt-roasting process, and sweet wort boiling, although in the latter case compounds with lower molecular weight and lighter color are formed (Woffenden et al., 2002). In addition to melanoidins, other substances derived from the caramelization of sugars may also be formed during malt or cereal roasting and boiling. The high temperatures applied in malt roasting lead to pyrolytic production of compounds with different chemical structures such as furfural and its derivatives, pyrazines, pyrroles, pyrrolidines, azepines, and maltoxazine. All these compounds contribute to beer flavor and color, and some of them such as maltoxazine are important due to their bitterness. The levels of these compounds in beer are low, and all brewing processes are controlled to ensure the final content of furfural and its derivatives, such as hydroxymethylfurfural, remains below permitted limits. The antioxidant effect of melanoidins has been studied for many years and demonstrated in model systems, foodstuffs, and in biological studies (Valls-Bellés et al., 2004; Langner and Rzeski, 2014). As such, melanoidins in beer also contribute to the overall antioxidant capacity of the diet (Pastoriza and Rufian-Henares, 2014).

15.3 ANTIOXIDANT PROPERTIES OF BEER AND HEALTH EFFECTS

Beer contains compounds with well-known antioxidant properties, which are of great interest to the oxidant/antioxidant balance required for a healthy lifestyle. The human body responds to oxidative stress by activating its endogenous antioxidant defenses. This response may be insufficient in itself and, therefore, the intake of exogenous bioactive antioxidant compounds play a key role in strengthening these defenses. As such, the consumption of foodstuffs containing natural antioxidants that contribute to maintaining the balance between oxidants and antioxidants, or even to tip this balance in favor of antioxidants, is of vital importance. Consequently, the consumption of fruit, vegetables, olive oil, and certain drinks, such as beer, coffee, and tea, has been inversely associated with different pathophysiological situations linked to oxidative damage (Codoñer-Franch et al., 2013).

Beer is a drink with a high antioxidant capacity, due to its content of polyphenolic compounds, vitamins, and melanoidins, among other components. This antioxidant capacity has been widely studied, thus suggesting the potential role of beer in the prevention of various diseases (Gerhäusser et al., 2003; Sohrabvandi et al., 2012).

The antioxidant capacity of beer has been studied in vitro by several authors using different methods (Vinson et al., 2003; Rivero et al., 2005; Tedesco et al., 2005). These studies obtained different values for the antioxidant capacity of beer, which tend to be higher for darker beers/stouts than for their lighter and alcohol-free counterparts (Rivero et al., 2005). These findings may be attributed to the different melanoidin and polyphenol contents of the above-mentioned beers, as a positive correlation has been observed between the polyphenol content of beer and its antioxidant capacity (Tedesco et al., 2005).

An in vitro study of the effect of beer on biomarkers of oxidative stress showed that beer exerts a protective effect against DNA damage, preventing chain rupture and the formation of modified bases such 8-hydroxy-2′–deoxyguanosine (Rivero et al., 2005). Other studies conducted in experimental animals showed that beer consumption protects against the oxidative stress induced by the potent anticancer drug adriamycin (Valls-Belles et al., 2005; Valls-Bellés et al., 2008, 2010). However, it should be noted that in vivo effects depend on the bioaccessibility and bioavailability of the compounds present in beer and thus on how they are metabolized. The bioavailability of the various polyphenols differs markedly, and phenolic compounds may undergo glucuronidation and sulfation, thereby modifying their biological activity (Manach et al., 2004; del Rio et al., 2013) Phenolic acids are the polyphenols most readily absorbed by humans, followed by catechins, flavonones, and flavonols such as quercetin glucosides, with proanthocyanidins tending to be the least readily absorbed (Manach et al., 2005). It should also be noted that some polyphenols are metabolized by the intestinal microbiota and phenolic metabolites play a key role in the biological activity of beer. Nardini et al. (2006) studied the absorption of phenolic acids from beer consumption in humans and found increased plasma concentrations of phenolic acids such as ferulic, vanillic, caffeic, and p-coumaric acid at 30 min postconsumption. These findings suggest that absorption takes place in the proximal part of the intestine. In addition, these authors found that ferulic, vanillic, and caffeic acid were preferably metabolized into sulfates rather than glucuronates.

With regard to the bioavailability and distribution of the prenylated flavonoids and chalcones found in beer, which come almost exclusively from hops, it has been found that these compounds are normally modified by glucuronidation. Studies with experimental animals have shown that xanthohumol was detected in plasma in the form of two mono-glucuronides after oral administration (50 mg/kg) (Yilmazer et al., 2001). Other bioavailability studies in experimental animals showed that xanthohumol and its metabolites can be detected in feces 24h after both oral and intravenous administration (Nookandeh et al., 2004; Avula et al., 2004). Recent bioavailability studies in humans found isoxanthohumol in the urine of beer drinkers and thus this compound can be considered as a good biomarker for beer consumption in humans and as a tool for epidemiological studies (Quifer-Rada et al., 2014).

Various studies have suggested that the bioavailability of phenolic compounds may be affected by the presence of alcohol. Some authors have reported that polyphenols are absorbed better and pass more rapidly into the blood when consumed in alcohol-containing products (Hollman and Katan, 1999; Williamson and Scarlbert, 2000). Similar findings have been reported in studies conducted on experimental animals (Valls-Belles et al., 2005), which showed that the peak-plasma polyphenol concentration occurred at 60 min after consumption of beer, whereas the peak concentration after consumption of alcohol-free beer occurred at 90 min and was somewhat lower. The findings published by Ghiselli et al. (2000), who found that the consumption of alcohol-free beer resulted in a slight tendency for higher plasma antioxidant activity values at 90 min, differ somewhat in this respect. However, no differences between the effect of consuming beer with or without alcohol were detected in the first 30 min.

15.4 CARDIOVASCULAR DISEASES AND BEER

According to the World Health Organization (WHO), cardiovascular diseases will be the leading cause of death worldwide by 2020 (Islam et al., 2014). Obesity, diabetes, hypertension, genetics, dyslipidemia, and unhealthy lifestyle are risk factors in cardiovascular diseases, with hyperlipidemia being one of the most important risk factors in atherosclerotic processes (Gonzalo-Calvo et al., 2014). There is evidence suggesting that oxidative stress lipid peroxidation and oxidative modification of LDL are involved in the onset of atherosclerosis (Csala et al., 2015).

Although alcohol consumption is harmful to health, it is well-known that moderate consumption is associated with a reduction in cardiovascular disease-related mortality (Gorinstein et al., 1989; Constanzo et al., 2010; Hernandez–Hernandez et al., 2015). The preventive effect of moderate beer consumption with respect to cardiovascular diseases may be due to the presence of ethanol or bioactive compounds such as polyphenols (Kondo, 2004). The mechanisms of action by which these compounds exert their cardioprotective effect include an increase in high-density lipoproteins (HDLs), a decrease in platelet aggregation, a decrease in cholesterol and triglyceride levels, antiinflammatory effects, an increase in antioxidant and anticoagulant capacity, etc. (Sohrabvandi et al., 2012).

The antioxidant activity of beer is related to its content of phenolic compounds and its bioavailability, and it is well-known that beer consumption increases the antioxidant capacity of human plasma and reduces atherosclerosis-related oxidative stress (Ghiselli et al., 2000; Vinson et al., 2003). Protective effects related to moderate beer consumption for 1 month were observed with regard to the lipid profile of young adults (25–50 years) (Romeo et al., 2008). Similar results concerning the lipid profile and oxidative-stress biomarkers were observed in an older population (58–73 years) (Martínez-Álvarez et al., 2009).

Although there is a consensus regarding the cardioprotective effects of moderate beer consumption, the contribution of ethanol and nonalcoholic components, mainly polyphenols, remains unclear. In a recent preclinical study, the protective effect of moderate intake of beer with and without alcohol on hypercholesterolemia-induced coronary endothelial dysfunction was attributed to nonalcoholic components (Vilahur et al., 2014). This investigation found that the protective effect of beer is associated with an increase in urinary isoxanthohumol levels, a decrease in coronary oxidative stress, and activation of the Akt/eNOS signaling pathways. In a similar study, Chiva-Blanch et al. (2015) also assessed the contribution of ethanol and phenolic compounds in humans at high CVD risk having a moderate consumption of beer with or without alcohol for 4 weeks and compared with a group consuming gin. These authors concluded that the phenolic content of beer reduces the level of molecules involved in leukocyte and biomarkers of adhesion such as E-selectin, IL-6r (interleukin), and IL-15, among others, whereas alcohol affects the lipid profile and reduces the plasma levels of some inflammatory biomarkers associated with cardiovascular disease.

Another positive effect of moderate beer consumption that has been reported is improvement in arterial function (Karatzi et al., 2013). Moreover, Chiva-Blanch et al. (2014) reported that beer exerts a beneficial health effect by regulating the number of circulating endothelial progenitor cells (EPCs).

It should also be noted that inflammation is involved in cardiovascular risk factors and metabolic syndrome (Geronikaki and Gavalas, 2006). Consequently, beer may have a beneficial effect by exerting various antiinflammatory mechanisms of action. Its main actions involve its ability to inhibit the enzyme inducible nitric oxide synthase (iNOS) and to inhibit enzymes such as cyclo-oxygenase (COX-1), in which different beer flavonoids, such as flavones or the chalcone xanthohumol, are implicated (Milligan et al., 2000; Arranz et al., 2012).

In line with the above, it has been reported that both alcohol and various individual polyphenols regulate the expression of fibrinolytic proteins (t-PA tissue-type plasminogen activator, uPA urokinase-type plasminogen, PAI-1, or plasminogen activator inhibitor, uPAR, or urokinase-type plasminogen receptor and annexin II) at a cellular, molecular, and genetic level, thereby increasing fibrinolytic activity (Boosyse et al., 2007; Arranz et al., 2012). The alteration and reduction of the fibrinolytic activity will increase the risk of thrombotic/atherothrombotic complications. However, systemic factors in circulation, such as alcohol or polyphenols, which are able to increase fibrinolytic activity, can significantly reduce the overall risk for acute thrombotic/atherothrombotic events associated with cardiovascular disease (Booyse et al., 2007).

15.5 ANTIOSTEOPOROSIS EFFECT OF BEER

Osteoporosis is a disease characterized by low bone mass and deterioration of bone tissue, thus resulting in increased bone fragility. It has high prevalence in postmenopausal women due to a deficit in calcitonin secretion. The relationship between beer consumption and bone density has been studied and it has been reported that moderate beer intake is associated with an increase in bone-mineral density in postmenopausal women, possibly as a result of the bioactive compounds present in beer (Tucker et al., 2009). One of these components is silicon, which plays a key role in bone and cartilage growth and development (Reffitt et al., 2003; Sripanyakorn et al., 2004). Indeed, silicon supplementation in postmenopausal women inhibits bone resorption and increases bone-mineral density (Eisinger and Clairet, 1993). Sripanyakorn et al. (2004) found that silicon levels in the serum and urine of healthy individuals increased markedly after drinking beer, with absorption of 50%, thus suggesting that the silicon in beer is bioavailable and that the main route of excretion is urinary.

Flavones, which are also found in beer, have estrogenic effects (Uthian, 1973) and stimulate calcitonin secretion (Harkness et al., 2004). Phytoestrogens are a group of substances with similar structure to the compounds that bind to estrogen receptors (ORs) and exert estrogenic and antioestrogenic effects, thus they have been considered as a therapeutic option for treating osteoporosis. Thus 8-prenylnaringenin, for example, is a prenylflavonoid found in beer that exhibits a powerful phytoestrogenic activity, even exceeding that of genistein and daidzein, thereby making a significant contribution to bone protection (Ming et al., 2013).

15.6 ANTIMUTAGENIC AND ANTICARCINOGENIC EFFECTS OF BEER

The studies performed to date permit neither a firm nor a direct relationship to be established between moderate beer consumption and protection against cancer. There is some evidence suggesting that high alcohol consumption is related to increased risk of cancer, especially cancer of the mouth, pharynx, esophagus, liver, and colorectal cancer. Thus the protective and positive effects that could be exerted by the bioactive components of beer should be only taken into consideration under situations of moderate consumption, in which alcohol intake is also reduced.

Studies in vitro and in experimental animals have suggested that beer consumption inhibited induced colonic carcinogenesis (Nozawa et al., 2004). The majority of these studies concerned phenolic compounds potentially responsible for the cancer-prevention effects. In this regard, it has been reported that, in addition to their antioxidant capacity, phenolic acids such as benzoic and cinnamic acid, flavon-3-ols such as catechin, and flavones exert a chemopreventive effect due to their ability to modulate detoxifying enzymes and to inhibit the enzymes involved in inflammatory processes such as inducible iNOS and COX-1 (Gerhäuser et al., 2003). Other phenols found in beer, such as quercetin and its metabolites, and ferulic acid, are powerful antioxidants and their anticarcinogenic effects have been described in both in vitro and in vivo studies (Chen et al., 2014). Similarly, it is well-known that kaempferol, another flavonol found in beer, has anticarcinogenic effects due to its ability to act as a powerful free-radical scavenger and to preserve the activity of various antioxidant enzymes and glutathione metabolism. It also has antiproliferative effects, induces apoptosis, and has an angiogenic effect, among others (Song et al., 2015).

Xanthohumol, 8-prenylnaringenin, isoxanthohumol, flavanones, humulones, and proanthocyanidins, all of which can be found in beer, have been reported to be anticarcinogenic compounds and, despite their low bioavailability in some cases, may exert their effect in the colon (Arranz et al., 2012). Although the dietary intake of xanthohumol is low, it has the advantage of being lipophilic, thus it has a higher bioactivity than other polyphenols. In vitro and in vivo studies have shown this compound to have antiinflammatory, antioxidant, and anticarcinogenic effects, thereby suggesting a preventive role. Its known mechanisms of action include the inactivation of procarcinogens and the induction of detoxifying enzymes, as well as the ability to inhibit tumor growth due to its antiinflammatory and angiogenic effects (Ming et al., 2013; Negrao et al., 2013; Chen et al., 2014). In colon cancer, xanthohumol has been shown to induce apoptosis significantly by regulation of Bcl-2 and to activate the caspase cascade. Its angiogenic and antiinflammatory effects have been associated with inhibition of the Akt/NF-kB pathway, and it is able to inhibit NO production by suppressing iNOS. Furthermore, humulones may also contribute to the anticarcinogenic effect as a result of their marked angiogenic activity, and both in vitro and in vivo studies have shown that they act by preventing tumor growth and metastasis (Shimamura et al., 2001).

An indirect protective effect of beer with respect to mutagenesis has been associated with the inhibition of heterocyclic amine formation processes due to inhibition of DNA adduct formation (Nozawa et al., 2004; Arimoto-Kobayashi et al., 2005). In addition, it has been reported that beer compounds such as xanthohumol or 8-prenylnaringenin inhibit formation of the procarcinogens typically found in cooked meats (Miranda et al., 2000).

15.7 BEER AND HYDRATION

Hydration in general, and particularly rehydration after physical effort or activity, is essential to preserve health and to ensure well-being. Fluid requirements under normal temperature conditions and with average exercise are estimated to be around 2.5 L/day, with this value increasing to 3.5 L/day in hot weather or when exercising. Some of the established guidelines on minimal fluid consumption to restore homeostatic balance (Sawka et al., 2007) note that ideal drinks made for such purposes should contain low quantities of sodium and potassium (to replace sweat-related loss) and around 6–8% of carbohydrates (to replace the glycogen that may have been used up during exercise).

The main ingredient of beer is water and its preferred characteristic for consumers is its ability to refresh. In addition to water, beer also contains substances that allow heat- and exercise-related water/mineral losses to be replaced, as well as amino acids, B-group vitamins, and antioxidants that contribute to replenish those lost during exercise.

The presence of carbohydrates in water reduces the rate of gastric emptying as sugars have a retarding effect. The major carbohydrates are maltodextrins, which help to attenuate the osmotic effect of glucose and sucrose in the intestine and are metabolized more slowly. Studies performed by Jiménez-Pavón et al. (2009) to evaluate the suitability of beer for rehydrating athletes concluded that moderate consumption of beer after exercise has no negative effects and does not hinder recovery or negatively affect the psychokinetic qualities of athletes who frequently consume this drink.

15.8 EFFECTS OF BEER SUPPLEMENTATION IN BREASTFEEDING MOTHERS

Energy requirements are much higher in pregnancy. Indeed, the average additional energy cost of pregnancy has been estimated at around 80,000 kcal for the 9-month period, thus implying that an overdose of free radicals and, therefore, greater oxidative stress, occurs even during normal pregnancies. In addition, the newborn passes from an intrauterine to an extrauterine environment in which the partial oxygen pressure is higher, thus also implying an induction of free radicals. Consequently, both the mother and child are subjected to high oxidative stress, with hyperproduction of reactive-oxygen species during birth and the first few days of the child's life (Roblers et al., 2001).

The most natural food for newborns is mother's milk and its quality depends on numerous factors, especially nutritional ones. An antioxidant-rich diet favors the reduction of oxidative stress in both the mother and the breastfed infant. In the light of this, in an attempt to improve the oxidative stress situation in both the mother and the newborn child, the diet of breastfeeding mothers was supplemented with alcohol-free beer. After a period of 30 days, it was found that milk from mothers on the supplemented diet had higher antioxidant content, which coincided with higher antioxidant activity or levels of endogenous antioxidants and with less oxidative damage (as measured by biomarkers in different organs). Moreover, analysis of urine samples from the breastfed infants highlighted the fact that the decrease in oxidative damage typical of the infants' progress was much greater and faster for babies from mothers supplemented with alcohol-free beer. This study, therefore, concluded that mothers and babies from the group supplemented with alcohol-free beer presented lower levels of oxidative damage (Codoñer-Franch et al., 2013).

15.9 OTHER HEALTH EFFECTS OF BEER

Homocysteine is an amino acid obtained by demethylation of dietary methionine. High concentrations of this amino acid in the blood, known as hyperhomocysteinemia, are associated with cardiovascular diseases. Although alcohol intake may increase blood homocysteine levels, various studies have associated beer consumption with lower homocysteine levels, possibly due to the folic acid and vitamin B6 content, thus contributing to the protective effect against cardiovascular disease (Mayer et al., 2001; Van der Gaag et al., 2000).

Metabolic syndrome comprises a group of high-risk diseases that include hyperglycemia abdominal fat, dyslipemia, and hypertension. It has been observed that xanthohumol can attenuate some aspects of metabolic syndrome. Thus among other functions, its ability to reduce weight and plasma glucose levels in obese rats fed with xanthohumol has been studied (Legette et al., 2013).

Neurodegenerative disorders such as Alzheimer's disease are some of the most common causes of age-related dementia, with the number of cases increasing steadily in the past few years. The antioxidant properties of the phenolic acids found in beer suggest that it may exert a protective effect against the oxidative stress commonly observed in neurodegenerative diseases (Ebrahimi and Schluesener, 2012). Silicon, another component of beer, has also been related to the prevention of Alzheimer's disease due to its ability to reduce toxic levels of aluminum (Granero et al., 2004). Aluminum is one of the environmental factors that can contribute to the onset of this disease, as it is responsible for an increase in oxidative stress in the brain. In this regard, moderate consumption of beer as a source of silicon has been shown to reduce aluminum-related toxicity by decreasing the bioavailability thereof in the digestive tract and reducing renal resorption of aluminum, thus preventing its accumulation in the brain (González-Muñoz et al., 2007).

Immunomodulatory effects have been described after moderate beer consumption in a population of healthy Spanish adults. It was found that, after consumption of 330 mL of beer in women and 660 mL in men, CD3+ levels in women increased, along with the levels of immunoglobulins IgG, IgM, and IgA, interleukins IL-2, IL-4, IL-10, and IFN-γ (Romeo et al., 2007). Moreover, studies in rats showed that beer consumption stimulates the nonspecific immune system (Sohrabvandi et al., 2012).

15.10 CONCLUDING REMARKS

Beer is a drink rich in different bioactive compounds and its moderate consumption is related to potential health benefits. Its high water content together with moderate salt content means beer can be as beneficial as a sports drink. Furthermore, its fiber content can contribute to lowering cholesterol and can prevent diseases of the colon, and the content of highly bioavailable silicon plays an important role in bone resorption and increases bone-mineral density. Beer shows significant levels of antioxidant capacity, which is mainly due to the presence of different phenolic compounds. These compounds exert preventive effects against cardiovascular disease due to their hypercholesterolemic effect, their capacity to reduce the levels of inflammatory agents, and due to their effect on fibrinolytic activity. Beer phenolics also showed a preventive effect on cancer alterations due to their antimutagenic, anticarcinogenic, and angiogenic properties. New studies about the preventive effect of beer on other dysfunctions and diseases such as those related to brain disease and the potential use of beer to enhance the hydration in elder people that reject water could be future research lines. Furthermore, from a technological point of view, it may be beneficial to look for new recipes and types of beers that show higher levels of interesting bioactive compounds and special sensory characteristics suited for specific segments of populations with specific disease, dysfunction, or nutritional needs.

REFERENCES

Arimoto-Kobayashi, S., Takata, J., Nakandakari, N., Fujioka, R., Okamoto, K., Konuma, T., 2005. Inhibitory effects of heterocyclic amine-induced DNA adduct formation in mouse liver and lungs by beer. Journal of Agricultural and Food Chemistry 53, 812–815.

Arranz, S., Chiva-Blanch, G., Valderas-Martínez, P., Medina-Remón, A., Lamuela-Raventós, R.M., Estruch, R., 2012. Wine, beer, alcohol and polyphenols on cardiovascular disease and cancer. Nutrients 4, 759–781.

Avula, B., Ganzera, M., Warnick, J.E., Feltenstein, M.W., Sufka, K.J., Khan, I.A., 2004. High-performance liquid chromatographic determination of xanthohumol in rat plasma, urine, and fecal samples. Journal of Chromatographic Science 42, 378–382.

Booyse, F.M., Pan, W., Grenett, H.E., Parks, D.A., Darley-Usmar, V.M., Bradley, K.M., Tabengwa, E.M., 2007. Mechanism by which alcohol and wine polyphenols affect coronary heart disease risk. Annals of Epidemiology 17, S24–S31.

Bramforth, C.W., 2004. Beer Health and Nutrition. Blackwell Science Ltd., Oxford, UK.

Chen, W., Becker, T., Qian, F., Ring, J., 2014. Beer and beer compounds: physiological effects on skin health. Journal of the European Academy of Dermatology and Venereology 28, 142–150.

Chiva-Blanch, G., Condines, X., Magraner, E., Roth, I., Valderas-Martínez, P., Arranz, S., Casas, R., Martínez-Huélamo, M., Vallverdú-Queralt, A., Quifer-Rada, P., Lamuela-Raventos, R.M., Estruch, R., 2014. The nonalcoholic fraction of beer increases stromal cell derived factor 1 and the number of circulating endothelial progenitor cells in high cardiovascular risk subjects: a randomized clinical trial. Atherosclerosis 233, 518–524.

Chiva-Blanch, G., Magraner, E., Condines, X., Valderas-Martínez, P., Roth, I., Arranz, S., Casas, R., Navarro, M., Hervas, A., Sisó, A., Martínez-Huélamo, M., Vallverdú-Queralt, A., Quifer-Rada, P., Lamuela-Raventos, R.M., Estruch, R., 2015. Effects of alcohol and polyphenols from beer on atherosclerotic biomarkers in high cardiovascular risk men: a randomized feeding trial. Nutrition, Metabolism and Cardiovascular Diseases 25, 36–45.

Codoñer-Franch, P., Hernández-Aguila, M.T., Navarro-Ruiz, A., López-Jaén, A.B., Borja, C., Valls-Bellés, V., 2013. Nonalcoholic beer increases antioxidant properties of breast milk. Breastfeeding Medicine 8, 164–169.

Costanzo, S., di Castelnuovo, A., Donati, M.B., Lacoviello, L., De Gaetano, G., 2010. Alcohol consumption and mortality in patients with cardiovascular disease: a meta-analysis. Journal of the American College of Cardiology 55, 1339–1347.

Csala, M., Kardon, T., Legeza, B., Lizák, B., Mandl, J., Margittai, É., Puskás, F., Száraz, P., Szelényi, P., Bánhegyi, G., 2015. On the role of 4-hydroxynonenal in health and disease. Biochimica et Biophysica Acta 1852, 826–832.

De Bree, A., Verschuren, W.M., Blom, H.J., Kromhout, D., 2001. Alcohol consumption and plasma homocysteine: what's brewing? International Journal of Epidemiology 30, 626–627.

Del Rio, D., Rodriguez-Mateos, A., Spencer, J.P., Tognolini, M., Borges, G., Crozier, A., 2013. Dietary (poly)phenolics in human health: structures, bioavailability, and evidence of protective effects against chronic diseases. Antioxidant Redox Signaling 18, 1818–1892.

Denke, M.A., 2000. Nutritional and health benefits of beer. American Journal of the Medical Sciences 320, 320–326.

Di Castelnuovo, A., Costanzo, S., Bagnardi, V., Donati, M.B., Iacoviello, L., De Gaetano, G., 2006. Alcohol dosing and total mortality in men and women: an updated meta-analysis of 34 prospective studies. Archives of Internal Medicine 166, 2437–2445.

Diaz, L.E., Montero, A., Gonzalez-Gross, M., Vallejo, A.I., Romeo, J., Marcos, A., 2002. Influence of alcohol consumption on immunological status: a review. European Journal of Clinical Nutrition 56 (Suppl. 3), S50–S53.

Ebrahimi, A., Schluesener, H., 2012. Natural polyphenols against neurodegenerative disorders: potentials and pitfalls. Ageing Research Reviews 11 (2), 329–345.

Eisinger, J., Clairet, D., 1993. Effects of silicon, fluoride, etidronate and magnesium on bone mineral density: a retrospective study. Magnesium Research 6, 247–249.

Gallen, I.W., Rosa, R.M., Esparaz, D.Y., Young, J.B., Robertson, G.L., Batlle, D., Epstein, F.H., Landsberg, L., 1998. On the mechanism of the effects of potassium restriction on blood pressure and renal sodium retention. American Journal of Kidney Diseases 31, 19–27.

Gerhäuser, C., Becker, H., 2009. Phenolic compounds in beer. In: Preedy, V.R. (Ed.), Beer in Health and Disease Prevention. Academic Press, pp. 124–144.

Gerhäuser, C., Klimo, K., Heiss, E., Neumann, I., Gamal-Eldeen, A., Knauft, J., Liu, G.Y., Sitthimonchai, S., Frank, N., 2003. Mechanism-based in vitro screening of potential cancer chemopreventive agents. Mutation Research 523–524, 163–172.

Geronikaki, A.A., Gavalas, A.M., 2006. Antioxidants and inflammatory disease: synthetic and natural antioxidants with anti-inflammatory activity. Combinatorial Chemistry High Throughput Screening 9, 425–442.

Ghiselli, A., Natella, F., Guidi, A., Montanari, L., Fantozzi, P., Scaccini, C., 2000. Beer increases plasma antioxidant capacity in humans. Journal of Nutritional Biochemistry 11, 76–80.

Goñi, I., Diaz-Rubio, M.E., Saura-Calixto, F., 2009. Dietary fiber in beer: content, composition, colonic fermentability, and contribution to the diet. In: Preedy, V.R. (Ed.), Beer in Health and Disease Prevention. Elsevier, pp. 299–307.

González-Muñoz, M.J., Peña, A., Meseguer, I., 2007. Role of beer as a possible protective factor in preventing Alzheimer's disease. Food and Chemical Toxicology 46, 49–56.

Gonzalo-Calvo, D., Llorente-Cortés, V., Orbe, J., Páramo, J.A., Badimon, L., 2014. Altered atherosclerotic-related gene expression signature in circulating mononuclear leukocytes from hypercholesterolemic patients with low HDL cholesterol levels. International Journal of Cardiology 173, 337–338.

Gorinstein, S., Zemser, M., Weisz, M., Halevy, S.H., Martin-Belloso, O., Trakhtenberg, S., 1989. The influence of alcohol-containing and alcohol-free beverages on lipid levels and lipid peroxides in serum of rats. Journal of Nutritional Biochemistry 9, 682–686.

Gorinstein, S., Caspi, A., Zemser, M., Trakhtenberg, S., 2000. Comparative contents of some phenolics in beer red and white wines. Nutrition Research 20, 131–139.

Granero, S., Vicente, M., Aguilar, V., Martínez-Para, M.C., Domingo, J.L., 2004. Effects of beer as a source of dietary silicon on aluminium absorption and retention in mice. Trace Elements and Electrolytes 21, 28–32.

Harkness, L.S., Fiedler, K., Sehgal, A.R., Oravec, D., Lerner, E., 2004. Decreased bone resorption with soy isoflavone supplementation in postmenopausal women. Journal of Women's Health 13, 1000–1007.

Hernandez-Hernandez, A., Gea, A., Ruiz-Canela, M., Toledo, E., Beunza, J.J., Bes-Rastrollo, M., Martinez-Gonzalez, M.A., November 5, 2015. Mediterranean alcohol-drinking pattern and the incidence of cardiovascular disease and cardiovascular mortality: the SUN Project. Nutrients 7 (11), 9116–9122.

Hollman, P.C., Katan, M.B., 1999. Health effects and bioavailability of dietary flavonol. Free Radical Research 31S, 75–80.

Hough, J.S., Briggs, D.E., Stevens, R., Young, T.W., 1982. Malting and Brewing Science. Chapman and Hall, London.

Islam, S.M., Purnat, T.D., Phuong, N.T., Mwingira, U., Schacht, K., Fröschl, G., 2014. Non-communicable diseases (NCDs) in developing countries: a symposium report. Global Health 11 (10), 81.

Jiménez-Pavón, D., Cervantes, M., Castillo, M.J., Romeo, J., Marcos, A., 2009. Idoneidad de la cerveza en la recuperación del metabolismo de los deportistas. Centro de Información Cerveza y Salud (CICS), Madrid.

Jugdaohsingh, R., Anderson, S.H., Tucker, K.L., Elliott, H., Kiel, D.P., Thompson, R.P., Powell, J.J., 2002. Dietary silicon intake and absorption. The American Journal of Clinical Nutrition 75, 887–893.

Karatzi, K., Rontoyanni, V.G., Protogerou, A.D., Georgoulia, A., Xenos, K., Chrysou, J., Sfikakis, P.P., Sidossis, L.S., 2013. Acute effects of beer on endothelial function and hemodynamics: a single-blind, crossover study in healthy volunteers. Nutrition 29, 1122–1126.

Kondo, K., 2004. Beer and health: preventive effects of beer components on lifestyle-relate disease. Biofactors 22, 303–310.

Langner, E., Rzeski, W., 2014. Biological properties of melanoidins: a review. International Journal of Food Properties 17, 344–353.

Legette, L.L., Luna, A.Y., Reed, R.L., Miranda, C.L., Bobe, G., Proteau, R.R., Stevens, J.F., 2013. Xanthohumol lowers body weight and fasting plasma glucose in obese male Zucker fa/fa rats. Phytochemistry 91, 236–241.

Manach, C., Scalbert, A., Morand, C., Rémésy, C., Jiménez, L., 2004. Polyphenols: food sources and bioavailability. The American Journal of Clinical Nutrition 79 (5), 727–747.

Manach, C., Williamson, G., Morand, C., Scalbert, A., Rémésy, C., 2005. Bioavailability and bioefficacy of polyphenols in humans. I. Review of 97 bioavailability studies. The American Journal of Clinical Nutrition 81, 230S–242S.

Martínez Álvarez, J.R., Valls-Bellés, V., López-Jaén, A.B., VillarinoMarín, A., Codoñer-Franch, P., 2009. Effects of alcohol-free beer on lipid profile and parameters of oxidative stress and inflammation in elderly women. Nutrition 25, 182–187.

Mayer, O., Simon, J., Roslova, H., 2001. A population study of beer consumption on folate, and homocysteine concentrations. European Journal of Clinical Nutrition 55, 605–609.

McMurrough, I., Hennigan, G.P., Cleary, K., 1985. Interactions of proteins and polyphenols in worts, beers and model systems. Journal of the Institute of Brewing 91, 93–100.

Milligan, S.R., Kalita, J.C., Pocock, V., Van De Kauter, V., Stevens, J.F., Deinzer, M.L., Rong, H., de Keukeleire, D., 2000. The endocrine activities of 8-prenylnaringenin and related hop (*Humuluslupulus* L.) flavonoids. Journal of the Clinical Endocrinology and Metabolism 85, 4912–4915.

Ming, L.G., Lv, X., Ma, X.M., Ge, B.G., Zhen, P., Song, P., Zhou, J., Ma, H.P., Xian, C.J., Ming, C., 2013. The prenyl group contributes to activities of phytoestrogen 8-prenylnaringenin in enhancing bone formation and inhibiting bone resorption in vitro. Endocrinology 154, 1202–1214.

Miranda, C.L., Yang, Y.H., Henderson, M.C., Stevens, J.F., Santana-Rios, G., Deinzer, M.L., Buhler, D.R., 2000. Prenylflavonoids from hops inhibit the metabolic activation of the carcinogenic heterocyclic amine 2-amino-3-methylimidazo[4,5-f]quinoline, mediated by cDNA-expressed human CYP1A2. Drug Metabolism and Disposition 28, 1297–1302.

Moll, M., Fonknechten, G., Carnielo, M., Flayeux, R., 1984. Changes in polyphenols from raw materials to finished beer. Technical Quarterly of Master Brewers Association of the Americas 21, 79–87.

Murakami, A., Rader, S., Chicoye, E., Goldstein, H., 1989. Effect of hopping on the volatile composition of beer. Journal of the American Society of Brewing Chemists 47, 35–42.

Nardini, M., Natella, F., Scaccini, C., Ghiselli, A., 2006. Phenolic acids from beer are absorbed and extensively metabolized in humans. Journal of Nutritional Biochemistry 17, 14–22.

Negrão, R., Duarte, D., Costa, R., Soares, R., 2013. Isoxanthohumol modulates angiogenesis and inflammation via vascular endothelial growth factor receptor, tumor necrosis factor alpha and nuclear factor kappa B pathways. Biofactors 39, 608–612.

Nookandeh, A., Frank, N., Steiner, F., Ellinger, R., Schneider, B., Gerhäuser, C., Becker, H., 2004. Xanthohumol metabolites in faeces of rats. Phytochemistry 65, 561–570.

Nozawa, H., Yoshida, A., Tajima, O., Katayama, M., Sonobe, H., Wakabayashi, K., Kondo, K., 2004. Intake of beer inhibits azoxymethane-induced colonic carcinogenesis in male Fischer 344 rats. International Journal of Cancer 108, 404–411.

Ortega Herás, M., González-SanJosé, M.L., 2003. Beers: wort production. In: Caballero, B. (Ed.), Encyclopedia of Food Sciences and Nutrition. Elsevier Science, Amsterdam, pp. 429–434.

Pastoriza, S., Rufian-Henares, J.A., 2014. Contribution of melanoidins to the antioxidant capacity of the Spanish diet. Food Chemistry 164, 438–445.

Pennington, J.A.T., 1991. Silicon in foods and diets. Food Additives and Contaminants 8, 97–118.

Piendl, A., 1990. The role of beer in present-day nutrition. Brauwelt International III, 174–176.

Pulido, R., Hernandez-Garcia, M., Saura-Calixto, F., 2003. Contribution of beverages to the intake of lipophilic and hydrophilic antioxidants in the Spanish diet. European Journal of Clinical Nutrition 57, 1275–1282.

Quifer-Rada, P., Martínez-Huélamo, M., Chiva-Blanch, G., Jáuregui, O., Estruch, R., Lamuela-Raventós, R.M., 2014. Urinary isoxanthohumol is a specific and accurate biomarker of beer consumption. Journal of Nutrition 144, 484–488.

Reffitt, D.M., Ogston, N., Jugdaohsingh, R., Cheung, H.F., Evans, B.A., Thompson, R.P., 2003. Orthosilicic acid stimulates collagen type 1 synthesis and osteoblastic differentiation in human osteoblast-like cells in vitro. Bone 32, 127–135.

Rivero, D., Pérez-Magariño, S., González-Sanjosé, M.L., Valls-Belles, V., Codoñer, P., Muñiz, P., 2005. Inhibition of induced DNA oxidative damage by beers: correlation with the content of polyphenols and melanoidins. Journal of Agricultural and Food Chemistry 53, 3637–3642.

Roblers, R., Palomino, N., Robles, A., 2001. Oxidative stress in neonate. Early Human Development 65, S75–S81.

Romeo, J., Wärnberg, J., Nova, E., Díaz, L.E., González-Gross, M., Marcos, A., 2007. Changes in the immune system after moderate beer consumption. Annals of Nutrition and Metabolism 51, 359–366.

Romeo, J., González-Gross, M., Wärnberg, J., Dıaz, L.E., Marcos, A., 2008. Effects of moderate beer consumption on blood lipid profile in healthy Spanish adults. Nutrition, Metabolism and Cardiovascular Diseases 18, 365–372.

Sanchez-Martinez, M., Da Silva, E.G.,P., Perez-Corona, T., Camara, C., Ferreira, S.L.C., Madrid, Y., 2012. Selenite biotransformation during brewing. Evaluation by HPLC-ICP-MS. Talanta 88, 272–276.

Saura-Calixto, F., Goni, I., Martin Albarran, C., Pulido, R., 2002. In: Cerveza y Salud. Fibra dietetica en la cerveza: DietaryFiber in Beer -Contenido composición y evaluación nutricional. Centro de Información Cerveza y Salud, Madrid. www.cervezaysalud.com.

Sawka, M.N., Burke, L.M., Eichner, E.R., Maughan, R.J., Montain, S.J., Stachenfeld, N.S., 2007. American College of Sports Medicine position stand. Exercise and fluid replacement. Medicine and Science in Sports and Exercise 39, 377–390.

Sendra, J.M., Carbonell, J.V., 1999. Evaluación de las propiedades nutritivas, funcionales y sanitarias de la cerveza, en comparación con otras bebidas. Centro Información Cerveza y Salud, Madrid.

Sendra, J.M., Todo, V., Pinaga, F., Izquierdo, I., Carbonell, J.V., 1994. Evaluation of the effects of yeast and fermentation conditions on the volatile concentration profiles of pilot-plant lager beers. Monatsschrift fur Brauwissenschaft 47, 316–321.

Shimamura, M., Hazato, T., Ashino, H., Yamamura, Y., Iwasaki, E., Tobe, H., Yamamoto, K., Yamamot, S., 2001. Inhibition of angiogenesis by humulone, a bitter acid from beer. Biochemistry and Biophysics Research Communications 289, 220–224.

Sohrabvandi, S., Mortazavian, A.M., Rezaei, K., 2012. Health-related aspects of beer: a Review. International Journal of Food Properties 15, 350–373.

Song, H., Bao, J., Wei, Y., Chen, Y., Mao, X., Li, J., Yang, Z., Xue, Y., 2015. Kaempferol inhibits gastric cancer tumor growth: an in vitro and in vivo study. Oncology Report 33, 868–874.

Sripanyakorn, S., Jugdaohsingh, R., Elliott, H., Walker, C., Mehta, P., Shoukru, S., Thompson, R.P., Powell, J.J., 2004. The silicon content of beer and its bioavailability in healthy volunteers. British Journal of Nutrition 9, 403–409.

Stavri, M., Schneider, R., O'Donnell, G., Lechner, D., Bucar, F., Gibbons, S., 2004. The antimycobacterial components of hops (*Humulus lupulus*) and their dereplication. Phytotherapy Research 18, 774–776.

Tedesco, I., Nappo, A., Petitto, F., Iacomino, G., Nazzaro, F., Palumbo, R., Russo, G.L., 2005. Antioxidant and cytotoxic properties of lyophilized beer extracts on HL-60 cell line. Nutrition Cancer 52, 74–83.

Tucker, K.L., Kiel, D.P., Powell, J.J., Qiao, N., Hannan, M.T., Jugdaohsingh, R., 2001. Dietary silicon and bone mineral density: the Framingham Study. Journal of Bone and Mineral Research 16 (Suppl. 1), S510.

Tucker, K.L., Jugdaohsingh, R., Powell, J.J., 2009. Effects of beer, wine, and liquor intakes on bone mineral density in older men and women. American Journal of Clinical Nutrition 89, 1188–1196.

Uthian, W.H., 1973. Comparative trial of P1496, a new non-steroidaloestrogen. British Medical Journal 1, 579–581.

Valls-Bellés, V., Torres-Rodríguez, M.C., Muñiz, P., Boix, L., González-SanJose, M.L., Codoñer-Franch, P., 2004. The protective effects of melanoidins in adriamycin-induced oxidative stress in rat isolated hepatocytes. Journal of the Science of Food and Agriculture 80, 1701–1707.

Valls-Bellés, V., Codoñer-Franch, P., González-San José, M.L., Muñiz-Rodríguez, P., 2005. Biodisponibilidad de los flavonoides de la cerveza. Efecto antioxidante "in vivo". Centro de Información Cerveza y Salud, Madrid.

Valls-Belles, V., Torres, M.C., Boix, l., Muñiz, P., Codoñer-Franch, P., 2008. α-Tocopherol, MDA, HNE and 8-OHdG levels in liver and heart mitochondria of adriamycin-treated rats fed with alcohol-free beer. Toxicology 249, 97–101.

Valls-Bellés, V., Torres, M.C., Muñiz, P., Codoñer-Franch, P., 2010. Changes in mitochondrial rat liver and heart enzymes (complex I and complex IV) and coenzymes Q9 and Q10 levels induced by alcohol-free beer consumption. European Journal of Nutrition 49, 181–187.

Van der Gaag, M.S., Ubbink, J.B., Sillanaukee, P., Nikkari, S., Hendriks, H.F., 2000. Effect of consumption of red wine, spirits, and beer on serum homocysteine. Lancet 355, 1522.

Verzele, M., 1986. 100 years of hop chemistry and its relevance to brewing. Journal of the Institute of Brewing 92, 32–48.

Vilahur, G., Casani, L., Mendieta, G., Lamuela-Raventos, R.M., Estruch, R., Badimon, L., 2014. Beer elicits vasculoprotective effects through Akt/eNOS activation. European Journal of Clinical Investigation 44, 1177–1188.

Viñas, P., Aguinaga, N., López-García, I., Hernández-Córdoba, M., 2002. Determination of cadmium, aluminium, and copper in beer and products used in its manufacture by electrothermal atomic absorption spectrometry. Journal of AOAC International 87, 736–743.

Vinson, J.A., Mandarano, M., Hirst, M., Trevithick, J.R., Bose, P., 2003. Phenol antioxidant quantity and quality in foods: beers and effect of two types of beer on an animal model of atherosclerosis. Journal of Agricultural and Food Chemistry 51, 5528–5533.

WHO Williamson, G., Scarlbert, A., 2000. Dietary intake and bioavailability of polyphenols. Journal of Nutrition 130, 2073S–2085S.
Woffenden, H.M., Ames, J.M., Chandra, S., Anese, M., Nicoli, M.C., 2002. Effect of kilning on the antioxidant and pro-oxidant activities of palemalts. Journal of Agriculture and Food Chemistry 50, 4925–4933.
Yilmazer, M., Stevens, J.F., Deinzer, M.L., Buhler, D.R., 2001. In vitro biotransformation of xanthohumol, a flavonoid from hops (*Humuluslupulus*), by rat liver microsomes. Drug Metabolism and Disposition 29, 223–231.

Chapter 16

Fermented Pulses in Nutrition and Health Promotion

J. Frias, E. Peñas, C. Martinez-Villaluenga
Institute of Food Science, Technology and Nutrition (ICTAN-CSIC), Madrid, Spain

16.1 INTRODUCTION

Pulses are edible seeds of the family *Leguminosae* (*Fabaceae*) that include important grain species that rank second after cereals in their importance for human nutrition. According to the definition of the Food and Agricultural Organization of the United Nations, pulses include "*Leguminosae* crops harvested solely for dry grain, including dry bean, pea, lentil, chickpea and faba bean." The FAO (1994) includes 11 primary pulses (Table 16.1) and excludes oil-crop legume seeds (soybean and peanut) and those harvested green as fresh vegetables (green peas and green beans). Pulses are suited for sowing in a wide range of climate and soil conditions due to their unique capability to form nitrogen-fixing nodules harboring rhizobia, requiring low or no nitrogen fertilizers while providing the associated benefits to agriculture. In addition, their cultivation in rotation lowers the carbon footprint of other crops by decreasing the greenhouse gases released in comparison with nitrogen-fertilized systems, making them an elective sustainable crop available worldwide (Gan et al., 2011). Their production covers 78 million ha around the world for a total production of 70 million tons, contributing two-thirds to the world's food supply (FAOSTAT, 2014).

Pulses are characterized by their unique nutritional value as source of proteins, carbohydrates, fiber, vitamins, minerals, and phytochemicals with recognized health benefits. Pulses are consumed cooked, roasted, germinated, and fermented and constitute a significant part of the daily diet of most of the world's population. Pulse-derived fermented products vary widely in different regions of the world due to well-defined preferences. Table 16.2 lists some of the most popular legume-fermented foods produced in different countries.

Fermentation causes desirable biochemical changes in nutritional composition by the microorganisms involved: the indigenous microbiota present on the legume surface or the added starter cultures (Harlander, 1992). Fermentation improves pulse flavor, texture, appearance, shelf-life, nutrient digestibility, and nutritional quality. Furthermore, this process decreases nonnutritional

TABLE 16.1 Scientific Names of Pulses According to FAO (1994)

Pulses, Common Name	Botanic Name
1. Dry beans	
Kidney beans, haricot bean, pinto bean, navy bean	*Phaseolus vulgaris*
Adzuki bean	*Vigna angularis*
Butter bean, lima bean	*Phaseolus lunatus*
Mungo bean, golden bean, green gram	*Phaseolus aureus/Vigna radiata*
Black gram, urd	*Phaseolus mungo*
Scarlet runner bean	*Phaseolus coccineus*
Rice bean	*Vigna calcaratus*
Moth bean	*Vigna aconitifolius*
Tepary bean	*Phaseolus acutifolius*
2. Dry broad beans	
Horse bean	*Vicia faba* var. *equina*
Broad bean	*Vicia faba* var. *major*
Field bean	*Vicia faba* var. *minor*
3. Dry peas	*Pisum sativum, P. arvense*
4. Chickpeas, bengal gram, garbanzos	*Cicer arietinum*
5. Cowpea, blackeye pea/bean	*Vigna sinensis, Dolichos sinensis*
6. Pigeon pea, cajan pea	*Cajanus cajan*
7. Lentils	*Lens culinaris, L. esculenta*
8. Bambara groundnut	*Vigna subterranea*
9. Vetch	*Vicia sativa*
10. Lupins	*Lupinus* spp.
11. Minor pulses	
Jack bean, sword bean	*Canavalia* spp.
Guar bean	*Cyamopsis tetragonoloba*
Winged bean	*Psophocarpus tetragonolobus*
Velvet bean	*Stizolobium* spp.
Yam bean	*Pachyrhizus erosus*

TABLE 16.2 Most Representative Indigenous Pulse-Based Fermented Foods of Different Countries

Product	Country of Origin	Substrates	Fermentation Type	Main Microorganisms Involved	Recipe	Food Use
Dawadawa	West and Central Africa	Locust bean and local pulses	Solid-state fermentation	*Bacillus subtilis, B. pumilus, B. licheniformis, Bacillus* spp.	Boiled dhal are spread calabash tray, covered with a cloth, and fermented for 2–4 days	Meat substitute
Dhokla	India	Chickpeas, green gram, rice (2,2,1)	Liquid-state fermentation	*Lactobacillus fermentum, Leuconostoc mesenteroides, Streptococcus faecalis*	Ingredients are mixed and fermented naturally for 10–12 h	Steamed cake for breakfast or snack food
Doenjang	Korea	Black soybeans, local pulses	Liquid-state fermentation	*Leuconostoc mesenteroides, Enterococcus faecium, Tetragenococcus halophilus, Bacillus subtilis, Mucor plumbeus*	Mashed cooked seeds in ball shape are wrapped in rice straw and fermented for 1–6 months	Meat substitute
Cheonggukjang	Korea	Black soybeans, local pulses	Solid-state fermentation	*Bacillus subtilis, B. amyloliquefaciens, Rhizopus oligosporus*	Soaked and steamed seeds are inoculated with spores of *Bacillus* and fermented for 1–3 days	Main course as meat substitute

Continued

TABLE 16.2 Most Representative Indigenous Pulse-Based Fermented Foods of Different Countries—cont'd

Product	Country of Origin	Substrates	Fermentation Type	Main Microorganisms Involved	Recipe	Food Use
Dosa	India	Black gram and rice (1,1)	Liquid-state fermentation	*Leuconostoc mesenteroides, Lactobacillus delbrueckii, Lactobacillus fermentum, Streptococcus faecalis, Bacillus* spp., yeasts	Soaked dhal and parboiled rice fermented for 10–12 h	Steamed cake for breakfast or snack food
Idli	India, Sri Lanka	Black gram and rice (1,2)	Liquid-state fermentation	*Leuconostoc mesenteroides, Lactobacillus delbrueckii, Lactobacillus fermentum, Lactobacillus lactis, Pediococcus cerevisiae, Streptococcus lactis, Streptococcus faecalis,* yeasts	Soaked dhal and parboiled rice are fermented for 10–12 h	Breakfast food
Khaman	India	Chickpeas	Liquid-state fermentation	*Leuconostoc mesenteroides, Lactobacillus fermentum, Lactobacillus lactis, Pediococcus acidilactici, Bacillus* spp.	Ground dhal is fermented for 10–12 h	Breakfast food or snack

Fermented Pulses in Nutrition and Health Promotion Chapter | 16 389

Natto	Japan	Soybeans, local pulses	Solid-state fermentation	Bacillus natto	Soaked and steamed seeds are inoculated with spores of B. natto and fermented for 15–24 h	Main course as meat substitute
Tempeh	Indonesia, New Guinea, Surinam	Soybeans, chickpeas, groundnut, local pulses	Mold fermentation	Rhizopus oligosporus, Aspergillus oryzae	Soaked and steamed seeds are inoculated with spores of R. oligosporus and fermented for 24–48 h	Breakfast food or snack
Ugba	Nigeria	Locust bean and local pulses	Solid-state fermentation	Bacillus ssp., Staphylococcus, Micrococcus	Boiled dhal is sliced, wrapped in banana leaves, and fermented naturally for 4–5 days	Side dish
Wari	India, Pakistan	Black gram or bengal gram	Liquid-state fermentation	Saccharomyces cereviseae, Candida krusei, Lactobacillus ssp, Leuconostoc mesenteroides, Lactobacillus fermentum	Ground soaked dhal is backslopped for 1–3 days and ball molded	Fried balls as side dish

compounds present in legume seeds such as protease inhibitors, oligosaccharides, phytate, and lectins (Desphande et al., 2000).

Fermented legumes are currently receiving a great deal of attention for their health-promoting properties and disease-preventing effects. Although most of the human-controlled trials conducted thus far have been focused on soybean products (Sugano, 2005), there are many other fermented pulse products in which attributes are derived from their nutrient and phytochemical constituents and the probiotic features of microorganisms involved in fermentation. In this context, the objective of this chapter is to describe the nutritional composition of pulses, the biochemical changes occurring during fermentation, and the emerging evidence showing their potential effect on chronic disease prevention.

16.2 NUTRITIONAL AND PHYTOCHEMICAL COMPOSITION OF PULSES AND THEIR HEALTH BENEFITS

Pulses have a well-balanced nutritional composition consisting of considerable amounts of proteins, starch, fiber, vitamins, minerals, and phytochemicals (Tharanathan and Mahadevamma, 2003). The proximate composition and nutritive attributes of most representative pulses are given in Table 16.3.

16.2.1 Proteins

Grain legumes are considered to be one of the most important sources of low-cost proteins (~20–40% on a dry weight basis) and play an important role in the human diet. The protein quality of legumes is lower than in meat and dairy proteins due to their relatively low content in sulfur-containing amino acids, methionine and cysteine, as well as the aromatic amino acid tryptophan. However, they contain significant amounts of lysine, which is the limiting amino acid in cereals. Hence, when eaten together, they provide high-quality proteins (Boye et al., 2010). Moreover, legume proteins provide functional properties beneficial to human health and their consumption represents a healthier option in mitigation of obesity, inflammation, and diabetes (Ley et al., 2014; Wang and Beydoun, 2009). It has been reported that legume proteins are precursors of bioactive peptides exhibiting antimicrobial, anticancer, antihypertensive, hypocholesterolemic, and antioxidant activities (Mrudula et al., 2012; Roy et al., 2010; Zambrowicz et al., 2013), as was shown in Chapter 3.

Minor components of grain-legume proteins have been cataloged in the past as antinutrients since they negatively affect nutrient digestibility (Boye et al., 2010). However, the term antinutrient is often a misnomer as its involvement in health-promoting processes is gaining increasing attention. Protease inhibitors are capable of blocking trypsin and chymotrypsin activities in the gut, reducing protein digestibility (Muzquiz et al., 2012). However, the Bowman–Birk family of protease inhibitors has shown anticarcinogenic and anti-inflammatory effects in human colon cancer-cell lines (Clemente and Arques,

TABLE 16.3 Proximate Composition, Minerals, Fatty Acids, Vitamins, and Amino Acids of Some Pulses (USDA, National Nutrient Database for Standard Reference Rel. 27)

Nutrients	Dry Beans	Dry Peas	Chickpeas	Cowpeas	Pigeon Pea	Lentils	Bengal Gram
Proximates (g/100 g)							
Protein	23.58	23.82	20.47	23.85	21.70	24.63	23.86
Total lipids	0.83	1.16	6.04	2.07	1.49	1.06	1.15
Carbohydrates	60.01	63.74	62.95	59.64	62.78	63.35	62.62
Total dietary fiber	24.9	25.5	12.2	10.7	15.0	10.7	16.3
Water	11.75	8.62	7.68	11.05	10.59	8.26	9.05
Energy (Kcal/100 g)	333	352	378	343	343	352	347
Minerals (mg/100 g)							
Calcium, Ca	143	37	57	85	130	35	132
Iron, Fe	8.20	4.82	4.31	9.95	5.23	6.51	6.74
Magnesium, Mg	140	49	79	333	183	47	189
Phosphorus, P	407	321	252	438	367	281	367
Potasium, K	1406	823	718	1375	1392	677	1246
Sodium, Na	24	15	24	58	17	6	15
Zinc, Zn	2.79	3.55	2.76	6.11	2.76	3.27	2.68
Cupper, Cu	0.96	0.82	0.66	0.84	1.06	0.75	0.94

Continued

TABLE 16.3 Proximate Composition, Minerals, Fatty Acids, Vitamins, and Amino Acids of Some Pulses (USDA, National Nutrient Database for Standard Reference Rel. 27)—cont'd

Nutrients	Dry Beans	Dry Peas	Chickpeas	Cowpeas	Pigeon Pea	Lentils	Bengal Gram
Manganese, Mn	140	49	79	184	183	47	189
Selenium, Se	3.20	4.10	0.00	9.00	8.20	0.10	8.20
Fatty Acids (g/100 g)							
Fatty acids, total saturated	0.120	0.161	0.603	0.542	0.330	0.154	0.348
Fatty acids, total monounsaturated	0.064	0.242	1.377	0.173	0.012	0.193	0.161
Fatty acids, total polyunsaturated	0.457	0.495	2.731	0.889	0.814	0.526	0.384
Vitamins (mg/100 g)							
Vitamin C	4.5	1.8	4.0	1.5	0.0	4.5	4.8
Thiamin, B1	0.529	0.726	0.477	0.680	0.643	0.873	0.621
Riboflavin, B2	0.219	0.215	0.212	0.170	0.187	0.211	0.233
Niacin, B3	2.060	2.889	1.541	2.795	2.965	2.605	2.251
Piridoxin, B6	0.397	0.174	0.535	0.361	0.283	0.540	0.382
Folate	0.394	0.274	0.557	639	0.456	0.479	0.625
Vitamin E	0.22	0.09	0.82	0.63	0.77	0.49	0.51
Vitamin K	0.019	0.015	0.009	0.005	0.006	0.005	0.009

Amino Acids (g/100 g)								
Trp	0.28	0.28	0.20	0.29	0.21	0.22	0.26
Thr	0.99	0.87	0.77	0.90	0.77	0.88	0.78
Iso	1.04	1.01	0.88	0.96	0.78	1.06	1.01
Leu	1.74	1.76	1.46	1.80	1.55	1.79	1.85
Lys	1.59	1.77	1.38	1.59	1.52	1.72	1.66
Met	0.33	0.25	0.27	0.34	0.24	0.21	0.29
Cys	0.24	0.37	0.28	0.26	0.25	0.32	0.21
Phe	1.18	1.13	1.10	1.37	1.86	1.23	1.44
Tyr	0.62	0.71	0.51	0.76	0.54	0.67	0.71
Val	1.14	1.16	0.86	1.12	0.94	1.22	1.24
Arg	1.35	2.19	1.94	1.63	1.30	1.90	1.67
His	0.61	0.60	0.57	0.73	0.77	0.69	0.70
Ala	0.92	1.08	0.88	1.07	0.97	1.03	1.08
Asp	2.64	2.90	2.42	2.84	2.15	2.72	2.95
Glu	3.33	4.20	3.60	4.45	5.03	1.03	4.13
Gly	0.85	1.09	0.86	0.97	0.80	1.03	1.05
Pro	0.93	1.01	0.85	1.06	0.96	1.00	1.07
Ser	1.19	1.08	1.04	1.18	1.03	1.14	1.33

2014). α-Amylase inhibitors found in white beans bind irreversibly α-amylase enzyme decreasing basal and postpandrial circulating glucose, and hence insulin flow. These effects may contribute to weight control in humans (Celleno et al., 2007). Similarly, lectins are glycoproteins able to bind to the gut epithelial cells and interfere with nutrient digestion and absorption. In addition, they reduce the activities of mucosal intestinal maltase and invertase and thus alter glucose transportation (Vasconcelos and Oliveira, 2004). In contrasts to the reported negative effects, legume lectins can as act as antitumor agents causing cytotoxicity, apoptosis, and inhibition of tumor growth (De Mejía and Prisecaru, 2005). Several studies have proposed their application as nutraceuticals for weight management due to their ability to decrease nutrient absorption (Roy et al., 2010). Besides these biological activities, lectins also exert immunomodulatory, antimicrobial, and HIV-1 reverse transcriptase inhibitory activities (Hamid et al., 2013).

16.2.2 Carbohydrates

Approximately 55–65% of the total weight of pulses is carbohydrates, which is one of the most important sources of energy in the human diet. Pulse starch accounts for 22–45% of the seeds (Hoover and Ratnayake, 2002). Starch granular structure, formed by linear amylose and complex amylopectin, is difficult to digest (Hoover et al., 2010), which is related to the low glycemic index (GI) of pulses that are particularly beneficial to diabetic patients (McCrory et al., 2010). However, the GI depends on the type of pulse and the processing method (Atkinson et al., 2008).

Starch has been classified as rapidly digested starch (RDS), slowly digested starch (SDS), and resistant starch (RS) according to the rate of glucose release and its absorption throughout the course of the small intestine (Englyst et al., 1992). In addition, the higher amylose content in pulse starches is associated with its high capacity for retrodegradation, reducing the starch digestion rate (Hoover et al., 2010). Studies on the health benefits of legumes have linked SDS to diabetes management and promotion of satiation (Lehmann and Robin, 2007). In addition, health benefits of RS have an important role in digestive physiology, providing fermentable carbohydrates to colon microbiota and contributing to the prevention of colon cancer, providing a hypoglycemic effect and modulating fasting plasma triglyceride and cholesterol levels and absorption of minerals (Raigond et al., 2015).

Pulses also contain significant amounts of α-galactosides (0.5–12%), a significant group of nondigestible oligosaccharides represented by raffinose (a trisaccharide), stachyose (a tetrasaccharide), and verbascose (a pentasaccharide). These oligosaccharides resist digestion in the upper gut and pass to the large intestine to be fermented by the colon microbiota, decreasing the pH, releasing short-chain fatty acids (SCFA), and hence promoting gut health as prebiotics (Martínez-Villaluenga et al., 2008). Daily ingestion of 3 g of α-galactosides for

2 weeks has shown a bifidogenic effect without negative physiological discomfort (Hayakawa et al., 1990; Tomomatsu, 1994). In addition, animal studies have demonstrated that besides the prebiotic effect, α-galactosides are emerging as antioxidants, blood-glucose regulators, lipid-profile enhancers, and immune-stimulators in animal trials (Chen et al., 2010; Xie et al., 2012).

16.2.3 Lipids

Pulses contain a very low amount of lipids (2–6%) compared to those legumes used for oil extraction. Among pulses, chickpeas are the richest, followed by kidney beans, lentils, and peas. Legume seeds have low saturated fatty acids (SFA) content while unsaturated fatty acids (UFA) are predominant, with variable proportions of monounsaturated (MUFA), mostly palmitic (16:0) and stearic (18:0) acids, and polyunsaturated (PUFA), represented by oleic (18:1), linoleic (18:2), and linolenic (18:3) acids (Kalogeropoulos et al., 2010). Observational studies have shown an association between diets rich in α-linolenic acid and moderately lower risk of cardiovascular diseases (Mensink et al., 2003).

16.2.4 Dietary Fiber

Legumes are rich in dietary fiber (10–35%). Fiber fermentation in the colon enhances the growth of beneficial bacteria and the production of SCFA associated with the prevention of colon cancer (Ferrarelli, 2015). Other major health benefits linked to increased intake of dietary fiber include laxation, reduced risk of diabetes, cardiovascular diseases, and overweight (Fardet, 2010). Although the mechanisms are not fully understood, the beneficial effects of the consumption of dietary fiber seem to be due to their considerable structural diversity associated with water-holding capacity, viscosity, stool volume, ability to bind bile acids and fermentability, cholesterol-lowering activity, and modulation of blood-glucose and insulin levels (Phillips, 2013; Tosh and Yada, 2010).

16.2.5 Minerals and Vitamins

Pulses provide significant amounts of the B vitamins thiamine, riboflavin, niacin, pyridoxine, and folic acid (Prodanov et al., 1997), as well as fat-soluble vitamins A and E (Doblado et al., 2005). Pulses contain important amounts of essential minerals such as iron (Fe), zinc (Zn), calcium (Ca), potassium (K), and selenium (Se) and are low in sodium (Na) (Broughton et al., 2003). High levels of K increase insulin secretion, whereas K deficiency is associated with glucose intolerance and impaired insulin secretion (Chatterjee et al., 2011), while low Na content favors the maintenance of normal blood pressure (Miura et al., 2010). In pulses, the bioavailability of Ca and Zn is approximately 20%, while that of Fe is relatively low (Sandberg, 2002). This low availability of minerals is mainly due to phytic acid that impairs the full biological potential of divalent minerals by the strong chelating power associated with its six reactive phosphate groups (Urbano et al., 2000). Pulses provide considerable amounts of phytate, ranging from 0.2% in some *Phaseolus vulgaris* species to 2% in the soybean cultivars (Muzquiz et al., 2012) and may contribute to Fe deficiency when pulses are consumed as a staple food (Petry et al., 1992). Processes such as soaking, germination, and fermentation reduce the phytate content, enhancing the bioavailability of essential minerals (Luo et al., 2013).

16.2.6 Phytochemicals

Legumes are good sources of phytochemicals as phenolic compounds. Phenolics have proven to display antioxidant and antiinflammatory activities, thus protecting against oxidative stress, which is closely related to chronic diseases such as cardiovascular diseases and metabolic disorders (Andriantsitohaina et al., 2012; Soobrattee et al., 2005; Wang et al., 2011).

Dietary phenolic compounds in pulses vary from 102 mg/100 g in lentils (Xu and Chang, 2007) to 90 mg/g in dark-pigmented varieties of adzuki beans (Amarowicz and Pegg, 2008). As was recently reviewed (Vaz Patto et al., 2015), *trans*-ferulic, *trans-p*-coumaric, and syringic acids are in nearly all species and the hulls may contain gallic, syringic, *p*-hydroxybenzoic, protocatechuic, *p*-coumaric, vanillic, caffeic, sinapic, and ferulic acids. Pulses are a good source of flavonoids, represented by flavan-3-ols, flavanols, flavones, and anthocyanidins (Amarowicz and Pegg, 2008; Dueñas et al., 2015a). Isoflavones, a subclass of flavonoids widely studied in soybeans, have been only found in some other legumes including chickpeas and lupins (Ranilla et al., 2009; USDA, 2002). In pulses, flavonoids are mostly esterified to sugar moieties or bound to cell-wall polymers and their bioavailability is relatively low (Bouchenak and Lamri-Senhadji, 2013). A small percentage of phenolic compounds (5–10%) may be readily absorbed in the small intestine but most of them reach the colon almost intact evading the gastrointestinal enzymes (Appeldoorn et al., 2009). In the colon, they are hydrolyzed by colon microbiota enzymes to form more bioactive aglycones and metabolites that contribute to the maintenance of gut health (Dueñas et al., 2015b; Monagas et al., 2010).

Saponins are complex structures based on triterpene or nonpolar steroid nucleous (the aglycone) and polar glycosylations. Saponins are amphiphilic in nature and have been considered antinutritional factors due to their haemolytic activity. However, soyasaponins present in pulses are not toxic and also interact with bile acids and cholesterol-forming insoluble micelles, thus increasing fecal cholesterol excretion resulting in lower blood cholesterol levels (Milgate and Roberts, 1995). Other evidence suggests that legume saponins stimulate the immune system and may exert anticancer activity, decrease blood lipids, and inhibit platelet aggregation, playing a chemopreventive role against cardiovascular diseases (Shi et al., 2004). Soyasaponins have been found in many edible legumes such as lentils, chickpeas, and various species of beans and peas, ranging from traces to 60 g/kg (Khokhar and Chauhan, 1986; Ruiz et al., 1996; Shi et al., 2004).

Pulses are the main natural source of phytosterols, occurring as sterol glucosides and esterified sterol glucosides. They are associated with cholesterol-lowering properties, acting as competitive inhibitors in the intestinal cholesterol uptake (Gylling et al., 2014). β-Sitosterol, stigmasterol, campesterol, and stigmastanol have been found in lentils, chickpeas, and white beans, accounting for 1 g/kg (Jiménez-Escrig et al., 2006). Pulse consumption has been inversely associated with cardiovascular diseases, increasing hepatic secretion of cholesterol by partially interrupting the enterohepatic circulation of the bile acids, an effect attributed to phytosterols (Duane, 1997).

16.3 NUTRITIONAL CHANGES DURING FERMENTATION OF PULSE-BASED FOODS

Fermentation is an economic and well-suited process to convert and improve the nature of raw and poor digestible legumes into more acceptable, palatable, safe, nutritive, and healthy value-added edible products. The microorganisms involved in legume fermentation hydrolyze and metabolize seed constituents resulting in the production of derived-valuable products and have the ability to produce antimicrobial compounds and desirable organic acids that can preserve the food by the suppression of the growth and survival of undesirable microflora. Thus fermentation provides several advantages over other conventionally feasible methods of legume processing, in addition to being less expensive (Leroy and De Vuyst, 2004; Steinkraus, 1996).

16.3.1 Changes in Protein and Amino Acids During Fermentation

Fermented-food pulses contribute to the diet as an important source of proteins, eg, from 12–18% in *dosa* and *idli* products to 40% in *tempeh*-like derivatives (Krishnamoorthy et al., 2013; Starzyńska-Janiszewska et al., 2015). The positive changes in protein quality of pulses during lactic-acid fermentation have been reported by several researchers. In the production of *dosa*, fermentation of black gram led to a slight increase in the proteinase activity that brings about an increase in the total nitrogen, soluble proteins (Soni et al., 1985), and well-balanced amino acid batter with almost 50% of essential amino acids over total amino acids (Balasubramanian et al., 2015; Palanisamy et al., 2012). The replacement of black gram by mung bean produced more nutritious batters in terms of total nitrogen, protein, and free amino acids (Soni and Sandhu, 1989). *Idli* products provide larger limiting amino-acid scores and in vitro protein digestibility than unfermented seeds (Riat and Sadana, 2009). In addition, the active yeast fermentation of different varieties of cowpeas, peas, and kidney beans improved the protein chemical score and essential amino-acid index, contributing to the overall protein quality (Khattab et al., 2009). Lactic-acid fermentation of faba bean led to increased amounts of essential amino acids and the hypotensive γ-aminobutyric acid (GABA), and enhanced the in vitro protein digestibility (Coda et al., 2015). Similarly, during the production of *ugba* with local kidney beans, the content of essential amino acids was improved to meet FAO dietary requirements (Audu and Aremu, 2011). Similarly, chickpeas, common beans, and bambara groundnut *tempeh* foods exhibited higher in vitro protein digestibility, total amino-acid content, available Lys, as well as the calculated protein efficiency ratio than raw flour (Angulo-Bejarano et al., 2008; Bujang and Taib, 2014; Reyes-Bastidas et al., 2010; Reyes-Moreno et al., 2004). In addition, certain strains of *Rhizopus* exhibited higher proteolytic activity by releasing several times more amino acids than other strains (Baumann and Bisping, 1995).

Protein inhibitors are widely reduced after pulse fermentation, contributing to the enhancement of the overall protein quality. Trypsin inhibitory activity was mostly eliminated during the preparation of *tempeh*-like products from cowpeas, ground beans, and chickpeas, and a concomitant increase on protein digestibility was achieved (Abu-Salem and Abou-Arab, 2011; Hemalatha et al., 2007; Egounlety and Aworh, 2003). It has also been reported that lactic-acid bacteria (LAB) led to undetectable activity of trypsin inhibitors (Shimelis and Rakshit, 2008; Coda et al., 2015), while fermentation of different varieties of cowpeas, peas, and kidney beans with *Saccharomyces cerevisiae* caused only a decrease between 37% and 49% (Khattab and Arntfield, 2009).

16.3.2 Changes in Starch, Carbohydrates, and Dietary Fiber During Fermentation

Pulses have a protective, fibrous, indigestible hull that represents among 0.09–0.3% of dry seeds. Seed hull contains a considerable amount of insoluble fiber and this fraction can be removed when dehulled grains (*dhals*) are the starting material of fermented derived products. Hence, the dehulling process led to an increase in fiber solubility, palatability, digestibility and overall nutritive quality (Nalle et al., 2010). Soaking and cooking are usually previous treatments to fermentation. Soaking contributes to the hydration of the seeds and causes the leaching effect in soluble carbohydrates and heat treatment further provokes the starch gelatinization and enhances starch digestibility (Vidal-Valverde et al., 1998).

Fermented pulses provide an important source of carbohydrates (50–70%) (Abu-Salem and Abou-Arab, 2011; Sotomayor et al., 1999). Starch degradation is a complex biochemical process that is modulated by endogenous pulse enzymes and those provided by fermentative microorganisms (Sotomayor et al., 1999). Endogenous and microbial amylases play an important role during pulse fermentation and increased activity in the early stages was shown that declined gradually with the fermentation progress (Soni et al., 1985). During the hydrolysis of starch, amylose and amylopectin decrease gradually over the course of fermentation and reducing sugars are produced (Audu and Aremu, 2011;

Soni and Sandhu, 1989; Sotomayor et al., 1999). In this context, LAB glucose metabolism led to a pH drop associated with lactic-acid production (Leroy and De Vuyst, 2004).

Several reports have identified that lactic-acid fermentation notably increases the starch digestibility of pulses (Bhandal, 2008; Vidal-Valverde et al., 1993; Yadav and Khetarpaul, 1994). Similar findings have been reported in different pulse *tempeh*-like products. However, mixed-culture fermentation with *Rhizopus oligosporus* and *Aspergillus oryzae* brought about less in vitro bioavailability of sugars (Abu-Salem and Abou-Arab, 2011; Starzyńska-Janiszewska et al., 2012).

Fermented pulse products are recognized as a good source of RS (8–15%) (Granito and Álvarez, 2006; Veena et al., 1995). For instance, chickpea *tempeh* contains RS levels three-fold higher than raw flour (Angulo-Bejarano et al., 2008). *Idli* products have been identified as one of the major RS providers in Indian populations (Kavita et al., 1998). Different information has been reported about the content of total dietary fiber (TDF), insoluble (IDF), and soluble (SDF) fractions in fermented pulses. A decrease in SDF has been reported in fermented bengal grams, cowpeas, green grams, and black beans, while IDF showed a significant increase, contributing to the increase of TDF (Granito and Álvarez, 2006; Veena et al., 1995). In fermented lentils, the content of neutral dietary fiber (NDF), cellulose, and hemicellulose depleted, while lignin content increased twice (Vidal-Valverde et al., 1993).

There is a large body of information on the reduction of raffinose family oligosaccharides during the lactic-acid fermentation of pulses, making the final products more acceptable by relieving flatulence discomfort and intestinal cramps (Frias et al., 1996; Granito et al., 2003; Madodé et al., 2013; Martínez-Villaluenga et al., 2008; Shimelis and Rakshit, 2008). Similar results have been reported for cowpea *natto*, where *Bacillus subtilis* led to the total removal of raffinose, stachyose, and verbascose, and in vitro and in vivo studies exhibited considerable fermentability depletion (Madodé et al., 2013). However, fungal fermentation seems to be less effective in α-galactoside removal (Egounlety and Aworh, 2003; Starzyńska-Janiszewska et al., 2014), while yeast fermentation increased significantly the content of raffinose, stachyose, and verbascose of peas, cowpeas, and kidney beans (Khattab and Arntfield, 2009).

16.3.3 Changes in Lipids During Fermentation

Studies on the effect of the fermentation process on the lipid content and profile in pulses are relatively scarce. Prinyawiwatkul et al. (1996) investigated the effect of fungal fermentation in lipid content and fatty-acid composition of cowpea-like *tempeh*. Lipid content increased from 2.2% in unfermented flour to 2.8% after 24 h of fermentation and the major UFA were linoleic, palmitic, and linolenic acids. Niveditha et al. (2012) found that the lipid content of *Canavalia* varieties fermented with *Rhizopus oligosporus* underwent a slight increase and UFA predominated over the saturated ones. However, Reyes-Moreno et al. (2004) reported a 50% lipid decrease in chickpea *tempeh* in comparison with unfermented flour.

16.3.4 Changes in Phytic Acid and Mineral Bioavailability During Fermentation

One of the main constraints of grain legumes is the phytic acid-content, considered as mainly responsible for the low mineral bioavailability (Urbano et al., 2000). Many efforts have been made to reduce the amount of phytate in legumes, and fermentation has been identified as one of the most effective treatments. This is due to acidic conditions occurring during fermentation that increase the phytase activity leading to phytic-acid degradation and hence a larger mineral availability is achieved (Yadav et al., 2013). Phytic-acid degradation in legumes during fermentation depends on many processing parameters such as time and temperature (Kozlowska et al., 1996; Chitra et al., 1996; Khattab and Arntfield, 2009). In this sense, it has been noted that traditional Indian *idli* and *dosa* breakfasts provide a 69% reduction of phytic-acid content and an increase in Ca and Fe availability (Krishnamoorthy et al., 2013). The Zn bioaccesibility in those fermented products increased 71% and 50%, respectively, and to a greater extent the Fe bioaccessibility 277% and 127%, respectively (Hemalatha et al., 2007). These foods are well accepted for Indian children and are recommended for malnourishment (Dahiya et al., 2014). It has also been reported that solid-state fermentation contributes to the reduction in phytic-acid content (Reyes-Moreno et al., 2004), an effect that has been related with a higher in vitro protein digestibility (Abu-Salem and Abou-Arab, 2011).

16.3.5 Changes in Vitamins During Fermentation

Some microorganisms have the ability to synthetize water-soluble vitamins, making fermented pulses more nutritious. In general, natural lactic-acid fermentation of different lentils, kidney beans, and cowpeas led to larger thiamin and riboflavin content (Granito et al., 2002; Torres et al., 2006; Vidal-Valverde et al., 1997). Likewise, Indian *dosa* batters provide larger thiamin, riboflavin, and cobalamin content than unfermented products, and the replacement of black gram by mung beans exhibited higher B-group vitamin content (Soni and Sandhu, 1989). However, the fermentation of bambara groundnut with *Rhizopus oligosporus* reduced thiamin content, while riboflavin, folic acid, niacin, and biotin increased significantly (Fadahunsi, 2009). Additionally, the content of vitamin E increased during natural fermentation of cowpeas (Doblado et al., 2003), while decreased slightly in fermented pigeon peas and lupins (Frias et al., 2005; Torres et al., 2006).

16.3.6 Changes in Phenolic Compounds and Antioxidant Activity During Fermentation

In recent years, there has been increasing interest in investigating phenolic compounds due to their antioxidant activity related to protective health effects on oxidative stress-induced diseases (Morton et al., 2000; Newmark, 1996; Sharma et al., 2011). Pulses are an excellent source of phenolic compounds that are mainly accumulated in the hull (Dueñas et al., 2004). The dehulling process that usually is performed prior to fermentation led to a sharp reduction in tannin content (Dueñas et al., 2002, 2004; Gilani et al., 2012). A significant increase in (+)-catechin and hydroxybenzoi-acid content was found in naturally fermented lentils (Bartolomé et al., 1997). Fermentation of cowpeas with *Lactobacillus plantarum* resulted in a reduction of conjugated forms of ferulic and *p*-cumaric and hydroxycinnamic derivatives, the synthesis of tyrosol, and an increase in free quercetin due to hydrolysis of quercetin glucosides (Dueñas et al., 2005). Changes in the phenolic composition of pulses during fermentation are attributed to glycosidases and esterases from LAB releasing free aglycones and phenolic acids (Esteban-Torres et al., 2015; Ferreira et al., 2013; Jiménez et al., 2014; Limón et al., 2014), and less esterified proanthocyanidins and hydroxycinnamic acids contributing to their antioxidant properties (Dueñas et al., 2005; Esteban-Torres et al., 2015). Additionally, microbial phenolic acid decarboxylases and reductases allow the synthesis of phenolic metabolites with antioxidant activity (Landete et al., 2015; Rodríguez et al., 2009).

The antioxidant properties of fermented pulse-derived foods associated with phenolic compounds have been widely documented (Doblado et al., 2003; Moktan et al., 2011; Torino et al., 2013; Torres et al., 2006). *Dhokla* and *idli* products exhibited higher free-radical scavenging, metal-quelating, and lipid peroxidation inhibitory activities than their unfermented batters (Moktan et al., 2011). Lactic-acid fermentation of pigeon peas, bambara groundnuts, and kidney beans brought about an increase in the free-soluble polyphenols, diminished the content of bound polyphenols, and consequently enhanced antioxidant activity (Oboh et al., 2009; Pérez and Granito, 2012; Starzyńska-Janiszewska et al., 2014). Likewise, common bean and lentil *tempeh* products exhibited higher soluble phenolic concentration, radical scavenging, and antioxidant activities than unfermented grains (Reyes-Bastidas et al., 2010; Torino et al., 2013). Based on these results, the changes in content and composition of phenolic compounds in pulses suggest that fermentation can be considered as a feasible strategic process to promote health benefits and counteract oxidative stress.

16.3.7 Changes in Other Minor Bioactive Compounds During Fermentation

Several researchers have reported changes in some active components during fermentation of pulses. Fermentation of different pulses with LAB results in a noticeable increase in GABA content, a free nonprotein amino acid present in low amounts in grains, while fermentation with *Bacillus subtilis* was rather inferior (Liao et al., 2013b; Limón et al., 2014; Torino et al., 2013). These results suggest that glutamic acid decarboxylase, the enzyme responsible for the GABA biosynthesis, is more active in LAB than in the *Bacillus* strains involved in the fermentation process. Regarding saponins, Shimelis and Rakshit (2008) reported their elimination after natural and controlled lactic-acid fermentation of common beans. However, Fenwick and Oakenfull (1983) observed a 58% reduction in saponins of soybean tempeh fermented with *Rhizopus oligosporus*. Lactic-acid fermentation has been shown to decrease vicine and convicine, cyanogenic glycosides in faba beans (Coda et al., 2015), and the natural toxin β-ODAP from grass peas (Starzyńska-Janiszewska and Stodolak, 2011). In addition, the ability of LAB to degrade biogenic amines phenylethylamine, spermine, and spermidine has been described, although histamine and tyramine were found in fermented lupin products at levels lower than those causing adverse health effects (Bartkiene et al., 2015). Other minor compounds have also been described for pulses, but reports on their content in fermented seeds are scarce, and further studies are needed to better understand their transformations during fermentation.

16.4 ROLE OF FERMENTED PULSE FOODS IN HEALTH PROMOTION

Plant food-based diets are considered part of a healthy diet, providing constituents that slow down chronic diseases such as obesity, diabetes, and cardiovascular diseases among others (Pistollato and Battino, 2014). In general, it has been established that diets high in fiber, low in energy density and glycemic load, moderate in protein, low in fat, and rich in antioxidants promote health and well-being (García-Fernández et al., 2014). Fermented pulse foods fulfill these requirements and can have beneficial effects on the prevention and management of highly prevalent pathologies. In addition, fermented pulse products contain beneficial probiotic microorganisms that can improve gut health and related diseases. Although the nutritional benefits of fermented pulse constituents are widely documented, the scientific evidence demonstrating the role of fermented pulse consumption in disease prevention is scarce. This is probably due to the diversity of fermented pulses worldwide and to the lack of interventional studies in the populations consuming fermented pulses. Therefore this section summarizes the recent scientific evidence that supports the potential health benefits of fermented pulses.

16.4.1 Fermented Pulse Products and Weigh Management

Overweight and obese individuals are at risk for several medical conditions that contribute to morbidity and mortality, including diabetes, cardiovascular diseases, and other metabolic complications. A daily energy imbalance leads to weight gain over time and prevention of excess weight gain can be achieved with low energy-density foods. Replacing energy-dense foods with pulses can enhance satiety (Rebello et al., 2014).

To date, there are few published studies devoted to the effect of fermented-pulse products on appetite and satiety. Only one study has shown that *idli* received the best satiety score compared to other breakfast foods that did not include fermented pulses. Among the various factors examined for their influence on satiety scores, fiber content, energy density, and weight of the food item positively influenced satiety scores (Pai et al., 2005). In addition, fermented pulses provide a good source of proteins, peptides, and amino acids, which make them candidates to promote weight loss by sensation of fullness (Iwashita et al., 2006; Lejeune et al., 2006; McKnight et al., 2010). Although studies on fermented pulse-derived products are still forthcoming, their composition suggests that they can modulate biological processes that counteract obesity.

16.4.2 Fermented Pulse-Products and Diabetes

Pulses are elective foods in dietary strategies to manage blood glucose levels. Pulses provide high amounts of RS, generally defined as starch products not digested in the small intestine, which ultimately conducts to lower GI and lower insulin resistance (Messina, 2014). Lower GI and insulin resistance are contributing factors to reducing both the incidence and severity of type 2 diabetes, one of the factors involved in metabolic syndrome that is also associated with waistline adipose deposition, dyslipidemia, and hyperglycemia (Alberti et al., 2009). Among traditional pulses, mung beans have been recommended as a potential antidiabetic food for diabetic patients, and fermented foods also help to reduce the prevalence of diabetes in Asian populations (Yeap et al., 2012). Fermented mung bean products have been recommended to manage diabetes due to their low-GI, high-fiber content, and phenolic compounds, which improve oxidative stress-induced hyperglycemia (Atkinson et al., 2008; Landete et al., 2015; Maiti and Majumdar, 2012; Randhir and Shetty, 2007). In addition, the antihyperglycemic effect of fermented mung bean extracts observed in an alloxan-induced-diabetic mice model was attributed to its higher GABA content and free amino acids (Yeap et al., 2012). Fermented tropical legume-based diets also exhibited a modulatory effect on oxidative stress and protection of hepatic tissue damage in streptozotocin-induced diabetic rats, an effect attributed to higher intake in phenolic antioxidants (Ademiluyi and Oboh, 2012). While these results show preliminary evidence on the effect of fermented pulses on diabetes human clinical studies are encouraged to validate the results observed in preclinical studies.

16.4.3 Fermented Pulse-Derived Products and Cardiovascular Diseases

Naturally fermented cowpeas exhibited antioxidant and lipid-lowering properties that may contribute to lowering the risk of the development of cardiovascular diseases. Significant improvements in plasma antioxidant capacity and hepatic activity of antioxidant enzymes were observed in albino Wistar rats fed with cowpea-fermented flours for 14 days. In addition, liver weight and plasma cholesterol and triglyceride levels were positively affected (Kapravelou et al., 2015). These effects were attributed to dietary fiber components (Anderson and Major, 2002; Bazzano, 2008; Ma et al., 2008) as well as flavonoid intake (Cassidy et al., 2011). Other minor components such as saponins, oligosaccharides, and phytosterols can also contribute to inhibit the intestinal absorption of cholesterol.

Some fermented-legume products may contribute to lowering the risk of cardiovascular diseases due to their blood pressure-lowering effects. In a recent study, navy bean milk fermented by *L. bulgaricus* and *Lactobacillus plantarum* B1-6 exhibited angiotensin-I converting enzyme (ACE) inhibitory activity, an effect that was attributed to the presence of bioactive peptides (Rui et al., 2015). Similarly, fermentation of mung bean milk by *L. plantarum* B1-6 resulted in the release of smaller and more hydrophilic peptides with higher ACE inhibitory activity (Wu et al., 2015). Liquid-state fermentations of lentils and kidney beans with *L. plantarum* CECT 748 also showed ACE inhibitory activity (Limón et al., 2014; Torino et al., 2013). LAB strains have also been identified as major GABA producers, and hence legume-fermented products containing these amino acids can exert hypotensive effects (Dhakal et al., 2012). Fermented adzuki beans with *Lactococcus lactis* and *Lactobacillus rhamnosus* (Liao et al., 2013) and lactic-acid fermented lentils and kidney beans either naturally or by *L. plantarum* CECT 748 (Limón et al., 2014; Torino et al., 2013) exhibited noticeable accumulation of GABA, which may contribute to the antihypertensive effect associated with these fermented foods (Franciosi et al., 2015; Suwanmanon and Hsieh, 2014).

Antioxidants also contribute to the cardioprotective effect of fermented legumes as it has been shown during the oral administration of 50% ethanol extracts of red bean *natto* to Sprague–Dawley rats (Chou et al., 2008; Jhan et al., 2015). The presence of nattokinase in *natto* adds another cardiopreventive attribute to these products since this extracellular enzyme possesses fibrinolytic activity and thus has been considered effective against thrombolytic episodes. Other valuable advantages described for nattokinase include treatment of hypertension, Alzheimer's disease, and vitreoretinal disorders (Dabbagh et al., 2014).

16.4.4 Fermented Pulse-Derived Products and Cancer

Currently, the Food and Drug Administration (FDA), Canadian Cancer Society, and the World Cancer Research Fund (WCRF) recommend the consumption of pulses to reduce cancer risk. Moreover, they can contribute to the recommended daily intake of 25 g of nonstarch polysaccharide to help reach public health goals (WCRF/AICR, 2010). Fermented food consumption may confer a variety of important nutritional and therapeutic benefits to ameliorate the development of cancer (Kandasamy et al., 2011). The potential anticancer effect of *tempeh*-like extracts from *Canavalia cathatica* and *C. maritima* on colon cancer-cell lines MCF-7 and HT-29 has been demonstrated (Niveditha et al., 2013). Fermented mung bean extracts have shown chemopreventive action on 4T1 breast cancer cells through the stimulation of immunity, lipid peroxidation, and modulation of inflammation (Yeap et al., 2013).

Minor components of pulses such as soyasaponins are hydrolyzed by LAB releasing aglycones such as soyasapogenol A and B in which the lipophilic core have been identified as being responsible for cell death of colon-cancer cells (Gurfinkel and Rao, 2003). Bowman–Birk protease inhibitors have exhibited anticancer activity through inhibition of protease-mediated inflammation and growth of human colon-cancer cells (Clemente and Arques, 2014). Legume lectins are also considered as antitumor agents (De Mejía and Prisecaru, 2005), but no studies have been reported thus far on the anticancer activity of these components from fermented pulses.

16.4.5 Fermented Pulse-Derived Products in Healthy Aging and Stress

In addition to disease prevention, pulse intake has been associated with longevity and with potential enhancement of mental health in aging. A study of five cohorts of elderly people (≥ 70 years) identified high legume intake as a consistent and significant protective dietary component among nine major groups of food (Darmadi-Blackberry et al., 2004). Recently, it was reported that frequent pulse consumption is associated with reduced stress, anxiety emotional, distress, and somatic symptoms in older adults (Smith, 2012). Nevertheless, studies on fermented pulses are scarce and only one showed that a mung bean product fermented by *Rhizopus* sp. strain 5351 (aqueous extracts of 200 and 1000 mg/kg) exhibited potent antiinflammatory and antinociceptive activities in a dose-dependent manner in vitro (Ali et al., 2014). Although the mechanisms underlying these effects are not fully understood, the authors attributed them to the accumulation of GABA, total amino acids, and phenolic antioxidants. Additionally, fermented mung beans have recently been documented to contribute to the alleviation of heatstroke occurring under stress conditions in vivo (Yeap et al., 2014). Nevertheless, this is a preliminary study and further research should be conducted to establish the role of fermented pulses in healthy aging and well-being.

16.4.6 Fermented Pulse-Derived Products as Probiotic Vehicle

LAB play an essential role in the production of fermented-pulse foods. Their combination with large amounts of nonstarch polysaccharides provoke symbiotic potential benefits by the formation of various acidic compounds, such as acetate, lactate, butyrate, and propionate, and the release of short-chain fatty acids with further decrease of pH associated with favorable alteration in the gastrointestinal microecology (Parvez et al., 2006). Fermented mung bean milk prepared with *Lactobacillus plantarum* B1-6 was suggested as a probiotic carrier with health benefits (Wu et al., 2015). Likewise, *Rhizopus* fermentation leads to the decrease of pH, and it has been suggested that *tempeh*-like products can enhance the stability of intestinal beneficial bacteria (Dinesh Banu et al., 2009). The benefits of lower pH in the intestine include enhanced multiplication and survival of beneficial microorganisms while at the same time inhibits the growth of undesirable pathogens. Nevertheless, further studies should be encouraged to demonstrate the role of fermented-pulse foods on gastrointestinal health.

16.5 FINAL REMARKS

Fermentation has high potential to improve the nutritional quality of pulses, providing protein, starch, fiber, and other health-promoting compounds with physiological benefits contributing to the reduction of several risk factors associated with diabetes, cardiovascular diseases, colon cancer, stress, and aging. Moreover, fermented products derived from pulses can be considered as probiotic carriers and of benefit to gastrointestinal health. Current consumption of fermented pulses in North America and Europe is lower than in traditional markets in Asia and Africa, where processed pulses are consumed almost daily. Interestingly, per capita consumption in these traditional pulse markets has been declining in recent years and is associated with an increase in the same chronic disease issues facing developed countries (Curran, 2012). Well-informed consumers are becoming aware of the beneficial effects of fermented-pulse products and they should be considered as part of a nutritive and healthy diet. As governments and health organizations recognize the nutritional benefits of pulses globally, fermented pulses should be recommended to address malnutrition and be promoted as part of healthy eating. Hence, there is an increasing demand for scientific evidence to support future health claims for fermented pulses.

ACKNOWLEDGMENTS

This research was supported by the Spanish Ministry of Economy and Competitiveness and FEDER funding through the project number AGL2013-43247-R.E. Peñas is indebted to Spanish "Ramón y Cajal" Programme for financial support.

REFERENCES

Abu-Salem, F.M., Abou-Arab, E.A., 2011. Physicochemical properties of tempeh produced from chickpea seeds. Journal of American Science 7, 107–118.
Ademiluyi, A.O., Oboh, G., 2012. Attenuation of oxidative stress and hepatic damage by some fermented tropical legume condiment diets in streptozotocin-induced diabetes in rats. Asian Pacific Journal of Tropical Medicine 5, 692–697.

Alberti, K.G.M.M., Eckel, R.H., Grundy, S.M., Zimmet, P.Z., Cleeman, J.I., Donato, K.A., Fruchart, J.C., James, W.P.T., Loria, C.M., Smith, S.C., 2009. Harmonizing the metabolic syndrome, a joint interim statement of the international diabetes federation task force on epidemiology and prevention; National Heart, Lung, and Blood Institute; American Heart Association; World Heart Federation; International Atherosclerosis Society; and International Association for the Study of Obesity. Circulation 120, 1640–1645.

Ali, N.M., Mohd Yusof, H., Yeap, S.K., Ho, W.Y., Beh, B.K., Long, K., Koh, S.P., Abdullah, M.P., Alitheen, N.B., 2014. Antiinflammatory and antinociceptive activities of untreated, germinated, and fermented mung bean aqueous extract. Evidence-Based Complementary and Alternative Medicine 2014.

Amarowicz, R., Pegg, R.B., 2008. Legumes as a source of natural antioxidants. European Journal of Lipid Science and Technology 110, 865–878.

Anderson, J.W., Major, A.W., 2002. Pulses and lipaemia, short- and long-term effect, potential in the prevention of cardiovascular disease. British Journal of Nutrition 88, S263–S271.

Andriantsitohaina, R., Auger, C., Chataigneau, T., Étienne-Selloum, N., Li, H., Martínez, M.C., Schini-Kerth, V.B., Laher, I., 2012. Molecular mechanisms of the cardiovascular protective effects of polyphenols. British Journal of Nutrition 108, 1532–1549.

Angulo-Bejarano, P.I., Verdugo-Montoya, N.M., Cuevas-Rodríguez, E.O., Milán-Carrillo, J., Mora-Escobedo, R., Lopez-Valenzuela, J.A., Garzón-Tiznado, J.A., Reyes-Moreno, C., 2008. Tempeh flour from chickpea (*Cicer arietinum* L.) nutritional and physicochemical properties. Food Chemistry 106, 106–112.

Appeldoorn, M.M., Vincken, J.P., Gruppen, H., Hollman, P.C.H., 2009. Procyanidin dimers A1, A2, and B2 are absorbed without conjugation or methylation from the small intestine of rats. Journal of Nutrition 139, 1469–1473.

Atkinson, F.S., Foster-Powell, K., Brand-Miller, J.C., 2008. International tables of glycemic index and glycemic load values: 2008. Diabetes Care 31, 2281–2283.

Audu, S.S., Aremu, M.O., 2011. Effect of processing on chemical composition of red kidney bean (*Phaseolus vulgaris* L.) flour. Pakistan Journal of Nutrition 10, 1069–1075.

Balasubramanian, S., Jincy, M.G., Ramanathan, M., Chandra, P., Deshpande, S.D., 2015. Studies on millet idli batter and its quality evaluation. International Food Research Journal 22, 139–142.

Bartkiene, E., Krungleviciute, V., Juodeikiene, G., Vidmantiene, D., Maknickiene, Z., 2015. Solid state fermentation with lactic acid bacteria to improve the nutritional quality of lupin and soya bean. Journal of the Science of Food and Agriculture 95, 1336–1342.

Bartolomé, B., Estrella, I., Hernández, T., 1997. Changes in phenolic compounds in lentils (*Lens culinaris*) during germination and fermentation. Zeitschrift fur Lebensmittel-Untersuchung und-Forschung 205, 290–294.

Baumann, U., Bisping, B., 1995. Proteolysis during tempe fermentation. Food Microbiology 12, 39–47.

Bazzano, L.A., 2008. Effects of soluble dietary fiber on low-density lipoprotein cholesterol and coronary heart disease risk. Current Atherosclerosis Reports 10, 473–477.

Bhandal, A., 2008. Effect of fermentation on in vitro digestibilities and the level of antinutrients in moth bean [*Vigna aconitifolia* (Jacq.) Marechal]. International Journal of Food Science and Technology 43, 2090–2094.

Bouchenak, M., Lamri-Senhadji, M., 2013. Nutritional quality of legumes, and their role in cardiometabolic risk prevention: a review. Journal of Medicinal Food 16, 185–198.

Boye, J., Zare, F., Pletch, A., 2010. Pulse proteins, processing, characterization, functional properties and applications in food and feed. Food Research International 43, 414–431.

Broughton, W.J., Hernández, G., Blair, M., Beebe, S., Gepts, P., Vanderleyden, J., 2003. Beans (*Phaseolus* spp.) - model food legumes. Plant and Soil 252, 55–128.

Bujang, A., Taib, N.A., 2014. Changes on amino acids content in soybean, garbanzo bean and groundnut during pre-treatments and tempe making. Sains Malaysiana 43, 551–557.

Cassidy, A., O'Reilly, É.J., Kay, C., Sampson, L., Franz, M., Forman, J.P., Curhan, G., Rimm, E.B., 2011. Habitual intake of flavonoid subclasses and incident hypertension in adults. American Journal of Clinical Nutrition 93, 338–347.

Celleno, L., Tolaini, M.V., D'Amore, A., Perricone, N.V., Preuss, H.G., 2007. A dietary supplement containing standardized *Phaseolus vulgaris* extract influences body composition of overweight men and women. International Journal of Medical Sciences 4, 45–52.

Clemente, A., Arques, M.C., 2014. Bowman-Birk inhibitors from legumes as colorectal chemopreventive agents. World Journal of Gastroenterology 20, 10305–10315.

Coda, R., Melama, L., Rizzello, C.G., Curiel, J.A., Sibakov, J., Holopainen, U., Pulkkinen, M., Sozer, N., 2015. Effect of air classification and fermentation by *Lactobacillus plantarum* VTT E-133328 on faba bean (*Vicia faba* L.) flour nutritional properties. International Journal of Food Microbiology 193, 34–42.

Chatterjee, R., Yeh, H.C., Edelman, D., Brancati, F., 2011. Potassium and risk of type 2 diabetes. Expert Review of Endocrinology and Metabolism 6, 665–672.

Chen, H., Liu, L.j., Zhu, J.j., Xu, B., Li, R., 2010. Effect of soybean oligosaccharides on blood lipid, glucose levels and antioxidant enzymes activity in high fat rats. Food Chemistry 119, 1633–1636.

Chitra, U., Singh, U., Rao, P.V., 1996. Phytic acid, in vitro protein digestibility, dietary fiber, and minerals of pulses as influenced by processing methods. Plant Foods for Human Nutrition 49, 307–316.

Chou, S.T., Chao, W.W., Chung, Y.C., 2008. Effect of fermentation on the antioxidant activity of red beans (*Phaseolus radiatus* L. var. *Aurea*) ethanolic extract. International Journal of Food Science and Technology 43, 1371–1378.

Curran, J., 2012. The nutritional value and health benefits of pulses in relation to obesity, diabetes, heart disease and cancer. British Journal of Nutrition 108, S1–S2.

Dabbagh, F., Negahdaripour, M., Berenjian, A., Behfar, A., Mohammadi, F., Zamani, M., Irajie, C., Ghasemi, Y., 2014. Nattokinase, production and application. Applied Microbiology and Biotechnology 98, 9199–9206.

Dahiya, P.K., Nout, M.J.R., van Boekel, M.A., Khetarpaul, N., Grewal, R.B., Linnemann, A., 2014. Nutritional characteristics of mung bean foods. British Food Journal 116, 1031–1046.

Darmadi-Blackberry, I., Wahlqvist, M.L., Kouris-Blazos, A., Steen, B., Lukito, W., Horie, Y., Horie, K., 2004. Legumes, the most important dietary predictor of survival in older people of different ethnicities. Asia Pacific Journal of Clinical Nutrition 13, 217–220.

Desphande, S.S., Salunkhe, D.K., Oyewole, O.B., Azan-Ali, S., Battcock, M., Bressani, R., 2000. Fermented Grain Legumes, Seeds and Nuts, a Global Perspective. FAO Agricultural Services Bulletins. 142.

Dhakal, R., Bajpai, V.K., Baek, K.H., 2012. Production of GABA (γ-aminobutyric acid) by microorganisms: a review. Brazilian Journal of Microbiology 43, 1230–1241.

Dinesh Banu, P., Bhakyaraj, R., Vidhyalakshmi, R., 2009. A low cost nutritional food "tempeh"- a review. World Journal of Dairy & Food Sciences 4, 22–27.

Doblado, R., Frias, J., Muñoz, R., Vidal-Valverde, C., 2003. Fermentation of *Vigna sinensis* var. *carilla* flours by natural microflora and *Lactobacillus* species. Journal of Food Protection 66, 2313–2320.

Doblado, R., Zielinski, H., Piskula, M., Kozlowska, H., Muñoz, R., Frías, J., Vidal-Valverde, C., 2005. Effect of processing on the antioxidant vitamins and antioxidant capacity of *Vigna sinensis* var. *carilla*. Journal of Agricultural and Food Chemistry 53, 1215–1222.

Duane, W.C., 1997. Effects of legume consumption on serum cholesterol, biliary lipids and sterol metabolism in humans. Journal of Lipid Research 38, 1120–1128.

Dueñas, M., Estrella, I., Hernández, T., 2004. Occurrence of phenolic compounds in the seed coat and the cotyledon of peas (*Pisum sativum* L.). European Food Research and Technology 219, 116–123.

Dueñas, M., Fernández, D., Hernández, T., Estrella, I., Muñoz, R., 2005. Bioactive phenolic compounds of cowpeas (*Vigna sinensis* L). Modifications by fermentation with natural microflora and with *Lactobacillus plantarum* ATCC 14917. Journal of the Science of Food and Agriculture 85, 297–304.

Dueñas, M., Hernández, T., Estrella, I., 2002. Phenolic composition of the cotyledon and the seed coat of lentils (*Lens culinaris* L.). European Food Research and Technology 215, 478–483.

Dueñas, M., Martínez-Villaluenga, C., Limón, R.I., Peñas, E., Frias, J., 2015a. Effect of germination and elicitation on phenolic composition and bioactivity of kidney beans. Food Research International 70, 55–63.

Dueñas, M., Muñoz-González, I., Cueva, C., Jiménez-Girón, A., Sánchez-Patán, F., Santos-Buelga, C., Moreno-Arribas, M.V., Bartolomé, B., 2015b. A survey of modulation of gut microbiota by dietary polyphenols. BioMed Research International 2015.

Egounlety, M., Aworh, O.C., 2003. Effect of soaking, dehulling, cooking and fermentation with *Rhizopus oligosporus* on the oligosaccharides, trypsin inhibitor, phytic acid and tannins of soybean (*Glycine max* Merr.), cowpea (*Vigna unguiculata* L. *Walp*) and groundbean (*Macrotyloma geocarpa* Harms). Journal of Food Engineering 56, 249–254.

Englyst, H.N., Kingman, S.M., Cummings, J.H., 1992. Classification and measurement of nutritionally important starch fractions. European Journal of Clinical Nutrition 46, S33–S50.

Esteban-Torres, M., Landete, J.M., Reverón, I., Santamaría, L., de las Rivas, B., Muñoz, R., 2015. A *Lactobacillus plantarum* esterase active on a broad range of phenolic esters. Applied and Environmental Microbiology 81, 3235–3242.

Fadahunsi, I.F., 2009. The effect of soaking, boiling and fermentation with *Rhizopus oligosporus* on the water soluble vitamin content of bambara groundnut. Pakistan Journal of Nutrition 8, 835–840.

FAO, 1994. Definition and Classifications of Commodities, Pulses and Derived Products. Online. Available from: http://www.fao.org/es/faodef/fdef04ee.htm.

FAOSTAT, 2014. Food and Agriculture Organization of the United Nations Statistics Division. http://faostat3.fao.org.

Fardet, A., 2010. New hypotheses for the health-protective mechanisms of whole-grain cereals: what is beyond fiber? Nutrition Research Reviews 23, 65–134.

Fenwick, D.E., Oakenfull, D., 1983. Saponin content of food plants and some prepared foods. Journal of the Science of Food and Agriculture 34, 186–191.

Ferrarelli, L.K., 2015. Why a high-fiber diet prevents cancer. Science Signaling 8.

Ferreira, L.R., Macedo, J.A., Ribeiro, M.L., Macedo, G.A., 2013. Improving the chemopreventive potential of orange juice by enzymatic biotransformation. Food Research International 51, 526–535.

Franciosi, E., Carafa, I., Nardin, T., Schiavon, S., Poznanski, E., Cavazza, A., Larcher, R., Tuohy, K.M., 2015. Biodiversity and γ-aminobutyric acid production by lactic acid bacteria isolated from traditional alpine raw cow's milk cheeses. BioMed Research International 2015.

Frias, J., Miranda, M.L., Doblado, R., Vidal-Valverde, C., 2005. Effect of germination and fermentation on the antioxidant vitamin content and antioxidant capacity of *Lupinus albus* L. var. *Multolupa*. Food Chemistry 92, 211–220.

Frias, J., Vidal-Valverde, C., Kozlowska, H., Tabera, J., Honke, J., Hedley, C.L., 1996. Natural fermentation of lentils. Influence of time, flour concentration, and temperature on the kinetics of monosaccharides, disaccharide, and α-galactosides. Journal of Agricultural and Food Chemistry 44, 579–584.

Gan, Y., Liang, C., Wang, X., McConkey, B., 2011. Lowering carbon footprint of durum wheat by diversifying cropping systems. Field Crops Research 122, 199–206.

García-Fernández, E., Rico-Cabanas, L., Rosgaard, N., Estruch, R., Bach-Faig, A., Bach-Faig, A., 2014. Mediterranean diet and cardiodiabesity: a review. Nutrients 6, 3474–3500.

Gilani, G.S., Xiao, C.W., Cockell, K.A., 2012. Impact of antinutritional factors in food proteins on the digestibility of protein and the bioavailability of amino acids and on protein quality. British Journal of Nutrition 108, S315–S332.

Granito, M., Álvarez, G., 2006. Lactic acid fermentation of black beans (*Phaseolus vulgaris*), microbiological and chemical characterization. Journal of the Science of Food and Agriculture 86, 1164–1171.

Granito, M., Champ, M., Guerra, M., Frias, J., 2003. Effect of natural and controlled fermentation on flatus-producing compounds of beans (*Phaseolus vulgaris*). Journal of the Science of Food and Agriculture 83, 1004–1009.

Granito, M., Frias, J., Doblado, R., Guerra, M., Champ, M., Vidal-Valverde, C., 2002. Nutritional improvement of beans (*Phaseolus vulgaris*) by natural fermentation. European Food Research and Technology 214, 226–231.

Gurfinkel, D.M., Rao, A.V., 2003. Soyasaponins: the relationship between chemical structure and colon anticarcinogenic activity. Nutrition and Cancer 47, 24–33.

Gylling, H., Plat, J., Turley, S., Ginsberg, H.N., Ellegård, L., Jessup, W., Jones, P.J., Lütjohann, D., Maerz, W., Masana, L., Silbernagel, G., Staels, B., Borén, J., Catapano, A.L., De Backer, G., Deanfield, J., Descamps, O.S., Kovanen, P.T., Riccardi, G., Tokgözoglu, L., Chapman, M.J., 2014. Plant sterols and plant stanols in the management of dyslipidaemia and prevention of cardiovascular disease. Atherosclerosis 232, 346–360.

Hamid, R., Masood, A., Wani, I.H., Rafiq, S., 2013. Lectins, proteins with diverse applications. Journal of Applied Pharmaceutical Science 3, S93–S103.

Harlander, S., 1992. Food biotechnology. In: Lederberg, J. (Ed.), Encyclopedia of Microbiology. Academic Press, New York, pp. 191–207.

Hayakawa, K., Mizutani, J., Wada, K., Masai, T., Yoshihara, I., Mitsuoka, T., 1990. Effects of soybean oligosaccharides on human faecal flora. Microbial Ecology in Health and Disease 3, 293–303.

Hemalatha, S., Platel, K., Srinivasan, K., 2007. Influence of germination and fermentation on bioaccessibility of zinc and iron from food grains. European Journal of Clinical Nutrition 61, 342–348.

Hoover, R., Hughes, T., Chung, H.J., Liu, Q., 2010. Composition, molecular structure, properties, and modification of pulse starches: a review. Food Research International 43, 399–413.

Hoover, R., Ratnayake, W.S., 2002. Starch characteristics of black bean, chickpea, lentil, navy bean and pinto bean cultivars grown in Canada. Food Chemistry 78, 489–498.

Iwashita, S., Mikus, C., Baier, S., Flakoll, P.J., 2006. Glutamine supplementation increases postprandial energy expenditure and fat oxidation in humans. Journal of Parenteral and Enteral Nutrition 30, 76–80.

Jhan, J.K., Chang, W.F., Wang, P.M., Chou, S.T., Chung, Y.C., 2015. Production of fermented red beans with multiple bioactivities using co-cultures of *Bacillus subtilis* and *Lactobacillus delbrueckii* subsp. *bulgaricus*. LWT - Food Science and Technology 63, 1281–1287.

Jiménez-Escrig, A., Santos-Hidalgo, A.B., Saura-Calixto, F., 2006. Common sources and estimated intake of plant sterols in the Spanish diet. Journal of Agricultural and Food Chemistry 54, 3462–3471.

Jiménez, N., Esteban-Torres, M., Mancheño, J.M., De las Rivas, B., Muñoz, R., 2014. Tannin degradation by a novel tannase enzyme present in some *Lactobacillus plantarum* strains. Applied and Environmental Microbiology 80, 2991–2997.

Kalogeropoulos, N., Chiou, A., Ioannou, M., Karathanos, V.T., Hassapidou, M., Andrikopoulos, N.K., 2010. Nutritional evaluation and bioactive microconstituents (phytosterols, tocopherols, polyphenols, triterpenic acids) in cooked dry legumes usually consumed in the Mediterranean countries. Food Chemistry 121, 682–690.

Kandasamy, M., Bay, B.H., Lee, Y.K., Mahendran, R., 2011. Lactobacilli secreting a tumor antigen and IL15 activates neutrophils and dendritic cells and generates cytotoxic T lymphocytes against cancer cells. Cellular Immunology 271, 89–96.

Kapravelou, G., Martínez, R., Andrade, A.M., López Chaves, C., López-Jurado, M., Aranda, P., Arrebola, F., Cañizares, F.J., Galisteo, M., Porres, J.M., 2015. Improvement of the antioxidant and hypolipidaemic effects of cowpea flours (*Vigna unguiculata*) by fermentation, results of in vitro and in vivo experiments. Journal of the Science of Food and Agriculture 95, 1207–1216.

Kavita, V., Verghese, S., Chitra, G.R., Prakash, J., 1998. Effects of processing, storage time and temperature on the resistant starch of foods. Journal of Food Science and Technology 35, 299–304.

Khattab, R.Y., Arntfield, S.D., 2009. Nutritional quality of legume seeds as affected by some physical treatments 2. Antinutritional factors. LWT - Food Science and Technology 42, 1113–1118.

Khattab, R.Y., Arntfield, S.D., Nyachoti, C.M., 2009. Nutritional quality of legume seeds as affected by some physical treatments: part 1, protein quality evaluation. LWT - Food Science and Technology 42, 1107–1112.

Khokhar, S., Chauhan, B.M., 1986. Antinutritional factors in moth bean (*Vigna aconitifolia*), varietal differences and effects of methods of domestic processing and cooking. Journal of Food Science 51, 591–594.

Kozlowska, H., Honke, J., Sadowska, J., Frias, J., Vidal-Valverde, C., 1996. Natural fermentation of lentils, influence of time, concentration and temperature on the kinetics of hydrolysis of inositol phosphates. Journal of the Science of Food and Agriculture 71, 367–375.

Krishnamoorthy, S., Kunjithapatham, S., Manickam, L., 2013. Traditional Indian breakfast (Idli and Dosa) with enhanced nutritional content using millets. Nutrition and Dietetics 70, 241–246.

Lehmann, U., Robin, F., 2007. Slowly digestible starch - its structure and health implications, a review. Trends in Food Science and Technology 18, 346–355.

Lejeune, M.P.G.M., Westerterp, K.R., Adam, T.C.M., Luscombe-Marsh, N.D., Westerterp-Plantenga, M.S., 2006. Ghrelin and glucagon-like peptide 1 concentrations, 24-h satiety, and energy and substrate metabolism during a high-protein diet and measured in a respiration chamber. American Journal of Clinical Nutrition 83, 89–94.

Leroy, F., De Vuyst, L., 2004. Lactic acid bacteria as functional starter cultures for the food fermentation industry. Trends in Food Science and Technology 15, 67–78.

Ley, S.H., Sun, Q., Willett, W.C., Eliassen, A.H., Wu, K., Pan, A., Grodstein, F., Hu, F.B., 2014. Associations between red meat intake and biomarkers of inflammation and glucose metabolism in women1-3. American Journal of Clinical Nutrition 99, 352–360.

Liao, W.C., Wang, C.Y., Shyu, Y.T., Yu, R.C., Ho, K.C., 2013. Influence of preprocessing methods and fermentation of adzuki beans on γ-aminobutyric acid (GABA) accumulation by lactic acid bacteria. Journal of Functional Foods 5, 1108–1115.

Limón, R.I., Peñas, E., Torino, M.I., Martínez-Villaluenga, C., Dueñas, M., Frias, J., 2014. Fermentation enhances the content of bioactive compounds in kidney bean extracts. Food Chemistry 172, 343–352.

Luo, Y.W., Xie, W.H., Jin, X.X., Wang, Q., Zai, X.M., 2013. Effects of germination and cooking for enhanced in vitro iron, calcium and zinc bioaccessibility from faba bean, azuki bean and mung bean sprouts. CYTA - Journal of Food 11, 318–323.

Landete, J.M., Hernández, T., Robredo, S., Dueñas, M., De Las Rivas, B., Estrella, I., Muñoz, R., 2015. Effect of soaking and fermentation on content of phenolic compounds of soybean (*Glycine max* cv. Merit) and mung beans (*Vigna radiata* [L] Wilczek). International Journal of Food Sciences and Nutrition 66, 203–209.

De Mejía, E.G., Prisecaru, V.I., 2005. Lectins as bioactive plant proteins: a potential in cancer treatment. Critical Reviews in Food Science and Nutrition 45, 425–445.

Ma, Y., Hébert, J.R., Li, W., Bertone-Johnson, E.R., Olendzki, B., Pagoto, S.L., Tinker, L., Rosal, M.C., Ockene, I.S., Ockene, J.K., Griffith, J.A., Liu, S., 2008. Association between dietary fiber and markers of systemic inflammation in the Women's Health Initiative Observational Study. Nutrition 24, 941–949.

Madodé, Y.E., Nout, M.J.R., Bakker, E.J., Linnemann, A.R., Hounhouigan, D.J., van Boekel, M.A.J.S., 2013. Enhancing the digestibility of cowpea (*Vigna unguiculata*) by traditional processing and fermentation. LWT - Food Science and Technology 54, 186–193.

Maiti, D., Majumdar, M., 2012. Impact of bioprocessing on phenolic content & antioxidant activity of mung seeds to improve hypoglycemic functionality. International Journal of PharmTech Research 4, 924–931.

Martínez-Villaluenga, C., Frias, J., Vidal-Valverde, C., 2008. Alpha-galactosides: antinutritional factors or functional ingredients? Critical Reviews in Food Science and Nutrition 48, 301–316.

McCrory, M.A., Hamaker, B.R., Lovejoy, J.C., Eichelsdoerfer, P.E., 2010. Pulse consumption, satiety, and weight management. Advances in Nutrition 1, 17–30.

McKnight, J.R., Satterfield, M.C., Jobgen, W.S., Smith, S.B., Spencer, T.E., Meininger, C.J., McNeal, C.J., Wu, G., 2010. Beneficial effects of L-arginine on reducing obesity, potential mechanisms and important implications for human health. Amino Acids 39, 349–357.

Mensink, R.P., Zock, P.L., Kester, A.D.M., Katan, M.B., 2003. Effects of dietary fatty acids and carbohydrates on the ratio of serum total to HDL cholesterol and on serum lipids and apolipoproteins: a meta-analysis of 60 controlled trials. American Journal of Clinical Nutrition 77, 1146–1155.

Messina, V., 2014. Nutritional and health benefits of dried beans. American Journal of Clinical Nutrition 100, 437S–442S.

Milgate, J., Roberts, D.C.K., 1995. The nutritional & biological significance of saponins. Nutrition Research 15, 1223–1249.

Miura, K., Okuda, N., Turin, T.C., Takashima, N., Nakagawa, H., Nakamura, K., Yoshita, K., Okayama, A., Ueshima, H., 2010. Dietary salt intake and blood pressure in a representative Japanese population: baseline analyses of NIPPON DATA80. Journal of Epidemiology 20, S524–S530.

Moktan, B., Roy, A., Sarkar, P.K., 2011. Antioxidant activities of cereal-legume mixed batters as influenced by process parameters during preparation of dhokla and idli, traditional steamed pancakes. International Journal of Food Sciences and Nutrition 62, 360–369.

Monagas, M., Urpi-Sarda, M., Sánchez-Patán, F., Llorach, R., Garrido, I., Gómez-Cordovés, C., Andres-Lacueva, C., Bartolomé, B., 2010. Insights into the metabolism and microbial biotransformation of dietary flavan-3-ols and the bioactivity of their metabolites. Food and Function 1, 233–253.

Morton, L.W., Caccetta, R.A.A., Puddey, I.B., Croft, K.D., 2000. Chemistry and biological effects of dietary phenolic compounds: relevance to cardiovascular disease. Clinical and Experimental Pharmacology and Physiology 27, 152–159.

Mrudula, S., Apsana Begum, A., Ashwitha, K., Pindi, P.K., 2012. Enhanced production of alkaline protease by *Bacillus subtilis* in submerged fermentation. International Journal of Pharma and Bio Sciences 3, B619–B631.

Muzquiz, M., Varela, A., Burbano, C., Cuadrado, C., Guillamón, E., Pedrosa, M.M., 2012. Bioactive compounds in legumes: pronutritive and antinutritive actions. Implications for nutrition and health. Phytochemistry Reviews 11, 227–244.

Nalle, C.L., Ravindran, G., Ravindran, V., 2010. Influence of dehulling on the apparent metabolisable energy and ileal amino acid digestibility of grain legumes for broilers. Journal of the Science of Food and Agriculture 90, 1227–1231.

Newmark, H.L., 1996. Plant phenolics as potential cancer prevention agents. Advances in Experimental Medicine and Biology 401, 25–34.

Niveditha, V.R., Krishna Venkatramana, D., Sridhar, K.R., 2013. Cytotoxic effects of methanol extract of raw, cooked and fermented split beans of Canavalia on cancer cell lines MCF-7 and HT-29. IIOAB Journal 4, 20–23.

Niveditha, V.R., Krishna Venkatramana, D., Sridhar, K.R., 2012. Fatty acid composition of cooked and fermented beans of the wild legumes (Canavalia) of coastal sand dunes. International Food Research Journal 19, 1401–1407.

Oboh, G., Ademiluyi, A.O., Akindahunsi, A.A., 2009. Changes in polyphenols distribution and antioxidant activity during fermentation of some underutilized legumes. Food Science and Technology International 15, 41–46.

Pai, S., Ghugre, P.S., Udipi, S.A., 2005. Satiety from rice-based, wheat-based and rice-pulse combination preparations. Appetite 44, 263–271.

Palanisamy, B.D., Rajendran, V., Sathyaseelan, S., Bhat, R., Venkatesan, B.P., 2012. Enhancement of nutritional value of finger millet-based food (Indian dosa) by co-fermentation with horse gram flour. International Journal of Food Sciences and Nutrition 63, 5–15.

Parvez, S., Malik, K.A., Ah Kang, S., Kim, H.Y., 2006. Probiotics and their fermented food products are beneficial for health. Journal of Applied Microbiology 100, 1171–1185.

Pérez, S., Granito, M., 2012. Fermented and hydrolized concentrates of *Cajanus cajanus* and *Phaseolus vulgaris*: functional ingredients for food manufacture. Interciencia 37, 431–437.

Petry, C.D., Eaton, M.A., Wobken, J.D., Mills, M.M., Johnson, D.E., Georgieff, M.K., 1992. Iron deficiency of liver, heart, and brain in newborn infants of diabetic mothers. The Journal of Pediatrics 121, 109–114.

Phillips, G.O., 2013. Dietary fiber: a chemical category or a health ingredient? Bioactive Carbohydrates and Dietary Fiber 1, 3–9.

Pistollato, F., Battino, M., 2014. Role of plant-based diets in the prevention and regression of metabolic syndrome and neurodegenerative diseases. Trends in Food Science and Technology 40, 62–81.

Prinyawiwatkul, W., Beuchat, L.R., McWatters, K.H., Phillips, R.D., 1996. Changes in fatty acid, simple sugar, and oligosaccharide content of cowpea (*Vigna unguiculata*) flour as a result of soaking, boiling, and fermentation with *Rhizopus microsporus* var. *oligosporus*. Food Chemistry 57, 405–413.

Prodanov, M., Sierra, I., Vidal-Valverde, C., 1997. Effect of germination on the thiamine, riboflavin and niacin contents in legumes. European Food Research and Technology 205, 48–52.

Vaz Patto, M.C., Amarowicz, R., Aryee, A.N.A., Boye, J.I., Chung, H.J., Martín-Cabrejas, M.A., Domoney, C., 2015. Achievements and challenges in improving the nutritional quality of food legumes. Critical Reviews in Plant Sciences 34, 105–143.

Raigond, P., Ezekiel, R., Raigond, B., 2015. Resistant starch in food: a review. Journal of the Science of Food and Agriculture 95, 1968–1978.

Randhir, R., Shetty, K., 2007. Mung beans processed by solid-state bioconversion improves phenolic content and functionality relevant for diabetes and ulcer management. Innovative Food Science and Emerging Technologies 8, 197–204.

Ranilla, L.G., Genovese, M.I., Lajolo, F.M., 2009. Isoflavones and antioxidant capacity of Peruvian and Brazilian lupin cultivars. Journal of Food Composition and Analysis 22, 397–404.

Rebello, C., Greenway, F.L., Dhurandhar, N.V., 2014. Functional foods to promote weight loss and satiety. Current Opinion in Clinical Nutrition and Metabolic Care 17, 596–604.

Reyes-Bastidas, M., Reyes-Fernández, E.Z., López-Cervantes, J., Milán-Carrillo, J., Loarca-Piña, G.F., Reyes-Moreno, C., 2010. Physicochemical, nutritional and antioxidant properties of tempeh flour from common bean (*Phaseolus vulgaris* L.). Food Science and Technology International 16, 427–434.

Reyes-Moreno, C., Cuevas-Rodríguez, E.O., Milán-Carrillo, J., Cárdenas-Valenzuela, O.G., Barrón-Hoyos, J., 2004. Solid state fermentation process for producing chickpea (*Cicer arietinum* L) tempeh flour. Physicochemical and nutritional characteristics of the product. Journal of the Science of Food and Agriculture 84, 271–278.

Riat, P., Sadana, B., 2009. Effect of fermentation on amino acid composition of cereal and pulse based foods. Journal of Food Science and Technology 46, 247–250.

Rodríguez, H., Curiel, J.A., Landete, J.M., de las Rivas, B., de Felipe, F.L., Gómez-Cordovés, C., Mancheño, J.M., Muñoz, R., 2009. Food phenolics and lactic acid bacteria. International Journal of Food Microbiology 132, 79–90.

Roy, F., Boye, J.I., Simpson, B.K., 2010. Bioactive proteins and peptides in pulse crops: pea, chickpea and lentil. Food Research International 43, 432–442.

Rui, X., Wen, D., Li, W., Chen, X., Jiang, M., Dong, M., 2015. Enrichment of ACE inhibitory peptides in navy bean (*Phaseolus vulgaris*) using lactic acid bacteria. Food and Function 6, 622–629.

Ruiz, R.G., Price, K.R., Arthur, A.E., Rose, M.E., Rhodes, M.J.C., Fenwick, R.G., 1996. Effect of soaking and cooking on the saponin content and composition of chickpeas (*Cicer arietinum*) and lentils (*Lens culinaris*). Journal of Agricultural and Food Chemistry 44, 1526–1530.

Sandberg, A.S., 2002. Bioavailability of minerals in legumes. British Journal of Nutrition 88, S281–S285.

Sharma, G., Srivastava, A.K., Prakash, D., 2011. Phytochemicals of nutraceutical importance: their role in health and diseases. Pharmacology 2, 408–427.

Shi, J., Arunasalam, K., Yeung, D., Kakuda, Y., Mittal, G., Jiang, Y., 2004. Saponins from edible legumes, chemistry, processing, and health benefits. Journal of Medicinal Food 7, 67–78.

Shimelis, E.A., Rakshit, S.K., 2008. Influence of natural and controlled fermentations on α-galactosides, antinutrients and protein digestibility of beans (*Phaseolus vulgaris* L.). International Journal of Food Science and Technology 43, 658–665.

Smith, A.P., 2012. Legumes and well-being in the elderly, a preliminary study. Journal of Food Research 1, 165–168.

Soni, S.K., Sandhu, D.K., 1989. Nutritional improvement of Indian dosa batters by yeast enrichment and black gram replacement. Journal of Fermentation and Bioengineering 68, 52–55.

Soni, S.K., Sandhu, D.K., Vilkhu, K.S., 1985. Studies on dosa-an indigenous Indian fermented food, some biochemical changes accompanying fermentation. Food Microbiology 2, 175–181.

Soobrattee, M.A., Neergheen, V.S., Luximon-Ramma, A., Aruoma, O.I., Bahorun, T., 2005. Phenolics as potential antioxidant therapeutic agents: mechanism and actions. Mutation Research - Fundamental and Molecular Mechanisms of Mutagenesis 579, 200–213.

Sotomayor, C., Frias, J., Fornal, J., Sadowska, J., Urbano, G., Vidal-Valverde, C., 1999. Lentil starch content and its microscopical structure as influenced by natural fermentation. Starch/ Staerke 51, 152–156.
Starzyńska-Janiszewska, A., Stodolak, B., 2011. Effect of inoculated lactic acid fermentation on antinutritional and antiradical properties of grass pea (*Lathyrus sativus* 'Krab') flour. Polish Journal of Food and Nutrition Sciences 61, 245–249.
Starzyńska-Janiszewska, A., Stodolak, B., Duliński, R., Mickowska, B., 2012. The influence of inoculum composition on selected bioactive and nutritional parameters of grass pea tempeh obtained by mixed-culture fermentation with *Rhizopus oligosporus* and *Aspergillus oryzae* strains. Food Science and Technology International 18, 113–122.
Starzyńska-Janiszewska, A., Stodolak, B., Mickowska, B., 2014. Effect of controlled lactic acid fermentation on selected bioactive and nutritional parameters of tempeh obtained from unhulled common bean (*Phaseolus vulgaris*) seeds. Journal of the Science of Food and Agriculture 94, 359–366.
Starzyńska-Janiszewska, A., Stodolak, B., Wikiera, A., 2015. Proteolysis in tempeh-type products obtained with *Rhizopus* and *Aspergillus* strains from grass pea (*Lathyrus Sativus*) seeds. Acta Scientiarum Polonorum. Technologia Alimentaria 14, 125–132.
Steinkraus, K.H., 1996. Handbook of Indigenous Fermented Foods. Marcel Decker Inc., New York.
Sugano, M., 2005. Soy in Health and Disease Prevention. CRC Press, Boca Raton, FL.
Suwanmanon, K., Hsieh, P.C., 2014. Effect of γ-aminobutyric acid and nattokinase-enriched fermented beans on the blood pressure of spontaneously hypertensive and normotensive Wistar-Kyoto rats. Journal of Food and Drug Analysis 22, 485–491.
Tharanathan, R.N., Mahadevamma, S., 2003. Grain legumes - a boon to human nutrition. Trends in Food Science & Technology 14, 507–518.
Tomomatsu, H., 1994. Health effects of oligosaccharides. Food Technology 48, 61–65.
Torino, M.I., Limon, R.I., Martinez-Villaluenga, C., Makinen, S., Pihlanto, A., Vidal-Valverde, C., Frias, J., 2013. Antioxidant and antihypertensive properties of liquid and solid state fermented lentils. Food Chemistry 136, 1030–1037.
Torres, A., Frias, J., Granito, M., Vidal-Valverde, C., 2006. Fermented pigeon pea (*Cajanus cajan*) ingredients in pasta products. Journal of Agricultural and Food Chemistry 54, 6685–6691.
Tosh, S.M., Yada, S., 2010. Dietary fibers in pulse seeds and fractions, characterization, functional attributes, and applications. Food Research International 43, 450–460.
Urbano, G., López-Jurado, M., Aranda, P., Vidal-Valverde, C., Tenorio, E., Porres, J., 2000. The role of phytic acid in legumes, antinutrient or beneficial function? Journal of Physiology and Biochemistry 56, 283–294.
USDA. National Nutrient Database for Standard Reference Rel. 27 (Report day May 13, 2015).
USDA, 2002. USDA-Iowa State University Database on the Isoflavone Content of Foods. Agricultural Research Service, U.S. Department of Agriculture, Washington, DC.
Vasconcelos, I.M., Oliveira, J.T.A., 2004. Antinutritional properties of plant lectins. Toxicon 44, 385–403.
Veena, A., Urooj, A., Puttaraj, S., 1995. Effect of processing on the composition of dietary fiber and starch in some legumes. Die Nahrung 39, 132–138.
Vidal-Valverde, C., Frias, J., Prodanov, M., Tabera, J., Ruiz, R., Bacon, J., 1993. Effect of natural fermentation on carbohydrates, riboflavin and trypsin inhibitor activity of lentils. Zeitschrift für Lebensmittel-Untersuchung und-Forschung 197, 449–452.
Vidal-Valverde, C., Frias, J., Sotomayor, C., Diaz-Pollan, C., Fernandez, M., Urbano, G., 1998. Nutrients and antinutritional factors in faba beans as affected by processing. European Food Research and Technology 207, 140–145.

Vidal-Valverde, C., Prodanov, M., Sierra, I., 1997. Natural fermentation of lentils, influence of time, temperature and flour concentration on the kinetics of thiamin, riboflavin and niacin. European Food Research and Technology 205, 464–469.

Wang, S., Melnyk, J.P., Tsao, R., Marcone, M.F., 2011. How natural dietary antioxidants in fruits, vegetables and legumes promote vascular health. Food Research International 44, 14–22.

Wang, Y., Beydoun, M.A., 2009. Meat consumption is associated with obesity and central obesity among US adults. International Journal of Obesity 33, 621–628.

WCRF/AICR, 2010. Food, Nutrition, Physical Activity, and the Prevention of Cancer, a Global Perspective. AIRC, Washington, DC, USA. Availabre from: http://www.dietandcancerreport.org/cancer_resource_center/download/Second_Expert_Report_full.pdf.

Wu, H., Rui, X., Li, W., Chen, X., Jiang, M., Dong, M., 2015. Mung bean (*Vigna radiata*) as probiotic food through fermentation with *Lactobacillus plantarum* B1-6. LWT - Food Science and Technology 63, 445–451.

Xie, S., Zhu, J., Zhang, Y., Shi, K., Shi, Y., Ma, X., 2012. Effects of soya oligosaccharides and soya oligopeptides on lipid metabolism in hyperlipidaemic rats. British Journal of Nutrition 108, 603–610.

Xu, B.J., Chang, S.K.C., 2007. A comparative study on phenolic profiles and antioxidant activities of legumes as affected by extraction solvents. Journal of Food Science 72, S159–S166.

Yadav, M., Singh, P., Kaur, R., Gupta, R., Gangoliya, S.S., Singh, N.K., 2013. Impact of food phytic acid on nutritions, health and environment. Plant Archives 13, 605–611.

Yadav, S., Khetarpaul, N., 1994. Indigenous legume fermentation, effect on some antinutrients and in-vitro digestibility of starch and protein. Food Chemistry 50, 403–406.

Yeap, S.K., Beh, B.K., Ali, N.M., Mohd Yusof, H., Ho, W.Y., Koh, S.P., Alitheen, N.B., Long, K., 2014. In vivo antistress and antioxidant effects of fermented and germinated mung bean. BioMed Research International:694842.

Yeap, S.K., Mohd Ali, N., Mohd Yusof, H., Alitheen, N.B., Beh, B.K., Ho, W.Y., Koh, S.P., Long, K., 2012. Antihyperglycemic effects of fermented and nonfermented mung bean extracts on alloxan-induced-diabetic mice. Journal of Biomedicine and Biotechnology:285430.

Yeap, S.K., Mohd Yusof, H., Mohamad, N.E., Beh, B.K., Ho, W.Y., Ali, N.M., Alitheen, N.B., Koh, S.P., Long, K., 2013. In vivo immunomodulation and lipid peroxidation activities contributed to chemoprevention effects of fermented mung bean against breast cancer. Evidence-Based Complementary and Alternative Medicine 2013:708464.

Zambrowicz, A., Timmer, M., Polanowski, A., Lubec, G., Trziszka, T., 2013. Manufacturing of peptides exhibiting biological activity. Amino Acids 44, 315–320.

Chapter 17

Nonwheat Cereal-Fermented-Derived Products

Z. Ciesarová[1], L. Mikušová[2], M. Magala[2], Z. Kohajdová[2], J. Karovičová[2]
[1]NPPC National Agricultural and Food Centre, Food Research Institute, Bratislava, Slovak Republic; [2]Slovak University of Technology, Bratislava, Slovak Republic

17.1 INTRODUCTION

Today's growing emphasis on a healthy and balanced diet has created a demand for "functional foods," those that in addition to their original functions provide energy and nutrition and are able to favorably affect body functions and reduce the risk of chronic diseases. Foods with these positive health effects include cereal products naturally rich in biologically valuable constituents (eg, proteins, lipids, carbohydrates with favorable composition, high content of dietary fiber (DF), vitamins and minerals, and other micronutrients essential for health).

Cereals belong to the most important segments of plant production from an agronomic, economic, and consumer points of view. They contribute greatly to the nutritional balance of the world's population. In terms of consumption, cereals have a privileged position among other agricultural products. In addition to their nutritional importance, their health benefits are being increasingly highlighted. Cereals have a long history of consumption by humans in both developing and developed countries, being an important source of energy, carbohydrates, protein, fiber, and a wide range of vitamins (mainly B vitamins and vitamin E), minerals (iron, zinc, magnesium), and other phytochemicals. Regular consumption of cereals, especially in their whole-grain form, is associated with prevention of chronic diseases such as cardiovascular diseases, hypertension, type 2 diabetes, obesity, and some types of cancer (McKewith, 2004).

Many different types of cereals (eg, wheat, rice, rye, barley, oat, maize, sorghum, and millet) are grown worldwide and have some structural similarities. Wheat is considered the earliest field crop used for human processing. It became the leading grain used for consumption due to its nutritive profile and relatively easy harvesting, storing, transportation, and processing, as compared to other grains. Wheat together with rice are the most common cereals, reaching approximately 50% of the production worldwide, while maize is important mainly in Central and South America, and sorghum and millet in Africa (Poutanen, 2012).

On the other hand, amaranth, quinoa, and buckwheat create a group of grains referred to as pseudo-cereals, as their seeds resemble true cereals in function and composition, but botanically create a different group. Although the general requirements of the human diet are met mainly by dominantly grown grains (eg, wheat, rice, maize), minor crops support diversity and enrich the diet with natural valuable components taking into account traditional and geographical specifics. Last but not least, there is a visible trend toward a healthy beneficial diet and functional foods by consumers worldwide. Therefore this chapter is focused on:

- Nutritional aspects of nonwheat cereals
- Advantages and limitations of fermentation applied to nonwheat cereals
- Fermented nonwheat food products including nonwheat beverages and nonwheat fermented bakery products
- Health beneficial effects of fermented nonwheat products

17.2 NUTRITIONAL ASPECTS OF NONWHEAT CEREALS

Cereals and pseudo-cereals provide the majority of carbohydrates and fiber intake in many countries (Poutanen, 2012), although they are also sources of protein with a wide range of essential amino acids. The proximate nutritional composition of nonwheat cereals is summarized in Table 17.1.

The cereal kernel consists of endosperm rich in starch and proteins, which comprises the major part of the grain, bran (source of fiber, micronutrients, and other biologically active compounds), and germ (rich in lipids and other macro- and micronutrients), representing minor parts of the grain (Mikušová et al., 2011). Cereals with a high proportion of bran (up to 30%) are barley and oats, while the smallest bran proportion can be found in corn (3.5%). On the other hand, barley and oats have smallest germ fraction (3%), while the largest germ is present in corn (10%) and millet (17%) (Kamal–Eldin, 2008).

Rye (*Secale cereale* L.) is the second most commonly used cereal after wheat for preparation of bread and bakery products. Rye flour contains significantly more soluble protein and less gliadin and glutenin fractions than wheat. The rye starch is less resistant to amylolytic enzymes and has lower gelatinization temperature, a quality that favors the use of rye in production of sourdough. The other advantage of rye starch is slower starch retrogradation, which results in slower staling of rye and wheat–rye bakery products (Lorenz, 2000).

Oat (*Avena* spp. L.) is characterized by good taste and dietary, health, and metabolic stimulation properties. These properties are due to its high nutritional value, in particular due to its significant amount of biologically valuable proteins and lipids compared to other cereals. Moreover, oat possesses a favorable composition of carbohydrates and high vitamin and mineral content. Oat contains a significant amount of nondigestible polysaccharides-soluble and insoluble fiber, particularly β-glucans (2.2–10.0%), and has a high satiety effect. Continuous energy supply is ensured by gradual degradation of starch and,

TABLE 17.1 Informative Macronutrient Composition of Nonwheat Cereals Expressed as G/100 g Grain

	Carbohydrates	Proteins	Lipids	Dietary Fiber
Rye	73.3–80.1	9.2–13.4	1.7–2.0	13.2
Barley	71.1–80.7	10.1–13.4	1.7–2.1	12.5
Oats	65.6–69.8	11.6–13.0	5.2–5.8	7.5–10.6
Sorghum	62.2–72.9	9.9–10.9	2.7–3.4	7.3
Rice	66.9–86.8	6.7–9.4	0.6–3.6	1.5–2.9
Maize	64.2–81.2	8.9–10.4	3.9–4.0	9.2
Millet species	48.4–72.9	7.7–14.5	1.3–5.2	6.2
Buckwheat	66.0–73.8	9.7–12.3	2.1–2.3	12.2
Amaranth	59.2	13.4–16.6	6.4–7.2	11.3
Quinoa	48.1–69	9.8–16.5	5.3–6.3	10.4

Based on data from Charalampopoulos, D., Wang, R., Pandiella, S.S., Webb, C., 2002. Application of cereals and cereal components in functional foods: a review. International Journal of Food Microbiology 79, 131–141; Kocková, M., Valík, Ľ., 2011. Potential of cereals and pseudocereals for lactic acid fermentations. Potravinarstvo 5, 27–40; Miranda, M., Vega-Gálvez, A., Uribe, E., López, J., Martínez, E., Rodríguez, M.J., Quispe, I., Di Scala, K., 2011. Physico-chemical analysis, antioxidant capacity and vitamins of six ecotypes of chilean quinoa (Chenopodium quinoa Willd). Procedia Food Science 1, 1439–1446; Arendt, E.K., Emanuele, Z., 2013. Cereal Grains for the Food and Beverage Industries. Woohead Publishing Limited, pp. 220–243; Saleh, A.S.M., Zhang, Q., Chen, J., Shen, Q., 2013. Millet grains: nutritional quality, processing, and potential health benefits. Comprehensive Reviews in Food Science and Food Safety 12, 281–295; Franz, C.M., Huch, M., Mathara, J.M., Abriouel, H., Benomar, N., Reid, G., Galvez, A., Holzapfel, W.H., 2014. African fermented foods and probiotics. International Journal of Food Microbiology 190, 84–96; Nascimento, A.C., Mota, C., Coelho, I., Gueifão, S., Santos, M., Matos, A.S., Gimenez, A., Lobo, M., Samman, N., Castanheira, I., 2014. Characterisation of nutrient profile of quinoa (Chenopodium quinoa), amaranth (Amaranthus caudatus), and purple corn (Zea mays L.) consumed in the North of Argentina: proximates, minerals and trace elements. Food Chemistry 148, 420–426; and Compiled Food Composition Database (Food Research Institute, 1999–2002).

consequently, by slower increase of blood glucose and insulin levels, resulting in delayed hunger perception. Insoluble fiber helps to improve intestinal peristalsis, increases the rate of chymus passage that reduces the exposition of epithelium to toxins, and may reduce the risk of cancer of the digestive tract, especially the colon. Easily soluble fiber (especially β-glucans) contributes to lower total and LDL cholesterol and to increase HDL cholesterol, reducing the risk of cardiovascular diseases. An oat diet improves metabolic processes in diabetic patients and in patients suffering from acidosis and inflammation of the mouth, stomach, or intestines (Mårtensson et al., 2005). The most important production from oats is oat flakes, which have become more popular in recent years. Oatmeal is made only in a limited scale for incorporationin to special types of bread and has a positive impact on softness of breadcrumbs, similar to barley (McMullen, 2000; Ciesarova et al., 2014; Blažeková, 2015).

Barley (*Hordeum vulgare* L.) is the oldest known cereal and today only around 20–25% of the harvest crop is utilized for food production, in particular for malt production. Only a small part is processed for peeled barley production. Although the bakery quality of barley is lower than wheat, it is characterized by similar or even better nutritional properties. Barley has high-soluble fiber content, including β-glucans as one of the main constituents, whose beneficial effects were noted above (Hockett, 2000).

Interest in alternative crops, which include pseudo-cereals such as buckwheat, amaranth, and quinoa, has increased recently. The reason for their popularity may be their excellent nutritional and biological value, absence of gluten, and content of some health-promoting compounds (Barnhoorn, 2015). These pseudo-cereals can serve as a good source of energy in the diet and contribute to the improvement of nutrition across the world. Due to their strong resistance, pseudo-cereals represent a sustainable food source, even if potential food shortage occurs for different reasons. The advantage of these crops is their unassuming cultivation conditions. For amaranth, a hot climate is recommended, in contrast to hardy quinoa, while both are appropriate for high altitudes.

Buckwheat (*Fagopyrum esculentum*) is nutritionally interesting because of its high vitamin B1 and B2 content, proteins with significant essential amino-acid composition, and relatively large amount of lysine, which is often deficient in other cereals. Furthermore, it is rich source of flavonoids, phytosterols, soluble carbohydrates, and other substances such as D-chiro-inositol, fagopyritols, or thiamine-binding proteins. Buckwheat contains a higher amount of rutin (quercetin-3-rutinoside) compared to other crops and thus has significant antioxidant, antiinflammatory, and anticarcinogenic properties. Food products derived from buckwheat have many other biological effects, such as promotion of intestinal microbiota and growth support of colonies of lactic-acid bacteria in the gastrointestinal tract, glucose- and cholesterol-lowering effects, inhibition of proteases, and free radical scavenging ability (Giménez-Bastida and Zieliński, 2015).

Amaranth (*Amaranthus* L.) has higher protein content (17–18%) than conventional cereals with nearly optimal proportion of essential amino acids and is rich in lysine. Starch content is low, with a limited amount of amylose (about 10% of starch, while 90% is amylopectin). Amaranth is a good source of tocotrienols and flavonoids. Moreover, the lipid content in amaranth seeds is significant, containing 6–7% of squalene, a compound related to the reduction of cancer risk, antiaging effects on skin, regulation of lipid metabolism, and a positive effect on the human immune system. Amaranth is also rich in minerals such as magnesium, potassium, phosphorus, and zinc (Nascimento et al., 2014).

Quinoa (*Chenopodium quinoa* Willd.) has been recognized as a nutritionally valuable crop and FAO declared 2013 as "The International Year of Quinoa."

Compared to wheat, quinoa is richer in macronutrients, in particular proteins that are comparable to the quality of the casein. Proteins in quinoa lack prolamins, which ensures their gluten-free nature. Quinoa provides valuable amounts of heart-healthy lipids such as monounsaturated fat (as oleic acid) and small amounts of omega-3 fatty acids as alpha-linolenic acid. It also contains high levels of fiber, minerals, and has an appropriate composition of carbohydrates and polysaccharides ensuring a low glycemic index. In addition, phytochemicals and antioxidants such as tocopherols as well as flavonoids like quercetin and kaempferol have been found in quinoa (Nascimento et al., 2014; Nowak et al., 2015). Quinoa is also known to contain saponins, in addition to many other beneficial micro- and macronutrients (Alvarez-Jubete et al., 2010)

The protein content of pseudo-cereals is generally higher than wheat with a much more balanced amino-acid profile, higher amounts of essential amino acids, and greater protein bioavailability. Since they contain mainly globulins and albumins and almost no prolamin proteins, pseudo-cereals represent an important food source for people suffering from celiac disease (Alvarez-Jubete et al., 2010). Several cereals such as sorghum, millet, rice, and maize are also gluten-free. Interestingly, not all countries have approved all of the above-mentioned crops for production of gluten-free products (Kreisz et al., 2008). An overview of the most common nonwheat cereals and their gluten status is given in Table 17.2.

TABLE 17.2 Nonwheat Cereals and Pseudo-Cereals, Their Botanical Origin and Gluten Status

Crop	Cereal/Pseudo-Cereal	Genus	Gluten Status
Rye	Cereal	*Secale*	High
Barley	Cereal	*Hordeum*	High
Oats	Cereal	*Avena*	Low
Sorghum	Cereal	*Sorghum*	Gluten-free
Rice	Cereal	*Oryzopsis*	Gluten-free
Maize	Cereal	*Zea*	Gluten-free
Millet species	Cereal	*Various*	Gluten-free
Buckwheat	Pseudo-cereal	*Fagopyrum*	Gluten-free
Amaranth	Pseudo-cereal	*Amaranth*	Gluten-free
Quinoa	Pseudo-cereal	*Chenopodium*	Gluten-free

Adapted from Barnhoorn, R., February 2015. Ancient Grains for Modern Times. Patrick Mannion, Singapore.

17.3 ADVANTAGES AND LIMITATIONS OF FERMENTATION APPLIED TO NONWHEAT CEREALS

Fermented foods are part of the daily diet in most parts of the world (Durgadevi and Shetty, 2012), including various traditional products and recently discovered "novel" probiotic drinks, sourdough bakery products, etc. One of the reasons for the increase in consumption of fermented foods is the fact thta many people consider these products to be healthy and natural (Giraffa, 2004). The demand for dairy milk substitutes is also increasing. Cereal-based beverages have a huge potential to fulfill this expectation and provide functional compounds such as antioxidants, dietary fiber, minerals, and vitamins, as well as prebiotic sugars (Nionelli et al., 2014).

Fermented cereals have multiple advantages compared to native or cooked grains since they are better in terms of nutritional composition and digestibility, as well as shelf-life. Fermentation enhances the nutrient content of foods through the improvement of protein and carbohydrate digestibility, biosynthesis of vitamins, enhancement of micronutrient bioavailability, and degradation of antinutritional factors. Fermentation is also a relatively cost-effective and low-energy preservation process, which is essential for ensuring the shelf-life and microbiological safety of the product (Liu et al., 2011). Toxic compounds such as aflatoxins and cyanogens can be reduced during fermentation by protective factors such as lactic-acid, bacteriocins, carbon dioxide, hydrogen peroxide, and ethanol, which facilitate inhibition or elimination of foodborne pathogens and enhance food safety. In addition to its nutritive, safety, and preservative effects, fermentation enriches the diet through the production of a variety of flavors, textures, and aromas (Giraffa, 2004). Moreover, some studies have reported an increase in antioxidant activity of cereals during the fermentation process (Dordevic et al., 2010; Zaroug et al., 2014).

There are many traditional indigenous fermented foods and beverages based on cereal matrices or their combination with pulses or vegetables, but there is also a new category of novel cereal-fermented products with health beneficial properties. Another way of utilizing the benefits of fermented-cereal substrates is through the preparation of sourdough in bakery product development. The application of sourdough is of increasing interest for improvement of flavor, structure, but also stability of baked goods. In addition, sourdough can actively retard starch digestibility, leading to low glycemic response. Sourdough is especially beneficial in bran-rich products, since bioactive substances from the bran may be released and delivered into blood circulation. Sourdough may also influence the nutritional quality of bread by its fiber solubilization and gluten degradation. Moreover, new compounds such as prebiotic oligosaccharides, bioactive peptides, or other metabolites may be formed during the fermentative process (Poutanen et al., 2009).

In recent years, cereals have been investigated as a potential source of prebiotics. Oligosaccharides and different types of DF (eg, xylans, xylo-oligosaccharides, arabinoxylans, and β-glucans) present in grains have been reported to be utilized by various probiotic microorganisms, thus exhibiting great potential for development of symbiotic food products (Figueroa-Gonzalez et al., 2011).

17.3.1 Degradation of Antinutritive Compounds and Simultaneous Enhancement of Minerals Concentration

Fermentation is an important process that significantly lowers the content of antinutritive compounds such as phytates, tannins, and polyphenols in grains. Fermentation may have a diminishing effect on polyphenol concentration due to the activity of the polyphenol oxidase present in the food grain and microbiota (Sindhu and Khetarpaul, 2001). Tannin levels may be reduced as a result of lactic-acid fermentation, leading to an increased absorption of iron, except in some high-tannin cereals, where little or no improvement in iron availability has been observed (Nout and Ngoddy, 1997).

Fermentation is one of the most effective processes to reduce phytic-acid concentration (Bilgicli et al., 2006). Magala et al. (2015) studied the degradation rate of phytic acid during fermentation of cereal-based food products and found that 89% of phytic acid was degraded in *tarhana* cultured with *Lb. sanfrancisco* CCM 7699, and an 80% reduction was observed in the *tarhana* sample inoculated with *Lb. plantarum* CCM 7039 after 144h of fermentation. In general, lower pH and longer fermentation time resulted in more intensive degradation of phytic acid (Bilgicli et al., 2006; Bilgicli and Ibanoglu, 2007), which was attributed to the action of endogenous phytases of the grain flour as well as microbial phytases (Osman, 2004; Zotta et al., 2007).

As a consequence of phytic-acid degradation, mineral concentration has been shown to increase in *tarhana* samples, while its content was not modified in cereal-based fermented beverages (Magala et al., 2015). In line with this, an increase in calcium and magnesium of *tarhana* was reported (Bilgicli et al., 2006). The increase in mineral content was attributed, at least in part, to the degradation of phytic acid, leading to the release of the minerals from their complexes into a free form after fermentation (Toufeili et al., 1999; Albarracin, 2013).

17.3.2 Improvement of Nutritional Properties, Vitamins Bioavailability, Palatability, Sensorial Acceptation, and Textural Properties

Fermentation can have multiple effects on the nutritional value of food, which leads to a decrease in the level of carbohydrates as well as some nondigestible poly- and oligosaccharides. The latter reduces side effects such as abdominal distension and flatulence. Improvement in starch digestibility during fermentation is sometimes related to the enzymatic properties of fermenting microbiota breaking down starch and oligosaccharides. The enzymes convert amylose and amylopectin into dextrins, maltose, and glucose. Reduction in amylase inhibition activity may also be responsible for the enhancement in starch digestibility. Similarly, improvement in protein digestibility of fermented products is mainly associated with the proteolytic activity of the fermenting microbiota. Certain amino acids may be synthesized and the availability of the vitamin B group may be improved. It has been reported that fermentation of cereals by LAB increases free amino acids and their derivatives by proteolysis and/or by metabolic synthesis. The content of the essential amino acids lysine, methionine, and tryptophan has also been increased during the fermentation process (Kohajdova and Karovicova, 2007). In addition, the microbial mass can also supply low molecular weight nitrogenous metabolites by cellular lysis (Mugula et al., 2003).

While the traditional foods made from cereal grains usually lack flavor and aroma (Charalampopoulos et al., 2002), fermentation usually enhances the flavor of goods. During cereal fermentation, several volatile compounds are formed, which contribute to a complex blend of flavors. The presence of aromas represented by diacetyl, acetic acid, and butyric acid makes fermented cereal-based products more appetizing (Blandino et al., 2003). The proteolytic activity of fermenting microorganisms, often in combination with malt enzymes, may produce precursors of flavor compounds, such as amino acids that may be deamined and decarboxylated to aldehydes, which may be oxidized to acids or reduced to alcohols (Mugula et al., 2003).

Reduction of the viscosity of fermented cereal-based products can be achieved to enhance their attractiveness. The addition of germinated cereal grains to prepared porridge increases its nutrient density while keeping it sufficiently liquid to be swallowed by infants (Kohajdova and Karovicova, 2007). Fermented porridge prepared from rice flour, soya milk, and passion fruit DF by inoculation with amylolytic strains (LAB and bifidobacteria) led to a decrease in apparent viscosity and a probiotic yogurt-like consistency porridge was developed (Do Espirito-Santo et al., 2014).

17.4 FERMENTED NONWHEAT FOOD PRODUCTS

Traditional fermented foods play an important role in many societies worldwide. The major advantages of fermentation are its low cost and the fact that it is a "low-tech" procedure accessible to poorer rural societies. In African civilizations food fermentation plays a major role in combating food spoilage and foodborne disease (Franz et al., 2014). In addition, it enhances the nutritive value, appearance of the food, and reduces the energy required for cooking. Cereals and pseudo-cereals are usually consumed in the form of bread, breakfast cereals, or cereal bars in developed countries. In developing countries the consumption of fermented cereals in the form of beverages, cakes, or porridges is common. This kind of food could increase consumption of cereals and pseudo-cereals in developed countries. The manufacturing techniques, microorganisms responsible for the fermentation, raw materials, and processing after fermentation vary worldwide.

There are four main types of fermentation processes: alcoholic, lactic-acid, acetic-acid, and alkali fermentations. The most common fermentation for cereals is lactic-acid fermentation, which contributes to the safety and nutritional quality while achieving different taste and consistency. Alcoholic fermentation is mostly used for production of fermented-cereal alcoholic beverages. Typical fermented nonwheat cereal foods and beverages are traditionally produced from maize, sorghum, millet, rice, etc. The fermentation is mostly carried out spontaneously and involves mixed cultures of yeasts, bacteria, and fungi. Some microorganisms may participate in parallel, while others act sequentially, changing the dominant flora during the fermentation process. The type of bacterial flora developed in the particular food depends on the water activity, pH, salt concentration, temperature, and composition of the food matrix (Blandino et al., 2003).

The tropical developing countries have a long history of fermenting various raw materials including cereals. Naturally fermented cereals account for up to 80% of the total energy intake in many African countries and provide an important source of dietary protein (Guyot, 2012; Franz et al., 2014). In many regions, different names and raw materials may be used for the preparation of traditional food products. Fermented food products based on rice such as *burong isda* are typical in the Philippines, *idli* and *dosa* are traditionally consumed in India, maize is used to prepare *pozol* and *atole agrio* in Mexico and Guatemala, and teff (belonging to the millet family) is used for preparation of *injera* in Etiopia and Eritrea (Guyot, 2012). For more extensive information about traditional cereal-based fermented food, see Blandino et al. (2003), Guyot (2012), and Franz et al. (2014).

Kocková (2014) prepared probiotic products by inoculating buckwheat, dark buckwheat, barley, oat, soya, and chickpea in combination with oat with *Lactobacillus rhamnosus* GG and subsequent molding to eliminate water from the cooked grains. *Lb. rhamnosus* GG was able to grow in all substrates during fermentation and reached a cell density of 6.68–7.58 log CFU/g. The highest growth rate was found in the oat product (0.34 log CFU/g/h). After fermentation, the lowest pH value (4.52) was observed in the barley product and the greatest amount of lactic acid was exhibited in the oat-soya product (1977.8 mg/kg).

17.4.1 Fermented Nonwheat Beverages

Today many studies are focused on the preparation and analysis of traditional and novel cereal-based beverages fermented with various microorganisms providing potential probiotics and functional properties. The suitability of selected nonwheat cereals, pseudo-cereals, and legumes for new probiotic food development was recently monitored and it was shown that buckwheat, dark buckwheat, barley, oat, soya, and chickpea inoculated with *Lb. rhamnosus* performed acceptable cell growth, pH, organic-acid profiles, and sensory characteristics and could be considered probiotic products (Kocková, 2014). During the preparation of a yogurt-like beverage from a mixture of rice, barley, and oat fermented with selected probiotic LAB strains, Coda et al. (2012) found viable cell counts of 8.4 log CFU/ml after 30 days of storage. The addition of sugar substitutes in a symbiotic functional beverage based on oat substrate and probiotic strains did not affect the fermentation process and viability of LAB during storage, and the concentration of β-glucans remained unchanged during the 21 days estimated shelf-life (Angelov et al., 2006). In the same context, Rathore et al. (2012), fermenting malt and barley flour with probiotic LAB *Lb. plantarum* and *Lb. acidophilus*, found the minimum dose recommended for a probiotic product to confer a therapeutic effect (6 log CFU/ml, based on a 100 ml daily dose) after 6 h. A commercially available probiotic beverage based on oat and barley malt flour called *Proviva*, manufactured in Sweden since 1994, showed probiotic bacteria cell counts of 5 log CFU/ml and were recommended to be effective as probiotic functions (Nyanzi and Jooste, 2012). The preparation and analysis of fermented beverages based on various cereals fermented with different LAB strains has been widely reported by different authors (Gokavi et al., 2005; Gupta et al., 2010; Nionelli et al., 2014; Ghosh et al., 2015).

17.4.2 Fermented Nonwheat Bakery Products

In developed countries, increasing demand for nutritionally enhanced foods associated with a positive impact on human health is evident. In order to meet consumer expectations, great effort is being put toward the design of products, especially staple foods, with improved health-promoting attributes. However, bakery products made from nonwheat cereals usually do not attain the same quality as wheat bakery products, and their production in large quantities is unprofitable (Peñas et al., 2013). To take advantage of the favorable dietary characteristics of nonwheat cereals while maintaining the quality of traditional wheat bakery products, so-called blended flours based on wheat with limited portions of other cereals may be utilized.

Keeping this approach in mind, products fortified with oat, rich in β-glucan, are considered to be a valuable part of the human diet, considering their outstanding nutritional and healthy values and relatively standardized bakery quality. A new functional wheat–rye bread fortified with extruded wheat bran (10.0%), cereal β-glucan hydrogel (12.5%), and lactobacilli starter culture showed higher amounts of DF, flavonoid content, antioxidant activity, and exhibited reduced glucose levels in healthy males compared with the wheat control bread, but no statistical significant change in the insulin response was observed (Mikušová, 2013). Consumption of nutritionally enhanced wheat–oat bread naturally rich in β-glucan fermented by LAB improved digestibility and provided unambiguous health benefits associated with slower blood glucose release and higher subjective satiety and led to more attractive organoleptic quality than wheat bread (Ciesarova et al., 2014). Production of sourdough from a special fine wholegrain oat flour fraction and potentially probiotic *Lb. plantarum* achieved stable gel consistency suitable for its technological application as a starter culture in wheat–oat blends and provided high fiber and β-glucan contents as well as a low energy value. The rate of staling after 3 days of storage was comparable to that of the control sample without the addition of sourdough (Blažeková, 2015).

The demand for cakes with functional properties has resulted in efforts to develop new and innovative products such as muffins made of extra wholegrain fermented oat flour providing high nutritional parameters such as fiber, protein, β-glucan, and mineral content as well as positive sensory acceptability (unpublished data). A similar approach was provided in developed fermented buckwheat-based muffins (Wronkowska et al., 2015).

17.5 HEALTH BENEFICIAL EFFECTS OF NONWHEAT FERMENTED FOODS

In the past, fermentation was considered as a tool for food preservation, although over time fermented food became popular due to its ability to improve gastrointestinal health and other positive benefits. Fermented cereal beverages with reliable probiotic cultures can reduce diarrhea and malnutrition resulting from contaminated food, reduce fatalities, and improve the well-being of children in Africa. Unfortunately, most traditional fermented food products are poorly studied, with unsubstantiated claims linked with positive effects on human health. Ideally, the health benefits should be backed up by scientific evidence in the form of randomized controlled clinical trials; however, these studies are rare in the field of fermented cereal foods. Despite the need for definitive studies demonstrating the direct health benefits to consumers, in vitro and animal studies are encouraging (Marsh et al., 2014).

The symbiotic effect of fermented cereal pastes (a mixture of cereal grains, nuts, and seeds) caused a significant decrease in serum and hepatic cholesterol levels, lowered the ratio of LDL to HDL cholesterol, and increased the excretion of cholesterol, triglycerides (TG), and bile acids in faeces of hyperlipidemic hamsters. In addition, antioxidant activity levels and the population of LAB increased, while the count of coliform and *Clostridium perfringens* were significantly reduced in feces (Wang et al., 2012).

Fermented rice bran is considered as a preventative food in common diseases connected with oxidative stress such as cancer, cardiovascular disease, impaired glucose metabolism, and neurodegenerative disturbances due to its high amount of phenolic bound to DF and β-glucans with a wide variety of biological activities. Solid-state fermentation is presented as a strategic technology to potentially increase the content of bioactive compounds in rice bran and also to provide newly synthetized metabolites (Kim and Han, 2014). Likewise, fermented brown rice with LAB has been suggested as a supplement to reduce plasma and hepatic TG, total, and LDL cholesterol as well as very lowdensity lipoprotein cholesterol levels in rats (Hee et al., 2005). Similarly, brown rice fermented by *Aspergillus oryzae* may be a promising dietary agent for the prevention of human esophageal (Kuno et al., 2004), gastric (Tomita et al., 2008), and colon (Katyama et al., 2002) cancers, as well as to have an impact on the elimination of chemicals such as polychlorinated biphenyls in humans (Takasuga et al., 2004).

A blood pressure-lowering effect of buckwheat sprouts fermented by LAB was found in spontaneously hypertensive rats. Although buckwheat contains rutin and other phenolic compounds exhibiting vasorelaxant effects and angiotensin I-converting enzyme (ACE) inhibition activity, the antihypertensive effect was attributed more to lactic-acid fermentation than to sprouting, producing bioactive peptides and γ-aminobutyric acid (GABA), which could have caused additional ACE inhibition and vasodilatation (Nakamura et al., 2013).

Human studies with nonwheat cereal-fermented foods are relatively less common. In a randomized, double-blind, parallel group study of 62 volunteers with moderately increased plasma-cholesterol levels, the consumption of an oat beverage cofermented with a β-glucans-producing strain (ropy product) reduced their total cholesterol levels by 6% for 5 weeks and confirmed the prebiotic potential of the beverage by a significant increase of *Bifidobacterium* spp. in the feces compared to the control group (fermented dairy-based product). However, a fermented oat drink without the additional microbial β-glucans did not show similar effects (Mårtensson et al., 2005).

17.6 CONCLUSION

Cereals provide an important source of energy and are represented by carbohydrates, proteins, and fats, but also fiber, vitamins, minerals, and other phytochemicals essential to human health. Although wheat is the most exploited cereal for food production, other cereals and pseudo-cereals appear to have unambiguous advantages: more soluble proteins, higher content of vitamins and minerals, and a notable amount of nondigestible polysaccharides. Moreover, pseudo-cereals have excellent nutritional and biological value, do not contain gluten, and are rich in health-promoting substances. Fermentation of cereals brings multiple improvements to nutritional composition, digestibility, shelf-life, as well as flavor enhancement. Traditional fermented food and beverages are widely consumed all over the world due to their positive effects on gastrointestinal tract and ease preservation. There is also a promising trend in the production of novel fermented nonwheat cereal products aimed at the prevention of common diseases.

ACKNOWLEDGMENTS

This work resulted from the implementation of projects ITMS 26240220091 "Strategy of acrylamide elimination in food processing" and ITMS 26240120042 "The Centre of Excellence for Contaminants and Microorganisms in Foods" funded by the European Regional Development Fund and additionally supported by the grant VEGA no. 1/0453/13.

REFERENCES

Albarracin, M.G., Rolando, J., Drago, S.R., 2013. Effect of soaking process on nutrient bio-accessibility and phytic acid content of brown rice cultivar. LWT Food Science and Technology 53, 76–80.

Alvarez-Jubete, L., Arendt, E.K., Gallagher, E., 2010. Nutritive value of pseudocereals and their increasing use as functional gluten-free ingredients. Trends in Food Science & Technology 21, 106–113.

Angelov, A., Gotcheva, V., Kuncheva, R., Hristozova, T., 2006. Development of a new oat-based probiotic drink. International Journal of Food Microbiology 112, 75–80.

Arendt, E.K., Emanuele, Z., 2013. Cereal Grains for the Food and Beverage Industries. Woohead Publishing Limited, pp. 220–243.

Barnhoorn, R., February 2015. Ancient Grains for Modern Times. Patrick Mannion, Singapore.

Bilgicli, N., Ibanoglu, S., 2007. Effect of wheat germ and wheat bran on the fermentation activity, phytic acid content and colour of tarhana, a wheat flour-yoghurt mixture. Journal of Food Engineering 78, 681–686.

Bilgicli, N., Elgun, A., Turker, S., 2006. Effects of various phytase sources on phytic acid content, mineral extractability and protein digestibility of tarhana. Food Chemistry 98, 329–337.

Blandino, A., Al-Aseeri, M.E., Pandiella, S.S., Cantero, D., Webb, C., 2003. Cereal-based fermented foods and beverages. Food Research International 36, 527–543.

Blažeková, L., Polakovičová, P., Mikušová, L., Kukurová, K., Saxa, V., Ciesarová, Z., Šturdík, E., 2015. Development of innovative health beneficial bread using a fermented fiber-glucan product. Czech Journal of Food Sciences 33, 118–125.

Ciesarova, Z., Kukurova, K., Mikusova, L., Basil, E., Polakovicova, P., Duchonova, L., Vlcek, M., Sturdik, E., 2014. Nutritionally enhanced wheat-oat bread with reduced acrylamide level. Quality Assurance and Safety of Crops & Foods 6, 327–334.

Coda, R., Lanera, A., Trani, A., Gobbetti, M., Di Cagno, R., 2012. Yogurt-like beverages made of a mixture of cereals, soy and grape must: microbiology, texture, nutritional and sensory properties. International Journal of Food and Microbiology 155, 120–127.

Charalampopoulos, D., Wang, R., Pandiella, S.S., Webb, C., 2002. Application of cereals and cereal components in functional foods: a review. International Journal of Food Microbiology 79, 131–141.

Do Espirito-Santo, A.P., Mouquet-Rivier, C., Humblot, C., Cazevieille, C., Icard-Verniere, C., Soccol, C.R., Guyot, J.-P., 2014. Influence of cofermentation by amylolytic *Lactobacillus* strains and probiotic bacteria on the fermentation process, viscosity and microstructure of gruels made of rice, soy milk and passion fruit fiber. Food Research International 57, 104–113.

Dordevic, T.M., Siler-Marinkovic, S.S., Dimitrijevic-Brankovic, S.I., 2010. Effect of fermentation on antioxidant properties of some cereals and pseudocereals. Food Chemistry 119, 957–963.

Durgadevi, M., Shetty, P.H., 2012. Effect of ingredients on texture profile of fermented food, idli. In: Dan, Y. (Ed.), 3rd International Conference on Biotechnology and Food Science, vol. 2. pp. 190–198.

Figueroa-Gonzalez, I., Quijano, G., Ramirez, G., Cruz-Guerrero, A., 2011. Probiotics and prebiotics–perspectives and challenges. Journal of the Science of Food and Agriculture 91, 1341–1348.

Franz, C.M., Huch, M., Mathara, J.M., Abriouel, H., Benomar, N., Reid, G., Galvez, A., Holzapfel, W.H., 2014. African fermented foods and probiotics. International Journal of Food Microbiology 190, 84–96.

Ghosh, K., Ray, M., Adak, A., et al., 2015. Microbial, saccharifying and antioxidant properties of an Indian rice based fermented beverage. Food Chemistry 168, 196–202.

Giménez-Bastida, J.A., Zieliński, H., 2015. Buckwheat as a functional food and its effects on health. Journal of Agricultural and Food Chemistry 63, 7896–7913.

Giraffa, G., 2004. Studying the dynamics of microbial populations during food fermentation. Fems Microbiology Reviews 28, 251–260.

Gokavi, S., Zhang, L.W., Huang, M.K., Zhao, X., Guo, M.G., 2005. Oat-based symbiotic beverage fermented by *Lactobacillus plantarum*, *Lactobacillus paracasei* ssp. *casei*, and *Lactobacillus acidophilus*. Journal of Food Science 70, M216–M223.

Gupta, S., Cox, S., Abu-Ghannam, N., 2010. Process optimization for the development of a functional beverage based on lactic acid fermentation of oats. Biochemical Engineering Journal 52, 199–204.

Guyot, J.-P., 2012. Cereal-based fermented foods in developing countries: ancient foods for modern research. International Journal of Food Science and Technology 47, 1109–1114.

Hee, S., Park, B.S., Lee, H.G., 2005. Hypocholesterolemic action of fermented brown rice supplement in cholesterol-fed rats: cholesterol-lowering action of fermented brown rice. Journal of Food Science 70, s527–s531.

Hockett, E.A., 2000. In: Kulp, K., Ponte Jr., J.G. (Eds.), Barley. Handbook of Cereal Science and Technology. Marcel Dekker, Inc., New York, Basel, p. 790.

Kamal-Eldin, A., 2008. In: Hamaker, B.R. (Ed.), Micronutrients in Cereal Products: Their Bioactivities and Effects on Health. Woodhead Publishing Limited, Cambridge, England, pp. 86–111.

Katyama, M., Yoshimi, N., Yamada, Y., et al., 2002. Preventive effect of fermented brown rice and rice bran against colon carcinogenesis in male F344 rats. Oncology Reports 9, 817–822.

Kim, D., Han, G.D., 2014. Fermented Rice Bran Attenuates Oxidative Stress. pp. 467–480.

Kocková, M., Valík, Ľ., 2011. Potential of cereals and pseudocereals for lactic acid fermentations. Potravinarstvo 5, 27–40.

Kocková, M., Valík, Ľ., 2014. Development of new cereal-, pseudocereal-, and cereal-leguminous-based probiotic foods. Czech Journal of Food Sciences 32, 391–397.

Kohajdova, Z., Karovicova, J., 2007. Fermentation of cereals for specific purpose. Journal of Food and Nutrition Research 46, 51–57.

Kreisz, S., Arendt, E.K., Hübner, F., Zarnkov, M., 2008. 16-Cereal-based gluten-free functional drinks. In: Arendt, E.K., Bello, F.D. (Eds.), Gluten-free Cereal Products and Beverages. Academic Press, San Diego, pp. 373–392.

Kuno, T., Hirose, Y., Hata, K., et al., 2004. Preventive effect of fermented brown rice and rice bran on N-nitrosomethylbenzylamine-induced esophageal tumorigenesis in rats. International Journal of Oncology 25, 1809–1815.

Liu, S-n, Han, Y., Zhou, Z-j, 2011. Lactic acid bacteria in traditional fermented Chinese foods. Food Research International 44, 643–651.

Lorenz, K., 2000. Rye. In: Kulp, K., Ponte Jr., J.G. (Eds.), Handbook of Cereal Science and Technology. Marcel Dekker, Inc., New York, Basel, p. 790.

Magala, M., Kohajdova, Z., Karovicova, J., 2015. Degradation of phytic acid during fermentation of cereal substrates. Journal of Cereal Science 61, 94–96.

Marsh, A.J., Hill, C., Ross, R.P., Cotter, P.D., 2014. Fermented beverages with health-promoting potential: past and future perspectives. Trends in Food Science & Technology 38, 113–124.

Mårtensson, O., Biörklund, M., Lambo, A.M., Dueñas-Chasco, M., Irastorza, A., Holst, O., Norin, E., Welling, G., Öste, R., Önning, G., 2005. Fermented, ropy, oat-based products reduce cholesterol levels and stimulate the bifidobacteria flora in humans. Nutrition Research 25, 429–442.

McKewith, B., 2004. Nutrition aspects of cereals: briefing paper. Nutrition Bulletin 29, 111–142.

McMullen, M.S., 2000. Oats. In: Kulp, K., Ponte Jr., J.G. (Eds.), Handbook of Cereal Science and Technology. Marcel Dekker, Inc., New York, Basel, p. 790.

Mikušová, L., Šturdík, E., Holubková, A., 2011. Whole grain cereal food in prevention of obesity. Acta Chimica Slovaca 4, 95–114.
Mikušová, L., Gereková, P., Kocková, M., Šturdík, E., Valachovičová, M., Holubková, A., Vajdák, M., Mikuš, Ľ., 2013. Nutritional, antioxidant, and glycaemic characteristics of new functional bread. Chemical Papers 67, 284–291.
Miranda, M., Vega-Gálvez, A., Uribe, E., López, J., Martínez, E., Rodríguez, M.J., Quispe, I., Di Scala, K., 2011. Physico-chemical analysis, antioxidant capacity and vitamins of six ecotypes of chilean quinoa (*Chenopodium quinoa* Willd). Procedia Food Science 1, 1439–1446.
Mugula, J.K., Narvhus, J.A., Sorhaug, T., 2003. Use of starter cultures of lactic acid bacteria and yeasts in the preparation of togwa, a Tanzanian fermented food. International Journal of Food Microbiology 83, 307–318.
Nakamura, K., Naramoto, K., Koyama, M., 2013. Blood-pressure-lowering effect of fermented buckwheat sprouts in spontaneously hypertensive rats. Journal of Functional Foods 5, 406–415.
Nascimento, A.C., Mota, C., Coelho, I., Gueifão, S., Santos, M., Matos, A.S., Gimenez, A., Lobo, M., Samman, N., Castanheira, I., 2014. Characterisation of nutrient profile of quinoa (*Chenopodium quinoa*), amaranth (*Amaranthus caudatus*), and purple corn (*Zea mays* L.) consumed in the North of Argentina: proximates, minerals and trace elements. Food Chemistry 148, 420–426.
Nionelli, L., Coda, R., Curiel, J.A., Poutanen, K., Marco, G., Rizzello, C.G., 2014. Manufacture and characterization of a yogurt-like beverage made with oat flakes fermented by selected lactic acid bacteria. International Journal of Food Microbiology 185, 17–26.
Nout, M.J.R., Ngoddy, P.O., 1997. Technological aspects of preparing affordable fermented complementary foods. Food Control 8, 279–287.
Nowak, V., Du, J., Ruth Charrondière, U., 2015. Assessment of the nutritional composition of quinoa (*Chenopodium quinoa* Willd.). Food Chemistry 193.
Nyanzi, R., Jooste, P.J., 2012. Cereal-Based Functional Foods. In: Rigobelo, E. (Ed.), Probiotics. InTech, pp. 161–196.
Osman, M.A., 2004. Changes in sorghum enzyme inhibitors, phytic acid, tannins and in vitro protein digestibility occurring during Khamir (local bread) fermentation. Food Chemistry 88, 129–134.
Peñas, E., Martinez-Villaluenga, C., Vidal-Casero, C., Zieliński, H., Frias, J., 2013. Protein quality of traditional rye breads and ginger cakes as affected by the incorporation of flour with different extraction rates. Polish Journal of Food and Nutrition Sciences 63, 5–10.
Poutanen, K., 2012. Past and future of cereal grains as food for health. Trends in Food Science & Technology 25, 58–62.
Poutanen, K., Flander, L., Katina, K., 2009. Sourdough and cereal fermentation in a nutritional perspective. Food microbiology 26, 693–699.
Rathore, S., Salmeron, I., Pandiella, S.S., 2012. Production of potentially probiotic beverages using single and mixed cereal substrates fermented with lactic acid bacteria cultures. Food Microbiology 30, 239–244.
Saleh, A.S.M., Zhang, Q., Chen, J., Shen, Q., 2013. Millet grains: nutritional quality, processing, and potential health benefits. Comprehensive Reviews in Food Science and Food Safety 12, 281–295.
Sindhu, S.C., Khetarpaul, N., 2001. Probiotic fermentation of indigenous food mixture: effect on antinutrients and digestibility of starch and protein. Journal of Food Composition and Analysis 14, 601–609.
Takasuga, T., Senthilkumar, K., Takemori, H., Ohi, E., Tsuji, H., Nagayama, J., 2004. Impact of FEBRA (fermented brown rice with *Aspergillus oryzae*) intake and concentrations of PCDDs, PCDFs and PCBs in blood of humans from Japan. Chemosphere 57, 1409–1426.

Tomita, H., Kuno, T., Yamada, Y., et al., 2008. Preventive effect of fermented brown rice and rice bran on N-methyl-N'-nitro-N-nitrosoguanidine-induced gastric carcinogenesis in rats. Oncology Reports 19, 11–15.

Toufeili, I., Melki, C., Shadarevian, S., Robinson, R.K., 1999. Some nutritional and sensory properties of bulgur and whole wheatmeal kishk (a fermented milk-wheat mixture). Food Quality and Preference 10, 9–15.

Wang, C.-Y., Wu, S.-J., Fang, J.-Y., Wang, Y.-P., Shyu, Y.-T., 2012. Cardiovascular and intestinal protection of cereal pastes fermented with lactic acid bacteria in hyperlipidemic hamsters. Food Research International 48, 428–434.

Wronkowska, M., Christa, K., Ciska, E., Soral-Śmietana, M., 2015. Chemical characteristics and sensory evaluation of raw and roasted buckwheat groats fermented by *Rhizopus oligosporus*. Journal of Food Quality 38, 130–138.

Zaroug, M., Orhan, I.E., Senol, F.S., Yagi, S., 2014. Comparative antioxidant activity appraisal of traditional Sudanese *kisra* prepared from two sorghum cultivars. Food Chemistry 156, 110–116.

Zotta, T., Ricciardi, A., Parente, E., 2007. Enzymatic activities of lactic-acid bacteria isolated from Cornetto di Matera sourdoughs. International Journal of Food Microbiology 115, 165–172.

Chapter 18

Use of Sourdough Fermentation and Nonwheat Flours for Enhancing Nutritional and Healthy Properties of Wheat-Based Foods

C.G. Rizzello[1], R. Coda[2], M. Gobbetti[1]
[1]*University of Bari Aldo Moro, Bari, Italy;* [2]*University of Helsinki, Helsinki, Finland*

18.1 BACKGROUND

The opportunity to supplement or completely replace wheat with legumes, minor cereals, pseudo-cereals, and milling byproducts of higher nutritional value is a strategy that extends to the food chain, from farm to processed food, and that can be inherently beneficial to the public (Coda et al., 2014a).

After cereals, *Leguminosae* are the second most popular food crop worldwide. The global value for leguminous crops is thought to be approximately 2 billion US dollars per year (Duranti, 2006). Although legumes have been a part of most diets across the world for thousands of years, only recently have nutritional and functional value been rediscovered and deeply investigated (Coda et al., 2014a; Duranti, 2006; Rizzello et al., 2014). The *Leguminosae* family is the most important group of *Dicotyledonae* (Duranti, 2006), being one of the largest families of flowering plants with 18,000 species, which are classified into around 650 genera (Duranti, 2006). A large variety of legumes used for human consumption are cultivated extensively or locally (Duranti, 2006; Smartt and Nwokolo, 1996). The economic importance of the *Leguminosae* family is also increasing in marginal lands, since many species are able to grow under restrictive conditions (Duranti, 2006). Overall, legumes are an excellent source of proteins, carbohydrates, and dietary fibers (DF). They also provide many essential amino acids, vitamins, minerals, oligosaccharides, and phenolic compounds (Campos-Vega et al., 2010; Roy et al., 2010). Legumes are also considered effective at decreasing the risk of cardiovascular disease (Flight and Clifton, 2006), type 2 diabetes (Jenkins et al., 2012), some types of cancer (Feregrino-Perez et al., 2008), and obesity (Mollard et al., 2012). Nevertheless, legumes also contain several compounds that can lead to reduction of nutrient absorption, mineral availability, and protein digestibility (Díaz-Batalla et al., 2006) often referred to as antinutritional, namely raffinose, phytic acid, condensed tannins, alkaloids, lectins, pyrimidine glycosides (eg, vicine and convicine), and protease inhibitors (Coda et al., 2014a; Liener, 1990). Most of these antinutritional factors (ANF) are heat-labile (eg, protease inhibitors and lectins) and thus thermal treatments would remove potential negative effects during consumption (Muzquiz et al., 2012). On the other hand, phytic acid, raffinose, tannins, and saponins are heat stable, and various methods such as dehulling, soaking, air classification, extrusion, or cooking are used to decrease their negative impact on digestibility and bioavailability of different compounds (Coda et al., 2014a; Jezierny et al., 2010; Rizzello et al., 2014; Van Der Poel, 1990). Biological methods such as germination, enzyme treatments, and fermentation have also been proposed (Alonso et al., 2000; Coda et al., 2014a; Granito et al., 2002; Luo et al., 2009).

In spite of the huge amount produced, wheat is not a typical cereal in developing or emerging countries where other crops like sorghum (*Sorghum bicolor* L. Moench), millet (*Pennisetum glaucum*), acha or white fonio (*Digitaria exilis*), iburu or black fonio (*Digitaria iburua*), teff (*Eragrostis tef*), maize (*Zea mais*), and rice (*Oryza sativa*) comprise the local diet and play an essential role in providing healthy food for the poorest populations and regions around the world (Coda et al., 2014a). For this reason and because of their numerous useful properties, research and development on minor, ethnic, and ancient grains has received worldwide. Several ancient crops and/or minor cereals such as kamut, barley, spelt, rye, einkorn, millet, oat, and sorghum are often underutilized or only consumed locally; others are pseudo-cereals (eg, quinoa, amaranth, buckwheat), crops evolutionarily distant from cereals that produce grains (Coda et al., 2010a; Correia et al., 2010; Guyot, 2012; Sterr et al., 2009). The use of minor cereals and pseudo-cereals is of nutritional interest because of their nutrition profile, especially in regard to the minor components present in grains (eg, DF, resistant starch, minerals, vitamins, phenolic compounds). Furthermore, many of them, including spelt, emmer, teff, fonio, and other indigenous cereal grains, are low-input plants, suitable for growing without the use of pesticides in harsh ecological conditions and marginal areas of cultivation (Coda et al., 2010a; Jideani and Jideani, 2011; Moroni et al., 2010). In recent years, these grains have attracted interest, mainly as niche products advertised as healthier and more natural than modern wheat (Coda et al., 2014a). Pseudo-cereals such as buckwheat, amaranth, and quinoa have nutritional and textural features that make them suitable for replacing, at least in part, traditional cereal-based products (Coda et al., 2014a).

Nonwheat cereals and pseudo-cereals are often used for making fermented products. In particular, they are prepared as fermented beverages, gruels, porridges, soups, etc., and are called specific names (Beuchat, 1997). These fermented products play an important role because of improved shelf-life and the nutritional properties derived from fermentation. In Western countries, food fermentation is often integrated into marketing strategies in response to increased trends toward healthier lifestyles and to address specific organoleptic characteristics (Humblot and Guyot, 2008).

Refined cereal flours contain almost exclusively endosperm seed, while germ (embryo) and bran are discarded during milling despite the high DF content and functional compounds (α-tocopherol, vitamins, minerals, and phytochemicals). Wheat germ also contains proteins with high biological value. Although their use in the food and feed industry has increased in the past decade, the major part of milling byproducts is still used as livestock feed and only a small percentage is employed for food purposes (Coda et al., 2015a). In particular, wheat-germ consumption is limited by lipid oxidation during storage and by the presence of ANF such as raffinose and phytic acid (Rizzello et al., 2010b), whereas bran is underutilized because of the detrimental effects on the technological and organoleptic characteristics of wheat-based foods, such as baked goods and pasta (Rizzello et al., 2012). Nevertheless, the potential health benefits of consuming more whole-grain foods have been studied extensively in recent years and today the importance of cereal bran and germ is firmly recognized, due to their high DF content and antioxidant compounds (Vitaglione et al., 2008). Most consumers still prefer refined white-flour to whole-grain products, since the textural and flavor properties of bran are perceived as less attractive than products made with refined flour (Bakke and Vickers, 2007). Furthermore, despite the evidence of positive effects on health, the intake of DF is still less than the recommended 25 g/day, according to WHO/FAO guidelines (WHO, 2003).

Consumer demand for food products with improved nutritional value or health benefits has involved the cereal industry but also presented new challenges. Even though the market for new novel cereal-based products, which include legumes, alternative cereals, pseudo-cereals, and milling byproducts, is increasing, the use of such flours is restricted due to their low baking quality and reduced sensory quality of baked products (Gallagher et al., 2004).

It has been shown that sourdough fermentation of these alternative flours, through lactic-acid bacteria (LAB) metabolism, can improve both the sensory and baking qualities, providing wholesome food with attractive flavor and texture (Coda et al., 2014a). From a nutritional point of view, sourdough biotechnology offers the opportunity to increase the nutritional quality and decrease ANF (Coda et al., 2010b, 2011; Moroni et al., 2012).

18.2 USE OF LEGUMES, MINOR CEREAL, PSEUDO-CEREAL FLOURS, AND SOURDOUGH FERMENTATION FOR ENHANCING NUTRITIONAL AND FUNCTIONAL PROPERTIES OF WHEAT-BASED FOODS

18.2.1 Legume Flours and Sourdough Fermentation

Although the nutritional benefits of legumes are well recognized, a downward trend in their consumption has been noted (Vaz Patto et al., 2015). Therefore better tasting and more convenient legume food products are needed to increase consumer interest (Gómez et al., 2008; Schneider, 2002). Among the many options of food with increased nutritional value, legumes (eg, pea or chickpea) to bread (Dhinda et al., 2011; Kamaljit et al., 2010; Mohammed et al., 2012; Sadowska et al., 2003) and other bakery product formulas (eg, chapatti, cakes, biscuits, and crackers) (Eissa et al., 2007; Gómez et al., 2008; Kadam et al., 2012; Kohajdová et al., 2011; Tiwari et al., 2011) seem to offer great potential. The addition of legume flours to wheat bread is an excellent way to improve cereal nutritional properties since the high level of bioactive compounds, present in legumes (eg, DF, essential amino acids), may complement the nutritional and functional properties and/or deficiencies of a cereal-based diet (Angioloni and Collar, 2012). In bread-making applications, high amounts of legumes incorporated into baked products without any structuring agent are cost-effective and nutritionally advantageous but also technologically challenging (Angioloni and Collar, 2012). Thus when legume flours are used for bread making, several process parameters need to be adjusted to get the right structure and sensory qualities consumers demand (Kohajdová et al., 2013; Maninder et al., 2007).

In addition to the careful selection of legume flour, the use of adequate bioprocessing based on technology, nutritional, and healthy properties is considered key to optimal sensory and healthy features.

One option to improve the sensory and functional quality of breads containing legume flours is the use of sourdough fermentation. Currently, the literature is particularly rich with experimental data that show how sourdough fermentation enhances the nutritional and sensory quality of wheat breads. Sourdough fermentation may decrease the glycemic response of baked goods, improve the content of bioactive compounds, decrease the level of ANF, and increase the uptake of minerals (Gobbetti et al., 2014). Table 18.1 summarizes the advantages of sourdough applied to legume flour fermentation.

From a microbiological point of view, the effect of sourdough fermentation is related to the metabolic activities of two groups of microorganisms: yeasts and LAB. Nevertheless, LAB are mainly responsible for all the nutritional and functional advantages of sourdough fermentation, whereas yeasts are mostly related to leavening and aroma formation. For this reason, in most studies sourdough fermentation has been carried out with the sole use of starter LAB, especially if the improvement of the nutritional and functional features of food matrices is the main aim. Bacterial enzymatic activities have been shown to be responsible for an increase in functional compounds such as phenolic compounds and soluble fibers, whose consumption is recommended to prevent several diseases, to regulate energy intake and satiety, and to help with glycemic control. In a recent study, the suitability of several minor species and varieties of legumes for use in sourdough fermentation was assessed on the basis of improvement of nutritional characteristics (Curiel et al., 2015). The legume varieties belonging to *Phaseolus vulgaris*, *Cicer arietinum*, *Lathyrus sativus*, *Lens culinaris*,

TABLE 18.1 Main Nutritional Advantages Related to the Use of Sourdough Biotechnology in Legume Flours and Legume-Enriched Wheat Bread

Legume Flour	Effects	References
Effects of Sourdough Fermentation on Legume Flour		
Bean, chickpea, grass pea, lentil, pea (local cultivars)	Increase of phytase activity and free amino acids, GABA, soluble fibers, and total phenols concentrations. Decrease of raffinose and condensed tannins concentrations.	Curiel et al. (2015)
Grass pea	Decrease of phytic acid concentration and trypsin inhibitory activity	Starzyńska-Janiszewska et al. (2011)
Faba bean	Decrease of vicine and convicine concentration, trypsin inhibitor activity, starch hydrolysis index. Increase of protein digestibility, and free amino acids and GABA concentrations.	
Effects of Legume Flour Addition and Sourdough Fermentation in Wheat Bread		
Chickpea	Increase of free amino acid and GABA concentrations; decrease of the starch hydrolysis index (HI).	Coda et al. (2010b)
Soybean	Improved nutritional quality, as estimated by protein efficiency ratio (PER); net protein ratio (NPR); apparent digestibility (AD), true digestibility	Peñaloza-Espinosa et al. (2011)
Lupine	Decrease of the trypsin inhibitor activity; increase of protein digestibility.	Bartkiene et al. (2011)
Chickpea/lentil/bean (mixture)	Increase of free amino acid concentration; increase of the antioxidant and phytase activities.	Rizzello et al. (2014)

and *Pisum sativum* species, with high levels of protein (mainly grass pea and several lentil flours), free amino acids (FAA), phenolic compounds, DF, and ash were subjected to fermentation, according to protocols used for cereals, with the selected starters *Lactobacillus plantarum* C48 and *Lactobacillus brevis* AM7 (Curiel et al., 2015).

Legumes contain a variable concentration of α-galactosides of sucrose, namely raffinose, stachyose, and verbascose. These α-galactosides, also known as the "raffinose family of oligosaccharides" (RFOs), are not degraded in the upper gastrointestinal tract due to the lack of α-galactosidase (α-Gal) activity (Teixeira et al., 2012). RFOs are fermented in the large intestine by intestinal microbes, thus causing gastrointestinal symptoms (eg, abdominal discomfort, flatulence, and diarrhea). In contrast, moderate doses of RFOs favor the metabolism of beneficial intestinal microorganisms (eg, bifidobacteria). Treatment with α-Gal (EC 3.2.1.22) or with bacteria (eg, lactobacilli and *Leuconostoc*) capable of degrading RFOs markedly decreased the concentration of raffinose in legumes (Rizzello et al., 2010a; Teixeira et al., 2012). The partial hydrolysis of raffinose was also achieved through the fructansucrase (eg, levansucrase, EC 2.4.1.10) activity of lactobacilli (Teixeira et al., 2012). Curiel et al. (2015) reported that raffinose decreased by up to ca. 64% in traditional Italian legumes mainly due to sourdough fermentation. The pyrimidine glycosides, called vicine and convicine, are metabolites exerting an antinutritional effect present in faba bean (*Vicia faba* L.). Derivatives of vicine and convicine divicine and isouramil are toxic to human carriers of a genetic deficiency of the erythrocyte-located glucose-6-phosphate dehydrogenase, causing the hemolytic anemia disease known as favism (Crépon et al., 2010). In a recent study (Coda et al., 2015b) air classified faba bean flour was followed by fermentation with *Lactobacillus plantarum* VTT E–133328. Fermentation caused a decrease of vicine and convicine content by more than 91%, significantly reduced trypsin inhibitor activity, and condensed tannins by more than 40% (Coda et al., 2015b). Condensed tannins, mainly composed of flavonoid units, could have a marked effect on the nutritional value of legumes, since they can interfere with the digestive process by binding enzymes, other proteins, and/or minerals (Kosińska et al., 2011; Liener, 1990). The hydrolysis of condensed tannins follows different pathways, which involve enzymes such as decarboxylases and oxygenases. As recently shown for faba bean (Coda et al., 2014a) and many Italian legume flours (Curiel et al., 2015), microbial activities (Deschamps, 1989) and, especially, LAB fermentation, consistently decreases the levels of condensed tannins.

Phytic acid (*myo*inositol hexakisphosphate) is the main phosphate storage form in most cereals, legumes, and nuts that strongly binds minerals like iron and zinc (Lopez et al., 2002). Phytase, catalyzing the hydrolysis of phytic acid to myoinositol and phosphoric acid, makes phosphate available and leads to nonmetal chelator compound (Martinez et al., 1996). It has been shown that the optimal conditions for phytate degradation between different plant species vary. High native phytase activity was found in cereal and cereal byproducts, whereas lower activity was described for legumes (Steiner et al., 2007). The activity of grain endogenous phytase is stimulated under the acidic conditions promoted during sourdough fermentation, and a contribution to phytate degradation can also come from sourdough microflora (Gobbetti et al., 2014). The extent to which phytases are activated depends on the fermentation kinetics, which also depends on the raw materials used (Hammes et al., 2005). The optimal conditions for leguminous endogenous phytases are different from those of cereals (Gustafsson and Sandberg, 1995; Scott, 1991) but, depending on the case, fermentation can still efficiently decrease phytic-acid content in some legumes and favor the increase of phytase activity, especially that of kidney bean flours (Curiel et al., 2015; Doblado et al., 2003; Luo et al., 2009).

The partial or complete elimination of α-galactosides, tannins, phytic acid, and trypsin inhibitory activity was also obtained in kidney beans through spontaneous fermentation (Granito et al., 2002). It was also reported for grass pea that fermentation by *Lactobacillus plantarum* decreased the levels of phytic acid and trypsin inhibitory activity (Starzyńska-Janiszewska and Stodolak, 2011).

In addition to the reduction of antinutritional compounds, the literature largely confirms that LAB fermentation of legumes increases the amount of FAA, especially essential amino acids and γ-aminobutyric acid (GABA), enhances in vitro protein digestibility, and significantly lowers the starch hydrolysis index.

18.2.2 Legume Flours in Sourdough Wheat Bread Making

Improvements in digestibility, functionality, and suitability of legumes as ingredients in food applications can be obtained through LAB fermentation, able to modify the physicochemical, structural, and functional properties of legume flours, can be considered an efficient tool to improve its suitability to be used as ingredients in the food industry. The fortification of bread with legume ingredients has been the subject of several studies, as reviewed by Boye et al. (2010). In most of the studies legume was used as whole flour or fraction while in only some studies legume flour was incorporated into wheat bread after sourdough fermentation.

Aiming at the production of functional wheat-flour bread enriched in GABA, Coda et al. (2010b) fermented a blend of pseudo-cereals (quinoa and buckwheat) and chickpea flours using two selected LAB. This sourdough was added to wheat flour for bread making, obtaining a final product containing ca. 500 mg GABA/kg and concentration of total phenols and fibers higher than control wheat-flour bread. Moreover, the experimental bread was characterized by a lower starch hydrolysis index (HI), a parameter closely related to the glycemic index of the food matrix (Coda et al., 2010b). GABA is a four-carbon nonprotein amino acid that has several physiological functions, including being the major inhibitory neurotransmitter of the central nervous system, induction of antihypertensive, antidiabetic, diuretic, and tranquilizer effects (Coda et al., 2010b). Its use in pharmaceutical preparations is also extended to functional foods such as dairy products, gabaron tea, and shochu (Nomura et al., 1999; Sawai et al., 2001; Yokoyama et al., 2002). GABA attractiveness as a health-related active compound has driven the research, as reviewed by Diana et al. (2014). Numerous studies have reported the potential use of selected LAB on fermented sourdough, resulting in GABA-enriched sourdough that can increase the GABA content in the final bread making. However, testing and validation are important requirements in the prospects of future clinical trials of bioactive molecules (Diana et al., 2014).

The addition of legume (eg, chickpea, lentil, and bean) flours to wheat-flour bread was further investigated by Rizzello et al. (2014). In particular, type I sourdough containing a mixture of the three legumes (each added at 5% w/w of the wheat flour) was prepared and propagated (back-slopped) according to the traditional protocols used for making typical Italian breads.

Bread containing 15% (w/w) of mixed legume (chickpea, lentil, and bean) flours was produced using a wheat-legume sourdough (Fig. 18.1). Compared to bread produced without the addition of legume flours and fermented only with baker's yeast, marked increases of the level of total FAA, phytase, and antioxidant activities were found (Rizzello et al., 2014). The use of sourdough

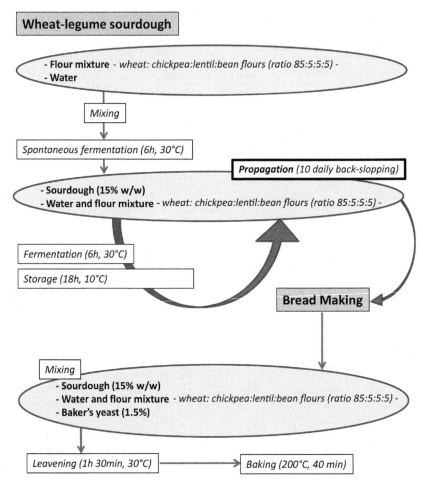

FIGURE 18.1 Flow chart of the process for wheat-legume sourdough making/propagation and for bread making. The dough yield (dough weight × 100/flours weight) of sourdough and dough for bread making was 160 (corresponding to 62.5% and 37.5% of flours and water, respectively).

fermentation allowed significant improvement of in vitro protein digestibility and a significant decrease of HI values compared to wheat-flour bread but still keeping a good sensory and technological quality.

Similar results were obtained when *Lupinus angustifolius* and *Lupinus luteus* whole-meal flours were fermented by *Pediococcus acidilactici*. The addition of lupin to wheat flour (10% w/w) decreased some technological properties of bread (volume and porosity). Nevertheless, fermentation with starter improved protein extraction, decreased trypsin inhibitor activity, and increased protein digestibility, without increasing the risk of biogenic amine formation (Bartkiene et al., 2011). Lupin is considered a valuable source of protein with good nutritional balance of essential amino acids, but also with DF higher than most other legumes, up to 27–31% of the kernel weight (Bartkiene et al., 2011; Nasar-Abbas and Jayasena, 2012). Bread prepared by adding lupin flour helped reduce blood pressure and cardiovascular risk (Lee et al., 2009).

The benefits of soybeans on human health have been the target of many studies in recent decades due to the particular proteins and phytochemicals (eg, isoflavones, phytosterols, bioactive peptides) found in soybeans, which may help prevent diseases such as diabetes, obesity, hypertension, cardiovascular disease, dyslipidemia, and cancer (Peñaloza-Espinosa et al., 2011). Soybean flour was used as the ingredient for making traditional sourdough bread, replacing wheat flour by up to 25% (w/w) and resulting in a product with lower volume, but with good acceptance by consumers and improved nutritional quality, as estimated by the protein biological indexes protein efficiency ratio (PER), net protein ratio (NPR), apparent digestibility (AD), and true digestibility (Peñaloza-Espinosa et al., 2011).

18.2.3 Minor Cereals and Pseudo-Cereals in Sourdough Wheat Bread Making

In recent years, the use of sourdough technology has been applied to many cereals other than wheat since wheat is not typical in developing or emerging countries where other crops like sorghum (Rooney and Awika, 2005), millet, maize, rice, acha or white fonio, iburu or black fonio, and teff (Joshi et al., 2008; Balasubramanian and Viswanathan, 2010) comprise the diet for human consumption and play an essential role in providing nutritional status for the poorest populations and regions (Jideani and Jideani, 2011; McKevith, 2004). Along with minor cereals, several other crops play a key role in nutrition in certain parts of the world. For example, pseudo-cereals such as buckwheat (*Fagopyrum esculentum* Moench), amaranth (*Amaranthus* spp.), and quinoa (*Chenopodium quinoa* Willd.) show nutritional and texture features, which make them suitable for replacing, at least in part, traditional cereal-based products. The literature reports many cases in which proper selection of authoctonous LAB starters made it possible to obtain better technological, nutritional, and sensory properties of the baked goods produced with these alternative flours (Coda et al.,

2010a, 2011). Overall, autochtonous LAB, adapted to the matrix, show better fermentative performance than strains isolated from other different environments. Indeed, microbial adaptability and performance are related to the abundance of different carbon sources and the concentration of phenolic compounds (Coda et al., 2014a).

There is increased interest in the consumption of locally available grains and pulses and the addition of a sourdough to these grains is a suitable way to enhance the overall quality of wheat bread (Gänzle, 2009). Bread is a staple food in most human diets and contributes up to 50% of dietary energy due to its significant carbohydrate content (Brites et al., 2011). Cereals like teff, sorghum, and barley, for instance, contain more minerals (iron, calcium, and zinc) than wheat and thus their incorporation into wheat bread seems like a good strategy for mineral enrichment. This could be particularly important for developing countries, since most foods consumed by young children are cereal based in the form of fermented drinks, porridge, or breads (Coda et al., 2014a). *Injera* bread is a good example of this staple food. It is an Ethiopian, thin flatbread processed from teff but also made of different cereal blends, according to geographical area. Sourdough fermentation makes this bread nutritionally valuable, promoting phytase activity. This effect was also confirmed in fermentation of barley–wheat and wheat–red sorghum mixtures. As a consequence of LAB fermentation, other antinutrients such as of α-galactosides were also removed (Baye et al., 2013). A very similar nutritional profile is shown by buckwheat, which is rich in micronutrients and antioxidants but also in phytate and tannins (Li and Zhang, 2001). Reduction of phytate content was obtained after sourdough fermentation of buckwheat used for the production of wheat bread. At the same time, a two-fold increase of polyphenols was also observed (Moroni et al., 2012). Recent studies on the use of nutrient-dense grains and sourdough bioprocessing have shown that this is a suitable alternative to enhancing the nutritional quality of wheat bread. This research trend, aimed at fulfilling increased consumer demand for healthy bread, has led to considerable efforts to develop breads that combine health benefits with good sensory properties (Rieder et al., 2012).

18.2.4 Nonwheat Flours in Sourdough Gluten-Free Bread Making

Flours derived from alternative grains have been used to improve the nutritional value and to diversify gluten-free bread formulations. The use of sorghum, millet, teff, amaranth, quinoa, and buckwheat can be a way to deliver nutritious gluten-free products containing cereal DF and other nutrients (Alvarez-Jubete et al., 2010; Capriles and Arêas, 2014). In fact, gluten-free bread made from refined flours and starches usually has inherently low levels of DF, vitamins, and minerals (Thompson, 2000); thus it is important to enhance the nutritional properties of this bread. Celiac disease (gluten-sensitive enteropathy) is a chronic gastrointestinal tract disorder where the ingestion of gluten from wheat, rye, and barley and their cross-related varieties leads to damage to the small intestinal mucosa by an autoimmune mechanism in genetically susceptible individuals (Tye-Din and Anderson, 2008). As gluten is essential for the structure of high-quality baked goods, obtaining high-quality gluten-free breads is a technological challenge that drives the search for ingredients and technologies able to improve bread-making performance.

Along with a variety of alternative raw materials, sourdough has also been used to improve the quality of gluten-free bread due to its positive effect on texture, aroma, nutritional properties, and shelf-life. Sorghum flour (Galle et al., 2012), mixtures of buckwheat, amaranth, chickpea, and quinoa flours (Coda et al., 2010b), or mixtures of brown rice, corn starch, buckwheat, and soy flours were successfully used for LAB fermentation, showing an improvement in bread nutritional properties and texture. Moreover, the proteolytic activities of LAB have shown great potential in the reduction of gluten contamination in gluten-free recipes (Di Cagno et al., 2002, 2004).

Using a pool of LAB selected for its strong proteolytic activity and long fermentation time, Di Cagno et al. (2004) obtained sourdough bread made by a flour mixture including wheat and gluten-free flours such as oat, millet, and buckwheat. The applied bioprocess allowed the almost complete hydrolysis of the wheat polypeptides involved in the celiac disease while still maintaining suitable technological performance for bread making. The in vivo acute challenge showed tolerance of celiac patients to the bread produced with this biotechnology (Di Cagno et al., 2004).

Since celiac disease is associated with high incidence of type I (insulin-dependent) diabetes, maintenance of good glycemic control is an important task. Developing low glycemic index gluten-free bread is therefore fundamental (Capriles and Arêas, 2014). Biological acidification as well as the DF content of many nonwheat cereals and pseudo-cereals have been reported to positively affect the in vitro glycemic index of gluten-free breads as well as of enriched wheat-based bread (Coda et al., 2010b; Novotni et al., 2012). To date, some gluten-free bread formulations have been promising means of enriching the daily diet with nutrient and bioactive compounds; therefore research on baking properties is needed to further develop high-quality gluten-free breads to meet the needs of celiac consumers.

18.3 USE OF WHEAT MILLING BYPRODUCTS AND SOURDOUGH FERMENTATION FOR ENHANCING NUTRITIONAL AND HEALTHY-RELATED PROPERTIES OF WHEAT-BASED FOODS

The milling industry is always looking for new value-added applications for wheat-milling byproducts, considering the large amount that is produced. The major byproduct of the milling process is wheat bran, followed by other valuable compounds such as wheat germ and parts of the endosperm.

Several researches have investigated the health potential of bran, concluding that bran is a key factor in determining whole-grain health benefits, as also reviewed by Prückler et al. (2014). Whole grains or the bran fraction itself were shown to modulate hunger and satiety moods, influence the glycemic, lipidic, and inflammatory status of consumers, and may exert prebiotic activity. In response to this health potential, the use of wheat bran in the food and feed industry has increased distinctly and visibly over the last decade, especially for the production of bakery and cereals (Prückler et al., 2014).

However, the use of bran and milling byproducts in food applications is more challenging and limited than refined grains due to some technological drawbacks. In bread, for instance, bran supplementation usually weakens the structure and baking quality of the dough and decreases the volume and elasticity of breadcrumbs, reducing the overall quality (Coda et al., 2015a). Notwithstanding the numerous studies showing the potential health benefits of consuming more whole-grain foods, the addition of bran to baked goods may confer difficulties in terms of processing, nutrition, or sensory acceptance by consumers. Therefore several strategies have been applied in order to overcome the technological and sensory drawback of bran implementation to food (Delcour et al., 2012). Pretreatment/processing of milling byproducts (alone or in combination with wheat flour) by sourdough fermentation has been shown to be successful in improving their incorporation into wheat-based foods, thanks to the positive effects of LAB on the technological and nutritional features. Bran has a very complex structure, which is formed by multiple layers in which arabinoxylans (AX) and β-glucans represent the most abundant cell-wall polysaccharides (Coda et al., 2015a). In these layers, bioactive compounds such as DF and phenolic acids are trapped in the strong cell-wall structures, resisting conventional milling and thus having low bioaccessibility (Delcour et al., 2012). From a nutritional point of view, the efforts made in recent years have been directed mostly toward the increase of in vitro and in vivo bioaccessibility of phenolic acids and of soluble DF (Coda et al., 2015a). For this purpose, LAB and yeasts were successfully used for bran fermentation, often in enzyme-combined bioprocessing (Coda et al., 2015a). Overall, fermentation of bran favored the enhancement of the bioactive potential and the increase of the concentration of folates, phenolic compounds, ferulic acid, and solubilized pentosans (Katina et al., 2012).

Baking technologies have been developed to incorporate bran as a functional ingredient. Its high-fiber content makes it suitable for bread with health-promoting effects, such as helping gut regulation and the control of blood glucose and cholesterol (Poutanen et al., 2009, 2014). The presence of DF in bread was shown to reduce the glycemic response by affecting starch state, water distribution in the matrix, and by increasing the viscosity of digesta (Poutanen et al., 2014). In particular, the preservation of starch from the access of human amylases, as well as the reduced pH during sourdough fermentation that may delay gastric emptying or create a barrier to starch digestion, seems to be more effective than DF per se in improving glucose metabolism (Scazzina et al., 2013). Sourdough technology is effective in reducing the starch digestibility and lowering the glycemic response when applied to the production of baked goods containing whole grain and bran as ingredients (Poutanen et al., 2009; Rizzello et al., 2012). The extent of these modifications and the metabolite profile can be modulated and tailored by changing the dimensions of the bran particles during milling prior to fermentation and by adjusting aeration conditions during fermentation (Coda et al., 2014a,b; Savolainen et al., 2014).

In addition to DF and phenolic compounds, wheat bran and germ are considered as important sources of minerals but also of phytic acid (Coda et al., 2015a). Bioprocessing of wheat bran with *Lactobacillus brevis* E95612 and *Kazachstania exigua* C81116 improved the antioxidant and phytase activities, the content of peptides, total FAA, and the in vitro digestibility of proteins (Coda et al., 2014c). Enzyme addition during fermentation further enhanced the positive effects of microbial starters (Coda et al., 2014c; Katina et al., 2012; Savolainen et al., 2014). Micronization, air fractionation, and fermentation were applied to durum wheat-bran fractions successively used for bread making (Rizzello et al., 2012). The addition of fermented micronized bran fractions has been shown to increase the concentration of FAA, total phenols, and DF as well as the phytase and antioxidant activities of dough and experimental breads produced at pilot plant scale (Rizzello et al., 2012).

Due to the rapid increase in global demand for protein consumption, wheat germ may be one of the most attractive and alternative sources of protein from vegetables (Rizzello et al., 2010a,b), having proteins with most of the essential amino acids in concentrations higher than in the reference egg protein pattern (Ge et al., 2001; Rizzello et al., 2010a). Wheat germ is a rich source of α-tocopherol, group B vitamins, proteins, DF, and minerals and is also rich in unsaturated fatty acids and phytochemicals (Rizzello et al., 2010a). Sourdough fermentation was employed to stabilize wheat germ and after fermentation with two autochthonous starters belonging to *Lactobacillus plantarum* and *Lactobacillus rossiae* species, the in vitro protein digestibility, phytase and antioxidant activities, and the concentration of phenols and FAA increased. Moreover, it was found that raffinose decreased during fermentation as a consequence of its consumption by the bacteria (Rizzello et al., 2010a). The application of sourdough fermentation to bran and wheat germ significantly decreased the in vitro and in vivo glycemic index of experimental breads containing percentages of these byproducts ranging from 4% to 10% (Rizzello et al., 2010b, 2012).

There is huge potential for the development and application of new technologies to milling byproducts to obtain bread and cereal-based foods high in DF, phytochemicals, and proteins. The use of selected LAB and sourdough biotechnology represents the most effective tool to maximize the technological, nutritional, functional, and sensory potential of leavened baked goods.

18.4 CONCLUSIONS

With the aim of obtaining novel ingredients for healthy wheat-based food production, the effect of sourdough fermentation on legumes, nonwheat cereals, pseudocereals, and cereal-milling byproducts is currently under intense investigation. In addition to developing technologies for producing bread with good nutritional profile and improved sensory quality, there have also been several studies aimed at improving bread physiological effects, in an effort to promote the long-term maintenance of health. In many cases, the addition of alternative flours to wheat-based food is a difficult task when the absence or interference of the gluten network decreases the technological performance and sensory profile of the final products.

Sourdough fermentation, carried out by the traditional back-slopping procedure, or enhanced by the use of selected LAB, has allowed the desired changes in the matrix, under several perspectives, to be obtained. The most recent findings show that sourdough bioprocessing is an important prerequisite for improving the nutritional and functional properties of alternative nonwheat flours, besides mere improvement of technological features. Thanks to LAB metabolism, the bioaccessibility and bioavailability of health-promoting compounds is improved, exploiting more of the alternative flours functional potential. Despite the growing evidence of this health potential, research is still needed to confirm effects on health that otherwise still remain speculative. In a recent Whole Grains Summit (2012), it was suggested that not only the chemical composition but also the food structure must be considered when studying the effects of whole grains, since distribution of water and air in the matrix also affects physiological responses (McKeown et al., 2013).

With sustainability as a global trend, the intake of plant protein will increase in the future and its use needs to be implemented by adequately biotechnological processes, such as sourdough fermentation, aiming at the development of novel products, such as leavened baked goods, with improved functionality and nutritional features.

REFERENCES

Alonso, R., Aguirre, A., Marzo, F., 2000. Effects of extrusion and traditional processing methods on antinutrients and in vitro digestibility of protein and starch in faba and kidney beans. Food Chemistry 68, 159–165.

Alvarez-Jubete, L., Arendt, E.K., Gallagher, E., 2010. Nutritive value of pseudo-cereals and their increasing use as functional gluten free ingredients. Trends in Food Science and Technology 21, 106–113.

Angioloni, A., Collar, C., 2012. High legume–wheat matrices: an alternative to promote bread nutritional value meeting dough viscoelastic restrictions. European Food Research and Technology 234, 273–284.

Bakke, A., Vickers, Z., 2007. Consumer liking of refined and whole wheat breads. Journal of Food Science 72, S473–S480.

Balasubramanian, S., Viswanathan, R., 2010. Influence of moisture content on physical properties of minor millets. Journal of Food Science and Technology 4, 279–284.

Bartkiene, E., Joudeikiene, G., Vidmantiene, D., Viskelis, P., Urbonaviciene, D., 2011. Nutritional and quality aspects of wheat sourdough bread using *L. luteus* and *L. angustifoliusflours* fermented by *Pediococcus acidilactici*. International Journal of Food Science and Technology 46, 1724–1733.

Baye, K., Mouquet-Rivier, C., Icard-Vernière, C., Rochette, I., Guyot, J.P., 2013. Influence of flour blend composition on fermentation kinetics and phytate hydrolysis of sourdough used to make injera. Food Chemistry 138, 430–436.

Beuchat, L.R., 1997. Traditional fermented foods. In: Doyle, M.P., Beuchat, L.R., Montville, T.J. (Eds.), Food Microbiology - Fundamentals and Frontiers. ASM Press, Washington, DC, pp. 629–648.

Boye, J., Zare, F., Pletch, A., 2010. Pulse proteins: processing, characterization, functional properties and applications in food and feed. Food Research International 43, 414–431.

Brites, C.M., Trigo, M.J., Carrapico, B., Alvina, M., Bessa, R.J., 2011. Maize and resistant starch enriched breads reduce postprandial glycemic responses in rats. Nutrition Research 31, 302–308.

Campos-Vega, R., Loarca-Piña, G., Oomah, B.D., 2010. Minor components of pulses and their potential impact on human health. Food Research International 43, 461–482.

Capriles, V.D., Arêas, J.A.G., 2014. Novel approaches in gluten-free breadmaking: interface between food science, nutrition, and health. Comprehensive Reviews in Food Science and Food Safety 13, 871–890.

Coda, R., Nionelli, L., Rizzello, C.G., De Angelis, M., Tossut, P., Gobbetti, M., 2010a. Spelt and emmer flours: characterization of the lactic acid bacteria microbiota and selection of mixed autochthonous starters for bread making. Journal of Applied Microbiology 108, 925–935.

Coda, R., Rizzello, C.G., Gobbetti, M., 2010b. Use of sourdough fermentation and pseudo-cereals and leguminous flours for the making of a functional bread enriched of γ-aminobutyric acid (GABA). International Journal of Food Microbiology 137, 236–245.

Coda, R., Di Cagno, R., Rizzello, C.G., Nionelli, L., Edema, M.O., Gobbetti, M., 2011. Utilization of African grains for sourdough bread making. Journal of Food Science 76, M329–M335.

Coda, R., Di Cagno, R., Gobbetti, M., Rizzello, C.G., 2014a. Sourdough lactic acid bacteria: exploration of nonwheat cereal-based fermentation. Food Microbiology 37, 51–58.

Coda, R., Kärki, I., Nordlund, E., Heiniö, R.-L., Poutanen, K., Katina, K., 2014b. Influence of particle size on bioprocess induced changes on technological functionality of wheat bran. Food Microbiology 37, 69–77.

Coda, R., Rizzello, C.G., Curiel, J.A., Poutanen, K., Katina, K., 2014c. Effect of bioprocessing and particle size on the nutritional properties of wheat bran fractions. Innovative Food Science & Emerging Technologies 25, 19–27.

Coda, R., Katina, K., Rizzello, C.G., 2015a. Bran bioprocessing for enhanced functional properties. Current Opinion in Food Science 1, 50–55.

Coda, R., Melama, L., Rizzello, C.G., Curiel, J.A., Sibakov, J., Holopainen, U., Pulkkinen, M., Sozer, N., 2015b. Effect of air classification and fermentation by *Lactobacillus plantarum* VTT-133328 on faba bean (*Vicia Faba* L.) flour nutritional properties. International Journal of Food Microbiology 193, 34–42.

Correia, A., Nunes, S., Guedes, A.S., Barros, I., Delgadillo, I., 2010. Screening of lactic acid bacteria potentially useful for sorghum fermentation. Journal of Cereal Science 52, 9–15.

Crépon, K., Marget, P., Peyronnet, C., Carrouee, B., Arese, P., Duc, G., 2010. Nutritional value of faba bean (*Vicia faba* L.) seeds for feed and food. Field Crops Research 115, 329–339.

Curiel, J.A., Coda, R., Centomani, I., Summo, C., Gobbetti, M., Rizzello, C.G., 2015. Exploitation of the nutritional and functional characteristics of traditional Italian legumes: the potential of sourdough fermentation. International Journal of Food Microbiology 196, 51–61.

Delcour, J.A., Rouau, X., Courtin, C.M., Poutanen, K., Ranieri, R., 2012. Technologies for enhanced exploitation of the health-promoting potential of cereals. Trends in Food Science & Technology 25, 78–86.

Deschamps, A., 1989. Microbial Degradation of Tannins and Related Compounds. Plant Cell Wall Polymers Biogenesis and Biodegradation. American Chemical Society, Washington, DC, pp. 559–566.

Dhinda, F., Lakshmi, J.A., Prakash, J., Dasappa, I., 2011. Effect of ingredients on rheological, nutritional and quality characteristics of high protein, high fibre and low carbohydrate bread. Food and Bioprocess Technology 5, 2998–3006.

Di Cagno, R., De Angelis, M., Lavermicocca, P., De Vincenzi, M., Giovannini, C., Faccia, M., Gobbetti, M., 2002. Proteolysis by sourdough lactic acid bacteria: effects on wheat flour protein fractions and gliadin peptides involved in human cereal intolerance. Applied and Environmental Microbiology 68, 623–633.

Di Cagno, R., De Angelis, M., Auricchio, S., Greco, L., Clarke, C., De Vincenzi, M., Giovannini, C., D'Archivio, M., Landolfo, F., Parrilli, G., Minervini, F., Arendt, E., Gobbetti, M., 2004. Sourdough bread made from wheat and nontoxic flours and started with selected lactobacilli is tolerated in celiac sprue patients. Applied and Environmental Microbiology 70, 1088–1096.

Diana, M., Quílez, J., Rafecas, M., 2014. Gamma-aminobutyric acid as a bioactive compound in foods: a review. Journal of Functional Foods 10, 407–420.

Díaz-Batalla, L., Widholm, J.M., Fahey Jr., G.C., Castaño-Tostado, E., Paredes-López, O., 2006. Chemical components with health implications in wild and cultivated Mexican common bean seeds (*Phaseolus vulgaris* L.). Journal of Agricultural and Food Chemistry 54, 2045–2052.

Doblado, R., Frias, J., Muñoz, R., Vidal-Valverde, C., 2003. Fermentation of *Vigna sinensis* var. *carilla* flours by natural microflora and *Lactobacillus* species. Journal of Food Protection 66, 2313–2320.

Duranti, M., 2006. Grain legume proteins and nutraceutical properties. Fitoterapia 77, 67–82.

Eissa, A., Hussein, A.S., Mostafa, B.E., 2007. Rheological properties and quality evaluation of Egyptian balady bread and biscuits supplemented with flours of ungerminated and germinated legume seeds or mushroom. Polish Journal of Food and Nutrition Sciences 57, 487–496.

Feregrino-Perez, A.A., Berumen, L.C., Garcia-Alcocer, G., Guevara-Gonzalez, R.G., Ramos-Gomez, M., Reynoso-Camacho, R., Acosta-Gallegos, J.A., Loarca-Piña, G., 2008. Composition of chemopreventive effect of polysaccharides from common beans (*Phaseolus vulgaris* L.) on azoxymethane-induced colon cancer. Journal of Agricultural and Food Chemistry 56, 8737–8744.

Flight, I., Clifton, P., 2006. Cereal grains and legumes in the prevention of coronary heart disease and stroke: a review of the literature. European Journal of Clinical Nutrition 60, 1145–1159.

Gallagher, E., Gormleya, T.R., Arendt, E.K., 2004. Recent advances in the formulation of gluten-free cereal-based products. Trends in Food Science and Technology 15, 143–152.

Galle, S., Schwab, C., Dal Bello, F., Coffey, A., Gänzle, M.G., Arendt, E.K., 2012. Influence of in-situ synthesized exopolysaccharides on the quality of gluten-free sorghum sourdough bread. International Journal of Food Microbiology 155, 105–112.

Gänzle, M.G., 2009. From gene to function: metabolic traits of starter cultures for improved quality of cereal foods. International Journal of Food Microbiology 134, 29–36.

Ge, Y., Sun, A., Ni, Y., Cai, T., 2001. Study and development of a defatted wheat germ nutritive noodle. European Food Research and Technology 212, 344–348.

Gobbetti, M., Rizzello, C.G., Di Cagno, R., De Angelis, M., 2014. How the sourdough may affect the functional features of leavened baked goods. Food Microbiology 37, 30–40.

Gòmez, M., Oliete, B., Rosell, C.M., Pando, V., Fernàndez, E., 2008. Studies on cake quality made of wheat-chickpea flour blends. LWT- Food Science and Technology 41, 1701–1709.

Granito, M., Frias, J., Doblado, R., Guerra, M., Champ, M., Vidal-Valverde, C., 2002. Nutritional improvement of beans (*Phaseolus vulgaris*) by natural fermentation. European Food Research and Technology 214, 226–231.

Gustafsson, E.L., Sandberg, A.S., 1995. Phytate reduction in brown beans (*Phaseolus vulgaris* L.). Journal of Food Science 60, 149–152.

Guyot, J.P., 2012. Cereal-based fermented foods in developing countries: ancient foods for modern research. International Journal of Food Science and Technology 47, 1109–1114.

Hammes, W.P., Brandt, M.J., Francis, K.L., Rosenheim, J., Seitter, M.F.H., Vogelmann, S.A., 2005. Microbial ecology of cereal fermentations. Trends in Food Science and Technology 16, 4–11.

Humblot, C., Guyot, J.P., 2008. Other fermentations. In: Cocolin, L., Ercolini, D. (Eds.), Molecular Techniques in the Microbial Ecology of Fermented Foods. Springer, New York, pp. 208–224.

Jenkins, D.J., Kendall, C.W., Augustin, L.S., Mitchell, S., Sahye-Pudaruth, S., Blanco Mejia, S., Chiavaroli, L., Mirrahimi, A., Ireland, C., Bashyam, B., Vidgen, E., de Souza, R.J., Sievenpiper, J.L., Coveney, J., Leiter, L.A., Josse, R.G., 2012. Effect of legume as part of low glycemic index diet on glycemic control and cardiovascular risk factors in type 2 diabetes mellitus: a randomized controlled trial. Archives of Internal Medicine 172, 1653–1660.

Jezierny, D., Mosenthin, R., Bauer, E., 2010. The use of grain legumes as a protein source in pig nutrition: a review. Animal Feed Science and Technology 157, 111–128.

Jideani, I.A., Jideani, V.A., 2011. Developments on the cereal grains *Digitaria exilis* (acha) and *Digitaria iburua* (iburu). Journal of Food Science and Technology 48, 251–259.

Joshi, A., Rawat, K., Karki, B., 2008. Millet as "Religious offering" for nutritional, ecological, and economical security. Comprehensive Reviews in Food Science and Food Safety 7, 369–372.

Kadam, M.L., Salve, R.V., Mehrajfatema, Z.M., More, S.G., 2012. Development and evaluation of composite flour for Missi roti/chapatti. Journal of Food Processing & Technology 3, 1000134.

Kamaljit, K., Baljeet, S., Amarjeet, K., 2010. Preparation of bakery products by incorporating pea flour as a functional ingredient. American Journal of Food Technology 5, 130–135.

Katina, K., Juvonen, R., Laitila, A., Flander, L., Nordlund, E., Kariluoto, S., Piironen, V., Poutanen, K., 2012. Fermented wheat bran as a functional ingredient in baking. Cereal Chemistry Journal 89, 126–134.

Kohajdová, Z., Karovičová, J., Magala, M., 2011. Utilisation of chickpea flour for cracker production. Acta Chimica Slovenica 4, 98–107.

Kohajdová, Z., Karovičová, J., Magala, M., 2013. Effect of lentil and bean flours on rheological and baking properties of wheat dough. Chemical Papers 67, 398–407.

Kosińska, A., Karamać, M., Penkacik, K., Urbalewicz, A., Amarowicz, R., 2011. Interactions between tannins and proteins isolated from broad bean seeds (*Vicia faba* Major) yield soluble and nonsoluble complexes. European Food Research and Technology 233, 213–222.

Lee, Y.P., Mori, T.A., Puddey, I.B., Sipsas, S., Ackland, T.R., Beilin, L.J., Hodgson, J., 2009. Effects of lupin kernel flour-enriched bread on blood pressure: a controlled intervention study. American Journal of Clinical Nutrition 89, 766–772.

Li, S., Zhang, Q.H., 2001. Advances in the development of functional foods from buckwheat. Critical Reviews in Food Science and Nutrition 41, 451–464.

Liener, I.E., 1990. Naturally-occurring toxic factors in animal feedstuffs. In: Wiseman, J., Cole, D.J.A. (Eds.), Feedstuff Evaluation. Elsevier Ltd., pp. 377–394.

Lopez, H.W., Leenhardt, F., Coudray, C., Remesy, C., 2002. Minerals and phytic acid interactions: is it a real problem for human nutrition? International Journal of Food Science and Technology 37, 727–739.

Luo, Y.W., Gu, Z.X., Han, Y.B., Chen, Z.G., 2009. The impact of processing on phytic acid, in vitro soluble iron and Phy/Fe molar ratio of faba bean (*Vicia faba* L.). Journal of the Science of Food and Agriculture 89, 861–866.

Maninder, K., Sandhu, K.S., Singh, N., 2007. Comparative study of the functional, thermal and pasting properties of flours from different field pea (*Pisum sativum* L.) and pigeon pea (*Cajanus cajan* L.) cultivars. Food Chemistry 104, 259–267.

Martinez, C., Ros, G., Periago, M.J., Lopez, G., Ortuno, J., Rincon, F., 1996. Phytic acid in human nutrition. Food Science and Technology International 2, 201–209.

McKeown, N.M., Jacques, P.F., Seal, C.J., de Vries, J., Jonnalagadda, S.S., Clemens, R., Webb, D., Murphy, L.A., van Klinken, J.-W., Topping, D., Murray, R., Degeneffe, D., Marquart, L.F., 2013. Whole grains and health: from theory to practice. Journal of Nutrition 143, 744S–758S.

McKevith, B., 2004. Nutritional aspects of cereals. Nutrition Bulletin 29, 111–142.

Mohammed, I., Ahmed, A.R., Senge, B., 2012. Dough rheology and bread quality of wheat–chickpea flour blends. Industrial Crops and Products 36, 196–202.

Mollard, R.C., Luhovyy, B.L., Panahi, S., Nunez, M., Hanley, A., Andersona, G.H., 2012. Regular consumption of pulses for 8 weeks reduces metabolic syndrome risk factors in overweight and obese adults. British Journal of Nutrition 108, 111–122.

Moroni, A.V., Iametti, S., Bonomi, F., Arendt, E.K., Dal Bello, F., 2010. Solubility of proteins from nongluten cereals: a comparative study on combinations of solubilising agents. Food Chemistry 121, 1225–1230.

Moroni, A.V., Zannini, E., Sensidoni, G., Arendt, E.K., 2012. Exploitation of buckwheat sourdough for the production of wheat bread. European Food Research and Technology 235, 659–668.

Muzquiz, M., Varela, A., Burbano, C., Cuadrado, C., Guillamón, E., Pedrosa, M., 2012. Bioactive compounds in legumes: pronutritive and antinutritive actions. Implications for nutrition and health. Phytochemistry Reviews 11, 227–244.

Nasar-Abbas, S.M., Jayasena, V., 2012. Effect of lupin flour incorporation on the physical and sensory properties of muffins. Quality Assurance and Safety of Crops and Foods 4, 41–49.

Nomura, M., Nakajima, I., Fujita, Y., Kobayashi, M., Kimoto, H., Suzukil, I., Aso, H., 1999. *Lactococcus lactis* contains only one glutamate decarboxylase gene. Microbiology 145, 1375–1380.

Novotni, D., Čukelj, N., N, Smerdel, B., Bituh, M., Dujmić, F., Ćurić, D., 2012. Glycemic index and firming kinetics of partially baked frozen gluten-free bread with sourdough. Journal of Cereal Science 2, 120–125.

Peñaloza-Espinosa, J., De La Rosa-Angulo, G., Mora-Escobedo, R., Chanona-Pérez, J., Farrera-Rebollo, R., Calderón-Domínguez, 2011. Sourdough and bread properties as affected by soybean protein addition. In: Ng, T.-B. (Ed.), Soybean - Applications and Technology (InTech).

Poutanen, K., Flander, L., Katina, K., 2009. Sourdough and cereal fermentation in a nutritional perspective. Food Microbiology 26, 693–699.

Poutanen, K., Sozer, N., Della Valle, G., 2014. How can technology help to deliver more of grain in cereal foods for a healthy diet? Journal of Cereal Science 59, 327–336.

Prückler, M., Siebenhandl-Ehn, S., Apprich, S., Höltinger, S., Haas, C., Schmid, E., Kneifel, W., 2014. Wheat bran-based biorefinery 1: composition of wheat bran and strategies of functionalization. LWT-Food Science and Technology 56, 211–221.

Rieder, A., Holtekjølen, A.K., Sahlstrøm, S., Moldestad, A., 2012. Effect of barley and oat flour types and sourdoughs on dough rheology and bread quality of composite wheat bread. Journal of Cereal Science 55, 44–52.

Rizzello, C.G., Nionelli, L., Coda, R., De Angelis, M., Gobbetti, M., 2010a. Effect of sourdough fermentation on stabilisation, and chemical and nutritional characteristics of wheat germ. Food Chemistry 119, 1079–1089.

Rizzello, C.G., Nionelli, L., Coda, R., Di Cagno, R., Gobbetti, M., 2010b. Use of sourdough fermented wheat germ for enhancing the nutritional, texture and sensory characteristics of the white bread. European Food Research and Technology 230, 645–654.

Rizzello, C.G., Coda, R., Mazzacane, F., Minervini, D., Gobbetti, M., 2012. Micronized byproducts from debranned durum wheat and sourdough fermentation enhanced the nutritional, textural and sensory features of bread. Food Research International 46, 304–313.

Rizzello, C.G., Calasso, M., Campanella, D., De Angelis, M., Gobbetti, M., 2014. Use of sourdough fermentation and mixture of wheat chickpea, lentil and bean flours for enhancing the nutritional, texture and sensory characteristics of white bread. International Journal of Food Microbiology 180, 78–87.

Rooney, L.W., Awika, J.M., 2005. Overview of products and health benefits of specialty sorghums. Cereal Foods World 50, 109–115.

Roy, F., Boye, J.I., Simpson, B.K., 2010. Bioactive proteins and peptides in pulse crops: pea, chickpea and lentil. Food Research International 43, 432–442.

Sadowska, J., Blaszczak, W., Fornnal, J., Vidal-Valverde, C., Frias, J., 2003. Changes of wheat dough and bread quality and structure as a result of germinated pea flour addition. European Food Research and Technology 216, 46–50.

Savolainen, O.I., Coda, R., Suomi, K., Katina, K., Juvonen, R., Hanhineva, K., Poutanen, K., 2014. The role of oxygen in the liquid fermentation of wheat bran. Food Chemistry 153, 424–431.

Sawai, Y., Yamaguchi, Y., Miyana, D., 2001. Cycling treatment of anaerobic and aerobic incubation increases the content of gamma-aminobutyric acid in tea shoots. Amino Acids 20, 331–334.

Scazzina, F., Siebenhandl-Ehn, S., Pellegrini, N., 2013. The effect of dietary fibre on reducing the glycaemic index of bread. British Journal of Nutrition 109, 1163–1174.

Schneider, A.V.C., 2002. Overview of the market and consumption of pulses in Europe. British Journal of Nutrition 88, S243–S250.

Scott, J.J., 1991. Alkaline phytase activity in nonionic detergent extracts of legume seeds. Plant Physiology 95, 1298–1301.

Smartt, J., Nwokolo, E., 1996. Food and Feed from Legumes and Oilseeds. Chapman and Hall, London.

Starzyńska-Janiszewska, A., Stodolak, B., 2011. Effect of inoculated lactic acid fermentation on antinutritional and antiradical properties of grass pea (*Lathyrus sativus* 'Krab') flour. Polish Journal of Food and Nutrition Sciences 61, 245–249.

Steiner, T., Mosenthin, R., Zimmermann, B., Greiner, R., Roth, S., 2007. Distribution of phytase activity, total phosphorus and phytate phosphorus in legume seeds, cereals and cereal byproducts as influenced by harvest year and cultivar. Animal Feed Science and Technology 133, 320–334.

Sterr, Y., Weiss, A., Schmidt, H., 2009. Evaluation of lactic acid bacteria for sourdough fermentation of amaranth. International Journal of Food Microbiology 136, 75–82.

Teixeira, J.S., McNeill, V., Gänzle, M.G., 2012. Levansucrase and sucrose phosphorylase contribute to raffinose, stachyose, and verbascose metabolism by lactobacilli. Food Microbiology 31, 278–284.

Thompson, T., 2000. Folate, iron, and dietary fiber contents of the gluten-free diet. Journal of the American Dietetic Association 100, 1389–1395.

Tiwari, B.K., Brennan, C.S., Jaganmohan, R., Surabi, A., Alagusundaram, K., 2011. Utilisation of pigeon pea (*Cajanus cajan* L.) byproducts in biscuit manufacture. LWT- Food Science and Technology 44, 1533–1537.

Tye-Din, J., Anderson, R., 2008. Immunopathogenesis of celiac disease. Current Gastroenterology Reports 10, 458–465.

Van Der Poel, A.F.B., 1990. Effect of processing on antinutritional factors and protein nutritional value of dry beans (*Phaseolus vulgaris* L.). A review. Animal Feed Science and Technology 29, 179–208.

Vaz Patto, M.C., Amarowicz, R., Aryee, A.N.A., Boye, J.I., Chung, H.J., Martín-Cabrejas, M.A., Domoney, C., 2015. Achievements and challenges in improving the nutritional quality of food legumes. Critical Reviews in Plant Sciences 34, 105–143.

Vitaglione, P., Napolitano, A., Fogliano, V., 2008. Cereal dietary fibre: a natural functional ingredient to deliver phenolic compounds into the gut. Trends in Food Science & Technology 19, 451–463.

WHO, 2003. Diet, Nutrition and the Prevention of Chronic Diseases. Technical Report Series 916.

Yokoyama, S., Hiramatsu, J.I., Hayakawa, K., 2002. Production of gamma-aminobutyric acid from alcohol distillery lees by *Lactobacillus brevis* IFO-12005. Journal of Bioscience and Bioengineering 93, 95–97.

Chapter 19

Tempeh and Other Fermented Soybean Products Rich in Isoflavones

V. Mani[1,2], L.C. Ming[3]
[1]Universiti Teknologi MARA (UiTM), Selangor, Malaysia; [2]Qassim University, Buraidah, Kingdom of Saudi Arabia; [3]University of Tasmania, Hobart, TAS, Australia

19.1 INTRODUCTION

A number of animal- and plant-based fermented foods constitute a part of the daily diet in many parts of the world. Food fermentation has been used for centuries as a method to preserve perishable food products. Many traditional Southeast Asian foods are fermented before consumption, reducing food costs and also promoting health. Fruits, cereals, honey, vegetables, milk, meat, and fish are fermented traditionally. The process of fermentation may facilitate the digestibility and bioavailability of proteins, carbohydrates, lipids, and minerals; enhance nutritional values; shorten cooking time; and increase microbial safety (McKeown, 2004). The microorganisms used for fermentation include bacteria, yeasts, and filamentous fungi (molds); among them, naturally occurring lactic-acid bacteria (LAB) play an important role (Lei et al., 2013).

Soybean cannot be eaten raw, thus it is usually processed to produce a variety of food products. The most common process is fermentation. Different countries in Asia have different names for fermented soybean food products, based on preparation and the type of microbial strain used, resulting in differences in consistency and forms of fermented products. In Japan, fermented soybean products are known as *natto* and *miso* (Yanaka et al., 2012); in Korea, *cheonggukjang* (Cho et al., 2011); and in China, *douchi* (Wang et al., 2008). *Tempeh* is a solid cake-like fermented soybean product produced with *Rhizopus oligosporus* (Haron et al., 2009). The main aim of this chapter is to summarize the evidence that shows *tempeh* as an affordable and rich source of nutrients for the prevention of various diseases.

19.2 SOYBEAN

Soybean (*Glycine max* L.) is the most cultivated plant in the world and contains a high amount of protein. Like most legume seeds, it contains several proteins, most of which contribute to nutrition and good health. In addition to proteins, they are notably rich in isoflavones, anthocyanins, saponins, lipids, and oligosaccharides (Barnes, 2010). Consumption of soybean and soybean-derived fermented products has been linked to many health benefits, including the reduction of cardiovascular disease incidence, reduced risk of ischemic stroke, and lower cholesterol levels that in turn reduce the incidence of atherosclerosis (Giordano et al., 2015; Liang et al., 2009; Nagarajan et al., 2008). The efficacy of soybean phytochemicals against several types of cancer such as breast, prostate, colorectal, ovarian, and also endometrial cancers has also been demonstrated (Carbonel et al., 2015; Karsli-Ceppioglu et al., 2015; Liu et al., 2015; Qu et al., 2014; Xiao et al., 2015). Soybean has also long been known to reduce menopausal symptoms in women and to reduce the risk of diabetes type 2 (Gilbert and Liu, 2013; Villa et al., 2009). Soybean is long been known to possess antioxidant activity (Devi et al., 2009). Among soybean phytochemicals, isoflavones have shown a wide range of health benefits. Soybean isoflavones are diphenolic compounds that are similar to estrogen and bind to estrogen receptors. Soybean isoflavones are categorized into several groups based on their chemical structures. The glycosides are daidzin and genistin and the corresponding aglycones are daidzein and genistein. Genistein, daidzein, glycitein, and their respective glucosidic conjugates are the major isoflavones found in soybean. Soybean isoflavones are also included since phytoestrogens bind to the estrogen receptors (ERα and ERβ) and affect estrogen-related physiology (da Silva et al., 2011). ERα is expressed in the endometrium, breast cancer cells, ovarian stromal cells, and the hypothalamus. ERβ has also been found in various tissues and organs, including ovarian granulosa cells, kidney, brain, bone, heart, lungs, intestinal mucosa, prostate, and endothelial cells. Among major isoflavones (Fig. 19.1), genistein has been reported to express higher affinities for ERβ than ERα and other derivatives (Russo et al., 2016). Furthermore, soybean isoflavones have inhibitory effects on tyrosine kinase, topoisomerase, and angiogenesis, which might reduce the risk of cancer (Danciu et al., 2012). Previous research found that malonyl genistin was the most abundant in soybean, followed by malonyl glycitin, genistin, daidzin, daidzein, genistein, and glycitin in decreasing order (Ho et al., 2002). Isoflavone glucosides are hydrolyzed to aglycones during soybean fermentation to produce other soybean food products (Ahn-Jarvis et al., 2013). The aglycones are absorbed at a higher rate than their counterpart, glucosides, in humans (Ahn-Jarvis et al., 2013; Izumi et al., 2000; Xu et al., 2015). The bioavailability of glycosides and aglycones is different. In the small intestine, aglycones are easily absorbed by passive diffusion due to its less complex structure and lower molecular weight (Haron et al., 2009). However, glycosides must be hydrolyzed to aglycones by β-glucosidases in the gut.

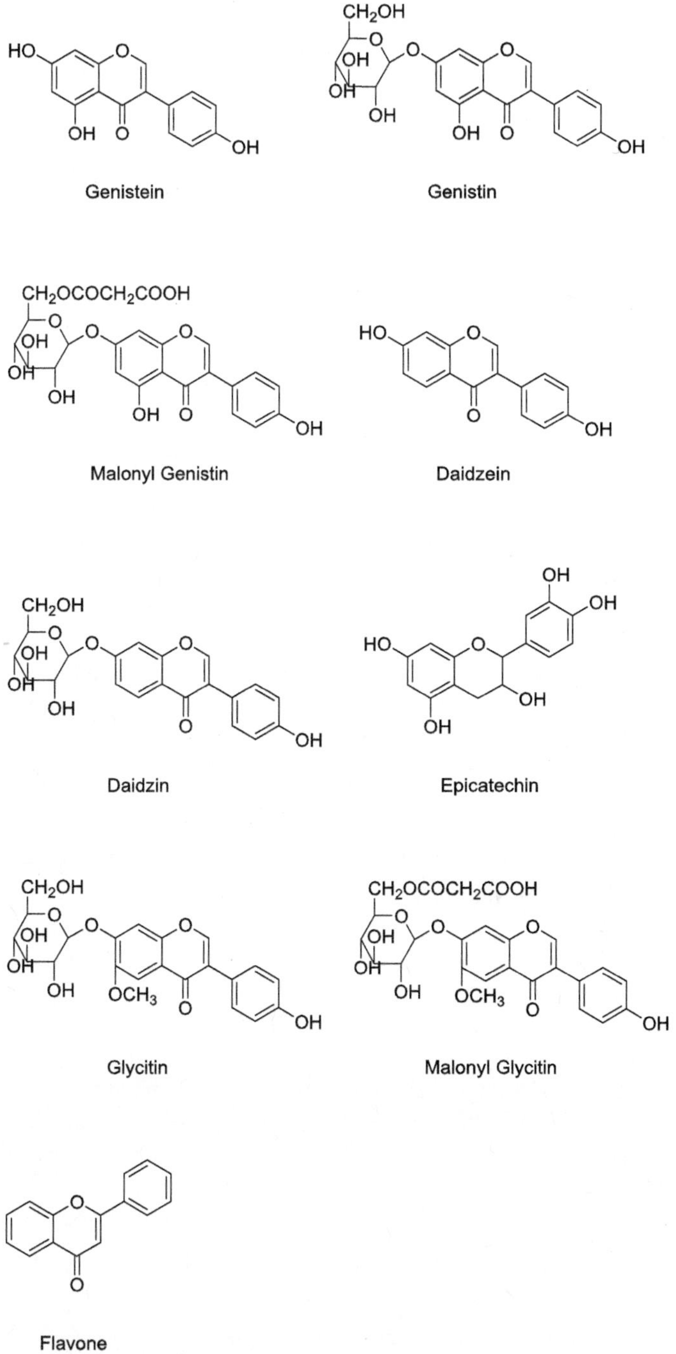

Genistein

Genistin

Malonyl Genistin

Daidzein

Daidzin

Epicatechin

Glycitin

Malonyl Glycitin

Flavone

FIGURE 19.1 Chemical structures of daidzein, genistein, and their glycosides.

Although recent investigations have noted that flavonoid glycosides, including genistin, phloretin, and quercetin, can be partially absorbed without previous hydrolysis of glucose moieties (Xiao et al., 2015).

19.3 TEMPEH

Tempeh was initially produced in the Central Java (Indonesia). Over the years, it has spread to Malaysia and Indonesia as a well-known traditional fermented soybean food. *Tempeh* has gained popularity in the world, especially in the diets of vegetarians and in developing countries, because it is easy to prepare and low in cost. *Tempeh* is the collective name for cooked and fermented beans, cereals, or food-processing byproducts, penetrated and bound together by the mycelium of a living mold (Nout and Kiers, 2005). In Malaysia, *tempeh* has been a staple source of proteins for several years. *Tempeh*, which has a firm texture and a nutty mushroom flavor, can also be used in different ways as an ingredient in soups, salads, and sandwiches. Moreover, it is also gaining attention as a complete dietary protein, providing essential and nonessential amino acids. The proteins and isoflavones from *tempeh* have been shown to contribute to gastrointestinal health benefits. Two recent studies demonstrated that *tempeh* consumption in Sprague–Dawley and *attus norvegicus* rats increased muscle mass and beneficial microbiota such as *Bacteroidetes* populations in the gut (Khasanah et al., 2015; Soka et al., 2014).

Traditional manufacture of *tempeh* includes two major processes, ie, soaking and mold fermentation initiated by the addition of a mold starter culture. Most commonly, the preferred raw material for *tempeh* preparation is yellow-seeded soybean. Other substrates such as barley, chickpea, ground bean, horse bean, lima bean, pea, oat, sorghum, and wheat have also been reported as a source for *tempeh* preparation (Nout and Kiers, 2005). The process of *tempeh* starts with the dehulling of the soybeans by soaking for 30–60 min. The dehulled beans are soaked for 6–24 h to increase the beans' moisture content, enable the microbial activity, render the beans edible, and reduce the amount of naturally occurring antimicrobial compounds (saponins) and bitter components. After draining, the beans are cooked for 20 min in fresh water. Then, the cooked beans are spread out on perforated trays to remove free water. The dried soybeans are inoculated using a starter containing sporangiospores of mainly *Rhizopus* spp. and rarely using *Mucor* spp. Traditionally, the beans are packed in punctured banana leaves, allowing a limited supply of air to the beans. The limited air supply restricts the formation of fungal sporangiospores. The packed inoculated beans are incubated for 1–2 days at 25–30°C. An attractive, creamy, white fresh *tempeh* cake forms after incubation and can be added to a variety of dishes (Nout and Kiers, 2005; Roubos-van den Hil and Nout, 2011).

19.3.1 Nutritional Enhancements of *Tempeh*

In general, the fermentation process is believed to improve the nutritional value of food by enhancing vitamin, essential amino acid, or fatty acid content, facilitating detoxification, and removing antinutritional factors. In *tempeh*, *Rhizopus* spp. produces a variety of amylases, lipases, and proteases, which in turn hydrolyze macronutrients into simpler, more water-soluble compounds. As a result, vitamins, phytochemicals, and antioxidative constituents are produced (Ahmad et al., 2015; Starzynska-Janiszewska et al., 2014). Interestingly, the soluble protein content in *tempeh* is increased as a consequence of fermentation (Murata et al., 1967, 1970). A comparative study showed higher protein content in *tempeh* produced from soybean as a raw material as compared to peanut, white kidney bean, and sesame (Chutrtong, 2013). Additionally, free amino acid levels also sharply increased in *tempeh* (Soni and Dey, 2014). Lipids are an important nutrient found in soybean. In *tempeh*, the hydrolysis of triacylglycerols into free fatty acids by the action of microbial lipases takes place during the fermentation process. These free fatty acids are used as a source of energy for the mold, resulting in lower lipid content in *tempeh*. During fermentation process, the lipid content decreased about 26% in *tempeh* as compared to unfermented soybeans, while the levels of trace minerals such as iron, calcium, and cuprum were not altered (Astuti et al., 2000). Total soluble iron is reported to increase in *tempeh* after fermentation. During the fermentation process, some of the proteins are broken down into free amino acids and peptides. As a result, the iron is liberated from the iron–protein complexes, thus increasing the total soluble iron (Astuti et al., 2000). In terms of vitamins, the level of vitamin B complexes increases except for thiamine. Vitamin B12 content was reported to increase from 0.15 µg/100 g in unfermented soybeans to 3.9 µg/100 g in *tempeh* (Liem et al., 1977). Ginting and Arcot (2004) reported that during *tempeh* fermentation, the folic-acid level increased from 71.6 µg/100 mg to 416.4 µg/100 mg. Recently, folic-deficient rats fed with *tempeh* for 5 weeks showed a two-fold increase in serum folic acid as compared to folic-deficient control (Khasanah, 2013). Tocopherol composition also changes during soybean fermentation. With the exception of alpha-tocopherol, the levels of beta-, gamma-, and delta-tocopherol increase in *tempeh* (Astuti et al., 2000). With respect to carbohydrates, the soybean consists of mainly cell-wall polysaccharides and small sugars such as fructose, stachyose, and raffinose. These small sugars are removed during soaking, cooking, and fermentation. Pectin, cellulose, and hemicelluloses are insoluble cell-wall polysaccharides degraded by the microbial enzymes during fermentation (Egounlety and Aworh, 2003). The level of glucose (possibly a product of the digestion of complex to simple carbohydrates) increases in *tempeh*, giving a hint of savory sweetness (van der Riet et al., 1987).

Antinutritional factors such as phytate, saponins, hemagglutinins, flatus factors, and antitrypsin are present in soybean, which are mainly leached out, hydrolyzed, or inactivated during soaking, cooking, and fermentation. These antinutritional factors have the ability to bind any divalent minerals thus lowering the mineral bioavailability. During *tempeh* fermentation process these antinutrional factors' levels declined about 65% due to activity of yeast enzyme, thus the fermentation process increases the minerals bioavailabilty (Nout and Kires, 2005).

Soybean isoflavones are also modified after fermentation, and while the amount of glycosides decreases, the amount of aglycones increases in *tempeh* (Ahmad et al., 2015; Hong et al., 2012).

19.3.2 Antioxidant Roles of *Tempeh*

Reactive-oxygen species (ROS) are involved in the progression of diseases such as cancer, diabetes, atherosclerosis, arthritis, and neurodegenerative disorders (Shinde et al., 2010). Free radicals are released during normal metabolism and oxidation processes. An excessive amount of free radicals in the body leads to oxidative damage in proteins, lipids, DNA, and RNA. Perpetual oxidative damage may disrupt the activity of a biological system and weaken the body's defense mechanism (Droge, 2002). Isoflavones are weak phytoestrogen antioxidants as they have the ability to trap singlet oxygen. Soybean and its fermentation products are known to possess a high amount of isoflavones (Lin et al., 2006). A comparative study on the antioxidant potency of genistein and daidzein with respect to their corresponding glycosides isolated from soybean seeds has been established. The results showed that the ferric reducing antioxidant power (FRAP) and 2,2-diphenylpicrylhydrazyl (DPPH) free-radical scavenging activity of genistin, daidzein, glycitin, malonyl glycitin, and malony genistein were similar to their corresponding aglycones genistein and daidzein (Lee et al., 2005a). Recent results have shown that isoflavones extract from *tempeh* had more potent DPPH free-radical scavenging activity ($IC_{50}=2.67$ mg/mL) than the soybean isoflavone extract ($IC_{50}=10$ mg/mL), while the ferrous-ion chelating ability was similar for both extracts ($IC_{50}=10.40$ mg/mL, soybean and 11.13 mg/mL, *tempeh*) (Ahmad et al., 2015). Soybean germ fermented with *Aspergillus niger* M46 exhibited a higher antioxidant activity against hydroxyl radicals and superoxide radicals ($IC_{50}=0.8$ and 6.15 µg/mL, respectively) than those of unfermented soybean germ ($IC_{50}=164.0$ and 290.48 µg/mL, respectively) (Sheih et al., 2014). Inoculation of *Lactobacillus* in the fungal fermentation procedure of *tempeh* production enhanced both DPPH•-scavenging and antioxidant activities (Starzynska-Janiszewska et al., 2014).

In vivo studies have demonstrated the positive effect of isoflavones consumption on antioxidant status. Dietary soybean isoflavones administrated for 30 days to male Wistar rats exhibited a positive effect on antioxidant status (Barbosa et al., 2011). Balb/c mice treated for 14 days with a nutrient-enriched soybean *tempeh* significantly enhanced the superoxide dismutase (SOD) level and FRAP and reduced the malondialdehyde (MDA) level in liver tissue (Mohd Yusof et al., 2013). Matsuo et al. (1997) isolated a new potent antioxidant compound 3-hydroxyanthranilic acid (HAA) from *tempeh* and the antioxidant mechanisms were established by using mouse lung tissue and HuH-7, a human hepatoma-derived cell line. Unpublished data from our laboratory showed that the different groups of rats treated with *tempeh* isoflavone extract significantly elevated the level of antioxidants catalase, SOD, glutathione reductase (GRD), and glutathione (GSH) in both the brain of normal and scopolamine-induced animals. Furthermore, the extract decreased the level of oxidative markers thiobarbituric acid reactive substances (TBARS) and nitric oxide (NO) in both models.

19.3.3 *Tempeh* and Dementia

Neurodegeneration is a series of neuronal dysfunctions due to the continual death of neurons. The loss of brain neurons leads to major age-related diseases such as dementia. The most deteriorating effect of neurodegeneration is memory loss. However, memory loss is more common in neurodegenerative disease or due to injury to the brain. Learning is an adaptive behavior influenced by experience while memory is the neuronal activity of experience storage (Yang et al., 2011). Animals and humans survive through the ability of learning and memory, which take place in the brain. There is growing interest in the physiological functions of soybean isoflavones on cognitive functions and other related neurodegenerative diseases. There is clinical evidence showing that soybean isoflavones may mimic the functions of estrogen in the brain and have positive effects on the cognitive functions in females (Duffy et al., 2003; Macready et al., 2009).

Studies of soybean isoflavone effects on spatial memory have not provided consistent results in males (Lee et al., 2005b). Studies of spatial learning and memory using the radial arm maze (RAM) in rats fed *tempeh* isoflavone extract for 15 days showed that the time taken for the animals to consume all baits was significantly reduced during the 8-day assessment against the scopolamine-induced memory impairment model (Ahmad et al., 2014). The reference memory error (RME) and working memory error (WME) are two variables that report the physiological status of the brain (Titus et al., 2007). Working memory is involved in temporarily maintaining cues of preciously experienced events, which can be called short-term memory. The reduction of WME by *tempeh* isoflavone extract explained the improvement in short-term memory in animal models (Ahmad et al., 2014). On the other hand, reference memory represents long-term memory that is based on well-learned responses in the presence of current stimulus or unmoved cues along the study (Titus et al., 2007). Moreover, the improvement of long-term memory was represented by a significant reduction in the RME of *tempeh* isoflavone extract (Ahmad et al., 2014). Previous studies showed that immediate recall, which is the ability to remember a small amount of information over a few seconds, is highly sensitive to dementia. A clinical survey that included rural community-dwelling elderly of Central Java highlighted that the regular consumption of *tempeh* food was associated with improvement of immediate recall in younger population (Hogervorst et al., 2011). The reported findings suggest that exposure to *tempeh* food among the elderly of 56–97 years old may not be helpful as found for young adults.

Cholinergic neurons of the brain play a vital role in cognitive deficits related to aging and neurodegenerative diseases. In the brain of senile dementia of the

Alzheimer's type, the common features observed included selective loss of cholinergic neurons, decrease in acetylcholine (ACh) level, and an increase of acetylcholinesterase (AChE) activity (Hsieh et al., 2009). According to the cholinergic hypothesis, memory impairment in patients with senile dementia is due to a selective and irreversible deficiency of the brain cholinergic functions (Park et al., 2012). AChE involves the regulation of ACh to proper levels. However, excessive AChE activity leads to constant ACh deficiency and cognitive impairments (Pepeu and Giovannini, 2010). Soybean isoflavones can influence the brain cholinergic system and cognitive function by increasing the level of ACh through the inhibition of AChE (Yang et al., 2011). Thus daidzein may play a vital role in ACh synthesis as it may act as an choline acetyltransferase activator (Heo et al., 2006). From our findings of Ahmad et al. (2014), the rats treated with *tempeh* isoflavone extract elevated the ACh level and declined the AChE activity in central nervous system against scopolamine, an antimuscarinic agent induced cholinergic deficit. The brain cholinergic activity is significantly higher than in rats treated with soybean isoflavone extract at the same dose levels (Ahmad et al., 2014). The higher cholinergic activity as a result of the fermentation is directly related to the enrichment of bioavailable aglycones by fermentation process.

Neuroinflammation is another pathogenic mechanism that leads to memory loss due to neurodegeneration. At the inflammation site, proinflammatory cytokines and ROS produced by activating microglia cells and astrocytes may result in apoptosis and necrosis (Glass et al., 2010). Moreover, proinflammatory mediators released from microglial and astrocytes activate each other to amplify inflammatory signals to neurons. The major proinflammatory cytokine is IL-1β, which has been reported to be upregulated in brain neurodegeneration. Among the antiinflammatory cytokines, IL-4, IL-10, and TGF-β were reported to be downregulated in Alzheimer's disease (AD) patients (Chatterjee et al., 2015; Rubio-Perez and Morillas-Ruiz, 2012).

Moreover, COX-1 and COX-2 are key enzymes that play central roles in the inflammation process by converting arachidonic acid (AA), which is released from membrane phospholipids by a phospholipase A_2, into bioactive prostaglandins and other lipid mediators. Our study explained that groups of rats treated with *tempeh* isoflavone extract declined both COX activities in the brain. Moreover, the extract attenuated the level of a proinflammatory cytokine IL-1β and enhanced an antiinflammatory cytokine IL-10 in the brain (Ahmad et al., 2014).

The insoluble amyloid plaques are predominantly aggregates of Aβ peptides of 39–43 amino acids formed from the sequential cleavage of amyloid precursor protein (APP). It is cleaved by β-site APP cleaving enzyme-1 (BACE-1) followed by subsequent cleavage by the γ-secretase complex to form Aβ peptides. Since the main culprit of Aβ aggregation is the β-secretase (BACE1), the inhibition of this enzyme is believed to play an important role in the prevention of AD (Park et al., 2004; Tresadern et al., 2011). The in vitro β-secretase inhibition of soybean and *tempeh* isoflavone extracts was determined using a BACE1 FRET assay kit that uses baculovirus-expressed BACE1 and a specific substrate

(Rh-EVNLDAEFK-quencher) based on the Swedish mutation of the APP. The IC_{50} values of soybean and *tempeh* isoflavones against BACE1 were 10.87 and 5.47 mg/mL, respectively (Ahmad et al., 2014).

The fermentation process of the *tempeh* not only elevated the isoflavones genistin and daidzein, but also increased the levels of folic acid, vitamin B12, magnesium, and other nutrients, which can potentially improve brain health. Moreover, the antioxidant activity of *tempeh* constituents also facilitates improvement of cognitive functions against free-radical production in the brain (Khasanah, 2013).

19.3.4 *Tempeh* and Hypocholesterolemia

Presently, mortality caused by cardiovascular diseases (CVD) are relative high among other noncommunicable diseases. CVD are considered as a severe problem in developed and developing countries. According to WHO reports, 17.3 million people died from CVD in 2008 and 23.6 million people will die in 2030. Atherosclerosis is believed to be a major risk factor and is related to hypercholesterolemia and low-density lipoprotein cholesterol (LDL-C); both factors are considered as major causes of the onset of the atherogenic process. Reducing elevated LDL-C is a key public health challenge. Recently, there is growing interest in defining the dietary approaches to the management of lipid disorders. Increasing dietary protein intake has been related to improved lipid profiles in humans and animals (El Khoury and Anderson, 2013). Dietary proteins are believed to regulate lipid metabolism in a manner dependent on quantity and composition. Additionally, there is a general consensus that proteins slow lipid absorption and synthesis and promote lipid excretion. Soybean protein has been intensively investigated and many studies show that its consumption reduces blood cholesterol (Pyo and Seong, 2009). Soybean protein in the diet reduces the concentration of total cholesterol (TC) and LDL-C in plasma (Carroll, 1991). Soy protein also reduced triglyceride (TG) concentrations in plasma and liver in experimental animal and human studies (Lin et al., 2004; Sirtori et al., 1995). Recently, 5-week continuous treatment with lactic acid-fermented soymilk showed significant reduction in liver weight and fat mass of rats fed a high cholesterol diet. The hepatic TG and cholesterol levels as well as plasma TC level were significantly decreased. The expression of SREBP-2, a cholesterol synthesis-related gene, was significantly decreased in liver tissue (Kobayashi et al., 2012). Another bioactive ingredient, isoflavone, has also shown lipid metabolism-modulating effects. A *meta*-analysis study of 11 randomized controlled trials in humans established that consumption of soybean isoflavones significantly reduced serum TC and LDL-C, but did not change HDL-C and triacylglycerol. Moreover, the reductions in LDL-C were larger in hypercholesterolemic than in normocholesterolemic subjects (Taku et al., 2007). Isoflavone extracted from soybean proteins also decreased plasma TC level in rodents (Demonty et al., 2003).

It has been shown that mean body weight (16.34 g ± 9.11) was reduced in rats fed *tempeh* ad libitum for 28 days compared with those fed buffalo meat, casein, and soybean protein (Babji et al., 2010). Consumption of γ-amino butyric acid (GABA) enriched-*tempeh* declined the level of blood plasma triacylglycerols as compared with soybean protein and casein. Increased HDL-C and decreased LDL-C levels in the GABA enriched-*tempeh* group favored the antiatherosclerosis effects of GABA-enriched *tempeh* (Watanabe et al., 2006). GABA enriched-*tempeh* can be considered as an antihypertensive food since it is rich in GABA (Watanabe et al., 2006). It was also shown that a reduction of serum cholesterol in hypercholesterolemic rabbits fed an alcoholic extract of *tempeh*. Furthermore, oleic and linolenic acids were identified in the lipid extracts of fermented soybean. These fatty acids inhibited the HMG-CoA reductase in vitro (Hermosilla et al., 1993). The effect of a *tempeh*-rich diet on cholesterol levels was reported by Mangkuwidjojo et al. (1985). *Tempeh* had a positive effect on cholesterol level and histopathological changes in the liver and arteries of rats after a 4-month feeding trial. *Tempeh* constituents inhibit the enzyme responsible for the biosynthesis of cholesterol and prevent the oxidation of LDL, thus minimizing the production of plaque in arteries. In addition, *tempeh* contains high antioxidant content including genistein, daidzein, and tocopherol, which can also prevent lipid-related diseases (Mangkuwidjojo et al., 1985).

19.3.5 *Tempeh* and Cancer Prevention

Recently, attention has also focused on the potential role of soybean products in reducing cancer risk. Epidemiological studies have revealed that high consumption of soy products is associated with low incidence of hormone-dependent cancers, including breast and prostate cancers. Breast cancer is the second most frequently diagnosed cancer and the leading cause of death among women in developed and developing countries. The constituents of soybeans have proven to be able to modulate carcinogenesis, namely initiation, promotion, and cancer progression. In particular, soybean isoflavones such as genistein and daidzein have attracted interest, as they exhibit a plethora of biological actions, including breast and prostate cancer chemopreventive activity (Taylor et al., 2009). It is well known that most breast cancers are hormone-receptor-positive. Soy phytoestrogens such as genistein and daidzein are closely related to human 17β-estradiol, but with lower estrogenic activity (Messina and Hilakivi-Clarke, 2009). Based on available experimental data, genistein is believed to possess pleiotropic molecular mechanisms of action including inhibition of tyrosine kinases, DNA topoisomerase II, 5α-reductase, galectin-induced G2/M arrest, protein histidine kinase, cyclin-dependent kinases, and modulation of different signaling pathways associated with the growth of cancer cells (Aggarwal and Shishodia, 2006; Li et al., 2008). The aglycone genistein is also reported as a potent inhibitor of angiogenesis. Genistein was found to inhibit angiogenesis through regulation of multiple pathways, such as regulation of vascular endothelial growth factor, matrix metalloproteinases, epidermal growth factor receptor expressions and nuclear factor kappa B, phosphatidylinositol-3-kinase and protein kinase B, extracellular signal-regulated kinases 1 and 2 signaling pathways (Varinska et al., 2015). Kiriakidis et al. (2005) isolated several isoflavones from the *tempeh* extract. The isolated isoflavones were identified as genistein, daidzein, 6,7,40 -trihydroxyisoflavone (factor 2), 7,8,40 -trihydroxyisoflavone (7,8,40 -TriOH), and 5,7,30,40 -tetrahydroxyisoflavone (orobol). The chicken chorioallantoic membrane assay was used to evaluate the angiogenesis effect of these isoflavones. From the results, all isolated isoflavones inhibited angiogenesis, genistein reduced angiogenesis by 75.09%, followed by orobol (67.96%), factor 2 (56.77%), daidzein (48.98%), and 7,8,40-TriOH (24.42%). These compounds also inhibited vascular endothelial growth factor-induced endothelial cell proliferation and expression of the Ets1 transcription factor, known to be implicated in the regulation of new blood-vessel formation (Kiriakidis et al., 2005).

Extensive clinical and preclinical evidence has reported the effect of beneficial dietary substances from soybean and *tempeh* on various types of cancers. Xu et al. (2002) isolated genistein, daidzein, glycitein, genistin, and daidzin from *tempeh* isoflavones and exposed them against three cancer cell lines including MCF-7 (breast cancer cell line), HeLa (immortal cervical cancer cell), and HO-8910 (ovarian cancer cell) to study anticancer activity. Among the isolated compounds, genistein significantly inhibited the tumor cell lines and daidzein had an effect on the HO-8910 cell line. Furthermore, *tempeh* isoflavones inhibited tumor growth with an inhibitory rate of 30.9% and significantly enhanced the thymus index and activity of macrophage in BALB/C mice implanted with S-180 sarcoma cancer cells. Overall, the *tempeh* isoflavones had stronger antitumor activity than soybean isoflavones (Xu et al., 2002). Genistein has shown to be adjuvant therapeutic agent to potentiate the antitumor effect of 5-fluorouracil against pancreatic cancer (Suzuki et al., 2014). Moreover, genistein was reported to inhibit pathways that regulate the metastatic transformation in human prostate tissue. Pavese et al. (2014) found that genistein inhibited cell detachment, protease production, cell invasion, and human prostate cancer metastasis at concentrations achieved in human dietary intake. Further, results from phase I and phase II clinical trials showed that concentrations of genistein associated with antimetastatic efficacy in preclinical models are achievable in humans, and treatment with genistein inhibits pathways that regulate the metastatic transformation in human prostate tissue. In addition, the combination of genistein and daidzein at the low dose level demonstrated a synergistic preventive effect on isogenic human prostate cancer cells, such as LNCaP and C4-2B cells, when compared with individual soybean isoflavone (Dong et al., 2013).

Colorectal cancer is one of the most common human malignancies and a leading cause of cancer-related deaths around the world. Recently, the antiproliferative effects of genistein on colorectal cancer cells HCT-116 and LoVo cells were assessed using MTT assay. Genistein inhibited cell proliferation and induced apoptosis of colorectal cancer cells. It also induced the mitochondrial pathway of apoptosis in HCT-116 cells by inhibiting phosphorylation of Akt (Qin et al., 2016). Interestingly, both isoflavones (genistein and daidzein) reduced the proliferation of the human colon adenocarcinoma grade II cell line (HT-29) at concentrations of 25 and 50–100 µM, respectively. Moreover, the effects of both isoflavones were studied on tumor development and progression by their regulation of cell proliferation. Genistein suppressed expression of β-catenin (CTNNBIP1) at a concentration of 50 µM. Neither genistein nor daidzein affected APC (adenomatous polyposis coli) or survivin (BIRC5) expression when cells were treated with concentrations of 10 or 50 µM (Lepri et al., 2014).

Recently, both types of isoflavones (genistein and daidzein) supported the beneficial mechanisms for the management of hepatocellular carcinoma. It is the fifth most common malignant disease in men and the eighth in women worldwide. Dei and colleagues reported the effects and mechanism of genistein on hepatocellular carcinoma. Results from cell scratch and transwell assays highlighted that genistein inhibited migration of cell lines using HepG2, SMMC-7721, and Bel-7402 cells. Further, genistein at both mRNA and protein levels enhanced E-cadherin and α-catenin, but reduced N-cadherin and Vimentin. Simultaneously, treatment with genistein suppressed epithelial–mesenchymal transition (EMT) induced by TGF-β. In HepG2 cells, genistein reduced mRNA and protein expressions of nuclear factor of activated T cells 1 (NFAT1), Abca3, Autotaxin, CD154, and Cox-2. The results showed that genistein inhibited hepatocellular carcinoma cell migration by reversing the EMT, which was partly mediated by NFAT1 (Dai et al., 2015). Daidzein was reported to be a potent inducer of apoptosis in hepatic cancer cells via mitochondrial pathway. Using hepatic cancer cells (SK-HEP-1), daidzein decreased cell proliferation by real-time cell electronic sensing analysis. Treatment of daidzein modulated redox homeostasis of cells by expression of peroxiredoxin-3 (Prdx-3). Additionally, SKHEP-1 cells treated with daidzein decreased the levels of ROS. The upregulation of Bak and down-regulation of Bcl-2 and Bcl-xL proteins confirmed daidzein-induced apoptosis in SK-HEP-1 cells. Moreover, daidzein treatment increased with the release of mitochondrial cytochrome C and activation of APAF-1, caspase 9, and caspase 3, indicating that daidzein is a potent inducer of apoptosis in hepatic cancer cells via mitochondrial pathway (Park et al., 2013). After the fermentation process of soybean and its end product *tempeh* is reported with enrichment of bioavailable aglycones genistein and daidzein; both can be a potential dietary aid for prevention and management of various cancers such as prostate, cervix, brain, breast, and colon cancers (Patisaul and Jefferson, 2010; Pudenz et al., 2014).

19.4 OTHER FERMENTED SOYBEAN PRODUCTS WITH HEALTH BENEFITS

19.4.1 Cheonggukjang

Cheonggukjang is a traditional Korean soybean paste made from cooked whole soybeans fermented with *Bacillus subtilis*, which usually occurs in the air or in rice straw, at about 40°C for 2–3 days (Cho et al., 2011). Fermentation by

Bacillus breaks down soybean protein and polysaccharides into peptides and monosaccharides, producing *cheonggukjang* with sticky gums and bioavailable nutrients. The elevation of total phenolic compounds, isoflavone-aglycone contents, and antioxidant activities has also been reported (Cho et al., 2011). According to the literature, *cheonggukjang* is known for its functionality, including antioxidant, immunostimulant, antiinflammatory, antihypertensive, neuroprotective, antiosteoporotic, antimicrobial, and antidiabetic activities, in addition to its nutritional value (Cho et al., 2015; Go et al., 2015). Additionally, a quantitative study of *cheonggukjang*, soybean fermented with *Bacillus subtilis* CS90, also showed a decrease in daidzin content, from 487.45 to 213.45 mg/kg while daidzein increased from undetected level to 330.09 mg/kg after 60 h of fermentation. A similar trend was observed for genistein, increasing from an undetected level to 25.62 mg/kg (Cho et al., 2011).

19.4.2 Doenjang

Doenjang is also a traditional fermented soybean food from Korea. *Meju* is the basis for *doenjang* fermentation, traditionally prepared by soaking, steaming, crushing the soybean, and fermentation with *Bacillus subtilis, Rhizopus spp.,* and *Aspergillus spp.* for 2–3 months. The fermented *Meju* is separated into two parts; the liquid part is filtered to make soybean sauce and the solid part is used for further ripening to make *doenjang* (Kim and Lee, 2014). *Doenjang* is rich in isoflavones and beneficial vitamins, minerals, and phytosterols that are reported to possess anticarcinogenic properties (Jeong et al., 2014). *Doenjang* seems to be effective at preventing cancer, heart disease, brain tumors, and lowering blood pressure, glucose, and cholesterol levels (Cha et al., 2014; Chung et al., 2014; Jeong et al., 2014). It has also been reported to promote gut health by regulating gut microbiota (Jang et al., 2014).

19.4.3 Doubanjiang

Doubanjiang is a spicy, salty paste of a Chinese traditional fermented soybean product obtained through the fermentation of soybeans by naturally occurring bacteria *Aspergillus oryzae* (Byun et al., 2013). The fermentation process takes about 2–5 months (Frisvad and Samson, 1991). Animal nutritional and human intervention studies have reported that isoflavones of *doubanjiang* reduce insulin resistance. Fermented soybeans exhibit a higher effect on the progression delay of type 2 diabetes than nonfermented soybeans (Kwon et al., 2010).

19.4.4 Douchi

Douchi is a traditional salt-fermented soybean food among Chinese communities worldwide made from black soybeans. The fermentation process involves fungal solid-state fermentation (prefermentation) with microorganisms (eg, *Aspergillus, Mucor, Rhizopus*, bacteria), followed by salting and maturation (postfermentation) (Chen et al., 2012). The *douchi* sauce is a rich source of protein and is also used as a flavored ingredient for cooking. Furthermore, it is used traditionally in relieving tiredness, weakness, insomnia, and poor appetite (Chen et al., 2015). It has also shown to have angiotensin-converting enzyme inhibitory activity as well as antioxidant activity (Iwaniak et al., 2014) and α-glucosidase inhibitory activity (Wang et al., 2008).

19.4.5 Gochujang

Gochujang is a traditional Korean fermented hot pepper-soybean paste. The characteristic flavor of *gochujang* is a combination of hot taste from red pepper, sweet taste from sugars, umami taste from amino acids, and salty taste from NaCl. *Gochujang* is produced by fermentation with *Aspergillus* sp. and *Bacillus* sp. for over years in large earthen pots by mixing glutinous rice powder, salt, and red pepper powder with *meju* powder (Kwon et al., 2010). It contains isoflavones and has been attributed with a number of biological activities such as anticancer effect, antihypertensive activity, hepatoprotective properties, and immune functions (Kim et al., 2014b).

19.4.6 Miso

Miso is a Japanese traditional paste produced by fermenting soybean with fungus *Aspergillus oryzae* and salt, and sometimes with rice, wheat, or oats. It contains vitamins, minerals, proteins, carbohydrates, isoflavones, and lecithin (Watanabe, 2013). A review by Watanabe (2013) recorded the effects of *miso* with reference to prevention of radiation injury, prevention of liver, breast, and intestinal tumors, and hypertension in animal models. Citizens of Okinawa (Japan) are known for their long life expectancy, high number of centenarians, and accompanying low risk of age-associated diseases (Willcox et al., 2009). The secret is "the Okinawa diet," in which miso soup is the epicenter. By decreasing oxidative stress in our body, the high antioxidant of miso reduces the risk of CVD, some forms of cancer, and chronic diseases (Zaheer and Akhtar, 2015).

19.4.7 Natto

Natto is a fermented traditional soybean product in Japan. When forming *natto*, the soybeans are first soaked and cooked, then mixed with the *Bacillus subtilis* to allow fermentation up to 24h at 40°C. Then the *natto* is cooled and aged in a refrigerator for up to 1 week to allow the development of stringiness. It contains vitamins, amino acids, proteins, sugars, fats, minerals, and dietary fibers (Hitosugi et al., 2015). *Natto* contains nattokinase, a polypeptide comprised of total 275 amino acid residues with anticoagulant, fibrinolytic, blood pressure-lowering effects, and antioxidant activity (Fujita et al., 2011; Hitosugi et al., 2015).

19.4.8 Soy Sauce

In East and Southeast Asian cuisines, soy sauce is a traditional ingredient used in cooking and as a condiment. It is made from a fermented paste of boiled soybeans, roasted grain, brine, and fermentation with *Aspergillus oryzae* or *Aspergillus sojae* molds. After the fermentation process, the soy sauce is separated by pressing the fermented paste and collecting liquid. Soy sauce contains many flavor compounds, including alcohols, aldehydes, esters, and acids, but amino acids form the main substance of the flavor of soy sauce (Zhao et al., 2015). Fermented soy sauce exhibited rich antioxidant activity and hypolipidemic and antiinflammatory effects in C57BL/6J mice (Kim et al., 2014a), neuroprotective action in albino rats (Abdelall et al., 2015), and antihypertensive activity in suppressing renin-angiotensin-aldosterone system in spontaneously hypertensive rats and Dahl salt-sensitive rats (Nakahara et al., 2011).

19.4.9 Tauchu (Tauco)

Tauchu is a preserved fermented yellow soybean paste found in *Chinese-Indonesian cuisine*. It is made by boiling yellow soybeans, grinding and mixing them with flour, and then fermenting them to make a soybean paste. Then the paste is soaked in salty water and sun-dried for several weeks, until the color of the paste turns yellow-reddish. *Tauchu* is used as a condiment and flavor for stir-fried dishes and in soups. Similar to *doubanjiang*, a recent study showed that *tauchu* may have antidiabetic properties (Kwon et al., 2010).

19.5 CONCLUSION

Among fermented soybean products, *tempeh* is considered as a good source of protein, vitamins, antioxidants, phytochemical, and other bioactive beneficial substances. Fermentation of soybean by the mold *Rhizopus* spp. induces a number of nutritional changes and thus makes it more beneficial to health. *Tempeh* enrichment with the bioavailable aglycones genistein and daidzein has been linked to various health benefits, making this product attractive to consumers. Due to its health and nutritional aspects, *tempeh* and other traditional fermented soybean products can be considered as functional foods to help in the prevention of various diseases and disorders.

REFERENCES

Abdelall, H.F., ElGhamrawy, T.A., Helmy, D., 2015. Morphological evaluation of the protective role of dark soy sauce against acrylamide induced neurotoxicity in albino rats. Folia Morphologica (Warsz) 74, 16–24.

Aggarwal, B.B., Shishodia, S., 2006. Molecular targets of dietary agents for prevention and therapy of cancer. Biochemical Pharmacology 71, 1397–1421.

Ahmad, A., Ramasamy, K., Jaafar, S.M., Majeed, A.B., Mani, V., 2014. Total isoflavones from soybean and *tempeh* reversed scopolamine-induced amnesia, improved cholinergic activities and reduced neuroinflammation in brain. Food and Chemical Toxicology 65, 120–128.

Ahmad, A., Ramasamy, K., Majeed, A.B., Mani, V., 2015. Enhancement of beta-secretase inhibition and antioxidant activities of *tempeh*, a fermented soybean cake through enrichment of bioactive aglycones. Pharmaceutical Biology 53, 758–766.

Ahn-Jarvis, J.H., Riedl, K.M., Schwartz, S.J., Vodovotz, Y., 2013. Design and selection of soy breads used for evaluating isoflavone bioavailability in clinical trials. Journal of Agricultural and Food Chemistry 61, 3111–3120.

Astuti, M., Meliala, A., Dalais, F.S., Wahlqvist, M.L., 2000. Tempe, a nutritious and healthy food from Indonesia. Asian Pacific Journal of Clinical Nutrition 9, 322–325.

Babji, A., Fatimah, S., Abolhassani, Y., Ghassem, M., 2010. Nutritional quality and properties of protein and lipid in processed meat products–a perspective. International Food Research Journal 17, 35–44.

Barbosa, A.C., Lajolo, F.M., Genovese, M.I., 2011. Effect of free or protein-associated soy isoflavones on the antioxidant status in rats. Journal of the Science of Food and Agriculture 91, 721–731.

Barnes, S., 2010. The biochemistry, chemistry and physiology of the isoflavones in soybeans and their food products. Lymphatic Research and Biology 8, 89–98.

Byun, B.Y., Bai, X., Mah, J.-H., 2013. Occurrence of biogenic amines in doubanjiang and tofu. Food Science and Biotechnology 22, 55–62.

Carbonel, A.A., Calio, M.L., Santos, M.A., Bertoncini, C.R., Sasso Gda, S., Simoes, R.S., et al., 2015. Soybean isoflavones attenuate the expression of genes related to endometrial cancer risk. Climacteric 18, 389–398.

Carroll, K.K., 1991. Review of clinical studies on cholesterol-lowering response to soy protein. Journal of the American Dietetic Association 91, 820–827.

Cha, Y.S., Park, Y., Lee, M., Chae, S.W., Park, K., Kim, Y., et al., 2014. Doenjang, a Korean fermented soy food, exerts antiobesity and antioxidative activities in overweight subjects with the PPAR-gamma2 C1431T polymorphism: 12-week, double-blind randomized clinical trial. Journal of Medicinal Food 17, 119–127.

Cha, Y.S., Yang, J.A., Back, H.I., Kim, S.R., Kim, M.G., Jung, S.J., et al., 2012. Visceral fat and body weight are reduced in overweight adults by the supplementation of Doenjang, a fermented soybean paste. Nutrition Research and Practice 6, 520–526.

Chatterjee, G., Roy, D., Khemka, V.K., Chattopadhyay, M., Chakrabarti, S., 2015. Genistein, the isoflavone in soybean, causes amyloid beta peptide accumulation in human neuroblastoma cell line: implications in Alzheimer's disease. Aging and Disease 6, 456–465.

Chen, C., Xiang, J.Y., Hu, W., Xie, Y.B., Wang, T.J., Cui, J.W., et al., 2015. Identification of key micro-organisms involved in Douchi fermentation by statistical analysis and their use in an experimental fermentation. Journal of Applied Microbiology 119, 1324–1334.

Chen, T., Jiang, S., Xiong, S., Wang, M., Zhu, D., Wei, H., 2012. Application of denaturing gradient gel electrophoresis to microbial diversity analysis in Chinese Douchi. Journal of the Science of Food and Agriculture 92, 2171–2176.

Cho, C.W., Han, C.J., Rhee, Y.K., Lee, Y.C., Shin, K.S., Shin, J.S., et al., 2015. Cheonggukjang polysaccharides enhance immune activities and prevent cyclophosphamide-induced immunosuppression. International Journal if Biological Macromolecules 72, 519–525.

Cho, K.M., Lee, J.H., Yun, H.D., Ahn, B.Y., Kim, H., Seo, W.T., 2011. Changes of phytochemical constituents (isoflavones, flavanols, and phenolic acids) during cheonggukjang soybeans fermentation using potential probiotics *Bacillus subtilis* CS90. Journal of Food Composition and Analysis 24, 402–410.

Chung, S.I., Rico, C.W., Kang, M.Y., 2014. Comparative study on the hypoglycemic and antioxidative effects of fermented paste (doenjang) prepared from soybean and brown rice mixed with rice bran or red ginseng marc in mice fed with high fat diet. Nutrients 6, 4610–4624.

Chutrtong, J., 2013. Acceptance of consumer on various *Tempeh* and protein content comparison. International Journal of Biological, Biomolecular, Agricultural, Food and Biotechnological Engineering 7, 183–186.

da Silva, L.H., Celeghini, R.M.S., Chang, Y.K., 2011. Effect of the fermentation of whole soybean flour on the conversion of isoflavones from glycosides to aglycones. Food Chemistry 128, 640–644.

Dai, W., Wang, F., He, L., Lin, C., Wu, S., Chen, P., et al., 2015. Genistein inhibits hepatocellular carcinoma cell migration by reversing the epithelial-mesenchymal transition: partial mediation by the transcription factor NFAT1. Molecular Carcinogenesis 54, 301–311.

Danciu, C., Soica, C., Csanyi, E., Ambrus, R., Feflea, S., Peev, C., et al., 2012. Changes in the anti-inflammatory activity of soy isoflavonoid genistein versus genistein incorporated in two types of cyclodextrin derivatives. Chemistry Central Journal 6, 58.

Demonty, I., Lamarche, B., Jones, P.J., 2003. Role of isoflavones in the hypocholesterolemic effect of soy. Nutrition Reviews 61, 189–203.

Devi, M.K.A., Gondi, M., Sakthivelu, G., Giridhar, P., Rajasekaran, T., Ravishankar, G.A., 2009. Functional attributes of soybean seeds and products, with reference to isoflavone content and antioxidant activity. Food Chemistry 114, 771–776.

Dong, X., Xu, W., Sikes, R.A., Wu, C., 2013. Combination of low dose of genistein and daidzein has synergistic preventive effects on isogenic human prostate cancer cells when compared with individual soy isoflavone. Food Chemistry 141, 1923–1933.

Droge, W., 2002. Free radicals in the physiological control of cell function. Physiological Reviews 82, 47–95.

Duffy, R., Wiseman, H., File, S.E., 2003. Improved cognitive function in postmenopausal women after 12 weeks of consumption of a soya extract containing isoflavones. Pharmacology Biochemistry and Behavior 75, 721–729.

Egounlety, M., Aworh, O.C., 2003. Effect of soaking, dehulling, cooking and fermentation with *Rhizopus oligosporus* on the oligosaccharides, trypsin inhibitor, phytic acid and tannins of soybean (Glycine max Merr.), cowpea (*Vigna unguiculata* L. Walp) and groundbean (*Macrotyloma geocarpa* Harms). Journal of Food Engineering 56, 249–254.

El Khoury, D., Anderson, G.H., 2013. Recent advances in dietary proteins and lipid metabolism. Current Opinion in Lipidology 24, 207–213.

Frisvad, J.C., Samson, R.A., 1991. Filamentous fungi in foods and feeds: ecology, spoilage and mycotoxins production. In: Arora, D.K., Mukerji, K.G., Marth, E.H. (Eds.), Handbook of Applied Mycology: Foods and Feeds. vol. 3. Marcel Dekker, New York, pp. 31–68.

Fujita, M., Ohnishi, K., Takaoka, S., Ogasawara, K., Fukuyama, R., Nakamuta, H., et al., 2011. Antihypertensive effects of continuous oral administration of nattokinase and its fragments in spontaneously hypertensive rats. Biological & Pharmaceutical Bulletin 34, 1696–1701.

Gilbert, E.R., Liu, D., 2013. Antidiabetic functions of soy isoflavone genistein: mechanisms underlying its effects on pancreatic beta-cell function. Food & Function 4, 200–212.

Ginting, E., Arcot, J., 2004. High-performance liquid chromatographic determination of naturally occurring folates during tempe preparation. Journal of Agricultural and Food Chemistry 52, 7752–7758.

Giordano, E., Davalos, A., Crespo, M.C., Tome-Carneiro, J., Gomez-Coronado, D., Visioli, F., 2015. Soy isoflavones in nutritionally relevant amounts have varied nutrigenomic effects on adipose tissue. Molecules 20, 2310–2322.

Glass, C.K., Saijo, K., Winner, B., Marchetto, M.C., Gage, F.H., 2010. Mechanisms underlying inflammation in neurodegeneration. Cell 140, 918–934.

Go, J., Kim, J.E., Kwak, M.H., Koh, E.K., Song, S.H., Sung, J.E., et al., 2015. Neuroprotective effects of fermented soybean products (Cheonggukjang) manufactured by mixed culture of *Bacillus subtilis* MC31 and *Lactobacillus sakei* 383 on trimethyltin-induced cognitive defects mice. Nutritional Neuroscience. http://dx.doi.org/10.1179/1476830515Y.0000000025.

Haron, H., Ismail, A., Azlan, A., Shahar, S., Peng, L.S., 2009. Daidzein and genestein contents in *tempeh* and selected soy products. Food Chemistry 115, 1350–1356.

Heo, H.J., Suh, Y.M., Kim, M.J., Choi, S.J., Mun, N.S., Kim, H.K., et al., 2006. Daidzein activates choline acetyltransferase from MC-IXC cells and improves drug-induced amnesia. Bioscience Biotechnology and Biochemistry 70, 107–111.

Hermosilla, J.A.G., Jha, H.C., Egge, H., Mahmud, M., Hermana, S., Rao, G.S., 1993. Isolation and characterization of hydroxymethylglutaryl coenzyme a reductase inhibitors from fermented soybean extracts. Journal of Clinical Biochemistry and Nutrition 15, 163–174.

Hitosugi, M., Hamada, K., Misaka, K., 2015. Effects of *Bacillus subtilis* var. natto products on symptoms caused by blood flow disturbance in female patients with lifestyle diseases. International Journal of General Medicine 8, 41–46.

Ho, H.M., Chen, R.Y., Leung, L.K., Chan, F.L., Huang, Y., Chen, Z.Y., 2002. Difference in flavonoid and isoflavone profile between soybean and soy leaf. Biomedicine & Pharmacotherapy 56, 289–295.

Hogervorst, E., Mursjid, F., Priandini, D., Setyawan, H., Ismael, R.I., Bandelow, S., et al., 2011. Borobudur revisited: soy consumption may be associated with better recall in younger, but not in older, rural Indonesian elderly. Brain Research 1379, 206–212.

Hong, G.E., Prabhat, K.M., Lim, K.W., Lee, C.H., 2012. Fermentation increases isoflavone aglycone contents in black soybean pulp. Asian Journal of Animal and Veterinary Advances 7, 502–511.

Hsieh, H.M., Wu, W.M., Hu, M.L., 2009. Soy isoflavones attenuate oxidative stress and improve parameters related to aging and Alzheimer's disease in C57BL/6J mice treated with D-galactose. Food and Chemical Toxicology 47, 625–632.

Iwaniak, A., Minkiewicz, P., Darewicz, M., 2014. Food-originating ACE inhibitors, including antihypertensive peptides, as preventive food components in blood pressure reduction. Comprehensive Reviews in Food Science and Food Safety 13, 114–134.

Izumi, T., Piskula, M.K., Osawa, S., Obata, A., Tobe, K., Saito, M., et al., 2000. Soy isoflavone aglycones are absorbed faster and in higher amounts than their glucosides in humans. Journal of Nutrition 130, 1695–1699.

Jang, S.E., Kim, K.A., Han, M.J., Kim, D.H., 2014. Doenjang, a fermented Korean soybean paste, inhibits lipopolysaccharide production of gut microbiota in mice. Journal of Medicinal Food 17, 67–75.

Jeong, J.K., Chang, H.K., Park, K.Y., 2014. Doenjang prepared with mixed starter cultures attenuates azoxymethane and dextran sulfate sodium-induced colitis-associated colon carcinogenesis in mice. Journal of Carcinogenesis 13, 9.

Karsli-Ceppioglu, S., Ngollo, M., Adjakly, M., Dagdemir, A., Judes, G., Lebert, A., et al., 2015. Genome-wide DNA methylation modified by soy phytoestrogens: role for epigenetic therapeutics in prostate cancer? OMICS 19, 209–219.

Khasanah, Y., 2013. Effect of Dietary Tempe Towards the Folate Nutritional Status in Sprague Dawley Rats (Ph.D. Thesis). Universitas Gadjah Mada, Yogyakarta, Indonesia. Available online- http://etd.repository.ugm.ac.id/index.php?mod=penelitian_detail&sub=PenelitianDetail&act= view&typ=html&buku_id=63598.

Khasanah, Y., Ratnayani, A.D., Angwar, M., Nuraeni, T., 2015. In vivo study on albumin and total protein in white rat (*Rattus Norvegicus*) after feeding of enteral formula from tempe and local food. Procedia Food Science 3, 274–279.

Kim, J.H., Jia, Y., Lee, J.G., Nam, B., Lee, J.H., Shin, K.S., et al., 2014a. Hypolipidemic and antiinflammation activities of fermented soybean fibers from meju in C57BL/6 J mice. Phytotherapy Research 28, 1335–1341.

Kim, M.K., Lee, K.G., 2014. Correlating consumer perception and consumer acceptability of traditional Doenjang in Korea. Journal of Food Science 79, S2330–S2336.

Kim, S.H., Kim, H.J., Shin, H.S., 2014b. Identification and quantification of antitumor thioproline and methylthioproline in Korean traditional foods by a liquid chromatography-atmospheric pressure chemical ionization-tandem mass spectrometry. Journal of Pharmaceutical and Biomedical Analysis 100, 58–63.

Kiriakidis, S., Hogemeier, O., Starcke, S., Dombrowski, F., Hahne, J.C., Pepper, M., et al., 2005. Novel *tempeh* (fermented soyabean) isoflavones inhibit in vivo angiogenesis in the chicken chorioallantoic membrane assay. British Journal of Nutrition 93, 317–323.

Kobayashi, M., Hirahata, R., Egusa, S., Fukuda, M., 2012. Hypocholesterolemic effects of lactic acid-fermented soymilk on rats fed a high cholesterol diet. Nutrients 4, 1304–1316.

Kwon, D.Y., Daily 3rd, J.W., Kim, H.J., Park, S., 2010. Antidiabetic effects of fermented soybean products on type 2 diabetes. Nutrition Research 30, 1–13.

Lee, C.H., Yang, L., Xu, J.Z., Yeung, S.Y.V., Huang, Y., Chen, Z.-Y., 2005a. Relative antioxidant activity of soybean isoflavones and their glycosides. Food Chemistry 90, 735–741.

Lee, Y.B., Lee, H.J., Sohn, H.S., 2005b. Soy isoflavones and cognitive function. Journal of Nutritional Biochemistry 16, 641–649.

Lei, X., Sun, G., Xie, J., Wei, D., 2013. Lactobacillus curieae sp. nov., isolated from stinky tofu brine. International Journal of Systematic and Evolutionary Microbiology 63, 2501–2505.

Lepri, S.R., Zanelatto, L.C., da Silva, P.B., Sartori, D., Ribeiro, L.R., Mantovani, M.S., 2014. Effects of genistein and daidzein on cell proliferation kinetics in HT29 colon cancer cells: the expression of CTNNBIP1 (beta-catenin), APC (adenomatous polyposis coli) and BIRC5 (survivin). Human Cell 27, 78–84.

Li, Z., Li, J., Mo, B., Hu, C., Liu, H., Qi, H., et al., 2008. Genistein induces cell apoptosis in MDA-MB-231 breast cancer cells via the mitogen-activated protein kinase pathway. Toxicology in Vitro 22, 1749–1753.

Liang, W., Lee, A.H., Binns, C.W., Huang, R., Hu, D., Shao, H., 2009. Soy consumption reduces risk of ischemic stroke: a case-control study in southern china. Neuroepidemiology 33, 111–116.

Liem, I.T., Steinkraus, K.H., Cronk, T.C., 1977. Production of vitamin B-12 in *tempeh*, a fermented soybean food. Applied and Environmental Microbiology 34, 773–776.

Lin, C.H., Wei, Y.T., Chou, C.C., 2006. Enhanced antioxidative activity of soybean koji prepared with various filamentous fungi. Food Microbiology 23, 628–633.

Lin, Y., Meijer, G.W., Vermeer, M.A., Trautwein, E.A., 2004. Soy protein enhances the cholesterol-lowering effect of plant sterol esters in cholesterol-fed hamsters. Journal of Nutrition 134, 143–148.

Liu, Y., Hilakivi-Clarke, L., Zhang, Y., Wang, X., Pan, Y.X., Xuan, J., et al., 2015. Isoflavones in soy flour diet have different effects on whole-genome expression patterns than purified isoflavone mix in human MCF-7 breast tumors in ovariectomized athymic nude mice. Molecular Nutrition & Food Research 59, 1419–1430.

Macready, A.L., Kennedy, O.B., Ellis, J.A., Williams, C.M., Spencer, J.P., Butler, L.T., 2009. Flavonoids and cognitive function: a review of human randomized controlled trial studies and recommendations for future studies. Genes & Nutrition 4, 227–242.

Mangkuwidjojo, S., Pranowo, D., Nitisuwirjo, S., Noor, Z., 1985. Observation on hypolipidemic activity of tempe. In: Hermana, H., Karyadi, D. (Eds.), Proceedings of the Symposium on the Utilization of Tempe in Health and Nutrition Development. Bogor Nutrition Research and Development Center, Department of Health of Republic of Indonesia, pp. 114–127.

Matsuo, M., Nakamura, N., Shidoji, Y., Muto, Y., Esaki, H., Osawa, T., 1997. Antioxidative mechanism and apoptosis induction by 3-hydroxyanthranilic acid, an antioxidant in Indonesian food *Tempeh*, in the human hepatoma-derived cell line, HuH-7. Journal of Nutritional Science and Vitaminology (Tokyo) 43, 249–259.

McKeown, N.M., 2004. Whole grain intake and insulin sensitivity: evidence from observational studies. Nutrition Reviews 62, 286–291.

Messina, M., Hilakivi-Clarke, L., 2009. Early intake appears to be the key to the proposed protective effects of soy intake against breast cancer. Nutrition and Cancer 61, 792–798.

Mohd Yusof, H., Ali, N.M., Yeap, S.K., Ho, W.Y., Beh, B.K., Koh, S.P., et al., 2013. Hepatoprotective effect of fermented soybean (nutrient enriched soybean *tempeh*) against alcohol-induced liver damage in mice. Evidence Based Complementary and Alternative Medicine 2013, 274274.

Murata, K., Ikehata, H., Miyamoto, T., 1967. Studies on the nutritional value of *tempeh*. Journal of Food Science 32, 580–586.

Murata, K., Miyamoto, T., Kokufu, E., Sanke, Y., 1970. Studies on the nutritional value of *tempeh*. 3. Changes in biotin and folic acid contents during *tempeh* fermentation. Journal of Vitaminology (Kyoto) 16, 281–284.

Nagarajan, S., Burris, R.L., Stewart, B.W., Wilkerson, J.E., Badger, T.M., 2008. Dietary soy protein isolate ameliorates atherosclerotic lesions in apolipoprotein E-deficient mice potentially by inhibiting monocyte chemoattractant protein-1 expression. Journal of Nutrition 138, 332–337.

Nakahara, T., Sugimoto, K., Sano, A., Yamaguchi, H., Katayama, H., Uchida, R., 2011. Antihypertensive mechanism of a peptide-enriched soy sauce-like seasoning: the active constituents and its suppressive effect on renin-angiotensin-aldosterone system. Journal of Food Science 76, H201–H206.

Nout, M.J.R., Kiers, J.L., 2005. Tempe fermentation, innovation and functionality: update into the third millenium. Journal of Applied Microbiology 98, 789–805.

Park, H.J., Jeon, Y.K., You, D.H., Nam, M.J., 2013. Daidzein causes cytochrome c-mediated apoptosis via the Bcl-2 family in human hepatic cancer cells. Food and Chemical Toxicology 60, 542–549.

Park, I.H., Jeon, S.Y., Lee, H.J., Kim, S.I., Song, K.S., 2004. A beta-secretase (BACE1) inhibitor hispidin from the mycelial cultures of *Phellinus linteus*. Planta Medica 70, 143–146.

Park, S.J., Kim, D.H., Jung, J.M., Kim, J.M., Cai, M., Liu, X., et al., 2012. The ameliorating effects of stigmasterol on scopolamine-induced memory impairments in mice. European Journal of Pharmacology 676, 64–70.

Patisaul, H.B., Jefferson, W., 2010. The pros and cons of phytoestrogens. Frontiers in Neuroendocrinology 31, 400–419.

Pavese, J.M., Krishna, S.N., Bergan, R.C., 2014. Genistein inhibits human prostate cancer cell detachment, invasion, and metastasis. American Journal of Clinical Nutrition 100 (Suppl. 1), 431S–436S.

Pepeu, G., Giovannini, M.G., 2010. Cholinesterase inhibitors and memory. Chemico-Biological Interactions 187, 403–408.

Pudenz, M., Roth, K., Gerhauser, C., 2014. Impact of soy isoflavones on the epigenome in cancer prevention. Nutrients 6, 4218–4272.

Pyo, Y.H., Seong, K.S., 2009. Hypolipidemic effects of Monascus-fermented soybean extracts in rats fed a high-fat and -cholesterol diet. Journal of Agricultural and Food Chemistry 57, 8617–8622.

Qin, J., Teng, J., Zhu, Z., Chen, J., Huang, W.J., 2016. Genistein induces activation of the mitochondrial apoptosis pathway by inhibiting phosphorylation of Akt in colorectal cancer cells. Pharmaceutical Biology 54, 74–79.

Qu, X.L., Fang, Y., Zhang, M., Zhang, Y.Z., 2014. Phytoestrogen intake and risk of ovarian cancer: a meta- analysis of 10 observational studies. Asian Pacific Journal of Cancer Prevention 15, 9085–9091.
Roubos-van den Hil, P.J., Nout, M.J.R., 2011. Antidiarrhoeal Aspects of Fermented Soya Beans. Rijeka, Croatia: INTECH.
Rubio-Perez, J.M., Morillas-Ruiz, J.M., 2012. A review: inflammatory process in Alzheimer's disease, role of cytokines. Scientific World Journal 2012, 756357.
Russo, M., Russo, G.L., Daglia, M., Kasi, P.D., Ravi, S., Nabavi, S.F., et al., 2016. Understanding genistein in cancer: the "good" and the "bad" effects: a review. Food Chemistry 196, 589–600.
Sheih, I.C., Fang, T.J., Wu, T.K., Chen, R.Y., 2014. Effects of fermentation on antioxidant properties and phytochemical composition of soy germ. Journal of the Science of Food and Agriculture 94, 3163–3170.
Shinde, A.N., Malpathak, N., Fulzele, D.P., 2010. Determination of isoflavone content and antioxidant activity in *Psoralea corylifolia* L. callus cultures. Food Chemistry 118, 128–132.
Sirtori, C.R., Lovati, M.R., Manzoni, C., Monetti, M., Pazzucconi, F., Gatti, E., 1995. Soy and cholesterol reduction: clinical experience. Journal of Nutrition 125, 598S–605S.
Soka, S., Suwanto, A., Sajuthi, D., Rusmana, I., 2014. Impact of *tempeh* supplementation on gut microbiota composition in Sprague-Dawley rats. Research Journal of Microbiology 9, 189–198.
Soni, S., Dey, G., 2014. Perspectives on global fermented foods. British Food Journal 116, 1767–1787.
Starzynska-Janiszewska, A., Stodolak, B., Mickowska, B., 2014. Effect of controlled lactic acid fermentation on selected bioactive and nutritional parameters of *tempeh* obtained from unhulled common bean (*Phaseolus vulgaris*) seeds. Journal of the Science of Food and Agriculture 94, 359–366.
Suzuki, R., Kang, Y., Li, X., Roife, D., Zhang, R., Fleming, J.B., 2014. Genistein potentiates the antitumor effect of 5-Fluorouracil by inducing apoptosis and autophagy in human pancreatic cancer cells. Anticancer Research 34, 4685–4692.
Taku, K., Umegaki, K., Sato, Y., Taki, Y., Endoh, K., Watanabe, S., 2007. Soy isoflavones lower serum total and LDL cholesterol in humans: a meta-analysis of 11 randomized controlled trials. American Journal of Clinical Nutrition 85, 1148–1156.
Taylor, C.K., Levy, R.M., Elliott, J.C., Burnett, B.P., 2009. The effect of genistein aglycone on cancer and cancer risk: a review of in vitro, preclinical, and clinical studies. Nutrition Reviews 67, 398–415.
Titus, A.D., Shankaranarayana Rao, B.S., Harsha, H.N., Ramkumar, K., Srikumar, B.N., Singh, S.B., et al., 2007. Hypobaric hypoxia-induced dendritic atrophy of hippocampal neurons is associated with cognitive impairment in adult rats. Neuroscience 145, 265–278.
Tresadern, G., Bartolome, J.M., Macdonald, G.J., Langlois, X., 2011. Molecular properties affecting fast dissociation from the D2 receptor. Bioorganic & Medicinal Chemistry 19, 2231–2241.
van der Riet, W.B., Wight, A.W., Cilliers, J.J.L., Datel, J.M., 1987. Food chemical analysis of *tempeh* prepared from South African-grown soybeans. Food Chemistry 25, 197–206.
Varinska, L., Gal, P., Mojzisova, G., Mirossay, L., Mojzis, J., 2015. Soy and breast cancer: focus on angiogenesis. International Journal of Molecular Sciences 16, 11728–11749.
Villa, P., Costantini, B., Suriano, R., Perri, C., Macri, F., Ricciardi, L., et al., 2009. The differential effect of the phytoestrogen genistein on cardiovascular risk factors in postmenopausal women: relationship with the metabolic status. Journal of Clinical Endocrinology & Metabolism 94, 552–558.
Wang, D., Wang, L-j, Zhu, F-x, Zhu, J-y, Chen, X.D., Zou, L., et al., 2008. In vitro and in vivo studies on the antioxidant activities of the aqueous extracts of Douchi (a traditional Chinese salt-fermented soybean food). Food Chemistry 107, 1421–1428.

Watanabe, H., 2013. Beneficial biological effects of miso with reference to radiation injury, cancer and hypertension. Journal of Toxicologic Pathology 26, 91–103.

Watanabe, N., Endo, Y., Fujimoto, K., Aoki, H., 2006. *Tempeh*-like fermented soybean (GABA-*tempeh*) has an effective influence on lipid metabolism in rats. Journal of Oleo Science 55, 391–396.

Willcox, D.C., Willcox, B.J., Todoriki, H., Suzuki, M., 2009. The Okinawan diet: health implications of a low-calorie, nutrient-dense, antioxidant-rich dietary pattern low in glycemic load. Journal of the American College of Nutrition 28 (Suppl.), 500S–516S.

Xiao, X., Liu, Z., Wang, R., Wang, J., Zhang, S., Cai, X., et al., 2015. Genistein suppresses FLT4 and inhibits human colorectal cancer metastasis. Oncotarget 6, 3225–3239.

Xu, D.P., Xia, G., Hh, J., Ck, Z., 2002. Effect of Tempe Isoflavones on Tumor, 25. J Nanjing Agricultural University, pp. 97–101.

Xu, L., Du, B., Xu, B., 2015. A systematic, comparative study on the beneficial health components and antioxidant activities of commercially fermented soy products marketed in China. Food Chemistry 174, 202–213.

Yanaka, K., Takebayashi, J., Matsumoto, T., Ishimi, Y., 2012. Determination of 15 isoflavone isomers in soy foods and supplements by high-performance liquid chromatography. Journal of Agricultural and Food Chemistry 60, 4012–4016.

Yang, H., Jin, G., Ren, D., Luo, S., Zhou, T., 2011. Mechanism of isoflavone aglycone's effect on cognitive performance of senescence-accelerated mice. Brain and Cognition 76, 206–210.

Zaheer, K., Akhtar, M.H., 2015. An updated review of dietary isoflavones: nutrition, processing, bioavailability and impacts on human health. Critical Reviews in Food Science and Nutrition. http://dx.doi.org/10.1080/10408398.2014.989958.

Zhao, G., Yao, Y., Wang, C., Tian, F., Liu, X., Hou, L., et al., 2015. Transcriptome and proteome expression analysis of the metabolism of amino acids by the fungus *Aspergillus oryzae* in fermented soy sauce. Biomed Research International 2015, 456802.

Section 3.4

Vegetables and Fruits Fermented Products

Chapter 20

Kimchi and Its Health Benefits

K.-Y. Park, H.-Y. Kim, J.-K. Jeong
Pusan National University, Busan, Korea

20.1 INTRODUCTION

Kimchi is a traditional fermented Korean vegetable food. Baechu cabbage (Chinese cabbage) is the major raw material for kimchi production, and kimchi prepared using baechu cabbage as the main ingredient is referred to as "baechu kimchi." However, many other vegetables, including radish, cucumber, leek, mustard leaf, and green onion are also used to prepare various types of kimchi (Cheigh and Park, 1994). There are more than 167 types of kimchi, with variations based on the main ingredient, region, season, and preparation methods (Lee et al., 1992). Nevertheless, baechu kimchi is the most popular and commonly consumed (>70%) in Korea.

During preparation of baechu kimchi, the cabbage is brined and then mixed with other ingredients such as sliced radish, green onion, red pepper powder, crushed garlic, crushed ginger, and fermented seafood (jeotgal). The kimchi is then stored at around 5°C to undergo lactic-acid bacteria (LAB) fermentation. The taste and quality of kimchi is dependent on the ingredients, fermentation conditions, and LAB involved in the fermentation (Park and Cheigh, 2004; Kwon and Kim, 2007).

The ingredients used to prepare kimchi contain high levels of vitamins, minerals, dietary fibers, and other functional components. Kimchi has previously been reported to have anticancer, antioxidative, antiatherosclerotic, antidiabetic, and antiobesity effects (Park, 1995; Islam and Choi, 2009; Kim et al., 2007a,b) derived from bioactive compounds contained in those ingredients. Moreover, the bioactivities of many of these ingredients have been found to improve during the fermentation process. Kimchi usually contains about 10^8–10^9 CFU (colony forming units)/g of LAB after fermentation under optimal conditions. Overall, these properties make kimchi consumption a good way to include more vegetables and probiotics in the diet and improve health.

The health benefits of kimchi have only recently been reported, which has led to increased interest in kimchi internationally, and today it is known worldwide as a representative food of Korea. In this chapter, a brief history of kimchi as well as a discussion of the processes used for its manufacture and fermentation are given. Moreover, the health benefits of kimchi such as antioxidative, antiaging, antimutagenic, anticancer, and antiobesity properties, as well as its safety are considered.

20.2 HISTORY OF KIMCHI

Kimchi is a generic term used to denote a highly varied group of salted and fermented vegetable foods that have been consumed for 2000 years in Korea. Many records indicate that kimchi has been prepared and consumed since the third or fourth century AD in Korea, but radish was likely the major raw material (Lee, 1975; Chang, 1975). The first record of kimchi is in the *Koguryoion* from the *Weizdongyizhuan* region of *Sangouzhi* China in AD 289. According to this book the Koguryo people (referring to Koreans) were skilled in preparation of fermented foods such as wine, soybean paste, and salted and fermented fish and vegetables, indicating that fermented foods in Korea were widely enjoyed at that time. During the Silla dynasty (around AD 720), fermented vegetables were prepared using a stone pickle jar (published in the Samkuksaki, AD 1145). Buddhism accepted vegetarian diets while rejecting meat consumption during the early Koryo dynasty (AD 918–1392). The addition of ingredients and preparation methods gained more diversity with time, and records indicate that kimchi was eventually prepared using garlic and spices such as Chinese pepper, ginger, and tangerine peels (Park and Cheigh, 2004). The *Kapyakyong*, from the *Dongkukisangkukjip* (AD 1241), states that kimchi is prepared for winter during a traditional custom known as kimjang, the fall kimchi preparation. Kimchi prepared during this time is similar to the broth containing radish kimchi (nabakji and dongchimi) that is found today.

Kimchi was originally referred to as "*dimchae*," "*dimchi*," and then kimchi (Park and Rhee, 2005). The word "kimchi" originated from "*chimchae*," which is the Chinese character meaning salted vegetables. This word appears for the first time in Yi Saek's *Mogunjip* (AD 1626). Radishes were the primary ingredient in kimchi until the Koryo dynasty, although cucumber, eggplant, and green onions were also used during this period.

Many foreign vegetable species were introduced into Korean foods during the early Choson dynasty (AD 1392–1600), during which time kimchi ingredients became more diversified, and the preparation methods more elaborate. The Chosun *Wangjoshillok*, the archive of Chosun Dynasty (AD 1409), describes a *chimjanggo* for the storage of kimchi, which was a designated place for the kimchi prepared during kimjang. The word kimjang originates from *chimjang*, and refers to the storage of kimchi for winter. Red peppers are first described in the *Jibongyusol* (AD·1613), and their use in kimchi was first recorded in the *Sallimkyongje* (AD 1715). The incorporation of red pepper as one of the main subingredients of kimchi resulted in a more harmonious taste. The number of vegetables used as ingredients of kimchi also grew during this period. Many types of salted and fermented fish came into use during this time as well, and animal meats, fishes, and vegetables were soon incorporated. Baechu cabbage and white radishes became the main ingredients of kimchi during this period, and baechu kimchi similar to the present kimchi evolved in the 19th century (Park and Cheigh, 2004). In 1776, *Jungbosallimkyoungje* introduced 41 types of kimchi, including varieties containing meat and fish ingredients. Many types of kimchi were introduced in written form for the first time in history in this book. By 1827, 97 varieties of kimchi were described (*Imwonshibyukji*, 1827), and the use of red peppers for kimchi was emphasized.

20.3 MANUFACTURING KIMCHI

The preparation of baechu kimchi is summarized as follows (Fig. 20.1). The cabbage is trimmed, washed, brined overnight in 10% brine, and then rinsed. Next, the excess water is drained and the raw materials are graded, washed, and cut (Cheigh and Park, 1994).

A mixture of chopped and sliced subingredients that typically consists of garlic, ginger, radish, red pepper powder, salt-fermented fish (anchovy, etc.), and other vegetables is then prepared and stuffed between each leaf of the cabbage. A standard composition of baechu kimchi ingredients is as follows: brined baechu cabbage (100%) mixed with 13% sliced radish, 3.5% red pepper powder, 1.4% garlic, 0.6% ginger, 2.2% fermented anchovy sauce, 1.0% sugar, and 2% green onion, then adjusted to a final salt level of 2.5% (Cho, 1999).

However, various other materials including watercress, mustard, pear, apple, pine nuts, chestnut, gingko nuts, cereals, fish, and crabs can be added to the

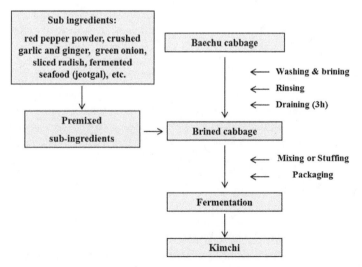

FIGURE 20.1 Process of manufacturing baechu kimchi.

premixture of subingredients depending on family tradition, economics, and seasonal and regional availability. Cabbage mixed with or stuffed with the premixture is then fermented at different temperatures, although 5°C is considered ideal for preparation of a good-tasting product. Recipes and fermentation conditions can be adjusted to improve various functionalities.

Our laboratory developed a kimchi recipe to improve anticancer properties by adding functional ingredients including mustard leaf (Kim et al., 2007b), Chinese pepper (Han et al., 2011), and Korean mistletoe extract (Kil, 2004). The kimchi containing Korean mistletoe extract had greater anticancer effects (80% inhibition ratio) than that without it (62% inhibition) (Kil, 2004). Additionally, the final salt concentration of anticancer functional kimchi (2.2%) was lower than that of standardized kimchi (2.5%), supplying a lower level of sodium. Pear, mushroom, and sea tangle juice were also included to enhance the taste. Other investigators have used different types of salt such as natural sea salt (Korean solar salt) without bittern (the saltwater that contains magnesium sulfate or magnesium chloride) or baked bamboo salt to increase the anticancer effects and taste of kimchi (Han et al., 2009).

20.4 FERMENTATION OF KIMCHI

Spontaneous kimchi fermentation without the sterilization of raw materials leads to the growth of various LAB, which usually results in variations in the quality of kimchi products (Park et al., 2010, 2012a; Jung et al., 2011). The production of organic acids from carbohydrates by LAB and the resulting reduction in pH maintain the vegetable freshness during storage. LAB produce various compounds in addition to organic acids, including CO_2, ethanol, mannitol, bacteriocins, γ-amino butyric acid, ornithine, conjugated linoleic acids, and oligosaccharides. These components contribute to the fermentation characteristics and health benefits of kimchi. The fermentation is markedly affected by environmental factors, such as temperature, salt concentration, and ingredients.

Kimchi fermentation is generally divided into four stages based on acidity; an initial stage (acidity <0.2%), immature stage (acidity 0.2–0.4%), optimum-ripening stage (acidity 0.4–0.9%), and overripening, or rancid stage (acidity >0.9%) (Chang and Chang, 2010). The LAB profile during kimchi fermentation fluctuates with pH and acidity, with *Leuconostoc mesenteroides* present during early fermentation (pH 5.64–4.27 and acidity 0.48–0.89%), and *Lactobacillus sakei* dominating later in the fermentation (pH 4.15 and acidity 0.98%). A distinct subset of LAB related to kimchi fermentation is greatly influenced by temperature, with *L. sakei* predominant in kimjang kimchi. This strain appears suitable for low fermentation temperatures (5–9°C) and storage temperatures (−2°C) (Lee et al., 2008).

Many previous studies have shown that LAB including members of the genera *Lactobacillus*, *Leuconostoc*, *Weissella*, *Lactococcus*, and *Pediococcus* are key players responsible for kimchi fermentation (Park and Kim, 2012; Lee

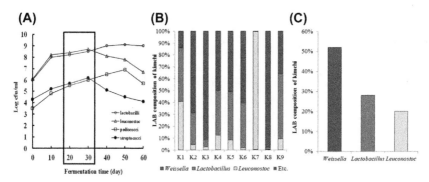

FIGURE 20.2 Microfloral change of LAB during kimchi fermentation at 5°C (A, Lee et al., 1992), LAB pyrosequences in nine Korean commercial kimchi at the genus level (B), and overall percentage of LAB genera in nine kimchi samples (C).

et al., 2008). According to a previous study (Lee et al., 1992), kimchi fermentation is generally dominated by *Leuconostoc* spp. and *Lactobacillus* spp. at 5°C, while other LAB such as *Pediococcus* spp. and *Streptococci* spp. are present at much lower levels (Fig. 20.2A). However, the strain and population dynamics appear to be markedly influenced by temperature and seasonal variations in raw materials.

The microorganisms in kimchi were first investigated in Korea in 1939, and many LAB have been isolated and subjected to taxonomic analysis (Lee et al., 1996, 1997). Due to the known limitations of culture methods, many recent studies have used culture-independent methods such as 16S rDNA-based PCR to determine the diversity and dynamics of the microbiota in kimchi and other fermented foods (Lee et al., 2005b).

Classical identification of bacterial isolates from kimchi has revealed that *Leuconostoc mesenteroides* and *Lactobacillus plantarum* were the predominant species (Lim et al., 1989). However, a variety of other species including *Leuconostoc citreum*, *Leuconostoc gasicomitatum*, *Lactobacillus sakei*, *Lactobacillus brevis*, *Weissella koreensis*, and *Weissella confusa* have been detected in kimchi samples using molecular approaches (Cho et al., 2006; Kim et al., 2000). Some recent studies using culture-independent methods confirmed these findings, and indicated that the majority of bacteria in kimchi are culturable (Kim and Chun, 2005; Lee et al., 2005b). Currently, detection, differentiation, and identification of bacteria including LAB is accomplished using numerous methods including phenotypic, biochemical, and immunological assays and molecular techniques (Settanni and Corsetti, 2007).

Results from our laboratory on microbial communities of nine representative types of commercial Korean kimchi at pH 4.2–4.4 using the pyrosequencing method showed that various LAB differed greatly among kimchi varieties (Fig. 20.2B). At the genus level, *Weissella* comprise 52% of the total population, followed by *Lactobacillus* (28%) and *Leuconostoc* (20%) (Fig. 20.2C). Previous studies have also shown that *Weissella* (especially *Weissella koreensis*) was the dominant species in kimchi. However, most investigations of the bacterial composition of kimchi published before 2000 did not mention the genus *Weissella*. This was likely because it was considered as *Leuconostoc* spp. before being reclassified by Collins et al. (1993). Many molecular methods (eg, pyrosequencing) have recently been widely applied for bacterial analysis, including LAB of kimchi.

To date, no rational approach to controlling the microbial community during kimchi fermentation has been established, which makes it difficult to produce high-quality and commercial kimchi products. However, LAB starters have been developed to solve this problem. Additionally, representative kimchi LAB have been isolated and tested for fermentation characteristics and probiotic activities to improve the quality and bioactivity of kimchi. *Leuconostoc citreum* (Chang and Chang, 2010), *Leuconostoc mesenteroides* (Jung et al., 2012), and mixed cultures of *L. mesenteroides* and *Lactobacillus plantarum* (Bong et al., 2013) that showed probiotic activities were employed as starters to ferment kimchi. Kimchi fermented with mixed starter cultures showed better overall acceptability based on taste preference and higher radical scavenging and anticancer effects than naturally fermented kimchi.

20.5 HEALTH BENEFITS OF KIMCHI

Baechu kimchi is a low-calorie food (18 kcal/100 g) that contains high levels of vitamins (vitamin C, β-carotene, vitamin B complex, etc.), minerals (Na, Ca, K, Fe, P, etc.), dietary fiber (24% on a dry basis), and other functional components including capsaicin, allyl compounds, gingerol, and chlorophyll. Kimchi also contains active compounds that have demonstrated anticancer, antiobesity, and antiatherosclerotic properties, such as benzyl isothiocyanate, indole compounds, thiocyanate, and β-sitosterol (Park and Rhee, 2005; Park and Kim, 2012). Kimchi is mainly prepared with yellow-green vegetables, which have been shown to prevent cancer, increase immune function, retard the aging process, and prevent constipation (Park, 1995). When kimchi is fermented, its taste and functionality are enhanced and it becomes a good probiotic food (Park and Kim, 2012).

20.5.1 Antioxidative and Antiaging Effects

Chlorophyll, phenol compounds, vitamin C, carotenoids, dietary fibers, and other phytochemicals from the raw materials or formed during the fermentation process, as well as LAB, exert antioxidative and antiaging activities. Antioxidative compounds in kimchi may scavenge free radicals formed in the body by acting as hydrogen donors (Hwang and Song, 2000). Additionally, a dichloromethane fraction of kimchi was found to inhibit Cu^{2+}-induced LDL oxidation, showing the greatest antioxidant effect against LDL oxidation by inhibiting thiobarbituric acid (TBARS) production by two-fold compared to the control (Hwang and Song, 2000).

Evaluation of the effects of kimchi intake on antiaging characteristics, free-radical production, and antioxidative enzyme activities in brains of senescence-accelerated mice (SAM) has shown that kimchi moderated the increase in free radical production due to aging (Kim et al., 2002). Among the kimchi-fed groups, those fed baechu kimchi containing both 30% mustard leaf and mustard leaf kimchi-fed groups showed greater inhibition of free-radical production in the brain than those that received standard baechu kimchi.

In a human skin cell study, keratinocytes (A431 human epidermoid carcinoma) and fibroblasts (CCD-986SK human normal, control) were cultured under oxidative stress induced by a superoxide anion generator, paraquat, and hydrogen peroxide in the absence and presence of kimchi extract. Optimally ripened kimchi extract led to a remarkable decrease in H_2O_2-induced cytotoxicity in keratinocyte cells (Ryu et al., 1997). Paraquat at concentrations greater than 1 mM exerted strong cell toxicity on keratinocyte cells, but the extracts of fermented kimchi for 1, 2, and 3 weeks at 8°C showed protective effects. Fibroblast cell viability was significantly affected by H_2O_2, but 2 week-fermented kimchi extract attenuated H_2O_2-induced cytotoxic effects (Ryu et al., 1997).

The antiaging activity of kimchi has also been evaluated using a stress-induced premature senescence (SIPS) model (Kim et al., 2011a). WI-38 human fibroblasts challenged with H_2O_2 showed decreased cell viability, increased lipid peroxidation, and reduced cell lifespan, indicating the induction of SIPS. However, treatment with kimchi attenuated cellular oxidative stress, as indicated by increased cell viability and inhibition of lipid peroxidation. The lifespan of WI-38 cells was also extended, suggesting that kimchi has the potential for use as an antiaging agent.

In an in vivo study, 10% freeze-dried baechu kimchi, leek kimchi, or mustard leaf kimchi was added to the AIN-76 diet and fed to hairless mice for 20 weeks. Morphological changes in the epidermis and dermis were observed at week 16, and the antioxidant effects against UV-induced photo aging and the free-radical scavenging effect were investigated at week 20 (Ryu, 2000). The epidermal thickness of hairless mice was found to be greater (22–37%) in the kimchi-fed groups, while the stratum corneum was thinner (62–58%) relative to the control group, indicating that kimchi promoted healthy skin. Collagen synthesis at the dermis increased with kimchi treatment as the activity of the rough endoplasmic reticulum of the fibroblast was greater than that of the control group (Ryu, 2000). Type IV collagen, which supports matrices at the basement membrane, was present in greater amounts in the kimchi-fed groups than the control group, especially the mustard leaf kimchi-fed group (52% higher vs. control). Lipid oxidation (based on the TBARS content) of the livers of the hairless mice fed a kimchi diet was retarded compared to the control, and the contents of superoxide anion, hydroxyl radical, and hydrogen peroxide were lower than in the control group, indicating greater free-radical scavenging capacity. The free-radical scavenging activity of kimchi was attributed to high levels of chlorophyll, vitamin C, carotenoids, and phenolics in kimchi.

The protective effects of kimchi against Aβ25-35-induced memory impairment were also investigated in an in vivo Alzheimer's mouse model (Choi et al., 2014). Aggregated Aβ25-35 was injected into mice brains (5 nmol/mouse), after which optimally ripened kimchi was orally administered for 2 weeks at 100 and 200 mg/kg of body weight. Subjection of mice to the Morris water maze test revealed that kimchi protected against cognitive impairment induced by Aβ25-35, suggesting that kimchi improves Aβ25-35-induced memory deficit and cognitive impairment.

20.5.2 Antimutagenic and Anticancer Effects

Kimchi has been shown to exert antimutagenic effects against aflatoxin B_1 and MNNG (N-methyl-Nitro-N-nitrosoguanidine) in the Ames test and the SOS chromotest in vitro (Park and Rhee, 2005; Park et al., 1995), with optimally ripened kimchi showing higher antimutagenicity than fresh and overripened kimchi (Park et al., 1995). Kimchi extract also exerted antimutagenic activity in the *Drosophila* wing test in vivo (Hwang et al., 1997), while kimchi exhibited anticlastogenic activity in mitomycin C-induced mice based on an in vivo supravital staining micronucleus assay (Ryu and Park, 2001).

Mouse embryonic C3H/10T1/2 cells form foci in culture media when exposed to carcinogens. In a previous study, foci that developed as type II and type III correlated well with tumor formations in 50% and 85% of C3H mice, respectively (Raznikoff et al., 1973). When 200 μg of methanol soluble fraction (MSF) from kimchi fermented for 3 weeks at 5°C were added to the cells with 3-methylcholanthrene, the number of type II and type III foci decreased significantly ($P < .5$) (Choi et al., 1997).

Several studies have proven the antiproliferative and cytotoxic effects of kimchi in a wide range of cancer cell lines. Sulforhodamine B assay, MTT (3-(4,5-dimethylthiazol-2-yl)-2,5-diphenyl tetrazolium bromide) assay, and a growth inhibition test indicated that kimchi extracts inhibited the survival or growth of human AGS gastric cancer, HT-29 colon cancer, MG-63 osteocaricinoma, HL-60 leukemia, and Hep 3B liver cancer cells (Cho, 1999; Park and Rhee, 2005). In addition, the dichloromethane fraction of kimchi inhibited ^{3}H thymidine incorporation into cancer cells (Hur et al., 1999).

The main active compounds that showed anticancer activity in kimchi were β-sitosterol and a linoleic-acid derivative. The kimchi dichloromethane fraction arrested tumor cells at the G2/M phase of the cell cycle and induced apoptosis of HL-60 human leukemia cells (Cho, 1999). Treatment of HIRc-B cells with the dichloromethane fraction of kimchi as well as the active kimchi compound (β-sitosterol) followed by microinjection of oncogenic H-ras^{v12} (Park et al., 2003) led to decreased DNA synthesis in the cells, indicating that active compounds of kimchi affected the signal transduction pathway via ras oncogene signaling to the nucleus (Park et al., 2003).

Our research group used sarcoma 180 cells transplanted into Balb/c mice that were subsequently treated with kimchi extracts. The MSF of the kimchi-treated group had smaller tumor weight (1.98 ± 1.9 g) than the control group (4.32 ± 1.5 g). The MSF from kimchi fermented for 3 weeks at 5°C also showed lower malondialdehyde formation than the control (Hur et al., 2000), as well as decreased hepatic cytosolic xanthine oxidase activity in sarcoma-180 treated Balb/c mice. However, MSF increased the hepatic cytosolic glutathione content and activities of glutathione S-transferase and glutathione reductase, indicating that kimchi might be involved in detoxification of xenotoxic materials in the liver. Kimchi extracts also enhanced natural killer cell and macrophage immune functions (Park and Rhee, 2005).

The concentration and type of salt used in kimchi preparation also influences its cancer-preventative effects. High levels of salt (8.5%) in kimchi have been shown to induce comutagenic activity (Park et al., 1990); however, kimchi containing 2.5% or less salt exhibited antimutagenic and anticancer effects. Moreover, baked salt has been shown to be very effective for preparation of cancer-preventive kimchi (Ha and Park, 1999). Experimental metastasis with colon 26-M3.1 cells revealed that subcutaneous administration of kimchi extract (0.05 to approximately 1.25 mg/mouse) 1 day after tumor-cell inoculation led to significant inhibition of lung metastasis. Moreover, functional kimchi significantly inhibited metastasis of colon 26-M3.1 cells in the lungs of the mice (Park and Rhee, 2005).

The anticancer effects of kimchi are primarily due to regulation of apoptosis of cancer cells and inhibition of inflammation. Optimally ripened kimchi led to significant induction of apoptosis in HT-29 human colon cancer cells and upregulated the expression of Bax protein and mRNA; however, it led to downregulation of the Bcl-2 protein and mRNA (Kim, 2009). An animal study using azoxymethane (AOM)- and dextran sulfate sodium (DSS)-induced colitis-associated colon cancer in mice also confirmed the antiinflammatory and anticancer properties of kimchi (Kim et al., 2014). Increased colon length and decreased tumor numbers in colon were observed in kimchi-treated groups, especially in response to anticancer kimchi prepared using organically cultivated baechu cabbage with mustard leaf and Chinese pepper. In that study, kimchi also regulated mRNA and protein expression of inflammation-related genes such as iNOS and COX-2 (Fig. 20.3).

Our group recently demonstrated that fecal pH, β-glucosidase, and β-glucuronidase activities decreased significantly in response to consumption of 210 g of kimchi for 28 days in a human study (Fig. 20.4). β-Glucosidase can form toxic aglycones from plant glycosides, while β-glucuronidase hydrolyzes glucuronic acid conjugates and increases the enterohepatic circulation of toxic compounds. Low pH inhibits the growth of many pathogens and reduces the intestinal absorption of potentially toxic compounds (Martineau and Laflamme, 2002). β-glucuronidase, nitroreductase, and azoreductase are the main human intestinal bacterial enzymes responsible for stimulation of the conversion of precarcinogens to carcinogens (Goldin and Gorbach, 1976). β-Glucuronidase and nitroreductase activities in the colons of Koreans and Germans eating kimchi and sauerkraut decreased significantly, as did fecal pH (Oh et al., 1993). Kimchi LAB can colonize the intestinal tract for long periods of time and generate short-chain

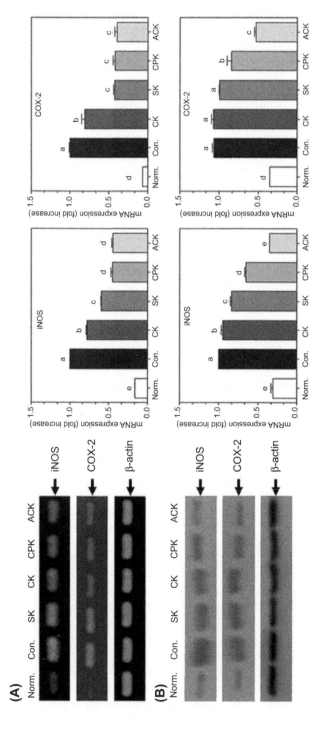

FIGURE 20.3 Effects of extracts from different types of kimchi (1.89 g/kg body weight) on the mRNA (A) and protein (B) levels of iNOS and COX-2 in colon tissue from BALB/c mice with AOM- and DSS-induced colitis associated colon cancer (Kim et al., 2014). *ACK*, Anticancer kimchi; *CK*, Commercial kimchi; *SK*, Standardized kimchi. Band intensities were measured with a densitometer and expressed as fold of the control. Fold increase = gene expression/β-actin × control value (control fold increase = 1). [a–e]Mean values with different letters on the bars are significantly different ($P < .05$) according to Duncan's multiple-range test.

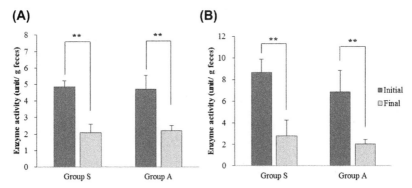

FIGURE 20.4 A clinical study for the effect of kimchi intake on colon health of Korean young adults. (A) Fecal β-glucosidase activity; (B) Fecal β-glucuronidase activity. Twenty-eight participants (age range 18–36) were assigned – group S: Standardized kimchi intake ($n=14$); group A: Anticancer kimchi (kimchi added *L. plantarum*, 3×bamboo salt, mistletoe extract, Chinese pepper, etc.) intake ($n=14$). All volunteers consumed kimchi 210 g/day for 28 days, and with usual meal, but except for antibiotics and probiotics. Food intake questionnaires were submitted once a week, and serum and stool were taken from volunteers on the first and last (28th) days. All data are means ±SD. ** Significantly different from initial value ($P<.01$) by paired *t*-test.

fatty acids from the dietary fiber of vegetables, which can induce apoptosis of colon cancer cells (Bengmark, 2001), and may help prevent colon cancer.

20.5.3 Antiobesity Effects

Kimchi has been found to help maintain normal body weight in rats (Choi, 2001; Choi et al., 2002; Yoon et al., 2004). Capsaicin in red pepper can cause body fat loss via stimulation of spinal nerves, which activates the release of catecholamines in the adrenal glands, thereby increasing metabolism and the expenditure of energy (Kim, 1998). Choi (2001) reported that the final body weight of rats decreased significantly when they were fed a diet containing red pepper powder with high-fat content relative to those fed only a high-fat diet. Although red pepper powder in kimchi may be responsible for its dietary effects, kimchi has been shown to exert a better lowering effect of body weight than red pepper powder alone. In a previous study, a group fed kimchi plus high-fat diet showed significantly lower body weight and perirenal fat pad weights than the red pepper powder plus high-fat diet group and the high-fat diet group (Choi, 2001). These effects might be due to red pepper powder, garlic, dietary fibers, and other ingredients. Fermented kimchi stimulates B-cell proliferation and lowers lipid accumulation in the epididymal fat pad in rats, and kimchi fermented for 6 weeks at 4°C (optimally fermented kimchi) was especially effective at reducing adipose cell numbers (Kim and Lee, 1997).

Kimchi inoculated with *Leuconostoc mesenteroides* starter cultures (LmK), when added to a high-fat diet at a concentration of 10%, showed better antiobesity effects than noninoculated kimchi in a high-fat diet-induced obesity mice model in our recent study (Cui et al., 2015). The body and adipose tissue weight decreased significantly in the kimchi-treated group relative to the HFD-treated group, with LmK treatment leading to significant reductions in the serum levels of triacylglycerol (TG), total cholesterol (TC), low-density lipoprotein-cholesterol (LDL-c), insulin and leptin, as well as marked increases in the serum levels of high-density lipoprotein-cholesterol (HDL-c) and adiponectin. Moreover, the mRNA levels of CCAAT/enhancer binding-protein alpha (C/EBP-α), peroxisome proliferator-activated receptor gamma (PPAR-γ), sterol regulatory element-binding proteins 1c (SREBP-1c), and fatty-acid synthase (FAS), which accelerate lipid formation in the liver, were decreased in liver tissues of mice fed kimchi, especially in the LmK-treated group (Fig. 20.5).

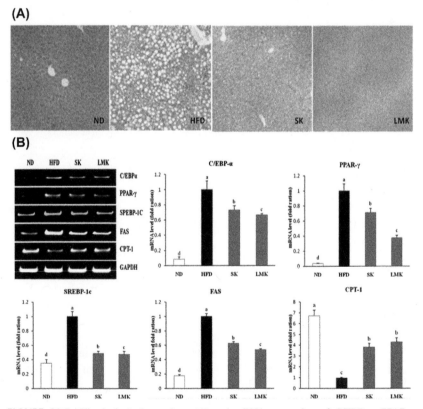

FIGURE 20.5 Histological observations (A) and mRNA expression of C/EBP-α, PPAR-γ, SREBP-1c, FAS, and CPT-1 (B) in liver tissue of DIO (diet-induced obesity) mice. *HFD*, Group received high-fat diet; *LmK*, group received high-fat diet containing 10% *Leuconostoc mesenteroides*–inoculated kimchi; *ND*, group received normal diet; *SK*, group received high-fat diet containing 10% standard kimchi. The intensity of the bands was measured with a densitometer and expressed as fold rate of control. Fold ratio: Gene expression/GAPDH × control numerical value (Control fold ratio: 1). a–dMeans with the different letters are significantly different ($P < .05$) by Duncan's multiple range test.

Conversely, mRNA expression of carnitine palmitoyltransferase 1 (CPT-1) was increased in kimchi-treated groups. Supplementation with kimchi also decreased the mRNA levels of monocyte chemoattractant protein-1 (MCP-1) and interleukin 6 (IL-6), resulting in reduced macrophage infiltration in epididymal fat tissue.

The antiobesity activity of kimchi has also been shown in humans. A group of obese women that supplemented their diets with kimchi capsules (3 g of freeze-dried kimchi/day) and exercised (1 h, once a week) exhibited a remarkable decrease in body weight and body mass index (BMI) compared to a group that did not receive the kimchi capsules. Moreover, the visceral fat area and serum triglycerides were significantly lower in the kimchi-supplemented group than the no-kimchi supplement group (Ahn, 2007). Kim et al. (2011b) reported that kimchi consumption leads to a reduction in body weight, BMI, and percentage of body fat in overweight and obese subjects, which may reduce the risk of cardiovascular disease and metabolic syndrome. Fermented kimchi intake also decreased waist-hip ratio (WHR), fasting blood glucose concentrations, fasting insulin, total cholesterol, MCP-1, and leptin levels. Furthermore, fermented kimchi led to greater decreases in percentage of body fat, systolic blood pressure, diastolic blood pressure, fasting glucose, and total cholesterol relative to the fresh kimchi group.

The effects of aerobic exercise and/or kimchi supplementation on changes in body composition and plasma lipids of obese middle-school girls were investigated (Baek et al., 2001). In this study, an exercise group (EG, eight obese girls) jogged and jumped rope for 60 min 4 times a week, while the kimchi group (KG, 12 obese girls) took 3 g of freeze-dried kimchi packed in 500 mg capsules (30 g of kimchi) daily. KG+EG led to a greater reduction in body fat than EG or KG. BMI, fat mass, abdominal fat, and TG concentration decreased, while HDL cholesterol increased in serum. Moreover, KG led to lower plasma cholesterol and LDL-cholesterol levels than EG. Overall, these findings indicate that kimchi supplementation in conjunction with exercise might decrease obesity and improve plasma lipids.

20.5.4 Other Health Benefits

Animal studies have investigated the hypolipidemic effects of dietary supplementation of kimchi. Plasma cholesterol and TG levels were found to be lower in rats fed a kimchi diet (supplemented at the concentration of 3%, 5%, or 10%) relative to a control (Kwon et al., 1997). Additionally, kimchi intake led to reduced very low-density lipoprotein (VLDL)-cholesterol levels, but significantly increased HDL-cholesterol levels in serum. Specifically, the concentration of LDL-cholesterol in subjects that received 10% kimchi diets was lower, as were the levels of hepatic cholesterol, TG, total lipids, and apolipoprotein B. However, fecal total fat, cholesterol, TG, and apolipoprotein A-1 levels were significantly increased (Kwon et al., 1997). Thus supplementation of kimchi may ameliorate hyperlipidemia.

Kim and Lee (1997) investigated the effects of kimchi on lipid metabolism and immune function in experiments using 63 male Sprague–Dawley (SD) rats fed six types of baechu kimchi for four weeks. Three types of freeze-dried baechu kimchi prepared using different fermentation periods (not fermented or fermented for 3 or 6 weeks at 4°C) were added at 5% and 10% to a diet containing 15% lard. The serum total lipids and TG levels were lower in all kimchi groups than in the control group. Fermented kimchi had greater suppressive effects on the total lipids, cholesterol, and TG levels of the epididymal fat pad than kimchi without fermentation. These findings indicated that kimchi fermentation increases the functional components of the raw materials. Moreover, the TG concentration of feces from the kimchi groups was higher than that of the control group, indicating that kimchi stimulates lipid mobilization from the epididymal fat pad as well as lipid excretion via feces. Accordingly, high levels of dietary fiber in kimchi may also contribute to the capture and excretion of cholesterol and TG via the feces.

Interestingly, LAB may be involved in the observed decreases in serum cholesterol after kimchi consumption. Indeed, Ray and Bhunia (2008) indicated that LAB metabolize dietary cholesterol and deconjugate bile salts in the colon, while also preventing their reabsorption in the liver, decreasing cholesterol levels in the serum.

Kwon et al. (1999) evaluated the hypolipidemic effects of daily kimchi consumption in 102 healthy Korean adult men aged 40–64 years that visited a hospital for physical examination. Specifically, they evaluated physical and biochemical parameters of blood, food record, preference for taste, and lifestyle habits. Dietary fiber and calcium intake were found to increase with increased kimchi intake. Moreover, kimchi consumption (0 to approximately 453 g) was positively correlated with HDL cholesterol and negatively correlated with LDL cholesterol. Additionally, a preference for spicy foods was negatively correlated with systolic blood pressure. Furthermore, Lee et al. (2012b) reported that kimchi consumption was positively correlated with HDL cholesterol and negatively correlated with body fat and BMI, suggesting that kimchi consumption is beneficial for elevating plasma HDL cholesterol and lowering body fat, plasma TG, and LDL cholesterol.

Kimchi has been shown to have blood glucose-lowering and antidiabetic effects (Islam and Choi, 2009; Lee et al., 2012b). Specifically, administration of freeze-dried baechu kimchi to high-fat (HF) diet-fed, streptozotocin (STZ)-induced type 2 diabetic rats resulted in increased serum insulin concentrations and pancreatic beta-cell function after 4 weeks, as well as decreased blood glycated hemoglobin levels. Moreover, lower fasting blood glucose and better glucose tolerance were observed in the kimchi-fed group. These findings suggest that dietary baechu kimchi has antidiabetic effects, even when administered in conjunction with a high-fat diet, although better effects are possible if it is consumed with normal or low-fat diets.

20.6 SAFETY OF KIMCHI

Kimchi is a salted and fermented vegetable product, and vegetables contain high levels of nitrate (NO_3). It has been suggested that the salt content in kimchi, as well as the NO_3, nitrite (NO_2^-), and nitrosamine levels are negative factors of kimchi, despite its various health beneficial effects. However, salt is essential to kimchi fermentation, acting as both a preservative and supporting LAB fermentation. In the past, the salt content of kimchi ranged from 2.5% to 5.0% based on region and season; however, the salt content of kimchi has considerably decreased and commercial kimchi today contains lower salt levels of 1.2–1.9%.

Salt (NaCl) may cause stomach cancer and elevated blood pressure in humans (D'Elia et al., 2012; O'Donnell and Mente, 2014). Although salt does not exhibit mutagenic or carcinogenic effects, it may exert comutagenic and cocarcinogenic effects when mutagens or carcinogens are also present (Joossens et al., 1996; Yoon and Chang, 2011; Choi and Park, 2002). Indeed, mutagenicities increased by about three-fold when 15% NaCl was added to MNNG compared to MNNG alone (Kim et al., 1995). However, typical kimchi contains around 2–3% salt, which induces antimutagenicity against MNNG. Moreover, vitamin C, which is present in high amounts in kimchi, decreased the comutagenicity of NaCl in MNNG-induced mutagenicity.

Korean solar salt used in kimchi contains various minerals that are not found in purified salt, including K, Ca, Fe, and Mg. Kimchi prepared using aged solar salt (1–4 years), especially without bittern showed increased organoleptic qualities such as hardness and taste, increased storage stability, and improved antioxidant activity and growth inhibitory effects against human cancer cells compared to kimchi made with purified salt and bittern nonaged solar salt. Baked salt or bamboo salt has also been shown to significantly improve kimchi quality and anticancer efficacy. Overall, these findings indicate that salt type is an important factor affecting the quality and health-promoting properties of kimchi (Choi and Park, 2002; Kim et al., 1995; Han et al., 2009; Chang et al., 2014; Yoon and Chang, 2011). Moreover, optimally fermented kimchi showed much higher inhibition of cancer-cell growth and lower comutagenicity than unfermented kimchi, indicating that fermentation can mitigate the negative effects of salt in kimchi.

The consumption of 60–100 g of kimchi/day containing 2.0% salt results only in a salt intake of 1.2–2.0 g of NaCl/day. Overall, this is a safe salt content, even if kimchi intake increases. Moreover, if a bamboo salt-KCl system is used during kimchi fermentation (Choi and Park, 2002), kimchi could be a good anticarcinogenic food without the risks associated with NaCl.

Park (1995) reported that kimchi contained 35–139 ppm NO_3. However, kimchi NO_3 levels decreased significantly by 70–80% during fermentation. NO_2^- levels in kimchi are very low, with only 1.6 ppm NO_2^- being detected after 5 weeks of fermentation at 5°C and 0.6 ppm after 6 weeks (Park and Cheigh, 1992).

Only nitrosamines such as N-nitrosodimethylamine (NDMA) were detected, while N-nitrosopyrolidine (NPYR) and N-nitrosodimethylamine (NDEA) were not detected in the fermented kimchi (Kim et al., 1984). In our study, nitrosamines were not detected with the exception of NDMA detected in kimchi fermented for 6 weeks at 5°C at a concentration of 0.044 ppb (Park and Cheigh, 1992). The presence of vitamin C (Lee and Woo, 1989), phenolic compounds, capsaicin, and their metabolites (Park et al., 1998) in kimchi can prevent formation of nitrosamines from NO_2^- and secondary amines. LAB, especially *Leuconostoc mesenteroides*, was found to deplete almost all NO_2^- during fermentation (Oh et al., 1997, 2004), as well as to significantly decrease the mutagenicity of nitrosamines. The levels of NO_2^- are too low to form nitrosamines during fermentation; therefore, kimchi contains little or no nitrosamines, although it does contain protective nutrients such as vitamin C, phenolic compounds, capsaicin, and other antioxidant metabolites, as well as LAB,, demonstrating that kimchi is safe from NO_3, NO_2^-, and nitrosamines contamination (Park, 1995).

Biogenic amines are nitrogenous, low molecular weight organic bases of aliphatic, aromatic, or heterocyclic structures that are primarily produced by microbial decarboxylation of amino acids in foods. These compounds are frequently found in fermented foods (Shruti et al., 2010; Kalac and Krausova, 2005). The ordinary consumption of these amines does not induce adverse reactions; however, they can pose a risk to humans if the amine-metabolizing capacity is saturated (Joosten and Nunez, 1996). Previous studies reported that the biogenic amine levels in kimchi were very low (0–150 mg/kg) and acceptable for human health (Lee et al., 2012a; Mah et al., 2004). Because kimchi is a fermented vegetable and biogenic amines are formed from amino acids, almost all of the biogenic amines in kimchi are from the subingredient jeotgal (fermented fish). Generally, the amount of jeotgal in kimchi is 0–3%; thus this fermented product is safe from biogenic amines.

20.7 HEALTH BENEFITS OF KIMCHI LAB

20.7.1 Antioxidative and Anticancer Effects

Antioxidative and anticancer effects are important probiotic functionalities of microorganisms in human health. The antioxidative activity of *Lactobacillus plantarum* KCTC 3099 from kimchi was evaluated by measuring its resistance to active oxygen species. Intact cell and cell-free extracts of *L. plantarum* showed potent antioxidative activity against lipid oxidation. Additionally, *L. plantarum* KCTC 3099 could survive after 8 h in the presence of 1 mM H_2O_2 and 0.4 mM hydroxyl radicals, as well as in the presence of superoxide anion generated using paraquat (Lee et al., 2005a).

Lactobacillus acidophilus KFRI342 from kimchi showed protective effects against cell damage via a comet assay (Chang et al., 2010). The intensity of tail DNA fluorescence of HT-29 cells subjected to 30 min of treatment without H_2O_2 differed significantly from that of cells subjected to treatment with H_2O_2.

Moreover, the cytoplasmic fraction of *L. acidophilus* KFRI342 showed strong antiproliferative effects in tumor cells. Although cytoplasmic fraction of LAB inhibited the growth of normal cells by less than 10%, it decreased SNY-C4 human colon cancer-cell growth by 38%. *L. acidophilus* KFRI342 doubled quinone reductase activity, while increasing immunostimulating activities and NO and IL-1α production in SNY-C4 human colon cancer cells (Chang et al., 2010).

Tumor formation in *Lactobacillus plantarum*- and *L. mesenteroides*-treated mice implanted with sarcoma-180 cells decreased by 57% and 39% versus control, respectively, in response to heat-treated and lyophilized LAB intraperitoneal administration (10^6 cells/mouse). Moreover, administration of *L. plantarum* and *Leuconostoc mesenteroides* to C57BL/6 mice with Lewis lung carcinoma upregulated tumor inhibition rates by 42% and 44%, respectively (Kim et al., 1991). In another study, the intake of kimchi ingredients, as well as a mixture of microorganisms from kimchi or pure culture of kimchi *L. plantarum* (cell lysate), decreased ascites tumor formation and increased the expected lifespan in mice (Shin et al., 1998). Taken together, these results indicate that feeding or intraperitoneal administration of LAB suppresses tumors implanted in rodents. This effect may occur due to stimulation of macrophages by muramyl dipeptide in the cell fractions, which would induce the release of superoxide anion and hydrogen peroxide, killing tumor and bacterial cells (Chang and Chang, 2010).

20.7.2 Immune-Boosting Effects

The cell-wall fraction of *Lactobacillus plantarum* from kimchi activated the phagocytic activity of macrophages in mice. Polysaccharides bound with phosphodiester bonds to the muramic acid in the cell wall, resulting in immunogenic activity (Chung, 1993). *L. plantarum* PS-21 from kimchi stimulated the proliferation of Peyers's patch cells, while cell-wall fractions exhibited strong mitogenic activity relative to the soluble cytoplasmic fraction. The peptidoglycan fraction was found to be an active mitogenic component when used in murine lymph nodes and spleen cell-test systems, as well as to enhance antibody production in lymph-node cells and production of TNF-α and IL-6 in RAW264.7 macrophage cells (Lee et al., 2006).

Cell lysates of *Lactobacillus plantarum* from kimchi fed to Balb/c mice exerted immunostimulatory effects (Chae et al., 1998). Specifically, they increased splenocyte and Peyer's patch-cell proliferation, as well as the production of intestinal secretory IgA, TNF-α, and IL-2 in blood. These results suggest that oral administration of the cell lysate enhanced enteric and systemic immune responses (Chae et al., 1998).

Orally administered *L. plantarum* HY7712 increased T-cell proliferation and the expression of IL-2 and IFN-γ in splenic cytotoxic T cells of cyclophosphamide (CP)-immunosuppressed mice (Jang et al., 2013). Moreover, this treatment restored CP-impaired phagocytosis in macrophages. As a result, *L. plantarum* HY7712 may accelerate recovery of CP-caused immunosuppression without adverse side effects (Jang et al., 2013).

20.7.3 Antiobesity Effects

In a study of SD rats fed a high-fat diet, kimchi LAB powder (KL) induced body weight- and lipid-lowering effects (Kwon et al., 2004). Specifically, the body weights of groups supplemented with 10% and 20% KL decreased by 13% and 15%, respectively, compared to control group. Visceral fats also decreased significantly by 42% and 48%, respectively versus control group. Moreover, plasma triglycerides, cholesterol, and LDL-c levels were reduced significantly, while triglycerides and cholesterol were excreted in the feces of the KL group (Kwon et al., 2004).

Administration of *L. acidophillus* supernatants (LS) into the brains of rats induced weight loss without a decrease in food consumption, but it also led to simultaneous increases in leptin expression in neurons and peripheral adipose tissue. Moreover, LS modulated the expression or release of other peptides involved in metabolism and body weight control (Sousa et al., 2008). Kimchi LAB plus kimchi raw materials and fermented products have also shown antiobesity effects, with optimally ripened kimchi leading to significantly decreased body fat and body weight in humans (Ahn, 2007).

Weissella koreensis OK1-6 isolated from kimchi showed antiobesity activity in differentiated 3T3-L1 adipocytes (Moon et al., 2012). Both cytoplasmic fraction of *W. koreensis* OK1-6 cells and its spent media led to significant decreases in the triglyceride concentration and intracellular lipid accumulation in treated 3T3-L1 adipocytes relative to the control. The mRNA expression levels of C/EBP-α, which is a major transcriptional factor involved in adipocyte differentiation, were significantly lower in 3T3-L1 adipocytes treated with the spent medium and cytoplasmic fraction of *W. koreensis* OK1-6, as were aP2, FAS, and SREBP1 gene expression. In addition, the in vivo antiobesity activity of kimchi improved when the strain was used as a starter to prepare kimchi (Park et al., 2012b). Taken together, these results will facilitate efforts by the nutraceutical and food industries to develop probiotic-based therapies for the treatment and prevention of obesity.

20.7.4 Other Effects

L. plantarum KCTC3928 from kimchi exerted hypocholesterolemic effects in C57BL/6 mice. Specifically, feeding double-coated *L. plantarum* with a high-fat diet to mice resulted in 42% and 32% decreases in plasma LDL cholesterol and triglyceride levels, respectively, while fecal bile-acid excretion increased by 45% (Jeun et al., 2010). Additionally, live *L. plantarum* exerted hypocholesterolemic effects in mice due to induction of fecal bile-acid secretion followed by increased degradation of hepatic cholesterol into bile acids (Jeun et al., 2010).

Numerous reports have shown that ingestion of specific strains of kimchi LAB reduces allergic skin symptoms and elevates serum IgE levels in mice (Segawa et al., 2008; Sunada et al., 2008; Watanebe et al., 2009). Moreover, their ability to inhibit allergic reactions may involve the enhancement of the

Th1/Th2 balance (Tobita et al., 2009). Won et al. (2011) also reported that *Lactobacillus plantarum* CJLP55, CJLP56, CJLP133, and CJLP136 strains isolated from kimchi had the capacity to inhibit atopic dermatitis (AD). *L. plantarum* CJLP55, CJLP133, and CJLP136 suppressed AD-like skin lesions, high serum IgE levels, and epidermal thickening in an NC/Nga mouse model (Fig. 20.6). Additionally, these strains diminished the accumulation of eosinophils and mast cells in topical inflammatory sites and the enlargement of axillary lymph nodes, which are responsible for dorsal dermatitis. These findings suggest that LAB isolated from kimchi inhibit AD, probably by altering the balance of the Th1/Th2 ratio or inducing IL-10 production.

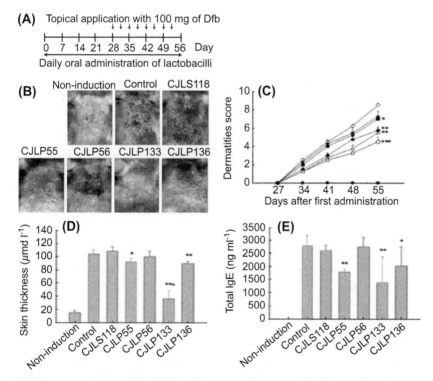

FIGURE 20.6 Experimental procedure and effects of the oral administration of *Lactobacillus* strains on dermatitis lesions, skin thickness, and serum IgE (Won et al., 2011). (A) Day 0 is defined as the first administration of *Lactobacillus plantarum* strains including CJLP55, CJLP56, CJLP133, or CJLP136 (all of these strain name stand for *L. plantarum* isolated from kimchi from CJ CheilJedang Co. in South Korea). (B) Aspects of Dfb-induced dermatitis in NC/Nga mice were taken on day 55. (C) Dermatitis scores were evaluated from day 27 to day 55 once a week. (D) The dorsal skins were excised, fixed with 10% formalin, embedded in paraffin, and stained with hematoxylin and eosin. (E) The concentration of total IgE in collected serum on day 41 was determined by ELISA. Data are shown as mean ± SD of changes in dermatitis score, skin thickness, and total IgE of eight mice ($n=8$). $*P<.05$; $**P<.01$; $***P<.001$ compared with control. (●) Noninduction; (○) Control; (▼) CJLS118; (△) CJLP55; (■) CJLP56; (□) CJLP133; and (◆) CJLP136.

Some LAB strains catalyze the decarboxylation of glutamate, producing γ-aminobutyric acid (GABA) and CO_2 (Higuchi et al., 1997); therefore fermented kimchi has the potential to contain high amounts of GABA, which is a nonprotein amino acid that is well-known as a major inhibitory neurotransmitter of the nervous system in animals (Jakobs et al., 1993; Wong et al., 2003). Cho et al. (2007) generated GABA using *Lactobacillus buchneri* isolated from kimchi, while Seok et al. (2008) produced GABA-enriched kimchi using *Lactobacillus sakei* as a starter. Direct GABA intake from kimchi is considered desirable because consuming foods with high GABA content as part of a regular diet may improve health outcomes.

20.8 CONCLUSION

Kimchi is a probiotic LAB-fermented vegetable-based food, baechu (cabbage) kimchi being the most popular product. The quality and bioactivity of kimchi is dependent on its raw ingredients, fermentation conditions, and the LAB involved in the fermentation. The health benefits and quality of kimchi can be increased by manipulating the types and amounts of ingredients, as well as the fermentation temperature, time, oxygen conditions, and starter cultures. Kimchi has antioxidative, antiaging, antimutagenic, anticancer, antiobesity, and antiatherosclerotic effects. Moreover, kimchi LAB may exert antioxidative, anticancer, immune-stimulatory, and antiobesity activities through a probiotic effect. Thus kimchi raw materials and fermentation with probiotic LAB are responsible for the increased bioactivity of kimchi. Moreover, the negative effects of salt in kimchi can be mitigated or removed by using different kinds of salt or reducing the level of salt during the manufacture of kimchi. This fermented product is also safe from NO_3, NO_2, nitrosamines, and biogenic amine contamination. Finally, the health-promoting, probiotic, and functional food properties of kimchi can be optimized or increased by manipulating the types and amounts of ingredients, as well as by using appropriate probiotic starters and kimchi preparation methods. Overall, optimally produced kimchi can be a very healthy food.

REFERENCES

Ahn, S.J., 2007. The Effect of Kimchi Powder Supplement on the Body Weight Reduction of Obese Adult Women (MS thesis). Pusan National University, Busan, Korea.

Baek, Y.H., Kwak, J.R., Kim, S.J., Han, S.S., Song, Y.O., 2001. Effects of kimchi supplementation and/or exercise training on body composition and plasma lipids in obese middle school girls. Journal of the Korean Society of Food Science and Nutrition 30, 906–912.

Bengmark, S., 2001. Use of prebiotics, probiotic and synbiotics in clinical immunonutrition. In: Proceedings of International Symposium on Food, Nutrition and Health for 21st Century. Seoul, pp. 187–213.

Bong, Y.J., Jeong, J.K., Park, K.Y., 2013. Fermentation properties and increased health functionality of kimchi by kimchi lactic acid bacteria starters. Journal of the Korean Society of Food Science and Nutrition 42, 1717–1726.

Chae, O.W., Shin, K.S., Chung, H., Choe, T.B., 1998. Immunostimulation effects of mice fed with cell lysate of *Lactobacillus plantarum* isolated from kimchi. Korean Journal of Biotechnology and Bioengineering 13, 424–430.

Chang, J.Y., Chang, H.C., 2010. Improvements in the quality and shelf life of kimchi by fermentation with the induced bacteriocin-producing strain, *Leuconostoc citreum* GJ7 as a starter. Journal of Food Science 75, M103–M110.

Chang, J.H., Shim, Y.Y., Cha, S.K., Chee, K.M., 2010. Probiotic characteristics of lactic acid bacteria isolated from kimchi. Journal of Applied Microbiology 109, 220–230.

Chang, J.Y., Kim, I.C., Chang, H.C., 2014. Effect of solar salt on kimchi fermentation during long-term storage. Korean Journal of Food Science and Technology 46, 456–464.

Chang, J.H., 1975. Studies on the origin of Korean vegetable pickles. Thesis collection. Sungsin Womens College 6, 149–154.

Cheigh, H.S., Park, K.Y., 1994. Biochemical, microbiological and nutritional aspect of kimchi. Critical Reviews in Food Science and Nutrition 43, 175–203.

Cho, J.H., Lee, D.Y., Yang, C.N., Jeon, J.I., Kim, J.H., Han, H.U., 2006. Microbial population dynamics of kimchi, a fermented cabbage product. FEMS Microbiology Letters 257, 262–267.

Cho, Y.R., Chang, J.Y., Chang, H.C., 2007. Production of γ-aminobutyric acid (GABA) by *Lactobacillus buchneri* isolated from kimchi and its neuroprotective effect on neuronal cells. Journal of Microbiology and Biotechnology 17, 104–109.

Cho, E.J., 1999. Standardization and cancer Chemopreventive Activities of Chinese Cabbage Kimchi (Ph.D. thesis). Pusan National University, Busan, Korea.

Choi, S.M., Park, K.Y., 2002. Effects of different kinds of salt on kimchi fermentation and chemopreventive functionality. Journal of Korean Association of Cancer Prevention 7, 192–199.

Choi, M.W., Kim, K.H., Kim, S.H., Park, K.Y., 1997. Inhibitory effects of kimchi extracts on carcinogen-induced cytotoxicity and transformation in C3H10T1/2 cells. Journal of Food Science and Nutrition 2, 241–245.

Choi, S.M., Jeon, Y.S., Rhee, S.H., Park, K.Y., 2002. Red pepper powder and kimchi reduce body weight and blood and tissue lipids in rats fed a high fat diet. Nutraceuticals and Food 7, 162–167.

Choi, J.M., Lee, S.H., Park, K.Y., Kang, S.A., Cho, E.J., 2014. Protective effect of kimchi against Aβ25-35-induced impairment of cognition and memory. Journal of the Korean Society of Food Science and Nutrition 43, 360–366.

Choi, S.M., 2001. Antiobesity and Anticancer Effects of Red Pepper Powder and Kimchi (Ph.D. thesis). Pusan National University, Busan, Korea.

Chung, H.K., 1993. The physiological characteristics and immunological functions of lactic acid bacteria from kimchi. Kimchi Science and Industry 2, 23–28.

Collins, M.D., Samelis, J., Metaxopoulos, J., Wallbanks, S., 1993. Taxonomic studies on some *leuconostoc*-like organisms from fermented sausages: description of a new genus *Weissella* for the *Leuconostoc paramesenteroides* group of species. Journal of Applied Bacteriology 75, 595–603.

Cui, M., Kim, H.Y., Lee, K.H., Jeong, J.K., Hwang, J.H., Yeo, K.Y., Ryu, B.H., Choi, J.H., Park, K.Y., 2015. Antiobesity effects of kimchi in diet-induced obese mice. Journal of Ethnic Foods 2, 137–144.

D'Elia, L., Rossi, G., Ippolito, R., Cappuccio, F.P., Strazzullo, P., 2012. Habitual salt intake and risk of gastric cancer: a meta-analysis of prospective studies. Clinical Nutrition 31, 489–498.

Goldin, B.R., Gorbach, S.L., 1976. The relationship between diet and rat fecal bacterial enzymes implicated in colon cancer. Journal of the National Cancer Institute 57, 371–375.

Ha, J.O., Park, K.Y., 1999. Comparison of auto-oxidation rate and comutagenic effect of different kinds of salt. Journal of Korean Association of Cancer Prevention 4, 44–51.

Han, G.J., Son, A.R., Lee, S.M., Jung, J.K., Kim, S.H., Park, K.Y., 2009. Improved quality and increased in vitro anticancer effects of kimchi by using natural sea salt without bittern and baked (Guwun) salt. Journal of Korean Society of Food Science and Nutrition 38, 996–1002.

Han, W., Hu, W., Lee, Y.M., 2011. Anticancer activity of human colon cancer (HT-29) cell line from different fraction of *Zanthoxylum schnifolium* fruits. Korean Journal of Pharmacology 42, 282–287.

Higuchi, T., Hayashi, H., Abe, K., 1997. Exchange of glutamate and γ-aminobutyrate in a *Lactobacillus* strain. Journal of Bacteriology 179, 3362–3364.

Hur, Y.M., Kim, S.H., Park, K.Y., 1999. Inhibitory effects of kimchi extracts on the growth and DNA synthesis of human cancer cells. Journal of Food Science and Nutrition 4, 107–112.

Hur, Y.M., Kim, S.H., Choi, Y.W., Park, K.Y., 2000. Inhibition of tumor formation and changes in hepatic enzyme activities by kimchi extracts in sarcoma-180 cell transplanted mice. Journal of Food Science and Nutrition 5, 48–53.

Hwang, J.W., Song, Y.O., 2000. The effects of solvent fractions of kimchi on plasma lipid concentration of rabbit-fed high-cholesterol diet. Journal of the Korean Society of Food Science and Nutrition 19, 204–210.

Hwang, S.Y., Hur, Y.M., Choi, Y.H., Rhee, S.H., Park, K.Y., Lee, W.H., 1997. Inhibitory effect of kimchi extracts on mutagenesis of aflatoxin B_1. Environmental Mutagens and Carcinogens 17, 133–137.

Islam, M.S., Choi, H., 2009. Antidiabetic effect of Korean traditional baechu (Chinese cabbage) kimchi in a type 2 diabetes model of rats. Journal of Medicinal Food 12, 292–297.

Jakobs, C., Jaeken, J., Gibson, K.M., 1993. Inherited disorders of GABA metabolism. Journal of Inherited Metabolic Disease 16, 704–715.

Jang, S.E., Joh, E.H., Lee, H.Y., Ahn, Y.T., Lee, J.H., Huh, C.S., Han, M.J., Kim, D.H., 2013. *Lactobacillus plantarum* HY7712 ameliorates cyclophosphamide-induced immunosuppression in mice. Journal of Microbiology and Biotechnology 23, 414–421.

Jeun, J., Kim, S., Cho, S.Y., Jun, H.J., Park, H.J., Seo, J.G., Chung, M.J., Lee, S.J., 2010. Hypocholesterolemic effects of *Lactobacillus plantarum* KCTC 3928 by increased bile acid excretion in C57BL/6 mice. Nurition 26, 321–330.

Joossens, J.V., Hill, M.J., Elliott, P., Stamler, R., Stamler, J., Lesaffre, E., Dyer, A., Nichols, R., Kesteloot, H., 1996. Dietary salt, nitrate and stomach cancer mortality in 24 countries. International Journal of Epidemiology 25, 494–502.

Joosten, H.M.L.J., Nunez, M., 1996. Prevention of histamine formation in cheese by bacteriocin-producing lactic acid bacteria. Applied and Environmental Microbiology 62, 1178–1181.

Jung, J.Y., Lee, S.H., Kim, J.M., Park, M.S., Bae, J.W., Hahn, Y., Madsen, E.L., Jeon, C.O., 2011. Metagenomic analysis of kimchi, a traditional Korean fermented food. Applied and Environmental Microbiology 77, 2264–2274.

Jung, J.Y., Lee, S.H., Lee, H.J., Seo, H.Y., Park, W.S., Jeon, C.O., 2012. Effect of *Leuconostoc mesenteroides* starter cultures on microbial communities and metabolites during kimchi fermentation. International Journal of Food Microbiology 153, 378–387.

Kalac, P., Krausova, P., 2005. A review of dietary polyamines: formation, implication for growth and health and occurrence in foods. Food Chemistry 77, 349–351.

Kil, J.H., 2004. Studies on Development of Cancer Preventive and Anticancer Kimchi and Its Anticancer Mechanism (Ph.D. thesis). Pusan National University, Busan, Korea.

Kim, M.J., Chun, J.S., 2005. Bacterial community structure in kimchi, a Korean fermented vegetable food, as revealed by 16S rRNA gene analysis. International Journal of Food Microbiology 103, 91–96.

Kim, J.Y., Lee, Y.S., 1997. The effects of kimchi intake on lipid contents of body and mitogen response of spleen lymphocytes in rats. Journal of the Korean Society of Food Science and Nutrition 26, 1200–1207.

Kim, S.H., Lee, E.H., Kawabata, T., Ishbashi, T., Endo, T., Matsui, M., 1984. Possibility of N-nitrosamine formation during fermentation of kimchi. Journal of the Korean Society of Food Science and Nutrition 13, 291–306.

Kim, H.Y., Bae, H.S., Baek, Y.J., 1991. *In vivo* antitumor effects of lactic acid bacteria on Sarcoma 180 and mouse Lewis lung carcinoma. Journal of the Korean Cancer Association 23, 188–196.

Kim, S.H., Park, K.Y., Suh, M.J., 1995. Comutagenic effect of sodium chloride in the *Salmonella/* mammalian microsome assay. Food Science and Biotechnology 4, 264–267.

Kim, B.J., Chun, J., Han, H.U., 2000. *Leuconostoc kimchii* sp. nov., a new species from kimchi. International Journal of Systematic and Evolutionary Microbiology 50, 1915–1919.

Kim, J.H., Ryu, J.D., Lee, H.G., Park, J.H., 2002. The effect of kimchi on production of free radicals and antioxidative enzyme activities in the brain of SAM. Journal of the Korean Society of Food Science and Nutrition 31, 117–123.

Kim, H.J., Lee, J.S., Chung, H.Y., Song, S.H., Suh, H.S., Noh, J.S., Song, Y.O., 2007a. 3-(4′-hydroxyl-3′,5′-dimethoxylphenyl) propionic acid, an active principle of kimchi, inhibits development of atherosclerosis in rabbits. Journal of Agricultural and Food Chemistry 55, 10486–10492.

Kim, Y.T., Kim, B.K., Park, K.Y., 2007b. Antimutagenic and anticancer effects of leaf mustard and leaf mustard kimchi. Journal of Food Science and Nutrition 12, 84–88.

Kim, B.K., Park, K.Y., Kim, H.Y., Ahn, S.C., Cho, E.J., 2011a. Antiaging effects and mechanisms of kimchi during fermentation under stress-induced premature senescence cellular system. Food Science and Biotechnology 20, 643–649.

Kim, E.K., An, S.Y., Lee, M.S., Kim, T.H., Lee, H.K., Hwang, W.S., Choe, S.J., Kim, T.Y., Han, S.J., Kim, H.J., Kim, D.J., Lee, K.W., 2011b. Fermented kimchi reduces body weight and improves metabolic parameters in overweight and obese patients. Nutrition Research 31, 436–443.

Kim, H.Y., Song, J.L., Chang, H.K., Kang, S.A., Park, K.Y., 2014. Kimchi protects against azoxymethane/dextran sulfate sodium–induced colorectal carcinogenesis in mice. Journal of Medicinal Food 17, 833–841.

Kim, K.M., 1998. Increases in Swimming Endurance Capacity of Mice by Capsaicin (Ph.D. thesis). Kyoto University, Kyoto, Japan.

Kim, B.K., 2009. Antiaging Effects and Anticancer Mechanisms of Kimchi During Fermentation (Ph.D. thesis). Pusan National University, Busan, Korea.

Kwon, E.A., Kim, M.H., 2007. Microbial evaluation of commercially packed kimchi products. Food Science and Biotechnology 16, 615–620.

Kwon, M.J., Song, Y.O., Song, Y.S., 1997. Effect of kimchi on tissue and fecal lipid composition and apoprotein and thyroxine levels in rats. Journal of Korean Society of Food Science and Nutrition 26, 507–513.

Kwon, M.J., Chun, J.H., Song, Y.S., Song, Y.O., 1999. Daily kimchi consumption and its hypolipidemic effect in middle-aged men. Journal of the Korean Society of Food Science and Nutrition 28, 1144–1150.

Kwon, J.Y., Cheigh, H.S., Song, Y.O., 2004. Weight reduction and lipid lowering effects of kimchi lactic acid powder in rats fed high fat diets. Korean Journal of Food Science and Technology 36, 1014–1019.

Lee, S.W., Woo, S.J., 1989. Effect of some materials on the contents of nitrate, nitrite and vitamin C in kimchi during fermentation. Korean Journal of Dietary Culture 4, 161–166.

Lee, C.W., Ko, C.Y., Ha, D.M., 1992. Microfloral changes of the lactic acid bacteria during kimchi fermentation and identification of the isolates. Korean Journal of Applied Microbiology and Biotechnology 20, 102–109.

Lee, J.S., Jung, M.C., Kim, W.S., Kim, H.J., Park, C.S., Joo, Y.J., Lee, H.J., Ahn, J.S., Park, W., Park, Y.H., Mheen, T.I., 1996. Identification of lactic acid bacteria from kimchi by cellular FAMEs analysis. Korean Journal of Microbiology and Biotechnology 24, 234–241.

Lee, J.S., Chun, C.O., Jung, M.C., Kim, W.S., Kim, H.J., Hector, M., Kim, S.B., Park, C.S., Ahn, J.S., Park, Y.H., Mheen, T.I., 1997. Classification of isolates originating from kimchi using carbon-source utilization patterns. Journal of Microbiology and Biotechnology 7, 68–74.

Lee, J., Hwang, K.T., Heo, M.S., Lee, J.H., Park, K.Y., 2005a. Resistance of *Lactobacillus plantarum* KCTC 3099 from kimchi to oxidative stress. Journal of Medicinal Food 8, 299–304.

Lee, J.S., Heo, G.Y., Lee, J.W., Oh, Y.J., Park, J.A., Park, Y.H., Pyun, Y.R., Ahn, J.S., 2005b. Analysis of kimchi microflora using denaturing gradient gel electrophoresis. International Journal of Food Microbiology 102, 143–150.

Lee, J.H., Kweon, D.H., Lee, S.C., 2006. Isolation and characterization of an immunopotentiating factor from *Lactobacillus plantarum* in kimchi: assessment of immunostimulatory activities. Food Science and Biotechnology 15, 877–883.

Lee, D.Y., Kim, S.J., Cho, J.H., Kim, J.H., 2008. Microbial population dynamics and temperature changes during fermentation of kimjang kimchi. Journal of Microbiology 46, 590–593.

Lee, G.I., Lee, H.M., Lee, C.H., 2012a. Food safety issues in industrialization of traditional Korean foods. Food Control 24, 1–5.

Lee, S.Y., Song, Y.O., Han, E.S., Han, J.S., 2012b. Comparative study on dietary habits, food intakes, and serum lipid levels according to kimchi consumption in college students. Journal of the Korean Society of Food Science and Nutrition 41, 351–361.

Lee, S.W., 1975. Studies on the movements and inter-changes of vegetable pickles in China, Korea and Japan. Journal of the Korean Society of Food Science and Nutrition 4, 71–76.

Lim, C.R., Park, H.K., Han, H.U., 1989. Reevaluation of isolation and identification of gram positive bacteria in kimchi. The Korean Journal of Microbiology 27, 404–414.

Mah, J.H., Kim, Y.J., No, H.K., Hwang, H.J., 2004. Determination of biogenic amines in kimchi, Korean traditional fermented vegetable products. Food Science and Biotechnology 13, 826–829.

Martineau, B., Laflamme, D.P., 2002. Effect of diet on markers of intestinal health in dogs. Research in Veterinary Science 72, 223–227.

Moon, Y.J., Soh, J.R., Yu, J.J., Sohn, H.S., Cha, Y.S., Oh, S.H., 2012. Intracellular lipid accumulation inhibitory effect of *Weissella koreensis* OK1-6 isolated from kimchi on differentiating adipocyte. Journal of Applied Microbiology 113, 652–658.

O'Donnell, M., Mente, A., 2014. Urinary sodium and potassium excretion, mortality, and cardiovascular events. New England Journal of Medicine 371, 612–623.

Oh, Y.J., Hwang, I.J., Leitzmann, C., 1993. Regular intake of kimchi prevents colon cancer. Kimchi Science and Industry 2, 9–22.

Oh, C.K., Oh, M.C., Hyon, J.S., Choi, W.J., Lee, S.H., Kim, S.H., 1997. Depletion of nitrite by lactic acid bacteria from kimchi (II). Journal of Korean Society of Food Science and Nutrition 26, 556–562.

Oh, C.K., Oh, M.C., Kim, S.H., 2004. The depletion of sodium nitrite by lactic acid bacteria isolated from kimchi. Journal of Medicinal Food 7, 38–44.

Park, K.Y., Cheigh, H.S., 1992. Kimchi and nitrosamines. Journal of Korean Society of Food Science and Nutrition 21, 109–116.

Park, K.Y., Cheigh, H.S., 2004. Kimchi. In: Hui, Y.H., Meunier-Goddik, L., Hansen, A.S., Josephsen, J., Nip, W.K., Stanfield, P.S., Toldra, F. (Eds.), Handbook of Food Science and Beverage Fermentation Technology. Marcel Dekker, New York, pp. 621–655.

Park, K.Y., Kim, B.K., 2012. Lactic acid bacteria in vegetable fermentation. In: Lahtnen, S., Ouwehand, A.C., Salminen, S., von Wright, A. (Eds.), Lactic Acid Bacteria. CRC Press, Boca Raton, FL, USA, pp. 187–211.

Park, K.Y., Rhee, S.H., 2005. Functional foods from fermented vegetable products: kimchi (Korean fermented vegetables) and functionality. In: Shi, J., Ho, C.T., Shahidi, F. (Eds.), Asian Functional Foods. CRC Press Inc., Boca Raton, FL, pp. 341–380.

Park, K.Y., Kim, S.H., Suh, M.J., 1990. Comutagenicity of high-salted kimchi in the Salmonella assay system. Journal of the College of Home Economics, Pusan National University 16, 45–50.

Park, K.Y., Baek, K.A., Rhee, S.H., Cheigh, H.S., 1995. Antimutagenic effect of kimchi. Foods and Biotechnology 4, 141–145.

Park, J.S., Park, K.Y., Yu, R., 1998. Inhibition of in vitro nitrosation by capsaicin and its metabolites. Journal of Korean Society of Food Science and Nutrition 27, 1015–1018.

Park, K.Y., Cho, E.J., Rhee, S.H., Jung, K.O., Yi, S.J., Jhun, B.H., 2003. Kimchi and an active component, beta-sitosterol, reduce oncogenic H-Ras^{v12}-induced DNA synthesis. Journal of Medicinal Food 6, 151–156.

Park, J.M., Shin, J.H., Lee, D.W., Song, J.C., Suh, H.J., Chang, U.J., Kim, J.M., 2010. Identification of the lactic acid acteria in kimchi according to initial and over-ripened fermentation using PCR and 16S rRNA gene sequence analysis. Food Science and Biotechnology 19, 541–546.

Park, E.J., Chun, J.S., Cha, C.J., Park, W.S., Jeon, C.O., Bae, J.W., 2012a. Bacterial community analysis during fermentation of ten representative kinds of kimchi with barcoded pyrosequencing. Food Microbiology 30, 197–204.

Park, J.A., Pichiah, P.B.T., Yu, J.J., Oh, S.H., Daily III, J.W., Cha, Y.S., 2012b. Anti-obesity effect of kimchi fermented with Weissella koreensis OK1-6 as starter in high-fat diet-induced obese C57BL/6J mice. Journal of Applied Microbiology 113, 1507–1516.

Park, K.Y., 1995. The nutritional evaluation, and antimutagenic and anticancer effects of kimchi. Journal of the Korean Society of Food Science and Nutrition 24, 169–182.

Ray, B., Bhunia, A., 2008. Fundamental Food Microbiology, fourth ed. CRC Press, Boca Raton, FL, USA, pp. 165–168.

Raznikoff, C.A., Bertram, J.S., Brankow, D.W., Heidelberger, C.L., 1973. Quantitative and qualitative studies of chemical transformation of cloned C3H mouse embryo cells sensitive to post confluence inhibition of cell division. Cancer Research 33, 3239–3429.

Ryu, J.C., Park, K.Y., 2001. Anticlastogenic effect of Baechu (Chinese cabbage) kimchi and Buchu (leek) kimchi in mitomycin C-induced micronucleus formations by supravital staining of mouse peripheral reticulocytes. Environmental Mutagens and Carcinogens 21, 51–56.

Ryu, S.H., Jeon, Y.S., Kwon, M.J., Moon, J.W., Lee, Y.S., Moon, G.S., 1997. Effect of kimchi extracts to reactive oxygen species in skin cell cytotoxicity. Journal of the Korean Society of Food Science and Nutrition 26, 814–821.

Ryu, B.M., 2000. Effect of Kimchi on Inhibition of Skin Aging of Hairless Mouse (Ph.D. thesis). Pusan National University, Busan, Korea.

Segawa, S., Hayashi, A., Nakakita, Y., Kaneda, H., Watari, J., Yasui, H., 2008. Oral administration of heat-killed Lactobacillus brevis SBC8803 ameliorates the development of dermatitis and inhibits immunoglobulin E production in atopic dermatitis model NC/Nga mice. Biological and Pharmaceutical Bulletin 31, 884–889.

Seok, J.H., Park, K.B., Kim, Y.H., Bae, M.O., Lee, M.K., Oh, S.H., 2008. Production and characterization of kimchi with enhanced levels of γ-aminobutyric acid. Food Science and Biotechnology 17, 940–946.

Settanni, L., Corsetti, A., 2007. The use of multiplex PCR to detect and differentiate food- and beverage-associated microorganisms: a review. Journal of Microbiological Methods 69, 1–22.

Shin, K.S., Chae, O.W., Park, I.C., Hong, S.K., Choe, T.B., 1998. Antitumor effects of mice fed with cell lysate of *Lactobacillus plantarum* isolated from kimchi. Korean Journal of Biotechnology and Bioengineering 13, 357–363.

Shruti, S., Park, H.K., Kim, J.K., Kim, M.H., 2010. Determination of biogenic amines in Korean traditional fermented soybean paste (Doenjang). Food Chemical Toxicology 48, 1191–1195.

Sousa, R., Halper, J., Zhang, J., Lewis, S.J., Li, W.I., 2008. Effect of *Lactobacillus acidophilus* supernatants on body weight and leptin expression in rats. BMC Complementary and Alternative Medicine 8, 1–8.

Sunada, Y., Nakamura, S., Kamei, C., 2008. Effect of *Lactobacillus acidophilus* strain L-55 on the development of atopic dermatitis-like skin lesions in NC/Nga mice. International Immunopharmacology 8, 1761–1766.

Tobita, K., Yanaka, H., Otani, H., 2009. Heat-treated *Lactobacillus crispatus* KT strains reduce allergic symptoms in mice. Journal of Agricultural and Food Chemistry 57, 5586–5590.

Watanabe, T., Hamada, K., Tategaki, A., Kishida, H., Tanaka, H., Kitano, M., Miyamoto, T., 2009. Oral administration of lactic acid bacteria isolated from traditional South Asian fermented milk 'dahi' inhibits the development of atopic dermatitis in NC/Nga mice. Journal of Nutritional Science and Vitaminology 55, 271–278.

Won, T.J., Kim, B., Lim, Y.T., Song, D.S., Park, S.Y., Park, E.S., Lee, D.I., Hwang, K.W., 2011. Oral administration of *Lactobacillus* strains from kimchi inhibits atopic dermatitis in NC/Nga mice. Journal of Applied Microbiology 110, 1195–1202.

Wong, C.G.T., Bottiglieri, T., Snead, O.C., 2003. GABA, γ-hydroxybutyric acid, and neurological disease. Annals of Neurology 54, S3–S12.

Yoon, H.H., Chang, H.C., 2011. Growth inhibitory effect of kimchi prepared with four year-old solar salt and Topan solar salt on cancer cells. Journal of the Korean Society of Food Science and Nutrition 40, 935–941.

Yoon, J.Y., Jung, K.O., Kim, S.H., Park, K.Y., 2004. Antiobesity effect of baek-kimchi (whitish baechu kimchi) in rats fed high fat diet. Journal of Food Science and Nutrition 9, 259–264.

Chapter 21

The Naturally Fermented Sour Pickled Cucumbers

H. Zieliński[1], M. Surma[2], D. Zielińska[3]
[1]*Institute of Animal Reproduction and Food Research of the Polish Academy of Sciences, Olsztyn, Poland;* [2]*University of Agriculture in Krakow, Krakow, Poland;* [3]*University of Warmia and Mazury in Olsztyn, Olsztyn, Poland*

21.1 ORIGIN OF CUCUMBER

Cucumber (*Cucumis sativus* L.) belongs to the *Cucurbitaceae* family. Of the 30 species of *Cucumis*, *C. sativus* has the greatest economic significance. The most important cucumber cultivars originate from Europe and America, the western part of India, China, and the Himalayas. Cucumber has enclosed dicotyledonous seeds and it develops from a flower, and therefore it is classified as a fruit (Mukherjee et al., 2013). Cucumbers are water-rich vegetables, with only 4% dry mass. Considering that fresh cucumbers expire relatively fast, they are widely consumed mainly fresh in salads or most frequently eaten in a preserved form, for example pickled or marinated.

21.2 CHEMICAL COMPOSITION AND BIOACTIVE COMPOUNDS IN CUCUMBER

Cucumber (*Cucumis sativus*) contains water (96.4%), carbohydrates (2.7–2.8%), protein (0.4–0.67%), fat (0.19–0.13%), dietary fiber (0.8%), minerals (0.3%), calcium (0.01%), phosphorus (0.03%), iron (1.5 mg/100 g), and group B vitamins (30 IU/100 g). The content of vitamin C may range from 7.5 up to 18.1 mg/100 g fresh weight (Chu et al., 2002). Fleming and McFeeters (2002) and Lu et al. (2002) found that during cucumber fruit growth malic acid content slightly decreases from 0.28% to 0.21% (wet basis); glucose and fructose contents increase from 0.80% to 1.16% and 0.95–1.25% (wet basis), respectively; and dry matter content declines from 5.01% to 4.35%.

Cucumbers are a particularly rich source of phenolic compounds (Murcia et al., 2009). Total phenolic content, flavonols, and proanthocyanidins were found to be 9.05 ± 0.83, 2.06 ± 0.09, and 55.66 ± 1.52 mg/100 g, respectively, in cucumber extracts (Melo et al., 2006). The free phenolics content in cucumber was found to be 14.37±1.48 mg/100 g, whereas bound phenolic compounds were found to be 2.92±0.07 mg/100 g fresh weight (Chu et al., 2002). More recently, about 73 phenolic compounds were identified in cucumber (Abu-Reidah et al., 2012).

Cucumbers contain softening enzymes, such as pectinesterase, exo-polygalacturonase, endo-polygalacturonase, cellulose, endo-xylanase, β-xylosidase, and endo-1,4-β-glucanase, which may have an impact on ripening and processing. Elaterase, oxidase, succinic, and malic dehydrogenase have also been reported in cucumber fruits (Maruvada and McFeeters, 2009).

The cucumber pulp and peel extracts contain a high concentration of lactic acid (~7–8%, w/w).

Cucumbers contain cucurbitacins (types A, B, C, D, E, and I), which are responsible for bitterness and toxicity. The amount of bitterness in cucumbers appears to vary from year to year and from location to location. The bitterness of cucumber is caused mainly by cucurbitacin C but it is present in very small amounts (lower than 1 mg/100 g fresh weight) even when selecting a strongly bitter cucumber fruit (threshold level of the bitterness is less than 0.1 mg/L) (Hideki et al., 2007; Mukherjee et al., 2013). Maturated cucumbers contain low levels of cucurbitacins because elaterase enzyme hydrolyzes them to its nonbitter principles. Therefore, cucurbitacins do not accumulate very heavily in the fruit (Mukherjee et al., 2013).

Volatiles from cucumber have been isolated by steam distillation at 70°C and comprise of 1-nonanol, *trans*-2-nonen-1-ol, *cis*-3-nonen-1-ol, *cis*-6-nonen-1-ol, *trans, cis*-2,6-nonadien-1-ol, *cis, cis*-3,6-nonadien-1-ol, *cis*-6-nonenal, *cis*-3-nonenal, and *cis, cis*-3,6-nonadienal (Kemp et al., 1974). Sotiroudis et al. (2010) isolated essential oil from three distinct cucumber cultivars and found 21 compounds including Z-6-nonenol (61.54%), E-2-nonenol (6.98%), E,Z-2,6-nonadienal (47.08%), E-2- nonenal (17.39%), Z-3-nonenol (14.79%), 3-nonenal (7.32%), pentadecanal (43.47%), 9,12,15-octadecatrienal (14.52%), and 9,17-octadecadienal (12.33%) (Sotiroudis et al., 2010; Mukherjee et al., 2013). Among them, *E,Z*-2, 6-nonadienal and *E*-2-nonenal (NE) showed antibacterial activity against human and foodborne pathogen bacteria, such as *Bacillus cereus*, *Escherichia coli*, *Listeria monocytogenes*, and *Salmonella typhimurium* (Cho et al., 2004).

21.3 LACTIC ACID FERMENTATION OF CUCUMBERS

Considering that fresh cucumbers expire relatively fast, they are most frequently eaten as fermented pickles. It is believed that cucumbers were first fermented around 2000 BC in the Middle East; however, early written records of cucumber pickles come from paper fragment remains of a play (*The Taxiarchs*) by the Greek writer Eupolis (429–412 BC) (Breidt et al., 2013). Nowadays, cucumber pickles are one of the most important primary retail fermented vegetable products produced in the United States and Europe. Fermented pickle belongs to the products stabilized with salt and lactic acid which are produced by lactic acid bacteria (LAB) fermentation. The word "pickle" by itself usually refers to a pickled cucumber. However, the products prepared in diluted acetic acid solution spiced with some flavor or aroma herbals are also included under this definition. Therefore, fermented pickles should be defined as a product obtained from fresh cucumbers with added aromatic and flavoring spices, covered with table salt solution, and subjected to lactic acid fermentation.

Fermentation, which is by definition an anaerobic process, allows to extend the shelf-life of cucumbers and enhances nutritional and sensory attributes including flavor, color, nutrient digestibility, and stability, while reducing toxicity and removing antinutritional factors (Di Cagno et al., 2013). At present, fermentation of cucumbers (Atul and Ramesh, 2008), cabbages, and olives has only an industrial significance due to their commercial importance (Montet et al., 2006). Most commercial cucumber fermentations rely upon growth of LAB that are naturally present on the surface of cucumbers. LAB are initially present on fresh vegetables in lower numbers, 10^2–10^3 colony-forming units (CFU)/g, compared with other mesophilic microorganisms.

Therefore, immersion and washing are two important steps in cucumber pretreatment before fermentation. Cucumbers of appropriate quality are placed in open containers with cold and clean running water. The length of immersion depends on the firmness of the flesh varying, between 1 and 4 h. The role of the immersion process is to remove all contaminants from the cucumbers and restore the firmness lost during transportation and storage. Immediately after immersion, cucumbers are cleaned by washers. Washing prior to fermentation reduces the initial microbial count, thus favoring the development of lactic acid flora (Ana et al., 2002).

Currently, the fermentation is typically carried out in brine equilibrated at about 5–7% of NaCl. There is also some research on reduction of the salt concentration (4% or less) and addition of calcium chloride to 0.4% to the cover brine, to keep the firm texture of the fermented cucumbers during fermentation and storage (Di Cagno et al., 2013; Breidt et al., 2013). The initial pH of brined cucumbers is about 6.5.

Lactobacillus spp., *Leuconostoc* spp., and *Pediococcus* spp. are the main LAB naturally present on the cucumber surface and responsible for fermentation. Spontaneous fermentation is still widely used in traditional household production of fermented pickles in Africa, Asia, Europe, and Latin America (Holzapfel, 2002; Steinkraus, 2002). On the other hand, the use of starter cultures, such as *L. plantarum*, *L. rhamnosus*, *L. gasseri*, and *L. acidophilus* provide reproducibility at industrial-scale production (Karovicova et al., 1999). During fermentation, LAB synthesize bacteriocins and release antimicrobial peptides that inhibit spoilage bacteria growth. Carbon dioxide (CO_2) may be synthesized through cucumber respiration and malolactic fermentation by *L. plantarum* (Di Cagno et al., 2013). A malolactic-deficient strain of *L. plantarum* (US Food Fermentation Laboratory culture collection, Raleigh, NC, USA) was recommended as a starter culture for industrial cucumber fermentation because

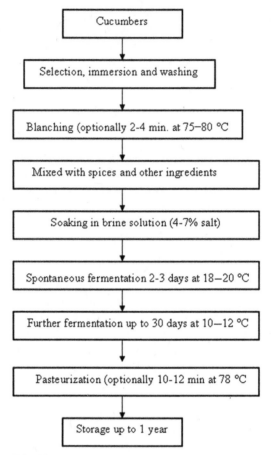

FIGURE 21.1 Basic outline of cucumber fermentation.

it does not utilize malic acid (a natural acid present in cucumbers) to produce CO_2 (Daeschel et al., 1984). Optimal temperature range from 15°C to 18°C during fermentation has been reported for pickled cucumber production. However, lower temperatures (7–8 °C) can also be applied as long as the fermentation process is extended. At the end of the fermentation, there may be 1.5% lactic acid, a pH of 3.1–3.5, and little or no residual sugar (Breidt et al., 2013). A basic outline of cucumber fermentation is given in Fig. 21.1.

After fermentation by *L. plantarum* or other LAB, cucumbers may be stored in fermentation tanks for 1 year or more. Fermented cucumbers are shown in Fig. 21.2.

Spoilage microbial growth during fermentation is prevented by organic acids, low pH, and lack of fermentable sugars. Some examples of traditional fermented cucumbers, produced through the various worldwide regions, are shown in Table 21.1.

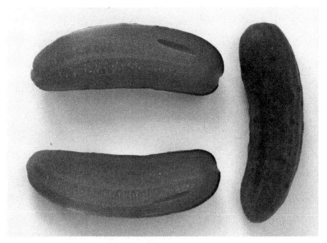

FIGURE 21.2 The fermented cucumber. *H. Zieliński [own photo collection].*

21.4 FACTORS AFFECTING CUCUMBER FERMENTATION

21.4.1 Effects of Cucumber Size and Enzyme Activity

Cucumber size may influence fermentation and final nutrient composition of fermented product. Fleming and McFeeters (2002) reported that small cucumbers with diameters lower than 27 mm after lactic acid fermentation may be spoiled by undesirable bacteria during storage and a final pH of 3.1–3.5 is required for microbial stability of the product, depending upon salt concentration (Fleming et al., 1996). Activity of enzymes involved in cucumber softening also varies according to fruit size (Fleming et al., 1978), being responsible for bloater damage in large-sized cucumbers (39–51 mm in diameter) at the early step of fermentation. Therefore, blanching prior to fermentation has been suggested by Etchells et al. (1964) to reduce bloater formation and to increase firmness retention of the fermented cucumbers (Fleming et al., 1978, 1996). It has been found that enzymatic tissue softening was associated with fungal polygalacturonases rather than cucumber endogenous enzymes (Maruvada and McFeeters, 2009). Empty spaces may appear in cucumbers, particularly at high temperatures due to the growth of gas-producing microorganisms, such as yeasts, bacteria from the *Aerobacter* family, and heterofermentative LAB.

21.4.2 Effect of Salt Concentration

Typically cucumbers are fermented in 5–8% NaCl. This high salt level creates anaerobic conditions, inhibits the growth of undesirable microorganisms, serves as a selection agent for acid-tolerant homofermentative LAB to ferment the sugars to lactic acid, and helps to retain the firm texture of the fruit (Ross et al., 2002; Viander et al., 2003). Moreover, high salt concentration inhibits potential

TABLE 21.1 Examples of Emerging and Traditional Fermented Cucumbers, Which Are Manufactured Through Various Worldwide Regions

Origin	Local Name	Ingredients	Additional Ingredients	Fermentation Time/ Temperature	Main Lactic Acid Bacteria Involved	References
Nepal	Khalpi	No	Salt, mustard oil, powdered chillies	3–5 days/room temperature	L. plantarum L. brevis Leuconostoc fallax P. pentosaceus	Dahal et al. (2005), Tamang et al. (2005)
Republic of China	Paocai	Salt	Ginger, sugar, hot red pepper	4–5 h/room temperature	L. pentosus L. plantarum L. brevis L. lactis L. fermentum L. mesenteroides	Yan et al. (2008)
Taiwan	Jiang-gua	Salt	Sugar, vinegar soy sauce	4–5 h/room temperature, then at least 1 day/6–10°C	L. plantarum, Weissella cibaria, W. hellenica, Leuconostoc lactis, Enterococcus casseliflavus	Chen et al. (2012)
USA, Central and East Europe	Pickled cucumber	Salt	Dill, horseradish roots, spices	3–5 days/room temperature then up to 30 days at 10–12°C	L. plantarum P. pentosaceus	Di Cagno et al. (2013)
Turkey	Tursu	Salt	Vinegar	4 weeks at ca. 20°C	L. plantarum L. mesenteroides L. brevis P. pentosaceus E. fecalis	Kabak and Dobson (2011)

pectinolytic and proteolytic enzymes that can cause vegetable softening and further putrefaction. The cucumbers absorb salt until the salt concentration in the cucumbers matches the brine salt concentration (about 3%) (Reina et al., 2005). It has been shown that addition of calcium to cucumbers helps to improve firmness retention and makes it unnecessary to increase salt levels during postfermentation storage (Maruvada and McFeeters, 2009). Guillou and Floros (1993) found that yeasts were almost absent from cucumber fermentations in the presence of NaCl and $CaCl_2$ mixtures with potassium sorbate added as a preservative. Yamani et al. (1999) fermented cucumbers in various NaCl concentrations and found that 4% NaCl was the lowest concentration to produce acceptable products, but 3% NaCl in combination with 0.5% KCl and 0.5% $CaCl_2$ was also acceptable. It should be mentioned that the addition of $CaCl_2$ used to improve structure may cause whitening of fermented cucumbers.

21.4.3 Effect of Sugar Addition

The sugars diffusing from cucumber are fermented sequentially by *L. mesenteroides*, *P. cerevisiae*, *L. brevis*, and *L. plantarum*. Depending on fermentation conditions, approximately 0.6–1.2% lactic acid is formed in 7–14 days. As pH decreases to 3.2, the metabolism of *L. plantarum* is inhibited (Marsili and Miller, 2000). Breidt et al. (2013) described that sugar addition to brine results in incomplete fermentation by LAB, leading to secondary fermentation by fermentative yeasts and subsequent bloater formation in the final product.

21.4.4 Effect of Spice or Aromatic Herb Addition

Spices or aromatic herbs are added to improve the flavor of the fermented cucumbers. Basic spices include: fresh dill (stems and the upper area) (addition 2.5%), horseradish roots (0.2%), horseradish leaves (0.8%), and garlic (15%). Secondary spices are added at 0.2–0.5% in the brine, and they include: blackcurrant leaves, mustard, tarragon, grapevine leaves, peppers, green marjoram, and bay leaves. Spices and herbs are sources of nontoxic inhibitors of pectinolytic and cellulolytic enzymes and the most commonly used are leaves of grapevine or blackcurrant.

21.4.5 Effect of Chemical Preservative Addition

Chemical preservatives, such as ascorbic or benzoic acid salts are sometimes used in industrial production of lactic acid-fermented cucumbers (Steinkraus, 2002). To promote the growth of LAB over yeasts, molds, and other pathogenic or unwanted bacterial strains, acids or buffer solutions are often added to the fermentation medium. A buffer solution of acetic acid and calcium acetate has been reported to improve lactic acid fermentation of cucumbers (Swain et al., 2014).

21.4.6 Biopreservation

Preservation and microbial safety of fermented cucumbers are extremely important due to the long storage period. In this context, LAB play a significant role in fermented cucumber quality. The antimicrobial activity of LAB has been attributed to production of organic acids, reuterin, proteinaceous compounds, and cyclic dipeptides. Organic acids in their undissociated, hydrophobic form diffuse across the cell membrane of the pathogens, neutralize the electrochemical potential, and increase its permeability, resulting in bacteriostasis and death (Divya et al., 2012). Reuterin, a broad-spectrum antimicrobial agent is considered a competitive inhibitor of the enzyme ribonucleotide reductase, thereby preventing growth of spoilage microorganisms and increasing the shelf-life of foods (Settanni and Corsetti, 2008).

21.5 HEALTH BENEFITS OF FERMENTED CUCUMBERS

Even though fermented cucumbers show health benefits, they are usually consumed in low amounts as appetizers, and therefore, their beneficial properties are limited.

21.5.1 Lactic Acid Bacteria (LAB) Isolated From Fermented Cucumbers as Probiotics

LAB isolated from fermented cucumbers are proposed as potential probiotics due to their potential therapeutic and prophylactic attributes. According to the Food and Agriculture Organization (FAO) probiotics are defined as "live microorganisms which provide health benefits to host by producing inhibitory substances, such as acid against pathogenic strains and also by competing with them for essential nutrients and adhesion sites."

Many studies have supported that maintenance of healthy gut microbiota provides protection against gastrointestinal disorders including gastrointestinal infections and inflammatory bowel diseases (IBDs) (Peres et al., 2012). LAB can modify the intestinal microbiota positively and prevent the colonization of other enteric pathogens. LAB strains also improve the digestive functions, enhance the immune system, reduce the risk of colorectal cancer, control the serum cholesterol levels, and eliminate the antinutritional compounds present in food materials (Divya et al., 2012). On the other hand, LAB can be used as an alternative to antibiotics to reduce the symptoms of antibiotic-associated diarrhea (Rastall et al., 2005). The two main IBD disorders, ulcerative colitis (UC) and Crohn disease (CD) are both characterized by chronic lesions of the intestinal mucosa. Current medicine lacks an adequate therapy to eradicate these diseases of unknown etiology. Efforts have been focused on isolation of LAB with potentially probiotic features from fermented cucumbers. It was found that plantaricin C19 was active against *Listeria grayi* IP 6818 (Atrih et al., 1993).

Shukla and Goyal (2014) isolated *Pediococcus pentosaceus* CRAG3 (GenBank accession number JX679020) from fermented cucumber, which displayed probiotic properties and expressed glucansucrase, an enzyme responsible for the biosynthesis of glucan, a compound with a potential role for functional food applications. Tamminen et al. (2012) studied the viability of a potentially probiotic *Lactobacillus paraplantarum* DSM 14,485 in the intestinal tract of 22 healthy test subjects in a randomized double-blind crossover study. The subjects were given vegetables fermented by *L. paraplantarum* DSM 14,485 as starter. The isolated strain has shown essential in vitro probiotic qualities, such as tolerance of bile salts, pancreatic juice, and acidity, and adhesion to human intestinal mucus.

21.5.2 Production of Bacteriocins

The term "bacteriocin" was introduced in 1953 to define colicin produced by *Escherichia coli*. As reported by Cotter et al. (2005) "bacteriocins can be used to confer a rudimentary form of innate immunity to foodstuffs." Bacteriocins are ribosomally synthesized, extracellularly released low-molecular-mass peptides or proteins which exert a bactericidal or bacteriostatic effect on other bacteria, either in the same species (narrow spectrum) or across genera (broad spectrum). Bacteriocins of less than 20 kDa cause depolarization of the target cell membrane and/or inhibit cell wall synthesis and those with more than 20 kDa degrade the murein layer (Settanni and Corsetti, 2008; Cotter et al., 2005). Bacteriocin production has been found in numerous species of bacteria, among which, due to their "generally recognized as safe" (GRAS) status, LAB have attracted great interest in terms of food safety (Singh and Ramesh, 2008). Therefore, substantial research has been done in recent years to isolate bacteriocin-producing LAB from natural sources and exploit them for production and preservation of fermented product.

The importance of bacteriocin-producing LAB is even more in relation to the microbial interaction observed in a natural fermentation process, wherein antagonistic cultures are expected to enjoy a distinct selective advantage as fermentation proceeds. Halami et al. (2005) found that fermenting cucumber was a potential source for the isolation of pediocin-like bacteriocin producers. Singh and Ramesh (2008) identified pediocin, plantaricin A, mesentericin, enterocin A, and nisin obtained at different time periods from fermented cucumbers in 4% saline solution in a static mode at 35°C. The bacteriocins produced by LAB belong to the class IIa category, which, like pediocins, are particularly known for their strong antilisterial activity, making them attractive candidates for application in food systems (Singh and Ramesh, 2008).

Recently bacteriocins found an application as active constituents in active packaging technology due to their cationic character and easy interaction with Gram-positive bacteria (Settanni and Corsetti, 2008).

21.5.3 Fermented Cucumber as a Source of Starter Cultures Able to Produce Oligosaccharides

Fermented cucumber can be a source of starter cultures with the ability to produce oligosaccharides. It was shown that *Leuconostoc* is the major bacterial genus in the initial phase of the lactate fermentation of vegetables. The dextransucrase produced by this bacterium is used to synthesize dextran polymers or prebiotic oligosaccharides (Eom et al., 2007). Dextrans are a type of α-glucans which are comprised of a polymeric chain of glucosyl units and that are synthesized by dextransucrase by acceptor reaction. The polysaccharides, especially α-glucans and β-glucans, serve as potential immunostimulator, immunomodulator and anticancer agents. The dextrans from *Leuconostoc* have been extensively studied, while those from *Weissella* and *Pediococcus* species have not been widely studied. Shukla and Goyal (2013) found porous branched dextran produced from *Pediococcus pentosaceus* CRAG3 isolated from fermented cucumbers. In vitro cytotoxicity analysis of dextran showed that it displayed anticancer activity against cervical cancer (HeLa) and colon cancer (HT29) cell lines which might be due to the ability of dextran to modify membrane surface proteins in tumor cells, leading to adhesion disturbances (Shukla and Goyal, 2013).

21.5.4 Immune Modulation

LAB have been shown to induce different mucosal immune responses, including both specific and nonspecific responses. Similarly, LAB strains from fermented vegetable pickles have been effective in ameliorating type-1 allergies, via improvement of Th1/Th2 balance (helper T cells) and decrease in IgE production (Masuda et al., 2010).

21.5.5 Macroelements From Fermented Cucumber

The partial substitution of sodium chloride with calcium, magnesium, and potassium chloride salts during fermentation may provide additional health benefits to fermented cucumber consumption. The K, Ca, and Mg are macroelements and are mentioned in the list of nutrients that can be included in the nutritional labeling in the USA and the EU (European Commission, 1990; Code of Federal Regulations, 2007). These elements have also permitted health claims within the current EU legislation (European Commission, 2012).

21.5.6 Removal of Antinutrient Compounds

Most vegetables contain toxins and antinutritional compounds. Such antinutrients include oxalate, protease, and α-amylase inhibitors, lectins, condensed tannins, and phytic acid. Polyphenols chelate metals, such as iron and zinc, and reduce the absorption of these nutrients, but they also inhibit digestive enzymes and may also precipitate proteins. LAB fermentation results in reduction of antinutritional factors and, therefore, their potential to enhance the absorption of certain nutrients.

21.6 FINAL REMARKS

Current industrial fermentation practices are greatly based on traditional practices that have been adapted to a larger scale. The naturally fermented sour pickled cucumbers belong to the products stabilized with salt and lactic acid. Fermented cucumbers are microbiologically safe, nutritious, and have appealing sensory characteristics, and can be conveniently stored for extended periods without refrigeration (Breidt et al., 2013). Factors affecting cucumber fermentation have an impact on shelf-life, the nutritional and sensory value, flavors, digestibility, nutrient retention, and reduction of toxicity and antinutrients. During fermentation, LAB synthesize several bacteriocins and liberate antimicrobial peptides, which are inhibitory to spoilage bacteria. LAB-fermented cucumber help to enhance human nutrition in several ways, such as the attainment of balanced nutrition, providing minerals, and contributing to disease prevention as a supplier of probiotic bacteria. Basic understanding about the relationship between food, beneficial microorganisms, and human health is important to improve the quality of fermented cucumbers and the prevention of several diseases.

REFERENCES

Abu-Reidah, I.M., Arráez-Román, D., Quirantes-Piné, R., Fernández-Arroyo, S., Segura-Carretero, A., Fernández Gutiérrez, A., 2012. HPLC–ESI-Q-TOF-MS for a comprehensive characterization of bioactive phenolic compounds in cucumber whole fruit extract. Food Research International 46, 108–117.

Ana, A.S., Azeredo, D.P., Costa, M., Macedo, V., 2002. Analysis of risks of minimal processing of vegetables. Higiene Alimentaria 16, 80–84.

Atrih, A., Rekhif, N., Milliere, J.B., Lefebvre, G., 1993. Detection and characterization of a bacteriocin produced by *Lactobacillus plantarum* C19. Canadian Journal of Microbiology 39, 1173–1179.

Atul, K.S., Ramesh, A., 2008. A succession of dominant and antagonistic lactic acid bacteria in fermented cucumber: insights from a PCR-based approach. Food Microbiology 25, 278–287.

Breidt, F., McFeeters, R.F., Perez-Diaz, I., Lee, Z.-H., 2013. Fermented vegetables. In: Doyle, M.P., Buchanan, R.L. (Eds.), Food Microbiology: Fundamentals and Frontiers. ASM Press, Washington DC, pp. 841–855.

Chen, Y.S., Wu, H.C., Lo, H.Y., 2012. Isolation and characterization of lactic acid bacteria from jiang-gua (fermented cucumbers), a traditional fermented food in Taiwan. Journal of the Science of Food and Agriculture 92, 2069–2075.

Cho, M.J., Buescher, R.W., Johnson, M., Janes, M., 2004. Inactivation of pathogenic bacteria by cucumber volatiles (E, Z)-2, 6- nonadienal and (E)-2-nonenal. Journal of Food Protection 67, 1014–1016.

Chu, Y.-F., Sun, J., Wu, X., Liu, R.H., 2002. Antioxidant and antiproliferative activities of common vegetables. Journal of Agricultural and Food Chemistry 50, 6910–6916.

Code of Federal Regulations, 2007. 21 CFR 101, Food Labeling. http://frwebgate3.access.gpo.gov/cgibin/waisgate.cgi?WAISdocID=69469928114+1+0+0&WAISaction=retrieve.

Cotter, P.D., Hill, C., Ross, R.P., 2005. Bacteriocins: developing innate immunity for food. Nature Reviews: Microbiology 3, 777–788.

Daeschel, M.A., McFeeters, R.F., Fleming, H.P., Klaenhammer, T.R., Sanozky, R.B., 1984. Mutation and selection of *Lactobacillus plantarum* strains that do not produce carbon dioxide from malate. Applied and Enviromental Microbiology 47, 419–420.

Dahal, N.R., Karki, T.B., Swamylingappa, B., Li, Q., Gu, G., 2005. Traditional foods and beverages of Nepal – a review. Food Reviews International 21, 1–25.

Di Cagno, R., Coda, R., De Angelis, M., Gobbetti, M., 2013. Exploitation of vegetables and fruits through lactic acid fermentation. Food Microbiology 1, 1–10.

Divya, J.B., Varsha, K.K., Nampoothiri, K.M., Ismail, B., Pandey, A., 2012. Probiotic fermented foods for health benefis. Engineering in Life Sciences 4, 377–390.

Eom, H.J., Seo, D.M., Han, N.S., 2007. Selection of psychrotrophic *Leuconostoc* spp. producing highly active dextransucrase from lactate fermented vegetables. International Journal of Food Microbiology 117, 61–67.

Etchells, J.L., Costilow, R.N., Anderson, T.E., Bell, T.A., 1964. Pure culture fermentation of brined cucumbers. Applied Microbiology 12, 523–535.

European Commission, 1990. Council Directive 90/496/EEC of 24 September on nutrition labeling for foodstuffs. Official Journal of the European Communities 276, 40–44.

European Commission, 2012. Commission Regulation (EU) 432/2012 of 16 May 2012 establishing a list of permitted health claims made of foods, other than those referring to the reduction of disease risks and to children's development and health. Official Journal of the European Union L136/1–L136/40.

Fleming, H.P., Thompson, R.L., Bell, T.A., Hontz, L.H., 1978. Controlled fermentation of sliced cucumbers. Journal of Food Science 43, 888–891.

Fleming, H.P., Thompson, R.L., McFeeters, R.F., 1996. Assuring microbial and textural stability of fermented cucumbers by pH adjustment and sodium benzoate addition. Journal of Food Science 61, 832–836.

Fleming, H.P., McFeeters, R.F., 2002. Effects of fruit size on fresh cucumber composition and the chemical and physical consequences of fermentation. Journal of Food Science 67, 2934–2939.

Guillou, A.A., Floros, J.D., 1993. Multiresponse optimization minimizes salt in natural cucumber fermentation and storage. Journal of Food Science 58, 1381–1389.

Halami, P.M., Ramesh, A., Chandrashekar, A., 2005. Fermenting cucumber, a potential source for the isolation of pediocin like bacteriocin producers. World Journal of Microbiology and Biotechnology 21, 1351–1358.

Hideki, H., Hidekazu, I., Katsunari, I., Keiko, A., Yoshiteru, S., Isamu, I., 2007. Cucurbitacin C-bitter principle in cucumber plants. JARQ-Japan Agriculture Research Quarterly 41, 65–68.

Holzapfel, W.H., 2002. Appropriate starter culture technologies for mall-scale fermentation in developing countries. International Journal of Food Microbiology 75, 197–212.

Kabak, B., Dobson, A.D.W., 2011. An introduction to the traditional fermented foods and beverages of Turkey. Critical Reviews in Food Science and Nutrition 51, 248–260.

Karovicova, J., Drdak, M., Greif, G., Hybenova, E., 1999. The choice of strains of Lactobacillus species for the lactic acid fermentation of vegetable juices. European Food Research and Technology 210, 53–56.

Kemp, T.R., Knavel, D.E., Stoltz, L.P., 1974. Identification of some volatile compounds from cucumber. Journal of Agricultural and Food Chemistry 22, 717–718.

Lu, Z., Fleming, H.P., McFeeters, R.F., 2002. Effects of fruit size on fresh cucumber composition and the chemical and physical consequences of fermentation. Journal of Food Science 67, 2934–2939.

Marsili, R.T., Miller, N., 2000. Determination of major aroma impact compounds in fermented cucumbers by solid-phase microextraction-gas chromatography-mass spectrometry-olfactometry detection. Journal of Chromatographic Science 38, 307–314.

Maruvada, R., McFeeters, R.F., 2009. Evaluation of enzymatic and non-enzymatic softening in low salt cucumber fermentations. International Journal of Food Science and Technology 44, 1108–1117.

Masuda, T., Kimura, M., Okada, S., Yasui, H., 2010. Pediococcus pentosaceus Sn26 inhibits IgE production and the occurrence of ovalbumin-induced allergic diarrhoea in mice. Bioscience, Biotechnology and Biochemistry 74, 329–335.

Melo, E.D.A., Lima, V.L.A.G., Maciel, M.I.S., Caetano, A.C.S., Leal, F.L.L., 2006. Polyphenol, ascorbic acid and total carotenoid contents in common fruits and vegetables. Brazilian Journal of Food Technology 9, 89–94.

Montet, D., Loiseau, G., Zakhia-Rozis, N., 2006. Microbial technology of fermented vegetables. In: Ray, R.C., Ward, O.P. (Eds.), Microbial Biotechnology in Horticulture, vol. 1. Science Publishers, Enfield, USA, pp. 309–343.

Mukherjee, P.K., Nema, N.K., Maity, N., Sarkar, B.K., 2013. Phytochemical and therapeutic potential of cucumber. Fitoterapia 84, 227–236.

Murcia, M.A., Jiménez, A.M., Martinez-Tomé, M., 2009. Vegetables antioxidant losses during industrial processing and refrigerated storage. Food Research International 42, 1046–1052.

Peres, C.M., Peres, C., Hernandez-Mendoza, A., Malcata, F.X., 2012. Review on fermented plant materials as carriers and sources of potentially probiotic lactic acid bacteria – with an emphasis on table olives. Trends in Food Science & Technology 26, 31–42.

Rastall, R.A., Gibson, G.R., Gill, H.S., Guarner, F., Klaenhammer, T.R., Pot, B., Reid, G., Rowland, I.R., Sanders, M.E., 2005. Modulation of the microbial ecology of the human colon by probiotics, prebiotics and synbiotics to enhance human health: an overview of enabling science and potential applications. FEMS Microbiology Ecology 52, 145–152.

Reina, L.D., Breidt Jr., F., Fleming, H.P., Kathariou, S., 2005. Isolation and selection of lactic acid bacteria as biocontrol agents for nonacidified, refrigerated pickles. Journal of Food Science 70, 7–11.

Ross, R.P., Morgan, S., Hill, C., 2002. Preservation and fermentation: past, present and future. International Journal of Food Microbiology 79, 3–16.

Settanni, L., Corsetti, A., 2008. Application of bacteriocins in vegetable food biopreservation. International Journal of Food Microbiology 121, 123–138.

Shukla, R., Goyal, A., 2013. Novel dextran from *Pediococcus pentosaceus* CRAG3 isolated from fermented cucumber with anti-cancer properties. International Journal of Biological Macromolecules 62, 352–357.

Shukla, R., Goyal, A., 2014. Probiotic potential of Pediococcus pentosaceus CRAG3: a new isolate from fermented cucumber. Probiotics and Antimicrobial Proteins 6, 11–21.

Singh, A.K., Ramesh, A., 2008. Succession of dominant and antagonistic lactic acid bacteria in fermented cucumber: insights from a PCR-based approach. Food Microbiology 25, 278–287.

Sotiroudis, G., Melliou Sotiroudis, E., Chinou, I., 2010. Chemical analysis, antioxidant and antimicrobial activity of three Greek cucumber (*Cucumis sativus*) cultivars. Journal of Food Biochemistry 34, 61–78.

Steinkraus, K.H., 2002. Fermentations in world food processing. Comprehensive Reviews in Food Science and Food Safety 1, 23–32.
Swain, M.R., Anandharaj, M., Ray, R.C., Rani, R.P., 2014. Fermented fruits and vegetables of Asia: a potential source of probiotics. Biotechnology Research Journal. http://dx.doi.org/10.1155/2014/250424.
Tamang, J.P., Tamang, B., Schillinger, U., Franz, C.M.A.P., Gores, M., Holzapfel, W.H., 2005. Identification of predominant lactic acid bacteria isolated from traditionally fermented vegetable products of the Eastern Himalayas. International Journal of Food Microbiology 105, 347–356.
Tamminen, M., Ouwehand, A.C., Mäki, M., Joutsjoki, T., Sjöblom, M., Nissinen, K., Ryhänen, E.-L., 2012. Viability of *Lactobacillus paraplantarum* DSM 14485 in human gastrointestinal tract and its molecular and biochemical identification after fermented vegetable consumption. Agricultural and Food Science 21, 182–196.
Viander, B., Mäki, M., Palva, A., 2003. Impact of low salt concentration, salt quality on natural large-scale sauerkraut fermentation. Food Microbiology 20, 391–395.
Yamani, M.I., Hammouth, F.G.A., Humeid, M.A., Robinson, R.K., 1999. Production of fermented cucumbers and turnips with reduced levels of sodium chloride. Tropical Science 39, 233–237.
Yan, P.-M., Xue, W.-T., Tan, S.-S., Zhang, H., Chang, X.-H., 2008. Effect of inoculating lactic acid bacteria starter cultures on the nitrite concentration of fermenting Chinese paocai. Food Control 19, 50–55.

Chapter 22

Role of Natural Fermented Olives in Health and Disease

C.M. Peres[1], C. Peres[2], F. Xavier Malcata[3,4]
[1]*Instituto de Tecnologia Química e Biológica, Universidade Nova de Lisboa, Oeiras, Portugal;*
[2]*INIAV, IP, Oeiras, Portugal;* [3]*University of Porto, Porto, Portugal;* [4]*LEPABE, Laboratory for Engineering of Processes, Environment, Biotechnology and Energy, Porto, Portugal*

22.1 GENERAL CONSIDERATIONS

Humans have been deliberately fermenting foods as a form of preservation since ancient times, and each nation has its own types of fermented food, that represent the staple diet and the raw ingredients available in that particular place. Today one-third of the human diet globally consists of fermented foods (Borresen et al., 2012). Food fermentation probably originated from natural microbial interactions that led to positive outcomes in terms of food preservation (Vogiatzakis et al., 2006). The nutritional value of a particular food depends on its digestibility and its content of essential nutrients. Both digestibility and the nutrient content may be improved by fermentation. Therefore, fermented foods can be more nutritious than their nonfermented counterparts due to the catabolic activity of microorganisms responsible for the breakdown of complex compounds releasing free fatty acids, amino acids, and simple sugars, but also due to the microbial synthesis of several vitamins and bioactive compounds (Parvez et al., 2006).

Plant-derived fermented products and, in general, fermented olives depend on lactic acid fermentation that typically consists of the conversion of carbohydrates to organic acids, catalyzed by yeasts, bacteria, molds, or a combination thereof (Guerzoni et al., 2011). Microbiota responsible for olive fermentation normally break down carbohydrates, proteins, and lipids present in the raw materials by secreting enzymes, which constitutes a nutritionally desirable event, as the food is partially digested prior to consumption. Besides, fermentation reduces the levels of certain antinutritional factors that interfere with digestion, so making them more efficiently utilized at the human digestive tract (Soomro et al., 2002).

Over recent decades, development and consumption of probiotic foods has increased, along with awareness of their beneficial effects in promoting health, as well as in disease prevention and therapy (D'Aimmo et al., 2007). Fermented olives are considered as a probiotic food, since some of the microorganisms present can survive during passage through the gastrointestinal tract (GIT), and they can exert antimicrobial effects against local pathogens in the colon, among other health effects (Madureira et al., 2005; Peres et al., 2014; Bautista-Gallego et al., 2013; De Bellis et al., 2010). On the other hand, probiotic olives provide plenty of vital nutrients and bioactive compounds that can affect a number of functions of the human body in a positive way. Many bioactive compounds in fermented olives, such as phenolic compounds, have been found to possess a protective effect against several diseases (Martinez and Poole, 2004). Moreover, the beneficial health effects of lactic acid bacteria (LAB) have been scientifically documented (Scott and Sullivan, 2008; Patel et al., 2014). This is a major reason for the current boom of health food demand by educated consumers, for whom a wider variability in sensory terms is also appealing as a sign of less processed and industrialized products.

22.2 PRODUCTION OF TRADITIONAL FERMENTED OLIVES

The olive tree (mainly *Olea europaea* L.) is an important crop in the Mediterranean Basin. Its cultivation probably started about 6000 years ago in present Turkey. About the year 3000 BC, olive trees could be found throughout the Middle East. Phoenicians introduced it to Greek islands from where it extended to all the Mediterranean area, influenced by the Greeks and Romans. Olive trees still have strong historical, symbolic, and economical relevance. Olives have traditionally been a dry extensive culture of the Mediterranean Basin, where they are still significant in the national agricultural trade. They are an essential element for the preservation of Mediterranean ethnicity and environment. More recently, this crop expanded to the United States of America and Australia, where it is drastically expanding (Vossen, 2007).

Worldwide there are about 850 million olive trees which cover more than 10 million hectares of land—but 98% of them are grown in the Mediterranean region, where they play a major role on the environment and the rural economy (FAO, 2004), containing 85% of table olive production in the world. Table olive production is set to rise about 4%, and generated more than 2,660,500 tons in 2013/2014 season, Spain being the world leading table olive producer, with a share of about 20% of global table olive production (IOOC, 2014).

The olive fruit is a drupe containing a bitter component (oleuropein), low sugar concentration, and high oil content, but its composition varies depending on the degree of maturity and the variety (Garrido-Fernández et al., 1997). Even at full maturation, olives are hardly edible as such because of their bitter-tasting compounds, primarily oleuropein, which must be eliminated prior to consumption. Processing also contributes to preserve olives from natural deterioration and, therefore, table olives can be stored and consumed gradually throughout the year. Besides being palatable, the processed olives should be safe upon ingestion and able to retain most of their nutritional attributes (Garrido-Fernández et al., 1997).

According to the International Olive Oil Council (IOOC), the designation of "natural olives" may be applied to fruits in various maturation stages (green, turned color, and black), which have not undergone any alkaline treatment. Natural olives are harvested when the fruits are green, or semi- or fully ripe. Depending on the region, such olives can be reddish black, violet black, deep violet, greenish black, or deep chestnut, with both skin and flesh being colored (IOOC, 2004). The fruits are kept in water until they lose (partially or totally) their natural bitterness (Gomez et al., 2006; Panagou et al., 2008; Oliveira et al., 2004; Silva et al., 2011). "Natural olives" are processed by submerging them directly in brine after harvesting, or repeatedly soaking in water, where fermentation takes place; the finished product retains some fruity and bitter flavors. As per market demand, olives are sorted, graded, and packed; and in some commercial presentations, they can be broken or cut along their longer longitudinal diameter, and/or seasoned with natural products or their flavors (IOOC, 2004). During processing, olives often lose their intense black purple pigments, thus resulting in pale to dark brown colored olives, but the original color can be (partially) restored by exposing them to air after processing. They are preserved in brine, following sterilization or pasteurization, or by addition of a preservative (Panagou et al., 2008).

22.3 OLIVE FERMENTATION

In natural olive fermentations the final product is obtained in brine by interactions between microorganisms, of which metabolism is essential for transformation of natural substrates (Martins et al., 2011) contributing to improve nutritional value, appearance and flavor, as well as favoring the degradation of undesirable factors, contributing to make safer products (Prajapati and Nair, 2003; Vogiatzakis et al., 2006).

Natural olive fermentation relies on a complex microbiota, including *Enterobacteriaceae*, LAB, and other Gram-positive bacteria, as well as yeasts and molds coexisting until the end of fermentation and dominating ecologically over spoilage and pathogenic microorganisms that would otherwise produce sensory and health hazards (Schnurer and Magnusson, 2005). Lactic acid fermentation is considered as the major contributor to the beneficial characteristics observed in fermented foods (Moslehi-Jenabian et al., 2010).

LAB involved in plant food fermentations have been investigated extensively with regard to the release of plant metabolites, such as phenolic compounds. These compounds are often glycosylated and have a low bioavailability. Oleuropein is a phenolic glucoside responsible for the bitterness of unprocessed olives, and partial degradation of oleuropein is required before table olives can be consumed (Peres et al., 2012). Data have demonstrated that α-rhamnosidase and β-galactosidase enzymes produced by LAB can effectively remove glycoside residues from olives (Beekwilder et al., 2009). The most important species for the fermentation of olives is *Lactobacillus plantarum*. However, *Lactobacillus pentosus*, *Lactobacillus brevis*, and *Pediococcus pentosaceus* have been isolated from fermented olives as well (Ghabbour et al., 2011). More recent studies have shown that certain LAB strains tolerate and degrade oleuropein by using oleuropein and X-gluc as substrates; however, it is not known whether LAB can breakdown oleuropein via any route besides acid hydrolysis (Peres et al., 2012, 2014). Recent studies have accordingly focused on selected LAB to bring about debittering of olives via degradation of oleuropein, to control microbial spoilage, and to improve fermentation performance, all with the final goal of product optimization (Peres et al., 2012, 2014). β-Glucosidase activity could be a promising way to design an eco-friendly approach in table olive processing, reducing the impact of lye (in Spanish type) by using microorganisms with a β-glucosidase activity (Tofalo et al., 2014).

The yeasts isolated are mainly of the species *Saccharomyces*, *Kluyveromyces*, *Debaryomyces*, and *Rhodotorola* spp. (Jones and Jew, 2007). Yeasts appear to be active in synthesizing vitamins, amino acids, and purines, and also hydrolyze complex carbohydrates that are essential for the growth of *Lactobacillus* spp. Thiamine (vitamin B_1), nicotinic acid, pyridoxine (vitamin B6), and pantothenic acid are among the vitamins and other enzyme cofactors accumulated and/or synthesized by yeasts (Arroyo-Lopez et al., 2008; Silva et al., 2011; Peres et al., 2012). On the other hand, certain molds contain cellulases that cannot be synthesized by human beings, and which will soften the texture of the food and release sugars that would otherwise be unavailable. Microbial cellulases hydrolyze cellulose into sugars, which are then readily digestible by humans; similarly, pectinases soften the texture of foods and release sugars for digestion (Odunfa et al., 2001).

As the raw materials become hydrolyzed, the environment also changes—and sometimes undergoes a pH drop. In addition, the extra peptides and amino acids thus formed may be further converted into smaller volatile molecules—that are odoriferous, and thus improve the flavor characteristics of the fermented olives (General et al., 2011).

22.4 LACTIC ACID BACTERIA OF OLIVE FERMENTATION AS PROBIOTICS

LAB have received considerable attention as probiotics over the past few years. They have "generally regarded as safe" (GRAS) status and also provide a number of good technological properties and clinically validated and documented health effects (Aslam and Qazi, 2010). The majority of microorganisms used as probiotics belong to the LAB and bifidobacteria. Bifidobacteria are microorganisms of paramount importance in the active and complex ecosystem of the intestinal tract of humans and other warm-blooded animals (Figueiroa-González et al., 2011). The indigenous microbiota of infants is dominated by bifidobacteria, which is established shortly after birth; their number decreases with increasing age of an individual and eventually becomes the third most abundant genus (Soccol et al., 2010).

Lactobacilli, distributed in various ecological niches throughout the gastrointestinal and genital tracts, are rarely associated with cases of gastrointestinal infection, and are regarded as nonpathogenic and safe microorganisms. On the contrary, they are known as health promoters, especially in these human tracts (Soccol et al., 2010). Probiotic foods have to date been restricted almost exclusively to dairy products, and have encompassed mainly *Bifidobacterium* spp. However, vegetable matrices, such as fermented olives, constitute a rich alternative source of probiotic microorganisms, mainly LAB and yeasts (Peres et al., 2014; Silva et al., 2011; Argyri et al., 2013; Blana et al., 2014). Within the group of LAB, *Lactobacillus* species are the most commonly utilized group of microorganisms for their potential beneficiary properties as probiotics (Pundir et al., 2013).

The skin of olives contains microscopic pores, called stomata, which are thought to be the primary portals for entrance and exit of solutes and gases. Their location at the interface between internal plant tissues and the environment makes them convenient gates for endophytic bacteria colonization; in this regard, LAB have been shown to enter through stomata and predominate in the intercellular space of the substomal cells (Nychas et al., 2002). These findings highlight the potential of table olives as biological vectors for selected probiotic strains Lavermicocca et al. (2005) assessed the ability of human-borne probiotic strains, belonging to *Lactobacillus rhamnosus*, *Lactobacillus paracasei*, *Bifidobacterium bifidum*, and *Bifidobacterium longum* to incorporate in, and survive on the olive surface. After a 3-month experiment, most strains showed an inconspicuous compatibility with olives, likely because they are not naturally involved in spontaneous fermentation and are hardly adapted to the prevailing environmental conditions. For instance, bifidobacteria and *L. rhamnosus* GG showed a moderate survival rate until 30 days (c. 10^6 colony-formimg units [CFU]/g), but the population remained unchanged afterwards for *L. rhamnosus* GG, while bifidobacteria declined by c. 1 log cycle. Conversely, *L. paracasei* IMPC2.1 showed the highest viability (over 10^7 CFU/g) throughout the whole experiment. In order to validate table olives as carrier matrix of bacteria into the human GIT, assays were performed in human volunteers fed probiotic olives and strain *L. paracasei* IMPC2.1 was recovered from fecal samples from four of the volunteers. The microstructure of the olive fruit surface apparently guarantees cell integrity during transit through the GIT (Lavermicocca et al., 2005) and international patents covering these aspects have been issued (Lavermicocca et al., 2007).

Despite the dual role of starter and probiotic of *Lactobacillus paracasei* IMPC2.1 for development of a new table olive-based probiotic product (Sisto and Lavermicocca, 2012), most strains proposed for "probiotication" of vegetables are of dairy origin, thus exhibiting a poor adaptation to the harsh conditions prevailing in the brine of fermented table olive (Bevilacqua et al., 2010). Therefore, research of table olives as a wild source of LAB with probiotic and technological features constitutes an emerging (and promising) field of work (Peres et al., 2012). Previous screening for such probiotic properties as hemolytic activity, acid and bile salt tolerance, inhibition of pathogenic bacteria, and antibiotic resistance have been performed on strains that had been isolated from fermented olives (Peres et al., 2014; Mourad and Nour-Eddine, 2006). In addition, *Lactobacillus plantarum* strains isolated from Bella de Cerignola table olives have confirmed the possibility of multifunctional starters with probiotic features, including their ability to adhere to mammalian jejunum epithelial cells IPEC-J2, coupled with desired technological properties (Bevilacqua et al., 2010).

Interest in the relationship between diet, health, and well-being has been growing exponentially in Europe, thus mimicking a trend observed also in the rest of the developed world. Probiotic bacteria are increasingly incorporated into food products as carriers for their delivery to the human body intended to confer health benefits in the human gut and beyond (Sanders and Marco, 2010). Food regulates their colonization and contains other functional ingredients, such as bioactive components, which may interact with probiotics to alter their functionality and efficacy. Tolerance to gut environment, adherence to gastrointestinal epithelium and the acid production of probiotics are affected by the food ingredients used in probiotic delivery (Ranadheera et al., 2010). The exopolysaccharide-producing ability of the probiotics could be an appropriate way to obtain strains with adequate viability, since these polymers can act as protectors of the producing bacteria, contributing to their viability (Ramchandran and Shah, 2010). Ten potential probiotic strains isolated from Portuguese olive fermentation are able to produce exopolysaccharides (Peres et al., 2014).

Furthermore, a shift of interest has been recently experienced toward strains possessing probiotic features (Bautista-Gallego et al., 2013; Abriouel et al., 2012; Argyri et al., 2013). Technological issues related to microorganisms in probiotic foods are usually complex and diverse. Processing of microbial systems aiming at manufacturing probiotic foods can be organized into processing of starter cultures, processing of products with desired functional characteristics, and processing of foods containing probiotics. In order to make the ideal culture for any particular food application, it is necessary to understand the function we demand of the culture, and to have tools to improve the function of the culture (Hansen, 2002). Technological challenges include the necessity to obtain high initial productivity and viability as starter cultures, as well as high stability, viability, and productivity as probiotic strains (Yerlikaya, 2014).

Several factors must be considered when using probiotic bacteria in fermented products, in general being applied to fermented olives (Blana et al., 2014). The probiotics must be viable and present in high counts ($6-7 \log \text{CFU/g}$)

at time of consumption to achieve the desired benefits. Probiotics have been consumed regularly in a quantity of higher than 100 g per day, in other words, at least 9 log CFU per day (Codex, 2003). This level could be strain-specific, and the number of surviving cells along the GIT is more important (Sarvari et al., 2014). However, many probiotics containing food products fail to maintain the recommended probiotic concentrations due to instability of probiotics in food matrices (Sadeghar et al., 2012). Taking into account table olives, the daily recommendation for a healthy adult is around 25 g of olives per day, or about seven olives carrying about 10^9–10^{10} viable cells of probiotic strains (Lavermicocca et al., 2005).

22.5 HEALTH EFFECTS OF OLIVE FERMENTATION PROBIOTICS

Probiotics, in general, are receiving a great deal of attention in different fields of microbial biotechnology applications including health improvement and therapeutic purposes. In particular, there are important advances on LAB and yeast characterization from fermented olives with a focus on their particular technological and probiotic features (Hurtado et al., 2012; Blana et al., 2014; Bautista-Gallego et al., 2013; Figeiroa-González et al., 2011; Soccol et al., 2010). Recent research effort has demonstrated that LAB strains isolated from olive fermentations are potential candidates for a probiotic culture, and consequently for future development of novel health-promoting foods from plant origin (Peres et al., 2014; Silva et al., 2011).

As already emphasized, lactic acid fermentation is the predominant biological preservation method contributing to unique flavor development, and considered as the major contributor to the beneficial characteristics observed in fermented foods including olive fermentation (Soomro et al., 2002; Peres et al., 2012, 2014; Blana et al., 2014; Bautista-Gallego et al., 2013; Botta and Cocolin, 2012; Giraffa, 2004). As a result of an increased number of investigations in the probiotic field, this concept has been expanded to include bacteria from traditionally fermented foods; they constitute the untapped source for a wide variety of beneficial probiotic microorganisms (Zeng et al., 2010). Probiotics are involved in all fermentations and LAB are the most common type of microorganisms used as probiotics, frequently in association with yeasts (Reid et al., 2003; Viljoen, 2006).

There has been a recent increase of interest in probiotics due to consumer demand for better therapies and problems, such as drug resistance and opportunistic infections due to inadvertent use of antimicrobial therapy (Peres et al., 2012). Recent scientific evidence has in addition proven that diet may modulate several functions in the human body—as well as delay, or even decrease, the incidence of several chronic diseases and health conditions (Marteau et al., 2001). In fact, when the intestinal microbiota is altered, administration of probiotic bacteria not only reestablishes its normal equilibrium, but also improves the microbial balance and properties of the endogenous microbiota (Cross, 2002). Health conditions that can benefit from probiotics therapy include diarrhea, gastroenteritis, irritable bowel syndrome, inflammatory bowel disease, colitis, and alcoholic liver disease; probiotic consumption will also reduce the risk of colon and liver cancer (Juan et al., 2006; Cross, 2002).

To understand how probiotics influence gastrointestinal function, it is important to know about the physiology, microbiology of GIT, and the digestive process. Probiotics increase intestinal mucin production, which prevents the attachment of enteropathogens by steric hindrance or through competitive inhibition for attachment sites on mucins, as well as augmenting production of antimicrobial peptides, and decreasing epithelial permeability to intraluminal pathogens and toxins (Quigley, 2010). Assays on antagonistic activity performed with some potential probiotic lactobacilli strains isolated from olive fermentation against 10 target gastrointestinal pathogens reduced their incidence; this suggests that they can act either directly when included in the olives, or after accommodation in the intestine of the host, after ingestion (Peres et al., 2014).

Verocytotoxigenic *Escherichia coli* (VTEC), *Salmonella enterica*, and *Listeria monocytogenes* are among the most important agents responsible for food outbreaks worldwide and the most important zoonotic agents in the European Union (EU). Most of these cases were caused by the serogroup O157 (EFSA & ECDC, 2015). Two *Lactobacillus* spp. strains, *Lactobacillus plantarum* LB95 and *Lactobacillus paraplantarum* LB13, previously isolated from spontaneously fermenting olive brines, and two reference probiotic strains, *Lactobacillus casei* Shirota and *Lactobacillus rhamnosus* GG, were investigated in the attenuation of the virulence of four strains of these three pathogenic species using animal cell culture assays. In competitive exclusion assays, the relative percentages of adhesion and invasion of *S. enteritidis* were significantly reduced when the human HT-29 cell line was previously exposed to *L. plantarum* LB95. Similarly, the relative percentage of invasion of *L. monocytogenes* was significantly reduced when the HT-29 cells were previously exposed to *L. plantarum* LB95. Results suggested that *L. plantarum* LB95 isolated from Portuguese olive fermentations may be able to attenuate the virulence of Gram-positive and Gram-negative foodborne pathogens, which together with other reported features of these strains point toward their potential use in probiotic foods with interesting potential in preventing enteric infections in humans (Dutra et al., 2016). Previous research done in Portuguese table olives indicated that several of the bacteria present during the fermentation process have the ability to inhibit the growth of *Helicobacter pylori*, a common human pathogen that is resistant to a growing number of antibiotics. This shows a potential for such probiotics to be used as an antibiotic alternative (unpublished results).

Cholesterol, a fatty substance known as a lipid vital for the normal functioning of the body, is a precursor to certain hormones and vitamins and a component of cells; it is mainly produced by the liver but can also be found in some foods (Nes, 2011). Hypercholesterolemia is considered a major risk factor for the development of coronary heart disease, with cellular cholesterol homeostasis being very important for the prevention of cardiovascular diseases (Reyes-Nava and Rivera-Espinoza, 2014). The main cause of the reduced flow in the heart is an accumulation of plaques in arteries, mainly in the intimae of arteries, causing atherosclerosis (Clark and Weiser, 2013). Regular consumption of probiotics could result in lower blood pressure, particularly, in people with hypocholesterolemic high blood pressure; consequently, scientific evidence indicated that there is potential for the derivation of health benefits providing a prophylactic effect in heart disease, through consumption of food containing probiotics (Parvez et al., 2006; Sobol, 2014). The cholesterol-lowering effects can be partially ascribed to bile salt hydrolase activity—other possible mechanisms include assimilation of cholesterol by the bacteria, binding of cholesterol to the bacterial cell walls, or physiological actions of the end products of short-chain fatty acid fermentation. One important transformation of bile acids is via deconjugation, a reaction that occurs before further modifications and it is catalyzed by bile salt hydrolase (BSH) enzymes (Aloglu and Oner, 2006; Begley et al., 2006). LAB with active BSH have been claimed to lower cholesterol levels via interaction with host bile salt metabolism; enzymes able to deconjugate bile salts to amino acids and cholesterol reduce the corresponding toxicity. This mechanism could be used in the control of cholesterol levels in the blood by conversion of deconjugated bile acids into secondary bile acids by probiotics (Kumar et al., 2012). The ability of probiotic strains to hydrolyze bile salts has often been included among the criteria for probiotic strain selection. However, no *Lactobacillus* isolated from spontaneous fermented olives have shown BSH activity so far (Peres et al., 2015).

Carcinogens can be ingested or generated by metabolic activity of microorganisms that live in the gastrointestinal system (Shinde, 2012). It has been hypothesized that probiotic cultures may decrease the exposure to chemical carcinogens by detoxifying ingested carcinogens, modulation of the environment of the intestine, and thereby decreasing populations or metabolic activities of bacteria that may generate carcinogenic compounds, reducing the production of carcinogenic metabolic products, production of anticarcinogenic compounds, or stimulation of the immune system (Shinde, 2012; Lee et al., 2000). Cases of microbial foodborne infections associated with carcinogen production have been reported in association with fermented olives, as well as risk involving microbial food intoxications due to mycotoxins from raw materials, production of bacterial toxins or possible mycotoxin production by fungi, toxins by *Escherichia coli* O157:H7, and other toxic fermentation products including biogenic amines. Mycotoxins occur in several fermented food products associated with specific species or subgenera of *Aspergillus* (Palencia et al., 2010), *Penicillium*, and *Fusarium* genera. The presence of mycotoxins has been recently reported in table olives (El Adlouni et al., 2006; Franzetti et al., 2011), mainly in the type "natural olives" where the environmental conditions are adequate for the presence of those mold mycotoxin-producers. Aflatoxins have been recently considered as an important safety problem (Jarvis, 2002). Once ingested and due to their low molecular weight, they are rapidly adsorbed in the GIT through a nondescribed passive mechanism, and then quickly appear as metabolites in the blood after just a few minutes (Moschini et al., 2006). They are hepatocarcinogenic and recent studies also suggest that the B aflatoxins may cause neural disturbances (Wild and Gong, 2010). The elucidation of metabolic pathways responsible for the production of aflatoxins by *A. flavus* group has progressed rapidly. Studies have identified aflatoxins as the toxicological agent that initiated mycotoxicology as a serious and complex problem of food safety (Hussein and Brasel, 2001). Biodegradation of aflatoxins by LAB offers an attractive alternative for their elimination in fermented olives, preserving their quality and safety. Potential probiotic *Lactobacillus* spp. strains from olive fermentation effectively removed aflatoxin B1, via binding and reduction of its bioaccessibility from artificially contaminated olive brines (in situ assays). These results support the suggestion that specific *Lactobacillus* strains from fermented olives can reduce the bioavailability of dietary carcinogens (unpublished data).

Escherichia coli O157:H7 toxins have been referred to as a problem occurring during olive technology (Argyri et al., 2013; Grounta et al., 2013). Enterohemorrhagic *E. coli* must survive during passage through the GIT and express virulence genes (Barnett Foster, 2013). Studies to assess the survival and toxin production by *E. coli* O157:H7 during its transit through the GIT were performed. Results indicate that, under the tested conditions, the pathogen was not able to produce Shiga toxins at detectable levels. The high survival rate of the enteropathogen in colon conditions suggests that toxin production would take place there but this phenomenon can be controlled by probiotics (Arroyo-Lopez et al., 2014).

Biogenic amines have been also detected in fermented olives and their occurrence is attributed to spoilage microorganisms possessing decarboxylase enzymes, which convert amino acids to amines. Putrescine seems to be produced during the active fermentation phase of olive fermentation, maintaining low levels in the final products. The effects of temperature and the debittering process on cadaverine and tyramine formation related to fermented olive "zapatera" spoilage have been reported (Arena et al., 2007; Garcia et al., 2004). Limits of toxicity of biogenic amines in a given product have not been established yet, because their effects do not depend on their presence alone (type of amine and levels present), but they are also influenced by other compounds (Linares et al., 2011).

Despite functional properties of phenolic compounds from table olives, as well as their promising performance as carriers of probiotic strains, studies focusing on the combined effect of phenolic compounds with such wild probiotic bacteria have been reported (Dutra et al., 2016; Peres et al., 2015). Studies to elucidate the relationship between probiotic strains originating in fermented table olives and the foodborne pathogen *Escherichia coli*, when in the presence of oleuropein and hydroxytyrosol at the recommended daily dose for table olives, were performed (Peres et al., 2015). Results showed that phenolic compounds can modify the intestinal microbiota positively and prevent the in vitro colonization of *E. coli* by adhesive reduction, which can be a potential approach to effectively control the pathogenic infections (Chapman et al., 2014; Dunne et al., 2001). Data on biofilm formation and biofilm composition (probiotics and *E. coli*) suggested that the synergic effect of probiotic and oleuropein and hydroxytyrosol can effectively affect the *E. coli* biofilm growth, explaining the hypothetical decreasing of the adhesion capacity of *E. coli* by the competition for sites of intestinal cells (Peres et al., 2015; Tahmourespour and Kermanshahi, 2011). The potential synergism between phenolic compounds and probiotic bacteria may be taken advantage of to selectively stimulate proliferation or activity of probiotics aimed at pathogen control.

22.6 HEALTHY BIOACTIVE MOLECULES AND METABOLITES FROM FERMENTED OLIVES

With table olives being one of the most popular fermented products consumed in the West, the vast majority of people there have an incipient knowledge on their nutritional and health value. Biochemical and epidemiological studies demonstrated that the Mediterranean dietary pattern with high intake of olive antioxidants is associated with a low incidence of chronic diseases—including cardiovascular diseases and cancer, and showed beneficial effects on *diabetes mellitus* and glucose metabolism in general (Martinez-Gonzalez et al., 2008). Many of the health benefits are attributed to monosaturated fatty acids (mainly oleic acid), and phenolic compounds (mainly oleuropein and hydroxytyrosol) (Garrido-Fernández et al., 1997), known by their important biological attributes, such as antioxidant, antiinflammatory, and antibacterial properties (Bianchi, 2003).

Southern European populations that consume a traditional Mediterranean Diet, where olive products are the primary source of fat, have lower rates of cancer than in North America, Northern Europe, and Australia (Trichopoulou et al., 2000). In general, cancer is caused by mutation or activation of abnormal genes that control cell growth and division. One of the mechanisms linking olive intake to cancer protection may involve genes. Most of the abnormal cells do not result in cancer as normal cells usually out-compete abnormal ones. The immune system also recognizes and destroys most abnormal cells (Shyu et al., 2014). Research on whole olives and cancer has often focused on breast cancer and stomach (gastric) cancer. In the case of breast cancer, special attention has been paid to the triterpene phytonutrients in olives, including erythrodiol, uvaol, and oleanolic acids that have been shown to help interrupt the life cycle of breast cancer cells. Interruption of cell cycles has also been shown in the case of gastric cancer, but with this second type of cancer, the exact olive phytonutrients involved are less clear (Allouche, 2011).

Oleic acid has been found to be particularly effective against breast, colon, and prostate cancer cells by interaction with the human genome and suppressing the overexpression of an oncogene, invasive progression, and metastasis (Menendez and Lupu, 2006). Research on the antioxidants of olive oil has also focused on their capacity to inhibit proliferation and promote apoptosis in several tumor cell lines (Visioli et al., 2004). Olives contains oleocanthal, an in vitro cyclooxygenase inhibitor with potential antiinflammatory and analgesic properties similar to the nonsterol antiinflammatory drug ibuprofen. Ibuprofen reduces the risks of some cancers and platelet aggregation. Oleocanthal may also offer special protection against Alzheimer's disease (Pitt et al., 2009).

Fermentation can positively affect the nutritional quality of food by inducing important physicochemical changes that improve the nutrient density and increase its amount and bioavailability. Fermentation may reduce the content of nondigestible material, such as cellulose, hemicellulose, and polygalacturonic and glucuronic acids hydrolyzed by the enzymatic action of LAB, either polysaccharides or other high-molecular-weight compounds (Furukawa et al., 2013). The acidic nature of the fermentation products enhances the activity of microbial enzymes. Enzyme involvement of cellulose, hemicellulose, and related polymers in the cell walls of plant tissues during fermentation release the nutrients locked into plant structures and cells and improve nutritionally the fermented product (Potter and Hotchkiss, 2006). In addition, fermentation reduces the levels of some antinutritional compounds present in vegetables, such as oxalate, protease and α-amylase inhibitors, lectins, tannins, and phytic acid (Swain et al., 2014), which generally interfere with the assimilation of some nutrients. The reduction of antinutrients leads to an increased bioavailability of minerals, such as iron, proteins, and simple sugars. Increased utilization of iron from fermented foods is due to breakaway of inorganic iron from complex substances under the influence of vitamin C (Akande et al., 2010).

Table olives are also rich in natural antioxidants, such as vitamins. They provide small amounts of B group vitamins as well as liposoluble vitamins, such as provitamin A and vitamin E, considered to have great antioxidant effects (Peres et al., 2012). Plants, yeasts, and some bacterial species from fermented food contain the folate biosynthesis pathway and produce natural folates, but mammals lack the ability to synthesize folate and they are therefore dependent on sufficient intake from the diet (Kariluoto et al., 2006). *Saccharomyces cerevisiae* is a rich dietary source of native folate and produces high levels of folate (Hjortmo et al., 2008) essential for LAB growth in olive fermentation (Peres et al., 2012). Vitamin B12 synthesized by propionibacteria is an important cofactor for the metabolism of fatty acids, amino acids, carbohydrates, and nucleic acids. Vitamin K, an essential cofactor, is synthesized by LAB and is involved in the posttranslational modification of glutamic acid residues in certain proteins, such as blood clotting proteins and proteins involved in tissue calcification (Subrota et al., 2013).

The content of nutritionally interesting antioxidant phenolic compounds is also noteworthy in table olives, as compared with other fruits (Visioli et al., 2002; Tripoli et al., 2005). Phenolic compounds may contribute to fruit quality in a number of ways, for instance by contributing to sensory attributes, such as color and flavor. This is the case for some specific phenolics, in particular oleuropein (Bianchi, 2003; Ozdemir et al., 2014; Soler-Rivas et al., 2000), given its intense bitterness. Other promising substances from olive tissues include hydroxytyrosol, tyrosol, 5-O-caffeoilquinic acid, verbascoside, luteolin 7-O-glucoside, rutin, and luteolin, as well as two secoiridoid glucosides of uncertain structure containing tyrosol, elenolic acid, and glucose moieties, and phospholipid complex, which work synergistically, and provide an effective year-round defense system for the olive tree (Blekas et al., 2002).

It is estimated that dietary intake of phenolics is approximately 1 g/day. Bioavailability of these bioactive components is dependent on food preparation processes, gastrointestinal digestion, absorption, and metabolism (Scalbert and Williamson, 2000). It has been estimated that more than 50% of phenolic compounds from olive oil and olives are absorbed after ingestion, primarily in the small intestine (Vissers et al., 2002). Oleuropein and hydroxytyrosol represent the molecules with major biological and pharmacological interest, and are among the most investigated antioxidant natural compounds (Omar, 2010; Bulotta et al., 2014). Olive fermentation plays an important role in phenolic compound degradation since they can be hydrolyzed and solubilized into the brine. Recent studies on product optimization have focused on selected LAB to bring about debittering of green olives via degradation of oleuropein, to control microbial spoilage and to improve fermentation performance (Peres et al., 2012). *L. plantarum* strains isolated from olive fermentation brines synthesize β-glucosidase enzyme that can hydrolyze oleuropein in vitro. First, the hydrolysis of the glycosidic linkage of the oleuropein by β-glucosidase synthesized by *L. plantarum* strains with formation of oleuropein-aglycone occurs; this aglycone is converted to elenolic acid and hydroxytyrosol by hydrolysis, from *L. plantarum* esterase action (Peres et al., 2014). Experimental studies encompassing the ecosystem olive/brine under commonly encountered conditions have indicated that special strains of adventitious LAB can use phenolic compounds as the only carbon source in the absence of sugars (Blana et al., 2014; Ghabbour et al., 2011; Ciafardini et al., 1994).

Hydroxytyrosol has been recognized as GRAS and this compound and other phenolic constituents of olives have shown potent biological activities in vitro, including antioxidant action related to their highly bioavailability (Visioli and Galli, 2002). Tyrosol and hydroxytyrosol are absorbed by humans after ingestion in a dose-dependent manner, and they are excreted in the urine as glucuronide conjugates (Visioli et al., 2000; Covas et al., 2006; Furneri et al., 2002). Hydroxytyrosol is endowed with interesting pharmacological activities, many of which have already been demonstrated in vivo (Romero et al., 2007; Bulotta et al., 2014). This last evidence may have important implications in protection from atherosclerosis (Visioli and Galli, 2002; Miles et al., 2005; Covas et al., 2006). In particular, hydroxytyrosol and oleuropein scavenge free radicals, such as superoxide (Rietjens et al., 2007). Hydroxytyrosol induces the synthesis of detoxifying enzymes that are involved in the protection against oxidative damage and mitochondrial biogenesis, critical pathways occurring in the fight against oxidative stress (Zhu et al., 2010).

Many diseases are associated with low-grade inflammation triggered and sustained by oxidative stress. A growing body of research supports that chronic inflammation is the root cause of major illnesses, such as neurological disorders (Parkinson and Alzheimer diseases), obesity, metabolic syndrome, rheumatoid arthritis, cancer, and cardiovascular diseases (Wahle et al., 2004). Several phenolic acids affect the expression and activity of enzymes involved in the production of inflammatory mediators of pathways thought to be important in the development of gut disorders. Olive phenolic compounds are also bioactive against specific cancer lines (Juan et al., 2006). Studies investigating the effect of oleuropein and hydroxytyrosol from olives on HT-29 cancer cell proliferation showed that there was a reduction in cell viability in a concentration-dependent manner, after exposure to both compounds; the effect was more evident for the last one (Peres et al., 2015), which can be due to the number and position of hydroxyl groups (Zhiping et al., 2014). Results suggest that these bioactive compounds may protect against cancer by preventing initiation of the chain of reactions that transform normal cells to cancer ones (Peres et al., 2015).

It has been reported that hydroxytyrosol is able to modulate an adaptive signaling pathway activated after stress and to ameliorate homeostasis (Giordano et al., 2014). Finally, hydroxytyrosol and related olive phenols have been tested, as supplements, in humans and it has been demonstrated that this compound retains its antioxidant activity after ingestion; the human metabolic pathway has been elucidated, and suggests an extensive glucuronidation and subsequent urinary excretion, with a protective role from second-hand smoke-induced oxidative damage, inhibition of platelet aggregation, increase of brain cell resistance to oxidation and mitochondrial membrane potential (Nakbi et al., 2010). The findings published in the EFSA Journal support that dietary consumption of hydroxytyrosol and related polyphenol compounds from olive fruit offers protection to the blood lipids from oxidative damage (BIOHAZ, 2011). Studies provided further substantiation of the cardiovascular benefits of olives and olive oil. After 3 months of consuming olive oil rich in phenolic compounds, the studied subjects had a downregulation in the expression of atherosclerosis-related genes in their peripheral blood mononuclear cells affecting coronary heart disease (Konstantinidou et al., 2010).

Other studies have confirmed the antithrombotic potential of hydroxytyrosol, coupled with its ability to ameliorate osteoarthritis, a common chronic degenerative- and aging-associated disease related to extracellular matrix degradation and mineralization (Aigner and Gerwin, 2007; Shen et al., 2012).

The contribution of excessive free radical formation to the onset of certain pathologies (eg, atherosclerotic heart disease and cancer) strongly recommends higher dietary intake of fruits and vegetables, foods with large contents of antioxidant vitamins, flavonoids, and phenolic compounds (Granados-Principal et al., 2010).

22.7 FERMENTED OLIVES CAN MODULATE THE DIGESTIVE MICROBIOTA

Fermentation can be considered almost like the beginning of digestion (Selhub et al., 2014). The gut includes a complex microbial ecosystem composed of resident commensals or ones passing through the GIT that form an intimate and beneficial association with the host playing protective roles in chronic disease and acute infection. With 1.5 kg of bacteria in the colon and a density of 10^{12} cells per gram of intestinal content, the gut microbiota has been recognized as an important metabolic organ comparable to the liver (Garrett et al., 2010). Due to its scale and its important role in maintaining health, the gut microbiota can be considered as a "new organ" inside the human body (Chassard and Lacroix, 2013).

The human GIT is one of the most complex ecosystems known, consisting of an assortment of bacteria that shape many important physiological and metabolic processes as well as the development of the immune system (Sekirov et al., 2010). The associated microbial communities play a critical role in human health and predisposition to disease, but the degree to which they also shape therapeutic interventions is not well understood.

There is a long array of beliefs on the medicinal properties of fermented food products—yet the beneficial health effects of LAB on the intestinal microbiota have been scientifically documented (Masood et al., 2011). Probiotics are able to alter the intestinal microbiota in order to improve the health of the host or the gut microbiota (Selhub et al., 2014). Modulation of the intestinal microbiota is one of the potential health-beneficial effects of probiotics (Chapman et al., 2011; Gareau et al., 2010; Gerritsen et al., 2011). In the light of the increased understanding of the impact of the gut microbiome on host health, the question then arises as to how food-ingested microbial communities modulate the indigenous microbiota when psychological stress or dietary changes induce changes in species and levels of microbiota (Knowles et al., 2008; Veiga et al., 2014).

The colon is an organ which is the preferred site for bacterial colonization. The effects of interplay between phenolic compounds from olives and specific intestinal microbiota functions remain largely uncharacterized (Laparra, 2010). Since the GIT is an important target for probiotics, some factors from olive composition, such as oleuropein and hydroxytyrosol that can be related to strain adhesion in intestine cells were investigated (Peres et al., 2015). The effect of oleuropein and hydroxytyrosol on potential probiotic lactobacilli upon adhesion to Caco-2 cells showed that it depends on the compound; in the presence of oleuropein adhesion ability of *E. coli* decreased significantly. The reduction of pathogen adhesion can be a potential approach to effectively control the pathogenic infections.

Fermented foods provide a dual benefit to the consumer's gut, promoting health and immune function; providing high fiber content, there are an increasing of beneficial microorganisms, such as LAB in gut (Hugenholtz, 2013). Of particular relevance is the high content in dietary fiber and vitamins of fermented olives, regardless of some differences in cultivar composition (Garrido-Fernández and Lopéz-Lopéz, 2008). In the fermentation processes, specific compounds are originated from either biotransformation reactions or biosynthesis, and these can affect the health of the consumer (Borresen et al., 2012). Gut microbiota composition and activity is controlled both by diet as well as through modulation by the host mucosal immune system. Evidence of benefits of gut microbiota suggest that it also plays a pivotal role in nutrient digestion and regulation of energy obtained from the diet (Kinross et al., 2011). Indeed, diet and hence fermented foods are important modulators of the gut microbiota activity that can be consumed to improve host health (Walker et al., 2011). Genetic and environmental factors influence the abundance and type of beneficial and pathogenic bacteria in the gut (Qin et al., 2010). Recent DNA techniques based on 16S rRNA gene quantification or sequencing allow for a more broad evaluation of the intestinal microbiota as mediated by diet (Moschen et al., 2012). These techniques showed that dietary patterns largely determine the main phyla of the gut microbial profile (Bested et al., 2013). The potential synergism between phenolic compounds from olives and probiotic bacteria may be taken advantage of selectively to stimulate proliferation or activity of probiotics aimed at pathogen control. Moreover, this combination may provide an opportunity to produce cells in biofilms, a barrier against pathogen-harmful antigens in the environment, and thus account (at least partially) for the mucosal barrier function. Results on the relation between probiotic strains originating in fermented table olives and the foodborne pathogen *E. coli* in the presence of oleuropein and hydroxytyrosol at the recommended daily dose for table olives, seem to show the possibility to control intestine infections by reduction of *E. coli* (Peres et al., 2015).

The phenolic constituents of olives have been associated with potent biological activities in vitro, besides antioxidant action; tyrosol and hydroxytyrosol are absorbed by humans after ingestion in a dose-dependent manner, and excreted in the urine as glucuronide conjugates and an increase in the dose of phenolics administered apparently increases the proportion of conjugation with glucuronide (Visioli et al., 2000). Phenolic compounds occur in olives mainly as esters, glycosides, and polymers which cannot be absorbed in these native forms; they require hydrolysis by digestive system enzymes or intestinal microbiota. Following their uptake with the diet, phenolic compounds not metabolized by fermentation may be absorbed to some extent, but they may also undergo transformation by intestinal bacteria that may result in an inactive compound or, alternatively, the activity may be enhanced as a result of bacterial transformation (Lewandowska et al., 2013). Hence, bacterial metabolites in the colon may have a different biological role than that demonstrated for the parent compound. There is little knowledge about the phenolic derivatives present in the colon, their physiological concentrations, and how the colonic content modulates the functions of the bioactives (Aura, 2008).

22.8 FUTURE TRENDS

Advances in food and medical sciences coupled with changing consumer demand have encouraged the growth of the fermented foods market. Nowadays, one-third of the human diet globally consists of foods prepared by fermentation and each nation has its own types of fermented food (Borresen et al., 2012).

Consumption of olives is recognized as a key factor supporting the beneficial effects of the Mediterranean Diet. Currently, due to continuing scientific evidence supported by numerous epidemiological and clinical experimental studies, the recognition of olives as a source of health-promoting compounds is much recognized. Therefore, the future looks encouraging for the demand for fermented foods, guaranteeing low environmental impact production and favorably contributing to consumer health; furthermore sociodemographic trends are also in favor for such functional foods (Siro et al., 2008).

Food fermentation is originated from natural microbial interactions that led to positive outcomes in terms of food preservation. Lactic acid fermentation is considered as the major contributor to the beneficial characteristics observed in fermented olives (Moslehi-Jenabian et al., 2010) and the indigenous enzymes of raw materials may also play a role in the associated nutritional value and elimination of undesirable compounds, providing safer products (Bamforth, 2005).

Fermented olives are still a vast and unexplored resource of potentially healthy microbial strains and bioactive components. Biotechnological innovations for table olives represent a great challenge, and producers require new solutions to design innovative products by the use of novel starters that mimic the natural microbiota with some other desired traits, traditional regions of fermented olives, and health-oriented olives. An important step on fermented olives research area will be to understand the functions of the uncharacterized microorganisms related to olive fermentation, particularly their roles in the breakdown of olive matrix and how the associated byproducts contribute to health and disease (Conlon and Bird, 2015; Saad et al., 2013).

There are still many gaps in the knowledge of the interactions between diet, lifestyle, gut microbiota, and health and many questions remain to be clarified. Dietary factors and habits are indeed a major tool to aid in the reduction of incidence of diet-related diseases and health conditions (Leon-Munoz et al., 2015). Particular composition of olives related to the presence of bioactive nutrients and probiotic bacteria constitutes an area where fundamental and applied research is still needed. Research regarding the bioavailability of olive phenolics has focused on the absorption and excretion of two major components, hydroxytyrosol and tyrosol, from olive oil. In this regard, significant efforts should also be focused on the isolation and structural elucidation of olive phenolics and other bioactive derivatives. In addition, there is still a lack of knowledge on the metabolic pathways as well as the relationship between specific compounds and their corresponding bioactivity, bioaccessibility, and bioavailability, which thus demands more research in table olives.

Future studies are in great need for collecting data to definitely assess, by in vivo studies, whether the living bacteria, as well as the table olive beneficial components, significantly improve the health of consumers. Health claims scientifically substantiated may help inform consumers of the potential benefits of table olives.

ACKNOWLEDGMENTS

This work was supported by FCT project NEW PROTECTION: NativE, Wild PRObiotic sTrain EffecCT In Olives in briNe, ref. PTDC/AGRALI/117658/2010, PhD fellowship ref. SFRH/BD/81997/2011 (both supervised by author FXM).

REFERENCES

Abriouel, H., Benomar, N., Cobo, A., Caballero, N., Fuentes, M.A.F., Perez-Pulido, R., Gálvez, A., 2012. Characterization of lactic acid bacteria from naturally-fermented Manzanilla Alorena green table olives. Food Microbiology 32 (2), 308–316.

Aigner, T., Gerwin, N., 2007. Growth plate cartilage as developmental model in osteoarthritis research – potentials and limitations. Current Drug Targets 8 (2), 377–385.

Akande, K.E., Doma, U.D., Agu, H.O., 2010. Major anti-nutrients found in plant protein sources: their effect on nutrition. Pakistan Journal of Nutrition 9, 827–832.

Allouche, Y., Warleta, F., Campos, M., Sanchez-Quesada, C., Uceda, M., Eltrán, G., Gaforio, J.J., 2011. Antioxidant, antiproliferative, and pro-apoptotic capacities of pentacyclic triterpens found in the skin of olive on MCF-7 human breast cancer cells and their effects on DNA damage. Journal of Agriculture of Food Chemistry 59, 121–130.

Aloglu, H., Oner, Z., 2006. Assimilation of cholesterol in broth, cream, and butter by probiotic bacteria. European Journal of Lipid Science and Technology 108 (9), 709–713.

Arena, E., Campisi, S., Fallico, B., Maccarone, E., 2007. Distribution of fatty acids and phytosterols as a criterion to discriminate geographic origin of pistachio seeds. Food Chemistry 104 (1), 403–408.

Argyri, A.A., Zoumpopoulou, G., Karatzas, K.-A.G., Tsakalidou, E., Nychas, G.-J., Panagou, E.Z., et al., 2013. Selection of potential probiotic lactic acid bacteria from fermented olives by in vitro tests. Food Microbiology 33 (2), 282–291.

Arroyo-Lopez, F.N., Querol, A., Bautista-Gallego, J., Garrido-Fernandez, A., 2008. Role of yeasts in table olive production. International Journal of Food Microbiology 128 (2), 189–196.

Arroyo-Lopez, F.N., Blanquet-Diot, S., Denis, S., Thevenot, J., Chalancon, S., Alric, M., Rodríguez-Gómez, F., Romero-Gil, V., Jiménez-Díaz, R., Garrido-Fernández, A., 2014. Survival of pathogenic and lactobacilli species of fermented olives during simulated human digestion. Frontiers in Microbiology 5, 9 pp.

Aslam, S., Qazi, J.I., 2010. Isolation of acidophilic lactic acid bacteria antagonistic to microbial contaminants. Pakistan Journal of Zoology 42 (5), 567–573.

Aura, A.M., 2008. Microbial metabolismo of dietary phenolic compouns in the colon. Phytochemistry Reviews 7, 407–429.
Bamforth, C.W., 2005. Food, Fermentation and Microrganisms. Blackwell Science Ltd. A Blackwell Publishing Company, Oxford.
Barnett Foster, D., 2013. Modulation of the enterohemorrhagic *E. coli* virulence program through the human gastrointestinal tract. Virulence 4 (4), 315–323.
Bautista-Gallego, J., Arroyo-Lopez, F.N., Rantsiou, K., Jimenez-Diaz, R., Garrido-Fernandez, A., Cocolin, L., 2013. Screening of lactic acid bacteria isolated from fermented table olives with probiotic potential. Food Research International 50 (1), 135–142.
Beekwilder, J., Marcozzi, D., Vecchi, S., de Vos, R., Janssen, P., Francke, C., van Hylckama, V.J., Hall, R.D., 2009. Characterization of rhamnosidases from *Lactobacillus plantarum* and *Lactobacillus acidophilus*. Applied and Environmental Microbiology 75 (11), 3447–3454.
Begley, M., Hill, C., Gahan, C.G.M., 2006. Bile salt hydrolase activity in probiotics. Applied and Environmental Microbiology 72 (3), 1729–1738.
Bested, A.C., Logan, A.C., Selhub, E.M., 2013. Intestinal microbiota, probiotics and mental health: from Metchnikoff to modern advances: part III – convergence toward clinical trials. Gut Pathogens 5, 5–21.
Bevilacqua, A., Altieri, C., Corbo, M.R., Sinigaglia, M., Ouoba, L.I.I., 2010. Characterization of lactic acid bacteriai solated from Italian Bella di Cerignola table olives: selection of potential multifunctional starter cultures. Journal of Food Science 75 (8), M536–M544.
Bianchi, G., 2003. Lipids and phenols in table olives. European Journal of Lipid Science and Technology 105 (5), 229–242.
(BIOHAZ) EFSA Panel on Biological Hazards, 2011. Scientific opinion on risk based control of biogenic amine formation in fermented foods. EFSA Journal 2393.
Blana, V.A., Grounta, A., Tassou, C.C., Nychas, G.J.E., Panagou, E.Z., 2014. Inoculated fermentation of green olives with potential probiotic *Lactobacillus pentosus* and *Lactobacillus plantarum* starter cultures isolated from industrially fermented olives. Food Microbiology 38, 208–218.
Blekas, G., Vassilakis, C., Harizanis, C., Tsimidou, M., Boskou, D.G., 2002. Biophenols in table olives. Journal of Agricultural and Food Chemistry 50 (13), 3688–3692.
Borresen, E.C., Henderson, A.J., Kumar, A., Weir, T.L., Ryan, E.P., 2012. Fermented foods: patented approaches and formulations for nutritional supplementation and health promotion. Recent Patents on Food, Nutrition & Agriculture 4 (2), 134–140.
Botta, C., Cocolin, L., 2012. Microbial dynamics and biodiversity in table olive fermentation: culture-dependent and -independent approaches. Frontiers in Microbiology 6 (3), 1–10.
Bulotta, S., Celano, M., Lepore, S.M., Montalcini, T., Pujia, A., Russo, D., 2014. Beneficial effects of the olive oil phenolic components oleuropein and hydroxytyrosol: focus on protection against cardiovascular and metabolic diseases. Journal of Translational Medicine 12, 1–9.
Chapman, C.M.C., Gibson, G.R., Rowland, I., 2011. Health benefits of probiotics: are mixtures more effective than single strains? European Journal of Nutrition 50 (1), 1–17.
Chapman, C.M.C., Gibson, G.R., Rowland, I., 2014. Effects of single- and multi-strain probiotics on biofilm formation and *in vitro* adhesion to bladder cells by urinary tract pathogens. Anaerobe 27, 71–76.
Chassard, C., Lacroix, C., 2013. Carbohydrates and the human gut microbiota. Current Opinion in Clinical Nutrition and Metabolic Care 16 (4), 453–460.
Ciafardini, G., Marsilio, V., Lanza, B., Pozzi, N., 1994. Hydrolysis of oleuropein by l'*Lactobacillus-plantarum* strains associated with olive fermentation. Applied and Environmental Microbiology 60 (11), 4142–4147.

Clark, S.E., Weiser, J.N., 2013. Microbial modulation of host immunity with the small molecule phosphorylcholine. Infection and Immunity 81 (2), 392–401.

Codex, 2003. Codex Standard for Fermented Milks (Vol. Codex Stan 243). Codex Alimentarius Commission.

Conlon, M.A., Bird, A.R., 2015. The impact of diet and lifestyle on gut microbiota and human health. Nutrients 7 (1), 17–44.

Covas, M.I., Nyyssonen, K., Poulsen, H.E., Kaikkonen, J., Zunft, H.J.F., Kiesewetter, H., Gaddi, A., de la Torre, R., Mursu, J., Baumler, H., Nascetti, S., Salonen, J.T., Fito, M., Virtanen, J., Marrugat, J., 2006. The effect of polyphenols in olive oil on heart disease risk factors – a randomized trial. Annals of Internal Medicine 145 (5), 333–341.

Cross, M.L., 2002. Microbes *versus* microbes: immune signals generated by probiotic lactobacilli and their role in protection against microbial pathogens. FEMS Immunology and Medical Microbiology 34 (4), 245–253.

D'Aimmo, M.R., Modesto, M., Biavati, B., 2007. Antibiotic resistance of lactic acid bacteria and *Bifidobacterium* spp. isolated from dairy and pharmaceutical products. International Journal of Food Microbiology 115 (1), 35–42.

De Bellis, P., Valerio, F., Sisto, A., Lonigro, S.L., Lavermicocca, P., 2010. Probiotic table olives: microbial populations adhering on olive surface in fermentation sets inoculated with the probiotic strain *Lactobacillus paracasei* IMPC2.1 in an industrial plant. International Journal of Food Microbiology 140 (1), 6–13.

Dunne, C., O'Mahony, L., Murphy, L., Thornton, G., Morrissey, D., O'Halloran, S., Feeny, M., Flynn, S., Fitzgerald, G., Daly, C., Kiely, B., O'Sullivan, G.C., Shanahan, F., Collins, J.K., 2001. In vitro selection criteria for probiotic bacteria of human origin: correlation with in vivo findings. American Journal of Clinical Nutrition 73 (2), 386S–392S.

Dutra, V., Silva, C., Cabrita, P., Peres, C., Malcata, F.X., Brito, L., 2016. *Lactobacillus plantarum* LB95 a potential probiotic strain against gram-positive and gram-negative food borne pathogens. Journal of Medical Microbiology 65, 28–35.

EFSA, ECDC, 2015. The European Union summary report on trends and sources of zoonoses, zoonotic agents and food-borne outbreaks in 2013. EFSA Journal 13 (1), 3991–4153.

El Adlouni, C., Tozlovanu, M., Naman, F., Faid, M., Pfohl-Leszkowicz, A., 2006. Preliminary data on the presence of mycotoxins (ochratoxin A, citrinin and aflatoxin B1) in black table olives "Greek style" of Moroccan origin. Molecular Nutrition & Food Research 50 (6), 507–512.

FAO, 2004. The State of Food Insecurity in the World.

Figueiroa-González, I., CRuz-Guerrero, A., Quijano, G., 2011. The benefits of probiotics on human health. Journal of Microbial & Biochemical Technology 1, 1–6.

Franzetti, L., Scarpellini, M., Vecchio, A., Planeta, D., 2011. Microbiological and safety evaluation of green table olives marketed in Italy. Annals of Microbiology 61 (4), 843–851.

Furneri, P.M., Marino, A., Saija, A., Uccella, N., Bisignano, G., 2002. In vitro antimycoplasmal activity of oleuropein. International Journal of Antimicrobial Agents 20 (4), 293–296.

Furukawa, S., Watanabe, T., Toyama, H., Morinaga, Y., 2013. Significance of microbial symbiotic coexistence in traditional fermentation. Journal of Bioscience and Bioengineering 116 (5), 533–539.

Garcia, P.G., Barranco, C.R., Quintana, M.C.D., Fernandez, A.G., 2004. Biogenic amine formation and "zapatera" spoilage of fermented green olives: effect of storage temperature and debittering process. Journal of Food Protection 67 (1), 117–123.

Gareau, M.G., Sherman, P.M., Walker, W.A., 2010. Probiotics and the gut microbiota in intestinal health and disease. Nature Reviews Gastroenterology & Hepatology 7 (9), 503–514.

Garrett, W.S., Gallini, C.A., Yatsunenko, T., Michaud, M., Du Bois, A., Delaney, M.L., Punit, S., Karlsson, M., Bry, L., Glickman, J.N., Gordon, J.I., Onderdonk, A.B., Glimcher, L.H., 2010.

Enterobacteriaceae act in concert with the gut microbiota to induce spontaneous and maternally transmitted colitis. Cell Host & Microbe 8 (3), 292–300.

Garrido-Fernández, A., Lopéz-Lopéz, A., 2008. Revalorisación nutricional de la aceituna de mesa. In: II Jornadas Internacionales de la Aceituna de Mesa. Seville. Spain.

Garrido-Fernández, A., Fernández-Díez, M.J., Adams, M.R., 1997. Physical and chemical characteristics of olive fruit. In: Garrido-Fernández, A., et al. (Ed.), Table Olives. Chapman & Hall, London, United Kingdom.

General, T., Koijam, K., Appaiah, K.A., 2011. Process improvement as influenced by inoculum and product preservation in the production of Hawaijar – a traditional fermented Soybean. African Journal Food Science 5, 63–68.

Gerritsen, J., Smidt, H., Rijkers, G.T., de Vos, W.M., 2011. Intestinal microbiota in human health and disease: the impact of probiotics. Genes and Nutrition 6 (3), 209–240.

Ghabbour, N., Lamzira, Z., Thonart, P., Peres, C., Markaoui, M., Asehraou, A., 2011. Selection of oleuropein-degrading lactic acid bacteria strains isolated from fermenting Moroccan green olives. Grasas Y Aceites 62 (1), 84–89.

Giordano, E., Davalos, A., Nicod, N., Visioli, F., 2014. Hydroxytyrosol attenuates tunicamycin-induced endoplasmic reticulum stress in human hepatocarcinoma cells. Molecular Nutrition & Food Research 58 (5), 954–962.

Giraffa, G., 2004. Studying the dynamics of microbial populations during food fermentation. FEMS Microbiology Reviews 28 (2), 251–260.

Gomez, A.H., Garcia, P., Navarro, L., 2006. Elaboration of table olives. Grasas Y Aceites 57 (1), 86–94.

Granados-Principal, S., Quiles, J.L., Ramirez-Tortosa, C.L., Sanchez-Rovira, P., Ramirez-Tortosa, M.C., 2010. Hydroxytyrosol: from laboratory investigations to future clinical trials. Nutrition Reviews 68 (4), 191–206.

Grounta, A., Nychas, G.J.E., Panagou, E.Z., 2013. Survival of food-borne pathogens on natural black table olives after post-processing contamination. International Journal of Food Microbiology 161 (3), 197–202.

Guerzoni, M.E., Gianotti, A., Serrazanetti, D.I., 2011. Fermentation as a tool to improve healthy properties of bread. In: Preedy, V.R., Watson, R.R., Patel, V.B. (Eds.), Flour and Breads and Their Fortification in Health and Disease Prevention. Academic Press, Elsevier, London, Burlington, San Diego, pp. 385–393.

Hansen, E.B., 2002. Commercial bacterial starter cultures for fermented foods of the future. International Journal of Food Microbiology 78 (1-2), 119–131.

Hjortmo, S., Patring, J., Andlid, T., 2008. Growth rate and medium composition strongly affect folate content in *Saccharomyces cerevisiae*. International Journal of Food Microbiology 123 (1-2), 93–100.

Hugenholtz, J., 2013. Traditional biotechnology for new foods and beverages. Current Opinion in Biotechnology 24 (2), 155–159.

Hurtado, A., Reguant, C., Bordons, A., Rozès, N., 2012. Lactic acid bacteria from fermented table olives. Food Microbiology 31 (1), 1–8.

Hussein, H.S., Brasel, J.M., 2001. Toxicity, metabolism, and impact of mycotoxins on humans and animals. Toxicology 167 (2), 101–134.

IOOC, 2004. Trade Standard Applying to Table Olives. IOC.

IOOC, 2014. Areas of Activity. Economy. World Olive Oil Figures. IOC.

Jarvis, B.B., 2002. Chemistry and toxicology of molds isolated from water-damaged buildings. Mycotoxins and Food Safety. Advances in Experimental Medicine & Biology 504, 43–52.

Jones, P.J., Jew, S., 2007. Functional food development: concept to reality. Trends in Food Science & Technology 18 (7), 387–390.

Juan, M.E., Wenzel, U., Ruiz-Gutierrez, V., Daniel, H., Planas, J.M., 2006. Olive fruit extracts inhibit proliferation and induce apoptosis in HT-29 human colon cancer cells. Journal of Nutrition 136 (10), 2553–2557.

Kariluoto, S., Aittamaa, M., Korhola, M., Salovaara, H., Vahteristo, L., Piironen, V., 2006. Effects of yeasts and bacteria on the levels of folates in rye sourdoughs. International Journal of Food Microbiology 106 (2), 137–143.

Kinross, J.M., Darzi, A.W., Nicholson, J.K., 2011. Gut microbiome-host interactions in health and disease. Genome Medicine 3, 12.

Knowles, S.R., Nelson, E.A., Palombo, E.A., 2008. Investigating the role of perceived stress on bacterial flora activity and salivary cortisol secretion: a possible mechanism underlying susceptibility to illness. Biological Psychology 77 (2), 132–137.

Konstantinidou, V., Covas, M.I., Munoz-Aguayo, D., Khymenets, O., de la Torre, R., Saez, G., Tormos, M., del, C., Toledo, E., Marti, A., Ruiz-Gutiérrez, V., Ruiz Mendez, M.V., Fito, M., 2010. In vivo nutrigenomic effects of virgin olive oil polyphenols within the frame of the Mediterranean diet: a randomized controlled trial. FASEB Journal 24 (7), 2546–2557.

Kumar, M., Nagpal, R., Kumar, R., Hemalatha, R., Verma, V., Kumar, A., et al., 2012. Cholesterol-lowering probiotics as potential biotherapeutics for metabolic diseases. Experimental Diabetes Research 1, 1–14.

Laparra, J.M., Sanz, Y., 2010. Interactions of gut microbiota with functional food components and nutraceuticals. Pharmacological Research 61, 219–225.

Lavermicocca, P., Valerio, F., Lonigro, S.L., De Angelis, M., Morelli, L., Callegari, M.L., Rizzello, C.G., Visconti, A., 2005. Study of adhesion and survival of lactobacilli and bifidobacteria on table olives with the aim of formulating a new probiotic food. Applied and Environmental Microbiology 71 (8), 4233–4240.

Lavermicocca, P., Lonigro, S.L., Visconti, A., De Angelis, M., Valerio, F., Morelli, L., 2007. Table Olives Containing Probiotic Microorganisms US2007086990 A1.

Lee, Y.K., Lim, C.Y., Teng, W.L., Ouwehand, A.C., Tuomola, E., Salminem, S., 2000. Qualitative approach in the study of adhesion of lactic acid bacteria on intestinal cells and their competition with enterobacteria. Applied and Environmental Microbiology 66, 3692–3697.

Leon-Munoz, L.M., Garcia-Esquinas, E., Lopez-Garcia, E., Banegas, J.R., Rodriguez-Artalejo, F., 2015. Major dietary patterns and risk of frailty in older adults: a prospective cohort study. BMC Medicine 13, 9 pp.

Lewandowska, U., Szewczyk, K., Hrabec, E., Janecka, A., Gorlach, S., 2013. Overview of metabolism and bioavailability enhancement of polyphenols. Journal of Agricultural and Food Chemistry 61 (50), 12183–12199.

Linares, D.M., Martin, M.C., Ladero, V., Alvarez, M.A., Fernandez, M., 2011. Biogenic amines in dairy products. Critical Reviews in Food Science and Nutrition 51 (7), 691–703.

Madureira, A.R., Pereira, C.I., Truszkowska, K., Gomes, A.M., Pintado, M.E., Malcata, F.X., 2005. Survival of probiotic bacteria in a whey cheese vector submitted to environmental conditions prevailing in the gastrointestinal tract. International Dairy Journal 15 (6-9), 921–927.

Marteau, P.R., de Vrese, M., Cellier, C.J., Schrezenmeir, J., 2001. Protection from gastrointestinal diseases with the use of probiotics. American Journal of Clinical Nutrition 73 (2), 430S–436S.

Martinez, M.G., Poole, N., 2004. The development of private fresh produce safety standards: implications for developing Mediterranean exporting countries. Food Policy 29 (3), 229–255.

Martinez-Gonzalez, M.A., de la Fuente-Arrillaga, C., Nunez-Cordoba, J.M., Basterra-Gortari, F.J., Beunza, J.J., Vazquez, Z., et al., 2008. Adherence to Mediterranean diet and risk of developing diabetes: prospective cohort study. British Medical Journal 336 (7657), 1348–1351.

Martins, D.A.B., do Prado, H.F.A., Leite, R.S.R., Ferreira, H., de Souza Moretti, M.M., da Silva, R., Gomes, E., 2011. Agroindustrial wastes as substrates for microbial enzymes production and source of sugar for bioethanol production. Integrated Waste Management 17 (2), 319–362.

Masood, M.I., Qadir, M.I., Shirazi, J.H., Khan, I.U., 2011. Beneficial effects of lactic acid bacteria on human beings. Critical Reviews in Microbiology 37 (1), 91–98.

Menendez, J.A., Lupu, R., 2006. Mediterranean dietary traditions for the molecular treatment of human cancer: anti-oncogenic actions of the main olive oil's monounsaturated fatty acid oleic acid (18:1n-9). Current Pharmaceutical Biotechnology 7 (6), 495–502.

Miles, E.A., Zoubouli, P., Calder, P.C., 2005. Differential anti-inflammatory effects of phenolic compounds from extra virgin olive oil identified in human whole blood cultures. Nutrition 21 (3), 389–394.

Moschen, A.R., Wieser, V., Tilg, H., 2012. Dietary factors: major regulators of the gut's microbiota. Gut and Liver 6 (4), 411–416.

Moschini, R.C., Sisterna, M.N., Carmona, M.A., 2006. Modelling of wheat black point incidence based on meteorological variables in the southern Argentinean Pampas region. Australian Journal of Agricultural Research 57 (11), 1151–1156.

Moslehi-Jenabian, S., Pedersen, L.L., Jespersen, L., 2010. Beneficial effects of probiotic and food borne yeasts on human health. Nutrients 2 (4), 449–473.

Mourad, K., Nour-Eddine, K., 2006. In vitro preselection criteria for probiotic *Lactobacillus plantarum* strains of fermented olives origin. International Journal of Probiotics and Prebiotics 1, 27–32.

Nakbi, A., Tayeb, W., Grissa, A., Issaoui, M., Dabbou, S., Ellouz, M., Miled, A., Hammami, M., 2010. Effects of olive oil and its fractions on oxidative stress and the liver's fatty acid composition in 2,4-Dichlorophenoxyacetic acid-treated rats. Nutrition & Metabolism 7 11 pp.

Nes, W.D., 2011. Biosynthesis of cholesterol and other sterols. Chemical Reviews 111 (10), 6423–6451.

Nychas, G.J.E., Panagou, E.Z., Parker, M.L., Waldron, K.W., Tassou, C.C., 2002. Microbial colonization of naturally black olives during fermentation and associated biochemical activities in the cover brine. Letters in Applied Microbiology 34 (3), 173–177.

Odunfa, S.A., Adeniran, S.A., Teniola, O.D., Nordstrom, J., 2001. Evaluation of lysine and methionine production in some lactobacilli and yeasts from Ogi. International Journal of Food Microbiology 63 (1–2), 159–163.

Oliveira, M., Brito, D., Catulo, L., Leitão, F., Gomes, L., Silva, S., Vilas-boas, L., Peito, A., Fernandes, I., Gordo, F., Peres, C., 2004. Biotechnology of olive fermentation of 'Galega' Portuguese variety. Grasas Y Aceites 55 (3), 219–226.

Omar, S.H., 2010. Oleuropein in olive and its pharmacological effects. Scientia Pharmaceutica 78 (2), 133–154.

Ozdemir, Y., Guven, E., Ozturk, A., 2014. Understanding the characteristics of oleuropein for table olive processing. Food Processing & Technology 5 (5), 328–344.

Palencia, E.R., Hinton, D.M., Bacon, C.W., 2010. The black *Aspergillus* species of maize and peanuts and their potential for mycotoxin production. Toxins 2, 399–416.

Panagou, E.Z., Schillinger, U., Franz, C., Nychas, G.J.E., 2008. Microbiological and biochemical profile of cv. Conservolea naturally black olives during controlled fermentation with selected strains of lactic acid bacteria. Food Microbiology 25 (2), 348–358.

Parvez, S., Malik, K.A., Kang, S.A., Kim, H.Y., 2006. Probiotics and their fermented food products are beneficial for health. Journal of Applied Microbiology 100 (6), 1171–1185.

Patel, A., Shah, N., Prajapati, J.B., 2014. Clinical application of probiotics in the treatment of *Helicobacter pylori* infection-A brief review. Journal of Microbiology Immunology and Infection 47 (5), 429–437.

Peres, C.M., Peres, C., Hernandez-Mendoza, A., Malcata, F.X., 2012. Review on fermented plant materials as carriers and sources of potentially probiotic lactic acid bacteria – with an emphasis on table olives. Trends in Food Science & Technology 26 (1), 31–42.

Peres, C.M., Alves, M., Hernandez-Mendoza, A., Moreira, L., Silva, S., Bronze, M.R., Villa-Boas, L., Peres, C., Malcata, F.X., 2014. Novel isolates of lactobacilli from fermented Portuguese olive as potential probiotics. LWT-Food Science and Technology 59 (1), 234–246.

Peres, C.M., Hernandez-Mendonza, A., Bronze, M.R., Peres, C., Malcata, F.X., 2015. Synergy of olive bioactive phytochemicals and potential probiotic strain in food-borne pathogens control. LWT-Food Science and Technology 64, 938–945.

Pitt, J., Roth, W., Lacor, P., Smith, A.B., Blankenship, M., Velasco, P., De Felice, F., Breslin, P., Klein, L., 2009. Alzheimer's-associated A beta oligomers show altered structure, immunoreactivity and synaptotoxicity with low doses of oleocanthal. Toxicology and Applied Pharmacology 240 (2), 189–197.

Potter, N.N., Hotchkiss, J.H., 2006. Food Science. CBS Publishers and Distributors, New Delhi, pp. 264–327.

Prajapati, J.B., Nair, B.M., 2003. The history of fermented foods. Handbook of Fermented Functional Foods 1–25.

Pundir, R.K., Rana, S., Kashyap, N., Kau, A., 2013. Probiotic potential of lactic acid bacteria isolated from food samples: an *in vitro* study. Journal of Applied Pharmaceutical Science 3, 85–93.

Qin, J.J., Li, R.Q., Raes, J., Arumugam, M., Burgdorf, K.S., Manichanh, C., Nielsen, T., Pons, N., Levenez, F., Yamada, T., 2010. A human gut microbial gene catalogue established by metagenomic sequencing. Nature 464 (7285), 59–70.

Quigley, E.M.M., 2010. Probiotics in gastrointestinal disorders. Hospital Practice (1995) 38 (4), 122–129.

Ramchandran, L., Shah, N.P., 2010. Characterization of functional, biochemical and textural properties of sinbiotic low fat yogurts during refrigerated storage. LWT-Food Science and Technology 43, 819–827.

Ranadheera, R., Baines, S.K., Adams, M.C., 2010. Importance of food in probiotic efficacy. Food Research International 43 (1), 1–7.

Reid, G., Sanders, M.E., Gaskins, H.R., Gibson, G.R., Mercenier, A., Rastall, R., Roberfroid, M., Rowland, I., Cherbut, C., Klaenhammer, T.R., 2003. New scientific paradigms for probiotics and prebiotics. Journal of Clinical Gastroenterology 37 (2), 105–118.

Reyes-Nava, L.A., Rivera-Espinoza, Y., 2014. Isolation sources of bile salt hydrolase-microorganisms. Herald Journal of Agriculture and Food Science Research 3, 49–54.

Rietjens, S.J., Bast, A., Haenen, G., 2007. New insights into controversies on the antioxidant potential of the olive oil antioxidant hydroxytyrosol. Journal of Agricultural and Food Chemistry 55 (18), 7609–7614.

Romero, C., Medina, E., Vargas, J., Brenes, M., De Castro, A., 2007. *In vitro* activity of olive oil polyphenols against *Helicobacter pylori*. Journal of Agricultural and Food Chemistry 55 (3), 680–686.

Saad, N., Delattre, C., Urdaci, M., Schmitter, J.M., Bressollier, P., 2013. An overview of the last advances in probiotic and prebiotic field. LWT-Food Science and Technology 50 (1), 1–16.

Sadeghar, Y., Mortazavian, A.M., Ehsani, A., 2012. Survival and activity of 5 probiotic lactobacilli strains in 2 types of flavoured fermented milk. Food Science Biotechnology 21, 151–157.

Sanders, E., Marco, L., Doyle, M.P., Klaenhammer, T.R., 2010. Food formats for effective delivery of probiotics. Annual Review of Food Science and Technology 1, 65–85.

Sarvari, F., Mortazavian, A.M., Fazelli, M.R., 2014. Biochemical characteristics and viability of probiotic and yogurt during the fermentation and refrigerated storage. Applied Food Biotechnology 1, 55–61.

Scalbert, A., Williamson, G., 2000. Dietary intake and bioavailability of polyphenols. Journal of Nutrition 130 (8), 2073S–2085S.

Schnurer, J., Magnusson, J., 2005. Antifungal lactic acid bacteria as biopreservatives. Trends in Food Science & Technology 16 (1–3), 70–78.

Scott, R., Sullivan, W.C., 2008. Ecology of fermented foods. Human Ecology Review 1, 25–31.

Sekirov, I., Russell, S.L., Antunes, L.C.M., Finlay, B.B., 2010. Gut microbiota in health and disease. Physiological Reviews 90 (3), 859–904.

Selhub, E.M., Logan, A.C., Bested, A.C., 2014. Fermented foods, microbiota, and mental health: ancient practice meets nutritional psychiatry. Journal of Physiological Anthropology 33, 1–12.

Shen, C.L., Smith, B.J., Lo, D.F., Chyu, M.C., Dunn, D.M., Chen, C.H., Shen, C.L., Smith, B.J., Lo, D.F., Chyu, M.C., Dunn, D.M., Chen, C.H., Kwun, I.S., 2012. Dietary polyphenols and mechanisms of osteoarthritis. Journal of Nutritional Biochemistry 23 (11), 1367–1377.

Shinde, P.B., 2012. Probiotic: an overview for selection and evolution. International Journal of Pharmaceutical Sciences 4, 14–21.

Shyu, P.T., Glenn, G.O., Cabrera, E.C., 2014. Citotoxycity of probiotics from Philippine commercial dairy products on cancer cells and the effect on expression of $cfos$ and $sjun$ early apoptotic-promoting genes and interleukin-1β and Tumor Necrosis Factor α proinflammatory cytokine genes. BioMed Research International 2014, 1–9.

Silva, T., Reto, M., Sol, M., Peito, A., Peres, C.M., Peres, C., Malcata, F.X., 2011. Characterization of yeasts from Portuguese brined olives, with a focus on their potentially probiotic behavior. LWT-Food Science and Technology 44 (6), 1349–1354.

Siro, I., Kapolna, E., Kapolna, B., Lugasi, A., 2008. Functional food. Product development, marketing and consumer acceptance-a review. Appetite 51 (3), 456–467.

Sisto, A., Lavermicocca, P., 2012. Suitability of a probiotic *Lactobacillus paracasei* strain as a starter culture in olive fermentation and development of the innovative patented product "probiotic table olives". Frontiers in Microbiology 3, 1–5.

Sobol, C.V., 2014. How microbioma impact on the cardiovascular system. Journal of Clinical Trials in Cardiology 1, 1–4.

Soccol, C.R., de Sousa Vandenberghe, L.P., Spier, M.R., Medeiros, A.B.P., yamaguishi, C.T., de Dea Lindner, J., Pandey, A., Soccol, V., 2010. The potential of probiotics: a review. Food Technology and Biotechnology 48, 413–443.

Soler-Rivas, C., Espin, J.C., Wichers, H.J., 2000. Oleuropein and related compounds. Journal of the Science of Food and Agriculture 80 (7), 1013–1023.

Soomro, A.H., Masud, T., Answer, K., 2002. Role of lactic acid bacteria in food preservation and human health – a review. Pakistan Journal of Nutrition 1, 20–24.

Subrota, H., Shilpa, V., Brij, S., Vandna, K., Surajit, M., 2013. Antioxidatif activity and polyphenol content in fermented soy milk supplemented with WPC-70 by probiotic lactobacilli. International Food Research Journal 20, 2125–2131.

Swain, M.R., Anandharaj, M., Ray, R.C., Parveen Rani, R., 2014. Fermented fruits and vegetables of Asia: a potential source of probiotics. Biotechnology Research International 2014, 1–19.

Tahmourespour, A., Kermanshahi, R.K., 2011. The effect of a probiotic strain (*Lactobacillus acidophilus*) on the plaque formation of oral Streptococci. Bosnian Journal of Basic Medical Sciences 11 (1), 37–40.

Tofalo, R., Perpetuini, G., Schirone, M., Ciarrocchi, A., Fasoli, G., Suzzi, G., Corsetti, A., 2014. *Lactobacillus pentosus* dominates spontaneous fermentation of Italian table olives. LWT-Food Science and Technology 57 (2), 710–717.

Trichopoulou, A., Lagiou, P., Kuper, H., Trichopoulos, D., 2000. Cancer and Mediterranean dietary traditions. Cancer Epidemiology Biomarkers & Prevention 9 (9), 869–873.

Tripoli, E., Giammanco, M., Tabacchi, G., Di Majo, D., Giammanco, S., La Guardia, M., 2005. The phenolic compounds of olive oil: structure, biological activity and beneficial effects on human health. Nutrition Research Reviews 18 (1), 98–112.

Veiga, P., Pons, N., Agrawal, A., Oozeer, R., Guyonnet, D., Brazeilles, R., Faurie, J.-M., van Hylckama Vlieg, J.E.T., Houghton, L.A., Whorwell, P.J., Ehrlich, S.D., Kennedy, S.P., 2014. Changes of the human gut microbiome induced by a fermented milk product. Scientific Reports 4, 1–9.

Viljoen, B.C., 2006. Yeast ecological interactions – yeast-yeast, yeast-bacteria, yeast-fungi interactions and yeasts as biocontrol agents. Yeasts in Food and Beverages 2, 83–110.

Visioli, F., Galli, C., 2002. Biological properties of olive oil phytochemicals. Critical Reviews in Food Science and Nutrition 42 (3), 209–221.

Visioli, F., Galli, C., Bornet, F., Mattei, A., Patelli, R., Galli, G., et al., 2000. Olive oil phenolics are dose-dependently absorbed in humans. FEBS Letters 468 (2–3), 159–160.

Visioli, F., Poli, A., Galli, C., 2002. Antioxidant and other biological activities of phenols from olives and olive oil. Medicinal Research Reviews 22 (1), 65–75.

Visioli, F., Grande, S., Bogani, P., Galli, C., 2004. The role of antioxidants in the Mediterranean diets: focus on cancer. European Journal of Cancer Prevention 13 (4), 337–343.

Vissers, M.N., Zock, P.L., Roodenburg, A.J.C., Leenen, R., Katan, M.B., 2002. Olive oil phenols are absorbed in humans. Journal of Nutrition 132 (3), 409–417.

Vogiatzakis, I.N., Mannion, A.M., Griffiths, G.H., 2006. Mediterranean ecosystems: problems and tools for conservation. Progress in Physical Geography 30 (2), 175–200.

Vossen, P., 2007. Olive oil: history, production, and characteristics of the world's classic oils. Hortscience 42 (5), 1093–1100.

Wahle, K.W.J., Caruso, D., Ochoa, J.J., Quiles, J.L., 2004. Olive oil and modulation of cell signaling in disease prevention. Lipids 39 (12), 1223–1231.

Walker, A.W., Ince, J., Duncan, S.H., Webster, L.M., Holtrop, G., Ze, X.L., et al., 2011. Dominant and diet-responsive groups of bacteria within the human colonic microbiota. ISME Journal 5 (2), 220–230.

Wild, C.P., Gong, Y.Y., 2010. Mycotoxins and human disease: a largely ignored global health issue. Carcinogenesis 31 (1), 71–82.

Yerlikaya, O., 2014. Starter cultures used in probiotic dairy product preparation and popular probiotic dairy drinks. Food Science and Technology 34 (2), 221–229.

Zeng, X.Q., Pan, D.D., Guo, Y.X., 2010. The probiotic properties of *Lactobacillus buchneri* P2. Journal of Applied Microbiology 108 (6), 2059–2066.

Zhiping, H., Rankin, G.O., Sakul, Y.R., Chen, Y.C., 2014. Selecting bioactive phenolic compounds as potential agents agents to inhibit proliferation and VEGF expression in human ovarian cancer cells. Oncology Letters 9 (3), 1444–1450.

Zhu, L., Liu, Z., Feng, Z., Hao, J., Shen, W.L., Li, X., Sharman, E., Wang, Y., Wert, K., Weber, P., Shi, X., Liu, J., 2010. Hydroxytyrosol protects against oxidative damage by simultaneous activation of mitochondrial biogenesis and phase II detoxifying enzyme systems in retinal pigment epithelial cells. Journal of Nutritional Biochemistry 21 (11), 1089–1098.

Chapter 23

Pulque

J.A. Gutiérrez-Uribe[1], L.M. Figueroa[1], S.T. Martín-del-Campo[2], A. Escalante[3]

[1]*Tecnologico de Monterrey, Campus Monterrey, Monterrey, Mexico;* [2]*Tecnologico de Monterrey, Campus Querétaro, Querétaro, Mexico;* [3]*National Autonomous University of Mexico, Cuernavaca, Mexico*

23.1 INTRODUCTION

Agave is one of the most important natural resources in Mexico from an economic, social, and agro-ecological view. Agave species most commonly used for pulque production include: *Agave americana, Agave atrovirens, Agave ferox, Agave mapisaga,* and *Agave salmiana* (Ortiz-Basurto et al., 2008). These species are also known as "maguey pulquero." Prehispanic mural paintings show images of maguey plants and possible pulque consumption, and new chemical evidence has been reported for the use of ceramic vessels from Teotihuacan (150 BC to 650 AD) to contain pulque (Correa-Ascencio et al., 2014).

Fermentation of the Agave sap or nectar, known as aguamiel, begins at a very slow rate in the collection cavity, where native microorganisms, such as yeasts, lactic acid bacteria (LAB), ethanol-producing bacteria, and exopolysaccharide-producing bacteria can be found. Most pulque production in Mexico is done by artisans. The principal variables determining nectar fermentation time are ambient temperature, microorganism type, and concentration of the seed pulque used to ferment the maguey nectar (Gómez-Aldapa et al., 2011).

Pulque is a nondistilled alcoholic beverage produced by spontaneous fermentation of aguamiel (Escalante et al., 2012), which is currently produced and consumed mainly in the central states of Mexico (Escalante et al., 2008; Gómez-Aldapa et al., 2012). Aguamiel is the slightly cloudy, thick, sweet, yellowish agave sap obtained from plants about 8–10 years old (Ortiz-Basurto et al., 2008). Immature flower stem is removed to lead to a circular hole in the central part of the plant, where the sap is stored and then collected twice a day for 2–6 months (Ortiz-Basurto et al., 2008). Collection of aguamiel is made by oral suction with a nut (*Lagenaria siceraria*) known as acocote (Lappe-Oliveras et al., 2008). After collecting nectar, the walls of the cavity are scraped with a sharp tool to remove approximately 0.5-cm wall thickness and stimulate nectar secretion (Gómez-Aldapa et al., 2011). Each agave plant produces approximately 1500 L of aguamiel in a period of 4–6 months (Santos-Zea et al., 2012).

Since nectar fermentation starts during aguamiel collection, the Mexican regulation considers several parameters in this substrate that may significantly affect the fermented beverage quality. Type I aguamiel is the freshest product with a lower degree of fermentation, whereas type II has a more advanced degree of fermentation (NMX-V-022-1972). According to Ramírez et al. (2004), type I aguamiel is the fresh sap collected during the first 60 days of flower stem removal. Type II aguamiel may have a pH as low as 4.5 with a maximum acidity of 4.00 mg/100 mL of aguamiel (as lactic acid) while type I is in the range of 6.6–7.5 and acidity between 0.90 and 1.03 mg/100 mL. Commercial pulque is fed with both types of aguamiel and might be enriched with aguamiel concentrates in different stages of production, this is referred to as type II by the Mexican regulation (NMX-V-037-1972).

On the other hand, pulque type I includes *semilla de pulque* (seed or starter) and *pulque de punta* (NMX-V-037-1972). *Semilla de pulque* is the product obtained from first-grade quality aguamiel that is prepared for optimal growth of natural microbiota. It can be added with other substances to promote better growth of indigenous microorganisms. *Pulque de punta* is the product obtained from the first production tanks in the fermentation process, and its preparation involves the seed and type I aguamiel. This pulque is used as the basis for subsequent production, explaining the differences in alcoholic grade between pulque types I and II.

23.2 BIOACTIVE CONSTITUENTS OF AGUAMIEL

Aguamiel contains water, sugars (glucose, fructose, and sucrose mainly), proteins, gums, and mineral salts as the most important components (Lappe-Oliveras et al., 2008). It has 11.5 wt% (percentage by weight) of dry matter, composed mainly of sugars (75 wt%) and among these sugars, 10 wt% were fructooligosaccharides (FOS). Other components include 0.3 wt% of free amino acids, 3 wt% of proteins, and 3 wt% of ash (Ortiz-Basurto et al., 2008).

Agave plants present a crassulacean acid metabolism (CAM) and their principal photosynthetic products are fructans (López and Urías-Silvas, 2007). Fructans are polymers of fructosyl units connected to the fructose residue of a sucrose molecule through β-(2, 1) and/or β-(2, 6) linkages. The degree of polymerization (DP) can vary from 3 to 60. If the DP is below 10, these polymers are known as fructooligosaccharides (FOS), but if it is higher than 10, then they are called inulins (Ortiz-Basurto et al., 2008). Vijn and Smeekens (1999) classified fructans into five major groups according to the majority structural units; those are inulin, levan, mixed levan, inulin neoseries, and levan neoseries. Fructans of the inulin type consist of lineal (2-1)-linked β-D-fructosyl units. Fructans are classed in the *graminans* group, as β-fructofuranosyl linkages are present, in addition to nonbranched fructofuranosyl moieties called agavins (Mancilla-Margalli and López, 2006). Inulin has been reported to be the principal storage carbohydrate in *Agave tequilana* (Sánchez-Marroquín and Hope, 1953) and *A. americana* (Bathia and Nandra, 1979). 1-Kestose (DP3), nystose

(DP4), neokestose (DP3), and fructans with DP5 have been reported in *A. deserti* (Wang and Nobel, 1998). Martinez del Campo-Padilla (1999) reported the presence of fructans in aguamiel from *A. atrovirens*.

Fructans have numerous nutraceutical properties, one of them being their prebiotic effect (López and Urías-Silvas, 2007). A prebiotic is a selectively fermented ingredient that allows specific changes, both in composition and/or activity in the gastrointestinal microbiota, that confer benefits upon host well-being and health (Gibson, 2004). Inulin-type fructans are an example of such carbohydrates. These compounds possess mainly β-(2-1) linkage that escape from the action of digestive enzymes and can be fermented by colonic microbiota producing short-chain fatty acids (SCFAs), mainly acetate, propionate, and butyrate as well as lactate. Fructans from the same *Agave* species, but from plants grown in different geographic zones, showed differences on their prebiotic effect (López and Urías-Silvas, 2007).

Evidence suggests that agave inulin-type fructans selectively stimulate the growth of bifidobacteria, both in vitro and in vivo (López and Urías-Silvas, 2007). Fermentation of the *A. tequilana* fructans increased *Bifidobacteria* and *Lactobacilli* counts in vitro, by 15.4% and 8.7%, respectively, compared with the use of cellulose, which decreased these microbial populations by 0.76% and 1.26%, respectively (Gómez et al., 2010). Fructans from *A. cantala* from Oaxaca have been shown to stimulate the growth of *Lactobacillus casei*. Fructans from *Agave angustifolia* spp. tequilana have been shown to increase the growth of *Lactobacillus acidophilus*, *L. casei*, and *Bifidobacterium lactis* in vitro, as well as production of SCFAs (Rendón-Huerta et al., 2011).

Studies in rats showed that fructans from *A. angustifolia* ssp. tequilana decreased blood glucose concentrations, fecal *Clostrodium* spp. counts, and liver steatosis, whereas blood HDL concentrations and fecal *Lactobacillus* spp. and *Bifidobacterium* spp. counts increased. These results were compared with fructans from *Cichorium intybus* L. Asteraceae and *Helianthus tuberosus* L. Asteraceae and it was demonstrated that agave fructans were more effective in decreasing blood glucose and cholesterol in obese subjects with diabetes (Rendón-Huerta et al., 2012). Fructans from *A. tequilana* have shown a reductive effect on food intake and body weight gain in mice, in comparison to those who were fed a standard diet. Mice fed agave fructans showed lower serum glucose and cholesterol (Urías-Silvas et al., 2008).

Aguamiel from *A. salmiana* has been shown to possess high antioxidant capacity, and to increase concentrations of hemoglobin, hematocrit, and mean corpuscular hemoglobin in rabbits (Tovar-Robles et al., 2011). Antioxidant activity has been reported as 904.8-µM gallic acid equivalents and it contained 0.14 mg/mL of ascorbic acid (Tovar-Robles et al., 2011).

Other components identified in aguamiel from *A. mapisaga* included free amino acids representing 0.26% of dry matter, corresponding to 0.3 g/L of aguamiel, including most of the essential amino acids except for methionine (Ortiz-Basurto et al., 2008). In addition, this kind of aguamiel has a γ-aminobutyric acid (GABA) concentration of 26 mg/L of aguamiel (Ortiz-Basurto et al., 2008).

Saponins, particularly glycosylated steroids, have been widely studied in agave leaves but they also are important components of aguamiel. The amount of these compounds depends on the agave species as well as the ripening stage (Leal-Díaz et al., 2015). Among the sapogenins identified by Leal-Díaz et al. (2015), hecogenin has shown important biological effects. Santos Cerqueira et al. (2012) reported a hecogenin dose-dependent gastro-protective effect against alcohol damage on experimental models of gastric ulcer even at a low concentration of 10.92 ± 1.17 mg/kg. Meanwhile, Brito Cortez Lima et al. (2012) mentioned the antidepressant effects in mice.

Saponins have shown a protective role against visceral leishmaniasis in CB hamsters (87.7%) and Balb/c mice (84%) when they are used as vaccine coadjuvants. Saponins from *Agave sisalana* have shown a hemolytic effect with a median lethal dose (LD_{50}) of 18.2 μg/mL (Santos et al., 1997)

Agave extracts have shown different biological activities. Methanolic extracts of *A. americana* showed a potent cytotoxic effect against MCF-7 cancer cell line with an IC_{50} of 545.9 and 1854 μg/mL against Vero cell line (Anajwala et al., 2010). On the other hand, *A. fourcroydes* methanolic extracts showed cytotoxic activity against HeLa cells (Ohtsuki et al., 2004). Aqueous *A. intermixta* L. extracts showed effect against Gram-positive and Gram-negative bacteria (García et al., 1999). The aqueous extract of *A. attenuata* showed a molluscicidal activity against the *Schistosoma haematobium* host, the snail species *Bulinus tropicus* and *B. africanus* with a LD_{90} of 0.015 and 0.027 g/L, respectively (Brackenbury and Appleton, 1997). This effect was attributed to saponin content in the *A. attenuata* extract. On the other hand, this extract did not show health hazards to communities that use the treated waterbody or to the people harvesting, preparing, and applying the molluscicide (Brackenbury and Appleton, 1997).

23.3 BIOACTIVE CONSTITUENTS OF PULQUE

The fermentation process starts in the maguey since native microorganisms, such as yeasts, LAB, ethanol-producing bacteria, and exopolysaccharide-producing bacteria ferment part of the available carbohydrates (Escalante et al., 2004; Lappe-Oliveras et al., 2008). For pulque production, fresh aguamiel is transported in wooden barrels or in bags made of young goat skins, and transferred to large barrels where fermentation takes place. The process is accelerated by adding a portion of previously produced pulque (*semilla de pulque*). The entire process is performed in nonaseptic conditions, hence the mixture of microorganisms involved in the fermentation process is naturally found in aguamiel and incorporated during collection, transport, and handling inoculation (Escalante et al., 2008). Most pulque production in Mexico is done by artisans. The principal variables determining nectar fermentation time are ambient temperature, microorganism type, and concentration of the seed pulque used to ferment the aguamiel (Gómez-Aldapa et al., 2011).

Traditionally, viscosity development due to the synthesis of exopolysaccharides (EPSs) has been one of the main criteria for determining the degree of fermentation (fresh or aged pulque). There is a wide diversity on the structures of these polymers produced during fermentation by LAB, particularly of the genus *Leuconostoc*, since they have a linear backbone of α-(1, 6)-linked D-glucopyranosyl units but with different branches (Torres-Rodríguez et al., 2014). In pulque, viscosity has been associated with EPS (dextran) production by *Leuconostoc mesenteroides* strains (Sánchez-Marroquín and Hope, 1953; Chellapandian et al., 1998). Chellapandian et al. (1998), identified *L. mesenteroides* IBT-PQ strain isolated from pulque, which showed the ability to produce a dextransucrose polymer. Escalante et al. (2008) reported the presence of one EPS after 6h of aguamiel fermentation. In addition, these polysaccharides are prebiotics but they also have been related with other bioactivities, such as antitumoral, immunoregulator, and cholesterol-lowering (Patel et al., 2012).

Besides the concentration of prebiotics, pulque consumption has also been related to ameliorate iron and zinc deficiencies by the presence of phytases derived from the microorganisms found in this drink (Tovar et al., 2008). Phytase activity is very relevant to enhance mineral availability in vegetable sources. Based on in vitro assays, fresh pulque phytases hydrolyzed 78% of the phytate found in corn tortilla that was used as a substrate model (Tovar et al., 2008).

Higher plasma ferritin was associated with greater intakes of ascorbic acid and nonheme iron in nonpregnant Mexican women. Pulque consumption was associated with a reduced risk of low ferritin and hemoglobin values in these women. This study showed that pulque was the primary source of ascorbate and the third most important source of nonheme iron after the phytate-rich sources of maize tortillas and legumes. Tovar et al. (2008) found an ascorbic acid content of 13.3 mg/500mL and iron of 0.35 mg/500mL in fresh pulque.

23.4 MICROORGANISMS IN AGUAMIEL AND PULQUE

Among microorganisms isolated from aguamiel and pulque, homo-fermentative and hetero-fermentative species of *Lactobacillus, L. mesenteroides, L. dextranicum, Saccharomyces carbajali*, and *Pseudomonas linderi* have been reported (Sanchez-Marroquin and Hope, 1953). Besides, the presence of other microorganisms in the pulque has been documented, including bacteria: *Acetobacter aceti, A. aceti* ssp. *xylinus, Bacillus simplex, Bacillus subtilis, Cellulomonas* sp., *Escherichia* sp., *Kokuria rosea, Lactobacillus delbrueckii, Lactobacillus vermiform, Macrococcus caseolyticus, Micrococcus luteus,* and *Sarcina* spp. Yeasts reported in pulque are *Cryptococcus* spp., *Candida parapsilosis, Clavispora lusitaniae, Debaryomyces carsonii, Hanseniaspora uvarum, Geotrichum candidum, Pichia* spp., *Pichia guilliermondii, Pichia membranifaciens, Rhodotorula* sp., *R. mucilaginous, Saccharomyces bayanus, S. pastorianus,* and *Torulaspora delbrueckii* (Lappe-Oliveras et al., 2008).

Analysis of 16S rDNA sequencing showed that bacterial diversity present in pulque is dominated by *Lactobacillus* species (80.97%). Particularly at the end of the fermentation, the bacterial population was mainly composed by homofermentative *L. acidophilus*, hetero-fermentative *L. mesenteroides, Lactococcus lactis* subsp. *lactis,* and proteobacterium *Acetobacter malorum* (Escalante et al., 2008). Based on this study, it can be inferred that the microbial population during the fermentation process of pulque undergoes changes, due to the environmental changes that these microorganisms produce into the aguamiel. Several authors have documented changes from aguamiel to commercial pulque but there is only one report that describes the changes during fermentation (Table 23.1).

In another study, yeast strains with the ability to inhibit the growth of other microorganisms were isolated from aguamiel and pulque. Yeasts isolated from aguamiel were *Candida lusitaneae, Kluyveromyces marxianus* var. *bulgaricus,* and *Saccharomyces cerevisiae (capensis). K. marxianus* var. *bulgaricus* inhibited the growth of other microorganisms. Other strains isolated from pulque included *Candida valida, S. cerevisiae (chevalieri), S. cerevisiae (capensis),* and *K. marxianus* var. *lactis. C. valida* and *K. marxianus* var. *lactis* showed potential to inhibit the growth of other microorganisms (Estrada-Godina et al., 2001). Besides, yeasts isolated from aguamiel and pulque have been evaluated for inulinase production. Two strains of *Kluyveromyces marxianus* isolated from aguamiel exhibited the ability to produce 2.5 times more enzyme than the control hyperproducing strain *K. marxianus* CDBB-L-278. One strain of *Kluyveromyces lactis* var. *lactis* isolated from pulque was also an excellent inulinase producer, being the first strain of this species reported as such (Cruz-Guerrero et al., 2006).

Microbial safety has been evaluated in pulque production (Gomez-Aldapa et al., 2011). Pulque production is an artisanal manufacturing process that takes place in an open-air environment without sanitary practices; therefore, pulque may have risk of pathogen contamination. Gómez-Aldapa et al. (2011) examined the behavior of pathogens, such as *Salmonella typhimurium, Staphylococcus aureus, Listeria monocytogenes, Shigella flexneri,* and *Shigella sonnei* when they were added to fresh aguamiel for pulque manufacturing. When starter pulque was added and the product was fermented, all pathogens were quickly deactivated since they competed with the indigenous microbiota, and none were detected in the final product. This demonstrates the effect of potential antimicrobials contained in this beverage. In another study, the behavior of *Escherichia coli* O157:H7 in aguamiel and pulque was evaluated. *E. coli* O157:H7 is a foodborne pathogen that can cause severe diseases, such as hemorrhagic colitis, and death. Under conditions of the study, *E. coli* O157:H7 survived during the

TABLE 23.1 Changes in Bacteria and Yeast Composition From Aguamiel to Pulque

Microorganism	Aguamiel	Pulque (at 0 h)	Pulque (at 3 h)	Pulque (at 6 h)	Commercial pulque
Bacteria					
Acetobacter aceti					X[a]
Acetobacter pomorium					X[b]
Acetobacterium malorum		X[c]	X[c]	X[c]	
Acetobacter orientales				X[c]	
Acinetobacter radioresistens	X[c]	X[c]	X[c]		
Bacillus spp.				X[c]	
Bacillus licheniformis		X[c]			
Bacillus simplex					X[a]
Bacillus subtilis					X[a]
Cellulomonas spp.					X[a]
Chryseobacterium spp.				X[c]	
Citrobacter spp.	X[c]				
Enterobacter spp.	X[c]				
Enterobacter agglomerans		X[c]	X[c]		
Erwinia rhapontici	X[c]				
Escherichia spp.					X[a]
Flavobacterium johnsoniae					X[b]
Gluconobacter oxydans					X[b]
Hafnia alvei					X[b]
Kluyvera ascorbata		X[c]	X[c]		
Kluyvera cochleae	X[c]				

Continued

TABLE 23.1 Changes in Bacteria and Yeast Composition From Aguamiel to Pulque—cont'd

Microorganism	Aguamiel	Pulque (at 0 h)	Pulque (at 3 h)	Pulque (at 6 h)	Commercial pulque
Kokuria rosea					X[a]
Lactobacillus spp.		X[c]	X[c]	X[c]	X[a]
Lactobacillus acetotolerans					X[b]
Lactobacillus acidophilus			X[c]	X[c]	X[b]
Lactobacillus delbrueckii					X[a]
Lactobacillus hilgardii		X[c]			X[b]
Lactobacillus kefir					X[b]
Lactobacillus paracollinoides			X[c]		
Lactobacillus plantarum					X[b]
Lactobacillus sanfranciscencis		X[c]			
Lactobacillus vermiforme					X[a]
Lactococcus spp.	X[c]				
Lactococcus lactis		X[c]	X[c]		
Leuconostoc spp.					X[a]
Leuconostoc citreum	X[c]	X[c]	X[c]		
Leuconostoc gasicomitatum		X[c]			
Leuconostoc kimchi	X[c]				
Leuconostoc mesenteroides	X[a,c,d]	X[c]	X[c]	X[c]	X[a]
Leuconostoc lactis		X[c]	X[c]	X[c]	
Leuconostoc pseudomesenteroides			X[c]		X[b]
Macrococcus caseolyticus					X[a]

TABLE 23.1 Changes in Bacteria and Yeast Composition From Aguamiel to Pulque—cont'd

Microorganism	Aguamiel	Pulque (at 0 h)	Pulque (at 3 h)	Pulque (at 6 h)	Commercial pulque
Microbacterium arborescens					X[b]
Micrococcus luteus					X[a]
Pediococcus urinaeequi			X[c]		
Providencia spp.				X[c]	
Sarcina spp.					X[a]
Serratia grimensis	X[c]	X[c]			
Streptococcus devriesei	X[c]				
Sterotrophomonas spp.		X[c]			
Zymomonas mobilis			X[c]		X[a]
Yeasts					
Cryptococcus spp.					X[a]
Candida lusitaneae	X[e]				
Candida parapsilosis					X[a]
Candida valida					X[e]
Clavispora lusitaniae					X[a]
Debaryomyces carsonii					X[a]
Hanseniaspora uvarum					X[a]
Kluyveromyces lactis					X[f]
Kluyveromyces marxianus	X[e,f]				X[e]
Geotrichum candidum					X[a]
Pichia spp.					X[a]
Pichia guilliermondii					X[a]

Continued

TABLE 23.1 Changes in Bacteria and Yeast Composition From Aguamiel to Pulque—cont'd

Microorganism	Aguamiel	Pulque (at 0 h)	Pulque (at 3 h)	Pulque (at 6 h)	Commercial pulque
Pichia membranifaciens					X[a]
Rhodotorula spp.					X[a]
Rhodotorula mucilaginosa					X[a]
Saccharomyces bayanus					X[a]
Saccharomyces cerevisiae	X[e]	X[c]			X[a,e]
Saccharomyces pastorianus					X[a]
Torulaspora delbrueckii					X[a]

[a]Lappe-Oliveras et al. (2008).
[b]Escalante et al. (2004).
[c]Escalante et al. (2008).
[d]Castro-Rodríguez, D., Hernández-Sánchez, H., Yáñez-Fernández, J., 2015. Probiotic properties of Leuconostoc mesenteroides isolated from Aguamiel of Agave salmiana. Probiotics and Antimicrobial Protien 7, 107–117.
[e]Estrada-Godina et al. (2001).
[f]Cruz-Guerrero et al. (2006).

pulque fermentation process, suggesting that this pathogen constitutes a high potential health risk to pulque consumers. Results suggest that *E. coli* O157:H7 can develop acid and alcohol tolerance in pulque. However, further research is required using other pulque production conditions and different inoculums and more strains of *E. coli* to better understand the real microbial risk to consumers from this traditional Mexican fermented beverage (Gómez-Aldapa et al., 2012).

Agave sap composition varies among regions, plant species, weather, collection time, and other factors that also affect the endogenous microbiota. Recently, it has been demonstrated that plant bacterial composition of *A. tequilana* and *A. salmiana* were affected by factors, such as soil, host species, and season (Desgarennes et al., 2014). This explains the microbiologic variability observed by Valadez-Blanco et al. (2012) when the aguamiel was obtained from the same region but different site (Fig. 23.1).

Interestingly, even if there were significant differences in the aguamiel microbial composition, particularly for anaerobic LAB, pulque had similar counts of these microorganisms (Fig. 23.2). This was not the case for *Zymomonas* spp.,

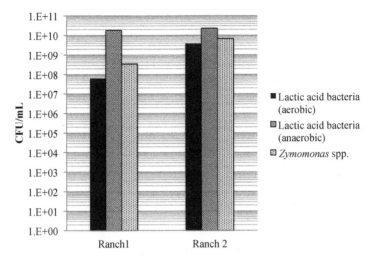

FIGURE 23.1 Comparison of the total counts of LAB and *Zymomonas* spp. of aguamiel collected in two different locations in Oaxaca, México. *Data reported by Valadez-Blanco, R., Bravo-Villa, G., Santos-Sánchez, N.F., Velasco-Almendarez, S.I., Montville, T.J., 2012. The artisanal production of pulque, a traditional beverage of the Mexican highlands. Probiotics and Antimicrobial Protiens 4, 140–144.*

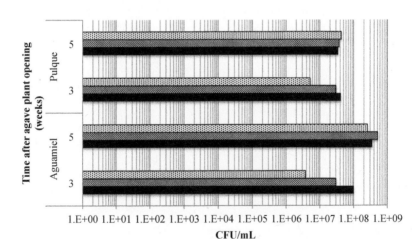

FIGURE 23.2 Comparison of the total counts of lactic acid bacteria and *Zymomonas* spp. of pulque and aguamiel collected 3 or 5 weeks after plant opening. *Data reported by Valadez-Blanco, R., Bravo-Villa, G., Santos-Sánchez, N.F., Velasco-Almendarez, S.I., Montville, T.J., 2012. The artisanal production of pulque, a traditional beverage of the Mexican highlands. Probiotics and Antimicrobial Protiens 4, 140–144.*

responsible for ethanol production in pulque, but the magnitude of the difference in pulque was not as high as that observed for aguamiel.

23.5 CONCLUSIONS

Pulque is an important source of prebiotics and probiotics that have diverse biological activities to promote health. The complex microbiota is affected by the manufacturing process as well as by the conditions of aguamiel collection. Its potential health benefits may be enhanced by the selection of agaves and microorganisms used for the manufacturing process. Pulque is not only consumed as a beverage but also as a food ingredient. Therefore, the stability of the metabolites produced during the fermentation, as well as the potential health benefits derived from its consumption, must be evaluated to overcome the disadvantages of ethanol intake.

REFERENCES

Anajwala, C., Patel, R., Dakhara, S., Jariwala, J., 2010. In vitro cytotoxicity study of *Agave americana, Strychnos nuxvomica* and *Areca catechu* extracts using MCF-7 cell line. Journal of Advanced Pharmaceutical Technology and Research 1, 245–252.

Bhatia, I.S., Nandra, K.S., 1979. Studies on fructosyl transferase from *Agave americana*. Phytochemistry 18, 923–927.

Brackenbury, T.D., Appleton, C.C., 1997. A comprehensive evaluation of *Agave attenuata*, a candidate plant molluscicide in South Africa. Acta Tropica 68, 201–213.

Brito Cortez Lima, N., Dantas Cavalcante Abreu, R.N., Barbosa Filho, J.M., Leal, L.K.A.M., Ribeiro Honorio Júnior, J.E., Camelo Chaves, E.M., et al., 2012. Evaluation of the action mechanism of hecogenin's antidepressant effect from *Agave sisalana* Perrine in mice. European Neuropsychopharmacology 22 (Suppl. 2), S275.

Campo-Padilla, M.D., 1999. Determinación, cuantificación e hidrólisis de inulina en el aguamiel del agave pulquero *Agave atrovirens* (Thesis). Facultad de Química. UNAM, México, DF.

Chellapandian, M., Larios, C., Sánchez-González, M., López-Munguía, A., 1988. Production and properties of a dextransucrase from *Leuconostoc mesenteroides* IBT-PQ isolated from 'pulque', a traditional Aztec alcoholic beverage. Journal of Industrial Microbiology and Biotechnology 21, 51–56.

Correa-Ascencio, M., Robertson, I.G., Cabrera-Cortés, O., Cabrera-Castro, R., Evershed, R.P., 2014. Pulque production from fermented agave sap as a dietary supplement in Prehispanic Mesoamerica. PNAS 111, 14223–14228.

Cruz-Guerrero, A.E., Olvera, J.L., García-Garibay, M., Gomez-Ruiz, L., 2006. Inulinase-hyperproducing strains of *Kluyveromyces* sp. isolated from aguamiel (agave sap) and pulque. World Journal of Microbiology & Biotechnology 22, 115–117.

Desgarennes, D., Garrido, E., Torres-Gomez, M.J., Pena-Cabriales, J.J., Partida-Martinez, L.P., 2014. Diazotrophic potential among bacterial communities associated with wild and cultivated Agave species. FEMS Microbiology Ecology 90, 844–857.

Escalante, A., Flores, M.E., Martínez, A., López-Munguía, A., Bolívar, F., Gosset, G., 2004. Characterization of bacterial diversity in pulque, a traditional Mexican alcoholic fermented beverage, as determined by 16S rDNA analysis. FEMS Microbiology Letters 235, 273–279.

Escalante, A., Giles-Gómez, M., Hernández, G., Córdova-Aguilar, M.S., López-Munguía, A., Gosset, G., Bolívar, F., 2008. Analysis of bacterial community during the fementation of pulque, a traditional Mexican alcoholic beverage, using a polyphasic approach. International Journal of Food Microbiology 124 (2), 126–134.

Escalante, A., Giles-Gómez, M., Esquivel-Flores, G., Matus-Acuña, V., Moreno-Terrazas, R., López-Munguía, A., Lappe-Oliveras, P., 2012. Pulque fermentation. In: Hui, Y.H. (Ed.), Handbook on Plant-based Fermented Food Beverage Technol, second ed. CRC Press, Boca Raton, FL, pp. 691–706.

Estrada-Godina, A.R., Cruz-Guerrero, A.F., Lappe, P., Ulloa, M., García-Garibay, M., Gómez-Ruíz, L., 2001. Isolation and characterization of killer-yeasts from agave sap (aguamiel) and pulque. World Journal of Microbiology and Biotechnology 17, 557–560.

García, M.D., Saenz, M.T., Puerta, R., Quilez, A., Fernandez, M.A., 1999. Antibacterial activity of *Agave intermixta* and *Cissus sicyoides*. Fitoterapia 70 (1), 71–73.

Gibson, G.R., Probert, H.M., Loo, J.V., Rastall, R.A., 2004. Dietary modulation of the human colonic microbiota: updating the concept of prebiotics. Nutrition Research Reviews 17, 259–275.

Gómez, E., Tuohy, K.M., Gibson, G.R., Klinder, A., Costabile, A., 2010. *In vitro* evaluation of the fermentation properties and potential prebiotic activity of Agave fructans. Journal of Applied Microbiology 108 (6), 2114–2121.

Gómez-Aldapa, C.A., Díaz-Cruz, C.A., Villarruel-López, A., Torres-Vitela, M.D.R., Añorve-Morga, J., Rangel-Vargas, E., Cerna-Cortes, J.F., Vigueras-Ramírez, J.G., Castro-Rosas, J., 2011. Behavior of *Salmonella typhimurium, Staphylococcus aureus, Listeria monocytogenes*, and *Shigella flexneri* and *Shigella sonnei* during production of pulque, a traditional mexican beverage. Journal of Food Protection 74 (4), 580–587.

Gómez-Aldapa, C.A., Díaz-Cruz, C.A., Villarruel-López, A., Torres-Vitela, M.D.R., Rangel-Vargas, E., Castro-Rosas, J.G., 2012. Acid and alcohol tolerance of *Escherichia coli* O157:H7 in pulque, a typical Mexican beverage. International Journal of Food Microbiology 154 (1–2), 79–84.

Lappe-Oliveras, P., Moreno-Terrazas, R., Arrizón-Gaviño, J., Herrera-Suárez, T., García-Mendoza, A., Gschaedler-Mathis, A., 2008. Yeasts associated with the production of Mexican alcoholic nondistilled and distilled Agave beverages. FEMS Yeast Research 8, 1037–1052.

Leal-Díaz, A.M., Santos-Zea, L., Martínez-Escobedo, H.C., Guajardo-Flores, D., Gutiérrez-Uribe, J.A., Serna-Saldivar, S.O., 2015. Effect of *Agave americana* and *Agave salmiana* ripeness on saponin content from aguamiel (Agave Sap). Journal of Agricultural and Food Chemistry 63 (15), 3924–3930.

López, M.G., Urías-Silvas, J.E., 2007. Agave fructans as prebiotics. In: Norio, S., Noureddine, B., Shuichi, O. (Eds.), Recent Advances in Fructooligosaccharides Research. Research Signpost, Kerala, India, pp. 296–310.

Mancilla-Margalli, N.A., López, M.G., 2006. Water-soluble carbohydrates and fructan structure patterns from Agave and Dasylirion species. Journal of Agricultural and Food Chemistry 54, 7832–7839.

NMX-V-022, 1972. Aguamiel. Hydromel. Dirección General de Normas, Normas Mexicanas.

NMX-V-037, 1972. Pulque Manejado a Granel. Pulque Handled in Bulk. Dirección General de Normas, Nomas Mexicanas.

Ohtsuki, T., Koyano, T., Kowithayakorn, T., Sakai, S., Kawahara, N., Goda, Y., et al., 2004. New chlorogenin hexasaccharide isolated from *Agave fourcroydes* with cytotoxic and cell cycle inhibitory activities. Bioorganic & Medicinal Chemistry 12 (14), 3841–3845.

Ortiz-Basurto, R.I., Pourcelly, G., Doco, T., Williams, P., Dornier, M., Belleville, M.P., 2008. Analysis of the main components of the aguamiel produced by the maguey-pulquero (*Agave mapisaga*) throughout the harvest period. Journal of Agricultural and Food Chemistry 56 (10), 3682–3687.

Patel, S., Majumder, A., Goyal, A., 2012. Potentials of exopolysaccharides from lactic acid bacteria. Indian Journal of Microbiology 52 (1), 3–12.

Ramírez, J.F., Sánchez-Marroquín, A., Álvarez, M.M., Valyasebi, R., 2004. Industrialization of mexican pulque. In: Steinkraus, K. (Ed.), Industrialization of Indigenous Fermented Foods, second ed. Marcel Deckker, New York, pp. 547–586.

Rendón-Huerta, J.A., Juárez-Flores, B.I., Pinos-Rodríguez, J.M., Aguirre-Rivera, J.R., Delgado-Portales, R.E., 2011. Effects of different kind of fructans on in vitro growth of *Lactobacillus acidophilus, Lactobacillus casei* and *Bifidobacterium lactis*. African Journal of Microbiology Research 5 (18), 2706–2710.

Rendón-Huerta, J.A., Juárez-Flores, B.I., Pinos-Rodríguez, J.M., Aguirre-Rivera, J.R., Delgado-Portales, R.E., 2012. Effects of different sources of fructans on body weight, blood metabolites and fecal bacteria in normal and obese non-diabetic and diabetic rats. Plant Foods for Human Nutrition 67, 64–70.

Sánchez-Marroquín, A., Hope, P.H., 1953. Agave juice: fermentation and chemical composition studies of some species. Journal of Agricultural and Food Chemistry 1, 246–249.

Santos Cerqueira, G., dos Santos e Silva, G., Rios Vasconcelos, E., Fragoso de Freitas, A.P., Arcanjo Moura, B., Silveira Macedo, D., Lopes Souto, A., Barbosa Filho, J.M., de Almeida Leal, L.K., de Castro Brito, G.A., Souccar, C., de Barros Viana, G.S., 2012. Effects of hecogenin and its possible mechanism of action on experimental models of gastric ulcer in mice. European Journal of Pharmacology 683, 260–269.

Santos, W.R., Bernardo, R.R., Peçanha, L.M.T., Palatnik, M., Parente, J., de Sousa, C.B.P., 1997. Haemolytic activities of plant saponins and adjuvants. Effect of *Periandra mediterranea* saponin on the humoral response to the FML antigen of *Leishmania donovani*. Vaccine 15 (9), 1024–1029.

Santos-Zea, L., Leal-Díaz, A.M., Cortés-Ceballos, E., Gutiérrez-Uribe, J.A., 2012. Agave (*Agave* spp.) and its traditional products as a source of bioactive compounds. Current Bioactive Compounds 8 (3), 218–231.

Torres-Rodríguez, I., Rodríguez-Alegría, M.E., Miranda-Molina, A., Giles-Gómez, M., Morales, R.C., López-Munguía, A., Bolívar, F., Escalante, A., 2014. Screening and characterization of extracellular polysaccharides produced by *Leuconostoc kimchii* isolated from traditional fermented pulque beverage. Springer Plus 3, 583.

Tovar, L.R., Olivos, M., Gutierrez, M.E., 2008. Pulque, an alcoholic drink from rural Mexico, contains phytase. Its *in vitro* effects on corn tortilla. Plant Foods Human Nutrition 63, 189–194.

Tovar-Robles, C.L., Perales-Segovia, C., Nava-Cedillo, A., Valera-Montero, L.L., Gómez-Leyva, J.F., Guevara-Lara, F., Hernández-Duque, J.L.M., Silos-Espino, H., 2011. Effect of aguamiel (agave sap) on hematic biometry in rabbits and its antioxidant activity determination. Italian Journal of Animal Science 10 (21), 106–110.

Urías-Silvas, J.E., Cani, P.D., Delmée, E., Neyrinck, A., López, M.G., Delzenne, N.M., 2008. Physiological effects of dietary fructans extracted from *Agave tequilana* Gto. and *Dasylirion* spp. British Journal of Nutrition 99, 254–261.

Valadez-Blanco, R., Bravo-Villa, G., Santos-Sánchez, N.F., Velasco-Almendarez, S.I., Montville, T.J., 2012. The artisanal production of pulque, a traditional beverage of the Mexican highlands. Probiotics and Antimicrobial Protiens 4, 140–144.

Vijn, I., Smeekens, S., 1999. Fructan: more than a reserve carbohydrate? Plant Physiology 120, 351–359.

Wang, N., Nobel, P.S., 1998. Phloem transport of fructans in the crassulacean acid metabolism species *Agave deserti*. Plant Physiology 116, 709–714.

Chapter 24

Sauerkraut: Production, Composition, and Health Benefits

E. Peñas, C. Martinez-Villaluenga, J. Frias
Institute of Food Science, Technology and Nutrition (ICTAN-CSIC), Madrid, Spain

24.1 BRIEF HISTORY OF SAUERKRAUT

Salted and fermented vegetables have a long history in human nutrition since ancient times, with sauerkraut being one of the most popular vegetable fermented products. Sauerkraut, a traditional dish in Central and Eastern Europe, the United States, and Asia, results from the lactic acid fermentation of chopped and salted white cabbage (*Brassica oleracea* var. *capitata*).

The word sauerkraut originated from the German term for "sour cabbage." The first mention of sauerkraut dates from 2000 years ago in China, where cabbage was fermented in rice wine, and was introduced 1000 years later in Europe by Genghis Khan after invading China (Wacher et al., 2010). Europeans substituted wine with salt in sauerkraut.

The health benefits of sauerkraut were well-known in early civilizations. Hippocrates, a Greek physician, recommended sauerkraut against overweight and Romans consumed it to prevent intestinal infections. Moreover, Captain James Cook, an English navigator and explorer, replenished the food supplies of his ships with sauerkraut on his lengthy voyages, as he noticed that it did not need refrigeration and prevented scurvy in sailors (Saloheimo, 2005).

24.2 SAUERKRAUT MANUFACTURE

For sauerkraut production, fresh cabbage heads are trimmed of outer leaves and their central cores are removed (Fig. 24.1). Cabbage is further shredded to 0.7–2-mm thick strips and salted with 0.7–2.5% sodium chloride (Adams and Moss, 1995; Holzapfel, 2003). The addition of salt is necessary for the development of anaerobic conditions during fermentation and for inhibiting the growth of spoilage microorganisms and activity of endogenous pectinolytic enzymes responsible for cabbage softening. The concentration of salt added affects the

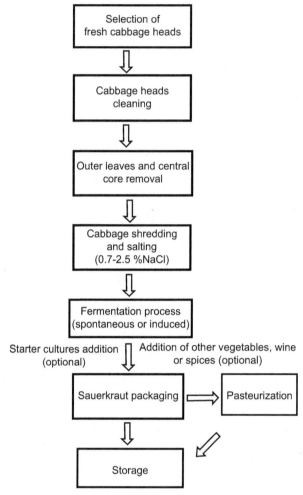

FIGURE 24.1 Schematic representation of sauerkraut production.

type of microbial population and the sensory quality of sauerkraut. The amount of sodium chloride used usually depends on fermentation temperature and market preference (Wolkers-Rooijackers et al., 2013). However, due to the new guidelines on sodium salt intake published by WHO in 2012, there is a trend towards the reduction of the salt level during sauerkraut manufacture. Sometimes, spices, herbs, carrot, and wine are added at this stage of fermentation to enhance sauerkraut flavor. After salting, cabbage is placed into fermentation vessels and it is tightly pressed to exclude air. Fermentation chambers are then covered with a lid to allow the development of anaerobic conditions and cabbage is left to ferment between 1 week and several months. After fermentation, sauerkraut is packaged in metal cans or glass jars, and it is consumed as fresh product or it is pasteurized to extend its shelf-life.

24.3 MICROBIAL CHANGES DURING SPONTANEOUS SAUERKRAUT FERMENTATION

Sauerkraut is usually produced by spontaneous fermentation that relies on populations of lactic acid bacteria (LAB) naturally present in raw cabbage. Before fermentation, raw cabbage harbors a variety of microorganisms, including aerobic spoilage bacteria, such as *Pseudomonas*, *Enterobacter*, yeasts, and molds (Nguyen-the and Carlin, 1994). The populations of these aerobic mesophilic microorganisms range from 10^4–10^6 colony forming units (CFU)/g, whereas the LAB population in raw cabbage is lower, 10^2–10^3 CFU/g (Peñas et al., 2010a; Breidt et al., 2013). When shredded cabbage is pressed during fermentation, the concentration of oxygen decreases in the vessel, leading to a reduction of dominant aerobic bacterial communities. These anaerobic conditions, however, favor the growth of LAB populations. It is known that during cabbage fermentation a rapid turnover of different heterofermentative and homofermentative LAB species occurs due to the modification of ecological conditions throughout the process. Fermentation is initiated by heterofermentative LAB, mainly *Leuconostoc mesenteroides*. This bacterium dominates the microbiota at the beginning of fermentation since it possess less acid tolerance and microaerophilic properties, and has a shorter generation time than other LAB over temperatures of 18–20°C and NaCl concentrations up to 5% (Lu et al., 2003; Breidt et al., 2013). Other LAB, such as *Leuconostoc fallax* seem to be also involved in the early fermentation period (Barrangou et al., 2002). These bacterial communities produce significant amounts of acetic and lactic acids that lead to a decrease of the pH, as well as carbon dioxide that provides an anaerobic environment. When the acid content increases to 0.7–1% and the pH decreases below 4.5, *Leuconostoc* species are replaced by more acid-tolerant homofermentative LAB (McDonald et al., 1990), such as *Lactobacillus plantarum* and *Lactobacillus brevis*. These populations produce lactic acid almost exclusively and dominate the late stage of the fermentation, when the pH reaches values ranging from 3.4 to 3.7 (Nguyen-the and Carlin, 1994; Plengvidhya et al., 2007). The correct succession of these bacterial communities during fermentation is essential to obtain sauerkraut with good sensory quality. At the end of the fermentation process, sauerkraut contains approximately 1% acetic acid and more than 2% lactic acid (Breidt et al., 2013).

24.4 INOCULATION OF STARTER CULTURES DURING SAUERKRAUT MANUFACTURE

Even though sauerkraut is generally produced by spontaneous fermentation, its flavor and quality markedly depend on the composition of the indigenous microbial community and on the quality of the cabbage used for fermentation. The microbial population of vegetables is subjected to fluctuations on the physical and nutritional conditions (Lindow and Brandl, 2003), and hence, changes in the autochthonous microbiota of raw cabbage will be reflected in modifications of sauerkraut quality.

The use of starter cultures during sauerkraut production has been proposed in recent years to minimize the variation of the quality and to ensure the uniformity of the product. At present, there is a lack of commercial starters suitable for cabbage fermentation. Therefore, the development of starter cultures targeted for sauerkraut manufacture is still a scientific challenge. A careful selection of the bacterial strain used as starter during fermentation should be performed in order to ensure an optimal sauerkraut quality. The selection criteria must be related to the adaptation of the strain to cabbage material, as well as to the tolerance of the strain to low pH values and high salt concentrations. Moreover, a rapid growth rate and the production of acids causing a rapid acidification of sauerkraut are important criteria to be considered for the starter culture selection. On the other hand, a heterofermentative metabolism, and the synthesis of antimicrobial compounds, such as bacteriocins and hydrogen peroxide are important features expected in starter cultures used for cabbage fermentation (Di Cagno et al., 2013). Starter cultures consisting of single or multiple strains can be used. Inoculation with sauerkraut obtained from a previous fermentation process has also been reported (Wiander and Ryhänen, 2005).

Recent studies have investigated the use of several starter cultures to obtain good-quality sauerkraut with enhanced levels of bioactive compounds or reduced concentrations of salt and biogenic amines. A nisin-resistant *Leuconostoc mesenteroides* strain alone (Breidt et al., 1995) or in combination with a nisin-producing *Lactococcus lactis* strain (Harris et al., 1992) have been applied for cabbage fermentation to inhibit the growth of naturally present LAB, thus improving the uniformity of sauerkraut quality. Tolonen et al. (2002) documented the production of good-quality sauerkraut containing a high amount of beneficial phytochemicals using a mixed starter culture containing *L. mesenteroides* and *Pediococcus dextrinicus*. The same group reported that the application of a *Lactobacillus sakei* culture conducted to sauerkrauts with antimicrobial activity and a concentration of bioactive compounds two to three times higher than in sauerkraut produced using another starter cultures (Tolonen et al., 2004). Moreover, the application of *L. mesenteroides* as starter culture in combination with mineral salt with low NaCl resulted in a mild-tasting sauerkraut juice with good and uniform sensory and microbiological qualities (Wiander and Ryhänen, 2005). Similarly, Johanningsmeier et al. (2007) observed that the addition of *L. mesenteroides* during fermentation resulted in a more uniform sauerkraut compared to noninoculated fermentations, while allowing 50% NaCl reduction. The fermentation of cabbage with *L. mesenteroides* and low salt concentration (0.5%) retained a large concentration of vitamin C and caused higher production of ascorbigen, a glucosinolate hydrolysis compound with recognized anticarcinogenic properties (Peñas et al., 2010b) and lower content of biogenic amines than cabbages fermented by *Lactobacillus plantarum* (Peñas et al., 2010c). A decrease of biogenic amine concentration has also been observed after application of *L. plantarum*, *Lactobacillus casei* subsp. *casei* 2763, or *Lactobacillus curvatus* 2771 and 2775 as starter cultures when compared with spontaneous fermentation (Halász et al., 1999; Kalač et al., 2000; Rabie et al., 2011).

Recent studies have shown potential benefits of using probiotic starter cultures for sauerkraut production. In this sense, the application of the probiotic strain *Lactobacillus plantarum* L4 in combination with *Leuconostoc mesenteroides* LMG 7954 positively influenced the fermentation process by reducing the NaCl concentration from 4.0% to 2.5% and accelerating fermentation by 14 days. The resultant sauerkraut showed an improved quality and can be considered a probiotic product since a viable probiotic cell count in the final product was higher than 10^6 CFU/g of product (Beganović et al., 2011). Starter cultures with probiotic properties offer additional advantages compared to classical starter cultures, representing a valuable approach to improving the quality, safety, and health-promoting properties of sauerkraut.

24.5 NUTRITIONAL AND PHYTOCHEMICAL COMPOSITION OF SAUERKRAUT

Cabbage is recognized as a health-promoting vegetable due to its high nutritional value and its large levels of bioactive compounds. The main constituents of white cabbage are carbohydrates (4.18–5.51 g/100 g) and dietary fiber (1.9–2.9 g/100 g) followed by proteins (1.27–1.37 g/100 g), minerals (0.3–0.7g/100 g), fat (0.06–0.20 g/100 g), and vitamins, especially vitamin C (0.03–0.04 g/100 g) (Souci et al., 2000; USDA, 2011). Cabbage also contains phytochemicals, mainly phenolic compounds and glucosinolates (GLS). GLS, a distinctive feature of *Brassica* vegetables, are a group of nitrogen- and sulfur-containing plant secondary metabolites that are responsible for the characteristic flavor and odor of these vegetables (Verkerk et al., 2009). GLS are among the most-studied bioactive compounds in *Brassica* vegetables and they are associated with their cancer-protective properties. The profile and content of GLS depends on the genotype, environmental conditions, geographical growing location, and postharvest processing (Rosa and Heaney, 1996; Ciska et al., 2000; Verkerk et al., 2009; Peñas et al., 2011). The main GLS present in white cabbage are sinigrin, an aliphatic GLS compound, and glucobrassicin, an indolic GLS (Peñas et al., 2011). GLS themselves are not biologically active, but during food processing they are hydrolyzed to a wide range of biologically active compounds.

During LAB fermentation, the chemical composition of cabbage changes, resulting in a fermented product rich in carbohydrates, proteins, fat, dietary fiber, minerals, and vitamins (Table 24.1). Sauerkraut also contains organic acids, mostly lactic and acetic acids (1–2%), but also propionic, malic, and succinic acids, as well as ethanol, ethyl acetate, acetaldehyde, and CO_2 (Trail et al., 1996). Sauerkraut is considered a particularly good source of antioxidants, such as vitamin C, whose content ranges from 14.7 to 75 mg/100g fresh weight (fw), and phenolic compounds (0.44–1.06 mg gallic acid equivalents/100 g fw) (Ciska et al., 2005; Podsędek, 2007; Peñas et al., 2011; USDA, 2011).

TABLE 24.1 Sauerkraut Composition

Nutrient	Units	Content (Per 100 g Fresh Weight)
Proximates		
Water	g	92.5
Energy	kcal	19
Protein	g	0.91
Fat	g	0.14
Carbohydrates	g	4.28
Total dietary fiber	g	2.9
Sugars (total)	g	1.78
Glucose	g	0.14
Fructose	g	0.04
Minerals		
Ca	mg	30
Fe	mg	1.47
Mg	mg	13
P	mg	20
K	mg	170
Na	mg	661
Zn	mg	0.19
Cu	mg	0.096
Mn	mg	0.15
Se	µg	0.6
F	µg	0.7
Vitamins		
Vitamin C	mg	14.7
Thiamin	mg	0.021
Riboflavin	mg	0.022
Niacin	mg	0.143
Pantothenic acid	mg	0.093

TABLE 24.1 Sauerkraut Composition—cont'd

Nutrient	Units	Content (Per 100 g Fresh Weight)
Vitamin B6	mg	0.130
Folic acid	µg	24
Vitamin A	µg	1
B-Carotene	µg	8
A-Carotene	µg	5
Vitamin E	µg	0.14
Vitamin K	µg	13.0
Lipids		
Saturated fatty acids	g	0.034
Monounsaturated fatty acids	g	0.013
Polyunsaturated fatty acids	g	0.067

Data obtained from USDA, December 7, 2011. National Nutrient Database for Standard Reference.

Moreover, this fermented vegetable presents high levels of GLS breakdown products with recognized health benefits. In intact vegetables, GLS are located in tissue compartments that are physically separated from those containing myrosinase (thioglucosidase EC 3.2.1.147), an enzyme responsible for GLS hydrolysis. During cabbage shredding and fermentation, a disruption of cabbage cells occurs, and GLS are completely hydrolyzed by myrosinase enzyme to a variety of GLS breakdown products including isothiocyanates (ITCs), thiocyanates, nitriles, and epithionitriles, compounds with recognized health-promoting properties (Peñas et al., 2010a; Nugrahedi et al., 2015). In particular, glucobrassicin is hydrolyzed into indol-3-carbinol (I3C) by myrosinase and, as the pH decreases during cabbage fermentation, this indole compound reacts nonenzymatically with ascorbic acid to yield ascorbigen (ABG) (Wagner and Rimbach, 2009). Studies have shown that ABG is the main GLS breakdown compound in sauerkraut, and it is present at levels between 3 and 18 µmol/100 g fw (Ciska and Pathak, 2004; Martinez-Villaluenga et al., 2009; Peñas et al., 2010b, 2012a, 2015; Palani et al., 2016). The anticarcinogenic properties of ABG due to its ability to induce activation of xenobiotic-metabolizing enzymes and apoptosis of tumoral cells have been well established (Stephensen et al., 1999; Bonnesen et al., 2001; Kravchenko et al., 2001). I3C and indol-3-acetonitrile (I3A) have also been detected in sauerkraut at concentrations of 0.13–0.94 µmol/100 g fw and 0.07–0.25 µmol/100 g fw, respectively (Tolonen et al., 2002; Ciska et al., 2009; Ciska and Honke, 2012; Peñas et al., 2012a,b; Palani et al., 2016). It should be highlighted that the addition of sodium selenite (0.3 mg/kg fw) has been found to be beneficial to enhance the concentration of both compounds in sauerkraut (Peñas et al., 2012a,b). Indole GLS hydrolysis products exhibit cancer-protective properties (Bonnesen et al., 2001). In fact, a plethora of studies have clearly shown the anticarcinogenic properties of I3C through regulation of inflammation, cell proliferation, and inhibition of tumor invasion in a variety of tumors (Perez-Chacon et al., 2014; Tin et al., 2014; Lin et al., 2015; Safa et al., 2015; Wang et al., 2015). Interestingly, clinical trials evaluating the effect of oral intake of I3C and its dimer 3,3-diindolylmethane (DIM) in the prevention or treatment of cancerous lesions have established the effectiveness of these compounds as chemopreventive agents in the early precancerous stages of cervix and larynx carcinogenesis (Auborn et al., 1998; Bell et al., 2000; Reed et al., 2005) Furthermore, antiinflammatory and protective properties against reactive oxygen species (ROS) of I3C have been demonstrated in animal models (El-Naga et al., 2014; Jayakumar et al., 2014).

Apart from indole GLS breakdown products, other GLS derivatives are formed during sauerkraut production. Sulforaphane (SFN), a compound derived from the GLS glucoraphanin, has been found in sauerkraut at concentrations of 28–32 µmol/100 g dw, whilst allyl isothiocyanate (AITC), formed by siningrin decomposition, was present at levels of 16–125 µmol/100 g dw, depending on the fermentation conditions and the starter culture used (Tolonen et al., 2002, 2004; Peñas et al., 2012a,b, 2015). Sauerkraut also contains iberin (1-isothiocyanate-3[methylsulfinyl]-propane) and iberin nitrile (4-[methylsulfinyl]-butane nitrile), both derived from glucoiberin, as well as allyl cyanide (3-butenenitrile), derived from sinigrin, at concentrations ranging between 29 and 39 µmol/100 g dw, 36–39 µmol/100 g dw, 15–72 µmol/100 g dw, respectively (Tolonen et al., 2002, 2004; Peñas et al., 2012a,b, 2015). All these GLS hydrolysis compounds have attracted attention as potential chemopreventive agents against various types of cancer.

24.6 HEALTH BENEFITS OF SAUERKRAUT

A large body of available data obtained from epidemiological and clinical studies has shown that the intake of *Brassica* vegetables reduces the risk of certain chronic diseases, such cancer and cardiovascular diseases (Verhoeven et al., 1996; Hayes et al., 2008; Tse and Eslick, 2014). Although the beneficial effects of sauerkraut on health have been less extensively studied, there is convincing evidence that sauerkraut possesses a wide range of health benefits that can be attributed to its high levels of phytochemicals.

24.6.1 Antioxidant Benefits

Oxidative stress, a situation that occurs when the production of ROS is greater than organisms' antioxidant protective ability, has been increasingly recognized as a contributing factor in aging and in the pathogenesis of many chronic health problems, such as cardiovascular diseases, neurodegenerative diseases, and cancer (Persson et al., 2014; Pisoschi and Pop, 2015). Sauerkraut contains high levels of vitamins C and E as well as phenolic compounds that act as potent free radical scavengers, protecting against oxidative stress (Podsędek, 2007). Vitamin C reduces the levels of C-reactive protein (CRP), involved in inflammation and atherothrombosis, and acts (together with phenolic compounds) as an electron donor for eight human enzymes, neutralizing superoxide and hydroxyl radicals (Ellulu et al., 2015). Vitamin E exhibits antioxidant activity due to its ability to donate a hydrogen atom, showing protecting effects against cardiovascular diseases due to inhibition of LDL oxidation (Podsędek, 2007). Moreover, some GLS hydrolysis compounds present in sauerkraut, such as allyl isothiocyanate and phenyl isothiocyanate, have been shown to scavenge hydroxyl radicals in vitro (Manesh and Kuttan, 2003).

The free radical scavenging potential of sauerkraut produced using different fermentation conditions has been widely investigated. In this sense, peroxyl radical scavenging activity in the range of 132–164 μmol trolox/g dry weight (dw) was observed by ORAC-FL in sauerkrauts obtained by natural or induced fermentation (Martinez-Villaluenga et al., 2012; Peñas et al., 2012a,b). On the other hand, TEAC values of 1.0–3.0 and 4–6 μmol/g dw were found by ABTS and DPPH assays for juices and methanolic extracts, respectively, obtained from naturally fermented sauerkrauts (Kusznierewicz et al., 2008, 2010). Ciska et al. (2005) reported an antioxidant activity of 31 μmol trolox/g in extracts from sauerkrauts obtained by natural fermentation (Ciska et al., 2005). The antioxidant activity observed for sauerkraut in all studies was higher than that observed in raw cabbage.

24.6.2 Anticarcinogenic Properties

A few recent studies have documented the protective effects of sauerkraut against cancer. The chemopreventive activity of sauerkraut is related to the induction of phase II detoxifying and antioxidant enzymes by ITCs and other GLS breakdown products, such as I3C and ABG formed during cabbage fermentation. It has been reported that oral administration of sauerkraut juice to rats (1.25 ml/kg body weight) for 28 days increased the expression and activity of glutathione S-transferase (GST) and NAD(P)H quinone oxidoreductase 1 (NQO1), two key detoxifying enzymes, in liver through activation of NF-E2-related transcription factor 2 (Nrf2) and aryl-hydrocarbon receptor (AhR) (Krajka-Kuźniak et al., 2011). These effects were organ-dependent since no effects of sauerkraut juice were noted on antioxidant enzymes in kidney. The same research group investigated the effect of sauerkraut juices on the expression and activity of cytochrome P450 enzymes (CYP1A1/1A2, 1B1, and CYP2B enzymes), phase I enzymes involved in detoxification of several environmental carcinogens and endogenous chemicals, in rat liver and kidney (Szaefer et al., 2012a). Results indicated that rats fed sauerkraut juice increase CYP1A1 and CYP1A2 activities in liver. These effects of sauerkraut juice intake on antioxidant and detoxifying enzymes were similar to those found in rats after oral administration of pure I3C and penethyl isothiocyanate (PEITC). Additional findings from these studies were that sauerkraut juices induced the expression of CYP1A1, CYP1A2, and CYP1B1, all involved in estrogen metabolism, in breast cell lines (Szaefer et al., 2012b). Estrogens are considered a major breast cancer risk factor and their metabolism by P450 enzymes substantially contributes to carcinogenic activity. Furthermore, sauerkraut juice showed a potent inhibitory activity of CYP19 expression in MCF10A breast cells (Licznerska et al., 2013). CYP19 encodes the enzyme aromatase that synthesizes estrogens and presents high levels in breast cancer tissue. These results suggest that sauerkraut could have potential breast-protective activity against breast, liver, and kidney cancer. These findings are consistent with those reported by other authors, who found that ABG itself, or in combination with I3C or SFN, modulates the activity of xenobiotic metabolizing enzymes, including the CYP1A1 family (Stephensen et al., 1999; Kravchenko et al., 2001).

The chemopreventive effects of processed cabbage and sauerkraut against different types of cancer have also been reported by other authors. A review summarizing the epidemiological studies performed on brassica vegetables and prostate cancer shows that cabbage consumption is associated with a lower risk of prostate cancer (Kristal and Lampe, 2002). These findings are in line with those found in a prospective study that observed a statistically significant lower risk of pancreatic cancer after consumption of cabbage (≥ 1 serving per week) (Larsson et al., 2006). On the other hand, the antiestrogenic activity of sauerkraut extracts at low concentrations (5–25 ng/mL) in estrogen-dependent human breast cancer (MCF-7) cells has been reported (Ju et al., 2000). These results are consistent with those of a case–control study performed in Polish migrant women showing that short-cooked cabbage and sauerkraut consumption (>3 servings per week) during adolescence and adulthood was linked to a 72% reduced risk of breast cancer compared with those who consumed 1.5 or fewer servings per week (Nelson, 2006).

Numerous studies performed in rodents and humans have shown the ability of AITC and SFN to inhibit different stages of cancer development in several organs (Cornblatt et al., 2007; Kim et al., 2014; Lenzi et al., 2014; Rajakumar et al., 2015). SFN exhibits a potent ability to induce phase II detoxification enzymes, such as quinone reductases, UDP-glucoronosyl transferases, and glutathione transferase (Abdull Razis et al., 2011) and to inhibit phase I enzymes of the cytochrome P450 system (Herr and Büchler, 2010). SFN also suppresses expression of the androgen receptor protein, the central signaling pathway in prostate cancer, by inactivating histone deacetylases proteins in vitro and in vivo (Myzak et al., 2006, 2007; Gibbs et al., 2009), and induce repression of cancer cell proliferation, stimulation of cancer cell apoptosis, inhibition of tumor progression (by its antiangiogenic effect), and metastasis (Kaminski et al., 2012; Lenzi et al., 2014). Recent experimental data pointed to SFN inhibition of intracellular signaling pathways activated by the epidermal growth factor receptor (EGFR), which play a central role in cancerous cell survival and proliferation (Chen et al., 2015). SFN also exerts an antiinflammatory effect since it reduces expression of genes encoding inflammatory cytokines, inducible nitric oxide synthase (iNOS) and cyclooxygenase 2 (COX-2), as well as secretion of tumor necrosis factor alpha (TNF-α) in cultivated macrophages (Heiss et al., 2001). Therefore, SFN exerts a cancer-protective effect through several mechanisms of action. Numerous studies in vitro and in animals have also demonstrated the cancer chemopreventive properties of AITC through the inhibition of cell proliferation, tumor growth, and induction of cell cycle arrest and apoptosis (Srivastava et al., 2003; Zhang et al., 2003; Chen et al., 2010; Lau et al., 2010; Rajakumar et al., 2015; Sávio et al., 2015).

Although less extensively studied, iberin has also been shown a great deal of attention since it exhibits anticarcinogenic properties by increasing the expression of tumor-suppressing genes and induction of cell cycle arrest and apoptosis (Verhoeven et al., 1996; Jakubikova et al., 2006; Jadhav et al., 2007a,b; Traka et al., 2009). Furthermore, epidemiological studies showed that dietary intake of isothiocyanates reduced the incidence of breast, lung, and colon cancer (Zhao et al., 2001; Seow et al., 2002). It has been reported that doses between 53 and 150 μmol of ITCs are enough to display anticarcinogenic effects. Taking into account that the content of ITC in sauerkraut is in the range 22 μmol/100 g fw (Peñas et al., 2012a,b), it could be assumed that a weekly consumption of 200–250 g of sauerkraut would provide effective ITC doses to exert cancer chemopreventive effects.

24.6.3 Protection Against Oxidative DNA Damage

It is well known that an elevated concentration of ROS can induce oxidative DNA damage producing, consequently, DNA mutations. DNA mutation is a crucial step in carcinogenesis. In fact, the role of increased levels of oxidized base lesions in DNA in the initiation, promotion, and progression stages of carcinogenesis has been well established (Cooke et al., 2003; Scott et al., 2014). The ability of sauerkraut juice to prevent DNA damage in human colon tumor HT29 cells exposed to an oxidative insult (300-μM H_2O_2) has been investigated (Kusznierewicz et al., 2010). The authors demonstrated that sauerkraut juices prevented DNA damage when applied at doses regarding normal dietary consumption to HT29 cells concomitantly with H_2O_2. ITCs and indoles present in sauerkraut could be responsible for these beneficial effects, since the ability of both groups of phytochemicals to protect against chemicals, such as benzo(*a*)pyrene and H_2O_2 that damage DNA in LS-174 colon cells has been reported (Bonnesen et al., 2001). Chemoprevention by these compounds, however, required the colon cells to be treated with both phytochemicals before exposure to genotoxic compounds. Moreover, pretreatment of LS-174 cells with both ITCs and indoles conferred significantly greater protection against

DNA fragmentation than did each compound alone. The protection of DNA from chemical insults by ITCs, such as SFN, in human hepatoma, mammary epithelial, and liver epithelial cells has been confirmed by other investigations (Barceló et al., 1998; Singletary and MacDonald, 2000; Jiang et al., 2003).

24.6.4 Antiinflammatory Effects

Chronic inflammation plays a critical role in various chronic diseases, such as cancer, cardiovascular diseases, type II diabetes, and obesity (Zhao, 2013; Donath, 2014; Fernandes et al., 2015; Ueland et al., 2015). The complex process of inflammation is stimulated by endotoxins, such as LPS. Macrophages are the most important immune cells involved in the inflammatory response. The activated macrophages can produce inflammatory mediators, such as nitric oxide (NO), eicosanoids, TNF-α, and cytokines. The inhibition of the production of these inflammatory mediators is an important target in the treatment of inflammatory diseases (Ghayor et al., 2015). Hence, LPS-induced macrophages have usually been used as a model for evaluating the antiinflammatory effects of phytochemicals contained in foods.

The ability of raw cabbage and sauerkraut to prevent NO production by LPS-induced macrophages has been recently investigated. Both cabbage and sauerkraut, obtained by natural or induced fermentations, suppressed NO production in a dose-dependent manner (Martinez-Villaluenga et al., 2012). Sauerkraut presented higher NO production inhibitory potency than raw cabbage. These authors studied whether the potential NO production inhibitory effect of sauerkraut could be attributed to ABG, by treating LPS-activated macrophages with synthetic ABG (Peñas et al., 2012a,b). ABG showed a weak inhibitory activity of NO production, suggesting that other bioactive compounds present in sauerkraut would be responsible for the observed biological activity of this fermented vegetable. Data available in the literature indicate that I3C and SFN inhibit production of cytokine and expression of proinflammatory enzymes, such as inducible NO synthase (Heiss and Gerhäuser, 2005; Cho et al., 2008), suggesting that these phytochemicals could contribute to the antiinflammatory activity of sauerkraut. Moreover, current data clearly indicate that AITC also exhibits antiinflammatory activity through downregulation of inflammatory gene expression in cultured macrophages in vitro (Woo et al., 2007; Wagner et al., 2012). The addition of selenite (1.6 mg Se/kg dw) during cabbage fermentation notably improved the NO production inhibitory effects of sauerkraut (Peñas et al., 2012a,b).

24.6.5 Sauerkraut as a Source of Probiotic Bacteria

Sauerkraut can be considered as a vehicle of probiotic bacteria. Probiotics are defined as live microorganisms that, when administered in adequate amounts, confer a health benefit on the host (Sanders, 2008). Many studies reported that LAB isolated from sauerkraut are potential probiotics. Beganović et al. (2014)

isolated two LAB strains, *Lactobacillus paraplantarum* SF9 and *Lactobacillus brevis* SF15, from sauerkraut and demonstrated their ability to survive in the acid gastrointestinal environment as well as their competitive exclusion of enteropathogens. Furthermore, these bacteria showed adhesion capacity to Caco-2 cells, an important prerequisite for exerting probiotic effects. Similarly, different *Lactobacillus* strains isolated from sauerkraut effectively survived in simulated gastrointestinal conditions and inhibited growth of six pathogens. These strains also exhibit high ability to adhere to Caco-2 cells (Feng et al., 2015). It should be emphasized that these probiotic bacterial strains could be used as starter cultures in other vegetable fermentation processes for producing functional foods.

24.7 CONCLUDING REMARKS

Sauerkraut is a nutritious fermented vegetable food, highly appreciated for its particular sensory characteristics. There is strong scientific evidence that sauerkraut provides numerous health benefits, such as antioxidant and anticarcinogenic effects, but also by attenuating inflammation and DNA damage. Moreover, sauerkraut is a natural unexplored source of probiotic bacteria that can be potentially used as a starter culture in other vegetable fermentation processes. The health-promoting properties of sauerkraut are attributed to its high levels of bioactive constituents, especially glucosinolate breakdown products. The data pointing to health promotion of sauerkraut are currently coming from in vitro and epidemiological studies. Unfortunately, clinical data supporting the potentially health benefits of sauerkraut remain scarce. Therefore, intervention studies are mandatory to confirm the beneficial properties of sauerkraut. Such studies would allow the demonstration of the disease-prevention properties of this fermented vegetable, of great importance since health care costs continue to rise.

ACKNOWLEDGMENTS

This research was cofunded by the Ministry of Economy and Competitiveness (MINECO, Spain) and the European Union through the project number AGL2013-43247-R and FEDER programme, respectively. E. Peñas in indebted to the Spanish "Ramón y Cajal" Programme for financial support.

REFERENCES

Abdull Razis, A.F., Iori, R., Ioannides, C., 2011. The natural chemopreventive phytochemical *R*-sulforaphane is a far more potent inducer of the carcinogen-detoxifying enzyme systems in rat liver and lung than the *S*-isomer. International Journal of Cancer 128, 2775–2782.

Adams, M.R., Moss, M.O., 1995. Food Microbiology, third ed. The Royal Society of Chemistry, Cambridge.

Auborn, K., Abramson, A., Bradlow, H.L., Sepkovic, D., Mullooly, V., 1998. Estrogen metabolism and laryngeal papillomatosis: a pilot study on dietary prevention. Anticancer Research 18, 4569–4573.

Barceló, S., Macé, K., Pfeifer, A.M.A., Chipman, J.K., 1998. Production of DNA strand breaks by *N*-nitrosodimethylamine and 2-amino- 3-methylimidazo[4,5-*f*]quinoline in THLE cells expressing human CYP isoenzymes and inhibition by sulforaphane. Mutation Research—Fundamental and Molecular Mechanisms of Mutagenesis 402, 111–120.

Barrangou, R., Yoon, S.S., Breidt Jr., F., Fleming, H.P., Klaenhammer, T.R., 2002. Identification and characterization of *Leuconostoc fallax* strains isolated from an industrial sauerkraut fermentation. Applied and Environmental Microbiology 68, 2877–2884.

Beganović, J., Pavunc, A.L., Gjuračić, K., Špoljarec, M., Šušković, J., Kos, B., 2011. Improved sauerkraut production with probiotic strain *Lactobacillus plantarum* L4 and *Leuconostoc mesenteroides* LMG 7954. Journal of Food Science 76, M124–M129.

Beganović, J., Kos, B., Leboš Pavunc, A., Uroić, K., Jokić, M., Šušković, J., 2014. Traditionally produced sauerkraut as source of autochthonous functional starter cultures. Microbiological Research 169, 623–632.

Bell, M.C., Crowley-Nowick, P., Bradlow, H.L., Sepkovic, D.W., Schmidt-Grimminger, D., Howell, P., Mayeaux, E.J., Tucker, A., Turbat-Herrera, E.A., Mathis, J.M., 2000. Placebo-controlled trial of indole-3-carbinol in the treatment of CIN. Gynecologic Oncology 78, 123–129.

Bonnesen, C., Eggleston, I.M., Hayes, J.D., 2001. Dietary indoles and isothiocyanates that are generated from cruciferous vegetables can both stimulate apoptosis and confer protection against DNA damage in human colon cell lines. Cancer Research 61, 6120–6130.

Breidt, F., Crowley, K.A., Fleming, H.P., 1995. Controlling cabbage fermentations with nisin and nisin-resistant *Leuconostoc mesenteroides*. Food Microbiology 12, 109–116.

Breidt, F., McFeeters, R.F., Perez-Diaz, I., Lee, C.-H., 2013. Fermented Vegetables. ASM Press, Whasington, DC.

Chen, N.G., Chen, K.T., Lu, C.C., Lan, Y.H., Lai, C.H., Chung, Y.T., Yang, J.S., Lin, Y.C., 2010. Allyl isothiocyanate triggers G2/M phase arrest and apoptosis in human brain malignant glioma GBM 8401 cells through a mitochondria-dependent pathway. Oncology Reports 24, 449–455.

Chen, C.Y., Yu, Z.Y., Chuang, Y.S., Huang, R.M., Wang, T.C.V., 2015. Sulforaphane attenuates EGFR signaling in NSCLC cells. Journal of Biomedical Science 22.

Cho, H.J., Seon, M.R., Lee, Y.M., Kim, J., Kim, J.K., Kim, S.G., Park, J.H., 2008. 3,3′-Diindolylmethane suppresses the inflammatory response to lipopolysaccharide in murine macrophages. The Journal of Nutrition 138, 17–23.

Ciska, E., Honke, J., 2012. Effect of the pasteurization process on the contents of ascorbigen, indole-3-carbinol, indole-3-acetonitrile, and 3,3′-diindolylmethane in fermented cabbage. Journal of Agricultural and Food Chemistry 60, 3645–3649.

Ciska, E., Pathak, D.R., 2004. Glucosinolate derivatives in stored fermented cabbage. Journal of Agricultural and Food Chemistry 52, 7938–7943.

Ciska, E., Martyniak-Przybyszewska, B., Kozlowska, H., 2000. Content of glucosinolates in cruciferous vegetables grown at the same site for two year under different climatic conditions. Journal of Agricultural and Food Chemistry 48, 2862–2867.

Ciska, E., Karamac, M., Kosinska, A., 2005. Antioxidant activity of extracts of white cabbage and sauerkraut. Polish Journal of Food and Nutrition Sciences 14/55, 367–373.

Ciska, E., Verkerk, R., Honke, J., 2009. Effect of boiling on the content of ascorbigen, indole-3-carbinol, indole-3-acetonitrile, and 3,3′-diindolylmethane in fermented cabbage. Journal of Agricultural and Food Chemistry 57, 2334–2338.

Cooke, M.S., Evans, M.D., Dizdaroglu, M., Lunec, J., 2003. Oxidative DNA damage: mechanisms, mutation, and disease. FASEB Journal 17, 1195–1214.

Cornblatt, B.S., Ye, L., Dinkova-Kostova, A.T., Erb, M., Fahey, J.W., Singh, N.K., Chen, M.S.A., Stierer, T., Garrett-Mayer, E., Argani, P., Davidson, N.E., Talalay, P., Kensler, T.W., Visvanathan, K., 2007. Preclinical and clinical evaluation of sulforaphane for chemoprevention in the breast. Carcinogenesis 28, 1485–1490.

Di Cagno, R., Coda, R., De Angelis, M., Gobbetti, M., 2013. Exploitation of vegetables and fruits through lactic acid fermentation. Food Microbiology 33, 1–10.

Donath, M.Y., 2014. Targeting inflammation in the treatment of type 2 diabetes: time to start. Nature Reviews Drug Discovery 13, 465–476.

Ellulu, M.S., Rahmat, A., Patimah, I., Khaza'Ai, H., Abed, Y., 2015. Effect of vitamin C on inflammation and metabolic markers in hypertensive and/or diabetic obese adults: a randomized controlled trial. Drug Design, Development and Therapy 9, 3405–3412.

El-Naga, R.N., Ahmed, H.I., Abd Al Haleem, E.N., 2014. Effects of indole-3-carbinol on clonidine-induced neurotoxicity in rats: impact on oxidative stress, inflammation, apoptosis and monoamine levels. Neurotoxicology 44, 48–57.

Feng, J., Liu, P., Yang, X., Zhao, X., 2015. Screening of immunomodulatory and adhesive *Lactobacillus* with antagonistic activities against *Salmonella* from fermented vegetables. World Journal of Microbiology and Biotechnology 31, 1947–1954.

Fernandes, J.V., Cobucci, R.N.O., Jatobá, C.A.N., de Medeiros Fernandes, T.A.A., de Azevedo, J.W.V., de Araújo, J.M.G., 2015. The role of the mediators of inflammation in cancer development. Pathology and Oncology Research 21, 527–534.

Ghayor, C., Gjoksi, B., Siegenthaler, B., Weber, F.E., 2015. N-methyl pyrrolidone (NMP) inhibits lipopolysaccharide-induced inflammation by suppressing NF-kB signaling. Inflammation Research 64, 527–536.

Gibbs, A., Schwartzman, J., Deng, V., Alumkal, J., 2009. Sulforaphane destabilizes the androgen receptor in prostate cancer cells by inactivating histone deacetylase 6. Proceedings of the National Academy of Sciences of the United States of America 106, 16663–16668.

Halász, A., Baráth, Á., Holzapfel, W.H., 1999. The influence of starter culture selection on sauerkraut fermentation. European Food Research and Technology 208, 434–438.

Harris, L.J., Fleming, H.P., Klaenhammer, T.R., 1992. Novel paired starter culture system for sauerkraut, consisting of a nisin- resistant *Leuconostoc mesenteroides* strain and a nisin-producing *Lactococcus lactis* strain. Applied and Environmental Microbiology 58, 1484–1489.

Hayes, J.D., Kelleher, M.O., Eggleston, I.M., 2008. The cancer chemopreventive actions of phytochemicals derived from glucosinolates. European Journal of Nutrition 47, 73–88.

Heiss, E., Gerhäuser, C., 2005. Time-dependent modulation of thioredoxin reductase activity might contribute to sulforaphane-mediated inhibition of NF-κB binding to DNA. Antioxidants and Redox Signaling 7, 1601–1611.

Heiss, E., Herhaus, C., Klimo, K., Bartsch, H., Gerhäuser, C., 2001. Nuclear factor κB is a molecular target for sulforaphane-mediated anti-inflammatory mechanisms. Journal of Biological Chemistry 276, 32008–32015.

Herr, I., Büchler, M.W., 2010. Dietary constituents of broccoli and other cruciferous vegetables: implications for prevention and therapy of cancer. Cancer Treatment Reviews 36, 377–383.

Holzapfel, 2003. Sauerkraut. In: Farnworth, E.R. (Ed.), Handbook of Fermented Functional Foods. CRC Press, Boca Raton, FL, USA, pp. 343–360.

Jadhav, U., Ezhilarasan, R., Vaughn, S.F., Berhow, M.A., Mohanam, S., 2007a. Dietary isothiocyanate iberin inhibits growth and induces apoptosis in human glioblastoma cells. Journal of Pharmacological Sciences 103, 247–251.

Jadhav, U., Ezhilarasan, R., Vaughn, S.F., Berhow, M.A., Mohanam, S., 2007b. Iberin induces cell cycle arrest and apoptosis in human neuroblastoma cells. International Journal of Molecular Medicine 19, 353–361.

Jakubikova, J., Bao, Y., Bodo, J., Sedlak, J., 2006. Isothiocyanate iberin modulates phase II enzymes, posttranslational modification of histones and inhibits growth of Caco-2 cells by inducing apoptosis. Neoplasma 53, 463–470.

Jayakumar, P., Pugalendi, K.V., Sankaran, M., 2014. Attenuation of hyperglycemia-mediated oxidative stress by indole-3-carbinol and its metabolite 3,3′-diindolylmethane in C57BL/6J mice. Journal of Physiology and Biochemistry 70, 525–534.

Jiang, Z.Q., Chen, C., Yang, B., Hebbar, V., Kong, A.N.T., 2003. Differential responses from seven mammalian cell lines to the treatments of detoxifying enzyme inducers. Life Sciences 72, 2243–2253.

Johanningsmeier, S., McFeeters, R.F., Fleming, H.P., Thompson, R.L., 2007. Effects of *Leuconostoc mesenteroides* starter culture on fermentation of cabbage with reduced salt concentrations. Journal of Food Science 72, M166–M172.

Ju, Y.H., Carlson, K.E., Sun, J., Pathak, D., Katzenellenbogen, B.S., Katzenellenbogen, J.A., Helferich, W.G., 2000. Estrogenic effects of extracts from cabbage, fermented cabbage, and acidified brussels sprouts on growth and gene expression of estrogen-dependent human breast cancer (MCF-7) cells. Journal of Agricultural and Food Chemistry 48, 4628–4634.

Kalač, P., Špička, J., Křížek, M., Pelikánová, T., 2000. The effects of lactic acid bacteria inoculants on biogenic amines formation in sauerkraut. Food Chemistry 70, 355–359.

Kaminski, B.M., Steinhilber, D., Stein, J.M., Ulrich, S., 2012. Phytochemicals resveratrol and sulforaphane as potential agents for enhancing the anti-tumor activities of conventional cancer therapies. Current Pharmaceutical Biotechnology 13, 137–146.

Kim, H.N., Kim, D.H., Kim, E.H., Lee, M.H., Kundu, J.K., Na, H.K., Cha, Y.N., Surh, Y.J., 2014. Sulforaphane inhibits phorbol ester-stimulated IKK-NF-κB signaling and COX-2 expression in human mammary epithelial cells by targeting NF-κB activating kinase and ERK. Cancer Letters 351, 41–49.

Krajka-Kuźniak, V., Szaefer, H., Bartoszek, A., Baer-Dubowska, W., 2011. Modulation of rat hepatic and kidney phase II enzymes by cabbage juices: comparison with the effects of indole-3-carbinol and phenethyl isothiocyanate. British Journal of Nutrition 105, 816–826.

Kravchenko, L.V., Avren'eva, L.I., Guseva, G.V., Posdnyakov, A.L., Tutel'yan, V.A., 2001. Effect of nutritional indoles on activity of xenobiotic metabolism enzymes and T-2 toxicity in rats. Bulletin of Experimental Biology and Medicine 131, 544–547.

Kristal, A.R., Lampe, J.W., 2002. Brassica vegetables and prostate cancer risk: a review of the epidemiological evidence. Nutrition and Cancer 42, 1–9.

Kusznierewicz, B., Śmiechowska, A., Bartoszek, A., Namieśnik, J., 2008. The effect of heating and fermenting on antioxidant properties of white cabbage. Food Chemistry 108, 853–861.

Kusznierewicz, B., Lewandowska, J., Kruszyna, A., Piasek, A., Śmiechowska, A., Namieśnik, J., Bartoszek, A., 2010. The antioxidative properties of white cabbage (*Brassica oleracea* var. *capitata f. alba*) fresh and submitted to culinary processing. Journal of Food Biochemistry 34, 262–285.

Larsson, S.C., Håkansson, N., Näslund, I., Bergkvist, L., Wolk, A., 2006. Fruit and vegetable consumption in relation to pancreatic cancer risk: a prospective study. Cancer Epidemiology Biomarkers and Prevention 15, 301–305.

Lau, W.S., Chen, T., Wong, Y.S., 2010. Allyl isothiocyanate induces G2/M arrest in human colorectal adenocarcinoma SW620 cells through down-regulation of Cdc25B and Cdc25C. Molecular Medicine Reports 3, 1023–1030.

Lenzi, M., Fimognari, C., Hrelia, P., 2014. Sulforaphane as a promising molecule for fighting cancer. Cancer Treatment and Research 207–223.

Licznerska, B.E., Szaefer, H., Murias, M., Bartoszek, A., Baer-Dubowska, W., 2013. Modulation of CYP19 expression by cabbage juices and their active components: indole-3-carbinol and 3,3′-diindolylmethene in human breast epithelial cell lines. European Journal of Nutrition 52, 1483–1492.

Lin, H., Gao, X., Chen, G., Sun, J., Chu, J., Jing, K., Li, P., Zeng, R., Wei, B., 2015. Indole-3-carbinol as inhibitors of glucocorticoid-induced apoptosis in osteoblastic cells through blocking ROS-mediated Nrf2 pathway. Biochemical and Biophysical Research Communications 460, 422–427.

Lindow, S.E., Brandl, M.T., 2003. Microbiology of the phyllosphere. Applied and Environmental Microbiology 69, 1875–1883.

Lu, Z., Breidt, F., Plengvidhya, V., Fleming, H.P., 2003. Bacteriophage ecology in commercial sauerkraut fermentations. Applied and Environmental Microbiology 69, 3192–3202.

Manesh, C., Kuttan, G., 2003. Anti-tumour and anti-oxidant activity of naturally. Occurring isothiocyanates. Journal of Experimental and Clinical Cancer Research 22, 193–199.

Martinez-Villaluenga, C., Peñas, E., Frias, J., Ciska, E., Honke, J., Piskula, M.K., Kozlowska, H., Vidal-Valverde, C., 2009. Influence of fermentation conditions on glucosinolates, ascorbigen, and ascorbic acid content in white cabbage (*Brassica oleracea* var. *capitata* cv. Täler) cultivated in different seasons. Journal of Food Science 74, C62–C67.

Martinez-Villaluenga, C., Peñas, E., Sidro, B., Ullate, M., Frias, J., Vidal-Valverde, C., 2012. White cabbage fermentation improves ascorbigen content, antioxidant and nitric oxide production inhibitory activity in LPS-induced macrophages. LWT—Food Science and Technology 46, 77–83.

McDonald, L.C., Fleming, H.P., Hassan, H.M., 1990. Acid tolerance of *Leuconostoc mesenteroides* and *Lactobacillus plantarum*. Applied and Environmental Microbiology 56, 2120–2124.

Myzak, M.C., Hardin, K., Wang, R., Dashwood, R.H., Ho, E., 2006. Sulforaphane inhibits histone deacetylase activity in BPH-1, LnCaP and PC-3 prostate epithelial cells. Carcinogenesis 27, 811–819.

Myzak, M.C., Tong, P., Dashwood, W.M., Dashwood, R.H., Ho, E., 2007. Sulforaphane retards the growth of human PC-3 xenografts and inhibits HDAC activity in human subjects. Experimental Biology and Medicine 232, 227–234.

Nelson, N.J., 2006. Migrant studies aid the search for factors linked to breast cancer risk. Journal of the National Cancer Institute 98, 436–438.

Nguyen-the, C., Carlin, F., 1994. The microbiology of minimally processed fresh fruits and vegetables. Critical Reviews in Food Science and Nutrition 34, 371–401.

Nugrahedi, P.Y., Verkerk, R., Widianarko, B., Dekker, M., 2015. A mechanistic perspective on process-induced changes in glucosinolate content in *Brassica* vegetables: a review. Critical Reviews in Food Science and Nutrition 55, 823–838.

Palani, K., Harbaum-Piayda, B., Meske, D., Keppler, J.K., Bockelmann, W., Heller, K.J., Schwarz, K., 2016. Influence of fermentation on glucosinolates and glucobrassicin degradation products in sauerkraut. Food Chemistry 190, 755–762.

Peñas, E., Frias, J., Gomez, R., Vidal-Valverde, C., 2010a. High hydrostatic pressure can improve the microbial quality of sauerkraut during storage. Food Control 21, 524–528.

Peñas, E., Frias, J., Sidro, B., Vidal-Valverde, C., 2010b. Chemical evaluation and sensory quality of sauerkrauts obtained by natural and induced fermentations at different NaCl levels from *Brassica oleracea* Var. *capitata* Cv. Bronco grown in eastern Spain. Effect of storage. Journal of Agricultural and Food Chemistry 58, 3549–3557.

Peñas, E., Frias, J., Sidro, B., Vidal-Valverde, C., 2010c. Impact of fermentation conditions and refrigerated storage on microbial quality and biogenic amine content of sauerkraut. Food Chemistry 123, 143–150.

Peñas, E., Frias, J., Martínez-Villaluenga, C., Vidal-Valverde, C., 2011. Bioactive compounds, myrosinase activity, and antioxidant capacity of white cabbages grown in different locations of Spain. Journal of Agricultural and Food Chemistry 59, 3772–3779.

Peñas, E., Martinez-Villaluenga, C., Frias, J., Sánchez-Martínez, M.J., Pérez-Corona, M.T., Madrid, Y., Cámara, C., Vidal-Valverde, C., 2012a. Se improves indole glucosinolate hydrolysis products content, Se-methylselenocysteine content, antioxidant capacity and potential anti-inflammatory properties of sauerkraut. Food Chemistry 132, 907–914.

Peñas, E., Pihlava, J.M., Vidal-Valverde, C., Frias, J., 2012b. Influence of fermentation conditions of *Brassica oleracea* L. var. *capitata* on the volatile glucosinolate hydrolysis compounds of sauerkrauts. LWT—Food Science and Technology 48, 16–23.

Peñas, E., Martínez-Villaluenga, C., Pihlava, J.M., Frias, J., 2015. Evaluation of refrigerated storage in nitrogen-enriched atmospheres on the microbial quality, content of bioactive compounds and antioxidant activity of sauerkrauts. LWT—Food Science and Technology 61, 463–470.

Perez-Chacon, G., De Los Rios, C., Zapata, J.M., 2014. Indole-3-carbinol induces cMYC and IAP-family downmodulation and promotes apoptosis of Epstein-Barr virus (EBV)-positive but not of EBV-negative Burkitt's lymphoma cell lines. Pharmacological Research 89, 46–56.

Persson, T., Popescu, B.O., Cedazo-Minguez, A., 2014. Oxidative stress in Alzheimer's disease: why did antioxidant therapy fail? Oxidative Medicine and Cellular Longevity 2014.

Pisoschi, A.M., Pop, A., 2015. The role of antioxidants in the chemistry of oxidative stress: a review. European Journal of Medicinal Chemistry 97, 55–74.

Plengvidhya, V., Breidt Jr., F., Lu, Z., Fleming, H.P., 2007. DNA fingerprinting of lactic acid bacteria in sauerkraut fermentations. Applied and Environmental Microbiology 73, 7697–7702.

Podsędek, A., 2007. Natural antioxidants and antioxidant capacity of *Brassica vegetables*: a review. LWT—Food Science and Technology 40, 1–11.

Rabie, M.A., Siliha, H., El-Saidy, S., El-Badawy, A.A., Malcata, F.X., 2011. Reduced biogenic amine contents in sauerkraut via addition of selected lactic acid bacteria. Food Chemistry 129, 1778–1782.

Rajakumar, T., Pugalendhi, P., Thilagavathi, S., 2015. Dose response chemopreventive potential of allyl isothiocyanate against 7,12-dimethylbenz(a)anthracene induced mammary carcinogenesis in female Sprague-Dawley rats. Chemico-Biological Interactions 231, 35–43.

Reed, G.A., Peterson, K.S., Smith, H.J., Gray, J.C., Sullivan, D.K., Mayo, M.S., Crowell, J.A., Hurwitz, A., 2005. A phase I study of indole-3-carbinol in women: tolerability and effects. Cancer Epidemiology Biomarkers and Prevention 14, 1953–1960.

Rosa, E., Heaney, R., 1996. Seasonal variation in protein, mineral and glucosinolate composition of Portuguese cabbages and kale. Animal Feed Science and Technology 57, 111–127.

Safa, M., Tavasoli, B., Manafi, R., Kiani, F., Kashiri, M., Ebrahimi, S., Kazemi, A., 2015. Indole-3-carbinol suppresses NF-κB activity and stimulates the p53 pathway in pre-B acute lymphoblastic leukemia cells. Tumor Biology 36, 3919–3930.

Saloheimo, P., 2005. Captain cook used sauerkraut to prevent scurvy. Duodecim 121, 1014–1015.

Sanders, M.E., 2008. Probiotics: definition, sources, selection, and uses. Clinical Infectious Diseases 46, S58–S61.

Sávio, A.L.V., da Silva, G.N., Salvadori, D.M.F., 2015. Inhibition of bladder cancer cell proliferation by allyl isothiocyanate (mustard essential oil). Mutation Research 771, 29–35.

Scott, T.L., Rangaswamy, S., Wicker, C.A., Izumi, T., 2014. Repair of oxidative DNA damage and cancer: recent progress in DNA base excision repair. Antioxidants and Redox Signaling 20, 708–726.

Seow, A., Yuan, J.M., Sun, C.L., Van Den Berg, D., Lee, H.P., Yu, M.C., 2002. Dietary isothiocyanates, glutathione S-transferase polymorphisms and colorectal cancer risk in the Singapore Chinese Health Study. Carcinogenesis 23, 2055–2061.

Singletary, K., MacDonald, C., 2000. Inhibition of benzo[a]pyrene- and 1,6-dinitropyrene-DNA adduct formation in human mammary epithelial cells by dibenzoylmethane and sulforaphane. Cancer Letters 155, 47–54.

Souci, S.W., Fatchmann, W., Kraut, H., 2000. Food Composition and Nutritional Tables, sixth ed. MedPharm Scientific Publishers.

Srivastava, S.K., Xiao, D., Lew, K.L., Hershberger, P., Kokkinakis, D.M., Johnson, C.S., Trump, D.L., Singh, S.V., 2003. Allyl isothiocyanate, a constituent of cruciferous vegetables, inhibits growth of PC-3 human prostate cancer xenograft in vivo. Carcinogenesis 24, 1665–1670.

Stephensen, P.U., Bonnesen, C., Bjeldanes, L.F., Vang, O., 1999. Modulation of cytochrome P4501A1 activity by ascorbigen in murine hepatoma cells. Biochemical Pharmacology 58, 1145–1153.

Szaefer, H., Krajka-Kuźniak, V., Bartoszek, A., Baer-Dubowska, W., 2012a. Modulation of carcinogen metabolizing cytochromes P450 in rat liver and kidney by cabbage and sauerkraut juices: comparison with the effects of indole-3-carbinol and phenethyl isothiocyanate. Phytotherapy Research 26, 1148–1155.

Szaefer, H., Licznerska, B., Krajka-Kuniak, V., Bartoszek, A., Baer-Dubowska, W., 2012b. Modulation of CYP1A1, CYP1A2 and CYP1B1 expression by cabbage juices and indoles in human breast cell lines. Nutrition and Cancer 64, 879–888.

Tin, A.S., Park, A.H., Sundar, S.N., Firestone, G.L., 2014. Essential role of the cancer stem/progenitor cell marker nucleostemin for indole-3-carbinol anti-proliferative responsiveness in human breast cancer cells. BMC Biology 12.

Tolonen, M., Taipale, M., Viander, B., Pihlava, J.M., Korhonen, H., Ryhänen, E.L., 2002. Plant-derived biomolecules in fermented cabbage. Journal of Agricultural and Food Chemistry 50, 6798–6803.

Tolonen, M., Rajaniemi, S., Pihlava, J.M., Johansson, T., Saris, P.E.J., Ryhänen, E.L., 2004. Formation of nisin, plant-derived biomolecules and antimicrobial activity in starter culture fermentations of sauerkraut. Food Microbiology 21, 167–179.

Trail, A.C., Fleming, H.P., Young, C.T., McFeeters, R.F., 1996. Chemical and sensory characterization of commercial sauerkraut. Journal of Food Quality 19, 15–30.

Traka, M.H., Chambers, K.F., Lund, E.K., Goodlad, R.A., Johnson, I.T., Mithen, R.F., 2009. Involvement of KLF4 in sulforaphane- and iberin-mediated induction of p21waf1/cip1. Nutrition and Cancer 61, 137–145.

Tse, G., Eslick, G.D., 2014. Cruciferous vegetables and risk of colorectal neoplasms: a systematic review and meta-analysis. Nutrition and Cancer 66, 128–139.

Ueland, T., Gullestad, L., Nymo, S.H., Yndestad, A., Aukrust, P., Askevold, E.T., 2015. Inflammatory cytokines as biomarkers in heart failure. Clinica Chimica Acta 443, 71–77.

USDA, December 7, 2011. USDA National Nutrient Database for Standard Reference.

Verhoeven, D.T.H., Goldbohm, R.A., Van Poppel, G., Verhagen, H., Van Den Brandt, P.A., 1996. Epidemiological studies on *Brassica* vegetables and cancer risk. Cancer Epidemiology Biomarkers and Prevention 5, 733–748.

Verkerk, R., Schreiner, M., Krumbein, A., Ciska, E., Holst, B., Rowland, I., de Schrijver, R., Hansen, M., Gerhäuser, C., Mithen, R., Dekker, M., 2009. Glucosinolates in *Brassica* vegetables: the influence of the food supply chain on intake, bioavailability and human health. Molecular Nutrition and Food Research 53, 219–265.

Wacher, C., Díaz-Ruiz, G., Tamang, J.P., 2010. Fermented vegetable products. In: Tamang, J.P., Kailasapathi, K. (Eds.), Fermented Foods and Beverages of the World. CRC Press, Boca Raton, FL, pp. 151–190.

Wagner, A.E., Rimbach, G., 2009. Ascorbigen: chemistry, occurrence, and biologic properties. Clinics in Dermatology 27, 217–224.

Wagner, A.E., Boesch-Saadatmandi, C., Dose, J., Schultheiss, G., Rimbach, G., 2012. Anti-inflammatory potential of allyl-isothiocyanate - role of nrf2, nf-κB and microrna-155. Journal of Cellular and Molecular Medicine 16, 836–843.

Wang, X., He, H., Lu, Y., Ren, W., Teng, K.Y., Chiang, C.L., Yang, Z., Yu, B., Hsu, S., Jacob, S.T., Ghoshal, K., Lee, L.J., 2015. Indole-3-carbinol inhibits tumorigenicity of hepatocellular carcinoma cells via suppression of microRNA-21 and upregulation of phosphatase and tensin homolog. Biochimica et Biophysica Acta - Molecular Cell Research 1853, 244–253.

WHO, 2012. Guideline: Sodium Intake for Adults and Children. World Health Organization, Geneva.

Wiander, B., Ryhänen, E.L., 2005. Laboratory and large-scale fermentation of white cabbage into sauerkraut and sauerkraut juice by using starters in combination with mineral salt with a low NaCl content. European Food Research and Technology 220, 191–195.

Wolkers-Rooijackers, J.C.M., Thomas, S.M., Nout, M.J.R., 2013. Effects of sodium reduction scenarios on fermentation and quality of sauerkraut. LWT—Food Science and Technology 54, 383–388.

Woo, H.M., Kang, J.H., Kawada, T., Yoo, H., Sung, M.K., Yu, R., 2007. Active spice-derived components can inhibit inflammatory responses of adipose tissue in obesity by suppressing inflammatory actions of macrophages and release of monocyte chemoattractant protein-1 from adipocytes. Life Sciences 80, 926–931.

Zhang, Y., Tang, L., Gonzalez, V., 2003. Selected isothiocyanates rapidly induce growth inhibition of cancer cells. Molecular Cancer Therapeutics 2, 1045–1052.

Zhao, B., Seow, A., Lee, E.J.D., Poh, W.T., Teh, M., Eng, P., Wang, Y.T., Tan, W.C., Yu, M.C., Lee, H.P., 2001. Dietary isothiocyanates, glutathione S-transferase -M1, -T1 polymorphisms and lung cancer risk among Chinese women in Singapore. Cancer Epidemiology Biomarkers and Prevention 10, 1063–1067.

Zhao, L., 2013. Chronic inflammation of obesity, chronic inflammation: causes. Treatment Options and Role in Disease 143–155.

Chapter 25

Vinegars and Other Fermented Condiments

M.C. Garcia-Parrilla[1], M.J. Torija[2], A. Mas[2], A.B. Cerezo[1], A.M. Troncoso[1]
[1]Universidad de Sevilla, Sevilla, Spain; [2]Universitat Rovira i Virgili, Tarragona, Spain

25.1 INTRODUCTION/GENERAL OVERVIEW

Vinegar is defined as a liquid fit for human consumption and contains a specified amount of acetic acid and water. It is produced from raw materials of different agricultural origin containing starch and sugars, that are subjected to a process of double fermentation, alcoholic and acetous (Tesfaye et al., 2002). Vinegar is one of the oldest fermented products and its history dates back to around 2000 BC, having been considered for a long time as the poor relative among fermented food products (Solieri and Giudici, 2009). Despite not having nutritional value, its value in the context of healthy food habits seems outstanding, as well as its property to preserve foods due to its high acetic acid content (Murooka and Yamshita, 2008).

Vinegar is obtained by a two-step process (alcoholic fermentation plus acetification) in which bioconversion of different carbohydrate sources takes place (Mas et al., 2014). Different agricultural raw materials may be used as starting material containing an edible carbohydrate source (fermentable sugars). Yeasts are the first microorganisms that convert sugar into alcohol followed by acetic acid bacteria, which convert ethanol into acetic acid; the latter process is known as "acetification" instead of the more popular term of "acetous fermentation," due to its strict requirement for oxygen (Mas et al., 2015). The most common raw materials used to obtain vinegar are shown in Table 25.1. They are mainly plant materials with some exceptions, such as whey and honey, and encompass a wide variety of origins. Hence, typical retail varieties of vinegar include balsamic, cider, malt, rice, fruit, and white distilled alcohol as well as white and red wine. Flavored vinegars and herbal specialty vinegars are made from wine or white distilled, with popular flavorings including garlic, basil, and tarragon. Specialty fruit vinegars are also made from wine or white distilled, and can be produced by adding fruit or fruit juice to create a sweet-and-sour

TABLE 25.1 Vinegars of the World: Substrate, Name and Region/Country of Origin

Substrate (Raw Material)	Name	Region/Country (Production and Distribution)
Grape	Wine vinegar	Global
	Balsamic vinegar	Global
	Red vinegar	Global
	White vinegar	Global
	Distilled white vinegar	Global
	Sherry vinegar	Global
	Traditional balsamic vinegar	Global
Apple	Cider vinegar	US, Canada
Different fruits (mango, kaki, berries)	Fruit vinegar	East and Southeast Asia
Date	Date vinegar	Middle East
Coconut	Coconut vinegar	Tropical Africa
Rice	Rice vinegar	China, Japan, Korea
	Kurosu	China, Japan, Korea
Malta	Malt vinegar	USA, Northern Europe
	Distilled malt vinegar	USA, Northern Europe
Whey (dairy by products)	Whey vinegar	Europe
Honey	Honey vinegar	Global

taste. Apple- and raspberry-flavored vinegars remain widely popular, among others. Moreover, the variety of raw materials used in the production of vinegar is outstanding, ranging from byproducts and agricultural surpluses to high-quality substrates for the most unique and prized vinegars, such as sherry vinegar (Spain) and aceto balsamico tradizionalle (Italy) (Mas et al., 2014). However, most of the vinegar produced is "white" vinegar, that is vinegar produced directly from diluted alcohol to be used by the food industry in different sauces, dressings, and formulas (Solieri and Giudici, 2009).

Although acetic acid bacteria are feared among oenologists because they spoil wine, they are the main agents in the production of vinegar. Acetic acid

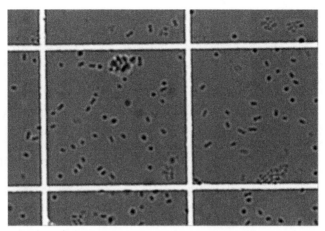

FIGURE 25.1   Acetic acid bacteria from a laboratory wine acetification process.

bacteria are Gram-negative or Gram-variable, ellipsoidal or cylindrical, and can be observed under the microscope alone, in pairs, or in aggregates and chains (Fig. 25.1) (Sengun and Karabiyikli, 2011). Acetic acid bacteria have aerobic respiratory metabolism, and oxygen is generally used as the final electron acceptor; however, other compounds may occasionally act as final electron acceptors, allowing the bacteria to survive under nearly anaerobic conditions, such as those present during wine fermentation (Drysdale and Fleet, 1988). The main genera involved in acetification are *Acetobacter*, *Gluconacetobacter*, and *Komagataeibacter*, which are represented by more than 40 species, although less than half of them have been described in vinegars. Acetic acid bacteria are normally inoculated in wines as "mother of vinegar," an undefined mixture of different species that are present in the production of vinegar. Several attempts have been done to have single, well-defined species of acetic acid bacteria for the production of vinegar, although it has been concluded that a mixture of at least two species (one of them as "starter" and the other as "finisher," with different acetic acid sensitivities) is the most appropriate to be used as inoculum for the production of vinegar, especially those above 5% (w/v) acetic acid (Gullo et al., 2009; Hidalgo et al., 2010). In addition to the production of vinegar, acetic acid bacteria are also of importance in the fermentation of cocoa and coffee (Sengun and Karabiyikli, 2011).

There are two main elaboration processes to produce vinegar, as follows: (1) traditional methods encompassing the transformation of ethanol into acetic acid by static surface culture of acetic acid bacteria placed at the interface between the liquid and air; and (2) submerged culture methods involving the use of bioreactors with a continuous air flow, thus the conversion of alcohol into acetic acid is much faster and takes place in the air–liquid interfaces of the air bubbles. Optimization of parameters governing the submerged process is a priority challenge for vinegar industry (Gullo et al., 2014). In general, traditional methods provide higher-quality vinegars due to an outstanding organoleptic complexity.

Vinegar was known by most ancient civilizations, and its use as a seasoning or preserving agent is as ancient as the use of wine. Although it is a spontaneous process which takes place in wines and musts in contact with air, vinegar is far from being the simple spoilage of wine. Hence, as a food side-product from alcoholic fermented liquids, it has lately acquired an important role as a salad dressing and ingredient for sauces and other food products, such as marinades. Its processing is an ancient technology (Egypt, 8000 BC) that follows wine production and it was used by Greeks as a medicine, while by Romans it was commonly utilized as a beverage (called oxicrat, composed by vinegar, water, and eggs), flavoring (the famous Apicio recipes were based on vinegar), medicine, and cosmetic. Several vinegar therapeutical applications are mentioned in historical documents; as an example, it was recommended for disinfection during cholera diffusion (Mazza and Yoshikatsu, 2009). Today, vinegar is used exclusively for gastronomy and food technology, for dressing and pickled food, and as a sauce ingredient. At the same time, new and innovative goods with marketable health benefits and multiple uses are appearing globally since there has been an increase in consumer demand for natural and organic foods that have nutritional value.

Black vinegars from China or "kurosu" from Japan are obtained from rice and other cereals (including sorghum, wheat, and others) through a slow acetification process with an outstanding aging process and subsequent concentration to yield a dense, dark product. This vinegar is produced by a solid-state fermentation in which yeasts and bacteria grow in the absence of water (Wu et al., 2010). Since kurosu is produced from unpolished rice, it is characterized by higher levels of amino acids and organic acids (Murooka and Yamshita, 2008). These vinegars may even be consumed as a health beverage in small quantities in Japan. However, the direct use of vinegar as a drink is rare (Mas et al., 2015).

The main component of vinegar is acetic acid, which gives a sour taste and pungent aroma. The acetic acid content of the different vinegars present in the global market usually ranges from four to seven acetic degrees (w/v; 4–7 g acetic acid/100 ml of vinegar). Other minor constituents of vinegar include other organic acids (eg, citric, malic, and lactic acids), esters, alcohols, vitamins, mineral salts, amino acids, as well as compounds with antioxidant properties (mainly polyphenols).

25.2 ANTIMICROBIAL EFFECTS

Although vinegar is used mainly for its organoleptic properties, its role as a preserving agent is well known due to its antimicrobial effects, which make it useful for various industrial uses and applications. Hence, it has been traditionally used for the relief of a number of infectious diseases, such as nail fungus or otitis (Dohar, 2003; Budak et al., 2014).

The organic acids in vinegar (mainly acetic acid) are capable of passing through cell membranes entering the bacteria and leading to cell death. When the killing effects of different organic acids were compared, acetic acid was reported to be the most effective to inactivate *Escherichia coli* O157:H7 (Ryu et al., 1999). Lately, acetic acid has been shown to be an effective and economical bactericidal agent for *M. tuberculosis* and nontuberculous mycobacteria, although exposure time required for optimal killing is longer (20–30 min) than the recommended for some commercial bactericides (5 min). Exposure to 6% acetic acid for 30 min resulted in an 8 log reduction of viable *M. tuberculosis* bacteria, including extensively drug-resistant (XDR) and multidrug-resistant (MDR) strains (Cortesia et al., 2014).

Different studies have reported the capability of vinegar to inhibit the proliferation of pathogenic bacteria in different foods, such as raw vegetables (Sengun and Karapinar, 2004, 2005) including carrots and leaf vegetables, among others. Experiments conducted by Chang and Fang (2007) proved the antimicrobial effect of rice vinegar on lettuce inoculated with *E. coli* O157:H7, obtaining excellent microbial reductions (3 log units) for treatments with vinegar containing 5% acetic acid for 5 min at 25°C.

A 2% acetic acid treatment was effective at reducing *Escherichia coli* and *Salmonella typhimurium* viable counts over time in refrigerated and frozen storage of beef products, and this treatment did not cause adverse sensory changes when evaluated by a consumer panel (Harris et al., 2006). Birk et al. (2010) assessed the viability of *Listeria monocytogenes* on uncured turkey breast containing buffered vinegar solutions and demonstrated the suppressed outgrowth of *L. monocytogenes* during an extended refrigerated shelf-life. Wine vinegar either alone or in combination with soy sauce, has been demonstrated to be an active antibacterial agent against *Campylobacter jejuni*, the main microorganism involved in foodborne outbreaks due to the consumption of chicken meat. Moreover, a blend containing sugar and vinegar was demonstrated to be effective on *C. jejuni* and *S. typhimurium* without affecting the texture of chicken breasts (Park et al., 2014). Furthermore, a recent study shows that a combination of 1.0% vinegar with CO_2 packaging can extend the shelf-life from 12 to 20 days for chicken retail cuts without negatively affecting the quality and sensory properties of the broiler meat (Desai et al., 2014).

25.3 BIOACTIVE COMPOUNDS AND ANTIOXIDANT ACTIVITY

25.3.1 Phenolic Compounds

This group of compounds encompasses an enormous variety of chemical structures widespread in nature. Indeed, thousands of phenolic structures have been determined so far in natural sources with their occurrence strongly related to the botanical origin. Moreover, food elaboration processes do affect them and fermentation is not an exception. Particularly in vinegars, the main factors influencing their presence and concentration include: the nature of the raw material used in the elaboration, the type of acetification process involved in their production, and contact with wood during aging, if any.

Apart from the differentiation of vinegars from different sources (wine, apple, cider, honey) their relevance to ascertain the botanical origin has been proposed to discriminate origins. Thus, phenolic compounds have been formerly studied as markers for vinegar authenticity purposes, mainly to differentiate vinegars included in the European Union's Protected Designation of Origin (PDO) framework as sherry wines (García-Parrilla et al., 1997) and aceto balsamico, either aceto balsamico tradizionale (ABT), regulated by two different PDO labels (ABT di Modena or ABT di Reggio Emilia), and aceto balsamico di Modena, which has a Protected Geographical Indication (PGI) status (Masino et al., 2008; Verzelloni et al., 2007).

Furthermore, their ability to scavenge free radicals and the additional metal-chelating properties determine that they possess antioxidant activity with a putative health effect, preventing damage by reactive oxygen species.

On the other hand, phenolic compounds are related to sensory properties of food and beverages, in particular color, astringency, and flavor, attracting the interest of scientific studies in their determination for these aims.

25.3.2 Phenolic Composition of Vinegars

The concentration of phenolic compounds in vinegars is reported in literature and compiled in databases of bioactive compounds as, such Phenol Explorer (http://phenol-explorer.eu/contents/food/141) summarizing values given by different authors. Vinegars contain hydroxybenzoic acids [gallic acid (2.59 mg/L), protocatechuic acid (0.81 mg/L), *p*-hydroxybenzoic acid (0.20 mg/L), vanillic acid (0.09 mg/L), syringic acid (0.3 mg/L)]; hydroxycinnamic acids [caffeic acid (0.28 mg/L), *p*-coumaric acid (0.15 mg/L), ferulic acid (0.11 mg/L)] and the tartaric esters of the hydroxycinnamic acids caffeoyl tartaric acid (1.1 mg/L), *trans*-coumaroyltartaric acid (0.52 mg/L), *cis*-coumaroyltartaric acid (0.18 mg/L), feruloyl tartaric acid (0.1 mg/L), aldehydes [vanillin (0.1 mg/L), protocatechualdehyde (0.11 mg/L), syringaldehyde (0.07 mg/L), *p*-hydroxybenzaldehyde (0.16 mg/L)], the flavanols [(+)-catechin (1.23 mg/L), (−)-epicatechin (0.55 mg/L)] and the esthers caffeic ethyl ester (0.03 mg/L) and coumaric ethyl ester (0.03 mg/L).

The decrease in total phenolic content after the acetification process occurs regardless of the source used to elaborate the vinegar, for instance 40% in cider vinegar (Andlauer et al., 2000), 13% in red wine vinegars (Cerezo et al., 2008), 8% in white wine vinegars (Andlauer et al., 2000; García-Parrilla et al., 1998), or 13–60% in strawberry vinegars (Ubeda et al., 2013), with anthocyanin compounds present in red wine and strawberry vinegars accounting for a major decrease. Not only does the acetification diminish the phenolic compounds but also the rate that it occurs at has an important role. Indeed, submerged culture provokes a more acute change than surface culture acetification as it supplies more oxygen and these antioxidants are more affected.

Contact with wood can influence the phenolic content as stated before, increasing the concentration of certain polyphenols, such as gallic acid and aldehydes produced by alcoholysis of the wood lignin (vanillin, p-hydroxybenzaldehyde, and protocatechualdehyde) (Tesfaye et al., 2009). Traditionally, vinegars are aged in wood barrels but the use of shavings or chips is an increasing tendency for less expensive vinegars. This technique provides the product with desired characteristics that can be perceived by the senses (Tesfaye et al., 2004) but also modifies their phenolic profile in 15 days of contact. Normally, barrels or chips are toasted before the contact with vinegar to provoke the formation of phenolic aldehydes (vanillin, protocatechualdehyde, benzaldehyde, etc.) released from lignin, providing a much appreciated flavor to the liquid. More recently, experiences with nontoasted wood have been carried out revealing that some compounds, such as dihydrorobinetin or (+)-taxifolin, markers of acacia and cherry wood respectively, clearly diminish after toasting (Cerezo et al., 2009, 2014). As these flavonoids possess antioxidant properties, this fact has to be taken into account when designing innovative products with meant health properties. Vinegars can be aged in wood barrels made of oak, chestnut, cherry, or acacia among others for variable periods of time. The type of wood, degree of toasting, surface of contact, and time of aging determine the phenolic concentration in the vinegar.

25.3.3 Antioxidant Properties of Vinegars

Due to their bioactive content, mainly phenolic compounds, vinegars have exhibited antioxidant properties as determined by in vitro methods (ORAC, DPPH) (Dávalos et al., 2005). In addition, melanoidins present in traditional vinegars contribute to antioxidant values (Verzelloni et al., 2010). The natural source used in vinegar production has a great influence on its antioxidant activity. For instance, antioxidant activity values for wine vinegar were higher than those reported for apple cider vinegar (Budak et al., 2011) but lower than those determined for persimmon vinegar (Ubeda et al., 2011).

25.4 HEALTH EFFECTS

Vinegar has been traditionally considered an essential component of a healthy diet, mainly because it is used as a condiment in many vegetable-based recipes. Moreover, due to its properties as a flavoring and acidifying agent, it may enable less use of salt, thus reducing the risk of hypertension.

The use of vinegar to fight infections and other acute conditions dates back to Hippocrates, but recent research suggests that vinegar ingestion favorably influences biomarkers for heart disease, cancer, and diabetes (Johnston, 2009).

Literature concerning properties attributed to vinegar is mostly related to the antidiabetic and antiobesity properties of acetic acid (Budak et al., 2014). It is worth mentioning that although many health effects have been claimed as a result of vinegar consumption, only a few of these effects are based on clear evidence (Mas et al., 2015).

Developed countries are experiencing a major overweight and obesity epidemic. Worldwide obesity has duplicated since 1980 and according to WHO 39% of adults aged ≥18 years were overweight in 2014, and 13% were obese (www.who.int/mediacentre/factsheets/fs311/en/). Most of the world's population lives in countries where overweight and obesity are responsible for more deaths than underweight and this excessive weight is associated with the leading causes of death, such as cardiovascular diseases, cancer, stroke, and type II diabetes.

25.4.1 Effect on Glucose Metabolism and Insulin Resistance

The rising epidemic of diabetes worldwide is of significant concern. In type II diabetes, insulin is secreted but tissues are resistant to insulin and therefore, blood glucose increases. The consumption of vinegar with meals was used as a home remedy for diabetes before the advent of pharmacologic glucose-lowering therapy (O'Keefe et al., 2008).

Scientific evidence supporting an antiglycemic effect of vinegar at mealtimes was first reported in 1988 (Johnston, 2009). Vinegar ingestion can reduce the glycemic effect of a meal, a phenomenon that has been related to satiety and reduced food intake (Johnston et al., 2004). The acetic acid in vinegar may reduce the glycemic response to high glycemic load foods by inhibiting disaccharidases in the small intestinal epithelium (Ogawa et al., 2000) or by stimulating glucose uptake and utilization in peripheral tissues (Fushimi et al., 2001). Recent studies in animals have shown that vinegar may be used for diabetic treatment (Gu et al., 2012). Indeed, a dietary intervention study using different types of vinegar for 6 weeks proved to have beneficial effects on diabetic rats and also showed a hypocholesterolemic effect. Moreover, the vinegar did not affect the stomach histopathological structure and protected the pancreas from damage (undesirable change in B cells) (Soltan and Shehata, 2012).

Human studies, both in healthy subjects and in patients with insulin resistance or diabetes mellitus, have provided limited evidence of beneficial effects of vinegar/acetic acid on glucose metabolism (Johnston and Buller, 2005; Johnston et al., 2010, 2013; Liatis et al., 2010; Östman et al., 2005). Johnston et al. (2004) reported that vinegar significantly improved postprandial insulin sensitivity in insulin-resistant subjects. In addition, Liatis et al. (2010) concluded that the addition of vinegar reduces postprandial glycemia in patients with type II diabetes only when it is added to a high glycemic index meal. Furthermore, a vinegar extract from *Panax* ginseng, moderately improved HbA1c level and it was well tolerated by type II diabetic patients with inadequate glycemic control

(Yoon et al., 2012). Conversely, another human intervention concluded that vinegar coingestion does not attenuate the postprandial rise in plasma glucose or insulin concentrations following glucose ingestion in well-controlled type II diabetic patients (Van Dijk et al., 2012).

Probably these effects are related to acetic acid which suppressed disaccharidase activity and raised glucose-6-phosphate concentrations in skeletal muscle. Recent research suggests that the addition of vinegar to meals has antiglycemic effects in adults at risk for type II diabetes. These effects are possibly related to carbohydrate maldigestion (Johnston et al., 2013), since postpandrial glycemia improves when vinegar is administered together with polysaccharides and no effect is observed with simple sugars administration (Johnston et al., 2010). Indeed, a dose–response effect for acetic acid intake was observed (Östman et al., 2005). Moreover, in individuals with impaired glucose tolerance, vinegar ingestion before a mixed meal results in an enhancement of muscle blood flow, an improvement of glucose uptake by the forearm muscle, and a reduction of postprandial hyperinsulinemia and hypertriglyceridemia. Hence, vinegar may be considered beneficial for improving insulin resistance and metabolic abnormalities in the atherogenic prediabetic state (Mitrou et al., 2015).

Although still unclear, several mechanisms have been proposed to explain the effects of vinegar on glucose metabolism (Petsiou et al., 2014), such as the delay in gastric emptying that may lead to a reduced postpandrial blood glucose level (Lin et al., 1990). Another possible mechanism would be the suppression of enteral carbohydrate absorption, more specifically of the disaccharidase activity in enterocytes, thus decreasing the digestion of di- and oligosaccharides.

Vinegar ingestion at bedtime moderates waking glucose concentrations in adults with well-controlled type II diabetes (White and Johnston, 2007), suggesting an effect on endogenous release of glucose by the liver, but the mechanisms are still unknown. Sakakibara et al. (2010) observed an increased blood flow following vinegar consumption, which could lead to improved vascular reactivity and endothelial function and, finally, to improved insulin action. Finally, recent experiments with animals (Seok et al., 2012; Gu et al., 2012) have shown that treatment with either balsamic vinegar or rice vinegar may improve beta cell function and insulin secretion.

25.4.2 Effect on Lipid Metabolism

The effect of vinegar on lipid metabolism has been mostly studied in animals, with the information from human trials very limited. Hence, Fushimi et al. (2006) have reported that chronic administration of acetic acid reduces total serum cholesterol and triglyceride levels in rats fed a cholesterol-rich diet. These effects were also found in obese and type II diabetic rats submitted to chronic treatments with acetic acid (Soltan and Shehata, 2012). Human studies (Kondo et al., 2009; Beheshti et al., 2012; Panetta et al., 2013) have so far been scarce and encompass important limitations.

The effect of acetate on high-density lipoprotein (HDL) cholesterol is somewhat contradictory since animal studies show a beneficial effect, whereas human studies do not (Petsiou et al., 2014). This discrepancy could be related to the small sample size and lack of a control group, among other factors.

Another claimed effect, although weakly substantiated, is related to body weight loss. Several animal studies have shown a reduction in body weight following the administration of vinegar but human studies are still scarce. It is worth mentioning a double-blind, placebo-controlled human intervention trial in 155 obese Japanese during 12 weeks (Kondo et al., 2009). The results indicate a moderate decrease in body weight, body fat mass, and waist circumference at the fourth week, in both vinegar groups (15 and 30 ml of apple vinegar), in a dose-dependent manner. This study also suggests that continuous administration may be necessary to maintain the effects. Several mechanisms may explain this effect on body weight and visceral fat accumulation (Petsiou et al., 2014). Reduced lipogenesis could be one explanation since acetate may affect transcription factors involved in the conversion of glucose to fatty acids in the liver. Other proposed referred mechanisms could be increased lipolysis, increase in oxygen consumption or in energy expenditure. Indeed, if vinegar is consumed with a high-carbohydrate meal, the glycemic index of that meal decreases. This effect has been observed in both patients with type II diabetes and in healthy individuals, with a notable reduction of insulin resistance in the diabetic patients. This reduction could be one of the reasons for claims that the direct consumption of vinegar can result in weight loss (Mas et al., 2015). Anyway, the scientific evidence for these effects is scarce and more controlled trials are necessary as well as a deeper insight into the biological mechanisms involved.

25.4.3 Other Effects and Safety Issues

Studies on the cancer-protective effects of vinegar are limited and involve the use of several cell lines. In this sense, the effect of Kurosu vinegar on the proliferation of a variety of human cancer cell lines has been studied, including colon adenocarcinoma, lung carcinoma, breast adenocarcinoma, bladder carcinoma, and prostate carcinoma cells. Kurosu inhibited the proliferation of all tested cell lines in a dose-dependent manner (Nanda et al., 2004). Kibizu, sugar cane vinegar produced in Japan, inhibited the growth of typical human leukemia cells (Mimura et al., 2004). More recently, it has been proved that it inhibits the proliferation of human squamous cell carcinoma cells via programmed necrosis (necroptosis) (Baba et al., 2013).

Another health effect of vinegar is the enhancement of calcium ion absorption. It is well known that the reduction of calcium absorption can be a problem, especially in the elderly, and many components of the diet result in reduced absorption. The combination of calcium ions with acetate in the intestine is facilitated by the change of pH between the stomach and intestine and, thus, the formed compound, calcium acetate, is easily absorbed, resulting in an increased absorption of calcium (Mas et al., 2015).

The moderate consumption of vinegar during meals as a dressing seems to be safe. However, it is important to emphasize the high risk of the direct consumption of vinegar as a drink due to the caustic effects on the pharynx and esophagus and intestinal inflammation (Mas et al., 2015).

25.5 OTHER FERMENTED CONDIMENTS

In the food industry, acetic acid bacteria can also be used to produce other products different from vinegar. Recently, gluconized beverages have been considered (and even commercialized) as health food products. These beverages are based on the capacity of acetic acid bacteria to oxidize glucose into gluconic acid. Gluconic acid is a sweet-and-sour acid that can be pleasant in low concentration and due to its "fixed acidity" can enhance taste and buffer low pHs. Gluconic fermentation has an additional advantage from a nutritional perspective, since *Gluconobacter* strains convert glucose into mostly gluconic acid, without fermenting fructose (Attwood et al., 1991). Gluconic acid health effects are still poorly understood. It is hardly absorbed in the small intestine and thus it reaches the lower gut, where it can stimulate the intestinal microbiota and human epithelial cells, as it is an easily metabolized compound (Anderson et al., 2008). Gluconic acid is also used in the pharmaceutical industry as a counter ion in calcium and iron supplementation (Ramachandran et al., 2006). A recent research has shown that gluconic fermentation of strawberry puree can produce beverages with a low glucose content and a high content of nonanthocyanin polyphenol compounds, maintaining the antioxidant potential of the strawberry raw material as compared to alcoholic and acetic acid fermentations, which clearly diminished its bioactive potential (Álvarez-Fernández et al., 2014). Thus, this application could be of interest for the production of a new line of healthy, nonalcoholic, fermented beverages.

REFERENCES

Álvarez-Fernández, M.A., Hornedo-Ortega, R., Cerezo, A.B., Troncoso, A.M., García-Parrilla, M.C., 2014. Non-anthocyanin phenolic compounds and antioxidant activity of beverages obtained by gluconic fermentation of strawberry. Innovative Food Science & Emerging Technologies 26, 469–481.

Anderson, K., Yu, Z., Chen, J., Jenkins, J., Courtney, P., Morrison, M., 2008. Analyses of Bifidobacterium, Lactobacillus, and total bacterial populations in healthy volunteers consuming calcium gluconate by denaturing gradient gel electrophoresis and real-time PCR. International Journal of Probiotics & Prebiotics 3, 31–36.

Andlauer, W., Stumpf, C., Fürst, P., 2000. Influence of the acetification process on phenolic compounds. Journal of Agricultural and Food Chemistry 48, 3533–3536.

Attwood, M.M., van Dijken, J.P., Pronk, J.T., 1991. Glucose metabolism and gluconic acid production by *Acetobacter diazotrophicus*. Journal of Fermentation and Bioengineering 72, 101–105.

Baba, N., Higashi, Y., Kanekura, T., 2013. Japanese black vinegar "izumi" inhibits the proliferation of human squamous cell carcinoma cells via necroptosis. Nutrition and Cancer 65, 1093–1097.

Beheshti, Z., Huak Chan, Y., Sharif Nia, H., Hajihosseini, F., Nazari, R., Shaabani, M., Salehi Omran, M.T., 2012. Influence of apple cider vinegar on blood lipids. Life Science Journal 9, 2431–2440.

Birk, T., Grønlund, A.C., Christensen, B.B., Knøchel, S., Lohse, K., Rosenquist, H., 2010. Effect of organic acids and marination ingredients on the survival of *Campylobacter jejuni* on meat. Journal of Food Protection 73, 258–265.

Budak, N.H., Aykin, E., Seydim, A.C., Greene, A.K., Guzel-Seydim, Z.B., 2014. Functional properties of vinegar. Journal of Food Science 79, R757–R764.

Budak, N.H., Kumbul Doguc, D., Savas, C.M., Seydim, A.C., Kok Tas, T., Ciris, M.I., Guzel-Seydim, Z.B., 2011. Effects of apple cider vinegars produced with different techniques on blood lipids in high-cholesterol-fed rats. Journal of Agricultural and Food Chemistry 59, 6638–6644.

Cerezo, A.B., Álvarez-Fernández, M.A., Hornedo-Ortega, R., Troncoso, A.M., García-Parrilla, M.C., 2014. Phenolic composition of vinegars over an accelerated aging processu different wood species (acacia, cherry, chestnut, and oak): effect of wood toasting. Journal of Agriculture and Food Chemistry 62, 4369–4376.

Cerezo, A.B., Espartero, J.L., Winterhalter, P., Garcia-Parrilla, M.C., Troncoso, A.M., 2009. (+)-Dihydrorobinetin: a marker of vinegar aging in acacia (*Robinia pseudoacacia*) wood. Journal of Agriculture and Food Chemistry 57, 9551–9554.

Cerezo, A.B., Tesfaye, W., Torija, M.J., Mateo, E., García-Parrilla, M.C., Troncoso, A.M., 2008. The phenolic composition of red wine vinegar produced in barrels made from different woods. Food Chemistry 109, 606–615.

Chang, J.-M., Fang, T.J., 2007. Survival of *Escherichia coli* O157:H7 and *Salmonella enterica* serovars *Typhimurium* in iceberg lettuce and the antimicrobial effect of rice vinegar against *E. coli* O157:H7. Food Microbiology 24, 745–751.

Cortesia, C., Vilchèze, C., Bernut, A., Contreras, W., Gómez, K., de Waard, J., Jacobs Jr., W.R., Kremer, L., Takiff, H., 2014. Acetic acid, the active component of vinegar, is an effective tuberculocidal disinfectant. mBio 5 (2), 1–4.

Dávalos, A., Bartolomé, B., Gómez-Cordovés, C., 2005. Antioxidant properties of commercial grape juices and vinegars. Food Chemistry 93, 325–330.

Desai, M.A., Kurve, V., Smith, B.S., Campano, S.G., Soni, K., Schilling, M.W., 2014. Utilization of buffered vinegar to increase the shelf life of chicken retail cuts packaged in carbon dioxide. Poultry Science 93, 1850–1854.

Dohar, J.E., 2003. Evolution of management approaches for otitis externa. Pediatric Infectious Disease Journal 22, 299–306.

Drysdale, G.S., Fleet, G.H., 1988. Acetic acid bacteria in winemaking: a review. American Journal of Enology and Viticulture 39, 143–154.

Fushimi, T., Suruga, K., Oshima, Y., Fukiharu, M., Tsukamoto, Y., Goda, T., 2006. Dietary acetic acid reduces serum cholesterol and triacylglycerols in rats fed a cholesterol-rich diet. British Journal of Nutrition 95, 916–924.

Fushimi, T., Tayama, K., Fukaya, M., Kitakoshi, K., Nakai, N., Tsukamoto, Y., Sato, Y., 2001. Acetic acid feeding enhances glycogen repletion in liver and skeletal muscle of rats. Journal of Nutrition 131, 1973–1977.

García-Parrilla, M.C., González, G.A., Heredia, F.J., Troncoso, A.M., 1997. Differentiation of wine vinegars based on phenolic composition. Journal of Agricultural and Food Chemistry 45, 3487–3492.

García-Parrilla, M.C., Heredia, F.J., Troncoso, A.M., 1998. The influence of the acetification process on the phenolic composition of wine vinegars. Sciences des Aliments 18, 211–221.

Gu, X., Zhao, H.-L., Sui, Y., Guan, J., Chan, J.C.N., Tong, P.C.Y., 2012. White rice vinegar improves pancreatic beta-cell function and fatty liver in streptozotocin-induced diabetic rats. Acta Diabetologica 49, 185–191.

Gullo, M., De Vero, L., Guidici, P., 2009. Succession of selected strains of *Acetobacter pasteurianus* and other acetic acid bacteria in traditional balsamic vinegar. Applied and Environmental Microbiology 75, 2585–2589.

Gullo, M., Verzelloni, E., Canonico, M., 2014. Aerobic submerged fermentation by acetic acid bacteria for vinegar production: process and biotechnological aspects. Process Biochemistry 49, 1571–1579.

Harris, K., Miller, M.F., Loneragan, G.H., Brashears, M.M., 2006. Validation of the use of organic acids and acidified sodium chlorite to reduce *Escherichia coli* O157 and *Salmonella typhimurium* in beef trim and ground beef in a simulated processing environment. Journal of Food Protection 69, 1802–1807.

Hidalgo, C., Vegas, C., Mateo, E., Tesfaye, W., Cerezo, A.B., Callejón, R.M., Poblet, M., Guillamon, J.M., Mas, A., Torija, M.J., 2010. Effect of barrel design and the inoculation of *A. pasteurianus* in wine vinegar production. International Journal of Food Microbiology 141, 56–62.

Johnston, C.S., 2009. Medicinal uses of vinegar. In: Complementary and Alternative Therapies and the Aging Population. Elsevier, Amsterdam, pp. 433–443.

Johnston, C.S., Buller, A.J., 2005. Vinegar and peanut products as complementary foods to reduce postprandial glycemia. Journal of the American Dietetic Association 105, 1939–1942.

Johnston, C.S., Kim, C.M., Buller, A.J., 2004. Vinegar improves insulin sensitivity to a high-carbohydrate meal in subjects with insulin resistance or type 2 diabetes. Diabetes Care 27, 281–282.

Johnston, C.S., Quagliano, S., White, S., 2013. Vinegar ingestion at mealtime reduced fasting blood glucose concentrations in healthy adults at risk for type 2 diabetes. Journal of Functional Foods 5, 2007–2011.

Johnston, C.S., Steplewska, I., Long, C.A., Harris, L.N., Ryals, R.H., 2010. Examination of the antiglycemic properties of vinegar in healthy adults. Annals of Nutrition and Metabolism 56, 74–79.

Kondo, T., Kishi, M., Fushimi, T., Ugajin, S., Kaga, T., 2009. Vinegar intake reduces body weight, body fat mass, and serum triglyceride levels in obese Japanese subjects. Bioscience, Biotechnology, and Biochemistry 73, 1837–1843.

Liatis, S., Grammatikou, S., Poulia, K.-A., Perrea, D., Makrilakis, K., Diakoumopoulou, E., Katsilambros, N., 2010. Vinegar reduces postprandial hyperglycaemia in patients with type II diabetes when added to a high, but not to a low, glycaemic index meal. European Journal of Clinical Nutrition 64, 727–732.

Lin, H.C., Doty, J.E., Reedy, T.J., Meyer, J.H., 1990. Inhibition of gastric emptying by acids depends on pH, titratable acidity, and length of intestine exposed to acid. American Journal of Physiology 259, G1025–G1030.

Mas, A., Torija, M.J., García-Parrilla, M.C., Troncoso, A.M., 2014. Acetic acid bacteria and the production and quality of wine vinegar. The Scientific World Journal 2014, 1–6.

Mas, A., Troncoso, A.M., García-Parrilla, M.C., Torija, M.J., 2015. Vinegar. In: Caballero, B., Finglas, P., Toldrá, F. (Eds.), Encyclopedia of Food and Health, vol. 5. Academic Press, Oxford, pp. 418–423.

Masino, F., Chinnici, F., Bendini, A., Montevecchi, G., Antonelli, A., 2008. A study on relationships among chemical, physical, and qualitative assessment in traditional balsamic vinegar. Food Chemistry 106, 90–95.

Mazza, S., Yoshikatsu, M., 2009. Vinegars through the ages. In: Solieri, L., Giudici, P. (Eds.), Vinegars of the World. Springer-Verlag, Milan, pp. 17–39.

Mimura, A., Suzuki, Y., Toshima, Y., Yazaki, S., Ohtsuki, T., Ui, S., Hyodoh, F., 2004. Induction of apoptosis in human leukemia cells by naturally fermented sugar cane vinegar (kibizu) of Amami Ohshima Island. BioFactors 22, 93–97.

Mitrou, P., Petsiou, E., Papakonstantinou, E., Maratou, E., Lambadiari, V., Dimitriadis, P., Spanoudi, F., Raptis, S.A., Dimitriadis, G., 2015. The role of acetic acid on glucose uptake and blood flow rates in the skeletal muscle in humans with impaired glucose tolerance. European Journal of Clinical Nutrition 69, 734–739.

Murooka, Y., Yamshita, M., 2008. Traditional healthful fermented products of Japan. Journal of Industrial Microbiology and Biotechnology 35, 791–798.

Nanda, K., Miyoshi, N., Nakamura, Y., Shimoji, Y., Tamura, Y., Nishikawa, Y., Uenakai, K., Kohno, H., Tanaka, T., 2004. Extract of vinegar "Kurosu" from unpolished rice inhibits the proliferation of human cancer cells. Journal of Experimental and Clinical Cancer Research 23, 69–75.

Ogawa, N., Satsu, H., Watanabe, H., Fukaya, M., Tsukamoto, Y., Miyamoto, Y., Shimizu, M., 2000. Acetic acid suppresses the increase in disaccharidase activity that occurs during culture of Caco-2 cells. Journal of Nutrition 130, 507–513.

O'Keefe, J.H., Gheewala, N.M., O'Keefe, J.O., 2008. Dietary strategies for improving post-prandial glucose, lipids, inflammation, and cardiovascular health. Journal of the American College of Cardiology 51, 249–255.

Östman, E., Granfeldt, Y., Persson, L., Björck, I., 2005. Vinegar supplementation lowers glucose and insulin responses and increases satiety after a bread meal in healthy subjects. European Journal of Clinical Nutrition 59, 983–988.

Panetta, C., Jonk, Y., Shapiro, A., 2013. Prospective randomized clinical trial evaluating the impact of vinegar on lipids in non-diabetics. World Journal of Cardiovascular Diseases 3, 191–196.

Park, N.Y., Hong, S.H., Yoon, K.S., 2014. Effects of commercial marinade seasoning and a natural blend of cultured sugar and vinegar on *Campylobacter jejuni* and *Salmonella typhimurium* and the texture of chicken breasts. Poultry Science 93, 719–727.

Petsiou, E.I., Mitrou, P.I., Raptis, S.A., Dimitriadis, G.D., 2014. Effect and mechanisms of action of vinegar on glucose metabolism, lipid profile, and body weight. Nutrition Reviews 72, 651–661.

Phenol explorer, http://phenol-explorer.eu/contents/food/141.

Ramachandran, S., Fontanille, P., Pandey, A., Larroche, C., 2006. Gluconic acid: properties, applications and microbial production. Food Technology and Biotechnology 44, 185–195.

Ryu, J.-H., Deng, Y., Beuchat, L.R., 1999. Behavior of acid-adapted and unadapted *Escherichia coli* O157:H7 when exposed to reduced pH achieved with various organic acids. Journal of Food Protection 62, 451–455.

Sakakibara, S., Murakami, R., Takahashi, M., Fushimi, T., Murohara, T., Kishi, M., Kajimoto, Y., Kitakaze, M., Kaga, T., 2010. Vinegar intake enhances flow-mediated vasodilatation via upregulation of endothelial nitric oxide synthase activity. Bioscience, Biotechnology, and Biochemistry 74, 1055–1061.

Sengun, I.Y., Karabiyikli, S., 2011. Importance of acetic acid bacteria in food industry. Food Control 22, 647–656.

Sengun, I.Y., Karapinar, M., 2004. Effectiveness of lemon juice, vinegar and their mixture in the elimination of *Salmonella typhimurium* on carrots (*Daucus carota* L.). International Journal of Food Microbiology 96, 301–305.

Sengun, I.Y., Karapinar, M., 2005. Effectiveness of household natural sanitizers in the elimination of *Salmonella typhimurium* on rocket (*Eruca sativa* Miller) and spring onion (*Allium cepa* L.). International Journal of Food Microbiology 98, 319–323.

Seok, H., Lee, J.Y., Park, E.M., Park, S.E., Lee, J.H., Lim, S., Lee, B.-W., Kang, E.S., Lee, H.C., Cha, B.S., 2012. Balsamic vinegar improves high fat-induced beta cell dysfunction via beta cell ABCA1. Diabetes and Metabolism Journal 36, 275–279.

Solieri, L., Giudici, P., 2009. Vinegars of the world. In: Solieri, L., Giudici, P. (Eds.), Vinegars of the World. Springer-Verlag, Milan, pp. 1–16.

Soltan, S.S.A., Shehata, M.M.E.M., 2012. Antidiabetic and hypocholesrolemic effect of different types of vinegar in rats. Life Science Journal 9, 2141–2151.

Tesfaye, W., Morales, M.L., Benítez, B., García-Parrilla, M.C., Troncoso, A.M., 2004. Evolution of wine vinegar composition during accelerated aging with oak chips. Analytica Chimica Acta 513, 239–245.

Tesfaye, W., Morales, M.L., García-Parrilla, M.C., Troncoso, A.M., 2002. Wine vinegar: technology, authenticity and quality evaluation. Trends in Food Science & Technology 13, 12–21.

Tesfaye, W., Morales, M.L., García-Parrilla, M.C., Troncoso, A.M., 2009. Jerez vinegar. In: Solieri, L., Giudici, P. (Eds.), Vinegars of the World. Springer-Verlag, Milan, pp. 188–190.

Ubeda, C., Callejón, R.M., Hidalgo, C., Torija, M.J., Troncoso, A.M., Morales, M.L., 2013. Employment of different processes for the production of strawberry vinegars: effects on antioxidant activity, total phenols and monomeric anthocyanins. LWT—Food Science and Technology 52, 139–145.

Ubeda, C., Hidalgo, C., Torija, M.J., Mas, A., Troncoso, A.M., Morales, M.L., 2011. Evaluation of antioxidant activity and total phenols index in persimmon vinegars produced by different processes. LWT—Food Science and Technology 44, 1591–1596.

Van Dijk, J.-W., Tummers, K., Hamer, H.M., Van Loon, L.J.C., 2012. Vinegar co-ingestion does not improve oral glucose tolerance in patients with type 2 diabetes. Journal of Diabetes and its Complications 26, 460–461.

Verzelloni, E., Tagliazucchi, D., Conte, A., 2007. Relationship between the antioxidant properties and the phenolic and flavonoid content in traditional balsamic vinegar. Food Chemistry 105, 564–571.

Verzelloni, E., Tagliazucchi, D., Conte, A., 2010. From balsamic to healthy: traditional balsamic vinegar melanoidins inhibit lipid peroxidation during simulated gastric digestion of meat. Food and Chemical Toxicology 48, 2097–2102.

White, A.M., Johnston, C.S., 2007. Vinegar ingestion at bedtime moderates waking glucose concentrations in adults with well-controlled type 2 diabetes. Diabetes Care 30, 2814–2815.

WHO, 2015. Obesity and Over Weight. Fact Sheet N°311. http://www.who.int/mediacentre/factsheets/fs311/en/.

Wu, J., Gullo, M., Cheng, F., Giudicci, P., 2010. Diversity of *Acetobacter pasteurianus* strains isolated from solid-state fermentation of cereal vinegars. Current Microbiology 60, 280–286.

Yoon, J.W., Kang, S.M., Vassy, J.L., Shin, H., Lee, Y.H., Ahn, H.Y., Choi, S.H., Park, K.S., Jang, H.C., Lim, S., 2012. Efficacy and safety of ginsam, a vinegar extract from *Panax ginseng*, in type 2 diabetic patients: results of a double-blind, placebo-controlled study. Journal of Diabetes Investigation 3, 309–317.

Chapter 26

Wine

I. Fernandes, R. Pérez-Gregorio, S. Soares, N. Mateus, V. De Freitas
Universidade do Porto, Porto, Portugal

26.1 PREAMBLE

Wine, the product of grape juice fermentation, has played an important role in the development of human culture. The earliest chemical evidence for wine production was found in the Middle East in well-preserved ancient jars dated from 5400 to 5000 BC (McGovern et al., 1996). The multitude of ancient and historical images of wines and grapes found as decorative elements on ancient coins, temples, ritual potteries, and mosaic sculptures clearly demonstrated the importance of grape and its products in ancient societies across the globe. For example, in the Balkans and the Eastern Mediterranean, grape and wine were considered divine and dedicated to various deities: "Zagreus" by the Thracians, "Dionysus" by the Greeks, and "Bacchus" by the Romans (Riedel et al., 2012; McGovern et al., 1996). The mystical powers of grape and wine in the social life and cultural traditions of ancient people were not futile. Modern science continues to decipher the benefits of grape and wine as rich sources of valuable phytonutrients with remarkable positive effects on human health.

Red wine is a fermented beverage of high chemical complexity in continuous evolution, changing both organoleptic and biological properties, which constitutes a true challenge for wine chemists, companies, and consumers. Chemically, wine is a hydroalcoholic solution (~78% ethanol) that comprises a wide array of chemical components that include sugars, soluble proteins, minerals, lipids, vitamins, organic acids, aldehydes, ketones, esters, and phenolics.

The main organoleptic attributes of wine, such as color, flavor, and aroma, are due to the presence of a variety of organic compounds extracted from grapes, and are affected by the chemical transformations that they undergo during winemaking and wine aging.

The health-promoting and disease-prevention properties associated with red wine consumption have been correlated to the wine chemistry, and especially to some classes of phenolic compounds. To unravel the full wine chemistry that is behind all biological events remains a challenging task for researchers worldwide.

Presently, red wine is one of the most important sources of dietary polyphenols. During the last decade, not only has the knowledge on the phenolic content of wine increased considerably (El Darra et al., 2016; Ferreira et al., 2016; Olejar et al., 2015) but also numerous scientific studies have revealed the in vitro biological properties of some of these phenolic compounds.

Phenolic compounds are commonly divided into two major groups: flavonoids and nonflavonoids. The main flavonoid compounds present in wine comprise several classes, such as flavanols, flavonols, and anthocyanins, while nonflavonoid compounds present in wine include phenolic acids, phenols, and stilbenes. Many of the reported health properties associated with red wine consumption have been positively correlated with some of these phenolic classes. However, due to the remaining insufficient epidemiological and in vivo evidences on this subject, the presence of a high number of variables such as human age, metabolism, the presence of alcohol, the complex wine chemistry, and the wide array of in vivo biological effects of many of these compounds suggest that only cautious conclusions must be drawn from studies focusing on the direct effect of wine on a specific health issue.

Still, there are several reports on the health-promoting properties of wine phenolics on several diseases like cardiovascular diseases, some cancers, obesity, neurodegenerative diseases, diabetes, allergies, and osteoporosis. The mechanisms by which many of the biological events lead to disease prevention are still a deep matter of research. Not only must cellular events be targeted, but also other issues preceding these actions such as bioavailability issues. Still, some questions are imposed: which compounds are actually involved in these biological events? Which amounts of those are required? Indeed, it is currently acknowledged that the bioactive forms in vivo are not necessarily the same that occur in food. A major drawback of most of the epidemiological studies and human trials made to date is the use of foods or extracts that are not chemically well defined, which makes it difficult to clearly comprehend any positive results or establish structure–activity relationships.

After drinking a glass of a nice red wine, there is a long journey before its components may exert a health-promoting property. They must pass through the oral cavity, the gastrointestinal tract, undergo metabolism events, pass cellular barriers, and eventually trigger a biological event (Fig. 26.1). This chapter aims at reviewing some of these issues and revisiting some of the health-promoting effects of red wine reported so far.

26.2 DISEASE-PROTECTIVE/PREVENTIVE EFFECT OF WINE OR ITS PHENOLIC COMPOUNDS

Antioxidants are substances that delay, prevent, or reverse oxidative damage to a target molecule (Gutteridge and Halliwell, 2010). Among them, wine phenolic compounds have gained special interest, since the antioxidant power of some of its constituents, such as the proanthocyanidins, showed that it was 50

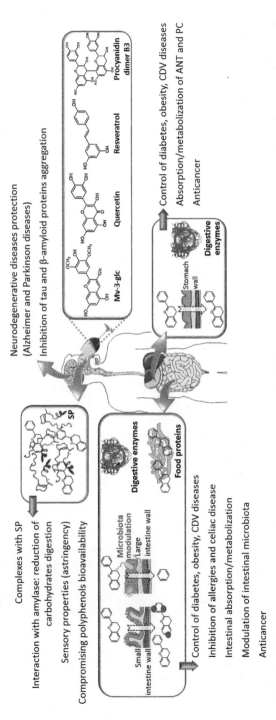

FIGURE 26.1 Schematic representation of the major biological effects of red wine polyphenols. The text refers to the most significant biological activities attributed to polyphenols in the different organs. The schemes highlight some of the major activities in the assigned organ. *ANT*, anthocyanins; *CDV*, cardiovascular; *Glc*, glucose; *Mv-3-glc*, malvidin-3-glucoside; *PC*, procyanidins; *SP*, salivary proteins.

times greater than vitamin C and 20 times greater than vitamin E (Karjalainen et al., 2009). Oxidative stress in biological systems is a condition when there is an imbalance between the production of reactive oxygen species (ROS) and the capacity of antioxidants (endogenous or exogenous) to neutralize the ROS produced. Consumption of sufficient amounts of dietary antioxidants, with bioprotective capacity, from wine or other food sources has been shown to reduce oxidative stress and thus reduce the risk of developing some diseases.

In vitro chemical assays are frequently used to assess the wine antioxidant activity such as ABTS (2,2'-azinobis[3-ethylbenzothiazoline 6-sulfonate]), DPPH (2,2-diphenyl-1-picrylhydrazyl), FRAP (ferric reducing antioxidant potential), and ORAC (oxygen radical absorption capacity) assays (Alam et al., 2013). These assays measure the ability of natural antioxidants to scavenge free radicals using different chemical reaction mechanisms, either via electron transfer or hydrogen atom transfer. However, very often, the free radical scavenging capacity of wine does not necessarily reflect their health benefits in vivo since different bioactive compounds may act in vivo through different mechanisms under biological conditions such as in cell signaling and even in the interaction with biological proteins (further reviewed in Section 26.3) (López-Alarcón and Denicola, 2013). For this reason, the protective/preventive action of wine phenolic compounds is supported not only by the measurement of its in vitro antioxidant capacity but also by in vivo experiments. Based on human epidemiological data, several studies highlighted an inverse relationship between the occurrence of cardiovascular diseases, certain cancers, and age-related degenerative diseases and the consumption of polyphenols. Hence, the biological effect of wine has been evaluated since it is a huge source of a large diversity of phenolic compounds and in varying amounts. The biological activity of wine on different organs and systems was illustrated by either experimental or clinical studies. Fig. 26.2 summarizes the biological effects of wine polyphenols and their relationship with the prevention of several diseases. The impact of wine consumption on disease has been widely investigated but it is important to note that some studies included wine consumption as a part of generalized alcohol intake, which limits the detection of the specific effects of wine. In addition, human clinical trials have shown that the potential benefits of wine intake are not always detected, possibly because the concentrations of the phenolic metabolites that reach the human body are always very low.

Reactive oxygen and nitrogen species are considered major determinants of some degenerative pathology. These compounds can act by damaging biological structures, but they can affect cell signaling by changing the redox cellular status or by participating in intracellular signaling (Fig. 26.2). The radical scavenging process was defined as one of the most important mechanisms of polyphenols to prevent diseases. The free-radical scavenging activity could be associated with specific structural features such as their chemistry; substitution pattern, especially hydroxyl groups; presence of double bonds; methylations;

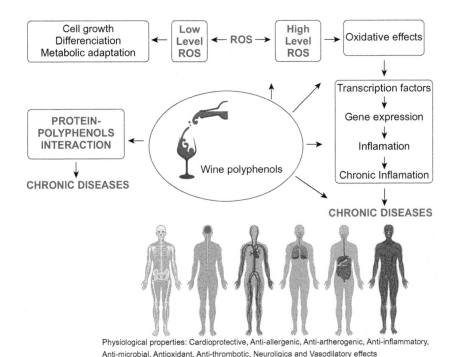

FIGURE 26.2 Influence of wine polyphenols in human health.

glycosylation; degree of polymerization, etc. (Seyoum et al., 2006) and also the quantity of antioxidants. Thus, the different phenolic compounds present in wine could act in a different way. Synergies and antagonisms have also been demonstrated (Kurin et al., 2013).

Although the antioxidant properties of phytonutrients have been a major focus of research, there is an emerging view that they may act not only by scavenging reactive oxygen and nitrogen species or suppressing their production but also by enhancing the endogenous antioxidant capacity of cells/tissues or by influencing signaling pathways through interaction with cellular receptors and enzymes (Bellaver et al., 2015; Testai, 2015).

In this section, the achievements to date in the potential of wine as a preventive agent for certain diseases derived from its phenolic content will be described.

26.2.1 Cardiovascular Diseases

Cardiovascular diseases (CVDs) are the leading cause of death globally (WHO, 2015). Diet and lifestyle have been considered modifiable risk factors for CVD for a long period of time. Although a positive effect of alcohol on atherosclerosis was considered as early as 1904, it was not until 1979 that St Leger et al.

observed a strong and specific negative association between ischemic heart disease deaths and wine (Cordova and Sumpio, 2009). The fact that ischemic mortality rate was lower in France than in other countries with a similar consumption of saturated fats, such as the United States and the United Kingdom, was attributed to the high consumption of red wine in the French population and coined the term "French Paradox." This association was wholly attributable to wine consumption and not to alcohol, as plasma HDL cholesterol levels, which are increased by alcohol, were similar in France and in other countries with higher rates of CVD. Despite the interpretation of the French data, there is a debate as to whether the reported benefits of wine on ischemic heart disease are mainly due to its alcohol content, or whether wine provides specific protection. The cardiovascular protection of wine phenolic compounds has been described as the result of multiple biological activities independent of their antioxidant capacity, including improvement of endothelial functionality, decrease in LDL uptake, diminution in LDL oxidation and aggregation, reduction of blood pressure, and inhibition of platelet aggregation. Some studies have associated alcohol with favorable changes in factors impacting atherosclerosis. For example, the consumption of alcohol has been followed by an increase in HDL-cholesterol and ApoA-I, inhibition of platelet aggregation, and production of a decreased procoagulant state, improvement in flow-mediated dilation and induction in the release of nitric oxide (NO) from the endothelium due to activation and expression of eNOS (Chiva-Blanch et al., 2015).

26.2.2 Cancer

The chemopreventive activity of wine phenolic compounds has also been suggested to be associated with the ability to block the progression of tumors. Different mechanisms of phenolics as carcinopreventive agents have also been described.

Phenolic compounds can modulate the action either of proteins in charge of their extracellular export and of enzymes that detoxify carcinogens (Jose et al., 2014), such as the platelet-derived growth factor receptor (eg, anthocyanins) (Lamy et al., 2008). In particular, *trans*-resveratrol promotes apoptosis and cell death of tumor cells, vascularizes the microtumors by blocking metalloproteases, and reduces inflammation by inhibiting cyclooxygenase 2 and was also proved as a promoter of A549 cells (human lung adenocarcinoma epithelial cells) apoptosis (Lucas and Kolodziej, 2015). In vitro and clinical studies have shown that red wine inhibits the proliferation of cells from different human cancers—such as lung, oral squamous carcinoma, and prostate (Erdrich et al., 2015; Shrotriya et al., 2015; Song et al., 2015; Yu et al., 2015). Zell et al. (2008) found that regular wine consumption had favorable effects on stage at presentation and survival in familial colorectal cancer cases (Pierini et al., 2008). Protection of moderate wine consumption was also detected against renal cancer in a metaanalysis (Bassig et al., 2012).

26.2.3 Metabolic Diseases

There is some literature showing the benefits of wine consumption against metabolic diseases like diabetes or obesity. It is well known that postprandial oxidative stress contributes to the development of diabetes since hyperglycemia and glycemic oscillations are produced (Haxhi et al., 2015).

Recent studies have demonstrated that many dietary phenolic compounds can act as mimics of caloric restriction and activate AMPK (AMP-activated protein kinase is a conserved sensor of cellular energy), thus suppressing hepatic gluconeogenesis, inducing hepatic fatty-acid β-oxidation, and stimulating glucose transporters in muscle and adipose tissues, with an overall reduction of the glucose level in the blood and the lipid content of the liver as well as an improvement in insulin sensitivity (Dragan et al., 2015; Kamakura et al., 2015).

Resveratrol, quercetin, and anthocyanins present in wine, extend the lifespan in different species. In particular, resveratrol directly inhibits cAMP-degrading phosphodiesterases, leading to an increase in cAMP and subsequent activation of AMPK. Thus, AMPK pathway activation has been suggested as a strategy for treating metabolic syndrome and delaying aging. Because diabetes mellitus is a metabolic disease often associated with aging, the effect of wine phenolics on this pathology is also interesting. An in vivo study on rats has demonstrated that a dietary intake of wine phenolic compounds may protect the pancreatic islets from morphological and functional decline. In fact, this consumption inhibited two enzymes required for starch digestion to maltodextrins and thereby conversion into absorbable monosaccharides. The inhibition of these enzymes improved metabolic homeostasis, delaying the age-related decline in basal insulin, ameliorating pancreatic β-cell glucose responsiveness, and increasing β-cell mass in normal-aged rats.

Relative to the influence of wine phenolic compounds in obesity, two mechanisms of action were proposed (1) decreasing absorption of lipids and proteins by wine polyphenols in the intestine, thus reducing calorie intake (further explained in Section 26.3.2); and (2) activating AMP-activated protein kinase by wine phenolics that are bioavailable in the liver, skeletal muscle, and adipose tissues. The relative importance of these two mechanisms depends on the wine consumed by individuals. The activated AMP-activated protein kinase would decrease gluconeogenesis and fatty acid synthesis and increase catabolism, leading to body weight reduction and metabolic syndrome alleviation (Bhaswant et al., 2015).

26.2.4 Neurological Diseases

Neuroinflammation constitutes a beneficial process involved in the maintenance of organ homeostasis and the brain response to infection or injury (Glass et al., 2010). However, sustained neuroinflammatory processes may contribute to the cascade of events leading to the progressive neuronal damage observed in aging

(Barrientos et al., 2015) and age-related cognitive disorders, such as Parkinson disease (Herrero et al., 2015) and Alzheimer disease (Heneka et al., 2015; Heppner et al., 2015). Brain damage and neuroinflammatory process begin years before substantial neurodegeneration appears. Hence, there is a crucial need for the development of new strategies capable to prevent, delay the onset or treat brain dysfunction and associated cognitive decline. Growing evidence sheds light on the use of dietary polyphenols to improve cognitive performance and reduce the neuroinflammatory and oxidative stress responses occurring with age and neurodegenerative pathologies.

The effect of moderate alcohol intake on neurological diseases is unclear. A prospective study in the Netherlands, however, observed that the consumption of red wine, but no other type of alcoholic beverages, was inversely associated with the progression of cognitive decline in a population of middle-aged subjects. These findings support the hypothesis that nonalcoholic substances found in red wine are involved in a favorable effect in cognitive function (Fenwick et al., 2015; Nooyens et al., 2014). There is some experimental support for these clinical observations. Resveratrol has been proven to provide neuroprotective effects in in vitro and in vivo models for Alzheimer disease. Resveratrol supplementation lowers β-amyloid protein (Aβ) peptide levels by promoting non-amyloidogenic cleavage of amyloid precursor protein, thus reducing neuronal damage (Barrientos et al., 2015; Bastianetto et al., 2015; Heneka et al., 2015). There are also initiatives exploring the effect on Parkinson disease (Pavlin et al., 2015; Virmani et al., 2013). The mechanisms underlying these effects are still poorly understood (Glass et al., 2010). The underlying mechanisms of neuronal degeneration associated with motor and cognitive decline remain elusive, although it is thought that several cellular and molecular events are involved which are sensitive to oxidative stress and chronic neuroinflammation. Resveratrol and proanthocyanidins from grape seed extract have shown to complex with β-amyloid protein (Aβ) (Smid et al., 2012) and tau protein (Santa-Maria et al., 2012; Wang et al., 2010), respectively, inhibiting aggregate formation as further explained in Section 26.3.

Resveratrol was also related to the reduction of prostaglandin and isoprostane in rat activated microglia. In a microglial–neuronal coculture, resveratrol reduced neuronal death induced by activated microglia (Wendeburg et al., 2009). Some investigators have claimed that the urinary excretion of isoprostanes, a class of eicosanoids derived from fatty acid oxidation in arachidonic acid, meets enough reliability for measuring oxidative stress in vivo (Deon et al., 2015). The antioxidant effect of red and white wine was investigated in healthy volunteers by measuring urinary content of prostaglandin-F2-III (PR2-III), an indicator of the isoprostane production, after 15 days of supplementation. A greater decrease in urinary PF2-III was observed in those who were given red wine as compared to white wine, which was attributed to the higher polyphenol content of red wine. Of interest, dealcoholized wine was capable of maintaining antioxidant benefits, supporting the role of polyphenols on the oxidative stress process evaluated by isoprostane concentrations. Furthermore, resveratrol mediates the downregulation of other neuroinflammatory biomarkers such as interleukins (Wendeburg et al., 2009).

26.2.5 Osteoporosis

The impact of alcohol and wine on skeletal health has received attention since the 1990s. Much of the focus has concentrated on osteoporosis, with less evidence in relation to cartilage degradation in osteoarthritis or other skeletal diseases. Osteoporosis may be investigated at the level of changes on bone mineral density or bone turnover biochemical markers, the information being meagre in relation to fractures. Besides the greater difficulty of investigating fragility fractures as an endpoint, the interference of an increased risk for falls creates additional difficulty when considering significant amounts of wine, or alcohol in general. Studies in the latter years have included more refined analyses, which have tried to clarify whether there are differences between the different types of alcoholic beverages, or between men and women. Ethanol itself seems to have an effect. Although there is not absolute unanimity in experimental studies, most of the evidence has shown that resveratrol induces proliferation and differentiation of preosteoblastic murine cell lines and, interestingly, interferes with the receptor activator of NF-κB ligand (RANKL) and the bone-specific transcription factor Runx2, two crucial determinants in the control of bone resorption and bone formation (Tsai et al., 2015).

26.2.6 Wine and Mortality: Impact on Longevity

Since the findings of St Leger et al. (1979) several epidemiological studies have found an association between wine consumption and a reduced risk of cardiovascular disease-dependent mortality (Hernandez–Hernandez et al., 2015). Also notable are findings from other studies, which found that wine consumption was associated with a higher reduction of cardiovascular risk and total mortality than consumption of beer and spirits. However, in comparison with other alcoholic beverages wine consumption is associated with higher socioeconomic status, healthier behavior, and less risk factors for major illnesses (Herttua et al., 2015). Taking into account that this kind of epidemiological study did not analyze all possible confounding factors, the healthier status of consumers might influence the apparent unique effect of wine. In this vein, a study of wine consumption among late-middle-aged adults found that failure to control for confounding factors produced a spurious effect of a mortality advantage for drinkers who prefer wine compared with those who preferred other alcoholic beverages. There is no consensus on whether the type of wine has any influence on the benefits to longevity as some epidemiological studies noted an association with red wine, while others have not found differences between red and white wine.

26.3 HEALTH-PROMOTING ACTIVITY OF POLYPHENOLS RESULTING FROM THEIR INTERACTION WITH BIOLOGICAL PROTEINS

As already referred to, the positive health effects in preventing some diseases (such as cancer, allergies, cardiovascular, Parkinson and Alzheimer diseases) have been related to the strong antioxidant capacity of phenolic compounds but also to much more complex effects related to cell signaling and even to the interaction with biological proteins. Some red wine polyphenols, in particular the ones that belong to the tannin group, have a special ability to interact with proteins. This interaction involving some biological proteins may be at the origin of positive and negative nutritional and health effects. On the other hand, these interactions could compromise the absorption of phenolic compounds, their metabolism and following bioactivities, as further discussed later. Therefore, the study of these interactions has been the scope of numerous research groups and the recent literature has given great advances on understanding this interaction and its impact on human health.

26.3.1 Phenolic Compound Journey Starts in the Oral Cavity

Ingestion of red wine phenolic compounds starts in the oral cavity, where the interaction with some salivary proteins (SPs) is particularly significant. In fact, during red wine consumption, phenolics interact with SPs, especially with proline-rich proteins (PRPs) (acidic, basic, and glycosylated) and statherins forming (in)soluble aggregates, which is majorly related to astringency perception (Brandão et al., 2014; Perez-Gregorio et al., 2014; Soares et al., 2011a,b). Astringency is a tactile sensation described as dryness, tightening, and puckering sensations perceived in the oral cavity, well-known for red wine consumers and highly linked to red wine quality. In general, the nature of tannin–protein interactions is mainly noncovalent through hydrophobic and hydrogen bonds, despite some covalent interactions that could also occur. The major difference between covalent and noncovalent interactions is based on whether the molecules are irreversibly bound to each other or not, respectively.

Among the phenolic compound group, tannins are the major compounds known to interact with SPs as well as with other proteins. However, recently an interaction between the major red wine anthocyanin (malvidin-3-glucoside) with salivary acidic PRPs was also reported (Ferrer-Gallego et al., 2015).

The interaction of tannins with SPs could also impart the bioavailability and so the succeeding biological effects of polyphenols. A recent work found that PRP/grape seed procyanidin precipitates were resistant to stomach digestion and phenolic compounds were not released from complexes to be further metabolized and absorbed (Soares et al., 2015). It was observed that the complexes of the more polymerized tannins were more highly resistant than the smaller ones.

26.3.2 Interaction With Digestive Enzymes

Besides the interaction with SPs, phenolics such as flavonoids, phenolic acids, tannins (proanthocyanidins and ellagitannins), and stilbenes (resveratrol) also have the ability to interact with and inhibit several digestive enzymes.

Potential efficacy of phenolic compounds on carbohydrate metabolism and glucose homeostasis has been well investigated in in vitro, animal models and some clinical trials (Hui and Xiaoqing, 2012; Johnston et al., 2005; McDougall et al., 2005; Tadera et al., 2006). Amylases and α-glucosidase are the key enzymes responsible for digestion of dietary carbohydrates. These enzymes are present in the oral cavity and are also secreted by the pancreas to the small intestine, being responsible for the digestion of carbohydrates by reducing starch polymers to oligosaccharides and glucose. Disorders of carbohydrate metabolism may cause severe health problems such as diabetes, obesity, or dental caries. Therefore, these enzymes are targets of drug design in attempts to treat the referred diseases.

It is well reported that phenolic compounds have the ability to inhibit these enzymes. An interesting review summarized several structural features that affect the inhibitory capacity of phenolic compounds on these enzymes (Hui and Xiaoqing, 2012).

Regarding condensed tannin compounds, results in the literature about their effect on digestive enzyme activities are not consensual. Barrett et al. (2013) found that larger and more complex proanthocyanidins (trimers and tetramers) isolated from fruits (such as grapes) inhibited more effectively amylase and glucosidase than did less polymerized tannins (monomers). However, Lee et al. (2007) found that the effect of proanthocyanidin polymerization degree is dependent on the enzyme. These authors observed that polymers showed a strong inhibitory activity against α-amylase, while oligomers had a relatively weak effect. On the other hand, against α-glucosidase activity, oligomers exerted a stronger protective effect than polymers. Gonçalves et al. (2011) observed the same trend for amylase inhibition. The most polymerized procyanidins (from pentamers to hexamers) were the most effective inhibitors of amylase activity rather than less polymerized structures (trimers).

Therefore, phenolic compounds could in theory slow starch digestion by inhibition of these digestive enzymes and increase satiety by modulation of glucose "spiking" and depletion that occurs after carbohydrate-rich meals, which could make possible the use of these natural compounds as antidiabetic agents.

Besides their influence on carbohydrate metabolism, phenolic compounds have also been reported to affect lipid metabolism being highly linked to weight gain, obesity, and cardiovascular protection. Several mechanisms have been reported by which they have these actions, including reduction of dietary lipid digestion and absorption from the gut. Once again, this inhibition occurs by interaction and inhibition of a digestive enzyme, lipase.

Before fat can be absorbed, triglycerides from foods must be hydrolyzed, and the pancreatic lipase plays a key role in the efficient digestion of these compounds. Grape seed extracts (Moreno et al., 2003) and oligomeric procyanidins (higher than trimer) (Gonçalves et al., 2010; Sugiyama et al., 2007) inhibit the activity of pancreatic lipase, thus suggesting limited dietary triglycerides absorption (Sugiyama et al., 2007). Inhibitory effects of procyanidins increased according to the degree of polymerization from dimer to pentamer. On the other hand, pentamer or greater procyanidins showed maximal inhibitory effects on pancreatic lipase. These results suggested that with respect to in vitro pancreatic lipase inhibition, the degree of polymerization is an important factor. Kurihara et al. (2006) also found that small structural differences in phenolic compounds lead to great differences in lipase activity inhibition. For instance, while epicatechin gallate, a common flavan-3-ol present in red wine, inhibits lipase significantly its counterpart epigallocatechin does not. In addition, Sbarra et al. (2005) found that cholesterol esterase from humans and rats, an enzyme involved in the duodenal digestion of lipids, was inhibited both by resveratrol and a red wine polyphenol extract.

Similar results were obtained in vivo. A lower weight gain and a lower adipogenesis were observed in mice fed a high-fat diet supplemented with epigallocatechingallate (EGCG), suggested to be due in part to EGCG's inhibitory effect on pancreatic lipase activity resulting in reduced fat absorption from the gut as indicated by the presence of a high fecal fat content (Ikeda et al., 2005; Nakai et al., 2005). Bladé and collaborators made an extensive revision comprising in vivo studies on proanthocyanidin effects on lipid metabolism (Bladé et al., 2010).

Other authors support that the hypolipidemic effects of procyanidins are related to inhibition of other key enzymes in lipid biosynthesis pathways. Ardévol et al. (2000) also found a significant inhibition of glycerol-3-phosphate dehydrogenase activity, an enzyme which plays an important role in the synthesis of lipids in the adipocyte, by procyanidin extracts from grape and wine. Once again the most efficient extracts were the most polymerized ones.

Procyanidins also present the ability to interact with lipids reducing their availability. The micellar solubility of cholesterol in vitro is reduced significantly more by procyanidin dimers B2 and B5, trimer C1, and tetramer A2 than by catechin and epicatechin (Yasuda et al., 2008), suggesting that procyanidins inhibit the absorption of cholesterol and bile acids by decreasing micellar cholesterol solubility.

26.3.3 Interaction With Serum Proteins

Besides interaction and inhibition of digestive enzymes, there are several studies about the interaction of phenolic compounds with plasma proteins which could also affect delivery of phenolics and their metabolites to cells and tissues. The reversible binding of phenolic compounds to blood serum proteins, such as serum albumin, α1-acid glycoprotein, and lipoproteins has been reported (Diniz et al., 2008; Urpí-Sardà et al., 2005).

Several authors report the interaction of human serum albumin and several red wine phenolic compounds, namely resveratrol (Lu et al., 2007; N' soukpoé-Kossi et al., 2006; Pantusa et al., 2012), quercetin (Sengupta and Sengupta, 2002), catechin, epicatechin, epicatechin gallate, and procyanidin dimers B3 and B4 (Li and Yan, 2015; Soares et al., 2007). Besides, it was found that the interaction of gallic acid, epigallocatechin, catechin, and epicatechin with albumin weaken the phenolics' antioxidant capacity (Arts et al., 2002). Similarly, Riedl and Hagerman (2001) found that the rate of radical scavenging by procyanidin dimer B1 was decreased in the presence of protein (model albumin and gelatin) because procyanidin and protein formed substantial amounts of perceptible complexes. A similar trend was observed for quercetin bound to serum albumin (Rohn et al., 2004). These authors showed that the quercetin–albumin interaction decreased the total antioxidant activity in comparison to an equivalent amount of free quercetin.

Urpí-Sardà et al. (2005) have reported binding of resveratrol and its metabolites to human LDL, suggesting that these compounds may exert their antioxidant effect to LDL.

26.3.4 Interaction With Platelets

Several red wine phenolic compounds have also been proved to inhibit platelet aggregation (Beretz et al., 1982; Hubbard et al., 2003; Renaud and de Lorgeril, 1992; Struck et al., 1994). Platelets are important elements of blood because they are active both in the initiation of thrombosis as well as within the development of atherosclerosis. It is also well known that patients with diabetes, hypertension, and patients who smoke have hyperactive blood platelets. Therefore, this activity of natural red wine phenolics to avoid platelet aggregation is also of great importance to prevent atherosclerosis. Beretz et al. (1982) found that quercetin and catechin, as well as other naturally occurring phenolics, inhibited platelet aggregation. These authors studied a wide library of compounds and identified several structural features of phenolic compounds that diminished to a great extent this effect, such as saturation of the C-2, C-3 double bond, lack of the C-4 carbonyl, glycosylation at C-3, and a high number of hydroxyl substituents.

Ruf et al. (1995) were able to reproduce the protective effect of red wine that they previously observed on platelets by condensed tannins (procyanidins) extracted from grape seeds or red wine and added to 6% ethanol.

26.3.5 Interaction With Neurotoxic Proteins

Another very important neuroprotective activity has been attributed to some red wine phenolics due to interaction with proteins in the brain. With an emphasis on neurodegenerative conditions, such as Alzheimer disease, there are some polyphenols, such as resveratrol, capable of directly interfering with the Alzheimer disease hallmark toxic β-amyloid protein (Aβ), thereby inhibiting fibril and aggregate formation (Smid et al., 2012).

Some works that studied the effect of two unrelated red wines, a Cabernet Sauvignon and a Muscadine wine, observed that Cabernet Sauvignon wine was effective in reducing the generation of Aβ peptides, while polyphenols from the Muscadine wine attenuate Aβ aggregation (Bitan et al., 2001; Ho et al., 2009; Pasinetti, 2012; Wang et al., 2006). The authors identified anthocyanin polyphenolic components as those responsible for Cabernet Sauvignon. Regarding the Muscadine wine, they pointed out that its polyphenolics significantly interfere with Aβ protein-to-protein interactions that are critical for the initial assembly of monomeric Aβ peptides into increasingly large aggregated species.

The same authors found that grape seed phenolic extracts enriched in proanthocyanidins are also capable of interfering with tau-mediated toxicity by interfering with the abnormal aggregation of tau (Santa-Maria et al., 2012; Wang et al., 2010).

Guéroux et al. (2015) studied the mode of action upon the tau protein aggregation of 10 different flavan-3-ols, including several red wine polyphenols that cover their natural structural diversity (epicatechin, epigallocatechin, procyanidin dimers B1 to B4, the C4–C8 trimer of catechin, and the B3 galloylated dimers) by using NMR and dynamic molecular modeling. The results showed that the fixation to protein occurs in the peptide region where phosphorylations usually take place and the affinity has been shown to depend on the presence of some procyanidin structural elements (galloyl moiety and the degree of polymerization were shown to increase the affinity).

26.3.6 Interaction With Allergy Proteins

Another possible biological action of phenolic compounds is related to antiallergic effect. The interaction of phenolic compounds with proteins can modulate the process of allergic sensitization and their direct effect on allergic symptoms. The ability of some phenolic compounds such as caffeic, ferulic, and gallic acid to form insoluble complexes with potentially allergenic proteins can render the protein hypoallergenic (Plundrich et al., 2015; Singh et al., 2011; Zuercher et al., 2010).

26.4 BIOAVAILABILITY OF RED WINE

Besides all the biological effects described for red wine or its main phenolic components, there are now strong evidences that the molecules responsible for those effects are not the original ones but rather their metabolites arising after absorption takes place.

This discovery confers no physiological relevance to some of the in vitro and animal studies performed with the original molecules and also to studies in which the concentration ranges are above the ranges detected in human plasma, urine, and fecal samples.

Due to the almost absence of human studies concerning red wine bioavailability, this topic will include some data with the main red wine polyphenols as purified compounds or included in other food matrixes.

26.4.1 Absorption and Metabolism of Red Wine Anthocyanins and Derivatives

In 2001, Bub et al. performed a comparative study on changes in plasma malvidin-3-glucoside (Mv3glc), the main red wine anthocyanin, and its urinary excretion after ingestion of red wine, dealcoholized red wine, and red grape juice (Bub et al., 2001). All subjects consumed each beverage with different Mv3glc quantities: red wine 68 mg, dealcoholized red wine 58 mg, and red grape juice 117 mg Mv3glc was found in plasma and urine after ingestion of all the beverages studied and no metabolite or aglycone forms were observed in plasma or urine. Increases in plasma Mv3glc concentrations were not significantly different after the consumption of either red wine or dealcoholized red wine and were two times less than those measured after consumption of red grape juice. This difference may be caused by the about two times higher Mv3glc concentration present in red grape juice (Bub et al., 2001).

Later, due to their low bioavailability, any significant contribution of anthocyanins to health-protecting properties of red wine or red grape juice after consumption by healthy volunteers was questioned (Frank et al., 2003). The intestinal absorption of anthocyanins of red grape juice seemed to be improved compared to red wine, suggesting a possible synergistic effect of the glucose content of the juice (Bitsch et al., 2004).

The key difference compared to the other flavonoid glycosides, is that anthocyanins undergo rearrangements in response to pH and temperature (Brouillard and Delaporte, 1977). Actually, anthocyanins have been reported to have one of the lowest bioavailabilities of all of the dietary flavonoid subclasses (Manach et al., 2005) with typical urinary excretion of the total amount of anthocyanins and their derivatives well below 1% of the ingested amount.

Upstream of gastrointestinal absorption, a variety of binding processes can take place, namely interaction with food proteins or with salivary proteins and digestive enzymes, as previously described (Matsui et al., 2001; Walle et al., 2005; Wiese et al., 2009). Some studies have been conducted aiming at understanding the fast kinetics of plasma appearance of anthocyanins (Cao and Prior, 1999; Milbury et al., 2002; Murkovic et al., 2001; Tsuda et al., 1999). Passamonti et al. (2003) suggested the possible involvement of the gastric barrier in the rapid detection of anthocyanins in plasma. Moreover, a high proportion of pelargonidin-3-glucoside was found to be rapidly absorbed from rat stomach (Felgines et al., 2007).

Recently an in vitro gastric model was developed and showed that anthocyanins could be efficiently transported (Fernandes et al., 2012a; Oliveira et al., 2015).

A different picture appears with radiolabeled anthocyanins and the detection of anthocyanin metabolites or catabolites.

A current study using ^{13}C-labeled anthocyanins carried out in humans by Czank et al. (2013) concluded that these compounds could be more bioavailable than previously reported. The accumulation of multiple phenolic metabolites might ultimately be responsible for the reported bioactivity of anthocyanins, with the gut microbiota apparently playing an important role in the biotransformation process. Although phase II conjugates of cyanidin-3-O-glucoside and cyanidin (cyanidin-glucuronide, methyl cyanidin, and methyl-cyanidin-glucuronide) were detected, the most important metabolites corresponded to products from anthocyanin degradation (ie, benzoic, phenylacetic, and phenylpropenoic acids, phenolic aldehydes, and hippuric acid) and their phase II conjugates, which were found at 60- and 45-fold higher concentrations than their parent compounds in urine and plasma, respectively (Czank et al., 2013). An interesting observation was the rapid appearance of cyanidin-3-O-glucoside degradation products and their phase II conjugates within the serum, which suggested that some degradation likely occurred in the small intestine (either pre- or postabsorption) and, therefore, anthocyanin C-ring cleavage may not require the action of colonic microflora. Estimates made by Czank et al. (2013) of the half-lives of elimination for ^{13}C-labeled metabolites were between 12.44 ± 4.22 and 51.62 ± 22.55 h, which suggested a relatively slow urinary clearance of some metabolites. Long elimination half-lives may be the result of a combination of hepatic recycling, enterohepatic circulation, and prolonged colonic production and absorption.

After anthocyanin-rich grape juice consumption the most abundant anthocyanins found in plasma and urine were malvidin and peonidin as native anthocyanins and as glucuronidated metabolites as well as 3,4-dihydroxybenzoic acid; minor anthocyanins (delphinidin, cyanidin, and petunidin) were only detected as native glycosides (Kuntz et al., 2015).

All in all, these data suggest that anthocyanins would be as bioavailable as other flavonoid subclasses, such as flavan-3-ols and flavones, which have relative bioavailabilities between 2.5% and 18.5% (Manach et al., 2005).

Chemical reactions of phenolic compounds are particularly important in wine because they are responsible for the color and taste changes that occur during aging. Grape phenolic compounds, namely, anthocyanins, flavonols, hydroxycinnamic acids, and flavanols including catechins and proanthocyanidins, represent only approximately one half of the polyphenol content of a 2-year-old red wine polyphenol extract (Cheynier, 2005). The other half consists of unknown phenolic species derived from grape polyphenol reactions during winemaking and aging. Various chemical reactions involving anthocyanins and/or flavanols have been demonstrated to occur during red wine aging (Oliveira et al., 2014). Anthocyanin pyruvic-acid adducts can rapidly reach rat plasma after oral administration of malvidin-3-glucoside-pyruvic acid adduct (Faria et al., 2009). In addition, flavanol-anthocyanin pigments were found to be absorbed in the intestinal caco-2 cell model (Fernandes et al., 2012b).

Based on the reported studies, a new field of interest that is often overlooked arises: the anthocyanin absorption as anthocyanin-derived pigments (dimeric anthocyanins or flavonol–anthocyanin dimers) that could act as "proanthocyanins" (delivered systems) or could have their own biological impact in the organism.

Earlier data indicated that formation of anthocyanin adducts with pyruvic acid decreases the hydroxyl and superoxide anion scavenging and thus could decrease the antioxidant potential of these compounds (Garcia-Alonso et al., 2005). Recently, several pyranomalvidin-3-glucosides were suggested as good candidates as antioxidant compounds because they easily donate an H atom to the free radicals, originating stable species (Azevedo et al., 2014) supporting the fact that the antioxidant potential arising from anthocyanins is not impaired by some of their transformations during red wine aging.

Considering their biological activity few in vitro studies have been performed demonstrating that extracts of blueberry anthocyanin–pyruvic acid adducts possess anticancer properties by inhibiting breast cancer cell proliferation and by acting as cell antiinvasive factors and chemoinhibitors (Faria et al., 2010).

26.4.2 Absorption of Catechins and Proanthocyanidins

Early in the stomach, flavan-3-ols and procyanidins were found to cross the MKN-28 gastric cell barrier model with similar percentages, of 6% and 2% after 180 min, respectively (Fernandes et al., 2015).

Both human and animal studies indicated that (+)-catechin and (−)-epicatechin were rapidly absorbed from the upper portion of the small intestine. (+)-Catechin was detected in humans at 1.4 h after intake of dealcoholized red wine (Bell et al., 2000). Many cell, animal, and human studies have shown that flavanol monomers are also extensively metabolized to O-methylated forms and/or conjugated to glucuronides and sulfates during absorption into the circulation (Delgado et al., 2014; Lambert et al., 2007; Natsume et al., 2008). One such metabolite, $3'$-O-methylepicatechin, has been shown to exert protective effects against oxidative stress-induced cell death (Spencer et al., 2001b).

Major conjugates of (−)-epicatechin in human plasma, bile, and urine were (−)-epicatechin $3'$-O-sulfonate, and (−)-epicatechin $3'$-O-β-glucuronide after ingestion of 50 mg of (−)-epicatechin by volunteers have been described (Romanov-Michailidis et al., 2012). Conjugation of (−)-epicatechin with glucuronic acid was also already observed in a human BBB cell line model (Faria et al., 2011).

Similarly, human studies have shown unconjugated procyanidin B1, B2, and B5 in plasma and serum within 30 min of consumption of the test material (Holt et al., 2002; Sano et al., 2003). Unconjugated procyanidin B2 concentration in the systemic circulation reaches a peak at approximately 2 h after administration in humans (Holt et al., 2002; Sano et al., 2003). However, the levels of intact procyanidin B2 detected in human plasma (~10–40 nM) after high doses of proanthocyanidins are lower, sometimes by several orders of magnitude, than the concentrations observed to be effective in various in vitro tests.

No transporters have been identified for proanthocyanidins. Proanthocyanidins are absorbed through passive diffusion. Proanthocyanidins are not likely to pass the lipid bilayer via the transcellular pathway due to its large number of hydrophilic hydroxyl groups. Paracellular diffusion was thought to be a preferential absorption mechanism (Deprez et al., 2001). The cleavage of higher procyanidin oligomers to mixtures of monomer and dimer in the stomach may act to enhance the potential for their absorption in the small intestine as higher oligomers have very limited absorption (Spencer et al., 2000). In contrast, a later study suggested that procyanidin dimers and trimers were highly stable under gastric and duodenal digestion conditions (Serra et al., 2010). Accordingly, a recent study reported that no (−)-epicatechin detected in human plasma and urine was derived from ingested procyanidin oligomers and polymers (Ottaviani et al., 2012).

26.4.3 Absorption and Metabolism of Resveratrol

At the current state of the art, the only known source of bioavailable resveratrol is resveratrol-rich wine, with average amounts between 1.9 and 3.6 mg resveratrol per liter (Vang et al., 2011).

One aspect that is in favor of the active role of resveratrol in the cardioprotection afforded by wine is the fact that it is absorbed only when it is ingested in wine (Alberto, 2005). When wine containing resveratrol is ingested, bioavailability reaches 100 nM up to a maximum of 1 μM (Walle, 2011), whereas the same does not occur when pure resveratrol is ingested in a tablet form (Brown et al., 2010; Landete, 2012; Patel et al., 2010).

This phenomenon has recently been explained by the formation of glucuronides and sulfates that occur in the organism, with the purpose of eliminating the polyphenol with urine (Delmas et al., 2011; Vang et al., 2011; Vitaglione et al., 2005). The finding in human serum of *trans*-resveratrol glucuronides, rather than the free form of the compound, raises some doubts about the health effects of dietary resveratrol consumption in the original form. The oral absorption of resveratrol in humans is about 75%, although extensive metabolism in the intestine and liver results in an oral bioavailability considerably less than 1%. Major metabolites include glucuronides and sulfates of resveratrol (Walle, 2011).

26.4.4 Microbiota Impact on Red Wine Availability and Bioactivity

Wine components that reach the microbiota are those not absorbed in the upper GI level and, also, their metabolites excreted in the bile and/or from the enterohepatic circulation.

In the microbiota, red wine phenolic compounds are extensively metabolized, especially by genera and species with enzymes necessary to catalyze these reactions (hydrolyzing and conjugating enzymes) (Rechner et al., 2002; Spencer et al., 2001a; Tapiero et al., 2002). Therefore, the colon is being considered as an active site for metabolism rather than a simple excretion route and has been receiving much attention from the scientific community (Aura, 2008).

This conversion may contribute to the absorption and modulation of the biological activity of red wine phenolics, which is different from the original compounds (Landete, 2012; Vendrame et al., 2011).

In a human intervention study, modulation of the gut microbiota by using red wine was already described as an effective strategy for managing metabolic diseases associated with obesity (Moreno-Indias et al., 2015).

An intervention study with red wine revealed significant changes in eight metabolites: 3,5-dihydroxybenzoic acid, 3-O-methylgallic acid, p-coumaric acid, phenylpropionic acid, protocatechuic acid, vanillic acid, syringic acid, and 4-hydroxy-5-(phenyl)valeric acid (Jimenez-Giron et al., 2013), without any effect of ethanol on microbial action. It is important to highlight that O-methyl benzoic acids such as syringic and vanillic acids could arise from methoxylated anthocyanin catabolism (Aura et al., 2005; Williamson and Clifford, 2010) and protocatechuic acid from cyanidin. p-Coumaric acid may also be a product from p-coumaroyl-acylated anthocyanins (Monagas et al., 2005). In addition, since anthocyanin phenolic acids can be further absorbed in colon, it is possible that these phenolic acids are additionally metabolized by hepatic cells (Woodward et al., 2011). Health benefits associated with red wine anthocyanin intake may also be explained by slow and continuous release of phenolic compounds through the gut into the bloodstream.

Similar to what was reported for anthocyanins, proanthocyanidins can also be metabolized by microbial enzymes. Twenty-four "dimeric" metabolites with molecular weight greater than 290 were detected after procyanidin B2 was incubated with human fecal microbiota (Stoupi et al., 2010). Microbiota may also cleave the interflavan bond to convert procyanidin B2 into two (−)-epicatechin, after procyanidin B2 incubation with human fecal microbiota for up to 12h (Stoupi et al., 2010). The majority of proanthocyanidins reach the colon intact and are degraded into phenylvalerolactones and phenolic acids by colon microbiota (Appeldoorn et al., 2009; Stoupi et al., 2010).

In vivo studies showed that microbial-derived phenylvalerolactone and phenolic acids were the predominant metabolites of procyanidins in blood and urine (Ottaviani et al., 2012; Urpi-Sarda et al., 2009). Urinary excretion of p-coumaric acid, vanillic acid, 3-hydroxybenzoic acid, and ferulic acid increased by over twofold in humans after consumption of 40g of procyanidin-rich cocoa powder (Urpi-Sarda et al., 2009). There was an increase in urinary excretion of microbial catabolites after human consumption of procyanidins and catechin monomers (Rios et al., 2003).

Recent studies unraveled the pivotal role of gut microbiota on human health (Robles Alonso and Guarner, 2013). Increasing evidence suggests that anthocyanins and proanthocyanidins have the potential to confer health benefits via modulation of the gut microbiota and by exerting prebiotic-like effects (Cueva et al., 2013; Queipo-Ortuno et al., 2012; Tzounis et al., 2008).

REFERENCES

Alam, M.N., Bristi, N.J., Rafiquzzaman, M., 2013. Review on in vivo and in vitro methods evaluation of antioxidant activity. Saudi Pharmaceutical Journal 21, 143–152.

Alberto, A.E.B., 2005. Pharmacokinetics and Metabolism of Resveratrol, Resveratrol in Health and Disease. CRC Press, pp. 631–642.

Appeldoorn, M.M., Vincken, J.P., Aura, A.M., Hollman, P.C., Gruppen, H., 2009. Procyanidin dimers are metabolized by human microbiota with 2-(3,4-dihydroxyphenyl)acetic acid and 5-(3,4-dihydroxyphenyl)-gamma-valerolactone as the major metabolites. Journal of Agricultural Food Chemistry 57, 1084–1092.

Ardévol, A., Bladé, C., Salvadó, M.J., Arola, L., 2000. Changes in lipolysis and hormone-sensitive lipase expression caused by procyanidins in 3T3-L1 adipocytes. International Journal of Obesity and Related Metabolic Disorders 24, 319–324.

Arts, M.J.T.J., Haenen, G.R.M.M., Wilms, L.C., Beetstra, S.A.J.N., Heijnen, C.G.M., Voss, H.-P., Bast, A., 2002. Interactions between flavonoids and proteins: effect on the total antioxidant capacity. Journal of Agricultural and Food Chemistry 50, 1184–1187.

Aura, A.-M., 2008. Microbial metabolism of dietary phenolic compounds in the colon. Phytochemistry Reviews 7, 407–429.

Aura, A.M., Martin-Lopez, P., O'Leary, K.A., Williamson, G., Oksman-Caldentey, K.M., Poutanen, K., Santos-Buelga, C., 2005. In vitro metabolism of anthocyanins by human gut microflora. European Journal of Nutrition 44, 133–142.

Azevedo, J., Oliveira, J., Cruz, L., Teixeira, N., Bras, N.F., De Freitas, V., Mateus, N., 2014. Antioxidant features of red wine pyranoanthocyanins: experimental and theoretical approaches. Journal of Agricultural Food Chemistry 62, 7002–7009.

Barrett, A., Ndou, T., Hughey, C.A., Straut, C., Howell, A., Dai, Z., Kaletunc, G., 2013. Inhibition of α-amylase and glucoamylase by tannins extracted from cocoa, pomegranates, cranberries, and grapes. Journal of Agricultural and Food Chemistry 61, 1477–1486.

Barrientos, R.M., Kitt, M.M., Watkins, L.R., Maier, S.F., 2015. Neuroinflammation in the normal aging hippocampus. Neuroscience 309, 84–99.

Bassig, B.A., Lan, Q., Rothman, N., Zhang, Y., Zheng, T., 2012. Current understanding of lifestyle and environmental factors and risk of non-hodgkin lymphoma: an epidemiological update. Journal of Cancer Epidemiology.

Bastianetto, S., Ménard, C., Quirion, R., 2015. Neuroprotective action of resveratrol. Biochimica et Biophysica Acta (BBA) - Molecular Basis of Disease 1852, 1195–1201.

Bell, J.R., Donovan, J.L., Wong, R., Waterhouse, A.L., German, J.B., Walzem, R.L., Kasim-Karakas, S.E., 2000. (+)-Catechin in human plasma after ingestion of a single serving of reconstituted red wine. The American Journal of Clinical Nutrition 71, 103–108.

Bellaver, B., Souza, D.G., Bobermin, L.D., Souza, D.O., Gonçalves, C.A., Quincozes-Santos, A., 2015. Resveratrol protects hippocampal astrocytes against LPS-induced neurotoxicity through HO-1, p38 and ERK pathways. Neurochemical Research 40, 1600–1608.

Beretz, A., Cazenave, J.-P., Anton, R., 1982. Inhibition of aggregation and secretion of human platelets by quercetin and other flavonoids: structure-activity relationships. Agents and Actions 12, 382–387.

Bhaswant, M., Fanning, K., Netzel, M., Mathai, M.L., Panchal, S.K., Brown, L., 2015. Cyanidin 3-glucoside improves diet-induced metabolic syndrome in rats. Pharmacological Research 102, 208–217.

Bitan, G., Lomakin, A., Teplow, D.B., 2001. Amyloid beta-protein oligomerization: prenucleation interactions revealed by photo-induced cross-linking of unmodified proteins. The Journal of Biological Chemistry 276, 35176–35184.

Bitsch, R., Netzel, M., Frank, T., Strass, G., Bitsch, I., 2004. Bioavailability and biokinetics of anthocyanins from red grape juice and red wine. Journal of Biomedicine and Biotechnology 2004, 293–298.

Bladé, C., Arola, L., Salvadó, M.-J., 2010. Hypolipidemic effects of proanthocyanidins and their underlying biochemical and molecular mechanisms. Molecular Nutrition & Food Research 54, 37–59.

Brandão, E., Soares, S., Mateus, N., de Freitas, V., 2014. In vivo interactions between procyanidins and human saliva proteins: effect of repeated exposures to procyanidins solution. Journal of Agricultural and Food Chemistry 62, 9562–9568.

Brouillard, R., Delaporte, B., 1977. Chemistry of anthocyanin pigments. 2. Kinetic and thermodynamic study of proton transfer, hydration, and tautomeric reactions of malvidin 3-glucoside. Journal of the American Chemical Society 99, 8461–8468.

Brown, V.A., Patel, K.R., Viskaduraki, M., Crowell, J.A., Perloff, M., Booth, T.D., Vasilinin, G., Sen, A., Schinas, A.M., Piccirilli, G., Brown, K., Steward, W.P., Gescher, A.J., Brenner, D.E., 2010. Repeat dose study of the cancer chemopreventive agent resveratrol in healthy volunteers: safety, pharmacokinetics, and effect on the insulin-like growth factor axis. Cancer Research 70, 9003–9011.

Bub, A., Watzl, B., Heeb, D., Rechkemmer, G., Briviba, K., 2001. Malvidin-3-glucoside bioavailability in humans after ingestion of red wine, dealcoholized red wine and red grape juice. European Journal of Nutrition 40, 113–120.

Cao, G., Prior, R.L., 1999. Anthocyanins are detected in human plasma after oral administration of an elderberry extract. Clinical Chemistry 45, 574–576.

Cheynier, V., 2005. Polyphenols in foods are more complex than often thought. The American Journal of Clinical Nutrition 81, 223S–229S.

Chiva-Blanch, G., Magraner, E., Condines, X., Valderas-Martínez, P., Roth, I., Arranz, S., Casas, R., Navarro, M., Hervas, A., Sisó, A., Martínez-Huélamo, M., Vallverdú-Queralt, A., Quifer-Rada, P., Lamuela-Raventos, R.M., Estruch, R., 2015. Effects of alcohol and polyphenols from beer on atherosclerotic biomarkers in high cardiovascular risk men: a randomized feeding trial. Nutrition, Metabolism and Cardiovascular Diseases 25, 36–45.

Cordova, A.C., Sumpio, B.E., 2009. Polyphenols are medicine: is it time to prescribe red wine for our patients? The International Journal of Angiology: Official Publication of the International College of Angiology, Inc. 18, 111–117.

Cueva, C., Sanchez-Patan, F., Monagas, M., Walton, G.E., Gibson, G.R., Martin-Alvarez, P.J., Bartolome, B., Moreno-Arribas, M.V., 2013. In vitro fermentation of grape seed flavan-3-ol fractions by human faecal microbiota: changes in microbial groups and phenolic metabolites. FEMS Microbiology Ecology 83, 792–805.

Czank, C., Cassidy, A., Zhang, Q., Morrison, D.J., Preston, T., Kroon, P.A., Botting, N.P., Kay, C.D., 2013. Human metabolism and elimination of the anthocyanin, cyanidin-3-glucoside: a ^{13}C-tracer study. The American Journal of Clinical Nutrition 97, 995–1003.

Delgado, L., Fernandes, I., Gonzalez-Manzano, S., de Freitas, V., Mateus, N., Santos-Buelga, C., 2014. Anti-proliferative effects of quercetin and catechin metabolites. Food & Function 5, 797–803.

Delmas, D., Aires, V., Limagne, E., Dutartre, P., Mazue, F., Ghiringhelli, F., Latruffe, N., 2011. Transport, stability, and biological activity of resveratrol. Annals of the New York Academy of Sciences 1215, 48–59.

Deon, M., Sitta, A., Faverzani, J.L., Guerreiro, G.B., Donida, B., Marchetti, D.P., Mescka, C.P., Ribas, G.S., Coitinho, A.S., Wajner, M., Vargas, C.R., 2015. Urinary biomarkers of oxidative stress and plasmatic inflammatory profile in phenylketonuric treated patients. International Journal of Developmental Neuroscience 47, 259–265.

Deprez, S., Mila, I., Huneau, J.-F., Tome, D., Scalbert, A., 2001. Transport of proanthocyanidin dimer, trimer, and polymer across monolayers of human intestinal epithelial Caco-2 cells. Antioxidants & Redox Signaling 3, 957–967.

Diniz, A., Escuder-Gilabert, L., Lopes, N.P., Villanueva-Camañas, R.M., Sagrado, S., Medina-Hernández, M.J., 2008. Characterization of interactions between polyphenolic compounds and human serum proteins by capillary electrophoresis. Analytical and Bioanalytical Chemistry 391, 625–632.

Dragan, S., Andrica, F., Serban, M.C., Timar, R., 2015. Polyphenols-rich natural products for treatment of diabetes. Current Medicinal Chemistry 22, 14–22.

El Darra, N., Turk, M.F., Ducasse, M.A., Grimi, N., Maroun, R.G., Louka, N., Vorobiev, E., 2016. Changes in polyphenol profiles and color composition of freshly fermented model wine due to pulsed electric field, enzymes and thermovinification pretreatments. Food Chemistry 194, 944–950.

Erdrich, S., Bishop, K.S., Karunasinghe, N., Han, D.Y., Ferguson, L.R., 2015. A pilot study to investigate if New Zealand men with prostate cancer benefit from a Mediterranean-style diet. PeerJ.

Faria, A., Pestana, D., Monteiro, R., Teixeira, D., Azevedo, J., de Freitas, V., Mateus, N., Calhau, C., 2009. Bioavailability of Anthocyanin-Pyruvic Acid Adducts in Rat. International Conference on Polyphenols and Health. Yorkshire, Leeds, pp. 170–171.

Faria, A., Pestana, D., Teixeira, D., Couraud, P.-O., Romero, I., Weksler, B., de Freitas, V., Mateus, N., Calhau, C., 2011. Insights into the putative catechin and epicatechin transport across blood-brain barrier. Food & Function 2, 39–44.

Faria, A., Pestana, D., Teixeira, D., de Freitas, V., Mateus, N., Calhau, C., 2010. Blueberry anthocyanins and pyruvic acid adducts: anticancer properties in breast cancer cell lines. Phytotherapy Research 24, 1862–1869.

Felgines, C., Texier, O., Besson, C., Lyan, B., Lamaison, J.-L., Scalbert, A., 2007. Strawberry pelargonidin glycosides are excreted in urine as intact glycosides and glucuronidated pelargonidin derivatives in rats. The British Journal of Nutrition 98, 1126–1131.

Fenwick, E.K., Xie, J., Man, R.E.K., Lim, L.L., Flood, V.M., Finger, R.P., Wong, T.Y., Lamoureux, E.L., 2015. Moderate consumption of white and fortified wine is associated with reduced odds of diabetic retinopathy. Journal of Diabetes and Its Complications 29, 1009–1014.

Fernandes, I., de Freitas, V., Reis, C., Mateus, N., 2012a. A new approach on the gastric absorption of anthocyanins. Food & Function 3, 508–516.

Fernandes, I., Faria, A., de Freitas, V., Calhau, C., Mateus, N., 2015. Multiple-approach studies to assess anthocyanin bioavailability. Phytochemistry Reviews 1–21.

Fernandes, I., Nave, F., Gonçalves, R., de Freitas, V., Mateus, N., 2012b. On the bioavailability of flavanols and anthocyanins: flavanol–anthocyanin dimers. Food Chemistry 135, 812–818.

Ferreira, V., Fernandes, F., Pinto-Carnide, O., Valentão, P., Falco, V., Martín, J.P., Ortiz, J.M., Arroyo-García, R., Andrade, P.B., Castro, I., 2016. Identification of *Vitis vinifera* L. grape berry skin color mutants and polyphenolic profile. Food Chemistry 194, 117–127.

Ferrer-Gallego, R., Soares, S., Mateus, N., Rivas-Gonzalo, J., Escribano-Bailón, M.T., de Freitas, V., 2015. New anthocyanin–human salivary protein complexes. Langmuir 31, 8392–8401.

Frank, T., Netzel, M., Strass, G., Bitsch, R., Bitsch, I., 2003. Bioavailability of anthocyanidin-3-glucosides following consumption of red wine and red grape juice. Canadian Journal of Physiology and Pharmacology 81, 423–435.

Garcia-Alonso, M., Rimbach, G., Sasai, M., Nakahara, M., Matsugo, S., Uchida, Y., Rivas-Gonzalo, J.C., De Pascual-Teresa, S., 2005. Electron spin resonance spectroscopy studies on the free radical scavenging activity of wine anthocyanins and pyranoanthocyanins. Molecular Nutrition and Food Research 49, 1112–1119.

Glass, C.K., Saijo, K., Winner, B., Marchetto, M.C., Gage, F.H., 2010. Mechanisms underlying inflammation in neurodegeneration. Cell 140, 918–934.

Gonçalves, R., Mateus, N., de Freitas, V., 2010. Study of the interaction of pancreatic lipase with procyanidins by optical and enzymatic methods. Journal of Agricultural and Food Chemistry 58, 11901–11906.

Gonçalves, R., Mateus, N., de Freitas, V., 2011. Inhibition of α-amylase activity by condensed tannins. Food Chemistry 125, 665–672.

Guéroux, M., Pinaud-Szlosek, M., Fouquet, E., de Freitas, V., Laguerre, M., Pianet, I., 2015. How wine polyphenols can fight Alzheimer disease progression: towards a molecular explanation. Tetrahedron 71, 3163–3170.

Gutteridge, J.M.C., Halliwell, B., 2010. Antioxidants: molecules, medicines, and myths. Biochemical and Biophysical Research Communications 393, 561–564.

Haxhi, J., Leto, G., Di Palumbo, A.S., Sbriccoli, P., Guidetti, L., Fantini, C., Buzzetti, R., Caporossi, D., Di Luigi, L., Sacchetti, M., 2015. Exercise at lunchtime: effect on glycemic control and oxidative stress in middle-aged men with type 2 diabetes. European Journal of Applied Physiology 1–10.

Heneka, M.T., Carson, M.J., Khoury, J.E., Landreth, G.E., Brosseron, F., Feinstein, D.L., Jacobs, A.H., Wyss-Coray, T., Vitorica, J., Ransohoff, R.M., Herrup, K., Frautschy, S.A., Finsen, B., Brown, G.C., Verkhratsky, A., Yamanaka, K., Koistinaho, J., Latz, E., Halle, A., Petzold, G.C., Town, T., Morgan, D., Shinohara, M.L., Perry, V.H., Holmes, C., Bazan, N.G., Brooks, D.J., Hunot, S., Joseph, B., Deigendesch, N., Garaschuk, O., Boddeke, E., Dinarello, C.A., Breitner, J.C., Cole, G.M., Golenbock, D.T., Kummer, M.P., 2015. Neuroinflammation in Alzheimer's disease. The Lancet Neurology 14, 388–405.

Heppner, F.L., Ransohoff, R.M., Becher, B., 2015. Immune attack: the role of inflammation in Alzheimer disease. Nature Reviews. Neuroscience 16, 358–372.

Hernandez-Hernandez, A., Gea, A., Ruiz-Canela, M., Toledo, E., Beunza, J.J., Bes-Rastrollo, M., Martinez-Gonzalez, M.A., 2015. Mediterranean alcohol-drinking pattern and the incidence of cardiovascular disease and cardiovascular mortality: the SUN project. Nutrients 7, 9116–9126.

Herrero, M.-T., Estrada, C., Maatouk, L., Vyas, S., 2015. Inflammation in Parkinson's disease: role of glucocorticoids. Frontiers in Neuroanatomy 9, 32.

Herttua, K., Mäkelä, P., Martikainen, P., 2015. Minimum prices for alcohol and educational disparities in alcohol-related mortality. Epidemiology.

Ho, L., Chen, L.H., Wang, J., Zhao, W., Talcott, S.T., Ono, K., Teplow, D., Humala, N., Cheng, A., Percival, S.S., Ferruzzi, M., Janle, E., Dickstein, D.L., Pasinetti, G.M., 2009. Heterogeneity in red wine polyphenolic contents differentially influences Alzheimer's disease-type neuropathology and cognitive deterioration. Journal of Alzheimer's Disease 16, 59–72.

Holt, R.R., Lazarus, S.A., Sullards, M.C., Zhu, Q.Y., Schramm, D.D., Hammerstone, J.F., Fraga, C.G., Schmitz, H.H., Keen, C.L., 2002. Procyanidin dimer B2 [epicatechin-(4beta-8)-epicatechin] in human plasma after the consumption of a flavanol-rich cocoa. The American Journal of Clinical Nutrition 76, 798–804.

Hubbard, G.P., Wolffram, S., Lovegrove, J.A., Gibbins, J.M., 2003. The role of polyphenolic compounds in the diet as inhibitors of platelet function. The Proceedings of the Nutrition Society 62, 469–478.

Hui, C., Xiaoqing, C., 2012. Structures required of flavonoids for inhibiting digestive enzymes. Anti-Cancer Agents in Medicinal Chemistry 12, 929–939.

Ikeda, I., Tsuda, K., Suzuki, Y., Kobayashi, M., Unno, T., Tomoyori, H., Goto, H., Kawata, Y., Imaizumi, K., Nozawa, A., Kakuda, T., 2005. Tea catechins with a galloyl moiety suppress postprandial hypertriacylglycerolemia by delaying lymphatic transport of dietary fat in rats. The Journal of Nutrition 135, 155–159.

Jimenez-Giron, A., Queipo-Ortuno, M.I., Boto-Ordonez, M., Munoz-Gonzalez, I., Sanchez-Patan, F., Monagas, M., Martin-Alvarez, P.J., Murri, M., Tinahones, F.J., Andres-Lacueva, C., Bartolome, B., Moreno-Arribas, M.V., 2013. Comparative study of microbial-derived phenolic metabolites in human feces after intake of gin, red wine, and dealcoholized red wine. Journal of Agricultural and Food Chemistry 61, 3909–3915.

Johnston, K., Sharp, P., Clifford, M., Morgan, L., 2005. Dietary polyphenols decrease glucose uptake by human intestinal Caco-2 cells. FEBS Letters 579, 1653–1657.

Jose, R., Sajitha, G.R., Augusti, K.T., 2014. A review on the role of nutraceuticals as simple as Se^{2+} to complex organic molecules such as glycyrrhizin that prevent as well as cure diseases. Indian Journal of Clinical Biochemistry 29, 119–132.

Kamakura, R., Son, M.J., de Beer, D., Joubert, E., Miura, Y., Yagasaki, K., 2015. Antidiabetic effect of green rooibos (*Aspalathus linearis*) extract in cultured cells and type 2 diabetic model KK-A^y mice. Cytotechnology 67, 699–710.

Karjalainen, R., Anttonen, M., Saviranta, N., Hilz, H., Törrönen, R., Stewart, D., McDougall, G.J., Mattila, P., 2009. A review on bioactive compounds in black currants (*Ribes nigrum* L.) and their potential health-promoting properties. Acta Horticulturae 301–307.

Kuntz, S., Rudloff, S., Asseburg, H., Borsch, C., Frohling, B., Unger, F., Dold, S., Spengler, B., Rompp, A., Kunz, C., 2015. Uptake and bioavailability of anthocyanins and phenolic acids from grape/blueberry juice and smoothie in vitro and in vivo. The British Journal of Nutrition 113, 1044–1055.

Kurihara, H., Shibata, H., Fukui, Y., Kiso, Y., Xu, J.-K., Yao, X.-S., Fukami, H., 2006. Evaluation of the hypolipemic property of *Camellia sinensis* Var. *ptilophylla* on postprandial hypertriglyceridemia. Journal of Agricultural and Food Chemistry 54, 4977–4981.

Kurin, E., Fakhrudin, N., Nagy, M., 2013. ENOS promoter activation by red wine polyphenols: an interaction study. Acta Facultatis Pharmaceuticae Universitatis Comenianae 60, 27–33.

Lambert, J.D., Sang, S., Yang, C.S., 2007. Biotransformation of green tea polyphenols and the biological activities of those metabolites. Molecular Pharmaceutics 4, 819–825.

Lamy, S., Beaulieue, É., Labbé, D., Bédard, V., Moghrabi, A., Barrette, S., Gingras, D., Béliveau, R., 2008. Delphinidin, a dietary anthocyanidin, inhibits platelet-derived growth factor ligand/receptor (PDGF/PDGFR) signaling. Carcinogenesis 29, 1033–1041.

Landete, J.M., 2012. Updated knowledge about polyphenols: functions, bioavailability, metabolism, and health. Critical Reviews in Food Science Nutrition 52, 936–948.

Lee, Y.A., Cho, E.J., Tanaka, T., Yokozawa, T., 2007. Inhibitory activities of proanthocyanidins from persimmon against oxidative stress and digestive enzymes related to diabetes. Journal of Nutritional Science and Vitaminology 53, 287–292.

Li, X., Yan, Y., 2015. Probing the binding of procyanidin B3 to human serum albumin by isothermal titration calorimetry. Journal of Molecular Structure 1082, 170–173.

López-Alarcón, C., Denicola, A., 2013. Evaluating the antioxidant capacity of natural products: a review on chemical and cellular-based assays. Analytica Chimica Acta 763, 1–10.

Lu, Z., Zhang, Y., Liu, H., Yuan, J., Zheng, Z., Zou, G., 2007. Transport of a cancer chemopreventive polyphenol, resveratrol: interaction with serum albumin and hemoglobin. Journal of Fluorescence 17, 580–587.

Lucas, I.K., Kolodziej, H., 2015. *Trans*-resveratrol induces apoptosis through ROS-triggered mitochondria-dependent pathways in A549 human lung adenocarcinoma epithelial cells. Planta Medica 81, 1038–1044.

Manach, C., Williamson, G., Morand, C., Scalbert, A., Remesy, C., 2005. Bioavailability and bioefficacy of polyphenols in humans. I. Review of 97 bioavailability studies. The American Journal of Clinical Nutrition 81, 230–242.

Matsui, T., Ueda, T., Oki, T., Sugita, K., Terahara, N., Matsumoto, K., 2001. α-Glucosidase inhibitory action of natural acylated anthocyanins. 1. Survey of natural pigments with potent inhibitory activity. Journal of Agricultural and Food Chemistry 49, 1948–1951.

McDougall, G.J., Shpiro, F., Dobson, P., Smith, P., Blake, A., Stewart, D., 2005. Different polyphenolic components of soft fruits inhibit α-amylase and α-glucosidase. Journal of Agricultural and Food Chemistry 53, 2760–2766.

McGovern, P.E., Glusker, D.L., Exner, L.J., Voigt, M.M., 1996. Neolithic resinated wine. Nature 381, 480–481.

Milbury, P.E., Cao, G., Prior, R.L., Blumberg, J., 2002. Bioavailablility of elderberry anthocyanins. Mechanisms of Ageing and Development 123, 997–1006.

Monagas, M., Bartolome, B., Gomez-Cordoves, C., 2005. Updated knowledge about the presence of phenolic compounds in wine. Critical Reviews in Food Science and Nutrition 45, 85–118.

Moreno, D.A., Ilic, N., Poulev, A., Brasaemle, D.L., Fried, S.K., Raskin, I., 2003. Inhibitory effects of grape seed extract on lipases. Nutrition 19, 876–879.

Moreno-Indias, I., Sánchez-Alcolado, L., Pérez-Martínez, P., Andrés-Lacueva, C., Cardona, F., Tinahones, F., Queipo-Ortuño, M.I., November 24, 2015. Red wine polyphenols modulate fecal microbiota and reduce markers of the metabolic syndrome in obese patients. Food & Function. http://dx.doi.org/10.1039/C5FO00886G.

Murkovic, M., Mülleder, U., Adam, U., Pfannhauser, W., 2001. Detection of anthocyanins from elderberry juice in human urine. Journal of the Science of Food and Agriculture 81, 934–937.

N' soukpoé-Kossi, C.N., St-Louis, C., Beauregard, M., Subirade, M., Carpentier, R., Hotchandani, S., Tajmir-Riahi, H.A., 2006. Resveratrol binding to human serum albumin. Journal of Biomolecular Structure and Dynamics 24, 277–283.

Nakai, M., Fukui, Y., Asami, S., Toyoda-Ono, Y., Iwashita, T., Shibata, H., Mitsunaga, T., Hashimoto, F., Kiso, Y., 2005. Inhibitory effects of oolong tea polyphenols on pancreatic lipase in vitro. Journal of Agricultural and Food Chemistry 53, 4593–4598.

Natsume, M., Osakabe, N., Yasuda, A., Osawa, T., Terao, J., 2008. Inhibitory effects of conjugated epicatechin metabolites on peroxynitrite-mediated nitrotyrosine formation. Journal of Clinical Biochemistry and Nutrition 42, 50–53.

Nooyens, A.C.J., Bueno-de-Mesquita, H.B., van Gelder, B.M., van Boxtel, M.P.J., Verschuren, W.M.M., 2014. Consumption of alcoholic beverages and cognitive decline at middle age: the Doetinchem Cohort Study. The British Journal of Nutrition 111, 715–723.

Olejar, K.J., Fedrizzi, B., Kilmartin, P.A., 2015. Influence of harvesting technique and maceration process on aroma and phenolic attributes of sauvignon blanc wine. Food Chemistry 183, 181–189.

Oliveira, H., Fernandes, I.L., Brás, N.F., Faria, A., De Freitas, V., Calhau, C., Mateus, N., 2015. Experimental and theoretical data on the mechanism by which red wine anthocyanins are transported through human MKN-28 gastric cell model. Journal of Agricultural and Food Chemistry 63 (35).

Oliveira, J., Mateus, N., de Freitas, V., 2014. Previous and recent advances in pyranoanthocyanins equilibria in aqueous solution. Dyes and Pigments 100, 190–200.

Ottaviani, J.I., Kwik-Uribe, C., Keen, C.L., Schroeter, H., 2012. Intake of dietary procyanidins does not contribute to the pool of circulating flavanols in humans. The American Journal of Clinical Nutrition 95, 851–858.

Pantusa, M., Sportelli, L., Bartucci, R., 2012. Influence of stearic acids on resveratrol-HSA interaction. European Biophysics Journal 41, 969–977.

Pasinetti, G.M., 2012. Novel role of red wine-derived polyphenols in the prevention of Alzheimer's disease dementia and brain pathology: experimental approaches and clinical implications. Planta Medica 78, 1614–1619.

Passamonti, S., Vrhovsek, U., Vanzo, A., Mattivi, F., 2003. The stomach as a site for anthocyanins absorption from food. FEBS Letters 544, 210–213.

Patel, K.R., Brown, V.A., Jones, D.J., Britton, R.G., Hemingway, D., Miller, A.S., West, K.P., Booth, T.D., Perloff, M., Crowell, J.A., Brenner, D.E., Steward, W.P., Gescher, A.J., Brown, K., 2010. Clinical pharmacology of resveratrol and its metabolites in colorectal cancer patients. Cancer Research 70, 7392–7399.

Pavlin, M., Repič, M., Vianello, R., Mavri, J., 2015. The chemistry of neurodegeneration: kinetic data and their implications. Molecular Neurobiology 53 (5).

Perez-Gregorio, M.R., Mateus, N., De Freitas, V., 2014. New procyanidin B3–human salivary protein complexes by mass spectrometry. Effect of salivary protein profile, tannin concentration, and time stability. Journal of Agricultural and Food Chemistry 62, 10038–10045.

Pierini, R., Gee, J.M., Belshaw, N.J., Johnson, I.T., 2008. Flavonoids and intestinal cancers. The British Journal of Nutrition 99, ES53–ES59.

Plundrich, N.J., White, B.L., Dean, L.L., Davis, J.P., Foegeding, E.A., Lila, M.A., 2015. Stability and immunogenicity of hypoallergenic peanut protein-polyphenol complexes during in vitro pepsin digestion. Food & Function 6, 2145–2154.

Queipo-Ortuno, M.I., Boto-Ordonez, M., Murri, M., Gomez-Zumaquero, J.M., Clemente-Postigo, M., Estruch, R., Cardona Diaz, F., Andres-Lacueva, C., Tinahones, F.J., 2012. Influence of red wine polyphenols and ethanol on the gut microbiota ecology and biochemical biomarkers. The American Journal of Clinical Nutrition 95, 1323–1334.

Rechner, A.R., Kuhnle, G., Bremner, P., Hubbard, G.P., Moore, K.P., Rice-Evans, C.A., 2002. The metabolic fate of dietary polyphenols in humans. Free Radical Biology & Medicine 33, 220–235.

Renaud, S., de Lorgeril, M., 1992. Wine, alcohol, platelets, and the French paradox for coronary heart disease. The Lancet 339, 1523–1526.

Riedl, K.M., Hagerman, A.E., 2001. Tannin-protein complexes as radical scavengers and radical sinks. Journal of Agricultural and Food Chemistry 49, 4917–4923.

Riedel, H., Saw, N.M.M.T., Akumo, D.N., Kütük, O., Smetanska, I., 2012. In: Valdez, B. (Ed.), Wine as Food and Medicine, Scientific, Health and Social Aspects of the Food Industry.

Rios, L.Y., Gonthier, M.P., Remesy, C., Mila, I., Lapierre, C., Lazarus, S.A., Williamson, G., Scalbert, A., 2003. Chocolate intake increases urinary excretion of polyphenol-derived phenolic acids in healthy human subjects. The American Journal of Clinical Nutrition 77, 912–918.

Robles Alonso, V., Guarner, F., 2013. Linking the gut microbiota to human health. The British Journal of Nutrition 109 (Suppl. 2), S21–S26.

Rohn, S., Rawel, H.M., Kroll, J., 2004. Antioxidant activity of protein-bound quercetin. Journal of Agricultural and Food Chemistry 52, 4725–4729.

Romanov-Michailidis, F., Viton, F., Fumeaux, R., Lévèques, A., Actis-Goretta, L., Rein, M., Williamson, G., Barron, D., 2012. Epicatechin B-ring conjugates: first enantioselective synthesis and evidence for their occurrence in human biological fluids. Organic Letters 14, 3902–3905.

Ruf, J.-C., Berger, J.-L., Renaud, S., 1995. Platelet rebound effect of alcohol withdrawal and wine drinking in rats: Relation to tannins and lipid peroxidation. Arteriosclerosis, Thrombosis, and Vascular Biology 15, 140–144.

Sano, A., Yamakoshi, J., Tokutake, S., Tobe, K., Kubota, Y., Kikuchi, M., 2003. Procyanidin B1 is detected in human serum after intake of proanthocyanidin-rich grape seed extract. Biosciences, Biotechnology, and Biochemistry 67, 1140–1143.

Santa-Maria, I., Diaz-Ruiza, C., Ksiezak-Reding, H., Chen, A., Ho, L., Wang, J., Pasinetti, G.M., 2012. GSPE interferes with tau aggregation in vivo: implication for treating tauopathy. Neurobiology of Aging 33, 2072–2081.

Sbarra, V., Ristorcelli, E., Petit-Thévenin, J.L., Teissedre, P.-L., Lombardo, D., Vérine, A., 2005. In vitro polyphenol effects on activity, expression and secretion of pancreatic bile salt-dependent lipase. Biochimica et Biophysica Acta (BBA) - Molecular and Cell Biology of Lipids 1736, 67–76.

Sengupta, B., Sengupta, P.K., 2002. The interaction of quercetin with human serum albumin: a fluorescence spectroscopic study. Biochemical and Biophysical Research Communications 299, 400–403.

Serra, A., Macia, A., Romero, M.P., Valls, J., Blade, C., Arola, L., Motilva, M.J., 2010. Bioavailability of procyanidin dimers and trimers and matrix food effects in in vitro and in vivo models. The British Journal of Nutrition 103, 944–952.

Seyoum, A., Asres, K., El-Fiky, F.K., 2006. Structure–radical scavenging activity relationships of flavonoids. Phytochemistry 67, 2058–2070.

Shrotriya, S., Agarwal, R., Sclafani, R.A., 2015. A perspective on chemoprevention by resveratrol in head and neck squamous cell carcinoma. Advances in Experimental Medicine and Biology 815, 333–348.

Singh, A., Holvoet, S., Mercenier, A., 2011. Dietary polyphenols in the prevention and treatment of allergic diseases. Clinical and Experimental Allergy 41, 1346–1359.

Smid, S.D., Maag, J.L., Musgrave, I.F., 2012. Dietary polyphenol-derived protection against neurotoxic [small beta]-amyloid protein: from molecular to clinical. Food & Function 3, 1242–1250.

Soares, S., Brandao, E., Mateus, N., de Freitas, V., 2015. Interaction between red wine procyanidins and salivary proteins: effect of stomach digestion on the resulting complexes. RSC Advances 5, 12664–12670.

Soares, S., Mateus, N., de Freitas, V., 2007. Interaction of different polyphenols with bovine serum albumin (BSA) and human salivary α-amylase (HSA) by fluorescence quenching. Journal of Agricultural and Food Chemistry 55, 6726–6735.

Soares, S., Sousa, A., Mateus, N., de Freitas, V., 2011a. Effect of condensed tannins addition on the astringency of red wines. Chemical Senses 37, 191–198.

Soares, S., Vitorino, R., Osório, H., Fernandes, A., Venâncio, A., Mateus, N., Amado, F., de Freitas, V., 2011b. Reactivity of human salivary proteins families toward food polyphenols. Journal of Agricultural and Food Chemistry 59, 5535–5547.

Song, H., Jung, J.I., Cho, H.J., Her, S., Kwon, S.H., Yu, R., Kang, Y.H., Lee, K.W., Park, J.H.Y., 2015. Inhibition of tumor progression by oral piceatannol in mouse 4T1 mammary cancer is associated with decreased angiogenesis and macrophage infiltration. The Journal of Nutritional Biochemistry 26, 1368–1378.

Spencer, J.P., Schroeter, H., Kuhnle, G., Srai, S.K., Tyrrell, R.M., Hahn, U., Rice-Evans, C., 2001a. Epicatechin and its in vivo metabolite, 3'-O-methyl epicatechin, protect human fibroblasts from oxidative-stress-induced cell death involving caspase-3 activation. Biochem J 354, 493–500.

Spencer, J.P.E., Chaudry, F., Pannala, A.S., Srai, S.K., Debnam, E., Rice-Evans, C., 2000. Decomposition of cocoa procyanidins in the gastric milieu. Biochemical and Biophysical Research Communications 272, 236–241.

Spencer, J.P.E., Schroeter, H., Rechner, A.R., Rice-Evans, C., 2001b. Bioavailability of flavan-3-ols and procyanidins: gastrointestinal tract influences and their relevance to bioactive forms in vivo. Antioxidants and Redox Signaling 3, 1023–1039.

St Leger, A.S., Cochrane, A.L., Moore, F., 1979. Ischæmic heart-disease and wine. The Lancet 313, 1294.

Stoupi, S., Williamson, G., Drynan, J.W., Barron, D., Clifford, M.N., 2010. A comparison of the in vitro biotransformation of (−)-epicatechin and procyanidin B2 by human faecal microbiota. Molecular Nutrition & Food Research 54, 747–759.

Struck, M., Watkins, T., Tomeo, A., Halley, J., Bierenbaum, M., 1994. Effect of red and white wine on serum lipids, platelet aggregation, oxidation products and antioxidants: a preliminary report. Nutrition Research 14, 1811–1819.

Sugiyama, H., Akazome, Y., Shoji, T., Yamaguchi, A., Yasue, M., Kanda, T., Ohtake, Y., 2007. Oligomeric procyanidins in apple polyphenol are main active components for inhibition of pancreatic lipase and triglyceride absorption. Journal of Agricultural and Food Chemistry 55, 4604–4609.

Tadera, K., Minami, Y., Takamatsu, K., Matsuoka, T., 2006. Inhibition of α-glucosidase and α-amylase by flavonoids. Journal of Nutritional Science and Vitaminology 52, 149–153.

Tapiero, H., Tew, K.D., Ba, G.N., Mathe, G., 2002. Polyphenols: do they play a role in the prevention of human pathologies? Biomedicine & Pharmacotherapy 56, 200–207.

Testai, L., 2015. Flavonoids and mitochondrial pharmacology: a new paradigm for cardioprotection. Life Sciences 135, 68–76.

Tsai, Y.M., Chong, I.W., Hung, J.Y., Chang, W.A., Kuo, P.L., Tsai, M.J., Hsu, Y.L., 2015. Syringetin suppresses osteoclastogenesis mediated by osteoblasts in human lung adenocarcinoma. Oncology Reports 34, 617–626.

Tsuda, T., Horio, F., Osawa, T., 1999. Absorption and metabolism of cyanidin 3-O-β-D-glucoside in rats. FEBS Letters 449, 179–182.

Tzounis, X., Vulevic, J., Kuhnle, G.G., George, T., Leonczak, J., Gibson, G.R., Kwik-Uribe, C., Spencer, J.P., 2008. Flavanol monomer-induced changes to the human faecal microflora. The British Journal of Nutrition 99, 782–792.

Urpí-Sardà, M., Jáuregui, O., Lamuela-Raventós, R.M., Jaeger, W., Miksits, M., Covas, M.-I., Andres-Lacueva, C., 2005. Uptake of diet resveratrol into the human low-density lipoprotein. Identification and quantification of resveratrol metabolites by liquid chromatography coupled with tandem mass spectrometry. Analytical Chemistry 77, 3149–3155.

Urpi-Sarda, M., Monagas, M., Khan, N., Llorach, R., Lamuela-Raventos, R.M., Jauregui, O., Estruch, R., Izquierdo-Pulido, M., Andres-Lacueva, C., 2009. Targeted metabolic profiling of phenolics in urine and plasma after regular consumption of cocoa by liquid chromatography-tandem mass spectrometry. Journal of Chromatography A 1216, 7258–7267.

Vang, O., Ahmad, N., Baile, C.A., Baur, J.A., Brown, K., Csiszar, A., Das, D.K., Delmas, D., Gottfried, C., Lin, H.Y., Ma, Q.Y., Mukhopadhyay, P., Nalini, N., Pezzuto, J.M., Richard, T., Shukla, Y., Surh, Y.J., Szekeres, T., Szkudelski, T., Walle, T., Wu, J.M., 2011. What is new for an old molecule? Systematic review and recommendations on the use of resveratrol. PLoS One 6, e19881.

Vendrame, S., Guglielmetti, S., Riso, P., Arioli, S., Klimis-Zacas, D., Porrini, M., 2011. Six-week consumption of a wild blueberry powder drink increases bifidobacteria in the human gut. Journal of Agricultural and Food Chemistry 59, 12815–12820.

Virmani, A., Pinto, L., Binienda, Z., Ali, S., 2013. Food, nutrigenomics, and neurodegeneration - neuroprotection by what you eat!. Molecular Neurobiology 48, 353–362.

Vitaglione, P., Sforza, S., Galaverna, G., Ghidini, C., Caporaso, N., Vescovi, P.P., Fogliano, V., Marchelli, R., 2005. Bioavailability of trans-resveratrol from red wine in humans. Molecular Nutrition & Food Research 49, 495–504.

Walle, T., 2011. Bioavailability of resveratrol. Annals of the New York Academy of Sciences 1215, 9–15.

Walle, T., Browning, A.M., Steed, L.L., Reed, S.G., Walle, U.K., 2005. Flavonoid glucosides are hydrolyzed and thus activated in the oral cavity in humans. The Journal of Nutrition 135, 48–52.

Wang, J., Ho, L., Zhao, Z., Seror, I., Humala, N., Dickstein, D.L., Thiyagarajan, M., Percival, S.S., Talcott, S.T., Pasinetti, G.M., 2006. Moderate consumption of Cabernet Sauvignon attenuated beta-amyloid neuropathology in a mouse model of Alzheimer's disease. FASEB Journal 20, 2313–2320.

Wang, J., Santa-Maria, I., Ho, L., Ksiezak-Reding, H., Ono, K., Teplow, D.B., Pasinetti, G.M., 2010. Grape derived polyphenols attenuate tau neuropathology in a mouse model of Alzheimer's disease. Journal of Alzheimer's Disease 22, 653–661.

Wendeburg, L., de Oliveira, A.C.P., Bhatia, H.S., Candelario-Jalil, E., Fiebich, B.L., 2009. Resveratrol inhibits prostaglandin formation in IL-1β-stimulated SK-N-SH neuronal cells. Journal of Neuroinflammation 6, 26.

WHO, 2015. Cardiovascular Diseases (CVDs). Fact Sheet N°317.

Wiese, S., Gärtner, S., Rawel, H.M., Winterhalter, P., Kulling, S.E., 2009. Protein interactions with cyanidin-3-glucoside and its influence on α-amylase activity. Journal of the Science of Food and Agriculture 89, 33–40.

Williamson, G., Clifford, M.N., 2010. Colonic metabolites of berry polyphenols: the missing link to biological activity? The British Journal of Nutrition 104 (Suppl. 3), S48–S66.

Woodward, G.M., Needs, P.W., Kay, C.D., 2011. Anthocyanin-derived phenolic acids form glucuronides following simulated gastrointestinal digestion and microsomal glucuronidation. Molecular Nutrition & Food Research 55, 378–386.

Yasuda, A., Natsume, M., Sasaki, K., Baba, S., Nakamura, Y., Kanegae, M., Nagaoka, S., 2008. Cacao procyanidins reduce plasma cholesterol and increase fecal steroid excretion in rats fed a high-cholesterol diet. BioFactors 33, 211–223.

Yu, X.D., Yang, J.L., Zhang, W.L., Liu, D.X., 2015. Resveratrol inhibits oral squamous cell carcinoma through induction of apoptosis and G2/M phase cell cycle arrest. Tumor Biology 37 (3).

Zell, J.A., McEligot, A.J., Ziogas, A., Holcombe, R.F., Anton-Culver, H., 2007. Differential effects of wine consumption on colorectal cancer outcomes based on family history of the disease. Nutrition and Cancer 59 (1), 36–45.

Zuercher, A.W., Holvoet, S., Weiss, M., Mercenier, A., 2010. Polyphenol-enriched apple extract attenuates food allergy in mice. Clinical and Experimental Allergy 40, 942–950.

Section 4

Hazardous Compounds and Their Implications in Fermented Foods

Chapter 27

Biogenic Amines in Fermented Foods and Health Implications

L. Simon Sarkadi
Szent István University, Budapest, Hungary

27.1 CLASSIFICATION, BIOSYNTHESIS, AND METABOLISM OF BIOGENIC AMINES

Biogenic amines (BA) are generally classified on the basis of chemical structure, the number of amine groups, biosynthesis, or physiological functions.

According to their chemical structures there are three groups of BA: aliphatic amines, aromatic amines, and heterocylic amines. The main representatives of aliphatic biogenic amines are putrescine (Put; 1,4-diaminobutane), cadaverine (Cad; 1,5-diaminopentane), and agmatine (Agm; 1-(4-aminobutyl)guanidine); aromatic amines are tyramine (Tym; 4-(2-aminoethyl)phenol) and phenylethylamine (Phem; 2-phenylethylamine); and heterocyclic amines are histamine (Him; 2-(1H-imidazol-4-yl)ethanamine) and tryptamine (Trm; 2-(1H-indol-3-yl) ethanamine). The structures of the major BAs are shown in Fig. 27.1.

Depending on the number of amine groups, biogenic amines can be monoamines (Tym, Phem, Trm, Him), diamines (Put, Cad), or polyamines (Spm, Spd). On the basis of the biosynthetic pathway and physiological functions, amines are classified as natural polyamines (spermidine [Spd; N-(3-aminopropyl)-1,4-diaminobutane] and spermine [Spm; N,N'-bis(3-aminopropyl)-1,4-diaminobutane] and Put) or natural BA (Him) formed during de novo biosynthesis.

Polyamines include Spd, Spm, and Put. Put, a diamine classified in both groups, is widely formed by the decarboxylation of amino acid ornithine, but it is also an intermediate in Spd and Spm biosynthesis. Nevertheless, diamine Cad produced by enzymatic decarboxylation of lysine does not rank among physiological polyamines, similar to Agm, formed by enzymatic decarboxylation of arginine.

The biosynthesis of Put involves the decarboxylation of either ornithine by ornithine decarboxylase (ODC) or arginine by arginine decarboxylase (ADC) and subsequent conversion of the Agm into Put by agmatinase. Put is then converted into Spm and Spd by sequential transfer of aminopropyl groups donated

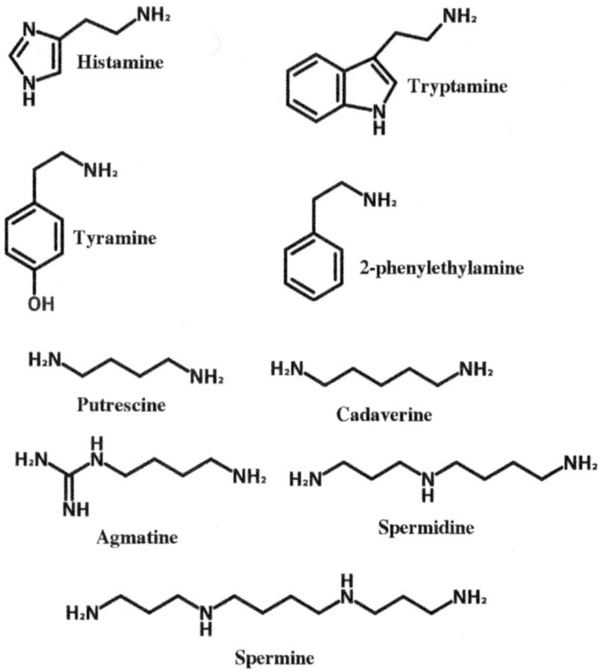

FIGURE 27.1 Structures of the major biogenic amines.

by decarboxylated S-adenosylmethionine (SAM), in reactions catalyzed by spermidine and spermine synthase, respectively. Decarboxylated S-adenosylmethionine (SAM) is produced from SAM by the enzyme S-adenosylmethionine decarboxylase (SAMDC). In the return direction of the interconversion pathway, Spm and Spd can be acetylated by spermidine/spermine N^1-acetyltransferase (SSAT) to produce compounds suitable for oxidation by polyamine oxidase (PAO), finally yielding Spd from Spm and Put from Spd (Medina et al., 2003). Him can be either natural (formed during de novo biosynthesis, stored in mast cells or basophils) or biogenic derived from histidine by decarboxylation. Tym is formed from tyrosine, Phem is derived from phenylalanine, and Trm from tryptophan.

Low amounts of BA do not constitute a risk for healthy consumers. They are easily metabolized in the gut to physiologically less active degradation products during the food intake process. The intestinal detoxifying system includes specific enzymes, such as monoamine oxidase (MAO), diamino oxidase (DAO), and polyamine oxidase (PAO). However, upon intake of high amounts of BA in foods, the detoxification system is unable to eliminate BA sufficiently or when it is genetically deficient. Moreover, the amine metabolism could be inhibited by simultaneous ingestion of alcoholic beverages, certain drugs, or monoamine oxidase inhibitors.

In the degradation of monoamines, but not for polyamines, there are methylation reactions using S-adenosylmethionine as the high-energy methyl donor. There are two ways of Him metabolism in the human body. The major process is when nitrogen in the imidazole cycle is methylated by histamine N-methyltransferase (HNMT) at the formation of N-methylhistamine, which is further oxidized by MAO to N-methylimidazolylacetic acid. This enzyme is very selective for Him detoxification and involves S-adenosylmethionine as donor of methyl group. The other possibility is the oxidation of Him by DAO to imidazolylacetic acid (Medina et al., 2003).

27.2 MICROBIAL PRODUCTION OF BIOGENIC AMINES IN FOODS

In foods, BA are mainly formed as a result of microbial decarboxylation of amino acids. There are specific key microbial enzymes with amino acid decarboxylase activity: ODC for Put, ADC for Agm, histidine decarboxylase (HDC) for Him, and aromatic amino acid decarboxylase (DDC) for Trm.

The formation of BA in food requires the availability of free amino acids, the presence of decarboxylase-positive microorganisms, and favorable conditions for bacterial growth, and decarboxylase activity. Free amino acids either occur as such in foods or may be liberated by proteolysis during processing or storage. Decarboxylase-producing microorganisms may be part of the associated microbiota of a particular food or may be introduced by contamination before, during, or after processing of the food. In the case of fermented foods and beverages, the applied starter cultures may also affect the production of BA (Simon Sarkadi, 2009).

Lactic acid bacteria (LAB) are the microorganisms mainly responsible for BA formation in fermented food products. Two different metabolic routes have been described in LAB for the biosynthesis of Put. The ODC pathway is a typical decarboxylation system consisting of an ODC and an ornithine/Put exchanger (Coton et al., 2010a; Marcobal et al., 2006b; Romano et al., 2012, 2014). In contrast, the agmatine deiminase (AgDI) pathway is a more complex system, comprising AgDI, a putrescine transcarbamylase, a carbamate kinase, and an Agm/Put exchanger (Griswold et al., 2004; Ladero et al., 2011b; Lucas et al., 2007). It has been shown that the prevalence of either pathway in the accumulation of Put depends on the type of food. For example, in cider and cheese, the AgDI pathway has a predominant role (Ladero et al., 2011a, 2012a), whereas in wine, Put is mainly produced through the ODC pathway (Nannelli et al., 2008).

The BA content of foods has been widely studied because of their adverse health implications. There are significant differences between the BA composition of foods from plant and animal origins. Plant-derived foods contain high amounts of Put, Spm, and Spd but significantly lower amount of Him than animal-derived foods. Generally, the plant-derived foods may be considered as low-risk products with regard to the presence of BA, while the food products undergoing microbial fermentation (wine, beer, meat, cheese) may contain relatively high amounts of BA (Simon Sarkadi, 2009).

Despite the fact that BA may cause several problems for susceptible consumers, there is a general absence of specific regulation on the BA content of food. The European Union (European Council Directive, 1991) established legislative limit values only for Him in fish, since Him is implicated in the most frequent foodborne intoxications. Some countries have regulated the maximum amounts of Him in different foods at a national level. Generally, upper limits of 100 mg Him/kg in food and 2 mg/L in beverages have been suggested. There are recommendations for Tym (100–800 mg/kg) and for Phem (30 mg/kg) in food (Brink et al., 1990).

27.3 FACTORS AFFECTING BIOGENIC AMINE CONTENT IN FERMENTED FOODS

Knowledge of BA in fermented foods is necessary to make an assessment of the health hazards arising from the consumption of these products and it can also provide information to improve food quality with respect to BA content. Several studies have monitored the BA formation and occurrence in food. The first general monograph on BA was published by Guggenheim in 1920 (Guggenheim, 1920). Information on recent developments of this topic is given in valuable reviews (Kalac and Krausova, 2005; Kalac, 2006, 2014; Alvarez and Moreno-Arribas, 2014).

Many factors contribute to the presence and accumulation of BA in fermented foods, such as the availability of free amino acids, pH, water activity, salt content, temperature, and the presence of microorganisms possessing amino acid decarboxylase activity, such as lactobacilli, enterococci, micrococci, and many strains of the genus Enterobacteriaceae. Knowledge concerning the origin and factors involved in BA production in fermented foods is well documented (Ancín-Azpilicueta et al., 2008; Linares et al., 2012; Spano et al., 2010). The most important BA occurring in fermented foods are Him, Tym, Put, Cad, Trm, Phem, Spm, Spd, and Agm.

27.3.1 Dairy Products

Cheese results from a lactic acid fermentation of milk. Proteolysis of casein during cheese ripening leads to an increase of free amino acids. Cheese represents an ideal environment for BA production because of the great availability of amino acids and the presence of bacteria. Several factors may contribute to BA formation in cheese, such as the type of raw milk, the use of starter cultures, and the conditions and time of the ripening process. Higher ripening temperature and pH and low salt concentration may contribute to the ability of the microorganisms to produce biogenic amines (Simon Sarkadi, 2009).

The microbial population of raw milk is one of the main factors that influence BA formation in cheese, even when thermal treatments are applied. Numerous bacteria may possess amino acid decarboxylase activity. The genera *Enterococcus*, *Lactobacillus*, *Leuconostoc*, and *Streptococcus* include some strains that are endowed with high decarboxylating potential. LAB are the main Him and Tym producers and Enterococci have been described as the most efficient Tym producers in fermented foods (Ladero et al., 2012b; Marcobal et al., 2012). The presence of efficient histaminogenic strains of *Streptococcus thermophilus* has also been reported (Calles-Enríquez et al., 2010; Rossi et al., 2011; Tabanelli et al., 2012). Helinck et al. (2013) demonstrated in model cheeses that *Debaryomyces hansenii*, *Proteus vulgaris*, *Psychrobacter* sp., and *Microbacterium foliorum* were able to produce BA.

The effect of milk quality (unpasteurized, pasteurized) and the type of milk (cow, ewe, goat) on BA content of cheese has also been extensively studied. Higher BA contents are usually detected in cheese made with raw milk than in those made with pasteurized milk (Novella Rodriguez et al., 2004). Enterococci are the predominant bacteria in ripened raw milk cheese (Foulquié Moreno et al., 2006). Cheese produced from ewes' or goat milk showed lower amounts of BA in comparison with cows' milk cheese. Other milk products, such as yogurt and kefir made from pasteurized milk have little or no detectable amounts of Tym (Guzel Seydim et al., 2003).

The content of BA in several kinds of cheese has been extensively studied and their amounts varied greatly even within the same variety. Him (38%), Tym (26%), and Put (9%) were the major amines in ripened Pecorino Abruzzese cheeses made from raw milk (batch A), whereas in cheeses prepared with starter cultures (batch B) Phem (28%), Tym (23%), and Put (15%) were the most abundant BA. The Him concentration in cheese B (76 mg/kg) was much lower than in cheese A (261 mg/kg), whereas Phem (305 mg/kg) and Put (163 mg/kg) were higher in the batch B cheese (Martuscelli et al., 2005).

Bunkova et al. (2013) have tested 40 different cheeses made in small-scale farms. Low Spm levels were detected in two-thirds of cheeses. Him was found only in four cheese samples containing up to 25 mg/kg. Twenty-four samples contained between 7.2 and 207 mg/kg of Tym. Six cheeses contained more than 100 mg/kg Tym. Twenty products contained between 7.0 and 149.0 mg/kg of Cad. Put was found in 17 cheeses at the levels of 12.2–230 mg/kg. Four cheeses contained over 100 mg/kg Put (Bunkova et al., 2013).

Data on BA content of some cheeses have been summarized in Table 27.1.

High levels of BA in cheese indicate that starter cultures should be carefully checked for their potential to form BA during cheese processing conditions. Regarding the safety of starters, the European Food Safety Authority (EFSA) has introduced a system for a premarket safety assessment of selected taxonomic groups of microorganisms leading to a "Qualified Presumption of Safety" (QPS) (the European equivalent of the Generally Recognized as Safe [GRAS] status) (EFSA, 2007). *Lactobacillus* associated with food, including

TABLE 27.1 Content of Biogenic Amines in Different Cheeses

	Biogenic Amines							
Cheese	Him (mg/kg)	Tym (mg/kg)	Trm (mg/kg)	Put (mg/kg)	Cad (mg/kg)	Spd (mg/kg)	Spm (mg/kg)	References
Castelmagno	645.81±30.15	1009.07±48.75	1048.72±51.47		310.44±14.21	0.42±0.02	449.56±20.54	Gosetti et al. (2007)
Raschera	452.36±19.36	153.94±7.56	389.91±19.55		118.89±5.12	10.59±0.47	352.65±16.23	
Toma Piemontese	587.63±29.01	282.32±14.16	255.46±12.44		1.25±0.06	6.58±0.23	193.85±9.14	
Dutch-type hard cheese (FD 30% fat)	10.7±5.9	15.7±0.4	1.8±0.8	6.0±1.0	0.8±0.1	0.3±0.0	0.1±0.0	Komprda et al. (2007)
Dutch-type hard cheese (CH 30% fat)	17.1±0.1	41.2±1.2	0.6±0.5	6.7±0.1	0.7±0.1		0.2±0.0	
Dutch-type hard cheese (45% fat)	1.8±0.4	299.8±4.8		60.8±2.2	2.0±0.1	0.3±0.1	0.2±0.1	
Appenzeller	51.9	375.5		8.2	8.3	6.2	16.4	Mayer et al. (2010)
Parmigiano	10.9	6.4		1.8	3.2		4.4	
Roquefort	9.9	4.6		18.3	8.9		18.1	
Emmental	23.5	52.2		38	98.3	16.8		
Cheddar	25.4			4.8		18.2	8.5	
Gouda		2.43				1.73	0.5	
Grana	249.0	18.0						
Edam	3.2							

FD, (Flora Danica Normal) microbial culture (Chr. Hansen, Germany); *CH*, (CH-N-11) microbial culture (Chr. Hansen, Germany).

Lactobacillus buchneri, *Lactobacillus brevis*, and *Lactobacillus hilgardii* have obtained QPS status, although some strains of these species have been described as BA producers (Lucas et al., 2005; Martín et al., 2005; Coton and Coton, 2009). This could raise the question of the addition of "absence of BA production and BA production associated genes" as qualification criteria in the QPS context (Spano et al., 2010).

27.3.2 Fermented Fish Products

Fermented fish sausages are popular in Oriental countries and also in some Western countries for their special characteristics. The fermentation process for fish-derived products may provide the conditions required for abundant production of BA.

The BA content in fermented fish products depends on many factors, such as raw materials, good handling practices, and storage conditions. Many studies on fish have reported that fresh fish contains low amounts of BA, but considerably increased amounts have been observed when raw materials are handled under poor hygienic conditions or stored inappropriately. In a study about the effect of storage temperature on BA formation in fish it was established that the elevated accumulation of Him and other BA occur at higher temperatures (Krizek et al., 2004). However, some other studies have also demonstrated that Him and other BA can be accumulated in fish stored at low temperatures (Hernandez-Herrero et al., 2002).

The microbial population of raw materials is also a relevant factor that influences the BA content of fermented fish products. The major Him-producing bacteria in fish are Gram-negative mesophilic enteric and marine bacteria (Bjornsdottir-Butler et al., 2010; Kim et al., 2003). Strains of *Morganella morganii*, *Enterobacter aerogenes*, *Raoultella planticola*, *Raoultella ornithinolytica*, and *Photobacterium damselae* can secrete ≥1000 ppm Him during optimal in vitro culture conditions. Strains of other species, including *Hafnia alvei*, *Citrobacter freundi*, *Vibrio alginolyticus*, and *Escherichia coli*, are weak Him producers (or non-Him producers), yielding concentrations <500 ppm under similar in vitro culture conditions (Kim et al., 2001; Takahashi et al., 2003).

The main BA found in fish sauce are Him, Put, Cad, and Tym. Trm and Phem are occasionally present at a low level, while Spm, Spd, and Agm are trace amines in this fermented product (Zaman et al., 2009). Production of BA was also reported in dried and fermented fish (Tsai et al., 2006), and vacuum-packaged and cold-smoked fermented fish products (Tome et al., 2008).

The European Union regulations have fixed a maximum of 100 mg Him/kg as the average value of nine samples of fresh or canned fish and 200 mg Him/kg for fermented fish or other enzymatically ripened products (European Commission, 2005). The US Food and Drug Administration (FDA) have set a limit of 50 mg/kg Him for most fish products (Food and Drug Administration, 2011). In contrast to Him, no tolerable levels for Put and Cad in foods have been elaborated yet.

27.3.3 Meat and Meat Products

Fermented meat products and, especially, dry fermented sausages are one of the most common sources of BA. During fermentation, maturation, and storage of dry fermented sausages, suitable environmental conditions take place favoring the activity of microorganisms bearing decarboxylase enzymes and, thus, the accumulation of BA. Besides the presence of microorganisms, BA accumulation is affected by a number of factors, such as the raw material (meat composition, size, and formulation of sausage), additives (salt, sugar, nitrites), fermentation time, and storage conditions. These factors strongly influence the microbial growth and interaction among microbial communities as well as acidification and proteolysis, determining the decarboxylase enzyme activity (Latorre-Moratalla et al., 2012).

The role of the microbiota in fermented meat products is fundamental in the final characteristics of these products. The leading bacterial group during the sausage-ripening process is LAB. Among lactobacilli, *Lactobacillus sakei*, *Lactobacillus curvatus*, and/or *Lactobacillus plantarum* generally constitute the predominant microbiota during traditional sausage ripening (Curiel et al., 2011; Rivas et al., 2008). The initial population of LAB is usually low in the raw material (3–4 log colony-forming units [CFU]/g), but it becomes dominant during the fermentation step (8 log CFU/g) in traditional sausages (Leroy et al., 2006). LAB are the main producers of BA. Starter cultures are frequently used in sausage manufacture in order to shorten the ripening time, ensure color development, enhance flavor, and improve product safety.

Content of BA in fermented sausages has shown great variation and the quality of raw materials influences the composition and the concentration of BA produced during the ripening of sausages (Bover Cid et al., 2006b). There are other factors, such as sausage diameter, pH, water activity, and NaCl that may influence the formation of BA in dry fermented sausage. A larger diameter might lead to a more favorable environment (lower NaCl concentration and drying level, higher water activity) for the growth of microorganisms and for the development of BA (Komprda et al., 2004). Similarly, Latorre-Moratalla et al. (2012) found that control of fermentation temperatures and relative humidity together with a small diameter can contribute in preventing the formation of high levels of BA in fermented sausages.

Several studies have showed that Tym is the most abundant BA found in dry sausage, followed by Put and Cad (Suzzi and Gardini, 2003; Latorre-Moratalla et al., 2008; De Mey et al., 2014). Tym is mainly related to the activity of fermentative LAB while Put and Cad are usually the result of the action of nonfermentative strains. Drosinos et al. (2007) and Pircher et al. (2006) found only tyrosine decarboxylase activity in the lactobacilli species isolated from traditionally fermented sausages.

The contents of BA were higher in the dry-ripened sausages than in soft sausages; however, during storage, in spite of the low storage temperature, traditional soft sausages showed BA levels comparable to those found in dry-ripened sausages (González-Tenorio et al., 2013).

The control of fermentation by introducing competitive LAB starter strains is an important method suggested to retard the formation of BA by preventing the growth of amine-producing bacteria in meat products, which leads to health-related benefits (Ammor and Mayo, 2007).

One of the novel approaches to reduce BA content in fermented meat products is the addition of wine to fermented sausages. The positive effects of wine on fermented sausage aroma profiles were confirmed (Gardini et al., 2013). Furthermore, the addition of wine showed reduced concentrations of Put in fermented sausages and did not negatively affect the ripening time (Coloretti et al., 2014). Current tendencies in the production of reduced- and low-fat dry fermented sausages with partial replacement of pork backfat by konjac gel modifies the BA profile without affecting relevant microbial development. Fat reduction linked to the presence of konjac gel favors production of certain BA (Tym, Put, and Spm) and reduces production of Agm in dry fermented sausage. The design of healthier meat products, reducing fat content and promoting inclusion of konjac gel, is a promising avenue of research, especially as the safety of these products in relation to the presence of BA seems to be guaranteed (Ruiz-Capillas et al., 2012).

27.3.4 Alcoholic Beverages

Alcoholic beverages constitute another category of fermented food products that sometimes bear substantial quantities of BA.

27.3.4.1 Wine

Wine is known to contain many biologically active compounds. Their amounts and compositions depend on the type of grapes and their degree of ripeness, climate, and soil of the viticulture area, as well as vinification techniques applied. Amino acids represent the main substrate for BA production in wine. The amines are formed primarily during and after the spontaneous malolactic fermentation process by decarboxylation of the precursor free amino acids. The presence of BA in wines is well documented in the literature (Heberger et al., 2003; Landete et al., 2005b; Bover Cid et al., 2006a). Surveys made on wines showed that winemaking technology had greater effect on BA formation in wines than geographical origin, grape variety, and year of vintage (Heberger et al., 2003; Landete et al., 2005b). Predominant BA in wine are Him, Tym, Put, and Agm (Marcobal et al., 2006a; Ferreira and Pinho, 2006; Ancín-Azpilicueta et al., 2008; Smit et al., 2008).

It is known that *Pediococcus, Lactobacillus, Leuconostoc*, and *Oenococcus* spp. are implicated in BA production in wine (Landete et al., 2005a). Some LAB strains are responsible Him, Tym, and Phem formation during winemaking.

Put, Agm and Spd originate mainly from the grape juices (Landete et al., 2005b). Different strains of *Lactobacillus hilgardii, Lactobacillus buchneri, Lactobacillus brevis,* and *Lactobacillus mali* produce a variety of BA in wine (Moreno-Arribas et al., 2003; Martín et al., 2005; Constantini et al., 2006; Landete et al., 2007a). *Leuconostoc mesenteroides* has a high potential to produce Tym or Him in this fermented beverage (Moreno-Arribas et al., 2003; Landete et al., 2007a).

Him has been extensively studied in wine for many years. In general, white wines exhibit very low Him concentrations, while red wines have often provoked physiological distress owed to the higher amount of Him. These differences may be due to the different fermentation processes. Some studies have reported relatively low levels of Him in wine. In Chinese red wines, the Him content varied between 0 and 10.51 mg/L, with 86% of wines containing between 0 and 5.0 mg/L (Zhijun et al., 2007). The average Him level found in Spanish red wines was 2.97 mg/L, with a maximum of 15.7 mg/L (Izquierdo Cañas et al., 2009). A mean of 8.5 mg/L and a maximum concentration of up to 27 mg/L of Him were recently found in Austrian wines (Konakovsky et al., 2011).

Recommendations on acceptable amounts of Him in wine established in some European countries affect the import and export of wines. The recommended upper limit for Him in wine has been reported to be 10 mg/L in Austria, 5–6 mg/L in Belgium, 8 mg/L in France, 2 mg/L in Germany, 10 mg/L in Hungary, 3.5 mg/L in the Netherlands, and 10 mg/L in Switzerland (Simon Sarkadi, 2009).

27.3.4.2 Beer

Beer is defined as an alcoholic beverage from starch-containing raw materials serving as a source for maltose and glucose, which are fermented by brewers' yeast.

Beers are classified into two groups, top- and bottom-fermented, based on whether yeast floats or sinks by the end of fermentation. Besides *Saccharomyces cerevisiae* (top fermenting) and *Saccharomyces carlsbergensis* (bottom fermenting), various wild yeasts together with LAB are involved in the brewing process of special local beers.

The BA content of beer is influenced by the barley variety used in the brewing process, malting technology, wort processing, and the conditions during fermentation. The presence of higher amounts of Him and Tym in beers has been associated with microbial contamination during brewing. In contrast, Put, Agm, Spd, and Spm are considered natural beer constituents, primarily originated from malt (Kalac and Krizek, 2003; Romero et al., 2003), while Tym and Phem could be present in hop (Slomkowska and Ambroziak, 2002).

The total average content of BA measured in Polish beers ($n=27$) was 16.15 ± 2.89 mg/L. Spm (mean content 8.43 ± 3.61 mg/L), Spd (3.37 ± 2.07 mg/L), and Put (1.75 ± 0.79 mg/L) were the major amines detected in these beers. Other BA were found at concentrations less than 2.0 mg/L (Slomkowska and

Ambroziak, 2002). Put and Agm were the most predominant amines in Belgian beer samples ($n=297$), produced with different brewing processes. In spontaneously fermented beers, Put concentration (14.0±10.4 mg/L) was significantly higher than that of other fermentation types. These beer samples also exhibited the highest levels of Cad (10 mg/L mean value), Him (11.9±8.61 mg/L), and Tym (28.7±17.3 mg/L). In top-fermented beers two other amines, Phem (0.85–1.45 mg/L) and Trm (2.96–2.03 mg/L), were also detected (Loret et al., 2005). Put (2.09–12.78 mg/L) and Tym (0.39–5.92 mg/L) were the main amines in Portugese beers ($n=22$) followed by Cad (0.19–1.38 mg/L). Him was present in 10 of the 22 samples, with levels ranging from 0.02 to 0.34 mg/L (Almeida et al., 2012).

27.3.4.3 Cider

Cider is an alcoholic beverage made from apple juice that is consumed after a minimum 3-month fermentation period. There are two main types of cider, sparkling and natural, depending on the addition of sugars and CO_2 (Picinelli et al., 2000).

During the elaboration of natural cider, two types of spontaneous fermentation occur, alcoholic (AF) and malolactic (MLF) fermentations. They are driven by the indigenous yeast and LAB microbiota (for the AF and the MLF, respectively) since there is no inoculation with starters. *Lactobacillus* and *Oenococcus* are the most abundant LAB genera in cider (Coton et al., 2010b). Some strains from these genera have been identified as BA producers in cider, ie, *Lactobacillus hilgardii, Lactobacillus collinoides, Lactobacillus dioliovorans*, and *Oenococcus oeni* as Him producers; *L. dioliovorans* and *O. oeni* as Tym producers (Garai et al., 2007). More recently, strains isolated from French ciders, i,e *L. collinoides, Lactobacillus brevis, Lactobacillus mali, Leuconostoc mesenteroides,* and *O. oeni*, were identified as potential Put producers (via the Agm deiminase pathway) and *L. brevis* and *Sporolactobacillus* sp. strains were identified as potential Tym producers (Coton et al., 2010b).

A different profile and BA concentration were observed in 43 commercial Spanish and French ciders depending on cider origin (Ladero et al., 2011a). The average total BA detected in Spanish ciders was 19.21 mg/L with a range from 0 to 67 mg/L, while in French ciders was 6.41 mg/L and ranged from 1 to 14 mg/L. Overall, Cad and Put were detected at the highest levels (34 mg/L) with a mean concentration and standard deviation of 2.74±8.11 and 3.49±8.89 mg/L, respectively, followed by Him (maximum 16 mg/L, mean concentration and standard deviation of 1.28±3.07 mg/L) and Tym (maximum 14 mg/L, mean concentration and standard deviation of 3.30±4.63 mg/L).

A good correlation between the BA content and the presence of BA-producing microorganisms has been observed. The agmatine deiminase pathway seems to be the main route for Put accumulation in cider. The agmatine deiminase pathway had been previously identified in a *Lactobacillus brevis* strain isolated from wine (Lucas et al., 2007). Ladero et al. (2011a) found that *Lactobacillus collinoides, Lactobacillus mali, Leuconostoc mesenteroides,* and *Oenococcus oeni* strains isolated from cider also presented this metabolic pathway. Moreover they reported for the first time new potential Him- and Put-producing *Lactobacillus paracollinoides* strains.

27.3.5 Fermented Plant Products

The following vegetables are commonly fermented at household and industrial scales: cabbage, cucumber, green pepper, red beet, cauliflower, green tomato, and onion. The fermentation process can be carried out using either spontaneous fermentation (which relies on LAB occurring naturally on vegetables) or controlled fermentation (using starter culture strains of *Lactobacillus* species).

27.3.5.1 Sauerkraut

Sauerkraut has been very popular in many European countries due to its sensorial properties and favorable nutritional value. The fermentation process can be carried out spontaneously or by adding starter cultures (controlled fermentation). Among microorganisms contributing to sauerkraut production, *Leuconostoc mesenteroides*, *Lactobacillus plantarum*, *Lactobacillus brevis*, *Pediococcus*, and *Enterococccus* are of special importance.

The BA content of sauerkraut is highly influenced by cabbage variety, fermentation conditions (temperature, pH value change, oxygen access, or sodium chloride content), microbial starters used for fermentation and bacterial contamination. The main amines found in sauerkraut are Put, Him, Tym, and Cad, while Spm and Spd occurred only in small amounts (Kalac et al., 2000a, 2000b; Spicka et al., 2002).

Kalac et al. (1999) tested more than 120 sauerkraut products from Czech and Austrian manufacturers and the concentrations of BA varied greatly (Put 2.8–529.0 mg/kg, Cad 0–293.0 mg/kg, Tym 0–37.5 mg/kg, Spd 0–47.0 mg/kg, Him 0–229.0 mg/kg). In spontaneously fermented sauerkraut, Tym (85–578 mg/kg) was present at the highest levels after 12 months of storage, followed by Put (10–233 mg/kg) and Cad (2–31 mg/kg). Spd concentrations were below 14 mg/kg (Kalac et al., 2000a). A high level of Put (265–446 mg/kg), Tym (85–212 mg/kg), and Cad (60–122 mg/kg) in control sauerkraut variants was significantly suppressed by *Lactobacillus plantarum* and Microsil (Kalac et al., 2000b). The accumulation of BA in spontaneously fermented Chinese sauerkraut exhibited Put (24–45 mg/kg), Cad (10–35 mg/kg), and Tym (30–38 mg/kg) concentrations that increased with the fermentation time. However, in sauerkraut inoculated with *L. plantarum* and *Zygosaccharomyces rouxii*, BA contents decreased significantly (Wu et al., 2014).

27.3.5.2 Soybean Products

Soybean has been one of the main plant protein sources in Eastern countries since ancient times and is used worldwide in a variety of traditional fermented products, such as miso, natto (Japanese traditional fermented soybean pastes), Doenjang, Cheonggukjang, Gochujang (Korean traditional fermented soybean pastes), Doubanjiang, Dajiang (Chinese traditional fermented soybean pastes), tempeh (Indonesian traditional fermented soybean cake), stinky tofu, and sufu (traditional cheese-like fermented soybean Chinese food product) and soy sauce (Mah, 2015).

Biogenic amines have been reported to occur in many fermented soybean products, such as tempeh (Nout et al., 1993), miso (Kung et al., 2007b; Shukla et al., 2010), natto (Tsai et al., 2007a), douchi (Tsai et al., 2007b), soy sauce (Lu et al., 2009), and sufu (Kung et al., 2007a).

Cho et al. (2006) reported the presence of Him and Tym in Doenjang at a level of 952 and 1430 mg/kg. Tsai et al. (2007a) tested BA levels in seven soybean and 11 black bean douchi, among which four soybean douchi products had Him levels greater than 5 mg/100 g, and four black bean douchi samples contained Him between 56 and 81 mg/100 g. Fermented soybean products analyzed by Toro-Funes et al. (2015) showed high BA contents. The mean Tym content in tempeh was 3.6 ± 6.1 mg/kg, in tamari 63 ± 20 mg/kg, in soy pasta 52 ± 89 mg/kg, in sufu 1190 ± 458 mg/kg, and the mean Him content found in tempeh was 3.6 ± 6.1 mg/kg, in miso 3.4 ± 1.7 mg/kg, in soy pasta 17 ± 30 mg/kg, and in sufu 448 ± 303 mg/kg.

Guan et al. (2013) analyzed different types of sufu samples. The mean total BA content in gray sufu (570.5 ± 386.4 mg/kg, wet weight basis) was higher than that in white (153.6 ± 160.2 mg/kg) and red (10.9 ± 11.6 mg/kg) sufu. Put and Cad were the most commonly found BA in all sufu samples. The ranges for each of the detected BA (wet weight basis) in sufu samples were as follows: for Put, 0.5–316.9 mg/kg; for Cad, 0.6–85.8 mg/kg; for Spd, nd (not detected)–4.0 mg/kg; for Spm, nd–0.9 mg/kg; for Trm, nd–104.1 mg/kg; for Phem, nd–36.3 mg/kg; for Him, nd–196.9 mg/kg; and for Tym, nd–446.6 mg/kg. The high amount of BA could be explained by the fact that sufu is produced under an open-type fermentation environment and stored at ambient temperatures.

Most soybean foods are fermented or contaminated by *Bacillus* spp., particularly *Bacillus subtilis*, and some strains have strong decarboxylating amino acid capability and, thereby are potential BA producers (Bai et al., 2013; Kim et al., 2012). Different species isolated from fermented soybean products of genera of *Clostridium, Bacillus, Enterococci, Enterobacter,* and *Pseudomonas* have been reported as Tym and Him producers. In miso, tyrosine decarboxylase bacteria have been identified in *Enterococcus faecium* and *Lactobacillus bulgaricus,* and histidine decarboxylase has been associated with *Lactobacillus* species and *Lactobacillus sanfrancisco* (Ibe et al., 1992). Tsai et al. (2007a) identified some Him-producing bacteria belonging to *Lactobacillus* species in natto products. Moon et al. (2010) reported that a *Clostridium* strain, isolated from traditional soybean pastes, was a potent Him producer among the tested cultures.

27.3.6 Summary on Occurrence Data on the Major Biogenic Amines in Food and Beverages

The EFSA published in June 2010 a public call for data on the presence of BA in food and beverages. Member states, research institutions, academia, and other stakeholders were specifically invited to submit data. Table 27.2 summarizes the main data on the presence of BA in food and beverages collected by EFSA.

The food categories showing the highest mean values of Him were: dried anchovies (348 mg/kg), fish sauce (196–197 mg/kg), fermented vegetables (39.4–42.6 mg/kg), cheese (20.9–62 mg/kg), other fish and fish products (26.8–31.2 mg/kg), and fermented sausages (23.0–23.6 mg/kg); for Tym they were fermented sausages (136 mg/kg), fish sauce (105–107 mg/kg), cheese (68.5–104 mg/kg), fermented fish (47.2–47.9 mg/kg), and fermented vegetables (45–47.4 mg/kg); for Put they were fermented vegetables (264 mg/kg), fish sauce (98.1–99.3 mg/kg), fermented sausages (84.2–84.6 mg/kg), cheese (25.4–65 mg/kg), and fermented fish (13.4–17 mg/kg); and for Cad they were fish sauce (180–182 mg/kg), cheese (72–109 mg/kg), fermented sausages (37.4–38 mg/kg), fermented vegetables (26–35.4 mg/kg), and fermented fish (14–17.3 mg/kg) (EFSA, 2011).

27.4 BIOGENIC AMINES AND HUMAN HEALTH

The BA content of some foods has been widely studied due to their adverse health effects and potential toxicity. Low levels of BA in food do not constitute a risk for healthy consumers. However, large amounts are ingested and/or the natural metabolism of amines is impaired or inhibited, acute toxic symptoms may occur in sensitive individuals.

High amounts of exogenous BA, especially Him and Tym, in the human diet may contribute to a wide variety of toxic effects. These amines are categorized as psychoactive or vasoactive. Psychoactive amines act on the neural transmission in the central nervous system, while vasoactive amines act in the vascular system. The severity of the intoxication is influenced by many factors, such as the presence of other amines (Put and Cad) in the diet and consumption of alcohol or use of drugs inhibiting amine oxidase activity as well as diseases of the gastrointestinal system.

Him is one of the BA with the highest biological activity. Him poisoning symptoms include headache, nausea, vomiting, diarrhea, itching, oral burning sensation, red rash, and hypotension. Symptoms can be reduced by a Him-free diet or be eliminated by antihistamines. However, because of the multifaceted nature of the symptoms, the existence of Him intolerance has been underestimated or its symptoms are misinterpreted. Clinical symptoms and their provocation by certain foods and beverages appear similar in different diseases, such as food allergy and intolerance of sulfites, or other BA (eg, Tym) (Maintz and Novak, 2007).

A relationship between Tym content of foods and illnesses after ingestion has also been established. These illnesses include headache, migraine, neurological disorders, nausea, vomiting, respiratory disorders, and hypertension (Ladero et al., 2010). Intolerance of Tym that has vasoconstrictive properties

TABLE 27.2 Mean Data on the Presence of Biogenic Amines in Food and Beverages

Food Class	Subcategory	n	Biogenic Amines			
			Him (mg/kg)	Tym (mg/kg)	Put (mg/kg)	Cad (mg/kg)
Alcoholic beverages	Beer	188	1.4	6.1	3.3–3.5	1.3–1.5
	Fortified and liqueur wines	28	1.1	6	1.4	0.1
	Wine, white, sparkling	45	1	4.9	5.2	0.1
Condiments	Fish sauce	71	198–199	105–107	98.1–99	180–182
	Other savoury sauces	27	0.5–10.1	1.5–10	6–13.6	3–12.7
Fish products	Fermented fish products	68	7.7–11.4	45.5–47	12.2–15	14.4–17
Meat products	Fermented sausages	369	23	136	84	37.4–38
	Other ripened meat products	92	6–6.2	44–44.2	32.8	17
	Other meat products	75	3.9–4.4	16	17	6.7–6.8
Dairy products	Cheese	2136	20.6–61	59–98	25.4–65	72.2–109
	Fresh cheese	98	3.2–38	12.8–48	5.5–41	10.7–45
	Hard cheese	1062	25–65	67.1–103	26.6–65	47.8–83
	Washed rind cheese	676	8.5–54	31.6–76	32.3–72	147–186
	Blue cheese	296	21.3–63	63.2–100	20.9–62	83.1–120
	Acid curd cheese	4	51.3–55	335	449	628
	Yogurt	7	0.5	1.9	0.7	3.2
Vegetables	Fermented vegetables	9	39.4–42	45–47.4	264	26–35.4
	Other vegetables	14	5.4	1.8	37.2	17

n, number of samples.
Data from European Food Safety Authority, 2011. Scientific opinion on risk based control of biogenic amine formation in fermented foods. EFSA Journal 9(10), 1–93, 2393.

that lead to hypertensive crisis and headache has been known mostly in patients taking monoamine oxidase (MAO)-inhibiting drugs.

Tym and Phem have been suggested as the initiators of hypertensive crisis and dietary-induced migraine in certain patients. Aged cheeses are the most serious and frequently reported. The phenomenon is the so-called cheese reaction caused by high amount of Tym in cheese, as has been revised by McCabe Sellers et al. (2006).

Polyamines (PA), such as Spd, and Spm, although they do not exert direct toxic effect, can potentiate the toxic effects of Tym and Him. PA may increase the intestinal permeability to food allergens and facilitate the induction of food allergy (Sugita et al., 2007).

The knowledge of the main roles of PA in health, disease, and aging was reviewed by Larque et al. (2007) and more recently Kalac (2014) set up an excellent review on the role of PAs in human health.

Determination of the exact toxicity threshold of BA in individuals is extremely difficult. The toxic dose is strongly dependent on the efficiency of the detoxification mechanisms of different individuals. Leuschner et al. (2013) reported that, according the Rapid Alert System for Food and Feed (RASFF) database, the current Him situation continues to be of public health importance considering the number of human cases and concentrations reported. In the period 2002–2010, nearly 300 notifications were related to fish and products thereof. Seventeen were related to soups, broths, and sauces. Him concentrations in soya sauce were between 430 and 750 mg/kg and in fish sauce between 268 and 413 mg/kg. One notification was concerned with grated cheese (850 mg/kg Him). A high content of BA in organic salami was notified for information with one affected person in 2010.

There is a lack of scientific data for a complete BA risk assessment. Recently, the European Food Safety Authority (2011) published a report on risk assessment based on the monitoring of BA formation in fermented foods. The report states that Him and Tym are the most toxic amines and encourages further research to estimate safe levels of BA in foods.

However, national limits can be established, based on the principles of risk assessment. This may be achieved by relating the amounts of orally administered amines that do not cause health effects to amounts of food consumed, as our group has already demonstrated for Him, Tym, and Phem in fermented sausages, fish, and cheese (Rauscher-Gabernig et al., 2009; Paulsen et al., 2012).

27.5 REDUCTION OF BIOGENIC AMINES IN FERMENTED FOOD

Numerous efforts have been made in food science and in the food industry to reduce and to prevent formation of BA in food. There are some approaches to reduce BA in foods: handling and processing under sanitary conditions, utilizing some amine-negative starter cultures, adding some probiotic bacterial strains alone or in combination with the starter cultures, high-pressure processing, or low-dose gamma irradiation (Simon Sarkadi, 2009).

Using high-quality fresh meat and good manufacturing practices greatly reduces the risk of BA formation in processed meat products (Bover Cid et al., 2006b; Bauer, 2006). Several studies have reduced BA content in foods by using different bacteria. The presence of a selected starter culture lacking decarboxylase activity is one of the ways to reduce the development of BA in fermented foods (Fernández et al., 2007; Landete et al., 2007b; Casquete et al., 2011b). The use of starter cultures nonproducing BA has been effective in inhibiting BA formation in fermented meat sausages (Gençcelep et al., 2007; Lu et al., 2010). Samples of different cheeses were screened for the presence of BA-degrading LAB and 17 isolate strains identified by 16S rRNA sequencing as *L. casei* were found able to degrade Tym and Him (Herrero-Fresno et al., 2012).

The use of selected pure or mixed starter cultures is reported to reduce BA in fish products (Hu et al., 2007; Zhong-Yi et al., 2010). Mah and Hwang (2009) successfully reduced BA in a salted and fermented anchovy (Myeolchijeot) using starter cultures of *Staphylococcus xylosus* during ripening. Application of *Bacillus amyloliquefaciens* and *Staphylococcus carnosus* was able to reduce Him, and *Staphylococcus intermedius* and *Bacillus subtilis* were able to degrade Put and Cad in Malaysian fish sauce (Zaman et al., 2010). It was also found that *B. amyloliquefaciens* and *St. carnosus* isolates were able to reduce the accumulation of Him in laboratory-scale fish sauce fermentation (Zaman et al., 2011). Nie et al. (2014) demonstrated that a mixed starter culture (*Lactobacillus plantarum* ZY40 plus *Saccharomyces cerevisiae* JM19) had very great effects on BA contents in silver carp sausages. The addition of the mixed starter culture inhibited the growth of undesirable bacteria, such as *Pseudomonas* and *Enterobacteriaceae* and, consequently, reduced the accumulation of Put and Cad, whereas Tym accumulation was enhanced due to the slow pH decline.

Mixed starter cultures (*Lactobacillus sakei*, *Staphylococcus carnosus*, and *St. xylosus*) greatly reduced (about 90%) the presence of Put, Cad, and Tym in fermented Spanish sausages (Bover Cid et al., 2000). Similar results were found by Xie et al. (2015) using *St. xylosus* and *Lactobacillus plantarum*. These mixed starter cultures effectively reduced Trm, Phem, Put, Cad, Him, and Tym by nearly 100%, 100%, 86%, 63%, 82%, and 43%, respectively, in fermented Chinese sausages. Mixed starter culture (*L. sakei* and *St. equorum*) was capable of reducing tyraminogenesis by nearly 50% in Spanish fuet (Latorre-Moratalla et al., 2010). Likewise, significant reductions of Cad, Put, Him, and Tym contents were observed in Nham using *L. plantarum* (Tosukhowong et al., 2011). Mangia et al. (2013) suggested a suitable combination of starter cultures (*Lactobacillus curvatus* and *Staphylococcus xylosus*) in the manufacture of salsiccia sarda sausage that are able to reduce the BA contents of ripened sausages.

Different combinations of *Pediococcus acidilactici* and *St. vitulus* autochthonous selected strains showing desirable technological properties of proteolytic and lipolytic activities and low BA production have been used in fermented products with different ripening processes (Casquete et al., 2011a).

There are some works reporting successful degradation of BA in wine. Some LAB strains isolated from fermented foods have been proven to degrade BA through the production of amine oxidase enzymes (García-Ruiz et al., 2011). Callejón et al. (2014) reported that LAB strains isolated from the winemaking process were able to remove BA in wine and *Lactobacillus plantarum* J16 and *Pediococcus acidilactici* CECT 5930 strains were identified as enzymes degrading amine producers.

Utilization of starter cultures enables sauerkraut production with a standardized quality. Wu et al. (2014) demonstrated that a combination of *Lactobacillus plantarum* and *Zygosaccharomyces rouxii* resulted in a reduction of BA in Chinese sauerkraut.

Several papers deal with the effects of high hydrostatic pressure (HHP) on the changes of BA in foods. HHP processing is one of the most encouraging alternatives to traditional thermal treatment for food preservation. Its effectiveness in the deactivation of pathogenic and spoilage microorganisms is well documented (Wuytack et al., 2002; Kheadr et al., 2002; Lanciotti et al., 2007). Latorre-Moratalla et al. (2007) reported that pressure treatment of 200 MPa for 10 min strongly inhibited Put and Cad production in meat batter, but did not affect Spd, Spm, and Tym content. Similarly, HHP applied to chorizo (Ruiz-Capillas et al., 2007) and vacuum-packaged cooked sliced ham (Ruiz-Capillas and Jimenez-Colmenero, 2010) had only a limited effect on Spd and Spm changes during cold storage, while Put content increased with the prolonged storage. The HHP treatment at 500 MPa for 10 min showed an inhibitory effect on Put and Cad formation in most of cases, while it showed activation of Tym and Spm formation in dry sausage during storage at 8°C up to day 28, compared with the control (Simon Sarkadi et al., 2012).

27.6 CONCLUSIONS

BA formation in fermented food is a complex process affected by multiple factors and their interactions are variable and product-specific. Estimating safe levels of the total amounts of ingested BA is the key issue to understand health effects to consumers. Currently information on existence, distribution, and concentration of BA in fermented foods is not included in food composition databases, however, it may be useful for the food industry and health professionals, as well as consumers.

REFERENCES

Almeida, C., Fernandes, J.O., Cunha, S.C., 2012. A novel dispersive liquid-liquid microextraction (DLLME) gas chromatography-mass spectrometry (GC-MS) method for the determination of eighteen biogenic amines in beer. Food Control 25, 380–388.

Alvarez, M., Moreno-Arribas, M., 2014. The problem of biogenic amines in fermented foods and the use of potential biogenic amine-degrading microorganisms as a solution. Trends in Food Science and Technology 39, 146–155.

Ammor, M.S., Mayo, B., 2007. Selection criteria for lactic acid bacteria to be used as functional starter cultures in dry sausage production: an update. Meat Science 76, 138–146.

Ancín-Azpilicueta, C., González-Marco, A., Jiménez-Moreno, N., 2008. Current knowledge about the presence of amines in wine. Critical Reviews in Food Science and Nutrition 48, 257–275.

Bai, X., Byun, B.Y., Mah, J.H., 2013. Formation and destruction of biogenic amines in Chunjang (a black soybean paste) and Jajang (a black soybean sauce). Food Chemistry 141, 1026–1031.

Bauer, F., 2006. Assessment of process quality by examination of the final product. 1. Assessment of the raw material. Fleischwirtschaft 867, 106–107.

Björnsdóttir-Butler, K., Bolton, G.E., Jaykus, L.A., McClellan-Green, P.D., Green, D.P., 2010. Development of molecular-based methods for determination of high histamine producing bacteria in fish. International Journal of Food Microbiology 139, 161–167.

Bover-Cid, S., Izquierdo-Pulido, M., Vidal-Carou, M.C., 2000. Mixed starter cultures to control biogenic amine production in dry fermented sausages. Journal of Food Protection 63, 1556–1562.

Bover-Cid, S., Izquierdo-Pulido, M., Mariné-Font, A., Vidal-Carou, M.C., 2006a. Biogenic mono-, di- and polyamine contents in Spanish wines and influence of a limited irrigation. Food Chemistry 96, 43–47.

Bover-Cid, S., Miguelez-Arrizado, M.J., Latorre-Moratalla, L.L., Vidal-Carou, M.C., 2006b. Freezing of meat raw materials affects tyramine and diamine accumulation in spontaneously fermented sausages. Meat Science 72, 62–68.

Bunkova, L., Adamcova, G., Hudcova, K., Velichova, H., Pachlova, V., Lorencova, E., Bunka, F., 2013. Monitoring of biogenic amines in cheeses manufactured at small-scale farms and in fermented dairy products in the Czech Republic. Food Chemistry 141, 548–551.

Callejón, S., Sendra, R., Ferrer, S., Pardo, I., 2014. Identification of a novel enzymatic activity from lactic acid bacteria able to degrade biogenic amines in wine. Applied Microbiology and Biotechnology 98, 185–198.

Calles-Enríquez, M., Eriksen, B.H., Andersen, P.S., Rattray, F.P., Johansen, A.H., Fernández, M., Ladero, V., Alvarez, M.A., 2010. Sequencing and transcriptional analysis of the streptococcus thermophilus histamine biosynthesis gene cluster: factors that affect differential hdcA expression. Applied and Environmental Microbiology 76, 6231–6238.

Cañas, P.M, I., Gómez Alonso, S., Ruiz Pérez, P., Seseña Prieto, S., García Romero, E., Palop Herreros, M.L.L., 2009. Biogenic amine production by *Oenococcus oeni* isolates from malolactic fermentation of Tempranillo wine. Journal of Food Protection 72, 907–910.

Casquete, R., Benito, M.J., Martín, A., Ruiz-Moyano, S., Hernández, A., Córdoba, M.G., 2011a. Effect of autochthonous starter cultures in the production of "salchichón", a traditional Iberian dry-fermented sausage, with different ripening processes. LWT – Food Science and Technology 44, 1562–1571.

Casquete, R., Benito, M.J., Martín, A., Ruiz-Moyano, S., Córdoba, J.J., Córdoba, M.G., 2011b. Role of an autochthonous starter culture and the protease EPg222 on the sensory and safety properties of a traditional Iberian dry-fermented sausage "salchichón". Food Microbiology 28, 1432–1440.

Cho, T.Y., Han, G.H., Bahn, K.N., Son, Y.W., Jang, M.R., Lee, C.H., Kim, S.H., Kim, D.B., Kim, S.B., 2006. Evaluation of Biogenic amines in korean commercial fermented foods. Korean Journal of Food Science and Technology 38 (6), 730–737.

Coloretti, F., Tabanelli, G., Chiavari, C., Lanciotti, R., Grazia, L., Gardini, F., Montanari, C., 2014. Effect of wine addition on microbiological characteristics, volatile molecule profiles and biogenic amine contents in fermented sausages. Meat Science 96, 1395–1402.

Costantini, A., Cersosimo, M., Del Prete, V., Garcia-Moruno, E., 2006. Production of biogenic amines by lactic acid bacteria: screening by PCR, thin-layer chromatography, and high-performance liquid chromatography of strains isolated from wine and must. Journal of Food Protection 69, 391–396.

Coton, E., Coton, M., 2009. Evidence of horizontal transfer as origin of strain to strain variation of the tyramine production trait in *Lactobacillus brevis*. Food Microbiology 26, 52–57.

Coton, E., Mulder, N., Coton, M., Pochet, S., Trip, H., Lolkema, J.S., 2010a. Origin of the putrescine-producing ability of the coagulase-negative bacterium staphylococcus epidermidis 2015B. Applied and Environmental Microbiology 76, 5570–5576.

Coton, M., Romano, A., Spano, G., Ziegler, K., Vetrana, C., Desmarais, C., Lonvaud-Funel, A., Lucas, P., Coton, E., 2010b. Occurrence of biogenic amine-forming lactic acid bacteria in wine and cider. Food Microbiology 27, 1078–1085.

Curiel, J.A., Ruiz-Capillas, C., de Las Rivas, B., Carrascosa, A.V., Jiménez-Colmenero, F., Muñoz, R., 2011. Production of biogenic amines by lactic acid bacteria and enterobacteria isolated from fresh pork sausages packaged in different atmospheres and kept under refrigeration. Meat Science 88, 368–373.

de las Rivas, B., Ruiz-Capillas, C., Carrascosa, A.V., Curiel, J.A., Jiménez-Colmenero, F., Muñoz, R., 2008. Biogenic amine production by Gram-positive bacteria isolated from Spanish drycured "chorizo" sausage treated with high pressure and kept in chilled storage. Meat Science 80, 272–277.

De Mey, E., De Klerck, K., De Maere, H., Dewulf, L., Derdelinckx, G., Peeters, M.C., Fraeye, I., Vander Heyden, Y., Paelinck, H., 2014. The occurrence of N-nitrosamines, residual nitrite and biogenic amines in commercial dry fermented sausages and evaluation of their occasional relation. Meat Science 96, 821–828.

Drosinos, E.H., Paramithiotis, S., Kolovos, G., Tsikouras, I., Metaxopoulos, I., 2007. Phenotypic and technological diversity of lactic acid bacteria and staphylococci isolated from traditionally fermented sausages in Southern Greece. Food Microbiology 24, 260–270.

European Commission, 2005. Commission Regulation (EC) No 2073/2005 on microbiological criteria for foodstuffs. In: Official Journal of the European UnionL 338, pp. 1–26.

European Council Directive, 1991. 91/493/EEC the Health Conditions for the Production and the Placing on the Market of Fishery Products.

European Food Safety Authority, 2007. Opinion of the Scientific Committee on a request from EFSA on the introduction of a Qualified Presumption of Safety (QPS) approach for assessment of selected microorganisms referred to EFSA. EFSA Journal 587, 1–16.

European Food Safety Authority, 2011. Scientific opinion on risk based control of biogenic amine formation in fermented foods. EFSA Journal 9 (10), 1–93 2393.

Fernández, M., Linares, D.M., Rodríguez, A., Alvarez, M.A., 2007. Factors affecting tyramine production in *Enterococcus durans* IPLA 655. Applied Microbiology and Biotechnology 73, 1400–1406.

Ferreira, I.M.P.L.V.O., Pinho, O., 2006. Biogenic amines in Portuguese traditional foods and wines. Journal of Food Protection 69, 2293–2303.

Food and Drug Administration, 2011. Fish and Fishery Products Hazards and Controls Guidance, fourth ed. FDA, Center for Food Safety and Applied Nutrition, Washington, DC.

Foulquié Moreno, M.R., Sarantinopoulos, P., Tsakalidou, E., De Vuyst, L., 2006. The role and application of enterococci in food and health. International Journal of Food Microbiology 106, 1–24.

Garai, G., Dueñas, M.T., Irastorza, A., Moreno-Arribas, M.V., 2007. Biogenic amine production by lactic acid bacteria isolated from cider. Letters in Applied Microbiology 45, 473–478.

García-Ruiz, A., González-Rompinelli, E.M., Bartolomé, B., Moreno-Arribas, M.V., 2011. Potential of wine-associated lactic acid bacteria to degrade biogenic amines. International Journal of Food Microbiology 148, 115–120.

Gardini, F., Tabanelli, G., Lanciotti, R., Montanari, C., Luppi, M., Coloretti, F., Chiavari, C., Grazia, L., 2013. Biogenic amine content and aromatic profile of Salama da sugo, a typical cooked fermented sausage produced in Emilia Romagna Region (Italy). Food Control 32, 638–643.

Gençcelep, H., Kaban, G., Kaya, M., 2007. Effects of starter cultures and nitrite levels on formation of biogenic amines in sucuk. Meat Science 77, 424–430.

González-Tenorio, R., Fonseca, B., Caro, I., Fernández-Diez, A., Kuri, V., Soto, S., Mateo, J., 2013. Changes in biogenic amine levels during storage of mexican-style soft and spanish-style dry-ripened sausages with different aw values under modified atmosphere. Meat Science 94, 369–375.

Gosetti, F., Mazzucco, E., Gianotti, V., Polati, S., Gennaro, M.C., 2007. High performance liquid chromatography/tandem mass spectrometry determination of biogenic amines in typical Piedmont cheeses. Journal of Chromatography A 1149, 151–157.

Griswold, A.R., Chen, Y.Y.M., Burne, R.A., 2004. Analysis of an agmatine deiminase gene cluster in *Streptococcus mutans* UA159. Journal of Bacteriology 186, 1902–1904.

Guan, R.F., Liu, Z.F., Zhang, J.J., Wei, Y.X., Wahab, S., Liu, D.H., Ye, X.Q., 2013. Investigation of biogenic amines in sufu (furu): a Chinese traditional fermented soybean food product. Food Control 31, 345–352.

Guggenheim, M., 1920. Die biogenen Amine und ihre Bedeutung fur die Physiologie und Pathologie des pflanzlichen und tierischen Stoffwechsels. Springer, Berlin.

Guzel Seydim, Z.B., Seydim, A.C., Greene, A.K., 2003. Comparison of amino acid profiles of milk, yogurt and Turkish kefir. Milchwissenschaft – Milk Science International 58, 158–160.

Héberger, K., Csomós, E., Simon-Sarkadi, L., 2003. Principal component and linear discriminant analyses of free amino acids and biogenic amines in Hungarian wines. Journal of Agricultural and Food Chemistry 51, 8055–8060.

Helinck, S., Perello, M.C., Deetae, P., De Revel, G., Spinnler, H.E., 2013. *Debaryomyces hansenii*, *Proteus vulgaris*, *Psychrobacter* sp. and *Microbacterium foliorum* are able to produce biogenic amines. Dairy Science and Technology 93, 191–200.

Hernández-Herrero, M.M., Duflos, G., Malle, P., Bouquelet, S., 2002. Amino acid decarboxylase activity and other chemical characteristics as related to freshness loss in iced cod (*Gadus morhua*). Journal of Food Protection 65, 1152–1157.

Herrero-Fresno, A., Martínez, N., Sánchez-Llana, E., Díaz, M., Fernández, M., Martin, M.C., Ladero, V., Alvarez, M.A., 2012. *Lactobacillus casei* strains isolated from cheese reduce biogenic amine accumulation in an experimental model. International Journal of Food Microbiology 157, 297–304.

Hu, Y., Xia, W., Liu, X., 2007. Changes in biogenic amines in fermented silver carp sausages inoculated with mixed starter cultures. Food Chemistry 104, 188–195.

Ibe, A., Nishima, T., Kasai, N., 1992. Bacteriological properties of and amine-production conditions for tyramine- and histamine-producing bacterial strains isolated from soybean paste (miso) starting materials. Japanese Journal of Toxicology and Environmental Health 38, 403–409.

Kalač, P., 2006. Biologically active polyamines in beef, pork and meat products: a review. Meat Science 73, 1–11.

Kalač, P., 2014. Health effects and occurrence of dietary polyamines: a review for the period 2005-mid 2013. Food Chemistry 161, 27–39.

Kalač, P., Krausová, P., 2005. A review of dietary polyamines: formation, implications for growth and health and occurrence in foods. Food Chemistry 90, 219–230.

Kalač, P., Křížek, M., 2003. A review of biogenic amines and polyamines in beer. Journal of the Institute of Brewing 109, 123–128.

Kalač, P., Špička, J., Křížek, M., Steidlová, S., Pelikánová, T., 1999. Concentrations of seven biogenic amines in sauerkraut. Food Chemistry 67, 275–280.

Kalač, P., Špička, J., Křížek, M., Pelikánová, T., 2000a. Changes in biogenic amine concentrations during sauerkraut storage. Food Chemistry 69, 309–314.

Kalač, P., Špička, J., Křížek, M., Pelikánová, T., 2000b. The effects of lactic acid bacteria inoculants on biogenic amines formation in sauerkraut. Food Chemistry 70, 355–359.

Kheadr, E.E., Vachon, J.F., Paquin, P., Fliss, I., 2002. Effect of dynamic high pressure on microbiological, rheological and microstructural quality of Cheddar cheese. International Dairy Journal 12, 435–446.

Kim, S.H., Field, K.G., Chang, D.S., Wei, C.I., An, H.J., 2001. Identification of bacteria crucial to histamine accumulation in pacific mackerel during storage. Journal of Food Protection 64, 1556–1564.

Kim, S.H., Barros-Velazquez, J., Ben-Gigrey, B., Eun, J.B., Jun, S.H., Wie, C.I., An, H.J., 2003. Identification of the main bacteria contributing to histamine formation in seafood to ensure product safety. Food Science and Biotechnology 12, 451–460.

Kim, B., Byun, B.Y., Mah, J.H., 2012. Biogenic amine formation and bacterial contribution in Natto products. Food Chemistry 135, 2005–2011.

Komprda, T., Smělá, D., Pechová, P., Kalhotka, L., Štencl, J., Klejdus, B., 2004. Effect of starter culture, spice mix and storage time and temperature on biogenic amine content of dry fermented sausages. Meat Science 67, 607–616.

Komprda, T., Smělá, D., Novická, K., Kalhotka, L., Šustová, K., Pechová, P., 2007. Content and distribution of biogenic amines in Dutch-type hard cheese. Food Chemistry 102, 129–137.

Konakovsky, V., Focke, M., Hoffmann-Sommergruber, K., Schmid, R., Scheiner, O., Moser, P., Jarisch, R., Hemmer, W., 2011. Levels of histamine and other biogenic amines in high-quality red wines. Food Additives and Contaminants, A 28, 408–416.

Kung, H.-F., Lee, Y.-H., Chang, S.C., Wei, C.-I., Tsai, Y.-H., 2007a. Histamine contents and histamine-forming bacteria in sufu products in Taiwan. Food Control 18, 381–386.

Kung, H.-F., Tsai, Y.-H., Wei, C.I., 2007b. Histamine and other biogenic amines and histamine-forming bacteria in miso products. Food Chemistry 101, 351–356.

Křížek, M., Vácha, F., Vorlová, L., Lukášová, J., Cupáková, Š., 2004. Biogenic amines in vacuum-packed and non-vacuum-packed flesh of carp (*Cyprinus carpio*) stored at different temperatures. Food Chemistry 88, 185–191.

Ladero, V., Calles-Enriquez, M., Fernandez, M., Alvarez, M, A., 2010. Toxicological effects of dietary biogenic amines. Current Nutrition and Food Science 6, 145–156.

Ladero, V., Coton, M., Fernández, M., Buron, N., Martín, M.C., Guichard, H., Coton, E., Alvarez, M.A., 2011a. Biogenic amines content in Spanish and French natural ciders: application of qPCR for quantitative detection of biogenic amine-producers. Food Microbiology 28, 554–561.

Ladero, V., Rattray, F.P., Mayo, B., Martín, M.C., Fernández, M., Alvarez, M.A., 2011b. Sequencing and transcriptional analysis of the biosynthesis gene cluster of putrescine-producing *Lactococcus lactis*. Applied and Environmental Microbiology 77, 6409–6418.

Ladero, V., Cañedo, E., Pérez, M., Martín, M.C., Fernández, M., Alvarez, M.A., 2012a. Multiplex qPCR for the detection and quantification of putrescine-producing lactic acid bacteria in dairy products. Food Control 27, 307–313.

Ladero, V., Fernández, M., Calles-Enríquez, M., Sánchez-Llana, E., Cañedo, E., Martín, M.C., Alvarez, M.A., 2012b. Is the production of the biogenic amines tyramine and putrescine a species-level trait in enterococci? Food Microbiology 30, 132–138.

Lanciotti, R., Patrignani, F., Iucci, L., Guerzoni, M.E., Suzzi, G., Belletti, N., Gardini, F., 2007. Effects of milk high pressure homogenization on biogenic amine accumulation during ripening of ovine and bovine Italian cheeses. Food Chemistry 104, 693–701.

Landete, J.M., Ferrer, S., Pardo, I., 2005a. Which lactic acid bacteria are responsible for histamine production in wine? Journal of Applied Microbiology 99, 580–586.

Landete, J.M., Ferrer, S., Polo, L., Pardo, I., 2005b. Biogenic amines in wines from three Spanish regions. Journal of Agricultural and Food Chemistry 53, 1119–1124.

Landete, J.M., Ferrer, S., Pardo, I., 2007a. Biogenic amine production by lactic acid bacteria, acetic bacteria and yeast isolated from wine. Food Control 18, 1569–1574.

Landete, J.M., de las Rivas, B., Marcobal, A., Muñoz, R., 2007b. Molecular methods for the detection of biogenic amine-producing bacteria on foods. International Journal of Food Microbiology 117, 258–269.

Larqué, E., Sabater-Molina, M., Zamora, S., 2007. Biological significance of dietary polyamines. Nutrition 23, 87–95.

Latorre-Moratalla, M.L., Bover-Cid, S., Aymerich, T., Marcos, B., Vidal-Carou, M.C., Garriga, M., 2007. Aminogenesis control in fermented sausages manufactured with pressurized meat batter and starter culture. Meat Science 75, 460–469.

Latorre-Moratalla, M.L., Veciana-Nogués, T., Bover-Cid, S., Garriga, M., Aymerich, T., Zanardi, E., Ianieri, A., Fraqueza, M.J., Patarata, L., Drosinos, E.H., Lauková, A., Talon, R., Vidal-Carou, M.C., 2008. Biogenic amines in traditional fermented sausages produced in selected European countries. Food Chemistry 107, 912–921.

Latorre-Moratalla, M.L., Bover-Cid, S., Talon, R., Garriga, M., Zanardi, E., Ianieri, A., Fraqueza, M.J., Elias, M., Drosinos, E.H., Vidal-Carou, M.C., 2010. Strategies to reduce biogenic amine accumulation in traditional sausage manufacturing. LWT – Food Science and Technology 43, 20–25.

Latorre-Moratalla, M.L., Bover-Cid, S., Bosch-Fusté, J., Vidal-Carou, M.C., 2012. Influence of technological conditions of sausage fermentation on the aminogenic activity of *L. curvatus* CTC273. Food Microbiology 29, 43–48.

Leroy, F., Verluyten, J., De Vuyst, L., 2006. Functional meat starter cultures for improved sausage fermentation. International Journal of Food Microbiology 106, 270–285.

Leuschner, R.G.K., Hristova, A., Robinson, T., Hugas, M., 2013. The Rapid Alert System for Food and Feed (RASFF) database in support of risk analysis of biogenic amines in food. Journal of Food Composition and Analysis 29, 37–42.

Linares, D.M., Del Río, B., Ladero, V., Martínez, N., Fernández, M., Martín, M.C., Álvarez, M.A., 2012. Factors influencing biogenic amines accumulation in dairy products. Frontiers in Microbiology 3, 180.

Loret, S., Deloyer, P., Dandrifosse, G., 2005. Levels of biogenic amines as a measure of the quality of the beer fermentation process: data from Belgian samples. Food Chemistry 89, 519–525.

Lu, Y., Xiaohong, C., Mei, J., Xin, L., Rahman, N., Mingsheng, D., Yan, G., 2009. Biogenic amines in Chinese soy sauce. Food Control 20, 593–597.

Lu, S., Xu, X., Zhou, G., Zhu, Z., Meng, Y., Sun, Y., 2010. Effect of starter cultures on microbial ecosystem and biogenic amines in fermented sausage. Food Control 21, 444–449.

Lucas, P.M., Wolken, W.A.M., Claisse, O., Lolkema, J.S., Lonvaud-Funel, A., 2005. Histamine-producing pathway encoded on an unstable plasmid in *Lactobacillus hilgardii* 0006. Applied and Environmental Microbiology 71, 1417–1424.

Lucas, P.M., Blancato, V.S., Claisse, O., Magni, C., Lolkema, J.S., Lonvaud-Funel, A., 2007. Agmatine deiminase pathway genes in *Lactobacillus brevis* are linked to the tyrosine decarboxylation operon in a putative acid resistance locus. Microbiology 153, 2221–2230.

Mah, J.H., 2015. Fermented soybean foods: significance of biogenic amines. Austin Journal of Nutrition and Food Sciences 3, 1058–1060.

Mah, J.H., Hwang, H.J., 2009. Inhibition of biogenic amine formation in a salted and fermented anchovy by *Staphylococcus xylosus* as a protective culture. Food Control 20, 796–801.

Maintz, L., Novak, N., 2007. Histamine and histamine intolerance. American Journal of Clinical Nutrition 85, 1185–1196.

Mangia, N.P., Trani, A., Di Luccia, A., Faccia, M., Gambacorta, G., Fancello, F., Deiana, P., 2013. Effect of the use of autochthonous *Lactobacillus curvatus*, *Lactobacillus plantarum* and *Staphylococcus xylosus* strains on microbiological and biochemical properties of the Sardinian fermented sausage. European Food Research and Technology 236, 557–566.

Marcobal, A., Martín-Alvarez, P.J., Polo, M.C., Muñoz, R., Moreno-Arribas, M.V., 2006a. Formation of biogenic amines throughout the industrial manufacture of red wine. Journal of Food Protection 69, 397–404.

Marcobal, A., de las Rivas, B., Moreno-Arribas, M.V., Munoz, R., 2006b. Evidence for horizontal gene transfer as origin of putrescine production in *Oenococcus oeni* RM83. Applied and Environmental Microbiology 72, 7954–7958.

Marcobal, A., de las Rivas, B., Landete, J.M., Tabera, L., Muñoz, R., 2012. Tyramine and phenylethylamine biosynthesis by food bacteria. Critical Reviews in Food Science and Nutrition 52, 448–467.

Martín, M.C., Fernández, M., Linares, D.M., Alvarez, M.A., 2005. Sequencing, characterization and transcriptional analysis of the histidine decarboxylase operon of *Lactobacillus buchneri*. Microbiology 151, 1219–1228.

Martuscelli, M., Gardini, F., Torriani, S., Mastrocola, D., Serio, A., Chaves-López, C., Schirone, M., Suzzi, G., 2005. Production of biogenic amines during the ripening of Pecorino Abruzzese cheese. International Dairy Journal 15, 571–578.

Mayer, H.K., Fiechter, G., Fischer, E., 2010. A new ultra-pressure liquid chromatography method for the determination of biogenic amines in cheese. Journal of Chromatogrraphy A 1217, 3251–3257.

McCabe-Sellers, B.J., Staggs, C.G., Bogle, M.L., 2006. Tyramine in foods and monoamine oxidase inhibitor drugs: a crossroad where medicine, nutrition, pharmacy, and food industry converge. Journal of Food Composition and Analysis 19, S58–S65.

Medina, M.A., Urdiales, J.L., Rodríguez-Caso, C., Ramírez, F.J., Sánchez-Jiménez, F., 2003. Biogenic amines and polyamines: similar biochemistry for different physiological missions and biomedical applications. Critical Reviews in Biochemistry and Molecular Biology 38, 23–59.

Moon, J.S., Cho, S.K., Choi, H.I., Kim, J.E., Kim, S.Y., Cho, K.J., Han, N.S., 2010. Isolation and characterization of biogenic amine-producing bacteria in fermented soybean pastes. Journal of Microbiology 48 (2), 257–261.

Moreno-Arribas, M.V., Polo, M.C., Jorganes, F., Muñoz, R., 2003. Screening of biogenic amine production by lactic acid bacteria isolated from grape must and wine. International Journal of Food Microbiology 84, 117–123.

Nannelli, F., Claisse, O., Gindreau, E., De Revel, G., Lonvaud-Funel, A., Lucas, P.M., 2008. Determination of lactic acid bacteria producing biogenic amines in wine by quantitative PCR methods. Letters in Applied Microbiology 47, 594–599.

Nie, X., Zhang, Q., Lin, S., 2014. Biogenic amine accumulation in silver carp sausage inoculated with *Lactobacillus plantarum* plus *Saccharomyces cerevisiae*. Food Chemistry 153, 432–436.

Nout, M.J.R., Ruikes, M.M.W., Bouwmeester, H.M., Beljaars, P.R., 1993. Effect of processing conditions on the formation of biogenic amines and ethyl carbamate in soybean tempe. Journal of Food Safety 13, 293–303.

Novella-Rodríguez, S., Veciana-Nogués, M.T., Roig-Sagués, A.X., Trujillo-Mesa, A.J., Vidal-Carou, M.C., 2004. Comparison of biogenic amine profile in cheeses manufactured from fresh and stored (4 degrees C, 48 hours) raw goat's milk. Journal of Food Protection 67, 110–116.

Paulsen, P., Grossgut, R., Bauer, F., Rauscher-Gabernig, E., 2012. Estimates of maximum tolerable levels of tyramine content in foods in Austria. Journal of Food and Nutrition Research 51, 52–59.

Picinelli, A., Suárez, B., Moreno, J., Rodríguez, R., Caso-García, L.M., Mangas, J.J., 2000. Chemical characterization of Asturian cider. Journal of Agricultural and Food Chemistry 48, 3997–4002.

Pircher, A., Bauer, F., Paulsen, P., 2006. Formation of cadaverine, histamine, putrescine and tyramine by bacteria isolated from meat, fermented sausages and cheeses. European Food Research and Technology 226, 225–231.

Rauscher-Gabernig, E., Grossgut, R., Bauer, F., Paulsen, P., 2009. Assessment of alimentary histamine exposure of consumers in Austria and development of tolerable levels in typical foods. Food Control 20, 423–429.

Romano, A., Trip, H., Lonvaud-Funel, A., Lolkema, J.S., Lucas, P.M., 2012. Evidence of two functionally distinct ornithine decarboxylation systems in lactic acid bacteria. Applied and Environmental Microbiology 78, 1953–1961.

Romano, A., Ladero, V., Alvarez, M.A., Lucas, P.M., 2014. Putrescine production via the ornithine decarboxylation pathway improves the acid stress survival of Lactobacillus brevis and is part of a horizontally transferred acid resistance locus. International Journal of Food Microbiology 175, 14–19.

Romero, R., Bagur, M.G., Sánchez-Viñas, M., Gázquez, D., 2003. The influence of the brewing process on the formation of biogenic amines in beers. Analytical and Bioanalytical Chemistry 376, 162–167.

Rossi, F., Gardini, F., Rizzotti, L., Tabanelli, G., 2011. Features of the histidine decarboxylase activity of *Streptococcus thermophilus* PRI60: quantitative analysis of hdcA transcription and factors influencing histamine production. Applied and Environmental Microbiology 77, 2817–2822.

Ruiz-Capillas, C., Jiménez-Colmenero, F., 2010. Effect of an argon-containing packaging atmosphere on the quality of fresh pork sausages during refrigerated storage. Food Control 21, 1331–1337.

Ruiz-Capillas, C., Jiménez Colmenero, F., Carrascosa, A.V., Muñoz, R., 2007. Biogenic amine production in Spanish dry-cured "chorizo" sausage treated with high-pressure and kept in chilled storage. Meat Science 77, 365–371.

Ruiz-Capillas, C., Triki, M., Herrero, A.M., Jiménez-Colmenero, F., 2012. Biogenic amines in low- and reduced-fat dry fermented sausages formulated with konjac gel. Journal of Agricultural and Food Chemistry 60, 9242–9248.

Shukla, S., Park, H.K., Kim, J.K., Kim, M., 2010. Determination of biogenic amines in Korean traditional fermented soybean paste (Doenjang). Food and Chemical Toxicology 48, 1191–1195.

Simon-Sarkadi, L., 2009. Biogenic amines. In: Stadler, R.H., Lineback, D.R. (Eds.), Process-induced Food Toxicants. Occurrence, Formation, Mitigation, and Health Risks. John Wiley & Sons, Inc., Hoboken, pp. 321–361.

Simon-Sarkadi, L., Pásztor-Huszár, K., Dalmadi, I., Kiskó, G., 2012. Effect of high hydrostatic pressure processing on biogenic amine content of sausage during storage. Food Research International 47, 380–384.

Slomkowska, A., Ambroziak, W., 2002. Biogenic amine profile of the most popular Polish beers. European Food Research and Technology 215, 380–383.

Smit, A.Y., du Toit, W.J., du Toit, M., 2008. Biogenic amines in wine: understanding the headache. South African Journal of Enology and Viticulture 29, 109–127.

Spano, G., Russo, P., Lonvaud-Funel, A., Lucas, P., Alexandre, H., Grandvalet, C., Coton, E., Coton, M., Barnavon, L., Bach, B., Rattray, F., Bunte, A., Magni, C., Ladero, V., Alvarez, M., Fernández, M., Lopez, P., de Palencia, P.F., Corbi, A., Trip, H., Lolkema, J.S., 2010. Biogenic amines in fermented foods. European Journal of Clinical Nutrition 64, S95–S100.

Špička, J., Kalač, P., Bover-Cid, S., Křížek, M., 2002. Application of lactic acid bacteria starter cultures for decreasing the biogenic amine levels in sauerkraut. European Food Research and Technology 215, 509–514.

Sugita, Y., Takao, K., Toyama, Y., Shirahata, A., 2007. Enhancement of intestinal absorption of macromolecules by spermine in rats. Amino Acids 33, 253–260.

Suzzi, G., Gardini, F., 2003. Biogenic amines in dry fermented sausages: a review. International Journal of Food Microbiology 88, 41–54.

Tabanelli, G., Torriani, S., Rossi, F., Rizzotti, L., Gardini, F., 2012. Effect of chemico-physical parameters on the histidine decarboxylase (HdcA) enzymatic activity in Streptococcus thermophilus PRI60. Journal of Food Science 77, M231–M237.

Takahashi, H., Kimura, B., Yoshikawa, M., Fujii, T., 2003. Cloning and sequencing of the histidine decarboxylase genes of gram-negative, histamine-producing bacteria and their application in detection and identification of these organisms in fish. Applied and Environmental Microbiology 69, 2568–2579.

ten Brink, B., Damink, C., Joosten, H.M.L.J., Huis in't Veld, J.H.J., 1990. Occurrence and formation of biologically active amines in foods. International Journal of Food Microbiology 11, 73–84.

Tomé, E., Pereira, V.L., Lopes, C.I., Gibbs, P.A., Teixeira, P.C., 2008. In vitro tests of suitability of bacteriocin-producing lactic acid bacteria, as potential biopreservation cultures in vacuum-packaged cold-smoked salmon. Food Control 19, 535–543.

Toro-Funes, N., Bosch-Fuste, J., Latorre-Moratalla, M.L., Veciana-Nogues, M.T., Vidal-Carou, M.C., 2015. Biologically active amines in fermented and non-fermented commercial soybean products from the Spanish market. Food Chemistry 173, 1119–1124.

Tosukhowong, A., Visessanguan, W., Pumpuang, L., Tepkasikul, P., Panya, A., Valyasevi, R., 2011. Biogenic amine formation in Nham, a Thai fermented sausage, and the reduction by commercial starter culture, Lactobacillus plantarum BCC 9546. Food Chemistry 129, 846–853.

Tsai, Y.H., Lin, C.Y., Chien, L.T., Lee, T.M., Wei, C.I., Hwang, D.F., 2006. Histamine contents of fermented fish products in Taiwan and isolation of histamine-forming bacteria. Food Chemistry 98, 64–70.

Tsai, Y.H., Chang, S.C., Kung, H.F., 2007a. Histamine contents and histamine-forming bacteria in natto products in Taiwan. Food Control 18 (9), 1026–1030.

Tsai, Y.H., Kung, H.F., Chang, S.C., Lee, T.M., Wei, C.I., 2007b. Histamine formation by histamine-forming bacteria in douchi, a Chinese traditional fermented soybean product. Food Chemistry 103, 1305–1311.

Wu, C., Zheng, J., Huang, J., Zhou, R., 2014. Reduced nitrite and biogenic amine concentrations and improved flavor components of Chinese sauerkraut via co-culture of *Lactobacillus plantarum* and *Zygosaccharomyces rouxii*. Annals of Microbiology 64, 847–857.

Wuytack, E.Y., Diels, A.M.J., Michiels, C.W., 2002. Bacterial inactivation by high-pressure homogenisation and high hydrostatic pressure. International Journal of Food Microbiology 77, 205–212.

Xie, C., Wang, H.-H., Nie, X.-K., Chen, L., Deng, S.-L., Xu, X.-L., 2015. Reduction of biogenic amine concentration in fermented sausage by selected starter cultures. CyTA – Journal of Food 13 (4), 491–497.

Zaman, M.Z., Abdulamir, A.S., Bakar, F.A., Selamat, J., Bakar, J., 2009. A review: microbiological, physicochemical and health impact of high level of biogenic amines in fish sauce. American Journal of Applied Sciences 6, 1199–1211.

Zaman, M.Z., Bakar, F.A., Selamat, J., Bakar, J., 2010. Occurrence of biogenic amines and amines degrading bacteria in fish sauce. Czech Journal of Food Sciences 28, 440–449.

Zaman, M.Z., Bakar, F.A., Jinap, S., Bakar, J., 2011. Novel starter cultures to inhibit biogenic amines accumulation during fish sauce fermentation. International Journal of Food Microbiology 145, 84–91.

Zhijun, L., Yongning, W., Gong, Z., Yunfeng, Z., Changhu, X., 2007. A survey of biogenic amines in Chinese red wines. Food Chemistry 105, 1530–1535.

Zhong-Yi, L., Zhong-Hai, L., Miao-Ling, Z., Xiao-Ping, D., 2010. Effect of fermentation with mixed starter cultures on biogenic amines in bighead carp surimi. International Journal of Food Science and Technology 45, 930–936.

Chapter 28

Occurrence of Aflatoxins in Fermented Food Products

S. Shukla[1], D.-H. Kim[2], S.H. Chung[3], M. Kim[1]
[1]*Yeungnam University, Gyeongsan, Republic of Korea;* [2]*National Agricultural Products Quality Management Service, Gimcheon, Republic of Korea;* [3]*Korea University, Seoul, Republic of Korea*

28.1 INTRODUCTION

Aflatoxins represent a group of extremely toxic fungal metabolites produced by some species of the genus *Aspergillus*, including *Aspergillus flavus*, *Aspergillus parasiticus*, and *Aspergillus nomius*, during their growth on foods and feeds (Moss, 2002; Reddy et al., 2010b). *Aspergillus flavus* produces only aflatoxin B, while the two other species (*Aspergillus parasiticus* and *Aspergillus nomius*) produce both aflatoxins B and G, which are considered to be significant threats to both humans and animals due to potent carcinogenic and mutagenic activities (Pitt et al., 2001), immunotoxic, as well as cause growth retardation in animals (Karunaratne et al., 1990). In addition to their hazardous effects on animals, they cause major health issues in humans (IARC, 1993). These aflatoxins are consumed by humans through intake of plants and animal-derived foods, such as dairy products contaminated either directly or indirectly with these aflatoxigenic fungi (Smith, 2001). Mycotoxins are produced directly by molds present in foods, whereas they indirectly enter via cross-contamination in food chain (Arab et al., 2012; Smith, 2001).

Aflatoxin contamination has caused huge economic losses to food commodities in the last few decades (Benblesa et al., 2004). Major economic losses caused by aflatoxin contamination can either be direct in the form of losses to crop, livestock, and dairy productivity or indirect with adverse effects on quality control programs, research, education, foreign exchange earnings, as well as storage and packaging costs of vulnerable commodities (Benblesa et al., 2004). The diseases and health hazards caused by aflatoxins have severe consequences of human illness which have led to the monitoring of these toxins in various food commodities by several international regulatory authorities. As reported previously, most foods are susceptible to aflatoxigenic fungi during production, processing, transportation, and storage. The first outbreak of aflatoxicosis reported in England in 1960 caused the death of a huge population of livestock (Blount, 1961), leading to the discovery of aflatoxins in groundnut meal contaminated by *Aspergillus flavus* (Mohammadi, 2011). Other types of aflatoxins have been discovered in feeds, especially maize (Chakrabarty, 1981) and cottonseed meal (Sharma et al., 1994).

In general, only aflatoxins B1 (AFB1), B2 (AFB2), G1 (AFG1), G2 (AFG2), and M1 (AFM1) are found normally (Scudamore, 1998). Specifically, *Aspergillus flavus* produces AFB1 and AFB2, whereas *Aspergillus parasiticus* produces AFB1, AFB2, AFG1, and AFG2 (Calvo, 2005; Fallah, 2010). Among the reported aflatoxins, AFB1 is considered to be the most toxic (Hernandez-Mendoza et al., 2009; Torkar and Vengust, 2008) and frequent, and it is majorly responsible for aflatoxin food contamination in foods and feeds (Hernandez-Mendoza et al., 2009).

Generally, aflatoxins show both acute and chronic poisoning effects (Turner et al., 2000). AFB1 and AFM1 have been classified as a Class 1A agent (carcinogenic to humans) and Class 2B agent (possibly carcinogenic to humans), respectively, by the World Health Organization (WHO). Long-term exposure to even extremely low levels of aflatoxins in the diet can induce very severe health complications in humans. The European Union (EU) and the United States of Food and Drug Administration (US FDA) have set maximum admissible aflatoxin levels of 0.05 and 0.50 g/L in milk, respectively (Arab et al., 2012; Govaris et al., 2002).

Aflatoxins are considered to be the greatest contributing factors to crop contamination worldwide from a food safety point of view (Reddy et al., 2010a). A number of strategies, such as application of chemical, physical, and biological control measures, are used to control contamination of aflatoxins in foods. Among these applied strategies, biological control appears to be the most promising way of controlling the occurrence of aflatoxins. In addition, a number of bacterial genera, including *Lactobacillus*, *Bacillus*, *Pseudomonas*, and *Burkholderia*, have shown significant antagonistic effects with great potential for inhibiting growth of aflatoxigenic fungi in vitro, leading to possible alternatives for controlling aflatoxin contamination, as reported by our research group (Kim, 2007).

Recently, it was reported that several types of aflatoxins have shown remarkable stability in various fermented foods and are fairly resistant to degradation, which is dependent on formulation and process factors, such as protein content, pH, temperature, length of heat treatment, and the nature of starter microorganisms (Scudamore, 1998; Shukla et al., 2014). Therefore, the purpose of this review is to consider the occurrence, process and storage factors, control strategies, and detection methods of aflatoxins in various fermented food products.

28.2 STRUCTURES OF AFLATOXINS

Aflatoxins are a group of chemically similar mycotoxins produced by a few mold species. Their names are derived from the fungus *Aspergillus flavus*, on which much of the early work related to aflatoxins was performed (Asao, 1965). Furthermore, it was discovered that aflatoxins represent a large number of distinct but structurally related compounds, and the four most commonly reported and designated ones are B1, B2, G1, and G2. The B and G designations of aflatoxins are originated from compounds which exhibit blue and yellow-green fluorescence under ultraviolet light, respectively (Asao, 1965). On the other hand, aflatoxins designated by M are considered to be hydroxylated derivatives of aflatoxins B and are most commonly found in milk, milk products, or meat, which explains the M designation (Asao, 1965). AFM1 and AFM2 are generally formed in animals by metabolism of B aflatoxins, that occur by subsequent absorption of contaminated feed. However, AFB1 has been reported to be the most frequent aflatoxin present in contaminated food samples, followed by AFB2, AFG1, and AFG2 (Asao, 1965). Chemical structures of the most prevalent aflatoxins are presented in Fig. 28.1.

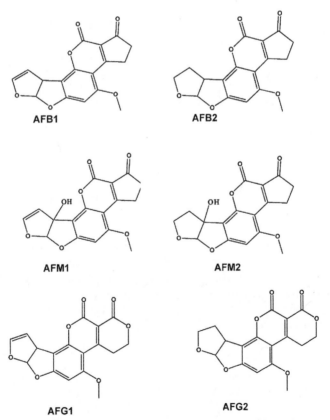

FIGURE 28.1 Chemical structures of major aflatoxin (AF) types (B1, B2, M1, M2, G1, G2) in fermented food products.

AFM1 produces blue-violet fluorescence, while AFM2 produces violet fluorescence (Goldblatt, 1969). Aflatoxins are highly similar in structure, as a unique group of highly oxygenated, naturally occurring heterocyclic compounds, and show natural fluorescence due to their pentaheterocyclic structure. Aflatoxins AFB2 and AFG2 exhibit 10-fold higher fluorescence capacity than other aflatoxin compounds, such as AFB1 and AFG1, possibly due to structural differences conferring a double bond in the furanic ring (Cepeda et al., 1996). This double bond plays a very crucial role in the photophysical properties of a number of aflatoxin derivatives (Cepeda et al., 1996). The fluorescence excitation efficiency of aflatoxins AFB1 and AFG1 can be promoted in number of ways, including postcolumn iodination (Tuinstra and Haasnoot, 1983), postcolumn bromination (Kok, 1994; Versantroort et al., 2005), and application of cyclodextrin compounds (Chiavaro et al., 2001) and trifluoroacetic acid (Nieduetzki et al., 1994).

28.3 OCCURRENCE OF AFLATOXINS IN DIFFERENT FERMENTED FOODS

Tropical and subtropical regions are major incubation sites for aflatoxins since these areas offer ideal conditions, such as temperature and humidity for the growth of fungal pathogens or molds, which produce mycotoxins, such as aflatoxins. Among the reported substrates, maize and peanuts are considered to be the most favorable substrates for growth of molds and subsequent production of aflatoxins (Farombi, 2006). Among them, AFB1 is responsible for contamination of a variety of important fermented foods and feed products, including soybean-based fermented foods, dairy-based fermented foods, alcoholic beverages, and fermented juices or wine (Kim, 2007).

28.3.1 Aflatoxins in Soybean Fermented Food Products

Although trypsin inhibitors present in soybeans are thought to prevent microbial pathogen growth, studies of aflatoxigenic strains of *Aspergillus flavus* and *Aspergillus parasiticus* have confirmed that these inhibitors do not affect the growth of these fungi or aflatoxin production. In addition, increased growth of aflatoxigenic fungi and their corresponding aflatoxins directly following cooking of soybeans, soybean flour, or soybean-based medium implies that heat-labile seed constituents (eg, proteinase inhibitors) might have an antifungal effect, thus reducing aflatoxin production (Gupta and Venkitasubramanian, 1975). Various important factors can be considered as essential for the degradation of aflatoxins during fermentation, such as ammonia production, light, microbial competition with *Bacillus* spp., and addition of charcoal, which has been shown to reduce aflatoxins in soybean products (Kang et al., 2000; Park et al., 2003). *Meju* has been used in Korea as a starter culture of *Doenjang*, a traditional Korean soybean fermented paste, in which *Bacillus* and *Aspergillus* species (mainly *Aspergillus oryzae* and *Aspergillus niger*) participate in fermentation (Kwon et al., 2010; Shukla et al., 2014). Although no report has investigated aflatoxins produced by these fungal strains, *Meju* might be contaminated by some other aflatoxigenic *Aspergillus* species during fermentation such as *Aspergillus flavus*, *Aspergillus parasiticus*, and *Aspergillus nomius* (Kwon et al., 2010). Aflatoxins in *Meju* cannot be degraded during processing of fermented foods and can thus pose potential risks to consumers (Kim et al., 2011).

Recently, we investigated aflatoxin levels in traditionally in-house prepared *Doenjang* (Kim and Kim, 2012). One of the *Doenjang* samples had a maximum aflatoxin level of $42.2 \pm 9.1\,\mu g/kg$, which is above the safety limit approved by the Codex Alimentarius Commission (CAC) regulatory agency (Kim and Kim, 2012). These findings reinforced the suggestion that further research is necessary to determine how *Doenjang* safety can be improved via elimination/reduction of microbial contamination during fermentation and storage as well as whether or not the use of microbial starter cultures for fermentation is feasible. Based on such findings, another strategy was adopted for the production of *Doenjang* samples with reduced aflatoxin levels using various starter cultures (Shukla et al., 2014). The total aflatoxin amount in *Doenjang* samples was under the standard limit defined by the Food and Agricultural Organization (FAO) (Shukla et al., 2014). A recent survey in China showed that local soy sauce products contain a maximum level of aflatoxins, which is still below the limit set by the European Commission (Qi and Che, 2010). The presence of AFB1 has been reported in fermented soybean products from Thailand (Charoenpornsook and Kavisarasai, 2014). In this respect, an interesting finding to reduce aflatoxin levels in soy fermented products was found by Petchkongkaew et al. (2008). These researchers confirmed that two *Bacillus* strains, *Bacillus licheniformis* and *Bacillus subtillis*, isolated from Thai fermented soybeans, were able to inhibit the growth of toxic fungi *Aspergillus flavus* and effectively degrade AFB1 and ochratoxin A (OTA).

28.3.2 Aflatoxins in Dairy Fermented Food Products

The yogurt and fermented milk market currently generates 46 billion dollars in North America, Europe, and Asia, representing a total of 77% of the market (Marsh et al., 2014). According to CAC, fermented milk may have been manufactured from products obtained from milk with and without compositional modification by the action of suitable microorganisms and resulting in reduction of pH with or without coagulation (Arab et al., 2012). These starter culture microorganisms are viable, active, and abundant in the final product (Anon, 2011). Safety, hygiene, detection, and control of toxins are considered very important parameters to define the quality of fermented milk and milk products. Confirmation of even extremely low levels of aflatoxins in milk samples can have unsuitable consequences on health and technological aspects. As reported previously, the occurrence of AFM1 in milk samples is of particular concern for humans due to its chronic and acute side effects (Turner et al., 2000).

To protect consumers, especially children, permissible levels of AFM1 in milk and dairy products have been reviewed and are regulated internationally (Fallah, 2010). The USFDA has set a limit of 0.5 mg/L for AFM1 in milk (USFDA, 1996). On the other hand, the Institute of Standards and Industrial Research of Iran and the European Commission have set an AFM1 limit in milk of 0.05 mg/L (EC, 2001; ISIRI, 2002). In addition, the maximum admissible levels of AFM1 in other dairy products, such as dried milk, butter/butter milk, and cheese have been defined as 0.5, 0.02, and 0.25 mg/kg, and for yogurt 0.05 mg/L, respectively (ISIRI, 2002; Fallah, 2010). Several studies have confirmed the presence of AFM1 in milk and dairy products (Carvajal et al., 2003; Fallah, 2010; Kim et al., 2010a). Dairy products, including cheese, yogurt, and dairy drinks, produced in Brazil have been found to contain AFM1 (Iha et al., 2011). Torkar and Vengust (2008) observed AFM1 contamination above 50 ng/kg in 10% of the total Slovenian cheese samples analyzed. In a survey by Khoury et al. (2011) administered in Lebanon, AFM1 was found to be present in 40.62% and 32.81% of the total milk and yogurt samples analyzed, respectively. However, contamination levels varied and were higher in milk and yogurt products of Lebanon according to European regulation limits (Khoury et al., 2011). It was reported that toxins can be detected if dairy products are manufactured from AFM1-contaminated milk (Bakirci, 2001). However, the content of AFM1 was found to be relatively stable during processing and storage of various dairy products. Reports have confirmed that AFM1 shows resistance to thermal processing, such as pasteurization and ultra high temperature (Fallah, 2010).

28.3.3 Aflatoxins in Alcoholic Beverage Products

One of the most toxic aflatoxins, AFB1, produced by *Aspergillus carbonarius* has been found to be present in various fermented food products as well as in beer and wine (Iamanaka et al., 2007). Pure AFB1 is a pale-white to yellow crystalline, odorless solid with solubility in various solvents, including methanol, chloroform, acetone, acetonitrile, and water (Rasch et al., 2010). Different aflatoxins and/or mycotoxins originated from malted grains or from food additives may be transmitted from contaminated grains into beer during the brewing process (Scott, 1996). Initially, occurrence of AFB1 in wine was attributed to fungal pathogens, which can grow on grapes during their processing, such as crushing as well as grape juice preparation, and are thus carried over into the wine (Zaki et al., 2008). Since mycotoxins show heat resistance and remain on foods indefinitely during the production process, they can only be eliminated through harsh treatment. Nowadays, there is an increased focus on diminishing mycotoxin contamination in wine through pre- and postharvest treatments.

OTA, which is produced by a number of fungal species, has garnered a considerable amount of attention due to its potential ability to colonize a variety of food products. Major concerns have been raised over OTA in terms of its severe health complications, which include nephrotoxic, teratogenic, and immunosuppressive effects in humans and animals (Pitt et al., 2001). In addition, reports have confirmed the contamination of alcoholic beverages, such as beer and wine, by OTA (Cabanes et al., 2002; Khoury et al., 2006). A number of reports carried out on wine, grape juice, and dried wine fruits in different countries have confirmed the presence of OTA due to contamination of grapes by aflatoxigenic fungi that produce OTA (Abarca et al., 2003; Bejaoui et al., 2006). Among OTA-producing fungi, *A. carbonarius* is considered to be the major OTA producer species for wine grapes (Cabanes et al., 2002). It has been reported that wines from southern regions of Europe and North Africa contain high amounts of OTA due to a climate characterized by high humidity and temperatures (Blesa et al., 2004; Khoury et al., 2006). However, only these few reports are available on wine contamination by aflatoxins, and no further reports have investigated the contamination of grapes by aflatoxins.

28.4 VARIOUS METHODS TO CONTROL AFLATOXINS

28.4.1 Physical Methods to Control Aflatoxins in Fermented Foods

Information related to mycotoxin contamination and their prevention in foods, such as physical, chemical, and biological control measures, has been provided by various international organizations, including the WHO, FAO, and United Nations. As reported previously, physical and vigorous methods are typically used to prevent the growth of fungi capable of producing aflatoxins, eventually leading to control or reduction of aflatoxin formation in various fermented foods (Thanaboripat, 2002). Although seeds contaminated by aflatoxigenic fungal pathogens can be removed physically by means of picking and photoelectric tools, these techniques are expensive, labor-intensive, and time-consuming (Thanaboripat, 2002). Since aflatoxins have a thermostable nature, physical heat treatment can result in the reduction of aflatoxin levels to a very small quantity (Tripathi and Mishra, 2010). However, physical treatments, such as heating and cooking under pressure, are able to reduce aflatoxin levels in rice by 70%, which is a significant result compared to a 50% reduction by atmospheric pressure (Coomes et al., 1966). In addition, physical parameters such as drying and roasting have shown sufficient aflatoxin reduction ability, specifically 50–70% reduction of AFB1 (Feuell, 1966). Moreover, prolonged cooking and overheating may result in the loss of various nutritional components of fermented food products since aflatoxins show higher temperature resistance.

28.4.2 Biochemical Methods to Control Aflatoxins in Fermented Foods

Biochemical methods have also shown significant importance in controlling aflatoxin contamination in various fermented foods. Among them, use of organic solvents has shown potential efficacy for removal of such compounds from fermented foods with less toxic effects on nutritional properties. In addition, decontamination of fermented foods and feeds has been achieved by application of an ammoniation process (Allameh et al., 2005). Use of ozone has gained USFDA approval for the safe decontamination of fermented foods and feeds containing aflatoxin compounds (Zorlugenc et al., 2008). Application of chemical inhibitors, such as fungicides, antifungal drugs like chloramphenicol, has also significantly contributed to controlling the growth of aflatoxigenic fungi in fermented foods and feeds (Moreno-Martinez et al., 2000).

As aflatoxins exert adverse health effects in humans, natural antifungal agents of plant origin have gained wide attention as a natural alternative in order to minimize aflatoxin contamination by aflatoxigenic fungi in various fermented foods and food products (Yin and Cheng, 1998). Application of plant-based natural compounds as potent antimicrobials for food preservation to inhibit growth of aflatoxigenic fungi has been previously carried out (Krishnamurthy and Shashikala, 2006; Thanaboripat et al., 2004; Wilson and Wisniewski, 1992). As reported previously, a number of plant-based organic extracts and essential oils have shown potent ability to inhibit growth of aflatoxigenic fungi as well as aflatoxin formation (Razzaghi-Abyaneh et al., 2008; Reddy et al., 2010a). Plant-based compounds have been shown to be better alternatives than physical and chemical control measures (Mishra and Das, 2003).

28.4.3 Microbiological Methods to Control Aflatoxins in Fermented Foods

Biological degradation of mycotoxins has been proven to be the best strategy for the effective elimination of these compounds from fermented foods and food products. Biological control, which includes the use of microorganisms for controlling hazardous toxins, has been considered as an important approach against aflatoxins (Reddy et al., 2010a). Palumbo et al. (2006) reported that a number of microorganisms, including *Bacillus*, *Pseudomonas*, *Ralstonia*, and *Burkholderia*, can effectively inhibit growth of *Aspergillus flavus*. Furthermore, *Bacillus subtilis* and *Pseudomonas solanacearum* bacterial strains isolated from the maize rhizosphere were shown to inhibit aflatoxin accumulation (Palumbo et al., 2006). A brief description of the microbial community able to inhibit aflatoxins under laboratory conditions is summarized in Table 28.1. Kim (2007) studied the combined effect of starter cultures on growth reduction and aflatoxin production in Korean fermented foods by coinoculating *Aspergillus* species with several bacterial strains, including *Bacillus subtilis*. In addition, other factors such as ammonia production during fermentation, light, microbial competition with *Bacillus* species, and addition of charcoal or vitamin C may also lead to degradation of aflatoxins in soybean products during fermentation (Lee et al., 2012; Kim et al., 2010b). El-Nezami et al. (2002) reported that the lactic acid bacteria *Lactobacillus rhamnosus* is able to remove AFB1 from artificially contaminated liquid media with about 80% toxin removal efficiency. Application of competitive aflatoxigenic strains, such as *Aspergillus flavus* and *Aspergillus parasiticus* is known to result in the biological control of aflatoxin contamination in different crops to a greater extent (Yin et al., 2008).

TABLE 28.1 A List of Bacterial and Fungal Starter Cultures That Inhibit Growth of Aflatoxin-Producing Fungi

Starter Culture Strain	Inhibited Fungi	References
Bacterial Species		
Bacillus subtilis	Aspergillus flavus	Reddy et al. (2009) and Zhang et al. (2008)
Bacillus pumilis	Aspergillus flavus, Aspergillus parasiticus	Cho et al. (2009)
Lactobacillus species	Aspergillus flavus, Aspergillus parasiticus	Karunaratne et al. (1990)
Stenotrophomonas sp.	Aspergillus flavus, Aspergillus parasiticus	Jermnak et al. (2013)
Lactobacillus casei	Aspergillus flavus	Chang and Kim (2007)
Bacillus cereus	Aspergillus species	Kumar et al. (2014)
Pseudomonas putida	Aspergillus flavus, Aspergillus parasiticus	Haggag et al. (2014)
Streptomyces aureofaciens	Aspergillus flavus, Aspergillus parasiticus	Haggag et al. (2014)
Fungal Species		
Trichoderma viride	Aspergillus flavus	Reddy et al. (2009)
Trichoderma harzianum	Aspergillus flavus	Anjaiah et al. (2006)
Aspergillus flavus	Aspergillus flavus, Aspergillus parasiticus	Dorner (2004, 2008) and Hruska et al. (2014)
Nannocystic exedens	Aspergillus flavus, Aspergillus parasiticus	Taylor and Draughon (2001)
Pichia anomala	Aspergillus flavus, Aspergillus parasiticus	Haggag et al. (2014)

Nowadays, a considerable amount of research is being focused on the biological control or detoxification of OTA in grape and wine, which are the second highest sources of OTA after cereals (Halasz et al., 2009). A number of microbes, including fungi and bacteria, have been effectively used as a feasible biological control approach to significantly degrade OTA up to more than 95% (Abunrosa et al., 2006). Based on these findings, it can be hypothesized that microorganisms can have real applicability for aflatoxin decontamination in foods and feeds. These findings are also supported by results showing that microorganisms, such as yeasts and bacteria, show potent efficacy in controlling aflatoxin decontamination, although variations in degradation rate were detected for both (Halasz et al., 2009).

28.5 LEGAL VALIDATION FOR AFLATOXIN LIMITS

Toxigenic fungi are ubiquitously distributed in nature and occur regularly in worldwide food supplies due to mold infestation of susceptible agricultural products, such as cereal grains, nuts, and fruits. Although higher numbers of mycotoxins occur in nature, only a few constitute food safety challenges. Hence, control measures are necessary to prevent hazardous outbreaks in humans and animals caused by aflatoxins, along with their tolerance levels by international authorities. Regulations concerning aflatoxin tolerance levels vary from country to country, as industrialized countries have set lower tolerance levels than developing countries where most susceptible commodities are produced. For instance, tolerance levels for aflatoxins in foods from Sweden, Japan, and Brazil have been set at 5, 10, and 30 μg/kg, respectively (Mazumder and Sasmal, 2001). The first legislative act was enacted in 1965 by the USFDA, which proposed a tolerance level of 30 μg/kg for total aflatoxins in foods. However, it was further lowered to 20 μg/kg due to the potent toxicity of aflatoxins.

According to the Korean Food and Drug Administration (KFDA), a sample containing 10 ppb (10 μg/kg) of aflatoxins is assumed to be hazardous (KFDA, 2009). However, the CAC Joint FAO/WHO food standards program adopted a consumption limit of 15 ppb (15 μg/kg) for total aflatoxins (CAC, 2001). A description of hazardous limits of various aflatoxins announced by various international agencies is summarized in Table 28.2. Regulation parameters for

TABLE 28.2 Description of Hazardous Limits of Different Aflatoxins in Various Fermented Foods According to Different International Agencies

Food/Feed	Aflatoxin	Level	Agency
All foods	Total aflatoxin	20 ng/kg	USFDA
Milk	Total aflatoxin	0.5 ng/kg	
Cotton seed meal (as a feed)	Total aflatoxin	300 ng/kg	
Corn for dairy cattle	Total aflatoxin	20 ng/kg	
Feed for dairy cattle	Aflatoxin B1	5 μg/kg	FAO/WHO
Milk	Aflatoxin M1	0.05 μg/kg	
Raw peanuts for humans (all foods)	Total aflatoxin	15 μg/kg	
All food materials	Aflatoxin B1	10 μg/kg	KFDA
	Aflatoxin M1	0.5 μg/kg	
	Total aflatoxin	10 μg/kg	

setting tolerance limits of aflatoxins in different countries are based on carcinogenic risk for humans exposed to aflatoxins.

In developing countries, aflatoxin occurs especially after harvesting and storage, since crops are usually sundried in open fields and then stored under humid conditions, leading to formation of aflatoxins in agricultural products. For instance, Pakistan as the sixth largest exporter of chilies previously detected aflatoxins in exported dried chili materials as confirmed by EU Food Authorities (EC, 2001).

28.6 GENERAL DETECTION METHODS TO ANALYZE AFLATOXINS IN FERMENTED FOODS

Application of chromatographic techniques, such as gas chromatography (GC), thin-layer chromatography (TLC), and high-performance liquid chromatography (HPLC) has been proven to be an effective approach for analysis of mycotoxins, including aflatoxins in various fermented foods or food products, followed by their fluorescence-based detection (Cavaliere et al., 2006; Vosough et al., 2010). Various conventional methods have been used for the detection of different aflatoxins in fermented foods and food products, but these have been proven to be less effective due to poor separation ability, inaccuracy, low sensitivity, prolonged operation times, as well as pre- and postderivatization for complete detection (Huang et al., 2010). However, application of HPLC coupled with fluorescence detection has achieved significant results for the accurate detection of aflatoxins from a variety of fermented foods (Shephard, 2009). Moreover, application of more advanced chromatographic techniques such as ultra-high-performance liquid chromatography has made accurate detection of desired aflatoxins possible in less time (Huang et al., 2010). Recently, Huang et al. (2010) performed quantitative determination of aflatoxins from fermented products with a quantitative limit of 0.012–0.073 µg/kg. On the other hand, detection of aflatoxins in fermented maize analyzed by TLC and spectrophotometric measurements resulted in a detection limit of 2.20–2.55 µg/mL (Adegoke et al., 2010). Chromatographic methods have been proven to be advantageous for quantitative detection of aflatoxins due to their better sensitivity and specificity.

28.7 RAPID DETECTION METHODS TO ANALYZE AFLATOXINS IN FERMENTED FOODS

Due to increasing interest in the beneficial health effects of fermented foods in humans, it is important to ensure the food safety of fermented food products from a consumer point of view (Lee et al., 2012). Several methods have been used to detect aflatoxins, such as the traditional dilution plating method, TLC, GC, HPLC, and chemical analysis, in foods and feeds. Although these methods have been proven to be effective, they have several drawbacks. Thus, there is a need to develop more rapid and advanced methods for easy and rapid detection of aflatoxins. Hence, in this chapter, we summarize the importance and significance of various advanced and rapid detection methods for the detection of aflatoxins in fermented food products. A list of various recent adopted methods for the analysis of aflatoxins in various food samples is presented in Table 28.3.

28.7.1 Polymerase Chain Reaction (PCR)-Based Molecular Detection Methods

Molecular approaches, such as PCR-based methods, can serve as good alternatives for the detection of aflatoxins due to their simplicity as well as specific and sensitive detection (Niessen, 2008). According to Levin (2012), there is no specific PCR assay for the detection of any of the four aflatoxins produced biologically. The late stages of AFB1 synthesis are carried out by two enzymes, a methyltransferase encoded by the aflP (omtA) gene that converts sterigmatocystin to O-methylsterigmatocystin and an oxidoreductase encoded by the aflQ (ordA) gene that converts O-methylsterigmatocystin to AFB1 (Bhatnagar et al., 1991; Levin, 2012). It was confirmed that these late stages of AFB1 and AFB2 synthesis are catalyzed by common enzymes that use separate precursors as substrates for each toxin (Bhatnagar et al., 1991).

28.7.2 Enzyme-Linked Immunosorbent Assay (ELISA)-Based Detection Methods

A number of ELISA techniques have been used for the detection of aflatoxins in foods, including solid-phase radioimmunoassay, monoclonal affinity column immunoassay, and ELISA (Pestka et al., 1980). Based on the results obtained using ELISA or affinity column techniques for aflatoxin detection, it was confirmed that these techniques are more appropriate than solid-phase radioimmunoassay for toxin detection and should thus be used and developed more extensively (Pittet, 2005). The major advantages of ELISA and affinity column methods include speed, ease of sample preparation, ease of use, and potentially low cost per analysis (Pittet, 2005). However, variations in antibody specificities for B1 and cross-reactivity with other aflatoxins have been noted as major disadvantages. In addition, ELISA techniques are qualitative or semiquantitative, as well as temperature-sensitive (Pittet, 2005). Nowadays, ELISA techniques are widely used for easy screening of AFB1 even at low concentrations. The color developed by the enzyme-mediated reaction allows the determination of the AFB1 amount present in the sample.

Recently, Xie et al. (2014) developed a method for the easy detection of AFB1 in soybean sauce using an immunomagnetic bead system for pretreatment and ELISA for quantification. The pretreatment method using immunomagnetic beads showed better performance than conventional extraction and

TABLE 28.3 Description of Occurrence of Different Aflatoxins in Various Fermented Foods

Food	Aflatoxin	Country	Detected Level	Detection Method	References
White dent maize (Zea mays)	Aflatoxin B1	Nigeria	1.05–2.55 µg/mL	Spectrophotometry	Adegoke et al. (2010)
Beer and wine	Aflatoxin B1	Germany	<43 ppb	Photon-induced fluorescence	Rasch et al. (2010)
Fermented alcoholic beverages	Aflatoxin B1	Thailand	0.3–2.15 µg/kg	ELISA	Charoenpornsook and Kavisarasai (2014)
Blue cheese	Aflatoxin B1	Thailand	0.5–1.25 µg/kg	ELISA	Charoenpornsook and Kavisarasai (2014)
Fermented soybean product	Aflatoxin B1	Thailand	0.2–3.20 µg/kg	ELISA	Charoenpornsook and Kavisarasai (2014)
Finish red wine (grapes)	AFB1	France	0.012–0.126 µg/L	Reversed-phase HPLC	Khoury et al. (2006)
Traditional Korean Doenjang	Total aflatoxin	Korea	0.00–42.2 µg/kg	ELISA	Kim and Kim (2012)
Cheese	Aflatoxin M1	Slovenia	50 µg/kg	ELISA	Torkar and Vengust (2008)
Lighvan cheese	Aflatoxin M1	Iran	0.85 µg/kg	TLC	Fallah (2010)
Traditional Iranian yogurt	Aflatoxin M1	Iran	0.018 µg/kg	TLC	Fallah (2010)
Commercial Iranian yogurt	Aflatoxin M1	Iran	0.038 µg/kg	TLC	Fallah (2010)

ELISA, Enzyme-Linked Immunosorbent Assay; TLC, Thin layer chromatography.

immunoaffinity column methods. As a result, ELISA exhibited a good linear relationship for AFB1 with a concentration range of 0.05–0.3 µg/kg, and average recoveries across spike levels varied from 0.5 to 7 µg/kg or 83.6–104%. This proposed method facilitates rapid and easy detection of AFB1 with a high recovery rate, suggesting it can be used for the trace determination of AFB1 in soybean fermented foods.

28.7.3 Biosensor-Based Detection Methods

Biosensors are small, portable, and analytical devices based on a combination of recognition biomolecules with an appropriate transducer, and they are able to detect chemical or biological materials selectively with high sensitivity (Paddle, 1996). The detection principle of biosensor devices is based on specific binding of the analyte of interest to the complementary biorecongnition element immobilized on a suitable support medium (Liu et al., 2006). Biosensor devices can be divided into different groups, such as electrochemical, optical, thermometric, piezoelectric, or magnetic biosensors, based on the method of signal transduction (Liu et al., 2006). Among these devices, electrochemical and optical biosensors are the most commonly used methods for detection of aflatoxins (Velasco-Garcia and Mottram, 2003). Detection of aflatoxins is coupled with reduction of acetylcholinesterase activity, which is measured using a choline oxidase amperometric biosensor (Nayak et al., 2009). On the other hand, amperometric methods have been proven to be very effective for measuring even low amounts of aflatoxins, which has become a burning issue for classical spectrophotometry due to omission of the dilution step (Nayak et al., 2009).

Moreover, electrochemical immunosensor devices, which use antibodies incorporated into a biorecognition layer to produce electroactive signals detectable by transducers (amplifiers), are used for analytical purposes (Morgan et al., 1996; Wacoo et al., 2014). Signal measurement is achieved through differential pulse voltammetry, cyclic voltammetry, chronoamperometry, electrochemical impedance spectroscopy, or linear sweep voltammetry (Valimaa et al., 2010; Wacoo et al., 2014). A number of electrochemical immunosensors involving immobilization of antibodies onto the electrode surface have been developed for the quantitative estimation of aflatoxins (Liu et al., 2006; Masoomi et al., 2013; Wacoo et al., 2014).

28.7.4 Immunochromatographic Test Strip Detection Methods

Nowadays, there is an increasing demand for quick immunochromatographic tests for the rapid and direct detection of aflatoxins using a disposable device. Lateral flow immunoassay, in which carrier membranes (usually polyvinylidene difluoride, nylon, or nitrocellulose) are used to immobilize either the antibody or antigen, has become a popular and improved alternative to ELISA. Based on the format pattern, results can be obtained in just one to three working steps (Krska et al., 2005). The immunochromatographic assay device determines the concentration of the target analyte with simple and rapid performance (Tanaka et al., 2006). In addition, immunochromatographic assays have several advantages over other detection methods, such as rapid on-site detection within a few minutes, only one sample extraction step before use, and detection of concentration levels of target analytes using only the naked eye (Tanaka et al., 2006), making these assays popular choices for a wide range of applications. Zhang et al. (2011) developed a selective immunochromatographic assay for the detection of AFB1 with a visual detection limit of 1 ng/mL and detected no cross-reactivity with other aflatoxins in peanuts, pu-erh tea, vegetable oil, and feedstuff, confirming its suitability for selective detection of AFB1 in agro-products.

Moreover, other immunochromatographic assays utilizing colloidal gold particles and antibodies have been developed for the effective detection of AFB1 in various food products, with a complete analysis time of 10 min as well as a twofold higher detection limit (Xiulan et al., 2005). Huang et al. (2014) also successfully developed a novel method for detection of AFM1 by immunochromatography coupled with enrichment based on immunomagnetic nanobeads in raw milk sample with an AFM1 detection limit of 0.1 ng/mL, which is lower than the regularly defined limit of AFM1 in foods as defined by Chinese legislation. These methods also showed no crossreactivity with any other aflatoxin, suggesting that the proposed method can be used for the rapid and on-site detection of AFM1 in various food samples (Huang et al., 2014).

28.8 CONCLUSIONS

Aflatoxins are highly poisonous secondary metabolites produced by fungal strains of *Aspergillus* species such as *Aspergillus flavus* and *Aspergillus parasiticus*. Aflatoxins have been found in moldy human foods and animal feeds and are implicated in various health complications in humans and animals. Aflatoxins can also pass through the food chain to humans through consumption and accumulation in animals. Disease control departments worldwide should create a healthy environment for maintaining social stability and national security as well as improving public health through prevention and control of diseases resulting from aflatoxin contamination. In order to prevent health risks, a number of methods (storage at proper moisture and temperature, standardization of production methods, improvement of production conditions, use of starter cultures, microwave treatments, packaging, etc.) can be applied to reduce/eliminate molds from various fermented food products, leading to formation of reduced levels of aflatoxins.

ACKNOWLEDGMENTS

This research was supported by the Ministry of Trade, Industry and Energy (MOTIE) and Korea Institute for Advancement of Technology (KIAT) through the Research and Development for Regional Industry (Grant No. R0003533) in 2014.

REFERENCES

Abarca, M.L., Accensi, F., Bragulat, M.R., Castella, G., Cabanes, F.J., 2003. *A. carbonarius* as the main source of ochratoxin A contamination in dried vine fruits from the Spanish market. Journal of Food Protection 66, 504–506.

Abunrosa, L., Santos, L., Venancio, A., 2006. Degradation of ochratoxin-A by proteases and crude enzyme extract of *Aspergillus niger*. Food Biotechnology 20, 231–236.

Adegoke, A.S., Akinyanju, J.A., Olajide, J.E., 2010. Fermentation of aflatoxin contaminated while dent Maize (*Zea mays*). Research Journal of Medical Sciences 4, 111–115.

Akerstrand, K., Molander, A., Andersson, A., Nilsson, G., 1976. Occurrence of molds and mycotoxins in frozen blueberries. Var Foda 28, 197–200.

Allameh, A., Safamehr, A., Mirhadi, S.A.M., Shivazad, M., Razzaghi-Abyaneh, M., Afshar-Naderi, A., 2005. Evaluation of biochemical and production parameters of broiler chicks fed ammonia treated aflatoxin contaminated maize grains. Animal Feed Science and Technology 22, 289–301.

Anjaiah, V., Thakur, R.P., Koedam, N., 2006. Evaluation of bacteria and *Trichoderma* for biocontrol of pre-harvest seed infection by *Aspergillus flavus* in groundnut. Biocontrol Science and Technology 16, 431–436.

Anon, 2011. CAC/RCP 243: 2003, Codex Standard for Fermented Milks.

Arab, M., Sohrabvandi, S., Mortazavian, A.M., Mohammadi, R., Rezaei, T.M., 2012. Reduction of aflatoxin in fermented milks during production and storage. Toxin Reviews 31, 44–53.

Asao, T., 1965. Structures of aflatoxins B1 and G1. Journal of the American Chemical Society 87, 882–886.

Bakirci, I., 2001. A study on the occurrence of aflatoxin M1 in milk and milk products produced in Van province of Turkey. Food Control 12, 47–51.

Bejaoui, H., Mathieu, F., Thailladier, P., Lebrihi, A., 2006. Biodegradation of ochratoxin-A by *Aspergillus* section *Nigri* species isolated from French grapes: a potential means of ochratoxin-A decontamination in grape juices and musts. FEMS Microbiology Letters 255, 203–206.

Benblesa, J., Soriano, J.M., Molto, J.C., Manes, J., 2004. Limited survey for the presence of aflatoxins in foods from local markets and supermarkets in Valencia, Spain. Food Additives & Contaminants 21, 165–171.

Bhatnagar, D., Cleveland, T., Kingston, D., 1991. Enzymological evidence for separate pathways for aflatoxin B1 and B2 biosynthesis. Biochemistry 30, 4343–4350.

Blesa, J., Soriano, J.M., Molto, J.C., Manes, J., 2004. Concentration of ochratoxin A in wines from supermarkets and stores of Valencian Community (Spain). Journal of Chromatography A 1, 397–401.

Blount, W.P., 1961. Turkey "X" disease. Journal of the British Turkey Federation 9, 52–54.

Cabanes, F.J., Accensi, F., Bragulat, M.R., Abarca, M.L., Castella, G., Minguez, S., Pons, A., 2002. What is the source of ochratoxin A in wine. International Journal of Food Microbiology 79, 213–215.

CAC - Codex Alimentarius Commission, 2001. Joint FAO/WHO Food Standards Programme, Codex Committee on Food Additives and Contaminants. Thirty-Third Session. CODEX, Hague, Netherlands.

Calvo, A.M., 2005. Mycotoxins. In: Darbrowski, W.M. (Ed.), Toxins in Food. CRC Press, Boca Raton, FL, USA, pp. 216–225.

Carvajal, M., Bolaños, A., Rojo, F., Méndez, I., 2003. Aflatoxin M1 in pasteurized and ultra-pasteurized milk with different fat content in Mexico. Journal of Food Protection 66, 1885–1892.

Cavaliere, C., Foglia, P., Pastorini, E., Samperi, R., Lagana, A., 2006. Liquid chromatography/tandem mass spectrometric confirmatory method for determining aflatoxin M1 in cow milk: comparison between electrospray and atmospheric pressure photoionization sources. Journal of Chromatography A 1101, 69–78.

Cepeda, A., Franco, C.M., Fente, C.A., Vazquez, B.I., Rodriguez, J.L., Prognon, P., Mahuzier, G., 1996. Post-column excitation of aflatoxins using cyclodextrins in liquid chromatography for food analysis. Journal of Chromatography A 1, 69–74.

Chakrabarty, A.B., 1981. Detoxification of aflatoxin in corn. Journal of Food Protection 44, 173–176.

Chang, I., Kim, J.D., 2007. Inhibition of aflatoxin production of *Aspergillus flavus* by *Lactobacillus casei*. Mycobiology 35, 76–81.

Charoenpornsook, K., Kavisarasai, P., 2014. Determination of aflatoxin B1 in food products in Thailand. African Journal of Biotechnology 13, 4761–4765.

Chiavaro, E., Asta, D., Galaverna, G., Biancardi, A., Gambarelli, E., Dossena, A., Marchelli, R., 2001. New reversed-phase liquid chromatography method to detect aflatoxins in food and feed with cyclodextrins as fluorescence enhancers added to the eluent. Journal of Chromatography A 937, 31–40.

Cho, K.M., Math, R.K., Hong, S.Y., Islam, S.M.A., Mandanna, D.K., Cho, J.J., Yun, M.J., Kim, J.M., Yun, H.D., 2009. Iturin produced by *Bacillus pumilus* HY1 from Korean soybean sauce (*kanjang*) inhibits growth of aflatoxin producing fungi. Food Control 20, 402–406.

Coomes, T.J., Crowther, P.C., Feuell, A.J., Francis, B.J., 1966. Experimental detoxification of groundnut meals containing aflatoxin. Nature 290, 406–407.

Dorner, J.W., 2004. Biological control of aflatoxin contamination of crops. Journal of Toxicology Toxin Reviews 23, 425–450.

Dorner, J.W., 2008. Management and prevention of mycotoxins in peanuts. Food Additives & Contaminants 25, 203–208.

El-Nezami, H., Polychronaki, N., Salminen, S., Mykkanen, H., 2002. Binding rather than metabolism may explain the interaction of two food-grade *Lactobacillus* strains with zearalenone and its derivative α-zearalenol. Applied and Environmental Microbiology 68, 3545–3549.

European Commission, 2001. Regulation (EC) No. 466/2001 of 8 March 2001, Setting maximum levels for certain contaminants in foodstuffs. Official Journal of European Commission 77, 1–13.

Fallah, A.A., 2010. Aflatoxin M1 contamination in dairy products marketed in Iran during winter and summer. Food Control 21, 1478–1481.

Farombi, E.O., 2006. Aflatoxin contamination of foods in developing countries: implications for hepatocellular carcinoma and chemopreventive strategies. African Journal of Biotechnology 5, 1–14.

Feuell, A.J., 1966. Aflatoxin in groundnuts, problems of detoxification. Tropical Science 8, 61–70.

Goldblatt, L.A., 1969. Aflatoxin. Academic Press, New York, NY, USA.

Govaris, A., Roussi, V., Koidis, P.A., Botsoglou, N.A., 2002. Distribution and stability of aflatoxin M1 during production and storage of yoghurt. Food Additives & Contaminants 19, 1043–1050.

Gupta, S.K., Venkitasubramanian, T.A., 1975. Production of aflatoxin on soybeans. Applied Microbiology 29, 834–836.

Haggag, W., El Habbasha, E.F., Mekhail, M., 2014. Potential biocontrol agents used for management of aflatoxin contamination in corn grain crop. The Research Journal of Pharmaceutical, Biological and Chemical Sciences 5, 521–527.

Halasz, A., Lasztity, R., Abony, T., Bata, A., 2009. Decontamination of mycotoxin-contaminating food and feed by biodegradation. Food Reviews International 25, 284–298.

Hernandez-Mendoza, A., Garcia, H.S., Steele, J.L., 2009. Screening of *Lactobacillus casei* strains for their ability to bind aflatoxin B1. Food and Chemical Toxicology 47, 1064–1068.

Hruska, Z., Rajasekaran, K., Yao, H., Kincaid, R., Darlington, D., Brown, R., Bhatnagar, D., Cleveland, T.E., 2014. Co-inoculation of aflatoxigenic and non-aflatoxigenic strains of *Aspergillus flavus* to study fungal invasion, colonization and competition in maize kernels. Frontiers in Microbiology 5, 1–7.

Huang, B., Han, Z., Cai, Z., Wu, Y., Ren, Y., 2010. Simultaneous determination of aflatoxins B1, B2, G1, G2, M1 and M2 in peanuts and their derivative products by ultra- high-performance liquid chromatography-tandem mass spectrometry. Analytica Chimica Acta 662, 62–68.

Huang, Y.M., Liu, D.F., Lai, W.H., Xiong, Y.H., Yang, W.C., Liu, K., Wang, S.Y., 2014. Rapid detection of aflatoxin M1 by immunochromatography combined with enrichment based on immunomagnetic nanobead. Chinese Journal of Analytical Chemistry 42, 654–659.

Iamanaka, B.T., de Menezes, H.C., Vicente, E., Leite, R.S.F., Taniwaki, M.H., 2007. Aflatoxigenic fungi and aflatoxins occurrence in sultanas and dried figs commercialized in Brazil. Food Control 18, 454–457.

Iha, M.H., Barbosa, C.B., Okada, I.A., Trucksess, M.W., 2011. Occurrence of aflatoxin M1 in dairy products in Brazil. Food Control 22, 1971–1974.

International Agency for Research on Cancer (IARC), 1993. Monographs on the Evaluation of Carcinogenic Risks to Human. Some Naturally Occurring Substances: Food Items Constituents, Heterocyclic Aromatic Amines Mycotoxins, vol. 56, pp. 362–395.

ISIRI - Institute of Standard and Industrial Research of Iran, 2002. Maximum Tolerated Limits of Mycotoxins in Foods and Feeds. National Standard No. 5925.

Jermnak, U., Chinaphuti, A., Poapolathep, A., Kawai, R., Nagasawa, H., Sakuda, S., 2013. Prevention of aflatoxin contamination by a soil bacterium of *Stenotrophomonas* sp. that produces aflatoxin production inhibitors. Microbiology 159, 902–912.

Kang, K.J., Park, J.H., Cho, J.I., 2000. Control of aflatoxin and characteristics of the quality in *Doenjang* (soybean paste) prepared with antifungal bacteria. Korean Journal of Food Science and Technology 32, 1258–1265.

Karunaratne, A., Wezenberg, E., Bullerman, L.B., 1990. Inhibition of mold growth and aflatoxin production by *Lactobacillus* spp. Journal of Food Protection 53, 230–236.

KFDA - Korea Food and Drug Administration, 2009. Outbreak Food Poisoning. http://e-start.kfda.go.kr.

Khoury, A.E., Rizk, T., Lteif, R., Azouri, H.l., Delia, M.L., Lebrihi, A., 2006. Occurrence of ochratoxin A- and aflatoxin B1-producing fungi in Lebanese grapes and ochratoxin A content in musts and finished wines during 2004. Journal of Agricultural and Food Chemistry 54, 8977–8982.

Khoury, A.E., Atoui, A., Yaghi, J., 2011. Analysis of aflatoxin M1 in milk and yogurt and AFM1 reduction by lactic acid bacteria used in Lebanese industry. Food Control 22, 1695–1699.

Kim, J.G., 2007. Anti-aflatoxigenic activity of some bacteria related with fermentation. In: Méndez-Vilas, A. (Ed.), Communicating Current Research and Educational Topics and Trends in Applied Microbiology. Formatex, Badajoz, Spain, pp. 322–328.

Kim, H.J., Lee, J.E., Kwak, B.M., Ahn, J.H., Jeong, S.H., 2010a. Occurrence of aflatoxin M1 in raw milk from South Korea winter seasons using an immunoaffinity column and high performance liquid chromatography. Journal of Food Safety 30, 804–813.

Kim, T.W., Kim, Y.H., Kim, S.E., Lee, J.H., Park, C.S., Kim, H.Y., 2010b. Identification and distribution of Bacillus species in Doenjang by whole cell protein patterns and 16S rRNA gene sequence analysis. Journal of Microbiology and Biotechnology 20, 1210–1214.

Kim, D.M., Chung, S.H., Chun, H.S., 2011. Multiplex PCR assay for the detection of aflatoxigenic and non-aflatoxigenic fungi in *Meju*, a Korean fermented soybean food starter. Food Microbiology 28, 1402–1408.

Kim, M., Kim, Y.S., 2012. Detection of foodborne pathogens and analysis of aflatoxin levels in home-made *Doenjang* samples. Preventive Nutrition and Food Science 17, 172–176.

Kok, W.T., 1994. Derivatization reactions for the determination of aflatoxins by liquid chromatography with fluorescence detection. Journal of Chromatography B 659, 127–137.

Krishnamurthy, Y.L., Shashikala, J., 2006. Inhibition of aflatoxin B1 production of *Aspergillus flavus* isolated from soybean seeds by certain natural plants products. Letters in Applied Microbiology 43, 469–474.

Krska, R., Welzig, E., Berthiller, F., Molinelli, A., Mizaikoff, B., 2005. Advances in the analysis of mycotoxins and its quality assurance. Food Additives & Contaminants 22, 345–353.

Kumar, S.N., Sreekala, S.R., Chandrasekaran, D., Nambisan, B., Anto, R.J., 2014. Biocontrol of *Aspergillus* species on peanut kernels by antifungal diketopiperazine producing *Bacillus cereus* associated with entomopathogenic nematode. PLoS One 9, e106041.

Kwon, D.Y., Hong, S.M., Ahn, I.S., Kim, M.J., Yang, H.J., Park, S.M., 2010. Isoflavonoids and peptides from *Meju*, long-term fermented soybeans, increase insulin sensitivity and exert insulinotropic effects in vitro. Nutrition 27, 244–252.

Lee, G.I., Lee, H.M., Lee, C.H., 2012. Food safety issues in industrialization of traditional Korean foods. Food Control 24, 1–5.

Levin, R.E., 2012. PCR detection of aflatoxin producing fungi and its limitations. International Journal of Food Microbiology 156, 1–6.

Liu, Y., Qin, Z., Wu, X., Jiang, H., 2006. Immune-biosensor for aflatoxin B1 based bio-electrocatalytic reaction on micro-comb electrode. Biochemical Engineering Journal 32, 211–217.

Marsh, M.A.J., Hill, C., Ross, R.P., Cotter, P.D., 2014. Fermented beverages with health-promoting potential: past and future perspectives. Trends in Food Science & Technology 38, 113–124.

Masoomi, L., Sadeghi, O., Banitaba, M.H., Shahrjerdi, A., Davarani, S.S.H., 2013. A non-enzymatic nanomagnetic electroimmunosensor for determination of aflatoxin B1 as a model antigen. Sensors and Actuators B: Chemical 177, 1122–1127.

Mazumder, P.M., Sasmal, D., 2001. Mycotoxins – limits and regulations. Ancient Science of Life 1, 1–9.

Mishra, H.N., Das, C., 2003. A review on biological control and metabolism of aflatoxin. Critical Reviews in Food Science and Nutrition 43, 245–264.

Mohammadi, H., 2011. A review of aflatoxin M1, milk, and milk products, In: Guevara-Gonzalez, R.G. (Ed.), Aflatoxins - Biochemistry and Molecular Biology. InTech Publication, 395-398.

Moreno-Martinez, E., Vazquez-Badillo, M., Facio-Parra, F., 2000. Use of propionic acid salts to inhibit aflatoxin production in stored grains of maize. Agrociencia 34, 477–484.

Morgan, C.L., Newman, D.J., Price, C.P., 1996. Immunosensors: technology and opportunities in laboratory medicine. Clinical Chemistry 42, 193–209.

Moss, M.O., 2002. Risk assessment for aflatoxins in foodstuffs. International Biodeterioration & Biodegradation 50, 137–142.

Nayak, M., Kotian, A., Marathe, S., Chakravortty, D., 2009. Detection of microorganisms using biosensors–a smarter way towards detection techniques. Biosensors and Bioelectronics 25, 661–667.

Nieduetzki, G., Lach, G., Ceschwill, K., 1994. Determination of aflatoxin in food by use of an automatic work station. Journal of Chromatography A 661, 175–180.

Niessen, L., 2008. PCR-based diagnosis and quantification of mycotoxin-producing fungi. Advances in Food and Nutrition Research 54, 81–138.

Paddle, B., 1996. Biosensors for chemical and biological agents of defense interest. Biosensors and Bioelectronics 11, 1079–1113.

Palumbo, J.D., Baker, J.L., Mahoney, N.E., 2006. Isolation of bacterial antagonists of *Aspergillus flavus* from almonds. Microbial Ecology 52, 45–52.

Park, K.Y., Jung, K.O., Rhee, S.H., Choi, Y.H., 2003. Antimutagenic effects of doenjang (Korean fermented soy paste) and its active compounds. Mutation Research 523, 43–53.

Pestka, J.J., Gaur, P.K., Chu, F.S., 1980. Quantitation of aflatoxin B1 and B1 antibody by an enzyme-linked immunosorbent microassay. Applied and Environmental Microbiology 40, 1027–1031.

Petchkongkaew, A., Taillandier, P., Gasaluck, P., Lebrihi, A., 2008. Isolation of *Bacillus* spp. from Thai fermented soybean (Thua-nao): screening for aflatoxin B1 and ochratoxin A detoxification. Journal of Applied Microbiology 104, 1495–1502.

Pitt, D.B., Plestina, J.I., Shepard, R., Solfrizzo, G., Verger, M., Walker, P.J.P., 2001. Safety evaluation of certain mycotoxins in food. In: Joint FAO/WHO Expert Committee on Food Additives (JECFA). Food and Agriculture Organization, Rome Italy, pp. 281.

Pittet, A., 2005. Modern methods and trends in mycotoxin analysis. Mitteilungen aus Lebensmitteluntersuchung und Hygiene 96, 424–444.

Qi, X., Che, Z., 2010. Introduction of the detection and coercion of aflatoxin in fermentation products. China Condiments 12, 22–28.

Rasch, C., Bottcher, M., Kumke, M., 2010. Determination of aflatoxin B1 in alcoholic beverages: comparison of one- and two-photon-induced fluorescence. Analytical and Bioanalytical Chemistry 397, 87–92.

Razzaghi-Abyaneh, M., Shams-Ghahfarokhi, M., Yoshinari, T., Rezaee, M.B., Jaimand, K., Nagasawa, H., Sakuda, S., 2008. Inhibitory effects of *Satureja hortensis* L. essential oil on growth and aflatoxin production by *Aspergillus parasiticus*. International Journal of Food Microbiology 123, 228–233.

Reddy, K.R.N., Nurdijati, S.B., Salleh, B., 2010a. An overview of plant derived products on control of mycotoxigenic fungi and mycotoxins. Asian Journal of Plant Sciences 9, 126–133.

Reddy, K.R.N., Reddy, C.S., Muralidharan, K., 2009. Potential of botanicals and biocontrol agents on growth and aflatoxin production by *Aspergillus flavus* infecting rice grains. Food Control 20, 173–178.

Reddy, K.R.N., Salleh, B., Saad, B., Abbas, H.K., Abel, C.A., Sheir, W.T., 2010b. An overview of mycotoxin contamination in foods and its implications for human health. Toxin Reviews 29, 3–26.

Scott, P.M., 1996. Mycotoxins transmitted into beer from contaminated grains during brewing. Journal of AOAC International 79, 875–882.

Scudamore, K.A., 1998. Mycotoxins. In: Watson, D.H. (Ed.), Natural Toxicants in Food. Sheffield Academic Press Ltd, Sheffield, UK, pp. 147–154.

Sharma, R.S., Trivedi, K.R., Wadodkar, U.R., Murthy, T.N., Punjarath, J.S., 1994. Aflatoxin B1 content in deoiled cakes, cattle feeds and damaged grains during different seasons in India. Journal of Food Science and Technology 31, 244–246.

Shephard, G., 2009. Aflatoxin analysis at the beginning of the twenty-first century. Analytical and Bioanalytical Chemistry 395, 1215–1224.

Shukla, S., Park, H.K., Lee, J.S., Kim, J.K., Kim, M., 2014. Reduction of biogenic amines and aflatoxins in *Doenjang* samples fermented with various *Meju* as starter cultures. Food Control 42, 181–187.

Smith, J.E., 2001. Mycotoxins. In: Watson, D.H. (Ed.), Food Chemical Safety. Woodhead Publishing Limited, Abington, MA, USA, pp. 238–245.

Tanaka, R., Yuhi, T., Nagatani, N., Endo, T., Kerman, K., Takamura, Y., Tamiya, E., 2006. A novel enhancement assay for immunochromatographic test strips using gold nanoparticle. Analytical and Bioanalytical Chemistry 385, 1414–1420.

Taylor, W.J., Draughon, F.A., 2001. *Nannocystis exedens*: a potential biocompetitive agent against *Aspergillus flavus* and *Aspergillus parasiticus*. Journal of Food Protection 64, 1030–1034.

Thanaboripat, D., Mongkontanawut, N., Suvathi, Y., Ruangrattametee, V., 2004. Inhibition of aflatoxin production and growth of *Aspergillus flavus* by citronella oil. King Mongkut's Institute of Technology Ladkrabang Science Journal 4, 1–8.

Thanaboripat, D., 2002. Importance of aflatoxin. King Mongkut's Institute of Technology Ladkrabang Science Journal 2, 38–45.

Torkar, K.G., Vengust, A., 2008. The presence of yeasts, moulds and aflatoxin M1 in raw milk and cheese in Slovenia. Food Control 19, 570–577.

Tripathi, S., Mishra, H.N., 2010. Enzymatic coupled with UV degradation of aflatoxina B1 in red chili powder. Journal of Food Quality 33, 186–203.

Tuinstra, L.G.M., Hassnott, W., 1983. Rapid determination of aflatoxin B1 in Dutch feeding stuffs by high performance liquid chromatography and post-column derivatization. Journal of Chromatography 282, 457–462.

Turner, P.C., Mendy, M., Whittle, H., Fortuin, M., Hall, A.J., Wild, C.P., 2000. Hepatitis B infection and aflatoxin biomarker levels in Gambian children. Tropical Medicine & International Health 5, 837–841.

USFDA, 1996. Sec. 527.400 Whole Milk, Low Fat Milk, Skim Milk-Aflatoxin M1 (CPG7106.210). FDA Compliance Policy Guides. FDA, Washington, DC, USA.

Valimaa, A.L., Kivisto, A.T., Leskinen, P.I., Karp, M.T., 2010. A novel biosensor for the detection of zearalenone family mycotoxins in milk. Journal of Microbiological Methods 80, 44–48.

Velasco-Garcia, M., Mottram, T., 2003. Biosensor technology addressing agricultural problems. Biosystems Engineering 84, 1–12.

Versantroort, C.H.M., Oomen, A.G., Sips, A.J.A.M., 2005. Applicability of an in vitro digestion model in assessing the bioaccessibility of mycotoxins from food. Food and Chemical Toxicology 43, 31–40.

Vosough, M., Bayat, M., Salemi, A., 2010. Matrix-free analysis of aflatoxins in pistachio nuts using parallel factor modeling of liquid chromatography diode array detection data. Analytica Chimica Acta 663, 11–18.

Wacoo, A.P., Wendiro, D., Vuzi, P.C., Hawumba, J.F., 2014. Methods for detection of aflatoxins in agricultural food crops. Journal of Applied Chemistry e2014:706291.

Wilson, C.L., Wisniewski, M.E., 1992. Further alternatives to synthetic fungicides for control of postharvest diseases. In: Tjamos, E.T. (Ed.), Biological Control of Plant Diseases. Plenum Press, New York, NY, USA, pp. 133–138.

Xie, F., Lai, W.H., Saini, J., Shan, S., Cui, X., Liu, D.F., 2014. Rapid pretreatment and detection of trace aflatoxin B1 in traditional soybean sauce. Food Chemistry 150, 99–105.

Xiulan, S., Xiaolian, Z., Jian, T., Zhou, J., Chu, F., 2005. Preparation of gold-labeled antibody probe and its use in immunochromatography assay for detection of aflatoxin B1. International Journal of Food Microbiology 99, 185–194.

Yin, M.C., Cheng, W.S., 1998. Inhibition of *Aspergillus niger* and *Aspergillus flavus* by some herbs and spices. Journal of Food Protection 61, 123–125.

Yin, Y., Yan, L., Jiang, J., Ma, Z., 2008. Biological control of aflatoxin contamination of crops. Journal of Zhejiang University Science B 9, 789–792.

Zaki, M.S., Nevin, E.S., Hend, R., Susan, O.M., Olfat, M.F., 2008. Diminution of aflatoxicosis in *Tilapia nilotica* fish by dietary supplementation with fix in toxin and *Nigella sativa* oil. American-Eurasian Journal of Agricultural & Environmental Science 3, 211–215.

Zhang, T., Shi, Z.Q., Hu, L.B., Cheng, L.G., Wang, F., 2008. Antifungal compounds from *Bacillus subtilis* BFS06 inhibiting the growth of *Aspergillus flavus*. World Journal of Microbiology and Biotechnology 24, 783–788.

Zhang, D., Li, P., Yang, Y., Zhang, Q., Zhang, W., Xiao, Z., Ding, X., 2011. A high selective immunochromatographic assay for rapid detection of aflatoxin B1. Talanta 85, 736–742.

Zorlugenc, B., Zorlugenc, F.K., Oztekin, S., Evliya, I.B., 2008. The influence of gaseous ozone and ozonated water on microbial flora and degradation of aflatoxina B1 in dried figs. Food and Chemical Toxicology 46, 3593–3597.

Chapter 29

Antibiotic Resistance Profile of Microbes From Traditional Fermented Foods

H. Abriouel[1], C.W. Knapp[2], A. Gálvez[1], N. Benomar[1]
[1]*Universidad de Jaén, Jaén, Spain;* [2]*University of Strathclyde, Glasgow, Scotland, United Kingdom*

29.1 OVERVIEW OF HISTORY OF ANTIBIOTIC RESISTANCE

The antibiotic era began very earlier in history. Several scientific evidences found large doses of tetracycline embedded in the bones of ancient Sudanese Nubian mummies (350–550 CE) (Armelagos et al., 1969; Bassett et al., 1980). Recently, the tetracycline-laden bone samples were analyzed by Nelson et al. (2010) by acid extraction and mass spectrometry; results obtained showed that the tetracycline in the bone samples was not a result of contamination from the environment, rather from the Sudanese Nubian population dietary regimen. Thus, a diet of fermented beer or gruel containing large amounts of tetracycline was consumed by these populations, providing them at the same time protection against infectious diseases without knowing the concept of bacteria. The discovery of antibiotics as potential chemotherapeutic agents against pathogenic bacteria was in 1928 by Sir Alexander Fleming, who discovered the first antibiotic, "penicillin." Since then, many antibiotics have been discovered and revolutionized human history; these antibacterial substances were introduced to medicine as successful and powerful agents capable of saving many lives.

The indiscriminate use of antibiotics in human medicine, animal husbandry, aquaculture, and agriculture for several decades has increased the appearance of resistant and opportunistic pathogens, thus resulting in an important public-health risk (Dixon, 2000; Feinman, 1999; Knapp et al., 2010; Magee et al., 1999; Munsch-Alatossava and Alatossava, 2007). However, antibiotic resistance (AR) is an old and natural phenomenon used by bacteria to guarantee their survival against several biotic and abiotic factors in a changing environment. The interconnected microbial ecosystem complicates the problem due to the interaction of bacteria, antibiotics, environment, and humans; resistant bacteria flow from animals to the food chain and, thereafter, amongst the community reservoir. Exposure to antibiotics known as "societal drugs" (Levy, 1997) over several decades has bestowed bacteria with several biochemical and physiological mechanisms responsible for resistance. The selective pressure exerted by antibiotics in the environment, the food chain, and even within the gastrointestinal tract results in the emergence and spread of resistant bacteria and their resistance genes—causing increased cases of disease, treatment failures and, consequently, more deaths and higher therapy costs to society. Thus, the use of antibiotics has had not only an impact on human and animal life but also on bacterial life, since antibiotics have throughout evolution selected the most robust bacteria to survive under the pressure of antibiotics.

Owing to the emergence of AR, a search for new strategies to overcome this growing challenge has been the central subject of several research works, governmental programs, and recommendations by several international organizations, such as the Scientific Committee for Animal Nutrition (SCAN, 1996, 1998), the World Health Organization (WHO, 2001), the Food and Agriculture Organization (FAO), and the World Organization for Animal Health (OIE) (FAO/OIE/WHO, 2003). Furthermore, WHO proclaimed AR a critical, international human-health challenge without geographical limitations, with about 25,000 deaths being recorded in the European Union each year from infections caused by antibiotic-resistant bacteria (WHO, 2011). Those international organizations emphasize the importance of antimicrobial resistance surveillance and antibiotic-usage monitoring as a global strategy for containment of antimicrobial resistance. Among those strategies to reduce the AR gene pool, restriction of both therapeutic and prophylactic uses of antibiotics in clinical settings and food animal production was the first measure adopted, in parallel to the discovery of new antimicrobials with less resistance and modification of the other existing ones, creating a new group of synthetic antibiotics with improved activity and additional characteristics. However, resistance emergence continues and new targets are detected. Thus, restriction of antibiotic use will not be enough, rather knowledge about "how" to use antibiotics, "when" and "for what" is crucial to avoid the spread of resistance. In this sense, many questions arise about why antibiotic resistance is not yet controlled, why we are looking for stronger drugs, etc. To understand this complicated issue and address these questions, we should fill knowledge gaps concerning the spread of AR in the food chain, gastrointestinal tract, and the environment, and also consider consumer habits, which may affect dissemination and persistence of AR all over the world. From a food safety perspective, several reports were published in recent years by the European Food Safety Authority (EFSA, 2008) regarding foods of animal origin being an important reservoir of AR due to several causes, such as historic use of antibiotics for growth promotion in food animals nowadays banned in Europe (SCAN, 1996, 1998), or as therapeutic and prophylactic treatments. Consequently, AR assessment in starter cultures and probiotics remains one of the main criteria to be evaluated.

As antibiotic resistance is recognized as a major public health problem that challenges health care globally, an understanding of avenues for resistance transfer and development in nonclinical environments becomes paramount. In this chapter, we assess the antibiotic resistance profile of microbes present in traditional fermented foods (starter-free) due to their dietary importance, especially in developing countries.

29.2 ANTIBIOTIC RESISTANCE IN TRADITIONAL FERMENTED FOODS

Food fermentation is a natural process, which mainly involves lactic acid bacteria (LAB) and/or yeasts present in the raw material (traditional fermented foods) or added as starter cultures in controlled fermentation. The wide variety of traditional fermented foods is a result of differences in the most represented bacterial species naturally found in raw ingredients of each geographical region, local environmental conditions, and traditional processing procedures. Thus, the wide range of fermented end products is characterized by a diverse microbiota and a variety of organoleptic properties as reported by several authors (Abriouel et al., 2008, 2011, 2012). Traditional fermented foods are not only attractive by their nutritional value (proteins, minerals, fats, and vitamins), distinct flavors, and consistencies, but they also present livelihood opportunities for farmers, processors, and sellers. Microorganisms present in each traditional fermented food could act as resistance superbugs since several reports indicated that fermented foods could be considered as vehicles of antibiotic-resistant bacteria and, thus, AR genes may be transferred to other bacteria including pathogens and commensals through the food chain and into the gastrointestinal tract (Forslund et al., 2013; Wang et al., 2012). In general, characterization of fermentative bacteria in traditional foods of their AR profile and also their virulence properties has not been routinely accomplished, and in many cases, culture-dependent methods leave 99% of uncultured bacteria unexplored. In fact, the uncharacterized indigenous microbiota present in traditional fermented foods may represent a major risk since many AR genes may be present, but not detectable, in cultivable bacteria; furthermore, their sequences may have been modified by mutation, or even new resistance genes may be present, thus making detection by conventional PCR methods a difficult task. Recently, genome sequencing has provided new information about AR genes in cultivable bacteria, and metagenomic analysis of complex bacterial communities remains a strategy for analyzing the global resistance of each fermented product to inform about its safety (Devirgiliis et al., 2014).

29.2.1 Phenotypic and Genotypic Antibiotic Resistance Profile of Lactic Acid Bacteria

LAB are the main guilds responsible for the lactic acid fermentation of different products, resulting in end products, such as lactic acid alone, or in combination with ethanol, acetate, formate, succinate, and carbon dioxide. During the fermentation process, besides lactic acid, an arsenal of antimicrobial substances are produced, such as bacteriocins, hydrogen peroxide, and diacetyl, which are active against pathogens, thus improving the safety of the fermented product and increasing its shelf-life. However, the microbiota present in the raw material depending on contamination with exogenous bacteria and environmental conditions, can result in several microbial successions in the different food-related "ecosystems," such as dairy products, fermented vegetables, and meat (Giraffa, 2004; Weckx et al., 2010; Wimpenny et al., 1995), thus many different variations may result in the microbial community. In the literature, we found several reports on the different traditional foods manufactured all over the world, each involving different genera, species, and even strains of LAB among similarly fermented products. Here, we describe the phenotypic and genotypic AR profile of LAB isolated from traditional fermented foods on the basis of the raw material used, geographical region, and LAB involved. Dominant AR genes can vary depending on the food ecosystem, geographical area, and manufacturing practices. Thus, traditional fermented foods ingested directly without further processing may contain a high level of antibiotic-resistant bacteria and then represent the most significant threats to public health associated with the intake of ethnic foods.

29.2.1.1 Dairy Products

Traditional dairy products were the first fermented foods known by humankind, being highly consumed all over the world due to their healthier and nutritional value. The microbiological quality of raw milk used in the manufacturing of various dairy products is important since many antibiotic-resistant bacteria present in the natural microbiota have been associated with the animal host, such as *Lactococcus* sp., *Leuconostoc* sp., *Lactobacillus* sp., and *Streptococcus uberis*, but also in the environment as contaminants, they may have a great impact on AR gene spread throughout the food chain. Wichmann et al. (2014) showed that by using functional metagenomics, the dairy cow microbiome acts as a reservoir for a great diversity of AR determinants (β-lactams, phenicols, aminoglycosides, and tetracyclines), thus spread of AR genes via horizontal gene transfer to other bacteria throughout the food chain and environment is highly possible and could aggravate the acquisition and spread of AR to pathogenic bacteria. Furthermore, Wang et al. (2006) reported that LAB and *Staphylococcus* sp. were the main AR gene carriers in dairy products. In this sense, LAB frequently isolated from traditional dairy products have belonged to different genera, such as *Lactococcus* sp., *Lactobacillus* sp., *Enterococcus* sp., *Leuconostoc* sp., and *Streptococcus* sp. (Table 29.1), however their prevalence and distribution may vary according to the animal species used as the source of raw milk, the dairy product, the environmental conditions, and the manufacturing practices (Kousta et al., 2010; Tannock et al., 1990). Among LAB, enterococci were the subject of many studies regarding AR due to their importance as opportunistic pathogens involved in serious infectious diseases in humans, such as endocarditis

TABLE 29.1 Phenotypic and Genotypic Antibiotic Resistance of Lactic Acid Bacteria Isolated From Traditional Dairy Products

Dairy Product	Raw Material	Geographical Zone	LAB	Phenotypic Antibiotic Resistance	Genotypic Antibiotic Resistance	References
Several dairy products	Milk	Egypt	Lactococcus spp., Lactobacillus spp., Streptococcus spp.	ERY, PEN, TET, VAN	erm(B), tet(M)	Gad et al. (2014)
Karish (cheese)	Raw milk	Egypt	E. faecalis, E. faecium	CHL, CIP, CLI, ERY, GEN, KAN, LNZ, STR, TET, VAN	tet(M), tet(L), tet(K), erm(B), aph(3')[a]	Hammad et al. (2015)
Nunu	Nonpasteurized cow milk	Nigeria	E. faecium, E. faecalis	CRO, CHL, CIP, CTZ, ERY, GEN, PEF, STR, VAN	ND	Oguntoyinbo and Okueso (2013)
Wara (soft cheese)	Nonpasteurized whole milk from cattle added with plant extract Calotropis procera	Nigeria	E. faecium, E. faecalis	CRO, CHL, CIP, CTZ, ERY, GEN, PEF, STR, VAN	ND	Oguntoyinbo and Okueso (2013)
Gariss (sour milk)	Camel milk	Sudan	E. faecalis, E. faecium, E. dispar, E. durans, E. hirae, E. mundtii, E. sanguinicola, Lb. plantarum, Lb. pentosus, S. acidominimus, S. thermophilus	AMP, CHL, CIP, ERY, PEN, STR, TET, VAN	ND	Ahmed et al. (2012)
Jben (soft white cheese)	Cow's raw milk	Morocco	Lb. brevis, Lb. paracasei, Lb. plantarum	AMP, CHL, GEN	ND	Jamly et al. (2011)

Continued

TABLE 29.1 Phenotypic and Genotypic Antibiotic Resistance of Lactic Acid Bacteria Isolated From Traditional Dairy Products—cont'd

Dairy Product	Raw Material	Geographical Zone	LAB	Phenotypic Antibiotic Resistance	Genotypic Antibiotic Resistance	References
Smen (fermented butter)	Cow's milk	Morocco	Lb. plantarum	—	ND	Jamly et al. (2011)
Riab (curd)	Cow's milk	Morocco	Lb. paracasei	—	ND	Jamly et al. (2011)
Dairy	Milk	China	Lb. salivarius, Lb. vaginalis	ERY, TET	erm(B), tet(M)	Nawaz et al. (2011)
Yogurt	Milk	China	Lb. acidophilus, Lb. brevis, Lb. fermentum, Lb. plantarum, S. thermophilus	ERY, TET	erm(B), tet(M), tet(S)	Nawaz et al. (2011)
Cheese	Goat milk	Brazil	Enterococcus sp., Lactococcus sp.	AMP, CHL, GEN, RIF, VAN	vanA	Perin et al. (2014)
Dairy products (dahi, buttermilk, butter, curd, khoa, yoghut, ice-cream)	Milk	India	E. casseliflavus, E. faecium, E. durans, E. lactis, Lb. fermentum, Lb. plantarum, P. pentosaceus, Ln. mesenteroides	ERY, TET	erm(B)[a], msrC, tet(K) tet(L), tet(M)[a] and tet(W)	Thumu and Halami (2012)
Cheese	Cow's raw milk	Croatia	Ln. mesenteroides subsp. cremoris, Ln. mesenteroides subsp. dextranicum	EFX, NIT, OXA, PMB, STR, SUL	ND	Zdolec et al. (2011)
Bryndza cheese	Sheep and cow's milk	Slovakia	E. faecalis, E. faecium	AMP, ERY, GEN	ND	Kročko et al. (2011)

Cabrales (cheese)	Cow's, goat, or sheep milk	Spain	*Enterococcus* sp., *Lactobacillus* sp., *Leuconostoc* sp., *Lc. lactis*	CHL, CLI, ERY, TET, VAN	*tet*(M)	Floréz et al. (2005, 2008)
Cheese	Cow's milk	Spain	*Lc. garviae*	TET	*tet*(M)[b]	Floréz et al. (2012)
De La Mesta (cheese)	Raw milk	Spain	LAB	ERY, TET	*erm*(B), *erm*(F), *tet*(L), *tet*(M), *tet*(K), *tet*(O), *tet*(S), *tet*(W)	Floréz et al. (2014)
Asiago (cheese)	Raw milk	Italy	LAB	ERY, TET	*erm*(B), *erm*(F), *tet*(M), *tet*(K), *tet*(O), *tet*(S), *tet*(W)	Floréz et al. (2014)
Cheese	Goat milk	Italy	*Lc. garvieae*	TET	*tet*(S)	Fortina et al. (2007)
Mozzarella di Bufala Campana	Buffalo milk	Italy	*E. faecalis*, *Lb. helveticus*, *Lb. plantarum*, *Lb. fermentum*, *Lb. paracasei*, *Lb. delbrueckii*, *Lc. lactis*, *S. bovis*, *S. salivarius*	AMP, ERY, KAN, TET	*erm*(B), *erm*(C), *tet*(M)[a], *tet*(S), ORFs (AMP, KAN)[c]	Devirgiliis et al. (2008, 2010, 2014)
Fiore Sardo PDO[d] (cheese)	Raw ewe's milk	Italy	*Lb. paracasei*	TET	*tet*(M)	Comunian et al. (2010)
Water buffalo Mozzarella PDO[d]	Buffalo milk	Italy	*Lb. paracasei*	TET	*tet*(M)	Comunian et al. (2010)
Pannerone cheese	Cow's milk	Italy	*Lb. paracasei*	ERY, TET	*erm*(B), *tet*(W)	Comunian et al. (2010)
Pannerone curd	Cow's milk	Italy	*Lb. paracasei*	ERY, TET	*erm*(B), *tet*(W)	Comunian et al. (2010)

Continued

TABLE 29.1 Phenotypic and Genotypic Antibiotic Resistance of Lactic Acid Bacteria Isolated From Traditional Dairy Products—cont'd

Dairy Product	Raw Material	Geographical Zone	LAB	Phenotypic Antibiotic Resistance	Genotypic Antibiotic Resistance	References
Parmigiano Reggiano (cheese)	Cow's raw milk	Italy	Lb. rhamnosus, Lb. paracasei; Lb. casei; Lb. harbinensis, Lb. fermentum	STR, VAN	ND	Solieri et al. (2014)
Cheese	Milk	Switzerland	Lc. Lactis subsp. lactis	CHL, STR, TET	tet(S)[a] Conjugative plasmid pK214	Perreten et al. (1997)
Cheese	Goat milk	Bulgaria	E. faecium	CRO, CTX, VAN	vanB	Favaro et al. (2014)
Handmade cheese	Milk	Turkey	Lactobacillus sp.	ERY, TET	erm(B), tet(M)	Çataloluk and Gogebakan (2004)
Dadih	Buffalo milk	Indonesia	Lb. fermentum, Lb.plantarum, P. acidilactici	CHL, ERY	erm(B), erm(C), cat, Tn554	Sukmarini et al. (2014)

Antibiotics: AMP, ampicillin; CHL, chloramphenicol; CIP, ciprofloxacin; CLI, clindamycin; CRO, ceftriaxone; CTX, Cefotaxime; CTZ, cotrimoxazole; EFX, enrofloxacine; ERY, Erythromycin; GEN, gentamycin; KAN, kanamycin; LNZ, linezolid; NIT, nitrofurantoin; OXA, oxacillin; PEF, pefloxacin; PEN, penicillin; PMB, polymyxin B; RIF, rifampicin; STR, streptomycin; SUL, sulfonamide; TET, tetracycline; VAN, vancomycin.
Bacteria: E, Enterococcus; Lb, Lactobacillus; Lc, Lactococcus; Ln, Leuconostoc; P, Pediococcus; S, Streptococcus.
[a]Tn916-like transposon.
[b]Tn6086.
[c]Predicted ORFs with a possible function in antibiotic resistance as revealed by metagenomic study.
[d]Protected Designation of Origin.

(Hummel et al., 2007). Recently, many reports described AR in lactobacilli, since the resistance profile is one of the main criteria for LAB intended to be used as starter cultures or as probiotics, according to the European Food Safety Authority (EFSA, 2005). EFSA's proposed scheme in 2005 and the updated guidelines by the FEEDAP Panel in 2008 were based on the qualified presumption of safety (QPS) and involved the individual assessment and evaluation of acquired AR determinants in LAB with the aim to categorize the resistance type harbored by those strains and then determine their safety in food and feed. For example, lactobacilli have been implicated as etiological agents in some cases of endocarditis (Salvana and Frank, 2006), thus phenotypic and molecular determination of AR is now required to help ensure their susceptibility to antibiotics. Resistance observed in LAB may be intrinsic (or natural), which is chromosomally encoded, mutational (mutation in different regions of the chromosome), or acquired by horizontal gene transfer. Thus, strains exhibiting acquired resistance must be regarded as unsafe and unacceptable to be used as feed additives due to their capacity to gain and transfer mobile genetic elements to other bacteria.

Resistance to tetracycline in bacteria isolated from foods of animal origin is common due to the widespread use of these antibiotics in veterinary practices (Ogier and Serror, 2008). The most common AR genes among LAB from traditional dairy products are *erm*(B) and *tet*(M) genes coding for resistance to erythromycin and tetracycline, respectively (Table 29.1); this fact was previously reported by Ammor et al. (2007). The presence of both *tet*(M) and *erm*(B) genes in *Lc. lactis*, *E. faecalis*, *Lactobacillus* sp., *Leuconostoc* sp., and *Streptococcus* sp. from various dairy products confirmed that both genes are linked, as reported by several authors (Devirgiliis et al., 2010; Walther et al., 2008), being frequently located on plasmids. Furthermore, the *tet*(M)-bearing Tn*916* transposon was described in *Lb. paracasei* isolated from natural whey starter cultures employed in the manufacturing process of Mozzarella di Bufala Campana (Devirgiliis et al., 2009); however, low conjugation frequency to *E. faecalis* was shown, indicating reduced risk of horizontal transfer to pathogenic species within the human gut microflora. The resistance patterns of LAB from traditional dairy products sometimes also showed the presence of other erythromycin and tetracycline resistance determinants, such as *tet*(S), which was demonstrated to be also linked to *erm*(B) on the same nonconjugative plasmid in *Lc. lactis* isolated from the Italian Protected Designation of Origin (PDO) dairy food product Mozzarella di Bufala Campana (Devirgiliis et al., 2010). In the literature, different tetracycline resistance genes [eg, tetracycline efflux genes *tet*(L) and *tet*(K), and ribosomal protection protein *tet*(W) gene] were routinely reported in enterococci isolated from traditional cheeses. In this sense, Hammad et al. (2015) reported that enterococcal (*E. faecium* and *E. faecalis*) strains isolated from Egyptian fresh cheese "karish" carried *tet*(L) and *tet*(M) associated with Tn*916* conjugative transposon, thus the genes synergistically increase the bacteria's MIC (minimum inhibitory concentration) value against tetracycline (Ammor et al., 2008). Furthermore, other resistance genes, such as aph(3′) and erm(B), were also associated with Tn916 in enterococci. Enterococci are intrinsically resistant to aminoglycosides; however, the acquisition of an aminoglycoside phosphotransferase [aph(3′)] that confers resistance to kanamycin by horizontal gene transfer resulted in higher level of aminoglycoside resistance. The presence of enterococci harboring the mobile element Tn916, associated with several AR genes, can undergo precise intra- and interspecies transfer through horizontal gene transfer events, thus Egyptian fresh raw milk cheese could act as a potential reservoir of AR enterococci. This may constitute a public health hazard due to the spread of resistance genes to pathogenic bacteria within the food chain and gastrointestinal tract. In general, traditional fermented foods harboring enterococci have been characterized by multidrug resistance profiles to be intrinsically resistant to several antibiotics (eg, β-lactams, cephalosporins, aminoglycosides, lincosamides, glycopeptides, and trimethoprim-sulfamethoxazole) and also capable of acquiring additional resistance. In fact, the high degree of multidrug resistance in enterococci (both intrinsic and acquired) from raw ingredients of Egyptian cheese and Indian dairy products may be due to the contamination of milk with fecal bacteria, milking equipment, or other environmental sources, such as contaminated water (Poznanski et al., 2004).

On the other hand, Thumu and Halami (2012) showed that the presence of erm(B) and tet(M) in *Pediococcus* sp. and other LAB (Table 29.1) is likely associated with mobile genetic elements, as revealed by their maximum identity with erm(B) and tet(M) associated with Tn916, Tn2010, and Tn5253-like transposons of pathogenic bacteria, such as *Streptococcus pneumoniae* and *Staphylococcus aureus*. Furthermore, Flórez et al. (2012) reported the presence of a transposon highly similar to conjugative Tn6086 from *E. faecalis*, which harbored an active tet(M) gene. The linkage between both AR genes with Tn916, or other transposons, may explain their widespread appearance among multiple genera (Roberts, 2008). The prevalence of acquired erythromycin resistance genes, such as msrC, was also shown in LAB isolated from naturally fermented dairy products (Thumu and Halami, 2012).

Traditional cheeses could be considered as potential reservoirs for large numbers of many types of AR determinants as was reported by Flórez et al. (2014) for traditional Spanish (De La Mesta) and Italian (Asiago) cheeses where they found a large diversity of tetracycline and erythromycin resistance determinants [erm(B), erm(F), tet(L), tet(M), tet(K), tet(O), tet(S) and tet(W)] (Table 29.1). Hence these cheeses may play a key role in the spread of AR genes via the food chain, especially if they are associated with mobile genetic elements (eg, plasmid, transposon, and integrons). The fact that LAB from some traditional dairy products showed less diversity in their AR genotypic patterns does not mean that they are resistant to few antibiotics. Some phenotypic resistances are frequently not associated with the commonly described genes, or they may have natural resistance—such as the intrinsic resistance to vancomycin, which is probably due to the presence of D-Ala-D-lactate in their peptidoglycan, instead of the normal dipeptide D-Ala-D-Ala to which vancomycin binds poorly (Deghorain et al., 2007). However, vancomycin resistance gene markers, such as the transferable *vanA* or *vanB* genes coding for resistance to vancomycin and teicoplanin or only teicoplanin, respectively, were recently detected, on both plasmids and chromosomes, in enterococci isolated from traditional cheeses made with raw goat's milk (Favaro et al., 2014; Perin et al., 2014). However, other phenotypic resistance traits were not related to intrinsic resistance, such as resistance to ampicillin, penicillin, and chloramphenicol. Such resistance may be due to the divergence in nucleotide sequences in those genes rendering them undetectable by described PCR methods, or there may be new genes involved in resistance as was shown by Devirgiliis et al. (2014), which determined the presence of new encoding ORFs responsible for resistance to ampicillin and kanamycin. Moreover, Bennedsen et al. (2011) showed that screening for AR genes via genome sequencing revealed the presence of *tet*(W) and *tet*(S) genes on plasmids in LAB used as probiotics.

According to the European Food Safety Authority (EFSA, 2007) guidelines, the presence of transmissible AR genes is considered undesirable, thus screening for resistance genes and also their location (ie, chromosome, plasmid, transposon, or integron) must be carried out to control the risk related to traditional dairy products containing multidrug-resistant LAB.

29.2.1.2 Fermented Meat

Antibiotics have been widely used to promote growth in livestock for over 60 years, thus pressure exerted by antibiotics has selected for several resistant bacteria in agriculture. Traditional fermented meat has carried several multidrug-resistant LAB, such as *Lactobacillus* sp., *Carnobacterium* sp., *Leuconostoc* sp., *Pediococcus* sp., and *Enterococcus* sp.; many of which have demonstrated resistance to erythromycin and tetracycline. Specifically, the most common AR genes have been the acquired *erm*(B) and *tet*(M) in LAB isolated from several fermented sausages and salami of different countries (Table 29.2); there is the possibility of the genes being spread to other bacteria present in the same ecological niche, such as coagulase-negative staphylococci involved in the final pigmentation of meat, lipolysis, and proteolysis (Iacumin et al., 2006). The presence of mobile genetic elements was reported, such as the conjugative plasmid pRE25 of *E. faecalis* carrying several resistance genes to be spread to other LAB in fermented sausages (Teuber et al., 2003), and also Tn*554* -from *Staphylococcus aureus*, which encodes for resistance to erythromycin and spectinomycin in LAB from Indonesian fermented meat (Sukmarini et al., 2014). Thus, naturally fermented meats could act as a vehicle for AR genes to be spread to other bacteria in the food system and also in the gastrointestinal tract.

TABLE 29.2 Phenotypic and Genotypic Antibiotic Resistance of Lactic Acid Bacteria Isolated From Fermented Meats

Fermented Meat	Raw Material	Geographical Zone	LAB	Phenotypic Antibiotic Resistance	Genotypic Antibiotic Resistance	References
Sausage	Meat	Croatia	Ln. mesenteroides subsp. mesenteroides, Lb. curvatus	EFX, MTZ, NAL, NM, NIT, OXA, PMB, STR, SUL, TMP, VAN	ND	Zdolec et al. (2011)
			Lb. brevis, Lb. fermentum			
			Lb. paracasei subsp. paracasei			
Ciauscolo salami	Meat	Italy	Carnobacterium sp., E. faecalis, Lactobacillus sp., Lactococcus sp., Ln. mesenteroides, P. pentosaceus, W. hellenica	AMP, CHL, CLI, ERY, GEN, STR, TET, VAN	erm(B), tet(M)	Federici et al. (2014)
Piacentino salami	Pork meat and lard	Italy	Lactobacillus sp., Lb. sakei, Lb. curvatus, Lb. plantarum	ERY, TET	erm(B), erm(C), tet(M), tet(S), tet(W)	Zonenschain et al. (2009)
Salame Piacentino PDO*	Pork meat	Italy	Lb. paracasei	ERY, TET	erm(B), tet(M)	Comunian et al. (2010)
Dry-cured sausages	Pork meat and pork back fat	Spain	E. faecium, Lb. coryniformis, Lb. paracasei, Lb. plantarum, Lb. sakei	AI, AMP, CIP, CLI, NIT, PEG, RIF, TET, VAN	ND	Landeta et al. (2013)
Sausages	Pork meat	China	Lb. plantarum	AMP, CHL, CIP, CLI, ERY, KAN, TET	aphA3, erm, erm(B), tet(M)	Pan et al. (2011)
Bekasam or tempoyak	Meat + rice or meat	Indonesia	Lb. fermentum, Lb. plantarum, P. acidilactici	CHL, ERY	erm(B), erm(C), cat, Tn554	Sukmarini et al. (2014)
Sausage	Meat	Switzerland	E. faecalis	AZT, CHL, CLA, CLI, ERY, KAN, LIN, NC, RXM, STR, TET, TYL	Conjugative plasmid pRE25	Teuber et al. (2003)

Antibiotics: AI, amikacin; AMP, ampicillin; AZT, azithromycin; CHL, chloramphenicol; CIP, ciprofloxacin; CLA, clarithromycin; CLI, clindamycin; EFX, enrofloxacin; ERY, Erythromycin; GEN, gentamycin; KAN, kanamycin; LIN, lincomycin; MTZ, metronidazol; NAL, nalidixan; NC, nourseothricin; NIT, nitrofurantoin; NM, Neomycin; OXA, oxacillin; PDO, Protected Designation of Origin; PEG, penicillin G; PMB, Polymyxin B; RIF, rifampicin; RXM, roxithromycin; STR, streptomycin; SUL, sulfonamides; TET, tetracycline; TMP, trimethoprim; TYL, tylosin; VAN, vancomycin. Bacteria: E, Enterococcus; Lb, Lactobacillus; Ln, Leuconostoc; P, Pediococcus; W, Weissella; *, Protected Designation of Origin.

29.2.1.3 Fermented Vegetables

Several traditional fermented vegetables have been consumed for centuries all over the world; however, little is known about the incidence of AR in LAB, which depends on the geographic region, the raw material, the production practices, and the antibiotics used in agriculture. Resistance assessment of LAB draws increased attention when the European Food Safety Authority (EFSA) in 2007 recommended that bacterial strains harboring transferable AR genes must be considered as unsafe and should not be used in animal feeds, or fermented and probiotic foods for humans (EFSA, 2007). So, when comparing with fermented products of animal origin (dairy and fermented meat), the incidence of antibiotic-resistant LAB is lower in fermented vegetables (Pan et al., 2011). Lactobacilli are the main guilds involved in the natural fermentation of vegetables. They are intrinsically resistant to many antibiotics, such as quinolones, trimethoprim, and sulfonamides, but some species are also resistant to glycopeptides. However, they are susceptible to all protein synthesis inhibitors except aminoglycosides. Several studies have been carried out on the assessment of phenotypic AR, and the results revealed multidrug-resistant LAB isolated from fermented cocoa beans (Ghana), koko (Ghana), dolo, and pito wort (Burkina Faso and Ghana); however, we could not evaluate whether the strains harbored transferable AR genes because of the lack of genotypic data of AR genes (Table 29.3). Recently, analysis of phenotypic and genotypic AR profile of lactobacilli from different traditional fermented vegetables revealed that the most common resistance genes found in lactobacilli were *erm*(B) and *tet*(M) being located on chromosome and plasmids. Thus, the presence of those genes, as previously explained, on mobile genetic elements may participate in the dissemination of AR genes to other bacteria within the same ecosystem or within the gastrointestinal tract. Furthermore, Pan et al. (2011) showed that besides the presence of *erm*(B) and *tet*(M) on both plasmid and chromosome in certain strains, the gene *aphA3* coding for kanamycin resistance was also detected on plasmid in LAB strains isolated from the Chinese pickles. Furthermore, Pan et al. (2011) suggested that *E. faecium* SZ109 might have had coresistance genes with lactobacilli since *erm*(B), *tet*(M) and in some cases *aphA3* were located on the same plasmid, which again raises concerns of horizontal gene transfer to other bacteria in the food chain. The acquired resistance gene *aphA3* was also detected in LAB isolated from Spanish wine, along with other aminoglycoside resistance genes (Rojo-Bezares et al., 2006). Nawaz et al. (2011) showed a high prevalence of multidrug-resistant LAB isolated from traditional fermented vegetables in China, which carried *erm*(B) and *tet*(M) genes on plasmid. Furthermore, the plasmid harboring *erm*(B) and *tet*(M) genes were successfully transferred to *E. faecalis* 181 by filter mating; thus, this acquired resistance raises concerns in Chinese foods since transfer to other genera has been demonstrated (Nawaz et al., 2011). In this sense, other acquired AR genes were also detected in LAB isolated from traditional fermented vegetables as reported by Thumu and Halami (2012), such as genes encoding for tetracycline [eg, *tet*(K), *tet*(L), *tet*(M)

TABLE 29.3 Phenotypic and Genotypic Antibiotic Resistance of Lactic Acid Bacteria Isolated From Fermented Vegetables

Fermented Vegetables	Raw Material	Geographical Zone	LAB	Phenotypic Antibiotic Resistance	Genotypic Antibiotic Resistance	References
Fermenting cocoa beans	Cocoa pulp	Ghana	Lb. plantarum, Lb. ghanensis, Ln. pseudomesenteroides	KAN, STR, TET, VAN	ND	Adimpong et al. (2012) and Nielsen et al. (2007)
Koko sour water (KSW)	Sorghum, maize, millet	Ghana	Lb. plantarum, Lb. salivarius, P. acidilactici	KAN, STR, TET, VAN	ND	Adimpong et al. (2012) and Lei and Jakobsen (2004)
			P. pentosaceus, W. confusa			
Koko	Millet	Ghana	Lb. fermentum, Lb. paraplantarum, Lb. salivarius, W. confusa	AM, CIP, COL, KAN, NAL, NEO, SMZ, SPC, STR, TET, TMP, VAN	—	Ouoba et al. (2008)
Dolo and pito wort	Sorghum, maize	Burkina Faso and Ghana	Lb. delbrueckii, Lb. fermentum	KAN, STR, VAN	ND	Adimpong et al. (2012) and Sawadogo-Lingani et al. (2006)
Jiang shui (Chinese drink)	Mixture of vegetables	China	Lb. plantarum	TET	tet(S)	Nawaz et al. (2011)
Pickle	Vegetables	China	Lb. salivarius	ERY, TET	erm(B), tet(M)	Nawaz et al. (2011)
Fermented vegetable	Vegetables	China	Lb. animalis, Lb. brevis, Lb. parabuchneri, Lb. plantarum, Lb. salivarius, Lb. vaginalis	ERY, TET	erm(B), tet(M), tet(S)	Nawaz et al. (2011)
Pickles	Vegetables	China	Lb. brevis, Lb. fermentum, Lb. helveticus, Lb. mali, L. namurensis	CHL, CIP, ERY, KAN	aphA3, erm, erm(B), tet(M)	Pan et al. (2011)

Food	Country	Bacteria	Antibiotic resistance	Resistance genes	Reference
Gilaburu (drink)	Cranberrybush	Turkey	Lb. plantarum, Lb. casei Lb. brevis, Lactobacillus sp., Ln. pseudomesenteroides	KAN, STR, VAN	ND	Sagdic et al. (2014)
Idli batter, dosa batter	Cereal	India	E. casseliflavus, E. faecium, E. durans, E. lactis, Lb. fermentum, Lb. plantarum, P. pentosaceus, Ln. mesenteroides	ERY, TET	erm(B), msrC, tet(K), tet(L), tet(M) and tet(W)	Thumu and Halami (2012)
Fermented vegetables	Vegetables	India	E. casseliflavus, E. faecium, E. durans, E. lactis, Lb. fermentum, Lb. plantarum, P. pentosaceus, Ln. mesenteroides	ERY, TET	erm(B), msrC, tet(K), tet(L), tet(M) and tet(W)	Thumu and Halami (2012)
Wine	Grapes	Spain	Lb. plantarum, P. acidilactici, P. parvulus, P. pentosaceus, O. oeni	ERY, GEN, KAN, STR, TET	aac(6')-aph(2''), ant(6), aph (3')-IIIa, erm(B), tet(L), tet(M)	Rojo-Bezares et al. (2006)
Aloreña table olives	olives	Spain	Lb. pentosus Ln. pseudomesenteroides	CFX, CIP, CLI, STR, SMZ, TET, TMP, TPL,VAN	acrA, mdeA, mepA, norA	Casado Muñoz et al. (2014)
Tape ketan	Rice + ragi	Indonesia	Lb. fermentum, Lb.plantarum, P. acidilactici	CHL, ERY	erm(B), erm(C), cat, Tn554	Sukmarini et al. (2014)

Antibiotics: *AM*, apramycin; *CHL*, chloramphanicol; *CFX*, cefuroxime; *CIP*, ciprofloxacin; *CLI*, clindamycin; *COL*, colistin; *ERY*, Erythromycin; *GEN*, gentamycin; *KAN*, kanamycin; *NAL*, nalidixan; *NEO*, neomycin; *SMZ*, sulfomethoxazol; *SPC*, spectinomycin; *STR*, streptomycin; *TET*, tetracycline; *TMP*, trimethoprim; *TPL*, teicoplanin; *VAN*, vancomycin.
Bacteria: *E*, Enterococcus; *Lb*, Lactobacillus; *Ln*, Leuconostoc; *O*, Oenococcus; *P*, Pediococcus; *W*, Weissella.

and *tet*(W) determinants] and erythromycin [eg, *erm*(B) and *msrC*] resistance; also *tet*(S) in LAB from Chinese fermented vegetables (Nawaz et al., 2011); and *erm*(C) in LAB from Tape ketan from Indonesia (Sukmarini et al., 2014). To highlight horizontal gene transfer between LAB and other genera, Tn*554* from *Staphylococcus aureus*, coding for resistance to erythromycin and spectinomycin, was found in LAB isolated from Indonesian Tape ketan (Sukmarini et al., 2014). It is alarming that the microorganisms associated with these foods are able to transfer AR genes to pathogens. However, other naturally fermented vegetables, such as Aloreña table olives have not shown the presence of any AR genes in LAB (*Lactobacillus pentosus* and *Leuconostoc pseudomesenteroides*), but only genes coding for efflux pumps (Casado Muñoz et al., 2014). The presence of multidrug resistance (MDR) systems, such as efflux pumps, which mediates resistance to several drugs was already reviewed by Gueimonde et al. (2009) for lactobacilli, and Lubelski et al. (2007) for other Gram-positive and Gram-negative bacteria, including pathogens. Among MDR transporters, the ABC-type MDR transporters (or homologs) have been detected in different LAB (*Enterococcus*, *Lactococcus*, and *Lactobacillus* among others) being involved in intrinsic and acquired drug resistance.

29.2.2 Antibiotic Resistance Profile of Pathogens

Traditional fermented foods rely on natural fermentation, thus the principal ingredient in dairy products may be used either raw or heated for only a few minutes to temperatures insufficient to kill many pathogenic bacteria. The contamination of traditional dairy products with pathogenic bacteria can be of endogenous origin (from the animal) or exogenous origin (personnel, water, or environment). Consequently, dairy products made with raw milk may pose a health risk to consumers from direct exposure to foodborne pathogens. Similarly, vegetables and meat are not subjected to heat treatment to conserve the natural microbiota involved in fermentation processes. Thus, the occurrence of different pathogens, such as *Escherichia coli*, *Salmonella*, *Staphylococcus aureus*, *Bacillus cereus*, *Listeria monocytogenes*, and *Pseudomonas* spp., are among the most common pathogens associated with traditional fermented foods. It has been extensively reported that pathogens present in traditional fermented foods carried virulence determinants, but they could also be carriers of AR genes. Thus, these products require a better control of food contamination sources and spread of antibiotic-resistant pathogens (Normanno et al., 2007).

29.2.2.1 Dairy Products

The most predominant pathogens associated with milk and traditional fermented dairy products are opportunistic pathogens like *Staphylococcus aureus* and *Escherichia coli* and also pathogenic *Listeria monocytogenes*, which can be reservoirs of AR genes. The natural ecological niches of *S. aureus* are the nasal cavity and the skin of warm-blooded animals and in the environment, thus raw milk could be contaminated with staphylococci from animals (skin, teats, and udders) and also from farmers' hands, thus products made with raw milk may be a vehicle of these opportunistic pathogens. Iranian Feta cheese, which has an annual consumption per capita of 5.4 kg (Alizadeh et al., 2005), was reported by Arefi et al. (2014) as the potential source of the multidrug-resistant meticillin-resistant *S. aureus* (MRSA) (Table 29.4). However, the multidrug resistance profile of MRSA may vary depending on the region where different antibiotics were used and exerted a selective pressure. Thus, the prevalence of MRSA in developing countries is high (Arefi et al., 2014) despite varying AR profiles among different regions; however, this occurrence is lower in European countries and is due to the differences in production processes, farming hygienic conditions, hygienic levels of farmers and personnel carrying out dairy production, and also on the antibiotics used in therapy. Furthermore, the use of antibiotics in growth promotion of animals has not been allowed in Europe since 2006, thus the risk of developing resistant *S. aureus* was lowered by reducing their selective pressure. MRSA is a global health threat because it is considered the most prevalent cause of gastroenteritis worldwide. Its acquisition of meticillin-resistance gene (*mecA*), which makes them resistant to all β-lactam antibiotics, and other resistance genes has also been detected in *S. aureus*; multiple antibiotic-resistant strains have started to emerge. In this sense, the traditional Italian cheese analyzed by Spanu et al. (2014) showed the presence of multidrug-resistant *S. aureus*, especially those resistant to β-lactams and tetracyclines [e.g., *blaZ*, *tet*(L), *tet*(M) and *tet*(S)] (Table 29.4). Similarly, Can and Çelik (2012) indicated the presence of multidrug-resistant MRSA in traditional white and Tulum cheeses from Turkey as a potential risk for consumer health, however no genotypic data were reported. On the other hand, the Iranian traditional cheese analyzed by Jamali et al. (2015) was a carrier of multidrug-resistant MRSA, which harbored an arsenal of AR traits involved in erythromycin, tetracycline, aminoglycoside, and lincosamide resistance. Due to the importance of traditional dairy products as an integral part of the diet, especially in developing countries, they could act as a vehicle to spread MRSA. The presence of enterotoxigenic and multidrug-resistant *S. aureus* strains in several traditional dairy products constitutes a potential risk for public health, thus strict hygienic measures are necessary for food safety due to the fact that MRSA strains are a global health concern.

On the other hand, antibiotic-resistant enteric bacteria were isolated from the most popular type of Egyptian cheese (Hammad et al., 2009), such as *Escherichia coli* and *K. pneumonia* (also other species of *Klebsiella* and *Citrobacter freundii*) which were shown to be resistant to a large pool of antibiotics (Table 29.4). Screening for resistance by genotypic methods revealed the presence of several AR genes, class 1 and class 2 integrons, and plasmid-mediated quinolone resistance genes (Hammad et al., 2009). The high antimicrobial resistance contained in Domiati cheese samples is alarming, especially because Hammad et al. (2015) found for the first time new resistance genes (genes coding for two novel SHV β-lactamases, "SHV-110 and SHV-111," and a novel AmpC β-lactamase,

TABLE 29.4 Prevalence of Phenotypic and Genotypic Antibiotic Resistance in Pathogens Isolated From Traditional Fermented Dairy Products

Traditional Dairy Product	Raw Material	Geographical Zone	Pathogen	Phenotypic Antibiotic Resistance	Genotypic Antibiotic Resistance	References
Cheese	Goat or sheep's milk	Iran	S. aureus	MET	mecA	Shanehbandi et al. (2014)
White cheese Feta cheese	Cow's and sheep's milk	Iran	S. aureus	AMP, CF, CHL, CLX, ERY, MET, OXA, PEG, STR, TET	mecA	Arefi et al. (2014)
Cheese	Sheep's milk	Italy	S. aureus	AMP, CLX, PEN, TET	blaZ, tet(L), tet(M), tet(S)	Spanu et al. (2014)
Cheeses	Milk	Turkey	S. aureus	AMP, CLI, ERY, MET, OXA, TET	mecA	Can and Çelik (2012)
Tulum cheese	Sheep's milk	Turkey	S. aureus	AMC, FOX, OXA, PEN	mecA	Özpinar and Gümüşsoy (2013)
Cheese	Milk	Iran	S. aureus	CLI, CXT, ERY, GEN, KAN, LIN, MET, OXA, PEG, STR, TET	aacA-aphD, ant(4′)-la, ant(6)-la, blaZ,, cfr, ermA, ermB, ermC, ermT, inuA, inuB, mecA, mphC, msrA, msrB, tet(K), tet(L), tet(M)	Jamali et al. (2015)

Yogurt-type products	Milk	Iran	E. coli	AMP, CF, CHL, CIP, ERY, GEN, LIN, NFX, NIT, PEN, SMZ, STR, TET, TMP	aac (3′)-IV, aadA1, blaSHV, cat1, CITM, cmlA, dfrA1, ereA, qnr, Sul1, tetA, tetB	Dehkordi et al. (2014)
Soft cheese, butter, and ice cream	Milk	Iran	E. coli	AMP, CF, CHL, CIP, GEN, LIN, NFX, NIT, PEN, SMZ, STR, TET, TMP	aadA1, cat1, blaSHV, CITM, cmlA, dfrA1, qnr, tetB	Momtaz et al. (2012)
Domiati cheese	Milk	Egypt	Citrobacter freundii	AMP, ATM, CAZ, CF, CFP, CPD, CPO, CRO, CTT, CTX, FEP, FOX, GEN, IMP, KAN, NAL, NOR, OXA, STR, SMZ, TET	aac(6′)-Ib, aadA1a, aadA1, aadA2, aadA22, bla$_{CMY4}$, bla$_{CMY-41}$, bla$_{CTX-M-14}$, bla$_{OXA-1}$, bla$_{OXY-1}$, bla$_{SHV-1}$, bla$_{SHV-25}$, bla$_{SHV-26}$, bla$_{SHV-110}$, bla$_{SHV-111}$, bla$_{TEM-1}$, dfrA1-aadA1a, dfrA12-aadA2, dfrA1, dfrA5, dfrA7, dfrA15, qnrS, qnrB9-like	Hammad et al. (2009)
			E. coli			
			Klebsiella sp.			
Cheese	Sheep or cow	Iran	Listeria sp.	AMP, CHL, CIP, ERY, NAL, PEN, TET	ND	Rahimi et al. (2012)
Cheese, butter	Sheep or cow	Iran	Campylobacter jejuni	AMP, CIP, ERY, NAL, NFX, TET	ND	Rahimi et al. (2013)
			Campylobacter coli			
			Campylobacter sp.			

Antibiotics: AMC, amoxicillin-clavulanic; AMP, ampicillin; ATM, aztreonam; CAZ, ceftazidime; CF, cephalothin; CFP, cefoperazone; CXT, cefoxitin; CHL, chloramphanicol; CIP, ciprofloxacin; CLI, clindamycin; CLX, cloxacillin; CPD, cefpodoxime; CPO, cefpirome; CRO, ceftriaxone; CTT, cefotetan; CTX, cefotaxime; ERY, Erythromycin; FEP, cefepime; FOX, cefoxitin; GEN, gentamycin; IPM, imipenem; KAN, kanamycin; LIN, lincomycin; MET, methicillin; NAL, nalidixan; NFX, enrofloxacin; NIT, nitrofurantoin; NOR, norfloxacin; OXA, oxacillin; PEN, penicillin; PEG, penicillin G; SMZ, sulfomethoxazol; STR, streptomycin; TET, tetracycline; TMP, trimethoprim.
Bacteria: E, Escherichia; S, Staphylococcus.

"CMY-41") in food samples, along with mobile genetic elements (integrons and plasmids). Thus, those emerging AR genes could be spread throughout the food chain and also within the gastrointestinal tract, which may aggravate the antibiotic resistance problem. Other dairy products, such as yogurt-type products (Dehkordi et al., 2014), soft cheese, and butter (Momtaz et al., 2012) from Iran were also reported as vehicles of multidrug-resistant *E. coli* carriers of a pool of AR genes (Table 29.4). Furthermore, Rahimi et al. (2013) reported the presence of multidrug-resistant *Listeria* sp. and *Campylobacter* sp. in the same products, which was correlated with the antibiotics used to treat animal infections in Iran.

Overall, surveillance and control of the microbiological quality of raw milk used in the production of traditional fermented products is necessary all over the world with the aim to avoid morbidity and mortality associated with food intoxication. However, also decreasing the spread of opportunistic pathogens carrying a pool of AR genes may limit the spread and emergence of new types of resistance within the microbiota of the fermented product, and also to other products and the gastrointestinal tract.

29.2.2.2 Fermented Meat and Vegetables

Fermented foods with a pH value below 4 are usually safe, as most pathogens are unable to survive under these conditions. However, postcontamination with pathogens able to develop acid tolerance and cause food poisoning may still occur. There are several reports describing the prevalence of pathogenic bacteria in traditional fermented foods, such as *Staphylococcus aureus*, *Bacillus cereus*, *Escherichia coli*, *Campylobacter*, *Vibrio cholerae*, *Aeromonas*, *Klebsiella*, *Shigella* sp., and *Salmonella* among others. In this sense, the incidence of virulence genes of pathogens was investigated in depth rather than the prevalence of AR genes (Ahmed Mohammed et al., 2014; Rantsiou et al., 2012). Furthermore, many reports were based on the characterization of desirable microorganisms with the aim to select potential starter cultures or probiotics. Overall, the presence of acquired tetracycline and erythromycin resistance genes was found in staphylococci from fermented sausages (Table 29.5), although the presence of coagulase-negative staphylococci is required in these types of fermented product due to their involvement in the final pigmentation of meat, lipolysis, and proteolysis (http://www.sciencedirect.com/science/article/pii/S0309174014001508; Albano et al., 2009; Iacumin et al., 2006).

29.3 NEW INSIGHTS INTO ANTIBIOTIC RESISTANCE

The raw ingredients employed for artisanal fermented food production may contain several bacteria (LAB, commensals, and pathogens), and in the absence of selected starter cultures able to predominate and guarantee a product with desirable properties and acceptable safety, several products were described in the literature containing a variable microbial community, which makes them attractive for the consumer. However, here we report that several bacteria present in

TABLE 29.5 Prevalence of Phenotypic and Genotypic Antibiotic Resistance in Pathogens Isolated From Traditional Fermented Meat

Fermented Meat	Raw Material	Geographical Zone	Pathogen	Phenotypic Antibiotic Resistance*	Genotypic Antibiotic Resistance	References
Salame Piacentino PDO* (Dontorou et al., 2003; Marty et al., 2012; Rebecchi et al., 2015; Zdolec et al., 2013)	Pork	Italy	Staphylococcus sp.	ERY, KAN, TET	erm(A), erm(C), tet(K), tet(L), tet(M), aacA, aphD, aphA3 and aadD	Rebecchi et al. (2015)
Dry fermented sausage kulen	Pork	Croatia	S. epidermis Staphylococcus sp.**	AMP, ERY, GEN, TET	Tet(K), tet(M)	Zdolec et al. (2013)
Slavonian fermented sausage	Pork	Croatia	S. epidermis Staphylococcus sp.**	ERY, GEN, TET	Tet(K), tet(M)	Zdolec et al. (2013)
Swiss sausages and spontaneously fermented Swiss meat products	Meat of several animals	Switzerland	S. warneri, S. capitis, S. epidermidis, S. carnosus, S. equorum	AMP, AMX, CLX, ERY, FA, OXA, PEN, STR, TET, TMP	blaZ, str, tet(K), lnu(A), mph(C)	Marty et al. (2012)
Fresh sausages	Pork	Greece	E. coli O157:H7	–	–	Dontorou et al. (2003)

* Protected Designation of Origin.
** Coagulase-negative staphylococci (CNS).
Antibiotics: AMP, ampicillin; AMX, amoxicillin; CLX: cloxacillin, ERY, erythromycin; FA, fusidic acid; GEN, gentamycin; KAN, kanamycin; OXA, oxacillin; PDO, Protected Designation of Origin; PEN, penicillin; STR, streptomycin; TET, tetracycline; TMP, trimethoprim.
Bacteria: E, Escherichia; S, Staphylococcus.

traditional fermented foods harbored acquired AR genes, although the organoleptic properties of the end products were not affected by the presence of those bacteria. In this sense, assessment of antibiotic susceptibility in a complex bacterial community present in the fermented products is necessary in safety assessment of fermented foods. However, many culture-dependent methods are based on screening of the most common and known resistance genes described in the literature, and in many cases, we are unable to detect those genes occurring at low frequencies since >99% of microorganisms present in an ecosystem remain uncultured. To resolve those limitations, high-throughput technologies based on culture-independent methods, such as metagenomics, are key to identify uncultured bacteria (Streit and Schmitz, 2004), sensitively detect novel AR genes and gene transfer mechanisms, and evaluate (in depth) the safety of traditional fermented foods. Devirgiliis et al. (2014) applied this method for screening of AR genes in traditional Mozzarella di Bufala Campana cheese microbiota to identify underrepresented species carrying specific genes of interest. Those authors revealed the presence of AR genes coding for resistance to ampicillin and kanamycin, which were poorly characterized in LAB (Devirgiliis et al., 2014), and they also detected many transposons and insertion sequence (IS) elements responsible for horizontal gene transfer. The metagenome-based strategy was already applied in the analysis of AR genes in a large cohort of human gut microbiota (Hu et al., 2013). On the other hand, culture-dependent methods may provide the opportunity to screen for strains with interesting functional traits, but assessment of AR should not be limited to PCR detection of known AR genes, since genome sequencing is crucial to detect and describe novel AR genes and transfer mechanisms. Furthermore, to evaluate the safety of traditional fermented foods, metagenomics should be carried out since it could reveal AR in a complex system without limitations, since several traditional fermented foods, especially those from developing countries, have been shown to be carriers of multidrug-resistant bacteria. Indeed, selection of starters or probiotic strains from those fermented foods should be carried out with caution, because they could also be a vehicle for AR genes to be spread to human.

The acquired status "generally regarded as safe" (GRAS) by LAB strains should be used with caution, since many species and even strains within the same species may carry transferable AR genes, which could aggravate the resistance problem. In this sense, analysis in silico for AR genes in LAB isolated from traditional fermented foods revealed the presence of an arsenal of uncharacterized AR genes and mobile genetic elements, such as transposons. In fact, screening for *Tn*916 in genomic NCBI database revealed the presence of this mobile genetic element associated with AR genes located on plasmids in bacteria, such as *Lc. lactis* subsp. *lactis* isolated from an artisanal cheese (Flórez et al., 2008) carrying transposon *Tn*916 and *tet*(M) genes on pAA211 and pAA291 plasmids. Furthermore, *tet*(S) gene found in *Lc. lactis* subsp. *lactis* K214 isolated from a raw milk soft cheese (Perreten et al., 1997) was joined by sequences similar to a region of transposon Tn*916* from *E. faecalis*, which is involved in

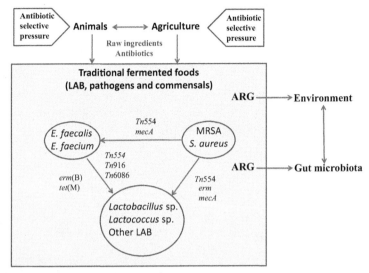

FIGURE 29.1 Dissemination of antibiotic-resistant bacteria and the most common genes into and out of traditional fermented food and other ecosystems highlighting the role of mobile genetic elements.

tet(M) expression, and also by insertion-sequence element from *E. faecium*, suggesting horizontal gene transfer between genera. Indeed, Tn554 originally from *Staphylococcus aureus* was also detected in *P. acidilactici* isolated from Indonesian Tape Ketan (Sukmarini et al., 2014). In a complex system, as is a traditional fermented food, many genetic exchanges resulting from bacterial interactions (Fig. 29.1) may occur between bacterial communities (LAB, commensals, and pathogens), thus the presence of a selected starter bacteria, free of transferable AR genes, could predominate and eliminate pathogens as they are a potential vehicle of AR genes and also virulence.

29.4 CONCLUSIONS

The most common AR profiles of bacteria present in traditional fermented foods pointed out the ubiquity of erythromycin and tetracycline resistance genes, especially in LAB, which may be due in part to their common use as growth promoters and in therapy, but also may be due to the scarcity of data about other antibiotic resistance. Thus, sequencing tools provide analysis in depth of AR in cultured and noncultured bacteria present in traditional fermented foods with the aim to adopt preventive measures in resistance spread throughout the food chain. In this sense, metagenomics provides a complete picture of AR in a complex microbiota. More attention should be given to traditional fermented foods in developing countries due to the imprudent use of antibiotics in animal husbandry and also for the limitations in hygienic conditions during production, especially at household scale. Thus, selection of probiotics and starter cultures in those types of products should be carried out considering all safety aspects including the presence of transferable AR genes, since we reported here that some traditional products represent a huge reservoir of AR genes.

ACKNOWLEDGMENTS

The authors acknowledge research grants AGL2009-08921 (Ministerio de Economía y Competitividad, MINECO), AGL2013-43571-P (Ministerio de Economía y Competitividad, MINECO, FEDER), and UJA2014/07/02 (Plan Propio UJA).

REFERENCES

Abriouel, H., Martín-Platero, A., Maqueda, M., Valdivia, E., Martínez-Bueno, M., 2008. Biodiversity of the microbial community in a Spanish farmhouse cheese as revealed by culture-dependent and culture-independent methods. International Journal of Food Microbiology 127, 200–208.

Abriouel, H., Benomar, N., Lucas, R., Gálvez, A., 2011. Culture-independent study of the diversity of microbial populations in brines during fermentation of naturally fermented Aloreña green table olives. International Journal of Food Microbiology 144, 487–496.

Abriouel, H., Benomar, N., Cobo, A., Caballero, N., Fernández Fuentes, M.A., Pérez-Pulido, R., Gálvez, A., 2012. Characterization of lactic acid bacteria from naturally-fermented Manzanilla Aloreña green table olives. Food Microbiology 32, 308–316.

Ahmed Mohammed, T., Elsubki Khalid, A., Mohamed Saadabi, A., 2014. PCR detection of Staphylococcal Enterotoxin A and B Genes in *Staphylococcus aureus* isolated from Salted Fermented Fish. Microbiology Journal 4, 51–56.

Albano, H., Van Reenen, C.A., Todorov, S.D., Cruz, D., Fraga, L., Hogg, T., 2009. Phenotypic and genetic heterogeneity of lactic acid bacteria isolated from "Alheira", a traditional fermented sausage produced in Portugal. Meat Science 82, 387–398.

Alizadeh, M., Hamedi, M., Khosroshahi, A., 2005. Optimizing sensorial quality of Iranian White Brine cheese using response surface methodology. Journal of Food Science 70, 299–303.

Ammor, M.S., Florez, A.B., Mayo, B., 2007. Antibiotic resistance in non-enterococcal lactic acid bacteria and bifidobacteria. Food Microbiology 24, 559–570.

Ammor, M.S., Gueimonde, M., Danielsen, M., Zagoree, M., van Hoek, A.H., Gavilan de los Reyes, C.G., Mayo, B., Margolles, A., 2008. Two different tetracycline mechanisms, plasmid carried *tet*(L) and chromosomally located transposon-associated *tet*(M), coexist in *Lactobacillus sakei* rits 9. Applied and Environmental Microbiology 74, 1394–1401.

Arefi, F., Mohsenzadeh, M., Razmyar, J., 2014. Isolation, antimicrobial susceptibility and *mecA* gene analysis of methicillin-resistant *Staphylococcus aureus* in Iranian white cheeses. Iranian Journal of Veterinary Research 15, 127–131.

Armelagos, G.J., 1969. Disease in ancient Nubia. Science 163, 225–258.

Ahmed, A.I., Mohamed, B.E., Yousif, N.M.E., Faye, B., Loiseau, G., 2012. Antimicrobial activity and antibiotic resistance of LAB isolated from Sudanese traditional fermented camel (*Camelus dromedarius*) milk gariss. International Journal of Biosciences 2, 129–136.

Adimpong, D.B., Nielsen, D.S., Sørensen, K.I., Derkx, P.M.F., Jespersen, L., 2012. Genotypic characterization and safety assessment of lactic acid bacteria from indigenous African fermented food products. BMC Microbiology 12, 75.

Bassett, E.J., Keith, M.S., Armelagos, G.J., Martin, D.L., Villanueva, A.R., 1980. Tetracycline-labeled human bone from ancient Sudanese Nubia (A.D. 350). Science 209, 1532–1534.

Bennedsen, M., Stuer-Lauridsen, B., Danielsen, M., Johansen, E., 2011. Screening for antimicrobial resistance genes and virulence factors via genome sequencing. Applied and Environmental Microbiology 77, 2785–2787.

Can, Y.H., Celik, H.T., 2012. Detection of enterofaxigenic and antimicrobial resistant *Staphylococcus aureus* in Turkish cheese. Food Control 24, 100–103.

Casado Muñoz, M.C., Benomar, N., Lavilla Lerma, L., Gálvez, A., Abriouel, H., 2014. Antibiotic resistance of *Lactobacillus pentosus* and *Leuconostoc pseudomesenteroides* isolated from naturally-fermented Aloreña table olives throughout fermentation process. International Journal of Food Microbiology 172, 110–118.

Çataloluk, O., Gogebakan, B., 2004. Presence of drug resistance in intestinal lactobacilli of dairy and human origin in Turkey. FEMS Microbiology Letters 236, 7–12.

Comunian, R., Daga, E., Dupre, I., Paba, A., Devirgiliis, C., Piccioni, V., Perozzi, G., Zonenschain, D., Rebecchi, A., Morelli, L., De Lorentiis, A., Giraffa, G., 2010. Susceptibility to tetracycline and erythromycin of *Lactobacillus paracasei* strains isolated from traditional Italian fermented foods. International Journal of Food Microbiology 138, 151–156.

Deghorain, M., Goffin, P., Fontaine, L., Mainardi, J.L., Daniel, R., Errington, J., Hallet, B., Hols, P., 2007. Selectivity for D-lactate incorporation into the peptidoglycan precursors of *Lactobacillus plantarum*: role of Aad, a VanX-like D-alanyl-D-alanine dipeptidase. Journal of Bacteriology 189, 4332–4337.

Dehkordi, F.S., Yazdani, F., Mozafari, J., Valizadeh, Y., 2014. Virulence factors, serogroups and antimicrobial resistance properties of *Escherichia coli* strains in fermented dairy products. BMC Research Notes 7, 217.

Devirgiliis, C., Coppola, D., Barile, S., Colonna, B., Perozzi, G., 2009. Characterization of the Tn*916* conjugative transposon in a food-borne strain of *Lactobacillus paracasei*. Applied and Environmental Microbiology 75, 3866–3871.

Devirgiliis, C., Barile, S., Caravelli, A., Coppola, D., Perozzi, G., 2010. Identification of tetracycline- and erythromycin-resistant Gram-positive cocci within the fermenting microflora of an Italian dairy food product. Journal of Applied Microbiology 109, 313–323.

Devirgiliis, C., Zinno, P., Stirpe, M., Barile, S., Perozzi, G., 2014. Functional screening of antibiotic resistance genes from a representative metagenomic library of food fermenting microbiota. Biomed Research International 2014, 1.

Dixon, B., 2000. Antibiotics as growth promoters: risks and alternatives. ASM News 66, 264–265.

Devirgiliis, C., Caravelli, A., Coppola, D., Barile, S., Perozzi, G., 2008. Antibiotic resistance and microbial composition along the manufacturing process of Mozzarella di Bufala Campana. International Journal of Food Microbiology 128, 378–384.

Dontorou, C., Papadopoulou, C., Filioussis, G., Economou, V., Apostolou, I., Zakkas, G., Salamoura, A., Kansouzidou, A., Levidiotou, S., 2003. Isolation of *Escherichia coli* O157:H7 from foods in Greece. International Journal of Food Microbiology 82, 273–279.

European Commission (SCAN), 1996. Report of the Scientific Committee for Animal Nutrition (SCAN) on the Possible Risk for Humans on the Use of Avoparcin as Feed Additive. Office for EC Publications, Luxemburg. Opinion expressed May 21, 1996.

European Commission (SCAN), July 10, 1998. Opinion of the Scientific Committee for Animal Nutrition (SCAN) on the Immediate and Long-term Risk to the Value of Streptogramins in Human Medicine Posed by the Use of Virginiamycin as an Animal Growth Promoter. Office for EC Publications, Luxemburg.

European Food Safety Authority (EFSA), 2005. Opinion of the Scientific Committee on a request from EFSA related to a generic approach to the safety assessment by EFSA of microorganisms used in food/feed and the production of food/feed additives. The European Food Safety Authority Journal 226, 1–16.

European Food Safety Authority (EFSA), 2007. Opinion of the Scientific Committee on a request from EFSA on the introduction of a qualified presumption of safety (QPS) approach for assessment of selected microorganisms referred to EFSA. EFSA Journal 187, 1–16.

European Food Safety Authority (EFSA), 2008. Scientific opinion of the panel on biological hazards on a request from the European Food Safety Authority on food borne antimicrobial resistance as a biological hazard. EFSA Journal 765, 1–87.

FAO/OIE/WHO, 2003, December 2003. First Joint FAO/OIE/WHO Expert Workshop on Non-human Antimicrobial Usage and Antimicrobial Resistance: Scientific Assessment, Geneva, Switzerland, pp. 1–5. http://www.who.int/foodsafety/micro/meetings/nov2003/en/.

Favaro, L., Basaglia, M., Casella, S., Hue, I., Dousset, X., Franco, B.D.G.M., Todorov, S.D., 2014. Bacteriocinogenic potential and safety evaluation of non-starter *Enterococcus faecium* strains isolated from home-made white brine cheese. Food Microbiology 38, 228–239.

Feinman, S.E., 1999. Antibiotics in animal feeds-drug resistance revisited. ASM News 64, 24–29.

Flórez, A.B., Ammor, M.S., Mayo, B., 2008. Identification of *tet*(M) in two *Lactococcus lactis* strains isolated from a Spanish traditional starter-free cheese made of raw milk and conjugative transfer of tetracycline resistance to lactococci and enterococci. International Journal of Food Microbiology 121, 189–194.

Flórez, A.B., Reimundo, P., Delgado, S., Fernández, E., Alegría, A., Guijarro, J.A., Mayo, B., 2012. Genome sequence of *Lactococcus garvieae* IPLA 31405, a bacteriocin-producing, tetracycline-resistant strain isolated from a raw-milk cheese. Journal of Bacteriology 194, 5118–5119.

Flórez, A.B., Alegría, A., Rossi, F., Delgado, S., Felis, G.E., Torriani, S., Mayo, B., 2014. Molecular identification and quantification of tetracycline and erythromycin resistance genes in Spanish and Italian Retail cheeses. BioMed Research International 2014, 746859.

Forslund, K., Sunagawa, S., Kultima, J.R., Mende, D.R., Arumugam, M., Typas, A., Bork, P., 2013. Country-specific antibiotic use practices impact the human gut resistome. Genome Research 23, 1163–1169.

Flórez, A.B., Delgado, S., Mayo, B., 2005. Antimicrobial susceptibility of lactic acid bacteria isolated from a cheese environment. Canadian Journal of Microbiology 51, 51–58.

Fortina, M.G., Ricci, G., Foschino, R., Picozzi, C., Dolci, P., Zeppa, G., Cocolin, L., Manachini, P.L., 2007. Phenotypic typing, technological properties and safety aspects of *Lactococcus garvieae* strains from dairy environments. Journal of Applied Microbiology 103, 445–453.

Federici, S., Ciarrocchi, F., Campana, R., Ciandrini, E., Blasi, G., Baffone, W., 2014. Identification and functional traits of lactic acid bacteria isolated from Ciauscolo salami produced in Central Italy. Meat Science 98, 575–584.

Giraffa, G., 2004. Studying the dynamics of microbial populations during food fermentation. FEMS Microbiology Reviews 28, 251–260.

Gueimonde, M., Flórez, A.B., de los Reyes-Gavilán, C.G., Margolles, A., 2009. Intrinsic resistance in lactic acid bacteria and bifidobacteria: the role of multidrug resistance transporters. International Journal of Probiotics and Prebiotics 4, 181–186.

Gad, G.F.M., Abdel-Hamid, A.M., Farag, Z.S.H., 2014. Antibiotic resistance in lactic acid bacteria isolated from some pharmaceutical and dairy products. Brazilian Journal of Microbiology 45, 25–33.

Hammad, A.M., Ishida, Y., Shimamoto, T., 2009. Prevalence and molecular characterization of ampicillin-resistant *Enterobacteriaceae* isolated from traditional Egyptian Domiati cheese. Journal of Food Protection 72, 624–630.

Hammad, A.M., Hassan, H.A., Shimamoto, T., 2015. Prevalence, antibiotic resistance and virulence of *Enterococcus* spp. in Egyptian fresh raw milk cheese. Food Control 50, 815–820.

Hu, Y., Yang, X., Qin, J., Lu, N., Cheng, G., Wu, N., Pan, Y., Li, J., Zhu, L., Wang, X., Meng, Z., Zhao, F., Liu, D., Ma, J., Qin, N., Xiang, C., Xiao, Y., Li, L., Yang, H., Wang, J., Yang, R., Gao, G.F., Wang, J., Zhu, B., 2013. Metagenome-wide analysis of antibiotic resistance genes in a large cohort of human gut microbiota. Nature Communications 4, 2151.

Hummel, A.S., Hertel, C., Holzapfel, W.H., Franz, C.M.A.P., 2007. Antibiotic resistances of starter and probiotic strains of lactic acid bacteria. Applied and Environmental Microbiology 73, 730–736.

Iacumin, L., Comi, G., Cantoni, C., Cocolin, L., 2006. Ecology and dynamics of coagulase-negative cocci isolated from naturally fermented Italian sausages. Systematic Applied Microbiology 29, 470–486.

Jamali, H., Paydar, M., Radmehr, B., Ismail, S., Dadrasnia, A., 2015. Prevalence and antimicrobial resistance of *Staphylococcus aureus* isolated from raw milk and dairy products. Food Control 54, 383–388.

Jamaly, N., Benjouad, A., Bouksaim, M., 2011. Probiotic potential of *Lactobacillus* strains isolated from known popular traditional Moroccan dairy products. British Microbiology Research Journal 1, 79–94.

Knapp, C.W., Dolfing, J., Ehlert, P.A.I., Graham, D.W., 2010. Evidence of increasing antibiotic resistance gene abundances in archived soils since 1940. Environmental Science and Technology 44, 580–587.

Kousta, M., Mataragas, M., Skandamis, P., Drosinos, E.H., 2010. Prevalence and sources of cheese contamination with pathogens at farm and processing levels. Food Control 21, 805–815.

Kročko, M., Čanigová, M., Ducková, V., Artimová, A., Bezeková, J., Poston, J., 2011. Antibiotic resistance of Enterococcus Species Isolated from Raw Foods of Animal Origin in South West Part of Slovakia. Czech Journal of Food Sciences 6, 654–659.

Levy, S.B., 1997. Antibiotic resistance: an ecological imbalance. Ciba Foundation Symposium 207, 1–9.

Lubelski, J., Konings, W.N., Driessen, A.J.M., 2007. Distribution and physiology of ABC-type transporters contributing to multidrug resistance in bacteria. Microbiology and Molecular Biology Reviews 71, 463–476.

Landeta, G., Curiel, J.A., Carrascosa, A.V., Muñoz, R., de las Rivas, B., 2013. Technological and safety properties of lactic acid bacteria isolated from Spanish dry-cured sausages. Meat Science 95, 272–280.

Lei, V., Jakobsen, M., 2004. Microbiological characterization and probiotic potential of koko and koko sour water, African spontaneously fermented millet porridge and drink. Journal of Applied Microbiology 96, 384–397.

Magee, J.T., Pritchard, E.L., Fitzgerald, K.A., Dunstan, F.D.J., Howard, A.J., 1999. Antibiotic prescribing and antibiotic resistance in community practice: retrospective study, 1996–8. British Medical Journal 319, 1239–1240.

Momtaz, H., Farzan, R., Rahimi, E., Dehkordi, F.S., Souod, N., 2012. Molecular characterization of Shiga toxin-producing *Escherichia coli* isolated from ruminant and donkey raw milk samples and traditional dairy products in Iran. Scientific World Journal 2012, 231342.

Munsch-Alatossava, P., Alatossava, T., 2007. Antibiotic resistance of raw milk associated psychrotrophic bacteria. Microbiological Research 162, 115–123.

Marty, E., Bodenmann, C., Buchs, J., Hadorn, R., Eugster-Meier, E., Lacroix, C., Meile, L., 2012. Prevalence of antibiotic resistance in coagulase-negative staphylococci from spontaneously fermented meat products and safety assessment for new starters. International Journal of Food Microbiology 159, 274–283.

Nawaz, M., Wang, J., Zhou, A., Ma, C., Wu, X., Moore, J.E., Millar, B.C., Xu, J., 2011. Characterization and transfer of antibiotic resistance in lactic acid bacteria from fermented food products. Current Microbiology 62, 1081–1089.

Nelson, M.L., Dinardo, A., Hochberg, J., Armelagos, G.J., 2010. Brief communication: mass spectroscopic characterization of tetracycline in the skeletal remains of an ancient population from Sudanese Nubia 350–550 CE. American Journal of Physical Anthropology 143, 151–154.

Normanno, G., La Salandra, G., Dambrosio, A., Quaglia, N.C., Corrente, M., Parisi, A., Santagada, G., Firinu, A., Crisetti, E., Celano, G.V., 2007. Occurrence, characterization and antimicrobial resistance of enterotoxigenic *Staphylococcus aureus* isolated from meat and dairy products. International Journal of Food Microbiology 115, 290–296.

Nielsen, D.S., Teniola, O.D., Ban-Koffi, L., Owusu, M., Andersson, T.S., Holzapfel, W.H., 2007. The microbiology of Ghanaian cocoa fermentations analysed using culture-dependent and culture-independent methods. International Journal of Food Microbiology 114, 168–186.

Ogier, J.C., Serror, P., 2008. Safety assessment of dairy microorganisms: the *Enterococcus* genus. International Journal of Food Microbiology 126, 291–301.

Oguntoyinbo, F.A., Okueso, O., 2013. Prevalence, distribution and antibiotic resistance pattern among enterococci species in two traditional fermented dairy foods. Annals of Microbiology 63, 755–761.

Ouoba, L.I., Lei, V., Jensen, L.B., 2008. Resistance of potential probiotic lactic acid bacteria and bifidobacteria of African and European origin to antimicrobials: determination and transferability of the resistance genes to other bacteria. International Journal of Food Microbiology 121, 217.

Özpinar, N., Gümüşsoy, K.S., 2013. Phenotypic and genotypic determination of antibiotic resistant and Biofilm forming *Staphylococcus aureus* isolated in Erzincan tulum cheese. Kafkas Universitesi Veteriner Fakultesi Dergisi 19, 517–521.

Pan, L., Hu, X., Wang, X., 2011. Assessment of antibiotic resistance of lactic acid bacteria in Chinese fermented foods. Food Control 22, 1316–1321.

Perin, L.M., Miranda, R.O., Todorov, S.D., Franco, B.D.G.M., Nero, L.A., 2014. Virulence, antibiotic resistance and biogenic amines of bacteriocinogenic lactococci and enterococci isolated from goat milk. International Journal of Food Microbiology 185, 121–126.

Perreten, V., Schwarz, F., Cresta, L., Boeglin, M., Dasen, G., Teuber, M., 1997. Antibiotic resistance spread in food. Nature 389, 801–802.

Poznanski, E., Cavazza, A., Cappa, F., Cocconcelli, P., 2004. Indigenous raw milk microbiota influences the bacterial development in traditional cheese from an alpine natural park. International Journal of Food Microbiology 92, 141–151.

Rahimi, E., Sepehri, S., Momtaz, H., 2013. Prevalence of *Campylobacter* species in milk and dairy products in Iran. Revue de Médecine Vétérinaire 164, 283–288.

Rantsiou, K., Mataragas, M., Alessandria, V., Cocolin, L., 2012. Expression of virulence genes of *Listeria monocytogenes* in food. Journal of Food Safety 32, 161–168.

Roberts, M.C., 2008. Update on macrolide-lincosamide-streptogramin, ketolide, and oxazolidinon resistance genes. FEMS Microbiology Letters 282, 147–159.

Rojo-Bezares, B., Sáenz, Y., Poeta, P., Zarazaga, M., Ruiz-Larrea, F., Torres, C., 2006. Assessment of antibiotic susceptibility within lactic acid bacteria strains isolated from wine. International Journal of Food Microbiology 111, 234–240.

Rahimi, E., Momtaz, H., Sharifzadeh, A., Behzadnia, A., Ashtari, M.S., Zandi Esfahani, S., Riahi, M., Momeni, M., 2012. Prevalence and antimicrobial resistance of *Listeria* species isolated from traditional dairy products in Chahar Mahal & Bakhtiyari, Iran. Bulgarian Journal of Veterinary Medicine 15, 115–122.

Rebecchi, A., Pisacane, V., Callegari, M.L., Puglisi, E., Morelli, L., 2015. Ecology of antibiotic resistant coagulase-negative staphylococci isolated from the production chain of a typical Italian salami. Food Control 53, 14–22.

Salvana, E.M.T., Frank, M., 2006. *Lactobacillus* endocarditis: case report and review of cases reported since 1992. Journal of Infection 53, 5–10.

Spanu, V., Scarano, C., Cossu, F., Pala, C., Spanu, C., De Santis, E.P., 2014. Antibiotic resistance traits and molecular subtyping of *Staphylococcus aureus* isolated from raw sheep milk cheese. Journal of Food Science 79, M2066–M2071.

Streit, W.R., Schmitz, R.A., 2004. Metagenomics–the key to the uncultured microbes. Current Opinion in Microbiology 7, 492–498.

Sukmarini, L., Mustopa, A.Z., Normawati, M., Muzdalifah, I., 2014. Identification of antibiotic-resistance genes from lactic acid bacteria in Indonesian fermented foods. HAYATI Journal of Biosciences 21, 144–150.

Solieri, L., Bianchi, A., Mottolese, G., Lemmetti, F., Giudici, P., 2014. Tailoring the probiotic potential of non-starter *Lactobacillus* strains from ripened Parmigiano Reggiano cheese by *in vitro* screening and principal component analysis. Food Microbiology 38, 240–249.

Sagdic, O., Ozturk, I., Yapar, N., Yetim, H., 2014. Diversity and probiotic potentials of lactic acid bacteria isolated from gilaburu, a traditional Turkish fermented European cranberrybush (*Viburnum opulus* L.) fruit drink. Food Research International 64, 537–545.

Sawadogo-Lingani, H., Lei, V., Diawara, B., Nielsen, D.S., Moller, P.L., Traore, A.S., Jakobsen, M., 2006. The biodiversity of predominant lactic acid bacteria in dolo and pito wort for the production of sorghum beer. Journal of Applied Microbiology 103, 765–777.

Shanehbandi, D., Baradaran, B., Sadigh-Eteghad, S., Zarredar, H., 2014. Occurrence of methicillin resistant and enterotoxigenic *Staphylococcus aureus* in traditional cheeses in the North West of Iran. ISRN Microbiology 2014, 129580.

Tannock, G.W., Fuller, R., Pedersen, K., 1990. *Lactobacillus* succession in the piglet digestive tract demonstrated by plasmid profiling. Applied and Environmental Microbiology 56, 1310–1316.

Teuber, M., Schwarz, F., Perreten, V., 2003. Molecular structure and evolution of the conjugative multiresistance plasmid pRE25 of *Enterococcus faecalis* isolated from a raw fermented sausage. International Journal of Food Microbiology 88, 325–329.

Thumu, S.C., Halami, P.M., 2012. Presence of erythromycin and tetracycline resistance genes in lactic acid bacteria from fermented foods of Indian origin. Antonie Van Leeuwenhoek 102, 541–551.

Walther, C., Rossano, A., Thomann, A., Perreten, V., 2008. Antibiotic resistance in *Lactococcus* species from bovine milk: presence of a mutated multidrug transporter *mdt*(A) gene in susceptible *Lactococcus garvieae* strains. Veterinary Microbiology 131, 348–357.

Wang, H.H., Manuzon, M., Lehman, M., Wan, K., Luo, H., Wittum, T.E., Yousef, A., Bakaletz, L.O., 2006. Food commensal microbes as a potentially important avenue in transmitting antibiotic resistance genes. FEMS Microbiology Letters 254, 226–231.

Wang, H., McEntire, J.C., Zhang, L., Li, X., Doyle, M., 2012. The transfer of antibiotic resistance from food to humans: facts, implications and future directions. Revue Scientifique et Technique (International Office of Epizootics) 31, 249–260.

Weckx, S., Van der Meulen, R., Maes, D., Scheirlinck, I., Huys, G., Vandamme, P., De Vuyst, L., 2010. Lactic acid bacteria community dynamics and metabolite production of rye sourdough fermentations share characteristics of wheat and spelt sourdough fermentations. Food Microbiology 27, 1000–1008.

Wichmann, F., Udikovic-Kolic, N., Andrew, S., Handelsman, J., 2014. Diverse antibiotic resistance genes in dairy cow manure. mBio 5, e01017–13.

Wimpenny, J.W.T., Leistner, L., Thomas, L.V., Mitchell, A.J., Katsaras, K., Peetz, P., 1995. Submerged bacterial colonies within food and model systems: their growth, distribution and interactions. International Journal of Food Microbiology 28, 299–315.

World Health Organization (WHO), 2001. WHO Global Strategy for Containment of Antimicrobial Resistance. Geneva, Switzerland. , pp. 1–99.

World Health Organization (WHO), 2011. Tackling Antibiotic Resistance from a Food Safety Perspective in Europe. Copenhagen, Denmark. , pp. 1–88.

Zdolec, N., Filipović, I., Fleck, Ž.C., Marić, A., Jankuloski, D., Kozačinski, L., Njari, B., 2011. Antimicrobial susceptibility of lactic acid bacteria isolated from fermented sausages and raw cheese. Veterinarski ARHIV 81, 133–141.

Zdolec, N., Račic, I., Vujnović, A., Zdelar-Tuk, M., Matanović, K., Filipović, I., Dobranić, V., Cvetnić, Ž., Spicic, S., 2013. Antimicrobial resistance of coagulase-negative staphylococci isolated from spontaneously fermented sausages. Food Technology and Biotechnology 51, 240–246.

Zonenschain, D., Rebecchi, A., Morelli, L., 2009. Erythromycin- and tetracycline-resistant lactobacilli in Italian fermented dry sausages. Journal of Applied Microbiology 107, 1559–1568.

Section 5

Revalorization of Food Wastes by Fermentation into Derived Outcomes

Chapter 30

Fermentation of Food Wastes for Generation of Nutraceuticals and Supplements

S. Patel[1], S. Shukla[2]
[1]San Diego State University, San Diego, CA, United States; [2]Yeungnam University, Gyeongsan, Republic of Korea

30.1 INTRODUCTION

The food industry generates a huge amount of waste, as consumer preference and adaptation towards only selective parts of crops or animals is ingrained. The phenomenal amounts of residues are incinerated or they end up as landfill, fertilizer, animal feed, fishmeal, etc. However, this unsustainable disposal convention is undesirable due to environmental concerns (carbon dioxide and methane emission aggravating global warming, foul odor, sanitation issues), and in the wake of looming food insecurity. Compliance with ecological balance is as important as ensuring nutritious food. Innovative utilization of the wastes might address both these issues. The endeavor to use the wastes for biogas, biodiesel, and bioethanol production is soaring. The evaluation of the wastes as a source of nutrients is rather pioneering, yet obligatory for the current scenario of depleting resources and rampant population growth. In this chapter, the feasibility of utilizing these wastes as food and additives will be explored. Fermentation is a well-established food preservation and nutrition improvement strategy. How this microbial-driven process might release functional ingredients will be discussed here by visiting different discards from agro, fruits, dairy, bakery, brewery, and fishery sectors. Antioxidant, antimicrobial, immunostimulatory, antidiabetic, anticancer, antihypertensive, anticoagulant, calcium-binding, hypocholesterolemic and appetite suppression effects of the nutrients will be summarized. Though the pace for repurposing of wastes has definitely intensified in recent times, the extent of exploitation is far less than feasible. This chapter strives to present a current state of affairs in a comprehensive manner.

Several interesting and insightful reviews shedding light on various facets of waste utilization for food valorization have been published in recent times. Bioconversion of wastes for bioactive mining hinges to a great extent on solid-state fermentation. The knowledge of optimal operation of fermentation, the substrates used and microorganisms employed play a decisive role in the extent of functional component recovery. An overview addressing these aspects of solid-state fermentation could be useful in this regard (Martins et al., 2011). Further, this fermentation technique as an alternative bioprocess technology for agro wastes was appraised (Martinez-Avila et al., 2014). Solid-state fermentation of grape skin and olive pomace with *Aspergillus niger* 113N and *Aspergillus fumigatus* 3 for production of lignocellulolytic enzymes such as xylanase, cellulase, β-glucosidase, and pectinase was reviewed (Romo Sánchez et al., 2014). Opportunities and challenges in deriving nutraceuticals from agricultural discards have been reviewed (Galanakis, 2013). Another review summarizes the protein-derived bioactive peptides identified in marine processing waste, fish, crustaceans, and molluscs (Harnedy and FitzGerald, 2012). Animal byproducts like blood, liver, lungs, kidney, brain, spleen, and tripe have good nutritive value. In addition, collagen, chitosan, and pigments promise myriad uses. Utilization of these waste materials has been reviewed (Jayathilakan et al., 2012). Fishery waste formulations were tested as growth substrates for food-grade microbial enzymes such as protease, lipase, and chitinolytic and ligninolytic enzymes. The efficacy of this strategy has been reviewed (Ben Rebah and Miled, 2012). Fish processing encompasses deheading, washing, scaling, gutting, cutting of fins, meat bone separation, etc. These steps produce a huge amount of waste (20–80%) depending upon the level of processing and type of fish. It has been observed that fish waste can be used to recover bioactive peptides, collagen and gelatin proteins, oil, amino acids, minerals, enzymes, etc. Fish oil as an omega-3 fatty acid-rich dietary supplement and fish sauce as pharmaceutical raw material have been reviewed (Ghaly et al., 2013). Fruit processing generates huge amounts of garbage in the form of peels, seeds, pomace, etc. The potency of date palm wastes has been reviewed. Resorting to fermentation and enzyme processing of date palm stones for the production of biosurfactants, organic acids, antibiotics, biofuels, biopolymers, industrial enzymes, etc. has been discussed (Chandrasekaran and Bahkali, 2013). Fermented apple processing wastes such as skin, pulp, and seeds as a source of functional ingredients, enzymes, and organic acids has been extensively reviewed (Dhillon et al., 2013a,b). The nutraceutical and pharmaceutical implications of the food manufacturing recovered nutrients have been outlined (Mirabella et al., 2014). The recyclable food wastes, the legislations associated with them, and the traditional and innovative techniques have been explored (Baiano, 2014). Fermentation of lignocellulosic wastes as a substrate for XOS (xylo-oligosaccharides), a prebiotic and sweetener, has been reviewed (Samanta et al., 2015). Apart from direct consumption of the fermented form of the waste food, it can be used as a growth medium for the cultivation of dietary supplements. This review recounts the major progress in this field over the last few years.

30.2 FUNCTIONAL ENHANCEMENT OF WASTES BY FERMENTATION

Nutrient-grade wastes can be categorized as agro, fruits, dairy, bakery, brewery, and fishery (Table 30.1). Fig. 30.1 shows the common, nutritionally exploitable wastes.

The biochemical makeup and potential utility of each of them varies widely. They serve as substrates for different microorganisms. These aspects and their nutraceutical relevance are discussed below. Various nutrient-laden organic wastes are generated from the dairy, agro, fruits, brewery, bakery, and fishery industries. Fig. 30.2 presents the types of wastes, fermenting microorganisms, and the recovered nutraceutical and pharmaceutical compounds.

30.2.1 Agro-Wastes

Agricultural harvest and subsequent processing generates a huge quantity of lignocellulose wastes in the form of straw, stalks, empty fruit bunches, legume hulls, cereal brans, soy pulp, corn cob, inferior-grade raw materials, molasses, waste syrup, vegetable pulp, seeds, etc.

30.2.1.1 Stalks, Straws, and Other Lignocellulosic Biomass

The solid-state fermentation of lignocellulosic biomass by *Pleurotus eous* and *Lentinus connatus* has been carried out (Rani et al., 2008). Both species of mushroom degraded lignocellulosic biomass of paddy straw and sorghum stalk by secreting cellulase and laccase. The use of laccase in the food industry for stabilization of beverages, promotion of pectin gelation, and improvement of food color, odor, and taste has been well-reviewed (Osma et al., 2010). The microalga *Aurantiochytrium* sp. KRS101 liberated a high concentration of docosahexaenoic acid (DHA) when cellulosic biomass, palm oil empty fruit bunches were utilized (Hong et al., 2012). Xylitol production by fermentation of mung bean hulls using *Candida tropicalis* was assessed. The obtained xylitol ameliorated serum glucose, cholesterol, and triglyceride levels in diabetic rats (Mushtaq et al., 2014).

30.2.1.2 Cereal Brans

The worth of the nutrient-dense outer layer of cereal kernels, the brans, has started to be appreciated with the discovery of beneficial phytochemicals. The phenolic acids and oryzanol-laden rice bran are being integrated in many foods such as bars, beverages, snacks, cereals, and baked products. A study was undertaken to estimate the phenolics from heat-stabilized, defatted rice bran (Webber et al., 2014). *Bacillus subtilis* fermented rice bran liberated phenolic of 26.8 mg ferulic acid equivalents/g. The phenolic acid profile comprised of (−)-epicatechin, syringic acid, gentistic acid, caffeic acid, *p*-courmaric acid, ferulic acid, and sinapic acid. Nukadoko is a fermented rice bran paste traditionally used for pickling vegetables in Japan. The microorganisms

TABLE 30.1 Waste Materials, Nutrients Tapped From Them, and the Fermentation Microbes

Waste Material	Nutrients/Food Ingredients	Fermentation Microbes	References
Paddy straw Sorghum stalk Mung bean hulls Palm oil empty fruit bunches	Cellulose Polyphenol oxidase Laccase Xylitol Docosahexaenoic acid	Pleurotus eous Lentinus connotus Candida tropicalis Aurantiochytrium sp. KRS101	Rani et al. (2008) Mushtaq et al. (2014) Hong et al. (2012)
Rice bran	Phenolic acids	Bacillus subtilis	Webber et al. (2014)
Wheat bran	Nigerloxin Rennet Fibrinolytic enzyme	Aspergillus niger MTCC-5166 Bacillus amyloliquefaciens D4 Bacillus cereus IND1	Chakradhar et al. (2009) Zhang et al. (2013) Vijayaraghavan and Vincent (2014)
Sugarcane waste	Inulinase	Kluyveromyces marxianus NRRL Y-7571	Bender et al. (2006)
Soy waste	Citric acid Antioxidant Iturin A Growth medium Prebiotic Polysaccharide	Aspergillus sp. B. subtilis Monascus anka Ganoderma lucidum Lactobacillus plantarum B16	Khare et al. (1995) Zhu et al. (2008) Khan (2012) Wong-Leung et al. (1993) Villanueva-Suárez et al. (2013) Shi et al. (2014) Xiao et al. (2015)
Corn cob	Xylanase Xylitol Ethanol Monascus pigment Xylooligosaccharide	Bacillus mojavensis A21 C. tropicalis W103 Monascus sp. Trichoderma koeningi	Haddar et al. (2012) Cheng et al. (2014) Bandikari et al. (2014)

Tea processing	Polysaccharides Polyphenols Free amino acids Proteins Caffeine Volatile components	*Grifola frondosa* *Ganoderma lucidum*	Bai et al. (2013)
Molasses	Omega-3 fatty acids	*Aurantiochytrium* sp. KRS101	Hong et al. (2011)
Syrup waste	Docosahexaenoic acid Astaxanthin	*Aurantiochytrium* sp. KH105	Iwasaka et al. (2013)
Sugar beet pulp	Pectin-derived oligosaccharides Edible oil	Fecal microbiota	Leijdekkers et al. (2014) Yagoub et al. (2004) Ouoba et al. (2008)
Carrot seeds Roselle seed	Essential oil	*Mucor circinelloides* *B. subtilis* *Bacillus licheniformis* *B. cereus* *Bacillus pumilus*	Śmigielski et al. (2014)
Orange peel Mango peel Pineapple peel Persimmon peel Litchi pericarp	Citric acid Pectic oligosaccharides Polygalacturonase Lactic acid Growth medium Vinegar Prebiotics Low-molecular-weight polysaccharides Antioxidants	*Aspergillus niger* NRRL 599 Bifidobacteria and Lactobacilli *B. licheniformis* SHG10 Indigenous microbes *Pediococcus pentosaceus* *Saccharomyces cerevisae* *Acetobacter aceti* Gut flora *Aspergillus awamori*	Torrado et al. (2011) Gómez et al. (2014) Embaby et al. (2014) Jawad et al. (2013) Diaz-Vela et al. (2013) Hwang et al. (2013) Da Silva et al. (2014) Lin et al. (2013)

Continued

TABLE 30.1 Waste Materials, Nutrients Tapped From Them, and the Fermentation Microbes—cont'd

Waste Material	Nutrients/Food Ingredients	Fermentation Microbes	References
Olive waste	Astaxanthin	Xanthophyllomyces dendrorhous ATCC 24202	Dursun et al. (2014)
Spent osmotic solution of tamarillo	Fructooligosaccharides	Aspergillus oryzae N74	Ruiz et al. (2014)
Apple pomace	Polyphenol Ethanol Acetic acid Citric acid Chitosan	Phanerochaete chrysosporium S. cerevisiae A. aceti Aspergillus niger Congronella butleri	Ajila et al. (2011) Parmar and Rupasinghe (2013) Dhillon et al. (2013a,b) Vendruscolo and Ninow (2014)
Grape pomace	Xylanase and pectinase Anthocyanins Trans-resveratrol Quercetin Gallic acid Catechin and epicatechin, proanthocyanidins	Lactobacillus S. cerevisiae Aspergillus awamori	Botella et al. (2007) Brazinha et al. (2014) Guchu et al. (2015)
Date palm juice	Curdlan	Rhizobium radiobacter ATCC 6466™	Ben Salah et al. (2011)
Dairy waste	Branched chain amino acids Whey-based beverages Peptide-enriched powder Antimicrobial compounds Galactooligosaccharide	Streptococcus thermophilus CRL 804 Lactobacillus delbrueckii subsp. bulgaricus CRL 454 Lactobacillus helveticus 209 Aspergillus oryzae L. plantarum 299v	Pescuma et al. (2008) Burns et al. (2010) Londero et al. (2011) Golowczyc et al. (2013)

Source	Products	Organisms	References
Brewery wastes (barley husks, spent grains, grain fragments, brewery cake, malt extract)	Xylooligosaccharides Anticomplementary polysaccharides Polyphenolic compounds Phenolic-rich beverage Prebiotics	Aspergillus sp. Trichoderma sp. Ganoderma lucidum	Gullón et al. (2011) Bae et al. (2012) Gupta et al. (2013) Reis et al. (2014) Aziizi et al. (2012)
Bakery wastes	Glucoamylase Omega-3 fatty acid Growth medium	Aspergillus awamori Thraustochytrium sp. AH-2 Schizochytrium sp. SR21 Lipomyces starkeyi DSM 70296	Lam et al. (2013) Thyagarajan et al. (2014) Tsakona et al. (2014)
Fishery wastes	Antioxidant Antimicrobial Fish oil Gelatin Astaxanthin Lutein and β-carotene Antioxidant peptides Amino acids Chitinase and protease Polysaccharides	Pediococcus acidilactici NCIM5368 Pediococcus acidilactici FD3 Enterococcus faecium NCIM5335 Lactic acid bacteria Serratia sp. TKU017 B. licheniformis OPL-007 Exiguobacterium sp.	Ruthu et al. (2014) Wang et al. (2010) Mao et al. (2013) Anil Kumar and Suresh (2014)
Underutilized resources (bamboo shoot, opuntia peel, rose hips, low-profile fish species)	Nutritious fermented food Vinegar substitute Colorant betanin Antioxidant-rich drink Fish sauce	B. subtilis Lactobacillus brevis and Lactobacillus plantarum Saccharomyces cerevisiae var. bayanus AWRI 796 Saccharomyces bayanus Aspergillus oryzae	Jeyram et al. (2010) Chavan et al. (2013) Czyzowska et al. (2014) Taira et al. (2007)

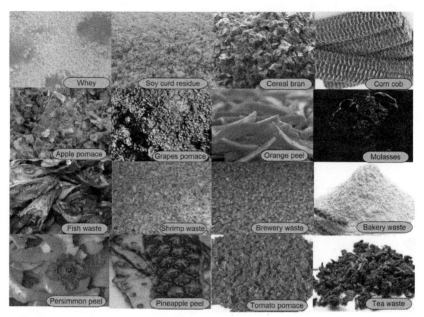

FIGURE 30.1 Various nutrient-laden organic wastes generated from agro, fruits, dairy, brewery, bakery, and fishery industries.

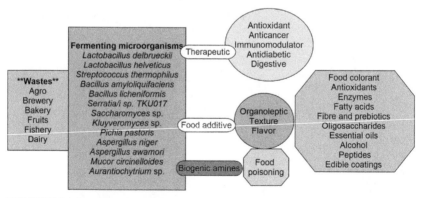

FIGURE 30.2 Overview of types of wastes, fermenting microorganisms, nutraceutical, and pharmaceutical applications.

associated with it were identified to be fast-growing lactic acid bacteria strains as well as slow-growing *Lactobacillus plantarum*. Formulation of a starter culture with the above species to boost initial fermentation of nukadoko was proposed (Ono et al., 2014). Solid-state fermentation of wheat bran by *Aspergillus niger* MTCC-5166 generated nigerloxin, a lipoxygenase inhibitor with an ability to improve the antioxidant defense system. A twofold increase in the production of nigerloxin was obtained by supplementing the bran with 10% sweet lemon peel and methanol (Chakradhar et al., 2009). Submerged fermentation of wheat bran with *Bacillus amyloliquefaciens* D4 liberated milk-clotting enzyme. Statistical optimization further improved the rennet yield (Zhang et al., 2013). Wheat bran fermentation with *Bacillus cereus* IND1 produced a potent fibrinolytic enzyme. The pH stable enzyme exhibited potency to activate plasminogen and significantly degraded the fibrin reticulum of blood clot (Vijayaraghavan and Vincent, 2014).

30.2.1.3 Soy Waste

Soybean has consolidated its status as a staple legume crop. The fermentation process has played a vital role in rendering a multitude of bioactive compounds and intensifying the unique flavor of soy food products such as tofu, soy sauce, miso, natto, and tempeh. In Korea, a number of soybean-based fermented products with multifarious biological benefits such as doenjang, ganjang, and cheonggukjang have been developed with novel starter cultures (Shukla et al., 2014). However, recent times have seen a surge in bioconversion of industrial soy wastes into value-added products with profound biological efficacy. Tofu byproducts called soybean-curd lees or tofu-cake are generated in tonnes. Due to their high moisture and protein content, they are prone to deterioration. These fermented wastes have shown better digestibility and palatability compared to original raw waste. Furthermore, the fermentation of soybean curd residue by *Aspergillus* sp. and *Bacillus subtilis* resulted in the production of citric acid (Khare et al., 1995), antioxidant (Zhu et al., 2008), and an potent antifungal substance iturin A (Khan, 2012). In addition, the potential role of soybean waste as vital substrate for the growth of *Monascus anka* mold during α-galactosidase production has been confirmed (Wong-Leung et al., 1993). Exploitation of microorganisms for deamination and hydrolysis of soy waste to liberate low-cost vegetable protein for human consumption seems to be an attractive approach for nutraceutical recovery (Kasai et al., 2004; O'Toole, 1999). Okara, a byproduct of soymilk production consisting of insoluble parts of the soybean which remain after pureed soybeans are filtered, has high nutritive value, but poor shelf-life. Okara, when treated with enzymes yielded a soluble dietary fiber and an enhanced ratio of soluble to insoluble fiber. Attributes of the modified product, such as water and oil retention, and swelling were improved. Polysaccharides were isolated from the treated okara. Fermentation of the product with *Bifidobacterium bifidus* synthesized short-chain fatty acids (SCFAs) such as acetic acid, propionic and butyric acids, while lowering the pH. The bifidogenic property of modified okara was confirmed (Villanueva-Suárez et al., 2013). The effect of inulin and okara flour on texture and taste of probiotic soy yogurt during 28 days storage at 4°C was evaluated. Nutritional and functional properties were improved. It was inferred that okara is suitable for incorporation into the probiotic product (Bedani et al., 2014). Soybean curd residue is rich in fiber, fat, protein, and vitamins. Soybean curd residue was subjected to solid-state fermentation for extraction of polysaccharides by *Flammulina velutipes* (enoki mushroom). Response surface methodology (RSM)-optimized medium of 74.5% moisture content, 9.69 inoculum size, and 30.27 carbon/nitrogen produced a polysaccharide content of 59.15 mg/g. The polysaccharide possessed multiple biological roles such as antioxidant and immunomodulatory (Shi et al., 2012). In addition, soybean curd residue was used to grow *Ganoderma lucidum* (reishi mushroom). A therapeutic polysaccharide was extracted from fermented curd residue with the aid of ultrasonic-assisted extraction. The polysaccharide exhibited antioxidant (by DPPH and ABTS assay) and immunomodulatory (stimulation of the proliferation of macrophages, production of nitric oxide, and phagocytosis) effects (Shi et al., 2014). *Lactobacillus plantarum* B16-fermented soy whey was richer in antioxidants (total phenolic content, isoflavone aglycone) and showed potential to be developed into a functional soy drink (Xiao et al., 2015).

30.2.1.4 Corn Cob

Corn cob is a good source of xylan. The cob-extracted xylan was subjected to fermentation by *Bacillus mojavensis* A21. Statistical optimization of the medium resulted in 7.45 U/mL xylanase yield. This enzyme might possess potential to be used for XOS production (Haddar et al., 2012). In addition, when corn cob was hydrolyzed with *Trichoderma koeningi* xylanase, xylose and its XOS were liberated (Bandikari et al., 2014). Probiotic strains such as *Bifidobacterium adolescentis*, *Bifidobacterium bifidum*, *Lactobacillus fermentum* and *Lactobacillus acidophilus* ferment these oligosaccharides (Chapla et al., 2012). The biological role of XOS in gut health restoration is shown in Fig. 30.3A. Acid-pretreated corn cob was successfully fermented with *Candida tropicalis* W103 to produce integrated xylitol (17.1 g/L) and ethanol (Cheng et al., 2014). The fungus degraded the inhibitors acetate, furfural, and 5-hydromethylfurfural and metabolite xylose to xylitol under aerobic conditions. Then it produced ethanol by anaerobic saccharification and fermentation. Corn cob was also soaked in aqueous ammonia to generate inhibitor-free acid hydrolysate. Fermentation of the hydrolysate by *C. tropicalis* generated xylitol. Detoxification with ammonia pretreatment seemed desirable as it enhanced xylitol yield to 60 g/L compared to 30 g/L for two-step adsorption detoxification (Deng et al., 2014). Monascus pigments produced by *Monascus* sp. are natural food colorants and therapeutic molecules eliciting anticancer, antiobesity, and antimicrobial potencies. Corn cob hydrolysate submerged fermentation was verified to produce more *Monascus* pigments than conventional glucose substrate. In addition, these pigments were free of citrin mycotoxin (Zhou et al., 2014).

30.2.1.5 Fermented Tea

Tea processing generates an enormous quantity of waste. The top leaves discarded after picking the tender ones were used for solid-state fermentation and culture of medicinal mushrooms *Grifola frondosa* and *Ganoderma lucidum* to obtain a therapeutic tea. The resultant beverage was rich in polysaccharides, polyphenols, free amino acids, proteins, caffeine, and volatile components

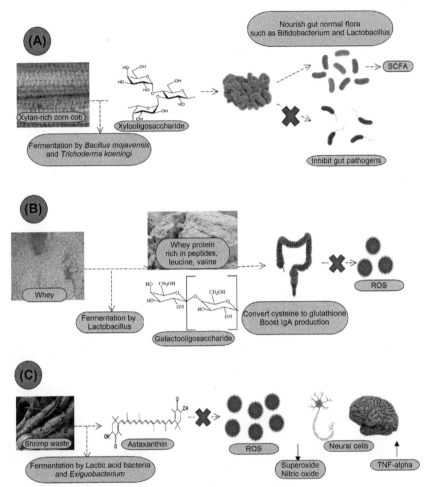

FIGURE 30.3 Mechanistic of the metabolic intervention caused by waste-recovered fermented products. (A) Xylooligosaccharide production from xylan-rich corn cob. (B) Biological ingredients such as whey protein and galactooligosaccharide production from whey. (C). Utilization of shrimp waste for antioxidant astaxanthin mining.

(alcohols, esters, aldehydes, and ketones) and, hence, fermentation of the tea leaves improved its health benefits (Bai et al., 2013).

30.2.1.6 Sugarcane Waste, Molasses, and Syrup Waste

Inulinase production by solid-state fermentation of bagasse supplemented with molasses and corn steep liquor was carried out. *Kluyveromyces marxianus* NRRL Y-7571 was used to liberate the enzyme capable of fructooligosaccharides (FOS) elaboration. An enzymatic activity of 445 U/g of dry substrate was achieved (Bender et al., 2006). Artificial neural network (ANN)-based algorithm predicted the amount of inulinase production when the above substrates and yeast strain were used to carry out the fermentation (Mazutti et al., 2009).

Thraustochytrid (marine protists) are now recognized candidates for commercial production of the omega-3 fatty acids. A microalgal strain *Aurantiochytrium* sp. KRS101, containing a high content of DHA was isolated from a Malaysian mangrove. The strain produced a high concentration of DHA, even when cheap substrate like molasses was used as the carbon source (Hong et al., 2011).

Syrups loaded with sugar are often drained as effluents or incinerated, while sustainable options exist. The possibility of utilizing waste syrup as a carbon source for production of DHA and astaxanthin by microalga *Aurantiochytrium* sp. KH105 was investigated (Iwasaka et al., 2013). Both the desired bioactive concentrations were high. Furthermore, medium optimization by RSM and use of activated charcoal enhanced the yield (Iwasaka et al., 2013).

30.2.1.7 Vegetable Pulp and Seeds

Sugar beet pulp is the residue of the beet sugar industry. The pulp shows potential to be transformed into pectin-derived oligosaccharides. These oligosaccharides were composed of partially acetylated rhamnogalacturonan and partially methyl-esterified/acetylated homogalacturonan. The oligosaccharides were completely fermented by human and pig fecal microbiota (Leijdekkers et al., 2014). The effect of an enzyme preparation sourced from *Mucor circinelloides* on the extraction of essential oil from waste carrot seeds was investigated. Statistical optimization of hydrodistillation yielded essential oil rich in carotol, sabinene, α-pinene, and daucol, which exhibited inhibition of *Bacillus subtilis*, *Staphylococcus aureus*, *Escherichia coli*, *Pseudomonas aeruginosa*, *Candida* sp., *Aspergillus niger*, and *Penicillium expansum* (Śmigielski et al., 2014). Roselle (*Hibiscus sabdariffa* L) seed is a byproduct during calyces harvesting. Though discarded, the seeds have been shown to be a reserve of a good quantity (20%) edible health-promoting oil. In Sudan, furundu is a preparation of cooked roselle seeds fermented for 9 days. The fermentation caused biochemical changes in the seed nutrients and promoted in vitro digestibility of the proteins (Yagoub et al., 2004). Bikalga is another roselle seed fermented condiment popular in Burkina Faso. Phenotyping and genotyping of the isolates from bikalga revealed the dominance of potential probiotic strains *Bacillus subtilis*, *Bacillus licheniformis*, *Bacillus cereus*, *Bacillus pumilus*, etc. (Ouoba et al., 2008).

30.2.2 Fruit Wastes

Fruit processing generates a huge quantum of refuges as peels, pomace, and spent solution. It has been realized that enzymes, antioxidants, pigments, organic acids, oligosaccharides, and prebiotics can be produced by proper processing of these wastes.

30.2.2.1 Peels

Orange peels were subjected to the production of citric acid by solid-state fermentation with *Aspergillus niger* NRRL 599. Under optimal conditions, 193 mg citric acid/g dry orange peel was obtained (Torrado et al., 2011). Orange peel waste was fermented with *Bacillus licheniformis* SHG10 to obtain polygalacturonase. RSM optimization implied the major parameters to be orange peel waste (1.76%), pH of medium (8.0), and incubation temperature (37.8°C). When these conditions were satisfied, polygalacturonase activity was maximized, manifested in the produced galacturonic acid level. Furthermore, the enzyme was pH- and temperature-stable (Embaby et al., 2014). Prebiotic potency of pectic oligosaccharides (POS) obtained by hydrothermal treatment of orange peels was investigated (Gómez et al., 2014). These oligosaccharides increased the population of bifidobacteria and lactobacilli in human fecal slurries. The generation of SCFAs proved the potential prebiotic efficacy of POS. The production of lactic acid by fermentation of mango peels was investigated. At an optimal condition of pH 10, incubation time of 6 days and temperature 35°C, maximum lactic acid production of 17.484 g/L was reported (Jawad et al., 2013). The fiber-rich passion fruit peel was envisioned to be a potential prebiotic. The ability of the peel to sustain probiotic growth by generation of SCFAs was explored. Two weeks of feeding with the peel flour revealed the production of butyrate and acetate in the cecal content of treated rats. It indicated the possible role of passion fruit peel flour in bolstering gut health and restoring homeostasis (da Silva et al., 2014). Fermentation of persimmon peel for vinegar production was investigated. Hydrolysis of the peel with viscozyme was followed by fermentation. The hydrolysate contained reducing sugars and soluble solids. On RSM-determined optimal fermentation settings (enzyme concentration of 2.38 g/100 mL, reaction temperature of 49.19°C, and reaction time of 6.32 h), vinegar contained acetic acid, citric acid, oxalic acid, and succinic acid (Hwang et al., 2013). *Aspergillus awamori* was effective in transforming litchi pericarp polysaccharides into a bioactive mixture with low-molecular-weight polysaccharides. The elaborated polysaccharides promoted the growth of *Lactobacillus bulgaricus* and *Streptococcus thermophilus* in the fermented medium. In addition, these polysaccharides demonstrated antioxidant activity verified by DPPH assay (Lin et al., 2013). Cactus pear and pineapple peel flours were evaluated as alternative carbon sources during fermentation with potential probiotic strains (*Pediococcus pentosaceus*). The total fiber content of both flours was over 60% and total soluble carbohydrate content was around 20%, indicating a good carbon source for lactic acid bacteria. Both the flours hold potential to be fermented and consequently employed as functional ingredients (Diaz-Vela et al., 2013).

30.2.2.2 Pomace

Grape pomace represents massive waste from grape processing. Usage of the pomace as substrate for xylanase and pectinase production by solid-state fermentation with *Aspergillus awamori* was evaluated and an economical yield was achieved (Botella et al., 2007). Grape pomace has been recognized as a substantial source of phenolic compounds, such as anthocyanins, *trans*-resveratrol, quercetin, and proanthocyanidins (Brazinha et al., 2014). After fermentation, over 90% of the total amount of gallic acid, catechin, and epicatechin and approximately 50% of the quercetin-3-*O*-glucoside were found in the pomace of Grenache noir and Airén variety grapes (Guchu et al., 2015).

Apple pomace is rich in moisture, cellulose, hemicellulose and lignin, glucose, and fructose. The pomace was subjected to solid-state fermentation by the fungus *Phanerochaete chrysosporium*. The polyphenol content in the acetone extract was found to be in the range of 5.78–16.12 mg gallic acid equivalents/g of pomace, depending on the solvent, extraction time, and temperature (Ajila et al., 2011). Enzymatic hydrolysis (using cellulase, pectinase, and β-glucosidase) of cellulose in apple pomace released glucose. After saccharification, the sugars were fermented with *Saccharomyces cerevisae* which yielded 19.0 g ethanol/100 g and *Acetobacter aceti*, which produced 61.4 g acetic acid/100 g (Parmar and Rupasinghe, 2013). The possibility of fermenting apple pomace with *Aspergillus niger* NRRL 2001 for citric acid production was evaluated. After RSM optimization, 312.32 g/kg dry weight was achieved at 75% (v/w) moisture and 3% (v/w) methanol after a 144 h incubation period (Dhillon et al., 2013a,b). In addition, apple pomace was assessed for chitosan production by the fungus *Gongronella butleri*. Addition of ammonium sulfate and setting the aeration rate to 0.6 vvm significantly enhanced deacetylation and elevated chitosan yield (Vendruscolo and Ninow, 2014).

Fortification of milk with phenolic extracts from olive and grape pomace was evaluated. Near-infrared spectroscopy characterized the properties of the fermentative media, which reflected stability, antioxidants, and sensorial acceptability (Aliakbarian et al., 2014). A yeast strain *Xanthophyllomyces dendrorhous* ATCC 24202, isolated from olive waste, was subjected to solid-state fermentation for astaxanthin production. The fermentation spanning 10 days produced 5.72–8.67 mg/L astaxanthin (Dursun et al., 2014).

30.2.2.3 Spent Solution and Juice Waste

Curdlan is a safe food and pharmaceutical additive with wide application as a gelling agent. The possibility of using date palm juice waste for curdlan production using *Rhizobium radiobacter* ATCC 6466 was investigated. Under optimized parameters of optimal medium pH 7, ammonium sulfate concentration 2 g/L, date glucose juice concentration 120 g/L, operating at 30°C with an inoculum ratio of 5 mL/100 mL, an agitation speed of 180 rpm, and a fermentation period of 51 h, a curdlan yield of 22.83 g/L was achieved. The curdlan had a molecular weight of 180 kDa and a linear structure of β-1,3 glucosidic linkages (Ben Salah et al., 2011). Fructooligosaccharide (FOS) production from spent osmotic solution of tamarillo (*Cyphomandra betacea*) on *Aspergillus oryzae* N74 fructosyltransferase treatment was assessed, which on scale-up with bioreactor fermentation resulted in a maximum FOS yield of 58.51% (Ruiz et al., 2014).

30.2.3 Dairy Wastes

Whey and buttermilk are the discards of the dairy industry. Whey, a byproduct of the cheese industry, is disposed of due to ignorance of its nutritional abundance. In recent times, whey proteins have been regarded as a source of healthy compounds. Unprecedented numbers of publications agree on the importance of whey proteins. An in-depth review on whey reports its interesting bioactive profile spanning lactoferrin, β-lactoglobulin, α-lactalbumin, glycomacropeptide, and immunoglobulins. Furthermore, its versatile ameliorative attributes as an antioxidant, antihypertensive, hypolipidemic, antitumor, antimicrobial, endurance improvement, and food packaging agent has been recounted (Marshall, 2004). The feasibility of employing lactic acid bacteria to formulate functional whey-based drink was evaluated. When fermented at 37–42°C for 24 h in reconstituted whey powder, the tested strains showed different results. *Streptococcus thermophilus* CRL 804 consumed substantial lactose and produced the maximum lactic acid amounts. *L. delbrueckii* subsp. *bulgaricus* CRL 454 exerted most proteolysis and released the branched-chain amino acids leucine and valine. These two strains might be used to develop fermented whey-based beverages (Pescuma et al., 2008). A proteolytic strain of lactic acid bacteria *Lactobacillus helveticus* 209 was used to ferment buttermilk to produce functional peptide-enriched powder. The product, when administered to rodents, stimulated proliferation of IgA-producing cells in the gut (Burns et al., 2010). The possibility of using kefir grains to ferment whey in order to obtain a natural acidic drink with microbial inhibitory properties was carried out. The fermented acidic liquor containing probiotic microorganisms was capable of suppressing *Salmonella enterica* serovar Enteritidis 2713 and *Escherichia coli* 2710 (Londero et al., 2011). The growth and proteolytic activity of various lactic acid bacteria strains in whey were evaluated to develop a starter culture for whey-based beverage formulations. Fermentation was performed at 37°C for 24 h in 10% and 16% (w/v) reconstituted whey powder. Most of the lactobacilli released amino acids and small peptides during the first 6 h incubation. Allergenic whey protein β-lactoglobulin, was degraded the most by *Lactobacillus acidophilus* CRL 636 and *L. delbrueckii* subsp. *bulgaricus* CRL 656 (Pescuma et al., 2012). Whey permeate was used to produce galactooligosaccharides (GOS) and to develop a culture medium for probiotics. β-Galactosidase from *Aspergillus oryzae* was used to liberate galactose from whey. The enzyme-treated medium was used to culture *Lactobacillus plantarum* 299v. The concomitant production of GOS and dehydrated viable probiotic microorganisms indicated functional food potential of whey (Golowczyc et al., 2013). Fig. 30.3B presents the schematic diagram of the functional ingredients recovered from whey.

30.2.4 Malting and Brewing Wastes

Breweries generate piles of solid wastes in the form of barley husks, spent grains, and grain fragments. Recent times have seen many interesting uses for these discards. Hydrothermal and endoxylanase treatment of the wastes generates liquor from which XOS can be isolated. XOS were fermented in vitro to assess their prebiotic potential. The presence of SCFAs such as succinate, lactate, formiate, acetate, propionate, and butyrate confirmed the prebiotic potential of these oligosaccharides (Gullón et al., 2011). Makgeolli is a traditional Korean wine brewed from rice, with an aromatic and sweet-sour taste. The byproducts during the alcoholic fermentation comprise rice bran and brewery cake. The crude polysaccharide isolated from cytolase hydrolysates of Makgeolli contained glucose and mannose, and exhibited anticomplementary (immune-manipulation) activities (Bae et al., 2012). Spent grain was evaluated for production of a phenolic compound-rich beverage. RSM optimization improved the release of polyphenolic compounds (268.6 mg gallic acid equivalents/mL of phenolic compounds and 135 mg quercetin equivalents/mL of flavonoid compounds). The beverage was shelf-stable for 15 days (Gupta et al., 2013). Arabinoxylans extracted from spent grain showed good fermentability and produced SCFAs when metabolized by human fecal slurries. Increment in the gut bifidobacteria density implied its prebiotic potential (Reis et al., 2014). Amended malt extract was explored for feasibility of growing medicinal mushroom *Ganoderma lucidum*. The highest mycelia growth rate (10.6 mm/day) was obtained in a combination of beech sawdust with 2.5% malt extract and 10% wheat bran (Azizi et al., 2012).

30.2.5 Bakery Waste

Bakeries and confectionery generate a high quantity of wastes from the flour processing and bread, pastry, and cake manufacture. The feasibility of using pastry waste as substrate for glucoamylase (an enzyme capable of saccharification of partially processed starch/dextrin to glucose) was evaluated by solid-state fermentation. A 10 days fermentation of 100 g pastry waste with *Aspergillus awamori* produced about 53- g glucose (Lam et al., 2013). Bread crumbs were used as fermentation medium for *Thraustochytrium* sp. AH-2 and *Schizochytrium* sp. SR21, to biosynthesize ω-3 fatty acid. Though the yield was low, necessary adjustments might improve the production (Thyagarajan et al., 2014). Wheat milling byproducts were subjected to solid-state fermentation by *Aspergillus awamori* for the production of glucoamylase and protease. The enzyme-rich hydrolysate was further used to grow oleaginous yeast *Lipomyces starkeyi* DSM 70296 for oil (oleic acid and linoleic acid) harvest. By fed-batch bioreactor fermentation, a culture dry weight of 109.8 g/L and oil content of 57.8% was obtained (Tsakona et al., 2014).

30.2.6 Fishery Byproducts

Fishery comprises a major component of the food sector. In addition, the seafood harvest is rising. Together they generate an enormous quantity of head, viscera, skin, exoskeletons, bones, scales, etc. Bioactive peptides contained in them are healthy and can be revalorized into functional foods and nutraceuticals. The feasibility of extracting these peptides from muscle protein of fish has been reviewed and the angiotensin-1-converting enzyme (ACE)-inhibitory and antioxidative peptides have been emphasized (Ryan et al., 2011).

After oil extraction, the discarded head part of salmon still contains protein and high-value fatty acids. Fermentation of the byproduct with lactic acid bacteria improves its shelf-life (Bower and Hietala, 2010). The possibility of lactic acid fermentation for recovery of lipids and proteins from freshwater fish head was evaluated. Proteolytic lactic acid bacteria *Pediococcus acidilactici* NCIM5368, *Pediococcus acidilactici* FD3, and *Enterococcus faecium* NCIM5335 were isolated from the fish processing wastes. The fermented slurry demonstrated antioxidant activity as determined by DPPH assay. In addition, it exhibited inhibition of pathogens such as *Listeria monocytogenes*, *Salmonella itridicus*, *Aspergillus ochraceus*, and *Penicillium chrysogenum*. Both the oil and protein hydrolysate-rich slurry held potential to be implicated in food formulations (Ruthu et al., 2014). Wastes from cultured tuna, sardine, eel, catfish, and milkfish showed possibility of mining fish oil. Fish skin and scales can be enzymatically treated to obtain gelatin.

The part of the shrimp often thrown away can be a valuable food ingredient. Astaxanthin is a robust carotenoid antioxidant (more potent than α-tocopherol, lutein, zeaxanthin, canthaxanthin, and β-carotene) and found in abundance in the shell of shrimp, crab, and other crustaceans. Astaxanthin ingestion has immense physiological benefits as it lowers the production of superoxides, nitric oxide, and enhanced tumor necrosis factor-α (TNF-α). Shrimp shell fermented with lactic acid bacteria at 27–36°C yielded astaxanthin (Pacheco et al., 2009). Shrimp waste was fermented by lactic acid bacteria and the generated slurry was lyophilized. The fermentation product contained antioxidant peptides and amino acids. The bioactive protein derivatives along with carotenoid content could be developed as fish feed (Sachindra and Bhaskar, 2008). Shrimp shell medium fermented with enterobacteriaceae member *Serratia* sp. TKU017 produced chitinase and protease. Both enzymes are useful for myriad food processing, so the shrimp shell waste could be effectively purposed (Wang et al., 2010). Furthermore, the fermentation of shrimp head waste by *Bacillus licheniformis* OPL-007 generated total phenols, polysaccharides, reducing sugars, free amino acids (eight essential amino acids), and organic acids. In addition, the broth exhibited antioxidant activity as verified by DPPH assay (Mao et al., 2013). Supercritical fluid extraction of tiger shrimp shell with ethanol yielded astaxanthin, lutein, and β-carotene (Radzali et al., 2014). Submerged fermentation shrimp waste medium with marine bacterium *Exiguobacterium* sp. concomitantly liberated protease and antioxidants. RSM optimization predicted that shrimp waste (21.2%), sugar (10.5%), and phosphate (2.3%) are ideal for production of protease, whereas shrimp waste (28.8%), sugar (12%) and phosphate (0.32%) are suitable for production of antioxidants (Anil Kumar and Suresh, 2014). Fig. 30.3C illustrates the antioxidant and neuroprotection mechanism of astaxanthin.

30.3 FOOD WASTES AS CULTURE MEDIUM OF FUNCTIONAL FOODS

In addition, multiple studies have shown that assorted liquid and solid wastes can be used as cheap sources of functional food-grade microorganisms (microalga, lactic acid bacteria, and yeasts). Food waste was hydrolyzed by *Aspergillus awamori* and *Aspergillus oryzae* and the hydrolysate was used to grow heterotrophic microalgae *Schizochytrium mangrovei* and *Chlorella pyrenoidosa* (Pleissner et al., 2013). *Chlorella vulgaris* cultivation in food waste hydrolyzate was investigated. The microalgal biomass very efficiently utilized the nutrients in the medium as 400 mg carbohydrates/g, 200 mg proteins/g, and 200 mg lipids/g conversion was reported (Lau et al., 2014). The growth of *Saccharomyces cerevisiae*, *Kluyveromyces marxianus*, and kefir by solid-state fermentation of various food industry waste mixtures was investigated. The fat- and protein-rich product could be used to add value to feeds (Aggelopoulos et al., 2014).

30.4 VALORIZATION OF UNDERUTILIZED FOOD SOURCES THROUGH FERMENTATION

Fermentation can be harnessed as a strategy to improve nutritional status of many obscure, ill-explored, and perishable food sources. This innovative approach is akin to and as crucial as rescuing nutrients from wastes. Fermentation of tender bamboo shoot tip results in a flavorful acidic product "Soidon", popular in Northeast India. The starter mixture "soidon mahi" was identified to be dominated by *Bacillus subtilis, Lactobacillus brevis*, and *Lactobacillus plantarum* (Jeyaram et al., 2010). The extract and paste of bamboo could be used as a substitute for chemical preservative vinegar in the preparation of pork pickle products. This food had a shelf-life of 90 days at room temperature (Chavhan et al., 2013). No adverse physicochemical and microbial spoilage was noticed during that time period. *Opuntia* grows luxuriantly in arid parts of the world. Some species of this genus are commercially used for fruits, while most of them languish in obscurity. Fermentation of the pigment-rich peels with *Saccharomyces cerevisiae* var. *bayanus* AWRI 796 resulted in red food colorant betanin (Castellar et al., 2008). The rose hips from wild species, namely *Rosa canina* L. and *Rosa rugosa* Thunb., have been established as potential functional foods. Food preparations like teas, jellies, jams, and beverages have become popular. However, experimental outcomes have corroborated the nutritional augmentation of the rose hips on fermentation. The fruit pulp fermented with *Saccharomyces bayanus* produced ascorbic acid of 1200 and 600 mg/L, from *Rosa rugosa* Thunb. and *Rosa canina* L., respectively. The antioxidant-rich rose hip wines exerted antimutagenicity as tested with *Salmonella typhimurium* TA98 and *S. typhimurium* TA100 (Czyzowska et al., 2014). To prepare fish sauce from underutilized fish species, microbial fermentation was adopted. Flying fish, small dolphin fish, and deep sea smelt were fermented using salt and *koji* mold. During the 180 days of fermentation, the pH decreased while total nitrogen and free amino acids increased. The microbial transformation reduced the smell, saltiness, and bitterness, while augmenting the sweetness and umami taste (Taira et al., 2007). The cited findings are only a fraction of the immense valorization capability of fermentation process.

30.5 POSSIBLE HARMS AND HURDLES

Some microorganisms produce biogenic amines (eg, putrescine, cadaverine, histamine, tyramine, tryptamine, spermidine, and spermine) during fermentation. Most of the biogenic amines generated by amino acid decarboxylation compromise the quality and safety of food. In Yulu, a Chinese fermented fish, seven biogenic amines were determined. This finding calls for strict monitoring of biogenic amine content in other fermented foods as well (Jiang et al., 2014). Cost-effective microbial fermentation depends on the utilization of cheap substrates and processes. Waste lignocellulosic biomass is among the most attractive raw materials because of its high abundance, low price, and renewability. However, current biomass conversion technologies are still economically inefficient.

30.6 FUTURE DIRECTIONS AND BARRICADES TO SURMOUNT

In the pursuit to ensure higher nutrient recovery, bioprospecting for novel microbial strains, metabolic engineering of microorganisms, fermentation process optimization, and novel processing techniques are important keys. So far, the microorganisms harnessed for the fermentation are lactic acid bacteria (wine, dairy products), some yeasts, some species of *Bacillus*, and a few microalgae. However, a host of other microorganisms with probable bioconversion efficiency is languishing in obscurity, which must be prospected. Manipulation of the microbial genome has already shown encouraging results.

Citrus peel is rich in pectin with D-galacturonic acid as the main monomer. *Aspergillus niger* was engineered to circumvent its D-galacturonic acid consumption trait to L-galacturonic acid conversion, a precursor for L-ascorbic acid (vitamin C) (Kuivanen et al., 2014).

Sometimes inherent compounds in the wastes inhibit fermentation and compromise desirable nutrient recovery. Limonene is a terpene abundant in most citrus peels which exerts antimicrobial action. On the other hand, D-limonene is used as a flavor compound in food processing to mask bitterness of alkaloids (Moraes et al., 2009). Extraction of this compound before subjecting the peels to fermentation has been suggested. Pretreatment before the fermentation process might enhance nutrient recovery. Hot acid impregnation, steam explosion, and aqueous ammonia soaking have been recognized as effective pretreatment modalities. Nutrient extraction techniques must be effective for maximum recovery. The advent of pressurized fluid extraction, supercritical fluid extraction, ultrasound-assisted extraction, microwave-assisted extraction, pulsed electric field extraction, and enzyme-assisted extraction appear promising in this regard.

Many major regional discards have not been evaluated for nutritive enhancement through fermentation. Oil palm plantations produce tonnes of empty fruit bunches, kernel shells, and fronds. Phytochemical investigations have shown their abundance in carotenoids, phenolic compounds, sterols, tocols, etc. (Ofori-Boateng and Lee, 2013). Oyster shells are discarded in tonnes in Taiwan. Algae have not been explored much for their fermentative ability. It does not mean they cannot be utilized, it just implies any breakthrough has not taken place so far.

Of course, fermentation of agro and food industry wastes to tap nutrients can be extended to proper utilization of perishable foods like fruits and vegetable. Tonnes of these undergo spoilage due to their acid and moisture content. As the popularity of sauerkraut (fermented cabbage) expands, other vegetables are being treated with "generally regarded as safe (GRAS)" microorganisms. such as cucumbers, beets, turnips, cauliflower, celery, radishes, and carrots (Swain et al., 2014). Almost all vegetables can be fermented to avoid rotting. Though fruits contain higher sugar amounts and require controlled fermentation, peach, apple, and cranberry have been fermented successfully to prolong their shelf-life. Several endemic and exotic tropical fruits are excellent sources of nutrients, yet their perishability poses a hurdle to their global popularity. Fermentation might be recruited to prolong their shelf-lives.

Exploration of the innovative uses of recyclable wastes is a rather nascent field and much needs to be investigated. Many start-up companies have initiated this commendable job of surveying the potential use of wastes. Converting the waste into food as well as extracting nutraceutical compounds from them is being assessed.

30.7 CONCLUSIONS

The impending food insecurity necessitates sustainability in food production. The agro-industrial wastes are laden with nutrients. They just need the right technology to extract them. Fermentation has proved beneficial in getting rid of undesirable traits while augmenting the desirable ones. Despite their immense potential, the food-grade wastes have not acquired much emphasis. Nutrient tapping by fermentation will resolve the issues of environmental pollution and boost sustainability. Research input in mining the nutrients optimally must be intensified. Innovative processing techniques must be harnessed in this pursuit. There goes a saying "wastes are resources that we are not intelligent enough to utilize yet". This practice of nutrient recovery from discarded organic sources will not only reduce production costs, it will ensure sustainability and minimize the anthropogenic footprint on an already-threatened environment. When millions of people are suffering or dying prematurely, especially in resource-poor countries, due to inadequate nutrient intake, it is sheer imprudence to dispose of nutrient sources. Fortunately, the old science of fermentation is emerging to be the pertinent technology to address this issue, which needs research input to make it cost-effective and streamlined.

REFERENCES

Aggelopoulos, T., Katsieris, K., Bekatorou, A., Pandey, A., Banat, I.M., Koutinas, A.A., 2014. Solid state fermentation of food waste mixtures for single cell protein, aroma volatiles and fat production. Food Chemistry 145, 710–716. http://dx.doi.org/10.1016/j.foodchem.2013.07.105.

Ajila, C.M., Brar, S.K., Verma, M., Tyagi, R.D., Valéro, J.R., 2011. Solid-state fermentation of apple pomace using *Phanerocheate chrysosporium*—liberation and extraction of phenolic antioxidants. Food Chemistry 126, 1071–1080. http://dx.doi.org/10.1016/j.foodchem.2010.11.129.

Aliakbarian, B., Casale, M., Paini, M., Casazza, A.A., Lanteri, S., Perego, P., 2014. Production of a novel fermented milk fortified with natural antioxidants and its analysis by NIR spectroscopy. LWT—Food Science and Technology. http://dx.doi.org/10.1016/j.lwt.2014.07.037.

Anil Kumar, P.K., Suresh, P.V., 2014. Biodegradation of shrimp biowaste by marine *Exiguobacterium* sp. CFR26M and concomitant production of extracellular protease and antioxidant materials: production and process optimization by response surface methodology. Marine Biotechnology 16, 202–218. http://dx.doi.org/10.1007/s10126-013-9531-2.

Azizi, M., Tavana, M., Farsi, M., Oroojalian, F., 2012. Yield performance of Lingzhi or Reishi medicinal mushroom, *Ganoderma lucidum* (W.Curt.:Fr.) P. Karst. (higher Basidiomycetes), using different waste materials as substrates. International Journal of Medicinal Mushrooms 14, 521–527.

Bae, S.H., Choi, J.W., Ra, K.S., Yu, K.-W., Shin, K.-S., Park, S.S., Suh, H.J., 2012. Anti-complementary activity of enzyme-treated traditional Korean rice wine (Makgeolli) hydrolysates. Journal of the Sciene of Food and Agriculture 92, 1765–1770. http://dx.doi.org/10.1002/jsfa.5543.

Bai, W.-F., Guo, X.-Y., Ma, L.-Q., Guo, L.-Q., Lin, J.-F., 2013. Chemical composition and sensory evaluation of fermented tea with medicinal mushrooms. Indian Journal of Microbiology 53, 70–76. http://dx.doi.org/10.1007/s12088-012-0345-0.

Baiano, A., 2014. Recovery of biomolecules from food wastes – a review. Molecules 19, 14821–14842. http://dx.doi.org/10.3390/molecules190914821.

Bandikari, R., Poondla, V., Obulam, V.S.R., 2014. Enhanced production of xylanase by solid state fermentation using *Trichoderma koenigi* isolate: effect of pretreated agro-residues. 3 Biotech 4, 655–664. http://dx.doi.org/10.1007/s13205-014-0239-4.

Bedani, R., Campos, M.M.S., Castro, I.A., Rossi, E.A., Saad, S.M.I., 2014. Incorporation of soybean by-product okara and inulin in a probiotic soy yoghurt: texture profile and sensory acceptance. Journal of the Science of Food and Agriculture 94, 119–125. http://dx.doi.org/10.1002/jsfa.6212.

Ben Rebah, F., Miled, N., 2012. Fish processing wastes for microbial enzyme production: a review. 3 Biotech 3, 255–265. http://dx.doi.org/10.1007/s13205-012-0099-8.

Ben Salah, R., Jaouadi, B., Bouaziz, A., Chaari, K., Blecker, C., Derrouane, C., Attia, H., Besbes, S., 2011. Fermentation of date palm juice by curdlan gum production from *Rhizobium radiobacter* ATCC 6466™: purification, rheological and physico-chemical characterization. LWT—Food Science and Technology 44, 1026–1034. http://dx.doi.org/10.1016/j.lwt.2010.11.023.

Bender, J.P., Mazutti, M.A., de Oliveira, D., Di Luccio, M., Treichel, H., 2006. Inulinase production by *Kluyveromyces marxianus* NRRL Y-7571 using solid state fermentation. Applied Biochemistry and Biotechnology 129–132, 951–958.

Botella, C., Diaz, A., de Ory, I., Webb, C., Blandino, A., 2007. Xylanase and pectinase production by *Aspergillus awamori* on grape pomace in solid state fermentation. Process Biochemistry 42, 98–101. http://dx.doi.org/10.1016/j.procbio.2006.06.025.

Bower, C., Hietala, K., 2010. Stabilizing smoked salmon (*Oncorhynchus gorbuscha*) tissue after extraction of oil. Journal of Food Science 75, C241–C245. http://dx.doi.org/10.1111/j.1750-3841.2010.01521.x.

Brazinha, C., Cadima, M., Crespo, J.G., 2014. Optimization of extraction of bioactive compounds from different types of grape pomace produced at wineries and distilleries. Journal of Food Science 79, E1142–E1149. http://dx.doi.org/10.1111/1750-3841.12476.

Burns, P., Molinari, F., Beccaria, A., Páez, R., Meinardi, C., Reinheimer, J., Vinderola, G., 2010. Suitability of buttermilk for fermentation with *Lactobacillus helveticus* and production of a functional peptide-enriched powder by spray-drying. Journal of Applied Microbiology 109, 1370–1378. http://dx.doi.org/10.1111/j.1365-2672.2010.04761.x.

Castellar, M.R., Obón, J.M., Alacid, M., Fernández-López, J.A., 2008. Fermentation of *Opuntia stricta* (Haw.) fruits for betalains concentration. Journal of Agriculture and Food Chemistry 56, 4253–4257. http://dx.doi.org/10.1021/jf703699c.

Chakradhar, D., Javeed, S., Sattur, A.P., 2009. Studies on the production of nigerloxin using agroindustrial residues by solid-state fermentation. Journal of Industrial Microbiology and Biotechnology 36, 1179–1187. http://dx.doi.org/10.1007/s10295-009-0599-7.

Chandrasekaran, M., Bahkali, A.H., 2013. Valorization of date palm (*Phoenix dactylifera*) fruit processing by-products and wastes using bioprocess technology—review. Saudi Journal of Biological Sciences 20, 105–120. http://dx.doi.org/10.1016/j.sjbs.2012.12.004.

Chapla, D., Pandit, P., Shah, A., 2012. Production of xylooligosaccharides from corncob xylan by fungal xylanase and their utilization by probiotics. Bioresource Technology 115, 215–221. http://dx.doi.org/10.1016/j.biortech.2011.10.083.

Chavhan, D.M., Hazarika, M., Brahma, M.L., Hazarika, R.A., Rahman, Z., 2013. Effect of incorporation of fermented bamboo shoot on physicochemical and microbial quality of pork pickle. Journal of Food Science and Technology. http://dx.doi.org/10.1007/s13197-013-1082-z.

Cheng, K.-K., Wu, J., Lin, Z.-N., Zhang, J.-A., 2014. Aerobic and sequential anaerobic fermentation to produce xylitol and ethanol using non-detoxified acid pretreated corncob. Biotechnology for Biofuels 7, 166. http://dx.doi.org/10.1186/s13068-014-0166-y.

Czyzowska, A., Klewicka, E., Pogorzelski, E., Nowak, A., 2014. Polyphenols, vitamin C and antioxidant activity in wines from *Rosa canina* L. and *Rosa rugosa* Thunb. Journal of Food Composition and Analysis. http://dx.doi.org/10.1016/j.jfca.2014.11.009.

da Silva, J.K., Cazarin, C.B.B., Bogusz Junior, S., Augusto, F., Maróstica Junior, M.R., 2014. Passion fruit (*Passiflora edulis*) peel increases colonic production of short-chain fatty acids in Wistar rats. LWT—Food Science and Technology 59, 1252–1257. http://dx.doi.org/10.1016/j.lwt.2014.05.030.

Deng, L.-H., Tang, Y., Liu, Y., 2014. Detoxification of corncob acid hydrolysate with SAA pretreatment and xylitol production by immobilized *Candida tropicalis*. The Scientific World Journal 2014, 214632. http://dx.doi.org/10.1155/2014/214632.

Dhillon, G.S., Brar, S.K., Kaur, S., Verma, M., 2013a. Screening of agro-industrial wastes for citric acid bioproduction by *Aspergillus niger* NRRL 2001 through solid state fermentation. Journal of the Science of Food and Agriculture 93, 1560–1567. http://dx.doi.org/10.1002/jsfa.5920.

Dhillon, G.S., Kaur, S., Brar, S.K., 2013b. Perspective of apple processing wastes as low-cost substrates for bioproduction of high value products: a review. Renewable and Sustainable Energy Reviews 27, 789–805. http://dx.doi.org/10.1016/j.rser.2013.06.046.

Diaz-Vela, J., Totosaus, A., Cruz-Guerrero, A.E., de Lourdes Pérez-Chabela, M., 2013. In vitro evaluation of the fermentation of added-value agroindustrial by-products: cactus pear (*Opuntia ficus-indica* L.) peel and pineapple (*Ananas comosus*) peel as functional ingredients. International Journal of Food Science and Technology 48, 1460–1467. http://dx.doi.org/10.1111/ijfs.12113.

Dursun, D., Altınçiçek, E.A., Boğusoğlu, M., Dalgıç, A.C., 2014. Astaxanthin pigment production by ATTC 24202 from fruit wastes using solid state fermentation method. Journal of Biotechnology 185, S24. http://dx.doi.org/10.1016/j.jbiotec.2014.07.081.

Embaby, A.M., Masoud, A.A., Marey, H.S., Shaban, N.Z., Ghonaim, T.M., 2014. Raw agro-industrial orange peel waste as a low cost effective inducer for alkaline polygalacturonase production from *Bacillus licheniformis* SHG10. SpringerPlus 3, 327. http://dx.doi.org/10.1186/2193-1801-3-327.

Galanakis, C.M., 2013. Emerging technologies for the production of nutraceuticals from agricultural by-products: a viewpoint of opportunities and challenges. Food and Bioproduct Processing 91, 575–579. http://dx.doi.org/10.1016/j.fbp.2013.01.004.

Ghaly, A.E., Ramakrishnan, V.V., Brooks, M.S., Budge, S.M., Dave, D., 2013. Fish processing wastes as a potential source of proteins, amino acids and oils: a critical review. Journal of Microbial and Biochemical Technology 5, 4.

Golowczyc, M., Vera, C., Santos, M., Guerrero, C., Carasi, P., Illanes, A., Gómez-Zavaglia, A., Tymczyszyn, E., 2013. Use of whey permeate containing in situ synthesised galacto-oligosaccharides for the growth and preservation of *Lactobacillus plantarum*. The Journal of Dairy Research 80, 374–381. http://dx.doi.org/10.1017/S0022029913000356.

Gómez, B., Gullón, B., Remoroza, C., Schols, H.A., Parajo, J.C., Alonso, J.L., 2014. Purification, characterization and prebiotic properties of pectic oligosaccharides from orange peel wastes. Journal of Agriculture and Food Chemistry. http://dx.doi.org/10.1021/jf503475b.

Guchu, E., Ebeler, S.E., Lee, J., Mitchell, A.E., 2015. Monitoring selected monomeric polyphenol composition in pre- and post-fermentation products of *Vitis vinifera* L. cv. Airén and cv. Grenache noir. LWT—Food Science and Technology 60, 552–562. http://dx.doi.org/10.1016/j.lwt.2014.09.019.

Gullón, P., González-Muñoz, M.J., Parajó, J.C., 2011. Manufacture and prebiotic potential of oligosaccharides derived from industrial solid wastes. Bioresource Technology 102, 6112–6119. http://dx.doi.org/10.1016/j.biortech.2011.02.059.

Gupta, S., Jaiswal, A.K., Abu-Ghannam, N., 2013. Optimization of fermentation conditions for the utilization of brewing waste to develop a nutraceutical rich liquid product. Industrial Crops and Products 44, 272–282. http://dx.doi.org/10.1016/j.indcrop.2012.11.015.

Haddar, A., Driss, D., Frikha, F., Ellouz-Chaabouni, S., Nasri, M., 2012. Alkaline xylanases from *Bacillus mojavensis* A21: production and generation of xylooligosaccharides. International Journal of Biological Macromolecules 51, 647–656. http://dx.doi.org/10.1016/j.ijbiomac.2012.06.036.

Harnedy, P.A., FitzGerald, R.J., 2012. Bioactive peptides from marine processing waste and shellfish: a review. Journal of Functional Foods 4, 6–24. http://dx.doi.org/10.1016/j.jff.2011.09.001.

Hong, W.-K., Kim, C.H., Rairakhwada, D., Kim, S., Hur, B.-K., Kondo, A., Seo, J.-W., 2012. Growth of the oleaginous microalga *Aurantiochytrium* sp. KRS101 on cellulosic biomass and the production of lipids containing high levels of docosahexaenoic acid. Bioprocess and Biosystems Engineering 35, 129–133. http://dx.doi.org/10.1007/s00449-011-0605-0.

Hong, W.-K., Rairakhwada, D., Seo, P.-S., Park, S.-Y., Hur, B.-K., Kim, C.H., Seo, J.-W., 2011. Production of lipids containing high levels of docosahexaenoic acid by a newly isolated microalga, *Aurantiochytrium* sp. KRS101. Applied Biochemistry and Biotechnology 164, 1468–1480. http://dx.doi.org/10.1007/s12010-011-9227-x.

Hwang, I.-W., Chung, S.-K., Jeong, M.-C., Chung, H.-S., Zheng, H.-Z., 2013. Optimization of enzymatic hydrolysis of persimmon peels for vinegar fermentation. Journal of the Korean Society for Applied Biological Chemistry 56, 435–440. http://dx.doi.org/10.1007/s13765-013-3036-6.

Iwasaka, H., Aki, T., Adachi, H., Watanabe, K., Kawamoto, S., Ono, K., 2013. Utilization of waste syrup for production of polyunsaturated fatty acids and xanthophylls by *Aurantiochytrium*. Journal of Oleo Science 62, 729–736.

Jawad, A.H., Alkarkhi, A.F.M., Jason, O.C., Easa, A.M., Nik Norulaini, N.A., 2013. Production of the lactic acid from mango peel waste – factorial experiment. Journal of King Saud University – Science 25, 39–45. http://dx.doi.org/10.1016/j.jksus.2012.04.001.

Jayathilakan, K., Sultana, K., Radhakrishna, K., Bawa, A.S., 2012. Utilization of byproducts and waste materials from meat, poultry and fish processing industries: a review. Journal of Food Science and Technology 49, 278–293. http://dx.doi.org/10.1007/s13197-011-0290-7.

Jeyaram, K., Romi, W., Singh, T.A., Devi, A.R., Devi, S.S., 2010. Bacterial species associated with traditional starter cultures used for fermented bamboo shoot production in Manipur state of India. International Journal of Food Microbiology 143, 1–8. http://dx.doi.org/10.1016/j.ijfoodmicro.2010.07.008.

Jiang, W., Xu, Y., Li, C., Dong, X., Wang, D., 2014. Biogenic amines in commercially produced Yulu, a Chinese fermented fish sauce. Food Additives and Contaminants. Part B, Surveillance 7, 25–29. http://dx.doi.org/10.1080/19393210.2013.831488.

Kasai, N., Murata, A., Inui, H., Sakamoto, T., Kahn, R.I., 2004. Enzymatic high digestion of soybean milk residue (okara). Journal of Agricultural and Food Chemistry 52, 5709–5716. http://dx.doi.org/10.1021/jf035067v.

Khan, A.W., 2012. Production of iturin A through glass column reactor (GCR) from soybean curd residue (okara) by *Bacillus subtilis* RB14-CS under solid state fermentation (SSF). Advances in Bioscience and Biotechnology 03, 143–148. http://dx.doi.org/10.4236/abb.2012.32021.

Khare, S.K., Jha, K., Gandhi, A.P., 1995. Citric acid production from Okara (soy-residue) by solid-state fermentation. Bioresource Technology 54, 323–325. http://dx.doi.org/10.1016/0960-8524(95)00155-7.

Kuivanen, J., Dantas, H., Mojzita, D., Mallmann, E., Biz, A., Krieger, N., Mitchell, D., Richard, P., 2014. Conversion of orange peel to L-galactonic acid in a consolidated process using engineered strains of *Aspergillus niger*. AMB Express 4, 33. http://dx.doi.org/10.1186/s13568-014-0033-z.

Lam, W.C., Pleissner, D., Lin, C.S.K., 2013. Production of fungal glucoamylase for glucose production from food waste. Biomolecules 3, 651–661. http://dx.doi.org/10.3390/biom3030651.

Lau, K.Y., Pleissner, D., Lin, C.S.K., 2014. Recycling of food waste as nutrients in *Chlorella vulgaris* cultivation. Bioresource Technology 170C, 144–151. http://dx.doi.org/10.1016/j.biortech.2014.07.096.

Leijdekkers, A.G.M., Aguirre, M., Venema, K., Bosch, G., Gruppen, H., Schols, H.A., 2014. In vitro fermentability of sugar beet pulp derived oligosaccharides using human and pig fecal inocula. Journal of Agricultural and Food Chemistry 62, 1079–1087. http://dx.doi.org/10.1021/jf4049676.

Lin, S., Wen, L., Yang, B., Jiang, G., Shi, J., Chen, F., Jiang, Y., 2013. Improved growth of *Lactobacillus bulgaricus* and *Streptococcus thermophilus* as well as Increased antioxidant activity by biotransforming litchi pericarp polysaccharide with *Aspergillus awamori*. BioMed Research International 2013, 413793. http://dx.doi.org/10.1155/2013/413793.

Londero, A., Quinta, R., Abraham, A.G., Sereno, R., De Antoni, G., Garrote, G.L., 2011. Inhibitory activity of cheese whey fermented with kefir grains. Journal of Food Protection 74, 94–100. http://dx.doi.org/10.4315/0362-028X.JFP-10-121.

Mao, X., Liu, P., He, S., Xie, J., Kan, F., Yu, C., Li, Z., Xue, C., Lin, H., 2013. Antioxidant properties of bio-active substances from shrimp head fermented by *Bacillus licheniformis* OPL-007. Applied Biochemistry and Biotechnology 171, 1240–1252. http://dx.doi.org/10.1007/s12010-013-0217-z.

Marshall, K., 2004. Therapeutic applications of whey protein. Alternative Medicine Review 9, 136–156.

Martinez-Avila, G.C.G., Aguilera, A.F., Saucedo, S., Rojas, R., Rodriguez, R., Aguilar, C.N., 2014. Fruit wastes fermentation for phenolic antioxidants production and their application in manufacture of edible coatings and films. Critical Reviews in Food Science and Nutrition 54, 303–311. http://dx.doi.org/10.1080/10408398.2011.584135.

Martins, S., Mussatto, S.I., Martínez-Avila, G., Montañez-Saenz, J., Aguilar, C.N., Teixeira, J.A., 2011. Bioactive phenolic compounds: production and extraction by solid-state fermentation. A review. Biotechnology Advances 29, 365–373. http://dx.doi.org/10.1016/j.biotechadv.2011.01.008.

Mazutti, M.A., Corazza, M.L., Maugeri Filho, F., Rodrigues, M.I., Corazza, F.C., Treichel, H., 2009. Inulinase production in a batch bioreactor using agroindustrial residues as the substrate: experimental data and modeling. Bioprocess and Biosystems Engineering 32, 85–95. http://dx.doi.org/10.1007/s00449-008-0225-5.

Mirabella, N., Castellani, V., Sala, S., 2014. Current options for the valorization of food manufacturing waste: a review. Journal of Cleaner Production 65, 28–41. http://dx.doi.org/10.1016/j.jclepro.2013.10.051.

Moraes, T.M., Kushima, H., Moleiro, F.C., Santos, R.C., Rocha, L.R.M., Marques, M.O., Vilegas, W., Hiruma-Lima, C.A., 2009. Effects of limonene and essential oil from *Citrus aurantium* on gastric mucosa: role of prostaglandins and gastric mucus secretion. Chemico – Biological Interaction 180, 499–505. http://dx.doi.org/10.1016/j.cbi.2009.04.006.

Mushtaq, Z., Imran, M., Salim-ur-Rehman, Z.T., Ahmad, R.S., Arshad, M.U., 2014. Biochemical perspectives of xylitol extracted from indigenous agricultural by-product mung bean (*Vigna radiata*) hulls in a rat model. Journal of the Science of Food and Agriculture 94, 969–974.

O'Toole, D.K., 1999. Characteristics and use of okara, the soybean residue from soy milk production – a review. Journal of Agricultural and Food Chemistry 47, 363–371.

Ofori-Boateng, C., Lee, K.T., 2013. Sustainable utilization of oil palm wastes for bioactive phytochemicals for the benefit of the oil palm and nutraceutical industries. Phytochemistry Reviews 12, 173–190. http://dx.doi.org/10.1007/s11101-013-9270-z.

Ono, H., Nishio, S., Tsurii, J., Kawamoto, T., Sonomoto, K., Nakayama, J., 2014. Monitoring of the microbiota profile in nukadoko, a naturally fermented rice bran bed for pickling vegetables. Journal of Bioscience and Bioengineering 118, 520–525. http://dx.doi.org/10.1016/j.jbiosc.2014.04.017.

Osma, J.F., Toca-Herrera, J.L., Rodríguez-Couto, S., 2010. Uses of laccases in the food industry. Enzyme Research 2010, 918761. http://dx.doi.org/10.4061/2010/918761.

Ouoba, L.I.I., Parkouda, C., Diawara, B., Scotti, C., Varnam, A.H., 2008. Identification of *Bacillus* spp. from Bikalga, fermented seeds of *Hibiscus sabdariffa*: phenotypic and genotypic characterization. Journal of Applied Microbiology 104, 122–131. http://dx.doi.org/10.1111/j.1365-2672.2007.03550.x.

Pacheco, N., Garnica-González, M., Ramírez-Hernández, J.Y., Flores-Albino, B., Gimeno, M., Bárzana, E., Shirai, K., 2009. Effect of temperature on chitin and astaxanthin recoveries from shrimp waste using lactic acid bacteria. Bioresource Technology 100, 2849–2854. http://dx.doi.org/10.1016/j.biortech.2009.01.019.

Parmar, I., Rupasinghe, H.P.V., 2013. Bio-conversion of apple pomace into ethanol and acetic acid: enzymatic hydrolysis and fermentation. Bioresource Technology 130, 613–620. http://dx.doi.org/10.1016/j.biortech.2012.12.084.

Pescuma, M., Hébert, E.M., Bru, E., Font de Valdez, G., Mozzi, F., 2012. Diversity in growth and protein degradation by dairy relevant lactic acid bacteria species in reconstituted whey. The Journal of Dairy Research 79, 201–208. http://dx.doi.org/10.1017/S0022029912000040.

Pescuma, M., Hébert, E.M., Mozzi, F., Font de Valdez, G., 2008. Whey fermentation by thermophilic lactic acid bacteria: evolution of carbohydrates and protein content. Food Microbiology 25, 442–451. http://dx.doi.org/10.1016/j.fm.2008.01.007.

Pleissner, D., Lam, W.C., Sun, Z., Lin, C.S.K., 2013. Food waste as nutrient source in heterotrophic microalgae cultivation. Bioresource Technology 137, 139–146. http://dx.doi.org/10.1016/j.biortech.2013.03.088.

Radzali, S.A., Baharin, B.S., Othman, R., Markom, M., Rahman, R.A., 2014. Co-solvent selection for supercritical fluid extraction of astaxanthin and other carotenoids from *Penaeus monodon* waste. Journal of Oleo Science 63, 769–777.

Rani, P., Kalyani, N., Prathiba, K., 2008. Evaluation of lignocellulosic wastes for production of edible mushrooms. Applied Biochemistry and Biotechnology 151, 151–159. http://dx.doi.org/10.1007/s12010-008-8162-y.

Reis, S.F., Gullón, B., Gullón, P., Ferreira, S., Maia, C.J., Alonso, J.L., Domingues, F.C., Abu-Ghannam, N., 2014. Evaluation of the prebiotic potential of arabinoxylans from brewer's spent grain. Applied Microbiology and Biotechnology 98, 9365–9373. http://dx.doi.org/10.1007/s00253-014-6009-8.

Romo Sánchez, S., Gil Sánchez, I., Arévalo-Villena, M., Briones Pérez, A., 2014. Production and immobilization of enzymes by solid-state fermentation of agroindustrial waste. Bioprocess and Biosystems Engineering. http://dx.doi.org/10.1007/s00449-014-1298-y.

Ruiz, Y., Klotz, B., Serrato, J., Guio, F., Bohórquez, J., Sánchez, O.F., 2014. Use of spent osmotic solutions for the production of fructooligosaccharides by *Aspergillus oryzae* N74. Food Science and Technology International 20, 365–372. http://dx.doi.org/10.1177/1082013213488611.

Ruthu, Murthy, P.S., Rai, A.K., Bhaskar, N., 2014. Fermentative recovery of lipids and proteins from freshwater fish head waste with reference to antimicrobial and antioxidant properties of protein hydrolysate. Journal of Food Science and Technology 51, 1884–1892. http://dx.doi.org/10.1007/s13197-012-0730-z.

Ryan, J.T., Ross, R.P., Bolton, D., Fitzgerald, G.F., Stanton, C., 2011. Bioactive peptides from muscle sources: meat and fish. Nutrients 3, 765–791. http://dx.doi.org/10.3390/nu3090765.

Sachindra, N.M., Bhaskar, N., 2008. In vitro antioxidant activity of liquor from fermented shrimp biowaste. Bioresource Technology 99, 9013–9016. http://dx.doi.org/10.1016/j.biortech.2008.04.036.

Samanta, A.K., Jayapal, N., Jayaram, C., Roy, S., Kolte, A.P., Senani, S., Sridhar, M., 2015. Xylooligosaccharides as prebiotics from agricultural by-products: production and applications. Bioactive Carbohydrates and Dietary Fibre. http://dx.doi.org/10.1016/j.bcdf.2014.12.003.

Shi, M., Yang, Y., Guan, D., Zhang, Y., Zhang, Z., 2012. Bioactivity of the crude polysaccharides from fermented soybean curd residue by *Flammulina velutipes*. Carbohydrate Polymers 89, 1268–1276. http://dx.doi.org/10.1016/j.carbpol.2012.04.047.

Shi, M., Yang, Y., Hu, X., Zhang, Z., 2014. Effect of ultrasonic extraction conditions on antioxidative and immunomodulatory activities of a *Ganoderma lucidum* polysaccharide originated from fermented soybean curd residue. Food Chemistry 155, 50–56. http://dx.doi.org/10.1016/j.foodchem.2014.01.037.

Shukla, S., Park, H.-K., Lee, J.-S., Kim, J.-K., Kim, M., 2014. Reduction of biogenic amines and aflatoxins in *Doenjang* samples fermented with various *Meju* as starter cultures. Food Control 42, 181–187. http://dx.doi.org/10.1016/j.foodcont.2014.02.006.

Śmigielski, K.B., Majewska, M., Kunicka-Styczyńska, A., Szczęsna-Antczak, M., Gruska, R., Stańczyk, Ł., 2014. The effect of enzyme-assisted maceration on bioactivity, quality and yield of essential oil from waste carrot (*Daucus carota*) seeds. Journal of Food Quality 37, 219–228. http://dx.doi.org/10.1111/jfq.12092.

Swain, M.R., Anandharaj, M., Ray, R.C., Parveen Rani, R., 2014. Fermented fruits and vegetables of Asia: a potential source of probiotics. Biotechnology Research International 2014, 250424. http://dx.doi.org/10.1155/2014/250424.

Taira, W., Funatsu, Y., Satomi, M., Takano, T., Abe, H., 2007. Changes in extractive components and microbial proliferation during fermentation of fish sauce from underutilized fish species and quality of final products. Fisheries Science 73, 913–923. http://dx.doi.org/10.1111/j.1444-2906.2007.01414.x.

Thyagarajan, T., Puri, M., Vongsvivut, J., Barrow, C.J., 2014. Evaluation of bread crumbs as a potential carbon source for the growth of thraustochytrid species for oil and omega-3 production. Nutrients 6, 2104–2114. http://dx.doi.org/10.3390/nu6052104.

Torrado, A.M., Cortés, S., Manuel Salgado, J., Max, B., Rodríguez, N., Bibbins, B.P., Converti, A., Manuel Domínguez, J., 2011. Citric acid production from orange peel wastes by solid-state fermentation. Brazilian Journal of Microbiology 42, 394–409. http://dx.doi.org/10.1590/S1517-83822011000100049.

Tsakona, S., Kopsahelis, N., Chatzifragkou, A., Papanikolaou, S., Kookos, I.K., Koutinas, A.A., 2014. Formulation of fermentation media from flour-rich waste streams for microbial lipid production by *Lipomyces starkeyi*. Journal of Biotechnology 189, 36–45. http://dx.doi.org/10.1016/j.jbiotec.2014.08.011.

Vendruscolo, F., Ninow, J.L., 2014. Apple pomace as a substrate for fungal chitosan production in an airlift bioreactor. Biocatalysis and Agricultural Biotechnology 3, 338–342. http://dx.doi.org/10.1016/j.bcab.2014.05.001.

Vijayaraghavan, P., Vincent, S.G.P., 2014. Statistical optimization of fibrinolytic enzyme production using agroresidues by *Bacillus cereus* IND1 and its thrombolytic activity in vitro. BioMed Research International 2014, 725064. http://dx.doi.org/10.1155/2014/725064.

Villanueva-Suárez, M.J., Pérez-Cózar, M.L., Redondo-Cuenca, A., 2013. Sequential extraction of polysaccharides from enzymatically hydrolyzed okara byproduct: physicochemical properties and in vitro fermentability. Food Chemistry 141, 1114–1119. http://dx.doi.org/10.1016/j.foodchem.2013.03.066.

Wang, S.-L., Li, J.-Y., Liang, T.-W., Hsieh, J.-L., Tseng, W.-N., 2010. Conversion of shrimp shell by using *Serratia* sp. TKU017 fermentation for the production of enzymes and antioxidants. Journal of Microbiology and Biotechnology 20, 117–126.

Webber, D.M., Hettiarachchy, N.S., Li, R., Horax, R., Theivendran, S., 2014. Phenolic profile and antioxidant activity of extracts prepared from fermented heat-stabilized defatted rice bran. Journal of Food Science 79, H2383–H2391. http://dx.doi.org/10.1111/1750-3841.12658.

Wong-Leung, Y.L., Fong, W.F., Lam, W.L., 1993. Production of α-galactosidase by *Monascus* grown on soybean and sugarcane wastes. World Journal of Microbiology and Biotechnology 9, 529–533. http://dx.doi.org/10.1007/BF00386288.

Xiao, Y., Wang, L., Rui, X., Li, W., Chen, X., Jiang, M., Dong, M., 2015. Enhancement of the antioxidant capacity of soy whey by fermentation with *Lactobacillus plantarum* B1-6. Journal of Functional Foods 12, 33–44. http://dx.doi.org/10.1016/j.jff.2014.10.033.

Yagoub, A.E.G.A., Mohamed, B.E., Ahmed, A.H.R., El Tinay, A.H., 2004. Study on furundu, a traditional Sudanese fermented roselle (*Hibiscus sabdariffa* L.) seed: effect on in vitro protein digestibility, chemical composition, and functional properties of the total proteins. Journal of Agricultural and Food Chemistry 52, 6143–6150. http://dx.doi.org/10.1021/jf0496548.

Zhang, W., He, X., Liu, H., Guo, H., Ren, F., Wen, P., 2013. Statistical optimization of culture conditions for milk-clotting enzyme production by *Bacillus amyloliquefaciens* using wheat bran-an agro-industry waste. Indian Journal of Microbiology 53, 492–495. http://dx.doi.org/10.1007/s12088-013-0391-2.

Zhou, Z., Yin, Z., Hu, X., 2014. Corncob hydrolysate, an efficient substrate for *Monascus* pigment production through submerged fermentation. Biotechnology and Applied Biochemistry 61, 716–723. http://dx.doi.org/10.1002/bab.1225.

Zhu, Y.P., Fan, J.F., Cheng, Y.Q., Li, L.T., 2008. Improvement of the antioxidant activity of Chinese traditional fermented okara (Meitauza) using *Bacillus subtilis* B2. Food Control 19, 654–661. http://dx.doi.org/10.1016/j.foodcont.2007.07.009.

Index

'*Note*: Page numbers followed by "f" indicate figures and "t" indicate tables.'

A

AADC. *See* Aromatic L-amino acid decarboxylase (AADC)
ABG. *See* Ascorbigen (ABG)
ABT. *See* Aceto balsamico tradizionale (ABT)
ABTS. *See* 2,2′-Azinobis[3-ethylbenzothiazoline 6-sulfonate] (ABTS)
ACE. *See* Angiotensin I converting enzyme (ACE)
Acetic acid, 367, 509, 580
 bacteria, 578–579, 579f
Acetification, 577–578
Aceto balsamico tradizionale (ABT), 582
Acetobacter spp., 6
Acetous fermentation, 577–578
N1-Acetyl-5-methoxykynuramine (AMK), 110
N-Acetyl-5-methoxytryptamine. *See* Melatonin (MEL)
N1-Acetyl-N2-formyl-5-methoxykynuramine (AFMK), 110
Acetylcholine (Ach), 459–460
Acetylcholinesterase activity (AChE activity), 459–460
N-Acetylserotonin, 108
Ach. *See* Acetylcholine (Ach)
Acha fonio. *See* White fonio (*Digitaria exilis*)
AChE activity. *See* Acetylcholinesterase activity (AChE activity)
Acidification, 278–280
Acocote, 543
Actinomucor elegans (*A. elegans*), 182
AD. *See* Alzheimer's disease (AD); Apparent digestibility (AD); Atopic dermatitis (AD)
ADC. *See* Arginine decarboxylase (ADC)
Adenomatous polyposis coli (APC), 463–464
S-Adenosyl-L-methionine (SAM), 108, 179, 189–190, 625–626
S-Adenosylmethinine decarboxylase (SAMDC), 625–626
Adhesion sites, 510
Adipose tissue, health benefits on, 39
Adult lymphoblastic leukemia, 346–347

Aerobacter family, 507
AF. *See* Aflatoxins (AF); Alcoholic fermentations (AF)
Aflatoxins (AF), 13–14, 653–654
 AFB1, 654, 655f, 656, 658
 AFB2, 654–656, 655f
 AFG1, 654–656, 655f
 AFG2, 654–656, 655f
 AFM1, 654–656, 655f
 AFM2, 654–656, 655f
 bacterial and fungal starter cultures, 661t
 contamination, 653–654
 control methods
  biochemical methods in fermented foods, 659–660
  microbiological methods in fermented foods, 660–661
  physical methods in fermented foods, 659
 general detection methods to analysis, 663
 legal validation for aflatoxin limits, 662–663, 662t
 occurrence in fermented foods, 656, 665t
  alcoholic beverage products, 658–659
  dairy fermented food products, 657–658
  soybean fermented food products, 656–657
 poisoning effects, 654
 rapid detection methods to analysis, 663–667
 structures, 654–656, 655f
AFMK. *See* N1-Acetyl-N2-formyl-5-methoxykynuramine (AFMK)
African civilizations food fermentation, 424
Agave, 543
Agavins, 544–545
AgDI pathway. *See* Agmatine deiminase pathway (AgDI pathway)
Aglycones, 63–64, 454–456
 genistein, 462–463
Agmatine (Agm), 625, 633
Agmatine deiminase pathway (AgDI pathway), 627

Agro-wastes, 709–718. *See also* Fruits wastes
  cereal brans, 709–715
  corn cob, 716
  fermented tea, 716–717
  metabolic intervention, 717f
  nutrient-laden organic wastes, 714f
  soy waste, 715–716
  stalks, straws, and other lignocellulosic biomass, 709
  sugarcane waste, molasses, and syrup waste, 717–718
  vegetable pulp and seeds, 718
  waste materials, nutrients and fermentation microbes, 710t–713t
Agroclavine, 295
Aguamiel, 543
  bioactive constituents of, 544–546
  microorganisms in, 547–554, 549t–552t, 553f
AhR. *See* Aryl-hydrocarbon receptor (AhR)
AITC. *See* Allyl isothiocyanate (AITC)
Akkermansia muciniphila (*A. muciniphila*), 167
Alcohol consumption, 373
Alcoholic beverages, 633
  AF in, 658–659
  beer, 634–635
  cider, 635–636
  wine, 633–634
Alcoholic fermentations (AF), 425, 635
Aliphatic amines, 625
Allergy proteins, phenolic compounds interaction with, 606
Allyl isothiocyanate (AITC), 564
Aluminum, 378
Alzheimer's disease (AD), 112, 378, 460, 605–606
  protection against, 112
Amaranth (*Amaranthus* L.), 420
Amaranthus L. *See* Amaranth (*Amaranthus* L.)
Amino acids, 86–87, 186, 204–207, 205t–206t
  changes during fermentation, 397–398
1-(4-Aminobutyl)guanidine. *See* Agmatine (Agm)
2-Aminoethane sulfonic acid, 204–207
4-(2-Aminoethyl)phenol. *See* Tyramine (Tym)
N,N'-bis(3-Aminopropyl)-1, 4-diaminobutane. *See* Spermine (Spm)
N-(3-Aminopropyl)-1,4-diaminobutane. *See* Spermidine (Spd)
AMK. *See* N1-Acetyl-5-methoxykynuramine (AMK)
AMP-activated protein kinase (AMPK), 599

AMPK. *See* AMP-activated protein kinase (AMPK)
Amyloid precursor protein (APP), 460–461
β-Amyloid protein (Aβ), 600
Anaerobic fermentation of dietary fibers, 165–166
Anaerobic process, 505
Andrastins, 294–295
ANF. *See* Antinutritional factors (ANF)
Angiotensin I converting enzyme (ACE), 10–11, 30–31, 210, 247–253, 308–309, 323, 350–351, 404, 428, 723
  ACE-2, 36
  ACE-inhibitory activity, 188
Animal proteins, bioactive peptides in, 24t–26t
Animal studies, 9, 11–12
ANN. *See* Artificial neural network (ANN)
Anserine, 207
Antiaging effects, 111, 482–484
Antibiotic resistance (AR), 675–677
  dissemination of antibiotic-resistant bacteria, 697f
  new insights, 694–697
  in traditional fermented foods, 677–694
    AR profile of pathogens, 690–694
    phenotypic and genotypic AR in LAB, 677–690, 679t–682t, 686t, 688t–689t
Antibiotics, 685
Anticancer
  effects, 110–111, 484–487, 492–493
  properties, 58
Anticarcinogenic effects
  of beer, 375–376
  effects of kefir, 346–347
Anticoagulant activity, 189
Antidiabetic effects of kefir, 351–352
Antihypertensive
  activity of traditional fermented seafood, 188–189
  effects of kefir, 350–351
  peptides, 247–253, 248t–252t
Antiinflammatory effects
  of kefir, 348–349
  of Sauerkraut, 568
Antimicrobial
  activity
    of kefir, 351
    of microorganisms, 183–185
  effects of vinegar, 580–581
  peptides, 256–257
Antimutagenic effects
  of beer, 375–376
  of kefir, 346–347
  of kimchi, 484–487

Antinutrient compound removal, 512–513
Antinutritional factors (ANF), 433–434, 457
Antinutritive compounds, degradation of, 423
Antiobesity
  effects
    of kimchi, 487–489, 494
    of MEL, 112
  properties of vinegar, 583–584
Antiosteoporosis effects
  of beer, 374–375
  of soymilk, 31
Antioxidant(s), 63, 65–67, 69, 75, 186, 209, 404, 594–596
  activity, 186–188, 373
    antioxidant properties of vinegars, 583
    changes, 401
    phenolic composition of vinegars, 582–583
    phenolic compounds, 581–582
  addition, 222, 223t–224t
  capacity of beer, 371
  compounds, 222
  effects
    of beer, 370–371
    of MEL, 110
  mechanisms, 458–459
  peptides, 210, 253–254
  properties
    of beer, 371–372
    of EPSs, 57–58
    of vinegars, 583
  roles of *Tempeh*, 458–459
Antioxidative effects of kimchi, 482–484, 492–493
Antiproliferative peptides, 255–256
Antitrypsin, 457
AOM. *See* Azoxymethane (AOM)
APC. *See* Adenomatous polyposis coli (APC)
aph(3′) gene, 683–684
APP. *See* Amyloid precursor protein (APP)
Apparent digestibility (AD), 441
Apple pomace, 720
AR. *See* Antibiotic resistance (AR)
Arabinoxylans (AX), 444, 722
Arginine decarboxylase (ADC), 625–627
Aroma compound formation, 285–286
Aromatic amines, 625
Aromatic herb addition, 509
Aromatic L-amino acid decarboxylase (AADC), 107–108
Artificial neural network (ANN), 717–718
Aryl-hydrocarbon receptor (AhR), 565–566
Ascorbic acid, 509

Ascorbigen (ABG), 561–564, 568
Aspergillus spp., 6, 656–657, 660
  A. oryzae, 399, 427–428, 465–466
  A. sojae, 466–467
  A. awamori, 719
  A. flavus, 653–657
  A. fumigates 3, 707–708
  A. niger, 33
    A. niger 113N, 707–708
    A. niger M46, 458
  A. nomius, 653
  A. parasiticus, 653, 656–657
Astringency, 602
Atole agrio (*A. agrio*), 425
Atopic dermatitis (AD), 494–495
Attus norvegicus rats, 456
Avena spp. L. *See* Oat (*Avena* spp. L.)
AX. *See* Arabinoxylans (AX)
2,2′-Azinobis[3-ethylbenzothiazoline 6-sulfonate] (ABTS), 596
Azoxymethane (AOM), 485
Aβ. *See* β-Amyloid protein (Aβ)
Aβ aggregation, 460–461

B

B-group vitamins, 369
B. uniformis CECT 7771, 167–168
BA. *See* Biogenic amines (BA)
BACE-1. *See* β-Site APP cleaving enzyme-1 (BACE-1)
Bacillus subtilis (*B. subtilis*), 399, 464–465, 709–716
Bacterial EPSs, 55t
Bacterial genomics
  fermented milk, 159–160
    as source of beneficial bacteria, 161–164
  gut ecosystem as source of beneficial bacteria, 166–168
  in probiotic health benefits identification, 164–166
Bacteriocins production, 511
Bacteroides uniformis (*B. uniformis*), 167
Bacteroidetes, 456
Baechu cabbage, 477
"Baechu kimchi", 477
Bakery waste, 722. *See also* Agro-wastes; Fruits wastes
Baking technologies, 444–445
Bamboo salt-KCl system, 491
Barley (*Hordeum vulgare* L.), 420
BCAAs. *See* Branched-chain amino acids (BCAAs)

BCMs. *See* β-Casomorphins (BCMs)
Beer, 121–122, 365, 376, 634–635. *See also* Wine
  antimutagenic and anticarcinogenic effects, 375–376
  antiosteoporosis effect, 374–375
  antioxidant properties of, 371–372
  bioactive components, 365–371
  brewing, 365
  cardiovascular diseases and, 373–374
  health effects, 371–372, 377–378
  hydration, 376
  supplementation effect in breastfeeding mothers, 377
Beneficial bacteria
  fermented milk as source, 161–164
  gut ecosystem as source, 166–168
Benzoic acid, 509
Beta-phenyl-gamma-aminobutyric acid, 85
Beverages, 96
Bifidobacteria, 34, 159–164, 520–521
Bifidobacterium spp, 428
  B. bifidus, 715–716
  B. breve, 58–59
Bikalga, 718
Bile salt hydrolase (BSH), 524–525
Bioactive components, 3–4
  of beer, 365–371
Bioactive compounds, 7–8, 86, 373, 378, 517–518
  antioxidant properties of vinegars, 583
  in cucumber, 503–504
  phenolic composition of vinegars, 582–583
  phenolic compounds, 581–582
  strategies for optimization, 209–225
    addition of vitamins and antioxidants, 222, 223t–224t
    bioactive peptide formation, 210
    curing agents, 225–226
    dietary fiber incorporation, 214–216
    improving fat content, 210–211, 212t–214t
    mineral addition, 216–222
    probiotics, 211–214, 215t
    reduction of biogenic amine formation, 222–224
Bioactive constituents
  of aguamiel, 544–546
  of pulque, 546–547
Bioactive peptides, 23, 239, 257. *See also* Phenolic compounds
  cheese, 246–264
  in fermented foods, 30–32
  from animal proteins, 24t–26t
  evidence for health effects, 35–40
  from plant proteins, 27t–29t
  production of, 32–33
  strategies to increasing production, 33–35
  formation, 210
Bioactive properties of yogurt, 310–318
  effect
    on cholesterol metabolism, 315–316
    on immunity, 311–312
    on lactose intolerance, 316–317
    on mineral absorption, 317–318
    on transit time/digestion, 317
  modulation of gut microbiota, 312–313
  yogurt as probiotic vector, 313–315
Bioactive proteins, 246–264
Bioavailability, 255
  of red wine, 606–612
    absorption and metabolism of resveratrol, 610
    absorption of catechins and proanthocyanidins, 609–610
    microbiota impact on red wine availability and bioactivity, 610–612
    red wine anthocyanins and derivatives, 607–609
Biochemical methods to control AF in fermented foods, 659–660
Bioconversion of wastes, 707–708
Biogenic amines (BA), 13, 190–192, 287–289, 288t, 492, 526, 625, 637
  affecting factors in fermented foods, 628–638
  classification, biosynthesis, and metabolism, 625–627
  in different cheeses, 630t
  in food and beverages, 638
    mean data, 639t
  groups, 625
  and human health, 638–640
  microbial production, 627–628
  reduction
    of BA formation, 222–224
    in fermented food, 640–642
  structures, 626f
Biological activity in traditional fermented seafood, 185–190
  anticoagulant activity, 189
  antihypertensive activity, 188–189
  antioxidant activity, 186–188
  fibrinolytic activity, 189
  SAM, 189–190

Biological control, 660
Biological effect
  of cheese vitamins, 246
  of wine, 595f, 596
Biopreservation, 510
Biosensor-based detection methods, 666
Biosynthesis of MEL, 107–109
Biotransformation
  hydroxybenzoic acid-derived compounds by
    L. plantarum, 69–73
  hydroxycinnamic acid-derived compounds
    by *L. plantarum*, 73–77
Black fonio (*Digitaria iburua*), 434
Black vinegars, 580
Blood pressure regulation, 86
Blue cheese. *See also* Cheese(s)
  lactic-acid bacteria, 286–289
  manufacturing, 275–286, 276t
    acidification, 278–280
    aroma compound formation, 285–286
    blue cheese ripening, 280–282
    coagulation, 278–280
    curd formation, 278–280
    cutting, 280
    lipolysis, 283–285
    milk preparation, 277–278
    molding, 280
    proteolysis, 282–283
    salting, 280
  preparation and maturation, 275–286
  ripening, 280–282
  secondary metabolites by *P. roqueforti*, 293–296
Blue-veined cheeses, 289–290
Blueberries, 10–11
Body mass index (BMI), 319–320, 489
Bone effects on MEL, 112
Bones, health benefits on, 39–40
Botulism, 13
Bowman–Birk protease inhibitors, 405
Brain damage, 599–600
Branched-chain amino acids (BCAAs), 308–309
Bread, 123, 440–441
Breast cancer, 325, 462–463
Breastfeeding mothers, beer supplementation
  effect in, 377
Brettanomyces, 6
Breweries, 722
Brewing, 365
  wastes, 722
Broad-spectrum antimicrobial agent, 510
BSH. *See* Bile salt hydrolase (BSH)
Buckwheat (*Fagopyrum esculentum*), 420, 442
Burong isda (*B. isda*), 425
Buttermilk, 721

C

C-reactive protein (CRP), 311, 564–565
C/EBP-α. *See* CCAAT/enhancer binding-
  protein alpha (C/EBP-α)
c3OH M. *See* Cyclic 3-hydroxymelatonin
  (c3OH M)
Cabbage, 557–558, 561
CAC. *See* Codex Alimentarius Commission
  (CAC)
CAC Joint FAO/WHO food standards program, 662–663
Cad. *See* Cadaverine (Cad)
Cadaverine (Cad), 625
p-Caffeic acid, 74
Calcium (Ca), 221–222, 306, 310, 320, 395–396
  acetate, 509
Camellia sinensis (*C. sinensis*), 6
Canavalia cathatica (*C. cathatica*), 404–405
Cancer, 204, 324–326, 598
  breast, 325, 462–463
  colon, 58
  colorectal, 324–325
  prevention, 462–464
Capsaicin, 487
Capsular polysaccharides (CPSs), 50
Captopril, 247–253
Carbohydrates, 178, 309–310, 394–395
  changes during fermentation, 398–399
Carbon dioxide (CO_2), 505–506
Carcinogens, 525–526
CARDIA. *See* Coronary Artery Risk
  Development in Young Adults
  (CARDIA)
Cardiometabolic diseases
  cancers, 324–326
  cardiovascular disease, 321–322
  hypertension, 322–323
  osteoporosis, 323–324
  type 2 diabetes, 320–321
Cardiovascular disease (CVD), 56, 204, 240,
  305, 321–322, 373–374, 461, 597–598
  protection against, 111
Cardiovascular system, health benefits on,
  35–37
Carnitine palmitoyltransferase 1 (CPT-1),
  487–489
Carnosine, 207

740 Index

Carotenoids, 222
Casein phosphopeptides (CPPs), 246, 257
Caseins, 38, 308–309
β-Casomorphins (BCMs), 254
Catalase (CAT), 110
Catechin, 78
Catechins absorption, 609–610
CCAAT/enhancer binding-protein alpha (C/EBP-α), 487–489, 494
CD. *See* Crohn disease (CD)
Celiac disease, 442–443
Cell envelope-associated proteinases (CEP), 32–33
Cereal-fermented products, 422
Cereal(s), 417–418
   brans, 709–715
   cereal-based products, 91–94
   fiber, 210–211
   kernel, 418
CFU. *See* Colony forming units (CFU)
CHD. *See* Coronary heart disease (CHD)
Cheese reaction, 640
Cheese(s), 94–95, 239, 628–629. *See also* Blue cheese; Kefir; Yogurt
   BA content, 630t
   bioactive proteins and peptides, 246–264
   biological effects of cheese vitamins, 246
   cheese-associated fungi, secondary metabolites in, 292–293, 292t, 293f
   cheese-derived peptide effects on nervous system, 254–255
   fat effect on health, 240–244, 241t–243t
   filamentous fungi in, 289–293
   minerals and impact on health, 244–246
   yeasts in, 289–293
Chemical composition
   in cucumber, 503–504
   kefir, 343–344, 345t
Chemical preservatives, 509
   addition effect, 509
Chemopreventive effects, 566
Chenopodium quinoa Willd. *See* Quinoa (*Chenopodium quinoa* Willd.)
Cheonggukjang, 464–465
Chimchae, 478
Chimjanggo, 478–479
Chlorella vulgaris cultivation, 724
Chocolate, 5–6
Cholesterol, 524–525
   effect on cholesterol metabolism, 315–316
   lowering effect, 56
Cholinergic neurons of brain, 459–460
"Chorizo" sausages, 210

CHP. *See* Cyclo(His–Pro) (CHP)
Chromatographic techniques, 113–114
Chronic disease control and prevention, 8–12
   legumes, 9–10
   specialty foods, 10–12
Chronic inflammation, 568
Chunghookjang (soybean product), 31
Cider, 635–636
Cinnamic acid esterases. *See* Feruloyl esterases
Cinnamoyl esterases. *See* Feruloyl esterases
Citrobacter rodentium (*C. rodentium*), 58–59
Citrus peel, 725
CLA. *See* Conjugated linoleic acid (CLA)
Classical probiotic candidates, 164–165
Clostridium botulinum (*C. botulinum*), 12–13
Clostridium perfringens, 427
Coagulation, 278–280
Cobalamin, 150–151, 209, 456–457
Cocoa beans, 5–6
Codex Alimentarius Commission (CAC), 657
Coenzyme Q (CoQs), 187–188
Coffee, 5–6
Colon, 531–532
   cancer, 58
Colony forming units (CFU), 305–306, 477, 505, 521, 559, 632
Colorectal cancer, 324–325
Compliance with ecological balance, 707
Conjugated linoleic acid (CLA), 208, 239–240, 309
Controlled fermentation, 636
Copper, 368
CoQs. *See* Coenzyme Q (CoQs)
Corn cob, 716
Coronary Artery Risk Development in Young Adults (CARDIA), 322–323
Coronary heart disease (CHD), 214–216, 322
m-Coumaric acids, 74
p-Coumaric acid. *See* Hydroxycinnamic acid
COX-1. *See* Cyclooxygenase-1 (COX-1)
CP. *See* Cyclophosphamide (CP)
CPPs. *See* Casein phosphopeptides (CPPs)
CPSs. *See* Capsular polysaccharides (CPSs)
CPT-1. *See* Carnitine palmitoyltransferase 1 (CPT-1)
Crohn disease (CD), 510–511
CRP. *See* C-reactive protein (CRP)
Cucumber (*Cucumis sativus* L.), 503
   chemical composition and bioactive compounds in, 503–504
   factors affecting cucumber fermentation, 507–510
      biopreservation, 510

effect of chemical preservative addition, 509
effects of cucumber size and enzyme activity, 507
effect of salt concentration, 507–509
effect of spice or aromatic herb addition, 509
sugar addition effect, 509
health benefits of fermented cucumbers, 510–513
lactic acid fermentation, 504–506
origin, 503
size effects, 507
Curd formation, 278–280
Curdlan, 720–721
Curing agents, 225–226
Cutting, 280
CVD. *See* Cardiovascular disease (CVD)
Cyclic 3-hydroxymelatonin (c3OH M), 110
Cyclo(His–Pro) (CHP), 187
Cyclooxygenase-1 (COX-1), 374
Cyclooxygenase 2 (COX-2), 566–567
Cyclophosphamide (CP), 493
Cynara cardunculus (*C. cardunculus*), 253–254

D

D.O. *See* Designations of origin (D.O.)
Daidzein, 454–456, 455f, 464
Dairy fermented food products, AF in, 657–658
Dairy products, 94–95, 628–631, 678–685, 690–694
Dairy wastes, 721. *See also* Agro-wastes; Bakery waste; Fruits wastes; Malting wastes
Dairy-derived food product, 7
DAO. *See* Diamino oxidase (DAO)
Dark teas, 6
DASH. *See* Dietary Approaches to Stop Hypertension (DASH)
DBP. *See* Diastolic blood pressure (DBP)
Decarboxylase (DDC), 627
Degree of polymerization (DP), 544–545, 604
Dehulling process, 398
Dementiai, 459–461
"Denomination of Origin", 122
Designations of origin (D.O.), 114–115
Dextran sulfate sodium (DSS), 485
Dextrans, 53, 512
Dextrins, 366–367
DF. *See* Dietary fibers (DF)

DGA. *See* Dietary guidelines for Americans (DGA)
DHA. *See* Docosahexaenoic acid (DHA)
Dhals, 398
Dhokla, 401
Diabetes mellitus, 599
Diamino oxidase (DAO), 626
1,4-Diaminobutane. *See* Putrescine (Put)
1,5-Diaminopentane. *See* Cadaverine (Cad)
Diastolic blood pressure (DBP), 35–36
Dicotyledonae, 433–434
Diet, 597–598
quality, 318–319
Dietary Approaches to Stop Hypertension (DASH), 322
Dietary fibers (DF), 395, 428, 433–434, 444–445
changes during fermentation, 398–399
incorporation, 214–216
Dietary guidelines for Americans (DGA), 318
Dietary soybean isoflavones, 458
Dietetic Products, Nutrition, and Allergies (NDA), 160
Digestive enzymes, interaction with, 603–604
Digestive microbiota, fermented olives modulating, 531–533
Digitaria exilis. See White fonio (*Digitaria exilis*)
Digitaria iburua. See Black fonio (*Digitaria iburua*)
3,3-Diindolylmethane (DIM), 561–564
DIM. *See* 3,3-Diindolylmethane (DIM)
"*Dimchae*". *See* Kimchi
"*Dimchi*". *See* Kimchi
Dimethylmenaquinone (DMK), 139
2,2-Diphenyl-1-picrylhydrazyl (DPPH), 596
Disaccharides, 367
Disease prevention, 339
yogurt in
cardiometabolic diseases, 320–326
diet quality, 318–319
weight management, 319–320
yogurt, 319–320
DMK. *See* Dimethylmenaquinone (DMK)
DNA
mutation, 567–568
sequencing techniques, 164–165
Docosahexaenoic acid (DHA), 186, 211, 709
Doenjang, 465, 657
Dosa, 425
production, 397–398
Doubanjiang, 465
Douchi, 465–466

DP. *See* Degree of polymerization (DP)
DPPH. *See* 2,2-Diphenyl-1-picrylhydrazyl (DPPH)
DPPH free-radical scavenging activity, 458
Drosophila wing test, 484
Dry fermented sausages, 203
Dry weight (DW), 114, 565
Dry-cured sausages, 203
DSS. *See* Dextran sulfate sodium (DSS)
DW. *See* Dry weight (DW)

E

E-2-nonenal (NE), 504
EAR. *See* Estimated average requirements (EAR)
EFSA. *See* European Food Safety Authority (EFSA)
EGCG. *See* Epigallocatechingallate (EGCG)
EGFR. *See* Epidermal growth factor receptor (EGFR)
Eicosapentaenoic (EPA), 186, 211
Electrochemical immunosensor devices, 666
Electrodialysis, 193
Elevated blood pressure, 491
ELISA. *See* Enzyme-linked immunosorbent assay (ELISA)
ELISA-based detection methods, 664–666
EMT. *See* Epithelial–mesenchymal transition (EMT)
Enalapril, 247–253
Endothelial progenitor cells (EPCs), 374
Energy-dense diets, 204
Enteric pathogens, interactions with, 58–59
Enterococcus faecalis (*E. faecalis*), 72
Enterotoxigenic *Escherichia coli* (ETEC), 58–59
Enzymatic hydrolysis, 720
Enzyme activity effects, 507
Enzyme-linked immunosorbent assay (ELISA), 113–114, 664–666
EPA. *See* Eicosapentaenoic (EPA)
EPCs. *See* Endothelial progenitor cells (EPCs)
EPIC. *See* European Prospective Investigation into Cancer and Nutrition (EPIC)
Epidermal growth factor receptor (EGFR), 566–567
Epigallocatechingallate (EGCG), 604
Epithelial–mesenchymal transition (EMT), 464
EPSs. *See* Exopolysaccharides (EPSs)
Eragrostis tef. See Teff (*Eragrostis tef*)
Eremofortins, 294
erm(B) gene, 683–684, 687–690
ERα, 454–456
ERβ, 454–456

Escherichia coli (*E. coli*), 524
  E. coli O157, 526
  E. coli O157:H7, 548–552, 581
ESE. *See* Ethanol-soluble extracts (ESE)
Estimated average requirements (EAR), 310
Estrogen receptors, 454–456
ETEC. *See* Enterotoxigenic *Escherichia coli* (ETEC)
Ethanol-soluble extracts (ESE), 247–253
EU. *See* European Union (EU)
European Commission, 160
European Food Safety Authority (EFSA), 105–106, 160, 287, 308, 629–631, 638, 687–690
European Prospective Investigation into Cancer and Nutrition (EPIC), 324–325
European Union (EU), 191, 524, 654
Ex vivo study, 36
Exopolysaccharides (EPSs), 50–53, 547
  and applications, 55t
  bacterial EPSs, 55t
  in food industry, 53–54
  health benefits, 54–59
    anticancer properties, 58
    antioxidant properties, 57–58
    cholesterol lowering effect, 56
    immunomodulation, 56–57
    interactions with enteric pathogens, 58–59
Extensively drug-resistant (XDR), 581
Extra-virgin olive oil, 114–115
Extraction process, 365

F

FAAs. *See* Free amino acids (FAAs)
FAD. *See* Flavin adenine dinucleotide (FAD)
Faecalibacterium prausnitzii (*F. prausnitzii*), 167–168
Fagopyrum esculentum. See Buckwheat (*Fagopyrum esculentum*)
FAO. *See* United Nations Food and Agriculture Organization (FAO)
FAS. *See* Fatty-acid synthase (FAS)
Fat, 207–208, 239
  content, 207–208, 210–211, 212t–214t
  replacers, 211
Fatty acid composition, 207–208
Fatty-acid synthase (FAS), 487–489
Favism, 437–438
Fermentation, 3, 49, 79, 177–178, 397, 422–423, 456–457, 477, 505, 528, 531, 546–547, 559, 707
  of cabbage, 560–561
  food wastes, 707–708

as culture medium of functional foods, 724
    functional enhancement, 709–724
  of kimchi, 480–482
  lipids changes during, 399–400
  phenolic compounds transformation, 65–67
  phytic acid and mineral bioavailability
      changes during, 400
  protein and amino acids changes during,
      397–398
  starch, carbohydrates, and dietary fiber
      changes during, 398–399
  valorization of underutilized food sources,
      724–725
  vitamins changes during, 400
Fermented apple processing wastes, 707–708
Fermented *Brassica* product, 95–96
Fermented cereals, 422
Fermented condiments, 587
Fermented cucumbers
  health benefits, 510–513
    fermented cucumber, 512
    immune modulation, 512
    LAB isolation, 510–511
    macroelements, 512
    production of bacteriocins, 511
    removal of antinutrient compounds,
        512–513
  as source of starter cultures to produce
      oligosaccharides, 512
Fermented dairy products, 211–214, 312
Fermented fish, 177–178
  and microorganisms, 180–185
    antimicrobial activity of microorganisms,
        183–185
    microorganisms as starters, 180–183
  products, 631
Fermented foods, 3–7, 49, 422, 478, 532. *See
    also* Fermented seafood products;
    Traditional fermented foods
  AF analysis
    general detection methods, 663
    rapid detection methods, 663–667
  AF control
    biochemical methods, 659–660
    microbiological methods, 660–661
    physical methods, 659
  AF occurrence, 656, 665t
    alcoholic beverage products, 658–659
    dairy fermented food products, 657–658
    soybean fermented food products,
        656–657
  BA reduction in, 640–642
  bioactive peptides in, 30–32
    from animal proteins, 24t–26t

evidence for health effects, 35–40
    from plant proteins, 27t–29t
    production of, 32–33
    strategies to increasing production, 33–35
  fermented foods, 5t
  food safety and quality control, 12–14
  health benefits, 7–12
    bioactive compounds, 7–8
    chronic disease control and prevention,
        8–12
  in MEL, 115–123, 116t–118t
    beer, 121–122
    orange juice, 122–123
    Saccharomyces species, 123
    wine, 115–121
Fermented legumes, 390
Fermented meat, 685, 694
  sausages, 203
    nutritional value, and health implications,
        204–209
    strategies for optimizing of bioactive
        compounds, 209–225
Fermented milk, 159–160
  products, 339–340
  as source of beneficial bacteria, 161–164
  traits and potential health benefits for LAB,
      162f
Fermented nonwheat food products, 424–427
  bakery products, 426–427
  beverages, 425–426
Fermented olives
  healthy bioactive molecules and metabolites
      from, 527–531
  modulating digestive microbiota, 531–533
Fermented papaya preparation (FPP), 11
Fermented plant products, 636
  sauerkraut, 636
  soybean products, 636–637
Fermented pulses, 385. *See also* Fermented
    soybean products
  and cancer, 404–405
  and cardiovascular diseases, 403–404
  and diabetes, 403
  foods role in health promotion, 402
  in healthy aging and stress, 405
  indigenous pulse-based fermented foods,
      387t–389t
  nutritional and phytochemical composition
      of pulses, 390–397
  nutritional changes during pulse-based foods
      fermentation, 397–402
  as probiotic vehicle, 405–406
  scientific names of pulses, 386t
  and weigh management, 402–403

Fermented rice bran, 427–428
Fermented seafood products, 177–179
  around world, 178t
  biological activity in traditional, 185–190
  fermented fish and microorganisms, 180–185
  and health, 179–180
  health risk, 190–194
    biogenic amines and strategies, 190–192
    parasite control, 194
    strategies to decreasing salt content, 192–193
    strategies to removing heavy metals, 193–194
Fermented small cyprinid fish (*Puntius sophore*), 189
Fermented soy permeate (FSP), 9–10
Fermented soybean products, 453. *See also* Fermented pulses
  Cheonggukjang, 464–465
  Doenjang, 465
  Doubanjiang, 465
  Douche, 465–466
  Gochujang, 466
  Miso, 466
  Natto, 466
  soy sauce, 466–467
  Tauchu, 467
Fermented tea, 716–717
Fermented vegetables, 687–690, 694
Ferric reducing-antioxidant power (FRAP), 458–459, 596
Ferulic acid esterases. *See* Feruloyl esterases
Feruloyl esterases, 75, 76t
Festuclavine, 295
FFA. *See* Free fatty acid (FFA)
Fibrinolytic activity, 189
Fibroblasts, 483
Filamentous fungi in cheese, 289–293
  P. roqueforti
    and fungi in blue cheese, 290–291
    strains in local nonindustrial blue cheeses, 291–292
  secondary metabolites in cheese-associated fungi, 292–293, 292t, 293f
Fish, 181
  fermentation, 177–178
  fermentation, 177–178
  fish-type products, 181
  oil, 707–708
  sauce, 179
Fishery byproducts, 723–724
Fishery waste formulations, 707–708

Flavin adenine dinucleotide (FAD), 148–149
Flavin mononucleotide (FMN), 148–149
Flavones, 375
Flavonoids, 63–64
  glycosides, 63–64
FMN. *See* Flavin mononucleotide (FMN)
Folate, 132–138
  concentrations in fermented foods, 134t–135t
  conjugase, 135–136
  content of fermented food, 133–136
  production by microorganisms, 136–138
Folic acid, 222
Food
  bioprocesses, 65
  EPSs in food industry, 53–54
  fermentation, 434–435, 517, 677
  safety, 12–14
Food waste fermentation, 707–708
  as culture medium of functional foods, 724
  functional enhancement, 709–724
    agro-wastes, 709–718
    bakery waste, 722
    dairy wastes, 721
    fishery byproducts, 723–724
    fruits wastes, 718–721
    malting and brewing wastes, 722
  future directions and barricades to surmount, 725–726
  harms and hurdles, 725
  valorization of underutilized food sources, 724–725
Food-processing techniques, 180
Foodborne pathogen bacteria, 504
FOS. *See* Fructooligosaccharides (FOS)
FPP. *See* Fermented papaya preparation (FPP)
FRAP. *See* Ferric reducing-antioxidant power (FRAP)
Free amino acids (FAAs), 179–180, 436–437, 439
Free fatty acid (FFA), 283, 284t
Free radicals, 458
Free-radical-scavenging activity, 186
"French Paradox", 597–598
French Roquefort cheese, 275
Fresh fruits, 122
Fructans, 545
Fructooligosaccharides (FOS), 544–545, 717–718, 720–721
Fruits, 216
Fruits wastes, 718–721. *See also* Agro-wastes
  peels, 719
  pomace, 719–720
  spent solution and juice waste, 720–721

FSP. *See* Fermented soy permeate (FSP)
Fu Zhuan tea, 6
Functional foods, 50, 428
  food wastes as culture medium, 724
Furundu, 718

G

GABA. *See* Gamma-aminobutyric acid (GABA)
GABA-enriched yogurt, 95
GAD. *See* Glutamic acid decarboxylase (GAD)
α-Gal. *See* α–Galactosidase (α-Gal)
Galactooligosaccharides (GOS), 721
α–Galactosidase (α-Gal), 437–438
β-Galactosidase, 161–163, 721
α-Galactosides. *See* Raffinose family of oligosaccharides (RFOs)
Gallic acid, 69
Gallotannin, 64
Gamma-aminobutyric acid (GABA), 85, 163, 350–351, 428, 439, 462, 496, 545–546
  decarboxylation reaction, 86f
  GABA-enriched fermented foods, 91–96, 92t–94t
    beverages, 96
    cereal-based products, 91–94
    dairy products, 94–95
    legumes, 95–96
    meat, 95–96
    vegetables, 95–96
  GABA-like substances, 85
  mechanisms of action, 89–90
  physiological functions, 86–87, 88t–89t
  production by LAB, 90–91
  side effects and toxicity, 96
Ganoderma lucidum (*G. lucidum*), 716–717
Gas chromatography (GC), 113–114, 663
Gastrointestinal system, health benefits on, 38
Gastrointestinal tract (GIT), 87, 310–311, 517–518
GC. *See* Gas chromatography (GC)
Genealogical concordance–phylogenetic species recognition (GC-PSR), 290–291
Generally recognized as safe (GRAS), 151, 511, 520–521, 629–631, 696–697
Generally regarded as safe. *See* Generally recognized as safe (GRAS)
Genistein, 454–456, 455f, 462–463
Genotypic AR in LAB, 677–690
  from fermented meats, 686t
  prevalence, 695t
  from fermented vegetables, 688t–689t
  from traditional dairy products, 679t–682t
  prevalence, 692t–693t
GI. *See* Glycemic index (GI)
GIT. *See* Gastrointestinal tract (GIT)
GLS. *See* Glucosinolates (GLS)
β-Glucans, 444
Gluconic acid, 587
Glucose metabolism, effect on, 584–585
β-Glucosidase, 67–68
β-Glucosides, 66–67
Glucosinolates (GLS), 561
Glutamic acid decarboxylase (GAD), 85
Glutathione (GSH), 207, 458–459
Glutathione reductase (GRD), 458–459
Glutathione S-transferase (GST), 565–566
Gluten-free products, 421
Gluten-sensitive enteropathy. *See* Celiac disease
Glycated hemoglobin (HbA1c), 9
Glycemic index (GI), 394
Glycine max L. *See* Soybean (*Glycine max* L.)
Glycosides, 454–456, 455f
Gochujang, 466
GOS. *See* Galactooligosaccharides (GOS)
Grain legumes, 390, 400
Gram-negative cheese-related bacteria, 289
Gram-positive bacteria, 511
Gram-positive cheese-related bacteria, 289
Grape phenolic compounds, 608–609
Grape pomace, 719–720
Grape wines, 32
Grapevine (*Vitis vinifera* L.), 119
GRAS. *See* Generally recognized as safe (GRAS)
GRD. *See* Glutathione reductase (GRD)
Grifola frondosa (*G. frondosa*), 716–717
GSH. *See* Glutathione (GSH)
GST. *See* Glutathione S-transferase (GST)
Gut, 531
  ecosystem as source of beneficial bacteria, 166–168
  microbiota modulation, 312–313

H

3-HAA. *See* 3-Hydroxyanthranilic acid (3-HAA)
Halophilic bacteria, 181
HbA1c. *See* Glycated hemoglobin (HbA1c)
HCT-116 cells, 463–464
HDC. *See* Histidine decarboxylase (HDC)

746 Index

HDL. *See* High-density lipoprotein (HDL)
HDL-c. *See* High-density lipoprotein-cholesterol (HDL-c)
Health, 385
  health-promoting process, 390–394
  healthier fermented sausage, 204
  healthy bioactive molecules, 527–531
  promoters, 521
Health benefits, 527
  EPSs, 54–59
    anticancer properties, 58
    antioxidant properties, 57–58
    cholesterol lowering effect, 56
    immunomodulation, 56–57
    interactions with enteric pathogens, 58–59
  of fermented cucumbers, 510–513
    fermented cucumber, 512
    immune modulation, 512
    LAB isolation, 510–511
    macroelements, 512
    production of bacteriocins, 511
    removal of antinutrient compounds, 512–513
  of kefir, 353
    anticarcinogenic effects, 346–347
    antidiabetic effects, 351–352
    antihypertensive effects, 350–351
    antiinflammatory effects, 348–349
    antimicrobial activity, 351
    antimutagenic effects, 346–347
    effects on immune system, 347–348
    hypocholesterolemic effects, 349–350
    kefir, 352–353
    lactose intolerance, 352–353
    osteoporosis, 353
  MEL, 110–113
  by phenolics interaction, 78–79
  of sauerkraut, 564–569
    anticarcinogenic properties, 565–567
    antiinflammatory effects, 568
    antioxidant benefits, 564–565
    protection against oxidative DNA damage, 567–568
    sauerkraut as source of probiotic bacteria, 568–569
Health effects, 583–587
  of beer, 371–372, 377–378
  of bioactive peptides, 35–40
    adipose tissue, health benefits on, 39
    bones, health benefits on, 39–40
    cardiovascular system, health benefits on, 35–37
    gastrointestinal system, health benefits on, 38
    immune system, health benefits on, 38–39
    nervous system, health benefits on, 37–38
  on glucose metabolism and insulin resistance, 584–585
  on lipid metabolism, 585–586
  and safety issues, 586–587
Health-promoting activity of polyphenols, 602–606
  interaction
    with allergy proteins, 606
    with digestive enzymes, 603–604
    with neurotoxic proteins, 605–606
    with platelets, 605
    with serum proteins, 604–605
  phenolic compound in oral cavity, 602
Heavy metals, strategies to removing, 193–194
HeLa cell line, 463
Hepatic cancer cells, 464
Hepatocellular carcinoma, 464
Hepta-peptide sequence HFGNPFH, 187
Heterocylic amines, 625
Heteropolysaccharide, 53
HF. *See* High-fat (HF)
HFD. *See* High-fat diet (HFD)
HHP. *See* High hydrostatic pressure (HHP)
HI. *See* Hydrolysis index (HI)
Hibiscus sabdariffa L. *See* Roselle seed (*Hibiscus sabdariffa* L)
High hydrostatic pressure (HHP), 642
High-density lipoprotein (HDL), 207–208, 349–350, 373, 586
High-density lipoprotein-cholesterol (HDL-c), 315–316, 487–489
High-fat (HF), 490
High-fat diet (HFD), 163
High-performance liquid chromatography (HPLC), 113–114, 663
High-performance liquid chromatography fluorimetric detection (HPLC-F detection), 113–114
High-performance liquid chromatography with electrochemical detection (HPLC-EC detection), 113–114
High-quality gluten-free breads, 442–443
Him. *See* Histamine (Him)
HIOMT. *See* Hydroxyindole-O-methyl transferase (HIOMT)
Histamine (Him), 628, 631, 633–634, 637
  intolerance, 638
Histamine *N*-methyltransferase (HNMT), 627
Histidine decarboxylase (HDC), 627

HNMT. *See* Histamine *N*-methyltransferase (HNMT)
HO-8910 cell line, 463
Homocysteine, 377
Homopolysaccharides, 50
Hops, 370
Hordeum vulgare L. *See* Barley (*Hordeum vulgare* L.)
Houji tea, 9–10
HPA axis. *See* Hypothalamus-pituitary adrenocortical axis (HPA axis)
HPLC. *See* High-performance liquid chromatography (HPLC)
HPLC-EC detection. *See* High-performance liquid chromatography with electrochemical detection (HPLC-EC detection)
HPLC-F detection. *See* High-performance liquid chromatography fluorimetric detection (HPLC-F detection)
HuH-7, 458–459
Human clinical trials, 36
Human studies, 9, 38
Hydration, 376
Hydrolysis index (HI), 439
3-Hydroxyanthranilic acid (3-HAA), 458–459
Hydroxycinnamic acids, 73–74, 79
   compounds, 64
   decarboxylation, 77
Hydroxyindole-O-methyl transferase (HIOMT), 107–108
Hydroxytyrosol, 529–530
Hypercholesterolemia, 208, 524–525
Hyperglycemia, 351–352
Hyperhomocysteinemia, 377
Hypertension, 10–11, 35–36, 247–253, 322–323
Hypocholesterolemia, 459–461
Hypocholesterolemic effects, 56, 349–350
Hypothalamus-pituitary adrenocortical axis (HPA axis), 37–38

I

I3A. *See* Indol-3-acetonitrile (I3A)
IAA. *See* Indole-3-acetic acid (IAA)
IAAld. *See* Indole-3-acetaldehyde (IAAld)
IAM. *See* Indole-3-acetamide (IAM)
IAN. *See* Indole-3-acetonitrile (IAN)
IBA. *See* Indole-3-butyric acid (IBA)
IBDs. *See* Inflammatory bowel diseases (IBDs)
IBS. *See* Irritable bowel syndrome (IBS)
Iburu fonio. *See* Black fonio (*Digitaria iburua*)
IDF. *See* Insoluble dietary fiber (IDF)
Idli, 425
   products, 397–399, 401
IFN-γ. *See* Interferon gamma (IFN-γ)
IGF-I. *See* Insulin-like growth factor I (IGF-I)
IL. *See* Interleukin (IL)
Immediate recall, 459
Immune system
   health benefits on, 38–39
   kefir effects on, 347–348
Immune-boosting effects, 493
Immune-manipulation, 722
Immunity, effect on, 311–312
Immunochromatographic test strip detection methods, 666–667
Immunomodulation, 56–57
Immunomodulatory effects, 347–348, 378
Immunoregulatory effect, 164
In vitro chemical assays, 596
Indol-3-acetonitrile (I3A), 561–564
Indole-3-acetaldehyde (IAAld), 108
Indole-3-acetamide (IAM), 108
Indole-3-acetic acid (IAA), 106, 108
Indole-3-acetonitrile (IAN), 108
Indole-3-butyric acid (IBA), 108
Indole-3-pyruvic acid (IPA), 108
Inducible nitric oxide synthase (iNOS), 374, 566–567
"Inflamm-aging", 111
Inflammation, 38–39
Inflammatory bowel diseases (IBDs), 510–511
Ingestion of LAB, 311–312
Injera, 425, 442
iNOS. *See* Inducible nitric oxide synthase (iNOS)
Insertion sequence (IS), 694–696
Insoluble dietary fiber (IDF), 399
Institute of Standards and Industrial Research of Iran (ISIRI), 658
Insulin resistance, 39
   vinegar effect on, 584–585
Insulin-like growth factor I (IGF-I), 325
Interferon gamma (IFN-γ), 38–39
Interleukin (IL), 168
   IL-6, 38–39, 311, 347–348, 487–489
Internal transcribed spacer (ITS1), 290–291
International Olive Oil Council (IOOC), 519
International Scientific Association for Probiotics and Prebiotics (ISAPP), 159–160
"International Year of Quinoa", 420–421
Inulin, 210–211, 544–545
   inulin-type fructans, 545

748 Index

Inulinase production, 717–718
Iodine (I), 221–222
IOOC. *See* International Olive Oil Council (IOOC)
IPA. *See* Indole-3-pyruvic acid (IPA)
Iron (Fe), 208, 395–396
Irradiation, 191–192
Irritable bowel syndrome (IBS), 317
IS. *See* Insertion sequence (IS)
ISAPP. *See* International Scientific Association for Probiotics and Prebiotics (ISAPP)
ISIRI. *See* Institute of Standards and Industrial Research of Iran (ISIRI)
Isoflavones, 396, 454–456
Isogenic human prostate cancer cells, 463
Isothiocyanates (ITCs), 561–564
ITCs. *See* Isothiocyanates (ITCs)
ITS1. *See* Internal transcribed spacer (ITS1)

J

Japanese fermented skipjack tuna (*Katsuwonus pelamis*), 31
Juice waste, 720–721

K

K1. *See* Phylloquinone (PK)
K2. *See* Menaquinones (MKs)
Katsuobushi, 31
Katsuwonus pelamis. *See* Japanese fermented skipjack tuna (*Katsuwonus pelamis*)
Kazachstania exigua C81116, 445
Kefir, 339, 352–353. *See also* Cheese; Pulque; Yogurt
  chemical composition, 343–344, 345t
  grains, 340–342, 340f
  health benefits, 346–353
  nutritional characteristics, 344–346
  production, 342–343, 342f
Kefiran, 56–57
Kefram-kefir, 351–352
KEGG. *See* Kyoto Encyclopedia of Genes and Genomes (KEGG)
Keratinocytes, 483
KFDA. *See* Korean Food and Drug Administration (KFDA)
Kimchi, 7, 477–479
  fermentation, 480–482
  health benefits, 482–490
    antiaging effects, 482–484
    anticancer effects, 484–487
    antimutagenic effects, 484–487
    antiobesity effects, 487–489
    antioxidative effects, 482–484
  manufacturing, 479–480
  safety, 491–492
Kimchi LAB, 482, 485–487
  health benefits, 492–496
    anticancer effects, 492–493
    antiobesity effects, 494
    antioxidative effects, 492–493
    immune-boosting effects, 493
    other effects, 494–496
Kitaibelia vitifolia (*K. vitifolia*), 224–225
Kluyveromyces marxianus NRRL Y-7571, 717–718
Koji, 9–10
Kombucha, 6
Korean Food and Drug Administration (KFDA), 662–663
Korean solar salt, 491
Kyoto Encyclopedia of Genes and Genomes (KEGG), 164–165

L

L. acidophillus supernatants (LS), 494
LAB. *See* Lactic acid bacteria (LAB)
Lactic acid bacteria (LAB), 5–6, 13, 31, 50, 67–68, 85, 159–160, 181, 210, 275–279, 305, 308, 339, 398, 405–406, 435–437, 441–443, 453, 477, 504–505, 517–520, 543, 559, 627, 633, 642, 677
  in blue cheeses, 286–289
    production of undesirable compounds, 287–289
  fermentation process, 308–309
  GABA production by, 90–91
  isolation from fermented cucumbers as probiotics, 510–511
  LAB-producing bacteriocins, 191–192
  of olive fermentation as probiotics, 520–523
  phenotypic and genotypic AR in, 677–690, 679t–682t, 686t, 688t–689t
Lactic acid fermentation, 425, 523
  of cucumbers, 504–506, 506f–507f, 508t
Lactobacilli, 521
Lactobacillus acidophilus (*Lb. acidophilus*), 425–426
Lactobacillus plantarum (*L. plantarum*), 67–68, 401, 425–426, 481, 709–715
  biochemical pathways, 70f, 74f
  health benefits by phenolics interaction with, 78–79

hydroxybenzoic acid-derived compounds
  biotransformation, 69–73
hydroxybenzoyl esterases, 71t
hydroxycinnamic acid-derived compounds
  biotransformation, 73–77
hydroxycinnamoyl esterases, 76t
L. plantarum B1–6, 404
L. plantarum CECT 748, 404
L. plantarum HY7712, 493
L. plantarum KCTC 3099, 492
L. plantarum PS-21, 493
as model bacteria for plant foods
  fermentation, 67–68
Lactobacillus spp., 115, 161, 458, 633–634
  L. bulgaricus, 312–313
  L. lactis, 404
  L. acidophilus KFRI342, 492–493
  L. brevis, 481
    L. brevis E95612, 445
  L. buchneri, 496
  L. bulgaricus, 137, 719
  L. curvatus, 210
  L. paraplantarum, 510–511
  L. rhamnosus, 244, 404, 660
    L. rhamnosus GG, 425
  L. sakei, 210, 480
Lactose, 309–310, 352–353
  intolerance, 352–353
    Yogurt effect on, 316–317
  maldigestion, 161–163
Lagenaria siceraria (L. siceraria), 543
LC-MS/MS. See Liquid chromatography-
  tandem mass spectrometry (LC-MS/MS)
LD_{50} tests. See Lethal dose$_{50}$ tests (LD_{50} tests)
LDL. See Low-density lipoprotein (LDL)
LDL-C. See Low-density lipoprotein
  cholesterol (LDL-C)
Learning, 459
Legumes, 9–10, 95–96, 216, 433–434, 437–438
  flours
    and sourdough fermentation, 435–439
    in sourdough wheat bread making, 439–441
  lectins, 405
Leguminosae, 385, 433–434
Lethal dose$_{50}$ tests (LD_{50} tests), 96
Leuconostoc spp., 512, 633–634
  L citreum, 481
  L gasicomitatum, 481
  L. fallax, 559
  L. mesenteroides, 278–279, 480, 492, 547
Lignocellulosic biomass, 709
Limonene, 725–726
Lipid(s), 309, 395

changes during fermentation, 399–400
  compounds, 207–208
  metabolism, 283
  vinegar effect on lipid metabolism, 585–586
Lipolysis, 283–286
Lipopolysaccharides (LPSs), 50
Liquid chromatography-tandem mass
  spectrometry (LC-MS/MS), 113–114
Long-term fermented-fish sauces, 188
Long-term memory, 459
Longevity, wine impact on, 601
LoVo cells, 463–464
Low bone mineral density, 353
Low-density lipoprotein (LDL), 207–208,
  315–316, 349–350
Low-density lipoprotein cholesterol (LDL-C),
  461, 487–489
Low-molecular-weight tannins, 69
"Low–tech" procedure, 424
LPSs. See Lipopolysaccharides (LPSs)
LS. See L. acidophilus supernatants (LS)
Lupin, 441
Lycopene, 222
Lysine, 207, 390

M

Macroelements from fermented cucumber, 512
MAD. See Malondialdehyde (MAD)
Magnesium (Mg), 208, 221–222
"Maguey pulquero", 543
Maillard reaction products (MRPs), 179
Maillard reactions, 5–6
Maize (Zea mais), 417–418, 425, 434
Malolactic fermentations (MLF), 635
Malolactic-deficient strain, 505–506
Malondialdehyde (MAD), 57–58, 458–459
Malting wastes, 722. See also Agro-wastes;
  Bakery waste; Fruits wastes
MAO. See Monoamine oxidase (MAO)
Mashing process, 365
Mass spectrometry (MS), 113–114
Maturation of blue cheeses, 275–286
MCF-7 cell line, 463
MCP-1. See Monocyte chemoattractant
  protein-1 (MCP-1)
MDR systems. See Multidrug resistance
  systems (MDR systems)
Meat, 95–96
  factor, 208
  and meat products, 632–633
  Meat products, 208
mecA. See Meticillin-resistance gene (mecA)

Mediterranean region, 518
Meju, 465, 656–657
MEL. *See* Melatonin (MEL)
Melanoidins, 370–371
Melatonin (MEL), 105
  biosynthesis, 107–109, 109f
  chains A and M, 107f
  EFSA, 105–106
  in fermented foods, 115–123, 116t–118t
  health benefits, 110–113
    antiaging effects, 111
    anticancer effects, 110–111
    antiobesity effects, 112
    antioxidant effects, 110
    effects on bone, 112
    migraine prevention, 113
    protection against Alzheimer's disease, 112
    protection against cardiovascular diseases, 111
    protection against UVR, 112–113
  mechanisms of action, 109–110
  in plant foods, 113–115
  structure and physicochemical properties, 106
Melatonin isomers (MIs), 106
Menadione, 4-naphtoquinone ring. *See* 2-methyl-1
Menaquinones (MKs), 138–139, 161–163
  concentrations of and variations in, 142t–147t
  menaquinone-producing bacterial species, 149t
Meta-analysis, 311, 321–322, 324–325
Metabolic
  diseases, 599
  syndrome, 377
Metabolites from fermented olives, 527–531
Methanol soluble fraction (MSF), 484
Methionine, 207
5-Methoxytryptamine (5-MT), 110
Methyl ketones, 284–286
2-Methyl-1,4-naphtoquinone ring, 138–139
N-Methyl-Nitro-N-nitrosoguanidine (MNNG), 484
N-Methylhistamine, 627
N-Methylimidazolylacetic acid, 627
5-Methyltetrahydrofolate, 132
Meticillin-resistance gene (*mecA*), 690–691
Microbial
  amylases, 398–399
  changes during spontaneous sauerkraut fermentation, 559
  community, 275

enzymes, 66–67
EPSs
  and linkages, 51t
  recovery, purification, and characterization, 52f
  fermentation, 724–725
  safety, 548–552
Microbial genomics, 166. *See also* Bacterial genomics
Microbial polysaccharides, 50. *See also* Exopolysaccharides (EPSs)
Microbiological methods to control AF in fermented foods, 660–661
Microbiota
  impact on red wine availability and bioactivity, 610–612
  for olive fermentation, 517
Microorganisms, 119, 308, 453, 677
  in aguamiel and pulque, 547–554, 549t–552t, 553f
  antimicrobial activity, 183–185
  folate production by, 136–138
  as starters, 180–183
Migraine prevention, 113
Milk, 133
  components, 313
  preparation, 277–278
  proteins, 350–351
Millet (*Pennisetum glaucum*), 434
Milling industry, 443
Mineral(s), 208–209, 310, 344–346, 368, 395–396
  addition, 216–222
    Ca, 221–222
    iodine enrichment, 221–222
    Mg, 221–222
    Se, 221–222
    strategies for sodium reduction, 216–220
  bioavailability changes, 400
  in cheese and impact on health, 244–246, 245t
  effect on absorption, 317–318
  simultaneous enhancement of concentration, 423
Minor bioactive compounds changes, 401–402
Minor cereals in sourdough wheat bread making, 441–442
MIs. *See* Melatonin isomers (MIs)
Miso, 453, 466
Mixed starter cultures, 641
MK-9 (4H). *See* MK-9 and tetrahydromenaquinone-9 (MK-9 (4H))
MK-9 and tetrahydromenaquinone-9 (MK-9 (4H)), 148

MKs. *See* Menaquinones (MKs)
MLF. *See* Malolactic fermentations (MLF)
MNNG. *See* N-Methyl-Nitro-N-nitrosoguanidine (MNNG)
Modulatory peptides of mineral absorptions, 257–264, 258t–263t
Molasses, 717–718
Molding, 280
Molds, 182
Monoamine oxidase (MAO), 626, 638–640
Monocyte chemoattractant protein-1 (MCP-1), 487–489
Monosaccharides, 367
Monounsaturated fatty acids (MUFAs), 207–208, 316, 395
Mortality, 601
Mouse embryonic C3H/10T1/2 cells, 484
Mouse lung tissue, 458–459
MPA. *See* Mycophenolic acid (MPA)
MRPs. *See* Maillard reaction products (MRPs)
MRSA. *See* Multidrug-resistant meticillin-resistant *S. aureus* (MRSA)
MS. *See* Mass spectrometry (MS)
MS/MS. *See* Tandem mass spectrometry (MS/MS)
MSF. *See* Methanol soluble fraction (MSF)
5-MT. *See* 5-Methoxytryptamine (5-MT)
Mucor spp., 456
MUFAs. *See* Monounsaturated fatty acids (MUFAs)
Multidrug resistance systems (MDR systems), 581, 687–690
Multidrug-resistant meticillin-resistant *S. aureus* (MRSA), 690–691
Multifunctional peptides, 30
Mung beans (*Vigna radiata*), 10
Mycophenolic acid (MPA), 295
Mycotoxins, 289–290, 294, 656
  biological degradation of, 660
Myoinositol hexakisphosphate. *See* Phytic acid

N

n-3 fatty acids, 207–208
n-3 PUFAs, 207–208, 211
N-terminal amino-acid sequence of DPYEEPGPCENLQVA, 189
NaCl, 491
NAD(P)H quinone oxidoreductase 1 (NQO1), 565–566
National Health and Nutrition Examination Survey (NHANES), 320–321
Natto, 453, 466

Nattokinase, 404
Natural killer cells (NK cells), 11–12
"Natural olives", 519. *See also* Olive tree (*Olea europaea L.*)
Natural polyamines, 625
NDA. *See* Dietetic Products, Nutrition, and Allergies (NDA)
NDF. *See* Neutral dietary fiber (NDF)
NDMA. *See* N-Nitrosodimethylamine (NDMA)
NE. *See* E-2-nonenal (NE)
Nectar fermentation, 543
Nervous system
  cheese-derived peptide effects on, 254–255
  health benefits on, 37–38
Net protein ratio (NPR), 441
Neurodegeneration, 459
Neurodegenerative disorders, 378
Neuroinflammation, 460, 599–600
  neuroinflammatory process, 599–600
Neurological diseases, 599–601
Neurotoxic proteins, interaction with, 605–606
Neutral dietary fiber (NDF), 399
Next-generation sequencing (NGS), 164–165
NF-E2-related transcription factor 2 (Nrf2), 565–566
NFAT1. *See* Nuclear factor of activated T cells 1 (NFAT1)
NGS. *See* Next-generation sequencing (NGS)
NHANES. *See* National Health and Nutrition Examination Survey (NHANES)
Niacin, 209
Nicotinoyl-GABA, 85
Nitrate (NO_3), 224–225, 491
Nitric oxide (NO), 36–37, 568, 597–598
Nitrite, 224–225
  reduction, 225–226
Nitrogen compounds, 204–207, 205t–206t
Nitrogen species, 596–597
N-Nitrosodimethylamine (NDMA), 492
N-Nitrosopyrolidine (NPYR), 492
NK cells. *See* Natural killer cells (NK cells)
NO. *See* Nitric oxide (NO)
Nondairy-derived food product, 7
NonNewtonian behavior of EPS, 54
Nonstarter lactic acid bacteria (NSLAB), 277–278, 286
Nonwheat cereals, 434–435
  advantages and limitations, 422–424
    degradation of antinutritive compounds, 423
  nutritional properties, 423–424
  palatability, 423–424

Nonwheat cereals (*Continued*)
    sensorial acceptation, 423–424
    simultaneous enhancement of minerals concentration, 423
    textural properties, 423–424
    vitamins bioavailability, 423–424
  fermented nonwheat food products, 424–427
  health beneficial effects of nonwheat fermented foods, 427–428
  nonwheat cereal–fermented–derived products, 417
  nutritional aspects of, 418–421
Nonwheat fermented foods, health beneficial effects of, 427–428
Nonwheat flours in sourdough gluten-free bread making, 442–443
NPR. *See* Net protein ratio (NPR)
NPYR. *See* N-Nitrosopyrolidine (NPYR)
NQO1. *See* NAD(P)H quinone oxidoreductase 1 (NQO1)
Nrf2. *See* NF-E2-related transcription factor 2 (Nrf2)
NSLAB. *See* Nonstarter lactic acid bacteria (NSLAB)
Nuclear factor of activated T cells 1 (NFAT1), 464
Nukadoko, 709–715
Nutraceuticals, 707–708, 714f, 723
Nutrient(s), 320
  nutrient-grade wastes, 709
  profile, 306–308, 307t
  recovery, 725–726
Nutritional
  aspects of nonwheat cereals, 418–421
    informative macronutrient, 419t
    nonwheat cereals and pseudo-cereals, 421t
  changes during pulse-based foods fermentation, 397–402
  characteristics of kefir, 344–346
  and phytochemical composition of sauerkraut, 561–564
  properties, 423–424
  studies, 9–10
Nutritional composition of pulses, 390–397
  carbohydrates, 394–395
  dietary fiber, 395
  lipids, 395
  minerals, 395–396
  phytochemicals, 396–397
  proteins, 390–394
  proximate composition, 391t–393t
  vitamins, 395–396

Nutritional value, 204–209
  amino acids, 204–207, 205t–206t
  antioxidants, 209
  fat content, 207–208
  fatty acid composition, 207–208
  minerals, 208–209
  other lipid compounds, 207–208
  other nitrogen compounds, 204–207, 205t–206t
  peptides, 204–207, 205t–206t
  proteins, 204–207, 205t–206t
  vitamins, 209
Nutritive attributes, 390

O

Oat (*Avena* spp. L.), 418–420
Oatmeal, 418–420
Obesity, 204
Ochratoxin A (OTA), 657–659, 661
OCTT. *See* Orocecal transit time (OCTT)
ODC. *See* Ornithine decarboxylase (ODC)
Oenococcus spp., 633–634
OGTT. *See* Oral glucose tolerance test (OGTT)
OIE. *See* World Organization for Animal Health (OIE)
Okara, 715–716
"Okinawa diet", 466
Olea europaea L. *See* Olive tree (*Olea europaea* L.)
Oleic acid, 528
Oleuropein, 518–519, 529
Oligosaccharide(s), 165–166, 422–423, 718
  fermented cucumber as source of starter cultures to produce, 512
Olive tree (*Olea europaea* L.), 518
  fermentation, 519–520
    health effects of probiotics, 523–527
    LAB, 520–523
  fermented olives modulating digestive microbiota, 531–533
  fruit, 518–519
  healthy bioactive molecules and metabolites from fermented olives, 527–531
  production of traditional fermented, 518–519
Omega-3 fatty acids, 420–421
Opioid peptides, 254
Opisthorchis, 194
ORAC. *See* Oxygen radical absorption capacity (ORAC)
Oral cavity, phenolic compound in, 602
Oral glucose tolerance test (OGTT), 321
Orange juice, 122–123

Orange peels, 719
Organic acids, 581
Organoleptic attributes, 593
Ornithine decarboxylase (ODC), 625–627
Orocecal transit time (OCTT), 317
Oryza sativa. *See* Rice (*Oryza sativa*)
Osteoporosis, 39–40, 323–324, 353, 374, 601
OTA. *See* Ochratoxin A (OTA)
Oxicrat, 580
Oxidative DNA damage, protection against, 567–568
Oxidative stress, 564–565, 594–596
Oxygen radical absorption capacity (ORAC), 596
Oyster shells, 726

P

P5P. *See* Pyridoxal 5′; phosphate (P5P)
PA. *See* Polyamines (PA)
PAD enzyme. *See* Phenolic acid decarboxylase enzyme (PAD enzyme)
Palatability, 423–424
Paneth cells, 38
PAO. *See* Polyamine oxidase (PAO)
Parasite control, 194
Paste, 178, 178t
Pathogens, AR profile, 690–694
  dairy products, 690–694
  fermented meat and vegetables, 694
PCR-based molecular detection methods, 664
PDO. *See* Protected designation of origin (PDO)
Pectic oligosaccharides (POS), 719
Pediococcus spp., 633–634
  P. parvulus, 56
  P. pentosaceus, 510–511
Peels, 719
Penethyl isothiocyanate (PEITC), 565–566
Penicillium roqueforti (*P. roqueforti*), 275–280, 282–283, 285, 289–291
  electron microscopy images, 281f
  and fungi in blue cheese, 290–291
  secondary metabolites by, 293–296
    agroclavine, 295
    andrastins, 294–295
    eremofortins, 294
    festuclavine, 295
    MPA, 295
    P. roqueforti metabolites, 296
    PR-toxin, 294
    roquefortines, 293–294
  strains in local nonindustrial blue cheeses, 291–292

Pennisetum glaucum. *See* Millet (*Pennisetum glaucum*)
PEP. *See* Phosphoenolpyruvate (PEP)
Peptides, 186, 204–207, 205t–206t
PER. *See* Protein efficiency ratio (PER)
Peroxidases (POX), 110
Peroxiredoxin-3 (Prdx-3), 464
Peroxisome proliferator-activated receptor gamma (PPAR-γ), 487–489
Perpetual oxidative damage, 458
PGI. *See* Protected geographical indication (PGI)
Phaseolus vulgaris species, 395–396
Phem. *See* Phenylethylamine (Phem)
Phenibut, 85
Phenolic acid decarboxylase enzyme (PAD enzyme), 77
Phenolic acids, 63, 371–372
Phenolic composition of vinegars, 582–583
Phenolic compounds, 529, 532–533, 581–582, 594, 598. *See also* Bioactive peptides
  changes, 401
  and derivatives, 64f
  disease-protective/preventive effect of wine, 594–601
    cancer, 598
    CVD, 597–598
    metabolic diseases, 599
    mortality, 601
    neurological diseases, 599–601
    osteoporosis, 601
    wine, 601
    wine polyphenols in human health, 597f
  in fermented foods, 63
  LAB and enzymatic activities, 66t
  in oral cavity, 602
  phytochemicals, 64
  transformation by fermentation, 65–67
Phenolics, 603
Phenols, 369–370, 375
Phenotypic AR in LAB, 677–690
  from fermented meats, 686t
    prevalence, 695t
  from fermented vegetables, 688t–689t
  from traditional dairy products, 679t–682t
    prevalence, 692t–693t
Phenylethylamine (Phem), 625
2-Phenylethylamine. *See* Phenylethylamine (Phem)
Phosphoenolpyruvate (PEP), 165–166
Phosphopeptides, 257–264, 258t–263t
Phylloquinone (PK), 138
Physical methods to controlling AF in fermented foods, 659

Phytase, 438–439
Phytic acid, 438–439
  changes, 400
  degradation, 400
Phytochemical(s), 7, 396–397
  composition of pulses, 390–397
    carbohydrates, 394–395
    dietary fiber, 395
    lipids, 395
    minerals, 395–396
    phytochemicals, 396–397
    proteins, 390–394
    proximate composition, 391t–393t
    vitamins, 395–396
Phytoestrogens, 66–67, 375
Picamilon, 85
PK. *See* Phylloquinone (PK)
Plant foods
  MEL in, 113–115
  plant food-based diets, 402
Plant proteins, bioactive peptides in, 27t–29t
Plant-derived fermented products, 517
Platelets, interaction with, 605
"Plombage", 281–282
Poly-unsaturated fatty acids (PUFAs), 165, 207–208, 211, 316, 395
Polyalcohols, 367
Polyamine oxidase (PAO), 625–626
Polyamines (PA), 625, 640
Polyphenols, 583
Polysaccharide(s), 165–166, 715–716. *See also* Exopolysaccharides (EPSs)
Pomace, 719–720
Pomegranate wines, 120
POS. *See* Pectic oligosaccharides (POS)
Potassium (K), 395–396
Potassium chloride (KCl), 220
POX. *See* Peroxidases (POX)
Pozol, 425
PPAR-γ. *See* Peroxisome proliferator-activated receptor gamma (PPAR-γ)
PR-toxin, 294
PR2-III. *See* Prostaglandin-F2-III (PR2-III)
Prdx-3. *See* Peroxiredoxin-3 (Prdx-3)
Prebiotic effect, 545
Proanthocyanidins absorption, 609–610
Proanthocyanins, 609
Probiotic(s), 7, 49–50, 159–160, 182–183, 211–214, 215t, 477, 568–569
  bacterial genomics in, 164–166
  health effects of olive fermentation, 523–527

LAB
  isolation from fermented cucumbers as, 510–511
  olive fermentation as, 520–523
  sauerkraut as source of probiotic bacteria, 568–569
  yogurt as probiotic vector, 313–315
Procyanidins, 604
Proinflammatory cytokine, 460
Proline-rich proteins (PRPs), 602
Propionibacterium strains, 148, 151
Prostaglandin-F2-III (PR2-III), 600–601
Protease inhibitors, 390–394
Protected designation of origin (PDO), 275, 582
Protected geographical indication (PGI), 275, 582
Protein efficiency ratio (PER), 441
Protein(s), 204–207, 205t–206t, 246–247, 308–309, 367–368, 390–394
  changes during fermentation, 397–398
  inhibitors, 398
  protein-derived bioactive peptides, 707–708
Proteinase (Prt), 34
Proteolysis, 204–207, 282–283
Proteomic analysis, 78–79
Proviva, 425–426
PRPs. *See* Proline-rich proteins (PRPs)
Prt. *See* Proteinase (Prt)
Pseudo-cereals, 417–418, 420, 434–435
  in sourdough wheat bread making, 441–442
Pseudoalteromonas alienates (*P. alienates*), 185
Psychoactive amines, 638
Pu-erh tea, 6
PUFAs. *See* Poly-unsaturated fatty acids (PUFAs)
Pulque, 543. *See also* Kefir
  bioactive constituents, 546–547
  bioactive constituents of aguamiel, 544–546
  fermentation, 544
  microorganisms in aguamiel and, 547–554, 549t–552t, 553f
  pulque type I, 544
  type I, 544
Pulque de punta, 544
Pulse-based foods fermentation, nutritional changes during, 397–402
  lipids changes, 399–400
  minor bioactive compounds changes, 401–402
  phenolic compounds and antioxidant activity changes, 401

phytic acid and mineral bioavailability changes, 400
protein and amino acids changes, 397–398
starch, carbohydrates, and dietary fiber changes, 398–399
vitamins during fermentation changes, 400
Pulses, 385, 394–395
Puntius sophore. *See* Fermented small cyprinid fish (*Puntius sophore*)
Putrescine (Put), 625, 633
Pyridoxal 5′ phosphate (P5P), 85
Pyridoxine, 209

Q

QA23 strain, 120–121
Qualified presumption of safety (QPS), 629–631, 678–683
Quality control, 12–14
Quercetin, 78
Quercetin-3-rutinoside, 420
Quinoa (*Chenopodium quinoa* Willd.), 420–421

R

Radial arm maze (RAM), 459
Radioimmunoassay (RIA), 113–114
Raffinose, 433–434
Raffinose family of oligosaccharides (RFOs), 437–438
RAM. *See* Radial arm maze (RAM)
Random amplified polymorphic DNA (RAPD), 290
Randomized controlled trials (RCT), 319–320
RANKL. *See* Receptor activator of NF-κB ligand (RANKL)
RAPD. *See* Random amplified polymorphic DNA (RAPD)
Rapid Alert System for Food and Feed (RASFF), 640
Rapidly digested starch (RDS), 394
RASFF. *See* Rapid Alert System for Food and Feed (RASFF)
Raw meats, 207
RCT. *See* Randomized controlled trials (RCT)
RDS. *See* Rapidly digested starch (RDS)
Reactive nitrogen species (RNS), 110
Reactive oxygen species (ROS), 57–58, 109–110, 253, 458, 460, 561–564, 594–597
Receptor activator of NF-κB ligand (RANKL), 601
Recyclable food wastes, 707–708

Red wine, 593
bioavailability, 606–612
absorption and metabolism of resveratrol, 610
absorption of catechins and proanthocyanidins, 609–610
anthocyanins and derivatives, 607–609
microbiota impact on availability and bioactivity, 610–612
polyphenols, 602
Refined cereal flours, 435
Regulatory T cells (Treg cells), 56–57, 168
Relative risk (RR), 321
Resistance genes, 675–677, 683–685, 687–691, 694–696
Resistant starch (RS), 394, 399
Response surface methodology (RSM), 718
Resveratrol, 600–601
absorption, 610
metabolism, 610
Reuterin, 510
RFOs. *See* Raffinose family of oligosaccharides (RFOs)
Rhizobium radiobacter ATCC 6466, 720–721
Rhizopus oligosporus (*R oligosporus*), 399, 453
Rhizopus spp., 397–398, 456
RIA. *See* Radioimmunoassay (RIA)
Riboflavin, 148–150, 209
Rice (*Oryza sativa*), 434
Ripening, 280–282
RNS. *See* Reactive nitrogen species (RNS)
Roots, 216
Roquefortines, 293–294
ROS. *See* Reactive oxygen species (ROS)
Rosa canina L., 724–725
Rosa rugosa Thunb., 724–725
Roselle seed (*Hibiscus sabdariffa* L), 718
RR. *See* Relative risk (RR)
RS. *See* Resistant starch (RS)
RSM. *See* Response surface methodology (RSM)
Rumenic acid, 240
Rye (*Secale cereale* L.), 418

S

Saccharomyces boulardii (*S. boulardii*), 8
Saccharomyces cerevisiae (*S. cerevisiae*), 119, 398, 528
Saccharomyces species, 6, 108, 123
Salivary proteins (SPs), 602

Salmonella, 13
Salt concentration effect, 507–509
Salt content, strategies to decreasing, 192–193
Salting, 280
SAM. *See* S-Adenosyl-L-methionine (SAM); Senescence-accelerated mice (SAM); Sympathetic adrenomedullary system (SAM)
SAMDC. *See* S-Adenosylmethinine decarboxylase (SAMDC)
Saponins, 396–397, 546
Saturated fatty acids (SFA), 204, 316, 395
Sauerkraut, 133, 636
　health benefits, 557, 564–569
　manufacturing, 557–558
　　inoculation of starter cultures during, 559–561
　microbial changes during fermentation, 559
　nutritional and phytochemical composition, 561–564, 562t–563t
　production, 558f
　salted and fermented vegetables, 557
SBP. *See* Systolic blood pressure (SBP)
SCFAs. *See* Short-chain fatty acids (SCFAs)
Scientific Committee for Animal Nutrition, 676
SD rats. *See* Sprague–Dawley rats (SD rats)
SDF. *See* Soluble dietary fiber (SDF)
SDS. *See* Slowly digested starch (SDS)
Seaweed, 216
Secale cereale L. *See* Rye (*Secale cereale* L.)
Secondary metabolites by *P. roqueforti*, 293–296
　agroclavine, 295
　andrastins, 294–295
　eremofortins, 294
　festuclavine, 295
　MPA, 295
　P. roqueforti metabolites, 296
　PR-toxin, 294
　roquefortines, 293–294
Seeds, 718
Selenium (Se), 208, 221–222, 395–396
Semilla de pulque, 544
Senescence-accelerated mice (SAM), 483
Sensorial acceptation, 423–424
Serotonin, 107–108
Serotonin-N-acetyl transferase (SNAT), 107–108
Serum proteins, interaction with, 604–605
SFA. *See* Saturated fatty acids (SFA)
SFN. *See* Sulforaphane (SFN)
Shellfish sauces, 188
Short-chain fatty acids (SCFAs), 165, 244, 394–395, 545, 719

Short-term memory, 459
SHRs. *See* Spontaneously hypertensive rats (SHRs)
Side effects of GABA, 96
Silicon, 374, 378
SIPS. *See* Stress-induced premature senescence model (SIPS)
β-Site APP cleaving enzyme-1 (BACE-1), 460–461
SKVYP peptide, 36
Slowly digested starch (SDS), 394
SNAT. *See* Serotonin-N-acetyl transferase (SNAT)
Soaking, 398
Societal drugs, 675–676
SOD. *See* Superoxide dismutase (SOD)
Sodium
　content, 216
　reduction, 216–220
Sodium chloride (NaCl), 216–220
Softening enzymes, 504
"Soidon mahi", 724–725
Solid-state fermentation, 427–428, 580
Soluble dietary fiber (SDF), 399
Sorghum (*Sorghum bicolor* L. Moench), 434
Sorghum bicolor L. Moench. *See* Sorghum (*Sorghum bicolor* L. Moench)
Sorghum flour, 443
Sourdough, 31–32
　fermentation, 442
　　legume flours and, 435–439
　　wheat milling byproducts and, 443–445
　gluten-free bread making, nonwheat flours in, 442–443
　wheat bread making
　　legume flours in, 439–441
　　minor cereals and pseudo-cereals in, 441–442
Soyasaponins, 405
Soybean (*Glycine max* L.), 66–67, 453–456
　fermented food products, AF in, 656–657
　flour, 441
　isoflavones, 454–456
　milk, 6–7
　phytoestrogens, 462–463
　products, 636–637
　sauce, 466–467
　soy-fermented products, 31
　soybean-curd lees, 715–716
　waste, 715–716
Spd. *See* Spermidine (Spd)
Specialty foods, 10–12
Spent solution, 720–721

Spermidine (Spd), 222–224, 625
Spermidine/spermine N1-acetyltransferase (SSAT), 625–626
Spermine (Spm), 222–224, 625
Spice herb addition effect, 509
Spm. *See* Spermine (Spm)
Spontaneously hypertensive rats (SHRs), 31, 95, 188
Sprague–Dawley rats (SD rats), 490
SPs. *See* Salivary proteins (SPs)
SREBP-1c. *See* Sterol regulatory element-binding proteins 1c (SREBP-1c)
SSAT. *See* Spermidine/spermine N1-acetyl-transferase (SSAT)
Stalks, 709
Staphylococcus spp., 181–182
Starch
    changes during fermentation, 398–399
    degradation, 398–399
Starter(s)
    cultures, 641
        inoculation during sauerkraut manufacture, 559–561, 562t–563t
    microorganisms as, 180–183
Sterol regulatory element-binding proteins 1c (SREBP-1c), 487–489
Stomach cancer, 491
Stomata, 521
Straws, 709
Streptococcus thermophilus (*S. thermophilus*), 308–309, 313–314, 719
    CRL 804, 721
Streptozotocin (STZ), 490
Stress-induced premature senescence model (SIPS), 483
STZ. *See* Streptozotocin (STZ)
Sugar addition effect, 509
Sugarcane waste, 717–718
Sulforaphane (SFN), 564
Superoxide dismutase (SOD), 57–58, 110, 458–459
Supplementation effect in breastfeeding mothers, 377
Sympathetic adrenomedullary system (SAM), 37–38
Syrup waste, 717–718
Systolic blood pressure (SBP), 35–36

T

T2D. *See* Type 2 diabetes (T2D)
T5H. *See* Tryptamine 5-hydroxylase (T5H)
Tandem mass spectrometry (MS/MS), 247–253

Tannase, 66–67, 69–73, 71t
Tannic acid, 64
Tannin(s), 64, 69
    acyl hydrolase, 69
    tannin–protein interactions, 602
Tarhana, 423
TAS. *See* Total antioxidant status (TAS)
Tauchu, 467
Taurine, 188, 204–207
Taxiarchs, 504–505
TBARS. *See* Thiobarbituric acid (TBARS)
TC. *See* Total cholesterol (TC)
TDA. *See* Tryptamine deaminase (TDA)
TDC. *See* Tryptophan decarboxylase (TDC)
TDF. *See* Total dietary fiber (TDF)
"Tea fungus", 6
Teff (*Eragrostis tef*), 434
Tempeh, 453, 456–464
    antioxidant roles of, 458–459
    and cancer prevention, 462–464
    and dementia, 459–461
    and hypocholesterolemia, 459–461
    nutritional enhancements of, 456–458
tet(K) gene, 683–684
tet(L) gene, 683–684
tet(M) gene, 683–684, 687–690, 696–697
tet(S) gene, 683–684, 696–697
Textural properties, 423–424
TG. *See* Triacylglycerol (TG); Triglycerides (TG)
Thiamine, 209
Thin-layer chromatography (TLC), 663
Thiobarbituric acid (TBARS), 482
Thrombosis, 37
Thymbra spicata oil, 222–224
TLC. *See* Thin-layer chromatography (TLC)
TNBS. *See* 2,4,6-Trinitrobenzenesulfonic acid (TNBS)
TNF-α. *See* Tumor necrosis factor alpha (TNF-α)
Tocols, 222
Tocopherols, 209, 417, 420–421
Tofu-cake, 715–716
Total antioxidant status (TAS), 122
Total cholesterol (TC), 315–316, 461, 487–489
Total dietary fiber (TDF), 399
Total volatile base nitrogen (TVB-N), 192
Toxic compounds, 422
Toxicity of GABA, 96
Toxigenic fungi, 662
TPH. *See* Tryptophan 5-hydroxylase (TPH)

Traditional fermented foods. *See also* Fermented foods
  AR, 677–694
    profile of pathogens, 690–694
    phenotypic and genotypic AR in LAB, 677–690
    from fermented meats, 686t
    from fermented vegetables, 688t–689t
    from traditional dairy products, 679t–682t
Traditional Specialties Guaranteed (TSG), 275
Trans-resveratrol, 598
Transit time/digestion, effect on, 317
Treg cells. *See* Regulatory T cells (Treg cells)
Triacylglycerol (TG), 487–489
Triglycerides (TG), 315–316, 427, 461
2,4,6-Trinitrobenzenesulfonic acid (TNBS), 168
TRP. *See* Tryptophan (TRP)
Trypsin inhibitor, 398
Tryptamine 5-hydroxylase (T5H), 108
Tryptamine deaminase (TDA), 108
Tryptophan (TRP), 106
Tryptophan 5-hydroxylase (TPH), 107–108
Tryptophan decarboxylase (TDC), 108
TSG. *See* Traditional Specialties Guaranteed (TSG)
Tubers, 216
Tumor necrosis factor alpha (TNF-α), 38–39, 311, 346–347, 566–567, 723–724
TVB-N. *See* Total volatile base nitrogen (TVB-N)
Tym. *See* Tyramine (Tym)
Type 2 diabetes (T2D), 305, 320–321
Type I aguamiel, 544
Type II aguamiel, 544
Tyramine (Tym), 625, 632–634, 637
  intolerance, 638–640
Tyrosol, 529

U

Ubiquinones, 187–188
UC. *See* Ulcerative colitis (UC)
UFA. *See* Unsaturated fatty acids (UFA)
Ugba production, 397–398
Ulcerative colitis (UC), 510–511
Ultra-performance liquid chromatography-high-resolution mass spectrometer (UPLC-HR-MS), 113–114
Ultra-performance liquid chromatography-tandem mass spectrometry (UPLC-MS/MS), 113–114
Ultrahigh-performance liquid chromatography (UPLC), 663

Ultraviolet (UV), 148
Uncontrolled free radicals, 186
Undesirable compound production, 287–289
United Nations Food and Agriculture Organization (FAO), 159–160, 385, 510, 657, 676
United Nations World Health Organization (WHO), 159–160
United States Department of Agriculture (USDA), 139–141, 306
United States Food and Drug Administration (USFDA), 191, 404–405, 631, 654, 658
Unsaturated fatty acids (UFA), 395
Unsustainable disposal convention, 707
UPLC. *See* Ultrahigh-performance liquid chromatography (UPLC)
UPLC-HR-MS. *See* Ultra-performance liquid chromatography-high-resolution mass spectrometer (UPLC-HR-MS)
UPLC-MS/MS. *See* Ultra-performance liquid chromatography-tandem mass spectrometry (UPLC-MS/MS)
USDA. *See* United States Department of Agriculture (USDA)
USFDA. *See* United States Food and Drug Administration (USFDA)
UV. *See* Ultraviolet (UV)
UV radiation (UVR), 112–113
  protection against, 112–113
UVR. *See* UV radiation (UVR)

V

Valorization of underutilized food sources, 724–725
Value-added products, 715–716
Valyl-prolyl-proline peptide (VPP peptide), 39–40
Vascular dysfunction, 36–37
Vegetable(s), 95–96
  pulp, 718
Verocytotoxigenic *Escherichia coli* (VTEC), 524
Vertebrates, 86
Very low-density lipoprotein (VLDL), 489
Vigna radiata. *See* Mung beans (*Vigna radiata*)
Vinegars, 577–578, 578t, 580
  acetic acid, 580
  acetic acid bacteria, 579f
  antimicrobial effects, 580–581
  bioactive compounds and antioxidant activity, 581–583
  fermented condiments, 587

health effects, 583–587
solid-state fermentation, 580
Vinegars phenolic composition, 582–583
4-Vinyl derivatives, 77
Viscosity, 53
Vitamin B, 417, 456–457
Vitamin B1. *See* Thiamine
Vitamin B2. *See* Riboflavin
Vitamin B3. *See* Niacin
Vitamin B6. *See* Pyridoxine
Vitamin B9. *See* Folate
Vitamin B12. *See* Cobalamin
Vitamin C, 491
Vitamin D, 209
Vitamin E, 417
Vitamin K, 138, 161–163
  comparison of vitamin K2 concentrations, 140t
  concentrations of and variations in menaquinones, 142t–147t
  content of fermented food, 139–141
  deficiency, 138
  vitamin K2 in fermented food, 148
Vitamin K3. *See* 2-Methyl-1,4-naphtoquinone ring
Vitamins, 131, 209, 239, 246, 310, 344–346, 395–396. *See also* Phenolic compounds
  addition, 222, 223t–224t
  bioavailability, 423–424
  changes during fermentation, 400
  folate, 132–138
  LAB fermentation, 152
  riboflavin, 148–150
  vitamin B12, 150–151
  vitamin K, 138–148
Vitis vinifera L. *See* Grapevine (*Vitis vinifera* L.)
VLDL. *See* Very low-density lipoprotein (VLDL)
Volatiles, 504
VPP peptide. *See* Valyl-prolyl-proline peptide (VPP peptide)
VTEC. *See* Verocytotoxigenic *Escherichia coli* (VTEC)

W

Waist-hip ratio (WHR), 489
Water-soluble extracts (WSE), 247–253
WCFS1 strain, 75
WCRF. *See* World Cancer Research Fund (WCRF)
Weight management, 319–320
Weisella confuse (*W. confuse*), 481
Weissella koreensis (*W. koreensis*), 481, 494

Wheat, 417–418, 434
  germ, 445
  Milling byproducts, 443–445
  nutritional advantages, 437t
  sourdough fermentation
    legumes, minor cereal, pseudo-cereal flours, and, 435–443
    wheat milling byproducts and, 443–445
  wheat-based foods, 435
  wheat-legume sourdough making/propagation, 440f
Whey, 721
  proteins, 343
White fonio (*Digitaria exilis*), 434
WHO. *See* United Nations World Health Organization (WHO); World Health Organization (WHO)
Whole grains, 444
WHR. *See* Waist-hip ratio (WHR)
WI-38 cells, 483
Wine, 115–121, 593, 601, 633–634. *See also* Beer
  bioavailability of red wine, 606–612
  disease-protective/preventive effect, 594–601
  health-promoting activity of polyphenols, 602–606
  polyphenols, 596
  preamble, 593–594
Working memory error (WME), 459
World Cancer Research Fund (WCRF), 404–405
World Health Organization (WHO), 216, 373, 654, 676
World Organization for Animal Health (OIE), 676
WSE. *See* Water-soluble extracts (WSE)

X

Xanthan gum, 53–54
Xanthophyllomyces dendrorhous ATCC 24202, 720
XDR. *See* Extensively drug-resistant (XDR)
XOS. *See* Xylooligosaccharide (XOS)
Xylitol production, 709
Xylooligosaccharide (XOS), 716

Y

Yeast(s), 347–348, 520
  in cheese, 289–293
    P. roqueforti and fungi in blue cheese, 290–291

Yeast(s) (*Continued*)
  P. roqueforti strains in local nonindustrial blue cheeses, 291–292
  secondary metabolites, 292–293, 292t, 293f
  nucleic acid, 135–136
  starters, 119
Yogurt, 94–95, 305, 308–309, 319. *See also* Cheese; Kefir
  bioactive properties, 310–318
  composition
    carbohydrates, 309–310
    codex alimentarius, 305–306
    lipids, 309
    microorganisms, 308
    minerals, 310
    nutrient profile, 306–308, 307t
    proteins, 308–309
    vitamins, 310
  in disease prevention, 318–326
  manufacture, 54
YPVEPFTE multifunctional peptide, 30–31
YQEPVLGPVRGPFPIIV peptide, 37

Z

Zea mais. *See* Maize (*Zea mais*)
Zinc (Zn), 209, 395–396